FUNDAMENTALS OF CHEMISTRY

SECOND EDITION

ROD O'CONNOR
TEXAS A&M UNIVERSITY

HARPER & ROW, PUBLISHERS
New York   Hagerstown   San Francisco   London

Dedicated to my wife, Shirley, and to Mark, Kara, Shanna, and Timothy Patrick; to my parents, Jay H. and Flora O'Connor; to my favorite teachers, Jim Cason, Shirley Gaddis, H. A. Mangan, Leroy Mason, and Bob Smith; and to my students in freshman chemistry.

Sponsoring Editor: John A. Woods/Wayne E. Schotanus
Project Editor: Lois Lombardo
Designer: Rita Naughton
Production Supervisor: Will C. Jomarrón
Compositor: York Graphic Services, Inc.
Printer and Binder: R. R. Donnelley & Sons Company
Art Studio: Eric G. Hieber E. H. Technical Services

Fundamentals of Chemistry, Second Edition.
Copyright © 1977 by Rod O'Connor

Library of Congress Cataloging in Publication Data

O'Connor, Rod, 1934–
  Fundamentals of chemistry.

  Includes index.
  1. Chemistry. I. Title.
QD31.2.025 1977   540   76-50549
ISBN 0-06-044874-1

Contents

Preface to the Student

"Chemistry, to me, is the most fascinating field of study in all the sciences. Not only is it an area of stimulating intellectual exercise, but . . . ."

That's how I started to write the preface for this book, when I realized that if any of my former students ever read it, they'd think the old man had lost his marbles. Let's be honest. Chemistry is *not* "fascinating," except to chemists, and even then it's mostly a job with only a few really great experiences—when something just goes beautifully. For everybody else it's usually something they *have* to take to get where they want to go.

I guess what I try to do in my classes is to present chemistry in such a way that students are sometimes surprised to find themselves *interested* and are usually surprised that the subject isn't as "impossible" as they thought it would be. That's what I've tried to do with this book, and it's a lot harder to write that way, than to do it "live." I would much rather sit down with you and talk about chemistry—that's when things sparkle—than to write about it, but I hope the book will have some of that flavor.

You may find that this book alone is not sufficient for your learning needs, although, hopefully, it can play a significant role. Many concepts and problem situations require more extensive discussion than any textbook can realistically provide. The extent to which additional discussion is necessary depends, to a large extent, on your own background, interests, and study patterns. Books can't really "teach," because teaching requires some give-and-take communication. This book was designed to be part of a total *learning system* that includes alternative resources, not the least of which is the interaction with an instructor. What this book *can* do is to facilitate your study of chemistry by helping you to identify and to assimilate important information, concepts, and approaches to problem solving. In addition, by means of short introductions and special topic sections in each unit, we can illustrate some of the applications of chemistry to other fields, showing that chemistry can be worth knowing even if you aren't going to be a professional chemist.

The content and level of this book have been designed for the student in science or a science-related field. While presenting an overview of chemistry for those students who will not take additional courses in the field, careful attention has been given to providing important background for additional studies in chemistry. Many students have commented that they have found the previous edition to be useful as a reference during subsequent courses and for review in preparing for various professional school admission exams.

Each unit begins with a set of *objectives* to direct your primary attention to the kinds of things you should learn and be able to do. There is more information in each unit than is required to satisfy these objectives so that you may see some of the background and applications associated with various topics. On the other hand, a single unit does not usually complete the development of a topic. Unless there is some later expansion and reinforcement of early ideas, these ideas tend to "slip away." You will, therefore, see many of the unit topics considered again and again in subsequent units, as they apply to and are amplified by later information. For those areas in which you find special interest or a need for additional study, supplementary readings are suggested at the end of each unit.

Each unit also concludes with a set of exercises designed to offer practice in recalling and applying the information presented in the unit, as identified by the objectives. For those wishing or needing additional practice, an extensive set of problems, *practice for proficiency,* is provided as well. Answers are given in Appendix I so that you can check your work. You may need to refer back to the information in the unit while working the exercises; you should do so freely.

Finally, a *self-test* is provided to permit you to evaluate your progress in satisfying the objectives and to pinpoint areas requiring further study or discussion with your instructor. Self-tests should be taken without referring to any information other than that indicated in the test heading. Answers to self-tests are also given in Appendix I.

The text is divided into *sections,* each containing two or more related *units.* At the end of each *section* is a *section overview test* designed to help you evaluate your comprehension of the broad area covered by the complete set of *units* in that *section.* Answers to these tests are given in Appendix J.

Your own instructor may wish to modify or expand on the objectives, exercises, and self-tests as dictated by the particular goals of your class.

The most efficient procedure for using this text may be summarized as follows:

(1) Read the unit objectives to determine the areas for study emphasis.
(2) Read through the unit once in its entirety.
(3) Participate in lectures, audiovisual programs, or discussions of the topic involved.
(4) Refer back to the objectives and study carefully the information and examples that apply to these guidelines.
(5) Work the exercises, referring back to unit information as often as necessary. Check your answers, then work as many of the problems in the practice for proficiency section as necessary. Discuss any difficulties with your instructor.
(6) Take the self-test without referring to unit information. If you miss any questions, review the appropriate topics carefully and consult with your instructor for additional help.
(7) When you have completed all *units* in a *section,* try the *section overview test* to evaluate your progress and to identify topics still requiring review.

I've been teaching chemistry for a long time. I've had more than 20,000 students. One former student is a well-known professional football player; hundreds are doctors or dentists, nurses or engineers; some are legislators, mechanics, or housewives; others are chemists and teachers. For most of them, chemistry was a set of hurdles to jump in the race for their own particular goal. A teacher can lower the hurdles by covering only the "easy stuff," or a teacher can make the hurdles so high that not many can jump them at all. Neither route appeals to me very much. I like to think that a teacher has the chance to help with a boost over the tougher hurdles. I've tried to do that in my teaching, and I think *Fundamentals of Chemistry, Second Edition,* can remain true to this intention.

This book is the result of working with students as colleagues in learning. It reflects hundreds of their suggestions. I would be most pleased to hear yours.

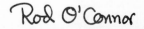
Rod O'Connor

Preface to the Instructor

Fundamentals of Chemistry, Second Edition, reflects my own experience in teaching chemistry to more than 20,000 students, and, to a large extent, it reflects student opinion on what a textbook of chemistry should offer. In earlier editions it has been class tested with many thousands of students and more than a hundred different faculty members. The major modifications made in this edition reflect the suggestions of students, other teachers who have used the book, and several special reviewers.

At the time that I wrote the preliminary version and the first edition, I was the only one using the text and I designed it to fit within my own particular needs, recognizing the extent to which it would be supplemented by lectures, audiovisual programs, and other resources. Now that I am directing a multisection program serving about 5000 students per semester, it is easy to see the need for the textbook to contain a more extensive treatment of many areas and thus to be more useful to other teachers whose approaches differ from my own.

This edition is much more than a minor revision. It is essentially a new book in which I have tried to preserve the better features of the first edition, while rewriting and expanding the more important sections, such as bonding and structure, and reducing the coverage of some areas of lesser importance. The text now contains more material than anyone would probably want to cover in a one-year course, but this is a deliberate effort to provide flexibility of coverage. In our own case, for example, we can now provide some of our sections (such as those serving certain engineering majors) with a fairly extensive coverage of appropriate topics in inorganic chemistry, while spending more time on organic and biochemistry topics with sections serving students in agriculture and other biology-related fields.

The emphasis on objectives and self-tests is in direct response to the obvious need for some way of helping students use their study time more efficiently without the frequent "wheel-spinning" of trying to guess what to study. If you want your students to use these, you will probably need to restrict examination questions to those related to specifications provided by objectives and self-tests. Should you wish to modify the text's learning specifications, you may wish to provide your students with suitable exercises and self-tests to correlate with the new objectives you define for them.

We have had a considerable success with the *learning systems approach* in our own program. More than half of our students now receive A's (90% or better achievement of objectives) or B's, and of these students, the ones going on to further courses in chemistry have demonstrated a superior comprehension of background material. Our program is the first on campus to receive a "commendation for excellence" from the University's Academic Council and student evaluations reflect the growing popularity of the program, while noting that it requires a great deal of work.

Because *Fundamentals of Chemistry, Second Edition,* places a major responsibility on the student, I find that my lectures play a somewhat different role in the learning process than they used to. I spend about the first third of "lecture" time working out exercises and answering questions on the preceding unit, and the remaining time on the development of selected new unit concepts and problem situations—in particular those best illustrated by demonstrations or audiovisual aids. Since the text's approach generates a desire for the student to talk with an instructor periodically, the staff at Texas A&M provides students with a *professor-tutorial* system whereby regular office hours are reserved for students in freshman chemistry. Students can therefore locate a senior faculty member at any hour of the academic day to discuss concepts or problems. The chemistry department also provides an *autotutorial center* with access to the supplementary readings suggested in the text as well as to a variety of audiovisual aids for remedial and enrichment needs.

Although many of the units are interrelated, others are essentially independent and may be omitted easily if not important to the goals of a particular course.

I would like to acknowledge all those of my students and colleagues who have made useful suggestions for this text, but the entire list would need a book of its own. Among the students, I particularly appreciate the suggestions for revisions of objectives, glossary, and appendixes made by Allen Zchiesche and the extensive help in manuscript preparation and proofreading by Mark and Kara O'Connor, both of whom have "suffered" through their father's course. For faculty comments, I am most grateful to Darell Axtell, Barry Barnhart, Ray Bogucki, Roy Caton, Bill Hutton, Paul Javora, Neil Kestner, Stanley Marcus, Ed Mercer, Bill Mooney, Don Nyberg, Larry Peck, Tim Rose, Jerry Sarquis, Spencer Seager, Jacob Seaton, Leonard Spicer, Yi-Noo Tang, Dave Torgerson, Catherine Travaglini, and Harry Zeitlin. In addition, John Woods, Chemistry Editor, and Lois Lombardo, Project Editor, of Harper & Row were of immense help on this project.

Finally, a very special thanks goes to the faculty, staff, and students in the first-year chemistry programs at Texas A&M University. They are, indeed, "the salt of the earth."

If you plan to use this text and would be interested in any of the details of our program, or if you have suggestions for improvement, I would be most pleased to hear from you.

Rod O'Connor

Mathematical skills for introductory chemistry

Students rarely find chemical concepts difficult to master, but the mathematical applications of chemical principles are often troublesome. Sometimes the difficulty lies in the area of mathematical manipulations, which were once familiar but have grown "rusty" through disuse. More frequently, it is the solving of "word problems" that is the major hindrance to learning chemistry.

The level of this text does not require a high degree of mathematical sophistication, but reasonable skill in problem solving is necessary. The following sections will illustrate the type of mathematics required and will enable you to identify those areas requiring some review or further study.

The general areas of competence may be roughly classified as follows:

A. Algebraic manipulations
B. Dimensional conversions (using the "unity factor" method)
C. Significant figures
D. Manipulation of exponents
E. Scientific notation
F. Logarithms

G. Problem solving (converting verbal problems into mathematical formulations for solution)

In addition, some facility with using a slide rule or a simple electronic calculator is highly desirable.

Try the following Pre-Test, and check your answers at the end of this chapter. If you miss any questions, review the topics indicated, then try the Post-Test. If you are still having trouble, consult your instructor for additional help.

PRE-TEST

(Check your answers in the list at the end of the "Math Skills" Review.)

A. Algebraic manipulations

1. $x + 6 = 10$
   $x =$ _____

2. $7 - x = 24$
   $x =$ _____

3. $2x + 8 = 4$
   $x =$ _____

4. $\dfrac{x}{6} = \dfrac{1}{4}$
   $x =$ _____

5. $\dfrac{3}{x} = \dfrac{2}{7}$
   $x =$ _____

6. $\dfrac{5x}{12} = \dfrac{9}{16}$
   $x =$ _____

B. Dimensional conversions

1. 4 ft = _____ in.
2. 264 ft = _____ mile
3. 2 in. = _____ yard
4. 16 tons = _____ lb
5. 15 pints = _____ gal
6. 1 ml = _____ liter
7. 1 cm = _____ m
8. 1 dm = _____ m
9. 1 kg = _____ g
10. 1 megaton = _____ tons
11. 1 in. = _____ cm
12. 1.00 lb = _____ g
13. 1.00 qt = _____ liter
14. 1 cm = _____ Å
15. 212°F = _____ °C
16. 100°C = _____ °K
17. 2.0 ft = _____ dm
18. 5.0 km = _____ miles
19. 125 lb = _____ kg
20. 0.0020 in. = _____ Å
21. 37.0°C = _____ °F
22. −40°F = _____ °K

C. Significant figures

Report each answer with the proper number of significant figures:

1. 0.20 lb × 125 ft = _____ ft lb
2. 0.1364 ft × 12 in./1 ft × 2.54 cm/1 in. = _____ cm
3. 2 × 13.7 mg = _____ mg

D. Manipulation of exponents

1. $10^2 \times 10^5 =$ _____

2. $\dfrac{10^5}{10^2} =$ _____

3. $\dfrac{10^4}{10^7} =$ _____

4. $10^{-2} \times 10^5 =$ _____

5. $\dfrac{10^4}{10^{-7}} =$ _____

6. $(10^4)^3 =$ _____

7. $\dfrac{(10^4)^2}{(10^4)^5} =$ _____

8. $\sqrt{10^{16}} =$ _____

E. Scientific notation

Write each quantity in scientific notation:

1. 125 cm = _____ cm
2. 0.125 cm = _____ cm
3. 1000.0 cm = _____ cm
4. 0.0020 cm = _____ cm
5. 1,000,000 cm = _____ cm

F. Logarithms (base 10) (see log table, Appendix B)

1. log 100 = _____
2. log(0.001) = _____
3. log 2.00 = _____
4. log 20.0 = _____
5. log 0.0200 = _____
6. log(2.0)² = _____
7. log(0.2)² = _____
8. $\log \sqrt{2.0} =$ _____
9. $\log \sqrt[3]{9.0} =$ _____
10. $\log \left(\dfrac{13 \times 0.47}{0.028} \right) =$ _____
11. log _____ = 6.000
12. log _____ = −4.000
13. log _____ = 0.6990
14. log _____ = 3.6990
15. log _____ = −0.6990
16. log _____ = −1.3980

G. Problem solving

1. A gross of assorted nails weighs 24 oz. What is the average weight of a nail in this assortment? _____[1]

2. A manufacturing concern plans to distribute nut/bolt/washer sets. Each set will consist of one 1.85-oz nut, one 3.10-oz bolt, and two 0.550-oz washers. The sets will be sold in 1-gross packages. What will be the net weight of each package? _____

3. What is the maximum weight of nut/bolt/washer sets, as described in Problem 2, that could be prepared from 8.00 lb of nuts, 18.2 lb of bolts, and 12.2 lb of washers? _____

4. The nut/bolt/washer set manufacturer described in Problem 2 finds that his assembly line is not entirely efficient. Some components are lost and some sets are improperly assembled and must be discarded at the inspection site. On the average, when the assembly process begins with 50.0 lb of nuts and the equivalent numbers of bolts and washers, only 138 lb of complete sets pass the final inspection. What percentage of the theoretical production actually passes inspection? _____

---

[1] A gross is 144 units.

REVIEW

The following discussions represent simple pragmatic approaches to the types of mathematical operations most frequently encountered in an introductory chemistry course. They are not intended as mathematical theory, but rather as a review or introduction to the how-to-do-it of solving simple problems.

SECTION A: ALGEBRAIC MANIPULATIONS

The key to solving simple algebraic equations containing a single unknown (e.g., $x + 6 = 10$) is to realize that the equation represents a statement that two quantities are equal. Thus any mathematical operation can be applied to one side of the equation if it is also applied to the other. It is then necessary only to determine what operations are required for the side containing the unknown (to "isolate the unknown") and to apply these operations to both sides of the equation. This principle is illustrated by the following solutions to the problems in part A of the Pre-Test.

Example 1

$$x + 6 = 10$$

Solution
To isolate x, it is necessary to *subtract* 6:

$$(x + 6) - 6 = (10) - 6$$
$$x = 4$$

Example 2

$$7 - x = 24$$

Solution
To isolate x, it is necessary to *subtract 7, then multiply by* -1:

$$(7 - x) - 7 = (24) - 7$$
$$-x = 17$$
$$(-x)(-1) = (17)(-1)$$
$$x = -17$$

Example 3

$$2x + 8 = 4$$

Solution
To isolate x, it is necessary to *subtract 8*, then *divide* by 2:

$$(2x + 8) - 8 = (4) - 8$$
$$2x = -4$$
$$\frac{(2x)}{2} = \frac{(-4)}{2}$$
$$x = -2$$

Example 4

$$\frac{x}{6} = \frac{1}{4}$$

Solution

To isolate x, it is necessary to *multiply* by 6:

$$\left(\frac{x}{6}\right)(6) = \left(\frac{1}{4}\right)(6)$$

$$x = \frac{6}{4}$$

$$x = 1\frac{1}{2}$$

Example 5

$$\frac{3}{x} = \frac{2}{7}$$

Solution

First, convert the unknown term from a denominator to a numerator position (by multiplying by x):

$$\left(\frac{3}{x}\right)x = \left(\frac{2}{7}\right)x$$

$$3 = \frac{2x}{7}$$

Then, to isolate x, it is necessary to *multiply by 7* and *divide by 2*, that is, multiply by $\frac{7}{2}$:

$$(3)\left(\frac{7}{2}\right) = \left(\frac{2x}{7}\right)\left(\frac{7}{2}\right)$$

$$\frac{21}{2} = x$$

$$x = 10.5$$

Example 6

$$\frac{5x}{12} = \frac{9}{16}$$

Solution

To isolate x, it is necessary to *multiply* by 12 and *divide* by 5, that is, multiply by $\frac{12}{5}$:

$$\left(\frac{5x}{12}\right)\left(\frac{12}{5}\right) = \left(\frac{9}{16}\right)\left(\frac{12}{5}\right)$$

$$x = \frac{108}{80}$$

$$x = 1.35$$

Table 1 COMMON CONVERSION FACTORS

ENGLISH-METRIC EQUIVALENTS

EXACT

1 inch (in.) = 2.54 centimeters (cm)

APPROXIMATE (To three-place accuracy)

1.00 pound (lb) = 454 grams (g)
1.00 quart (qt) = 0.946 liters (l)

EXACTLY DEFINED EQUIVALENTS

ENGLISH

1 mile = 5280 feet (ft)
1 yard = 3 feet (ft)
1 foot (ft) = 12 inches (in.)
1 ton = 2000 pounds (lb)
1 pound (lb) = 16 ounces (oz)
1 gallon (gal) = 4 quarts (qt)
1 quart (qt) = 2 pints (pt)
1 pint (pt) = 16 ounces (fluid) (oz)

METRIC

1 cubic centimeter (cm^3) = 1 milliliter (ml)
1 angstrom unit (Å) = 10^{-8} centimeter (cm)
1 micron (μ) = 10^{-6} meter (m)

SECTION B: DIMENSIONAL CONVERSIONS

Table 2 COMMON METRIC SYSTEM PREFIXES

mega = million (10^6)
kilo = thousand (10^3)
deci = tenth (10^{-1})
centi = hundredth (10^{-2})
milli = thousandth (10^{-3})
nano = billionth (10^{-9})

Table 3 TEMPERATURE SCALE CONVERSIONS

$°F = (\frac{9}{5})(°C) + 32°$
$°C = (\frac{5}{9})(°F - 32°)$
$°K = °C + 273°$

Most scientific measurements are expressed in metric system units. Until these units are universally accepted, it is necessary to convert from one measuring system to another. A large number of conversion factors have been tabulated in such sources as *The Handbook of Chemistry and Physics* (Chemical Rubber Company), but most conversions required in an introductory study of chemistry can be made by using the simple factors given in Table 1 and the metric system prefixes in Table 2. Conversions among the Celsius (centigrade), Fahrenheit, and Kelvin ("absolute") temperature scales are given in Table 3. For additional units and conversions, including the International System (SI) of Units, see Appendix A.

Any quantity can be multiplied by unity (one) without changing its value. A *unity factor* employs this to change the dimensions of a quantity without changing the *value* of the quantity, although the *numbers* used to express the value may be changed. If someone asked you the number of inches in 2 yards, you would reply, 72. The *value* of the length was not changed; that is, it is the same length whether you call it 2 yards or 72 inches (or 6 feet). You probably made this conversion almost intuitively. All the unity factor method does is to provide a systemtic procedure for making such conversions in cases a bit too complicated for the "intuitive approach."

Since any fraction whose numerator and denominator are *equivalent* has the value of unity, you can make unity factors from any dimensional equivalents, such as those given in Table 1 or implied by the prefix meanings of Table 2, for example,

Equivalents	Corresponding unity factors
1 in. = 2.54 cm	$\dfrac{1 \text{ in.}}{2.54 \text{ cm}}$ or $\dfrac{2.54 \text{ cm}}{1 \text{ in.}}$
1 mile = 5280 ft	$\dfrac{1 \text{ mile}}{5280 \text{ ft}}$ or $\dfrac{5280 \text{ ft}}{1 \text{ mile}}$
1 kg = 1000 g (kilo = thousand)	$\dfrac{1 \text{ kg}}{1000 \text{ g}}$ or $\dfrac{1000 \text{ g}}{1 \text{ kg}}$

Unity factors are valuable in problems of dimensional conversion and, as we shall see later, in a variety of other types of problem situations. The method simply requires setting up the problem in fractional form, using dimensions for all quantities involved. The quantity to be converted is multiplied by appropriate unity factors until all dimensions can be canceled except for those desired in the answer. In multistep conversions it is well to have a systematic sequence; typically, numerator dimensions are converted first, then denominator units.

Example 7

Convert 60 miles per hour (60 miles hr^{-1}) to feet per second. (Remember that "per" means "divide by.")

Stepwise Solution

$$\frac{60 \text{ miles}}{1 \text{ hr}} = \underline{\hspace{1cm}} \text{ ft sec}^{-1}$$

First, convert the numerator unit (miles $\longrightarrow$ feet), using a unity factor expressing the feet-miles equivalence, in a form such that the "miles" will cancel:

$$\frac{60 \text{ \sout{miles}}}{1 \text{ hr}} \times \frac{5280 \text{ ft}}{1 \text{ \sout{mile}}} = \frac{(60 \times 5280) \text{ ft}}{1 \text{ hr}}$$

Next, convert the denominator unit (hour $\longrightarrow$ seconds), using a set of unity factors expressing hour $\longrightarrow$ minute $\longrightarrow$ second equivalences, in proper form for dimensional cancellation:

$$\frac{(60 \times 5280) \text{ ft}}{1 \text{ \sout{hr}}} \times \frac{1 \text{ \sout{hr}}}{60 \text{ \sout{min}}} \times \frac{1 \text{ \sout{min}}}{60 \text{ sec}} = \frac{(60 \times 5280) \text{ ft}}{(60 \times 60) \text{ sec}}$$

Combined Solution

$$\frac{60 \text{ \sout{miles}}}{\text{\sout{hr}}} \times \frac{5280 \text{ ft}}{\text{\sout{mile}}} \times \frac{1 \text{ \sout{hr}}}{60 \text{ \sout{min}}} \times \frac{1 \text{ \sout{min}}}{60 \text{ sec}} = \frac{60 \times 5280 \text{ ft}}{60 \times 60 \text{ sec}} = 88 \text{ ft sec}^{-1}$$

Note that if any factors had been omitted or included "upside down," the necessary cancellation of units would not have worked.

Once the dimensions desired for the answer are obtained, the numbers are placed in scientific notation for calculation of the final answer. *It is important that an "estimation" step be included* as a check on the reasonableness of an answer. The estimation step involves rounding all numerical quantities to the nearest whole number and performing indicated operations to determine the order of magnitude expected for the final answer.

Example 8

Convert 62.4 pounds per cubic foot (62.4 lb ft^{-3}) to units of grams per cubic centimeter.

Solution
First, convert all dimensions:

$$\frac{62.4 \text{ lb}}{1 \text{ ft}^3} \times \frac{454 \text{ g}}{1.00 \text{ lb}} \times \frac{1 \text{ ft}^3}{(12 \text{ in.})^3} \times \frac{(1 \text{ in.})^3}{(2.54 \text{ cm})^3} = \frac{(62.4 \times 454) \text{ g}}{[(12)^3 \times (2.54)^3] \text{ cm}^3}$$

Second, convert to scientific notation (see Section E):

$$\frac{(6.24 \times 10^1 \times 4.54 \times 10^2)\,g}{[(1.2 \times 10^1)^3 \times (2.54)^3]\,cm^3} \equiv \frac{(6.24 \times 4.54 \times 10^3)\,g}{[(1.2)^3 \times (2.54)^3 \times 10^3]\,cm^3}$$

Third, estimate the order of magnitude:

$$6.24 \approx 6$$
$$4.54 \approx 5$$
$$1.2 \approx 1$$
$$2.54 \approx 3$$

$$\frac{6 \times 5 \times 10^3}{(1)^3 \times (3)^3 \times 10^3} = \frac{30 \times 10^3}{27 \times 10^3} = \sim 1$$

Remember that 1 ft³/12 in.³ is *exact* and does not affect significant figures.

Hence, the final answer should be closer to 1 than to, for example, 0.1 or 10.

Finally, perform the calculation (e.g., using a slide rule):

$$\frac{(6.24 \times 4.54 \times 10^3)\,g}{[(1.2)^3 \times (2.54)^3 \times 10^3]\,cm^3} = 1.00\ g\ cm^{-3} \quad \text{(to three-place accuracy)}$$

SECTION C:
SIGNIFICANT FIGURES

The numerical value of a measured quantity is limited by both the precision and the accuracy of the measurements involved. Precision is concerned with how close together a series of measurements of the same quantity can be made, whereas accuracy is determined by how close a measured value is to the "true" value (usually compared to an exactly defined standard).

For example, suppose someone wishes to know the time at which an event took place. He might consult his wristwatch. However, if for some reason he wants to be more certain of the time, he might check his watch against his alarm clock or even use the telephone to dial "time and temperature." The chances are that he would arrive at three different values, and he might choose to average these values. However, if the three values were widely different (poor precision), he might wish to consult a number of other timepieces. If he can obtain several independent readings that are relatively close together, then he feels a certain confidence in accepting the average value. If the knowledge of the time were really important, then he might telephone the National Bureau of Standards and set ("calibrate") his watch against an accepted standard. This procedure would guarantee an improvement in *accuracy,* whereas any improvement in *precision* alone would simply increase the confidence in the measured value. Ten very close readings from ten different electric clocks during a power failure would give good precision, but poor accuracy.

There are two types of errors: *systematic* and *random.* Systematic errors may result from incorrect calibration of instruments, from incorrect instrument design or manufacture, or from use of a method not well suited to the problem. Such errors affect all measurements the same way so that results may be precise, but inaccurate. Systematic errors can often be minimized by improving experimental conditions or compensated for by appropriate "correction factors." Random errors result from the difficulty of exactly repeating a measuring procedure. Even though skill and practice can reduce random error, it is not possible to eliminate it completely. In some cases, random error occurs for reasons beyond the control of the experimenter, as in a line voltage fluctuation during the use of a sensitive electrical measuring device.

Some comparisons of precision and accuracy are represented schematically in the following diagram.

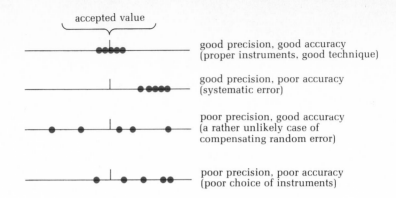

accepted value

good precision, good accuracy
(proper instruments, good technique)

good precision, poor accuracy
(systematic error)

poor precision, good accuracy
(a rather unlikely case of
compensating random error)

poor precision, poor accuracy
(poor choice of instruments)

Every result of a series of measurements should be reported in such a way as to indicate some idea of the error involved. There are various ways of doing this, some rather exotic. We will limit our work to reporting either the *range* or the average result with its *average deviation*. The average result is obtained by summing the individual results and dividing this sum by the number of individual values, for example for a series of five weighings with an analytical balance

1st weight	1.6752
2nd weight	1.6748
3rd weight	1.6754
4th weight	1.6747
5th weight	1.6753
sum:	8.3754

$$\text{average weight} = \frac{8.3754}{5} = 1.67508 \qquad \text{(rounded to 1.6751)}$$

The average deviation is determined by first finding the differences between each individual value and the average value, then dividing the sum of these differences (irrespective of algebraic sign) by the number of individual values. For example, for the preceding weighings,

$$1.6752 - 1.6751 = +0.0001$$
$$1.6748 - 1.6751 = -0.0003$$
$$1.6754 - 1.6751 = +0.0003$$
$$1.6747 - 1.6751 = -0.0004$$
$$1.6753 - 1.6751 = +0.0002$$

sum of differences: 0.0013

$$\text{average deviation} = \frac{0.0013}{5} = 0.00026 \qquad \text{(rounded to 0.0003)}$$

The result is then reported as

$$1.6751 \pm 0.0003 \text{ g}$$

All measured or calculated experimental results should indicate the error involved, as by the use of an average deviation. Results should be reported in such a way that the last figure given in the average value corresponds to the magnitude of "doubt" expressed by the average deviation. A result reported in this way is said to contain only *significant figures* (that is, those having experimental validity). For example, a weight obtained with an average deviation of 0.0003 g cannot be mean-

If a calculation yields more figures than are significant (as discussed later), the result is rounded to the proper number of significant figures. When the first nonsignificant digit is 0 through 4, omit it and all following digits. When the first nonsignificant digit is 5 through 9, add one unit to the preceding digit. For example,

$$\frac{1.6751}{5} = 0.33502$$

(rounded to 0.3350)

$$\frac{1.6751}{2} = 0.83755$$

(rounded to 0.8376)

When no average deviation is given, the last figure reported is assumed to be in doubt by ± 1 unit.

ingfully reported to 0.01-mg (0.00001 g). There is already a range of six units in the 0.1-mg place (from 1.6748 to 1.6754). An added figure in the 0.01-mg position does not add any *meaningful* information.

incorrect: 1.67508 ± 0.0003 g
correct: 1.6751 ± 0.0003 g

There is generally no problem in reporting the proper number of significant figures for the direct result of a series of experimental measurements. Additional figures sometimes appear, however, when calculations are performed, especially with one of the now popular "pocket electronic calculators." Although pure numbers (dimensionless) and exactly defined conversion factors (e.g., 1 ft = 12 in. or 1 in. = 2.54 cm) do not limit the number of meaningful figures in a computed quantity, other factors do. There is a method for determining the proper number of significant figures for any particular situation. It is, however, more sophisticated and complex than we really need at the level of our work. If we remember that no *mathematical operation* can increase the accuracy of an experimental result, we can get by with two simple rules:

1. If the operation involves multiplication or division, or both, round the result to the number of significant figures in the least accurately known quantity used; for example,

$$\frac{1.6751 \text{ g}}{0.332 \text{ ml}} = 5.0454819 \text{ g ml}^{-1} \qquad \text{(rounded to 5.05 g ml}^{-1}\text{)}$$
(calculator result)

2. If the operation involves only addition or subtraction, or both, round the result to the *decimal place* of the least accurately known quantity used; for example,

```
  16.5
  21.2
  68.34
  40.1
  10.327
 156.467    (rounded to 156.5)
```

Example 9

The diameter of a cylinder is estimated, using a yardstick, to be 1.1 in. The diameter of a second cylinder is measured carefully with a calibrated micrometer and found to be (0.8517 ± 0.0002) in. What value can meaningfully be reported as the sum of the diameters of the two cylinders?

Solution

```
 1.1
 0.8517
 1.9517
```

But the less accurately known quantity is measured only to the nearest 0.1 in., so the sum can be reported only as approximately 2.0 in., when properly rounded off.

Example 10

See Section D if you are unfamiliar with handling exponents.

How many samples of 25.0 mg each could be prepared from approximately 2 lb of a crystalline salt?

Solution

$$\frac{1 \text{ sample}}{25.0 \text{ mg}} \times \frac{10^3 \text{ mg}}{1 \text{ g}} \times \frac{454 \text{ g}}{1 \text{ lb}} \times \frac{2 \text{ lb}}{1} = \underline{\hspace{1cm}} \text{ samples}$$

Estimating:

$$\frac{10^3 \times 5 \times 10^2 \times 2}{2.5 \times 10^1} = 4 \times 10^4$$

Solving:

$$\frac{10^3 \times 4.54 \times 10^2 \times 2}{2.50 \times 10^1} = 3.63 \times 10^4$$

However, the number of significant figures is limited by the measured quantity "approximately 2 lb" (*note:* not by the exact figures "1 sample," "10^3 mg," "1 g," or "1 lb"). Thus the answer can be reported only as "approximately 4×10^4 samples."

The proper use of scientific notation clearly indicates the number of significant figures. (See also Sections D and E.)

Suggestion: To convince yourself that the solution to Example 10 is correct, remember that "approximately 2 lb" would generally indicate a weight between 1.5 and 2.4 lb (i.e., values in this range would "round off" to 2 lb). Calculate the number of samples if 1.5 lb were available and if 2.4 lb were available. Then explain why only one significant figure was reasonable for this example.

SECTION D: MANIPULATION OF EXPONENTS

Scientific calculations are frequently handled most conveniently by expressing quantities in scientific notation (see Section E). Such operations require simple manipulation of exponents, usually exponents of 10. When the same base (e.g., 10) is involved, the following rules apply.

1. When the operation involves multiplication, add the exponents algebraically; for example,

$$10^3 \times 10^2 = 10^{(3+2)} = 10^5$$
$$10^4 \times 10^8 \times 10^{-5} = 10^{[4+8+(-5)]} = 10^7$$

2. When the operation involves division, subtract the divisor exponent from the numerator exponent; for example,

$$\frac{10^5}{10^3} = 10^{(5-3)} = 10^2$$

$$\frac{10^7}{10^{11}} = 10^{(7-11)} = 10^{-4}$$

$$\frac{10^8}{10^{-2}} = 10^{[8-(-2)]} = 10^{10}$$

$$\frac{10^6 \times 10^4}{10^3 \times 10^9} = 10^{[6+4-(3+9)]} = 10^{-2}$$

3. When the operation involves roots or powers, divide the exponent by the root

To take "roots" for cases in which the exponent is not a multiple of the "root number" use logarithms (see Section F).

number or multiply the exponent by the power number, respectively; for example,

$$\sqrt{10^4} = (10^4)^{1/2} = 10^{(4/2)} = 10^2$$

$$\sqrt[5]{10^{20}} = (10^{20})^{1/5} = 10^{(20/5)} = 10^4$$

$$(10^2)^3 = 10^{(2 \times 3)} = 10^6$$

$$(10^{-7})^4 = 10^{(-7 \times 4)} = 10^{-28}$$

The operations described apply *only* to cases in which the same base number is involved. To illustrate this point, show that $2^3 \times 10^2$ is *not* equal to $(2 \times 10)^5$.

**SECTION E:
SCIENTIFIC NOTATION**

It is inconvenient to work with the standard form of very large or very small numbers. An exponential notation (see also Section D) is commonly employed to simplify calculations involving such numbers. The procedure for converting from standard form to scientific (exponential) notation is as follows:

In scientific notation the only zeros recorded are those that represent actual measured quantities. Zeros used in standard form merely to indicate numerical magnitude are omitted.

STEP 1 Reset the decimal point so that there is a single digit (not zero) to its left, for example,

0.0004.4

STEP 2 Count the number of digits between the original and new decimal point positions to determine the magnitude of the exponent of 10, for example,

123④
0.0004.4
10^4

STEP 3 Determine the sign for the exponent of 10 by the direction *from* the new *to* the original decimal point position (left is negative, right is positive); for example,

0.0004.4
10^{-4}

Thus $0.00044 = 4.4 \times 10^{-4}$.

Example 11

Assume that the approximate diameter of the solar system, to the nearest million miles, is 7,300,000,000. Write this in scientific notation.

Solution

STEP 1

7,.300,000,000. (understood)

STEP 2

$$1\ 23\ 4\ 56\ 78\ \textcircled{9}$$
$$7.300,000,00\ 0.$$
$$\uparrow\uparrow\uparrow\uparrow\uparrow\uparrow\uparrow\ \uparrow$$
$$10^9$$

STEP 3

$$7.300,000,000$$

(direction indicates *positive* exponent)

Since the measured quantity is known only to the nearest million, the last six zeros (indicating order of magnitude) are omitted. Thus 7,300,000,000 miles (to the nearest million) is written as 7.300×10^9 miles.

Example 12

Write 0.00020 in. in scientific notation.

Solution

STEP 1

$$0.0002.0$$

STEP 2

$$123\textcircled{4}$$
$$0.0002.0$$
$$\uparrow\uparrow\uparrow\ \uparrow$$
$$10^4$$

STEP 3

$$0.0002.0$$

(direction indicates *negative* exponent)

The four zeros preceding the 2 are not measured quantities, but only represent order of magnitude. The final zero does indicate a measurement to the nearest 0.00001 in. Thus 0.00020 in. is written as 2.0×10^{-4} in.

SECTION F: LOGARITHMS

Chemical calculations normally use base 10 logarithms. These logarithms are exponents of 10 and, as such, are manipulated as described in Section D. Logarithms consist of a whole number (characteristic) followed by a decimal number (mantissa) or zero. The mantissa is found from a logarithm table, such as the one in Appendix B.

With the advent of the inexpensive electronic calculator, it is unlikely that you will use logarithms in multiplication or division operations. However, we will encounter cases, such as the Nernst equation (Unit 24) and the pH scale (Unit 19), in which logarithms must be employed. It is therefore essential that you be able to use a logarithm table properly. The following discussion outlines the basic steps for working with base-10 logarithms.

1. To find the logarithm of a number:
   a. Write the number in scientific notation, that is, as the product of some number and some power of 10. For example,

   $$343 = 3.43 \times 10^2$$

Note that *mantissas* would be the same for numbers such as 3.43, 34.3, and 343, but *characteristics* would differ.

   b. Look up the logarithm (mantissa) of the number preceding the power of 10 in a log table. (Note: All such mantissas are decimal numbers, although most log tables do not show the decimal point.) For example,

   $$\log 3.43 = 0.5353$$

   c. Add this to the exponent of 10. For example,

   $$\log 343 = 0.5353 + 2 = 2.5353$$

In some cases the log will be a negative number. *This is perfectly legitimate,* although other forms are more frequently used. Operational uses of logs in many chemical calculations are simpler using negative logs. For example,

$$\log 0.0200 = ?$$

STEP a: $0.0200 = 2.00 \times 10^{-2}$

STEP b: $\log 2.00 = 0.3010$

STEP c: $\log 0.0200 = 0.3010 + (-2) = -1.6990$
         (entire logarithm is negative)

Technically referred to as the antilog.

2. To find a number* from its logarithm:
   a. Be sure the log is written in a form such that the decimal part (mantissa) is positive. (That is how the logs are conventionally tabulated.) For example, what number has the logarithm -4.3010?
   Add the log to the next largest whole number, then subtract that number. For example,

   $$(-4.3010 + 5) - 5 = 0.6990 - 5 \quad \text{(to be *used* in this form)}$$

If any of the numbers in the calculation are measured quantities or other than exactly defined conversion units, round off the final answer to the proper number of significant figures (see Section C).

   b. Look up the mantissa in the logarithm part of the table and find the number to which it most closely corresponds. For example, the number 5.00 corresponds to 0.6990 in the mantissa part of the table.
   c. Write this number as the first part of a scientific notation. Then use the whole number in the log as the exponent of 10. For example,

   0.6990 $- 5$

   from log table

   5.00 $\times 10^{-5}$

If the original log is positive, follow steps b and c above. For example, what number has the logarithm 11.0828?

11.**0828**

from log table

1.21 $\times 10^{11}$

3. To multiply or divide, using logs:

   a. Write the operation in fractional form. For example,

   $$\frac{0.0071 \times 148 \times 2.73}{8.1 \times 6.02 \times 10^{23}}$$

   b. Write all numbers in scientific notation. For example,

   $$\frac{(7.1 \times 10^{-3}) \times (1.48 \times 10^2) \times 2.73}{8.1 \times 6.02 \times 10^{23}}$$

   c. Combine all powers of 10 into a single power of 10 to appear in the numerator (see Section D). For example,

   $$\frac{7.1 \times 1.48 \times 2.73}{8.1 \times 6.02} \times 10^{(-3+2-23)} = \frac{7.1 \times 1.48 \times 2.73}{8.1 \times 6.02} \times 10^{-24}$$

   d. Obtain the sum of the logs of all numbers in the numerator. Subtract from this the sum of the logs of all numbers in the denominator. For example,

Numerator	Denominator	
log 7.1 = 0.8513	log 8.1 = 0.9085	1.4578
log 1.48 = 0.1703	log 6.02 = 0.7796	−1.6881
log 2.73 = 0.4362	sum: 1.6881	−0.2303
sum: 1.4578		

   e. Find the number that has this logarithm (Section F, 2). Combine it with the power of 10 found in step c above. For example,

   $$(-.2303 + 1) - 1 = 0.7697 - 1$$

   $$\underset{\text{5.88} \times 10^{-1}}{\overset{\textbf{0.7697} - 1}{\text{from log table}}}$$

   Then

   (step c)

   $$5.88 \times 10^{-1} \times 10^{-24} = 5.88 \times 10^{-25} \qquad (5.9 \times 10^{-25} \text{ to two place accuracy})$$

4. To find roots and powers, using logs: Since logs are exponents of 10, roots and powers are found in the same way as in any other exponential situation (Section D).

   Roots

   $$\sqrt[3]{1728} = ?$$

   a. Find the log of the number.

   $$\log(1728) = \log(1.728 \times 10^3) = 3.2375$$

   b. Divide the log by the "root number."

   $$\log \sqrt[3]{1728} = \tfrac{1}{3}(3.2375) = 1.0792$$

   c. Find the antilog (the number that has this logarithm).

   $$\sqrt[3]{1728} = \text{antilog } 1.0792 = 1.2 \times 10^1$$

Powers

$(25)^2 = ?$

a. Find the log of the number.

$\log(25) = \log(2.5 \times 10^1) = 1.3979$

b. Multiply the log by the power number.

$\log(25)^2 = 2 \times 1.3979 = 2.7958$

c. Find the antilog.

$(25)^2 = \text{antilog } 2.7958 = 6.25 \times 10^2$

SECTION G:
PROBLEM SOLVING

Some people seem to have an intuitive grasp of how to approach a problem to obtain a rapid and efficient solution. Although there is little doubt that some are more gifted than others in this respect, it is a talent that can be acquired. Proficiency in this skill, as in all skills, improves with practice.

When a mechanically inept customer takes his car to a mechanic with the complaint that it is getting poor gas mileage and makes a "funny rumbling sound," this poses a problem for the mechanic. Although the talented mechanic might not express it this way, his approach to this problem involves certain distinguishable phases:

1. Identification of what he wants to find out.
2. Recognition of what information is given that might be useful.
3. Identification of other information needed, but not given (i.e., information that must be obtained from memory and from other sources).
4. Selection of a method of attack on the problem.
5. Manipulation of information and tools to achieve a solution.
6. Checking that the solution is probably satisfactory.

For the mechanic, these steps might involve:

1. Find out what is causing the poor gas mileage and unusual engine noise.
2. Given: poor mileage + "funny rumbling sound."
3. Other information: memory of previous experience with similar problems + application of various suitable test instruments and other pertinent observations.
4. Selecting parts to be replaced, adjustments to make, and tools to be used.
5. Fixing the engine.
6. Test driving.

This system is a general one for solving all kinds of problems. It is the basis for most scientific research, and it is the key to solving all sorts of mathematical problems. The more problems you solve, the easier it becomes, until you, like the competent mechanic, can solve problems without formally thinking out each step of the method. Until the stage of "intuitive" proficiency is reached, however, it is worthwhile to proceed slowly and systematically through the various phases of problem solving.

Let us illustrate the approach by a "case study" of the first three problems given in Section G of the Pre-Test.

Problem 1

A gross of assorted nails weighs 24 oz. What is the average weight of a nail in this assortment?

Solution

STEP 1 Find *weight per nail* (in ounces)

STEP 2 Given *24 oz per gross* (of nails)

STEP 3 Needed, not given: *How many items per gross?*

Answer (from memory): 144 per gross

STEP 4 Method of attack
Divide weight by number to find average weight of each. Use unity factor method.

STEP 5 Manipulation

$$\frac{24 \text{ oz}}{\text{gross}} \times \frac{1 \text{ gross}}{144} = \frac{24 \text{ oz}}{144}$$

Estimate

$$\frac{2.4 \times 10^1}{1.44 \times 10^2} \approx \frac{2 \times 10^1}{1 \times 10^2} \qquad (\sim 0.2)$$

Solve

$$\frac{(2.4 \times 10^1) \text{ oz}}{1.44 \times 10^2} = 1.7 \times 10^{-1} \text{ oz} \qquad (0.17 \text{ oz})$$

STEP 6 Check

a. Dimensions canceled properly.
b. Math check:

$$0.17 \text{ oz} \times 144 = \sim 24 \text{ oz}$$

Problem 2

A manufacturing concern plans to distribute nut/bolt/washer sets. Each set will consist of one 1.85-oz. nut, one 3.10-oz bolt, and two 0.550-oz washers. The sets will be sold in 1-gross packages. What will be the net weight of each package?

Solution

STEP 1 Find *net weight of a 1-gross package of the sets.*

STEP 2 Given 1 nut/set @ 1.85 oz/nut; 1 bolt/set @ 3.10 oz/bolt; 2 washers/set @ 0.550 oz/washer; 1 gross per package.

STEP 3 Needed, not given:
How many items per gross?

Answer (from memory): 144 per gross

What does net weight mean?
Answer (from memory): weight of package content (i.e., 1 gross of the sets)
STEP 4 Method of attack

Find the weight of each set and multiply by the number of sets per package. Use unity factor method where appropriate.

(*Note:* Often there is no unique method of attack. Here, for example, we might have decided to find the weights of 1 gross of nuts, 1 gross of bolts, and 2 gross of washers and then added these together.)

STEP 5 Manipulation

$$\left(\frac{1.85 \text{ oz}}{1 \text{ nut}} \times \frac{1 \text{ nut}}{1 \text{ set}}\right) + \left(\frac{3.10 \text{ oz}}{1 \text{ bolt}} \times \frac{1 \text{ bolt}}{1 \text{ set}}\right) + \left(\frac{0.550 \text{ oz}}{1 \text{ washer}} \times \frac{2 \text{ washers}}{1 \text{ set}}\right)$$

$$= \frac{[1.85 + 3.10 + (2)(0.550)] \text{ oz}}{1 \text{ set}}$$

$$= \frac{6.05 \text{ oz}}{1 \text{ set}}$$

$$\frac{6.05 \text{ oz}}{1 \text{ set}} \times \frac{1 \text{ gross set}}{1 \text{ pkg}} \times \frac{144}{1 \text{ gross}} = \frac{(6.05 \times 144) \text{ oz}}{1 \text{ pkg}}$$

Estimate

$$6.05 \times 1.44 \times 10^2 \approx 6 \times 1 \times 10^2 \qquad (\sim 600)$$

Solve

$$\frac{(6.05 \times 1.44 \times 10^2) \text{ oz}}{1 \text{ pkg}} = 8.71 \times 10^2 \text{ oz/pkg} \qquad (871 \text{ oz/pkg})$$

STEP 6 Check
a. Dimensions canceled properly.
b. Arithmetic rechecked for accuracy.

Problem 3

What is the maximum weight of nut/bolt/washer sets, as described in Problem 2, that could be prepared from 8.00 lb of nuts, 18.2 lb of bolts, and 12.2 lb of washers?

Solution

STEP 1 Find *maximum weight of sets from given weights of components.*

STEP 2 Given *8.00 lb of nuts, 18.2 lb of bolts, 12.2 lb of washers*

STEP 3 Needed, not given:

Composition of a set
Answer: (Given in Problem 2)

1 nut + 1 bolt + 2 washers

Weights of components
Answer: (Given in Problem 2)

1.85 oz/nut, 3.10 oz/bolt, 0.550 oz/washer

Weight of a set
Answer: (Found during solution to Problem 2)

6.05 oz/set

STEP 4 Method of attack

The word "maximum" in the problem suggests that there is some limitation other than just the sum of the weights of the components given. It must be recognized that we can no longer make sets when we run out of any one component. Thus the *maximum* weight of sets will correspond to the weight obtainable from the limiting component (i.e., the smallest of the weights of sets obtainable from the weights of each component available). Since we are dealing with *weights* (rather than numbers), some appropriate equivalents would be useful in constructing unity factors.

$$1.85 \text{ oz nuts} = 6.05 \text{ oz sets}$$
$$3.10 \text{ oz bolts} = 6.05 \text{ oz sets}$$
$$(2 \times 0.550) \text{ oz washers} = 6.05 \text{ oz sets}$$

Parts-by-weight unity factors:

$$\frac{6.05 \text{ oz sets}}{1.85 \text{ oz nuts}} = \frac{6.05 \text{ sets}}{1.85 \text{ nuts}} \quad \text{(by weight)}$$

$$\frac{6.05 \text{ oz sets}}{3.10 \text{ oz bolts}} = \frac{6.05 \text{ sets}}{3.10 \text{ bolts}} \quad \text{(by weight)}$$

$$\frac{6.05 \text{ oz sets}}{1.10 \text{ oz washers}} = \frac{6.05 \text{ sets}}{1.10 \text{ washers}} \quad \text{(by weight)}$$

Then we must calculate the weight of sets obtainable by using up all of each component available and select the smallest of these values as the limit on the weight of sets obtainable.

STEP 5 Manipulation

From nuts

$$\frac{6.05 \text{ sets}}{1.85 \text{ nuts}} \times \frac{8.00 \text{ lb nuts}}{1} = 26.2 \text{ lb sets}$$

From bolts

$$\frac{6.05 \text{ sets}}{3.10 \text{ bolts}} \times \frac{18.2 \text{ lb bolts}}{1} = 35.5 \text{ lb sets}$$

From washers

$$\frac{6.05 \text{ sets}}{1.10 \text{ washers}} \times \frac{12.2 \text{ lb washers}}{1} = 67.1 \text{ lb sets}$$

Since our calculations show that we shall run out of nuts when 26.2 lb of sets have been constructed, the maximum weight of sets obtainable is 26.2 lb.

STEP 6 Check

a. Dimensions cancel.
b. Arithmetic rechecked for accuracy.
c. Logic rechecked.

Each new problem poses its own unique challenges, and there is no "magic formula" for problem solving. The stepwise approach suggested should prove useful in establishing a systematic technique. Facility will improve with practice.

POST-TEST

(Answers to Post-Test follow the Pre-Test answers at the end of this section.)

A. Algebraic manipulations

1. $x - 5 = 9$
$x =$ _____

2. $17 - x = 4$
$x =$ _____

3. $3x + 2 = -11$
$x =$ _____

4. $\dfrac{x}{5} = \dfrac{2}{9}$
$x =$ _____

5. $\dfrac{7}{x} = \dfrac{28}{9}$
$x =$ _____

6. $\dfrac{3x}{7} = \dfrac{9}{5}$
$x =$ _____

B. Dimensional conversions

1. 9.00 in. = _____ cm
2. 0.122 qt = _____ ml
3. 196 kg = _____ ton
4. 1.00 g cm^{-3} = _____ lb ft^{-3}
5. $-40°F$ = _____ °C
6. 70°C = _____ °K

C. Significant figures

Report each answer with the proper number of significant figures:

1. $\dfrac{12.72 \text{ g}}{3.0 \text{ cm}^3}$ = _____ g cm^{-3}

2. 0.0050 mile × 5280 ft mile^{-1} = _____ ft
3. 5 × 90.800 g × 1.00 lb/454 g = _____ lb

D. Manipulation of exponents

1. $10^3 \times 10^5 \times 10^{-1}$ = _____
2. $\dfrac{10^2}{10^8}$ = _____

3. $\dfrac{10^{-3}}{10^{-7}}$ = _____
4. $(10^{-3})^5$ = _____

5. $\sqrt[3]{10^6}$ = _____
6. $\dfrac{10^{-7} \times 10^4}{10^{-2} \times 10^5}$ = _____

E. Scientific notation

Write each quantity in scientific notation:

1. 1063 cm = _____ cm
2. 0.017 cm = _____ cm
3. 195.0 cm = _____ cm
4. 0.020 cm = _____ cm

F. Logarithms (base 10)

1. log 264 = _____
2. log 0.264 = _____ or _____
3. $\log(26.4)^3$ = _____
4. log $\sqrt[3]{0.0264}$ = _____ or _____
5. log _____ = 4.9030
6. log _____ = -5.6020

G. Problem solving

1. If three dozen eggs cost $1.70, what is the cost per egg, in cents? _____

2. A particular solvent mixture requires three parts acetic acid, two parts water, and two parts n-butyl alcohol (by weight). What weight of acetic acid is required for preparation of 140 g of the mixture? _____

3. The speed of light is approximately 3.0×10^{10} cm sec^{-1} through a vacuum. How long, in minutes, would it take for light traveling at this speed to go a distance of 193,000 miles? _____

4. Baking soda can be prepared from table salt, ammonia, and carbon dioxide. If experience shows that in a particular plant operation 1.00 g of table salt will produce 1.39 g of baking soda, using appropriate quantities of ammonia and carbon dioxide, what weight, in pounds, of table salt would be needed to form 5.00 lb of baking soda? _____

ANSWERS TO PRE-TEST

A. 1. 4
2. -17
3. -2
4. 1.5
5. 10.5
6. 1.35

If you missed any, review Section A and then try part A on the Post-Test.

B. 1. 48
2. 0.05
3. 0.0556
4. 32,000
5. 1.875
6. 0.001
7. 0.01
8. 0.1
9. 1000
10. 1,000,000
11. 2.54
12. 454
13. 0.946
14. 10^8
15. 100
16. 373
17. 6.1
18. 3.1
19. 56.8
20. 5.1×10^5
21. 98.6
22. 233

If you missed any, review Section B and then try part B on the Post-Test.

C. 1. 25 (limited by two significant figures in 0.20 lb.)
2. 4.157 (limited by the measured quantity 0.1364 ft, not by the exact equivalents 2.54 cm/1 in. or 12 in./1 ft.)
3. 27.4 (not limited by the pure number 2)

If you missed any, review Section C and then try part C on the Post-Test.

D. 1. 10^7
2. 10^3
3. 10^{-3}
4. 10^3
5. 10^{11}
6. 10^{12}
7. 10^{-12}
8. 10^8

If you missed any, review Section D and then try part D on the Post-Test.

E. 1. 1.25×10^2
2. 1.25×10^{-1}
3. 1.0000×10^3
4. 2.0×10^{-3}

5. 10^6 or 1.000000×10^6 (number of significant figures uncertain)

If you missed any, review Section E and then try part E on the Post-Test.

F. **1.** 2   **2.** -3
    3. 0.3010   **4.** 1.3010
    5. -1.6990 (or $\overline{2}.3010$)   **6.** 0.6020
    7. -1.3980 (or $\overline{2}.6020$)   **8.** 0.1505
    9. 0.3181
    10. $[1.1139 + (-0.3279) - (-1.5528) = 2.3388]$

11. 10^6 **12.** 10^{-4}
13. 5.00   **14.** 5.00×10^3 (5000)
15. 2.00×10^{-1} (0.200)   **16.** 4.00×10^{-2} (0.0400)

If you missed any, review Section F and then try part F on the Post-Test.

G. **1.** 0.17 oz   **2.** 871 oz/pkg
    3. 26.2 lb   **4.** 84.5%

If you missed any, review Section G and then try part G on the Post-Test.

ANSWERS TO POST-TEST

A. 1. 14   **2.** 13
    3. -4.33 **4.** 1.11
    5. 2.25   **6.** 4.2
B. 1. 22.9   **2.** 115
    3. 0.216   **4.** 62.4
    5. -40 **6.** 343
C. 1. 4.2   **2.** 26
    3. 1.00
D. 1. 10^7 **2.** 10^{-6}
    3. 10^4 **4.** 10^{-15}
    5. 10^2 **6.** 10^{-6}

E. 1. 1.063×10^3 **2.** 1.7×10^{-2}
    3. 1.950×10^2 **4.** 2.0×10^{-2}
F. 1. 2.4216   **2.** -0.5784 or $\overline{1}.4216$
    3. 4.2648   **4.** -0.5261 or $\overline{1}.4739$
    5. 8.00×10^4 **6.** 2.50×10^{-6}
G. 1. 4.72   **2.** 60
    3. 0.017   **4.** 3.60

If you missed any, except for a few accidental errors you can correct, consult your instructor for suggestions for additional study.

Section
ONE
ATOMIC STRUCTURE

The realm of our common experience encompasses only a narrow region of the dimension spectrum of the universe. We have a pretty good "feel" for measurements as large as a few thousand miles or as small as a speck of dust, but the distance to the nearest star or the size of a bacterial cell seems a little unreal to most of us. Scientists work routinely with numbers whose orders of magnitude are well outside the range of direct measurement, but which represent useful "reality" in scientific investigations.

Scientists "believe," for example, in the concept of very tiny atoms and subatomic particles—not because they have ever *seen* them, but because this concept provides a satisfactory model for explanations and predictions of the properties and behavior of matter.

In this section we will develop a relatively simple atomic model and illustrate some of the ways this model can be useful in studying chemical phenomena. More sophisticated models have been proposed and must be used for advanced theoretical studies or more quantitative investigations. It is generally satisfactory to employ the simplest model consistent with the level of study involved, so we shall leave the finer points of atomic theory for future use should you choose to continue your studies to more advanced areas of atomic sciences. As we pursue our discussions of the fundamentals of chemistry, we shall encounter many useful applications of our simple atomic model.

CONTENTS

Introduction

Toward a model for the atom

The development of the atomic theory of matter is a case history in the evolution of ideas. The present concepts of atomic structure did not spring full-blown from the minds of a handful of physicists and chemists at the beginning of the twentieth century, but emerged from the work of a troupe of philosophers, alchemists, and scientists whose lives span some 2600 years. Their goal was to fashion a "thought model"—a set of ideas that would explain and predict the behavior and properties of matter in terms of familiar systems.

By its very nature, a model is makeshift and temporary, useful only so long as its explanations and predictions correspond to observations of the real world. When it fails in this purpose, it must be modified or replaced. Mathematical models can be very complex and abstract. Physical models, such as the familiar ball-and-stick representations of chemical compounds, tend to be simpler and more concrete. It is not necessary to use complex models where simple ones suffice, but we must constantly bear in mind that all models are only constructs: representations of reality, not reality itself. One chemist observed that a ball-and-stick model is less like an actual molecule than a plaster mannequin is like a man.

Modern atomic theory is a model that provides a reasonably satisfactory explanation of the properties of matter, the mechanisms of chemical change, and the interactions of matter and energy. But this model is no more final than the latest model of automobile. The next generation is almost certain to improve it or discard it in favor of a better model.

Atomic theory is a product of the convergence of two separate lines of thought: on electricity and on atoms. Western historians trace both ideas to the ancient Greeks: electricity to Thales of Miletus (640–546 B.C.) and atomic theory to Democritus (546–460 B.C.). These two theories developed separately until the end of the nineteenth century. They merged in the first decade of this century—a marriage of ideas that gave birth to the modern "electrical" model of the atom.

The story begins with electricity. Thales experimented with amber, which in Greek is called elektron, and found that it acquires an electric charge when rubbed with fur. The implications of this observation were not recognized for more than two millennia. In fact, the few discoveries that emerged from more than 25 centuries of work on electricity—that is, the work up to about the end of the eighteenth century—can be mastered in a few afternoons in a high school physics laboratory. The discoveries that led to the modern theory concern the properties of electric charges at rest and in motion—in the language of the physicist, the fields of electrostatics and electrodynamics.

0.1 ELECTROSTATICS

Late in the eighteenth century, Benjamin Franklin observed that there were two kinds of electric charge, which he named positive and negative. These terms persist today. The basic rule that governs the behavior of two charged objects can easily be demonstrated in the laboratory. Two metal spheres are suspended from threads. If the spheres are uncharged, they will hang parallel, as shown in Figure 0.1. If we impart a negative charge to both spheres, they will move apart, because similar charges repel each other. If we impart a positive charge to both spheres, the result will be the same. If we charge one sphere negatively and the other positively, the two will move closer together, because unlike charges attract each other.

The French physicist Charles Coulomb (1736–1806) found that the electrostatic force between two opposite charges depends on the strength of the charges and on the square of the distance between them. *Coulomb's law* is summarized by the equation

$$F = k\frac{q_1 q_2}{d^2}$$

A laboratory humorist defined a constant as a number that, multiplied by the experimental result, gives the correct answer.

where F is the symbol for force and d is the symbol for distance; q is the usual symbol for electric charge, and if there is more than one, we call one of them q_1, the next q_2, and so on; k is a constant.

Note the inverse-square relationship between force and distance. This means that if the distance between two forces is doubled, the force between them decreases to one fourth its previous strength. Theoretically, one charged object exerts a force on all other charged objects in the universe; actually, because of the inverse-square relationship, the force becomes insignificant a relatively short distance away.

Students of physics will recognize a similarity between the mathematical form of

Figure 0.1   Charge interaction.

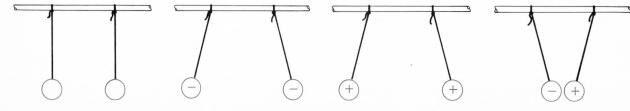

Figure 0.2 Magnetic field.

Coulomb's law and *Newton's law of gravitation,*

$$F = G \frac{m_1 m_2}{d^2}$$

where F is the symbol for the gravitational attraction between two bodies, G is the gravitational constant, and m_1 and m_2 are the respective masses of the two bodies. Electrical and gravitational forces share an important property: They can act on an object without touching it. In other words, an electric charge and a physical mass alter the space around them by creating a field of force. A magnet also has this property (Figure 0.2). Theoretical physicists have long sought to establish some relationship among these three types of force. So far gravity is the stumbling block: Einstein died without being able to incorporate gravity into his universal field equations. But the relationship between electricity and magnetism has been established and has been worked out in considerable detail.

There is one important difference between an electric field and a gravitational one: Electrostatic force can either push objects apart or pull them together, depending on their charges, but gravity always tends to pull objects together.

0.2 ELECTRODYNAMICS

Two effects of charges in motion have special bearing on atomic theory. Perhaps the most familiar example of charges in motion, the flow of electrons through a wire, led the Danish physicist Hans Christian Oersted to the accidental discovery of the relationship between electricity and magnetism. Almost 150 years ago, while preparing a lecture demonstration for his students, he noticed that the needle of a compass pointed toward a wire connected to a battery instead of toward magnetic north. If we pass a wire through a piece of cardboard with iron filings sprinkled on it, we can demonstrate that a current creates a magnetic field of force at right angles to the wire (Figure 0.3).

The other effect involves the motion of electric charges within a force field. Objects caught in a force field tend to move, and the motion is said to follow lines of force. The lines do not actually exist; they are merely imaginary contours that represent the strength of the field, much as the lines on a geologist's map represent the height of the land. Figure 0.4 shows the lines of force between two nearby electric charges, one positive and one negative. Lines close together indicate the strongest regions of the field; lines farther apart, the weaker regions. These lines enable us to predict the motions of charged particles.

The particles tend to move "down" the field, that is, in the direction that requires the least energy. A space vehicle approaches the earth at high velocity; once captured

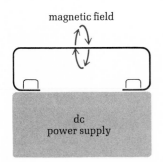

Figure 0.3 Current and magnetic field.

Figure 0.4 An electric force field.

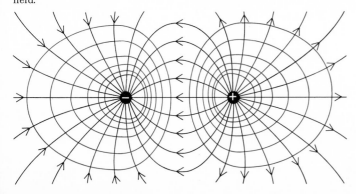

Figure 0.5 Vehicle in gravitational field.

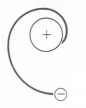

Figure 0.6 Charged particle in opposite electric field.

by the earth's gravitational field, it tends to spiral downward toward impact (Figure 0.5). A moving charged particle may follow the same spiral trajectory (Figure 0.6). In the process, both the space vehicle and the charged object lose energy. A moment's reflection will confirm this logic. An input of energy is required to remove a space vehicle from the earth's gravity or to separate two particles of opposite charge; when the objects fall back toward each other, they liberate energy. The space vehicle loses some of its energy as heat (in friction with the atmosphere) and the rest as momentum, which it transfers to the earth upon impact, much as a moving automobile gives up its energy when it collides with a wall. A charged particle moving down an electric field gives up energy by emitting radiation, particularly light. Electric charges traveling through magnetic fields also emit radiation.

The modern model of the atom is governed by the principles of electrostatics and electrodynamics, but it evolved from a series of models based largely on mechanical principles, beginning with the atoms of Democritus.

0.3 DEMOCRITUS TO DALTON

Democritus suggested that matter is not continuous, but is made up of tiny indivisible particles (the word "atom" means "indivisible" in Greek). A man of his time, he accepted the ancient idea of four elements: earth, air, fire, and water. He postulated that all varieties of matter result from combinations of atoms of these elements. Democritus based his model on intuition and logic, but it was discarded only a generation later by one of the greatest logicians of all time, the philosopher Aristotle (384–322 B.C.). A student of Plato and a teacher of Alexander the Great, Aristotle revived and strengthened the model of continuous matter—matter that is "all of a piece." His arguments were so persuasive that they stood virtually unchallenged until the Renaissance.

The ultimate superiority of Democritus' model was not recognized until the dawning of the age of science, an age that began when thinkers agreed on a rule that has governed scientific thought ever since: Any model must be not only logical but also consistent with experiment.

In the seventeenth century, workers such as Francis Bacon and Isaac Newton in England and Pierre Gassendi in France were troubled by experimental evidence that some substances behave in ways that are inconsistent with the idea of continuous matter. Aristotle's model collapsed when tested by experiment.

Newton explained the behavior of gases in terms of the motion of finite particles, and the Irish chemist William Higgins adapted these ideas in a limited way to explain the proportional combination of substances in chemical reactions. But it was not until 1808 that John Dalton, a Quaker schoolteacher in Manchester, England, proposed the unifying principle that all known properties of matter can be explained in terms of the behavior of unique, finite particles. By the beginning of the twentieth century, this principle was generally accepted, and today, after more than a century of modification, the atomic theory forms the foundation of modern chemistry.

0.4 SUBATOMIC PARTICLES

Recent evidence suggests that protons and neutrons may, themselves, have complex internal structures.[1]

Like Democritus, Dalton incorrectly believed the atom to be the ultimate particle, the smallest unit of matter. Later experiments demonstrated that atoms consist of still smaller units. So far physicists have cataloged more than 100 different *subatomic particles*—a collection humorously nicknamed the "particle zoo." The search for the ultimate particle continues, and so far no theory has satisfactorily explained the diversity within the "zoo." Fortunately, only three subatomic particles are important in the introductory study of chemistry: the electron, the proton, and the neutron.

[1] For a discussion of this idea, see "The Structure of the Proton and the Neutron," by H. W. Kendall, and W. K. H. Panofsky, in the June, 1971, issue of *Scientific American*.

Figure 0.7 Apparatus for determination of e/m.

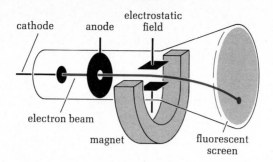

0.5 THE ELECTRON

The English physicist J. J. Thomson is generally credited with the first characterization of the electron, the "atom of electricity." Earlier investigators had described the emission of "rays" from the cathodes of electric discharge tubes. In 1897 Thomson, using an apparatus roughly similar to the picture tube in a television receiver, demonstrated that these rays could be interpreted as beams of negatively charged particles (Figure 0.7). The assignment of a negative charge was an arbitrary decision, based on the direction in which the beam was deflected by a magnetic field. In an elegant experiment, he determined the charge-to-mass ratio (e/m) of the electron: approximately 1.76×10^8 coulombs per gram ($C\ g^{-1}$). This result led Thomson to a powerful insight: Observing that the e/m for the cathode rays was independent of the type of gas in the discharge tube, he concluded that the electron must be a component of all matter.

Twelve years later the charge of the electron was precisely measured by the American physicist R. A. Millikan. Millikan's famous oil-drop experiment is depicted schematically in Figure 0.8. Basically, Millikan's apparatus was an air-filled chamber containing two electrically charged plates. The air molecules were ionized (stripped of some of their electrons) by exposure to x-rays. Millikan sprayed a fine mist of oil droplets into the top of the chamber, and as they settled toward the bottom their surfaces became charged as they adsorbed some of the free electrons. When a charged drop entered the electric field between the two plates, Millikan could suspend its fall by adjusting the field strength. The electrostatic force on the suspended droplet (the product of the field strength, E, and the charge on the droplet, q) exactly balanced the weight of the droplet (the product of the mass of the droplet, m, and the gravitational constant, G). By measuring the droplet's rate of fall with the field switched off, Millikan calculated its mass. Using the known value of the

total charge = unit charge (e) $\times$ number of unit charges, that is, $q = ne$

Figure 0.8 Millikan oil-drop experiment.

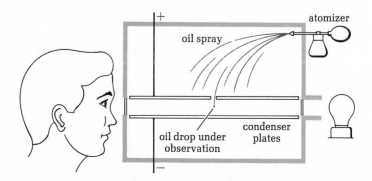

gravitational constant and the measured value of the field strength, he was then able to calculate the net charge on the droplet. He obtained a variety of values for q, but all were multiples of 1.60×10^{-19} C. Millikan correctly assumed that this value corresponded to the smallest possible unit of charge, the charge on the electron. From Thomson's value of the charge-to-mass ratio, it was then a simple task to calculate the mass of the electron: 9.1×10^{-28} g.

0.6 THE NUCLEUS

The experimental evidence for the neutron is described in Unit 35.

Because atoms are electrically neutral, the discovery of a negative subatomic particle suggested the presence of a particle of opposite charge. It was first observed in 1896 and later named the "proton." Its mass (1.7×10^{-24} g) proved to be almost 2000 times that of the electron, roughly the difference between the mass of a Ping-Pong ball and a cannonball. At one time chemists believed that all atoms consisted of combinations of hydrogen atoms (one proton and one electron). But studies of one of the products of radioactive decay, the α (Greek alpha) particle, revealed that many atoms also contain heavy particles that are electrically neutral. The missing particle—the neutron—was finally identified by very sophisticated experiments in 1932.

The term "neutron" had been suggested in 1921 by both Ernest Rutherford and the American chemist W. D. Harkins as a hypothesis to account for the difference between the charge-to-mass ratios of the proton and other subatomic particles, such as the α *particle* (now described as a stable combination of two protons and two neutrons). Experimental verification of the neutron as a particle of mass roughly equal to that of a proton, but having no electric charge, is credited to James Chadwick. Bothe and Becker, two German chemists, had demonstrated earlier that the bombardment of beryllium metal with α particles produced some form of penetrating radiation, which they believed to be γ (Greek gamma) rays (radiant energy similar to x-rays). Madame Curie and her husband were able to show that this radiation was capable of producing free protons when passed through paraffin. Chadwick, unable to conceive of proton formation by radiant energy, suggested that the radiation from the beryllium consisted of particles, and that it was the collision of these particles with the molecules of paraffin that "knocked loose" some of the protons of paraffin molecules. Chadwick's experiments were able to demonstrate the existence of such particles (the neutrons) and, indeed, to provide significant characterization of the particles.

How are subatomic particles arranged within the atom? In 1898, Thomson proposed a model of the atom as a sphere of positive electricity in which electrons are distributed rather uniformly, much like specks of sand embedded in a ball of lard (Figure 0.9). This model probably persisted as long as it did because of Thomson's well-earned prestige. In 1908, Geiger and Marsden, working under the direction of Lord Rutherford, tested the model by bombarding a thin metal foil with a beam of α particles (Figure 0.10). According to the model, the energetic, positively charged particles should have penetrated the diffuse atomic spheres like a bullet through a

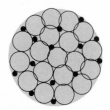

Figure 0.9 Thomson model for the atom.

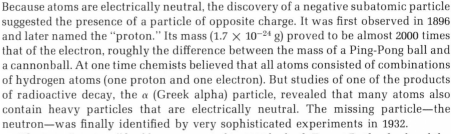

collimating slit (covered by thin metal foil for scattering experiments)

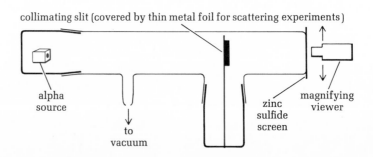

alpha source

to vacuum

zinc sulfide screen

magnifying viewer

Figure 0.10 Alpha-scattering apparatus.

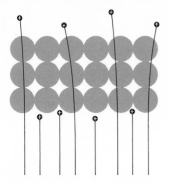

Figure 0.11 Alpha-scattering prediction by Thomson's model.

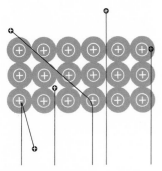

Figure 0.12 The *nuclear model* of the atom as interpretation of α-scattering observations.

0.7 PLANETARY MODEL

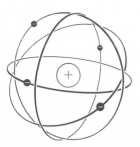

Figure 0.13 The planetary model of the atom.

haystack. Surprisingly, some of the particles were deflected through wide angles, and a few bounced back from the foil.

Rutherford's reasoning is easy to follow. Alpha (α) particles, having twice the charge and four times the mass of protons, should be repelled by the like-charged protons and attracted by the negatively charged electrons in the atoms of the metal foil. Because the α particles are massive (nearly 8000 times as heavy) compared to electrons, any electron–α-particle collisions should be negligible in affecting the α particle's path. Had the protons been uniformly distributed throughout the atom, as suggested by Thomson, collisions should have been rare and, when coupled with electrostatic repulsions, should have produced minor deflections of the α particles. The larger deflections observed could be attributed only to tightly grouped collections of protons. Indeed, the observations in later stages of the experiments, using detectors arranged to observe backscattering of α particles bouncing back out of the metal foil, must require a very dense "nucleus" of high positive charge, corresponding to an atomic center containing all the protons of the atom. Although the nuclear model of the atom is familiar to all of us today, Rutherford's postulate was a major surprise to the scientists of his time. He once described his original reaction to the experimental results by saying that he could not have been more surprised if he had fired a cannon at a sheet of paper and seen the cannonball bounce back.

Geiger tried to interpret the large degree of scattering in terms of successive collisions with protons diffused throughout the atoms of the metal foils (Figure 0.11). To Rutherford, however, the scattering suggested that "the diameter of the sphere of positive electricity is minute compared with the diameter of the sphere of influence of the atom" (Figure 0.12). Rutherford's paper, published in 1911, marked the discovery of the atomic nucleus.

The diameter of the nucleus is roughly 10^{-13} cm, or 1/100,000 the diameter of the atom. If an atom were magnified to the size of the Los Angeles Coliseum, the nucleus would be smaller than a walnut. Despite its almost negligible size, the nucleus contains virtually the entire mass of the atom.

Rutherford's conclusions led to the planetary model of the atom, in which the electrons were described as moving around the nucleus in curved orbits much as planets move around the sun (Figure 0.13). In the solar system the enormous gravitational attraction of the sun is precisely balanced by the centrifugal force of the planet as it travels in its orbit. In the model the powerful electrostatic pull of the nucleus was postulated as being similarly balanced by the motion of the electron.

The planetary model suffers from two major flaws. Experiments on macroscopic systems indicated that a sphere of negative electrostatic charge set in motion around a stationary sphere of positive charge would spiral toward it and eventually collide with it. In other words, according to the laws of classical (Newtonian) physics, the atom would collapse. Other experiments indicated that charged particles moving in a field of opposite charge lose energy by emitting radiation. But it was obvious that atoms in their normal state neither collapsed nor emitted radiation.

While physicists grappled with the implications of these results, chemists were gradually becoming aware that the properties of elements were somehow related to the number of electrons in the outer regions of the atom. Evidence was accumulating that molecules and polyatomic ions existed as collections of atoms arranged in uniquely characteristic shapes. If, as chemists suggested, these atoms are bound together by shared electrons, it was difficult to see how continuously orbiting electrons could be consistent with fixed relative positions of bonded atoms. Clearly, the planetary model of the atom required some modifications.

0.8 ATOMIC SPECTRA

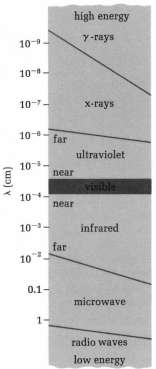

Figure 0.15 Electromagnetic spectrum.

By the second decade of the twentieth century the atomic theory was generally accepted, subatomic particles had been characterized, the periodic relationship between atomic number and the properties of elements was recognized, and evidence had accumulated that the electron is somehow involved in chemical reactions between elements. At about the same time, physicists suggested that the electron is involved in the emission of light from a glowing solid. Studies of atomic spectra led to a new model of the atom, the birth of the quantum theory, and some powerful insights into the nature of chemical reactions.

As early as the seventeenth century, Newton had observed that sunlight could be dispersed by a prism into a spectrum of its component wavelengths. Newton obtained a continuous spectrum: one in which the colors shade into one another almost imperceptibly (Figure 0.14, facing p. 32). We now know that visible light comprises only a rather small band in the broad spectrum of electromagnetic radiation, which includes radio waves, microwaves, light, and x-rays (Figure 0.15). The electromagnetic spectrum ranges from radio waves, with wavelengths of a kilometer or more, to γ rays, whose wavelengths may be shorter than 1 Å (10^{-8} cm).

The energy of electromagnetic radiation depends on its wavelength,

$$E = \frac{hc}{\lambda} \tag{0.1}$$

where E is the energy of the radiation (in calories), h is Planck's constant (1.58×10^{-34} cal sec), c is the speed of light (3×10^{10} cm sec^{-1}), and λ is the wavelength of the radiation (in centimeters).

Early in the nineteenth century physicists observed that when they heated certain elements in a flame or when they fired an electric discharge through gases in a tube, the glowing material emitted not a continuous spectrum but one made up of light of certain characteristic wavelengths. Passed through a prism, such a spectrum appears as a series of sharp, bright lines (Figure 0.16, facing p. 32). The German physicists Gustav Kirchhoff and Robert Bunsen emphasized that the line spectrum of an element is as characteristic as a fingerprint, and proceeded to demonstrate the power of spectroscopy as a tool for chemical analysis by discovering two previously unknown elements, cesium and rubidium. The emergence of atomic theory led physicists to associate the emission of light with the behavior of the electron and specifically with changes in the distance between the electron and the nucleus.

0.9 THE BOHR MODEL

In 1913 the great Danish physicist Niels Bohr advanced a revolutionary idea that not only explained why atoms emit spectral lines at characteristic wavelengths but also drastically modified the solar system model of the atom. Bohr argued that subatomic particles are not entirely governed by the laws of electrodynamics that apply to macroscopic bodies. An electron in an atom may possess only certain specific energies, and each of these energies corresponds to a particular orbit. In general, the higher the energy of the electron, the farther its orbit from the nucleus. The electron can acquire or lose energy only in discrete units called *quanta*. An appropriate quantum of heat energy, for example, can kick the electron into a new orbit slightly farther from the nucleus. By radiating away a particular quantum of energy, the electron can drop into an orbit closer to the nucleus, but it cannot drop below its normal stable orbit (the *ground state*). The energy differences between the various orbits correspond to the light energy associated with the spectral lines emitted by the atom (Figure 0.16).

The Bohr model was, in fact, developed as an attempt to explain, among other things, an empirical relationship (i.e., one based on correlation of experimental observations without a theoretical model) proposed in 1890 by Johannes Rydberg.

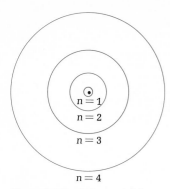

Figure 0.17 Bohr model for the hydrogen atom. Scale: 0.50 cm = 1.00 Å (nucleus not to scale). Note that radii vary with n^2.

The wavelengths in the visible region of the hydrogen spectrum (Figure 0.17) can be expressed by a rather simple equation. The same equation gives wavelengths found in the infrared and ultraviolet regions of the hydrogen spectrum. This *Rydberg equation* may be formulated as

$$\frac{1}{\lambda} = R\left(\frac{1}{n_a{}^2} - \frac{1}{n_b{}^2}\right)$$

in which λ is the wavelength of a spectral line in angstroms (1 Å = 10^{-8} cm); n_a is an integer (1, 2, 3, 4, . . .), n_b is equal to $n_a + 1$, $n_a + 2$, $n_a + 3$, . . ., and R is an empirical constant (1.09737 × 10^{-3} Å^{-1}).

Bohr set out to provide a model that would adapt the "nuclear atom" of Rutherford to an explanation of the Rydberg equation. His most important postulate was the *quantum condition*. In the normal condition of lowest energy (the *ground state*), the electron of the hydrogen atom follows a fixed orbit around the nucleus. The electron cannot gradually increase its distance from the nucleus but can, given the proper energy "boost," jump to a new orbit farther from the nucleus (an *excited state*). The energy difference between two possible orbits corresponds exactly to the energy of a specific wavelength in the hydrogen spectrum. Bohr's idea of quantized energy states was not related to any familiar system, but was postulated as a necessary condition "to make the model fit the facts." The electron does not spiral into the nucleus, radiating energy, *because it has an inherent property preventing this* (not an entirely satisfying "explanation").

The equation used by Bohr as a mathematical statement of his model can be derived from some basic principles of physics.

1. The electron, of mass m and charge e, circles around the nucleus (of charge e) with a velocity v in an orbit of radius r.
2. By virtue of its motion, the electron is "held" away from the nucleus by a centrifugal force expressed by

$$\frac{mv^2}{r}$$

3. Because of the opposite charges involved, the electron is pulled toward the nucleus by a "coulombic" force expressed by

$$\frac{e^2}{r^2}$$

4. The *quantum condition* imposed by Bohr was that the only possible orbits are those for which the radii result in angular momenta (given by Newtonian physics as mvr) equal to $nh/2\pi$, in which n is an integer (1, 2, 3, . . .) and h is Planck's constant, an empirical constant derived from earlier studies of the quantum characteristics of electromagnetic radiation.
5. The total energy of the system is given by the kinetic energy (energy of motion) plus the potential energy (the energy available under the static condition of the system); that is,

$$E_{\text{total}} = E_{\text{kinetic}} + E_{\text{potential}}$$

The kinetic energy for the moving electron is

$$E_{\text{kinetic}} = \frac{mv^2}{2}$$

and the potential energy of two particles of opposite charge separated by a distance r is

$$E_{potential} = \frac{-e^2}{r}$$

6. Now, the total energy is

$$E_{total} = \frac{mv^2}{2} + \left(\frac{-e^2}{r}\right) = \frac{mv^2}{2} - \frac{e^2}{r}$$

7. However, to maintain a stable radius r, the centrifugal force, mv^2/r, must equal the coulombic force, e^2/r^2, so

$$\frac{mv^2}{r} = \frac{e^2}{r^2}$$

from which

$$mv^2 = \frac{e^2}{r}$$

8. Substituting, then, in step 6 gives

$$E_{total} = \frac{e^2}{2r} - \frac{e^2}{r} = \frac{e^2}{2r} - \frac{2e^2}{2r} = \frac{-e^2}{2r}$$

9. The magnitude of e was known from the Millikan oil-drop experiments ($e = 1.6 \times 10^{-19}$ C), but r could not be measured experimentally. However, r could be expressed in terms of other quantities as required by Bohr's quantum condition (step 4) and the Newtonian requirements for stability (step 7). From step 4 ($mvr = nh/2\pi$),

$$v^2 = \frac{n^2h^2}{4\pi^2m^2r^2}$$

from step 7 ($mv^2/r = e^2/r^2$),

$$v^2 = \frac{e^2}{mr}$$

Thus

$$\frac{e^2}{mr} = \frac{n^2h^2}{4\pi^2m^2r^2}$$

from which

$$r = \frac{n^2h^2}{4\pi^2e^2m}$$

10. Substituting, then, in step 8,

$$E_{total} = \frac{2\pi^2me^4}{n^2h^2}$$

or

$$E_{total} = \frac{2\pi^2me^4}{h^2}\left(\frac{1}{n^2}\right)$$

11. Then, for the energy difference between any two orbits, n_a (of radius r_a) and n_b (of radius r_b),

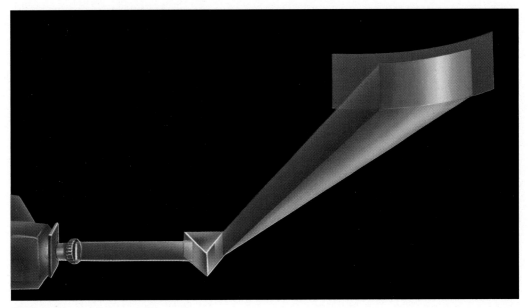

Figure 0.14 Continuous spectrum.

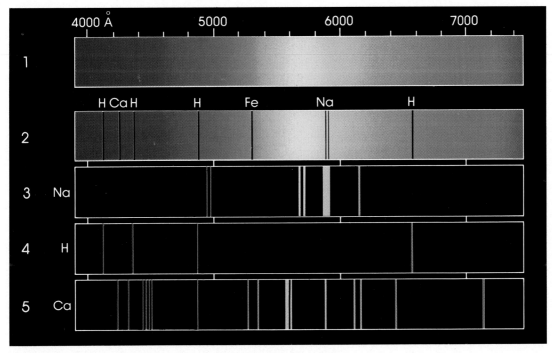

Figure 0.16 Discontinuous spectra. (1) Continuous spectrum from an incandescent solid. (2) Solar spectrum showing absorption (Fraunhofer) lines. (3-5) Discontinuous emission spectra of sodium, hydrogen, and calcium.

The Greek symbol Δ (delta) is used to represent a difference between two values, in this case an energy difference (ΔE).

$$E_a - E_b = \Delta E = \frac{2\pi^2 m e^4}{h^2}\left(\frac{1}{n_a{}^2} - \frac{1}{n_b{}^2}\right)$$

12. From equation (0.1), if ΔE corresponds to the energy of a particular wavelength in the spectrum, then for a wavelength associated with an "electron jump" between orbit n_a and orbit n_b,

$$\frac{1}{\lambda} = \frac{\Delta E}{hc} = \frac{2\pi^2 m e^4}{h^3 c}\left(\frac{1}{n_a{}^2} - \frac{1}{n_b{}^2}\right)$$

The derivation of the Bohr expression, although interesting to the more mathematically inclined, is of lesser importance than the end result. The quantities m, e, h, and c were all known with considerable precision and could be used to calculate the value of the term $2\pi^2 m e^4/h^3 c$. The *calculated value was almost exactly that of the empirical constant appearing in the* Rydberg equation, thus permitting accurate correlations of the simple Bohr model of the hydrogen atom with experimental values of wavelengths in the hydrogen spectrum.

As an important by-product of the Bohr model, it was also possible to calculate the ionization potential (Unit 1) for the hydrogen atom, that is, the energy required to remove the electron completely from the ground state of the hydrogen atom (n = 1). Again there was excellent agreement with the experimental value.

The beauty of the Bohr model of the hydrogen atom was that it allowed a rather straightforward calculation of the complete spectral characteristics of hydrogen, a calculation that could be verified experimentally. The model, however, did not adequately explain the spectra of the more complex atoms of other elements. Even after it was modified to include noncircular orbits, there was generally poor correlation between calculated and observed spectral characteristics.

Chemists, who are interested primarily in elements other than hydrogen, were generally less enthusiastic about the Bohr model than their colleagues in physics. Moreover, chemists were still troubled by an apparent conflict between the Bohr model and the results of the most recent work on the shape of molecules. Each type of molecule seemed to have a geometry that was characteristic and stable. This fixed geometry could hardly be reconciled with the idea of electrons in constantly shifting orbits.

The study of chemistry is primarily a study of matter and energy. The models used in modern explanations for chemical phenomena are significant modifications of the original "Dalton atom" and bear little resemblance to the planetary model so commonly invoked by the nonscientist even today. Various degrees of sophistication are found in the chemist's use of atomic models of matter, and the pragmatic approach of using the simplest model description appropriate to a given situation will become apparent during the further study of the science of chemistry.

SUGGESTED READING

Holiday, L. 1970. "Early Views on Forces between Atoms", *Sci. Amer.*, May.
Ihde, Aaron J. 1964. *The Development of Modern Chemistry.* New York: Harper & Row.
Young, Louise B. (Ed.). 1965. *The Mystery of Matter.* New York: Oxford University Press.
Zafiratos, C. D. 1972. "The Texture of the Nuclear Surface," *Sci. Amer.*, October.

Unit 1

Electrons in atoms

Objectives

(a) After studying the appropriate definitions in Appendix E and the terms' usage in this Unit, be able to demonstrate your understanding of the following terms by matching them with corresponding definitions, descriptions, or specific examples:

Bohr atom	node
Compton scattering	orbit
continuous spectrum	orbital
degenerate	orbital quantum number
discontinuous spectrum	orientation quantum number
electron diffraction	Pauli exclusion principle
energy level	principal quantum number
excited state	probability plot
ground state	quantum mechanics
ion	spin quantum number
ionization potential	sublevel
isoelectronic	Zeeman effect

(b) Be able to identify inadequacies of the "solar system" and Bohr models of the atom (Sections 1.1 and 1.2).

(c) Be able to draw

(1) an energy-level diagram suitable for assignment of electron configurations of simple atoms or ions having up to 56 electrons (Figure 1.10)

(2) boundary surface (outline) representations for s, p, and d orbitals (Figure 1.5)

(d) Given the name of an element or monatomic ion containing from 1 to 56 electrons be able to write both expanded and abbreviated electron configurations for the atomic or ionic ground state (Section 1.6).

(e) Be able to suggest plausible electron configurations for simple atomic or ionic excited states (Section 1.8).

OPTIONAL

Be able to use the Bohr equation, the Rydberg constant, equation (0.1) (Introduction), and appropriate conversion factors (Appendix A) to calculate energy or spectral wavelength corresponding to a specified electronic transition in the hydrogen atom.

With the rebirth of an atomic model for matter, chemists and physicists could begin the exciting game of explaining the mysteries of the submicroscopic world and, perhaps more important, predicting new properties and phenomena requiring ever-increasing sophistication for experimental testing.

It was not long until experimental results had outpaced the theoretical utility of a "hard-sphere" atomic model and, as is invariably the case, modifications of the model became necessary. The first such modifications resulted from the discoveries of the electrical nature of matter and the α (Greek alpha)-scattering results of Rutherford's group. Interestingly enough, the 1911 solar system model has persisted as the layman's view of the atom with a popularity undiminished by its scientific shortcomings, which were well recognized by the early years of this century.

The mathematical development of the Bohr Model was described in Section 0.9 of the Introduction.

In 1913 a young physicist, Niels Bohr, postulated some unique properties for electrons in atoms. To explain the observations of characteristic line spectra of chemical species, Bohr suggested that the energy of an electron in the atom is quantized: In Bohr's model the electron could revolve in certain specific orbits around the nucleus, but only orbits of particular radii were possible. This unique departure from the classical (Newtonian) laws of motion was a startling idea to the scientific world, but Bohr's mathematical model offered such close agreement to experimental observations of the hydrogen atom spectrum that the idea of quantized energy states found ready acceptance. However, this model was not as satisfactory as might be desired. Correlations and predictions for other than the simplest atom, hydrogen, were completely inadequate and, even more important, the quantum assumption seemed to have no parallel in familiar systems and appeared as a postulate having no real justification except "to make the theory fit the data."

The key to the modern model for electrons in atoms was suggested in 1925 by

Erwin Schrödinger. There were familiar systems having quantized energy conditions: musical instruments. The wave equations descriptive of the sound of a plucked guitar string could, with rather sophisticated extensions, serve to explain the quantum characteristics of electrons in atoms. The quantum mechanical model of the atom was established—a model that has proved invaluable in explaining and predicting a vast number of properties of matter.

Nature is very complex. Even the current models for the atom must be viewed as transient and imperfect descriptions of the intricate patterns of natural phenomena.

1.1 THE ATOM AS A SOLAR SYSTEM

The popular "planetary" model of the atom, evolved from Rutherford's studies of α scattering, is of little value to the scientist. It offers no clues to the periodic properties of the elements (Unit 2); it fails to explain the apparent exclusion of atoms from the laws governing other charge systems; and it appears inconsistent with experimental studies of molecular geometry (Unit 8).

By the early twentieth century, experimental and theoretical developments were appearing at a rapid pace (Figure 1.1). Three separate lines of investigation were converging: Experimental physicists were learning more about the properties of light; laboratory chemists were accumulating masses of data showing distinct correlations among numerous chemical properties; and theoreticians were developing increasingly sophisticated mathematical formulations for physical relationships. The genius of many investigators contributed to a giant step in the evolution of the model for the atom.

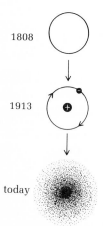

1808

1913

today

Figure 1.1 Time perspective of models for the atom.

1808   John Dalton
Atoms as simplest components of matter; all atoms of an element identical; atoms viewed as indestructible, tiny hard spheres.

1814   J. J. Berzelius ⎫ determinations
                              of relative
1819   Pierre Dulong and Alexis Petit ⎭ atomic weights.

1874–1891   G. Johnstone Stoney
Suggestion that electricity exists in discrete units associated with atoms; suggested the name *electron* for units of moving electric charge (current).

1895   Jean Perrin
"Cathode rays" in electric discharge tubes consist of particles, defined as "negative."

1896   Henri Becquerel
Discovery of radioactivity.
Atomic model modified: Atoms are not indestructible; they can decompose, forming atoms of different elements.

1897   J. J. Thomson
"Cathode rays" are really electrons; suggested electrons as components of all atoms.
Atomic model modified: Atom consists of smaller particles, positive and negative, mixed together like a plum pudding.

1900   Max Planck
Quantized characteristics of light.

1905   Albert Einstein
Particle characteristics of light (explanation for photoelectric effects) and relationship between mass and energy.

1906   R. A. Millikan
Value of unit charge = 1.6×10^{-19} C.

1911   Ernest Rutherford and co-workers

Interpretation of scattering of α particles by metal foil as evidence for the atomic nucleus. *Atomic model modified: The positive charge and most of the mass of the atom is centered in a tiny "nucleus."*

1912   J. J. Thomson

Discovery of "isotopes" of neon.

Atomic model modified: Atoms of an element all contain the same number of protons, but may differ in relative weights.

1913   Niels Bohr

Suggestion of quantized energy restrictions for electrons in atoms.

Atomic model modified: Electrons in atoms are limited to certain specific orbits around the nucleus.

1924   Louis de Broglie

Suggestion that electrons possess certain wave properties.

1925   C. J. Davisson and L. H. Germer

Experimental evidence for wave characteristics of electrons.

1925   G. E. Uhlenbeck and Samuel Goudsmit

Evidence that electrons have magnetic properties consistent with the model of a charged particle spinning around its axis.

1925–1927   Werner Heisenberg and Erwin Schrödinger

Quantum mechanical (wave equation) description for electrons in atoms; determination of electron trajectory (orbit) theoretically impossible.

Atomic model modified: Electrons in atoms are best described by wave equations. The idea of a definite orbit must be replaced by that of an orbital (region in space in which electron spends most of its time).

1927–present

Extensions and applications of quantum mechanics as a unifying model for electron behavior, dramatically aided by high-speed computers for the complex mathematical relationships involved.

1.2 PROBLEMS WITH THE BOHR MODEL

You will recall that the Bohr model for a hydrogen atom, as discussed in the Introduction, presented both a simple pictorial representation for electrons traveling around the nucleus and a mathematical representation that could be related to the experimental spectrum of hydrogen (Figure 1.2).

The excellent agreement between experiment and theory supported the validity of the model, as applied to the simple hydrogen spectrum. The problems, however, arose in attempting to adapt this simple model to a wider variety of applications. In 1896, the Dutch physicist Pieter Zeeman had observed that simple spectra became more complex when the emitting species were in a strong magnetic field. The original lines of the spectra were split into two or more new lines. The Bohr model, even when modified to include the possibility of noncircular orbits, was unable to account for this *Zeeman effect*. Of even greater importance, the Bohr model could not be extended to predictions of the complex spectra of multielectron atoms (Figure 0.17). The problem was apparent. Electrical interactions among the electrons contributed significantly to the energy states involved, but these interactions could not be evaluated within the context of the Bohr model.

There were conceptual problems, too. Scientists were familiar with the idea of energy quantization as applied to wave phenomena, such as light, but Bohr's suggestion of quantum states for *particles* had no parallel in the familiar world. It was

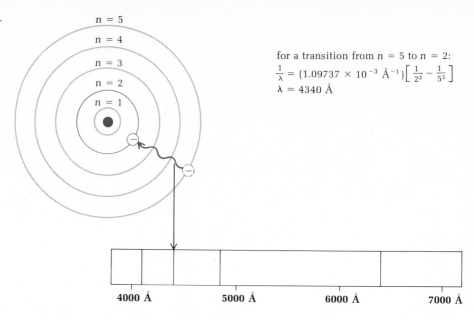

Figure 1.2 Bohr model correlation with hydrogen spectrum.

for a transition from $n = 5$ to $n = 2$:

$$\frac{1}{\lambda} = (1.09737 \times 10^{-3}\ \text{Å}^{-1})\left[\frac{1}{2^2} - \frac{1}{5^2}\right]$$

$$\lambda = 4340\ \text{Å}$$

apparent that although Bohr's model was more useful than the simple planetary model of the atom, some more satisfactory concepts of the behavior of electrons in atoms were needed.

1.3 MATTER WAVES

Photoelectric effect refers to the properties of certain materials that permit them to release electrons when exposed to light of appropriate wavelength. The familiar "electric eye" operates on this principle.

Note that the theory of relativity suggests that the mass of a particle varies with its velocity. The mass of a photon at zero velocity (rest mass) is believed to be zero.

For appropriate values of h and c, see Appendix A.

A clue to the problem was suggested in 1924 by the French physicist Louis de Broglie. Certain properties of light, such as its dispersion by a prism, can best be explained by the idea that light travels in waves. Other properties, such as the photoelectric effect, indicate that light is comprised of particles called "photons." de Broglie reconciled the two views with his concept of "matter waves": the idea that all matter really behaves in both ways. This duality can be expressed by a combination of equation (0.1) with the famous Einstein equation,

$$E = mc^2 \tag{1.1}$$

to give

$$\lambda = \frac{h}{mc} \tag{1.2}$$

Classical (Newtonian) physics was concerned primarily with the particle nature of matter. The wave nature, according to equation (2.2), would become significant only in the case of a particle whose mass is very small, such as the electron.

Experimental verification of de Broglie's suggestion followed just a year later. C. J. Davisson and L. H. Germer, at Bell Laboratories, were studying metal samples bombarded by high-energy electron beams. A sample of nickel was accidentally converted into a crystalline form and, when subjected to the electron beam, produced totally unexpected diffraction patterns (Figure 1.3) similar to those observed in x-ray diffraction by crystals (Unit 11). Such behavior indicated that electrons, like electromagnetic radiation, possess wave characteristics.

It is interesting to note that evidence for particle characteristics of light had been accumulated only a few years earlier. A. H. Compton found that x-rays could be "scattered" by interaction with electrons and that this scattering, analogous to the collision of a cue ball with a numbered billiard ball, could be interpreted as a

Figure 1.3 Diffraction pattern. Visible light passed through a tiny opening produces this diffraction pattern of alternate light and dark rings, similar to the wave pattern formed when a stone is dropped into still water.

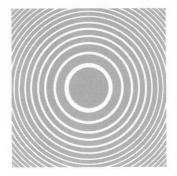

transfer of momentum from the x-rays to the electrons. Momentum, expressed as $p = mv$, requires a mass for the x-ray photon, and mass is a property of particles.

1.4 QUANTUM MECHANICS:[1] THE MUSICAL SCORE FOR THE ELECTRON

One of the objectionable features of the Bohr model of the atom was the *assumption* of quantized energy states for the electron. Such an assumption did not fit well with the whole idea of a *model*—an attempt to explain a complex system in terms of familiar phenomena. With the discovery of the wave characteristics of electrons, however, the key was provided. There *were* familiar systems having wave properties that showed quantized energy characteristics. A guitar string (Figure 1.4) is an example.

The Bohr description for the energy of electronic states in the hydrogen atom can be formulated in various equivalent mathematical statements. One of these may be expressed as

$$E = \frac{p^2}{2m} - \frac{e^2}{r}$$

Figure 1.4 When a single guitar string of length *l* is plucked, the wavelengths of the sound produced are related to the length of the vibrating string by integers, *n*, (*n* = 1, 2, 3, ...):

$$\lambda = \frac{2l}{n}$$

The sound itself actually consists of more than one wavelength. The musical terminology refers to a *fundamental* (for *n* = 1) and a series of *overtones* (for *n* = 2, 3, 4, . . .). The "richness" (timbre) of the sound results from the combination of these. An automobile could be accelerated uniformly so that the energy of the moving car showed a steady *continuous* increase, but the energy (the "energy" associated with the frequency of the sound is different from the *intensity* of the sound, which is energetically related to the square of the *amplitude* of the vibrating string) of the sound produced by the guitar string, as typically played, is limited to certain discrete values (i.e., *discontinuous*) mathematically related to the length of the string as determined by its "open" position or the various "finger positions" used. Thus the guitar string represents a system having wave properties and *quantized* energy states and provides a model for the wave descriptions of electrons in atoms. This "guitar analogy" will prove useful in our further discussion of quantum mechanics.

[1] The detailed development of quantum mechanics is beyond the scope of this text. The interested student will find an elegant and stimulating treatment in *Chemical Bonding Clarified Through Quantum Mechanics,* by George C. Pimentel and Richard D. Spratley (San Francisco: Holden-Day, 1969).

where p represents the momentum of the electron ($p = mv$), m is the mass, e is the charge of the electron, and r is the radius of its orbit. These terms, of course, refer to the electron simply as a particle, but with the assumption of quantum energy restrictions. Erwin Schrödinger, a theoretical physicist, suggested around 1927 that some characteristics of Bohr's treatment could be salvaged if the equation for electronic energy states could be expressed in terms of wave, rather than particle, considerations. A simplistic statement of the Schrödinger equation can be formulated, then, as

$$E\psi = \left(\frac{p^2}{2m} - \frac{e^2}{r}\right)\psi$$

where the Greek letter ψ (psi) represents a very complex wave function employed to convert the kinetic and potential energy terms from those representative of "particle" properties to their equivalents in wave equation forms.

1.5 QUANTUM NUMBERS

Inherent in the wave description of the electron are four *quantum numbers* used to designate different electron characteristics contributing to an overall energy state. The *principal quantum number* (n) can have any integer value ($1, 2, 3, \ldots$), like the n used in the Bohr equation or in the guitar analogy. This principal quantum number is related to the average distance of the electron from the positive nucleus and is thus of primary importance in determining the energy state involved.

The *orbital quantum number* (symbolized by l) can have any integer values from zero to $n - 1$. This term is used to designate an orbital (a geometric volume) in which the electron is moving. Some common orbital representations are given in Figure 1.5.

Figure 1.5 Simplified representations for some types of orbitals (Imaginary x, y, and z axes are shown to indicate *relative* orientations. The x,y plane is perpendicular to the page.)

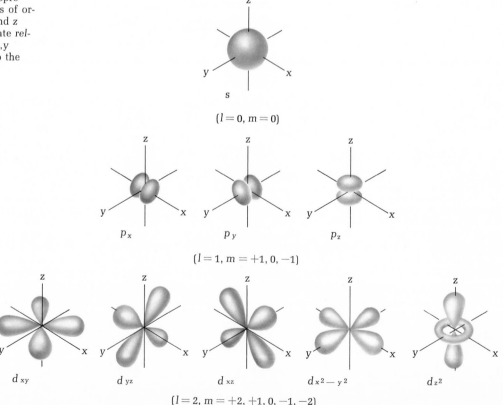

s

$(l = 0, m = 0)$

p_x p_y p_z

$(l = 1, m = +1, 0, -1)$

d_{xy} d_{yz} d_{xz} $d_{x^2-y^2}$ d_{z^2}

$(l = 2, m = +2, +1, 0, -1, -2)$

ORBITAL NOTATION

There are seven orientational distinctions in a set of f orbitals, corresponding to $l = 3$ with $m = +3, +2, +1, 0, -1, -2, -3$. The origin of the letters s, p, d, and f lies, historically, in descriptions used for lines in complex spectra. Early studies described four types of spectral lines, and later interpretations of these types in terms of related energy levels led to the designation of these levels (and the corresponding orbitals) by the first letter of the descriptive term. From single *sharp* lines, we get the symbol s. Sets of closely spaced *prominent* lines correspond to the p, and broader, more *diffuse* sets gave the d designation. The f term arises from certain lines referred to by early investigators as *fundamental*.

The *orientation* number m is also referred to as the *magnetic* quantum number.

Note distinctions among the different pictures for the three p and five d orbitals. The different relative orientations are designated by the *orientation quantum number* (m), which may have any integral value from $-l$ to $+l$, including zero. The letters used to define the orbitals (s, p, d, f) are now universally accepted in lieu of numerical values, as are the subscripts x, y, z, xy, and so on.

The fourth quantum number, s, may have values of $+\frac{1}{2}$ or $-\frac{1}{2}$. This *spin quantum number* is associated with the rotation of the electron around its own axis [Uhlenbeck and Goudsmit (1925), Figure 1.1]. In this respect the quantum mechanical model still considers some aspects of the particle nature of the electron. However, it should be noted that our current theories do not interpret the spin quantum number *literally* in terms of a spinning particle, but rather as an indication that two electronic energy states exist that interact differently with a magnetic field.

Since the negative electron and the positive nucleus are attracted by an electrostatic force, the energy an electron must possess to remain away from the nucleus should represent a major fraction of its total energy state. Hence the *principal quantum number, n,* is of considerable significance in defining possible energy levels (states) for electrons in atoms. Bohr's model for fixed electron orbits at increasing distances from the nucleus corresponded closely to the idea of different principal energy levels (Figure 1.6). In our present model, when applied to multielectron atoms, the maximum number of electrons that may "occupy" a particular principal energy level is given by $2n^2$.

Beyond this point the modern quantum mechanical treatment represents significant modification of the Bohr model for electrons in atoms. The wave equations used to define three-dimensional orbitals permit the refinement of energy *sublevels* within a given principal energy level (Figure 1.7). These sublevels (sets of similar orbitals, e.g., $2p_x$, $2p_y$, $2p_z$) permit a more detailed set of electronic transitions to be described and thus a more sophisticated treatment of atomic spectra. For an isolated atom (i.e., one not influenced by interaction with neighboring systems), the orbitals corresponding to a particular sublevel are of equivalent energy. (Another term for energy equivalence is *degeneracy*.) Further splitting of sublevels will be discussed later in Unit 27. The number of sublevels theoretically possible for a particular principal

Sublevels beyond f may be arbitrarily designated by g, h, i, ... We shall not consider the use of sublevels beyond d, leaving the more esoteric cases for advanced treatises.

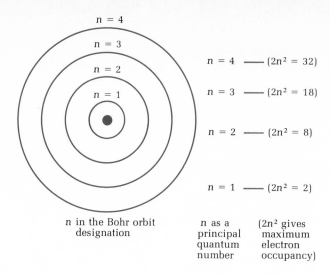

Figure 1.6 Bohr model and principal energy levels.

n as a principal quantum number	$(2n^2$ gives maximum electron occupancy$)$
$n = 4$ ——	$(2n^2 = 32)$
$n = 3$ ——	$(2n^2 = 18)$
$n = 2$ ——	$(2n^2 = 8)$
$n = 1$ ——	$(2n^2 = 2)$

n in the Bohr orbit designation

energy level is given by n. We shall use a numerical designation only for the quantum number n, employing the s, p, d notation (with appropriate subscripts) for sublevels (and individual orbitals).

Within a particular energy sublevel we describe a set of orbitals (one s, three p's, five d's, or seven f's). The *Pauli exclusion principle,* based on experimental observation, states that no orbital may contain more than two electrons and that this maximum can occur only when the two electrons are of opposite spin (i.e., the spin quantum number for one is $+\frac{1}{2}$, and for the other $-\frac{1}{2}$). Thus, in a multielectron atom, no two electrons have the same complete set of four quantum numbers. Although a rigorous treatment of the Pauli exclusion principle is beyond the scope of this text, we may consider some simplified logic in support of the concept.

Since electrons are identical in charge, electrostatic repulsion should suggest a lower energy for two electrons in separate orbitals than for a pair of electrons restricted to the same geometric region in space. Such is, in fact, the observation when degenerate orbitals are available. The lowest energy state for the isolated carbon atom, for example, corresponds to having two of the electrons in separate $2p$ orbitals. However, there is a significant energy difference between *different* sublevels. The pairing of electrons of *opposite spins* offers a weak magnetic attraction counter-

We may think, in simplistic terms, of electrons of opposite spin as similar to a "north" magnetic pole and a "south" magnetic pole, respectively.

Figure 1.7 Energy level diagram for a simple multielectron atom, showing sublevels and corresponding sets of degenerate orbitals. (Levels beyond $n = 3$ are not shown here. The picture is less simple at higher levels. For example, the $4s$ sublevel for simple atoms is of slightly lower energy than the $3d$.)

$(n = 3)$ $\begin{cases} (l = 2) \\ (l = 1) \\ (l = 0)\ 3s \end{cases}$

$3d$ $\overline{3d_{xy}}$ $\overline{3d_{yz}}$ $\overline{3d_{xz}}$ $\overline{3d_{x^2-y^2}}$ $\overline{3d_{z^2}}$

$3p$ $\overline{3p_x}$ $\overline{3p_y}$ $\overline{3p_z}$

$3s$ ——

$(n = 2)$ $\begin{cases} (l = 1) \\ (l = 0)\ 2s \end{cases}$

$2p$ $\overline{2p_x}$ $\overline{2p_y}$ $\overline{2p_z}$

$2s$ ——

increasing energy

$(n = 1, l = 0)$ $1s$ ——

acting, to some extent, the electrostatic repulsion. The resulting energy state is lower than that of two separated electrons in different sublevels. The magnetic attraction is, however, quite insufficient to overcome the electrostatic repulsion of more than two electrons in the same orbital. Thus a *pair* of electrons represents maximum orbital occupancy.

When each electron in an atom has been assigned its own unique set of four quantum numbers, then the electronic energy states have been described qualitatively. It is important to note that this does not indicate that two electrons in atoms of different elements (or in two different ions) having the same quantum numbers have the same *energies*. Again, a musical analogy may prove helpful. We can describe a saxophone and a trumpet as both playing "middle C." The "note" is the same, but the sound is decidedly different.

1.6 ELECTRON CONFIGURATION

A complete set of orbitals (e.g., $2p_x$, $2p_y$, and $2p_z$) has spherical symmetry; that is, when the complete set is drawn together around a common center, the representation appears as a sphere.

We shall arbitrarily use an arrow pointing up to represent positive spin and an arrow pointing down for negative spin. No difference in energies exists between "positive" and "negative" spins except in the presence of a magnetic field.

The lowest energy configuration is referred to as the *ground state*. Higher energy configurations are called *excited states*. Our "rules" are suitable only for *ground state* descriptions, although we may be able to make some reasonable guesses about simple *excited states*.

It is relatively easy to draw a picture for a single orbital, or for the simplest vibration of a guitar string (Figure 1.8), but it is not so easy to draw a single meaningful representation for the combination of all the orbitals of a many-electron atom, or for all the vibrations for the chord played on a multistring guitar. However, the chord may be *represented* by a *set* of musical notes, and the orbitals of a many-electron atom may be indicated by an electron configuration (Figure 1.9). The electron configuration is simply a notation indicating, by a simple convention, the quantum number assignments for all electrons in an atom.

The rules for writing electron configurations are relatively simple and involve, basically, assigning four quantum numbers to each electron involved. The principal quantum number n is indicated by an Arabic numeral (1, 2, 3, ...); the orbital quantum number l is designated by one of the letters s, p, d, or f (f orbitals will not be considered in this text). To distinguish between the various p and d orbitals within a given sublevel, the subscripts shown in Figure 1.5 are used rather than numerical values of the quantum number m, and electron spins (quantum number s) are indicated by vertical arrows. A maximum of two electrons (of opposite spin) are assigned to each orbital. Orbitals having the same n and l designations (e.g., $2p_x$, $2p_y$, $2p_z$), that is, degenerate orbitals, are only assigned one electron each until the complete set is half filled; then a second electron is assigned to each as necessary to account for the total number of electrons. Orbitals are filled in order of increasing energy (Figure 1.10). Two examples should be sufficient to illustrate the assignment of electron configurations.

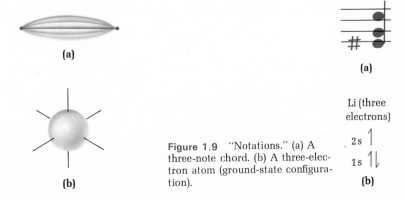

(a)

(b)

(a)

Li (three electrons)

2s ↑

1s ↑↓

(b)

Figure 1.8 "Representations." (a) Simplest vibration of a guitar string. (b) Simplest atomic orbital (1s).

Figure 1.9 "Notations." (a) A three-note chord. (b) A three-electron atom (ground-state configuration).

Figure 1.10 Energy-level diagram suitable for electron configuration assignments of up to 56 electrons (through 6s), for *representative* atoms and ions. (For exceptions to this scheme see Unit 27.)

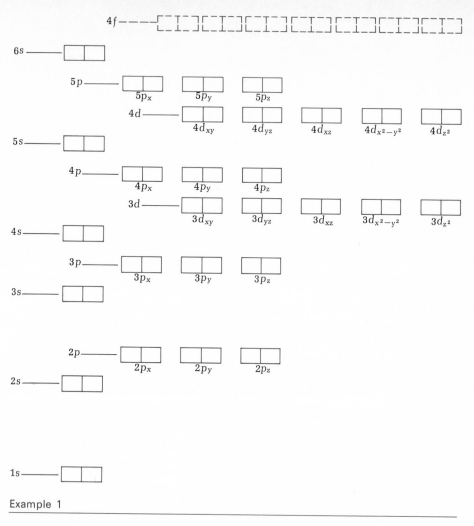

Example 1

Write the ground-state configuration for sodium (Na, 11 electrons).

Solution
Following the sequence of energy levels and sublevels shown in Figure 1.10, the first electron is assigned to the 1s state. Successive assignments are indicated by numbers above the "orbital boxes."

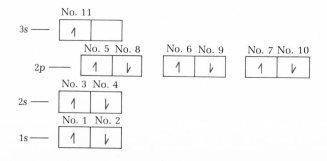

MEMORY AID TO ENERGY LEVELS

The following simplified chart is easy to remember and will be helpful in recalling the energy-level diagram. Remember that each p sublevel contains three orbitals and each d sublevel contains five. The "Aufbau" (building-up) sequence is indicated by the dashed line.

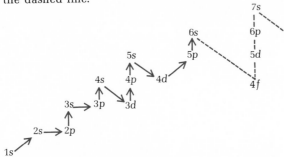

To reproduce the chart, simply fill out the complete sequence of each principal level on a separate horizontal row, indenting each row by one sublevel space.

The noble gas elements (Unit 2) are

helium   (He, 2 electrons)
neon   (Ne, 10 electrons)
argon   (Ar, 18 electrons)
krypton (Kr, 36 electrons)
xenon   (Xe, 54 electrons)
radon   (Rn, 86 electrons)

An ion is a charged particle. A negative charge (anion) indicates the number of extra electrons beyond those of the neutral particle, whereas a positive charge (cation) indicates the number of electrons less than those of the neutral particle.

The expanded electron configuration for sodium (or for any other 11-electron species) may be written as $1s\uparrow\downarrow$, $2s\uparrow\downarrow$, $2p_x\uparrow\downarrow$, $2p_y\uparrow\downarrow$, $2p_z\uparrow\downarrow$, $3s\uparrow$. Two abbreviated forms may also be used. The first omits spin designations ($1s^2$, $2s^2$, $2p^6$, $3s^1$), and the second indicates only the configuration beyond that of the noble gas element having the number of electrons closest to, but smaller than, the number for the species in question (Ne, $3s^1$, or (Ne), $3s\uparrow$).

Example 2

Write the ground-state electron configuration for the oxide ion (O^{2-}, 10 electrons).

Solution
The oxygen atom has 8 electrons (Z = 8), so the 2− ion must have 10 electrons. Then, following the procedure in Example 1, the electron configuration may be written as

$1s\uparrow\downarrow$, $2s\uparrow\downarrow$, $2p_x\uparrow\downarrow$, $2p_y\uparrow\downarrow$, $2p_z\uparrow\downarrow$

or

$1s^2$, $2s^2$, $2p^6$

or

(Ne)

that is, the O^{2-} has the same electron configuration as the neon atom. Such species are said to be *isoelectronic*.

1.7 ORBITALS VERSUS ORBITS (TRAJECTORIES)

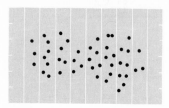

Figure 1.11   A probability plot.

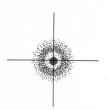

Figure 1.12   A two-dimensional probability plot (2s orbital).

Figure 1.13   A three-dimensional probability plot (2s orbital).

Figure 1.14   Nodal regions. (a) One type of vibration of a guitar string showing three nodal *points* (regions of zero string movement). (b) A drawing of a *p* orbital showing a nodal plane through the nucleus (the plan represents a surface of zero electron density).

The energies of electrons in atoms are of the same order of magnitude as x-rays of appropriate wavelengths. X-rays can be used to "map" the locations of regions of varying electron densities—the principle of x-ray crystallography (Excursion 1). Interactions between electrons and x-rays, however, are such as to disturb the electron's "normal" motion in the atom, so that trajectories (orbits) cannot be determined.

The problem of trying to locate an electron with x-rays is analogous to trying to find a vase in a darkened room by throwing baseballs at it. When one hears a crash, one knows where the vase *was* but not where it is. Of course the electron, unlike the vase, is a moving target, and the object of the search is to determine its path, not its position at a frozen instant in time.

A better analogy might be the task of trying to determine the path of a Volkswagen being driven on a football field in complete darkness. Its route is known only to its driver, who is navigating by compass. Imagine the difficulty of locating it by driving another car back and forth across the field until you hit it. Equipped with a grid map and a compass, you could mark your position on the grid after each collision and then back off and continue the search. The chances are excellent that as a result of your collision, the Volkswagen would be at least temporarily bumped from its normal path. After scores of collisions, your grid might look like Figure 1.11. You would have no way of knowing the path the Volkswagen was trying to follow. Some encounters may have occurred while the Volkswagen was deflected by a collision from its normal path. Others may have occurred while it was backing into position to start again. In most instances it would be difficult to determine which way the Volkswagen was headed at the moment of impact. Although you could not determine the Volkswagen's normal path by this type of detection, you could determine the area on the field in which collisions occurred most frequently and hence the region in which the Volkswagen spent most of its time. Similarly, locating an electron within an atom involves the use of energies large enough to disturb the electron's normal motion. In other words, the trajectory of an electron within an atom cannot be defined experimentally. The *Heisenberg uncertainty principle* states this as: "It is impossible to determine simultaneously the position and momentum of an electron in an atom."

However, it is possible to determine the relative probability of finding the electron in a certain region around the nucleus, just as we could determine the probability of finding the Volkswagen in a certain region of the football field. On a two-dimensional graph, the probability of finding the electron in a given region in the hydrogen atom looks like the plot in Figure 1.12. Because the electron moves in three dimensions, it is more informative to use a three-dimensional plot, like the one in Figure 1.13. Such three-dimensional probability plots of electron distribution are used in representing *orbitals*. They have replaced electron orbits in the current model of the atom and have provided a basis for explaining the properties of many elements and compounds. The orbitals can be thought of as "clouds of charge" created by moving electrons. Their sizes and shapes depend on the energy state of the electron. The probability description of an orbital is a mathematical result of Schrödinger's

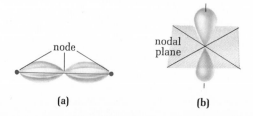

(a)   (b)

Figure 1.15 Probability-plot representations for some simple hydrogenlike orbitals.

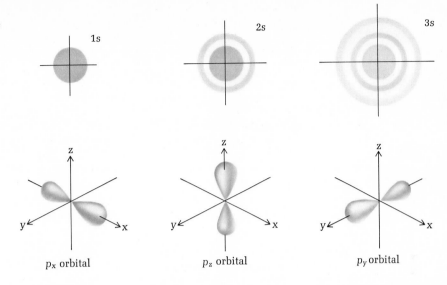

p_x orbital p_z orbital p_y orbital

wave-equation treatment for electrons in atoms. Certain orbitals (e.g., p and d orbitals) are characterized by *nodal surfaces* (Figure 1.14) representing regions of zero probability of finding the electron. Except for these restrictions, there is a *finite probability* of finding the electron at *any* distance from the nucleus. Thus typical orbital drawings represent regions of *maximum* probability (Figure 1.15), whereas the simpler outline drawings (Figure 1.5) indicate only boundary surfaces for maximum probability regions.

**1.8 EXCITED STATES:
SPECTRA**

Bohr's model of the hydrogen atom was related to wavelengths in the hydrogen atom spectrum by energies corresponding to differences between possible electron *orbits*. The quantum mechanical model employs the same basic approach, except that energy differences correspond to those between different principal energy levels, sublevels, or orbitals. Both models give excellent agreement with experimental results in the case of the simplest atom, hydrogen, but the Bohr model proved completely unsatisfactory for more complex situations. Quantum mechanics has been able to approximate experimental values closely for spectra of many different species, although the mathematical intricacies and the necessity of making various approximations require high-speed computers. The wave properties treated by quantum mechanics are perfectly appropriate for interpretation of discontinuous spectra (Figure 1.16).

A special advantage of the quantum mechanical model applies to the interpretation of the so-called Zeeman effect, described earlier. (In the presence of a magnetic field, the hydrogen spectrum appears more complex than that shown in Figure 1.2. In place of each of the "normal" lines, the spectrum shows a series of regularly spaced lines; that is, a magnetic field causes a "splitting" of spectral lines.) The Bohr model could not account for this observation, but the quantum mechanical model does: Whereas "normal" lines can be attributed to differences between *principal* energy states, the Zeeman effect can be explained by the smaller energy differences between *sublevels* (orbitals with the same principal quantum number) and by removal of the degeneracy of orbitals in the same sublevel.

Figure 1.16 Wave properties and spectra.

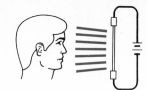

You hear a total sound.

You see a total color.

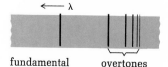

The guitar analogy used so frequently in these discussions is just an analogy and is by no means an exact parallel of electronic characteristics.

The total sound actually consists of a discontinuous spectrum (fundamental + overtones).

The total color actually consists of a discontinuous (line) spectrum.

The origin of the spectrum can be represented by a musical notation (the dark note is the fundamental and the "ghost" notes are overtones).

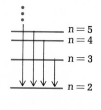

The origin of the spectrum can be represented by an energy-level diagram showing the ground state and various excited states.

In our use of the terms *qualitative* and *quantitative* here, we refer to *qualitative* correlations as those that can be described by simple conceptual (pictorial) models, whereas *quantitative* correlations would require related numerical data such as energy-to-wavelength correlations.

Noteworthy exceptions are observed with some of the *d*-transition elements (Unit 27).

We can employ electron configurations to describe excited states and, by comparing these with ground state or other excited state configurations, to relate electronic transitions to observed spectral lines. In Units 27 and 30 we will extend these applications in a qualitative way to the absorption spectra of complex ions and molecules. We will also find the use of excited state configurations valuable in the development of a simple model for chemical bonding (Units 7 and 8). Without some quantitative data on the relative energies of various electronic states, in particular with respect to the contributions of electron-electron interactions, we cannot determine the particular electron configurations necessary to account for specific spectral observations. This is still a major problem in quantum mechanics. We can, however, make some reasonable guesses about simple excited states by assuming that transitions from one sublevel to another, within the same principal energy level, should usually be of lower energy than transitions between principal energy levels.

Example 3

Suggest a plausible electron configuration for the simplest excited state of the carbon atom (C, six electrons).

Solution
Based on the energy-level diagram (Figure 1.10), the ground-state configuration should be

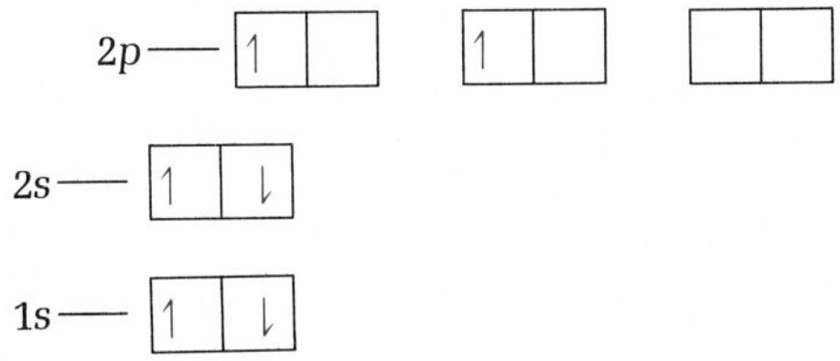

$2p$ — ☐☐ ☐☐ ☐☐

$2s$ — ☐☐

$1s$ — ☐☐

C(6 electrons): $1s\updownarrow$, $2s\updownarrow$, $2p_x\uparrow$, $2p_y$
or [He] $2s^2$, $2p^2$

Assuming that the simplest excited state would correspond to a transition within the second principal energy level, we might expect one of the $2s$ electrons to be excited to the $2p$ sublevel. Within that sublevel, lowest energy would correspond to one electron per orbital, giving

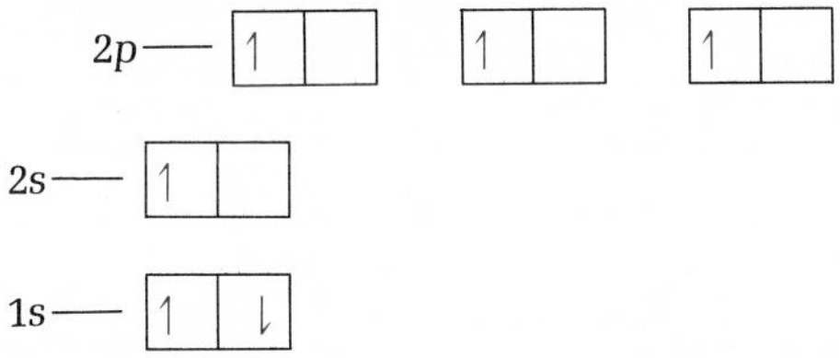

$2p$ — ☐☐ ☐☐ ☐☐

$2s$ — ☐☐

$1s$ — ☐☐

We could, then, represent a plausible configuration by

C(6 electrons): $1s\updownarrow$, $2s\uparrow$, $2p_x\uparrow$, $2p_y\uparrow$, $2p_z\uparrow$
or [He] $2s^1$, $2p^3$

Example 4

Write plausible electron configurations for the ground state and simplest excited state of the gallium atom (Ga, 31 electrons).

Solution

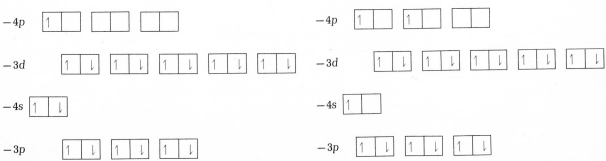

Begin with 1s, see p. 50.

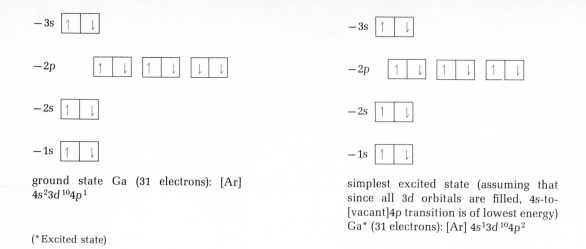

ground state Ga (31 electrons): [Ar] $4s^2 3d^{10} 4p^1$

(*Excited state)

simplest excited state (assuming that since all $3d$ orbitals are filled, 4s-to-[vacant]4p transition is of lowest energy) Ga* (31 electrons): [Ar] $4s^1 3d^{10} 4p^2$

1.9 IONIZATION POTENTIALS

The ionization potential refers to the energy required to remove an electron completely from the influence of the nucleus of a gaseous atom or ion. Since an electron in the highest energy level already possesses more energy than electrons in lower levels, less additional energy must be added to remove such an electron. Another way of looking at it is that the larger the average distance an electron is from the positive nucleus, the easier it is to pull it completely away. The removal of an outermost electron, then, is the lowest-energy ionization. Removal of successive electrons should be increasingly more difficult, since each remaining electron is affected by a larger *net* positive field. Major differences in successive ionization potentials would be expected as new principal energy levels are involved. These expectations are confirmed by experimental observations (Figure 1.17). Thus the quantum mechanical

Figure 1.17 Correlation of successive ionization potentials with electron configuration for sodium (Na, eleven electrons).

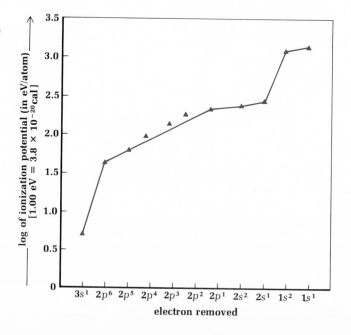

model for the atom has a useful application in terms of an important atomic property. We shall examine this property in more detail, along with other applications of our atomic model, in the context of periodic relationships among the chemical elements (Unit 2).

Although quantum mechanics provides us with a more useful treatment of electrons in atoms than was possible with earlier models, our present view of electronic structure is far from complete. We still have many unanswered questions, and quantitative predictions are still quite limited. Like the models of Dalton, Rutherford, and Bohr, the quantum mechanical description is *only* a model and, as such, awaits the modifications that must follow from continued research.

SUGGESTED READING

Berry, R. S. 1966. *"Atomic Orbitals," J. Chem. Ed.*, June.

Hochstrasser, R. M. 1964. *Behavior of Electrons in Atoms: Structure, Spectra, and Photochemistry of Atoms.* Menlo Park, Calif.: W. A. Benjamin.

Pimentel, George C., and Richard D. Spratley. 1969. *Chemical Bonding Clarified Through Quantum Mechanics.* San Francisco: Holden-Day. (Highly recommended.)

EXERCISES

1. Select the term best associated with each definition, description, or specific example.

    TERMS: (A) Bohr atom (B) discontinuous spectrum (C) ground state (D) ionization potential (E) node (F) orientation quantum number (G) Pauli exclusion principle (H) spin quantum number (I) sublevels

    a. for the hydrogen atom, energy corresponding to $H \longrightarrow H^+$

    b. for the hydrogen atom, energy state represented by $1s\uparrow$ (or $1s\downarrow$)

    c. 3s, 3p, 3d

    d. experimental evidence for quantized energy of electrons in atoms

    e. suggests that no two electrons in an atom can have the same set of four quantum numbers

    f. suggested that electrons in atoms were restricted to certain specific orbits around the nucleus

    g. for an orbital, a region of zero electron probability

    h. symbolized by m, or in orbital notation by subscripts such as x, y, z, xy

    i. symbolized in electron configurations by $\uparrow$ or $\downarrow$

2. Identify which of the following could not be explained in terms of (A) the simple "solar system" model for the atom, (B) the Bohr model for the atom:

    a. the Rutherford experiments indicating a nucleus containing most of the mass and all of the positive charge of the atom

    b. the simple emission spectrum of hydrogen

    c. the Zeeman effect

    d. the loss of an electron by an atom to form a positive ion

    e. the emission spectrum of mercury (Hg, 80 electrons)

3. Draw simple representations of

    a. an energy-level diagram suitable for up to 36 electrons

    b. an s orbital

    c. a p_x orbital

    d. a $d_{x^2-y^2}$ orbital

4. Write, in both expanded and abbreviated forms, the ground-state electron configurations expected for

    a. calcium atom (Ca, 20 electrons)

    b. phosphorus atom (P, 15 electrons)

    c. xenon atom (Xe, 54 electrons)

    d. iodide ion (I⁻, iodine atom has 53 electrons.)

    e. barium ion (Ba²⁺, barium atom has 56 electrons.) (Which of the preceding species are isoelectronic?)

5. Suggest plausible electron configurations, in expanded and abbreviated forms, for the simplest excited state of

    a. boron atom (B, 5 electrons)

    b. magnesium atom (Mg, 12 electrons)

OPTIONAL

6. Calculate, to three significant figures:[2]

    a. the wavelength of infrared radiation (the Paschen region) of the hydrogen atom spectrum corresponding to an electronic transition from $n = 5$ to $n = 3$

    b. the wavelength of visible radiation (the Balmer region) of the hydrogen atom spectrum corresponding to an electronic transition from $n = 5$ to $n = 2$

    c. the wavelength of ultraviolet radiation (the Lyman region) of the hydrogen atom spectrum corresponding to an electronic transition from $n = 5$ to $n = 1$.

    d. the ionization potential (in calories per atom) for the hydrogen atom (corresponding to a transition from $n = 1$ to $n = \infty$, that is, from $1/n^2 = 1$ to $1/n^2 = 0$). The experimental value for the ionization potential of the hydrogen atom is 13.6 eV (electron volts) per atom (eV atom⁻¹). Use the conversion factor given with Figure 1.17 to see how the calculated and experimental values compare.

[2] Refer to the *Rydberg equation* in Section 0.9 of the Introduction.

PRACTICE FOR PROFICIENCY

Objective (a)

1. Select the term best associated with each definition, description, or specific example.

   TERMS: (A) Compton scattering (B) continuous spectrum (C) degenerate (D) electron diffraction (E) energy levels (F) excited state (G) ion (H) isoelectronic (I) orbit (J) orbital (K) orbital quantum number (L) principal quantum number (M) quantum mechanics (N) Zeeman effect

   a. experimental evidence for wave properties of electrons
   b. magnetic field splitting of spectral lines
   c. for lithium atom (Li, 3 electrons), symbolized by $1s^2 2s^0 2p^1$
   d. Na^+ (10 electrons), F^- (10 electrons)
   e. a fixed trajectory (e.g., a definite path for an electron revolving around an atomic nucleus)
   f. a three-dimensional region in which an electron travels outside an atomic nucleus
   g. experimental evidence for particle properties of light
   h. modern treatment of electrons in atoms by use of wave equations
   i. light from an incandescent source
   j. symbolized by n, with whole-number assignments
   k. symbolized by l, conventionally designated by letters such as s, p, d, or f
   l. Na^+ or S^{2-}
   m. of equivalent energies, for example, $2p_x$, $2p_y$, $2p_z$ for an isolated atom
   n. method of describing successive energy states

Objective (b)

2. Identify which of the following could not be explained in terms of (A) the simple "solar system" model for the atom, (B) the Bohr model for the atom:
   a. the line spectrum of hydrogen in the ultraviolet region
   b. the line spectrum of strontium (Sr, 38 electrons) in the visible region
   c. the existence of molecules made of atoms combined in fixed geometric patterns
   d. that fluoride ion (F^-) is a fluorine atom with one additional electron
   e. the effect of a powerful magnetic field on the emission spectrum of hydrogen

Objective (c)

3. Draw simple representations of
   a. an energy-level diagram suitable for up to 56 electrons
   b. three "different" p orbitals
   c. five "different" d orbitals

Objective (d)

4. Write, in both expanded and abbreviated forms, the ground-state electron configurations expected for
   a. magnesium atom (Mg, 12 electrons)
   b. silicon atom (Si, 14 electrons)
   c. argon atom (Ar, 18 electrons)
   d. germanium atom (Ge, 32 electrons)
   e. cadmium atom (Cd, 48 electrons)
   f. potassium ion (K^+, Potassium atom has 19 electrons.)
   g. chloride ion (Cl^-, Chlorine atom has 17 electrons.)
   h. tin(II) ion (Sn^{2+}, Tin atom has 50 electrons.)

Objective (e)

5. Suggest plausible electron configurations, in expanded and abbreviated forms, for the simplest excited state of
   a. beryllium atom (Be, 4 electrons)
   b. silicon atom (Si, 14 electrons)
   c. gallium atom (Ga, 31 electrons)
   d. tin atom (Sn, 50 electrons)
   e. sodium ion (Na^+, 10 electrons)
      (Hint: Are *sublevel* transitions possible for Na^+? Would you estimate that simplest excitation of Na^+ would require more energy than that for Mg atom?)

OPTIONAL

6. The Bohr (or Rydberg) equation may be expressed as

$$\frac{1}{\lambda} = (1.097 \times 10^{-3}\ \text{Å}^{-1})\left[\frac{1}{n_a{}^2} - \frac{1}{n_b{}^2}\right]$$

   Calculate, to three significant figures:
   a. the wavelength in the hydrogen spectrum corresponding to an electron transition from $n = 4$ to $n = 2$ (To what region of the electromagnetic spectrum does this correspond?)
   b. the energy, in cal atom^{-1}, corresponding to the electron transition from $n = 4$ to $n = 2$ ($E = hc/\lambda$)
      (See Appendix A for constants and conversion factors.)

BRAIN TINGLERS

(Consult a periodic table of the elements for Questions 7–10.)

7. Identify in question 4 two sets of isoelectronic species. From the periodic table of the elements, find an atom from which a simple ion could be formed to include in one of the isoelectronic sets identified.

8. Compare electron configurations for species having 7, 12, 15, 20, 33, 38, 51, and 56 electrons, respectively. Locate elements in the periodic table having these "atomic numbers" (shown in the table as whole numbers above the element symbols). What trends do you note? What type of orbitals would you expect for highest energy occupancy in the ground states of radium (Ra) and bismuth (Bi), respectively?

9. The only chemically stable ion of rubidium is Rb^+. The most stable monatomic ion of bromine is Br^-. Krypton (Kr) is among the least reactive of the chemical elements. Compare the electron configurations of Rb^+, Br^-, and Kr. Then predict the most stable monatomic ions of strontium (Sr) and selenium (Se).

10. Neither atomic chromium nor atomic copper has an electron configuration predicted by the simple energy-level diagram given in Figure 1.10. The chromium atom has six *unpaired* electrons in the ground state; the copper atom has one (in a $4s$ orbital). Write the ground-state electron configurations for chromium (Cr) and copper (Cu). What do these configurations suggest in terms of the energy of filled or half-filled $3d$ orbitals? (This topic is pursued in more detail in Unit 27.)

Unit 1 Self-Test

1. Select the term best associated with each definition, description, or specific example.
   TERMS: (A) Bohr atom (B) Compton scattering (C) discontinuous spectra (D) ground state (E) isoelectronic (F) node (G) orbital
   a. experimental evidence for quantized energies of electrons in atoms
   b. experimental evidence for particle properties of light
   c. model suggesting quantized orbits for electrons in atoms
   d. $2p_x$
   e. for helium (He, 2 electrons), $1s^2$
   f. for an orbital, a region of zero electron probability
   g. helium (2 electrons), lithium ion (Li$^+$, lithium atom has 3 electrons), and hydride ion (H$^-$)

2. Which of the following cannot be explained in terms of *either* the simple "solar system" or the Bohr model for the atom?
   a. the α-scattering experiments of Rutherford's group
   b. the description of a magnesium ion (Mg^{2+}) as a magnesium atom that has lost two electrons
   c. the Zeeman effect
   d. the observed line spectrum of hydrogen in the infrared region
   e. the observed line spectrum of barium (Ba, 56 electrons) in the visible region
   f. the unique geometry of two hydrogen atoms combined with an oxygen atom to form a water molecule

3. Draw simple representations of
   a. an energy-level diagram sufficient to represent a 15-electron atom
   b. a $2p_y$ orbital, using a conventional set of x, y, z axes
   c. a $3d_{z^2}$ orbital

4. Write ground-state electron configurations:
   a. in expanded form for arsenic (As, 33 electrons)
   b. in an abbreviated form for cesium (Cs, 55 electrons)

5. Suggest a plausible electron configuration, in expanded form, for the simplest excited state of aluminum atom (Al, 13 electrons).

OPTIONAL

6. For an electron transition from $n = 7$ to $n = 2$ in a hydrogen atom, calculate
   a. the associated wavelength in the hydrogen spectrum
   b. the corresponding energy, in cal atom^{-1}

$$\frac{1}{\lambda} = (1.097 \times 10^{-3} \text{ Å}^{-1})\left(\frac{1}{n_a{}^2} - \frac{1}{n_b{}^2}\right)$$

$$E = \frac{hc}{\lambda}$$

(For constants and conversion factors, see Appendix A.)

Unit 2

. . . the ingredients of which all those perfectly mixt bodies are immediately compounded and into which they are ultimately resolved.
The Sceptical Chymist (1661)
ROBERT BOYLE

The chemical elements

Objectives

(a) Given the names or symbols of common elements, be able to write the corresponding symbols or names, respectively (Table 2.1).

(b) After studying the appropriate definitions in Appendix E and the terms' usage in this Unit, be able to demonstrate your understanding of the following terms by matching them with corresponding definitions, descriptions, or specific examples:

abundance	electron affinity
alkali metal	element
alkaline earth metal	group
anion	halogen
atomic mass unit (amu)	heterogeneous
cation	homogeneous
chalcogen	ionic
chemical family	ionization potential
chemical property	isotopes
compound	metal
covalent	metalloid

noble gas
nonmetal
period
physical property

representative element
transition element
transuranium element
valence electrons

(c) Given the atomic number and mass number of a neutral atom and the charge of a monatomic ion, in standard symbolism, be able to determine the numbers of protons, neutrons, and electrons in the atom or ion represented (Section 2.2).
(d) Be able to use the concepts of chemical periodicity and the "quantum" model of the atom to predict or explain some simple physical or chemical properties of specified representative elements (Sections 2.3–2.5).
(e) Given a periodic table and the names or symbols of some common elements, be able to
  (1) report the atomic number and atomic weight of each element (Figure 2.4)
  (2) indicate the charge expected for the most stable monatomic ion, or that no stable monatomic ion is expected (Figure 2.10)
  (3) arrange the names or symbols in order of relative sizes of the atoms of the elements (Section 2.7)
  (4) predict, on a qualitative basis, the size changes expected for the formation of specified cations or anions from the atoms of the elements (Section 2.7)
  (5) estimate relative magnitudes of ionization potentials or electron affinities for neutral gaseous atoms of the elements (Figures 2.8 and 2.9)

OPTIONAL

Given the isotopic abundances and isotopic masses (in atomic mass units) of an element, be able to calculate the atomic weight of the element (Example 1).

The Greek notion of four elements was challenged in the seventeenth century by the English physicist Robert Boyle, but was not finally toppled until the late eighteenth century, when the French chemist August-Laurent Lavoisier utilized chemical techniques to distinguish between elements and compounds. Lavoisier dazzled his contemporaries by compiling a list of 23 elements. A generation later, the list had grown to 34, and a handbook of chemistry published in 1852 listed 52.

In the same year, Julius Lothar Meyer independently suggested a similar periodic arrangement.

By this time (Figure 2.1) it was obvious that there were "families" of elements with similar properties, such as the "halogen triad": chlorine, bromine, and iodine. In 1869, the Russian chemist Dimitri Mendeleev published his periodic table of elements. In a revised version, published three years later, he arranged the 63 known elements in eight groups; equally important, he predicted the existence of 27 undiscovered elements. His predictions pointed the way for future research, and by 1900 the number of known elements had reached 84. In the spring of 1969, the American Chemical Society celebrated the centennial of Mendeleev's work in a program highlighted by a discussion of the synthesis of a new man-made element, number 104.

Figure 2.1 *Time perspective of the elements.*

ca. 500 B.C. Empedocles
All matter consists of combinations of four basic elements: earth, air, fire, and water.

"middle ages" Alchemists
The chemical substances consist of various combinations of three principles (elements): sulfur (principle of burning), salt (principle of earthiness), and mercury (principle of perfect metalness).

midseventeenth century Robert Boyle
The Sceptical Chymist (1661); Elements combine to make other substances but are not themselves made from any simpler species.
First serious attempts to classify elements and compounds by laboratory experimentation.

mideighteenth century Henry Cavendish
Prepared elemental hydrogen from metals and acids (1766).
mideighteenth century Joseph Priestley
Prepared elemental oxygen from various oxide compounds (1774).
mideighteenth century Carl Scheele
Proved that air is a mixture containing elemental oxygen and nitrogen (1777).

1789 Antoine Lavoisier
A first "textbook" of chemistry, listing twenty-three chemical elements (hydrogen, oxygen, nitrogen, carbon, sulfur, phosphorus, and 17 metals).

1808 John Dalton
The concept of atomic weights (weights of atoms relative to each other).

1817 J. W. Dobereiner
The existence of triads, families of three elements with similar properties and characteristic atomic weight relationships (e.g., iron, cobalt, and nickel with nearly the same atomic weights; chlorine, bromine, and iodine with the atomic weight of bromine nearly the average of the atomic weights of chlorine and iodine).
First significant evidence of "periodic" relationships.

1860s A. E. de Chancourtois
Periodic properties indicated by arranging elements in order of atomic weights in a helical pattern around a cylinder (1862).
1860s J. A. R. Newlands
Law of octaves—when elements are arranged in order of atomic weights, their properties are similar to those of the eighth preceding and following elements (1863).

1869 Dmitri Mendeleev and J. Lothar Meyer
Periodic charts showing elements of similar properties (e.g., lithium, sodium, potassium, rubidium, cesium) in vertical columns (1869).

The "periodic charts" developed independently by Meyer and Mendeleev bore considerable resemblance to the modern periodic table. Based on detailed experimental studies of many properties of the 63 known elements, these "charts" provided the basis for predicting the existence and properties of elements not yet known. Chemists were astounded at the accuracy of these predictions as the "missing" elements were discovered and characterized.

1914 H. G. J. Moseley
Bombardment of 42 solid elements with electron beams produced x-rays whose frequencies varied in an orderly periodic pattern as a function of *atomic numbers* rather than atomic weights.
Atomic numbers proved the key to periodic relationships among the elements.

1937 C. Perrier and E. Segrè
The first man-made element, technetium, number 43 (made by bombarding a molybdenum target with "heavy" hydrogen nuclei).

1940 D. R. Corson, K. R. MacKenzie, and E. Segrè
The last of the "preuranium" elements, astatine, number 85.

1940 G. T. Seaborg, E. M. McMillan, J. W. Kennedy, et al.
The first "transuranium" elements—neptunium (number 93) and plutonium (number 94).

Figure 2.1 (Continued)

1962   International Union of Pure and Applied Chemistry (IUPAC)
Agreement to select carbon-12 as reference standard for atomic weights.

1969   The Mendeleev Centennial Year
Marked by the description of a new element: atomic number 104.

future   ?
Elements beyond 104 have already been made. Research on "superheavy" elements may contribute greatly to the understanding of nuclear structure and stability.

2.1 ELEMENTS, COMPOUNDS, AND MIXTURES

One of the problems that plagued early chemists was how to tell an element from a compound. Today chemists characterize a substance as an element if it is chemically homogeneous, cannot be separated into simpler substances, and its nuclei all contain the same number of protons, a number unique to each element. These criteria bear a little closer examination.

In any of its three stable states—solid, liquid, or gas—matter can be heterogeneous or homogeneous. The composition of a heterogeneous material varies from one region to another within the sample; the composition of a homogeneous material does not. For example, a test tube full of water and iron filings comprises a heterogeneous sample. Sometimes a sample appears heterogeneous to the eye, but actually is not, in terms of *composition*. A jar of sulfur frequently contains large yellow lumps along with smaller fragments and a quantity of very pale yellow (nearly white) powder. Differences in particle size and color result from differing degrees of aggregation of the atoms of sulfur; the atoms themselves are chemically identical. A mixture of pure water and ice, although heterogeneous in terms of its physical states, is chemically homogeneous. Often rather sophisticated analysis is required to determine if a substance is truly homogeneous in a chemical sense. Purity, as the chemist uses the term, is a measure of the chemical homogeneity of a material in terms of a specified composition. For example, "pure" oxygen gas would contain only oxygen molecules (O_2, two atoms of oxygen connected by the sharing of electrons between the positive fields of their nuclei); "pure" sodium chloride would contain only ions (charged particles) of sodium and chlorine in equal numbers; a "pure" solution of ethyl alcohol in water would contain only molecules of these two compounds (although their relative numbers could be variable, since a solution is a mixture). In reality, "absolutely pure" materials represent an idealized condition, which can be approached more closely as more sophisticated purification techniques are devised, but which can never be attained on any large scale with currently available methods. The degree of purity of chemical products is usually indicated by their manufacturer. Such qualitative terms as "technical grade," "reagent grade," and (most recently) "chromatographically pure" are used to indicate increasing purity of material. Manufacturers also commonly list the impurities known to be present in a given sample as percentages of the total sample.

Most materials, both homogeneous and heterogeneous, can be separated into simpler components. Such separations (Figure 2.2) may involve physical techniques (sifting sand to remove gravel particles, filtering muddy water to remove large suspended particles, distilling water-alcohol mixtures to decrease the water content); in other cases chemical techniques may be required to effect the conversion of substances to simpler species. Much of the difficulty encountered by early chemists in their attempts to classify homogeneous substances as elements or compounds (chemical combinations of elements) resulted from the limitations imposed by the very few chemical procedures available. Thus mercuric oxide was classified quite early as a compound because it could be converted to simpler components, mercury

"Purity" is a term that is used in many different ways. A soap advertised as "$99\frac{44}{100}$ percent pure" may indicate something useful to the potential purchaser, but the chemist views the same soap as a rather complex mixture of chemical species and would not describe it as a chemically pure substance.

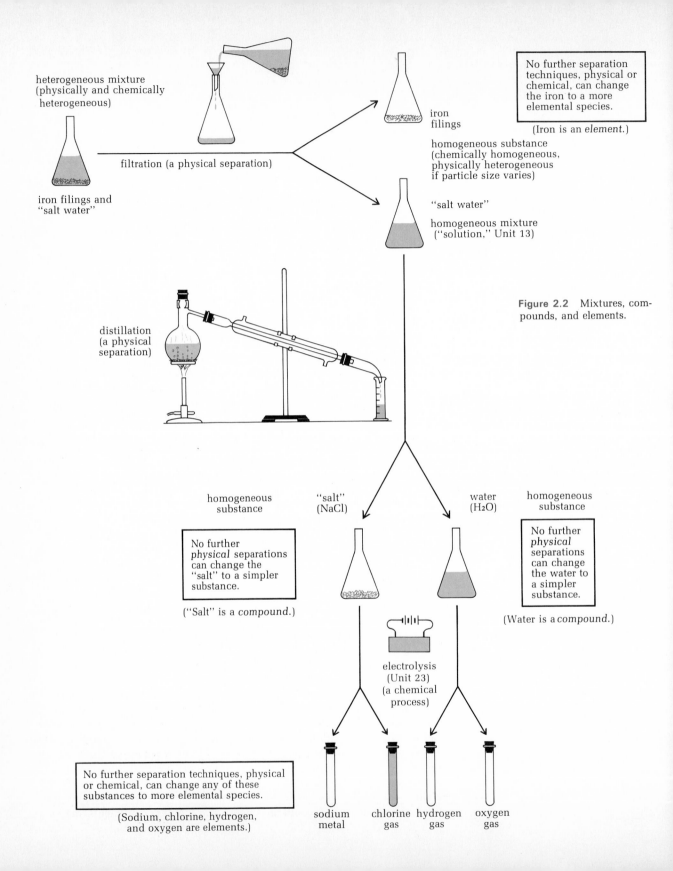

heterogeneous mixture
(physically and chemically
  heterogeneous)

iron filings and
"salt water"

filtration (a physical separation)

iron
filings

No further separation
techniques, physical or
chemical, can change
the iron to a more
elemental species.

(Iron is an *element*.)

homogeneous substance
(chemically homogeneous,
physically heterogeneous
if particle size varies)

"salt water"

homogeneous mixture
("solution," Unit 13)

distillation
(a physical
separation)

Figure 2.2 Mixtures, compounds, and elements.

homogeneous
substance

"salt"
(NaCl)

water
(H_2O)

homogeneous
substance

No further
physical separations
can change the
"salt" to a simpler
substance.

("Salt" is a *compound*.)

No further
physical
separations
can change
the water to
a simpler
substance.

(Water is a *compound*.)

electrolysis
(Unit 23)
(a chemical
process)

No further separation techniques, physical
or chemical, can change any of these
substances to more elemental species.

(Sodium, chlorine, hydrogen,
and oxygen are elements.)

sodium
metal

chlorine
gas

hydrogen
gas

oxygen
gas

and oxygen, by heating. These components, in turn, were classified as elements since even intense heating failed to change them to simpler species. Salt, on the other hand, presented a more complex problem. Table salt (sodium chloride) was not broken down into simpler components by the heating techniques available to early chemists. In fact, it was not until 1807, with the advent of electrochemical techniques, that Humphrey Davy was able to prove, by decomposition of salt with an electric current, that it is a compound of the elements sodium and chlorine.

The accepted criteria for classification of *homogeneous* substances, then, are as follows:

1. *Mixtures* may be separated into simpler components by *physical* techniques; that is, no charge in chemical composition of the components themselves is required.
2. *Compounds* cannot be separated into simpler components except by *chemical* techniques.
3. *Elements* cannot be separated into simpler components by either physical or chemical techniques.

Two of the three criteria for defining an element have been discussed. These were, in general, the sole criteria actually applied to chemical substances until fairly recent times.[1] As long as reasonably large samples of a substance could be obtained, its characterization was not unduly difficult, particularly with the techniques that had been developed by the early twentieth century. However, the third criterion—"an element consists of atoms whose nuclei contain the same number of protons, a number unique to each element"—has proved of critical importance in dealing with substances available only in minute quantities. In particular, such cases have occurred with several radioactive species (species whose nuclei tend to decompose). For example, original claims to the discovery of elements 43, 61, 85, and 87 have been successfully refuted by experiments showing that the species reported did not fit *all three* of the criteria listed.

The chemical identity of an atom ultimately depends on the structure of its nucleus and specifically on its *atomic number,* the number of protons in the nucleus. Customarily designated by the letter Z, the atomic number also indicates the number of electrons in an atom. (Because atoms are electrically neutral, the number of protons must equal the number of electrons.) Each atom of any one element has the same number of protons in its nucleus, and the structure of the nucleus cannot be disrupted by ordinary chemical or physical separation techniques.

2.2 ISOTOPES

In the strictest sense, even elements are not truly homogeneous. Most elements are a mixture of two or more *isotopes:* atoms whose nuclei contain the same number of protons but different numbers of neutrons. Hydrogen, for example, occurs as three isotopes: ordinary hydrogen, deuterium (heavy hydrogen), and tritium (radioactive hydrogen) (Figure 2.3). *Chemically,* isotopes of the same element are nearly identical. Chemists refer to isotopes by their *mass numbers* (the number of protons plus the number of neutrons). This convention has become familiar in the vocabulary of the twentieth century; we are frequently reminded of the presence of uranium-235, strontium-90, and cobalt-60. In chemical notation, the atomic number (Z) is usually written at the lower left of the symbol for the element and the mass number (A) at the

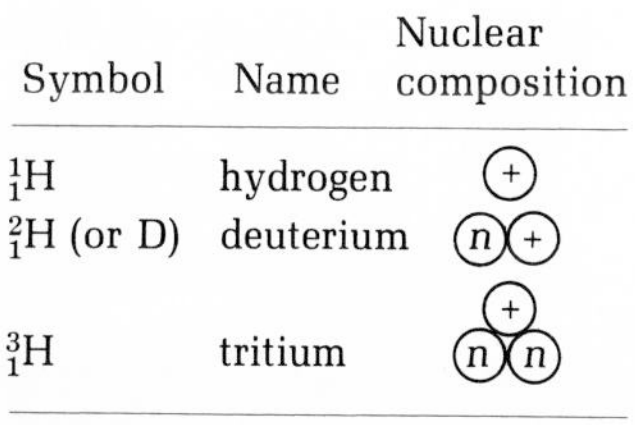

Symbol	Name	Nuclear composition
$^{1}_{1}$H	hydrogen	+
$^{2}_{1}$H (or D)	deuterium	n +
$^{3}_{1}$H	tritium	+ n n

Figure 2.3 Isotopes of hydrogen.

[1] For some interesting historical sidelights, see the series of articles by Bill Waggoner on "Spurious Elements" in *Chemistry,* beginning with the April, 1975, issue.

Table 2.1 NAMES AND SYMBOLS OF SOME COMMON ELEMENTS

SYMBOLS BASED ON CURRENT NAMES					
aluminum	Al	iodine	I	xenon	Xe
argon	Ar	krypton	Kr	zinc	Zn
arsenic	As	lithium	Li		
barium	Ba	magnesium	Mg		
beryllium	Be	manganese	Mn		
bismuth	Bi	neon	Ne	SYMBOLS BASED ON OLDER NAMES	
boron	B	nickel	Ni		
bromine	Br	nitrogen	N	antimony (L. stibium)	Sb
cadmium	Cd	oxygen	O	copper (L. cuprum)	Cu
calcium	Ca	phosphorus	P	gold (L. aurum)	Au
carbon	C	platinum	Pt	iron (L. ferrum)	Fe
cesium	Cs	radium	Ra	lead (L. plumbum)	Pb
chlorine	Cl	selenium	Se	mercury (L. hydrargyrum)	Hg
cobalt	Co	silicon	Si	potassium (L. kalium)	K
fluorine	F	strontium	Sr	silver (L. argentum)	Ag
germanium	Ge	sulfur	S	sodium (L. natrium)	Na
helium	He	titanium	Ti	tin (L. stannum)	Sn
hydrogen	H	uranium	U	tungsten (Ger. wolfram)	W

upper left: ${}^{235}_{92}U$, ${}^{90}_{38}Sr$, ${}^{60}_{27}Co$. Note that the symbols used are "chemical shorthand" representations for the element and are not just "abbreviations" of the element's current name (Table 2.1).

From standard symbolism, ${}^{A}_{Z}E^{(\text{charge})}$, it is possible to derive the numbers of protons, neutrons, and electrons contained in any neutral or ionic monatomic species:

The use of element symbols was first proposed, essentially in the present form, in 1811 by the Swedish chemist J. J. Berzelius.

$$\text{number of protons} = \text{atomic number}$$
$$\text{number of neutrons} = \text{mass number} - \text{atomic number}$$
$$\text{number of electrons} = \text{atomic number} - \text{ionic charge}$$

Example 1

Indicate the numbers of protons, neutrons, and electrons represented by ${}^{21}_{10}Ne$, ${}^{7}_{3}Li^{+}$, and ${}^{31}_{16}S^{2-}$.

Solution

For the Ne,

$$\text{number of protons} = 10$$
$$\text{number of neutrons} = 21 - 10 = 11$$
$$\text{number of electrons} = 10 - 0 = 10$$

For the Li^{+},

$$\text{number of protons} = 3$$
$$\text{number of electrons} = 7 - 3 = 4$$
$$\text{number of electrons} = 3 - 1 = 2$$

For the S^{2-},

$$\text{number of protons} = 16$$
$$\text{number of neutrons} = 31 - 16 = 15$$
$$\text{number of electrons} = 16 - (-2) = 18$$

Strictly speaking, of course, *weight* and *mass* are not equivalent terms. Common usage, however, frequently employs the term *weight* where *mass* would be more correct.

An alternative unit suggested as a replacement for *amu* is the *dalton* (D), in honor of John Dalton, originator of the atomic weight concept.

$$1D = 1\,amu = 1.660 \times 10^{-24}g$$

For a detailed discussion of "binding energy", see Unit 35.

It should be noted that the relative abundances of isotopes of an element vary somewhat with the geographical location of the element sample and, since the advent of nuclear devices, with the isotopic "pollution" created by man.

Although the mass number of an isotope and the atomic number of an element are always whole numbers, because they simply *count* particles, the *atomic weight* of an element is not (Figure 2.4). The atomic weight is found by a calculation involving experimentally determined isotopic masses and *abundances* (relative amounts), as illustrated in Example 2. Neither isotopic masses nor abundances are integers. The atomic weight may be expressed in "amu" (atomic mass units), defined as exactly 1/12 the mass of a carbon-12 atom (6 protons and 6 neutrons in the nucleus, plus 6 electrons outside the nucleus). Since naturally occurring carbon consists of a mixture of isotopes (Example 3), the atomic weight of carbon is not exactly 12 amu. (12.01115 ± 0.00005 is the presently accepted value). An additional factor contributing to decimal values in atomic weights arises from the fact that the isotopic mass is not exactly equal to the mass number. Protons and neutrons have slightly different masses, and a small difference exists between the masses of isolated protons and neutrons and their masses when combined in a nucleus. This mass difference is attributed to a conversion of some mass to energy ("binding energy of the nucleus") according to the Einstein equation, $E = mc^2$. The magnitude of this energy can be appreciated when one realizes that the reaction involved in nuclear fission (e.g., the atomic bomb) converts only a small fraction of the total mass to energy. Mass-energy conversion is discussed in Unit 35.

The atomic weight of an element can be calculated from the relative abundances (i.e., the fraction of each isotope in the total element sample) and the masses of the individual isotopes (in amu). Both types of data can now be determined with considerable accuracy by the technique of mass spectrometry, a method similar to that used by Thomson in determining the charge-to-mass ratio (e/m) of the electron (Introduction, Section 0.5). The necessary calculation may be formulated as

$$\text{atomic weight} = (\text{abundance}_1 \times \text{mass}_1) + (\text{abundance}_2 \times \text{mass}_2) + \cdots \tag{2.1}$$

Example 2

Calculate the atomic weight of oxygen from the following experimental data:

Isotope	Abundance	Mass (amu)
$^{16}_{8}O$	0.99759	15.995
$^{17}_{8}O$	0.00037	16.991
$^{18}_{8}O$	0.00204	17.991

Solution [based on equation (2.1)]

atomic weight = $(0.99759 \times 15.995) + (0.00037 \times 16.991) + (0.00204 \times 17.991)$

atomic weight = 15.999 amu

Example 3

Seven isotopes of carbon, from $^{10}_{6}C$ through $^{16}_{6}C$, have been detected. Only two of these, $^{12}_{6}C$ and $^{13}_{6}C$, are stable; the others are radioactive (Unit 35). If the accepted atomic weight value of 12.01115 for carbon is based only on the stable isotopes, what are the relative abundances of these two isotopes? (The mass of $^{13}_{6}C$ is 13.00335 amu.)

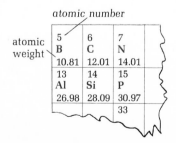

Figure 2.4 Segment of a simplified periodic table.

DIFFERENT ATOMIC WEIGHT SCALES

The first widely accepted use of relative atomic weights was based on John Dalton's suggestion of assigning an arbitrary value of 1 for the lightest element, hydrogen. A number of alternative reference standards have been employed from time to time, including Berzelius's suggestion (early nineteenth century) of 100 for oxygen. For almost a century, chemists used the standard of exactly 16 for the natural mixture of oxygen isotopes, as suggested by the Belgian chemist J. S. Stas. During this period, physicists, developing an interest in nuclear masses, changed to their own scale on which 16 was assigned for the single isotope $^{16}_{8}O$. The two scales were related by

(chemistry atomic weight) = 1.000272 × (physics atomic weight)

In 1962 the dual scales were abandoned, by international agreement, in favor of the present carbon-12 scale.

Because the accepted atomic weight of oxygen on the present scale is 15.9994 ± 0.0001, only very slight differences exist between current and pre-1962 data involving atomic weights. Scientists concerned with very accurate values must, however, be careful in using appropriate corrections when necessary.

Solution

If we represent the abundance of $^{12}_{6}C$ by y, then, since there are only two isotopes involved, the abundance of $^{12}_{6}C$ is $1 - y$. From equation (2.1),

$$12.01115 = [(1 - y) \times 12^*] + [y \times 13.00335]$$

$$12.01115 = 12 - 12y + 13.00335y = 12 + 1.00335y$$

$$y = \frac{0.01115}{1.00335} = 0.01111 \quad \text{(abundance of } ^{13}_{6}C\text{)}$$

and

$$1 - y = 1 - 0.01111 = 0.98889 \quad \text{(abundance of } ^{12}_{6}C\text{)}$$

* The mass of $^{12}_{6}C$ = exactly 12 amu, by definition.

2.3 CHEMICAL AND PHYSICAL PROPERTIES

Chemical properties of substances are those that are associated with chemical reactions: processes, we now recognize, in which electron distributions around nuclei of participating species are significantly altered without change in nuclear composition. Chemical reactions are discussed in Unit 3. Examples are such processes as the reaction of sodium metal with water to form hydrogen gas and sodium hydroxide; the conversion of lead and lead dioxide to lead sulfate (involving sulfuric acid) during the discharge of an automobile battery; and the electroplating of metallic copper from solutions of copper salts.

Physical properties do not involve major changes in electron distributions around nuclei; that is, the basic chemical composition of the species is not altered by physical changes. Such properties as the melting point (temperature at which a solid melts), electrical conductivity, surface luster ("shininess"), and hardness are considered physical properties.

The modern arrangement of elements in the form of a periodic table (Table 2.2) evolved historically (Figure 2.1) from careful observations of both physical and chemical properties, many of which showed repetitive patterns (periodicities) that proved to be functions of the atomic numbers of the elements. It is useful to examine some of these properties in order to establish the nature of chemical periodicity.

Before we do so, however, there are some important considerations we need to establish. Chemistry is an experimental science; our understanding of chemical phenomena must be based on observations and measurements. In attempting to distinguish between chemical and physical properties, or to characterize a unique substance by such properties, we must be aware of various factors that may influence our observations. One such factor is the *physical state* of the substance under investigation, that is, whether we are noting properties characteristic of the solid, the liquid, or the gaseous state of the substance. "Hardness," for example, is a property of water, but only for the solid state ("ice"). Iodine crystals appear blue-black on cursory examination, but the vapor (gas) is purple. Our common experience with oxygen as a gas would suggest that oxygen is colorless, but liquid oxygen is a beautiful blue (and will adhere to a magnet). The state of subdivision, that is, the particle size, of a sample can also be an important factor. Aluminum foil can be used to wrap food for cooking directly in a campfire, but aluminum wiring melts so easily that it is considered unsafe for many types of electric circuits and fine aluminum dust in a proper mixture with air may explode from a spark.

The environment in which observations are made may also have a significant influence. We know that temperature is a factor in the physical character of a substance. Water is a liquid only within a certain temperature range; at lower temperatures only the properties of the solid are observed and higher temperatures permit observation of only gas-phase properties. Other forms of energy must also be considered. Many chemical substances appear relatively stable under some conditions, but change drastically when exposed to light, as in the case of chemicals used in photographic film coatings. You are probably familiar with the recent attention given to the freons (compounds of carbon, fluorine, and chlorine) used in many aerosol sprays. Observations had characterized the freons as relatively unreactive chemicals and thus basically harmless until rather sophisticated studies suggested that they were *not* unreactive. Products from ultraviolet light-induced decomposition of freons react with the ozone of our upper atmosphere, thereby posing a very serious potential threat to the environment.

Most of the observations that we can make easily are of *bulk* properties, that is, those associated with very large numbers of atoms or atomic combinations. We will discuss several such properties throughout our study of chemistry. Of equal or greater importance, however, are "single-atom" properties, such as atomic size, ionization potential (Unit 1), or electronic structure. One of our goals in a study of chemistry is to relate "single-atom" properties to the more familiar bulk properties of matter. The keys to this relationship lie in an understanding of chemical bonding (Section Three) and the application of atomic theory to the investigation of the physical states of matter (Section Four).

For a discussion of "chemical hardness" in water, see Unit 6.

Many a novice prospector has followed the lure of "fool's gold," actually an iron salt.

Early studies of combustion (burning) provide a good example of the failure to recognize the influence of environment on observations. Many substances, such as magnesium, were observed to gain in weight when burned. The suggestion was made that such materials contained *phlogiston,* a substance of *negative weight.* Combustion released the phlogiston, allowing the material to become heavier. This theory was widely accepted until Antoine Lavoisier, in 1785, demonstrated that the weight gain resulted from combination with oxygen from the environment during combustion.

The smallest sample that we can see contains *many* more atoms than there are grains of sand on our largest public beach. A 1-mg sample of carbon, for example, consists of about 50,000,000,000,000,000,000 atoms of carbon.

Table 2.2 PERIODIC TABLE OF THE ELEMENTS

IA	IIA	IIIB	IVB	VB	VIB	VIIB	VIIIB	VIIIB	VIIIB	IB	IIB	IIIA	IVA	VA	VIA	VIIA	Zero (VIIIA)
e 1 **H** 1.0080																	2 **He** 4.0026
e 3 **Li** 6.94	4 **Be** 9.012											e 5 **B** 10.811	e 6 **C** 12.0115	e 7 **N** 14.0067	e 8 **O** 15.9994	e 9 **F** 18.9984	10 **Ne** 20.18
e 11 **Na** 22.9998	e 12 **Mg** 24.31											13 **Al** 26.9815	e 14 **Si** 28.086	e 15 **P** 30.974	e 16 **S** 32.06	e 17 **Cl** 35.453	18 **Ar** 39.948
e 19 **K** 39.102	e 20 **Ca** 40.08	21 **Sc** 44.96	22 **Ti** 47.90	e 23 **V** 50.94	e 24 **Cr** 51.996	e 25 **Mn** 54.938	e 26 **Fe** 55.847	e 27 **Co** 58.933	28 **Ni** 58.71	e 29 **Cu** 63.546	e 30 **Zn** 65.37	31 **Ga** 69.72	32 **Ge** 72.59	33 **As** 74.9216	e 34 **Se** 78.96	35 **Br** 79.909	36 **Kr** 83.80
37 **Rb** 85.47	38 **Sr** 87.62	39 **Y** 88.91	40 **Zr** 91.22	41 **Nb** 92.91	e 42 **Mo** 95.94	m 43 **Tc** (99)	44 **Ru** 101.07	45 **Rh** 102.91	46 **Pd** 106.4	47 **Ag** 107.868	48 **Cd** 112.40	49 **In** 114.82	e 50 **Sn** 118.69	51 **Sb** 121.75	52 **Te** 127.60	e 53 **I** 126.904	54 **Xe** 131.30
55 **Cs** 132.91	56 **Ba** 137.34	57 *****La** 138.91	72 **Hf** 178.49	73 **Ta** 180.95	74 **W** 183.85	75 **Re** 186.2	76 **Os** 190.2	77 **Ir** 192.22	78 **Pt** 195.09	79 **Au** 196.97	80 **Hg** 200.59	81 **Tl** 204.37	82 **Pb** 207.2	83 **Bi** 208.98	84 **Po** (210)	m 85 **At** (210)	86 **Rn** (222)
m 87 **Fr** (223)	88 **Ra** (226)	89 ‡**Ac** (227)	m 104 ‡**Rf** (261)	m 105 ‡**Ha** (260)	m 106 ‡?												

*Lanthanoid series

58 **Ce** 140.12	59 **Pr** 140.91	60 **Nd** 144.24	m 61 **Pm** (147)	62 **Sm** 150.4	63 **Eu** 151.96	64 **Gd** 157.25	65 **Tb** 158.9	66 **Dy** 162.50	67 **Ho** 164.93	68 **Er** 167.26	69 **Tm** 168.93	70 **Yb** 173.04	71 **Lu** 174.97
90 **Th** 232.0	91 **Pa** 231.0	92 **U** 238.03	m 93 **Np** 237.0	m 94 **Pu** (244)	m 95 **Am** (243)	m 96 **Cm** (247)	m 97 **Bk** (247)	m 98 **Cf** (251)	m 99 **Es** (254)	m 100 **Fm** (257)	m 101 **Md** (258)	m 102 **No** (255)	m 103 **Lr** (258)

‡Actinoid series

Key to Families of Elements

Group	Family	Group	Family
IA	Alkali	VB	Vanadium
IIA	Alkaline earth	VIB	Chromium
IIIA	Aluminum	VIIB	Manganese
IVA	Germanium	VIIIB	Iron/Platinum
VA	Nitrogen	IB	Copper
VIA	Chalcogens	IIB	Zinc
VIIA	Halogen	IIIB	Scandium
Zero (VIIIA)	Noble gas	IVB	Titanium

‡ Involved in disputed discovery claims (U.S.–U.S.S.R.).

= liquid

= solid

= gas   Reflects a normal temperature range of 15–30° C. Note that Cs, Fr, and Ga melt around 30° C.

= man-made elements

e = elements essential to living organisms

m = man-made elements

Table 2.3 APPROXIMATE MELTING AND BOILING POINTS (°C) OF SOME CHEMICAL ELEMENTS[a]

Key

	180
	Li
	1320

approximate b.p. at 1.00 atm pressure
symbol
approximate m.p.

	IA	IIA	IIIB	IVB	VB	VIB	IIIA	IVA	VA	VIA	VIIA	ZERO
m.p.	180	1280					2000	3730	−210	−220	−220	−250
sym	**Li**	**Be**					**B**	**C**	**N**	**O**	**F**	**Ne**
b.p.	1320	2510					2550	4200	−200	−180	−190	−250
m.p.	100	650					660	1410	40	120	−100	−190
sym	**Na**	**Mg**					**Al**	**Si**	**P**	**S**	**Cl**	**Ar**
b.p.	880	1110					2470	2480	280	440	−40	−190
m.p.	60	850	1540	1670	1890	1880	30	940	820	220	−10	−160
sym	**K**	**Ca**	**Sc**	**Ti**	**V**	**Cr**	**Ga**	**Ge**	**As**	**Se**	**Br**	**Kr**
b.p.	760	1480	2730	3260	3000	2200	2400	2700	610	690	60	−150
m.p.	40	770	1510	1850	2470	2610	160	230	630	450	110	−120
sym	**Rb**	**Sr**	**Y**	**Zr**	**Nb**	**Mo**	**In**	**Sn**	**Sb**	**Te**	**I**	**Xe**
b.p.	690	1380	3200	3580	4930	5560	2080	2270	1440	990	180	−110
m.p.	30	710	920	2230	3000	3410	300	330	270	250	⑦	−70
sym	**Cs**	**Ba**	**La**	**Hf**	**Ta**	**W**	**Tl**	**Pb**	**Bi**	**Po**	**At**	**Rn**
b.p.	710	1500	3470	5400	6100	5900	1460	1740	1630	960	⑦	−60

[a] The values tabulated here are rounded off to the nearest 10°C for simpler comparisons.

↑ too unstable (radioactive) for accurate data

2.4 PERIODICITY OF SOME PHYSICAL PROPERTIES

Most of the elements are solids at "room temperature" (∼25°C). A few—nitrogen, oxygen, fluorine, chlorine, hydrogen, and the noble gases (column 0, Table 2.2)—are gases; two of the common elements—mercury and bromine—are liquids. Cesium and gallium liquefy around 30°C.

Physical States at Room Temperature

Some trends are discernible in melting points and boiling points (Table 2.3), although these properties alone would not have suggested the present periodic system. Both melting and boiling points show a somewhat irregular increase from left to right in a horizontal row (*period*) until maximum values are reached, after which

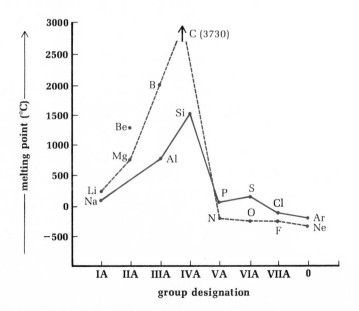

Figure 2.5 Some trends in melting points.

Table 2.4 DENSITIES (g cm^{-3}) OF SOME CHEMICAL ELEMENTS AT 20°C AND 1.0 ATM PRESSURE[a]

Li 0.53	Be 1.8				B 2.3	C 2.2	N 1.2 × 10^{-3}	O 1.3 × 10^{-3}	F 1.6 × 10^{-3}	Ne 0.83 × 10^{-3}
Na 0.97	Mg 1.7				Al 2.7	Si 2.4	P 2.2	S 2.0	Cl 3.0 × 10^{-3}	Ar 1.7 × 10^{-3}
K 0.87	Ca 1.6	Sc 3.0	Ti 4.5	V 6.0	Ga 5.9	Ge 5.4	As 5.7	Se 4.8	Br 3.1	Kr 3.5 × 10^{-3}
Rb 1.5	Sr 2.5	Y 5.5	Zr 6.4	Nb 8.4	In 7.3	Sn 7.2	Sb 6.7	Te 6.2	I 4.9	Xe 5.5 × 10^{-3}
Cs 1.9	Ba 3.5	La 6.2	Hf 13.3	Ta 16.6	Tl 11.8	Pb 11.4	Bi 9.7	Po (?)	At (?)	Rn (?)

no accurate data

[a]Many elements may exist in different forms in the solid state. Densities given are for the most common forms. For comparison purposes, water has a density of approximately 1 g cm^{-3} at 20°C. Dry "air" at 20°C and 1.0 atm pressure has a density of 1.21 × 10^{-3} g cm^{-3}.

there is a sharp decrease (Figure 2.5). Within vertical columns (*groups*), only the group VIIA and group 0 elements show simple, regular trends.

Densities A comparison of relative densities (ratio of mass to volume) is meaningful only for densities measured at the same temperature, since substances usually change volume as a function of temperature. As in the cases of melting and boiling points, densities of the elements show an increase from left to right in a given period up to some maximum value, followed by a sharp decrease. Densities are most conveniently measured around room temperature, so that those elements that are gases have very low densities in comparison to those of solid or liquid elements. Within a group, densities tend to increase with increasing atomic weight, with three notable exceptions—magnesium, potassium, and calcium (Table 2.4).

Metals, Metalloids, and Nonmetals Metals possess certain characteristic properties: malleability (ability to be flattened into thin sheets), ductility (ability to be drawn out into wires), surface luster ("shininess"), and—most important—electrical conductivity, *which generally decreases with increasing temperature*. Most of the chemical elements are metals. All elements to the left of the grouping B-Si–Ge–Sb–Po (Figure 2.6) are classified as metals, although this must be recognized in many cases as a somewhat arbitrary distinction. For example, one form of tin shows none of the typical metallic characteristics mentioned, while all are exhibited by one form of arsenic. Further studies will disclose that conductivity, as well as most other "bulk" properties, is often as much a reflection of the ways atoms are arranged in a particular sample as of the characteristics of the atoms themselves.

About 20 percent of the elements are classified as nonmetals, having essentially none of the typical metallic properties. These elements are found in the periodic table to the right of the grouping B-Si–As–Te–Po (Figure 2.6), although the group 0 elements are usually classified separately as "noble gases." Again, many cases of somewhat

Figure 2.6 Metallic ⟶ non-metallic trends.

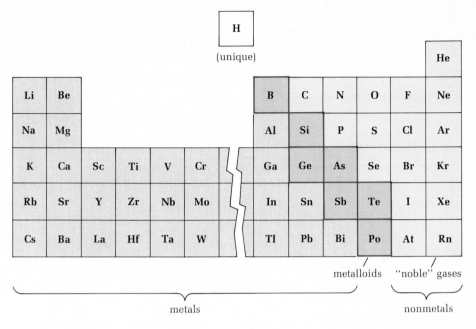

arbitrary classification could be mentioned, notably carbon which, in the form of graphite, has a significant electrical conductivity.

The remaining elements—boron, silicon, germanium, arsenic, antimony, tellurium, and polonium—are designated as *metalloids*. These have many properties typical of the nonmetals, but exhibit electrical conductivity which, unlike that of the metals, tends to *increase* with increasing temperature.

Physical Properties and the Atomic Model

Do the observed trends in properties such as melting point, density, and metallic character correlate with our simple "quantum" model for the atom? At first glance it would seem that there is little correlation, particularly with properties such as melting point where "trends" are irregular, at best. The only simple relationship appears in terms of the metallic ⟶ nonmetallic character. If we work out the electron configurations (Unit 1) of the first 20 elements, we note something rather interesting (Figure 2.7). Metals appear to have only a few electrons in the highest energy sublevels, whereas the highest sublevel is nearly filled for nonmetals (and completely filled for the group 0 elements). Metalloids, whose properties are intermediate between those of metals and nonmetals, have "intermediate" electron configurations. Metallic character, as we shall see in later discussions, is associated with the relatively easy removal of "outermost" electrons to form positive ions. If all such "outermost" electrons are to be removed, it seems reasonable in view of the trend in successive ionization potentials (Figure 1.17, Unit 1) that this is most likely to occur when the highest energy sublevels are sparsely occupied. Although the picture is more complex with heavier elements, an analogous relationship can be described (Unit 27).

Other physical properties are also explained well by our atomic model but because such properties as melting point and density are "bulk" properties, we must also consider the factors involved in the aggregation of atoms and atomic combinations. These factors will be explored in Section Four.

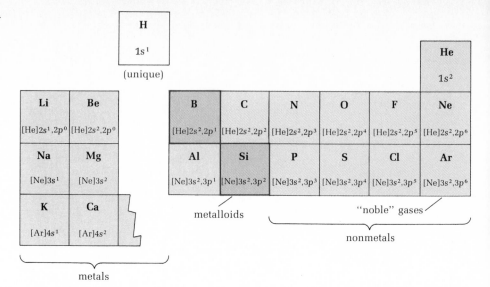

Figure 2.7 Metallic $\longrightarrow$ nonmetallic trends and electron configurations.

2.5 SOME CHEMICAL PROPERTIES

Although certain rough trends can be noted in many physical properties of the elements, these are not sufficiently periodic to form the basis of the highly ordered classification system of the periodic table. Many physical properties are more easily related to the type of atomic aggregations than to the nature of the atoms themselves.

Chemical properties, however, exhibit more definitive trends and can be related more directly to atomic characteristics.

In discussing chemical properties of elements, the chemist is considering changes in electron distribution around the atomic nuclei. Some terms used to refer to new chemical substances formed from atoms of the elements are as follows:

Formula: A collection of element symbols with subscript numbers indicating the number of atoms of each element in the chemical species described (e.g., chlorine molecule, Cl_2; barium fluoride, BaF_2).

Molecules: Characteristic groups of atoms held together by the sharing of electrons among two or more atomic nuclei (e.g., water, H_2O; molecular nitrogen, N_2; ammonia, NH_3).

Compound: A pure substance whose simplest components are molecules (covalent compound) or oppositely charged ions in ratios giving a zero net charge (ionic compound).

Cations: Ions having a *net* positive charge, resulting from the presence of fewer extranuclear electrons than nuclear protons (e.g., sodium ion, Na^+; calcium ion, Ca^{2+}; ammonium ion, NH_4^+).

Anions: Ions having a *net* negative charge, resulting from the presence of more extranuclear electrons than nuclear protons (e.g., chloride ion, Cl^-; sulfide ion, S^{2-}; bicarbonate ion, HCO_3^-).

(net ionic charge = total number of protons − total number of electrons)

A special case called the zwitterion (or dipolarion) in which equal numbers of positive and negative charge centers exist within atomic aggregates results in a zero *net* charge. Examples will be encountered in later discussions of amino acids and proteins (Unit 32).

Ionization Potentials and Electron Affinities

The atoms of certain elements, notably the metals, tend to lose electrons more or less readily to form cations. The ease with which electrons can be removed from the positive nuclear fields of gaseous atoms is indicated by the *ionization potentials* (Figure 2.8), which are related to the energy required to remove an electron from the influence of the positive nucleus. The lower the ionization potential, the easier it is

Figure 2.8 Trends in ionization potential (first ionization of neutral gaseous atom).

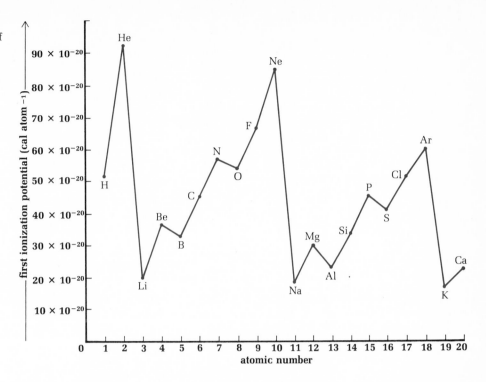

to convert the atom to a cation. Removal of a second, or third, electron from an atom requires considerably more energy than that for the first electron loss, since each successive electron is held by a larger *net* positive field (Unit 1).

Certain other elements, particularly those in groups VIA and VIIA (Table 2.2) form stable monatomic (one-atom) anions by gaining one or two additional electrons. Addition of one electron to a neutral atom might be expected to *release* energy, since the electron is moved into the positive field of the nucleus. Addition of more electrons, however, requires energy to force the negative electrons into an ion already having a net negative charge. *Electron affinity*[2] is a measure of the energy involved in the gain of electrons (Figure 2.9) by a gaseous neutral atom.

Some features of the ionization potential and electron affinity diagrams merit closer examination. The high energy required for ionization of the noble gas atoms, for example, indicates an unusual stability for these atoms which may be related to their electronic configurations. Species isoelectronic with these atoms are also relatively stable [e.g., Li^+, (He); Na^+, Mg^{2+}, Al^{3+}, N^{3-}, O^{2-}, F^-, (Ne)]. This stability is consistent with our atomic model, which suggests that such species have completely filled highest energy sublevels. Indeed, with helium and neon, and their isoelectronic species, the entire highest energy *level* is filled (all sublevels). We also observe that certain atoms, such as beryllium and nitrogen, or magnesium and phosphorus, appear "out of line" in both the ionization potential and electron affinity diagrams. Again, our atomic model is applicable. Beryllium and magnesium are ns^2 species

[2] For a description of how electron affinities may be determined, see "The Experimental Values of Atomic Electron Affinities," by E. C. M. Chen and W. E. Wentworth in the August 1975 issue of the *Journal of Chemical Education* (pp. 486–489).

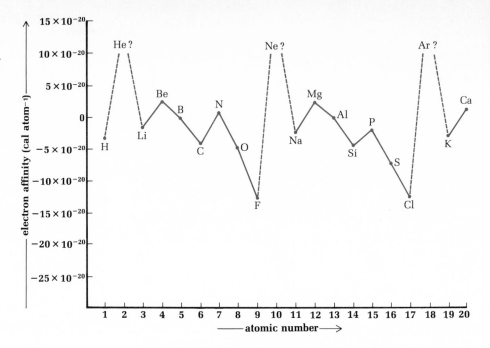

(with a filled sublevel) ($2s^2$ and $3s^2$, respectively), whereas nitrogen and phosphorus (ns^2, np^3) illustrate the stability associated with a half-filled sublevel (one electron in each $2p$ or $3p$ orbital, respectively).

Compound Formation

Chemical combination of atoms results in *compounds*, conveniently classified as either *ionic* (Unit 6) or *covalent* (Unit 7). The distinction between these two classes is somewhat arbitrary, since the differences are really a matter of degree rather than of type. In general, ionic compounds, in simple terms, are described as those consisting of aggregates of oppositely charged ions, whereas covalent compounds consist of electrically neutral molecules in which two or more atoms have combined through the "sharing" of orbital electrons by the positive fields of their respective nuclei. In either of the condensed states of matter (solid or liquid), some electron sharing among neighboring particles occurs, so the "100 percent ionic" compound is a considerable oversimplification, as we shall see in Units 6 and 7. Within the periodic table we find some trends in ion stability (Figure 2.10) that can, like earlier properties discussed, be related readily to our atomic model. The relationship, most simply noted with ions having noble gas electron configurations, becomes more complex with the heavier elements and with the "transition" elements (groups shown as B in the periodic table). We shall postpone the study of these more complicated cases, which involve *d*-orbital electron considerations, to later units.

The formation of chemical compounds is always associated with energy changes (Unit 5). If we compare the magnitudes of electron affinities (Figure 2.9) with those of ionization potentials (Figure 2.8), it would appear that the energy released when one atom gains an electron is considerably less than the energy required to remove that electron from another atom. This would seem to suggest that compound formation always requires an input of energy. Such is not, however, the case. Additional energy sources for ionic compound formation can be identified, such as the energy released by agglomeration of oppositely charged ions (Unit 6). Other, less easily visualized,

Figure 2.10 Some common stable monatomic ions. (The species B^{3+}, C^{4+}, C^{4-}, Si^{4+}, Si^{4-} are not chemically stable.)

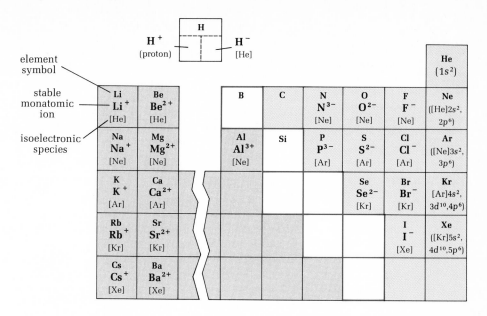

factors may contribute to the energy needs for covalent compound formation. Thus energy *may* be liberated as compounds are formed, although for many compounds this does not occur.

It is not always easy to predict whether two elements are more likely to form an ionic or a covalent compound. *As a first approximation, we will "guess" that an ionic compound will be formed from two elements when one can become isoelectronic with a noble gas by loss of no more than three electrons and the other can achieve noble gas configuration by gaining no more than three electrons.* If either of the elements under consideration fails to comply with these requirements, we will "guess" that covalent compound formation is more likely to occur. It should be apparent from Figure 2.10 that metal-nonmetal combinations are most likely to be ionic. It should be noted that although atoms of two different elements *may* combine in compounds, many such possibilities do not occur. This "first approximation" is a considerable simplification, but it will serve our needs until we can develop some more sophisticated concepts in later units.

Example 4

Which of the following pairs of elements will probably form ionic compounds, and which are more likely to form covalent compounds?

a. sodium and oxygen
b. nitrogen and oxygen
c. nitrogen and magnesium
d. carbon and chlorine

Many nitrogen-oxygen compounds are known and several of these have been identified as serious air pollutants from sources such as the automobile engine.

Solution

a. Sodium (a metal) forms the ion Na^+, isoelectronic with neon, by loss of one electron, while gain of two electrons by oxygen (a nonmetal) forms O^{2-}, isoelectronic with neon. These two elements would thus be expected to form an *ionic* compound.

The simplest carbon-chlorine compound, carbon tetrachloride (CCl_4), has been used in certain types of fire extinguishers and as a "cleaning fluid" for fabrics. Both uses have been generally discontinued because of evidence of health and safety hazards from CCl_4.

b. Although both nitrogen and oxygen can form ions isoelectronic with neon by a change involving no more than three electrons (N^{3-} and O^{2-}), both such species are anions. Neither element can form a stable cation by loss of three or fewer electrons. In such a case, *covalent* compound formation is predicted.

c. Both the criteria for *ionic* compound formation are met for combination of nitrogen (as N^{3-}) with magnesium (as Mg^{2+}).

d. Carbon may become isoelectronic with either helium (as C^{4+}) or neon (as C^{4-}), but either ion would require a four-electron change and this is unlikely. As a result the combination of carbon with chlorine is most likely to be covalent.

2.6 CHEMICAL FAMILIES

As more sophisticated techniques were developed, the isolation and characterization of the chemical elements began to indicate certain collections of elements with very similar properties. By the early nineteenth century, Dobereiner (Figure 2.1) had recognized many such groupings. Today it is common to refer to a number of *chemical families* (Figure 2.11) among the representative elements (those labeled A in the periodic table). Additional groupings often discussed are the *transition elements* (labeled B in the periodic table) and the *rare earth elements* (shown as the two lower, separate rows in Table 2.2).

Hydrogen is unique in having no other elements of similar properties.

The original arrangement of elements in vertical rows identified by group numbers was suggested by Mendeleev (Figure 2.1), on the basis of observations of

Figure 2.11 Chemical families.

alkali metals
group IA: Li, Na, K, Rb, Cs, Fr

Named for the strongly basic (alkaline) properties of their hydroxides. Characterized by low densities, unusual softness, violent reactions with water, formation of +1 cations as their only stable ions.

alkaline earth metals
group IIA: Be, Mg, Ca, Sr, Ba, Ra

Also form basic hydroxides, but these are significantly less soluble in water than group IA hydroxides. More brittle than the IA metals, react less readily with water, form +2 cations.

chalcogens (pronounced kal′-ko-jens)
group VIA: O, S, Se, Te, Po

Named from the Greek *chalkos* (copper) since many of these elements are found in copper ores, combined with the metal.

halogens
group VIIA: F, Cl, Br, I, At

Named for a Greek word meaning "salt-formers" since all form salts with group IA metals (e.g., sodium chloride). Highly reactive nonmetals, forming stable −1 anions. Except for fluorine, form many polyatomic (many-atom) anions with oxygen (e.g., ClO^-, ClO_2^-, ClO_3^-, ClO_4^-). These elements are very toxic in high concentrations.

noble gases
group 0: He, Ne, Ar, Kr, Xe, Rn

Named for their "disinterest in combining with the 'common' elements." Form no stable monatomic ions. Prior to 1960, considered "inert" or incapable of forming compounds. Compounds of some noble gas elements are now well known.

recurring (periodic) similarities which seemed, in most cases, to be a function of atomic weight. As more accurate methods for atomic weight determination were developed, it became apparent that any arrangement by *atomic weight* would produce anomolies from the periodic trends. For example, the positions of potassium (39.102) and argon (39.948) would be out of order. In 1914, the English physicist H. G. J. Moseley discovered that the wavelength of x-rays produced in an x-ray tube was a periodic function of the *atomic number* of the metal used as the anode in the x-ray tube. This discovery led to the revision of the periodic table, showing elements arranged in order of atomic numbers rather than weights. This revision is important in allowing us to relate the periodic properties of the elements to our atomic model.

Several different conventions have been suggested for designating the groups in the periodic table.[3] Some authors, for example, prefer VIIIA to zero, for the noble gases. Our convention uses B to indicate the transition elements (elements 21–30, 39–48, 57–80, and 89–?) and A for the other elements, commonly referred to as the *representative* elements. One advantage of this system is as a memory aid to the study of representative elements. As will be discussed in greater detail later, the chemical bonding in these elements involves the outermost (*valence*) electrons, and the group number tells us, for the A groups, the number of valence electrons of the neutral atom. For example, lithium, sodium, potassium, etc. (IA group) atoms have *one* valence electron each; oxygen, sulfur, etc. (VIA group) atoms each have *six* valence electrons.

2.7 ATOMIC AND IONIC DIMENSIONS

Modern experimental techniques permit rather accurate determinations of the sizes of atoms and ions. Such data correlate well with the positions of the elements in the periodic table (Figure 2.12). Atomic diameters show a regular increase with increasing

Figure 2.12 Trends in atomic sizes. The drawings indicate approximations of sizes for isolated atoms of the representative elements. Literature values vary considerably because of differences inherent in the various experimental methods employed and in definitions of "effective radii." Inclusion of the transition elements would have shown a size minimum near the middle of each period.

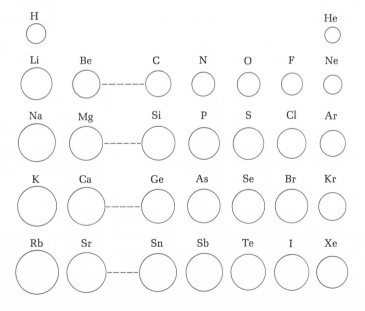

[3] For a brief historical discussion of group designations, see Fernelius, W. C., et al. in the September, 1971, issue of the *Journal of Chemical Education* (pp. 594–596).

Figure 2.13 Relative sizes and ionization potentials (eV atom^{-1}).

(1.0 eV = 3.8 × 10^{-20} cal)

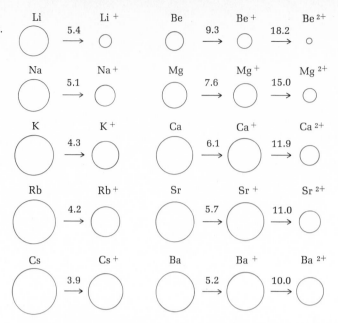

There is as yet no experimental method for direct measurements of sizes of *isolated* atoms. Since some degree of electron overlap occurs in atomic aggregates, we must deal with estimates of "effective" radii (Figure 2.12).

The valence (outermost) electrons of carbon, for example, sense an *effective* nuclear change of less than +6 because of the screening effect of the inner (1s^2) electrons.

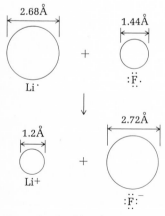

Figure 2.14 Size change in electron loss or gain.

atomic number within each group (vertical column) and a general decrease with increasing atomic number (from left to right) in a given period.

Two factors may be related, through our atomic model, to the periodic trends observed in atomic sizes. As we read from left to right within a given period of the atomic table, nuclear charge increases while electrons continue to fill a particular principal energy level. Within each *group,* however, nuclear charge increase is accompanied, for each successive element from top to bottom of the group, by the use of successively higher principal energy levels. Thus we may explain size decrease within a period by the increasing electrostatic attraction of successively more positive nuclei for electrons in the same principal level. The size increase moving down a group is attributed to the succession of higher energy levels for valence electrons and, in addition, the reduction (*shielding*) of effective nuclear charge by electrons in filled principal energy levels.

Stable monatomic ions range in diameter from about 0.6 Å (Be^{2+}) to around 4.4 Å (Te^{2-}). The proton (H$^+$) is a special case. Having no electrons, the proton is the smallest possible ion (diameter ~10^{-5}Å).

Trends in the relative sizes of ions of the same charge within a group of elements are the same as those observed with neutral atoms; that is, size increases with increasing atomic number.

When a positive ion is formed from an atom, there is a decrease in size (Figure 2.13). The most dramatic change is seen in the formation of the ion isoelectronic with the nearest noble gas atom. This is shown in the right half of Figure 2.13. Two effects can be suggested to account for the size decrease associated with electron loss. Since nuclear charge remains the same, electron loss results in fewer electrons to "share" the positive nuclear field; thus there is a net increase in nucleus-electron attraction. If all the electrons in the last principal energy level are lost (e.g., Li$^+$, Be^{2+}), the remaining electrons are in energy levels closer to the nucleus.

In contrast (Figure 2.14) the formation of a negative ion (electron gain) results in a size increase. Anions isoelectronic with the nearest noble gas atoms have gained an electron (or electrons) in a principal energy level already partly occupied. This effect

(continued on p. 76)

MAN-MADE ELEMENTS

When we think of man-made elements, we normally consider the relatively new radioactive elements of atomic number greater than 92 (the *transuranium* elements). However, three of the lighter elements predicted by Mendeleev's periodic chart are not found in nature. These correspond to "holes" in the early periodic tables for atomic numbers 43, 61, and 85. In addition, element 87 is more readily prepared by nuclear transformation than from natural sources.

Although many investigators had reported the "discovery" of these four elements prior to 1930, each successive claim was proved false when subjected to more careful testing. The first acceptable proof for one of these "missing" elements was for number 43, now called *technetium*. A sample of molybdenum that had been subjected to bombardment by deuterium nuclei ($^2_1H^+$) at the University of California was investigated in 1937 by C. Perrier and E. Segré in Italy. A radioactive isotope was isolated from the sample and demonstrated to have properties like those expected for an element in Group VIIB of the periodic table. Further studies definitely confirmed the isotope as a new element and, at Segré's suggestion in 1947, the name "technetium" was adopted to signify the "technical" (man-made) origin of the element. Other isotopes of technetium have since been prepared and enough of the pure element (a few milligrams) has been studied to indicate that its properties are quite similar to those of manganese, the first VIIB element. All isotopes prepared are radioactive (Unit 35) and it is doubtful if any detectable amount of element 43 occurs in nature.

The remaining "gaps" in the pre-uranium elements were filled by astatine (from a Greek word meaning "unstable"), promethium (named after the Greek god Prometheus, who supposedly stole fire from the gods to give to man), and francium (identified as a product of the radioactive decay of $^{227}_{89}Ac$ by the French scientist Marguerite Perey).

Three important discoveries led to the preparation of the first transuranium elements. In 1934, Marie Curie's group demonstrated the first "artificial" radioactivity by irradiating stable elements with α particles. Shortly thereafter James Chadwick identified the neutron and E. O. Lawrence, at the University of California, developed the cyclotron. The technology was now available to attempt the production of elements beyond uranium. Most prominent in the investigation of neutron bombardment of uranium was Enrico Fermi's group in Italy. Analysis of the products of Fermi's experiments showed a vast number of radioactive products rather than the expected one or two new isotopes. At first these were believed to be a whole series of new transuranium elements. Later investigations, however, proved the radioisotopes to be those of lighter elements. The uranium nucleus had been *split* by neutron bombardment. Experiments devised through pure scientific curiosity about heavier-than-uranium elements had resulted in the accidental discovery of *nuclear fission,* the process that was to culminate in the awesome detonation of the "atomic bomb."

It was not until 1940 that element 93 was discovered by the Americans E. M. McMillan and P. H. Abelson during their investigations of the exciting new fission process. They named the new species *neptunium,* because it was "beyond uranium just as the planet Neptune is beyond Uranus."

From 1940 to the Mendeleev Centennial Year (1969), 11 additional transuranium elements were prepared. Within a few years of the conclusion of World War II, the Soviet Union had developed a considerable nuclear technology. By the 1960s Soviet and American scientists were competing in the production of new elements and considerable controversy surrounded the discovery claims for element 104 and, more recently, elements 105 and 106. The American approach to heavy-element synthesis has involved bombardment of elements heavier than lead with ions of elements lighter than chromium. The present Soviet technique uses heavier ions for bombardment of lead or lighter elements, a method of considerable promise for the production of "superheavy" elements. Some theoreticians predict that elements such as 112 or 114 may be unusually stable, having half-lives (time for 50-percent decay) as long as several million years.

Although most of the transuranium elements are of purely theoretical interest at this time, some have found practical applications. Plutonium (94), for example, has been produced in large quantities as a fuel for nuclear power plants. What uses, if any, will be made of the more "exotic" transuranium elements will depend on continuing advances in research and technology.

alone would suggest no size change in formation of such anions. However, the nuclear charge is now shared among more electrons, and electron-electron repulsive forces are increased. These electrostatic effects appear to account for the size increase in anion formation.

SUGGESTED READING

Ahrens, Louis H. 1967. *Distribution of the Elements in Our Planet.* New York: McGraw-Hill.
Asimov, Isaac. 1966. *The Search for the Elements.* New York: Fawcett.
Pode, J. S. F. 1973. *The Periodic Table: Experiment and Theory.* New York: Halsted Press.
Seaborg, Glenn. 1963. *Man-Made Transuranium Elements.* Englewood Cliffs, N.J.: Prentice-Hall.
Sisler, Harry. 1973. *Electronic Structure, Properties, and the Periodic Law* (2nd ed.). New York: Van Nostrand Reinhold.
Weeks, Mary E., and Henry M. Leicester. 1965. *Discovery of the Elements* (7th ed.). Easton, Pa.: Chemical Education Publishing Company.
What's Happening in Chemistry? 1975. Washington, D.C.: American Chemical Society.

EXERCISES

1. Prepare a set of small cards (e.g., by cutting 3×5 inch cards into thirds). On the front of each card write the name of an element from Table 2.1 and on the back write the element's symbol. Shuffle the cards, then look at the front of a card and write on a sheet of paper what you believe is on the back of the card. Check your answer. If it is correct, place the card in a separate stack; if not, place the card on the bottom of the stack so you can try it again. Continue to use these "flash cards" until you have learned the names and symbols in Table 2.1. (You will find this method a useful technique for learning many items of information needed in your study of chemistry.) Save the cards for occasional review.

2. Select the term best associated with each definition, description, or specific example.

TERMS: (A) alkali metals (B) anion (C) chalcogens (D) chemical properties (E) covalent (F) group (G) homogeneous (H) isotopes (I) period (J) valence electrons

a. $^{12}_{6}C$, $^{13}_{6}C$, $^{14}_{6}C$

b. O, S, Se

c. Na, K, Rb

d. Li, Be, B, C, N, O, F, Ne

e. F, Cl, Br, I, At

f. the two 4s electrons of calcium

g. Br^- or O^{2-}

h. type of bonding expected between C and S

i. having the same properties throughout the sample

j. properties related to significant changes in electron distribution

3. Indicate the number of protons, neutrons, and electrons represented by
   a. $^{12}_{6}$C; b. $^{18}_{8}$O^{2-}; c. $^{23}_{11}$Na^{+}

4. Which of the following statements are incorrect? (Explain your choices.)
   a. Since sodium and potassium metals react violently with water, it is reasonable to expect that metallic cesium will react vigorously with water.
   b. Since potassium and calcium metals react violently with water, it is reasonable to expect that metallic chromium will react vigorously with water.
   c. Since fluorine and chlorine are gases at room temperature, but bromine is a liquid, it is unlikely that iodine is a gas at room temperature (and standard atmospheric pressure).
   d. Magnesium, aluminum, and phosphorus are expected to be good electrical conductors.
   e. It is probable that both the compound of potassium with bromine and the compound of silicon with chlorine are characterized as "ionic."

5. Using a simple periodic table:
   a. report the atomic numbers and atomic weights of aluminum, sodium, arsenic, boron, and lead.
   b. indicate the charge expected for the most stable monatomic ion (or that no stable monatomic ion is expected) for sulfur, potassium, fluorine, carbon, and strontium.
   c. arrange in order of atomic size, smallest first:

      aluminum; potassium; phosphorus; sodium

   d. arrange in order of ionic size, smallest first:

      Al^{3+}; Cl^{-}; Mg^{2+}; S^{2-}

   e. select the member of each element pair expected to have the highest first ionization potential and the member expected to have the largest *negative* value of the electron affinity. Give reasons for your choices.

      lithium/fluorine
      sodium/magnesium
      magnesium/aluminum
      nitrogen/oxygen

OPTIONAL

6. An element consists of four isotopes: 0.02 percent of mass 35.967 amu, 4.18 percent of mass 33.968 amu, 0.74 percent of mass 32.971 amu, and 95.06 percent of mass 31.972 amu. What is the atomic weight of the element? What is the name of this element?

PRACTICE FOR PROFICIENCY

Objective (a)

1. Supply the missing names or symbols:
   a. Al; b. antimony; c. Cd; d. copper; e. Ag; f. iron; g. Sn; h. potassium

Objective (b)

2. Select the term best associated with each definition, description, or specific example.
   TERMS: (A) abundance (B) alkaline earth metals (C) atomic mass unit (D) cations (E) chemical families (F) compounds (G) halogens (H) heterogeneous (I) ionic (J) metals (K) metalloids (L) noble gases (M) nonmetals (N) physical properties (O) transition elements (P) transuranium elements
   a. "alkali metals" and "chalcogens"
   b. melting point and density
   c. type of bonding expected between Sr and Br
   d. a mixture of sand and "salt water"
   e. the relative amounts of isotopes in an element sample
   f. new species formed when elements combine chemically
   g. defined as 1/12 the mass of a carbon-12 atom
   h. Bk, Cf
   i. Fe, Cu
   j. Sr, Ba
   k. Cl, Br
   l. Ne, Xe
   m. Mg^{2+}, Al^{3+}
   n. N, O, F
   o. B, Si, Sb
   p. Mg, Ag, Pb

Objective (c)

3. Indicate the number of protons, neutrons, and electrons represented by
   a. $^{24}_{12}$Mg   f. $^{26}_{13}$Al^{3+}
   b. $^{15}_{7}$N^{3-} g. $^{238}_{92}$U
   c. $^{3}_{1}$H^{+} h. $^{127}_{53}$I^{-}
   d. $^{31}_{15}$P   i. $^{90}_{38}$Sr^{2+}
   e. $^{36}_{17}$Cl^{-} j. $^{258}_{101}$Md

Objective (d)

4. Which of the following statements are incorrect? (Explain your choices.)
   a. The density of an element is a characteristic property, independent of temperature.
   b. In most cases the boiling points (at standard atmospheric pressure) of elements near the lower end of a group are higher than those of lighter elements in the same group.
   c. Since bromine and chlorine in the gaseous state react vigorously with liquid sodium, it is likely that fluorine gas will react readily with liquid potassium.
   d. Rubidium, calcium, and aluminum are expected to be good electrical conductors.
   e. It is probable that both the compound of sodium with fluorine and the compound of aluminum with sulfur are characterized as "ionic," whereas the compound of nitrogen with chlorine is more likely to be considered "covalent."

5. Using a simple periodic table:
   a. report the atomic numbers and atomic weights of arsenic, mercury, sodium, gold, beryllium, copper, tin, nickel, helium, and iron.
   b. indicate the charge expected for the most stable monatomic ion (or that no stable monatomic ion is expected) for aluminum, bromine, calcium, krypton, and silicon.
   c. arrange in order of atomic size, smallest first:

(1) calcium	**(2)** lithium	**(3)** sulfur
bromine	boron	selenium
potassium	neon	chlorine
selenium	sodium	tellurium

   d. arrange in order of ionic size, smallest first:

(1) Be^{2+}	**(2)** Ca^{2+}	**(3)** Li^+
F^-	Br^-	Na^+
H^+	K^+	Cl^-
N^{3-}	Se^{2-}	F^-

   e. for each of the following, suggest the most likely choice and propose an "atomic model" explanation for the choice given:
   (1) Which probably has the highest first ionization potential, lithium or beryllium?
   (2) Which probably has the highest first ionization potential, hydrogen or helium?
   (3) Which probably has the largest *negative* value of the electron affinity, magnesium or chlorine?
   (4) Which probably has a *positive* value of the electron affinity, beryllium or oxygen?
   (5) Which probably has the largest *difference* between first and second ionization potentials, calcium or potassium?

OPTIONAL

6. An element consists of two isotopes, 75.39 percent with mass 34.969 amu and 24.61 percent with mass 36.966 amu. What is the atomic weight of the element? What is the name of the element?

7.

Abundance (%)	Isotopic weight
48.89	63.93
27.81	65.93
4.11	66.93
18.57	67.93
0.62	69.93

The element for which these data are given is
   a. zinc   b. tin

   c. gallium   d. copper
   e. germanium

8.

Abundance (%)	Isotopic weight
1.22	105.91
0.88	107.90
12.39	109.90
12.75	110.90
24.07	111.90
12.26	112.90
28.86	113.90
7.58	115.90

The element for which these data are given is
   a. copper   b. palladium
   c. silver   d. tin
   e. cadmium

BRAIN TINGLERS

9. It has been pointed out that potassium and argon would be "anomolies" in an arrangement of elements by atomic weights. Identify two other elements among the transition elements whose positions in the periodic table would have been reversed in a "weight-sequence" arrangement. Which pair of elements would be most obviously "out of place" on the basis of chemical behavior? Explain your answer in terms of the atomic model, showing electron configurations for these elements. (You may need to consult Unit 27.)

10. Periodic trends within a "group" and within a "period" have been discussed. Another phenomenon, to be described in more detail in later units, is referred to as the "diagonal trend." For example, lithium has certain properties more like those of magnesium than those of the other alkali metals and boron is, in many respects, more like silicon than aluminum. From the following values of ionic radii, and the value of a unit charge (1.6×10^{-19} coulomb), calculate "charge densities" (charge per unit volume, in $C\,Å^{-3}$) for the ions indicated. Are there any correlations of "charge density" with "diagonal trends"?

Ion	Radius (Å)	Ion	Radius (Å)
Li^+	0.60	B^{3+}	0.20
Na^+	0.95	Al^{3+}	0.50
K^+	1.33	C^{4+}	0.15
Be^{2+}	0.31	Si^{4+}	0.41
Mg^{2+}	0.65	Ge^{4+}	0.53
Ca^{2+}	0.99		

Unit 2 Self-Test

(Refer to a periodic table when necessary.)

1. Supply the missing names and symbols:
   a. argon; **b.** Sn; **c.** gold; **d.** bromine; **e.** K; **f.** zinc; **g.** Mg; **h.** lead; **i.** sodium; **j.** Cl; **k.** copper; **l.** Hg

2. Select the term best associated with each definition, description, or scientific example.
   TERMS: (A) heterogeneous (B) covalent (C) cations (D) ionic (E) isotopes (F) abundance (G) transition elements (H) atomic mass unit (I) nonmetals (J) metalloids
   a. the reference standard is the assignment of exactly 12 for $_6^{12}C$
   b. relative isotopic composition
   c. composition varies in different portions of the sample
   d. the chemical bond in CCl_4
   e. the chemical bond in KCl
   f. $_5^{10}B$ and $_5^{11}B$
   g. B and Ge
   h. Cu and Ag
   i. K^+ and Mg^{2+}
   j. S and Br

3. The species formulated as $_{35}^{80}Br^-$ contains
   (a) _____ protons
   (b) _____ neutrons
   (c) _____ electrons

4. Which statement is incorrect?
   a. Liquid helium has a lower boiling point than liquid xenon.
   b. Since sulfur and selenium form simple compounds with hydrogen, formulated as H_2S and H_2Se, it is likely that tellurium forms the compound H_2Te.
   c. Since elements 11 through 17 will all burn in oxygen to form simple oxygen compounds, it is likely that element 18 will also burn readily in oxygen.
   d. It is expected that sodium, magnesium, and aluminum are good electrical conductors.

5. The atomic number of iron is _____ and its atomic weight is _____.

6. Which element is most likely to form a stable $+3$ monatomic ion?
   a. cesium; **b.** calcium; **c.** carbon; **d.** gallium

7. Which order of relative atomic sizes is incorrect?
   a. Li $<$ Na $<$ K; **b.** Al $<$ Mg $<$ Na; **c.** C $<$ Si $<$ Al; **d.** P $<$ As $<$ Se

8. Which order of relative ionic sizes is incorrect?
   a. $F^- <$ $Cl^- <$ Br^-; **b.** $Na^+ <$ $Mg^{2+} <$ Al^{3+}; **c.** $Na^+ <$ $F^- <$ O^{2-}; **d.** $Cl^- <$ $S^{2-} <$ Se^{2-}

9. Which element is "out of line" with the "normal" periodic trends in both ionization potentials and electron affinities?
   a. Si; **b.** P; **c.** S; **d.** Cl

OPTIONAL

10. An element consists of two isotopes, 93.10 percent with mass 38.963 amu and 6.90 percent with mass 40.973 amu. What is the atomic weight of the element? What is the name of the element?

1. Select the term best associated with each definition, description, or specific example.

TERMS: (A) alkali metals (B) chalcogens (C) degenerate (D) covalent (E) excited state (F) ground state (G) halogens (H) heterogeneous (I) homogeneous (J) ionic (K) isotopes (L) sublevels (M) transition elements (N) valence electrons

a. type of compound expected between Mg and Br
b. type of compound expected between Si and Br
c. uniform properties throughout the sample
d. varying properties in different parts of the sample
e. for example, $^{16}_{8}O$ and $^{18}_{8}O$
f. for example, Cu and Fe
g. for example, K and Cs
h. for example, Se and Te
i. for example, Br and I
j. for example, for calcium: [Ar]$4s^2$
k. for example, for calcium: [Ar]$4s^1$, $4p^1$
l. for example, for calcium: $4s^2$ electrons only
m. for example, $2p_x$, $2p_y$, $2p_z$
n. for example, $3s$, $3p$, $3d$

2. Which name/symbol combination is wrong?
a. argon/Ar; **b.** fluorine/F; **c.** tellurium/Te; **d.** selenium/Sn; **e.** none

3. Which best represents a p_x orbital?

4. The species represented as $^{27}_{13}Al^{3+}$ contains
a. 13p, 14n, 10e; **b.** 13p, 14n, 16e; **c.** 13p, 27n, 10e; **d.** 13p, 27n, 16e; **e.** 27n, 13p, 3e

5. The species represented as $^{31}_{16}S^{2-}$ contains
a. 16p, 31n, 2e; **b.** 16p, 15n, 14e; **c.** 16p, 15n, 18e; **d.** 31p, 16n, 33e; **e.** 31p, 16n, 29e

6. Which statement is incorrect?
a. The atomic number of potassium is 19.
b. The atomic weight of selenium is approximately 79.0.
c. A strontium atom is smaller than a calcium atom.
d. A strontium atom is smaller than a rubidium atom.
e. The compound CaF_2 is probably ionic.

7. The electron configuration for boron may be written as (He) $2s^2$, $2p_x{}^1$. Using this convention, the ground state configuration is, for sulfur,
a. (Ne) $3s^6$; **b.** (Ne) $3s^2$, $3p_x{}^2$, $3p_y{}^1$, $3p_z{}^1$; **c.** (Ne) $3s^2$, $3p_x{}^2$, $3p_y{}^2$, $3p_z{}^2$; **d.** (Ne) $3s^2$, $3p_x{}^2$, $3p_y{}^2$; **e.** none of these

8. The electron configuration for $^{120}_{50}Sn$ may be written as
a. (Ar) $4s^2 4p^6 4d^{10} 5s^2 5p^6 5d^6$
b. (Kr) $5s^2 5p^6 5d^{20} 6s^2 6p^4$
c. (Rn) $6s^2 5d^{10} 5f^{14} 6p^8$
d. (Kr) $5s^2 4d^{10} 5p^2$
e. (Kr) $5s^2 4d^8 5p^4$

9. Which of the following pairs on p. 81 is *not* isoelectronic?

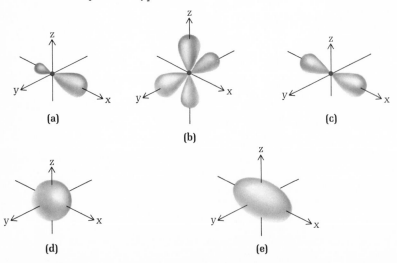

(a) **(b)** **(c)**

(d) **(e)**

a. Na^+ and Mg^{2+}
b. S^{2-} and Cl^-
c. O^{2-} and Na^+
d. Te and I^+
e. Na and Ne^+

10. Which anion is probably the least stable?
a. F^-; **b.** Cl^-; **c.** S^-; **d.** Se^{2-}; **e.** N^{3-}

11. Which cation is probably the least stable?
a. Li^+; **b.** K^+; **c.** Ca^{2+}; **d.** Sr^{2+}; **e.** Al^{2+}

12. Which indication of relative atomic or ionic sizes is incorrect?
a. $C^{4-} > C > C^{4+}$; **b.** $Mg > Al > B$; **c.** $Na^+ > Mg^{2+} > Al^{3+}$;
d. $F^- > O^{2-} > N^{3-}$; **e.** $As > P > N$

13. Which element is "out of line" in the "normal" periodic trend in first ionization potentials?
a. Na; **b.** Mg; **c.** Al; **d.** Si; **e.** S

14. Given that oxygen is a gas at room temperature, sodium is a solid at room temperature, and sodium reacts readily with oxygen at room temperature, periodic trends would suggest that
a. selenium is a gas at room temperature
b. lithium is a liquid at room temperature
c. sodium will react with nitrogen at room temperature
d. potassium will react with oxygen at room temperature
e. none of these should be expected

OPTIONAL

15.

Abundance (%)	Isotopic weight
7.76	75.92
36.53	73.92
7.76	72.92
27.42	71.92
20.53	69.92

The element for which these data are given is
a. germanium; **b.** gallium; **c.** gold; **d.** arsenic; **e.** tantalum

Section
TWO

CHEMICAL REACTIONS AND ENERGY CHANGES

With all the present attention given the "energy crisis" and the problems of environmental quality, it is hard to believe that just a few years ago these concerns were voiced by only a handful of scientists and environmentalists, often accused of "crying wolf" whenever they tried to focus public attention on the problems. Today most of us recognize what science has known for many years: If we are to survive, we must learn the limitations imposed by certain "natural laws".

One of these limitations may be expressed by the "mass conservation law": *Mass can be neither created nor destroyed by physical or chemical processes.* We have, of course, learned to destroy mass by conversion to energy, through nuclear reactions, but we have not learned to convert energy to mass. The restrictions imposed by the mass conservation law when coupled with the understanding that there are finite limits on every natural chemical source, warn us that we cannot continue to deplete our natural resources at the rate we have done in the past without suffering the most serious consequences. For every mile of copper wire produced, the available supply of copper ore must be further depleted; every gallon of gasoline consumed depletes still further the earth's dwindling oil reserves. There is a definite mathematical relationship among the masses of all substances involved in chemical reactions. Stoichiometry (Unit 4) is the study of this relationship.

Of equal, or greater, importance is the "energy conservation law": *Energy can be neither created nor destroyed by physical or chemical processes, but only changed in form.* This law is of particular concern in its warning of eventual depletion of many available energy sources. The study of energy conversions and the limitations on the efficiency of such conversions is the subject of our introduction to thermodynamics (Unit 5).

Before we can investigate the quantitative areas of stoichiometry and thermodynamics, we need to learn how the chemist describes chemical processes and how these may be represented by "chemical equations", the subjects of Unit 3.

CONTENTS

Unit 3 **Chemical Change**
How chemical change is defined and detected; writing and balancing simple chemical equations

Unit 4 **Stoichiometry**
Applications of mathematical analysis to mass relationships in chemical processes

Unit 5 **Thermodynamics**
Concepts and quantitative applications of energy conversions associated with chemical processes

Section Two Overview Test

Unit 3

Chemical change

Atoms move in the void and
catching each other up, jostle together, and some recoil. . . . and
others become entangled. . . .
SIMPLICIUS (5th century, A.D.)

Objectives

(a) Be able to demonstrate recall of the common names, symbols, and formulas
    listed in Tables 3.1, 3.2, and 3.3 by
    (1) writing formulas or symbols (including associated charges), given the names
        of the corresponding ions and molecules
    (2) writing correct chemical names for ions and molecules, given the corre-
        sponding symbols and formulas
(b) Given the names or formulas of simple salts, be able to write the corresponding
    formulas and names, respectively (Section 3.3).
(c) After studying the appropriate definitions in Appendix E and the terms' usage in
    this unit, be able to demonstrate your knowledge of the following terms by
    matching them with corresponding definitions, descriptions, or specific exam-
    ples:
    acid
    anhydrous
    base
    coefficient
    complete formula

empirical formula	reactant
hydrated	salt
net equation	strong (acid or base)
oxidation state	weak (acid or base)
product	''yields''

(d) Given a description of a chemical process, be able to write a qualitative chemical equation, in both ''complete'' and ''net'' forms, to represent the process (Sections 3.4 and 3.5).

(e) Be able to balance simple chemical equations (Section 3.6).

OPTIONAL

Be able to determine oxidation numbers for element symbols in simple compounds or polyatomic ions and to use oxidation numbers in classifying reactions as *metathetical* or *oxidation-reduction* (Section 3.7).

Most of us, from some prior exposure to science, have the general feeling that there is a very clear-cut distinction between physical changes and chemical changes. Certainly there is a sharply defined difference between what happens when ice melts and what happens when gasoline burns—or is there? Remember that chemistry is concerned with what happens at the atomic, ionic, or molecular level in matter. Consider a common definition of chemical change: "Chemical change involves a redistribution of electrons without a change in nuclear composition." In the light of this definition, let us now consider some simple phenomena that we normally think of as "physical changes." Look at the examples in Figures 3.1 and 3.2. Of course, we are really stretching the point a bit in considering these phenomena as chemical changes. After all, the "composition" of the substances has not been altered very

Figure 3.1 *Change of shape. A straight piece of wire is bent (a and b). No chemical change occurs, yet at the atomic level some atoms have been forced farther apart, some forced closer together (c and d). A crystal of sodium chloride (e) is cleaved into fragments. This is purely physical change, yet more ions are now exposed to the atmosphere in the fragments than in the original crystal (f and g); fewer ions are left in a symmetrical electrostatic field "inside" the crystal. Some change in electron distribution must occur. Pouring a liquid from one container to another hardly constitutes chemical change, since a liquid simply assumes the shape of its container (h and i). The ratio of surface particles to interior particles has changed in the liquid (j and k), and interactions of liquid particles with each other will be different for interior versus surface particles. Again, electronic distribution has changed (Photographs by Arden Literal, Washington State University.)*

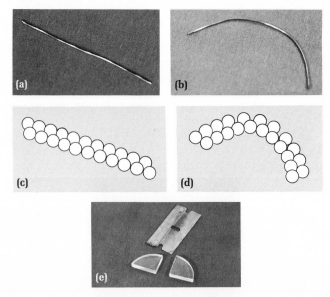

Figure 3.1 (Continued)

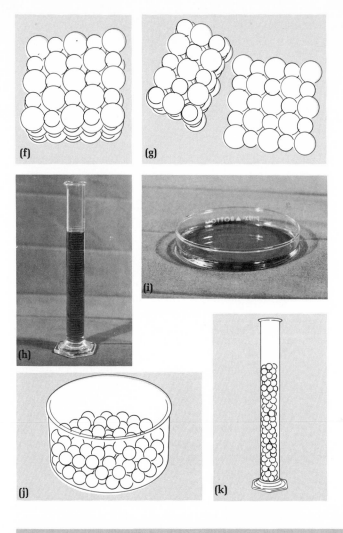

Figure 3.2 *Change of phase. Even in a case as simple as melting ice or boiling water, at the molecular level changes occur in interparticle distances, particle motions, and interparticle forces. These changes involve electrons.*

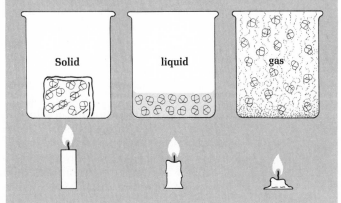

Solid   liquid   gas

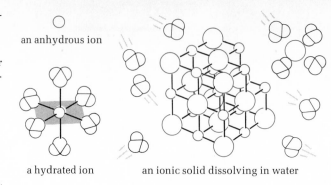

an anhydrous ion

a hydrated ion   an ionic solid dissolving in water

Figure 3.3 *Change of environment. Ionic solids consist of positive ions uniformly distributed among negative ions. In anhydrous substances no water is associated with the material. When ions are hydrated (associated with water molecules), either as solids or in aqueous solution, the water molecules strongly affect the electrons in the ions. The anhydrous Cu^{2+} ion, for example, is nearly colorless, whereas the hydrated ion $[Cu(H_2O)_6]^{2+}$ is a deep blue. The change of environment involved in the hydration of ions is certainly closer to what one normally thinks of as "chemical change" than were the examples given in Figures 3.1 and 3.2.*

much. Let us look, instead, at Figure 3.3, which illustrates some cases that are not quite so "obvious."

If one used "electron redistribution" as the definition of chemical change, it would be difficult to find any kind of change that failed to meet this criterion. Like so many "simple" definitions in science, the distinction between physical change and chemical change must be a rather arbitrary one, selected more for convenience than for all-encompassing accuracy.

3.1 CHEMICAL REACTIONS

Chemists are a fairly pragmatic group and, as such, are often less concerned with precise theoretical distinctions than with useful operational understandings. The most common working definition of a chemical change is one in which there is a change in *definable chemical species.* Such a statement is, of necessity, limited by the investigative tools and techniques available.

It is relatively easy to show that the species sulfuric acid and sodium hydroxide (formulated as H_2SO_4 and NaOH) combine, when mixed in the proper proportions, to form new species, water and sodium sulfate (H_2O and Na_2SO_4). The chemical characteristics of the original substances (called the *reactants*) are easily shown to be quite different from those of the new substances formed (the *products*). With situations in which more subtle changes occur, very tiny amounts are available to study, or extremely complex systems are involved—such as those in living organisms—it is much more difficult to detect chemical change.

The analgesic (pain killing) and narcotic properties of morphine have been known for several generations, but until the middle of this century there were only vague guesses as to the chemical behavior of morphine in the body. Two lines of thought had been proposed. One suggested that morphine retained its chemical identity in the body and acted simply as a physical blocking agent at certain sites in the chain of biochemical events. The other postulated that morphine was changed chemically during the course of its action in the body. Routine chemical investigations of body excretions from persons given morphine injections were inadequate to resolve the controversy. By the 1950s, new techniques had been developed to permit a much more sophisticated approach. Chromatographic methods (discussed in Unit 32 and Excursion 4) provided ways of separating and detecting chemical species on a very small scale, and radioactive isotopes (Unit 35) offered a way to "label" various parts of a molecule to determine if these were involved in chemical reactions. A team of investigators at the University of California at Berkeley, headed by Dr. Henry Rapoport, prepared morphine containing a radioisotope of carbon ($^{14}_{6}C$) in the so-called N-methyl group, starred in the structure that follows.

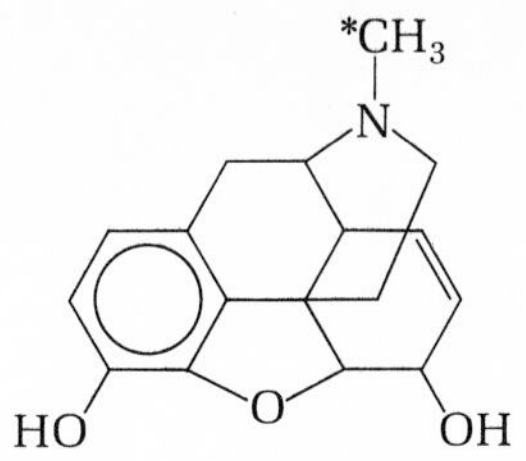

A group of volunteers received injections of the "labeled" morphine. Urine samples were collected and analyzed and, in later experiments, a very sophisticated instrument was used to collect and analyze exhaled carbon dioxide to detect traces of radioactive carbon associated with the removal of the labeled CH_3 group from the morphine molecule.

The investigative techniques used were able to show that although most of the morphine was excreted unchanged, a fraction had, indeed, undergone chemical reaction involving the particular group of atoms being studied.

3.2 EVIDENCE OF CHEMICAL CHANGE

Pure aspirin is odorless. Bottles of aspirins opened often allow enough atmospheric moisture to enter, especially in humidity, for detectable reaction with water. One of the products, acetic acid, is detected by its "vinegar" odor. If this evidence of chemical change is very "strong," the aspirin should be discarded.

The type of evidence of chemical change used by the chemist will vary with the particular reaction being studied. Some reactions are rather spectacular, and others are very subtle indeed. There are a number of techniques that can be employed.

The important thing to remember is that chemistry is an *experimental* science and the only meaningful way of showing that chemical change has occurred is by presenting acceptable experimental evidence that some chemical species (called *reactants*) have been consumed, and that different chemical species (called *products*) have been formed. This evidence may involve rather simple observations, such as the fact that mixing two odorless chemicals results in the evolution of a pungent gas. Few processes, however, are this simple and it is often necessary to utilize sophisticated analyses to demonstrate unambiguously that a chemical change has occurred. Some of the simpler types of evidence are:

Changes in Chemical Properties

Sodium bisulfite reacts with acid to evolve sulfur dioxide gas. Sodium bisulfate does not liberate gas when treated with acids such as sulfuric acid.

Suppose one wanted to know whether or not a sodium bisulfite sample left exposed to the air had been converted to sodium bisulfate by reaction with oxygen. He might take advantage of the different properties of the two substances and treat the sample with some acid. If no gas evolved, he might conclude that most of the sample was the bisulfate. (If some gas was formed, what could be safely concluded?)

Changes in Solubility

Reactions in aqueous solution (that is, with water as the solvent) leading to formation of insoluble precipitates or gases are easily detected. One can readily determine, for example, that Ba^{2+} reacts with sulfate ion by mixing solutions of a soluble barium salt and a soluble sulfate and observing the formation of a precipitate which can be shown to contain only barium and sulfate ions.

Changes in Color

Colored "indicators" are useful in detecting a wide variety of chemical changes. Many quantitative determinations depend on the use of such substances, which change color as some reaction progresses.

Instrumental Methods The modern chemist employs many instruments in his study of chemical changes. Ranging from inexpensive "make-it-yourself" gadgets to extremely expensive commercial devices, these instruments extend the chemist's studies and simplify his tasks to an extent undreamed of a few years ago. He may analyze complex reaction mixtures in minutes or detect new elements in samples as small as a few atoms. Many of the instruments used in modern research are discussed in later units.

3.3 SYMBOLS, FORMULAS, AND EQUATIONS

Mathematicians may say, "two plus two equals four," but it is simpler and more convenient to write this statement in the form of a mathematical equation:

$$2 + 2 = 4$$

In a similar way, it is convenient to write chemical equations, using chemical symbols and formulas to represent elements, ions, and compounds. Thus, the statement, "the reaction between sulfuric acid and sodium hydroxide *yields* (i.e., produces) water and sodium sulfate," may be expressed by the chemical equation:

$$\underbrace{H_2SO_4 + 2NaOH}_{\text{reactants (species consumed)}} \longrightarrow \underbrace{2H_2O + Na_2SO_4}_{\text{products (species formed)}}$$

Some of the more sophisticated methods employed for determining chemical formulas are described in Unit 30 and Excursion 5.

Such equations can be written only if the formulas of reactants and products are known. One of the games chemists play is the determination of formulas for new chemical species; some of the methods used are described in later sections. Simple formulas can be deduced from information given in Unit 2: *The simpler elements tend to react in ways resulting in electron configurations like those of the noble gas elements.* For example, sodium atoms can lose one electron each to form $+1$ ions, which are *isoelectronic* with (i.e., have the same electron configurations as) atoms of neon; chlorine atoms can gain one electron each to form -1 ions, which are isoelectronic with argon atoms. The simplest formula, then, for sodium chloride would be that showing a condition of electrical neutrality, namely, one ion each of Na^+ and

Table 3.1 COMMON MONATOMIC IONS

+	2+	3+	−	2−	3−
group IA (e.g., sodium, Na⁺)	group IIA (e.g., barium, Ba²⁺)	aluminum (Al³⁺)	hydride[b] (H⁻)	group VIA (e.g., oxide[b], O²⁻; sulfide[b], S²⁻;	nitride[b] (N³⁻)
hydrogen (as proton, H⁺)	cadmium (Cd²⁺)	scandium (Sc³⁺)	group VIIA (e.g., chloride[b], Cl⁻)	selenide[b], Se²⁻;	
silver (as Ag⁺)	zinc (Zn²⁺)	bismuth[a] (as Bi³⁺)		telluride[b], Te²⁻)	
copper[a] [as copper(I), or cuprous, Cu⁺]	cobalt[a] (as Co²⁺)	chromium[a] [as chromium(III), or chromic, Cr³⁺]			
	copper[a] [as copper(II), or cupric, Cu²⁺]	iron[a] [as iron(III), or ferric, Fe³⁺]			
	iron[a] [as iron(II), or ferrous, Fe²⁺]				
	lead (Pb²⁺)				
	manganese[a] (as Mn²⁺)				
	nickel (Ni²⁺)				
	tin[a] (as Sn²⁺)				
	mercury[a] [as mercury(II), or mercuric, Hg²⁺; or as mercury(I), or mercurous, Hg₂²⁺]				

[a] Many elements form more than one stable ion (e.g., Cu⁺, Cu²⁺ or Fe²⁺, Fe³⁺). Not all are shown in the table—only those encountered most frequently. In such cases, names indicate which ion is involved. Current terminology is, for example, copper(I) for Cu⁺. Older literature uses suffixes—*ous* for lower charge (e.g., Cu⁺, Fe²⁺), and *ic* for higher charge (e.g., Cu²⁺, Fe³⁺), often with Latin name prefixes (e.g., *cuprous, ferrous*, but *mercurous* rather than the Latin *hydrargyrous*).
[b] Note *ide* ending for monatomic anion names.

Cl^-: $NaCl$. In a similar fashion, magnesium should form a $+2$ ion and, since the net positive charge of $MgCl^+$ would still "attract" another Cl^- ion, the simplest formula for magnesium chloride should be $MgCl_2$. Note that *subscript* numbers are used to indicate the numbers of element symbols within the formula. Although this approach could be used, with some additional considerations, for most simple ionic compounds it is helpful to remember the charges of common monatomic ions (Table 3.1).

In many cases molecular substances were well known long before their chemical formulas were determined; "water" is the best example. As a result the common names for simple molecules rarely provide clues to their formulas, unlike the simpler "NaCl/sodium chloride" cases. A detailed system for naming organic compounds—compounds of carbon—has been developed and will be explored later, as will the systematic nomenclature for inorganic complexes such as $[Cu(NH_3)_4]^{2+}$. Until then, it is important that the names and formulas of common molecules (Table 3.2) and polyatomic ions (Table 3.3) be committed to memory.

Certain elements occur in nature as diatomic molecules. These should be remembered: H_2, N_2, O_2, F_2, Cl_2, Br_2, I_2.

An extensive discussion of inorganic and organic nomenclature is given in Appendix F and Appendix G, respectively. You may find it useful, at this point, to look over the parts of Appendix F dealing with the simpler cases.

Table 3.2 SOME COMMON MOLECULAR[a] SUBSTANCES

NAME	FORMULA	NAME	FORMULA
ACIDS[b]		NEUTRAL COMPOUNDS[c]	
acetic acid	CH_3CO_2H	arsine	AsH_3
hydrogen cyanide	HCN	benzene	C_6H_6
hydrogen fluoride	HF (anhyd)	butane	C_4H_{10}
hydrofluoric acid	HF (aq)	carbon dioxide	CO_2
hydrogen chloride	HCl (anhyd)	carbon monoxide	CO
hydrochloric acid	HCl (aq)	carbon	CCl_4
hydrogen bromide	HBr (anhyd)	tetrachloride	
hydrobromic acid	HBr (aq)	chloroform	$CHCl_3$
hydrogen iodide	HI (anhyd)	ethane	C_2H_6
hydroiodic acid	HI (aq)	ethanol	C_2H_5OH
hydrogen sulfide	H_2S (anhyd)	(ethyl alcohol,	
hydrosulfuric acid	H_2S (aq)	"grain alcohol")	
hypochlorous acid	$HClO$	methane	CH_4
	($HOCl$)	methanol	CH_3OH
chlorous acid	$HClO_2$	(methyl alcohol,	
chloric acid	$HClO_3$	"wood alcohol")	
perchloric acid	$HClO_4$	nitrous oxide	N_2O
sulfurous acid	H_2SO_3	nitric oxide	NO
sulfuric acid	H_2SO_4	propane	C_3H_8
nitrous acid	HNO_2	phosphine	PH_3
nitric acid	HNO_3	sulfur dioxide	SO_2
phosphoric acid	H_3PO_4	sulfur trioxide	SO_3
BASES[d]			
ammonia[e]	NH_3		
methylamine	CH_3NH_2		

[a] Certain acids (of those listed: HCl, HBr, HI, $HClO_3$, $HClO_4$, H_2SO_4, HNO_3) are ionized when dissolved in water. In fact, some of the oxygen-containing acids are ionized appreciably in the pure state (see Unit 19 for details).
[b] (anhyd) = anhydrous; (aq) = aqueous, i.e., in water solution. Acids are defined as species capable of donating (H^+) and are discussed in detail in Unit 19.
[c] Neutral compounds are those that do not have tendencies to donate or accept protons to an extent significantly greater than that of water. Note that there are alternative formulations for many molecular species. The ones given are those most commonly encountered in simple chemical equations.
[d] Bases are species capable of accepting protons; see Unit 19.
[e] Older literature often refers to aqueous ammonia as "ammonium hydroxide."

Table 3.3 COMMON POLYATOMIC IONS

NAME	FORMULA
ammonium	NH_4^+
methylammonium	$CH_3NH_3^+$
acetate	$CH_3CO_2^-$
bicarbonate[b]	HCO_3^-
carbonate	CO_3^{2-}
oxalate	$C_2O_4^{2-}$
cyanide	CN^-
thiocyanate	CNS^-
hypochlorite[a]	ClO^-
chlorite[a]	ClO_2^-
chlorate[a]	ClO_3^-
perchlorate[a]	ClO_4^-
bisulfite[b]	HSO_3^-
sulfite	SO_3^{2-}
bisulfate[b]	HSO_4^-
sulfate	SO_4^{2-}
thiosulfate	$S_2O_3^{2-}$
nitrite	NO_2^-
nitrate	NO_3^-
phosphate	PO_4^{3-}
arsenate	AsO_4^{3-}
chromate	CrO_4^{2-}
dichromate	$Cr_2O_7^{2-}$
permanganate	MnO_4^-
hydroxide	OH^-

[a] Note that other anions of halogens with oxygen are named like those of chlorine, e.g., bromate, BrO_3^-; periodate, IO_4^-. (See also Appendix F.)
[b] The current IUPAC nomenclature (Appendix F) uses "hydrogen carbonate" (for HCO_3^-), "hydrogen sulfite" for HSO_3^-), and "hydrogen sulfate" (for HSO_4^-). However, since most of the existing literature uses the older terminology, we will use "bicarbonate," etc. throughout this text.

The formulas of salts (ionic compounds having cations other than, or in addition to, H^+) may be written by combining the appropriate ionic symbols or formulas in the simplest ratios (indicated by subscripts) that will result in electrical neutrality. Ionic charges are not *written* in the formulas of salts, but must be kept in mind in determining the proper ratios. When a salt contains more than one of a particular polyatomic ion, the formula for that ion is enclosed in parentheses, with the subscript for the entire ion written outside the parentheses and to the lower right. By convention, cations are written—and named—before anions. A few examples will illustrate these points.

Example 1

Write the formula for potassium sulfate.

Solution
Potassium, a group IA element, forms the K^+ ion. Sulfate is formulated as SO_4^{2-}. For electrical neutrality, there must be two $+$ ions for each $2-$ ion:

K_2SO_4

Example 2

The Roman numeral refers to the oxidation state of the metal, as described in Section 3.7.

Write the formula for iron(III) nitrate (ferric nitrate).

Solution
The iron(III) ion is written as Fe^{3+}, and nitrate is formulated as NO_3^-. For electrical neutrality, there must be three $-$ ions for each $3+$ ion. Since the formula for nitrate appears three times, it must be enclosed in parentheses:

$Fe(NO_3)_3$

Example 3

Write the formula for aluminum carbonate.

Solution
Aluminum forms the Al^{3+} ion; carbonate is written as CO_3^{2-}. The smallest number containing both 3 and 2 is 6: two aluminum ions give 6 positive charges and three carbonate ions give 6 negative charges. Carbonate, a polyatomic ion appearing three times, is written in parentheses:

$Al_2(CO_3)_3$

Example 4

Write the formula for mercury(I) iodide (mercurous iodide).

Solution
It would be tempting to write HgI. This is, indeed, the *empirical* formula for this compound. Such a formula shows the *simplest* ratio of component elements. However, experimental evidence shows that the mercury(I) ion must be formulated as Hg_2^{2+} (Table 3.1). Although each mercury particle may have a $1+$ charge, two such particles are connected by a covalent bond. Thus the formula for this salt is

Hg_2I_2

The names of salts may be deduced from their formulas by a reverse of the process just described. For example, the formula Fe_2S_3 represents iron(III) sulfide (in older terminology, ferric sulfide). Since the sulfide ion has a 2− charge, three such ions give 6 negative charges. This is balanced by the charge of two ions of iron, so each of these ions is 3+.

3.4 WRITING CHEMICAL EQUATIONS

If we know the formulas of all reactants and products of a chemical reaction, we can write a *qualitative* equation to represent the chemical change involved. The most common convention places the formulas of reactants to the left of a horizontal arrow and the reactants to the right. Plus signs are used between successive reactant or product formulas. For purposes of convenience, the reactant and product formulas may be written in other relative locations, but the arrow used always points toward the products.

When it is desirable to indicate the physical state of reactants and products, each symbol or formula may be followed by a notation: (aq), in aqueous solution; (s), solid; (l), pure liquid; or (g), gas. For example, the reaction occurring when carbon dioxide gas is bubbled into a solution of calcium hydroxide may be formulated as

An aqueous solution is one in which the species are dissolved in water.

$$CO_2(g) + Ca(OH)_2(aq) \longrightarrow CaCO_3(s) + H_2O(l)$$

A *qualitative* equation tells us only the chemical species consumed and the new species formed, without any indication of the *amounts* of the different chemicals involved. For example, we can indicate that the burning of "natural gas" (methane) in air will usually produce a mixture of water vapor, carbon dioxide, and carbon monoxide by the equation

$$CH_4(g) + O_2(g) \longrightarrow H_2O(g) + CO(g) + CO_2(g)$$

Such an equation *does not* mean that equal numbers of CO and CO_2 molecules are produced. In fact, the relative amounts of these two species are determined by the conditions of the reaction (e.g., temperature and fuel-to-oxygen ratio) not by just the "chemical nature" of the reactants.

In many cases, especially with reactions of organic compounds, a wide variety of different products are formed and quite often we don't even know the formulas of all the minor products. It is common practice to write chemical equations showing only the major products, or even just the products of specific interest. For example, when methane reacts with a large excess of chlorine, the major products are CCl_4 and HCl, but some chloroform ($CHCl_3$), methylene chloride (CH_2Cl_2), and methyl chloride (CH_3Cl) are also produced. The *complete* qualitative equation would be

$$CH_4(g) + Cl_2(g) \longrightarrow CCl_4(g) + CHCl_3(g) + CH_2Cl_2(g)$$
$$+ CH_3Cl(g) + HCl(g)$$

We could, however, choose to discuss only the principal reaction, as expressed by

$$CH_4(g) + Cl_2(g) \longrightarrow CCl_4(g) + HCl(g)$$

or, if our interest was in describing the by-products (minor products) of the CCl_4 manufacturing process, we could write

$$CH_4(g) + Cl_2(g) \longrightarrow CHCl_3(g) + CH_2Cl_2(g) + CH_3Cl(g)$$

It is important to recognize the various ways that chemical equations may be employed so that we understand that a particular equation may not necessarily represent the *complete* chemical process.

Example 5

Write qualitative equations descriptive of the chemical changes occurring when

    a. a "milk of magnesia" tablet (solid magnesium oxide) neutralizes "stomach acid" (aqueous hydrochloric acid) by forming liquid water and aqueous magnesium chloride

    b. butane gas is burned in a camping stove, consuming oxygen and forming a gaseous mixture of carbon monoxide, carbon dioxide, and water vapor

    c. aqueous ethanol is metabolized in the body, forming, as *one* of the products, acetaldehyde ($CH_3CHO(aq)$), the chemical primarily responsible for the "alcohol hangover" (Assume that dissolved oxygen is consumed and liquid water is formed.)

Solution (Using formulas as described in Section 3.3)

    a. $MgO(s) + HCl(aq) \longrightarrow H_2O(l) + MgCl_2(aq)$

    b. $C_4H_{10}(g) + O_2(g) \longrightarrow CO(g) + CO_2(g) + H_2O(g)$

    c. $C_2H_5OH(aq) + O_2(aq) \longrightarrow CH_3CHO(aq) + H_2O(l)$

3.5 NET EQUATIONS

Salts (Unit 6) and many acids (Unit 19) exist in aqueous solution as separated ions. Chemical reactions involving such solutions often leave some of the ionic species unaffected; such ions are called *spectator ions* and need not be included in equations intended to represent only the chemical changes occurring. *Net* equations show only chemical species actually *participating* in the changes being described.

    For example, the reaction occurring when solutions of silver nitrate and sodium chloride are mixed may be represented by the *complete formula* equation

$$AgNO_3(aq) + NaCl(aq) \longrightarrow AgCl(s) + NaNO_3(aq)$$

or, since experimental evidence shows that the silver ions and chloride ions leave the

Table 3.4 SOME COMMON STRONG AND WEAK ACIDS AND BASES
(For a more extensive list, see Appendix C.)

NAME	TYPE	FORMULATE IN AQUEOUS SOLUTIONS AS
hydrochloric acid	strong acid	$H^+ + Cl^-$
hydrobromic acid	strong acid	$H^+ + Br^-$
nitric acid	strong acid	$H^+ + NO_3^-$
perchloric acid	strong acid	$H^+ + ClO_4^-$
sulfuric acid	strong acid (for first H^+ loss)	$H^+ + HSO_4^-$
hydroxide ion (from metal hydroxides)	strong base	(metal ion) $+ OH^-$
acetic acid	weak acid	CH_3CO_2H
hydrofluoric acid	weak acid	HF
hydrosulfuric acid	weak acid	H_2S
hypochlorous acid	weak acid	$HOCl$ (or $HClO$)
nitrous acid	weak acid	HNO_2
phosphoric acid	weak acid	H_3PO_4
bisulfate ion	weak acid	HSO_4^-
ammonia	weak base	NH_3
methylamine	weak base	CH_3NH_2

solution to form an insoluble solid (precipitate) of silver chloride while sodium and nitrate ions remain in solution, a *net equation* may be used:

$$Ag^+(aq) + Cl^-(aq) \longrightarrow AgCl(s)$$

In writing net equations, complete formulas are used for all covalent compounds (e.g., CO_2 or H_2O) and, *by convention,* for all *solid* ionic compounds (e.g., $CaCO_3(s)$ or $BaSO_4(s)$). Salts (e.g., NaCl or $AgNO_3$) or *strong* acids and bases (Table 3.4) that are dissolved in water are represented as separate ions, since such compounds are essentially 100 percent ionized in aqueous solutions (Unit 19). *Weak* acids and bases (Table 3.4), only slightly ionized in aqueous solution (Unit 19), are represented by complete formulas. We shall have to develop a more extensive understanding of chemistry before we can predict which salts will dissolve easily in water or which acids and bases are expected to be strong or weak. For now, we will rely on the use of information from Table 3.4 and the notation (s) or (aq) to help us identify species to be formulated as ions or by complete formulas.

To write the net equation, all reactants and products are first written as complete formulas, then converted to *conventional* form (i.e., as either complete formulas or as ions). Finally, any species represented by *identical* formulations in both reactants and products are deleted. Numerical coefficients (numbers in front of symbols or formulas) are ignored in developing *qualitative* net equations.

Example 6

Write net equations to represent the chemical changes occurring when

   a. aqueous sodium bicarbonate (baking soda) is used to neutralize spilled "battery acid" (aqueous sulfuric acid), forming water, carbon dioxide gas, and aqueous sodium bisulfate

   b. solid barium sulfate, used in making slurries for x-ray examinations of the gastrointestinal tract, is prepared by mixing aqueous solutions of barium chloride and sodium sulfate (Aqueous sodium chloride is the other product.)

   c. "vinegar" (aqueous acetic acid) is used to clean up a spilled drain cleaner consisting of aqueous sodium hydroxide, forming water and aqueous sodium acetate

Solution

  a. STEP 1 $NaHCO_3(aq) + H_2SO_4(aq) \longrightarrow H_2O(l) + CO_2(g) + NaHSO_4(aq)$

             a salt   a strong acid   covalent   a salt

    STEP 2 $Na^+ + HCO_3^- + H^+ + HSO_4^- \longrightarrow H_2O + CO_2 + Na^+ + HSO_4^-$

    STEP 3 $\cancel{Na^+} + HCO_3^- + H^+ + \cancel{HSO_4^-} \longrightarrow H_2O + CO_2 + \cancel{Na^+} + \cancel{HSO_4^-}$

    Answer: $HCO_3^-(aq) + H^+(aq) \longrightarrow H_2O(l) + CO_2(g)$

  b. STEP 1 $BaCl_2(aq) + Na_2SO_4(aq) \longrightarrow BaSO_4(s) + NaCl(aq)$

             a salt   a salt   a *solid* salt   a salt

    STEP 2 $Ba^{2+} + 2Cl^- + 2Na^+ + SO_4^{2-} \longrightarrow BaSO_4 + Na^+ + Cl^-$

                  (numerical coefficients ignored)

STEP 3 $Ba^{2+} + \cancel{Cl^-} + \cancel{Na^+} + SO_4^{2-} \longrightarrow BaSO_4 + \cancel{Na^+} + \cancel{Cl^-}$

Answer: $Ba^{2+}(aq) + SO_4^{2-}(aq) \longrightarrow BaSO_4(s)$

c. STEP 1 $CH_3CO_2H(aq) + NaOH(aq) \longrightarrow H_2O(l) + NaCH_3CO_2(aq)$

a weak acid ↓   a strong base ↓   covalent ↓   a salt ↓

STEP 2 $CH_3CO_2H + Na^+ + OH^- \longrightarrow H_2O + Na^+ + CH_3CO_2^-$

STEP 3 $CH_3CO_2H + \cancel{Na^+} + OH^- \longrightarrow H_2O + \cancel{Na^+} + CH_3CO_2^-$

Answer: $CH_3CO_2H(aq) + OH^-(aq) \longrightarrow H_2O(l) + CH_3CO_2^-(aq)$

Although net equations are useful in identifying the species actually participating in a chemical change, complete formulas are more often employed for equations needed to identify specific chemicals employed or to permit quantitative calculations (Stoichiometry, Unit 4).

3.6 BALANCED EQUATIONS

A chemical equation may simply indicate in a qualitative way what substances are consumed and what new substances are produced. However, a *balanced* equation may be used to obtain further information of a quantitative nature (Unit 4). Such an equation requires

1. the same number of each elementary symbol in both reactants and products, and
2. the same *net* charge for both reactants and products.

Simple equations can be balanced[1] by placing numerical coefficients for symbols and formulas as required, once all reactants and products are listed. Note that *subscript* numbers cannot be changed since they are part of the specific chemical formula. A systematic approach for certain types of more complicated equations is developed in Unit 22.

Example 7

Balance

$$H_2(g) + O_2(g) \longrightarrow H_2O(g)$$

(qualitative equation)

Solution

There are only three substances involved: hydrogen (H_2), oxygen (O_2), and water (H_2O). Only numerical coefficients, not additional chemical substances, may be introduced.

Since there are two O's ($\underline{O_2}$) in reactants and only one in products (H_2O), we can balance the O's by placing a 2 in front of the H_2O:

$$H_2 + \underline{O_2} \longrightarrow \underline{2H_2O}$$

Then there are four H's ($\underline{2H_2O}$) in products and only two (H_2) in reactants, so we now

---

[1] If you enjoy "mathematical games" and wish to learn a rather novel method for balancing all sorts of chemical equations, you may be interested in the article "An Algebraic Method for Balancing Chemical Equations" by David Greene in the March, 1975, issue of *Chemistry*.

place a 2 in front of the H_2:

$$2H_2 + O_2 \longrightarrow 2H_2O$$

The equation

$$2H_2(g) + O_2(g) \longrightarrow 2H_2O(g)$$

now fits both requirements for a balanced equation; that is,

symbols: 4H, 2O = 4H, 2O
and charge: 0 = charge 0

Example 8

Balance

$$Li(s) + H_2O(l) \longrightarrow Li^+(aq) + OH^-(aq) + H_2(g)$$

(qualitative equation)

Solution
Remember that for a qualitative equation (all reactants and products shown), only numerical coefficients may be introduced.

Balance the elementary symbols first: There are only two H's in reactants (H_2O) and three in products (OH^- + $\underline{H_2}$). As a trial-and-error attempt, place a 2 in front of H_2O and OH^- to balance the $\overline{H}$'s:

$$Li + \underline{2H_2O} \longrightarrow Li^+ + \underline{2OH^-} + \underline{H_2}$$

This also, by chance, balances the O symbols:
$(\underline{2H_2O} = \underline{2OH^-})$

The *net* charge is the algebraic sum of all ionic charges shown.

All symbols are now balanced. However, net charges are not. There are two $-$ and one $+$ for products for a net charge of -1, while there is a zero net charge for reactants. Placing a 2 in front of Li^+ will balance the charges, but this requires a 2 in front of Li to keep the symbols balanced. The resulting equation meets both criteria required:

$$2Li(s) + 2H_2O(l) \longrightarrow 2Li^+(aq) + 2OH^-(aq) + H_2(g)$$

symbols: 2Li, 4H, 2O = 2Li, 2O, 4H
and charges: 0 = (2+) + (2−)

Note that there are an infinite number of equations that would be balanced for this process, for example,

$$40Li + 40H_2O \longrightarrow 40Li^+ + 40OH^- + 20H_2$$

By convention, the *smallest* whole numbers sufficient to balance the equation are used.

3.7 TYPES OF CHEMICAL REACTIONS

Most chemical reactions can be classified as one of two general types: metathetical or oxidation-reduction. Oxidation-reduction reactions are characterized by the transfer of one or more electrons from one chemical species—atom, ion, or molecule—to another. In most cases such reactions can be used for the production of electric

Table 3.5 COMMON OXIDATION NUMBERS

TYPE OF SUBSTANCE	OXIDATION NUMBER ASSIGNMENT	EXAMPLES
"free" element	0 (zero)	$\overset{0}{Na}$, $\overset{0}{F_2}$, $\overset{0}{As_4}$, $\overset{0}{S_8}$
monatomic ion	same as the charge on the ion	$\overset{+I}{H^+}$, $\overset{+II}{Mg^{2+}}$, $\overset{+III}{Al^{3+}}$, $\overset{-I}{F^-}$, $\overset{-II}{S^{2-}}$
group IA elements in compounds	+I	$\overset{+I}{Na}Cl$, $\overset{+I}{Na_3}PO_4$
group IIA elements in compounds	+II	$\overset{+II}{Ca}SO_4$, $\overset{+II}{Ca_3}(PO_4)_2$
fluorine in compounds	−I	$H\overset{-I}{F}$, $Al\overset{-I}{F_3}$, $C\overset{-I}{F_4}$
oxygen in compounds or polyatomic ions	−II except as −I in peroxides (O—O bond) (as +II in OF₂)	$H_2\overset{-II}{O}$, $C\overset{-II}{O_2}$, $P\overset{-II}{O_4}{}^{3-}$ $H_2\overset{-I}{O_2}$, $Na_2\overset{-I}{O_2}$, $O\overset{+II}{F_2}$
hydrogen in compounds or polyatomic ions	+I except as −I when combined with metals	$\overset{+I}{H}Cl$, $\overset{+I}{H_2}O$, $\overset{+I}{N}H_3$, $O\overset{+I}{H^-}$ $Ca\overset{-I}{H_2}$, $LiAl\overset{-I}{H_4}$

current by arranging the system in such a way that electrons flow through a wire. Oxidation-reduction reactions, and the associated subject of electrochemistry, are considered in detail in Units 22–24. In metathetical reactions, no electron transfer occurs. A simple "bookkeeping" method may be used to classify reactions as one of these general types. *Oxidation numbers* are assigned for each element symbol appearing in the equation. These numbers, which may be written as Roman numerals with the appropriate algebraic signs, do not (except in the case of monatomic ions or "free elements") represent a "real" charge; they are only "bookkeeping" symbols.

For the simple cases to be considered at this time, oxidation numbers are assigned to common element symbols as indicated in Table 3.5. For other elements appearing with these in formulas, oxidation numbers are determined on the basis that *the algebraic sum of all oxidation numbers must equal the charge associated with the formula.* It is important to note that the oxidation number is counted each time an element symbol occurs; for example, in Al_2O_3, *each* oxygen is assigned an oxidation number of −II. Three oxygen symbols then represent −VI. The net charge on the formula is zero, so *each* aluminum is in a +III oxidation state.

This method is similar to that previously discussed for dealing with formulas of salts, except that it must be remembered that oxidation numbers are *assigned* for bookkeeping purposes, whereas ionic charges have experimental "reality."

The terms *oxidation number* and *oxidation state* are essentially equivalent, differing only slightly in their grammatical usage. For example, for the ion Cu^{2+}, we say that copper *is in* the +II oxidation state or that copper *has* an oxidation number of +II.

Example 9

What is the oxidation state of chromium in dichromate ion?

Solution

Dichromate ion is formulated as $Cr_2O_7{}^{2-}$. Oxygen is assumed to be in the −II oxidation state unless some indication is given that the species is a peroxide.

Let N be the oxidation number for chromium; then

$$2N + 7(-\text{II}) = -2$$

from which N is $+\text{VI}$.

You may find it simpler to use ordinary (Arabic) numerals for the algebraic work.

Example 10

Should the reaction formulated by the equation

$$\text{Ca(OH)}_2 + 2\text{H}^+ + \text{SO}_4{}^{2-} \longrightarrow \text{CaSO}_4 + 2\text{H}_2\text{O}$$

be classified as metathetical or oxidation reduction?

Solution
Assign oxidation numbers for each element symbol:

$$\overset{+\text{II}\ -\text{II}\ +\text{I}}{\text{Ca(O H)}_2} + \overset{+\text{I}}{2\text{H}^+} + \overset{+\text{VI}\ -\text{II}}{\text{S O}_4{}^{2-}} \longrightarrow \overset{+\text{II}\ +\text{VI}\ -\text{II}}{\text{Ca S O}_4} + \overset{+\text{I}\ -\text{II}}{2\text{H}_2\text{O}}$$

There is no change in oxidation states, so the reaction is classified as metathetical.

Figure 3.4 Choices. (Photograph by Arden Literal, Washington State University.)

Chemical changes can be controlled by knowledge of the processes involved. The use man will make of his knowledge of chemical reactions depends, in a free society, on the choices of scientists and nonscientists together (Figure 3.4).

SUGGESTED READING

Ferguson, L. N. 1972. *Organic Chemistry, A Science and an Art.* Boston: Willard Grant Press.
Fischer, R. B. 1971. *Science, Man, and Society.* Philadelphia: Saunders.
What's Happening in Chemistry? 1975. Washington, D.C.: American Chemical Society.

SOME PROBLEMS WITH "UNREACTIVE" CHEMICALS

Although many products of the chemical industry are designed to be used in reaction processes, some are planned specifically for purposes requiring "inert" substances. In other cases, little attention is given to the reactive character of a chemical product beyond the point of being sure that it will perform the particular task for which it was designed, although industry is now paying much more attention to a complete spectrum of chemical properties than was the case a few years ago. A few examples will illustrate some of the problems of "unreactive" chemicals.

Carbon tetrachloride is a relatively "inert" substance. Unlike most compounds of carbon, CCl_4 will not burn. This resistance to combustion, coupled with the fact that the vapors of CCl_4 are much more dense than air, suggested that this substance might be ideal for use as a "fire extinguisher." Initial testing confirmed this suggestion and glass bulbs containing carbon tetrachloride were soon on the market. When smashed near the base of a fire, the bulbs released the low-boiling liquid whose heavy vapors soon displaced the air around the blaze, thus stopping the combustion process. Unfortunately, testing had not been sufficiently extensive. Although CCl_4 did not burn, it did react rather rapidly at the temperatures around a fire area to form the extremely poisonous gas phosgene, $COCl_2$. It was not until a number of fatalities had been attributed to phosgene poisoning that the carbon tetrachloride extinguishers were banned for use against indoor fires.

Synthetic detergents provide a good example of chemical manufacture for specific uses without adequate consideration of long-range adverse effects. The original detergents were designed to be effective cleaning agents in "hard water." The fact that these chemicals were essentially inert to natural waste-destroying systems (primarily aquatic organisms) was not considered a problem until massive water pollution from accumulating waste detergents became very obvious. The molecular structures of the detergents were redesigned and all detergents now manufactured in the United States are "biodegradable."

A recent problem centers on the controversial use of Freons, such as Freon-12 (CF_2Cl_2). The gaseous Freons are relatively "inert" and nontoxic and have been employed for many years as compressor gases in refrigeration systems and as propellant gases in many aerosol sprays. It is the very "inertness" of the freons to reaction with "ordinary" atmospheric oxygen that has led to the problem now considered of great potential importance. Because, unlike many air pollutants, the Freons are quite unreactive with O_2 gas, they can survive and accumulate, eventually reaching the ozone (O_3) layer in the upper atmosphere. Freons are decomposed by ultraviolet light in the stratosphere, liberating highly reactive chlorine atoms. These atoms, in turn, react with ozone and if reaction is extensive enough to deplete the ozone layer significantly, this "protective O_3 shell" that shields us from excess ultraviolet radiation could be permanently damaged. How serious this threat is has not yet been fully determined, but it now is the subject of extensive research.

EXERCISES

1. Make and use a set of flash cards as described in Exercise 1, Unit 2, to learn the information given in Tables 3.1 3.2, 3.3, and 3.4.
2. Supply the missing names or formulas:
   a. $KHSO_4$; **b.** ammonium acetate; **c.** CuCN; **d.** silver chromate; **e.** Cr_2O_3; **f.** sodium sulfide; **g.** $Al_2(CO_3)_3$; **h.** iron(II) phosphate; **i.** $Co(OH)_2$; **j.** mercury(I) nitrate
3. Select the term best associated with each definition, description, or specific example.
   TERMS: (A) acid (B) empirical formula (C) hydrated (D) product (E) "yields"
   a. the meaning of the arrow in

   $$Ag^+ + Cl^- \longrightarrow AgCl$$

   b. the AgCl in the reaction represented by the equation in (a)
   c. HgCl (as compared to Hg_2Cl_2)
   d. HCl(aq)
   e. $CuSO_4(H_2O)_5$
4. Write qualitative equations (not necessarily balanced), in both "complete formula" and "net" forms, to represent the following processes:
   a. Newly installed ceramic floor tile is often discolored by white residues of calcium carbonate, formed when the tile was stored in an exposed area. The solid deposits can be removed by washing the tile with aqueous hydrochloric acid, forming water, carbon dioxide gas, and aqueous calcium chloride.
   b. Ammonium sulfate, a chemical fertilizer, may be prepared by mixing aqueous solutions of ammonia and sulfuric acid. The product remains dissolved in water, so the solid fertilizer is obtained by a subsequent physical process of evaporation.
   c. It is dangerous to mix many of the common household chemicals. For example, if aqueous hydrochloric acid tile cleaner is mixed with an aqueous bleaching solution of calcium hypochlorite, highly poisonous chlorine gas is generated. The other products are water and aqueous calcium chloride.
5. Balance both the "complete formula" and "net" equations written for the processes in Exercise 4.

OPTIONAL

6. Classify each of the following "qualitative" equations as metathetical (label *M*) or oxidation-reduction (label *O–R*).
   a. $NaHCO_3 + H_2SO_4 \longrightarrow Na_2SO_4 + H_2O + CO_2$
   b. $Zn + Cr_2O_7^{2-} + H^+ \longrightarrow Zn^{2+} + Cr^{2+} + H_2O$
   c. $ClO^- + Cl^- + H^+ \longrightarrow Cl_2 + H_2O$
   d. $Mg(OH)_2 + H_3PO_4 \longrightarrow Mg_3(PO_4)_2 + H_2O$
   e. $C_2O_4^{2-} + MnO_4^- + H^+ \longrightarrow CO_2 + Mn^{2+} + H_2O$

PRACTICE FOR PROFICIENCY

Objective (a)

1. Supply the missing names or formulas:
   a. acetic acid; **b.** PH_3; **c.** hydroxide ion; **d.** HNO_3; **e.** butane; **f.** HSO_3^-; **g.** chloroform; **h.** H_3PO_4; **i.** hydrogen sulfide; **j.** $Cr_2O_7^{2-}$; **k.** sulfuric acid; **l.** C_6H_6; **m.** hydrogen cyanide; **n.** CO_3^{2-} **o.** methanol; **p.** CH_3NH_2; **q.** perchloric acid; **r.** NH_4^+; **s.** thiosulfate ion; **t.** HOCl

Objective (b)

2. Supply the missing names or formulas:
   a. sodium bicarbonate; **b.** $KMnO_4$; **c.** aluminum oxide; **d.** Mg_2N_3; **e.** silver acetate; **f.** $Fe(NO_3)_3$; **g.** calcium chlorate; **h.** $MnSO_4$; **i.** ammonium dichromate; **j.** Cu_2S; **k.** zinc phosphate; **l.** CH_3NH_3Br; **m.** tin(II) fluoride ("stannous fluoride"); **n.** PbC_2O_4

Objective (c)

3. Select the term best associated with each definition, description, or specific example.
   TERMS: (A) anhydrous (B) base (C) coefficient (D) net equation (E) oxidation state (F) product (G) reactant (H) salt (I) strong acid (J) weak acid
   a. shows only the species participating in chemical change
   b. refers to species not associated with water
   c. a chemical substance capable of accepting a proton
   d. an ionic compound having a cation other than, or in addition to, H^+
   e. "bookkeeping" numbers assigned to element symbols (e.g., $+I$ for H in H_2O)
   f. HNO_3
   g. HNO_2
   h. (underlined): $\underline{2}H_2 + O_2 \longrightarrow 2H_2O$
   i. (underlined): $2H_2 + \underline{O}_2 \longrightarrow 2H_2O$
   j. (underlined): $2H_2 + O_2 \longrightarrow 2H_{\underline{2}}O$

Objective (d)

4. Write qualitative equations (not necessarily balanced), in both "complete formula" and "net" forms, to represent the following processes:
   a. Cases have been reported of housewives dying from inhalation of the poisonous gas chloramine, $ClNH_2$, released when household ammonia (aqueous) and an aqueous calcium hypochlorite bleaching solution were mixed in an attempt to make a "superpowerful" cleaning solution. The other reaction product is aqueous calcium hydroxide.
   b. "Chrome yellow," a commercially valuable paint pigment, can be prepared by mixing aqueous solutions of sodium chromate and lead nitrate, $Pb(NO_3)_2$. The pigment, lead chromate, precipitates as a solid and is isolated by filtration from the aqueous sodium nitrate solution.
   c. A commercial drain cleaner uses a mixture of solid sodium hydroxide and small chunks of aluminum. When the mixture is dropped into a clogged drain, the sodium hy-

droxide dissolves in the water present and this aqueous solution reacts with the aluminum to generate hydrogen gas. The bubbling gas helps dislodge debris clogging the drain. The other reaction product is an aqueous salt, sodium tetrahydroxoaluminate, formulated as $Na[Al(OH)_4]$.

d. Sulfur dioxide once found a use in public health for the "fumigation" of rooms or homes of scarlet fever victims. The gas could be generated by pouring hydrochloric acid into an aqueous solution of sodium sulfite. The other reaction products are water and aqueous sodium chloride.

e. New chemical techniques show a considerable promise for the production of "synthetic bone" for possible use in bone implant surgery. One such technique involves sealing a piece of coral (porous calcium carbonate) in a gold tube with ammonium hydrogen phosphate, $(NH_4)_2HPO_4$, dissolved in water. The tube is sealed and heated to 250°–300°C at high pressure. The reaction converts the calcium carbonate to "hydroxyapatite" $[Ca_5(OH)(PO_4)_3]$ having the same porous solid structure of the original coral, closely duplicating the structure of bone. The other products are dissolved in the water and can be completely washed away from the "synthetic bone."

Objective (e)

5. Balance both the "complete formula" and "net" equations written for the processes in question 4.

OPTIONAL

6. Assign oxidation numbers for each element symbol in
a. potassium permanganate; **b.** calcium oxalate;
c. $KAl(SO_4)_2$; **d.** $Cu_2C_2O_4$; **e.** $Fe(NO_3)_2$

7. Balance each of the following equations. Classify each as metathetical (label *M*) or oxidation-reduction (label *O–R*).
a. $Zn + H^+ \longrightarrow Zn^{2+} + H_2$
b. $Zn^{2+} + H_2S \longrightarrow ZnS + H^+$
c. $CaCO_3 + H_3PO_4 \longrightarrow Ca_3(PO_4)_2 + CO_2 + H_2O$
d. $NH_3 + O_2 \longrightarrow NO_2 + H_2O$

BRAIN TINGLERS

8. Since an acid may be defined as "a proton donor" and a base as a "proton acceptor", it follows that an acid should react with a base by a process we might call "proton exchange". Under these definitions, metal hydroxides themselves are not considered to be *bases*, but are *salts containing the base*

OH^-. Using these definitions and the information that all the ionic compounds formed from the following reactions remain in aqueous solution, predict the products for each reaction indicated and balance the resulting equations (in both "complete formula" and "net" forms). (Don't forget the distinction between "strong" and "weak".)
a. $NaOH(aq) + HCl(aq) \longrightarrow$
b. $KOH(aq) + HCN(aq) \longrightarrow$
c. $KOH(aq) + H_2SO_4(aq) \longrightarrow$
d. $NH_3(aq) + HNO_3(aq) \longrightarrow$
e. $Al(OH)_3(s) + HClO_4(aq) \longrightarrow$

9. All of the following useful ammonium salts dissolve easily in water. Suggest the appropriate aqueous acid for preparing each salt and describe each preparation by a balanced "complete formula" equation.
a. ammonium chloride: used in medicine as an expectorant and in certain soldering operations as a "flux"
b. ammonium nitrate: a valuable fertilizer, but potentially explosive; responsible for the famous Texas City disaster of 1947 (described in the Exercises for Unit 5)
c. ammonium sulfate: a common fertilizer often recommended for alkaline soils
d. ammonium phosphate: a particularly valuable fertilizer providing both nitrogen and phosphorus, essential elements for plant growth

10. Each of the following equations represents, qualitatively, a process occurring during the discharge of an electrochemical cell. All of these processes have something in common with *every* chemical reaction that could possibly be utilized for an electrochemical cell. Examine the equations carefully to see if you can find the common characteristic. Then see how many of the equations you can balance. (The topic of electrochemical cells is discussed in Units 22–24.)
a. (discharge reaction of the common automobile battery)

$$Pb(s) + PbO_2(s) + H_2SO_4(aq) \longrightarrow PbSO_4(s) + H_2O(l)$$

b. (discharge reaction, at low current, of a simple "dry cell")

$$Zn(s) + MnO_2(s) + NH_4Cl(aq) \longrightarrow ZnCl_2(aq) + Mn_2O_3(s) + NH_3(aq) + H_2O(l)$$

c. (discharge reaction of a rechargeable "Nicad" flashlight cell)

$$Cd(s) + NiO_2(s) + H_2O(l) \longrightarrow Cd(OH)_2(s) + Ni(OH)_2(s)$$

d. (discharge reaction of a methane-oxygen "fuel cell")

$$CH_4(g) + O_2(g) + KOH(aq) \longrightarrow K_2CO_3(aq) + H_2O(l)$$

Unit 3 Self-Test

1. Supply the missing names or formulas:
   a. CH_3CO_2H; **b.** nitric acid; **c.** C_2H_5OH; **d.** methylamine;
   e. H_3PO_4
2. Write formulas or names, respectively, for
   a. ammonium sulfate; **b.** copper(II) nitrate; **c.** trisodium
   phosphate; **d.** silver bromide; **e.** potassium chlorate;
   f. $Na_2Cr_2O_7$; **g.** $PbSO_4$; **h.** $KHCO_3$; **i.** CaO; **j.** $Fe(NO_3)_3$
3. Matching: Select the term best associated with each formula
   or equation.
   TERMS: (A) an acid (B) an anhydrous salt (C) a base (D) complete molecular formula for ethane (E) empirical formula for
   butane (F) a hydrated salt (G) a net equation (H) products
   (underlined) (I) reactants (underlined)
   a. NH_4NO_3
   b. $CuSO_4(H_2O)_5$
   c. C_2H_5
   d. C_2H_6
   e. HNO_3
   f. NH_3
   g. $H^+ + OH^- \longrightarrow H_2O$
   h. <u>NaOH + HCl</u> $\longrightarrow$ NaCl + H_2O
   i. LiOH + HBr $\longrightarrow$ <u>LiBr + H_2O</u>
4. Write qualitative equations (not necessarily balanced), in
   both "complete formula" and "net" forms, to represent the
   following processes:
   a. the preparation of solid "baking soda" (sodium bicarbonate) by the reaction of carbon dioxide gas with an aqueous
   solution of ammonia and sodium chloride (The other
   product is aqueous ammonium chloride.)
   b. the formation of solid silver bromide for use in photographic film by the reaction of aqueous silver nitrate with
   hydrobromic acid (The other product is aqueous nitric
   acid.)
   c. the reaction of solid zinc with hydrochloric acid to generate hydrogen gas (Aqueous zinc chloride is the other product.)
   d. the action of a solid aluminum hydroxide "antacid tablet"
   to neutralize "excess stomach acid" (hydrochloric acid)
   (The other product may be considered to be aqueous
   aluminum chloride.)
   e. the formation of crystals of "calcite" (calcium carbonate)
   in a limestone cavern as carbon dioxide gas from the
   atmosphere reacts with aqueous calcium hydroxide
   (Water is the other product.)
5. Balance both the "complete formula" and "net" equations
   written for question 4.

OPTIONAL

6. Supply appropriate numerical coefficients to balance each of
   the following equations. Classify each as M for metathetical or O–R for oxidation-reduction.
   a. ____Ag_2SO_4 + ____NaCl $\longrightarrow$ ____Na_2SO_4 + ____AgCl
   b. ____Al_2O_3 + ____H^+ $\longrightarrow$ ____Al^{3+} + ____H_2O
   c. ____CO + ____I_2O_5 $\longrightarrow$ ____CO_2 + ____I_2
   (a reaction used for detecting trace amounts of carbon
   monoxide in the atmosphere)
   d. ____Cu_2S + ____O_2 $\longrightarrow$ ____Cu + ____SO_2
   (one of the processes used in the smelting of copper)
   e. ____NH_4NO_3 $\longrightarrow$ ____N_2 + ____O_2 + ____H_2O
   (the reaction responsible for the explosion of a ship
   loaded with fertilizer off the coast of Texas a few years
   ago)

We have found the enemy—
he is us.
POGO POSSUM

Stoichiometry

Objectives

(a) After studying the appropriate definitions in Appendix E and the terms′ usage in this Unit, be able to demonstrate your knowledge of the following terms by identifying them with appropriate information provided in descriptions of chemical processes:
actual yield
Avogadro′s number
formula weight
gram-formula weight
limiting reagent
millimole
mole
percentage yield
theoretical yield

(b) Given a periodic table and the formula or name (Unit 3) of a chemical compound or polyatomic ion, be able to calculate the *formula weight* (Section 4.1).

(c) Given a description of a chemical reaction, by words or by an equation, be able to calculate the theoretical mass of any reactant or product from the given mass of another reactant or product by use of a stoichiometric ratio (Section 4.3).

(d) For a process using an excess of one or more reactants, be able to identify the *limiting reagent* for the reaction (Section 4.5).

(e) For a "nonstoichiometric" process:

(1) given the actual yield of a product and the mass of the limiting reagent, be able to calculate the *percentage yield* (Section 4.4).

(2) given the percentage yield of a product and appropriate mass data, be able to calculate the actual yield of the product or a minumum reactant mass (Section 4.4).

OPTIONAL

Given appropriate stoichiometric information and cost–profit factors, be able to calculate specified economic factors associated with a chemical process.

Science and technology have supplied the middle-class American with luxuries beyond the wildest dreams of the Roman emperors. This same science and technology—in answer to demands for more power, more luxuries, more things—now threatens to engulf us in the muck of foul air, polluted water, and "throw-away" trash, or to wipe us from the earth at the touch of the "nuclear pushbutton."

We are trapped by some inexorable laws. If we understand these laws, we shall see that they offer us some alternatives. Three of these laws lie in the realm of science:

1. *The useful energy of the universe is decreasing (Unit 5).*
2. *Many of the most critical properties of systems are functions of concentration (Unit 13).*
3. *Every chemical reaction is governed by the mathematical restrictions of conservation of energy and of mass.*

   (This is the fundamental basis for the subject of stoichiometry.)

An understanding of these laws—which tell us the limitations imposed on science and technology—is necessary, but not sufficient, for the survival of society as we know it. They define the broad alternatives we can select and delineate the predictable consequences of our choices. Which choices are actually made, however, will be governed by laws much more complex: the laws of "human nature."

4.1 STOICHIOMETRY   Stoichiometry refers to mathematical calculations based on chemical equations. *If any procedure can be considered essential to every area of chemistry and chemical engineering, this quantitative use of chemical relationships must be the one.* Through stoichiometric procedures the analyst can determine the detailed composition of samples; the organic chemist can estimate the efficiency of a new synthesis; the engineer can plan an economical process for large-scale production of new substances; the biochemist can follow the metabolic processes of an organism; the aerospace scientist can calculate the amount of fuel needed for a space vehicle.

Stoichiometric calculations are based on the fixed ratios among the species (atoms, ions, molecules) involved in chemical reactions. Such ratios are indicated by numerical subscripts appearing in formulas and numerical coefficients in balanced chemical equations (Unit 3). There is no significant loss of mass as a result of a chemical reaction. In a *nuclear* process, such as the explosion of an atomic bomb, it can be demonstrated that some matter actually disappears, reappearing as the equivalent ($E = mc^2$) in energy. The energy changes of chemical processes are too small to be detected as mass changes. As a result, a balanced chemical equation is associated with a *mass balance*.

Consider the reaction formulated by the equation

$$C + O_2 \longrightarrow CO_2$$

The average mass of carbon consumed when one atom combines with a molecule of oxygen is 12.0115 amu. The average mass of oxygen consumed is 31.9988 amu (2 atoms/molecule $\times$ 15.9994 amu/atom). The principal of *mass balance* then suggests that since there can be no significant mass loss, the carbon dioxide formed should "weigh" 44.0103 amu, that is, the sum of the masses of carbon and oxygen consumed.

The same answer is arrived at by calculating the *formula weight* of carbon dioxide. A formula weight is the summation of all atomic weights represented by the element symbols in a formula, considering that subscripts indicate the numbers of each element symbol in the total formula. Thus the formula weight of carbon dioxide (CO_2) is calculated as

C: 12.0115
O_2:   31.9988 (2 oxygens $\times$ 15.9994 each)
CO_2: 44.0103 amu

The units of formula weights, like those of atomic weights, are atomic mass units, although it is conventional to omit writing these units.

Only average masses (based on atomic weights) can be considered because of the existence of isotopes (Unit 2). Of course, if we could really work with a single atom, it would have its own characteristic isotopic mass.

Remember that we are using, as a matter of convenience, the terms *mass* and *weight* as though they were equivalent.

Example 1

Calculate the formula weight of sodium chromate to four significant figures.

Solution
The formula for sodium chromate is Na_2CrO_4 (Unit 3). Since we want the formula weight to four significant figures, we will use, as a convenience, atomic weights rounded to *five* figures, then round the answer to the desired four figures.

Na_2: (2 $\times$ 22.990) 45.980
Cr: (1 $\times$ 51.996) 51.996
O_4: (4 $\times$ 15.999) 63.996
                    161.972
    (rounded to 162.0)

Example 2

Calculate the formula weight of acetic acid, to three significant figures.

Solution
The formula for acetic acid is CH_3CO_2H (Unit 3). For purposes of formula weight calculation, it is convenient to count each element in the formula and rewrite the formula with each element symbol appearing only once. For example, since acetic

acid contains a total of two carbon atoms, four hydrogen atoms, and two oxygen atoms per molecule, we may write $C_2H_4O_2$. Then

C_2: (2×12.01) 24.02
H_4: (4×1.008) 4.032
O_2: (2×16.00) 32.00
_____
60.052

(rounded to 60.1)

A balanced equation represents not only the chemical change taking place with single atoms or molecules, but also the reaction occurring *on any scale*. This is relatively easy to see. For the formation of a molecule of CO_2 from its element, *each* molecule produced required one carbon atom and one oxygen molecule (as O_2, containing two oxygen atoms) as reactants. Then the formation of 10 CO_2 molecules, of necessity, must require 10 atoms of carbon and 10 molecules of oxygen. The *stoichiometric ratios* expressed by the numerical factors in the balanced equation determine the *relative* quantities of reactants and products at any level.

It is not generally convenient, experimentally, to *count* the numbers of atoms, ions, or molecules participating in a chemical reaction. It is much easier to *weigh* the quantities involved or, sometimes, to measure volumes of gases, liquids, or solutions. The stoichiometric ratios allow atomic or formula weights to be used in simple ratios for calculations of actual weight or volume quantities. The latter will be discussed for gases in Unit 10 and for liquids and solutions in Units 11 and 13.

Consider again the reaction formulated as

$$C + O_2 \longrightarrow CO_2$$

Suppose it is desirable to know the weight of carbon dioxide that could be formed by this reaction from 24 tons of carbon and sufficient oxygen.

From earlier discussion the weight of CO_2, from the *stoichiometric* amounts of reactants, *must* be 44.0103/12.0115 times the weight of carbon consumed in the reaction. Thus the weight of CO_2 formed from complete reaction of 24 tons of carbon is

$$\frac{44.0103}{12.0115} \times 24 \text{ tons} = 88 \text{ tons}$$

(Note that the answer is reported to only *two* significant figures because of limitations in the measured quantity "24 tons"; see Math Skills.)

A "Nuts and Bolts" Analogy

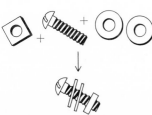

Figure 4.1 A "nuts and bolts" analogy.

Suppose that a manufacturing concern wishes to market nut/bolt/washer sets in packages of 1-gross units. Each set consists of one nut, one bolt, and two washers. We could formulate this relationship (Figure 4.1) as

$$\text{nut} + \text{bolt} + 2 \text{ washers} \longrightarrow \text{set}$$

It should be apparent that for each gross (144) of sets, 144 nuts, 144 bolts, and 288 (2×144) washers are required. It is generally easier to weigh large numbers of things than to count them. If it is found that 1 gross of nuts weighs 3.6 kg, 1 gross of bolts weighs 6.9 kg, and 1 gross of washers weighs 1.7 kg, then the weight of a gross of sets can be determined from

$$\left(\frac{3.6 \text{ kg}}{1 \text{ gross (nuts)}} \times \frac{1 \text{ nut}}{1 \text{ set}}\right) + \left(\frac{6.9 \text{ kg}}{1 \text{ gross (bolts)}} \times \frac{1 \text{ bolt}}{1 \text{ set}}\right)$$

$$+ \left(\frac{1.7 \text{ kg}}{1 \text{ gross (washers)}} \times \frac{2 \text{ washers}}{1 \text{ set}}\right) = 13.9 \text{ kg/gross (sets)}$$

These same basic relationships can be used in other situations. Suppose, for example, that our manufacturer receives an order from a construction company for 250 kg of set components, to be shipped in bulk, unassembled. Again, no *counting* operation is required. The weight ratios involved can be used as unity factors to determine the weights of the various components needed to fill the order.

From the preceding calculation, we can write a set of equivalents and use these to form appropriate unity factors:

$$3.6 \text{ kg (nuts)} = 13.9 \text{ kg (sets)}$$
$$6.9 \text{ kg (bolts)} = 13.9 \text{ kg (sets)}$$
$$(2 \times 1.7) \text{ kg (washers)} = 13.9 \text{ kg (sets)}$$

Then

$$\frac{250 \text{ kg (sets)}}{1} \times \underbrace{\frac{3.6 \text{ kg (nuts)}}{13.9 \text{ kg (sets)}}}_{\text{unity factor}} = 65 \text{ kg (nuts)}$$

$$\frac{250 \text{ kg (sets)}}{1} \times \underbrace{\frac{6.9 \text{ kg (bolts)}}{13.9 \text{ kg (sets)}}}_{\text{unity factor}} = 124 \text{ kg (bolts)}$$

$$\frac{250 \text{ kg (sets)}}{1} \times \underbrace{\frac{(2 \times 1.7) \text{ kg (washers)}}{13.9 \text{ kg (sets)}}}_{\text{unity factor}} = 61 \text{ kg (washers)}$$

Note that our "nuts and bolts" situation involved two important considerations:

1. a convenient counting unit [the gross (144 units)]
2. the ability to use weight relationships for convenient unity factors

We shall find these same two considerations in the stoichiometry of chemical processes.

4.2 A CONVENIENT COUNTING UNIT: THE MOLE

In arriving at a counting unit for chemical species, it was recognized that it would be convenient to select a number such that the weight of that many unit particles would correspond to the formula weight (or, for monatomic species, the atomic weight). The counting unit selected to correspond to the formula weight, *in grams,* of a substance is called the *mole.* Just as, in our analogy, 1 gross of nuts weighed 3.6 kg, 1 gross of bolts weighed 6.9 kg, and 1 gross of nut/bolt/washer sets weighed 13.9 kg, we can now state that

$$1 \text{ mole sodium atoms} = 22.99 \text{ g (1 gram-atomic weight)}$$
$$1 \text{ mole nitrogen atoms} = 14.01 \text{ g (1 gram-atomic weight)}$$
$$1 \text{ mole sodium nitrite } (NaNO_2) = [22.99 + 14.01 + (2 \times 16.00)]$$
$$= 69.00 \text{ g (1 gram-formula weight)}$$

The mole is, then, a very convenient unit in terms of weight relationships, but *how many* unit particles correspond to a mole? (That is, if a gross, for example, is 144 units, how many units constitute a mole?) To answer this question, we could employ any of a number[1] of experimentally determined values. Suppose we select the average mass of a single hydrogen atom (Unit 2) and the gram-atomic weight (gaw) of hydrogen. Then

$$\frac{1.008 \text{ g}}{\text{gaw}} \times \frac{1 \text{ gaw}}{1 \text{ mole}} \times \frac{1 \text{ atom}}{1.67 \times 10^{-24} \text{ g}} = {\sim}6.0 \times 10^{23} \text{ atoms/mole}$$

This number, whose value has now been determined experimentally as 6.02252 ($\pm$0.00028) $\times 10^{23}$, is known as Avogadro's number. It represents the number of fundamental units (atoms, molecules, ions, or groups of ions) in a *gram-formula weight* of any substance (i.e., the numerical value of the formula weight expressed in units of grams).

In the left margin: *In honor of Amadeo Avogadro who, in 1811, suggested the application of Dalton's atomic model to explain various aspects of gas behavior.*

As a counting unit, therefore, a mole is defined as the quantity of any species that contains Avogadro's number of unit particles of the species. The weight of a mole of any chemical substance is the formula weight (or, for monatomic species, the atomic weight) of the species when expressed in grams.

The term *mole* has gradually been modified through common usage to the point where it is now employed as synonymous with the term gram-formula weight (or gram-atomic weight). It should be recognized, however, that the mole is, like the gross, basically a counting unit.

Various terms based on the mole are often used for quantitative relationships more convenient in dealing with weight units other than grams. Thus a *millimole* (10^{-3} mole, abbreviated as mmol) represents a formula (or atomic) weight expressed in *milligrams,* and a *ton-mole* represents the same *numerical value,* but expressed in units of *tons.* It is important to realize that only the *mole* itself (i.e., the gram-mole) corresponds to Avogadro's number of unit particles.

In the left margin: *To convince yourself that this is correct, use appropriate unity factors to show that a ton-mole corresponds to approximately 5.5×10^{29} unit particles.*

Example 3

What mass of copper(II) sulfate pentahydrate would correspond to 2.50 moles of the $CuSO_4(H_2O)_5$?

Solution

STEP 1 Finding the formula weight:

$$
\begin{array}{lll}
\text{Cu:} & (1 \times 63.55) & 63.55 \\
\text{S:} & (1 \times 32.06) & 32.06 \\
O_4\text{:} & (4 \times 16.00) & 64.00 \\
(H_2)_5\text{:} & (10 \times 1.008) & 10.08 \\
(O)_5\text{:} & (5 \times 16.00) & 80.00 \\
& & \overline{249.69}
\end{array}
$$

where $(H_2O)_5$ brackets the $(H_2)_5$ and $(O)_5$ rows.

STEP 2 moles-to-grams conversion: (using unity factor method, Math Skills, Section B)

$$\frac{2.50 \text{ mole}}{1} \times \frac{249.69 \text{ g}}{1 \text{ mole}} = 624.23 \text{ g (rounded to 624 g)}$$

---

[1] For an interesting description of various experiments used, see *The Mole Concept in Chemistry,* by W. F. Kieffer (New York: Van Nostrand Reinhold, 1962).

Example 4

How many millimoles would correspond to 0.387 g of ammonium acetate?

Solution

STEP 1 The formula of ammonium acetate is $NH_4CH_3CO_2$ (Unit 3), rewritten as $C_2H_7NO_2$ for purposes of formula weight calculation (Example 2).

STEP 2 Formula weight:

C_2: (2×12.01) 24.02
H_7: (7×1.008) 7.056
N: (1×14.01) 14.01
O_2: (2×16.00) <u>32.00</u>
77.086

(rounded to 77.09)

STEP 3 Grams-to-mmol conversion:

$$\frac{0.387 \text{ g}}{1} \times \frac{1 \text{ mole}}{77.09 \text{ g}} \times \frac{10^3 \text{ mmol}}{1 \text{ mole}} = 5.020 \text{ mmol}$$

(rounded to 5.02 mmol)

4.3. STOICHIOMETRIC RATIOS AND UNITY FACTORS

The mole concept is exceedingly useful in a variety of chemical calculations. The quantitative relationships involving numbers of reactant and product species as expressed, for example, by *balanced* chemical equations (Unit 3) permit the application of the mole unit to a number of different situations. Relationships involving gas volumes are discussed in Unit 10, and those dealing with solution concentration and volume are introduced in Unit 13. For now we shall limit our discussion to weight-to-weight relationships.

The key to weight-to-weight stoichiometry is the realization that a *balanced* equation or any other means of expressing the numerical relationship of related chemical species indicates not only the single unit ratios, but also the mole ratios involved. For example, the equation

$$2SO_2 + O_2 \longrightarrow 2SO_3$$

means that one oxygen molecule combines with two molecules of sulfur dioxide to form two molecules of sulfur trioxide. It further implies, since the species must be related in the same numerical ratios on any scale, that *1 mole* of O_2 reacts with *2 moles* of SO_2 to produce *2 moles* of SO_3. The definition of the *weight* of a mole as the formula weight expressed in grams then permits the use of mole-to-weight equivalents in the form of unity factors to determine weights of reactants and products on any given scale of reaction.

Example 5

Following the general steps in problem solving as outlined in Math Skills, Section G.

What weight of oxygen is required to convert 196 g of SO_2 to SO_3?

Solution

1. What do we want to find?
Answer: the weight of O_2 required (presumably in grams, since no other unit was indicated).

2. What useful information is given?

Answer: the weight of SO_2 to be consumed (196 g).

3. What other information is needed?

Answer: (a) the mole ratio of O_2 to SO_2 (e.g., from the balanced equation $2SO_2 + O_2 \longrightarrow 2SO_3$), and (b) the formula weights of O_2 and SO_2 (the species involved in the problem) from atomic weights as given in the periodic table [for O_2, $2 \times 16.0 = 32.0$; for SO_2, $32.1 + (2 \times 16.0) = 64.1$].

4. What approach should be used?

Answer: Define and use appropriate unity factors:

$$1 \text{ mole } O_2 = 2 \text{ moles } SO_2$$
$$1 \text{ mole } O_2 = 32.0 \text{ g}$$
$$1 \text{ mole } SO_2 = 64.1 \text{ g}$$

5. Manipulations

$$\frac{196 \text{ g}(SO_2)}{1} \times \frac{1 \text{ mole } (SO_2)}{64.1 \text{ g } (SO_2)} \times \frac{1 \text{ mole } (O_2)}{2 \text{ moles } (SO_2)} \times \frac{32.0 \text{ g } (O_2)}{1 \text{ mole } (O_2)} = 48.9 \text{ g } (O_2)$$

$$[\text{given}] \times \begin{bmatrix} \text{g } (SO_2) \\ \downarrow \\ \text{moles } (SO_2) \end{bmatrix} \times \begin{bmatrix} \text{moles } (SO_2) \\ \downarrow \\ \text{moles } (O_2) \end{bmatrix} \times \begin{bmatrix} \text{moles } (O_2) \\ \downarrow \\ \text{g } (O_2) \end{bmatrix}$$

6. Check

1. Dimensions canceled properly.
2. Math check (estimate):

$$\frac{\sim 200 \times 32}{\sim 64 \times 2} = \sim \frac{200}{4} = \sim 50$$

In this and most other examples, we shall use the number of significant figures appropriate to typical slide rule accuracy.

The general approach, then, to weight-to-weight stoichiometry involves

1. establishing the required stoichiometric ratio
2. using tabulated atomic weights to find formula weights required
3. defining and using appropriate unity factors involving mole ratios and mole-to-weight equivalents

Although a balanced equation will always indicate mole ratios of reactants and products, such ratios may often be determined without the use of a complete equation.

Example 6

What weight of ammonium sulfate could be prepared from 51.0 g of ammonia and sufficient sulfuric acid?

Solution

1. To find
   Weight of ammonium sulfate.
2. Given
   51.0 g of ammonia and sufficient sulfuric acid.
3. Needed
   a. Formulas of species involved [from memory (Unit 3): ammonia is formulated

as NH_3, sulfuric acid as H_2SO_4, and ammonium sulfate as $(NH_4)_2SO_4$].

   b. Stoichiometric ratios for ammonia and ammonium sulfate (species involved) [the balanced equation could be written, but this is not necessary if it is observed that the formula $(NH_4)_2SO_4$ implies the requirement of two NH_3 units, that is,

$$2NH_3 + \cdots \longrightarrow (NH_4)_2SO_4$$

   c. Formula weights for NH_3 and $(NH_4)_2SO_4$ (from tabulated atomic weights):

<table>
<tr><td>for NH_3</td><td>for $(NH_4)_2SO_4$</td></tr>
<tr><td>N: $1 \times 14.0 = 14.0$</td><td>N_2: $2 \times 14.0 = 28.0$</td></tr>
<tr><td>H_3: $3 \times 1.0 = \underline{3.0}$</td><td>$H_8$: $8 \times 1.0 = 8.0$</td></tr>
<tr><td>NH_3: 17.0</td><td>S: $1 \times 32.1 = 32.1$</td></tr>
<tr><td></td><td>O_4: $4 \times 16.0 = \underline{64.0}$</td></tr>
<tr><td></td><td>$(NH_4)_2SO_4$: 132.1 ($\sim$132)</td></tr>
</table>

   4. Approach

$$1 \text{ mole } (NH_4)_2SO_4 = 2 \text{ moles } NH_3$$
$$1 \text{ mole } (NH_4)_2SO_4 = 132 \text{ g}$$
$$1 \text{ mole } NH_3 = 17.0 \text{ g}$$

   5. Manipulations

$$\frac{51.0 \text{ g } (NH_3)}{1} \times \frac{1 \text{ mole } (NH_3)}{17.0 \text{ g } (NH_3)} \times \frac{1 \text{ mole } [(NH_4)_2SO_4]}{2 \text{ moles } (NH_3)} \times \frac{132 \text{ g } [(NH_4)_2SO_4]}{1 \text{ mole } [(NH_4)_2SO_4]} = 198 \text{ g } [(NH_4)_2SO_4]$$

$$[\text{given}] \times \begin{bmatrix} \text{g } (NH_3) \\ \downarrow \\ \text{moles } (NH_3) \end{bmatrix} \times \begin{bmatrix} \text{moles } (NH_3) \\ \downarrow \\ \text{moles } [(NH_4)_2SO_4] \end{bmatrix} \times \begin{bmatrix} \text{moles } [(NH_4)_2SO_4] \\ \downarrow \\ \text{g } [(NH_4)_2SO_4] \end{bmatrix}$$

   Situations in which the weights of more than one species are to be determined are simply treated as two separate, but related, problems.

Example 7

In one of the more common processes for obtaining copper from sulfide ores, one stage involves roasting copper(I) sulfide in air. Oxygen is consumed and the copper(I) sulfide is converted to metallic copper and sulfur dioxide. What weight of oxygen is consumed and what weight of sulfur dioxide is formed for each 1.00 ton of metallic copper produced by this reaction?

Solution
Treat as two problems, both based on the equation (formulated from the description of the process)

$$Cu_2S + O_2 \longrightarrow 2Cu + SO_2$$

and given in both cases 1.00 ton of copper. (The following solutions are condensed from the detailed stepwise method, but be sure that you see how each step evolved.)

   A. Find the weight of O_2 consumed (given 1.00 ton of Cu formed).

From the equation

1 mole O_2 = 2 moles Cu

From atomic weights,

$\quad$ 1 mole O_2 = 2 × 16.0 g = 32.0 g
$\quad$ 1 mole Cu = 63.5 g

Note that we could have used ton-moles, but all weight units except the one given (ton) cancel anyway.

$$\frac{1.00 \text{ ton (Cu)}}{1} \times \frac{1 \text{ mole (Cu)}}{63.5 \text{ g (Cu)}} \times \frac{1 \text{ mole (O}_2)}{2 \text{ moles (Cu)}} \times \frac{32.0 \text{ g (O}_2)}{1 \text{ mole O}_2} = 0.252 \text{ ton (O}_2)$$

B. Find the weight of SO_2 formed (given 1.00 ton of Cu produced).
From the equation,

$\quad$ 2 moles Cu = 1 mole SO_2

From atomic weights,

$\quad$ 1 mole Cu = 63.5 g
$\quad$ 1 mole SO_2 = 32.1 + (2 × 16.0) = 64.1 g

$$\frac{1.00 \text{ ton (Cu)}}{1} \times \frac{1 \text{ mole (Cu)}}{63.5 \text{ g (Cu)}} \times \frac{1 \text{ mole (SO}_2)}{2 \text{ moles (Cu)}} \times \frac{64.1 \text{ g (SO}_2)}{1 \text{ mole (SO}_2)} = 0.505 \text{ ton (SO}_2)$$

AN ALTERNATIVE METHOD

There are several ways of approaching weight-to-weight stoichiometry problems, but one of the easiest uses a simple proportion method derived from the balanced chemical equation and appropriate formula weights. We may think of this as a series of steps:

1. Identify the two chemicals in the equation for which the weight-to-weight relationships are needed.
2. Below each of their formulas, write the corresponding formula weights multiplied by the coefficients of the respective formulas, as shown in the balanced equation.
3. Above each of their formulas, write the known weight and a symbol for the unknown weight (e.g., w), respectively.
4. Equate the two ratios established by "above" and "below" quantities and solve for the unknown term.

For example, suppose we wished to know the mass of sodium hydroxide required for neutralization of 116 g of sulfuric acid, according to the equation

$\quad$ $2NaOH + H_2SO_4 \longrightarrow Na_2SO_4 + 2H_2O$

Using the "steps" described, we would proceed as follows:

$$\boxed{2\text{NaOH}} + \boxed{\text{H}_2\text{SO}_4} \longrightarrow \text{Na}_2\text{SO}_4 + 2\text{H}_2\text{O}$$

STEP 2 $\underset{(2 \times 40.0)}{2\text{NaOH}} + \underset{(1 \times 98.1)}{\text{H}_2\text{SO}_4}\dots$

STEP 3 $\underset{(2 \times 40.0)}{\overset{w}{2\text{NaOH}}} + \underset{(1 \times 98.1)}{\overset{116\text{ g}}{\text{H}_2\text{SO}_4}}$

STEP 4 $\dfrac{w}{2 \times 40.0} = \dfrac{116\text{ g}}{98.1}$

from which

$$w = \frac{2 \times 40.0 \times 116\text{ g}}{98.1} = \textbf{94.6 g}$$

Note that this procedure does not employ any units for formula weights. Units that might have been used, such as amu or grams mole^{-1}, would have "cancelled" during the arithmetic of the process. Thus the units of the answer are determined by the units of weight (mass) employed for the "known" quantity.

Example 8

Chlorine for use in water purification systems may be obtained from industries using the electrolytic decomposition of sea water, for which the chemical change may be represented by

$$2\text{NaCl(aq)} + 2\text{H}_2\text{O(l)} \longrightarrow 2\text{NaOH(aq)} + \text{H}_2\text{(g)} + \text{Cl}_2\text{(g)}$$

What mass of sodium chloride would be consumed for the production of 25 tons of chlorine, assuming 100 percent efficiency.

Solution

$$\underset{(2 \times 58.4)}{\overset{w}{\boxed{2\text{NaCl}}}} + \dots \longrightarrow \dots + \underset{(70.9)}{\overset{25\text{ tons}}{\boxed{\text{Cl}_2}}}$$

$$\frac{w}{2 \times 58.4} = \frac{25\text{ tons}}{70.9}$$

$$w = \frac{25\text{ tons} \times 2 \times 58.4}{70.9} = \textbf{41 tons}$$

Whether you solve stoichiometry problems by the unity factor method or the "proportion" method is largely a matter of which procedure you find easier to use, although your instructor may suggest that you employ the method believed most generally useful to further applications. Because of its very general utility, the unity factor method will be employed throughout this text.[2]

4.4 "NONSTOICHIO-METRIC" PROCESSES

Strictly speaking, chemical *reactions* are always stoichiometric, that is, combinations must involve exact whole numbers of species in fixed, characteristic ratios. Our use of the term "nonstoichiometric" is purely with the intention of distinguishing between the simple processes for which calculations can be based satisfactorily on a single balanced equation and those for which such calculations apply only to *theoretical* amounts.

The term "nonstoichiometric" is also applied to compounds in which the component elements are present in other than whole number ratios. The existence of such "nonstoichiometric compounds" results from imperfections in crystals, as described in Excursion 1, not from any "flaws" in the chemical reaction itself.

Simple reactions between ions of opposite charge [e.g., $Ag^+(aq) + Cl^-(aq) \longrightarrow AgCl(s)$] are typically rapid and quantitative. Calculations based on the stoichiometric ratios for these reactions generally agree well with experimental determinations of quantities involved. Many of the more complex reactions are also stoichiometric. However, there are a large number of chemical processes for which a single balanced chemical equation represents an idealized situation or, perhaps, only one of several reactions involving one or more of the reactant species. For example, chloroform may be prepared by reaction of chlorine and methane gases. The simplified equation may be written as

$$CH_4 + 3Cl_2 \longrightarrow CHCl_3 + 3HCl$$

In actual practice, if 3 moles of chlorine are mixed with 1 mole of methane and the reaction is allowed to continue until all the methane is consumed, some chlorine will remain and the product mixture will contain chloroform, carbon tetrachloride (CCl_4), methylene chloride (CH_2Cl_2), methyl chloride (CH_3Cl), and hydrogen chloride. A calculation, based on the simplified equation given, of the amount of chloroform expected will yield a number larger than that actually obtained. For processes such as this, that is, "nonstoichiometric" processes, a calculation of the amount of a product expected from given quantities of reactants results in the *theoretical* yield, the amount that would have been obtained had the reaction proceeded quantitatively (100 percent) as expressed by the chemical equation written. In most cases, theoretical considerations are insufficient for prediction of *actual yields*; at best, only approximations are possible. Actual yields must be determined experimentally. The efficiency of a reaction in terms of a simplified equation may be expressed by the *percentage yield* for the process.

$$\text{percentage yield} = \frac{\text{actual yield}}{\text{theoretical yield}} \times 100\% \qquad \text{(4.1)}$$

This concept is of considerable importance in the chemistry of complex reactions, and much research is devoted to variations of reaction conditions (e.g., temperature, gas pressures) or reactant ratios in attempts to increase the percentage yields of synthetic reactions. In industry, for example, reaction efficiency is of considerable economic importance.

Example 9

Methyl salicylate (oil of wintergreen) can be prepared from salicylic acid and methanol in the presence of small amounts of sulfuric acid. The reaction may be

---

[2] If you are interested in further practice with the "proportion" technique, see *Solving Problems in Chemistry*, by Rod O'Connor and Charles Mickey (New York: Harper & Row, 1975).

formulated as

The circle inside the hexagonal ring represents a particular type of chemical bonding, as will be described in Units 7 and 8.

$$C_7H_6O_3 + CH_3OH \longrightarrow C_8H_8O_3 + H_2O$$

In a particular experiment, 60 g of salicylic acid and an equimolar amount of methanol produced 47 g of methyl salicylate. What was the percentage yield?

Solution

Since we are trying to find the theoretical yield, we must determine the amount of methyl salicylate that could have been produced if all the salicylic acid used (60 g) had been converted to this product.

From the equation given,

1 mole $C_7H_6O_3$ = 1 mole $C_8H_8O_3$

From atomic weights,

1 mole $C_7H_6O_3$ = $(7 \times 12.0) + (6 \times 1.0) + (3 \times 16.0) = 138$ g
1 mole $C_8H_8O_3$ = $(8 \times 12.0) + (8 \times 1.0) + (3 \times 16.0) = 152$ g

Let t represent the theoretical yield. Then

$$t = \frac{60 \text{ g } (C_7H_6O_3)}{1} \times \frac{1 \text{ mole } (C_7H_6O_3)}{138 \text{ g } (C_7H_6O_3)} \times \frac{1 \text{ mole } (C_8H_8O_3)}{1 \text{ mole } (C_7H_6O_3)} \times \frac{152 \text{ g } (C_8H_8O_3)}{1 \text{ mole } (C_8H_8O_3)}$$

$$= 66 \text{ g } (C_8H_8O_3)$$

Then, from equation (4.1)

$$\text{percentage yield} = \frac{47 \text{ g } (C_8H_8O_3)}{66 \text{ g } (C_8H_8O_3)} \times 100\% = 71\%$$

4.5 THE LIMITING REAGENT

Percentage yields in "nonstoichiometric" processes (incomplete reactions) can usually be increased by addition of an excess of one or more reactants. For example, an improved yield of chloroform results from mixing more than 3 moles of chlorine per mole of methane because some of the additional chlorine will convert part of the methyl chloride and methylene chloride to chloroform. However, the process is complicated by the fact that excess chlorine also favors formation of carbon tetrachloride. In this and similar cases, it is usually necessary to determine *experimentally* the optimum conditions and reactant ratios. In fact, an improved yield is not the only factor of importance. In many situations the additional cost of a process for maximum yield dictates the selection of less expensive conditions giving a lower yield, but also a lower cost per unit quantity of product formed.

In a chemical process in which an excess of one or more reactants is used, a reactant *not* in excess will, of course, determine the maximum theoretical yield of product obtainable. For the case of the chloroform synthesis, if 1 mole of methane is used with 3 *or more* moles of chlorine, the maximum theoretical yield of chloroform (assuming conditions could be found that avoid all other carbon-containing products) must be *1 mole,* no matter how large an excess of chlorine was used. Once the only carbon-containing reactant is completely consumed, there is no way of making more chloroform. The *limiting reagent* in a chemical reaction, then, is the reactant having the smallest *mole-to-coefficient ratio.* This ratio is obtained by dividing the number of moles of the reactant (*i.e., the weight used divided by the formula weight of the species*) by the coefficient for that species as it appears in the balanced chemical equation. *The actual quantity of limiting reagent is used in stoichiometric calculations of theoretical yields* (or theoretical amounts of other reactants consumed).

Example 10

Ammonia can be manufactured from the gas-phase reaction of molecular hydrogen and nitrogen. If 50.0 g of each of these gases are mixed, what is the theoretical yield of ammonia?

Solution

The reaction is expressed by the equation (not balanced)

$$H_2 + N_2 \longrightarrow NH_3$$

The equation can be balanced by placing a coefficient of 2 for NH_3 (to balance the two nitrogens represented by N_2) and then a coefficient of 3 for H_2 (to balance the six hydrogens represented by $2NH_3$):

$$3H_2 + N_2 \longrightarrow 2NH_3$$

The number of moles of each reactant will be for H_2 (formula weight 2×1.01):

$$\frac{1 \text{ mole}}{2.02 \text{ g}} \times 50.0 \text{ g} = 24.8 \text{ moles}$$

for N_2 (formula weight 2×14.0)

$$\frac{1 \text{ mole}}{28.0 \text{ g}} \times 50.0 \text{ g} = 1.79 \text{ moles}$$

Then, the mole-to-coefficient ratios are

for H_2:

$$\frac{24.8}{3} = 8.27$$

for N_2:

$$\frac{1.79}{1} = 1.79$$

Thus N_2 is the *limiting reagent* and must be used for calculation of the theoretical yield.

From the equation,

1 mole N_2 = 2 moles NH_3

From atomic weights,

1 mole $N_2 = 2 \times 14.0 = 28.0$ g
1 mole $NH_3 = 14.0 + (3 \times 1.0) = 17.0$ g

$$t = \frac{50.0 \text{ g (N}_2)}{1} \times \frac{1 \text{ mole (N}_2)}{28.0 \text{ g (N}_2)} \times \frac{2 \text{ moles (NH}_3)}{1 \text{ mole (N}_2)} \times \frac{17.0 \text{ g (NH}_3)}{1 \text{ mole (NH}_3)} = 60.7 \text{ g (NH}_3)$$

Example 11

How many tons of H_2SO_4 could be produced per day from a process using 38 tons per day of SO_2 with a 70 percent conversion efficiency, assuming SO_2 to be the limiting reagent?

Solution
Although the actual process involves several steps, the overall conversion may be represented by

$$2SO_2 + O_2 + 2H_2O \longrightarrow 2H_2SO_4 \quad \text{(see Figure 4.2)}$$

Then, for each day of production the theoretical yield may be determined from

$$SO_2 + \cdots \longrightarrow H_2SO_4$$

from which

$$1 \text{ mole } SO_2 = 1 \text{ mole } H_2SO_4$$

From atomic weights,

$$1 \text{ mole } SO_2 = 32.1 + (2 \times 16.0) = 64.1 \text{ g}$$
$$1 \text{ mole } H_2SO_4 = (2 \times 1.0) + 32.1 + (4 \times 16.0) = 98.1 \text{ g}$$

$$t = \frac{38 \text{ tons (SO}_2)}{1 \text{ day}} \times \frac{1 \text{ mole (SO}_2)}{64.1 \text{ g (SO}_2)} \times \frac{1 \text{ mole (H}_2SO_4)}{1 \text{ mole (SO}_2)} \times \frac{98.1 \text{ g (H}_2SO_4)}{1 \text{ mole (H}_2SO_4)}$$

$$= 58 \text{ tons (H}_2SO_4)/\text{day}$$

Since the process is only 70 percent efficient, Equation (4.1) suggests that

$$\text{actual yield expected} = \frac{58 \text{ tons (H}_2SO_4)}{\text{day}} \times \frac{70}{100}$$

$$= 41 \text{ tons (H}_2SO_4)/\text{day}$$

SUGGESTED READING

Kieffer, W. F. 1962. *The Mole Concept in Chemistry.* New York: Van Nostrand Reinhold.
Nash, Leonard K. 1966. *Stoichiometry.* Reading, Mass.: Addison-Wesley.
O'Connor, Rod, and Charles Mickey. 1975. *Solving Problems in Chemistry.* New York: Harper & Row.

ECONOMIC FACTORS

Every chemical process, whether it involves a massive industrial production or a "bench-scale" reaction in the more sheltered environment of academic research or an instructional laboratory, involves a cost analysis. The academic chemist must justify his costs to his administration, his funding agency, and, ultimately, to the taxpayer. The industrial chemist must "sell" his process to various group leaders, management, and, ultimately, the stockholder and consumer. The details of a single project tend increasingly to become buried in statistics for many other problems as they pass from the chemist to the man on the street who ultimately "pays the bills," but somewhere along the line decisions must be made on the economic feasibility of each proposed chemical process. The key to industrial chemistry is a combination of stoichiometry and economics. The sale price of a chemical product—whether it be an insecticide, a paint pigment, or a birth control pill—is determined by the costs of its production and the profits required by the industry.

Of the current U.S. production of H_2SO_4 (about 30 million tons annually), more than half is used for fertilizer production (mainly phosphates) and about 10% is used in petroleum refining. Less than 15% is sold for "nonindustrial" uses.

The importance of stoichiometry in deciding whether or not to manufacture a particular product can be illustrated by the production of sulfur dioxide, an important factor in air pollution. Fossil fuels and many metal ores contain significant quantities of sulfur or sulfur compounds that are converted to sulfur dioxide by traditional methods of combustion or smelting (compare Example 7). Some chemists have suggested removing the gaseous SO_2 from the stacks by converting it to sulfuric acid, a valuable commercial product used in automobile batteries, as a drain cleaner, and in the manufacture of synthetic detergents and many other products.

Suppose that a particular industry is considering installation of a plant capable of converting 70 percent of the sulfur dioxide it emits to sulfuric acid (Figure 4.2), and that the plant now emits 38 tons of SO_2 per day. How much sulfuric acid could be formed by the proposed new process? The answer to this question will influence such factors as storage and shipping requirements and, more important, the analysis of market potential. This is basically a problem of simple stoichiometry (Example 11).

If the industry now determines that the market for the expected output of H_2SO_4 is assured, then the next questions are the cost of new equipment, the wages of additional personnel, the increase in power requirements, warehouse and shipping costs, marketing discounts, advertising and legal fees, and taxes. After adding on all of these costs, can the sulfuric acid be produced profitably at a competitive price? Many of these factors can be estimated accurately; but even experienced economists

Figure 4.2 A sulfuric acid plant. The *net* process may be formulated as

$$2SO_2 + O_2 + 2H_2O \rightarrow 2H_2SO_4$$

(Photograph courtesy of ASARCO Incorporated.)

may be hard pressed to guess the impact of new sources of a chemical on the ever-changing commercial scene.

In many such cases, the economic facts of life make an industry unwilling to gamble. If the public demands a rapid decrease in industrial contaminants in the face of a potential loss to the industry, then it may be necessary for the government—the taxpayers—to subsidize the process by tactics such as tax benefits, price supports, or guaranteed minimum sales.

The following fictitious, but plausible, example illustrates some of the simpler economic realities of the chemical industry.

Example 12

A particular industry decides to use the process described in Example 11, having estimated a reasonable market potential for the product. The following economic factors are "fixed" by the competitive situation, profits essential to industrial survival, and long-range estimates of the economic picture:

1. Over a projected 10-year period the retail price of the product will average 2.80¢/lb.
2. The retailers will average a profit of 15 percent of sale price.
3. Industry-to-retailer costs will average 2.00¢/lb of product (e.g., packaging, labeling, storage, shipping, etc.).
4. Industry must make a profit of 12 percent of *its* selling price to remain in operation.

Now, the question is this: What is the maximum capital outlay that must be allowed for the projected 10-year period to cover all economic factors of the production beyond those listed (i.e., equipment, personnel, taxes, expendable supplies, power, etc.)?

Solution

a. If the retailer is to make a 15-percent profit, then his maximum purchase price would be (2.80¢ − 15% of 2.80¢) for each pound. Thus industry can sell the H_2SO_4 for no more than 2.38¢/lb.
b. Industry's profit of 12 percent of its selling price leaves (2.38¢ − 12% of 2.38¢) for each pound. Thus the total *cost* to industry per pound sold cannot exceed 2.09¢/lb.
c. Known costs of 2.00¢/lb then leave 0.09¢/lb as the maximum allowance for all other expenditures.
d. Over a 10-year period, then, the maximum expenditure must be limited to

$$\frac{\$0.0009}{lb} \times \frac{2000\ lb}{ton} \times \frac{41\ tons}{day} \times \frac{365\ days}{year} \times \frac{10\ years}{1} = \sim \$300,000$$

Thus the industry can determine that all costs of equipment, personnel, power, and so on cannot exceed $300,000 over a 10-year period. If ways can be found to hold costs at or below this figure, all is well. If, however, estimates indicate that the costs will be appreciably greater, the industry must conclude that the process is not economically feasible under the given conditions.

Dr. Barry Barnhart reports that in 1976 the Allegehny Power Company, in the Pittsburgh area, produced sulfuric acid from SO_2 emissions to the extent that 340 railroad cars, on annual lease, were in constant use. No market could be found close enough for economical shipping and the acid had to be dumped into deep wells for storage.

EXERCISES

1. Refer to Example 9 and use the information given in the problem and its solution to identify:
   a. the actual yield of methyl salicylate
   b. the formula weight of salicylic acid
   c. the weight of 1 mole of methyl salicylate
   d. the theoretical yield of methyl salicylate
2. Use tabulated values of atomic weights to calculate formula weights for
   a. CH_3CO_2H; **b.** ammonium sulfate; **c.** aluminum oxide;
   d. phosphate ion; **e.** nitric acid
3. Sodium carbonate reacts with hydrochloric acid to form sodium chloride, carbon dioxide, and water. Calculate the weights of carbon dioxide and water that could be formed by this process from 16.0 g of sodium carbonate, assuming complete reaction.
4. Ethyl acetate, a common solvent for certain glues and cements, can be prepared by reaction of acetic acid with ethanol in the presence of a small amount of sulfuric acid (not a reactant).

   If, in a particular experiment, a reaction mixture of 25.0 g each of acetic acid and ethanol produced 34.0 g of ethyl acetate, what was
   a. the limiting reagent?
   b. the percentage yield for the process described?

   The balanced equation for this process may be formulated as

   $$CH_3CO_2H + C_2H_5OH \longrightarrow CH_3CO_2C_2H_5 + H_2O$$

5. An industrial concern has decided to manufacture ethyl acetate by a process in which 50.0 kg each of ethanol and acetic acid are mixed under the conditions described in Exercise 4.
   a. Assuming the same percentage yield as that calculated for Exercise 4, what weight of ethyl acetate should be formed from 50.0 kg each of ethanol and acetic acid?

OPTIONAL

6. If ethanol costs $1.82/lb and acetic acid costs $2.79/kg, what should be the selling price per kilogram of the ethyl acetate to allow for a profit of 20 percent of the sale price, assuming the only other cost factors can be treated as "miscellaneous costs" of 35¢/kg of ethyl acetate?

PRACTICE FOR PROFICIENCY

Objective (a)

1. Refer back to Example 10 of this unit and use the information given in the problem and its solution to identify
   a. the limiting reagent
   b. the formula weight of ammonia
   c. the weight of one mole of nitrogen
   d. the theoretical yield of ammonia
   e. the gram-formula weight of H_2
   f. the number of *millimoles* of nitrogen used
   g. the percentage yield if a process using the reactant amounts shown formed 51.6 g of ammonia
   h. the actual yield corresponding to a process using the reactant amounts shown with a conversion efficiency of 70 percent
   i. the number of molecules of ammonia corresponding to the theoretical yield indicated (Remember Avogadro's number?)

Objective (b)

2. Use tabulated values of atomic weights to calculate formula weights for
   a. $(NH_4)_2CO_3$; **b.** sulfurous acid; **c.** copper(II) nitrate;
   d. chromate ion; **e.** ethanol; **f.** iron(III) sulfate; **g.** benzene;
   h. permanganate ion; **i.** phosphoric acid;
   j. $Ni(CH_3CO_2)_2(H_2O)_4$

Objective (c)

3. During discharge of a lead storage battery the chemical change can be represented by

   $$Pb + PbO_2 + 2H_2SO_4 \longrightarrow 2PbSO_4 + 2H_2O$$

   What weight of lead sulfate is formed by consumption of 41.4 g of lead in this process, assuming stoichiometric reaction?
4. Phosphates from detergents and agricultural wastes are recognized as serious water pollutants. In waste water treatment plants, phosphates can be removed by precipitating the insoluble aluminum phosphate:

   $$Al^{3+}(aq) + PO_4^{3-}(aq) \longrightarrow AlPO_4(s)$$

   a. What mass of *aluminum nitrate* would be needed, assuming complete reaction, to remove 68 kg of phosphate from waste water?
   b. What mass of aluminum phosphate would be formed?
5. Gases emitted during volcanic activity frequently contain high concentrations of both hydrogen sulfide and sulfur dioxide. These gases may react to produce deposits of sulfur in volcanic regions:

   $$2H_2S(g) + SO_2(g) \longrightarrow 3S(s) + 2H_2O(g)$$

   a. What mass of sulfur would result from complete reaction of 218 kg of hydrogen sulfide with excess sulfur dioxide?
   b. What mass of the SO_2 would be consumed by this process?
6. By May of 1974 vinyl chloride (CH_2CHCl) was recognized as a potent carcinogenic (cancer-causing) chemical. Vinyl chloride had found extensive use as a propellant in many aerosol sprays and in the manufacture of plastics (polyvinylchloride, PVC). More than 2.2×10^9 kg of vinyl chloride was produced in the United States in the single year 1970, but stringent health and safety restrictions now in effect may dramatically reduce industrial production of this chemical. Vinyl chloride is commonly prepared by the gas phase reaction of equi-

molar amounts of acetylene (C_2H_2) and hydrogen chloride. Assuming stoichiometric conversion, (a) how much acetylene and (b) how much hydrogen chloride (in kilograms) were consumed for the 1970 U.S. production of vinyl chloride?[3]

7. Each of the following processes, shown by qualitative equations (Unit 3), is used for the commercial production of agricultural fertilizers. Calculate the amount of each fertilizer that could be prepared from 25 kg of ammonia, assuming that ammonia is the limiting reagent and each process is 100 percent efficient? (Which fertilizer contains the highest percentage of nitrogen, by weight?)
   a. $NH_3 + HNO_3 \longrightarrow NH_4NO_3$
   b. $NH_3 + H_2SO_4 \longrightarrow (NH_4)_2SO_4$
   c. $NH_3 + H_3PO_4 \longrightarrow (NH_4)_3PO_4$

8. Trinitrotoluene (TNT) is prepared from toluene and nitric acid under carefully controlled conditions:

   a. What mass of TNT could be prepared from 2.2 kg of toluene with excess nitric acid?
   b. How much nitric acid would be consumed by this process?

9. Acetylene gas is generated as fuel for certain types of lanterns by the reaction of calcium carbide with water:

$$CaC_2(s) + 2H_2O(l) \longrightarrow Ca(OH)_2(s) + C_2H_2(g)$$

   a. How much calcium carbide would be needed to generate 75 g of acetylene, using excess water?
   b. How much calcium hydroxide would result?

10. In sealed environments, such as the cabins of space vehicles or submarine vessels, carbon dioxide "scrubbers" are required to prevent the accumulation of too much CO_2 in the environment. Lithium hydroxide can be used for this purpose, reacting with CO_2 to form lithium carbonate and water.
   a. What quantity of lithium hydroxide would be needed for the removal of 125 kg of CO_2 from the environment?
   b. What mass of lithium carbonate would result?

Objectives (d) and (e)

11. Baking soda can be prepared by adding carbon dioxide to an aqueous solution of sodium chloride and ammonia:

$$NaCl + NH_3 + CO_2 + H_2O \longrightarrow NaHCO_3 + NH_4Cl$$

[3] You may be interested in reading "The Vinyl Chloride Story" by John Moore in the June 1975 issue of *Chemistry*.

If in a particular experiment a reaction mixture of 25.0 g each of sodium chloride, ammonia, and carbon dioxide, in excess water, produced 24.0 g of sodium bicarbonate (baking soda), what was
   a. the limiting reagent?
   b. the percentage yield for the process described?

12. If 5.0 kg each of toluene and nitric acid were mixed for the process described in question 8 and 5.7 kg of TNT was isolated, what was
   a. the limiting reagent?
   b. the percentage yield?

13. If the process described in question 5 is only 45 percent efficient in producing sulfur, what amounts of (a) SO_2 and (b) H_2S must have actually been involved in the formation of a 2.9×10^3 kg deposit of sulfur, assuming that neither reactant was in excess?

OPTIONAL

14. An industrial operation is set up to produce baking soda by a process in which 100.0 lb each of the reactants are to be used under the conditions of question 11.
   a. Assuming the same percentage yield as that calculated for question 11, what weight of baking soda should be formed for each 100.0 lb of sodium chloride consumed?
   b. If sodium chloride costs 38¢ per pound, ammonia costs 61¢ per pound, and carbon dioxide costs 27¢ per pound, what should be the selling price per pound of baking soda to allow for a profit of 15 percent of the sale price, assuming all other cost factors are covered as "miscellaneous costs" of 14¢ per pound of baking soda?

BRAIN TINGLER

15. Many chemical processes involve a number of successive steps, each of which may be less than 100 percent efficient. Given the following information and assuming that all reactants are mixed in the mole ratios indicated by the balanced equations, calculate the amount of ammonia required for the production of 350 kg of nitric acid by the *Ostwald process*.

STEP 1 Gaseous ammonia is combined with oxygen at 1000°C to form nitric oxide and water vapor at an efficiency of 87 percent.

STEP 2 The nitric oxide is burned in oxygen to produce nitrogen dioxide gas at 72 percent efficiency.

STEP 3 Nitrogen dioxide is bubbled through water to form nitric acid and nitric oxide at 88 percent efficiency. One mole of nitric oxide is produced for each 2 moles of nitric acid formed.

Unit 4 Self-Test

(Refer to a periodic table when necessary.)

1. Use tabulated atomic weights to calculate the formula weights of
   a. C_2H_5OH; b. sulfuric acid, c. HCO_3^-; d. iron(II) phosphate; e. ammonium nitrate

2. What weight of zinc is required for formation of 50.0 g of chromium(II) sulfate, assuming stoichiometric reaction? The equation for this reaction is

$$4Zn + K_2Cr_2O_7 + 7H_2SO_4 \longrightarrow 7H_2O + K_2SO_4 + 4ZnSO_4 + 2CrSO_4$$

3. A 24.0-g sample of benzene was treated, dropwise, with 100.0 g of molecular bromine in the presence of a small amount of iron(III) bromide (not a reactant). A 28.0-g portion of bromobenzene (C_6H_5Br) was obtained. What was the percentage yield for this reaction, based on the limiting reagent? The equation is

$$C_6H_6 + Br_2 \longrightarrow C_6H_5Br + HBr$$

4. Use the information given in or derived from question 3 to identify
   a. the molecular formula of bromobenzene
   b. the actual yield of bromobenzene
   c. the formula weight of benzene
   d. the weight, in grams, of 1 mmol of bromobenzene
   e. the limiting reagent in the process described

OPTIONAL

5. If benzene costs $8.60/kg and bromine costs $11.40/kg, what should be the selling price per kilogram of bromobenzene formed (question 3) to permit a profit of 10 percent of sale price, assuming the only costs to be considered are those of the reactants specified?

If you think things are chaotic now, wait awhile and watch them get worse.

(*One statement of the second law of thermodynamics*)

Unit 5

Thermodynamics

Objectives

(a) After studying the appropriate definitions in Appendix E and the terms' usage in this unit, be able to demonstrate your familiarity with the following terms by matching them with corresponding symbols, descriptions, examples, or thermodynamic relationships:

calorimeter	free energy (Gibbs)
combustion	internal energy
efficiency (thermodynamic)	reaction rate
endothermic	reversible
enthalpy	spontaneous
entropy	standard state
exothermic	state function
formation	work

(b) Be able to identify statements involving thermodynamic concepts and definitions as correct or incorrect.

(c) Given appropriate standard heats of formation and a description of a chemical process, be able to calculate the standard heat of reaction (enthalpy change) for the process (Section 5.4).

(d) Given appropriate enthalpy information and reaction descriptions, be able to calculate the standard enthalpy change associated with a specified process (Example 2 and Section 5.5).

(e) Given ''input'' and ''output'' temperatures, be able to calculate the theoretical thermodynamic efficiency for heat-to-work conversion [equation (5.4)].

OPTIONAL

Be able to make qualitative estimates of the effect of temperature on process spontaneity (Section 5.7).

One of the most pressing problems of modern society is the ever-increasing demand for more energy—a demand that has caused us to use America's resources, and those of the rest of the world, as though there were an inexhaustible supply. There is not.

Thermodynamics is the study of energy: the forms it can take, the efficiency of its use, and the limitations on its availability for useful work. The laws of thermodynamics, based on repeated experimental testing, provide a key to the understanding of energy relationships for chemical processes. Thermodynamics can, for example, predict whether or not a chemical reaction can occur when two different substances are mixed. If a reaction can occur, thermodynamics can calculate the amount of energy theoretically required by or released from the process. Even the extent to which a reaction can proceed before reaching a condition of equilibrium (Unit 17) can be predicted by thermodynamic considerations.

The speed at which a given reaction proceeds, that is, the reaction rate, cannot be predicted on the basis of thermodynamics. This very important aspect of chemical processes depends on other factors (Unit 16).

As an introduction to a rather complex subject, let us consider a limited problem related to the production of electric energy.

Experts estimate that the demands for electric energy in the United States will double over the next 10 years. Some experts, more optimistic (if you own stock in the power companies) or pessimistic (if you are concerned with the environment) expect this effect within the next 5 years. In any event, realistic estimates of population growth and consumer demands suggest that ever greater quantities of electric energy must be provided.

There are several ways of increasing the U.S. output of electric energy. Hydroelectric plants, using the potential energy of large masses of water stored behind dams, offer a "clean" source of power. Such dams are, however, limited by the number of suitable locations for their construction—a limitation already closely approached. In addition, the problems of silt deposits have increased significantly within the last few years.

Geothermal plants, using the heat stored beneath the earth's crust to convert water to steam for turbines, have found increasing applications. However, this energy source is limited by the logistics of the operations required for its use, and

Additional discussions of thermodynamics will be found in subsequent units, including some important terms and considerations not covered in this unit. For a more detailed treatment, see Suggested Reading at the end of this unit.

even optimistic predictions do not expect geothermal plants to contribute much more than 10 percent of the nation's future power requirements.

Nuclear power plants, once hoped to be the ultimate answer to man's needs for electric energy, offer an alternative that has declined in popularity with the increasing public concern over disposal of radioactive wastes, over threats of "thermal pollution" from associated heat exchange processes, and over fears of "nuclear accidents." Whether these concerns are "overplayed" or not, the result has been to delay many of the original plans for developing this source of electric energy.

One of the most promising ideas for the future involves the utilization of solar energy (energy from the sun). Once limited to the small-scale applications of direct conversion to electric energy in so-called solar batteries, this enormous source of power is now being explored as a future means of large-scale electric energy production through the indirect process of storing the sun's heat and then converting this heat to electricity.

For the next few years, however, most electric energy will probably come from chemical processes. These may involve direct conversion of chemical to electric energy, as in electrochemical cells and batteries (Unit 23) or the large-scale indirect processes depending on the combustion of fossil fuels—coal, oil, or "gas." Most of the power plants now in operation, under construction, or in planning will depend on these combustion processes. All, unfortunately, are linked by the inherent laws of chemistry to the problems of pollution of the environment and the depletion of natural resources. As power demands continue to increase—tied inexorably to increases in population and in living standards—these demands will be met, at least for a while, by electric energy production dependent on increased consumption of fossil fuels.[1]

As an introduction to thermodynamic concepts, which will appear in greater detail further in our study of chemistry, let us consider a single type of power plant operation: one based on the use of methane as a fuel.

5.1 THE TOTAL ENERGY OF A METHANE MOLECULE

At room temperature and ordinary pressures, methane is a gas; this gas is described (Unit 10) as consisting of a very large number of methane molecules zooming around in constant, erratic motion (Figure 5.1).

If we select one of these moving molecules, it is possible to write a description for its total energy:

$$E_{\text{total}} = E_{\text{potential}} + E_{\text{kinetic}}$$

There are several factors contributing to the potential and kinetic energy terms for a moving molecule. Some contributions are appreciably larger than others. The total molecule has kinetic energy (KE) by virtue of its motion through space and of the movements of the component atoms relative to each other (Unit 9). In addition, all of the electrons in the molecule are moving in the positive fields of the atomic nuclei (Unit 1). Thus we can write

$$E_{\text{kinetic}} = KE_{\text{molecular motions}} + KE_{\text{electron motions}}$$

In a similar way, there are several factors contributing to the potential energy (PE). As long as the molecule is "up" it has some potential energy relative to whatever position is defined as "down," just as an automobile driving around a mountain road has some potential energy relative to a valley into which it might fall. Electrostatic

Figure 5.1 Molecules of methane in the gaseous state.

Kinetic energy may be thought of as "energy of motion," whereas potential energy refers to the comparison of one static system with another to which it might be changed.

[1]The entire September, 1971, issue of *Scientific American* is devoted to the subject of energy and power. An excellent analysis, using more recent data, is given in "An Overview of the Energy Crisis" by E. A. Walters and E. M. Wewerka in the May, 1975, issue of the *Journal of Chemical Education* (pp. 282–287).

forces within the molecule also contribute: electron–nucleus attractions, electron–electron repulsions, and nucleus–nucleus repulsions. Within the carbon nucleus there are strong forces holding the positive protons (and the neutrons) together. The total potential energy, then, may be expressed as

$$E_{potential} = PE_{molecule\ position} + PE_{electron-nucleus} \\ + PE_{electron-electron} + PE_{nucleus-nucleus} + PE_{atomic\ nucleus}$$

Some contributions to the total molecular energy can be estimated rather accurately; others are much more difficult. Even if the total energy of a single molecule could be calculated at a frozen instant of time, the calculation would have little practical value. Energies of molecules in gases are continuously changing as a result of collisions between molecules (or against the walls of a container). Indeed, the energy eventually appearing as electricity from a methane-fueled power plant can be related to "single-molecule" energies only in a very indirect sense. There are other considerations of much greater importance.

It is more profitable to deal with collections of large numbers of molecules, and thus with *average* energies, which can be rather informative for situations such as the consideration of methane gas as a fuel source for electric energy production.

5.2 INTERNAL ENERGY AND THE CONSERVATION LAW

The *first law of thermodynamics* states that energy can neither be created nor destroyed, but only changed in form. This must, of course, be understood in terms of the mass–energy equivalence expressed by $E = mc^2$. In ordinary chemical processes, mass–energy conversion is negligible, so for all practical purposes chemical reactions are bound by this law of *conservation of energy,* just as they were by the law of conservation of mass (Unit 4).

If we wish to know the maximum energy obtainable from the burning of a given amount of methane in air, the conservation of energy requires that this maximum will be *the difference between the total energy of all initial reactants and the total energy of all final products;* that is,

$$\Delta E_{reaction} = E_{products} - E_{reactants}$$

where ΔE (Greek delta) indicates the energy *difference.*

Chemists are therefore much more interested in energy *differences* between initial and final states of a process than in the absolute (total) energies of either state. Energy differences permit the elimination of some of the terms contributing to "total" energies. In a typical chemical reaction, for example, nuclei are not changed, so that contributions of nuclear energies "cancel out" of the ΔE expression. Although ΔE values can be determined experimentally, the absolute (total) energies cannot.

To simplify estimations of energy differences, some idealized conditions must be imposed. It is convenient to restrict the process under consideration to an *isolated system*—one that is completely insulated from its surroundings. Such a system, of course, cannot be achieved in practice, but it permits the estimation of maximum theoretical energy differences without the complications imposed by interactions with the surroundings. Such interactions are dependent on particular conditions of each specific situation. We shall find it convenient to refer to thermodynamic considerations in terms of three categories: an isolated *system,* the *surroundings,* and the *universe.* In terms of modern "set" notation, the *system* is a subset of the *surroundings* and the *surroundings* plus the *system* equals the *universe.* Although these categories represent idealized distinctions, they provide a useful way of visualizing many aspects of thermodynamics.

Although E is still widely used, the official symbol now accepted for *internal energy* is U. We shall maintain the older designation in our discussions because of the convenience of associating E with internal energy.

A second artificial simplification for this particular case will be the assumption that the reaction in question is stoichiometric and can be represented by

$$CH_4 + 2O_2 \longrightarrow CO_2 + 2H_2O$$

Thus the change in internal energy (i.e., the energy within the isolated system) can be formulated as

$$\Delta E_{reaction} = (E_{CO_2} + E_{H_2O}) - (E_{CH_4} + E_{O_2})$$

in which E_{CH_4} represents a summation of the energies of all the methane molecules in the system of initial reactants and E_{O_2}, E_{CO_2}, E_{H_2O} a similar summation for the species indicated.

The estimation of these absolute energy terms is exceedingly complex. It is treated by the very sophisticated techniques of statistical thermodynamics. Of greater interest to the problem of electric power are two aspects of the net energy *difference*, ΔE.

5.3 HEAT AND WORK

A generator to produce electric current can be built in a number of different ways. One way would use the combustion process to produce an increased volume of gases to expand against a movable piston. A series of these pistons could work together, as in the automobile engine, to produce mechanical energy. Some of this energy could function to turn a magnet within a coil of wire, thus inducing an electric current. The increased volume of product gases, as compared to that of reactant gases, could be obtained by selecting a reaction in which the number of moles of gaseous products is larger than that of gaseous reactants or by using the heat generated by the reaction to cause gas expansion (Unit 10). Sometimes both effects can be used, although the combustion of methane is restricted to the latter, as can be seen from the equation for the combustion process.

An alternative approach could use the heat from the combustion to convert water into steam. The steam, when directed through a turbine arrangement, would produce the mechanical energy necessary to run the generator.

Still other alternatives are possible, but in any of these the total energy from the combustion process is accounted for by two factors: heat and work.

$$\Delta E = q - w \tag{5.1}$$

The heat absorbed by the system is designated by q and acts to increase the total internal energy of the system. The internal energy is decreased by any work, w, performed by the system, such as the work done as expanding gases push a piston. Energy may be transferred either by heat flow or by work, and the relative amounts transferred by either route depend on the particular mechanism chosen (e.g., piston engine or steam generator). Remember, however, that ΔE itself is *independent of the mechanism*, being determined only by the difference between the initial and final states of the system. Such quantities are known in thermodynamics as *state functions*. This is an important concept. If the reaction in question is to be used as a process for a piston engine, the work done in moving the piston is *useful*. If the combustion process is merely a heat source, the work done as the gaseous products push back the atmosphere in escaping from the exhaust stack is *not* useful. Engineers can decide how they want to distribute total available energy between heat and work as they design a particular generating plant, but the molecules themselves "decide" how much total energy, ΔE, will be available for this distribution.

When a chemical reaction occurs, there are generally two ways in which energy appears as work: electrical work involving electron transfer through an external circuit (Unit 23) or pressure–volume work as in the expansion of gases. If the reaction

As a more familiar illustration of the concept of *state function,* consider the temperature of a room. If we note that the thermometer reading is 24.5°C, that tells us the temperature *state* and is independent of whether that temperature was arrived at by heating the room or cooling by air conditioning. The amount of work needed to achieve that temperature is *not* a state function, since it is determined, among other factors, by the particular "path" (method) employed for temperature control.

occurs within a constant volume and no electrical work is done, then the total energy change appears as heat; that is, when $w = 0$,

$$\Delta E = q_v$$

where q_v represents a particular condition of constant volume. Under these conditions, q_v is the maximum heat available from the particular chemical change. Neither of the methods suggested for the power plant process use electrical work *directly* involving the chemical change, but both require some pressure–volume work. Thus both involve less than the theoretical maximum heat energy.

Many chemical reactions occur under conditions of constant pressure, such as the combustion of methane in a furnace whose exhaust stack is open to the atmosphere. Under such conditions and when the only work is an expansion against a constant pressure, we can write

Simple reactions such as those in aqueous solutions may occur under conditions in which *both* pressure and volume are constant.

$$\Delta E = q_p - P\,\Delta V$$

in which q_p is the heat change and $P\,\Delta V$ is the work done by virtue of the volume change, ΔV. Then

By convention, all thermodynamic *state functions* are represented by capital letters.

$$q_p = \Delta E + P\,\Delta V$$

Since ΔE, P, and ΔV are all state functions, q_p must also be independent of the particular mechanism employed in the change from initial to final states. This heat term is of such importance in chemical processes that it is given a particular name and thermodynamic symbol: ΔH, representing the change in *heat content* or, more commonly, the *enthalpy*.

A more general definition, applicable to cases in which both pressure and volume may vary, can be formulated as

$$\Delta H = \Delta E + \Delta(PV)$$

$$\Delta H = \Delta E + P\,\Delta V \qquad \text{at constant pressure} \tag{5.2}$$

5.4 HEATS OF FORMATION

Since ΔH is a state function, it can be calculated for any chemical reaction from experimental values of ΔH for any other chemical changes that would theoretically result in the same overall process. Tabulations of experimental values of standard *heats of formation* (Table 5.1) for chemical compounds are particularly useful. Heat

Table 5.1   SOME STANDARD[a] **HEATS OF FORMATION**

COMPOUND	ΔH_f^0 (kcal mole^{-1})	COMPOUND	ΔH_f° (kcal mole^{-1})	COMPOUND	ΔH_f° (kcal mole^{-1})
acetic acid (l)	−116.4	hydrogen chloride (g)	−22.1	sodium bromide (s)	−86.0
ammonium nitrate (s)	−87.4	hydrogen fluoride (g)	−64.8	sodium carbonate (s)	−270.3
ammonia (g)	−11.0	hydrogen iodide (g)	+6.2	sodium chloride (s)	−98.4
benzene (l)	+11.7	hydrogen sulfide (g)	−4.8	sodium fluoride (s)	−136.5
carbon dioxide (g)	−94.0	methane (g)	−17.9	sodium hydroxide (s)	−102.0
carbon monoxide (g)	−26.4	methanol (l)	−57.0	sulfur dioxide (g)	−70.9
carbon tetra-		nitric acid (l)	−41.6	sulfur trioxide (g)	−94.6
chloride (l)	−32.4	nitric oxide (g)	+21.6	sulfuric acid (l)	−194.6
ethane (g)	−20.2	nitrous oxide (g)	+19.6	water (l)	−68.3
ethanol (l)	−66.4	propane (g)	−24.8	water (g)[b]	−57.8
hydrogen bromide (g)	−8.7	sodium bicarbonate (s)	−226.5		

[a] To facilitate calculations, values are tabulated for experimental quantities corrected to conditions of *standard states* (ΔH^0). These are defined as follows: temperature = 25°C, all gases at 1.00 atm pressure, and all solids in their most stable form. A negative sign indicates an exothermic process (heat evolved). A positive sign indicates an endothermic process (heat required). A temperature of 25°C (298°K) is not a requirement of the generalized *standard state* concept. Other temperatures may be selected for specific purposes, with ΔH_f^0 of the elements set at zero at these temperatures. We shall work exclusively with 25°C values.

[b] The value reported is "corrected" for differences between water *vapor* and the standard state (liquid) at 25°C.

The experimental discovery that the heat of reaction is independent of the reaction method is credited to Germain Hess (1840). As a result, calculations based on equation (5.3) are often referred to as "Hess's law calculations."

of formation refers to the change in enthalpy resulting from the formation of a compound *from its component elements*. For example, the enthalpy change associated with the burning of carbon (as graphite) with a stoichiometric amount of molecular oxygen to produce carbon dioxide corresponds to the *heat of formation* of CO_2.

Formation of CO_2

$$C(s) + O_2(g) \longrightarrow CO_2(g)$$

The production of carbon dioxide by some other route, such as the burning (*combustion*) of carbon monoxide, also results in an enthalpy change, but this process is not considered *formation* in the thermodynamic sense because reactants other than *elements* are involved.

Combustion of CO

$$2CO(g) + O_2(g) \longrightarrow 2CO_2(g)$$

It is necessary that we understand some important thermodynamic conventions. For example, the use of a *superscript zero* (as in ΔH^0 or ΔE^0) always refers to a value determined under specifically defined *standard state* conditions. These conditions may be defined in different ways for various special purposes, but we shall use a *thermodynamic standard state temperature* of 25°C (298°K) and *pressure* of 1.00 atm (760 torr) throughout this text as most appropriate to our needs. The use of a subscript f to specify *formation* of a species from its component elements is associated with a further convention: *By definition, the standard heat of formation of a free element in its most stable natural form is zero.* Finally, there is an algebraic sign convention: *A positive sign is used for energy transferred into a system and a negative sign is used for energy transferred out of a system.* For the specific case of *enthalpy* change, a process is said to be *exothermic* when ΔH is negative ($\Delta H < 0$) and *endothermic* when ΔH is positive ($\Delta H > 0$).

°K = °C + 273

Carbon, for example, exists in various forms such as *graphite* and *diamond*. Since graphite is the most stable form of carbon, it is *assigned* a standard heat of formation of zero. The ΔH_f^0 value for diamond is +0.45 kcal mole^{-1}.

The terminology for enthalpy change can easily be remembered as *endothermic* (heat enters) and *exothermic* (heat exits).

Note that equation (5.3) uses a specific convention, always employed for thermodynamic quantities, that the change (Δ) is defined as *final state minus initial state.*

Bond-energy calculations are made by the same methods used with ΔH calculations.

Experimental values of ΔH_f^0 are often calculated from reactions other than those of direct combination of the elements (Example 3). They may, in turn, be used to calculate enthalpy changes in other types of reactions:

$$\Delta H_{reaction}^0 = \Sigma \Delta H_f^0 \text{ (products)} - \Sigma \Delta H_f^0 \text{ (reactants)} \tag{5.3}$$

(The Greek symbol Σ represents "the sum of." Note that ΔH_f^0 values are on a "per mole" basis and must be multiplied by coefficients from the balanced equation; see Example 1.)

It is important to note that heats of formation refer to the net enthalpy change for the process of formation of a compound from its component elements in whatever form these elements occur naturally. The assignment of ΔH_f^0 values of zero to polyatomic elements (e.g., O_2 or P_4) does not imply that there is no enthalpy difference between the natural molecular elements and their free atoms. A related concept, the *bond energy*, refers to the enthalpy difference between free atoms and bonded atoms. The bond energies at 25°C for O_2 and N_2 are 118 kcal mole^{-1} and 225 kcal mole^{-1}, respectively. We shall restrict our considerations to heats of formation, leaving the discussions of bond energies to more advanced studies.

We will be working throughout this text with *calories* (or kilocalories) as units of energy. For those preferring the use of the joule, 4.184 J = 1.000 cal.

Since the formation of a compound from its elements is just the opposite of the decomposition of the compound to the component elements, the *magnitude* of the enthalpy change is the same for both processes. (Remember that ΔH depends only on the difference between the initial and final states.) However, the reversal of the selection of initial and final states (i.e., the reversal of the direction of reaction) will result in a change of sign for the ΔH involved; that is,

$$\Delta H_{formation}^0 = -\Delta H_{decomposition}^0$$

Example 1

Calculate the standard heat of combustion for methane in kilocalories per mole of methane.

A "Conceptual" Solution

The reaction is formulated as

$$CH_4(g) + 2O_2(g) \longrightarrow CO_2(g) + 2H_2O(l)$$

For the purpose of calculation, consider the reaction in terms of a series of *hypothetical* steps, each related to the formation of a compound from its elements:

STEP 1 $CH_4(g) \longrightarrow C(s) + 2H_2(g)$

STEP 2 $C(s) + O_2(g) \longrightarrow CO_2(g)$

STEP 3 $2H_2(g) + O_2(g) \longrightarrow 2H_2O(l)$

The first step is the reverse of the formation of methane from its elements. Since the formation *evolves* 17.9 kcal mole^{-1} (Table 5.1), the "decomposition" would *require* 17.9 kcal mole^{-1}. Step 2 evolves 94.0 kcal mole^{-1} (Table 5.1) and step 3 evolves 68.3 kcal mole^{-1}. The overall equation requires 2 moles of water per mole of methane, so step 3 corresponds to a total of $2 \times 68.3 = 136.6$ kcal (per mole of methane). Then the *net* enthalpy change is

$$
\begin{array}{ll}
\text{heat evolved:} & 94.0 + 136.6 = 230.6 \\
- \text{ heat required:} & \underline{\hspace{1.2em}17.9} \\
\text{net heat change:} & 212.7
\end{array}
$$

Since there is a net release of heat (i.e., the reaction is exothermic), the enthalpy change is expressed as

$$\Delta H^0_{\text{combustion}} = -212.7 \text{ kcal mole}^{-1} \text{ (CH}_4)$$

A Simple Solution Based on Equation (5.3)

$$\Delta H^0_{\text{reaction}} = \Sigma \, \Delta H_f^0 \text{ (products)} - \Sigma \, \Delta H_f^0 \text{ (reactants)}$$

for

$$CH_4 + 2O_2 \longrightarrow CO_2 + 2H_2O$$

Products (data from Table 5.1)

$$\left[\frac{1 \text{ mole (CO}_2)}{1 \text{ mole (CH}_4)} \times \frac{(-94.0) \text{ kcal}}{1 \text{ mole (CO}_2)} + \frac{2 \text{ moles (H}_2\text{O})}{1 \text{ mole (CH}_4)} \times \frac{(-68.3) \text{ kcal}}{1 \text{ mole (H}_2\text{O})} \right]$$

$$= [-230.6 \text{ kcal mole}^{-1} \text{ (CH}_4)]$$

Minus reactants (data from Table 5.1)

$$\left[\frac{1 \text{ mole (CH}_4)}{1 \text{ mole (CH}_4)} \times \frac{(-17.9) \text{ kcal}}{1 \text{ mole (CH}_4)} + \frac{2 \text{ moles (O}_2)}{1 \text{ mole (CH}_4)} \times \frac{(0) \text{ kcal}}{1 \text{ mole (O}_2)} \right]$$

$$= [-17.9 \text{ kcal mole}^{-1} \text{ (CH}_4)]$$

$$\Delta H^0_{\text{combustion}} = (-230.6) - (-17.9) = -212.7 \text{ kcal mole}^{-1} \text{ (CH}_4)$$

The Kirchhoff equation permits the calculation of ΔH at any temperature, $T(°K)$, $(T = 273° + °C)$ from appropriate use of experimentally based "correction" factors (Δa, Δb, Δc):

$$\Delta H_T = \Delta H° + \Delta a(T - 298)$$
$$+ 0.5\,\Delta b(T^2 - 298^2)$$
$$+ \Delta c\left(\frac{1}{298} - \frac{1}{T}\right)$$

Corrections allow for factors such as conversion of reactants and products to their stable forms at 700°C.

Hess's law is sometimes referred to as the law of constant heat summation, a more descriptive statement that reminds us that the "law" states that the enthalpy change for a process is determined only by the enthalpy difference in final and initial states, being independent of any particular process path.

It should be noted that methane does not burn in air at 25°C. The enthalpy change (ΔH), like other thermodynamic quantities, is a complex function of temperature. The value calculated in Example 1 includes the total heat that would have to be removed from the combustion products of 1 mole of methane to condense the water vapor to liquid water at 25°C and to cool the CO_2 gas to 25°C, as well as the heat required to warm the reactants from 25°C to the combustion temperature.

The ignition temperature of methane at atmospheric pressure is around 700°C, a temperature that varies somewhat with the construction details of the burner used. If we take this average as an approximation of the temperature in the combustion chamber of the power plant under consideration, then appropriate corrections give at 700°C

$$\Delta H_{combustion} \approx -191 \text{ kcal mole}^{-1}$$

If *all* of this heat were to be used to convert (Unit 12) water at 25°C to steam at 150°C, it would be sufficient for the generation of approximately 280 g of steam (at 150°C) for each mole (16 g) of methane burned. Alternatively, if the generator is to be of the piston-engine type rather than a steam turbine, the heat of combustion could be used for thermal expansion of the product gases (Unit 10) for the "piston-pushing" operation.

In either case it would appear that the combustion of methane might furnish a large quantity of energy for useful conversion to electricity. Such is indeed the case, but the amount of energy that can *actually* be utilized is appreciably less than the calculated value for ΔH of combustion.

Although ΔH calculations, as we shall see, are not sufficient to tell us the amount of *useful* energy we can obtain from a chemical process, such calculations do have many applications. We may, for example, use Hess's law calculations for the quantitative comparison of the theoretical amounts of heat available from different chemical reactions.

Example 2

In deciding on a particular vehicle fuel, it is necessary to consider, among other factors, both the energy aspects of fuel combustion and the weight requirements associated with a fuel load. Compare ΔH^0 combustion for methane with ΔH^0 combustion for benzene on this basis, in terms of kilocalories per gram, for complete combustion [$CO_2(g)$ and $H_2O(l)$ as the only products].

Solution

The data from Example 1 may be used for methane, employing an appropriate unity factor for conversion to kilocalories per gram.

$$\Delta H^0_{combustion}(CH_4) = \frac{-212.7 \text{ kcal}}{\text{mole } (CH_4)} \times \frac{1 \text{ mole } (CH_4)}{16.04 \text{ g } (CH_4)} = -13.26 \text{ kcal g}^{-1} (CH_4)$$

For benzene, it is first necessary to write the balanced equation (Unit 3) for complete combustion.

$$2C_6H_6(l) + 15O_2(g) \longrightarrow 12CO_2(g) + 6H_2O(l)$$

Since we wish our answer in units of benzene, the Hess's law calculation equation (5.3), is formulated as

Products (data from Table 5.1)

$$\Delta H^0_{combustion} = \left[\frac{12 \text{ moles } (CO_2)}{2 \text{ moles } (C_6H_6)} \times \frac{(-94.0) \text{ kcal}}{1 \text{ mole } (CO_2)} + \frac{6 \text{ moles } (H_2O)}{2 \text{ moles } (C_6H_6)} \times \frac{(-68.3) \text{ kcal}}{1 \text{ mole } (H_2O)} \right.$$

Reactants (data from Table 5.1)

$$\left[\frac{2 \text{ moles } (C_6H_6)}{2 \text{ moles } (C_6H_6)} \times \frac{(+11.7) \text{ kcal}}{1 \text{ mole } (C_6H_6)} + \frac{15 \text{ moles } (O_2)}{2 \text{ moles } (C_6H_6)} \times \frac{(0) \text{ kcal}}{1 \text{ mole } (O_2)}\right]$$

$$\Delta H^0_{\text{combustion}} = (-768.9) - (11.7) = -780.6 \text{ kcal mole}^{-1} (C_6H_6)$$

Or, for the comparison "per gram",

$$\Delta H^0_{\text{combustion}} = \frac{-780.6 \text{ kcal}}{\text{mole } (C_6H_6)} \times \frac{1 \text{ mole } (C_6H_6)}{78.11 \text{ g } (C_6H_6)} = -9.993 \text{ kcal g}^{-1} (C_6H_6)$$

Thus on a "heat-to-mass" comparison, methane would appear to be a more useful fuel than benzene. These are not, of course, the only factors that must be considered in fuel selection.

5.5 CALORIMETRY: EXPERIMENTAL DETERMINATION OF ΔH VALUES

How do we know the heats of formation of chemical compounds? Some processes, of course, can be observed rather easily and it is not difficult to visualize methods for direct experimental measurements of ΔH_f for such reactions as the formation of water from hydrogen and oxygen gases or the formation of HCl from H_2 and Cl_2. Many of the more interesting compounds, however, are not formed in any detectable amount by a direct combination of the component elements. For such cases it is necessary to *calculate* ΔH_f using enthalpy data for reactions that could, in reality or "on paper," be related through a series of steps to the desired formation from original elements.

A *calorimeter* (Figure 5.2) is an insulated device that can be used to determine the heat change associated with a chemical or physical (Unit 12) process. Thermal insulation ensures that any temperature change measured reflects the heat absorbed

Figure 5.2 A simple calorimeter.

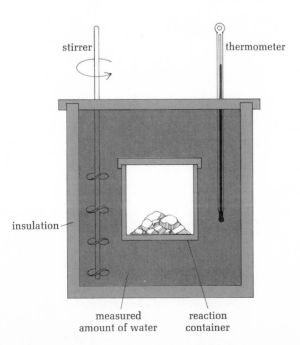

stirrer

thermometer

insulation

measured amount of water

reaction container

or released by the process under investigation, rather than by any heat exchanged with the surrounding environment. For a chemical process, measured amounts of reactants are typically mixed within a sealed container immersed in a measured amount of water and enclosed by the insulating calorimeter vessel. The initial temperature of the water is measured and stirring is continued during the course of reaction to maintain a uniform water temperature. From the temperature change and the mass of the water, with appropriate corrections for any heat transfer associated with the calorimeter components, the amount of heat released or absorbed by the chemical process can be determined. Depending on the accuracy desired and the complexity of the reaction under investigation, various modifications of the simple calorimeter may be required.

From calorimetric data for appropriate processes and mathematical "corrections" (if necessary) to standard state conditions, the heats of formation of the most complex substances can be determined with considerable accuracy.

Example 3

Ethanol is an example of a compound not formed in detectable amounts by direct combination of the component elements. Using the following reactions, for each of which the ΔH^0 values are based on direct calorimetric measurements (corrected to $25°C$), calculate the standard heat of formation for liquid ethanol.

Condensation (Unit 12), a physical process

$$C_2H_5OH(g) \longrightarrow C_2H_5OH(l) \qquad \Delta H^0_{condensation} = -9.7 \text{ kcal mole}^{-1} (C_2H_5OH)$$

Complete combustion of ethanol

$$C_2H_5OH(g) + 3O_2(g) \longrightarrow 2CO_2(g) + 3H_2O(g)$$
$$\Delta H^0_{combustion} = -304.7 \text{ kcal mole}^{-1} (C_2H_5OH)$$

Complete combustion of carbon

$$C(s) + O_2(g) \longrightarrow CO_2(g) \qquad \Delta H^0_{combustion} = -94.0 \text{ kcal mole}^{-1} (C)$$

Complete combustion of hydrogen

$$2H_2(g) + O_2(g) \longrightarrow 2H_2O(g) \qquad \Delta H^0_{combustion} = -57.8 \text{ kcal mole}^{-1} (H_2)$$

Solution

The *formation* (by definition, from the elements in their most stable natural forms) of liquid ethanol is represented by

$$4C(s) + 6H_2(g) + O_2(g) \longrightarrow 2C_2H_5OH(l)$$

Chemical equations may be added (or subtracted) in the same way as algebraic equations. All products are added together on one side of the arrow and all reactants on the other side. Species appearing in both reactants and products are deleted as in the writing of net ionic equations (Unit 3).

If we rearrange the equations given so that they can be added together to give the desired formation equation, Hess's law suggests that the ΔH terms can also be summed, *provided they are all expressed in the same units*. [We will do this by converting all units to kilocalories per mole (C_2H_5OH), based on stoichiometric ratios from the formation equation.]

To show 2 moles of liquid ethanol

$$2C_2H_5OH(g) \longrightarrow 2C_2H_5OH(l) \qquad \Delta H_1^0 = -9.7 \text{ kcal mole}^{-1} (C_2H_5OH)$$

To show the "origin" of the gaseous ethanol

$$6H_2O(g) + 4CO_2(g) \longrightarrow 2C_2H_5OH(g) + 6O_2(g)$$
$$\Delta H_2^0 = -(-304.7)$$
$$= +304.7 \text{ kcal mole}^{-1} (C_2H_5OH)$$

To show the formation of the $4CO_2$

$$4C(s) + 4O_2(g) \longrightarrow 4CO_2(g)$$

$$\Delta H_3{}^0 = \frac{-94.0 \text{ kcal}}{\text{mole (C)}} \times \frac{4 \text{ mole (C)}}{2 \text{ mole (C}_2\text{H}_5\text{OH)}}$$

$$= -188.0 \text{ kcal mole}^{-1} \text{ (C}_2\text{H}_5\text{OH)}$$

To show the formation of the $6H_2O$

$$6H_2(g) + 3O_2(g) \longrightarrow 6H_2O(g)$$

$$\Delta H_4{}^0 = \frac{-57.8 \text{ kcal}}{\text{mole (H}_2)} \times \frac{6 \text{ mole (H}_2)}{2 \text{ mole (C}_2\text{H}_5\text{OH)}}$$

$$= -173.4 \text{ kcal mole}^{-1} \text{ (C}_2\text{H}_5\text{OH)}$$

Net summation

$$4C(s) + 6H_2(g) + O_2(g) \longrightarrow 2C_2H_5OH(l) \quad \Delta H_f{}^0 = \Delta H_1{}^0 + \Delta H_2{}^0 + \Delta H_3{}^0 + \Delta H_4{}^0$$

$$\Delta H_f{}^0 = -66.4 \text{ kcal mole}^{-1} \text{ (C}_2\text{H}_5\text{OH)}$$

Example 4

Carbon dioxide can be removed from the atmosphere by reaction with any of a number of metal oxides or hydroxides. Calculate the standard heat of reaction of solid barium oxide with CO_2 gas to produce solid barium carbonate, in kilocalories per gram (CO_2), from the following data based on direct calorimetric determinations.

Neutralization of HCl by BaO

$$BaO(s) + 2HCl(g) \longrightarrow BaCl_2(s) + H_2O(g)$$

$$\Delta H^0_{\text{neutral}} = -85.7 \text{ kcal mole}^{-1} \text{ (BaCl}_2)$$

Reaction of $BaCO_3$ with HCl

$$BaCO_3(s) + 2HCl(g) \longrightarrow BaCl_2(s) + H_2O(g) + CO_2(g)$$

$$\Delta H^0_{\text{reaction}} = -22.4 \text{ kcal mole}^{-1} \text{ (BaCO}_3)$$

Solution
The desired equation is formulated as (Unit 3):

$$BaO(s) + CO_2(g) \longrightarrow BaCO_3(s)$$

Then, rearranging equations as necessary,

To show an "origin" of $BaCO_3$

$$BaCl_2(s) + H_2O(g) + CO_2(g) \longrightarrow BaCO_3(s) + 2HCl(g)$$

$$\Delta H_1{}^0 = -(-22.4)$$

$$= +22.4 \text{ kcal mole}^{-1} \text{ (BaCO}_3)$$

To show an "origin" of $BaCl_2$

$$BaO(s) + 2HCl(g) \longrightarrow BaCl_2(s) + H_2O(g)$$

$$\Delta H_2{}^0 = \frac{-85.7 \text{ kcal}}{\text{mole (BaCl}_2)} \times \frac{1 \text{ mole (BaCl}_2)}{1 \text{ mole (BaCO}_3)}$$

$$= -85.7 \text{ kcal mole}^{-1} \text{ (BaCO}_3)$$

Net summation

$$BaO(s) + CO_2(g) \longrightarrow BaCO_3(s) \quad \Delta H^0_{\text{reaction}} = H_1{}^0 + H_2{}^0$$

$$\Delta H^0_{\text{reaction}} = -63.3 \text{ kcal mole}^{-1} \text{ (BaCO}_3)$$

Then, using appropriate unity factors based on stoichiometric ratios and formula weights (Unit 3),

$$\Delta H^0_{\text{reaction}} = \frac{-63.3 \text{ kcal}}{\text{mole (BaCO}_3)} \times \frac{1 \text{ mole (BaCO}_3)}{1 \text{ mole (CO}_2)} \times \frac{1 \text{ mole (CO}_2)}{44.0 \text{ g (CO}_2)}$$

$$= -1.44 \text{ kcal g}^{-1} \text{ (CO}_2)$$

5.6 LIFE GETS COMPLICATED: ENTROPY

In Section 5.4, following Example 1, we indicated that calculations could be made showing that *all* of the heat generated by the combustion of 1 mole of methane at 700°C would be sufficient to convert 280 g of water at 25°C to steam at 150°C. Any good mechanical engineer could take the fuel-to-steam ratio calculated thus far and design a turbine generator. An electrical engineer could calculate how much electricity it should produce. Both would be sadly disappointed when they tried to make it work. All that heat is *not* going to prove "useful." *Work* is a manifestation of *ordered* energy: Something (a piston, a stream of superheated steam, etc.) is moved a particular distance in a specific direction. Heat, on the other hand, is energy corresponding to particles jostling around in *random* motion.

The *second law of thermodynamics* tells us that *any spontaneous change is associated with an increased randomness of the universe.* Experience confirms that this statement must apply to the *universe* (i.e., to a thermodynamic system *and* its total surroundings), since we can easily think of cases in which a spontaneous change in a *system* alone corresponds to increasing *order* (decreasing randomness). For example, when we place an ice cube tray full of liquid water in a freezing compartment, we fully expect to return later and find nice, regular ice cubes. As we shall see in subsequent discussions (Units 11 and 12), the molecules in liquid water are quite randomly arranged, whereas crystalline ice consists of a very orderly three-dimensional pattern of H_2O molecules. Here, indeed, is a case of a spontaneous *decrease* in randomness when water freezes. Remember, however, our distinctions of *system, surroundings,* and *universe.* If the second law of thermodynamics is correct, the spontaneous randomness decrease as ice freezes (in the system) must somehow be accompanied by a *larger* randomness *increase* in the surroundings. Careful experiments show this to be the case. In our freezing compartment example, it can be shown that the heat "pumped out" in order to freeze the water is sufficient to cause this larger randomness increase in the surroundings.

An alternative, but equivalent, statement of the *second law* is more easily related to the power plant problem: It is never possible to convert heat *completely* to other forms of energy. These statements are based on a massive accumulation of experimental information. In spite of repeated attempts, no instance has ever been observed to violate this second law.

The thermodynamic quantity that is related to this "randomness" of heat, and thus to the fraction of heat unavailable for conversion to useful work, is entropy (symbolized by S).

Let us return now to the problem of our generator plant. We shall decide to use the steam-turbine type of generator (a common choice) with a steam source as shown in Figure 5.3.

Entropy, like enthalpy, is a state function.[2] The heat energy (originally from the combustion process) that can be converted to useful work is limited by considerations of entropy, but also by another state function, temperature. The actual contri-

[2] For a more detailed treatment of entropy, see "Entropy: A Modern Discussion," by Otto Redlich, in the June, 1975, issue of the *Journal of Chemical Education* (pp. 374–376).

Figure 5.3 Simple steam source.

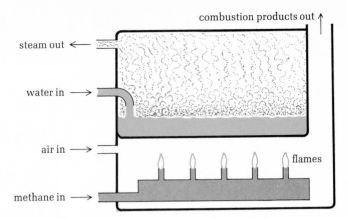

combustion products out

steam out ←

water in →

air in →

flames

methane in →

butions of entropy for an overall process—considering interactions with the surroundings—are difficult to estimate on an intuitive basis, but some idea of the efficiency of the turbine process can be gained from an older formulation developed during the days of the steam locomotive.

$$\text{efficiency} = \frac{T_{in} - T_{out}}{T_{in}} \tag{5.4}$$

By convention, a capital T refers to absolute temperature (°K),

°K = °C + 273

in which T_{in} is the temperature of the steam entering the turbine and T_{out} is the temperature of the cooler steam exhausted from the turbine after some of the energy has been converted to the work of turning the turbine blades.

Our example suggested the generation of steam at 150°C. Let us consider the case in which the exhaust temperature of the steam is 100°C (this avoids the complications of condensation to liquid water). Then, since T represents °K,

$$T_{in} = 150 + 273 = 423°K$$
$$T_{out} = 100 + 273 = 373°K$$
$$\text{efficiency} = \frac{423 - 373}{423} = 0.12$$

That is, under the conditions imposed we could expect a *maximum* efficiency of energy conversion of 12 percent. And this is only the *theoretical* limitation. In actual practice there are many additional factors that further limit the conversion of the heat of combustion to useful work. Some energy is lost as friction in the operation of the generator. Much of the original energy transferred to the steam is *still there* after the steam leaves the turbine (so it was "lost" insofar as useful work was concerned). In fact, not all of the heat of combustion was ever *really* useful in steam production, since much "went up the stack" as the hot combustion products escaped into the atmosphere. Engineering design factors can reduce many of these limitations, but they can never be fully eliminated.

There are ways of improving the efficiency of the overall process. One way is suggested by equation (5.3). The theoretical efficiency would be improved if the temperature of the steam exhausted could be lowered. (Ideally, an "output" temperature of 0°K, *absolute zero,* would give a theoretical efficiency of 100 percent.) There is a practical limitation on the output temperature. If we wish to exhaust the water as steam rather than as liquid water, the output temperature cannot be lower than 100°C (at 1 atm pressure).

DERIVATION OF EQUATION (5.4)

In describing the efficiency of conversion of heat to work, a useful ratio would be

$$e = \frac{w_{out}}{q_{in}}$$

That is, the efficiency (e) is defined as the ratio of work *done by* the system (w_{out}) to heat *absorbed by* the system (q_{in}). All experience has indicated that the *maximum* work available from any process is obtainable only when the process occurs in a *reversible* way, that is, in such a way that the system can be restored completely to its original condition without heat transfer to the surroundings. In practice, of course, a completely reversible process cannot be achieved because of inherent inefficiencies, such as frictional losses or imperfect insulation. The concept of reversibility is still a useful one in defining the *theoretical* limit of heat-to-work conversion efficiency. For an isolated system the heat input must all appear as work output plus heat output. (In the steam turbine, for example, the heat of the original steam equals the work done by the steam plus the heat of the "exhaust" steam.) That is,

$$q_{in} = w_{out} + q_{out}$$

from which

$$w_{out} = q_{in} - q_{out}$$

and

$$e = \frac{w_{out}}{q_{in}} = \frac{(q_{in} - q_{out})}{q_{in}}$$

In wishing to relate the efficiency of heat-to-work conversion, William Thomson (Lord Kelvin), decided to define a temperature scale such that

$$\left(\frac{q_{in}}{q_{out}}\right)_{reversible} = \frac{T_{in}}{T_{out}}$$

(This was the origin of the so-called "Kelvin temperature scale".) Now we can work with our efficiency equation as follows:

$$e = \frac{q_{in} - q_{out}}{q_{in}} = 1 - \frac{q_{out}}{q_{in}}$$

and, from the Kelvin temperature definition,

$$e = 1 - \frac{q_{out}}{q_{in}} = 1 - \frac{T_{out}}{T_{in}}$$

which may be rewritten as

$$e = \frac{T_{in} - T_{out}}{T_{in}} \qquad \text{[equation (5.4)]}$$

An alternative solution, also based on equation (5.3), would be to increase the temperature of the steam entering the turbine. This has a smaller effect on efficiency, since both the numerator and denominator are increased, but its effect can be demonstrated.

Example 5

What would be the theoretical efficiency of a steam turbine using incoming steam at 300°C and exhausting steam at 100°C?

Solution

$$T_{in} = 300 + 273 = 573°K$$
$$T_{out} = 100 + 273 = 373°K$$
$$\text{efficiency} = \frac{573 - 373}{573} = 0.35$$

Thus the efficiency is nearly three times as great with an input of steam at 300°C than at 150°C.

An alternative statement of the second law of thermodynamics is rather intuitively obvious: Heat can flow spontaneously only from a warmer to a cooler body. If the temperature of the burner used is 700°C, then this sets an upper limit on the input steam temperature and thus on the maximum theoretical efficiency.

What happens to the "waste" heat? The hot combustion gases and the exhaust steam lines transfer heat to the environment. (The *real* system was not, obviously, "isolated.") The increase in random motions of the air thus heated represents an *increase* in *entropy*.

5.7 FREE ENERGY (AT A PRICE)

Many spontaneous processes, such as combustion, are exothermic. This seems reasonable and might appear to imply the necessity for finding an external heat energy source to *force* a nonspontaneous process to occur. However, not all *exothermic* ($\Delta H < 0$) processes are spontaneous and many *endothermic* ($\Delta H > 0$) processes *do* occur spontaneously. The tendency to achieve a lower energy state by releasing heat may be opposed by the tendency to proceed spontaneously from an orderly to a less orderly condition, that is, by the spontaneity associated with entropy increase. Both the factors of enthalpy change (ΔH) and entropy change (ΔS) must be considered for a system. It is the balance between these two *system* factors that ultimately determines the spontaneity of a process. It is fairly easy to determine the enthalpy and entropy changes for an isolated *system*, but there are significant problems in measuring entropy changes of the *surroundings*. Since the use of ΔS as a criterion of spontaneity is restricted to $\Delta S_{universe}$, it would be most advantageous to define some new criterion based on the more readily determined factors of ΔH_{system} and ΔS_{system}. The *free energy* concept is defined to meet this need.

The symbol G is used for the "Gibbs free energy," a constant-pressure quantity. A similar relationship, known as the "Helmholtz free energy" (symbolized by A) applies to constant-volume processes,

$$\Delta A = \Delta E - T \Delta S$$

$$\Delta G_{system} = \Delta H_{system} - T \Delta S_{system} \tag{5.5}$$
$$\text{(at constant pressure)}$$

We shall explore the concept of *free energy* in more detail in Unit 23 (Electrochemistry), since the measurement of voltages for electrochemical cells is a particularly useful way of determining ΔG values experimentally. At that point we shall develop a derivation of equation (5.5) and of the following important relationship:

$$\Delta G_{system} < 0$$

for any *spontaneous* constant–pressure process.

For our present purposes, it is sufficient that we recognize ΔG as a useful criterion for spontaneity and as a measure of the maximum energy available for useful work (the difference between the total heat and the "unavailable heat," shown as $T \Delta S$). We shall find some important applications of the free-energy concept in later units and we may use equation (5.5) now to make some qualitative estimates of the effect of temperature on the spontaneity of a process (Example 6). As a matter of convenience, we shall omit subscripts of "system" on thermodynamic terms for our subsequent discussions, with the understanding that all such terms refer to *system* characteristics unless otherwise specified. We shall also assume *constant-pressure* conditions unless otherwise specified.

If you are interested at this point in a more detailed development of the *free-energy* concept, see Section 23.3 (Unit 23).

Example 6

Is an endothermic process associated with increasing entropy more likely to be spontaneous at a low temperature or at a very high temperature?

Solution

For spontaneity, $\Delta G < 0$. If a process is endothermic, $\Delta H > 0$, and increasing entropy means $\Delta S > 0$, thus

$$\Delta G = \underset{\text{(positive)}}{\Delta H} - \underset{\text{(positive)}}{T \Delta S}$$

(*Note:* T is always greater than zero on the Kelvin scale.) For ΔG to be less than zero the $T \Delta S$ term must be larger than the ΔH term. The higher the temperature, the larger the $T \Delta S$ term, so a process of this type is more likely to be spontaneous at a high temperature than at a low temperature.

Remember that our conventions imply *system* properties and constant-pressure conditions unless otherwise indicated.

It is important to note that free energy is not "free" in the economic sense. To obtain energy for useful work, we must pay the cost in terms of both the depletion of available energy sources and the increase in total entropy of the "universe" in which we live. It is also important to recognize that none of our thermodynamic considerations have offered any clues to the rate ("speed") of a process. Thermodynamics can allow us to determine whether or not a particular chemical reaction will occur spontaneously and to calculate various aspects of energy changes associated with chemical processes. Thermodynamic considerations do not, however, permit us to calculate how rapidly a reaction should proceed or at what rate energy is released or required. This is a significant limitation. The study of *reaction kinetics* (Unit 16) provides an understanding of the factors determining reaction rates.

5.8 SO WHAT? The net result of the methane-fueled, steam-turbine production of electricity involves, *for an idealized situation:*

1. Methane and oxygen—natural resources whose supply is *not* unlimited—are consumed.
2. Only a small percentage of the energy theoretically available from the combustion process eventually appears as electricity.

PROJECT SOLCHEM

Although any large-scale use of solar energy for power is probably many years in the future, the Naval Research Laboratory's project *Solchem* shows considerable promise for coupling the sun's energy with a simple chemical process to produce clean, economical power. A spherical reaction vessel is located at the focus of a paraboloidal reflector, permitting the focused solar rays to heat the chamber to around $800°C$. This heat is used to produce a chemical reaction, with the product gases diverted to a catalytic converter where they are recombined to the original reactant for recycling through the process. The heat released during the catalytic recombination may be used directly to generate steam for a turbine operation or stored (for example, by use to melt some solid salt for storage in an insulated reservoir) to permit continuous operation when sunlight is unavailable.

One of the chemical processes under investigation is the $SO_3(g)/(SO_2(g) + O_2(g)$ system:

$$\text{solar heat} + 2SO_3(g) \longrightarrow 2\underbrace{SO_2(g) + O_2(g)}$$

catalytic recombination
(exothermic)

Although most solar energy converters studied so far appear to have limited applications, it is believed that project *Solchem* could be designed on a scale quite reasonable for large, continuous power production. It is estimated that about ten of the SOLCHEM plants could match the power output of a large hydroelectric installation.

3. The "entropy of the universe" is increased; that is, there is now less energy that will ever be available for useful work.

In addition, it is important to realize that real processes impose problems in addition to those of the idealized case. We have already seen that one of our simplifying assumptions—an isolated system—is not truly realistic. The other assumption is no more valid. The real combustion process is *not* well represented by the equation

$$CH_4 + 2O_2 \longrightarrow CO_2 + 2H_2O$$

Pure oxygen is not used in power plant combustion processes; it is too expensive. Air is used instead, and air also contains nitrogen. Some of this nitrogen also "burns" during the combustion process, forming various oxides of nitrogen that escape into the atmosphere. In the presence of sunlight and organic residues from such sources as the automobile engine, these oxides react to form components of "photochemical smog," a mixture of substances whose hazardous properties are only beginning to be realized.

The methane is not completely oxidized in typical combustion processes. One of the products of its incomplete combustion is carbon monoxide, a highly dangerous substance. At the same time, incomplete combustion produces less energy per mole of methane consumed, requiring more fuel for production of a given amount of energy.

The implications of these problems are relatively clear. Scientists and engineers can do a few things to improve the efficiency of processes for converting chemical energy to electric energy. Some of the by-products can be reduced, if not eliminated. But the laws of chemical change and the laws of thermodynamics impose limitations that cannot be avoided.

These laws tell us that in terms of the availability of resources and energy and in terms of the increase of disorder, *things will get worse,* not better. These laws do not, however, offer any constraints to the *speed* at which this will occur. Man has the freedom to make some choices in this respect. All of us share in the responsibility for these choices.

SUGGESTED READING   **Angrist, Stanley W., and Loren G. Hepler.** 1967. *Order and Chaos: Laws of Energy and Entropy.* New York: Basic Books.

Holdren, John, and Philip Herrera. 1971. *Energy: A Crisis in Power.* San Francisco: Sierra Club.

Pimentel, George C., and Richard D. Spratley. 1969. *Understanding Chemical Thermodynamics.* San Francisco: Holden-Day. (Highly recommended.)

Spencer, J. N., David Gordon, and H. D. Schreiber, 1974. "Entropy and Chemical Reactions," *Chemistry.* December.

EXERCISES

1. Select the term best associated with each example, symbol, or relationship.
   TERMS: (A) combustion (B) formation (C) enthalpy (D) exothermic (E) internal energy
   a. E
   b. H
   c. $\Delta H < 0$
   d. $2C + 3H_2 \longrightarrow C_2H_6$
   e. $2C_2H_6 + 7O_2 \longrightarrow 4CO_2 + 6H_2O$

2. Indicate whether each of the following statements is true or false.
   a. The standard heat of formation of liquid oxygen $[O_2(l)]$ at $298°K$ is defined as zero.
   b. A defined condition of the standard state for gaseous substances is a pressure of 1.00 atm.
   c. For any spontaneous process, $\Delta S_{system} < 0$.
   d. If ΔH for reaction A is greater than ΔH for reaction B, then reaction A will proceed more rapidly than B.
   e. The theoretical thermodynamic efficiency is at its maximum for a reversible process.

3. For each of the following calculations, use appropriate ΔH_f^0 values from Table 5.1. (Assume all reactants and products to be in their stable forms at standard state conditions unless otherwise indicated.)
   a. Calculate the standard heat of combustion for propane (C_3H_8) in kilocalories per mole (C_3H_8) (to gaseous products).
   b. Calculate the standard heat of reaction for

   $$NaOH(s) + CO_2(g) \longrightarrow NaHCO_3(s)$$

   in kilocalories per mole ($NaHCO_3$).

   c. Calculate the net enthalpy change for production of sulfuric acid by the process:

   $$2H_2S(g) + 3O_2(g) \longrightarrow 2H_2O(g) + 2SO_2(g)$$
   $$2SO_2(g) + O_2(g) \longrightarrow 2SO_3(g)$$
   $$H_2O(g) \longrightarrow H_2O(l)$$
   $$H_2O(l) + SO_3(g) \longrightarrow H_2SO_4(l)$$

   Assume standard state conditions and express your answer in kilocalories per mole (H_2SO_4). Why is this value different from the standard heat of formation of sulfuric acid?

   d. Ammonium nitrate can be decomposed by *careful* heating to form laughing gas (nitrous oxide):

   $$NH_4NO_3(s) \longrightarrow N_2O(g) + 2H_2O(g)$$

   Calculate the standard heat of reaction for this process (i.e., ΔH at 25°C). Express your answer in kilocalories per mole of ammonium nitrate.

   e. The same ammonium nitrate can decompose by a different route. (The explosion of a ship loaded with ammonium nitrate fertilizer at Texas City in 1947 killed about 500 people and caused an estimated $50 million damage.) Calculate the standard heat of reaction for this process, formulated as

   $$2NH_4NO_3(s) \longrightarrow 2N_2(g) + O_2(g) + 4H_2O(g)$$

   Express your answer in kilocalories per mole of ammonium nitrate.

Compare answers from Exercises 3d and e. Does the difference account for the differences in "violence" of the two reactions? Explain your answer. (Hint: Is *rate* a thermodynamic consideration?)

4. Potassium superoxide may be thought of as a simple chemical analog of the "breathing cycle" of a living plant since, like a plant, KO_2 "takes up" carbon dioxide and releases oxygen:

$$4KO_2(s) + 2CO_2(g) \longrightarrow 2K_2CO_3(s) + 3O_2(g)$$

Calculate the standard enthalpy change for the consumption of 125 g of carbon dioxide by this process, using the following data based on calorimetric experiments. (Refer to Table 5.1 for any other data required.)

Combustion of potassium
$$K(s) + O_2(g) \longrightarrow KO_2(s)$$
$$\Delta H^0_{combustion} = -134.0 \text{ kcal mole}^{-1} (KO_2)$$

Reaction of CO_2 with KOH
$$KOH(s) + CO_2(g) \longrightarrow KHCO_3(s)$$
$$\Delta H^0_{reaction} = -33.5 \text{ kcal mole}^{-1} (KHCO_3)$$

Thermal decomposition of $KHCO_3$
$$2KHCO_3(s) \longrightarrow CO_2(g) + H_2O(g) + K_2CO_3(s)$$
$$\Delta H^0_{reaction} = +33.0 \text{ kcal mole}^{-1} (K_2CO_3)$$

Reaction of potassium with water vapor
$$2K(s) + 2H_2O(g) \longrightarrow 2KOH(s) + H_2(g)$$
$$\Delta H^0_{reaction} = -44.0 \text{ kcal mole}^{-1} (KOH)$$

5. Calculate (equation 5.4) the theoretical thermodynamic efficiency of the steam turbine discussed in this unit if the steam is exhausted at 100°C and the steam is introduced into the turbine at the maximum temperature possible (700°C) for the system described.

OPTIONAL

6. Is there any temperature at which a process can be spontaneous if $\Delta H > 0$ and $\Delta S < 0$? Explain. What does this mean in terms of the possible manufacture of synthetic diamonds (highly ordered arrangement of carbon atoms) from powdered charcoal?

PRACTICE FOR PROFICIENCY

Objective (a)

1. Select the term best associated with each example, symbol, description, or relationship.
   TERMS: (A) calorimeter (B) efficiency (C) endothermic (D) entropy (E) free energy (F) reaction rate (G) reversible (H) spontaneous (I) standard state (J) state functions (K) work
   a. $\Delta G < 0$ b. $\Delta H > 0$
   c. E, H, S, G   d. S
   e. G   f. w

g. $\dfrac{T_{in} - T_{out}}{T_{in}}$

h. speed of a process such as

$$2H_2(g) + O_2(g) \longrightarrow 2H_2O(g)$$

i. $H_2(g)$, or $O_2(g)$ at 298°K and 1 atm pressure
j. $2H_2(g) + O_2(g) \longrightarrow 2H_2O(g)$

k. insulated device used for experimental determination of enthalpy changes

Objective (b)

2. Indicate whether each of the following statements is true or false.
   a. ΔH^0 (reaction) $= \Sigma[\Delta H_f^0$ (products)$] - \Sigma[\Delta H_f^0$ (reactants)$]$.
   b. Work is not a state function.
   c. For all natural processes, $\Delta H > 0$.
   d. At constant pressure, $\Delta H = \Delta E + P \Delta V$.
   e. If ΔH for reaction A is greater than ΔH for reaction B, the speed of reaction A is not necessarily greater than that of B.
   f. The thermodynamic symbol for internal energy is E.
   g. When no work is done on or by a system, $\Delta E = q$ (constant pressure).
   h. ΔH_f^0 for $HCl(g) \neq \Delta H_f^0$ for $HCl(aq)$.
   i. One condition of the thermodynamic standard state defined requires that elements are in their most stable forms.
   j. The thermodynamic symbol for enthalpy is S.
   k. Internal energy is a state function.
   l. Entropy always decreases for a spontaneous process.
   m. Since neither q nor w is, in general, a state function, their difference $(q - w)$ can never be a state function.
   n. ΔH_f^0 for $N_2(g)$ is zero, by definition.
   o. Thermodynamic considerations alone do not predict reaction speeds.

Objective (c)

3. For each of the following calculations, use appropriate ΔH_f^0 values from Table 5.1. (Assume all reactants and products to be in their stable forms at standard state conditions unless otherwise indicated.)
   a. Calculate the standard heat of combustion for ethanol, a common fuel for simple laboratory "alcohol" burners.
   b. When wine "sours" to vinegar, the net process may be represented as

$$CH_3CH_2OH + O_2 \longrightarrow CH_3CO_2H + H_2O$$

   Calculate the standard enthalpy change for this process.
   c. Both hydrogen sulfide and sulfur dioxide are present in volcanic gases. The sulfur dioxide may originate from O_2 oxidation of hydrogen sulfide, with water as the other product. Calculate the standard enthalpy change for this oxidation, in kilocalories per mole (H_2S).

d. What is the standard enthalpy change for the reaction between sodium carbonate ("washing soda") and gaseous hydrogen bromide, in kilocalories per mole (Na_2CO_3)?

e. Benzene has been used as a fuel for certain types of internal combustion engines. As with most fuels, combustion may be complete (to $CO_2 + H_2O$) or incomplete (to $CO + CO_2 + H_2O$). What is the *difference* between the standard enthalpies of *complete* combustion and of combustion only to $CO + H_2O$, in kilocalories per mole (C_6H_6)?

f. Ammonium nitrate, for use as a fertilizer, may be prepared by bubbling gaseous ammonia through nitric acid. Calculate the standard heat of reaction for this process.

g. Some of the "fluoride" toothpastes use "stannous fluoride" (SnF_2), whereas others use a sodium fluoride additive. How much heat is generated, per mole of sodium fluoride formed, by reaction of gaseous HF with solid sodium hydroxide? (Water is the other product.)

h. Ammonia, one of the gaseous products of the decomposition of organic wastes, is slowly oxidized by atmospheric O_2 to nitric oxide and water. How much heat is released per mole of ammonia oxidized?

i. In 1953 Stanley Miller reported isolating simple organic molecules produced by an electric discharge through a mixture of gases believed to simulate the atmosphere of the primitive earth. Calculate the standard heat of reaction (in kilocalories per mole of glycine) for the hypothetical formation of glycine, a simple amino acid, from "primitive earth gases."

$$2CH_4(g) + NH_3(g) + 2H_2O(g) \longrightarrow$$
$$H_3\overset{+}{N}CH_2CO_2{}^-(s) + 5H_2(g)$$

(For glycine, $\Delta H_f^0 = -126.3$ kcal mole^{-1}.)

j. Some drugs are eventually metabolized in the body to simple chemical waste products. Calculate the standard heat of reaction (in kilocalories per mole) for the hypothetical decomposition of sulfanilamide, one of the "sulfa drugs."

$$H_2N(C_6H_4)SO_2NH_2 + 7O_2 \longrightarrow$$
$$6CO_2 + 2NH_3 + H_2O + SO_3$$

(For sulfanilamide, $\Delta H_f^0 = -1772$ kcal mole^{-1}.)

Objective (d)

4. Silicon carbide (Carborundum), a commercially important abrasive, is prepared by heating a mixture of coke and sand in an electric furnace. The process may be formulated as

$$SiO_2(s) + 3C(s) \longrightarrow SiC(s) + 2CO(g)$$

Calculate the standard enthalpy change for production of 1.8 kg of Carborundum by this process, using data from Table 5.1 and the following information based on calorimetric data.

Oxidation of silicon
$$Si(s) + O_2(g) \longrightarrow SiO_2(s)$$
$$\Delta H_{oxidation}^0 = -205.1 \text{ kcal mole}^{-1} (SiO_2)$$

Thermal decomposition of silicon carbide:
$$SiC(s) \longrightarrow Si(s) + C(s)$$
$$\Delta H_{decomposition}^0 = +26.7 \text{ kcal mole}^{-1} (SiC)$$

5. Nearly a million tons of urea are produced annually in the United States for agricultural fertilizers. Calculate the standard enthalpy change for the production of 5.0 kg of urea by the commercial process,

$$2NH_3(g) + CO_2(g) \longrightarrow CO(NH_2)_2(s) + H_2O(l)$$

The standard heat of combustion of urea (to nitric oxide, carbon dioxide, and water) is -151.6 kcal mole^{-1} (urea). Other useful data may be obtained from Table 5.1.

6. Acetylene is used as a fuel in metal-cutting torches. Acetylene lanterns, sometimes still employed for outdoor activities, used to find extensive use in mining operations before the advent of electric mining lamps. The combustion of acetylene is a highly exothermic process, resulting in a very hot flame. From the following information and appropriate data from Table 5.1, calculate the standard heat of combustion for acetylene (C_2H_2) and compare it with that found for the complete combustion of benzene (problem 3) on the basis of kilocalories per gram (fuel).

Generation of acetylene from calcium carbide
$$CaC_2(s) + 2H_2O(l) \longrightarrow Ca(OH)_2(s) + C_2H_2(g)$$
$$\Delta H_{reaction}^0 = -30.0 \text{ kcal mole}^{-1} (CaC_2)$$

Preparation of calcium carbide from lime and coke
$$CaO(s) + 3C(s) \longrightarrow CaC_2(s) + CO(g)$$
$$\Delta H_{reaction}^0 = +103.0 \text{ kcal mole}^{-1} (CaC_2)$$

Preparation of "slaked lime"
$$CaO(s) + H_2O(l) \longrightarrow Ca(OH)_2(s)$$
$$\Delta H_{reaction}^0 = -23.1 \text{ kcal mole}^{-1} (CaO)$$

Objective (e)

7. What is the theoretical efficiency of an internal combustion engine having an ignition chamber temperature of 850°C and an exhaust gas temperature of 150°C?

8. What is the limiting refrigeration efficiency for a freezer designed to maintain an internal temperature of -5°C in an environment of 25°C?

OPTIONAL

9. Which of the following processes should be spontaneous at *all* temperatures, assuming that ΔH and ΔS do not change sign as a function of temperature? Explain the reasons for your choice.

a. Process A: $\Delta H > 0$, $\Delta S < 0$
b. Process B: $\Delta H > 0$, $\Delta S > 0$
c. Process C: $\Delta H < 0$, $\Delta S < 0$
d. Process D: $\Delta H < 0$, $\Delta S > 0$

Remember the conventions of constant-pressure and *system* properties.

BRAIN TINGLER

10. The enthalpy changes associated with some processes cannot be determined from readily available calorimetric data. *Average bond energies*, defined in terms of the energy involved in breaking *one chemical bond* (but in units of kcal per *mole*) in a gaseous molecule to form two gaseous products, may be used to estimate heats of reaction when more accurate experimental data are unavailable. Average values must be employed since bond energy varies with the nature of the particular chemical species involved. Using the following values of average bond energies, estimate the standard heat of formation of cyclamic acid, a compound used in manufacturing the artificial "cyclamate" sweeteners. Report your answer as kilocalories per mole (cyclamic acid).

cyclamic
acid

Bond	Average bond energy (as $\Delta H°$)
C—H	99 kcal mole^{-1}
C—C	82 kcal mole^{-1}
C—N	70 kcal mole^{-1}
N—H	93 kcal mole^{-1}
N—S	84 kcal mole^{-1}
S=O	124 kcal mole^{-1}
S—O	69 kcal mole^{-1}
O—H	111 kcal mole^{-1}
O=O (for O_2)	118 kcal mole^{-1}
N≡N	225 kcal mole^{-1}
H—H	104 kcal mole^{-1}

for C(s) $\longrightarrow$ C(g)
$$\Delta H^0 = 171 \text{ kcal mole}^{-1}$$
for S(s) $\longrightarrow$ S(g)
$$\Delta H^0 = 66 \text{ kcal mole}^{-1}$$
for cyclamic acid(s) $\longrightarrow$ (g)
$$\Delta H^0 = 31 \text{ kcal mole}^{-1} \text{ (estimated)}$$

Now use the estimated ΔH_f^0 value, along with appropriate data from Table 5.1 to estimate the "food calorie" value of 1.00 g of cyclamic acid, closely approximated by the standard heat of combustion (in kilocalories) for complete combustion to CO_2, H_2O, NO, and SO_2.

Unit 5 Self-Test

(Refer to Table 5.1 when necessary.)

 1. Matching: Select the term designated by each symbol.
 TERMS: (A) free energy (B) internal energy (C) work (D) heat (E) enthalpy (F) entropy
 a. E; **b.** H; **c.** q; **d.** S; **e.** G; **f.** w

 2. Indicate whether each of the following statements is true (T) or false (F).
 a. Work is a state function.
 b. At 25°C water vapor [$H_2O(g)$] represents a standard state condition.
 c. A reaction for which ΔH is positive is said to be *endothermic*.
 d. The standard heat of formation of $O_2(g)$ is zero.
 e. Thermodynamics can predict the speed at which oxygen and hydrogen will react at 25°C.

 3. Calculate ΔH^0 for the reaction between hydrogen chloride gas and solid sodium hydroxide to form liquid water and solid sodium chloride. Express your answer in kilocalories per mole of HCl.

 4. A student at a southwestern university was rather severely burned in 1971 by a fire involving about 150 g of ethanol. Calculate the standard heat of combustion for ethanol (assuming complete conversion to CO_2 and H_2O *vapor*) and use this to estimate the heat released by the complete combustion of 150 g of ethanol. (Omit any corrections for temperature variation of ΔH.)

 5. Estimate the thermodynamic efficiency limit of a rocket engine using a combustion-chamber temperature of 1300°C and a gas-exhaust propulsion system of 550°C.

OPTIONAL

 6. Which of the following characteristics would best apply to a process more likely to be spontaneous at a low temperature than at a high temperature?
 a. $\Delta H > 0$, $\Delta S > 0$; **b.** $\Delta H > 0$, $\Delta S < 0$; **c.** $\Delta H < 0$, $\Delta S < 0$;
 d. $\Delta H < 0$, $\Delta S > 0$

Overview Test

Section TWO

1. Select the term best associated with each definition, description, symbolism, or specific example.
   TERMS: (A) acid (B) base (C) coefficient (D) combustion (E) endothermic (F) enthalpy (G) entropy (h) exothermic (I) formation (J) free energy (K) internal energy (L) oxidation state (M) product (N) reactant (O) state functions
   a. E
   b. G
   c. H
   d. S
   e. E, G, H, S, but not w
   f. (Underlined) $Ag_2SO_4 + 2NaCl \longrightarrow 2AgCl + Na_2SO_4$
   g. (Underlined) $Ag_2SO_4 + \underline{2NaCl} \longrightarrow 2AgCl + \overline{Na_2SO_4}$
   h. (Underlined) $\underline{Ag_2SO_4} + \underline{2NaCl} \longrightarrow 2AgCl + Na_2SO_4$
   i. HNO_3
   j. NH_3
   k. For H in HNO_3, it is $+I$
   l. $CH_4 + 2O_2 \longrightarrow CO_2 + 2H_2O$
   m. $2Na(s) + Br_2(l) \longrightarrow 2NaBr(s)$
   n. $\Delta H < 0$
   o. $\Delta H > 0$

2. Indicate whether each of the following statements is true (T) or false (F).
   a. The actual yield for a chemical reaction can exceed the theoretical yield only when the limiting reagent is in excess.
   b. The term *anhydrous*, from the Greek "with water," is used to describe substances dissolved in water.
   c. A calorimeter is used for the experimental determination of enthalpy changes associated with chemical or physical processes.
   d. A sample of ammonia having a mass of 17 g contains approximately 6×10^{23} ammonia molecules.
   e. All exothermic processes are spontaneous.
   f. No spontaneous processes are endothermic.
   g. Both nitric acid and hydrochloric acid are classified as "strong," but hydrofluoric acid is "weak."
   h. A 1.0-mmol sample of water has a mass of 0.018 g.
   i. The formula weight of sulfuric acid is 80.

j. The rate of a chemical reaction is not predicted by thermodynamic considerations alone.
k. $\Delta G = \Delta H - T \Delta S$
l. $\Delta G = \Delta H + \Delta w$
m. The thermodynamic efficiency of a process is a state function, independent of temperature.
n. At the defined standard state conditions, ΔH_f^0 is zero for liquid bromine.
o. Neither entropy nor enthalpy change alone can serve as the ultimate factor in determining process spontaneity.

3. Write correct formulas or names, respectively, for
   a. methylamine; b. $CHCl_3$; c. perchloric acid; d. NH_4NO_3; e. iron(II) sulfate; f. H_3PO_4; g. potassium hydroxide; h. C_2H_5OH; i. nitric oxide; j. $Na_2Cr_2O_7$

4. Write both "complete" and "net" forms of qualitative equations (not necessarily balanced) for
   a. the reaction of solid sodium carbonate with aqueous nitric acid to form carbon dioxide gas, aqueous sodium nitrate, and water
   b. the process in which carbon dioxide gas from the atmosphere dissolves in slightly alkaline "hard" water to form a sediment of calcite (Consider this to be a reaction between gaseous carbon dioxide and aqueous calcium hydroxide to produce solid calcium carbonate and liquid water.)
   c. the reaction occurring when "battery acid" is spilled on aluminum foil (Solid aluminum reacting with aqueous sulfuric acid generates hydrogen gas and an aqueous solution of aluminum bisulfate.)

5. Balance the equations for the processes in question 4.

6. Using the balanced equations from question 5, calculate the theoretical mass of
   a. sodium carbonate required for complete reaction with 85 g of nitric acid
   b. calcite formed from reaction of 17.6 mg of calcium hydroxide with a stoichiometric amount of carbon dioxide
   c. sulfuric acid consumed during the generation of 2.8 g of hydrogen by reaction with an appropriate amount of aluminum

7. In a test procedure for the synthesis of methanol by the

gas-phase reaction of carbon monoxide and hydrogen ($CO + 2H_2 \longrightarrow CH_3OH$), a mixture of 25 g of each reactant gas produced 23 g of methanol. For this process

a. What was the limiting reagent?

b. What was the percentage yield?

c. What mass of carbon monoxide, assuming the same reaction efficiency, would be required for production of 50 g of methanol?

8. Given:

Compound	ΔH_f^0 (kcal mole^{-1})
methanol (l)	-57.0
water (l)	-68.3
carbon dioxide (g)	-94.0

Calculate the standard heat of combustion for methanol (to CO_2 and H_2O) in kilocalories per mole (methanol).

9. The slow oxidation of pyrite (FeS_2) associated with coal dust in a mine may eventually build up enough heat to cause an explosion by spontaneous combustion. Calculate the heat released by the oxidation of 1.20 kg of pyrite, according to the equation

$$4FeS_2(s) + 11O_2(g) \longrightarrow 2Fe_2O_3(s) + 8SO_2(g)$$

Combustion of sulfur

$$S(s) + O_2(g) \longrightarrow SO_2(g)$$
$$\Delta H^0_{combustion} = -71.0 \text{ kcal mole}^{-1} \text{ (S)}$$

"Rusting" of iron

$$2Fe(s) + 3H_2O(l) \longrightarrow Fe_2O_3(s) + 3H_2(g)$$
$$\Delta H^0_{reaction} = +4.2 \text{ kcal mole}^{-1} \text{ (Fe)}$$

Formation of "pyrite"

$$Fe(s) + 2S(s) \longrightarrow FeS_2(s)$$
$$\Delta H_f^0 = -42.5 \text{ kcal mole}^{-1} \text{ (Fe)}$$

10. What is the thermodynamic efficiency of a steam engine using incoming steam at $627°C$, exhausted at $112°C$?

OPTIONAL

11. An example of an equation for an oxidation-reduction reaction is

a. $Ca(s) + H_2SO_4(aq) \longrightarrow CaSO_4(s) + H_2(g)$

b. $Ca(OH)_2(s) + H_2SO_4(aq) \longrightarrow CaSO_4(s) + 2H_2O(l)$

c. $Ag^+(aq) + Br^-(aq) \longrightarrow AgBr(s)$

d. $Al(OH)_3(s) + 3H^+(aq) \longrightarrow Al^{3+}(aq) + 3H_2O(l)$

e. none of these

12. What should be the selling price, per gram of methanol, to allow for a profit of 25% (of sale price) on the process described in question 7, assuming a cost of 3.0 cents per gram of hydrogen, 2.2 cents per gram of carbon monoxide, and remaining total costs equivalent to 5.8 cents per gram of methanol?

13. Which characteristics are most likely to be associated with a process found to be spontaneous at high temperatures but nonspontaneous at low temperatures?

a. $\Delta H < 0$, $\Delta S < 0$; **b.** $\Delta H < 0$, $\Delta S > 0$; **c.** $\Delta H > 0$, $\Delta S > 0$;

d. $\Delta H > 0$, $\Delta S < 0$

Section
THREE

CHEMICAL BONDING

Water is one of the simplest of chemical compounds, yet without it no living organism can function. The characteristics of the H_2O molecule make water an excellent solvent for a wide variety of substances, both ionic and covalent, which must be transported via a fluid medium in living systems. Chemical species ranging from simple monatomic ions to complex molecules, when dissolved in water, may be delivered as required to biochemical sites for various processes such as digestion, metabolism, or waste excretion. The properties of water also render it valuable in heat-exchange processes, as in the familiar phenomenon of cooling by evaporation of perspiration. Water may itself participate in biochemical reactions and, again, it is the unique characteristics of the water molecule that determine its role in such reactions as the digestion of proteins, carbohydrates, and fats.

How may we apply our atomic model to develop an understanding of the character of a water molecule or, indeed, of any particular chemical species? The first step in this development requires us to look at ways of describing relatively simple ionic and covalent species in terms of *chemical bonding*. This is the subject of Section Three. Within this discussion we shall find it useful to develop the concept of *electronegativity* as a measurement of the attraction of a bonded atom for the electrons that constitute the chemical bond. The *electronegativity difference* between two bonded atoms will provide us with a convenient way of classifying chemical bonds as *ionic*, *polar covalent*, or *nonpolar covalent*. Later we shall extend our development to more complex species and to the properties associated with *collections* of atoms, ions, and molecules.

CONTENTS

Unit 6

Ionic compounds

Objectives

(a) Be able to classify chemical species as "acids" or "bases" according to the Brønsted-Lowry concepts (Section 6.4).

(b) Be able to write and interpret simple equations illustrating the acid-base properties of amphoteric species (Table 6.4).

(c) Be able to predict the products of simple acid-base reactions and to write balanced equations for the reactions (Example 2).

(d) Be able to use "simple solubility rules (Table 6.5) to predict the products of "ion-exchange" (metathetical) reactions and to write balanced equations for these reactions (Section 6.5).

REVIEW

(e) Be able to apply the principles of stoichiometry (Unit 4) to ionic reactions.

*The term ion comes from a Greek word meaning "to move." It was originally chosen
because ions, charged particles, could be made to move by applying an electric field
to ionic compounds in solution or in the liquid state. The term is, perhaps, appropriate in a larger sense. Ionic compounds have played an important role in industrial–economic movements, in political–sociological movements, and, most recently,
in the movements undertaken to slow down man's destruction of the environment.*

*The economic importance of ionic compounds is significant in modern society
(Table 6.1).[1] The United States alone produces nearly 100 million tons of sodium
compounds annually, and thousands of ionic compounds of other elements are
important industrial products. The economic role of ions is not, however, unique to
the twentieth century. Ionic compounds—ores (impure salts) of silver and copper—played an important part in opening up the American West. Sources of borax
brought wealth from Death Valley beyond the wildest dreams of the early prospectors. In more primitive societies salt was so valuable that it often served as the
medium of exchange, and the salt trader was usually the most prosperous of the
early "traveling salesmen."*

*Wars have been fought for possession of rich sources of ionic compounds.
Deposits of silver ores on lands promised to the American Indian "as long as the sun
shall shine" led to treaty violations and massacres recorded in some of the darkest
pages of history. The Great Salt March led by Mahatma Gandhi in protest against
the British monopoly on salt was, like the Boston Tea Party, a beacon pointing the
way to eventual independence of a nation. In our own times we have seen how the
tremendous political influence of massive industrial and mining interests has often
been able to obtain legal shelters for the exploitation of the land, the water, and the
air.*

*Many ionic compounds formerly of interest only to the chemical industry or the
academic chemist have recently become of public concern. Mercury salts entering
natural waters as industrial waste or in runoff from agricultural lands using mer-
cury-based plant fungicides threaten the fishing industry and worry the housewife
when she opens a can of tuna or a package of frozen swordfish. Phosphates from
soluble fertilizers and detergents, increasing in concentration, favor the growth of
algae and certain types of amoebas in lakes and streams—rapidly making them
unfit for human use. Lead salts from automobiles using leaded gasoline (with
tetraethyl lead) have been linked to brain damage in children living near heavily
traveled freeways.*

*How are ionic compounds formed? What characteristics distinguish them from
the covalent compounds? What types of reactions do they undergo? Why does
production and use of the ionic compounds so essential to society seem to be
inevitably linked to the deterioration of the environment? These and other questions
are the concern of this unit.*

[1] For an interesting look at some aspects of industrial chemistry, see "Salt—A Pillar of the Chemical Industry,"
by James J. Leddy, in the May, 1970, issue of the Journal of Chemical Education (pp. 386–388).

Table 6.1 SOME IONIC COMPOUNDS

COMPOUND	TYPICAL USE	COMPOUND	TYPICAL USE
$AgBr$	used in photographic film (The silver bromide activated by light is selectively reduced in the developing process to free silver, appearing as the "black" in a black-and-white negative.)	$HgCl_2$ (only slightly ionized)	corrosive sublimate, a rodent poison (Like all mercury(II) salts, a cumulative poison to humans, destroying the kidneys.)
$Al(OH)_3$	an "antacid" ingredient in certain preparations for neutralizing "excess stomach acid."	$Hg(CNO)_2$	fulminate of mercury, used in detonators (blasting caps)
$Al_2(OH)_5Cl \cdot 2H_2O$	Aluminum chlorhydroxide, the active ingredient in many antiperspirants	KI	common additive in iodized salt
Na_3AsO_3	a common herbicide used against many types of weeds. Toxic to humans	KO_2	potassium superoxide, used in air "rebreather" systems
$Pb_3(AsO_4)_2$	an insecticide commonly used on certain fruit trees. Toxic to humans		$$(4KO_2 + 2CO_2 \longrightarrow 2K_2CO_3 + 3O_2)$$
$Na_2B_4O_7 \cdot 10H_2O$	borax, used in cleaning preparations, glass making, and in preparation of fireproof fabrics	$KClO_3$	the oxidizer component of matches
$NaBO_3 \cdot 4H_2O$	sodium peroxyborate, for bleaching	MgO	antacid (milk of magnesia, etc.)
BaO_2	a starting material for the preparation of hydrogen peroxide	$MgSO_4 \cdot 7H_2O$	epsom salts, laxative
$BaSO_4$	used in x-ray studies of the gastrointestinal tract	$NaCl$	table salt
CaO	lime (When heated by an oxyhydrogen flame, it emits a brilliant white light—"the limelight.")	$NaOH$	lye, used in certain types of drain cleaners (e.g., Drano)
$CaCl_2$	often sprinkled on icy roads or sidewalks to help melt the ice	$NaOCl$	an ingredient of some bleaching agents (e.g., Clorox)
$CaCO_3$	chalk	NaF	used as the fluoride additive in some toothpastes (Toxic if swallowed in moderate amounts.)
$(CaSO_4)_2 \cdot H_2O$	plaster of paris	$NaHCO_3$	baking soda, also used in dry-powder and "wet" fire extinguishers
$Ca(HSO_3)_2$	in the paper industry, used to dissolve lignins from wood, leaving cellulose fibers	Na_2SiO_3	sodium metasilicate, used with, or as substitute for, phosphate additives in detergents
Fe_2O_3	rust, "rouge"	NH_4NO_3 $(NH_4)_2SO_4$ $(NH_4)_3PO_4$	used in fertilizers
FeS_2	fool's gold (an iron pyrite)	NiO_2	a component of the nickel-cadmium flashlight cell (rechargeable)
Hg_2Cl_2	calomel, a stimulant to bile secretion	PbO_2	cathode reactant in the automobile battery
		Pb_3O_4	red lead, priming paint for iron and steel
		SnF_2	stannous fluoride, an alternative to NaF in fluoride toothpastes
		ZnO	used in various medicinal ointments
		ZnS	light emitter in fluorescent tubes

6.1 PROPERTIES OF IONIC COMPOUNDS

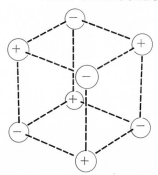

Figure 6.1 Idealized model for a crystalline ionic compound.

Ionic compounds are usually described as collections of discrete ions, even in the solid state (Unit 11). Many of the general properties reflect characteristics associated with the model of a regular three-dimensional array of charged particles (Figure 6.1).[2]

Ionic substances are typically high melting, have very high heats of fusion and vaporization (Unit 12), frequently dissolve in water (but rarely in other common solvents), and act as poor electrical conductors in the solid state. All of these properties are related to the high energy required to move ions from their very stable positions in the electrostatic fields of surrounding ions of opposite charge. Once the ordered arrangements of the solid state have been disrupted, such as by melting or dissolving the compound, ions can move more freely. Under the influence of an external electric field, positive ions drift rapidly to the cathode and negative ions move toward the anode (this is the basis for the names *cation* and *anion*). The moving

[2] As we shall see further in our study, the concept of discrete, spherical ions is a convenient simplification. For a more rigorous treatment of ionic solids, see *Chemical Bonds and Bond Energy*, by R. T. Sanderson (New York: Academic Press, 1971).

charged ions thus constitute an electric current, although different from the flow of *electrons* through a wire. Glass, a complex mixture of polymeric silicate salts, is an excellent electrical insulator—*unless* heat causes the glass to melt, in which case it becomes an effective ionic conductor.

The specific properties, both chemical and physical, of any particular ionic compound depend on the individual component ions and on their interactions with each other. Since ions are fairly "independent" species in solution, it is worth noting that "single-ion" properties are often of importance without regard to the particular compound from which the ions were obtained. If, for example, it is desired to precipitate some silver chloride from a solution containing silver ions, the *source* of the chloride ion is seldom important; any soluble chloride (e.g., NaCl, KCl, NH_4Cl, $BaCl_2$) will do. Similarly, there are many compounds suitable for introducing Mg^{2+} into growing plants (for inclusion in the chlorophyll needed for photosynthesis) or Fe^{2+} into humans (for incorporation into the hemoglobin of the blood).

A primitive tribe recently discovered in the jungles of the Philippines had never even seen salt.

On the other hand, it is often important to consider *all* the components of an ionic compound. Humans require a minimum regular intake of Na^+. Table salt is the most common source, but other compounds may serve instead. The anion associated in the sodium compound does have some obvious limitations—sodium cyanide would be a poor choice. The Ba^{2+} ion is fairly opaque to x-rays and, as such, provides a useful "shadow" to follow in x-ray examination of the gastrointestinal tract, *but* Ba^{2+} is exceedingly poisonous. No *soluble* barium salt could be used, but the insoluble $BaSO_4$ furnishes a concentration of Ba^{2+} well below the toxic limits.

Ions may be colorless. Most are. Others, such as Cu^{2+} (blue) or MnO_4^- (purple), have colors that may be changed by interaction with other ions, even though the latter are colorless. Thus copper(II) acetate is deep green, copper(II) chloride is yellow in the anhydrous form and green when hydrated, copper(II) nitrate is blue, copper(II) sulfite is white, and copper(II) sulfide is black. None of the anions of these salts is colored. The color of the salt results from electronic interactions of the component ions.

Not all ionic compounds are salts. Some acids (Unit 3) are ionic, or become so in aqueous solution (e.g., hydrochloric acid, nitric acid, sulfuric acid). Some others, such as acetic acid, and bases such as ammonia furnish some ions when dissolved in water but are not completely ionized. These "weak" acids and bases are discussed in Unit 19.

The most common test for the ionic character of a substance takes advantage of the movement of ions in an electric field. Substances that conduct an electric current, in the liquid state or in solution, by the movement of ions are called *electrolytes*. It should be noted that all ionic compounds are electrolytes, but not all electrolytes are *inherently* ionic compounds. Hydrogen chloride, for example, is a covalent substance. In aqueous solution, however, it forms ions. The change may be indicated as

The H_3O^+ ion is called the hydronium ion and represents a combination of a proton (H^+) and a water molecule. This is a convenient simplification which will be discussed further in later units.

$$HCl(g) + H_2O(l) \longrightarrow H_3O^+(aq) + Cl^-(aq)$$

We shall, for the present, restrict our considerations to the specific type of ionic compound called a salt, in which the cation is a metal ion or an "ammonium-type" ion (Unit 3).

As we have mentioned in Unit 3, ions are not accurately represented as *isolated* particles in chemical compounds. There is some degree of electron sharing (covalent bonding) between ions of opposite charge in even the simplest salts. We shall develop in Unit 7 the concept of *electronegativity,* which may be conveniently used in classifying compounds as ionic or covalent. On a simple experimental basis, however, we may consider a compound to be ionic if it is a nonconductor of electricity in

Electronegativity may be thought of as the attraction of one atom in a chemical bond for the bonding (shared) electrons.

Figure 6.2 Conductivity test for ionic character. (a) Solid ionic compound cannot conduct current because ions are not free to move. (b) Liquid salt shows ionic conductivity as charged ions migrate toward oppositely charged electrodes. Chemical reactions occur at electrodes (Unit 22).

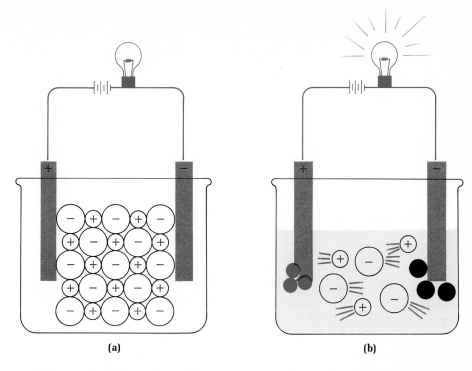

(a) (b)

the solid state, but a good electrical conductor when liquefied (Figure 6.2). This characterization is based on the idea that most solids do not undergo significant chemical change simply on melting, so that the observation of ionic conductivity in the liquid is indicative of the essential ionic character of the solid compound. This classification readily distinguishes ionic solids from metals (which conduct electricity in *both* the solid and liquid states) and further screens out "inherently covalent" species, such as HCl, which form ionic *solutions* only by reaction with an appropriate solvent. There are obvious cases, such as compounds that decompose on melting, not amenable to this simple experimental approach and the high melting points of many inorganic compounds require special techniques for conductivity studies.

A modification of this test is often used in introductory chemistry laboratories by measuring electrical conductivity of aqueous solutions. Ions in solution, like those in molten salts, are free to migrate under the influence of an electric field. Only water-soluble compounds can be investigated by this method, and covalent compounds such as HCl that ionize by reaction with water are not distinguished from "true" ionic compounds.

Because a characteristic of metals is the relatively easy loss of electrons to form positive ions (low ionization potentials, Unit 2) whereas nonmetals tend to form negative ions (Unit 2), we might expect metal-nonmetal compounds to be ionic. Such is, indeed, the case for many of the simpler compounds, particularly those of the lighter representative (A group) metals. When d orbitals become involved, however, as in the transition elements (Unit 27) or in the heavier representative metals (Unit 25), metal-nonmetal bonding assumes greater covalent character. Such compounds as $TiCl_4$ and $SnCl_4$, for example, do not conduct electricity in the liquid state.

IONS IN LIFE PROCESSES

All living systems require certain ionic compounds, usually in rather specific amounts. In humans many simple ionic species play critical roles in a variety of steps in the overall life process.

Carbonate and bicarbonate ions, along with dissolved CO_2, help to maintain the acidity of the blood within the narrow limits (Unit 20) required for healthy function of the body.

Sodium and potassium ions are found in a number of important biological systems, often in very specific concentration ratios (Unit 13). In human blood plasma, for example, sodium ion is present in much higher concentration than potassium ion, whereas their concentration ratio is reversed (i.e., $K^+ > Na^+$) in the fluid of muscle cells.

Many ions are required in only small amounts, often to act in concert with enzyme systems necessary to the complex chemical processes of living organisms (Units 32 and 34).

The detailed description of the functions of ionic compounds in the body is beyond the scope of this text. In fact, many aspects of ionic behavior in biochemical systems are still not fully understood. We shall, however, touch on some of the roles played by ions in the body as examples of concepts developed in later units.

6.2 IONIC CHARGE

An ionic bond refers to the electrostatic attraction between ions of opposite charge. The force between two charges separated by a distance d is given by Coulomb's law as

$$F = k\frac{q_1 \times q_2}{d^2}$$

For a discussion of "unit charge" and electrostatic force, see the Introduction: Toward a Model for the Atom.

For an *isolated* ion, the charge must be some multiple of the unit charge, 1.6×10^{-19} C. Thus all unipositive ions have the same charge, equal—but opposite in sign—to that of any uninegative ion. This might suggest that all ionic bonds of the same *type* (e.g., in LiCl, KI or in CaO, MgS) are of the same strength. Such is *not* the case. An approximation of relative strengths of ionic bonds of the same type can be made by comparisons of calculated enthalpy changes involved in the separation of ions from ion pairs in the gaseous state (Table 6.2).

The poor correlation of "bond strength" with simple electrostatic attraction is consistent with the idea that some degree of covalent bonding must be considered.

It is apparent that absolute values of ionic charges are not useful in explaining differences in enthalpy changes for dissociation of ion pairs. Coulomb's law, in its idealized form, deals with *point charges*. Isolated ions are better thought of as charged *spheres* rather than as point charges. In such cases interactions can best be expressed in terms of *charge density* rather than absolute charge. Although such values do not lend themselves readily to a simple Coulomb's law calculation, they do correlate in a qualitative way (Table 6.2) with trends in interionic attractions. The strengths of interionic forces show a decrease with decreasing charge density. For

Table 6.2 APPROXIMATIONS OF IONIC BOND STRENGTHS

$[MX(g) \longrightarrow M(g)^+ + X(g)^-]$

COMPOUND	CATION CHARGE $(\times 10^{19} C)$	ANION CHARGE $(\times 10^{19} C)$	$\Delta H_{separation}$ (kcal mole^{-1})	CATION CHARGE DENSITY^a $(\times 10^{21} C/Å^3)$	ANION CHARGE DENSITY^a $(\times 10^{21} C/Å^3)$
LiF	+1.6	−1.6	+175	+120	−16
LiCl	+1.6	−1.6	+148	+120	−6.4
LiBr	+1.6	−1.6	+142	+120	−5.1
LiI	+1.6	−1.6	+133	+120	−3.6
NaCl	+1.6	−1.6	+141	+42	−6.4
KI	+1.6	−1.6	+105	+16	−3.6

^{a} Charge density is obtained by dividing the ionic charge, in coulombs, by the ionic volume, in cubic angstroms (Å^3). The latter may be calculated from experimental estimations of ionic radii (Excursion 1).

example, the decreasing bond strengths in the lithium halides (LiF > LiCl > LiBr > LiI) correlates well with the decreasing charge densities (increasing volumes, at constant charge) of the halide ions.

The concept of a unique charge, some multiple of 1.6×10^{-19} C, or of a charge density as calculated for Table 6.2, is a useful simplification in discussing isolated monatomic ions. In collections of ions (ion pairs, liquid salts, ionic solids), in polyatomic ions, or in ions interacting with molecules (e.g., in aqueous salt solutions) the charge of a single ion is dissipated to some extent (Figure 6.3). In such cases the electron regions described as orbitals extend, to a varying degree depending on the nature of the substances involved, into the spheres of influence of additional nuclei. This *delocalization* of charge reduces the charge density significantly below that of an "isolated" ion.

Figure 6.3 Charge dissipation. (a) Isolated cation (charge = 1.6×10^{-19} C). (b) Ionic solid (charge dissipated by interaction with neighboring ions of opposite charge). (c) Acetate: a polyatomic ion (charge delocalized so that no single "atom" has a full unit charge). (d) Hydronium ion: hydrated proton (positive charge delocalized over entire H_3O^+ cluster).

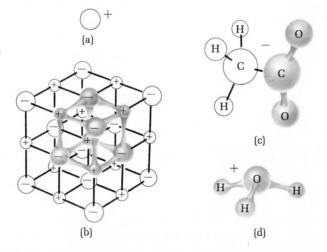

6.3 FORMATION OF IONIC COMPOUNDS

Lithium and fluorine are very reactive elements. If fluorine gas is introduced into a vessel containing some metallic lithium, a violent reaction occurs. The lithium bursts into flame and, when the reaction ceases, a white salt is observed—lithium fluoride. If the reactants are mixed in stoichiometric proportions, they are quantitatively converted to LiF with the evolution of a large amount of heat: about 144 kcal mole^{-1} of product.

It is fairly easy to determine the chemical composition of reaction products and reactants and the enthalpy change for a chemical process. The overall results can be summarized by a chemical equation and a ΔH term. For example, for lithium fluoride,

$$2Li(s) + F_2(g) \longrightarrow 2LiF(s)$$
$$\Delta H^0_{formation} \approx -144 \text{ kcal mole}^{-1}$$

We can describe the *net* process by saying that electrons have been transferred from lithium to fluorine and that the reaction is exothermic. Chemists, however, like to think in terms of behavior at the atomic and molecular level. What specific steps occur at this submicroscopic level to effect this electron transfer? The description of the steps by which a reaction proceeds is called the *reaction mechanism*; this problem is considered in Unit 16. A knowledge of reaction mechanisms is of more than theoretical interest. Frequently there are alternative mechanisms possible for a given reaction. By varying reaction conditions and reactant ratios, chemists can often favor one mechanism over another and thus control a reaction for the production of products of greater purity, for improving yields or reducing costs, or for eliminating or minimizing undesirable thermal (heat) or chemical side effects.

Because enthalpy is a state function (Unit 5), it is not necessary to know the actual mechanism of the reaction in order to offer some plausible suggestions for the origin of the heat of formation. Any logical series of steps for which ΔH values can be

Remember that enthalpy is a state function and, therefore, path independent.

Figure 6.4 Born–Haber cycle:

$2Li(s) + F_2(g) \longrightarrow 2LiF(s)$,
$\Delta H_F^0 = -144 \text{ kcal mole}^{-1}(LiF)$

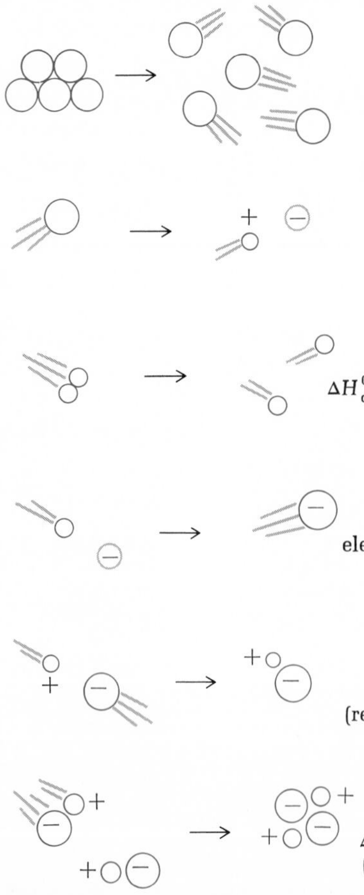

Li (s) $\longrightarrow$ Li (g)
(requires energy)
$\Delta H^0_{sublimation} = +37 \text{ kcal mole}^{-1}$
(see Unit 12)

Li (g) $\longrightarrow$ Li+ (g) $+ e^-$
(requires energy)
ionization potential =
$+124 \text{ kcal mole}^{-1}$
(see Unit 2)

F$_2$ (g) $\longrightarrow$ 2F (g)
(requires energy)
$\Delta H^0_{dissociation} = +36 \text{ kcal mole}^{-1}(\text{of } F_2)$
(see unit 7)

F (g) $+ e^- \longrightarrow$ F$^-$ (g)
(releases energy)
electron affinity $= -81 \text{ kcal mole}^{-1}$
(see Unit 2)

Li+ (g) $+$ F$^-$ (g) $\longrightarrow$ LiF (g)
(releases energy)
$\Delta H^0_{combination} = -175 \text{ kcal mole}^{-1}$
(reverse of ion separation, table 6.2)

n LiF (g) $\longrightarrow$ (LiF) n(s)
(releases energy)
$\Delta H^0_{condensation} = -67 \text{ kcal mole}^{-1}$
(reverse of $\Delta H^0_{sublimation}$, Unit 12)

measured or calculated can be used to offer clues to the net ΔH term. Such a *hypothetical* series of steps is involved in the *Born–Haber cycle* as a plausible explanation for the heat of formation of lithium fluoride (Figure 6.4). It can be seen that the principal contributions to the exothermic character of the reaction are the electron affinity of fluorine, the electrostatic attractions of the Li^+ and F^- ions, and the condensation of LiF sets.

Example 1

Use a Hess's law calculation (Unit 5) with the information from Figure 6.4 to demonstrate that the Born–Haber cycle shown is consistent with a standard heat of formation for lithium fluoride of -144 kcal mole^{-1}.

Solution

Following the method described in Unit 5, chemical equations and associated enthalpy changes are arranged for summation to the net equation and net ΔH required. Unity factors are used to convert all ΔH values to a common unit, in this case "kcal mole^{-1} (LiF)."

$$2Li(s) \longrightarrow 2Li(g) \qquad \Delta H^0 = \frac{+37 \text{ kcal}}{\text{mole (Li)}} \times \frac{2 \text{ mole (Li)}}{2 \text{ mole (LiF)}}$$
$$= +37 \text{ kcal mole}^{-1} \text{ (LiF)}$$

$$2Li(g) \longrightarrow 2Li^+(g) + 2e^- \qquad \Delta H^0 = \frac{+124 \text{ kcal}}{\text{mole (Li)}} \times \frac{2 \text{ mole (Li)}}{2 \text{ mole (LiF)}}$$
$$= +124 \text{ kcal mole}^{-1} \text{ (LiF)}$$

$$F_2(g) \longrightarrow 2F(g) \qquad \Delta H^0 = \frac{+36 \text{ kcal}}{\text{mole (F}_2)} \times \frac{1 \text{ mole (F}_2)}{2 \text{ moles (LiF)}}$$
$$= +18 \text{ kcal mole}^{-1} \text{ (LiF)}$$

$$2F(g) + 2e^- \longrightarrow 2F^-(g) \qquad \Delta H^0 = \frac{-81 \text{ kcal}}{\text{mole (F)}} \times \frac{2 \text{ mole (F)}}{2 \text{ mole (LiF)}}$$
$$= -81 \text{ kcal mole}^{-1} \text{ (LiF)}$$

$$2Li^+(g) + 2F^-(g) \longrightarrow 2LiF(g) \qquad \Delta H^0 = \frac{-175 \text{ kcal}}{\text{mole LiF}} \qquad = -175 \text{ kcal mole}^{-1} \text{ (LiF)}$$

$$2LiF(g) \longrightarrow 2LiF(s) \qquad \Delta H^0 = \frac{-67 \text{ kcal}}{\text{mole LiF}} \qquad = -67 \text{ kcal mole}^{-1} \text{ (LiF)}$$

$$2Li(s) + F_2(g) \longrightarrow 2LiF(s) \qquad \Delta H_f^0 = -144 \text{ kcal mole}^{-1} \text{ (LiF)}$$

6.4 IONIC REACTIONS Most ionic reactions in solution are fairly rapid, especially when they involve the combination of ions of opposite charge. Electrostatic forces, increasing as oppositely charged ions approach each other, help to accelerate the ions so that collisions are usually effective, even when surrounding water molecules must be "forced" out of the way (Figure 6.5).

Oxidation–reduction reactions (Unit 3) involving ions are of importance in electrochemistry, in analytical procedures, and in many industrial processes. "Kitchen chemistry" encounters such reactions on occasion: Bleaches are used because of their

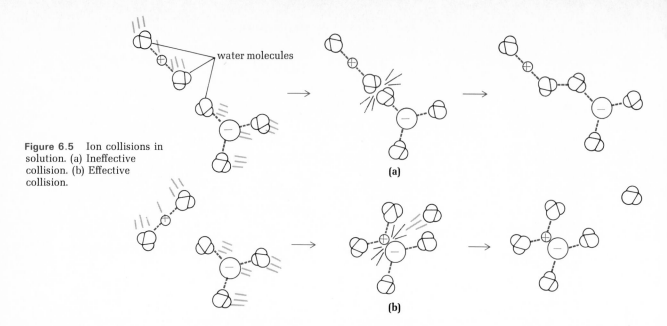

Figure 6.5 Ion collisions in solution. (a) Ineffective collision. (b) Effective collision.

(a)

(b)

Those reactions *not* involving "electron transfer" are classified as "metathetical" (Unit 3).

oxidizing action, and cases of poisoning from the toxic gas produced when bleaches are mixed with "household ammonia" are not uncommon. A detailed discussion of these electron-transfer reactions is found in Units 22–24.

Two types of metathetical reactions are of special significance, *proton exchange* and *ion exchange* (including precipitation reactions).

Proton-exchange reactions

Acids have been defined as species capable of donating protons (H^+) and bases as species capable of accepting protons (Unit 3). Acids or bases may be either ionic or covalent species, and the products of their reactions may be ionic or molecular substances, or a mixture of both (Table 6.3).

Table 6.3 SOME TYPICAL PROTON-EXCHANGE REACTIONS

$$\underset{\text{molecular}}{NH_3(g)} + \underset{\text{molecular}}{HCl(g)} \longrightarrow \underset{\text{ionic}}{NH_4Cl(s)}$$

Ammonia from bottles of aqueous ammonia combines in the gas phase with hydrogen chloride escaping from solutions of hydrochloric acid. The ammonium chloride formed appears as "smoke" or as the white deposit frequently seen on laboratory glassware.

$$\underset{\text{ionic}}{NaOH(s)} + \underset{\text{molecular}}{HCl(g)} \longrightarrow \underset{\text{ionic}}{NaCl\ (s + aq)} + \underset{\text{molecular}}{H_2O\ (l)}$$

Pellets of sodium hydroxide (containing the base, OH^-) are often used in traps for vacuum pumps for removal of acids such as HCl from the atmosphere to avoid damage to metal parts of the pumps.

$$\underset{\text{ionic}}{Ca_3(PO_4)_2(s)} + \underset{\text{partial ionic}^a}{4H_3PO_4(l)} \longrightarrow \underset{\text{ionic}}{3Ca(H_2PO_4)_2(s)}$$

"Triple superphosphate of lime" (calcium dihydrogen phosphate) is a soil additive useful in introducing phosphate into alkaline soil. It is prepared by the action of phosphoric acid on crushed phosphate rock [$Ca_3(PO_4)_2$, containing the base, PO_4^{3-}].

---

[a] Pure phosphoric acid consists of about 90 percent molecular H_3PO_4 and about 10 percent ions, including H^+, $H_2PO_4^-$, HPO_4^{2-}, and PO_4^{3-}. Of these, the last two constitute only minor components. The hydrogen ion (H^+) is associated with molecular phosphoric acid (see Unit 21).

Table 6.4 SOME COMMON IONIC ACIDS AND BASES

ION	FORMULA	REACTION
CATIONIC ACIDS		
ammonium ion	NH_4^+	$NH_4^+ + OH^- \longrightarrow NH_3 + H_2O$
methylammonium ion	$CH_3NH_3^+$	$CH_3NH_3^+ + OH^- \longrightarrow CH_3NH_2 + H_2O$
ANIONIC BASES		
fluoride ion	F^-	$F^- + H^+ \longrightarrow HF$
acetate ion	$CH_3CO_2^-$	$CH_3CO_2^- + H^+ \longrightarrow CH_3CO_2H$
carbonate ion	CO_3^{2-}	$CO_3^{2-} + H^+ \longrightarrow HCO_3^-$
sulfite ion	SO_3^{2-}	$SO_3^{2-} + H^+ \longrightarrow HSO_3^-$
phosphate ion	PO_4^{3-}	$PO_4^{3-} + H^+ \longrightarrow HPO_4^{2-}$
AMPHOTERIC SPECIES (MAY ACCEPT OR DONATE PROTONS)		
hydrazinium ion	$H_2NNH_3^+$	$H_2NNH_3^+ + H^+ \longrightarrow H_3NNH_3^{2+}$
		$H_2NNH_3^+ + OH^- \longrightarrow H_2NNH_2 + H_2O$
bicarbonate ion	HCO_3^-	$HCO_3^- + H^+ \longrightarrow CO_2 + H_2O$
		$HCO_3^- + OH^- \longrightarrow CO_3^{2-} + H_2O$
bisulfite ion	HSO_3^-	$HSO_3^- + H^+ \longrightarrow H_2SO_3$
		$HSO_3^- + OH^- \longrightarrow SO_3^{2-} + H_2O$
hydrogen phosphate ion	HPO_4^{2-}	$HPO_4^{2-} + H^+ \longrightarrow H_2PO_4^-$
		$HPO_4^{2-} + OH^- \longrightarrow PO_4^{3-} + H_2O$
dihydrogen phosphate ion	$H_2PO_4^-$	$H_2PO_4^- + H^+ \longrightarrow H_3PO_4$
		$H_2PO_4^- + OH^- \longrightarrow HPO_4^{2-} + H_2O$

Proton-exchange reactions may involve acids that are molecular [e.g., HCl(g)], anionic (e.g., $H_2PO_4^-$), or cationic (e.g., NH_4^+), furnishing protons to bases that may also be molecular (e.g., NH_3), anionic (e.g., OH^-), or (rarely) cationic (e.g., $H_2NNH_3^+$). The more common acids and molecular bases are listed in Table 3.2. Some ionic acids and bases are shown in Table 6.4. Present terminology considers compounds such as sodium hydroxide as salts that *contain* a base (OH^-). Early definitions (Unit 19) regarded bases as *compounds* that contain hydroxide ion. Fortunately, chemical species continue to function in their normal ways no matter how we choose to describe them, and they are most forgiving of chemists who switch definitions as dictated by the convenience of the situation.

Although we can delay the detailed considerations of acid-base theory for the present (Units 19–21), we need some familiarity with at least a set of "working definitions" of acid-base behavior. The most useful approach for our purposes was developed by Johannes Brønsted and Thomas Lowry in 1923:

An acid is any species *acting* to donate a proton (H^+).
A base is any species *acting* to accept a proton.

Amphoteric species (Table 6.4) may act as either acids or bases, depending on the nature of the other reactant involved.

The reaction of an acid with a base is a *dynamic competition* for the proton. The species that has furnished the proton now becomes a base, seeking (to some extent) to regain the lost proton. The species that has accepted the proton now becomes an acid, seeking (to some extent) to give away the captured proton.

For example, the reaction of hydrofluoric acid with aqueous ammonia may be formulated as

$$HF(aq) \ + \ NH_3(aq) \ \Longleftrightarrow \ NH_4^+(aq) + F^-(aq)$$

"original"   "original"   "new"   "new"
   acid   base   acid   base

The same competitive system would be generated by dissolving solid ammonium fluoride in water, in which case we might choose to write

$$NH_4{}^+(aq) + \quad F^-(aq) \quad \Longleftrightarrow \quad NH_3(aq) + HF(aq)$$

"original"   "original"   "new"   "new"
acid   base   base   acid

In either case, a system eventually develops in which the relative amounts of the different species in the solution reach constant values. Careful studies show that the dynamic exchange of a proton still continues, but the *rate* of the NH_3 + HF reaction has become equal to the *rate* of the $NH_4{}^+$ + F^- reaction, so that no further *net* change is observed. The particular condition characterized by constant composition and dynamic competition is called *chemical equilibrium,* a subject considered in detail in Units 17–21. The double arrows ($\Longleftrightarrow$) used in the preceding equations are used to represent an equilibrium process. We shall continue to use our convention of a single arrow ($\longrightarrow$) for chemical equations not requiring equilibrium considerations.

We need not concern ourselves at this point with equilibrium considerations. The important feature of the Brønsted-Lowry concepts, for our present purposes, is the requirement that acid-base *behavior* is expected only when *both* a proton donor and a proton acceptor are available. We will "pretend" (make a convenient simplification) that acid-base reactions proceed quantitatively, that is, with 100 percent efficiency of proton transfer. This will enable us to predict the products of proton-exchange reactions and to write "complete formula" or "net" equations for such processes. When coupled with solubility information on common salts (Section 6.5), this will allow us to develop quite a variety of "predictable" processes.

It will be useful at this point to review the rules for writing "net" equations, as discussed in Unit 3.

Example 2

Write balanced equations, in both *complete* and *net* forms, for each of the following processes for which a chemical reaction is expected. If no reaction is expected, indicate by NR.

a. Aqueous sodium hydroxide is mixed with hydrochloric acid. No insoluble solids are observed after mixing.

b. Aqueous ammonia is mixed with nitric acid. No insoluble solids are observed after mixing.

c. Aqueous calcium hydroxide is mixed with sulfuric acid, resulting in the formation of an insoluble white salt shown to contain sulfate ion.

d. Aqueous sodium chloride is mixed with nitric acid. No insoluble solids are observed after mixing.

e. Solid calcium carbonate is mixed with aqueous acetic acid. All solid material dissolves and carbon dioxide gas is evolved.

Solution (following methods described in Unit 3)

a. *Complete*

$$NaOH(aq) + HCl(aq) \longrightarrow H_2O(l) + NaCl(aq)$$

         ↑   ↑   ↑

contains the   acid   because no
base OH^- (strong,   insoluble solid
            Unit 3)   was observed

$$OH^- + H^+ \longrightarrow H_2O$$

$$Na^+(aq) + OH^-(aq) + H^+(aq) + Cl^-(aq) \longrightarrow H_2O(l) + Na^+(aq) + Cl^-(aq)$$

Net

$$H^+(aq) + OH^-(aq) \longrightarrow H_2O(l)$$

b. *Complete*

$$NH_3(aq) + HNO_3(aq) \longrightarrow NH_4NO_3(aq)$$

    ↑   ↑   ↑

   base   acid   because no

 (weak,   (strong,   insoluble solid

 Unit 3)   Unit 3)   was observed

$$NH_3 + H^+ \longrightarrow NH_4^+$$

$$NH_3(aq) + H^+(aq) + \cancel{NO_3^-}(aq) \longrightarrow NH_4^+(aq) + \cancel{NO_3^-}(aq)$$

Net

$$NH_3(aq) + H^+(aq) \longrightarrow NH_4^+(aq)$$

c. *Complete*

$$Ca(OH)_2(aq) + H_2SO_4(aq) \longrightarrow 2H_2O(l) + \quad CaSO_4(s)$$

    ↑   ↑   ↑   ↑

contains the   acid   (for   only possible

 base OH⁻   (strong, balance) insoluble salt

          Unit 3)   containing sulfate

$$OH^- + H^+ \longrightarrow H_2O$$

Net

$$Ca^{2+}(aq) + 2OH^-(aq) + H^+(aq) + HSO_4^-(aq) \longrightarrow 2H_2O(l) + CaSO_4(s)$$

d.

$$NaCl(aq) \quad + HNO_3(aq) \longrightarrow ? \text{ (probably NR)}$$

    ↑   ↑

neither acid   acid

nor base, so no

proton exchange

   possible

If Cl⁻ was considered a base, possible reaction would have to be written as

$$NaCl(aq) + HNO_3(aq) \longrightarrow NaNO_3(aq) + HCl(aq)$$

    ↑   ↑   ↑   ↑

 salt   strong   salt   strong

         acid   acid

$$\cancel{Na^+}(aq) + \cancel{Cl^-}(aq) + \cancel{H^+}(aq) + \cancel{NO_3^-}(aq) \longrightarrow$$

$$\cancel{Na^+}(aq) + \cancel{NO_3^-}(aq) + \cancel{H^+}(aq) + \cancel{Cl^-}(aq)$$

all shown as
(aq) because
no insoluble
solid observed

All species cancel, showing *no net* reaction, therefore NR.

e. *Complete*

$$CaCO_3(s) + 2CH_3CO_2H(aq) \longrightarrow CO_2(g) + \quad H_2O(l) \quad + Ca(CH_3CO_2)_2(aq)$$

↑	↑	↑	↑
(for balance)	acid (weak, Unit 3)	(given)	because no solid remains

Net

$$CaCO_3(s) + 2CH_3CO_2H(aq) \longrightarrow$$
$$CO_2(g) + H_2O(l) + Ca^{2+}(aq) + 2CH_3CO_2^-(aq)$$

(Remember, Unit 3, that weak acids are represented by complete formulas.)

Ion-exchange reactions

A more common term for reactions of this type is metathetical.

Some resins use H^+ rather than Na^+. In this case the water leaving the resin is acidic and must be neutralized by passage through an anion-exchange resin containing OH^-. The OH^- and H^+ combine to form water, and the binding sites in the anion-exchange resin are then occupied by the anion from the original "hard" water (e.g., Cl^-).

Precipitation may be thought of in the broad sense of the term ion exchange, since ion environments are changed.

Strictly speaking, any reaction (including proton exchange) that results in ions changing from one chemical environment to another could be called "ion exchange." The term is most commonly used for processes in which certain ions are removed from aqueous solution by replacing other ions held near charged groups that are parts of very large molecules (polymers). This is the basis for the most common type of "water softener" (Figure 6.6).

"Hard" water contains dipositive and, fairly often, tripositive ions leached from mineral deposits in the earth. The most common such ions are Ca^{2+}, Mg^{2+}, Fe^{3+}, and Al^{3+}. These ions form insoluble salts with ordinary soap (usually a sodium salt of a large molecular weight organic acid). These insoluble residues precipitate (e.g., "bathtub ring"), requiring an increased quantity of soap for cleaning purposes. Ion-exchange resins (polymers) can be employed in water softeners (Figure 6.6) installed in household water systems to remove the unwanted cations. The polymeric molecules of the cation-exchange resin contain ionic groups (e.g., $-CO_2^-$, $-SO_3^-$) which exert an electrostatic attraction on cations in solution. The strength of this attraction is a function of the charge density of the *hydrated* cation, that is, of the cation with associated water molecules which help to delocalize the cation charge. Competition for *binding sites* near negative groups in the resin is also a function of relative concentrations. Thus the electrostatic attraction of Mg^{2+} for a negative group may be greater than that of Na^+ but if the solution contains a very large number of sodium ions per unit volume and only a small concentration of magnesium ions, the sodium ions will probably capture most of the available binding sites.

In the operation of the water softener, the binding sites are originally occupied by unipositive ions, usually Na^+. As the hard water enters the resin, competition for binding sites occurs and the 2+ and 3+ ions are favored by their higher charge densities. Sodium ions displaced by this competition move on down the resin, along with anions from the entering water. Two sodium ions are displaced (exchanged) for each 2+ ion completely bound and three for each 3+ ion. The resin continues to function until a large fraction of the available binding sites are occupied by 2+ or 3+ ions. It must then be "recharged" by flushing with a concentrated saline (NaCl) solution. The concentration effects are sufficient to offset the unfavorable electrostatic competition, and the binding sites are recovered by sodium ions.

An alternative method for softening water uses *precipitation,* a different type of ion-exchange process. Unwanted ions are removed from solution in the form of an insoluble solid (precipitate).

"Temporary hardness" is associated with water that contains bicarbonate anions. Boiling favors the conversion of bicarbonate to carbonate and hydroxide ions because of the lower solubility of CO_2 in hot water:

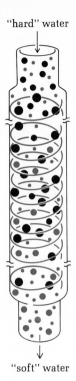

"hard" water

"soft" water

● anion group on polymer (resin)

● 2+ or 3+ ion
(e.g., Mg^{2+}, Ca^{2+}, Fe^{3+}, Al^{3+})

● unipositive ion
(usually Na^+ or H^+)

● mobile anion, unaffected by
the ion-exchange resin
(e.g., Cl^-, OH^-)

Figure 6.6 Ion-exchange
water softener.

$$2HCO_3^- \longrightarrow CO_3^{2-} + \left(CO_2(g)\uparrow\right) + H_2O$$

$$\downarrow H_2O$$

$$HCO_3^- + OH^-$$

When "temporarily hard" water is boiled, the increased concentrations of CO_3^{2-} and OH^- produce insoluble carbonates (e.g., $CaCO_3$) or hydroxides [e.g., $Mg(OH)_2$, $Fe(OH)_3$, $Al(OH)_3$]. On a large scale, such as a metropolitan water facility, boiling is impractical, but the same general effect can be achieved by the addition of carefully calculated amounts of lime water [aqueous $Ca(OH)_2$]. Precipitates formed can be removed by filtration systems.

Water having "permanent hardness" (i.e., HCO_3^- absent) cannot be "softened" by either of these methods. Precipitation of undesirable cations can, however, be effected by addition of appropriate amounts of lime water (to remove insoluble hydroxides) along with sodium carbonate ("washing soda") to remove Ca^{2+}. If the latter is the principal problem, washing soda alone is used to precipitate calcium carbonate. Of course, either "temporarily" or "permanently" hard water could have been "softened" by addition of excess soap to precipitate the insoluble calcium, magnesium, iron, or aluminum "soaps."

Still a third type of possibility exists for treatment of "hard" water. A chemical species may be added to the water to form a strong complex (Unit 33) which, although still in solution, effectively "ties up" the problem ion so that it can no longer react with added soap. Many detergents have used additives such as polyphosphates or silicates for this purpose, as well as to help in solubilizing inorganic materials such as clays.

Table 6.5 SIMPLE SOLUBILITY RULES[a] FOR SOME COMMON SALTS IN WATER

All common salts of ammonium ion and Group IA cations are soluble.
(LiF, Li_2CO_3, Li_3PO_4 only slightly soluble)

All common acetates and nitrates are soluble.
(Ag^+ and Hg_2^{2+} acetates only slightly soluble)

Most halides (chlorides, bromides, iodides) are soluble.
(Ag^+, Hg_2^{2+}, Pb^{2+} halides insoluble)

Most sulfates are soluble.
(Ca^{2+}, Ba^{2+}, Sr^{2+}, Pb^{2+} sulfates insoluble) [Ag_2SO_4 slightly soluble]

Few sulfides are soluble.
(groups IA and IIA, NH_4^+ sulfides soluble)

Very few oxides or hydroxides are soluble.
(group IA, Ba^{2+}, Sr^{2+} hydroxides soluble) [$Ca(OH)_2$ only slightly soluble] (Soluble oxides react with water to form hydroxides.)

Water solubility ↑

[a] "Solubility" is a rather imprecise term. A more quantitative treatment is given in Unit 18.

6.5 PREDICTING PRECIPITATION

We have already seen how net equations can be used to represent chemical processes when the reactants and products are known. It is worthwhile, in addition, to be able to predict whether or not given chemicals may react and if so, the products to be expected from their reaction. A recognition of acidic and basic species and application of the simple solubility "rules" listed in Table 6.5 can be employed to enable us to make predictions for a number of metathetical processes. Once these predictions have been made, the appropriate equations descriptive of the processes can be written (Unit 3).

It would be nice to be able to use some simple theoretical considerations to predict which salts are soluble in water and which would precipitate (be insoluble) if formed in an aqueous medium. However, there are too many competing factors involved in salt solubility (described in Units 13 and 18) to permit the use of any generally applicable simple model. Free energy change (Unit 5) is our criterion of spontaneity and we can identify both enthalpy and entropy factors associated with the dissolving of a salt in water. For example, energy is required to remove ions from the regular arrays of a crystalline salt (an endothermic process) and the resulting loss of "order" is associated with an entropy increase. Energy is released by the exothermic process of ion hydration (association of dissolved ions with water molecules). Other entropy factors are also involved. The concentration dependence of ΔG depends on the relative magnitudes of endothermic and exothermic parts of the overall process and on various contributions to the net entropy change. These factors cannot be reliably estimated on any simple basis and although some of the energy contributions can be determined experimentally, it is much easier to rely on experimental determinations of solubility in terms of the quantity of a salt dissolved by a particular volume of water. On this basis, compounds are generally considered to be *soluble* (Table 6.5) if they will dissolve to the extent of about 1 g or more per liter of water.

For our present purposes we will use the "simple solubility rules" of Table 6.5 to predict which salts should be indicated as solids (s) in chemical equations and which should be shown as aqueous (aq). Only those salts specified in the table as soluble should be indicated by (aq); others should be indicated by (s). It will be helpful to learn the information in Table 6.5, although you may wish to use the table of *solubility products* in Appendix C as a convenient reference source, since this table lists most of the more common *insoluble* salts. That is, most of the simple salts *not* listed in this table are water soluble.

To predict reaction products, we will simply "swap partners" in the reactant species by exchanging cation (H^+, NH_4^+, or metal ion) and anion combinations (as illustrated in Example 3) to formulate possible reaction products. *If this process results in the formation of new insoluble salts (or gases), the dissolving of solid reactants, or an acid-base reaction (Example 2), we will consider the resulting equation to represent a plausible reaction.* Another way of looking at it is to consider a possible net equation resulting from "exchange of ions". If the proper formulations of reactants and proposed products "cancel" (Unit 3), no reaction would be anticipated.

Example 3

Write balanced equations, in both "complete" and "net" forms for each of the following processes for which a chemical reaction is expected. If no reaction is expected, indicate by NR.

a. Aqueous ammonium chloride is mixed with aqueous silver nitrate.
b. Aqueous potassium hydroxide is mixed with aqueous copper(II) sulfate.
c. Aqueous sodium sulfide is mixed with aqueous magnesium acetate.
d. Solid zinc sulfide is mixed with hydrochloric acid. Hydrogen sulfide gas is evolved.
e. Solid lead carbonate is mixed with aqueous sulfuric acid. Carbon dioxide gas is evolved. The only ionic product contains sulfate.

Solution

a. Ammonium ion may be recognized as an acid (Table 6.4), but the mixture described contains no base capable of participating in a proton exchange. Exchange of other ions would result in

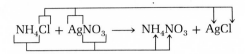

$$NH_4Cl + AgNO_3 \longrightarrow NH_4NO_3 + AgCl$$

All nitrates are water soluble, but silver chloride is insoluble (Table 6.5), so a reaction is expected.

Complete

$$NH_4Cl(aq) + AgNO_3(aq) \longrightarrow NH_4NO_3(aq) + AgCl(s)$$

$$NH_4^+(aq) + Cl^-(aq) + Ag^+(aq) + NO_3^-(aq) \longrightarrow$$
$$NH_4^+(aq) + NO_3^-(aq) + AgCl(s)$$

Net

$$Cl^-(aq) + Ag^+(aq) \longrightarrow AgCl(s)$$

b. Potassium hydroxide contains the base OH^-, but no acids are present. Exchange of ions would result in

$$KOH + Cu(NO_3)_2 \longrightarrow KNO_3 + Cu(OH)_2$$

Copper(II) hydroxide is insoluble in water (Table 6.5), so a reaction is expected.

Complete

$$2KOH(aq) + Cu(NO_3)_2(aq) \longrightarrow 2KNO_3(aq) + Cu(OH)_2(s)$$
$$\uparrow \qquad\qquad\qquad\qquad \uparrow$$
(for balance)   (for balance)

$$2K^+(aq) + 2OH^-(aq) + Cu^{2+}(aq) + 2NO_3^-(aq) \longrightarrow$$
$$2K^+(aq) + 2NO_3^-(aq) + Cu(OH)_2(s)$$

Net

$$2OH^-(aq) + Cu^{2+}(aq) \longrightarrow Cu(OH)_2(s)$$

c. No acid is present in the reactant mixture, so only other ions might be exchanged:

$$Na_2S + Mg(CH_3CO_2)_2 \longrightarrow 2NaCH_3CO_2 + MgS$$

All acetates are soluble and magnesium (a IIA element) forms a water-soluble sulfide, so no reaction is expected. This is confirmed by attempting to develop a net equation, in which all species are found to cancel:

$$2Na^+(aq) + S^{2-}(aq) + Mg^{2+}(aq) + 2CH_3CO_2^-(aq) \longrightarrow$$
$$2Na^+(aq) + 2CH_3CO_2^-(aq) + Mg^{2+}(aq) + S^{2-}(aq)$$

NR.

d. The information that hydrogen sulfide is evolved suggests an acid-base reaction in which sulfide acts as the proton acceptor.[3] Exchange of other ions would give

$$ZnS + HCl \longrightarrow ZnCl_2 + H_2S$$

We know that the original ZnS was solid and the H_2S was formed as a gas. Table 6.5 suggests that $ZnCl_2$ is soluble in water, so

Complete

$$ZnS(s) + 2HCl(aq) \longrightarrow ZnCl_2(aq) + H_2S(g)$$
$$\uparrow$$
$$\text{(for}$$
$$\text{balance)}$$

$$ZnS(s) + 2H^+(aq) + 2\cancel{Cl}^-(aq) \longrightarrow Zn^{2+}(aq) + 2\cancel{Cl}^-(aq) + H_2S(g)$$

Net

$$ZnS(s) + 2H^+(aq) \longrightarrow Zn^{2+}(aq) + H_2S(g)$$

e. The formation of CO_2 gas results from a two-step acid-base reaction (Table 6.4)

$$H^+ + CO_3^- \longrightarrow HCO_3^-$$
$$H^+ + HCO_3^- \longrightarrow H_2O + CO_2$$

Other ion exchange (noting that sulfate, rather than bisulfate, results) would give

$$PbCO_3 + H_2SO_4 \longrightarrow PbSO_4 + H_2O + CO_2$$

Lead sulfate is insoluble in water (Table 6.4):
Complete

$$PbCO_3(s) + H_2SO_4(aq) \longrightarrow PbSO_4(s) + H_2O(l) + CO_2(g)$$

Net

$$PbCO_3(s) + \underbrace{H^+(aq) + HSO_4^-(aq)} \longrightarrow PbSO_4(s) + H_2O(l) + CO_2(g)$$

(as described in Unit 3)

SUGGESTED READING

Benfey, O. T. 1963. *Classics in the Theory of Chemical Bonding.* New York: Dover.
Giddings, J. C., and M. Monroe. 1972. *Our Chemical Environment.* San Francisco: Canfield Press.
Lagowski, J. J. 1966. *The Chemical Bond.* Boston: Houghton Mifflin.
Robins, J. 1972. *Ions in Solution.* New York: Oxford University Press.
Sanderson, R. T. 1971. *Chemical Bonds and Bond Energy.* New York: Academic Press. (An advanced treatise, but highly recommended for those interested in an innovative approach to bonding.)

EXERCISES

1. Make flash cards and use them to help you learn the common ionic acids (as aqueous solutions), bases, and amphoteric species (Table 6.4) and the simple solubility rules (Table 6.5).

2. Identify the reactants in the following equations as *acid* or *base,* according to the Brønsted-Lowry system.

a. $HF(aq) + OH^-(aq) \longrightarrow H_2O(l) + F^-(aq)$
b. $NH_3(g) + HCl(g) \longrightarrow NH_4Cl(s)$
c. $NH_4^+(aq) + CN^-(aq) \longrightarrow NH_3(aq) + HCN(aq)$

d. $CO_3^{2-}(aq) + CH_3CO_2H(aq) \longrightarrow$
$$HCO_3^-(aq) + CH_3CO_2^-(aq)$$
e. $HSO_4^-(aq) + HPO_4^{2-}(aq) \longrightarrow SO_4^{2-}(aq) + H_2PO_4^-(aq)$

3. Write equations, in *net* form, for any reactions expected of the following with $H^+(aq)$ and with $OH^-(aq)$. If no reaction is expected, indicate by NR. Identify the amphoteric species.

a. $HCO_3^-(aq)$; **b.** $NH_4^+(aq)$; **c.** $H_2PO_4^-(aq)$; **d.** $H_2NNH_3^+(aq)$

(*Exercises continue on p. 170.*)

[3]Note that the formation of H_2S from HCl and S^{2-} illustrates that a strong acid may furnish a proton to the anion of a weaker acid.

IONIC COMPOUNDS AND THE ENVIRONMENT

Nature has many ways of destroying waste materials, including many of the common inorganic compounds. Most soaps and "biodegradable" detergents are destroyed by microorganisms in natural waters. Common ammonium salts can be utilized by plants of various types, including algae. Even such toxic species as Hg^{2+} and CN^- are metabolized by some organisms. Iron and steel cans oxidize on exposure to moist air, eventually forming oxide dusts that reenter the soil. (Aluminum cans, incidentally, are quite resistant to this process after formation of a fairly shallow surface oxide film.)

The problem is not that inorganic chemicals per se are a serious danger to the environment. The problem is more subtle than that. It is partly concerned with the *rate* at which society "dumps" chemicals on nature as compared to the rate at which natural processes can "eliminate" these pollutants.

Consider the following experiment (but do not try it past the third run!).

STEP 1 Eat one marshmallow per minute for 5 minutes.
STEP 2 Eat two marshmallows per minute for 5 minutes.
STEP 3 Eat three marshmallows per minute for 5 minutes.
STEP 4 Eat 10 marshmallows per minute for 5 minutes.
STEP 5 Eat 100 marshmallows per minute for 5 minutes.

Somewhere around step 3, you will discover that the marshmallows become available faster than they can be consumed.

In terms of inorganic chemical wastes and the natural environment, we are now around step 4.

The second part of the problem is concerned with the nature of the waste materials and the locations for their "dumping." If the wastes are "degradable," they can be dumped anywhere that has conditions suitable for the particular degradation (keeping in mind the *rate* problem). If the wastes are *not* degradable, then the location for their dumping must be one where waste accumulation poses no present *or future* problem.

Consider another analogy. You have the residues from a children's birthday party—cake crumbs and paper plates. You can toss the crumbs outside for the birds (if they can consume them before ants and flies are attracted), *but* it would not be very useful to toss the paper plates out for the birds. Those you can place in the trash can or pile on the kitchen table. The latter is a poor choice if you or anyone else is liable to object to a messy table.

You do not have to drive down many highways or visit more than a few rivers and lakes to see what choices we have been making in the analogous situation of the wastes, inorganic and organic, from our technological "party."

4. Write balanced net equations for the systems indicated. Predict reaction products on the basis of probable proton exchange or precipitation of insoluble salts. If no reaction is expected, indicate by NR.
   a. $CH_3CO_2H(aq) + OH^-(aq) \longrightarrow$
   b. aqueous sodium carbonate + hydrochloric acid $\longrightarrow$
   c. $BaCl_2(aq) + H_2SO_4(l) \longrightarrow$
   d. aqueous ammonia + aqueous nitric acid $\longrightarrow$
   e. $AgNO_3(aq) + Na_2SO_4(aq) \longrightarrow$
   f. aqueous sodium bicarbonate
      $+$ aqueous sodium hydroxide $\longrightarrow$
   g. $AgNO_3(aq) + HBr(aq) \longrightarrow$
   h. aqueous calcium chloride + aqueous nitric acid $\longrightarrow$
   i. $Sr(OH)_2(aq) + K_2SO_4(aq) \longrightarrow$
   j. solid barium sulfite + sulfuric acid $\longrightarrow SO_2(g) + \cdots$

5. What weight of phosphoric acid is required for formation of 5.0 tons of calcium dihydrogen phosphate from excess calcium phosphate? (See Table 6.3.)

OPTIONAL

6. Estimate ΔH for separation of the ions in the gaseous ion pair NaCl from the following data:

   $\Delta H^0_{formation}$ for $NaCl(s) = -99$ kcal mole^{-1}
   $\Delta H^0_{sublimation}$ for $Na(s) = +26$ kcal mole^{-1}
   $\Delta H^0_{dissociation}$ for $Cl_2(g) = +58$ kcal mole^{-1} [of Cl_2]
   ionization potential for $Na(g) = +118$ kcal mole^{-1}
   electron affinity for $Cl(g) = -83$ kcal mole^{-1}
   $\Delta H^0_{sublimation}$ for $NaCl(s) = +54$ kcal mole^{-1}

   What does this calculation suggest about the relative "strengths" of the NaCl and LiF bonds? (See Table 6.2.)

PRACTICE FOR PROFICIENCY

Objective (a)

1. Identify the reactants in the following equations as *acid* or *base* according to the Brønsted-Lowry system.
   a. $CH_3NH_2(g) + HBr(g) \longrightarrow CH_3NH_3Br(s)$
   b. $H_2PO_4^-(aq) + NH_3(g) \longrightarrow HPO_4^{2-}(aq) + NH_4^+(aq)$
   c. $H_2PO_4^-(aq) + H_2SO_3(aq) \longrightarrow H_3PO_4(aq) + HSO_3^-(aq)$
   d. $H_2O(l) + F^-(aq) \longrightarrow OH^-(aq) + HF(aq)$
   e. $H_2O(l) + HF(aq) \longrightarrow H_3O^+(aq) + F^-(aq)$
   f. $CO_3^{2-}(aq) + HCN(aq) \longrightarrow HCO_3^-(aq) + CN^-(aq)$
   g. $HCO_3^-(aq) + HSO_4^-(aq) \longrightarrow$
      $$H_2O(l) + CO_2(g) + SO_4^{2-}(aq)$$
   h. $HSO_3^-(aq) + HSO_4^-(aq) \longrightarrow H_2SO_3(aq) + SO_4^{2-}(aq)$
   i. $HSO_3^-(aq) + NH_3(aq) \longrightarrow SO_3^{2-}(aq) + NH_4^+(aq)$
   j. $NH_4^+(aq) + CO_3^{2-}(aq) \longrightarrow NH_3(aq) + HCO_3^-(aq)$

Objective (b)

2. Identify three different amphoteric species on the basis of reactions shown in question 1.

Objective (c)

3. Predict the products and write both *complete* and *net* balanced equations for all expected reactions. If no reaction is expected, indicate by NR. (No insoluble salts are formed.)
   a. $Ba(OH)_2(aq) + HNO_3(aq) \longrightarrow$
   b. $K_2HPO_4(aq) + NaOH(aq) \longrightarrow$
   c. $MgSO_4(aq) + CuBr_2(aq) \longrightarrow$
   d. $MgSO_4(aq) + HBr(aq) \longrightarrow$
   e. $(NH_4)_2CO_3(aq) + H_2SO_4(aq) \longrightarrow$
   f. $NH_4NO_3(aq) + LiOH(aq) \longrightarrow$
   g. $NH_3(aq) + CH_3CO_2H(aq) \longrightarrow$
   h. $CaCO_3(s) + KOH(aq) \longrightarrow$
   i. $BaSO_3(s) + HCl(aq) \longrightarrow$
   j. $CuO(s) + HNO_3(aq) \longrightarrow$

Objective (d)

4. Predict the products and write both *complete* and *net* balanced equations for all expected reactions. If no reaction is expected, indicate by NR. (You may wish to refer to the table of solubility products in Appendix C, or to Table 6.5.)
   a. $CdCl_2(aq) + H_2S(aq) \longrightarrow$
   b. $Na_2SO_4(aq) + Pb(NO_3)_2(aq) \longrightarrow$
   c. $FeSO_4(aq) + Ba(OH)_2(aq) \longrightarrow$
   d. $CuBr_2(aq) + AgNO_3(aq) \longrightarrow$
   e. $Zn(NO_3)_2(aq) + CH_3CO_2H(aq) \longrightarrow$
   f. aqueous calcium chloride
      $+$ aqueous sodium carbonate $\longrightarrow$
   g. aqueous ammonium nitrate
      $+$ aqueous sodium hydroxide $\longrightarrow$
   h. solid magnesium carbonate + aqueous sulfuric acid $\longrightarrow$
   i. solid iron(III) oxide + hydrochloric acid $\longrightarrow$
   j. aqueous calcium chloride + aqueous acetic acid $\longrightarrow$
   (Would precipitation be expected from mixing "vinegar" with "hard water"?)

Objective (e)

5. A commercially valuable pigment, cadmium yellow, may be prepared by the method shown in question 4a. How much cadmium chloride would be required for the production of 25 kg of cadmium yellow using excess hydrogen sulfide and assuming 100% efficiency?

6. Rust may be dissolved by hydrochloric acid according to the equation for question 4i. How many grams of HCl would be required to dissolve 18 g of rust, assuming 85% efficiency?

7. How much washing soda (sodium carbonate) would be required to "soften" 75 liters of water having a "hardness" of 37 ppm (mg per liter) of Ca^{2+} ion, assuming 100% efficiency? (See question 4f.)

OPTIONAL

8. Calculate ΔH^0 for separation of the ions in the gaseous ion pair NaI from the following data:

   $\Delta H^0_{formation}$ for $NaI(s) = -69$ kcal mole^{-1}
   $\Delta H^0_{sublimation}$ for $Na(s) = +26$ kcal mole^{-1}
   $\Delta H^0_{dissociation}$ for $I_2(g) = +36$ kcal mole^{-1} [per mole I_2]

$\Delta H^0_{\text{sublimation}}$ for $I_2(s) = +15$ kcal mole^{-1}
ionization potential for Na(g) $= +118$ kcal mole^{-1}
electron affinity for I(g) $= -71$ kcal mole^{-1}
$\Delta H^0_{\text{sublimation}}$ for NaI(s) $= +48$ kcal mole^{-1}

9. Using Table 5.1 (Unit 5) calculate the standard heat of reaction in kilocalories mole^{-1} (NaCl) for

$$Na_2CO_3(s) + 2HCl(g) \longrightarrow 2NaCl(s) + CO_2(g) + H_2O(g)$$

BRAIN TINGLER

10. *Electrolytes* may be defined as compounds that can conduct electricity *either* as the pure liquid *or* when dissolved in water.
    a. Discuss the statement, "Although all ionic compounds are electrolytes, not all electrolytes are inherently ionic."
    b. On the basis of the following information, identify all electrolytes and all ionic compounds among the substances described.

Substance (or mixture of substances)	Good electrical conductivity?
water (l)	no
sulfur dioxide (l)	no
carbon tetrachloride (l)	no
sodium iodide dissolved in $SO_2(l)$	yes
sodium iodide dissolved in $H_2O(l)$	yes
hydrogen iodide dissolved in $H_2O(l)$	yes
hydrogen iodide dissolved in $SO_2(l)$	yes
hydrogen iodide dissolved in $CCl_4(l)$	no
iron(III) chloride dissolved in $H_2O(l)$	yes
iron(III) chloride dissolved in $SO_2(l)$	no
urea [$CO(NH_2)_2$] dissolved in $H_2O(l)$	no

Unit 6 Self-Test

1. Identify the reactants in the following equations as *acid* or *base* according to the Brønsted-Lowry system.
   a. $HPO_4^{2-}(aq) + OH^-(aq) \longrightarrow PO_4^{3-}(aq) + H_2O(l)$
   b. $HPO_4^{2-}(aq) + HSO_4^-(aq) \longrightarrow H_2PO_4^-(aq) + SO_4^{2-}(aq)$
   c. $NH_3(aq) + HCO_3^-(aq) \longrightarrow NH_4^+(aq) + CO_3^{2-}(aq)$
   d. $CH_3CO_2H(aq) + HCO_3^-(aq) \longrightarrow$
   $\qquad\qquad\qquad CH_3CO_2^-(aq) + H_2O(l) + CO_2(g)$
   e. $CH_3CO_2H(g) + NH_3(g) \longrightarrow NH_4CH_3CO_2(s)$
2. Identify two amphoteric species shown in the equations for question 1.
3. Predict the products and write both *complete* and *net* balanced equations for all expected reactions. If no reactions are expected, indicate by NR. (You may wish to consult the table of solubility products in Appendix C.)
   a. aqueous ammonia + aqueous potassium bisulfate $\longrightarrow$
   b. $Al_2O_3(s) + HNO_3(aq) \longrightarrow$
   c. aqueous sodium carbonate
   $\qquad$ + aqueous strontium chloride $\longrightarrow$
   d. $CuCl_2(aq) + MgSO_4(aq) \longrightarrow$
   e. solid lead carbonate + aqueous sulfuric acid $\longrightarrow$
4. Zinc sulfide can be used as the fluorescent coating for cathode ray tubes. The white salt is precipitated if hydrogen sulfide gas is bubbled through an aqueous solution of zinc nitrate. What mass of zinc nitrate would be needed for the preparation of 26 g of zinc sulfide, assuming use of excess hydrogen sulfide under conditions corresponding to an 83% yield?

OPTIONAL

5. Calculate the standard enthalpy change for the reaction between aluminum oxide and nitric acid in kcal mole^{-1} (Al_2O_3). (See question 3b.)

Process	ΔH^0
Burning of aluminum	
$4Al(s) + 3O_2(g) \longrightarrow 2Al_2O_3(s)$	-199.5 kcal mole^{-1} (Al)
Reaction of aluminum with nitric acid	
$Al(s) + 4HNO_3(aq) \longrightarrow$ $Al(NO_3)_3(aq) + 2H_2O(l) + NO(g)$	-191.2 kcal mole^{-1} (Al)
Oxidation of nitric oxide	
$2NO(g) + O_2(g) \longrightarrow 2NO_2(g)$	-13.5 kcal mole^{-1} (NO)
Reaction of nitrogen dioxide with water	
$3NO_2(g) + H_2O(l) \longrightarrow$ $2HNO_3(aq) + NO(g)$	-11.0 kcal mole^{-1} (NO_2)

Covalent compounds

Atoms of various shapes, moving and combining in various ways, fall at last into certain arrangements of which the world of things is created.

De rerum natura
LUCRETIUS (first century B.C.)
[as translated in *Scientific American,* May 1970, page 117]

Objectives

(a) Be able to apply the principles of electron configuration (Unit 1) to the selection of appropriate atomic orbitals for use in simple molecular orbital descriptions of covalent bonds (Section 7.3).

(b) Be able to write electron-dot and line-bond formulations for simple molecules and polyatomic ions (Section 7.5) and to identify σ (*sigma*) and π (*pi*) bonds (Sections 7.4 and 7.5).

(c) Be able to use *electronegativities* for classification of chemical bonds as ionic, polar covalent, or nonpolar covalent (Section 7.6).

REVIEW

(d) Be able to apply the principles of stoichiometry (Unit 4) to reactions involving covalent compounds.

(e) Be able to calculate standard enthalpy changes (Unit 5) for processes involving covalent compounds.

Be able to draw complete sets of *resonance structures* (Example 6).

Covalent compounds, especially those of carbon, are among the most useful and interesting of all chemical substances (Table 7.1). In early societies, covalent compounds for medicines, dyes, cleaning agents, or fibers for making cloth were usually obtained from plant or animal sources. Today much of the chemical industry is involved in the production of synthetic compounds to supplement or replace those from natural sources—pigments, plastics, detergents, pharmaceuticals, and so on.

Table 7.1 SOME INTERESTING COVALENT COMPOUNDS

COMPOUND	COMMENT
hydrogen H_2	Formed by electrical decomposition of water or by action of acids (or water) on appropriate metals. Proper mixtures with oxygen are highly explosive. Accidents from such explosions are not uncommon among motorists checking their batteries or radiator water levels after dark by "matchlight."
carbon monoxide CO	A poisonous gas formed during incomplete combustion of hydrocarbon fuels. This substance complexes with the iron in hemoglobin, preventing its function as an oxygen carrier in the blood. A large-scale industrial process for preparing pure nickel uses the reaction $$Ni + 4CO(g) \longrightarrow Ni(CO)_4(g) \xrightarrow{\text{heat}} Ni + 4CO(g)$$ (impure) (pure) (pure)
carbon dioxide CO_2	One of the products of oxidation of most organic compounds; absorbed by plants for use in photosynthesis. This process is the best available for CO_2 removal and replenishment of atmospheric oxygen.
ammonia NH_3	In aqueous solution, used as "household ammonia" for cleaning purposes. Ammonia and its compounds are important "agrichemicals" for adding nitrogen to soils.
Acetylene $H—C\equiv C—H$ (C_2H_2)	Formed by reaction of calcium carbide (CaC_2) with water—as in the older mining lamps and carbide lanterns. Used in welding operations requiring high ignition temperatures.
Freon-12 $\overset{\displaystyle Cl}{\underset{\displaystyle F}{Cl—\overset{\mid}{\underset{\mid}{C}}—F}}$ (CF_2Cl_2)	Nontoxic gas used in refrigeration and air-conditioning systems.
ethanol $H—\overset{\displaystyle H}{\underset{\displaystyle H}{\overset{\mid}{\underset{\mid}{C}}}}—\overset{\displaystyle H}{\underset{\displaystyle H}{\overset{\mid}{\underset{\mid}{C}}}}—O—H$ (C_2H_5OH)	Often used to provide the "punch" for beverages served at various parties. An ingredient of gin, vodka, bourbon, rum, and various other "medicinal" drinks.
lactic acid $CH_3\overset{\displaystyle H}{\underset{\displaystyle OH}{\overset{\mid}{\underset{\mid}{C}}}}—CO_2H$	Formed when milk sours, this acid acts to destroy colloidal particles in milk, causing precipitation (curdling).

Table 7.1 (Continued)

COMPOUND	COMMENT

sucrose

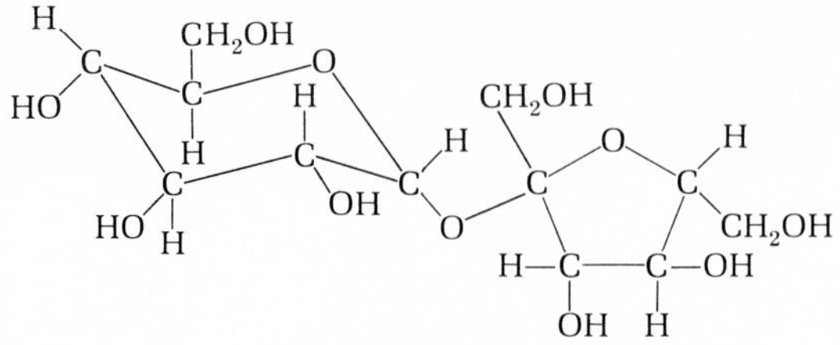

Ordinary table sugar.

$(C_{12}H_{22}O_{11})$

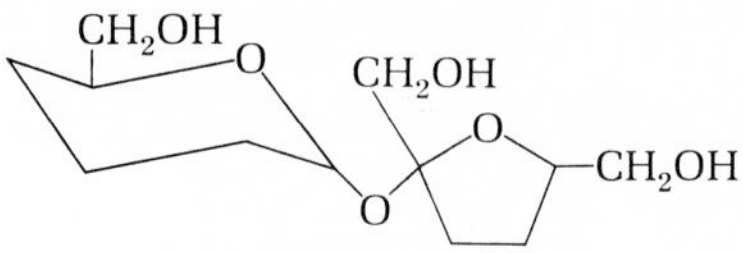

shorthand formulation

vitamin C (ascorbic acid)

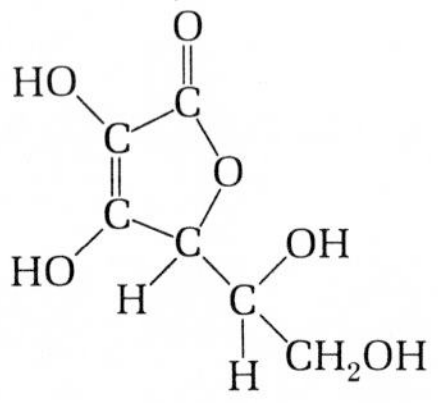

Found in many fresh vegetables and citrus fruits. A deficiency of this vitamin in the diet leads to scurvy, a debilitating diesase common to seamen in the days of the sailing ships. Involved in Linus Pauling's controversial suggestions (1970) for treatment and prevention of the common cold.

The hexagonal formula

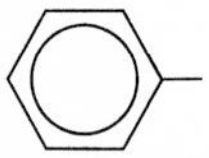

is "chemical shorthand" (Unit 8) to represent

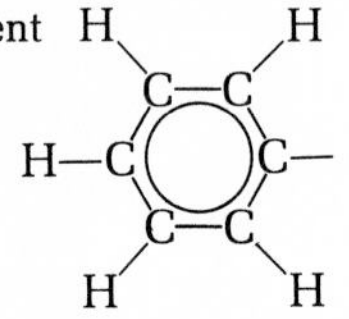

phenylalanine

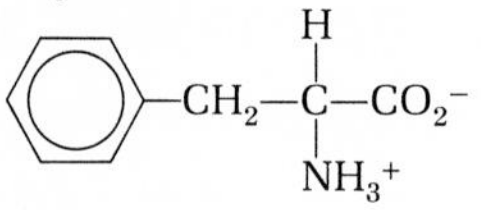

One of the first chemicals clearly linked to a "molecular disease," phenylketonuria, a cause of mental retardation in children. If detected early enough (by a very simple chemical test), the disease can be prevented by using a diet low in phenyl-alanine-containing proteins.

DDT

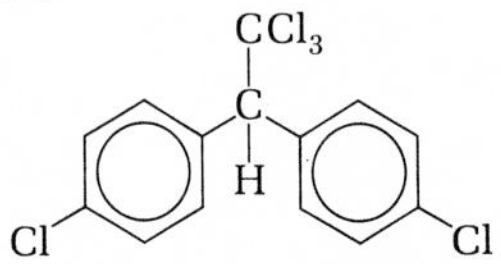

A very potent insecticide whose long residual life poses serious environmental problems, but whose continued use may be essential to insect control until satisfactory alternatives are developed.

aspirin (acetylsalicylic acid)

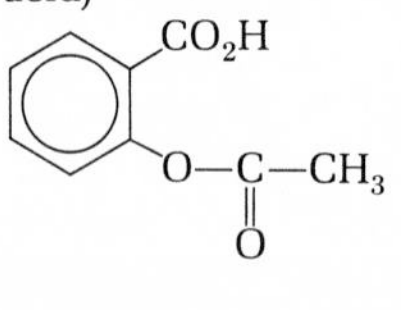

The most common treatment for those suffering from too much study of chemistry—or other "headaches." Commonly used as the sodium or calcium salt to increase solubility.

sulfathiazole

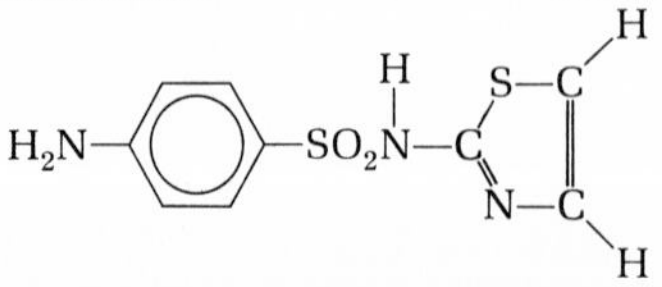

One of the more common "sulfa" drugs.

Table 7.1 (Continued)

COMPOUND	COMMENT
5-fluorouracil 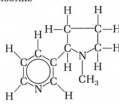	One of the earliest chemotherapeutic agents found effective against certain types of cancer.
nicotine 	The principal alkaloid in tobacco. In small quantities it acts as a stimulant and increases blood pressure. In higher concentrations it is a deadly poison and is a component of some insecticides.
3,4-benzpyrene	A very potent carcinogenic (cancer-causing) agent found in cigarette smoke.
phenobarbital	A tranquilizer ("downer"), one of the more common "barbiturate" drugs.
methamphetamine 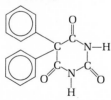	A common stimulant ("speed"), sometimes prescribed as an appetite depressant.
mescaline	A hallucinogenic drug found in certain cacti.

Table 7.1 (Continued)

COMPOUND	COMMENT
LSD (D-lysergic acid diethylamide)	The chemical fuel for a "bad trip."

The story of the growth of the synthetic chemical industries would make a book by itself. It is a story of business and politics, of the real advent of the "Madison Avenue" advertising man, and of the engineers and scientists who developed the processes used. Much of the scientific contribution was through carefully planned research—requiring detailed knowledge of chemical systems and plenty of long, hard work. Sometimes, however, a major advance can be traced to "serendipity," a lucky accident.

Several years ago, an industrial chemist opened a cylinder of tetrafluoroethylene (F_2CCF_2). He was studying the gas as a possibility for refrigeration systems (which now use other fluorine compounds, the Freons). To his surprise, no gas came out from the supposedly full cylinder. He might have discarded the cylinder or sent it back for replacement of a faulty valve. Instead, he decided to investigate the problem. He weighed the cylinder. The weight was right, so the gas apparently had not escaped through some crack or leaky valve. He shook the container and something rattled inside. The cylinder was cut open—carefully, to avoid a sudden release of gas. Inside was a white slippery solid. This polymer was found to have very useful properties. It was quite inert to the action of most chemicals, stable at high temperatures, and had a "self-lubricant" surface. The polymer was the first Teflon. Louis Pasteur was right: "Chance favors the prepared mind."[1]

Most of our further study of chemistry is concerned with species containing covalent bonds—simple molecules, polyatomic ions, and the complex molecules of importance in biochemistry.

As an introduction, this unit deals with some general characteristics of covalent species, some convenient ways of formulating their structures, and an application of modern atomic theory and thermodynamics to a description of simple molecules.

7.1 PROPERTIES OF COVALENT COMPOUNDS

As we shall see subsequently, there is no really clear-cut distinction between ionic and covalent bonds. The forces involved vary considerably, and differences are a matter of degree rather than kind. The terminology is largely a matter of arbitrary convenience.

Covalent bonds are responsible for atomic combinations in many elements (e.g., H_2, O_2, P_4, S_8, C_x; diamond is a "molecule" large enough to see), compounds (Table 7.1), and polyatomic ions. We shall restrict this introductory discussion primarily to simple molecules, reserving the more complex molecules and ions for later units.

Covalent bonds are usually (but not always) stronger than ionic bonds. The "pure" ionic bond is described as a simple electrostatic attraction of oppositely charged ions, whereas the covalent bond involves electrons sharing the positive fields of two or more nuclei.

[1] For some interesting case histories of chemical "serendipity," see *Chance Favors the Prepared Mind*, by Bernard E. Schaar, Reprint No. 80 (1968) from the American Chemical Society, Washington, D.C.

Table 7.2 ΔH **FOR ION SEPARATION**

REACTION	ΔH (kcal mole^{-1})
$HF(g) \longrightarrow H^+(g) + F^-(g)$	$+368$
$HCl(g) \longrightarrow H^+(g) + Cl^-(g)$	$+330$
$LiF(g) \longrightarrow Li^+(g) + F^-(g)$	$+175$
$LiCl(g) \longrightarrow Li^+(g) + Cl^-(g)$	$+148$

Table 7.3 **COMPARISONS OF APPROXIMATE MELTING AND BOILING POINTS**

SUBSTANCE	MELTING POINT (°C)	BOILING POINT (°C)
carbon (graphite)	3000	4830
silicon	1410	2290
germanium	960	2700
silver chloride	460	1560
potassium chloride	770	1410
sodium chloride	810	1470

As a rough approximation of the difference in strength of a 100-percent ionic bond (an idealized case) versus a 100-percent covalent bond, try the following experiment.

Run a comb through your hair several times. Use the charged comb to pick up a small piece of paper. This electrostatic attraction stimulates a "pure" ionic bond. Now see how much force you have to exert to pull the paper off the comb.

Compare this with the force required to pull a well-glued stamp off an envelope. The latter simulates a "pure" covalent bond.

It should be noted that the comb-and-paper and the glued stamp are only *analogies* representing some characteristics of chemical bonding.

The relative strengths of ionic and covalent bonds are reflected in a number of properties. For example, ΔH for ion separation is significantly greater in the covalent hydrogen halides than in the corresponding "ionic" alkali halides (Table 7.2; compare with data in Unit 6, Table 6.2).

In crystals of substances such as silicon (so-called covalent or network solids), unit particles are atoms; these are connected by covalent bonds in an intricate lattice network (Excursion 1).

The melting points and boiling points of covalent (network) substances (Unit 11) such as crystalline carbon (e.g., as graphite), silicon, or germanium also reflect the excess thermal energy needed to break the atoms loose from their normal covalent bonds as compared to the lower energies for disruption of ion clusters in salts (Table 7.3).

It is important to realize that the melting and boiling of substances such as silicon involve the disruption of covalent bonds. In molecular substances, such as carbon tetrachloride (melting point, $-23°C$; boiling point, $77°C$), melting and boiling *do not* break the covalent bonds of the individual molecules, but only the much weaker forces (Unit 9) between neighboring molecules.

Most molecular compounds do not conduct an electric current in the solid or liquid state or in solution; that is, they are not electrolytes (Unit 6). Some molecular compounds, such as many of the more common acids discussed in Unit 6, are partially ionized or become so when dissolved in solvents such as water (Units 13 and 19). These, of course, act as electrolytes in aqueous solution.

Additional comparisons of ionic and covalent substances are discussed in later units. The wide diversity of molecular substances makes generalizations somewhat risky, but it is reasonably safe to suggest that molecular substances are usually softer, lower melting, and less often water soluble than the typical ionic compounds.

7.2 FORMATION OF A COVALENT BOND

Only in completely isolated monatomic ions (a purely hypothetical situation) would one find an exact "full" charge (some multiple of the unit charge). In every other case, an ion's charge is affected by interaction with the electric fields of neighboring

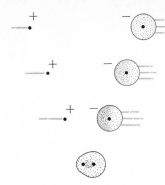

Figure 7.1 Polarization of Cl⁻ by H⁺, resulting in formation of a covalent bond.

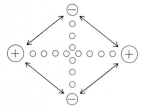

Figure 7.2 Electrostatic forces in the H_2 molecule.
↔ = attraction,
○○○○○ = repulsion

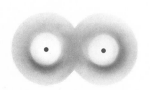

Figure 7.3 Simplified molecular orbital representation for H_2.

species—atoms, ions, or molecules. The extent to which charge is affected will vary depending on the nature of this interaction.

One way of looking at the formation of a covalent bond is in terms of the behavior of oppositely charged ions as they approach each other (Figure 7.1). Isolated monatomic ions are considered to be spherically symmetrical in terms of charge distribution. However, a negative ion approaching an ion of high positive *charge density* (charge per unit volume) will be distorted in shape (polarized) as electrons are pulled toward the positive field. In the case of H⁺ (very high positive charge density with no "shielding" electrons) and Cl⁻ (large negative ion), this distortion may result in a *molecule* of HCl; that is, *both nuclei are encompassed in the region of electron distribution.*

In general, negative ions are more readily polarized than positive ions, and large ions are more readily polarized than small ions. The higher the charge density of a positive ion, the greater effect it will have in polarizing a negative ion. On this basis, it is not surprising that molecules such as CCl_4 are more stable than $C^{4+} + 4Cl^-$. Polarization helps us to understand why some molecules are favored over their ionic counterparts. Additional considerations are necessary, however, to explain the formation and structure of molecular compounds. Among these considerations are comparisons of energy requirements for formation of possible ions (Unit 2). The energy needed for conversion of atomic carbon to C^{4+} ion, for example, suggests that such an occurrence is highly unlikely under normal conditions of chemical reactions.

The model of covalent bond formation based on ion polarization is quite reasonable for the combination of ions of opposite charge and is certainly consistent with the orbital model of a diffuse "charge cloud" of electrons which could easily "overlap" additional nuclei. Such an overlap in the case of a molecule such as HCl should result in a considerable energy change, purely on the basis of the electrostatic forces involved. The high negative ΔH of combination (opposite of "separation") of H⁺(g) with Cl⁻(g) (Table 7.2) is not surprising. However, the combination of two electrically *neutral* hydrogen atoms to form the H_2 molecule also releases energy (104 kcal mole⁻¹ of H_2 formed). This seems difficult to justify on the basis of simple electrostatic forces.

Electrostatic forces *do* exist in the H_2 molecule (Figure 7.2). At some optimum distance (the *bond length*), the attractive forces of the two electrons for the two nuclei exactly balance the electron–electron and proton–proton repulsions.

A similar situation could be postulated for simple species such as H_2^+ (one electron sharing the positive field of two nuclei), He_2^+ (three electrons), or He_2 (four electrons). Estimates of electrostatic forces might suggest that the stability of such species would reflect increasing attractions as the number of electron–proton interactions increased, that is, $He_2 > He_2^+, > H_2 > H_2^+$. Experimental investigations, however, show that the bond energy (enthalpy change for separation of the bonded species) for H_2 is nearly twice as large as that for H_2^+ or He_2^+ and that there is only a very weak force between two helium atoms. The unusual stability of *two* electrons in a positive nuclear field is reminiscent of that associated with two electrons, of opposite spin, in an atomic orbital (Unit 1). The quantum mechanical model for the atom can, indeed, be applied to molecules as well. From appropriate formulations of the Schrödinger equation, ψ^2 functions can be used to obtain three-dimensional probability plots of electron density for electrons sharing the positive fields of two or more nuclei (Figure 7.3). These *molecular orbitals,* restricted to a maximum of two electrons each, are now the most widely accepted models for the description of covalent bonds.

A useful generalization for the simpler elements is that atoms tend to combine in ways that result in species approximately isoelectronic with noble gas atoms (Unit 2).

SOME EARLY MODELS FOR THE COVALENT BOND

Figure 7.4 Interlocking-atom model (first century B.C.).

Long before any distinction between ionic and covalent bonding was recognized, philosophers and scientists were proposing models to account for ways in which atoms might "stick together" to form the multitude of substances found in the natural world. The "atomic" concept of Democritus (c. 500 B.C.) was the basis for the first attempts at describing atomic combination, most lucidly detailed by the Roman philosopher and poet Lucretius in the first century B.C., with an elaboration remarkably consistent with many aspects of modern theory. Indeed, the major difference between the primitive ideas and current hypotheses was in the *nature* of the atomic interaction. Lucretius thought of atoms as similar to the interlocking pieces of our present-day "picture puzzles" (Figure 7.4), with the unique atomic shapes determining which atoms could fit together in stable combinations. Although we do not recognize the validity of this idea today as a model for covalent bonding, we do employ some very similar descriptions in simple discussions of enzyme behavior (Units 32 and 34).

With the increasing acceptance of Aristotle's counter proposal of continuous (rather than atomic) matter and the generally antiscience attitude of major religious groups, interest in theoretical ideas about the nature of matter waned. It was not until the seventeenth century that scientific thought was again focused on the subject of atoms and their modes of combination. At first the "interlock" model was revived, but this was soon abandoned in favor of "attractive force" descriptions. In 1674, William Petty, an English physician, suggested magnetic atoms (Figure 7.5) with atomic combinations governed by the same laws of magnetic attraction observed in the macro world.

A major step toward our present concept of covalent bonding is credited to an

Figure 7.5 Magnetic-atom model (1674).

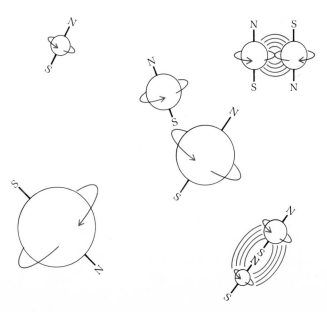

eighteenth-century Jesuit, Roger Boscovich. He proposed that both attractive and repulsive forces could exist between atoms, with stable atomic combination possible only at an interatomic distance corresponding to a balance between the opposing forces. When coupled with Boscovich's suggestion that atoms were *not* hard spheres, we have a model remarkably similar to that of the hydrogen molecule illustrated in Figures 7.2 and 7.3.

By the early nineteenth century, it was generally recognized that alternative theoretical considerations could only be resolved by more sophisticated experimental work, requiring technology not really developed until the twentieth century.[2] We have the necessary technology today and our current models for covalent bonding are based on massive accumulations of experimental evidence. What we have gained in sophistication, however, we have lost in simplicity, and our most advanced theoretical models are essentially incomprehensible to all but theoreticians. The simple "working models" we will find most useful are not really too far removed from those of our colleagues of the past.

[2] For a most interesting discussion of the historical perspective, see "Early Views on Forces between Atoms," by Leslie Holliday, in the May, 1970, issue of *Scientific American*.

Metal:nonmetal combinations often result in ionic compounds in which all the ions have noble gas electron configurations. Nonmetal:nonmetal combinations, on the other hand, usually result in covalent compounds in which the shared electron pair of the bond allows each bonded atom to *approach* the noble gas electronic structure.

7.3 A SIMPLE MOLECULAR ORBITAL MODEL

Covalent bonding, except in rare cases, appears to be most simply explained as the sharing of electron pairs among atoms in such a way that a "bond" results from the balance between internuclear and interelectron repulsions and the electrostatic attraction of the shared electron pair for the different positive nuclei (Figure 7.2). We can visualize this as happening in either of two ways. One atom (or ion) could furnish both electrons for the bond pair (e.g., Figure 7.1). Such a case is referred to as *coordinate* covalent bonding. Alternatively, as in the case of the H_2 molecule (Figure 7.2), each atom could furnish one electron for the bond pair ("simple" covalent bonding). It is important to note that the distinction between coordinate and simple bond formation is entirely one of convenience in classifying the *origin* of the bond pair. Once the bond is formed, the atoms "don't care" about this distinction and only a single bonding description is necessary. For example, we could visualize the formation of the HCl molecule as involving either coordinate bonding (from $H^+ + Cl^-$) or simple bonding (from atomic H and atomic Cl). The model chosen to describe the HCl molecule refers to the condition *after* bonding has occurred, so only a single model is required (Figure 7.6). Since we have found the *atomic orbital* to be a useful model for electrons in atoms (Units 1 and 2), a *molecular orbital* might prove equally pertinent to a description of the bond-pair electrons in molecules. A molecular orbital should have some of the characteristics described for an atomic orbital, such as a maximum occupancy of two electrons (of opposite spins, Unit 1). In addition, the molecular orbital should take into account the idea that the highest positive field would be between adjacent nuclei, so that bond-pair electrons would be expected to "spend a large fraction of their time" within the internuclear region. The optimum

It should appear reasonable that a "one-electron bond" might have insufficient electron-nuclei attraction to balance internuclear repulsion, and more than two electrons in the bonding region would increase the interelectron repulsions dramatically.

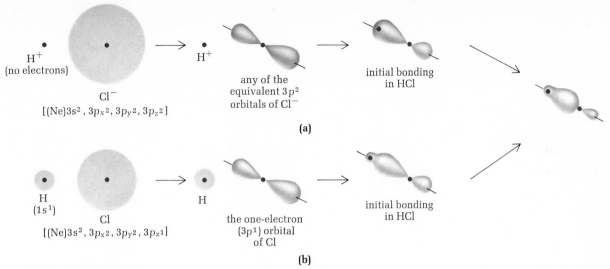

H⁺
(no electrons)

Cl⁻
$[(Ne)3s^2, 3p_x{}^2, 3p_y{}^2, 3p_z{}^2]$

H⁺

any of the
equivalent $3p^2$
orbitals of Cl⁻

initial bonding
in HCl

(a)

H
$(1s^1)$

Cl
$[(Ne)3s^2, 3p_x{}^2, 3p_y{}^2, 3p_z{}^1]$

H

the one-electron
$(3p^1)$ orbital
of Cl

initial bonding
in HCl

(b)

Figure 7.6 Molecular orbital model for HCl, showing maximum electron density within the general internuclear region. Simple molecular orbital description. (a) By *coordinate* bond formation with Cl⁻ furnishing both electrons for the bond. (Note that a *p* orbital was selected for bonding, since this provides a higher electron density along the line of H⁺ approach than would the spherical s orbital.) (b) By simple overlap of two one-electron atomic orbitals.

More sophisticated (and more accurate) molecular orbital theory requires a consideration of *all* electronic interactions, rather than just those involving the bond-pair electrons.

internuclear distance (*bond length*) should be determined by the balance between opposing electrostatic forces.

We shall consider two ways of approaching a molecular orbital model. In starting with species expected to combine by *coordinate* bond formation, we shall focus our attention on the particular atomic orbital containing the pair of electrons to be involved in bonding. This orbital should then be modified so that the covalent bond description permits the encompassing of *both* atomic nuclei within the general region of influence of the molecular orbital (Figure 7.6a). When two atoms combine by each furnishing one electron for the bond pair, we will approximate the molecular orbital by the *overlap* of the two appropriate one-electron atomic orbitals, with suitable modification to illustrate maximum electron probability within the general internuclear region.

Example 1

Suggest simple molecular orbital models for
  a. the fluorine molecule
  b. the hydrogen fluoride molecule
  c. the He₂²⁺ molecule-ion

Solution

a. The ground state electron configuration for atomic fluorine (Unit 1) may be represented as (He) $2s^2, 2p_x{}^2, 2p_y{}^2, 2p_z{}^1$. The combination of two fluorine atoms to form the F_2 molecule could then be represented, on a simplified basis, by the overlap of the two 1-electron *p* orbitals:

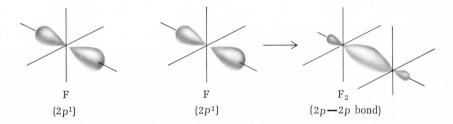

F
$(2p^1)$

F
$(2p^1)$

F_2
$(2p-2p$ bond$)$

b. Hydrogen fluoride could be considered as formed by either coordinate bonding of H^+ (no electrons) with F^- [(He) $2s^2$, $2p_x^2$, $2p_y^2$, $2p_z^2$] or by simple combination between atomic hydrogen ($1s^1$) and atomic fluorine [(He) $2s^2$, $2p_x^2$, $2p_y^2$, $2p_z^1$]:

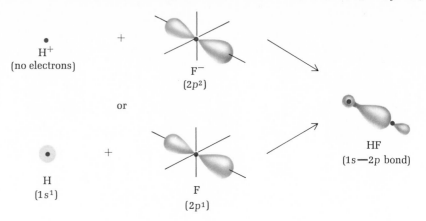

H^+
(no electrons)

F^-
$(2p^2)$

or

H
$(1s^1)$

F
$(2p^1)$

HF
$(1s\!-\!2p$ bond)

The He_2^{2+} ion is not very stable, because of the strong repulsion of the two nuclei. However, there is experimental evidence for the transient existence of this ion under appropriate conditions.

c. The He_2^{2+} molecule ion might form by coordinate bonding of He^{2+} (no electrons) with atomic helium ($1s^2$) or by combination of two He^+ ions (each $1s^1$):

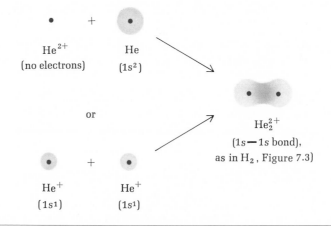

He^{2+}
(no electrons)

He
$(1s^2)$

or

He^+
$(1s^1)$

He^+
$(1s^1)$

He_2^{2+}
$(1s\!-\!1s$ bond),
as in H_2, Figure 7.3)

The simple model developed so far is satisfactory for approximating bond descriptions of many of the common diatomic molecules and can be extended, as we shall see, with appropriate modifications, to a broad spectrum of covalent species.

7.4 MULTIPLE BONDS   For many molecules and polyatomic ions, experimental evidence indicates that bonding involves more than one pair of electrons shared by adjacent nuclei. Because all experience suggests that two electrons represent the maximum occupancy of an atomic or molecular orbital, we need some bonding model that can account for *double bonds* (two shared electron pairs) and *triple bonds* (three shared pairs). The model should indicate, as in the case of the single bond, a maximum bond-pair electron density within the general internuclear region, but only one molecular orbital directly *between* the nuclei (along the internuclear axis). The directional characteristics of atomic *p* orbitals provide us with a useful approach to this model, by suggesting a side-to-side overlap of *p* orbitals (Figure 7.7) rather than an end-to-end overlap, as was illustrated in Example 1(a).

Figure 7.7 Simple model for double bond in C_2 molecule, by overlap of two 1-electron atomic p orbitals.

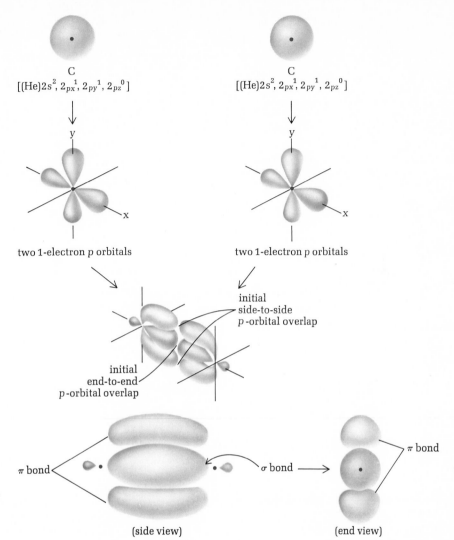

C
$[(He)2s^2, 2p_x^{\ 1}, 2p_y^{\ 1}, 2p_z^{\ 0}]$

C
$[(He)2s^2, 2p_x^{\ 1}, 2p_y^{\ 1}, 2p_z^{\ 0}]$

two 1-electron p orbitals

two 1-electron p orbitals

initial side-to-side p-orbital overlap

initial end-to-end p-orbital overlap

π bond

σ bond

π bond

(side view)

(end view)

The increased "degree of bonding" in multiple bonds has the effect of shortening the internuclear distance. A triple bond represents a shorter bond length than a double bond which, in turn, is shorter than a single bond. Bond-length measurement provides one type of experimental evidence for multiple bonding.

Hybrid orbitals, introduced in Unit 8, may be thought of as "averaged mixtures" of simple atomic orbital characteristics. For example, an sp hybrid has roughly 50% s-orbital character and 50% p-orbital character.

The side-to-side overlap is referred to as a *pi* (π) bond, and any bond along the internuclear axis is called a *sigma* (σ) bond. Our model then suggests that a *double bond consists of one σ and one π bond and a triple bond consists of one σ and two π bonds*. Since a triple bond (Figure 7.8) represents the maximum occupancy of the general internuclear region, no higher degree of multiple bonding can occur, that is, the combination of one σ and two π bonds occupies all of the available internuclear region.

There are only a few cases in which overlap of simple atomic orbitals can provide a reasonable approximation of covalent bonding. In fact the representations of the C_2 molecule (Figure 7.7) and the N_2 molecule (Figure 7.8) are only partially consistent with the observed properties of these molecules. As we shall see in Unit 8, the extension of our atomic orbital overlap model to include *hybrid* atomic orbitals provides a much more general and satisfying description of covalent bonding.

Figure 7.8 Simple model for triple bond in N_2 molecule, by overlap of three 1-electron atomic p orbitals.

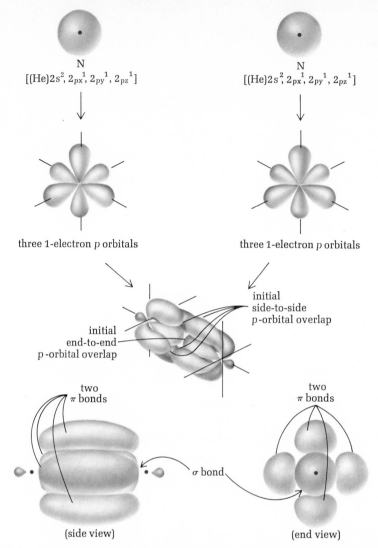

N
$[(He)2s^2, 2_{px}^{1}, 2_{py}^{1}, 2_{pz}^{1}]$

N
$[(He)2s^2, 2_{px}^{1}, 2_{py}^{1}, 2_{pz}^{1}]$

three 1-electron p orbitals

three 1-electron p orbitals

initial
side-to-side
p-orbital overlap

initial
end-to-end
p-orbital overlap

two
π bonds

two
π bonds

σ bond

(side view)

(end view)

7.5 ELECTRON-DOT AND LINE-BOND FORMULATIONS

If we are to gain any insight into the physical and chemical properties of polyatomic species, we must understand that these properties are determined by structural features of the atomic combinations.

There are a number of different ways of representing the formulas of polyatomic ions or molecules (Table 7.4), each having particular uses. In general, we employ the simplest formulation suitable for the purpose at hand. We have already noted, for example, the benefit of simplicity in using *molecular* formulas for stoichiometric calculations (Examples 2 and 4, Unit 4).

As a first step in expressing these structural characteristics, electron-dot or line-bond formulations will enable us to show the ways in which atoms are bonded together in molecules or polyatomic ions. The electron-dot formulas will show the distribution of *valence* electrons ("outermost" electrons, Units 1 and 2) involved in bonding (bond-pair electrons) or "free" (lone-pair electrons or single unpaired electrons). In line-bond representations, a line is used in place of an electron bond pair.

To write correct electron-dot formulas, we must first know the way in which the

Table 7.4 SOME CHEMICAL FORMULATIONS FOR ETHYLENEDIAMINE

FORMULATION	TYPE	DESCRIPTION	TYPICAL USE
CH_4N	empirical	only the simplest *ratio* of atoms in the molecule	simplest relationship between chemical formula and percent composition
$C_2H_8N_2$	molecular	total number of each atom per molecule	simplest for use in formula weight calculation
H:N:C:C:N:H (electron-dot structure with H atoms)	structural (expanded) (electron-dot)	Section 7.5	illustration of bond-pair and lone-pair electron locations
H—N—C—C—N—H (line-bond structure with H atoms)	structural (expanded) (line-bond)	Section 7.5	illustration of bonding arrangements (lone-pair "dots" optional)
$H_2NCH_2CH_2NH_2$ or $H_2N(CH_2)_2NH_2$	condensed structural	simple atomic groupings consistent with expanded structural formulas	simplest way of designating a unique formula (lone-pair "dots" optional)
(geometric structural diagram with N, C, H atoms)	geometric structural	Unit 8	illustration, in varying detail, of molecular geometry (lone-pair "dots" optional)

component atoms are connected. In simple cases only one logical bonding arrangement suggests itself (Example 2), but the number of possibilities increases with increasing complexity of the molecule. Structural formulas are determined by experimental methods and, for complex molecules, these methods may require quite sophisticated techniques (Excursion 5). As we learn more about the subject of chemistry, we will gradually acquire a familiarity with many of the common experimentally determined structural features. For now, to enable us to work with a reasonable variety of polyatomic species, we will have to rely on some simple "bonding rules" (Table 7.5), the experimental backgrounds for which will be developed in subsequent units.

With the information from Table 7.5, we can learn to write electron-dot and line-bond formulations for a large number of molecules and polyatomic ions. It was G. N. Lewis who suggested that bonding is most likely to occur in ways that permit atoms to achieve or approach the electronic structures of noble gas atoms. For example, if carbon forms four simple covalent bonds, the carbon atom then participates in the sharing of four pairs of electrons, thus approximating the "electron octet" (four pairs of electrons) of the valence electrons of neon [(He) $2s^2$, $2p^6$].

We may think of the method of writing electron-dot formulas as a "game" in which the "rules" may be stated as

1. The element symbol is used to represent the atomic nucleus and *kernel* electrons (all electrons except valence electrons).
2. Valence electrons (those in the last principal energy level) are represented by dots. These dots then represent electrons available for bonding.

Table 7.5 SOME SIMPLE "BONDING RULES"

1. For the representative elements (A groups), the number of valence electrons of the neutral atom is given by the group number. [For example, carbon (a IVA element) has four valence electrons.]

2. In most oxyanions (e.g., NO_2^-, CO_3^{2-}, PO_4^{3-}), oxygens are bonded to the "other element," not to each other. [The presense of an oxygen-oxygen bond in a molecule or polyatomic ion is indicated in the name of the species by "perox," as in hydrogen peroxide (H_2O_2, H—O̤—O̤—H)].

3. In most oxyacids (e.g., HClO, HNO_2, H_2SO_3), the "acidic" hydrogens are bonded to oxygen, not to the "other element" and oxygens are not bonded to each other except in "peroxy" acids. (Some oxyacids of phosphorus contain "nonacidic" hydrogens, Units 19 and 26.)

4. Within molecules or polyatomic anions, oxygen normally forms no more than two covalent bonds. ("Bonding" between neighboring molecules is introduced in Unit 9. Oxygen forms a triple bond in carbon monoxide.)

5. Only a few atoms form multiple bonds. The only common cases involve carbon, nitrogen, oxygen, or sulfur.

---

3. Dots are arranged so that each symbol has eight dots (an octet) around it as four *pairs,* except for hydrogen, which has one pair, or in cases for which insufficient electrons are available. (Example 5)

4. *For ions:* For each net positive charge, subtract one dot from the total available from component neutral atoms; for each net negative charge, add one dot.

5. Bonding must be consistent with experimentally determined structural features (e.g., Table 7.5), that is, the arrangement of element symbols is drawn according to these "rules".

Example 2

Draw electron-dot and line-bond representations for methane.

Solution
Carbon, a group IVA element, has four valence electrons, represented by dots:

·C·

Each hydrogen atom has a single electron:

H·

Methane is formulated as CH_4, so the hydrogen atoms must be arranged around the carbon to fit rule 3. Therefore, using ×'s and ∘'s to keep track of electron origins,

$$\overrightarrow{H^\circ} \times \underset{\underset{\uparrow \overset{\circ}{H}}{\times}}{\overset{\overset{H\downarrow}{\overset{\circ}{\times}}}{C}} \times {}^\circ\overleftarrow{H}$$

$$
\begin{array}{ccccc}
\text{H} & & \text{H} & & \text{H} \\
\text{H}\,{}^{×}_{○}\text{C}\,{}^{○}_{×}\,\text{H} & \text{or} & \text{H}:\overset{..}{\underset{..}{C}}:\text{H} & \text{or} & \text{H}—\overset{|}{\underset{|}{C}}—\text{H} \\
\text{H} & & \text{H} & & \text{H} \\
& & \text{(electron dot)} & & \text{(line bond)}
\end{array}
$$

Example 3

Draw electron-dot structures for products and reactants in

$$NH_3 \quad + \quad HCl \quad \longrightarrow \quad NH_4Cl$$

molecular   molecular   ionic

Solution

Hydrogen has one valence electron, nitrogen has five, and chlorine has seven. Therefore, using ×, ∘, and • to keep track of electron origins,

$$H \overset{\times\times}{\underset{\times\circ}{N}} \overset{\times}{\underset{}{}} H + H \overset{\bullet\bullet}{\underset{\bullet\bullet}{Cl}} \bullet \longrightarrow H \overset{\times\times}{\underset{\times\circ}{N}} H + \overset{\bullet\bullet}{\underset{\bullet\bullet}{Cl}} \bullet^- $$

$$\begin{array}{cc} & H \quad^+ \\ H\overset{\times\times}{N}\overset{\bullet}{}H & \bullet\bullet \\ \underset{H}{\underset{\times\circ}{}} & \bullet Cl\bullet^- \end{array}$$

or, without regard to electron origins,

$$H \overset{..}{N} H + H \overset{..}{Cl} \overset{..}{} \longrightarrow H \overset{..}{N} H + \overset{..}{Cl} \overset{..}{}^- $$

$$\begin{array}{cc} & H \quad^+ \\ H\!:\!\overset{..}{N}\!:\!H & \\ \underset{H}{} & :\overset{..}{\underset{..}{Cl}}:^- \end{array}$$

Example 4

Draw both electron-dot and line-bond formulations for nitrous acid.

Solution

The "available pieces" for HNO_2 may be represented as

$$H\cdot \quad \overset{\circ\circ}{\underset{\circ}{\circ N\circ}} \quad \overset{\times}{\underset{\times\times}{\times O}}\overset{}{}_{\times}^{} \quad \overset{\times}{\underset{\times\times}{\times O}}\overset{}{}_{\times}^{}$$

Because the "acidic" hydrogen is known to be bonded to oxygen (Table 7.5), we can write the two-atom piece

$$H\overset{\times}{\underset{\times\times}{\times O}}\overset{}{}_{\times}^{}$$

and because the oxygen atoms are not bonded to each other (Table 7.3), the structural sequence should be

     H   O   N   O   (same as O   N   O   H)

Now, to achieve "octets" for nitrogen and oxygen,

$$H\overset{\times\times}{\underset{\times\times}{\times O}}\overset{\circ\circ}{\underset{}{N}}\overset{\times\times}{\underset{\times\times}{\times O}}$$

or, without regard to electron origins,

$$\underset{\text{(electron dot)}}{H\!:\!\overset{..}{\underset{..}{O}}\!:\!\overset{..}{N}\!:\!:\!\overset{..}{\underset{..}{O}}} \quad \text{or} \quad \underset{\text{(line bond)}}{H\!-\!\overset{..}{\underset{..}{O}}\!-\!\overset{..}{N}\!=\!\overset{..}{\underset{}{O}}} \quad \overset{(\pi\text{ bond})}{}$$

No alternative structures "fit" the "octet rule" except for

$$H\!:\!\overset{..}{\underset{..}{O}}\!:\!:\!\overset{..}{N}\!:\!\overset{..}{\underset{..}{O}}:$$

and this is "forbidden" by rule 4, Table 7.5. It should be noted that the location of lone pairs with correct atoms is important, but *where* the dots are written is immaterial; that is, all the following formulations are equivalent.

$$H—\ddot{\underset{..}{O}}—N=\ddot{\underset{..}{O}} \equiv H—\ddot{\underset{..}{O}}—N=\ddot{\underset{..}{O}} \equiv H—\ddot{\underset{..}{O}}—N=\ddot{O}: \quad (etc.)$$

For many chemical species, the numbers of valence electrons involved require formulations other than the "octet" arrangement (Example 5). In other cases, experimental evidence is inconsistent with the "expected" electron-dot formula. For example, the blue color of liquid oxygen and the attraction of liquid oxygen for a magnet are interpreted as evidence for *unpaired* electrons in the oxygen molecule. The "expected" dot formula for O_2 ($\ddot{\underset{..}{O}}::\ddot{\underset{..}{O}}$) is inconsistent with this evidence. A simple formulation illustrating unpaired electrons may be written as $\cdot\ddot{\underset{..}{O}}:\ddot{\underset{..}{O}}\cdot$

It must be noted that the O_2 molecule is more complex than a simple electron-dot formula suggests. The bond length, for example, is shorter than that of a typical oxygen–oxygen single bond. Advanced molecular orbital theory provides the best bonding description for O_2.

Example 5

Write electron-dot and line-bond formulations for
  a. aluminum chloride
  b. nitric oxide
  c. sulfur hexafluoride

Solution

  a. Aluminum (Group IIIA) has three valence electrons and chlorine (VIIA) has seven. Since aluminum is not expected to form multiple bonds (Table 7.5), the only possible formulation is

:Cl:
Al:Cl:
:Cl:
(electron dot)

Cl
|
Al—Cl
|
Cl
(line bond, with lone-pair
electrons omitted)

  b. For nitric oxide (NO), we have available

$$\cdot\ddot{N}\cdot \quad and \quad \cdot\ddot{\underset{..}{O}}:$$

Oxygen could achieve an octet structure by forming a single bond (coordinate) with nitrogen:

$$\circ\,\underset{\circ}{\overset{\circ}{N}}\underset{\times\times}{\overset{\times\times}{O}}\times$$

This provides only five electrons for nitrogen. Nitrogen could more closely approach an octet by formation of a double bond:

$$\underset{\circ\circ}{\overset{\circ}{N}}\underset{\times\times}{\overset{\times\times}{O}}$$

Although nitrogen could form a triple bond, oxygen (with rare exceptions) does not (Table 7.5):

$$\ddot{N}::\ddot{\underset{..}{O}} \qquad N=\ddot{\underset{..}{O}} \;\;\overset{(\pi\ bond)}{\frown}$$

  (electron dot)   (line bond)

(In the absence of definitive information, we could have indicated the unpaired electron with oxygen, rather than nitrogen. The formula shown is consistent with experimental evidence.)

c. Sulfur (VIA) has six valence electrons, and fluorine (VIIA) has seven. The most plausible formulation is

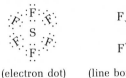

(electron dot)   (line bond, omitting lone-pair dots)

Figure 7.9 Representations for nitrate ion. (a) Complete set of contributors to the resonance hybrids. (b) Delocalized π-bond formulation. (Experimental evidence shows that all N—O bonds in nitrate ion are equivalent and intermediate in character between N=O and N—O.)

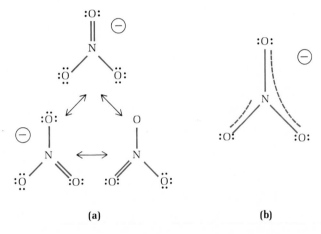

(a)   (b)

There are cases in which two or more formulations differing only in the location of double bonds are essentially equivalent. In such cases it is found experimentally that certain bond lengths are intermediate between those expected for double bonds and single bonds and that the structures are not well represented by single electron-dot or line-bond formulations. Structures of molecules or polyatomic ions of this type may be represented (Figure 7.9) by either a complete set of "contributor" formulations, called the *resonance hybrid*, or by a single formulation using a special representation for a *delocalized* π bond (one encompassing more than two atoms).

The concept of *resonance hybrids* may be used for more extensive analysis of "contributions" to a structural representation. We shall limit our use to the simple case of showing alternative locations for double bonds.

Example 6

Draw both resonance hybrid and delocalized π-bond formulations for
  a. nitrite ion
  b. nitric acid

Solution

a. For NO_2^- we have five valence electrons for nitrogen, six for each oxygen, and one "extra" electron (from the net negative charge). One acceptable formulation would be

$$-\overset{\times\times}{\underset{\times\times}{\times}}\!\overset{oo}{O}\!\overset{}{N}\!\overset{\times\times}{\underset{\times\times}{\times}}O$$

However, an equally plausible, and equivalent, formulation is

$$\overset{\times\times}{\underset{\times\times}{O}}\overset{\circ\circ}{\underset{\circ\circ}{\times}}N\overset{\times\times}{\underset{\times\times}{\circ}}O\times^-$$

The complete set (two) of resonance contributors is

$$^-\!:\!\overset{..}{\underset{..}{O}}\!:\!\overset{..}{N}\!:\!:\!\overset{..}{\underset{..}{O}} \longleftrightarrow \overset{..}{\underset{..}{O}}\!:\!:\!\overset{..}{N}\!:\!\overset{..}{\underset{..}{O}}\!:^- \qquad ^-\!:\!\overset{..}{\underset{..}{O}}\!-\!N\!=\!\overset{..}{\underset{..}{O}} \longleftrightarrow \overset{..}{\underset{..}{O}}\!=\!N\!-\!\overset{..}{\underset{..}{O}}\!:^-$$

$$\text{(electron dot)} \qquad\qquad\qquad \text{(line bond)}$$

The alternative "delocalized" formulation is

$$\overset{..}{\underset{..}{O}}\!=\!=\!N\!=\!=\!\overset{..}{\underset{..}{O}}{}^-$$

Dashed lines indicate the delocalized π bonding.

b. One formulation of nitric acid consistent with the "rules" (Table 7.5) is

$$H\!:\!\overset{..}{\underset{..}{O}}\!:\!N\!:\!:\!\overset{..}{\underset{..}{O}}$$
$$\underset{}{:\!\overset{..}{\underset{..}{O}}\!:}$$

Because resonance contributors, as we are employing them, show only alternative locations of double bonds and because oxygen is normally limited to a maximum of two covalent bonds within a molecule (Table 7.5), only two resonance hybrid contributors are required:

$$H\!-\!\overset{..}{\underset{..}{O}}\!-\!N\!=\!\overset{..}{\underset{..}{O}} \longleftrightarrow H\!-\!\overset{..}{\underset{..}{O}}\!-\!N\!-\!\overset{..}{\underset{..}{O}}\!:$$
$$\underset{:\overset{..}{\underset{..}{O}}:}{|} \qquad\qquad \underset{:\overset{..}{\underset{..}{O}}:}{\|}$$

The "delocalized" formulation may then be written as

$$H\!-\!\overset{..}{\underset{..}{O}}\!-\!N\!\underset{}{\overset{..}{=}}\!\overset{..}{O}\!:$$
$$\underset{:\overset{..}{\underset{..}{O}}}{\|}$$

7.6 ELECTRONEGATIVITY: POLAR BONDS

The simple overlap of atomic orbitals provides a very rough approximation of a covalent bond. Remembering that this "bond" is a description of a pair of electrons in the positive field of two different nuclei, we can modify the description, at least qualitatively, by considerations of electrostatic forces. The highest positive field would obviously be between the two nuclei, so the negative electrons should spend more time between the nuclei than in other regions. The orbital representation should be modified to show the higher probability of finding the electrons between the nuclei (Figure 7.10).

Different atoms in molecules will have different nuclei, different numbers of electrons, and hence different effects on the electron pair of a covalent bond. The attraction of a particular atom for the electron pair of a bond is referred to as *electronegativity,* and values for relative electronegativities of different elements have been estimated (Table 7.6).

Electronegativity values vary somewhat in different compounds, but a reasonable estimate may be made by averaging the values for first ionization potential and first electron affinity (Unit 2). Thus the electronegativity may be considered as representing the average of an atom's tendency to gain or lose electrons. Fluorine has the highest electronegativity of all the elements. Among the representative elements, electronegativity increases with atomic number within a period and decreases with increasing atomic number within a group. The periodic table, then, provides a useful guide to estimating, qualitatively, the relative electronegativities of common elements.

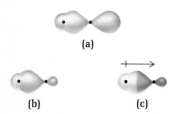

Figure 7.10 Modifications of the simple orbital representation for HF. (a) Overlapped atomic orbitals. (b) Higher electron density between nuclei. (c) Electron density displaced toward more electronegative atom.

Table 7.6   SEGMENT OF A PERIODIC TABLE SHOWING TRENDS IN RELATIVE ELECTRONEGATIVITIES

3 Li **1.0**	4 Be **1.5**											1 H **2.2**		5 B **2.0**	6 C **2.5**	7 N **3.0**	8 O **3.5**	9 F **4.0**
11 Na **0.9**	12 Mg **1.2**													13 Al **1.5**	14 Si **1.8**	15 P **2.1**	16 S **2.5**	17 Cl **3.0**
19 K **0.8**	20 Ca **1.0**	21 Sc **1.3**	22 Ti **1.5**	23 V **1.6**	24 Cr **1.6**	25 Mn **1.5**	26 Fe **1.8**	27 Co **1.8**	28 Ni **1.8**	29 Cu **1.9**	30 Zn **1.6**	31 Ga **1.6**	32 Ge **1.8**	33 As **2.0**	34 Se **2.4**	35 Br **2.8**
37 Rb **0.8**	38 Sr **1.0**								46 Pd **2.2**	47 Ag **1.9**	48 Cd **1.7**	49 In **1.7**	50 Sn **1.8**	51 Sb **1.9**	52 Te **2.1**	53 I **2.5**
55 Cs **0.7**	56 Ba **0.9**								78 Pt **2.2**	79 Au **2.4**	80 Hg **1.9**	81 Tl **1.8**	82 Pb **1.8**	83 Bi **1.9**	84 Po **2.0**	85 At **2.2**

The dipole moment of a polar molecule is equal to the partial charge of either end of the molecular dipole times the distance between the resultant centers of positive and negative charge density. The experimental measurement involves a determination of the voltage required to put a given amount of charge on the plates of a capacitor immersed in the substance being investigated. (For more details, see the Suggested Reading at the end of this unit.)

This concept provides another modification of the orbital picture (Figure 7.10) to show a higher *electron density* near the more electronegative atom in the bond. Any bond between atoms of different electronegativities will therefore be *polar* and can be represented as a *dipole,* using ↔ to indicate the direction of the dipole's electric field, that is, the bond *polarity.* (The arrow points toward the atom of higher electron density.)

The concept of electronegativity can be used as an arbitrary distinction between ionic and covalent bonds. Every chemical bond involves some degree of orbital overlap (Unit 6). Every bond between atoms of different electronegativities has an *unsymmetrical* overlap, with electron density higher near the more electronegative atom. Such asymmetry is referred to as *partial ionic character,* and this can be determined experimentally from measurements of bond length (internuclear distance) and dipole moment. The percent ionic character of the bond is then defined as the experimental value of the dipole moment divided by the calculated dipole moment for "complete" charges separated by a distance equal to the length of the molecular dipole (Figure 7.11).

This correlation is not highly accurate. For example the electronegativity difference between hydrogen and fluorine is 1.8 (Table 7.6), but the percent ionic character, based on experimental measurements, is only 45 percent (Figure 7.11). HF is thus a borderline case.

A rough correlation exists between percent ionic character and electronegativity differences (Table 7.7), that is, the numerical difference between tabulated values (Table 7.6) of the two atoms forming the bond. On the basis of this correlation, a bond that is more than 50 percent ionic (electronegativity difference > 1.7) is classified as ionic. Bonds having less than 50 percent ionic character, but with an electronegativity difference greater than zero, are said to be polar covalent. A zero electronegativity difference, then, corresponds to a nonpolar covalent bond (100 percent covalent). The distinctions are admittedly arbitrary, and calculations of electronegativity differences are most meaningful for the simple diatomic molecules. In polyatomic molecules the picture is complicated by the influence of additional atomic interactions. Molecules may be nonpolar even when they contain polar bonds. If polar bonds around a central atom (Figure 7.12) are directed at angles such that equal dipoles are mutually balanced, the molecule as a whole is nonpolar.

For a treatment of vectors, the interested student should consult the instructor for available references.

The dipole moment of a polyatomic molecule can be approximated by the *vector sum* of individual bond dipoles as calculated from considerations of their partial ionic character. Such a calculation requires some simple trigonometry, but for our

Figure 7.11 Percent ionic character of the HF molecule. To obtain the dipole moment in units of *debyes*, the charge is expressed in *electrostatic units* (1.000 esu $= 3.335 \times 10^{-10}$ C) and the bond length in centimeters (1 cm $= 10^8$ Å). Then the *debye* is defined as 10^{-18} esu-cm.

bond length $= 0.92$Å

experimental value of dipole moment $= 1.98$ debyes

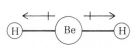

1.6×10^{-19} C 1.6×10^{-19} C

0.92Å

calculated value of dipole moment for unit charges separated by 0.92Å $= 4.42$ debyes

percent ionic character $=$

$$\frac{1.98}{4.42} \times 100\% = 45\%$$

Table 7.7 ELECTRONEGATIVITY DIFFERENCE AND PERCENT IONIC CHARACTER

PERCENT IONIC CHARACTER	ELECTRONEGATIVITY DIFFERENCE
0	0
1	0.2
10	0.6
25	1.1
50	1.7
75	2.4
90	3.1
100	3.5[a]

[a]This is an extrapolated value. The most electronegative element is fluorine (4.0), and the lease electronegative is cesium (0.7). Thus *no real* compound is 100 percent ionic.

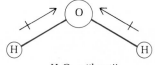

BeH$_2$, a linear molecule (Unit 8)

H$_2$O, a "bent" molecule (Unit 8)

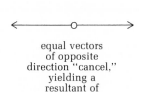

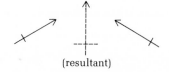

equal vectors of opposite direction "cancel," yielding a resultant of zero

(resultant)

vectors representing nonopposing bond dipoles yielding a resultant vector representing the *molecular dipole*

Figure 7.12 Bond dipoles, molecular shape, and polarity.

BeH$_2$ is a *nonpolar* molecule.

H$_2$O is a *polar* molecule.

purposes it is sufficient to note that *a molecule will be nonpolar if it contains equivalent bonds arranged in a symmetrical geometric pattern.*

Many properties of molecules depend on whether they are polar or nonpolar. Except for those molecules whose shapes can be estimated on the basis of simple atomic orbital overlap, any prediction of molecular polarity must await the development of a more useful model for predicting molecular geometry (Unit 8).

SUGGESTED READING

Gray, Harry. 1973. *Chemical Bonds.* Menlo Park, CA: Benjamin.

Griswold, Ernest. 1968. *Chemical Bonding and Structure.* Lexington, Mass.: Raytheon.

Pimentel, George C., and Richard D. Spratley. 1969. *Chemical Bonding Clarified Through Quantum Mechanics.* San Francisco: Holden-Day.

Sanderson, R. T. 1971. *Chemical Bonds and Bond Energy.* New York: Academic Press. (Advanced.)

EXERCISES

1. Suggest simple molecular orbital descriptions for:
   a. bromine molecule; **b.** hydroxide ion
2. Draw both electron-dot and line formulations for
   a. methylamine; **b.** methanol; **c.** hypochlorous acid;
   d. hydrogen sulfide; **e.** hydrogen cyanide (HCN)
3. Using the electronegativities given in Table 7.6, suggest appropriate classifications for
   a. the Al—Cl bond in $AlCl_3$
   b. the P—Cl bond in PCl_3
   c. the N—Cl bond in NCl_3
   d. the K—Cl bond in KCl
   e. the H—Cl bond in HCl
4. (*Review*) Methanol may be converted to formaldehyde (H_2CO), formic acid (HCO_2H), or carbon dioxide by selection of appropriate reagents and reaction conditions. Calculate the mass of each of the following products that could be produced from 96 g of methanol, assuming complete conversion to the product shown under conditions such that methanol is the limiting reagent:
   a. carbon dioxide; **b.** formic acid; **c.** formaldehyde
5. (*Review*) The standard heat of combustion of formaldehyde is -134.9 kcal mole^{-1} and that of formic acid is -62.8 kcal mole^{-1}. Calculate the standard enthalpy changes for oxidation of 96 g of methanol by each of the following reactions:
   a. $2CH_3OH + O_2 \longrightarrow 2H_2CO + 2H_2O$
   b. $CH_3OH + O_2 \longrightarrow HCO_2H + H_2O$
   c. $2CH_3OH + 3O_2 \longrightarrow 2CO_2 + 4H_2O$
   (Consult Table 5.1, Unit 5, for additional data.)

OPTIONAL

6. Draw a complete set of resonance contributors for carbonate ion.

PRACTICE FOR PROFICIENCY

Objective (a)

1. Draw simple molecular orbital representations for
   a. the hydrogen molecule
   b. the HS$^-$ ion
   c. the chlorine molecule
   d. lithium hydride molecule
   e. lithium iodide molecule

Objective (b)

2. Draw both electron-dot and line-bond formulations for
   a. ammonia; **b.** perchloric acid; **c.** ethanol; **d.** water;
   e. carbon dioxide; **f.** beryllium fluoride; **g.** nitrogen molecule;
   h. formic acid (HCO_2H); **i.** nitric oxide; **j.** carbon monoxide
   (In this molecule, oxygen forms three covalent bonds—a rare case.)
3. Formaldehyde may be represented as H_2CO. Draw a complete line-bond or electron-dot representation and then determine which statement is correct.

a. The formaldehyde molecule contains four σ bonds.
b. The formaldehyde molecule contains two lone pairs of valence electrons.
c. The formaldehyde molecule contains two π bonds.
d. The formaldehyde molecule contains one unpaired electron.
4. Which molecule-bond description is incorrect?
   a. ammonia: three σ bonds, one lone pair
   b. methanol: five σ bonds, no lone pairs
   c. carbon dioxide: two σ bonds, two π bonds, four lone pairs
   d. ethane: seven σ bonds, no lone pairs
   e. hydrogen fluoride: one σ bond, three lone pairs

Objective (c)

5. Using data from Table 7.6, suggest appropriate classifications (ionic, polar covalent, or nonpolar covalent) for
   a. the Si—Br bond in $SiBr_4$ **b.** the C—S bond in CS_2
   c. the Mg—F bond in MgF_2 **d.** the Be—H bond in BeH_2
   e. the N—O bond in NO_2
6. Which of the following is classified as nonpolar?
   a. the H—Cl bond in HCl
   b. the Ca—Cl bond in $CaCl_2$
   c. the N—Cl bond in NCl_3
   d. the C—O bond in CO_2
   e. the Cs—S bond in Cs_2S

Objective (d) [*Review*]

7. A prescription sleeping drug, chloral hydrate, can be prepared by the reaction of equimolar amounts of trichloroethanal and water:

$$\overset{\displaystyle H}{\underset{\displaystyle |}{Cl_3CC}}{=}O + H_2O \longrightarrow Cl_3CCH(OH)_2$$

If a particular process for this reaction is 72% efficient, how many grams of chloral hydrate could be prepared from 25 g of trichlorethanal?
   a. 15   **b.** 20   **c.** 25   **d.** 38   **e.** 50

Objective (e) [*Review*]

8. Illegal drugs confiscated by narcotics agents are usually destroyed by burning. Calculate the standard enthalpy change for the combustion of 1.00 kg of heroin, according to the equation,

$$4C_{21}H_{23}NO_5 + 97O_2 \longrightarrow 84CO_2 + 46H_2O + 2N_2$$

Compound	$\triangle H_f^0$ (kcal mole^{-1})
heroin	-73.7
water	-68.3
carbon dioxide	-94.0

   a. -7270 kcal; **b.** -6150 kcal; **c.** -5420 kcal; **d.** -4810 kcal;
   e. -2580 kcal

9. Draw complete sets of resonance contributors for
   a. nitrogen dioxide; **b.** bicarbonate ion; **c.** nitrous oxide (contains N—N bonding); **d.** formate ion (HCO_2^-)

BRAIN TINGLER

10. *Isomers* are defined as species having the same molecular formulas, but different structural formulas. For example,

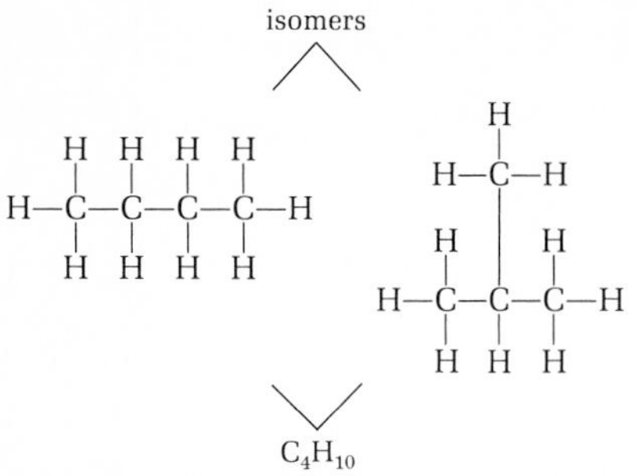

isomers

C_4H_{10}

It is important to note that isomeric differences are *real*, in terms of differences in structure, and do not represent just different ways of drawing the same structures. For example, the following drawings represent only *one* real structure:

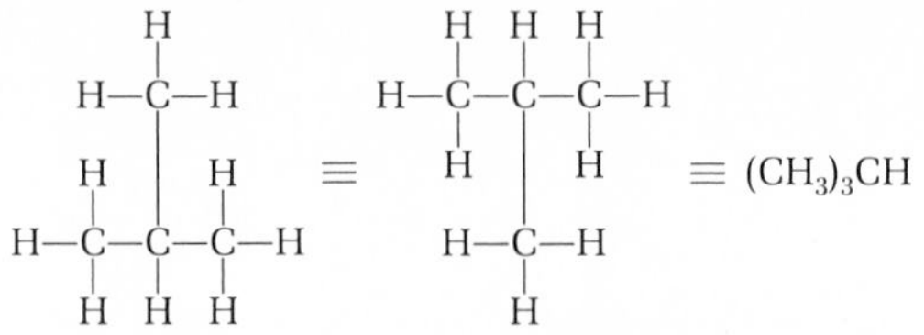

$\equiv (CH_3)_3CH$

a. Draw both electron-dot and line-bond representations for all isomeric compounds having the formula C_3H_8O.
b. One of these isomers is commonly marketed as "rubbing alcohol". This isomer contains two CH_3 groups and one O—H bond. Identify the structural formula of this compound.

Unit 7 Self-Test

1. Draw simple molecular orbital representations for
   a. hydrogen bromide; **b.** hypochlorite ion
2. Draw both electron-dot and line-bond formulations for each molecule or ion:
   a. methylammonium ion; **b.** ethane; **c.** ammonia; **d.** perchlorate ion; **e.** acetic acid
3. On the basis of electronegativity differences, classify the indicated bonds as probably ionic (I), polar covalent (PC), or nonpolar covalent (NP):

Element	Electronegativity	Element	Electronegativity
Al	1.5	F	4.0
Cl	3.0	N	3.0
H	2.2	Rb	0.8

   a. the Al—F bond in AlF_3
   b. the N—F bond in NF_3
   c. the N—Cl bond in NCl_3
   d. the H—N bond in NH_3
   e. the Rb—Cl bond in RbCl
4. (*Review*) Many years ago, during the gaslight era, a fuel called "water gas" was made by passing steam over heated coal. The carbon from the coal was converted to carbon monoxide, and the steam was reduced to molecular hydrogen. Assuming a stoichiometric conversion to an equimolar mixture of hydrogen and carbon monoxide ("water gas"), what weight of coal was required for formation of 12 tons of water gas for a city's gaslight district?
5. (*Review*) One of the problems with certain anesthetics, particularly cyclopropane or diethyl ether, is flammability. Mixtures of the gases or vapors with air may ignite explosively from a tiny spark. Calculate the standard enthalpy change for the complete combustion of 300 g of diethyl ether:

$$C_2H_5OC_2H_5 + 6O_2 \longrightarrow 4CO_2 + 5H_2O$$

Compound	$\triangle H_f^0$ (kcal mole^{-1})
diethyl ether	-88.6
carbon dioxide	-94.0
water	-68.3

   a. -5300 kcal; **b.** -4200 kcal; **c.** -3400 kcal; **d.** -2500 kcal; **e.** -2100 kcal

OPTIONAL

6. Draw the complete set of resonance contributors for the acetate ion.

Unit 8

It is perhaps too much to hope
that the geometrical arrangement
of primary particles will ever be
perfectly known.

W. H. WOLLASTON (1808)

Molecular geometry

Objectives

(a) Be able to draw representations of the geometric patterns described as *linear, bent* (V shape), *trigonal planar, pyramidal, square planar, tetrahedral, trigonal bipyramidal,* and *octahedral* (Figure 8.1).

(b) Be able to use VSEPR theory to predict the approximate geometries (including bond angles) of simple molecules and polyatomic ions (Sections 8.2 and 8.3).

(c) Be able to suggest appropriate simple molecular orbital models for molecules and polyatomic ions involving σ bonds from simple or hybrid atomic orbitals (Section 8.4).

(d) Be able to suggest simple molecular orbital models for molecules or polyatomic ions involving π-bond systems (Section 8.5).

(e) Be able to predict molecular polarity by considerations of bond dipoles, geometric features, and electronic symmetry or asymmetry (Section 8.6).

The shape of a molecule, or of a particular region within a molecule, is a critical factor in many aspects of chemical reactions in the laboratory. Molecular geometry is of even greater importance in biochemistry, the study of the chemical processes associated with living organisms.

Most chemical reactions of molecular species are relatively slow—far too slow to sustain the biochemical chain of events involved in a living system. Chemical reactions in cells and more complex organisms are usually speeded up (catalyzed) by complex macromolecules (giant molecules) called enzymes. These enzymes act only on certain specific molecules or molecular segments that will fit the catalytic region of the enzyme, the active site. Enzymes are therefore stereoselective in their action, requiring certain geometric features of atomic arrangements and certain positions of varying electron density in the reactants on which they operate. The special requirements for these reactants (the enzyme substrates) are associated with features of the shapes and polarities of molecules.

Many simpler processes also depend on characteristics of molecular geometry and varying regions of electron density.

During World War II, the Allied Armies feared the possibility of poison gas warfare, especially near the closing days of the conflict when Hitler's forces were contemplating desperate measures to postpone the inevitable defeat. In spite of international agreements banning the use of war gases, research had continued during the years following World War I, and much more sophisticated chemical warfare agents had been developed. One of these was an arsenic compound, lewisite, which caused severe burns on contact with the skin, rapid damage to lung and bronchial tissue, and—usually—death, apparently from irreversible destruction of vital sulfur-containing enzymes. The British, anticipating a possible exposure to lewisite gas, developed a specific chemical antidote, British antilewisite (BAL). The function of BAL was directly dependent on specific features of molecular geometry and electron density which permitted the molecule, 2,3-dimercaptopropanol [HSCH$_2$CH(SH)CH$_2$OH], to form a strong complex with arsenic.

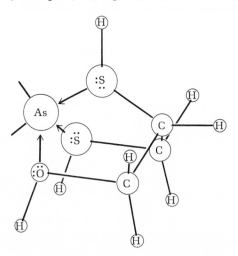

The lewisite threat never materialized, but BAL has proved most useful in treating cases of heavy metal poisoning from arsenic, mercury, copper, and nickel (but not lead) compounds.

An understanding of the geometries (Figure 8.1) and polarities of molecules is the key to any working relationship involving covalent compounds or polyatomic ions.

Figure 8.1 Common geometric patterns in simple molecules and polyatomic ions.

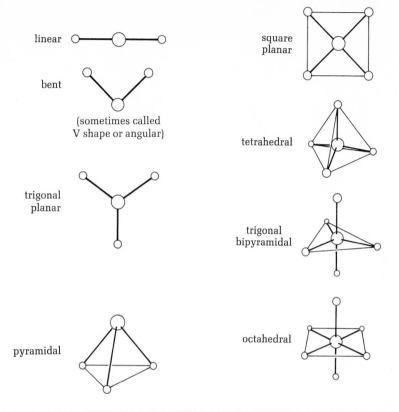

linear

bent

(sometimes called V shape or angular)

trigonal planar

pyramidal

square planar

tetrahedral

trigonal bipyramidal

octahedral

8.1 EXPERIMENTAL DETERMINATION OF MOLECULAR GEOMETRY

Reasonable guesses as to the shapes of molecules can be made on the basis of simple theories of electron repulsion (Section 8.2) and molecular orbitals (Unit 7). These theories have been available, at least in some rough form, since the 1920s. It has only been within the last few years, however, that it has been possible to obtain accurate experimental data for the determination of relative atomic locations within molecules.

Three-dimensional electron density maps, obtained by the techniques of x-ray diffraction (Figure 8.2), provide the most useful direct source of structural information. Accurate structural determinations are, unfortunately, limited to substances that form crystalline compounds, and heavy atoms are easier to locate with precision than lighter atoms. For molecules in the liquid or gaseous states, a somewhat analogous technique of electron diffraction can be employed. Neutron diffraction can be used to locate light atoms such as hydrogen. Of these methods, x-ray diffraction is by far the most widely used, and this technique has resulted in the accumulation of a vast amount of information related to molecular geometry.

X-ray diffraction requires the collection of a large body of data and its interpretation by a complex series of successive approximations. As a result, early x-ray studies were exceedingly laborious and time consuming. With the advent of high-speed computers and automated systems of data collection, this approach has become accessible to almost every research group involved in elucidation of molecular structure.

The successive approximations used in interpreting x-ray data are greatly sim-

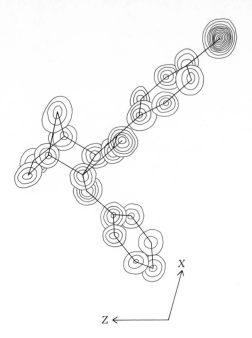

Figure 8.2 Electron density map from x-ray data. A projection down the y-axis of the final contoured electron density map for an organic compound containing bromine, nitrogen, oxygen, and carbon (hydrogens are not visible). The outline of the ring system has been superimposed over the contours. The extent of contouring readily identifies the bromine atom at upper right.
(Jon A. Kapecki, *J. Chem. Ed.* 49, 234 [1972], used by permission.)

plified if a major portion of the molecule's structure is already known. The sequence of atoms within a molecule, that is, the structural formula (Unit 7), can be determined by various chemical methods and by techniques of molecular spectroscopy. Many of the more common applications of these methods are discussed in later units. Indeed, many molecules can be related by chemical and physical investigations to substances whose three-dimensional structures have already been determined. Each piece of structural information accumulated aids in the ultimate determination of total molecular geometry.

It should be understood that any static description of molecular shape is only an approximation for real molecules, whose molecular motions (Unit 9) produce continuous distortions of bond angles and bond lengths. Interactions with other chemical species, such as those occurring in solutions, may involve dramatic changes from the molecular geometries determined for substances in pure crystals.

8.2 VSEPR THEORY: PREDICTING GEOMETRY

There are many different factors involved in the determination of molecular geometry. It must be admitted that we do not know all of these factors, nor do we have any highly accurate method for comparing the relative magnitudes of the factors that are understood. This means that we do not yet have a satisfactory theory for the accurate prediction of molecular geometry.

Fortunately, there is a simple method available for predicting the *approximate* shapes of molecules and polyatomic ions involving representative elements. Originating with the suggestions of G. N. Lewis in the 1920s, this *valence shell electron-pair repulsion* (VSEPR) theory owes its current popularity largely to the efforts of R. J. Gillespie of McMaster University (Hamilton, Ontario).[1] The basis for this theory is the Pauli exclusion principle (Unit 1), which states that any electron orbital is

---

[1] Students often form the impression that a workable theory is universally accepted by scientists. In fact, such is often not the case. For an enlightening look at alternative viewpoints, see the following articles; "The Electron-Pair Repulsion Model for Molecular Geometry" by R. J. Gillespie (*J. Chem. Ed.*, **47**, 18–23, 1970); "A Criticism of the Valence Shell Electron-Pair Repulsion Model as a Teaching Device" by R. S. Drago (*J. Chem. Ed.*, **50**, 244–245, 1973); "A Defense of the Valence Shell Electron-Pair Repulsion Model" by R. J. Gillespie (*J. Chem. Ed.*, **51**, 367–370, 1974).

limited to a maximum occupancy of two electrons of opposite spins. The VSEPR theory is an outgrowth of this principle and may be stated, in its simplest form as

The preferred arrangement of a given number of electron pairs in the valence shell of an atom is that which maximizes the distance between the pairs.

We must recognize, of course, the limitation on electron-pair arrangements imposed by multiple bonds (Unit 7). Although our orbital model of the atom restricts the *centralized* internuclear region to a single electron pair, the electron pairs of π bonds must still lie within the *general* internuclear region (surrounding the σ-bond orbital).

The first approximation of molecular shape is then made on the following basis:

1. *An electron-dot (or line-bond) formula of the species is drawn.* (For species requiring resonance hybrid representation, Unit 7, only a single drawing is required for this purpose.)
2. Any atom in the molecule (or polyatomic ion) bonded to more than one other atom is referred to as a *central atom.*
3. Each atom or lone pair of electrons shown on the central atom is referred to as a *group.*
4. Groups are arranged around the central atom in such a way as to provide the maximum separation between groups. Geometric designations (Figure 8.1) may then be applied to either group arrangements or, neglecting locations of lone-pair electrons, to atomic arrangements around the central atom. The latter practice is most common in describing *molecular* (or polyatomic) *geometry* (Table 8.1).
5. For molecules or polyatomic ions containing more than one "central atom", each is described separately, then "connected".

Table 8.1   SOME EXAMPLES OF GEOMETRIES FROM VSEPR THEORY

SPECIES	FORMULATION	CENTRAL ATOM	GROUPS	FIRST APPROXIMATION OF GEOMETRY "GROUP"	"MOLECULAR" (ATOMIC PATTERN)
BeH_2	H:Be:H	Be	2 (2 atoms)	H—Be—H (linear)	H—Be—H (linear)
CO_2	Ö::C::Ö	C	2 (2 atoms)	O=C=O (linear)	O=C=O (linear)
BF_3	:F:B:F: :F:	B	3 (3 atoms)	(trigonal planar)	(trigonal planar)
NO_2^-	-:Ö:N::Ö	N	3 (2 atoms + 1 lone pair)	(trigonal planar)	(bent, or V-shape)
CH_4	H:C:H H H	C	4 (4 atoms)	(tetrahedral)	(tetrahedral)

Table 8.1 (Continued)

SPECIES	FORMULATION	CENTRAL ATOM	GROUPS	FIRST APPROXIMATION OF GEOMETRY	
				"GROUP"	"MOLECULAR" (ATOMIC PATTERN)
NH_3	H:N̈:H H	N	4 (3 atoms + 1 lone-pair)	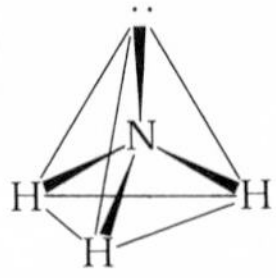 (tetrahedral)	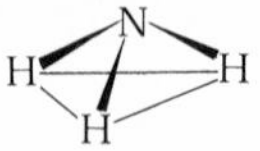 (pyramidal)
H_2O	H:Ö:H	O	4 (2 atoms + 2 lone-pairs)	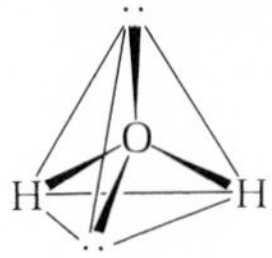 (tetrahedral)	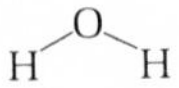 (bent, or V-shape)
PCl_5	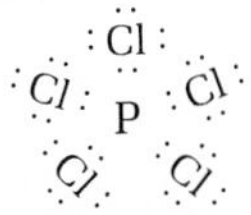 :Cl: :Cl: P :Cl: :Cl: :Cl:	P	5 (5 atoms)	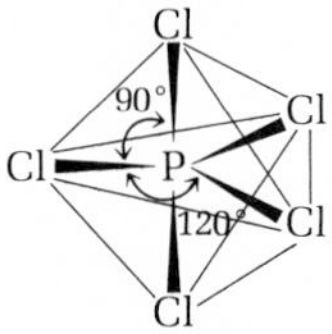 (trigonal bipyramidal)	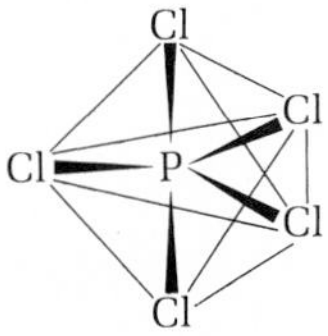 (trigonal bipyramidal)
SF_6	 :F: :F: :F: :F: S :F: :F: :F: :F:	S	6 (6 atoms)	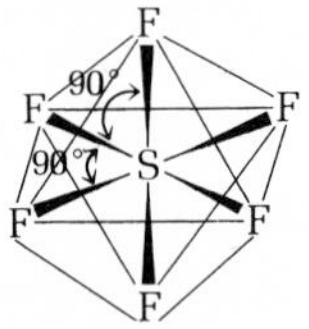 (octahedral)	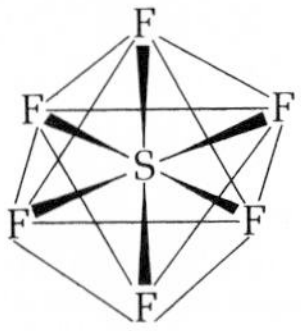 (octahedral)
XeF_4	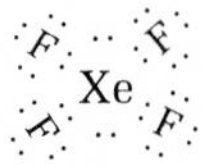 :F: :F: :F: Xe :F: :F: :F:	Xe	6 (4 atoms + 2 lone-pairs)	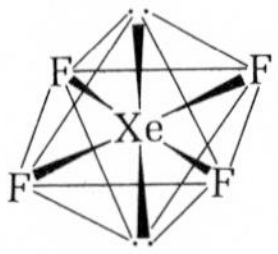 (octahedral)	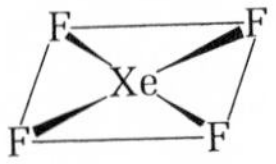 (square planar)

Example 1

Predict the geometries of
a. phosphorus trichloride
b. nitrate ion
c. acetone $\left(\begin{array}{c} O \\ \parallel \\ CH_3CCH_3 \end{array} \right)$

Solution

a. STEP 1 :Cl̈:P̈:Cl̈:
          :Cl̈:

STEP 2 Phosphorus is the central atom.

STEP 3 There are four groups (three atoms plus one lone pair).

STEP 4 Maximum separation of four groups is achieved by the tetrahedral pattern:

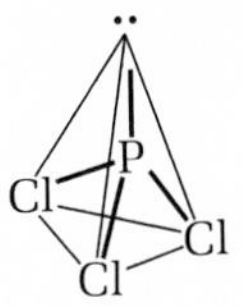

The *atomic* arrangement is

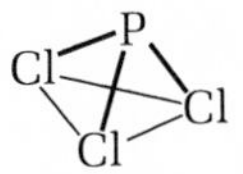

(pyramidal)

b. STEP 1 $^{-}:\ddot{O}:N::\ddot{O}$ (one of the three equivalent representations)

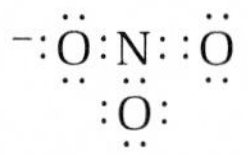

$:\underset{\cdot\cdot}{\overset{\cdot\cdot}{O}}:$

STEP 2 Nitrogen is the central atom.

STEP 3 There are three groups.

STEP 4 A trigonal planar arrangement provides maximum separation of three groups.

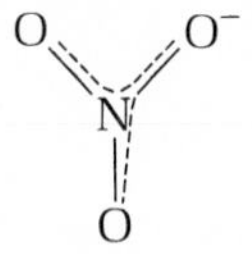

O ＼ ／ O⁻
N
‖
O   [delocalized representation (Unit 7)]

(trigonal planar)

c. STEP 1

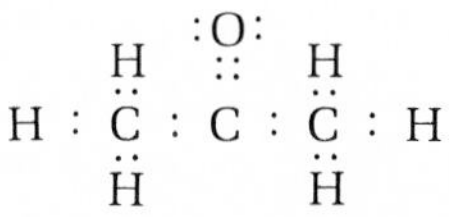

$$\begin{array}{ccc} & :O: & \\ H & :: & H \\ H:C:C:C:H & \\ H & H & \end{array}$$

STEP 2 Carbon is the central atom in each of three patterns:

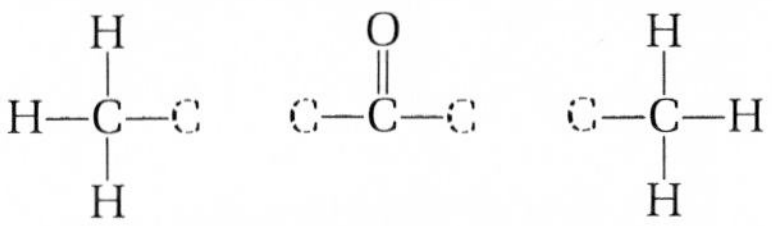

$$\begin{array}{ccc} H & O & H \\ | & \| & | \\ H-C-C & C-C-C & C-C-H \\ | & & | \\ H & & H \end{array}$$

STEP 3 In two of the "pieces" there are four groups; the other has three.

STEP 4

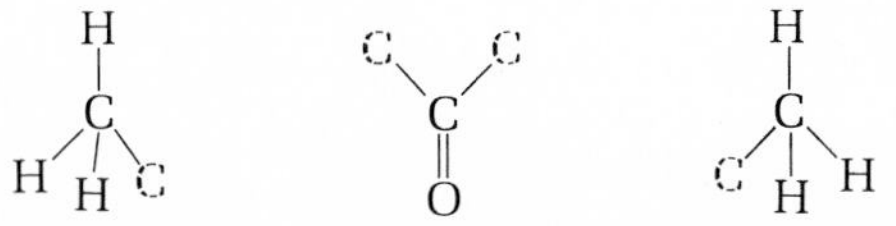

H | C H H C	C ＼ ／ C C ‖ O	H | C C H H
(tetrahedral region)	(trigonal planar region)	(tetrahedral region)

STEP 5 Total molecular geometry is approximated by "connecting the pieces."

(tetrahedral region)   (tetrahedral region)

(trigonal planar region)

8.3 NONEQUIVALENT "GROUPS": PREDICTING BOND-ANGLE VARIATIONS

The first approximations of molecular geometry by the VSEPR theory appear to suggest only a few possibilities for bond angles (180° with two groups; 120° with three groups; 109.5°, the tetrahedral angle, with four groups; 90° with six groups; and some 90° and some 120° with five groups). Experimentally, however, wide variations from these idealized angles are observed (Table 8.2).

The idealized geometries were logical extensions of treating all groups (e.g., lone pairs and various atoms) as though they were equivalent in the volume of space occupied and in the nature of group interactions. Although this was a convenient simplification for purposes of arriving at an approximation of molecular shape, it was not consistent with what we know about electrical interactions and atomic sizes (Unit 2). Perhaps we can use considerations of nonequivalence of groups to modify our suggestions of molecular geometry to bring them more into line with experimental observations.

We recognize that the electron pair of a σ bond (Unit 7) is localized within the central internuclear region. Lone-pair electrons, on the other hand, are not constrained to such a narrow region, so we might expect the "lone-pair volume" to be larger than a "bond-pair volume." The lone-pair region, then, would tend to "push away" neighboring bond-pair regions, forcing these closer together (decreasing the bond angle). This factor can account for the less-than-tetrahedral bond angles of molecules such as ammonia or water (Figure 8.3).

Table 8.2 SOME EXPERIMENTALLY DETERMINED BOND ANGLES

SPECIES	"IDEALIZED" BOND ANGLE (deg)	EXPERIMENTAL ANGLE (deg)
NH_3	109.5	107
NF_3	109.5	102
NCl_3	109.5	110
H_2O	109.5	105
H_2S	109.5	93
H_2Se	109.5	91
PH_3	109.5	93
PF_3	109.5	98
PCl_3	109.5	100
PBr_3	109.5	102
AsH_3	109.5	92
$\begin{array}{c} Cl \\ \diagdown \\ C=O \\ Cl \diagup \end{array}$	120	$\begin{array}{c} Cl \\ \diagdown \\ C = 118 \\ Cl \diagup \end{array}$
		$\begin{array}{c} Cl \\ \diagdown \\ C=O = 124 \end{array}$

Figure 8.3 Effect of nonequivalent groups on bond angle. Methane shows "perfect" tetrahedral angle. In ammonia, lone-pair to bond-pair repulsions exceed bond-pair to bond-pair repulsions, decreasing the bond angle. Further bond-angle decrease in water reflects additional space occupied by two lone-pair regions.

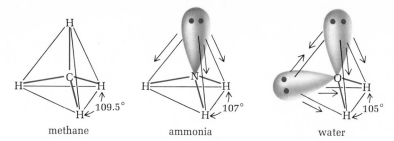

methane   ammonia   water

An even greater deviation from the idealized tetrahedral angle is noted in compounds of the heavier elements of Groups VA and VIA. For example, in PH_3 and H_2S the bond angles are about 93° (Table 8.2). We may rationalize this by suggesting that the lone-pair region increases with increasing size of the central atom, thus further compressing the bond angles. This suggestion would predict that bond angles are even smaller in AsH_3 and H_2Se. Such, indeed, is the case (Table 8.2).

That not all atomic groups are equivalent is obvious from the bond-angle differences in compounds such as PH_3 (93°), PF_3 (98°), PCl_3 (100°), and PBr_3 (102°). Here the trend is consistent with the differences in atomic size (Unit 2; H < F < Cl < Br). It would appear reasonable that bond-pair:bond-pair repulsions would increase with increasing size (increasing volume of electron regions) of "atomic groups." In most cases this is the observed trend. The hydrides of nitrogen and oxygen are notable exceptions (Table 8.2). If we remember the unique character of hydrogen (no electrons other than valence electrons), we should not be too surprised by apparent anomalies involving hydrogen in covalent molecules. We shall encounter others.

The influence of a multiple bond is consistent with the high electron density in the general internuclear region (Unit 7), as illustrated by phosgene ($Cl_2C{=}O$, Table 8.2).

Although these general considerations do not enable us to *predict* bond angles accurately, they do provide a qualitative method for "refining" the idealized approximations from first use of VSEPR theory.

Example 2

Predict idealized geometries, then suggest whether bond angles should be equal to (=), less than (<), or greater than (>) idealized angles for

  a. stibine (SbH_3)
  b. carbon tetrachloride
  c. iodoform (CHI_3)
  d. nitrous acid

Solution

  a. H:Sb:H   (four groups, three atoms plus one lone pair)
          ··
          H

The idealized geometry is a tetrahedral arrangement of groups (109.5° angles), with a pyramidal molecular geometry:

Since the lone-pair region on antimony (a large central atom) should occupy considerable space, bond angles should be

$\ll 109.5°$ (experimental value: 91.3 ± 0.3)

b. 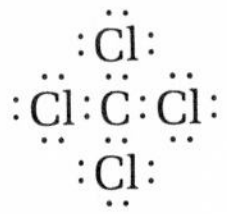
$$
\begin{array}{c}
:\ddot{C}l:\\
:\ddot{C}l:\ddot{C}:\ddot{C}l:\\
:\ddot{C}l:
\end{array}
$$
(four groups, all atoms)

The idealized geometry is tetrahedral and because all groups are equivalent, the bond angle should be

$= 109.5°$ (experimental value: $109.5°$)

c. 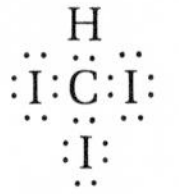
$$
\begin{array}{c}
H\\
:\ddot{I}:\ddot{C}:\ddot{I}:\\
:\ddot{I}:
\end{array}
$$
(four groups, all atoms, nonequivalent)

While recognizing the unique character of hydrogen, we might still expect the large iodine atoms to dominate, in terms of "repulsions" leading to distortion of the tetrahedral angles such that

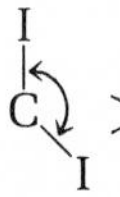

$> 109.5°$ and 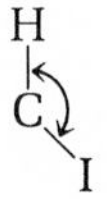 $< 109.5°$ (experimental value for : $113°$)

d.  $H:\ddot{O}:\ddot{N}::\ddot{O}$

For the H—$\ddot{O}$—N region we predict a tetrahedral arrangement of groups with a bent (V-shape) atomic pattern around oxygen. The $O:\dot{N}::O$ region should be trigonal planar with respect to groups. Idealized geometry could be represented as

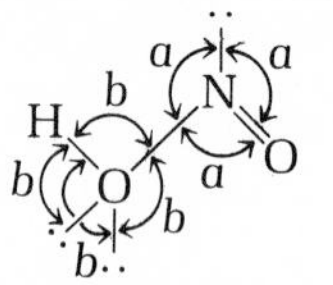

$a = 120°$
$b = 109.5°$

On the basis of the two lone-pair regions on oxygen, we can predict with reasonable assurance that

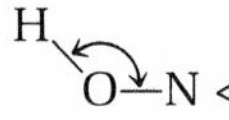

O—$N < 109.5°$ (experimental value: $108°$)

The remaining bond angle is not so easy to "guess," since we must weigh the relative influence of a lone-pair region versus a double-bond region. Because nitrogen is a small atom, we might expect the lone-pair region to be less important than the double-bond region. If such is the case, then

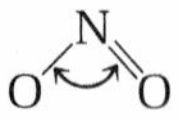

is $> 120°$

If on the other hand the lone-pair region dominates, the bond angle should be $< 120°$. We have no way of judging competing factors on a quantitative basis, but we can suggest that the two opposing "repulsive effects" should be rather close, so that the bond angle should not differ very much from the idealized angle. The experimental value is $116° \pm 2°$. Apparently the lone-pair region dominates. (This example illustrates the limitations of our theory and the fact that it is easier to rationalize an observation than to make an accurate prediction.)

8.4 SIMPLE MOLECULAR ORBITAL DESCRIPTIONS FOR SIGMA-BOND GEOMETRIES

Since multiple-bonded atoms are treated as simple groups by VSEPR theory, only σ-bond descriptions are required for designations of orbitals consistent with molecular geometries. The simple atomic p and d orbitals (Unit 1) limit possible descriptions to bond angles around 90°, since these orbitals are mutually perpendicular. Nondirectional (spherical) s orbitals provide no suitable descriptions of geometrical features.

In a few cases we can satisfactorily describe molecular geometries by the overlap of simple atomic p orbitals (Figure 8.4), but these provide no descriptions compatible with bond angles significantly larger than 90°.

Although simple molecular orbital theory is not sufficiently advanced to permit its use in the *prediction* of molecular shapes, the orbital model has considerable utility in helping us to visualize the arrangement of valence electrons in molecules and polyatomic ions and, as we shall see in later units, in helping to understand phenomena associated with electronic excitation. What we need are some atomic orbital descriptions consistent with linear, trigonal planar, tetrahedral, square planar, trigonal bipyramidal, and octahedral group geometries.

Hybrid orbitals represent a sort of "averaging" process to use the geometric features of simple orbitals in achieving new, *equivalent* orbital descriptions. The mathematical process involved is actually more complex than simply averaging the equations descriptive of atomic s, p, and d orbital characteristics, but we may think of it this way. The idea is to develop *equivalent* orbital descriptions consistent with known characteristics of bonds. Suppose, for example, that we need two colinear orbitals to account for bonding in a linear molecule, such as BeH_2. "Averaging" of any two p or d orbitals would have to result in something other than a linear shape, but averaging of a spherical s orbital with a single p orbital should result in two new orbitals whose directional characteristics resulted only from the p orbital (Figure 8.5).

In a similar way we can arrive at a set of trigonal planar orbitals. Here we need

The Salt Lake City Zoo had an animal called a liger, the product of the mating of a lion with a tiger. This *hybrid* had some tiger characteristics (e.g., stripes) and some lion characteristics (e.g., a mane). The process for its formation was also somewhat more complex than averaging.

Figure 8.4 Some σ-bond descriptions closely approximating molecular geometries, using simple atomic orbital overlap model.

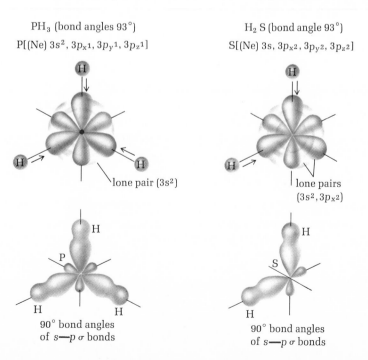

PH$_3$ (bond angles 93°)

P[(Ne) $3s^2$, $3p_x{}^1$, $3p_y{}^1$, $3p_z{}^1$]

lone pair ($3s^2$)

90° bond angles of s—p σ bonds

H$_2$S (bond angle 93°)

S[(Ne) $3s$, $3p_x{}^2$, $3p_y{}^2$, $3p_z{}^2$]

lone pairs ($3s^2$, $3p_x{}^2$)

90° bond angles of s—p σ bonds

Figure 8.5 Formation of *sp* hybrid orbitals, consistent with linear geometry.

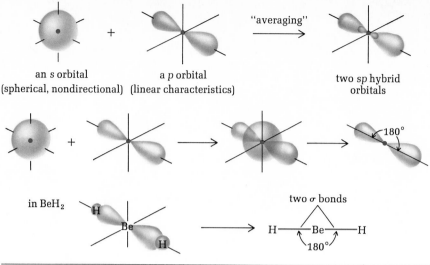

an *s* orbital
(spherical, nondirectional)

a *p* orbital
(linear characteristics)

"averaging"

two *sp* hybrid orbitals

180°

in BeH$_2$

H

Be

H

two σ bonds

H——Be——H

180°

Figure 8.6 Formation of *sp*^{2} hybrid orbitals, consistent with trigonal planar geometry.

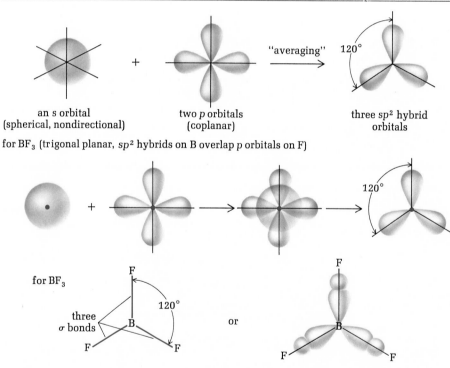

an *s* orbital
(spherical, nondirectional)

two *p* orbitals
(coplanar)

"averaging"

120°

three *sp*^{2} hybrid orbitals

for BF$_3$ (trigonal planar, *sp*^{2} hybrids on B overlap *p* orbitals on F)

120°

for BF$_3$

F

120°

three
σ bonds

B

F

F

or

F

B

F

F

three orbitals, none of which have directional characteristics outside a common plane. Any two *p* orbitals are coplanar and various coplanar combinations of *d* orbitals or *p*-with-*d* orbitals could be considered. The simplest hybridization would involve one *s* orbital with two *p* orbitals (Figure 8.6). The complex mathematics of the hybridization process reveals that only this set results in three new orbitals, *sp*^{2} (from "one part" *s* and "two parts" *p* character) and provides the unique 120° angle characteristic of the trigonal planar geometry. Note that the 120° angle is consistent with the requirement for *three equivalent* orbitals.

Square planar molecules or polyatomic ions are relatively rare. We shall reserve a discussion of these for the later study of transition element chemistry (Unit 27). It may be noted, however, that a dsp^2 hybrid orbital description is consistent with the square planar geometry.

For species having a "three-dimensional" arrangement of groups, the simplest possibility involves hybridization of an s orbital with a complete set (three) of p orbitals, resulting in four sp^3 hybrids of tetrahedral geometry (Figure 8.7). When all

Although sp^3 hybridization provides a convenient description for NH_3 and H_2O, experimental evidence reveals that lone-pair orbitals in water are *not* equivalent.[2]

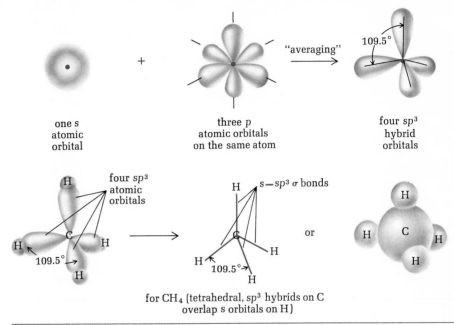

one s
atomic
orbital

three p
atomic orbitals
on the same atom

four sp^3
hybrid
orbitals

Figure 8.7 Formation of sp^3 hybrid orbitals, consistent with tetrahedral geometry.

four sp^3 atomic orbitals

s—sp^3 σ bonds

for CH_4 (tetrahedral, sp^3 hybrids on C overlap s orbitals on H)

Figure 8.8 Descriptions of sp^3 hybrid orbitals for ammonia (pyramidal) and water (bent).

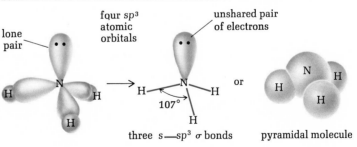

lone pair

four sp^3 atomic orbitals

unshared pair of electrons

three s—sp^3 σ bonds

pyramidal molecule

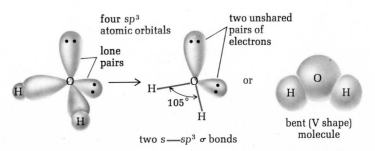

four sp^3 atomic orbitals

lone pairs

two unshared pairs of electrons

two s—sp^3 σ bonds

bent (V shape) molecule

[2] See: D. A. Sweigart, *J. Chem. Ed.,* **50,** 322, 1973).

Figure 8.9 Some possible hybrid orbital descriptions consistent with octahedral and trigonal bipyramidal geometries.

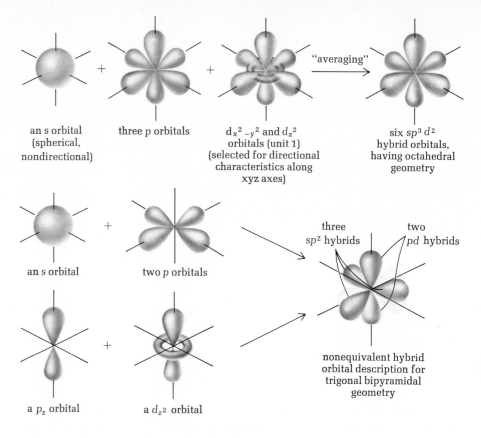

an s orbital (spherical, nondirectional)

three p orbitals

$d_{x^2-y^2}$ and d_{z^2} orbitals (unit 1) (selected for directional characteristics along xyz axes)

"averaging"

six $sp^3 d^2$ hybrid orbitals, having octahedral geometry

an s orbital

two p orbitals

a p_z orbital

a d_{z^2} orbital

three sp^2 hybrids

two pd hybrids

nonequivalent hybrid orbital description for trigonal bipyramidal geometry

four of these hybrids are used for σ-bond descriptions, tetrahedral molecules, such as methane, may be visualized. Assignment of lone pairs to one or two of the sp^3 hybrid orbitals provides reasonable descriptions for such molecules as ammonia or water (Figure 8.8).

Both octahedral and trigonal bipyramidal geometries require the inclusion of d orbitals in the hybridization process (Figure 8.9). Only a few of the representative elements form molecules having these geometries, but we shall find such descriptions (especially for the octahedral case) quite useful in the study of transition elements.

The use of simple molecular orbitals is, like the first approximations of VSEPR theory, a considerable simplification. Distortions from idealized geometries can be visualized on a qualitative basis from considerations of group repulsions (Section 8.3), while retaining the essential features of our orbital descriptions. However, we need to recognize that the molecular orbital model, although useful, is still just a *model* and not a "real picture" of molecular structure.

8.5 PI-BOND DESCRIPTIONS

We have already suggested (Unit 7) that the "side-to-side" overlap of atomic p orbitals can provide a simple molecular orbital description for π bonding. This can be applied to a variety of molecules and polyatomic ions and although such descriptions are unnecessary for our simple approach to molecular geometry, our orbital model for covalent bonding would be incomplete without some visualization of π bonds.

When p orbitals are available for overlap on only two σ-bonded atoms, *localized* π-bonding results. Our model for a double bond then looks much like a hotdog, with the frankfurter representing the σ bond and the bun analogous to the π bond (Figure 8.10). In a similar way we can represent a triple bond as "like a hotdog with an extra bun" (Figure 8.11). Again, valence electrons not required for the σ-bonding description are assigned to p orbitals.

Figure 8.10 Simple molecular orbital description for localized π bond in ethylene.

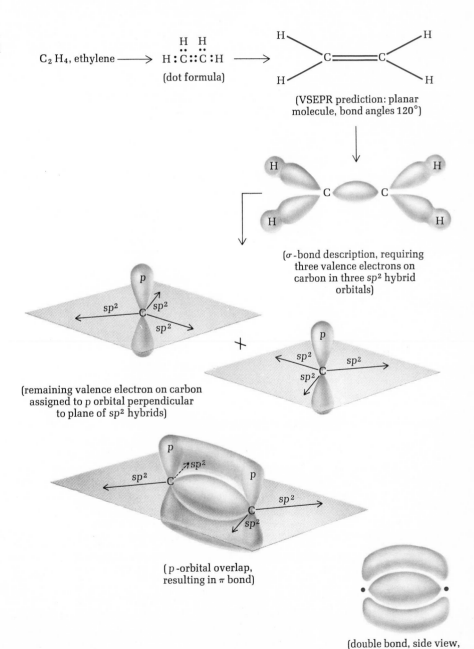

C_2H_4, ethylene $\longrightarrow$

H H
H:C::C:H
(dot formula)

$\longrightarrow$

(VSEPR prediction: planar molecule, bond angles 120°)

(σ-bond description, requiring three valence electrons on carbon in three sp^2 hybrid orbitals)

(remaining valence electron on carbon assigned to p orbital perpendicular to plane of sp^2 hybrids)

$+$

(p-orbital overlap, resulting in π bond)

(double bond, side view, showing "hot dog" similarity)

Figure 8.11 Simple molecular orbital description for two π bonds in the triple covalent bond of acetylene.

C_2H_2, acetylene $\longrightarrow$ H:C:::C:H $\longrightarrow$ H——C≡≡C——H
(dot formula) (VSEPR prediction: linear molecule)

(σ-bond description, requiring two valence electrons on carbon in two sp hybrid orbitals)

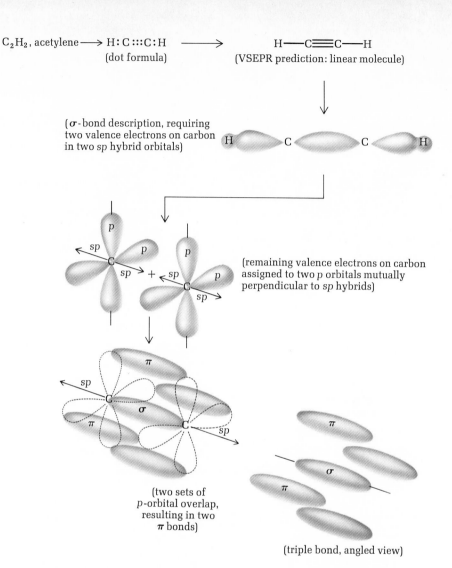

(remaining valence electrons on carbon assigned to two p orbitals mutually perpendicular to sp hybrids)

(two sets of p-orbital overlap, resulting in two π bonds)

(triple bond, angled view)

The concept of a delocalized π bond suggests orbital occupancy by more than two electrons. However, the system always "averages out" to two electrons per pair of bonded atoms.

Since the model of p-orbital overlap from σ-bonded atoms appears useful in describing π bonding, it would seem appropriate to suggest that π bonding *might extend throughout a polyatomic system in which p orbitals are available on contiguous atoms.* Remember that our orbital model suggests that electrons may "travel" throughout any given orbital. Thus electrons in an extensive π-bond system are no longer "localized" to the vicinity of their "parent atoms," and we now have an orbital description for *delocalized* bonding in polyatomic ions (Figure 8.12) or molecules (Figure 8.13, p. 212).

8.6 POLAR AND NONPOLAR MOLECULES

The concept of a molecular *dipole,* a model of a molecule as an unsymmetrical "charge cloud" with one end slightly positive (δ^+) with respect to the other (δ^-), is consistent with experimental observations of dielectric properties (Fig. 8.14, p. 212). Molecules having a net dipole moment greater than zero are said to be *polar,* whereas those of zero net dipole moment are *nonpolar.* The dipole moment of a molecule is a function of the magnitude of the effective charge difference of the two extremes of

Figure 8.12 Representation for delocalized π bond (and delocalized charge) in nitrate ion.

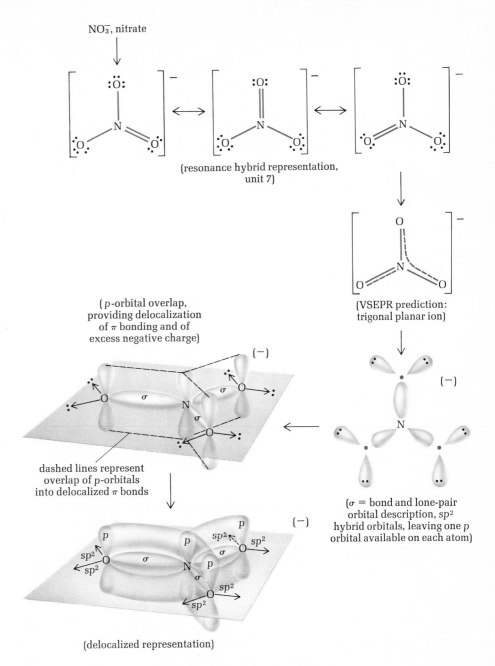

NO_3^-, nitrate

(resonance hybrid representation, unit 7)

(VSEPR prediction: trigonal planar ion)

(p-orbital overlap, providing delocalization of π bonding and of excess negative charge)

dashed lines represent overlap of p-orbitals into delocalized π bonds

(σ = bond and lone-pair orbital description, sp^2 hybrid orbitals, leaving one p orbital available on each atom)

(delocalized representation)

the dipole representation and of the effective distance of charge separation. This is most easily visualized for simple diatomic molecules, as has been described in Unit 7.

With diatomic molecules the molecular polarity is the same as the bond polarity, and this may be approximated from the difference in electronegativities of the bonded atoms (Unit 7). Polyatomic molecules, however, require additional consider-

Figure 8.13 Representation for delocalized π bonding in benzene.

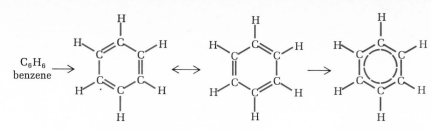

C_6H_6 benzene $\rightarrow$

(resonance–hybrid representation)

(remaining valence electron on each carbon assigned to p orbital, providing extensive overlap system)

(VSEPR prediction: planar molecule, all bond angles 120°)

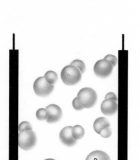

$\delta-$
$\delta+$

field off

(molecules in random orientations)

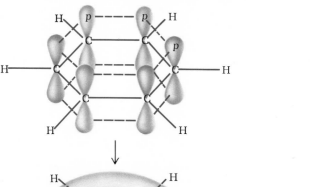

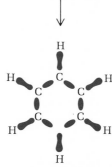

(σ-bond description using sp^2 hybrid orbitals on all carbon atoms)

(delocalized π-bond system)

(shorthand notation for benzene)

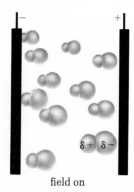

$\delta+$ $\delta-$

field on

(molecules aligned so that charged plates attract oppositely "charged" ends of dipole)

[$\delta+$ (Greek delta) represents region of lower electron density; $\delta-$ represents region of higher electron density.]

Figure 8.14 Experimental evidence for molecular polarity. Polar molecules located between charged plates reduce the effective voltage (potential difference) between the plates by an amount related to the dipole moment of the molecule. The experimental quantity determined by this magnitude is called the *dielectric constant*.

ations, in particular those of group geometry. We can state with considerable assurance that *any molecule containing equivalent bonds arranged in a symmetrical pattern will be nonpolar, whereas any molecule containing polar bonds arranged in an unsymmetrical pattern will be polar.* This is a logical consequence of the vector treatment of bond dipoles (Unit 7). Although it would seem that molecules having nonpolar bonds, such as NCl_3, would have a dipole moment of zero, this is not necessarily the case. The net molecular dipole is determined by the asymmetry of the total "electron cloud" of the molecule, so some consideration must be given to lone-pair electron regions. Thus although nitrogen and chlorine have the same electronegativity, 3.0 (Unit 7), the NCl_3 molecule *is* polar even though the dipole moment is quite small.

We are concerned with molecular polarity primarily because of the utility of this concept in helping us understand many aspects of the "bulk" properties of matter, such as boiling point, melting point, or solubility. As we explore these areas in later units, we will find that a rather simple approach to the prediction of molecular

Figure 8.15 Some polar and nonpolar molecules.

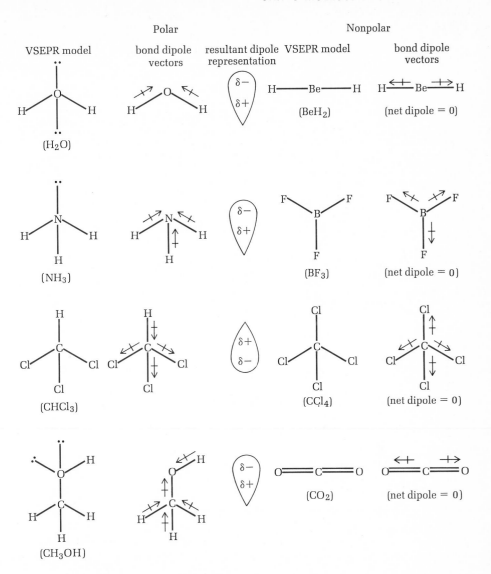

polarity will be quite adequate for our needs. It is only necessary to predict the group geometry by the first approximations of VSEPR theory. If the resulting prediction is one of complete symmetry, we will expect the molecule to be nonpolar; any unsymmetrical feature will suggest that the molecule is polar (Figure 8.15).

8.7 MOLECULAR GEOMETRY AND BIOCHEMISTRY

The shapes of molecules or of particular regions within molecules are of the utmost importance in many biochemical events. For example, starches and celluloses are complex polymers containing enormous numbers of glucose units (a simple sugar) connected by bond systems called *glycoside links*. The most important difference between starches and celluloses is concerned with the geometry of this linkage. Starches have the so-called α-glycoside configuration; celluloses are β (Greek *beta*)

THE STEREOCHEMICAL THEORY OF ODOR

Although many of the properties of covalent substances are determined primarily by the nature of particular atoms in the molecule and the relative atomic locations in the covalent bond system (Unit 29), there is experimental evidence that the *shape* of a molecule may be of special significance in certain "biochemical properties". Of particular interest is the theory that the odor we associate with a substance is determined, to a large extent, by the shape of the "odor" molecule that permits it to "fit" one of a few uniquely shaped cavities at our olfactory (odor-detecting) nerve endings.

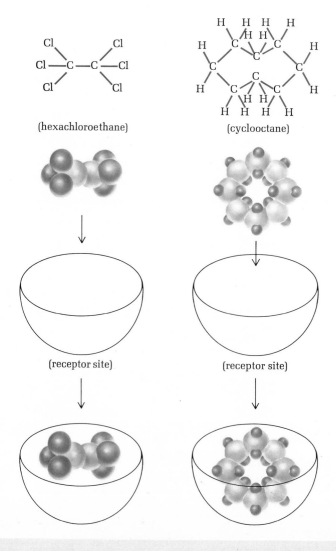

(hexachloroethane)   (cyclooctane)

(receptor site)   (receptor site)

It has been proposed that there are seven primary odors, each associated with an appropriately shaped nerve ending in our odor receptor system. The primary odors have been classified as camphoraceous (camphorlike), ethereal (etherlike), floral (flowerlike), musky, pepperminty, pungent, and putrid. Other odors are believed to result from the simultaneous stimulation of two or more of the primary receptors. Models for each of the receptor sites have been suggested, along with models for the corresponding shapes of "odorous molecules." As an example, such different molecules as hexachloroethane and cyclooctane have camphoraceous odors because their molecular shapes are such as to fit into a hemispherical receptor site.

"Complex" odors, resulting from multiple-receptor stimulation, may be associated with either mixtures of different molecules or with single-type molecules having a geometry that permits a "fit" with more than one of the primary receptor sites.

The molecular shape–receptor site model is not limited to humans. It has been found, for example, that spherical or oval-shaped molecules make more effective mosquito repellants than long, flat molecules, presumably because of the more effective blocking action of the "round" molecules at mosquito receptor sites.[3]

[3] If you find this aspect of molecular geometry of interest, you may wish to read "The Stereochemical Theory of Odor" by John Amoore, James Johnston, Jr., and Martin Rubin, in the February, 1964 issue of *Scientific American,* and "Why Mosquito Repellants Repel" by R. H. Wright, in the July, 1975 issue of *Scientific American.*

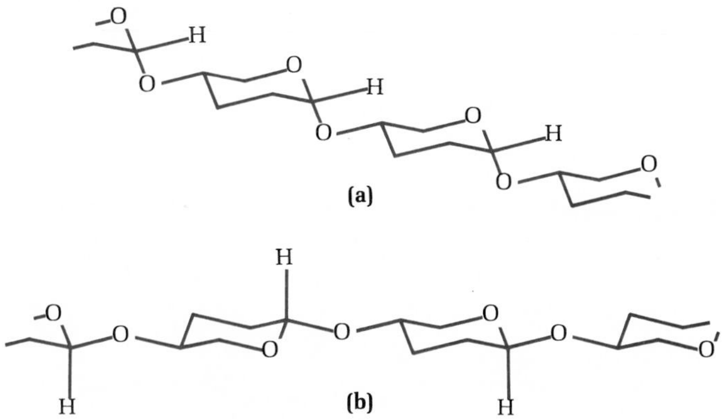

Figure 8.16 Alpha and beta glycosides. (a) Typical of starches. (b) Typical of celluloses. The "puckered hexagons" represent glucose structures. (Compare the formulation for sucrose given in Table 7.1, Unit 7.)

(Figure 8.16). Enzymes in the human digestive system are capable of rupturing only the α-glycoside links. Thus even though both starches and celluloses contain glucose, only the starches provide a source for humans of this important "chemical fuel."

One of the most fascinating stories concerning molecular geometry is connected with the work of Watson and Crick on the genetic code, culminating in the Nobel Prize for their studies of the genetic material.[4]

[4] Their research is summarized in the excellent book, *The Double Helix,* by James D. Watson (New York: New American Library, 1968).

SUGGESTED
READING

Bates, Robert B., and John P. Schaefer. 1971. *Research Techniques in Organic Chemistry.* Englewood Cliffs, N.J.: Prentice-Hall. (An excellent introduction to methods of structure determination.)

Coulson, C. A. 1973. *The Shape and Structure of Molecules.* New York: Oxford.

Ifft, James B. and John E. Hearst. 1974. *General Chemistry (Readings from Scientific American).* San Francisco: W. H. Freeman.

Pimentel, George C., and Richard D. Spratley. 1969. *Chemical Bonding Clarified Through Quantum Mechanics.* San Francisco: Holden-Day.

Ryschkewitsch, G. E. 1963. *Chemical Bonding and the Geometry of Molecules.* New York: Van Nostrand Reinhold.

EXERCISES

1. Using ball-and-stick models or styrofoam balls and toothpicks, build models of:

   a. two spheres attached to a central sphere in a colinear pattern.

   b. three spheres arranged in a trigonal planar pattern around a central sphere

   c. four spheres tetrahedrally spaced around a central sphere

   d. four spheres arranged in a square planar pattern around a central sphere

   e. five spheres arranged in a trigonal pyramidal pattern around a central sphere

   f. six spheres forming an octahedral arrangement around a central sphere

   g. a three-sphere V-shape (bent) pattern

   h. a four-sphere pyramidal pattern

   Now draw pictures of these models, using a convention acceptable to your instructor for (c), (e), (f), and (h).

2. Use VSEPR theory and "group-repulsion" considerations to predict the approximate shapes of the following species. Indicate any expected deviations from "idealized" bond angles (by > or <).

   a. boron trifluoride; **b.** carbon dioxide; **c.** phosphorus trichloride; **d.** carbonate ion; **e.** formate ion (HCO_2^-); **f.** sulfate ion; **g.** selenium hexafluoride

3. Suggest appropriate molecular orbital descriptions for σ bonding in

   a. boron trifluoride; **b.** arsine (AsH_3, bond angles ~90°); **c.** water; **d.** carbon tetrachloride; **e.** phosphorous pentachloride; **f.** selenium hexafluoride; **g.** beryllium hydride; **h.** ammonia

4. Draw simple representations for both σ-bond and π-bond systems, showing approximate atomic geometries, in

   a. carbon dioxide; **b.** bicarbonate ion; **c.** benzene; **d.** carbonate ion; **e.** acetylene

5. Draw representations of the molecular shapes of the following species. Indicate which are polar compounds.

   a. methanol; **b.** hydrogen cyanide; **c.** nitrous acid; **d.** chloroform; **e.** ethylene

PRACTICE FOR PROFICIENCY

Objective (a)

1. Draw representations for each of the following geometric patterns:

a. linear; **b.** bent; **c.** trigonal planar; **d.** pyramidal; **e.** square planar; **f.** tetrahedral; **g.** trigonal bipyramidal; **h.** octahedral

Objective (b)

2. Draw approximate representations for the shapes of the following species. Label bond angles with the expected "idealized" angles or, if deviation is probable, as greater than (>) or less than (<) the appropriate angle (in degrees).

   a. aluminum chloride; **b.** hydrogen cyanide; **c.** sulfur dioxide; **d.** nitric acid; **e.** hydroxylamine ($HONH_2$); **f.** nitrite ion; **g.** chlorate ion; **h.** perchlorate ion; **i.** phosphorus pentabromide; **j.** hexaaquoscandium(III) ion, $[Sc(H_2O)_6]^{3+}$ (*Hint:* The central scandium(III) ion is bonded to the oxygen in the water molecules.)

3. In which molecule are all atoms coplanar?

   a. CH_4; **b.** $SiCl_4$; **c.** PF_3; **d.** NH_3; **e.** BF_3

4. Which molecule has a pyramidal arrangement of component atoms?

   a. $AlCl_3$; **b.** $SiCl_4$; **c.** PCl_3; **d.** $BeCl_2$; **e.** BF_3

Objectives (c) and (d)

5. Write appropriate molecular orbital descriptions for each of the following species, indicating the nature of the orbitals used for σ bonding. Draw simple representations of both σ and π bonding, with appropriate atomic geometry, for each species. Indicate by an asterisk (*) any species in which delocalized π bonding occurs.

   a. hydrogen selenide (bond angle, 90°); **b.** hydrogen cyanide; **c.** formaldehyde (H_2CO); **d.** nitric acid; **e.** nitrobenzene **f.** nitrile ion; **g.** ammonium ion; **h.** acetate ion; **i.** oxalate ion; $(O_2CCO_2)^{2-}$; **j.** benzoate ion

6. Which statement is correct for the following molecule?

a. Carbon number 1 is described by sp^3 hybridization.
b. The molecule contains a total of 19 σ bonds.
c. Carbon number 2 is described by sp^2 hybridization.
d. The molecule contains a total of five π bonds.
e. Carbon number 3 is described by sp hybridization.

7. Which statement is correct for the following molecule?

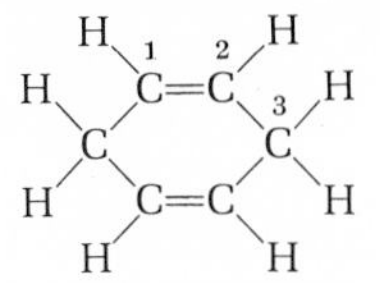

a. The molecule contains a total of 16 σ bonds.
b. Carbon number 1 is described by sp hybridization.
c. The molecule contains a total of four π bonds.
d. Carbon number 3 is described by sp^3 hybridization.
e. The molecule contains a completely delocalized π-bond system.

Objective (e)

8. Draw representations of the molecular shapes of the following species. Circle those representing polar molecules, then draw a simple "net dipole" representation for these molecules.
a. hydrogen sulfide; **b.** carbon disulfide; **c.** methylene chloride (CH_2Cl_2); **d.** chlorine oxide (Cl_2O); **e.** ethylene (H_2CCH_2); **f.** phosphine (PH_3); **g.** nitrosyl chloride ($ONCl$); **h.** aluminum chloride; **i.** para-dichlorobenzene 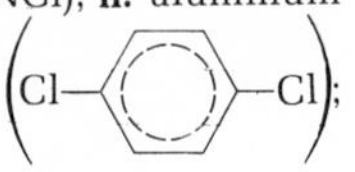 ;
j. phenol

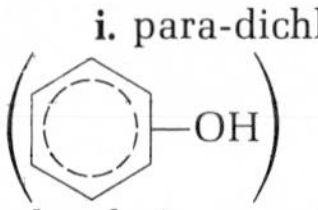

9. Which molecule is nonpolar?
a. BeH_2; **b.** $CHCl_3$; **c.** NH_3; **d.** ICl; **e.** none of these

BRAIN TINGLER

10. It has been suggested that "rounded" molecules (that would fit rather well in a spherical or oval "container") make more effective mosquito repellants than elongated molecules. Each of the following has been tested as a mosquito repellant and only three were found to be very effective. Construct models of each of these and predict which three are the effective

repellants. (Caution: Do not be deceived by the representations of molecular formulas. In constructing models, remember to keep delocalized π-bond systems coplanar. Molecules containing only σ bonds may be "twisted" into various shapes.)

a. dimethyl phthalate

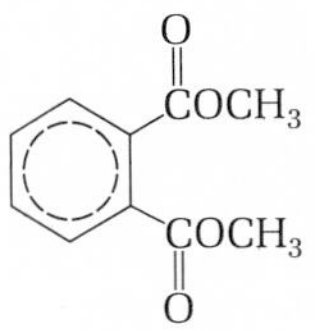

b. dimethyl terephthalate

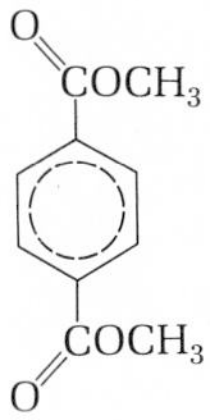

c. hexachlorophenol

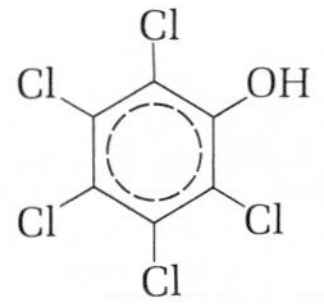

d. 2-ethyl-1,3-hexanediol

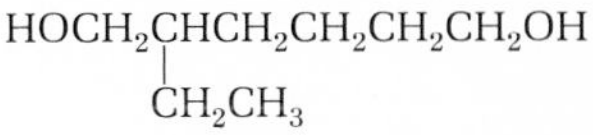

$$HOCH_2CHCH_2CH_2CH_2CH_2OH$$
$$\quad\quad\;\; |$$
$$\quad\quad\; CH_2CH_3$$

e. 1,3-propanediol monobenzoate

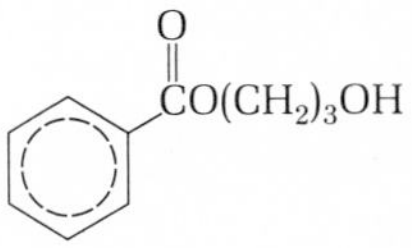

Unit 8 Self-Test

1. Match each of the following species with the appropriate description of molecular geometry (atomic pattern). (There is only one example of each geometric shape.)

   GEOMETRIES: (A) linear (B) bent (C) trigonal planar (D) pyramidal (E) tetrahedral (F) square planar

   a. BF_3; **b.** BeH_2; **c.** H_2Se; **d.** PF_3; **e.** $PtCl_4$; **f.** $SiCl_4$

2. Predict the bond angles for each of the following species (in degrees), indicating deviations from idealized angles: by $>$ or $<$.

   a. arsine (AsH_3); **b.** formate ion (HCO_2^-); **c.** nitrate ion; **d.** chloroform ($HCCl_3$); **e.** 1,1-dichloroethylene (Cl_2CCH_2)

3. Draw a complete line-bond or electron-dot formulation for carbon dioxide and then decide which statement is correct:

   a. The CO_2 molecule contains two lone pairs of valence electrons.

   b. The CO_2 molecule contains two σ bonds.

   c. The CO_2 molecule contains four π bonds.

   d. The carbon in CO_2 is described by sp^2 hybridization.

   e. The oxygen in CO_2 is described by sp^3 hybridization.

4. Which statement is correct for the following molecule?

   a. Carbon number 1 is described by sp^3 hybridization.

   b. The molecule contains a total of 10 σ bonds.

   c. Carbon number 2 is described by sp^2 hybridization.

   d. The molecule contains a total of two π bonds.

   e. Carbon number 3 is described by sp hybridization.

5. Predict which of the following molecules should be polar and draw a simple net dipole representation for each polar molecule:

   a. carbon dioxide; **b.** sulfur dioxide; **c.** hydrogen cyanide; **d.** cyanogen (NCCN); **e.** aluminum fluoride

1. In the following equations, underline each reactant species acting as an acid and circle each base (Bronsted-Lowry system):
   a. $HSO_3^-(aq) + OH^-(aq) \longrightarrow H_2O(l) + SO_3^{2-}(aq)$
   b. $HSO_3^-(aq) + HSO_4^-(aq) \longrightarrow$
   $$SO_2(g) + H_2O(l) + SO_4^{2-}(aq)$$
   c. $NH_4^+(aq) + CN^-(aq) \longrightarrow NH_3(aq) + HCN(aq)$
   d. $H_2PO_4^-(aq) + HSO_4^-(aq) \longrightarrow H_3PO_4(aq) + SO_4^{2-}(aq)$
   e. $H_2PO_4^-(aq) + HCO_3^-(aq) \longrightarrow$
   $$HPO_4^{2-}(aq) + H_2O(l) + CO_2(g)$$

2. Identify two amphoteric species on the basis of the reactions formulated in question 1.

3. Predict the products and write both *complete* and *net* balanced equations for all expected reactions. If no reaction is expected, indicate by NR. (You may wish to consult the table of solubility products in Appendix C.)
   a. aqueous acetic acid + aqueous sodium hydroxide $\longrightarrow$
   b. aqueous sulfuric acid + aqueous barium hydroxide $\longrightarrow$
   c. $CaCO_3(s) + KNO_3(aq) \longrightarrow$
   d. $Cu(NO_3)_2(aq) + Na_2S(aq) \longrightarrow$
   e. aqueous silver nitrate
   $$+ \text{ aqueous magnesium chloride} \longrightarrow$$

4. "Stannous fluoride" [tin(II) fluoride], a common additive in certain "fluoride" toothpastes, may be prepared by the reaction of tin(II) oxide with hydrofluoric acid:

   $$SnO + 2HF \longrightarrow H_2O + SnF_2$$

   If a particular process, using a ratio of 2 moles HF to 1 mole SnO, is 86% efficient, how many grams of tin(II) oxide would be required for production of 5.0 g of stannous fluoride, assuming the appropriate amount of HF was employed?
   a. 7.5; b. 5.0; c. 4.5; d. 3.0; e. 2.4

5. Given:

Compound	ΔH_f^0 (kcal mole^{-1})
sodium carbonate (s)	−270.3
sodium bromide (s)	−86.0
hydrogen bromide (g)	−8.7
carbon dioxide (g)	−94.0
water (l)	−68.3

Calculate the standard heat of reaction for conversion of sodium carbonate and hydrogen bromide to sodium bromide, H_2O, and CO_2 in kcal mole^{-1} (sodium carbonate).
a. −96.3; b. −114.9; c. −30.7; d. −46.6; e. −58.4

6. Draw simple molecular orbital representations for:
   a. hydrogen fluoride; b. fluorine molecule

7. Draw both electron-dot and line-bond formulations for:
   a. nitrous acid; b. hydrogen cyanide; c. ammonia; d. carbon dioxide; e. formic acid (HCO_2H)

8. On the basis of electronegativity differences, classify the indicated bonds as probably ionic (I), polar covalent (PC), or nonpolar covalent (NP):

Element	Electronegativity
Be	1.5
C	2.5
H	2.2
Br	2.8
K	0.8
S	2.5

   a. the H—S bond in H_2S; b. the H—Be bond in BeH_2; c. the K—Br bond in KBr; d. the C—S bond in CS_2; e. the C—Br bond in CBr_4

9. Draw simple representations for each of the following geometric patterns:
   a. pyramidal; b. trigonal planar; c. square planar; d. trigonal bipyramidal; e. octahedral

10. For each of the following species, give the expected molecular geometry (e.g., "bent"), the "idealized" bond angles, and any deviations expected from "idealized" angles (as > or < the angle, in degrees).
    a. NH_3; b. CH_4; c. NO_3^-; d. CO_2; e. SF_6

11. Which statement is correct for the following molecule?

    a. Carbon number 2 is described by sp^3 hybridization.

b. The molecule contains a total of 17 σ bonds.

c. Carbon number 4 is described by sp^2 hybridization.

d. The molecule contains a total of five π bonds.

e. Carbon number 6 is described by sp hybridization.

12. Which molecule is nonpolar?

   a. $CHCl_3$; **b.** H_2Se; **c.** NF_3; **d.** BF_3; **e.** none of these

OPTIONAL

13. Which is the correct *and complete* electron-dot structure for carbonate ion?

a. $\left[\ddot{O} ::C:: \ddot{O} \atop :\ddot{O}: \right]^{2-} \longleftrightarrow \left[:\ddot{O}:C:: \ddot{O} \atop :\ddot{O}: \right]^{2-} \longleftrightarrow \left[\ddot{O} ::C: \ddot{O}: \atop :\ddot{O}: \right]^{2-}$

b. $\left[\ddot{O} ::C: \ddot{O}: \atop :\ddot{O}: \right]^{2-} \longleftrightarrow \left[:\ddot{O}:C:: \ddot{O} \atop :\ddot{O}: \right]^{2-} \longleftrightarrow \left[:\ddot{O}:C: \ddot{O}: \atop :\ddot{O}: \right]^{2-}$

c. $\left[:\ddot{O}:C: \ddot{O}: \atop :\ddot{O}: \right] \longleftrightarrow \left[:\ddot{O}:C: \ddot{O}: \atop :\ddot{O}: \right] \longleftrightarrow \left[:\ddot{O}:C: \ddot{O}: \atop :\ddot{O}: \right]$

d. $\left[:\ddot{C}::O:\ddot{O}:\ddot{O}: \right] \longleftrightarrow \left[:\ddot{C}:O::\ddot{O}:\ddot{O}: \right] \longleftrightarrow \left[:\ddot{C}:O:O::\ddot{O} \right]$

e. $\left[\ddot{O} ::C:: \ddot{O} \atop :\ddot{O}: \right]^{2-} \longleftrightarrow \left[{:\ddot{O}: \atop \ddot{O} ::C:: \ddot{O}} \right]^{2-}$

Section FOUR

STATES OF MATTER

Up to this point we have concentrated on what we might call *single-particle* properties of matter, such as atomic size, ionization potential, ionic charge density, or molecular shape and polarity. Except for some of the more esoteric scientific studies, however, most of the observable characteristics of chemical systems represent *bulk* properties of matter: properties associated with *collections* of atoms, ions, or molecules. A diamond, for example, is a hard, transparent crystal, whereas charcoal is "soft" and black. Both are forms of carbon; the differences in properties are not characteristic of the individual carbon atom, but, rather, of the way in which the carbon atoms are held together in the "bulk" material.

Water may exist as ice, a brittle crystalline substance; as a liquid; or in the gaseous state, as *water vapor* or *steam*. When liquid water freezes, expansion occurs. (Ice is less dense than liquid water.) Gaseous water, like other gases, expands when heated at constant pressure, but steam does not behave, under many conditions, as simply as gases such as helium. In the solid, liquid, or gaseous state, water is still represented as H_2O. We are dealing with a single kind of molecule, but with many different characteristics. These characteristics depend, as we shall see, on the energies and types of aggregation of *collections* of H_2O molecules.

Molecules are not *static* (rigid); they are only roughly approximated by stick-and-ball models. Covalent bonds may stretch in various ways, resulting in changing bond lengths, and bond angles change as a result of "bending" vibrations within polyatomic molecules. These motions change molecular shapes and, in some cases, molecular polarities. Molecules may also move around and rotate in space. We must learn to think of molecules as dynamic species. Static molecular models represent molecules no better (and probably less accurately) than a modernistic statue represents a real, live, moving person.

Nevertheless, simple models for atoms, ions, and molecules are essential first steps to an understanding of the interactions and motions characteristic of the various kinds and states of matter. In this section, we shall use our simple single-particle models to develop the more extensive concepts required for explanations of such "bulk" properties as gas behavior, solid and liquid characteristics, and change of state.

CONTENTS

Unit 9

Moving and interacting particles

Objectives

(a) After studying the appropriate definitions in Appendix E and the terms'
usage in this Unit, be able to demonstrate your understanding of the following terms:

Boltzmann distribution	metallic bond
covalent radius	metallic radius
diffusion	momentum
dipole–dipole forces	rotation
dipole-induced dipole forces	solubility
dipole–ion forces	solute
effusion	solution
elastic collision	solvent
hydrogen bond	temperature
hydrophilic	translation
hydrophobic	translational kinetic energy
inelastic collision	van der Waals forces
ionic radius	van der Waals radius
London dispersion forces	vibration
mean free path	

(b) Be able to calculate the number and types of degrees of freedom for a molecule, to draw representations for simple molecular motions, and to identify degenerate vibrations (Sections 9.1 and 9.2).

(c) Given the name or formula of a compound or element, be able to identify the interparticle forces among unit particles (atoms, ions, or molecules) of the substance (Sections 9.5 and 9.6).

(d) On the basis of the relative strengths of interparticle forces, be able to make general predictions of the relative boiling points of simple substances (Section 9.8).

(e) Be able to identify the types of interparticle forces in mixtures and, on the basis of the relative strengths of these forces, to make general predictions of solubilities of substances in common solvents (Section 9.9).

REVIEW

Molecular geometries and polarities (Unit 8).

With the public attention focused on aerosols as possible environmental hazards, it will be interesting to see how long they survive as "common household items."

No home today is complete without its collection of aerosol cans—insecticides, deodorants, hair spray, and so on. Each can is carefully labeled, DO NOT PUNCTURE OR INCINERATE. The propellant gases, under pressure, can escape rapidly through a hole in the can or with explosive violence if the can is heated too much.

The familiar "pressure cooker" has some pressure-measuring device and a special safety seal that can pop if the steam pressure of an overheated cooker exceeds safe limits.

Red blood cells can be stored, at proper temperatures, in a solution of "physiological saline." The cells maintain their normal shapes by an osmotic pressure (Unit 15) of the cell fluids equal to that of the surrounding solution. If the cells are placed in pure water, this balance is disturbed. Water diffuses through the membrane into the cell, causing it to expand and burst (hemolysis). Placing the cells in a concentrated salt solution has the opposite effect; water diffuses irreversibly out of the cell, causing the cell to shrivel (crenation).

If a colloidal dispersion (Unit 14), such as a suspension of submicroscopic chalk particles in water, is viewed through a microscope with a properly lighted stage, tiny dots of light are observed in erratic jostling movements. This Brownian motion, a characteristic of colloidal dispersions, is not unlike the erratic jerking of a tin can under the impact of BBs fired by a group of surrounding children with BB guns.

Human hair consists of twisted strands of long protein molecules, much like the component fibers of a piece of hemp rope. Hair can be curled, or uncurled, by chemical substances causing changes in the shapes of protein molecules.

In forensic chemistry, a tiny sample of a pure chemical taken as legal evidence—such as methyl nitrate for manufacture of illicit explosives or an illegal drug—can be identified with certainty without destroying the sample. Infrared spectroscopy uses the absorption of specific wavelengths of infrared radiation to "fingerprint" chemical species. Each wavelength band absorbed is associated with vibrational and rotational motions of covalently bonded atoms in molecules or polyatomic ions. The complete infrared absorption spectrum for a pure compound, when compared with that of a known sample determined under the same conditions, can furnish unambiguous proof of the compound's identity.

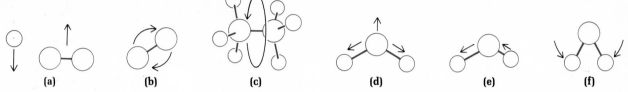

Figure 9.1 Particles in motion. (a) Translation. (b) Simple rotation. (c) Internal rotation. (d) Symmetric vibration. (e) Asymmetric vibration. (f) Bending vibration.

All of these seemingly unrelated situations have something in common. They all depend on characteristics of particles in motion—translational, rotational, and vibrational motions (Figure 9.1).

Water is beyond question the most thoroughly studied of all chemical substances. The water molecule is bent (Unit 8), containing two highly polarized O—H bonds and two lone pairs of valence electrons, at angles closely approximated by sp^3 hybrid orbitals on oxygen. Both oxygen and hydrogen are small, so the water molecule contains regions of high electron density associated with the oxygen atom and centers of relative positive character associated with the hydrogen atoms. The highly polar character (Unit 8) of the water molecule accounts for the strong attractive forces among water molecules, between water molecules and positive or negative ions, or between water molecules and highly polarized regions in other molecular substances. These interparticle forces account for many of the special properties of water which make it of unique importance.

Man requires about $2\frac{1}{2}$ quarts of water per day under "normal" conditions. Some of this is formed within the body as a product of the oxidation of certain chemicals taken in as food. Some is supplied as part of the food itself. Most must be consumed as liquid. Water is required by the human body for a number of different purposes. Liquid water forms a "transport medium" in body fluids for the flow of nutrients, bound oxygen, waste products, and many other substances in solutions or colloidal dispersions. Other, more isolated, aqueous systems provide "reaction media" for a host of essential biochemical processes; in many of these water itself is an important reactant, as in the hydrolysis of proteins and carbohydrates. In addition, water is of critical importance in controlling body temperature. In the process of perspiration, the high heat of vaporization of water permits the evaporation of a relatively small amount of water from the body surface to absorb a large quantity of heat.

Hydrolysis refers to any reaction in which water is a reactant.

The types of interparticle forces in which water may participate explain the behavior of water as a solvent for a wide variety of molecular and ionic species, as a continuous phase for many colloidal dispersions (Unit 14), and as an effective heat absorber in processes such as the melting of ice (Unit 12) or the vaporization of liquid water (Unit 12). These same interparticle forces promote the proper orientations of water molecules for the effective collisions necessary for hydrolysis reactions such as those involved in certain critical steps in the metabolism processes of living organisms.

Pure water is rapidly becoming a rare commodity. The same properties that make water a good solvent or dispersing agent in biological systems have contributed to the significant problems of water pollution. Much of the water in natural sources has always been unsuitable for humans to drink or for use in the irrigation of farm lands. About 70 percent of the earth's surface is covered by this "salt" water, whose content of Na^+, Cl^-, Mg^{2+}, SO_4^{2-}, and so on is far above the permissible limits for drinking or agricultural water. To this we can add the large quantities of "fresh" water now being contaminated by the addition of human and industrial waste

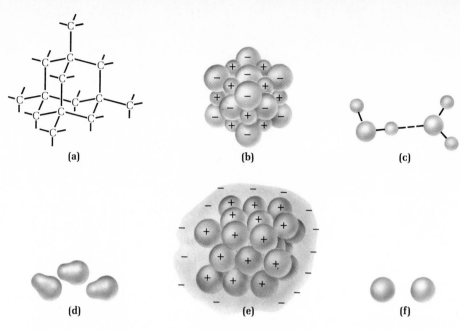

Figure 9.2 Interparticle forces in pure substances. (a) Covalent bonds in diamond. (b) Ionic bonds in rock salt. (c) Hydrogen bonds in water. (d) Dipole-dipole forces in hydrogen chloride. (e) Metallic bond in copper. (f) London dispersion forces in neon.

(a) (b) (c)

(d) (e) (f)

products. The removal of dissolved or suspended impurities requires energy, so "pure" water is not inexpensive. Research in desalinization processes (removal of salts from sea water) has progressed to the stage where, interestingly enough, it is often cheaper to prepare fresh water from sea water than to remove the contaminants that man has added to his "fresh" water systems.

In this unit we shall examine the nature of interparticle forces in pure chemical substances (Figure 9.2) and in mixtures of different species. These forces provide useful clues, not only to the properties of water, but to a host of related physical and chemical properties of all kinds of materials.

9.1 TYPES OF MOLECULAR MOTIONS

Kinetic energy is associated with movement. A bicycle with its rider pedaling along a road has kinetic energy by virtue of the movement of the bike and rider from one place to another (translation). This does not, however, represent the total kinetic energy. Wheels and pedals are going around (rotation); the rider's legs are pumping up and down (vibration). Even when the bike has stopped for a traffic light, the kinetic energy of the system (bike and rider) is not zero. The rider's chest may be heaving from the exertion of the ride, and he may be stretching his legs to avoid muscle cramps.

In an analogous manner, the kinetic energies of molecules encompass a number of different types of motions (Figure 9.1). When molecules are "stopped," as in crystalline compounds, some rotational and vibrational motions continue. Even at absolute zero ($-273°C$), molecular kinetic energies are greater than zero; even if all translational, rotational, and vibrational motions had ceased, electrons have minimum kinetic energies in their ground-state orbitals (Unit 1).

The average speeds of various types of molecular motions at room temperature may be calculated from equation (9.1), as illustrated in Section 9.3, and from theoretical considerations related to molecular spectra (Unit 30). These speeds vary with the characteristics of different molecules, as well as with temperature, physical state, and (for gases) pressure. The general orders of magnitudes, however, may be seen from the information in Table 9.1.

Average values of speeds discussed in this unit are not simple averages, but root-mean-square averages. The root-mean-square average is the square root of the average value of the square of the quantity. For *qualitative* purposes, this distinction may be disregarded.

Table 9.1 COMPARISONS OF MOLECULAR MOTIONS (IN CO_2 AT 25°C AND 1.00 atm)

translational velocity: ∼900 miles hr^{-1} (∼4.1 × 10^4 cm sec^{-1})
frequency of intermolecular collisions: ∼6.5 × 10^9 collisions sec^{-1}
frequency of symmetric stretching vibration: ∼4.0 × 10^{13} vibrations sec^{-1}
frequency of asymmetric stretching vibration: ∼7.0 × 10^{13} vibrations sec^{-1}
frequency of bending vibration: ∼2.0 × 10^{13} vibrations sec^{-1}
rotational frequency: ∼10^{11} rotations sec^{-1}

9.2 DEGREES OF FREEDOM

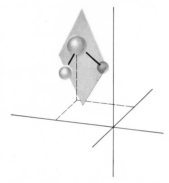

Figure 9.3 Degrees of freedom for a water molecule ($3n = 9$). Three translational degrees of freedom specify the distance from the center of mass to the x, y, and z axes. Three rotational degrees of freedom indicate the angle of the plane of the molecule with the xy, xz, and yz planes. The remaining three (vibrational) degrees of freedom specify the two O—H bond lengths (functions of stretching) and the bond angle (a function of bending).

Figure 9.4 Rotational degrees of freedom for CO_2. Only two angles are necessary to define the orientation of a straight line in space.

The number of degrees of freedom for a molecule is the number of Cartesian coordinates (x, y, z) needed to specify the locations of all atoms in the molecule (Figure 9.3). A molecule containing n atoms has $3n$ degrees of freedom, since three coordinates are needed to locate each atom. Three of these are always associated with the location in space of the center of mass of the molecule relative to some arbitrarily assigned reference point. Since these can be interpreted in terms of how far the molecule has moved from the reference position, they are called "translational degrees of freedom." The remaining degrees of freedom ($3n - 3$) are distributed between descriptions for vibrational and rotational motions. Linear molecules require only two terms to describe their rotation, leaving then $3n - 5$ ($3n - 3 - 2$) vibrational degrees of freedom. For nonlinear molecules the number of vibrational degrees of freedom is $3n - 6$. These values, ($3n - 5$) and ($3n - 6$), then predict the number of different fundamental vibrations (stretching and bending) possible for any molecule.

For better visualization of these ideas, we shall consider two different triatomic molecules, CO_2 (linear) and H_2O (bent). In both cases we could specify the exact position of a molecule in space by indicating distances to each atom in the molecule from a fixed reference point in terms of x, y, and z coordinates, using a set of conventional axes centered at the reference point. For any triatomic molecule, this would require nine items of information (three sets of three "distances").

There are, however, alternative ways of describing molecular position. Location of the center of mass of a molecule would require only three Cartesian coordinates and because the atoms are all connected, translational motion of the molecule could be followed by sets of these three translational degrees of freedom. Because the CO_2 molecule is linear (Unit 7), we can describe its orientation in space by *two* angles, those made by the intersections of an extension of the internuclear axis with two of the conventional reference planes (xy, yz, xz). Rotational motions could then be described in terms of changes of these angles, so the CO_2 molecule (or any *linear* molecule) has only two rotational degrees of freedom (Figure 9.4).

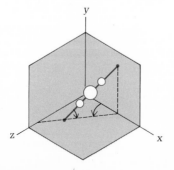

The conventional reference planes in a system of Cartesian coordinates are those defined by each pair of axes, that is, the xy-plane, the yz-plane, and the xz-plane.

Few molecules are linear. For any nonlinear polyatomic molecule, any three atoms can be used to define a plane and the orientation of this plane in space is described by *three* angles, those made by the intersections of this plane with the three conventional reference planes (Figure 9.5). Thus any nonlinear molecule, such as H_2O, has three rotational degrees of freedom.

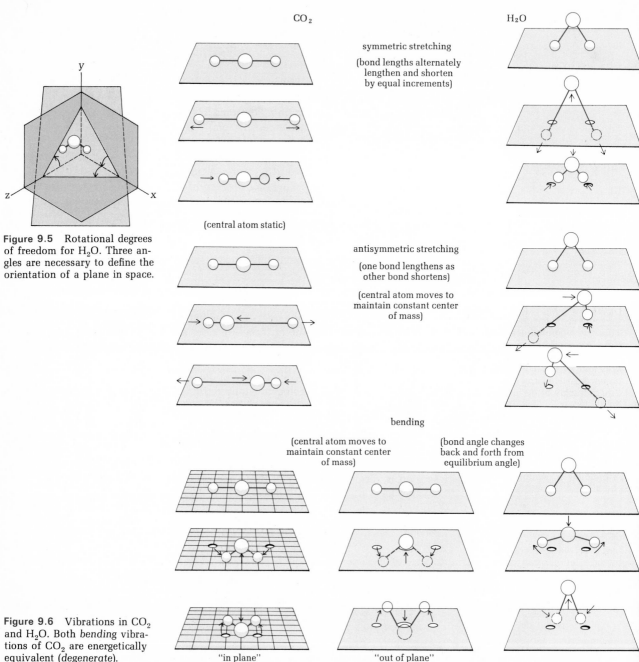

Figure 9.5 Rotational degrees of freedom for H_2O. Three angles are necessary to define the orientation of a plane in space.

CO_2

H_2O

symmetric stretching

(bond lengths alternately lengthen and shorten by equal increments)

(central atom static)

antisymmetric stretching

(one bond lengthens as other bond shortens)

(central atom moves to maintain constant center of mass)

bending

(central atom moves to maintain constant center of mass)

(bond angle changes back and forth from equilibrium angle)

"in plane"

"out of plane"

Figure 9.6 Vibrations in CO_2 and H_2O. Both *bending* vibrations of CO_2 are energetically equivalent (*degenerate*).

Since we have decided that a complete description of a triatomic molecule requires *nine* pieces of information, what can we use to supplement translational and rotational specifications? One obvious item is the distance of each atom in the molecule from the internal reference point located by "translational information." For the general case, these distances may be measured from the center of mass of the molecule to each component atom. For H_2O and CO_2, these distances are easily associated with the H—O and O—C bond lengths. Changes in these bond lengths are called *stretching vibrations* (Figure 9.6). The combination of translational and rotational degrees of freedom was five for CO_2 (or any linear molecule) and six for H_2O (or any nonlinear molecule). Since there are nine total degrees of freedom for a triatomic molecule, this leaves four (9 − 5) vibrational degrees of freedom for CO_2 and three for H_2O. Only two bond-length descriptions are needed for each molecule (hence, two possible stretching vibrations). The remaining vibrations must be thought of as *bending* (bond-angle changes) from the equilibrium bond angle described in our "static" model (Unit 8). Two such bending motions are possible for CO_2, "in plane" and "out of plane", and one bending motion ("in plane") is possible for H_2O (Figure 9.6). Note that any motion of an O—H bond outside the defined molecular plane would simply result in a new plane, so such a motion is rotation rather than vibration.

If we could see a real molecule, its vibrational motions would probably appear quite complex. All these motions could, however, be resolved mathematically into combinations of simple stretching and bending motions, referred to as *normal modes of vibration*. The description of these normal modes and the use of the concept of "degrees of freedom" to decide on the number of possible vibrations for a molecule have significant applications. As we shall see in Unit 30, the infrared spectrum of a chemical substance may be directly related to vibrational (and rotational) motions. On a more immediate basis, an analysis of the ways a molecule can move provides a method of looking at the possible distribution of a molecule's kinetic energy among translational, rotational, and vibrational motions. Although it is difficult to visualize all the possible motions of a complex molecule, it is easy to see that the number of possible vibrational motions increases with molecular complexity. This level of understanding is important in helping us "see" molecules as dynamic species.

Note that the center of mass of a molecule can change *only* with translational motion. For purely vibrational or rotational motions, the center of mass must remain constant. In some cases (Figure 9.6) this requires that *all* the atoms of a molecule move simultaneously.

Motions that involve equal energy are said to be *degenerate*.

Example 1

Both ammonia and acetylene (HCCH) contain four atoms per molecule.
  a. Predict the number of vibrational degrees of freedom for each.
  b. Draw representations for normal modes of vibration of an ammonia molecule and predict which are degenerate.

Solution
  a. Ammonia is a nonlinear molecule (pyramidal, Unit 8), whereas acetylene is linear (Unit 8). Each molecule has 12 ($3n = 3 \times 4$) total degrees of freedom. Three of these must always be translational. Linear molecules have two rotational degrees of freedom, whereas nonlinear molecules have three. The remaining degrees of freedom are rotational, thus,

NH_3 (nonlinear): 12 − (3 + 3) = 6 vibrations
HCCH (linear): 12 − (3 + 2) = 7 vibrations

b.

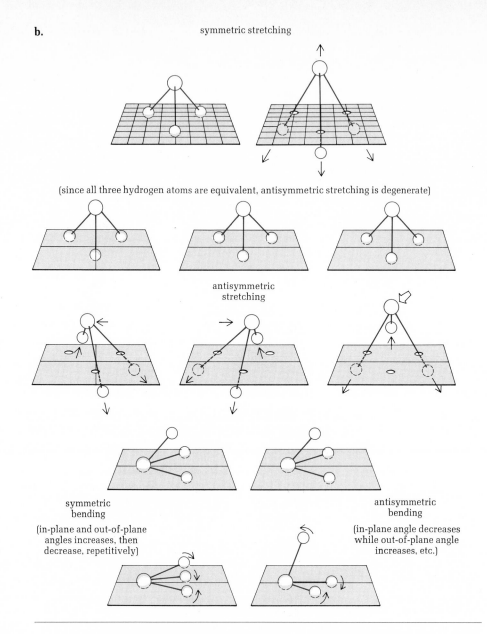

symmetric stretching

(since all three hydrogen atoms are equivalent, antisymmetric stretching is degenerate)

antisymmetric
stretching

symmetric
bending

(in-plane and out-of-plane
angles increases, then
decrease, repetitively)

antisymmetric
bending

(in-plane angle decreases
while out-of-plane angle
increases, etc.)

**9.3 ENERGIES OF
MOLECULAR MOTIONS**

The gaseous state (Unit 10) provides the simplest system for considering molecular motions because molecules or atoms in a gas are much farther apart, and thus more "independent," than in either of the "condensed" states (liquid or solid).

The pressure exerted by a gas is interpreted by the atomic model as resulting from collisions of gas particles with container walls, so gas pressure is related to molecular *translational* motion.

We can gain some insight into the nature of this motion from the simple *kinetic theory* of gas behavior.

KINETIC THEORY OF GASES

To simplify the model of a gas, we make the following assumptions:

1. Gaseous molecules or atoms may be approximated as tiny, incompressible spheres. These spheres are so small that the total volume of all spheres in a particular gas sample is negligible compared to the total gas volume (i.e., most of the volume of a gas is empty space).
2. Gaseous particles are always in random motion, with linear (translational) movement interrupted only by collisions with other particles or with container walls.
3. No forces, either of attraction or repulsion, exist among gaseous particles or between particles and container walls.
4. All particle collisions are perfectly *elastic*. This means that there is no *net* change in translational kinetic energy as a result of collision; energy lost by one particle is gained by another. Although different particles are moving at different speeds and in different directions, the total translational kinetic energy of the system remains constant (at any particular temperature).

Now, let us consider this gas model in terms of randomly moving particles in a cube-shaped container, of edge length l.

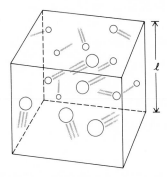

To further simplify matters, let us pretend that we can focus our attention on a single particle that manages to travel from the top of the box to hit the bottom without hitting another particle. (The mathematics involving collisions[1] is considerably more complex, but ultimately leads to the same final result.)

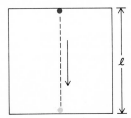

---

[1] The "collision problem" is discussed by Arthur M. Lesk in the February 1974 issue of the Journal of Chemical Education (p. 141).

STEP 1 The speed of the particle is constant, as required by the assumption of conservation of translational kinetic energy.

$$(KE = \tfrac{1}{2}ms^2$$

in which m is the mass of the particle and s is the speed).

STEP 2 Then the time required for the particle to travel from one wall to the other (a distance of l) is

$$\text{time } (t) = \frac{\text{distance}}{\text{speed}} = \frac{l}{s}$$

STEP 3 Sir Isaac Newton discovered that the *force* exerted by a particle in collision is equal to the rate of change of the particle's *momentum* with time:

$$F = \frac{\Delta(mv)}{t}$$

(Momentum is defined as the product of mass and *velocity*. Velocity, unlike "speed," is a directional quantity. Velocity may be represented by a *vector* for which the *magnitude* is "speed.")

STEP 4 The change of momentum, $\Delta(mv)$, of the particle when it hits the bottom of the box is

$$\Delta(mv) = \underset{\text{(final)}}{(-ms)} - \underset{\text{(initial)}}{(+ms)} = -2ms$$

(Since the particle changes direction, assumed to bounce directly back up, the *velocity* changes from $+s$ to $-s$) and, since "for every action there is an equal and *opposite* reaction," the momentum transferred to the box is $-(-2ms) = 2ms$.

STEP 5 From steps 2, 3, and 4, the force resulting from collision of the particle with the bottom of the box is

$$F = \frac{\Delta(mv)}{t} = \frac{2ms}{l/s} = \frac{2ms^2}{l}$$

STEP 6 By definition, *pressure* is force per unit area. Thus the pressure resulting from collision of our particle with the bottom of the box (area $= l^2$) is

$$P = \frac{F}{l^2} = \frac{2ms^2}{l} \times \frac{1}{l^2} = \frac{2ms^2}{l^3}$$

STEP 7 Now let us consider that there are many particles in the box, traveling in various directions at various speeds. We will represent the number of particles hitting container walls during the time interval t by x and their *average* speed by s_{av}. Since there are six equivalent walls (each of area l^2) for our cube-shaped box, the total area affected by collisions of the particles is $6l^2$ and the total force resulting by the x collisions is $2xms_{av}^2/l$, so that the total pressure is

$$P = \frac{2xms_{av}^2}{l} \times \frac{1}{6l^2} = \frac{xms_{av}^2}{3l^3}$$

STEP 8 If we now expand our treatment to the consideration of a very large number of particles, such that it is simpler to count them by *moles* (Unit 4) rather than by single particles, and allow the time interval to be such that the number of collisions with container walls is equal to the number of particles (nN_0, where n is the number of moles and N_0 is Avogadro's number).

$$P = \frac{nN_0 ms^2_{av}}{3l^3}$$

STEP 9 Since l^3 is the volume of the box,

$$P = \frac{nN_0 ms^2_{av}}{3V}$$

$$PV = \tfrac{1}{3}nN_0 ms^2_{av}$$

STEP 10 An experimentally verified relationship, known as the *general gas law*, is developed in Unit 10:

$$PV = nRT$$

in which R is a molar gas constant having the value of 8.313×10^7 g cm^2 sec^{-2} mole^{-1} °K^{-1}, and T is the Kelvin (absolute) temperature. (Other values of R in different units are given in Unit 10.)

STEP 11 From steps 9 and 10,

$$PV = nRT = \tfrac{1}{3}nN_0 ms^2_{av}$$

$$N_0 ms^2_{av} = 3RT$$

Since the molecular weight (or atomic weight, for monatomic species) is defined by $M = N_0 m$, then, from the definition of kinetic energy:

$$\text{average translational kinetic energy} = \tfrac{1}{2}Ms^2_{av} = \tfrac{3}{2}RT \tag{9.1}$$

in which M has the units of g mole^{-1}.

The intricacies of the kinetic theory are not particularly important. What is important is the demonstration that our kinetic particle model for gas behavior is consistent with experimental evidence, and two related implications of equation (9.1):

Translational kinetic energy is a simple function of temperature.
Temperature is a measurement of the average translational kinetic energy of the component particles of a system.

The use of *average* speeds suggests that some molecules travel more slowly than the average, and others move more rapidly. Such is indeed the case, as has been verified by experimental measurements (Figure 9.7).

These, and other, experimental measurements confirm the theoretical model for distribution of molecular speeds, the *Boltzmann distribution* (Figure 9.8).

This discussion of translational energies has adhered to classical mechanics; that is, no corrections for quantized energy states were introduced. The excellent agreement between classical theory and experiment supports the view that quantum considerations can be ignored without significant error in simple treatments of translational motion. Sophisticated considerations conclude that translational energy

Figure 9.7 Atomic speed distribution in gaseous cesium (method of Otto Stern, 1947). (a) Metallic cesium is vaporized in an oven. Atoms passing through a central slit impinge on a movable ionization detector, resulting in a current directly proportional to the number of atoms collected per unit time at a particular vertical position of the detector. The atoms follow a parabolic path determined by the gravitational force (constant) and the atomic speed. Slower atoms hit at a lower vertical position than faster atoms. (b) Current gives a measure of the number of atoms and detector height a measure of relative atomic speeds.

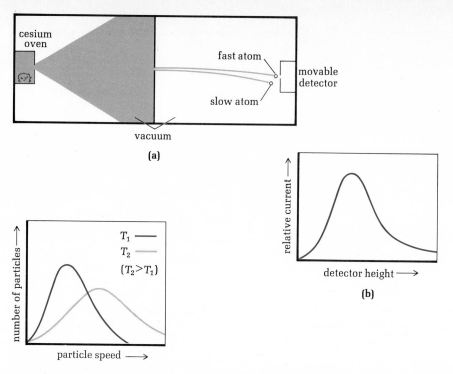

Figure 9.8 Boltzmann distribution of speeds. The most probable speed is represented by the maximum (peak) of the curve; that is, more particles have this speed than any other. Because the curve is unsymmetrical (to the right), the average (root-mean-square) speed is greater than the most probable speed; that is, there are slightly more fast molecules than slow molecules. Note the temperature dependence: At higher temperatures the *percentage* of fast molecules increases.

is quantized, but that energy levels are so closely spaced that they approximate a continuum except at quite low temperatures.

As an analogy, consider two staircases with equal slopes. One has many very tiny steps. The other has a few very large steps. If we wanted to calculate how long it would take a basketball to roll down the stairs, we could approximate this quite well for the tiny steps by equations of classical physics dealing with a smooth incline. These equations might not give a satisfactory answer for the large-step staircase. The ball might stop on the first large step or bounce occasionally as it traversed the stairs.

Energy levels involved with rotational and vibrational motions are sufficiently separated to require a quantum mechanical treatment, as introduced in Unit 30.

9.4 PARTICLES IN COLLISION

The average speed of molecules in the gaseous state may be estimated from equation (9.1). For carbon dioxide gas at 25°C (298°K),

$$s_{av} = \sqrt{\frac{3RT}{M}}$$

$$s_{av} = \sqrt{\frac{3 \times 8.313 \times 10^7 (\text{g cm}^2 \text{ sec}^{-2} \text{ mole}^{-1} {}^\circ\text{K}^{-1}) \times 298({}^\circ\text{K})}{44 \text{ g mole}^{-1}}}$$

$$s_{av} = 4.1 \times 10^4 \text{ cm sec}^{-1} \qquad (\sim 900 \text{ miles hr}^{-1}, \text{ Table 9.1})$$

Such a calculation suggests that the average gas molecule really zips across a room "faster than a speeding bullet." If this were the case, gases would diffuse at a fantastic rate. We know from experience that this does not happen. If someone lights a cigar in the still air of a classroom or opens a bottle of ammonia at the other end of the laboratory, we may detect some gaseous molecules by odor—but that odor does not zoom to us faster than a supersonic jet. A room full of air looks "empty," but it is not. Every cubic foot of that air contains *more than 10^{23}* molecules. A molecule cannot travel very far in that "crowd" without bumping into another molecule. The average molecular speed, then, applies to travel for only a very short distance in a straight line. The molecule then collides with another molecule. The frequency of such collisions is a function of the density of the gas (in terms of the number of molecules per unit volume) and the effective diameter of the gaseous molecule.

A calculation of collision frequency for a gas at 1.00 atm pressure at 25°C, containing approximately a mole of gas for each 24 liters (Unit 10), shows that such collisions happen very rapidly (Table 9.1).

The collision frequency calculation is beyond the scope of our treatment. If you are interested, see the Suggested Readings at the end of this unit.

The distance a molecule is allowed to travel unhindered at average speed is limited to the very short distance between collisions. In a collection of molecules, the average length of travel between intermolecular impacts is called the *mean free path*: It is very short ($\sim 10^{-5}$ cm in air at room temperature and 1 atm pressure).

What all this means is that the speed at which a gas will *diffuse* (travel through a group of other particles) is *very* much slower than the calculated average velocity. *Effusion* refers to a gas escaping from its original container through a tiny hole. The speed of effusion of a gas into a vacuum (where there are no particles to offer collision opportunities) approaches the calculated average molecular speed.

Actually the speed of effusion is the rate at which particles "hit" the hole in the container. Only for a very large number of particles does this rate happen to approach the average molecular speed.

The speed distribution in gases has already been described. It is important to note that it is extremely unlikely that a single molecule will retain any given velocity very long. As a result of collisions there is an exchange of momentum. The net effect of momentum transfer from molecular collisions with container walls is the *pressure* of a gas.

If the intermolecular collisions occur with a *conservation* of translational kinetic energy (KE) (the translational kinetic energy of one particle increased by exactly the translational kinetic energy loss of the other), the collision is said to be *elastic*. Many molecular collisions, however, are *inelastic*; that is, translational kinetic energy is not conserved. Some of the energy of translational motion may be converted to rotational energy. Changes in vibrational energy levels require considerably more energy. Except for high velocity collisions, translational energy is rarely converted to vibrational energy. Inelastic collisions are favored by strong interparticle forces, which tend to make the molecules "stick together."

If a covalent bond absorbs sufficient energy, the bond will break—like a rubber band that has been stretched too tightly. Bonds may break either *homolytically* (each component atom leaves with one of the bond-pair electrons) or *heterolytically* (ionization). Bond rupture requires a considerable amount of energy (Unit 7). At low temperatures very few collisions are sufficiently energetic to break chemical bonds, but as the temperature is increased, the percentage of total molecules having high velocities increases (Figure 9.8). This, as we shall see, is an important factor in the study and control of reaction rates.

A knowledge of molecular motions is of the utmost importance for any understanding of the behavior of gases (Unit 10), solutions (Unit 13), and colloids (Unit 14). Molecular rotations and vibrations provide valuable clues to the structures of molecules and polyatomic ions through techniques such as infrared spectroscopy (Unit 30). We cannot *see* the types of motions in which molecules engage, but we believe these motions occur because they provide models consistent with experimental observation and useful for predictions of phenomena yet to be studied.

9.5 INTERPARTICLE FORCES IN PURE SUBSTANCES

The types of interparticle forces discussed thus far, ionic bonds (Unit 6) and covalent bonds (Unit 7), have involved interactions among charged particles—oppositely charged ions or electrons and positive nuclei, in the case of a covalent bond. As we

rapidly moving propeller | orbital representation of moving electrons in helium

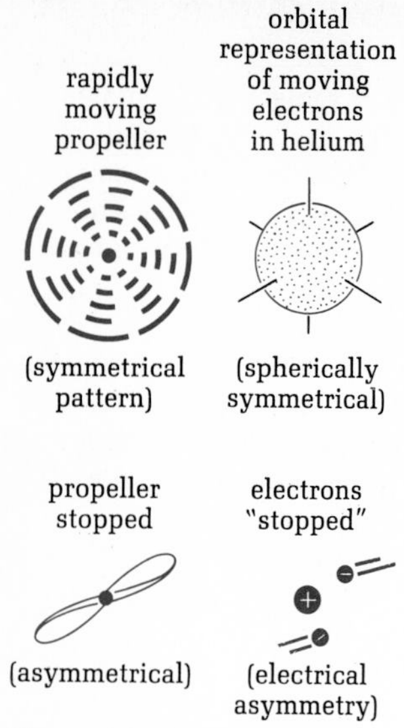

(symmetrical pattern) | (spherically symmetrical)

propeller stopped | electrons "stopped"

(asymmetrical) | (electrical asymmetry)

Figure 9.9 Asymmetry and apparent symmetry.

Atoms of metals in liquid or solid metals are held together by additional forces, the *metallic bond* (Section 9.6).

London dispersions

Fritz London suggested in 1928 a quantum mechanical model for the instantaneous charge fluctuation to produce a transient dipole.

shall see, a variety of interparticle forces may be described in terms of the electrical nature of matter. There is basically the same *kind* of interaction between unit particles in a crystalline salt as between neighboring atoms in a container of helium gas. The difference is a matter of degree, as we have already noted in the distinction between ionic and covalent bonds (Unit 7).

Consider the problem of classifying several bolts of cloth dyed by immersing the bolts for varying lengths of time in a vat of blue dye, then a vat of red dye. You are instructed to pack all the blue cloth in one box, the purple in another, and the red in a third. All the cloth has some red and some blue, depending on the relative lengths of time each bolt was immersed in each vat, but you could make some arbitrary decision as to which to call "red," which "purple," and which "blue." Someone else might have made a slightly different selection. The classification is, in any case, purely arbitrary and a matter of convenience.

Scientists often make the same types of distinction in their efforts to classify natural phenomena.

It is relatively easy to believe in electrostatic forces as the explanation for attractions between ions of opposite charge. It is less obvious that electrostatics might also account for attractions between electrically neutral particles, but such forces do exist. When helium atoms are slowed down by lowering the temperature and forced close together by increasing the pressure, atoms begin to collect into a separate phase: The helium begins to liquefy. Calculations of the magnitude of interparticle attractions sufficient to condense helium from a gas to a liquid show that the forces involved, though weak, are considerably greater than gravitational forces calculated on the basis of the masses of the atoms. These forces can, however, be explained in terms of electrical interactions.

The forces between electrically neutral molecules or between free atoms are broadly classified as *van der Waals forces,* in honor of the Dutch physicist Johannes van der Waals (1837–1923), who suggested a way of using such forces to account for deviations of real gases from ideal behavior (Unit 10). These forces vary considerably in magnitude, and this has led to some convenient subclassifications.

Atoms and nonpolar molecules (Unit 8) are electrically symmetrical as normally represented. However, it must be remembered that the orbital model for atoms or covalent bonds is a probability picture. The representation of an orbital, as commonly drawn, shows electron density *in a region of electron movement.* In this respect it is much like looking at the propeller of an airplane in rapid motion. If we stop the engine, we see that the propeller is asymmetric, although its motion made a symmetric pattern. In the same way, if we could "stop" the electrons, we might see a region of electrical asymmetry (Figure 9.9).

This is a simplified way of indicating that there is an instantaneous fluctuation of charge density at any given direction from the nucleus, resulting in a temporary dipole. If this transient dipole is near another atom or molecule, the electrical asymmetry may induce polarization (distortion of the "electron cloud") of the neighboring particle, much as a charged rod induces charge polarization on a neutral metal sphere (Figure 9.10). In solids or liquids, with the unit particles in close proximity, this effect may extend throughout a significant region, resulting in transient electrostatic attractions among neighboring particles. These forces, called *London dispersion forces,* are estimated to account for nearly all the attractions among nonpolar particles and for a significant fraction of other total interparticle forces (Table 9.3). It is found that the ease of polarization of a particle increases with increasing particle volume (Unit 7, Section 7.2). London dispersion forces show this general increase with increasing atomic or molecular volume. The relative magni-

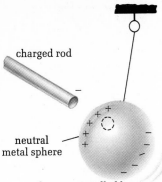

charged rod

neutral
metal sphere

electrons repelled by
charged rod leave
positive region near
rod so sphere swings
toward rod.

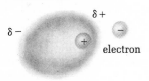

$\delta -$ $\delta +$

electron

polarized
atom

neutral
atom

Figure 9.10 Polarization of
neutral species by external
charge field.

**Table 9.2   SOME BOILING POINTS REFLECTING LONDON
DISPERSION FORCES**

SPECIES	APPROXIMATE MOLECULAR[a] (OR ATOMIC) VOLUME, IN Å³	BOILING POINT (°C AT 1.00 atm PRESSURE)
hydrogen (H_2)	0.4	−253
neon (Ne)	0.9	−246
nitrogen (N_2)	≪2.9[b]	−196
methane (CH_4)	2.7	−164
tetrafluoromethane (CF_4)	6.3	−128
tetrachloromethane (CCl_4)	18	+77
tetrabromomethane (CBr_4)	27	+190

[a] Based on summation of atomic volumes, with molecule approximated by "connected spherical atoms."
[b] In N_2 the triple covalent bond must reduce the molecular volume significantly from that calculated for two spherical nitrogen atoms.

tudes of these forces are reflected in the boiling points of nonpolar liquids (Table 9.2). Molecular volumes are difficult to estimate accurately, but rough approximations sufficient for our needs may be made by thinking of molecules as composed of connected spheres of atomic dimensions (Unit 2). It must be noted that the presence of multiple bonds in a molecule is associated with reduced bond lengths (Unit 7), so that the molecular volumes of molecules containing multiple bonds are expected to be significantly less than those predicted by the "connecting spheres" model.

Example 2

Match the given boiling points with the corresponding substances:

    a. chlorine, fluorine, neon (−246°C, −188°C, −35°C)
    b. ethylene (H_2CCH_2), ethane (H_3CCH_3), methane (CH_4) (−164°C, −104°C, −89°C)

Solution
    a. According to the periodic trends in atomic radii (Unit 2), we expect the relative molecular and atomic sizes to be

$$Cl_2 > F_2 > Ne$$

    The boiling points, then, should be

$$Ne(−246°C)$$
$$F_2(−188°C)$$
$$Cl_2(−35°C)$$

    (This is in agreement with experimental findings.)
    b. Methane should obviously have a smaller molecular volume than ethane or ethylene. Of these latter, ethylene should be smaller, both because of fewer hydrogen atoms and because of the influence of the double covalent bond. The boiling points, then, should be

      methane (−164°C)
      ethylene (−104°C)
      ethane (−89°C)

Although our model for predicting the relative boiling points (based on relative intermolecular forces) is moderately satisfactory, it is not sufficiently accurate to be considered completely reliable. We would, for example, have predicted that acetylene (H—C≡C—H) should be lower boiling than ethylene. It is not (boiling point = $-84°C$). In the absence of a better model, we shall have to be content with about a "75-percent predictability rating". Our simple model is, however, reasonably reliable in differentiating among relatively similar species (e.g., $CS_2 > CO_2$, $I_2 > Br_2$).

Polar molecules (Unit 8) act as dipoles that can be approximated by vector sums of component polar covalent bonds. The electric dipoles are often represented by δ^- and δ^+ to indicate that one end of the dipole has a slight negative charge and the other a slight positive charge as compared to the separated neutral atoms. These "partial charges" are always less than the unit charge (1.6×10^{-19} C) because of electron-sharing effects. The attractive forces between oppositely oriented dipoles add to the existing London dispersions to increase the total interparticle forces (Table 9.3). Even when permanent dipoles are not properly oriented for maximum attraction, one dipole may produce a transient distortion of the electron cloud of a nearby particle so that permanent dipole-induced dipole attractions further contribute to net forces. These are, however, only small contributions.

Hydrogen bonds

A very special case of dipole–dipole force occurs between molecules in which one contains hydrogen in a highly polarized bond and the other contains lone-pair valence electrons representing regions of high electron density. The attraction of the slightly positive hydrogen to the available lone pair is called a "hydrogen bond" (Figure 9.11). Although hydrogen bonds may occur in special cases involving other atoms, they are usually restricted to hydrogen compounds of fluorine, oxygen, and nitrogen. These atoms are quite electronegative (Unit 7) and, in most compounds, contain available unshared valence-electron pairs.

Hydrides of nitrogen, oxygen, and fluorine meet two criteria for strong intermolecular attractions. They contain highly polarized bonds (electronegativity differences from hydrogen are 0.8 for nitrogen, 1.3 for oxygen, and 1.8 for fluorine) *and* the electronegative atom is sufficiently small to allow neighboring molecules to approach each other closely before interatomic repulsions become significant. Remember that the electrostatic attraction between two particles is directly proportional to charge and inversely proportional to separation distance. Thus although the electronegativity difference between chlorine and hydrogen is as great as that between nitrogen and hydrogen (0.8), so that the polarity of the H—Cl bond is about the same as that of the H—N bond, the large chlorine atoms introduce Cl—Cl repulsions before two HCl molecules can approach closely enough for interaction to be strong enough to be considered effective hydrogen bonding (Figure 9.12). On the other hand a carbon

Bond polarity is only roughly approximated by electronegativity differences in complex molecules. Other bonds may influence the polarity of a given bond. In fluoroform (CHF_3), for example, the highly electronegative fluorine atoms increase the C—H bond polarity to the extent that hydrogen bonding may occur among neighboring CHF_3 molecules.

Table 9.3   RELATIVE CONTRIBUTIONS TO VAN DER WAALS FORCES

ATTRACTION	RELATIVE STRENGTHS OF NET FORCES	PERCENT CONTRIBUTION OF LONDON DISPERSIONS	PERCENT CONTRIBUTION OF DIPOLE-DIPOLE FORCES	PERCENT CONTRIBUTION OF PERMANENT DIPOLE-INDUCED DIPOLE FORCES
He · · · He	1	100	0	0
N_2 · · · N_2	90	100	0	0
HCl · · · HCl	200	83	13	4
HBr · · · HBr	330	96	2.4	1.6

Note that HCl is more polar than HBr (Unit 7), but total attractions are greater in HBr because of the higher London dispersion forces due to the larger bromine atom.

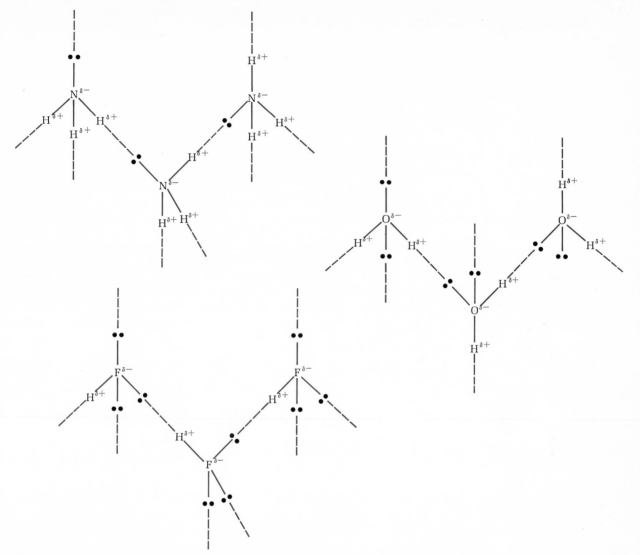

Figure 9.11 Hydrogen bonding in ammonia, water, and hydrogen fluoride.

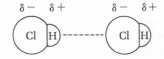

$\delta -$ $\delta +$ $\delta -$ $\delta +$

F↔F repulsion balances
H→F attraction at distance
short enough for effective
hydrogen bonding.

Figure 9.12 Hydrogen bonding (strong dipole-dipole force) in hydrogen fluoride and weaker dipole-dipole force in hydrogen chloride.

$\delta -$ $\delta +$ $\delta -$ $\delta +$

Higher Cl↔Cl repulsion
balances H→Cl attraction
at distance too large for
strong net attraction.

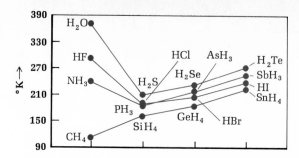

Figure 9.13 Hydrogen bonding and boiling points (at 1.00 atm). Hydrides of the group IVA elements indicate the expected trend when principal interparticle forces are London dispersions alone.

atom is small, but the electronegativity difference between carbon and hydrogen (0.3) is insufficient to produce a bond polarity high enough for effective hydrogen bonding.

Hydrogen bonds are longer and weaker than similar full covalent bonds. The strength of a hydrogen bond is seldom more than 10 percent of that of a typical covalent bond. In water the intermolecular O—H hydrogen bond is about 1.8 times as long as the intramolecular O—H covalent bond. Although hydrogen bonds are relatively weak, their addition to other intermolecular forces is appreciable, as can be noted from the anomalous boiling points of water, ammonia, and hydrogen fluoride (Figure 9.13).

We shall find the subject of intermolecular forces very useful in our studies of solids and liquids (Unit 11), solutions and colloids (Units 13 and 14), and organic and biochemistry (Units 28–34).

Terminology is by no means standardized in discussions of intermolecular forces. Hydrogen bonding is of such vast significance in topics ranging from simple interactions among water molecules to complex structural features of proteins and nucleic acids that it has been almost universally adopted as a separate classification. Other intermolecular forces are often lumped together as van der Waals forces. It is not uncommon to see the term "van der Waals forces" used to refer only to interactions among nonpolar species although, strictly speaking, these should be referred to as London forces or London dispersion forces. All of these types were included in the original ideas of van der Waals in his explanations for the nonideal behavior of gases.

9.6 METALLIC BOND

Metals are characterized by loosely held valence electrons occupying only a few of the possible orbitals of the outermost principal energy levels (Unit 2). In condensed phases (liquids or solids), metallic atoms are packed closely together. The valence electrons of one atom may spend a portion of their time in vacant valence orbitals of neighboring atoms. The close spacing of various possible electronic levels (sublevels) permits valence electrons in metals a considerable freedom of distribution. In metallic crystals and, to a lesser extent, in liquid metals, the valence electrons are quite mobile and are not closely associated with any particular atom. An appropriate model for metals in condensed phases is a collection of spherical cations in a "cloud" of fluid electrons (Figure 9.14). The metallic bond, then, is the net attraction of the positive metal ions for the total free electron cloud. The ease of electron flow (as in the electrical conductivity of metals), the malleability and ductility of metals (Unit 2), and the high surface luster of pure metals are explained in terms of this model for the metallic bond.

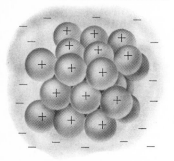

Figure 9.14 Metallic bond: Spherical cations in a "cloud" of mobile electrons.

9.7 INTERPARTICLE DISTANCES

There are a number of ways to determine atomic, ionic, and molecular dimensions. Each involves some degree of uncertainty in experimental measurement and some theoretical differences in interpretations of data. X-ray crystallography (Excursion 1) is perhaps the most common technique for the measurement of submicroscopic dimensions, especially those involving the heavier atoms. There are at least four kinds of dimensions commonly reported. *Ionic radii,* usually measured in crystalline salts, represent basically the distance from the center of a monatomic ion to the

Table 9.4 PARTICLE RADII

sodium in the crystalline metal:	*metallic radius* = 1.86 Å
iodine in the crystalline solid:	*covalent radius* = 1.33 Å
	van der Waals radius = 2.21 Å
sodium ion in NaI(s):	*ionic radius* = 0.97 Å
iodide ion in NaI(s):	*ionic radius* = 2.16 Å

Note the similarity between the ionic radius of I⁻ and the van der Waals radius of I. This is quite common in the nonmetals.

In some salts, anions (e.g., I⁻) are so large compared to the cations (e.g., Li⁺) that adjacent anions "touch" each other. In such cases, ionic radii are taken as one-half the closest internuclear distance for the "touching" ions (see Excursion 1).

region of minimum electron density between adjacent ions. Since this minimum is never zero (Unit 7), ionic radii in salts would only approximate the radii of theoretically "isolated" ions. *Metallic radii* are usually calculated to be one-half of the shortest observed internuclear distance in crystalline metals. Since crystalline metals appear to be really cations in an electron cloud, such measurements do not provide accurate evaluations of the size of an isolated metal atom. *Covalent radii* of the elements are also one-half the closest measured internuclear distance. *Van der Waals radii* apply to distances between nonbonded atoms. Such distances are, of course, meaningless unless certain parameters such as crystal form of solids or densities of gases are specified. Given this information, van der Waals radii are then taken as one-half the average distance between adjacent nonbonded atoms.

Since electrostatic forces are a function of distance between particles as well as a function of charge densities (Unit 6), it is important that correct values of appropriate radii be selected for estimation of any particular type of interparticle force. A comparison of different kinds of radius data is given in Table 9.4.

9.8 RELATIVE STRENGTHS OF INTERPARTICLE FORCES

Interparticle forces are so dependent on the individual characteristics of the particles concerned that it is hazardous to offer any sweeping generalizations. Some pure covalent bonds are very strong, such as the carbon–carbon bonds in diamond [ΔH^0 for C(diamond) $\longrightarrow$ C(g) is ~85 kcal mole⁻¹]. Others are quite weak [ΔH^0 for $I_2(g) \longrightarrow 2I(g)$ is ~18 kcal mole⁻¹ of I(g)]. Similar variations are observed in ionic and metallic bonding.

It may be safely suggested that covalent, ionic, and metallic bonds are stronger than the various interactions broadly classified as van der Waals forces, although it should be noted that the former imply forces in addition to, not instead of, London dispersions.

A useful rule of thumb (which still has numerous exceptions) can be applied to forces *involving particles of about the same size,* so that differences in London dispersion forces can be largely ignored. This rule is given in Figure 9.15.

In the absence of quantitative data, it is difficult to "weight" competing factors unless there are obviously large differences. For example, we might "guess" that intermolecular forces are stronger in nitrogen triiodide (very large iodine atoms, favoring strong London forces) than in ammonia (hydrogen bonding, but weak London forces). That guess is correct. A comparison of NH_3 with NF_3 is more difficult: Are the higher London forces for NF_3 sufficiently strong to outweigh the hydrogen bonding in ammonia? Without some numerical data, we really don't know. Experimentally, we find that the boiling point of ammonia ($-33°C$) is significantly higher than that of NF_3 ($-129°C$). Evidently in this case hydrogen bonding dominates. It is always easier to make predictions *after* we know the answer.

Unless otherwise specified, "boiling points" are for conditions of 1.00 atm pressure.

The strengths of interparticle forces are reflected in a number of "bulk" properties of matter. We shall consider many of these in later units. One of the simpler cases, as we have already noted, is the boiling point of a liquid. Since temperature is related to

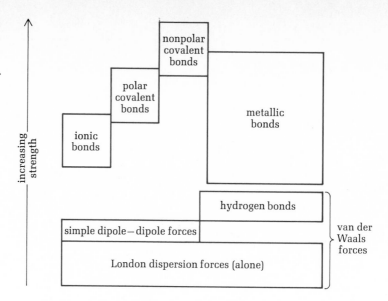

Figure 9.15 Relative strengths of interparticle forces for particles of essentially equivalent volume (so that London dispersion forces are about the same).

the average kinetic energy of the particles in a system, the boiling point reflects the energy a particle must have to escape from a collection of particles. This "escape energy," in turn, is a function of the strength of interparticle forces: Strong interparticle forces suggest higher boiling points than would weak forces. To predict relative boiling points, we must first identify the type of interparticle forces to be broken and then compare the relative strengths of these forces (Figure 9.15).

Example 3

Which member of each pair probably has the higher boiling point?
  a. lithium chloride or hydrogen chloride
  b. ethanol (CH_3CH_2OH) or dimethyl ether (CH_3OCH_3)
  c. carbon dioxide or formaldehyde (H_2CO)
  d. chloroform ($CHCl_3$) or iodoform (CHI_3)

Solution

(We can use the concepts of bond and molecular polarity, hydrogen bonding, and London dispersions to make these predictions.)
  a. Lithium chloride is considered to be ionic on the basis of an electronegativity difference of 2.0 (Unit 7), whereas hydrogen chloride is classified as polar covalent (electronegativity difference of 0.8). To boil lithium chloride, we must increase the translational kinetic energy of the ions sufficiently to permit them to escape from the electrostatic attractions of neighboring ions of opposite charge. It is essential to realize that boiling hydrogen chloride breaks the relatively weak dipole–dipole forces *between* neighboring HCl molecules, *not* the much stronger polar covalent bond *within* the molecule. According to Figure 9.15, ionic bonds are considerably stronger than simple dipole–dipole forces, so we predict that LiCl *will have a higher boiling point than* HCl. (London dispersion forces should be roughly equivalent for the two substances.)
    Experimental values are

  LiCl (boiling point = ~1350°C),   HCl (boiling point = −85°C)

b. Ethanol and dimethyl ether are isomers (Unit 8), so London dispersion forces are equivalent. Construction of geometric representations and bond dipoles (Unit 8) suggests that both molecules are polar.

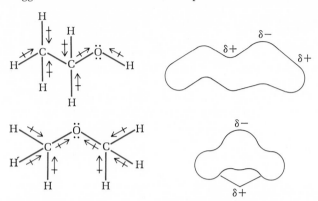

Ethanol contains the highly polarized O—H bond, whereas dimethyl ether does not. The boiling of ethanol, then, must overcome hydrogen bonding among neighboring molecules, whereas only simple dipole–dipole forces must be broken to boil dimethyl ether. We predict that *ethanol is higher boiling.*

Experimental values are

CH_3CH_2OH (boiling point = 79°C), CH_3OCH_3 (boiling point = −23°C)

c. London dispersion forces in carbon dioxide should be roughly equivalent to (probably slightly greater than) those in formaldehyde. Geometric and bond dipole considerations, however, indicate that CO_2 (linear) is nonpolar, whereas H_2CO (trigonal planar) is polar.

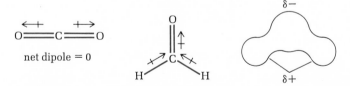

net dipole = 0

Forces to be broken among CO_2 molecules are only London dispersions, whereas those among H_2CO molecules are London dispersions *plus* dipole–dipole forces. Assuming that this added factor for formaldehyde offsets the slightly higher London forces expected for CO_2, we predict *formaldehyde to be higher boiling.*

Experimentally, formaldehyde boils at −21°C at 1.00 atm. The intermolecular forces in CO_2 are sufficiently weak so that molecules escape directly from *solid* CO_2 at −79°C in a process called sublimation (Unit 12).

d. Both $CHCl_3$ and CHI_3 should be polar (Unit 8) and neither should be involved in hydrogen bonding. The most significant difference is in molecular volume, due to the much larger size of iodine compared to chlorine. The higher London dispersion forces expected for CHI_3, then, suggest that *iodoform is higher boiling.*

Experimental values are

$CHCl_3$ (boiling point = 62°C), CHI_3 (boiling point = 218°C)

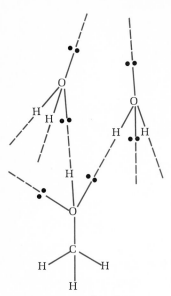

Figure 9.16 Methanol-water interactions. Note that a molecule of methanol may form a maximum of three hydrogen bonds, whereas a water molecule may form four.

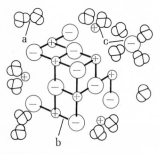

Figure 9.17 Interparticle forces involving salts and water. (a) Solvent-solvent forces (hydrogen bonds). (b) Solute-solute forces (ionic bonds). (c) Solute-solvent forces (dipole-ion interactions).

Every thinking person has an intuitive sense of interparticle forces, although you might be surprised to be told that. If you spill your soda pop on the sofa, you sponge it off with water—not the "cleaning fluid" you would have used to remove a grease stain.

The organic chemist expresses the same ideas when he says "like dissolves like." If he wishes to extract a nonpolar compound, he selects a nonpolar solvent. He uses water or simple alcohols to extract highly polar substances.

A *solution* is a homogeneous mixture. In order to predict whether or not two substances can form such an intimate mixture, it is necessary to consider the relative strengths of three different interactions: solute-solute forces, solvent-solvent forces, and solute-solvent forces.

The *solute* is the substance being dissolved. Strong forces among the unit particles of the solute material tend to hold these together, resisting solution formation.

The *solvent* is the dissolving agent. Forces among like particles of the solvent also tend to prevent mixing. Both solute-solute and solvent-solvent forces may be of any of the types described for pure substances.

The principal factor favoring solution formation, then, is the interaction between solvent particles and solute particles. This interaction must be strong enough to offset the combined opposing forces of solute-solute and solvent-solvent interactions. If solute and solvent particles are quite similar, the forces involved will be much like those in the pure substances themselves. In such cases mixing will usually occur—a situation favoring an increase in entropy (Unit 5). Thus small alcohol molecules mix readily with water by formation of hydrogen bonds (Figure 9.16).

Two forces must be considered for mixtures that were not described for pure substances: dipole-ion attractions and dipole-induced dipole forces. The former accounts, for example, for the solvent action of water, liquid ammonia, and various other polar liquids on many salts.

Because water is a very polar molecule, it is one of the best solvents for ionic compounds. The negative end of the water dipole (the lone-pair regions on oxygen) is attracted to cations, and the slightly positive hydrogens are attracted to anions (Figure 9.17). These attractions increase with increasing charge density of the ions (Unit 6). High charge density also increases interionic forces (solute-solute forces), so it is difficult to estimate the water solubilities of salts without considering additional factors (Units 6 and 13).

The dipole-induced dipole force has been mentioned earlier in connection with total forces among polar molecules (Table 9.3). We may now include the consideration that a polar molecule may polarize (induce polarity in) an otherwise nonpolar neighboring molecule or atom. Such interactions account in part, for example, for the relatively high solubility of HCl (polar) in CCl_4 (nonpolar). However, dipole-induced dipole interactions are relatively weak and London dispersion forces are undoubtedly more significant contributions to the mixing of polar and nonpolar species.

Water is such a uniquely important solvent that its character warrants some special comment. Because the water molecule is quite small, it has little to offer in the way of London dispersion forces. The highly polar character of a water molecule, coupled with its small size, makes it particularly "attractive" to ions and to molecules capable of forming hydrogen bonds. Such molecules may contain O—H or N—H bonds, in which case they may act as both "donors" *and* "acceptors" for hydrogen bond formation, as was the case with methanol (Figure 9.16). Molecules not having O—H or N—H bonds may still interact strongly with water if oxygen or nitrogen is present to offer the lone-pair electrons for bonding with the hydrogen of water (Figure 9.18). Note that water may dissolve even nonpolar molecules, such as CO_2 (Figure 9.18), if these molecules can hydrogen bond with water. The C—H bond is not

Figure 9.18 Interaction of water with acetone, trimethyl-amine, and carbon dioxide.

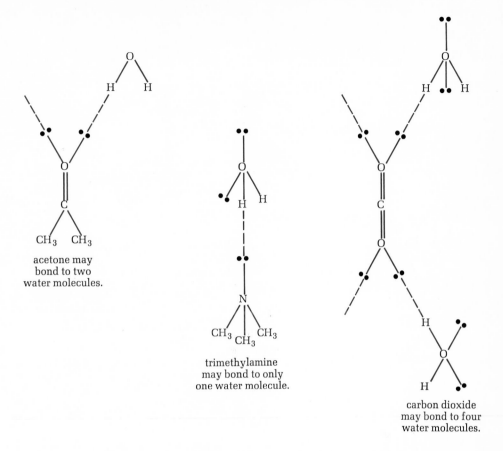

acetone may
bond to two
water molecules.

trimethylamine
may bond to only
one water molecule.

carbon dioxide
may bond to four
water molecules.

very polar, so water has little attraction for hydrogen atoms bonded to carbon and, of course, carbon has no lone-pair electrons in its compounds. As a result hydrocarbon regions (molecular segments containing hydrogen and carbon) do not interact significantly with water molecules.

Many molecules, especially the larger ones, contain regions that are quite polar and other regions that are nonpolar. Thus solvent–solute interactions may vary in different regions. For example, the long-chain alcohol,

1-decanol ($HOCH_2CH_2CH_2CH_2CH_2CH_2CH_2CH_2CH_2CH_3$)

Soaps and synthetic detergents, described in Unit 14, contain hydrophilic groups "attractive to" water and hydrophobic groups "attractive to" nonpolar substances such as oils or greases.

does not dissolve in water. The end containing the —OH group is immersed in the water, but the major portion of the hydrocarbon chain extends upward from the water surface. This compound has been used to retard evaporation from ponds by the formation of a molecular monolayer (a film one molecule thick) on the surface of the water. Groups having strong attractions for water are said to be *hydrophilic* ("water-loving"), whereas nonpolar regions having no strong attractions for water molecules are called *hydrophobic* ("water-fearing"). As a rough approximation, hydrophobic groups occupying more space than a linear three-carbon chain offset a single polar group sufficiently to cause a significant reduction in water solubility.

MOLECULES MAY NOT BE AS SIMPLE AS WE THINK

Formaldehyde, a "simple" polar molecule, is extremely soluble in benzene, a most "unlikely" solvent. Acetic acid, another "simple" molecule, capable of hydrogen bonding, is also soluble in benzene. Perhaps there is something about such molecules not readily apparent from their molecular formulas.

Careful investigations have established the reasons for the "unexpected" solubilities of formaldehyde and acetic acid in benzene. Formaldehyde molecules react readily with each other forming both *long-chain* and *cyclic* molecules:

$$CH_2OCH_2OCH_2OCH_2O$$

paraformaldehyde
(a long-chain combination
of many formaldehyde molecules)

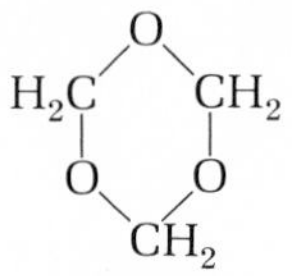

trioxane
(a cyclic combination of three
formaldehyde molecules)

Both of these forms, which decompose on heating to the "parent" formaldehyde, are, as might be expected from their relatively nonpolar character, quite soluble in benzene.

The explanation for the high solubility of acetic acid in benzene is even simpler. Two molecules of acetic acid, properly aligned, readily form a hydrogen-bonded *dimer* (two-molecule set), of a cyclic structure quite analogous in London forces to benzene.

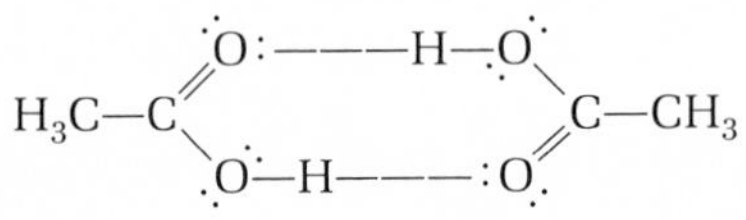

When "simple" molecules don't behave as expected, we don't have to consider them as either "magical" or "active nonconformists". There *is* an explanation and part of the fun of chemistry is finding out more about molecules that really aren't "simple".

***

Example 4

Which should be the best solvent for each of the following compounds, water, benzene, or chloroform ($CHCl_3$)?

a. ammonium chloride
b. iodine monochloride

c. pentane [$CH_3CH_2CH_2CH_2CH_3$]

d. formaldehyde (H_2CO)

Solution

a. Ammonium chloride is a salt and the highly polar water molecule should be more strongly attracted to ions than would the slightly polar chloroform or the essentially nonpolar benzene molecules. In addition, the ammonium ion should form hydrogen bonds with water. We predict, then, that *water is the best solvent for* NH_4Cl.

b. Iodine monochloride is polar (electronegativity difference, 0.5). It cannot, however, form hydrogen bonds with water. London dispersion forces between neighboring ICl molecules should be much stronger than London forces between H_2O and ICl molecules. We do not expect ICl to be very soluble in water. Stronger intermolecular London forces are offered by both benzene and chloroform molecules. Because the latter offer the additional factor of dipole–dipole forces, we predict that *chloroform should be the best solvent for* ICl.

Experimentally, we find that ICl will dissolve rather well in either benzene or chloroform but, as predicted, it is more soluble in chloroform.

c. The pentane molecule, being fairly symmetrical, is essentially nonpolar. Pentane is probably not appreciably soluble in water. Although both benzene and chloroform are possibilities, the stronger solvent–solvent forces in chloroform (large London forces plus dipole–dipole forces) should make it more resistant to mixing with pentane. *Benzene is predicted to be the best solvent for hexane.*

d. The formaldehyde molecule is rather small, so it should be only weakly attracted to benzene molecules (compared to benzene–benzene attractions). The polarity of H_2CO suggests that chloroform would be a better solvent than benzene, but the possibilities for hydrogen bonding suggests that *water is the best solvent.*

Experimentally, we find that formaldehyde is quite soluble in water. (The biological preservative "formalin" is a 40% solution of formaldehyde in water.) It is also soluble in chloroform and, surprisingly, it dissolves best in benzene.

Interparticle forces, among the like particles of pure substances and between different particles in mixtures, are essential to the understanding of chemistry. We shall see how these forces are used to describe the behavior of gases in Unit 10 and of solutions and colloids in Units 13 and 14. They are also of importance in explaining orientations of molecules for effective collisions in chemical reactions (Unit 16).

SUGGESTED
READING

Eisenberg, D., and W. Kauzmann. 1969. *The Structure and Properties of Water.* London: Oxford University Press.

Hildebrand, Joel H., and Robert L. Scott. 1964. *The Solubility of Nonelectrolytes.* New York: Dover.

Tabor, D. 1969. *Gases, Liquids, and Solids.* Baltimore, Md.: Penguin.

EXERCISES

1. Select the term best associated with each definition, description, or specific example.

TERMS: (A) covalent radius (B) diffusion (C) elastic collision (D) hydrogen bonds (E) hydrophilic (F) ionic radius (G) mean free path (H) metallic bonds (I) solvent (J) temperature

a. characterized by conservation of translational KE

b. average distance between successive collisions in a collection of moving particles

c. a function of the average kinetic energy of the particles in a system

d. principal forces among neighboring atoms in crystalline sodium

e. principal forces among neighboring water molecules in ice

f. the substance acting to dissolve another substance

g. movement of HCl molecules through air

h. a molecule or atomic grouping strongly attracted to water molecules

i. distance indicated:

 for I_2 molecule

j. distance indicated:

 for NaI(s)

2. Draw representations for
   a. translational motions of helium atoms in the gas phase
   b. symmetric stretching vibration in a carbon dioxide molecule
   c. asymmetric stretching vibration in a water molecule
   d. bending vibration in·a water molecule
   e. a rotational motion of a water molecule

3. Calculate the number of translational degrees of freedom (T), vibrational degrees of freedom (V), and rotational degrees of freedom (R) for
   a. water; **b.** carbon monoxide; **c.** beryllium hydride;
   d. aluminum bromide; **e.** acetylene; **f.** methanol
   (*Hint:* Predict molecular geometry first.)

4. Suggest the principal interparticle forces among
   a. neighboring particles in crystalline sodium chloride
   b. neighboring particles in crystalline copper
   c. neighboring molecules in liquid hydrogen
   d. neighboring carbon atoms in diamond
   e. neighboring molecules in liquid ethanol

5. On the basis of molecular geometries and polarities:
   a. Match the boiling points (at 1.00 atm) with the correct species in each pair. Explain.
   (1) CH_3OH, CH_3SH (7°C, 65°C)
   (2) CHF_3, $CHCl_3$ (-82°C, 61°C)

   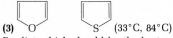
   (3)   O   S   (33°C, 84°C)

   b. Predict which should be the best solvent for each of the following compounds: water, chloroform, or benzene. Explain. (For formulas, see Table 7.1, Unit 7.)
   (1) acetylene; **(2)** sucrose; **(3)** ascorbic acid; **(4)** Freon-12; **(5)** methamphetamine

PRACTICE FOR PROFICIENCY

Objective (a)

1. Select the term best associated with each definition, description, or specific example.
   TERMS: (A) dipole–dipole forces   (B) dipole–ion forces
   (C) hydrogen bonds   (D) hydrophilic   (E) hydrophobic
   (F) inelastic collision (G) London dispersion forces (H) metallic radius (I) momentum (J) solute (K) solvent (L) translational kinetic energy (M) van der Waals radius
   a. characterized by net change in translational kinetic energy
   b. forces among unit particles in a solution of methanol in water
   c. forces between solvent particles and solute particles in a solution of sodium chloride in water
   d. underlined region: $\underline{CH_3CH_2CH_2CH_2}CH_2OH$
   e. underlined region: $CH_3CH_2CH_2CH_2CH_2\underline{OH}$
   f. forces among unit particles in liquid hydrogen
   g. forces among unit particles in liquid hydrogen chloride
   h. $\frac{1}{2}ms^2$
   i. mv
   j. distance indicated:

   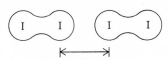

   for crystalline iodine

   k. distance indicated:

   for crystalline sodium

   l. water in an aqueous solution containing 5 g of sucrose in 100 g of water
   m. sucrose in an aqueous solution containing 5 g of sucrose in 100 g of water

2. Indicate whether each of the following statements is true (T) or false (F):
   a. The Boltzmann distribution of molecular speeds is a "normal bell-shaped curve" for which the percentage of "fast" molecules equals the percentage of "slow" molecules at all temperatures.
   b. The description of dipole-induced dipole forces may apply to both interactions between polar and nonpolar species and to interactions between two polar species.
   c. The movement of ammonia molecules through the air in a laboratory room is an example of *effusion*.
   d. Elastic collisions are characterized by no net change in translational kinetic energy.
   e. Although molecules of oxygen in air, at 1 atm pressure and 25°C, have a high average speed, they have very short mean free paths.
   f. Although translational motions of molecules are quantized, neither rotational nor vibrational motions are consistent with quantum descriptions.
   g. The solubility of one substance in another is determined primarily by the strength of solute–solvent forces relative to solute–solute and solvent–solvent forces.
   h. According to kinetic molecular theory, temperature is a function of the average kinetic energy of the particles in a system.

i. Van der Waals forces refer to the interactions between positive and negative ions and to the repulsions among ions of like charge.

Objective (b)

3. Calculate the maximum number of vibrational degrees of freedom for
   a. hydrogen fluoride; **b.** hydrogen cyanide; **c.** hydrogen sulfide; **d.** carbon tetrachloride; **e.** hydrazine (H_2NNH_2); **f.** carbon dioxide; **g.** formaldehyde (H_2CO); **h.** benzene (C_6H_6); **i.** nitrous acid (HONO); **j.** cyanogen (NCCN)
4. Draw representations for
   a. bending vibration in hydrogen sulfide
   b. symmetric stretching vibration in hydrogen sulfide
   c. asymmetric stretching vibration in hydrogen sulfide
   d. internal rotation in ethane
   e. all possible vibrations in carbon disulfide, identifying any degenerate vibrations
   f. all possible vibrations in nitrogen trichloride, identifying any degenerate vibrations

Objective (c)

5. Give the classification normally applied to each of the following interactions:

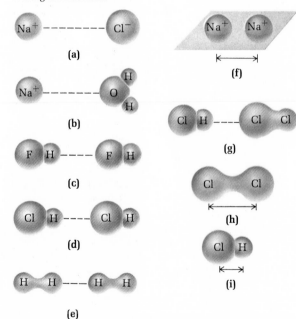

6. What is the maximum number of water molecules that could form hydrogen bonds directly with a single molecule of
   a. water; **b.** ammonia; **c.** methanol; **d.** acetone; **e.** hydrogen fluoride
   f. ethylene glycol ($HOCH_2CH_2OH$)
   g. diethyl ether ($CH_3CH_2OCH_2CH_3$)

h. glycerol ($HOCH_2CHCH_2OH$) with OH below
i. acetic acid (CH_3CO_2H)
j. trimethyl amine (CH_3NCH_3) with CH_3 below

Objective (d)

7. Match the boiling points (at 1.00 atm) with the correct member of each pair:
   a. $C_3H_7CH_2OH$, $C_3H_7C{=}O$ with H below   (74°C, 117°C)
   b. $C_2H_5OC_2H_5$, CCl_4 (34°C, 76°C)
   c. 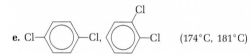 $-NO_2$, —CH_3 (108°C, 211°C)
   d. HF, HCl   (−85°C, 20°C)
   e. Cl—benzene—Cl, benzene with Cl, Cl   (174°C, 181°C)
   f. $CH_3CH_2CH_2NH_2$, $(CH_3)_3N$ (3°C, 48°C)
   g. HI, ICl   (−35°C, 97°C)
   h. benzene—F, benzene—I   (85°C, 188°C)
   i. CH_3CCH_3, CH_3COH (both with O)   (56°C, 118°C)
   j. CF_4, SiF_4 (−132°C, −86°C)

Objective (e)

8. Suggest the best solvent (water, chloroform, or benzene) for each of the following:
   (For formulas, see Table 7.1, Unit 7.)
   a. ammonium nitrate   **e.** nicotine
   b. DDT   **f.** ethanol
   c. lactic acid   **g.** 3,4-benzpyrene
                               h. mescaline
   d. toluene (benzene—CH_3)   **i.** phenylalanine
                               j. methyl bromide (CH_3Br)

9. Which is probably least soluble in water?
   a. calcium chloride
   b. 1-octanol ($CH_3CH_2CH_2CH_2CH_2CH_2CH_2CH_2OH$)
   c. dioxane $\left(O\begin{matrix} CH_2CH_2 \\ CH_2CH_2 \end{matrix}O \right)$
   d. glycerol ($HOCH_2CHCH_2OH$) with OH below

e. urea $\left(\begin{array}{c} O \\ \parallel \\ H_2NCNH_2 \end{array} \right)$

BRAIN TINGLER

10. There were certain assumptions made as the basis for the simple kinetic theory. Gases in which the unit particles exhibit behavior closely approximating these assumptions are referred to as "ideal gases." Steam at high pressure does not exhibit "ideal behavior." Suggest characteristics of the water molecule that might account for "nonideal behavior" of high-pressure steam.

Unit 9 Self-Test

1. Select the term best associated with each definition, description, or specific example.

TERMS: (A) covalent radius (B) dipole–dipole forces (C) hydrogen bonds (D) hydrophobic (E) inelastic collisions (F) London dispersion forces (G) metallic radius (H) solute (I) translation (J) van der Waals radius

a. movement of atoms in helium gas to new relative positions

b. characterized by a *net* change in translational KE

c. principal forces among neighboring molecules in liquid hydrogen

d. principal forces among neighboring molecules in liquid ammonia

e. principal forces among neighboring molecules in liquid hydrogen bromide

f. the substance being dissolved by another

g. molecule or atomic grouping not attracted to water molecules

h. one-half the internuclear distance in the hydrogen molecule

i. one-half the internuclear distance between adjacent atoms in liquid neon

j. one-half the internuclear distance between adjacent atoms in crystalline potassium

2. Draw representations for
a. translation of a helium atom
b. symmetric stretching vibration in phosphine

c. asymmetric stretching vibration in carbon disulfide
d. bending vibration in water

3. Calculate the number of vibrational degrees of freedom for **a.** hydrogen fluoride; **b.** hydrogen sulfide; **c.** ammonia; **d.** methane

4. Draw a representation for a hydrogen-bonded cluster involving one molecule each of methanol and acetone.

5. Match the boiling points with the correct species:

a. [benzene with OH] [benzene with SH] (170°C, 182°C)

b. $CH_3CH_2OCH_2CH_3$, $CH_3CH_2SCH_2CH_3$ (35°C, 92°C)
c. CH_3CH_2F, CH_3CH_2I (−38°C, 72°C)

6. Indicate the best solvent (W = water, C = chloroform, B = benzene) for each of the following:

a. *p*-xylene [benzene ring with CH$_3$ at top and CH$_3$ at bottom]

b. potassium nitrate
c. iodine monochloride (ICl)

Unit 10

Gases: real and ideal

Objectives

(a) Learn for recall and application:
   (1) the values defined for standard temperature and pressure (STP) as ap-
       plied to gas calculations (Section 10.2)
   (2) the value and meaning of the standard molar volume for gases (Sec-
       tion 10.2)
   (3) the general gas law in the forms given as Equations 10.1 and 10.2
(b) Be able to use the general gas law, or other suitable relationships assuming
    ideal behavior, to calculate from appropriate data:
   (1) changes in gas volume resulting from pressure or temperature changes
       (or both) (Example 1)
   (2) gas density (Example 2)
   (3) changes in gas pressure resulting from volume or temperature changes
       (or both) (Example 3)
   (4) molecular weight of a gas (Example 4)
(c) Be able to extend the principles of stoichiometry (Unit 4) to situations in-
    volving gas volume, assuming ideal behavior (Section 10.4).

(d) Be able to make calculations involving partial pressures or mole fractions (or both) for gas mixtures (Section 10.5).

OPTIONAL

Given the van der Waals equation and appropriate constants, be able to compare pressures calculated for real gases with those calculated under assumptions of ideal behavior (Section 10.7).

REVIEW

Kinetic theory (Unit 9) and stoichiometry (Unit 4)

Every small child has probably had the desire to float away in a balloon. There is something romantic and adventuresome about a balloon flight that has never been equaled by the sterile, controlled flight of a jet transport. Jules Verne used a balloon to start Phineas Fogg "around the world in eighty days" and to carry some passengers to a fantastic adventure.[1] Tom Sawyer and Huckleberry Finn[2] took a marvelous balloon trip and so did Tarzan[3]—to the center of the earth!

At one time huge passenger balloons, the zeppelins, offered serious competition to the luxury ocean liners for transatlantic traffic, but this ended shortly after the awesome disaster of the Hindenburg in 1937[4]—the hydrogen-filled gas chambers ignited as the craft was mooring, engulfing passengers and crew in a mass of flames. Today we see weather balloons on occasion or, sometimes, the Goodyear blimp (filled with inert helium, not hydrogen), but passenger balloons are very rare.

Balloons operate on the principle of Archimedes: A body immersed in a fluid is buoyed up by a force equal to the weight of the fluid displaced. Thus the lift of a balloon is the difference between its total weight and the weight of the displaced air—the volume occupied by the balloon times the density (mass per unit volume) of the air:

lift = (balloon volume × air density) − balloon weight

The lift of a balloon can be estimated by treating the balloon gas and the air by simple equations describing ideal gas behavior.

For many years, divers were troubled by problems of increased gas solubility at high pressures. Air is a mixture of several gases, principally oxygen and nitrogen. Neither is highly soluble in water, but the solubility of both is increased appreciably at higher pressures. Divers working at considerable depths in the ocean required high air pressures, resulting in increased solubility of the gases. The oxygen dissolved in the blood could be removed fairly rapidly by coordination with the iron in hemoglobin, but the nitrogen remained as a dissolved gas. As the diver returned to the surface, the decreasing pressure permitted nitrogen to escape from the blood

[1] Jules Verne, The Mysterious Island (New York: Scribner, 1918).
[2] Mark Twain, Tom Sawyer Abroad (New York: Harper & Row, 1878).
[3] Edgar Rice Burroughs, Tarzan at the Earth's Core (Tarzana, Calif.: Burroughs, 1930).
[4] A. A. Hoehling, Who Destroyed the Hindenburg? (New York: Popular Library, 1962).

back into the gas phase. If this process was done very slowly, the released gas could be exhaled with no significant problem. But if it was done too rapidly, nitrogen bubbles formed in the blood vessels themselves as the gas escaped from solution, and the bubbles expanded with decreasing pressure. The result was "the bends," a painful and often crippling experience—sometimes fatal. The only first aid measure available before the advent of the decompression chamber was to pack the victim in ice—reducing bubble volume and increasing nitrogen solubility.

A secondary problem was "nitrogen narcosis." Dissolved nitrogen in the blood acts slowly as a weak anesthetic, thus limiting the time a diver could work in high pressure air before gradually losing consciousness.

One modern-day solution to these problems involves substituting a mixture of oxygen and helium for the air. Helium atoms have very weak attractions for water molecules (Unit 9) and are thus essentially insoluble in the blood even at high pressures. The addition of helium to the breathing atmosphere has, however, had two interesting side effects. Workers in the Navy's Sealab programs, living in an atmosphere of low-density "air," have high, squeaky voices: The velocity of sound (and hence its frequency) is a function of the average velocity of the gas particles. In addition, the workers have trouble feeling warm enough: Helium has a higher thermal conductivity than nitrogen. Both properties are related to the low mass of the helium atom.

Many of the properties of gases—their densities, their rates of diffusion and effusion, their pressure-volume-temperature relationships—can be closely approximated by simple mathematical expressions for gas laws. Gases at high density, however, deviate significantly from this simple ideal behavior. The deviations can be understood on the basis of different types of molecular motions and interparticle forces (Unit 9).

We shall not develop Boyle's law or the law of Charles and Gay-Lussac separately, since we shall find a "combined" *general gas law* treatment to be more useful. If you are interested in more details of the "individual" gas laws, see the Suggested Reading at the end of this unit.

Table 10.1 SOME EARLY STUDIES OF GAS BEHAVIOR

1662   Robert Boyle
Gas volume varies inversely with pressure at a constant temperature.

Boyle's law
$(PV)_T = k$

   or

$P_1V_1 = P_2V_2$

1801   John Dalton
The total pressure of a mixture of gases equals the sum of the individual pressures each gas would exert if it occupied the entire container alone.

Dalton's law of partial pressures
$P_{total} = P_a + P_b + P_c + \cdots$

        or

$P_{total} = \Sigma\, P_{individual}$

1787   Jacques Charles
Gas volume varies directly with temperature at a constant pressure.
1802   Joseph Gay-Lussac
Gas volume changes 1/273 for each change of 1°C in temperature at constant pressure.

Charles' law (as made quantitative by Gay-Lussac)

$$V = V_0\left(1 + \frac{t}{273}\right) \qquad [V_0 = \textbf{volume at 0°C}]$$

Table 10.1 (Continued)

or

$$\frac{V_1}{T_1} = \frac{V_2}{T_2} \quad (t = \text{°C}, \ T = \text{°K})$$

1802   Joseph Gay-Lussac
Gas pressure changes 1/273 for each change of 1°C at constant volume.

Gay-Lussac's law

$$P = P_0\left(1 + \frac{t}{273}\right) \quad (P_0 = \text{pressure at 0°C})$$

or

$$\frac{P_1}{T_1} = \frac{P_2}{T_2}$$

1811   Amadeo Avogadro
Equal volumes of all gases at the same temperature and pressure contain equal numbers of molecules.

Avogadro's hypothesis
$$n_{T,P} = kV \quad n = \text{no. of moles}$$

1860   Stanislao Cannizzaro
Gas density will be proportional to molecular weight.

Cannizzaro's hypothesis
$$d = kM$$

Note: The k used simply represents some general proportionality constant. Thus k is a different *number* in each of the above formulations.

10.1 THE GENERAL GAS LAW

Each of the early gas laws (Table 10.1) dealt with a condition in which some quantity was held constant (P, V, or T). It would be useful to have a single expression relating all the variables concerned with gas behavior. Such an expression may be derived from a simple hypothetical experiment (Figure 10.1).

From experience with gases in balloons, automobile tires, pressure cookers, and so on, we all have some feeling for the relationships among gas volume, pressure, and temperature. Gas laws, in their various forms, simply provide useful quantitative formulations of these feelings. Let us apply some of the formulations to the situation illustrated in Figure 10.1.

We shall use the convention of capital T to represent temperature on the Kelvin scale (Unit 5) and lowercase t to indicate temperature on the Celsius (centigrade) scale. P and V represent pressure and volume, respectively.

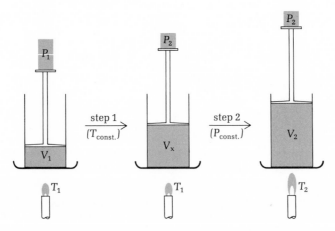

Figure 10.1   Two-step sequence for gas expansion.

STEP 1 A gas of volume V_1 under initial conditions P_1 and T_1 is expanded at constant temperature by reducing the pressure to P_2. The expanded gas volume is now V_x.

From Boyle's law (Table 10.1),

$$P_1 V_1 = P_2 V_x$$

Then

$$V_x = \frac{P_1 V_1}{P_2}$$

STEP 2 The pressure is kept constant (P_2) and the temperature is increased to T_2, causing the gas to expand to the final volume V_2.

From Charles' law (Table 10.1),

$$\frac{V_x}{T_1} = \frac{V_2}{T_2}$$

Then

$$V_x = \frac{V_2 T_1}{T_2}$$

From steps 1 and 2 of the hypothetical experiment, we find two quantities equal to the same quantity, V_z:

$$\frac{P_1 V_1}{P_2} = \frac{V_2 T_1}{T_2}$$

If we then multiply both sides of the equation by P_2/T_1,

$$\frac{P_1 V_1}{T_1} = \frac{P_2 V_2}{T_2} \qquad (10.1)$$

This is one form of the *general gas law*.

It follows from extension of equation (10.1) that

$$\frac{PV}{T} = \text{a constant}$$

As we shall see, real gases are well approximated by the general gas law only under conditions of moderately high temperatures and moderately low pressures.

and this constant can be evaluated. If we pick a specific volume as that containing Avogadro's number of molecules (Section 4.2), then for this *molar volume* V_m we shall designate the constant as R, the molar gas constant, such that

$$\frac{PV_m}{T} = R$$

Any two volumes are related, according to Avogadro's law, as

$$\frac{n_a}{n_b} = \frac{V_a}{V_b}$$

If V_a is the molar volume V_m, then n_a, the number of moles, is 1; then for any particular volume V containing n moles of gas,

$$\frac{1}{n} = \frac{V_m}{V}$$

$$V_m = \frac{V}{n}$$

Pressure measurements are reported most commonly either as atmospheres or as the height of a mercury column in a standard barometer: 1.00 atm = 760 mm.

A recent convention uses the unit *torr* as the equivalent of 1 mm *of mercury*. Thus 1.00 atm = 760 torr. (The *torr* was selected to honor Evangelista Torricelli, inventor of the barometer.)

In the new System Internationale Units, pressure is expressed in newtons per square meter: 1.00 atm = 1.01×10^5 N m^{-2}.

Substitution in the expression

$$\frac{PV_m}{T} = R$$

and rearranging terms gives another useful form of the *general gas law*:

$$PV = nRT \tag{10.2}$$

The molar gas constant R can then be evaluated by experimental measurements of P, V, T, and n. The units of R will depend on the units used to express the pressure, volume, temperature, and number of moles. The most common value reported for R requires pressure to be expressed in atmospheres, volume in liters, temperature in degrees Kelvin, and n simply as the number of moles:

$$R = 0.0821 \text{ liter atm mole}^{-1} \text{ deg}^{-1}(\text{K})$$

Alternative values for R, derived from appropriate conversion factors (Appendix A) are

62.4 liter torr mole^{-1} deg^{-1}(K)
8.31×10^3 liter N m^{-2} mole^{-1} deg^{-1}(K)
1.987 cal mole^{-1} deg^{-1}(K)
8.31 Joule mole^{-1} deg^{-1}(K)

10.2 STANDARD TEMPERATURE AND PRESSURE (STP)

Note that the *standard temperature for gases* (273°K) is different from the temperature of our defined *thermodynamic standard state* conditions (Unit 5).

It is convenient to have a set of reference conditions for comparison of gas properties. These have been arbitrarily set as

standard temperature: 273°K (0°C)
standard pressure: 1.00 atm (760 torr)

Under these conditions it is possible to calculate the *standard molar volume* (the volume of 1 mole of gas at STP) from equation (10.2):

$$V = \frac{nRT}{P}$$

$$\begin{matrix}\text{standard}\\\text{molar}\\\text{volume}\end{matrix} = \frac{1.00 \text{ mole} \times 0.0821 \text{ liter atm mole}^{-1} \text{ deg}^{-1} \times 273°}{1.00 \text{ atm}}$$

standard molar volume = 22.4 liters mole^{-1} at STP

10.3 SIMPLE GAS LAW PROBLEMS

Three general types of calculations are quite commonly applied to pure gases under conditions closely approximating "ideal gas behavior": the calculation (often referred to as "correction") of volume, or pressure, changes resulting from changes in temperature and pressure or volume, respectively; the calculation of gas density; and the calculation of the molecular weight of a gaseous compound. Each of these may utilize an appropriate formulation of the general gas law [equations (10.1) or (10.2)] or, perhaps more meaningfully, an analysis of the problem in terms of an understanding of gas behavior without "plugging numbers into an equation." These alternative approaches are illustrated by the following examples.

Example 1 (*P, V, T* "Corrections")

A balloon is filled with 4.00 liters of hydrogen at a pressure of 680 torr and a temperature of 25°C. What would be the approximate volume of the balloon at a high altitude at which the pressure is only 40 torr at −45°C, assuming free expansion of the balloon?

The use of temperature and pressure *ratios* is consistent with the *proportional* relationships expressed by Boyle's law and the law of Charles and Gay-Lussac (Table 10.1).

Solution (a) An Intuitive Analysis

The "intuitive" method is quite general and useful and avoids the necessity of relying on a memorized formula. If, for example, it is desired to correct a volume for a change in pressure and temperature, this intuitive approach suggests

$$V_{corr} = V_{orig} \times P \text{ ratio} \times T \text{ ratio}$$

The ratios are set so that the arithmetic result fits the intuitive conclusion. If the pressure has been increased, the volume will have decreased (compression), so the pressure ratio must be <1. If the temperature is increased, the volume increases (expansion), so the temperature ratio is >1. The same logic pattern is followed for either pressure decrease (P ratio >1) or temperature decrease (T ratio <1).

First make a table for the data to help keep track of the variations:

$$V_1 = 4.00 \text{ liters} \qquad V_2 = V_{corr}$$
$$P_1 = 680 \text{ torr} \qquad P_2 = 40 \text{ torr}$$
$$T_1 = (273 + 25) \qquad T_2 = (273 - 45)$$

A pressure change from 680 to 40 torr will permit a volume expansion, so the pressure ratio must be 680/40. A temperature change from 298° to 228°K will cause a volume contraction, so the temperature ratio must be 228/298. Then

$$V_{corr} = 4.00 \text{ liters} \times \frac{680}{40} \times \frac{228}{298}$$

$$V_{corr} = 52 \text{ liters}$$

(two significant figures, as limited by **40** torr.)

Solution (b) Using a Gas Law Equation

Because no gas escapes from the balloon, the number of moles (n) remains constant. For volume comparisons at constant n, at different temperatures and pressures, equation 10.1 is most appropriate, after converting temperatures to degrees Kelvin:

$$T_1 = 25 + 273 = 298°K$$
$$T_2 = -45 + 273 = 228°K$$

$$\frac{P_1 V_1}{T_1} = \frac{P_2 V_2}{T_2} \qquad \text{[equation (10.1)]}$$

$$\frac{680 \text{ torr} \times 4.00 \text{ liters}}{298°K} = \frac{40 \text{ torr} \times V_2}{228°K}$$

$$V_2 = \frac{228°K \times 680 \text{ torr} \times 4.00 \text{ liters}}{40 \text{ torr} \times 298°K}$$

$$V_2 = 52 \text{ liters}$$

(two significant figures, as limited by **40** torr)

Example 2 Gas Density Calculation

What is the density, in grams per cubic centimeter, of nitrogen gas at 27°C and 0.50 atm?

Solution (a) An Intuitive Analysis

Since the standard molar volume is 22.4 liters mole^{-1} (at STP), we can easily determine the density of any ("ideal") gas (at STP):

$$\text{density}_{(N_2)} = \frac{1 \text{ mole}}{22.4 \text{ liters}} \times \frac{28.0 \text{ g } (N_2)}{1 \text{ mole}}$$

Changes in pressure and temperature do not affect the amount (mass) of a gas in a sealed container, but the resulting volume change will affect the density. Any factors tending to increase volume will make the gas less dense, and density will increase if volume is decreased. Appropriate corrections by pressure and temperature ratios may then be applied.

$$d_{\text{corr}} = d_{\text{orig}} \times P_{\text{ratio}} \times T_{\text{ratio}}$$

Note that, since *mass* is independent of temperature, we could simply make a "volume correction," after which the density could be calculated by dividing the mass by the "corrected" volume.

$$d_{\text{orig}} = \frac{28.0 \text{ g}}{22.4 \text{ liters}} \qquad d_{\text{corr}} = ?$$

$$P_1 = 1.00 \text{ atm (standard)} \qquad P_2 = 0.50 \text{ atm}$$
$$T_1 = 273°\text{K (standard)} \qquad T_2 = (273 + 27)°\text{K}$$

A pressure decrease would cause a *volume* increase (with *lowered density*) and a temperature increase would have a similar effect. We may combine all arithmetic operations:

$$\text{density} = \underbrace{\frac{1 \text{ mole}}{22.4 \text{ liters}} \times \frac{28.0 \text{ g } (N_2)}{1 \text{ mole}}}_{\text{(at STP)}} \times \underbrace{\frac{0.50 \text{ atm}}{1.00 \text{ atm}} \times \frac{273°\text{K}}{300°\text{K}}}_{\substack{\text{(correction factors for} \\ \text{lowering density from that} \\ \text{at STP)}}} \times \frac{1 \text{ liter}}{10^3 \text{ cm}^3}$$

It is important to realize that the relationship among mass, volume, and density requires that both density and volume be considered *at the same conditions of temperature and pressure.*

$$= 5.7 \times 10^{-4} \text{ g cm}^{-3}$$

Solution (b) Using a Gas Law Equation

From equation (10.2), the density *in moles per liter* is

$$\frac{n}{V} = \frac{P}{RT} = \frac{0.50 \text{ atm}}{0.0821 \text{ liter atm mole}^{-1} \text{ deg}^{-1}(\text{K}) \times 300°\text{K}} = 2.03 \times 10^{-2} \text{ mole liter}^{-1}$$

Conversion to grams per cubic centimeter requires appropriate unity factors:

$$\text{density} = \frac{2.03 \times 10^{-2} \text{ mole}}{\text{liter}} \times \frac{28.0 \text{ g } (N_2)}{1 \text{ mole}} \times \frac{1 \text{ liter}}{10^3 \text{ cm}^3}$$

$$= 5.7 \times 10^{-4} \text{ g cm}^{-3}$$

(two significant figures, as limited by 0.50 atm)

Example 3 *P, V, T* Corrections

A Pyrex jar filled with air is sealed at 714 torr and 23°C. The jar is accidentally tossed into a trash incinerator. What pressure must the jar withstand to avoid exploding at the incinerator temperature of 680°C?

Solution (a) An Intuitive Analysis

We may assume a constant volume of gas (air), so we are dealing here with the variation of pressure with temperature. We know from experience (e.g., with pressure cookers or tires heated during a long, hot drive) that pressure increases with temperature. Thus

$$P_{680°\text{C}} = P_{23°\text{C}} \times \frac{(680 + 273)°\text{K}}{(23 + 273)°\text{K}}$$

$$= 714 \text{ torr} \times \frac{953}{296} = 2.30 \times 10^3 \text{ torr}$$

Solution (b) Using a Gas Law Equation

Equation (10.1) is more appropriate to this problem than would be equation (10.2). We assume the volume remains constant (if the jar does not explode), that is, $V_1 = V_2$:

$$\frac{P_1 V_1}{T_1} = \frac{P_2 V_2}{T_2}$$

$$\frac{714 \text{ torr} \times V_1}{(273 + 23)°K} = \frac{P_2 \times V_2}{(273 + 680)°K} = \frac{P_2 \times V_1}{(273 + 680)°K}$$

$$P_2 = \frac{953°K \times 714 \text{ torr} \times V_1}{296°K \times V_1} = 2.30 \times 10^3 \text{ torr}$$

Example 4 Molecular Weight

A 23.2-ml sample of a pure gas was found to weigh 0.028 g at 24°C and 692 torr. What is the approximate molecular weight of the gas?

Solution (a) An Intuitive Analysis

One value of avoiding equation (10.2) lies in the relief from memorizing (or looking up) the constant R in appropriate units, or from applying unit conversion factors. We may solve this problem without equation (10.2) simply by using the definition of molecular weight, the value of the molar volume at standard temperature and pressure, and appropriate correction factors for density (as in Example 2).

$$\text{molecular weight} = \underbrace{\frac{0.028 \text{ g}}{23.2 \text{ ml}} \times \frac{760 \text{ torr}}{692 \text{ torr}} \times \frac{297°K}{273°K}}_{\text{(density corrected to STP)}} \times \frac{10^3 \text{ ml}}{1 \text{ liter}} \times \underbrace{\frac{22.4 \text{ liters}}{1 \text{ mole}}}_{\text{(at STP)}}$$

$$= 32 \text{ g mole}^{-1}$$

Note again that gas volumes (or volumes and densities) can be compared only under the same conditions of temperature and pressure. This is a logical extension of Avogadro's hypothesis (Table 10.1).

Solution (b) Using a Gas Law Equation

From equation (10.2) we may calculate the number of moles (n) of gas in the sample. The mass given may then be used to calculate the molecular weight (as grams per mole).

$$PV = nRT$$

$$n = \frac{PV}{RT} = \frac{692 \text{ torr} \times 0.0232 \text{ liter}}{62.4 \text{ liter torr mole}^{-1} \text{ deg}^{-1}(K) \times 297°K}$$

$$n = 8.66 \times 10^{-4} \text{ mole}$$

$$\text{molecular weight} = \frac{0.028 \text{ g}}{8.66 \times 10^{-4} \text{ mole}} = 32 \text{ g mole}^{-1}$$

10.4 STOICHIOMETRY OF GASES

A large number of chemical processes involve gases as reactants, as products, or as both reactants and products. *Gas volumes* associated with chemical reactions can be calculated by a unity factor method similar to that used for weight per weight problems (Unit 4). Once the stoichiometric ratios of interest have been determined, the number of moles of gas can be calculated from molar ratios (Example 5) based on appropriate coefficients, and the gas volume can then be determined from equation (10.2). As an alternative, the *standard molar volume* can be used in the appropriate unity factor in place of the formula weight (Example 6), and the calculated volume can then be corrected for temperature or pressure (or both) different from standard temperature and pressure. Whichever method appears simpler for a particular problem is selected.

Example 5

What volume of gas, at 250°C and 1.00 atm, would be formed by the explosive decomposition of 5.0 g of ammonium nitrate, according to the equation

$$2NH_4NO_3(s) \longrightarrow 2N_2(g) + O_2(g) + 4H_2O(g)$$

Solution
The stoichiometric ratios show that for *each mole* of ammonium nitrate decomposed, 1 mole of N_2, $\frac{1}{2}$ mole of O_2, and 2 moles of H_2O are formed, that is, $3\frac{1}{2}$ moles of *gas* per mole of NH_4NO_3. Then

$$n = \frac{5.0 \text{ g } (NH_4NO_3)}{1} \times \frac{1 \text{ mole } (NH_4NO_3)}{80 \text{ g } (NH_4NO_3)} \times \frac{3.5 \text{ moles (gas)}}{1 \text{ mole } (NH_4NO_3)}$$

$$= 0.22 \text{ mole}$$

From equation (10.2),

$$V = \frac{nRT}{P}$$

$$= \frac{0.22 \times 0.0821 \times (273 + 250)}{1.00}$$

$$= 9.4 \text{ liters} \quad \text{(to two significant figures)}$$

A 5.0-g sample of ammonium nitrate occupies a volume of about 3 ml (0.003 liter). The volume change, in a "split second," involves more than a 3000-fold increase. It is not too surprising that this is classified as an explosion.

Example 6

What volume of dry carbon dioxide could be formed at 27°C and 730 torr by the reaction of 50 g of calcium bicarbonate with excess dilute hydrochloric acid?

Solution
The balanced equation, based on information from Unit 6, is

$$Ca(HCO_3)_2(s) + 2HCl(aq) \longrightarrow CaCl_2(aq) + 2H_2O(l) + 2CO_2(g)$$

The stoichiometric ratio of interest is

$$Ca(HCO_3)_2 + \cdots \longrightarrow \cdots + 2CO_2$$

An approach similar to that developed in Unit 4 can now be employed, with the addition of the important unity factor,

$$\frac{22.4 \text{ liters}}{1 \text{ mole}} \quad \text{(gas at STP)}$$

$$V_{\text{STP}} = \frac{50 \text{ g } [Ca(HCO_3)_2]}{1} \times \frac{1 \text{ mole } [Ca(HCO_3)_2]}{162 \text{ g } [Ca(HCO_3)_2]} \times \frac{2 \text{ moles } [CO_2]}{1 \text{ mole } [Ca(HCO_3)_2]} \times \frac{22.4 \text{ liters}}{1 \text{ mole } [CO_2]}$$

(at STP)

= 14 liters

Then, since the temperature given is higher than standard temperature and the pressure is lower than standard pressure, both factors act to increase the volume:

$$V_{\text{corr}} = 14 \text{ liters} \times \frac{(273 + 27)}{273} \times \frac{760}{730}$$

= 16 liters   (to two significant figures)

This represents an even greater volume change than in Example 5, but the decomposition of bicarbonates by acids is *very* much slower than the explosive decomposition of ammonium nitrate. However, what do you think would happen if this reaction had been attempted in a sealed liter bottle? (Estimate the gas pressure.)

10.5 GAS MIXTURES There are a number of ways of expressing the composition of mixtures. One that is particularly useful for gas mixtures is *mole fraction, f.*

Some chemists prefer the symbol x for mole fraction.

$$f_a = \frac{n_a}{n_{\text{total}}} \tag{10.3}$$

At low densities gas particles act independently as long as there is no chemical reaction possible. The total pressure of a gas mixture has been determined experimentally to be the sum of the individual pressures each component gas would exert if it were alone in the container. The individual pressure of gas a, called its partial pressure (P_a), is related to the mole fraction of the gas, as could be derived from the general gas law:

Dalton's law (Table 10.1), equation (10.5).

$$P_a = f_a P_{\text{total}} \tag{10.4}$$

The summation of a total number of moles of gas in a mixture has already been illustrated in Example 5. A similar summation of partial pressures is useful in a number of situations:

$$P_{\text{total}} = P_a + P_b + P_c + \cdots \tag{10.5}$$

One important application of Equation (10.5) is in situations involving a gas trapped above a liquid or solid contributing some vapor to the total gas mixture. In such a case the pressure of the gas itself is equal to the total pressure minus the equilibrium vapor pressure (Unit 11) of the solid or liquid phase. Tables of equilibrium vapor pressures for a number of common substances as a function of temperature are available and can be used to make appropriate corrections of pressures of trapped gases (e.g., see Table 10.2).

Example 7

A biology experiment was set up to monitor oxygen release by photosynthesis in certain algae. The released oxygen was passed through a series of traps, a final water trap, and was then measured by reading the gas volume in a constant-pressure

Table 10.2 EQUILIBRIUM VAPOR PRESSURES FOR WATER AND MERCURY

| TEMPERATURE (°C) | VAPOR PRESSURE (torr) | |
	WATER	MERCURY
0	4.6	0.000185
10	9.1	0.000490
20	17.4	0.001201
30	31.5	0.002777
40	54.9	0.006079
50	92.0	0.01267
60	148.9	0.02524
70	233.3	0.04825
80	354.9	0.08880
90	525.5	0.1582
100	760.0	0.2729

device. At a particular time in the experiment, a volume of 52.5 ml of oxygen had been collected over water at 30°C and a total pressure of 720 torr. How many moles of oxygen had been collected?

Solution

The pressure of the oxygen gas in the mixture of O_2 and water vapor is calculated from equation (10.5), using the vapor pressure of water at 30°C (Table 10.2) as the partial pressure of water:

$$P_{O_2} = P_{total} - P_{H_2O}$$
$$= 720 \text{ torr} - 32 \text{ torr} = 688 \text{ torr}$$

Then, from equation (10.2),

$$n_{O_2} = \frac{P_{O_2} V}{RT}$$

$$P_{O_2} = \frac{688}{760} \text{ (atm)}$$

$$V = 0.0525 \text{ (liter)}$$

$$T = 273 + 30 = 303°K$$

$$n_{O_2} = \frac{688 \times 0.0525}{760 \times 0.0821 \times 303}$$

$$= 1.91 \times 10^{-3} \text{ mole}$$

SAFE HYDROGEN-OXYGEN ATMOSPHERES

Contrary to popular belief, not all mixtures of hydrogen and oxygen are "dangerous." In fact, a mixture containing less than 5% (by volume) of oxygen will extinguish a flame.

This seems contrary to the basic principles of chemistry, which suggest that both reaction yields (Unit 4) and reaction rates (Unit 16) should be increased by use of an excess of one reactant. Nevertheless it is true that hydrogen–oxygen mixtures in which the mole fraction of hydrogen is 0.95 or greater remain unreactive. Various explanations have been proposed, but we are not yet certain of the reasons for this behavior.

The inert character of these low-oxygen mixtures has led to investigations of H_2/O_2 atmospheres as substitutes for He/O_2 mixtures in breathing systems for deep-sea diving. Dr. William Fife, Professor of Biology at Texas A & M University, has been studying H_2/O_2 atmospheres for several years and has himself made a number of deep dives using such systems. Test animals have survived very deep dives, with no apparent physiological changes, using breathing atmospheres containing as little as 0.6% O_2 (99.4% H_2).

Experiments of this type require application of Dalton's law of partial pressures. Oxygen-deficient atmospheres can, as we all know, lead to suffocation. An atmosphere, at normal pressure, containing less than 6% O_2 (by volume) will not support life. The secret lies in recognizing that the *partial pressure* of oxygen, not its percentage, is the essential factor. Oxygen is about 21% (by volume) of "normal" air. This means that, according to Dalton's law, the partial pressure of oxygen in the atmosphere we normally breathe is 0.21×760 torr $= 160$ torr. Biological experiments with hydrogen–oxygen breathing mixtures, typically using 3% O_2 (by volume) to stay well below the "safe" limits, can achieve the same partial pressure:

$$0.03 \times P_{total} = 160 \text{ torr} \qquad P_{total} = \frac{160 \text{ torr}}{0.03} = 5330 \text{ torr}$$

Thus a total gas pressure of 5330 torr (about 7 atm) will provide a partial pressure of oxygen equivalent to that of "normal" air. A pressure of 7 atm is quite suitable for deep-sea diving to around 200 feet.

At greater depths, the same approach is used. As total gas pressure must be increased, the percentage of oxygen is decreased to maintain P_{O_2} around 160 torr. Test animals appear to survive quite well in such high pressure–low oxygen atmospheres, in which oxygen is actually only a "trace component" ($<1\%$) of the breathing atmosphere.

An outgrowth of kinetic theory (Unit 9) is a comparison of average molecular speeds as a function of molecular weight. If we compare two different gases (in separate containers) under conditions such that molecular collisions can be neglected, such as effusion from a region of high molecular density into a vacuum chamber (or into a region of very low molecular density), at constant temperature, kinetic theory states

average translational kinetic energy (gas a) = average translational energy (gas b)

that is,

$$\tfrac{1}{2}m_a s_a^2 = \tfrac{1}{2}m_b s_b^2$$
$$\text{(av)(av)} \quad \text{(av)(av)}$$

Since the average mass of a gas particle is given by the atomic or molecular weight (M) divided by Avogadro's number, we may write

$$\frac{1}{2}\left(\frac{M_a}{6.02 \times 10^{23}}\right)s_a^2 = \frac{1}{2}\left(\frac{M_b}{6.02 \times 10^{23}}\right)s_b^2$$
$$\text{(av)} \qquad\qquad\qquad \text{(av)}$$

Multiplying both sides of the equation by $(2 \times 6.02 \times 10^{23})$ gives

$$M_a s_a^2 = M_b s_b^2$$
$$\text{(av)} \quad \text{(av)}$$

from which

$$\frac{s_a(\text{av})}{s_b(\text{av})} = \sqrt{\frac{M_b}{M_a}} \tag{10.6}$$

This equation is an expression of *Graham's law of effusion* and, within the limits of "ideal gas behavior," it provides an experimental verification for the kinetic theory of gases.

Graham's law indicates that at the same temperature lighter molecules move more rapidly than heavier molecules, on the average. This is, indeed, consistent with kinetic theory and implies that relative *diffusion* rates should also be more rapid for lighter molecules. Diffusion, however, involves the added complication of molecular collisions as the diffusing species moves through a region containing other particles. Because of this complication, equation (10.6) provides only a rough approximation of relative rates of *diffusion,* and the approximation decreases in accuracy with increasing molecular density of the gaseous mixture.

The Graham's law concept, even in approximate form, does have some practical applications. Helium-filled balloons, for example, gradually lose buoyancy as the more rapidly moving helium atoms escape through tiny pores in the balloon "skin" more frequently than slower O_2 or N_2 molecules from the surrounding atmosphere enter the balloon through the pores. (Of course, an "even exchange" of helium for a heavier gas would still have resulted in decreased buoyancy, but the buoyancy loss is more rapid than the even-exchange process would account for alone.)

A more dramatic example is the use of a gaseous effusion process for the separation of uranium isotopes (Unit 1) on the basis of slight differences in isotopic masses. In the isotopic separation, first performed on a large scale at the Oak Ridge (Tennessee) atomic energy facility, the uranium sample is converted into gaseous uranium hexafluoride. The gas is forced through a series of several thousand porous barriers, at each of which the UF_6 molecules containing the lighter uranium isotope gain a slight headway. Although the difference in isotopic masses provides an average speed for the lighter isotope of only 1.004 times that of the heavier isotope, the large number of effusion separations successively enriches the gas mixture in lighter UF_6 sufficiently for an adequate isolation of the desired product.

10.7 NONIDEAL GASES
The general gas law and the various results based on the kinetic molecular theory are found to be quite useful in approximating changes in gas volume, pressure, or temperature *under certain conditions*. Real gases, however, often deviate significantly from calculated behavior, especially at high pressures or low temperatures, or both, corresponding to high gas densities (Figure 10.2).

It is not difficult to think of cases in which real gases would deviate from "ideal" requirements. At very high molecular densities, for example, gas particles are packed so closely together that actual particle volume may constitute an appreciable fraction of total gas volume. Under such conditions larger particles (e.g., SiF_4) would produce greater deviations than smaller ones (e.g., helium) at a particular pressure. Polyatomic particles have more varieties of motions than do monatomic particles (Unit 9), so gases such as SF_6 would be expected to exhibit a larger fraction of inelastic collisions than gases such as helium. Interparticle forces (Unit 9) become appreciable at high densities (particles closer together) or low temperatures (particles passing each other slowly enough to permit interactions). Gases in which strong interparticle forces are possible will be "less ideal" than those in which only weak forces exist. Thus steam (containing hydrogen-bonded water molecule clusters) is far less "ideal" than helium under comparable conditions.

In spite of these difficulties, it is still very useful to be able to make simple approximations from general gas equations.

In 1873, Johannes van der Waals developed an equation for real gases that has the same general form as equation (10.2).

$$\left(P + \frac{n^2a}{V^2}\right)\left(V - nb\right) = nRT \qquad\qquad (10.7)$$

Figure 10.2 Nonideal behavior of Cl_2 and CH_4 gases under conditions of high molecular density. (Pressure is shown as a function of temperature for gases of density 1.00 mole liter^{-1}.)

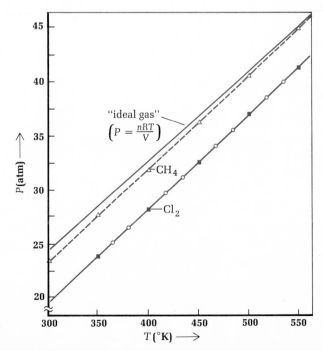

Table 10.3 SOME TYPICAL VAN DER WAALS CONSTANTS

SUBSTANCE	a (liter^2 atm mole^{-2})	b (liter mole^{-1})
He	0.034	0.024
Ne	0.211	0.017
Ar	1.35	0.032
H_2	0.24	0.027
N_2	1.39	0.039
Cl_2	6.48	0.056
CH_4	2.25	0.043
NH_3	4.17	0.037
H_2O	5.46	0.031
Hg	8.09	0.017

The terms a and b are actually "constant" only within certain definite limits of pressure and temperature.

In this equation, a and b are constants that must be determined experimentally for each particular gas (Table 10.3). The pressure is corrected for interparticle forces, related to the constant a, and for the opportunities for molecular interaction, a function of the square of the particle density, $(n/V)^2$. The correction for volume includes the constant b, which is related to the effective volume of the molecule—including repulsive forces—and the number of molecules in a given volume—a function of the number of moles, n.

The van der Waals equation is useful in obtaining approximations for gases under nonideal conditions, but it is still unsatisfactory in many situations. For example, steam engineers find steam tables, tabulations of precise experimental data, of more value in routine situations requiring a fair degree of accuracy.

For most simple calculations, the ideal gas laws are adequate, but one should be aware of their limitations.

SUGGESTED READING

Hildebrand, Joel H. 1963. *An Introduction to Molecular Kinetic Theory.* New York: Van Nostrand Reinhold.

Mendelsson, K. 1967. *The Quest for Absolute Zero.* New York: McGraw-Hill.

O'Connor, R., and C. Mickey. 1974. *Solving Problems in Chemistry.* New York: Harper & Row.

Parsonage, N. G. 1966. *The Gaseous State.* Elmsford, N.Y.: Pergamon.

EXERCISES

1. Fill in the blanks:
   a. When dealing with gases, calculations require temperatures to be expressed in degrees _____.
   b. By convention, for gases, *standard* temperature is _____ and *standard* pressure is _____ torr (_____ atm).
   c. The general gas law may be expressed as _____ = nRT, in which n represents _____.
   d. When the value 0.0821 liter atm mole^{-1} deg^{-1} is used for R, pressure must be expressed in units of _____ and volume in units of _____.

2. Calculate
   a. the volume occupied by a gas at STP if its volume was 28 ml at 127°C and 1900 torr
   b. the density, in grams per cubic centimeter, of oxygen gas at 37°C and 682 torr
   c. the pressure resulting from heating air at constant volume from 0°C to 100°C if the original air pressure was 1.00 atm
   d. the molecular weight of a gas for which a 26.0-ml sample weighed 0.037 g at 22°C and 710 torr

3. a. Calculate the volume of sulfur dioxide formed during the production of 25 metric tons of copper by the process

   $$Cu_2S + O_2 \longrightarrow 2Cu + SO_2$$

   Report your answer under conditions of 37°C and 650 torr.
   b. What is the maximum volume of ammonia gas that could be formed at 700°C at 5.0 atm by reaction of 8.0 kg of hydrogen with excess nitrogen. (Caution: Watch units.)

4. A 75-ml sample of nitrogen gas was collected over water at 20°C and a total pressure of 650 torr. Calculate
   a. the mole fraction of water vapor
   b. the partial pressure of nitrogen
   c. the weight of nitrogen collected (in grams)

OPTIONAL

5. Calculate the pressure that is exerted by 54 g of water vapor in a 5.00-liter vessel at 546°C:

a. from the general gas law
b. from van der Waals equation
Explain any difference in the two calculated values.

PRACTICE FOR PROFICIENCY

(Consult Tables 10.2 and 10.3 when necessary.)

Objective (a)

1. Fill in the blanks:
   a. Under ideal conditions, 1 mole of a gas occupies a volume of _____ liters at STP
   b. The "gas constant" R has the value of approximately _____ liter atm mole^{-1} deg^{-1} (K).
   c. $°K = °C +$ _____ $°$
   d. The general gas law may expressed as

   $PV =$ _____   or as $\dfrac{P_1 V_1}{T_1} =$ _____

Objective (b)

2. Calculate:
   a. the volume occupied by 4.2 g of nitrogen at STP
   b. the volume resulting from cooling 25 ml of nitrogen from 127°C to 0°C while decreasing the pressure from 702 torr to 351 torr
   c. the density of ammonia gas at 268°C and 1120 torr
   d. the molecular weight of a gas for which a 23.8 ml sample weighed 0.057 g at 25°C and 683 torr
3. A gas occupied a volume of 84 ml at -73°C and 380 torr. At 27°C and 570 torr, its volume would be approximately _____ ml.
4. The density of nitrous oxide (N_2O) gas at 27°C and 380 torr, in grams per liter, is approximately _____.
5. What pressure would be exerted by 30 g of argon gas in a 2.00-liter vessel at -20°C?
6. What is the molecular weight of a pure gaseous compound having a density of 1.15 g liter^{-1} at 127°C and 650 torr?
7. A gas occupies a volume of 7.5 ml at 70°C and 900 torr. At 10°C and 300 torr, its volume would be approximately _____ ml.
8. The density of benzene (C_6H_6) vapor at 127°C and 900 torr, in grams per liter, is approximately _____.
9. What pressure would be exerted by 60 g of argon gas in a 5.00-liter vessel at -30°C?
10. What is the molecular weight of a pure gaseous compound having a density of 1.53 g liter^{-1} at -23°C and 700 torr?

Objective (c)

11. The glassware in a chemical laboratory often accumulates a coating of white ammonium chloride, resulting from the combination of ammonia and hydrogen chloride vapors in the laboratory atmosphere:

    $NH_3(g) + HCl(g) \longrightarrow NH_4Cl(s)$

    What volume of gaseous ammonia, measured at 23°C and

120 torr, must have been consumed to form 25 g of ammonium chloride?
12. In a commercial process for the manufacture of baking soda, a solution of sodium chloride in aqueous ammonia is saturated with carbon dioxide gas:

    $NaCl(aq) + NH_3(aq) + CO_2(aq) + H_2O(l) \longrightarrow$
    $NaHCO_3(s) + NH_4Cl(aq)$

    What volume of carbon dioxide gas, measured at 20°C and 780 torr, would be required for the production of 5.00 lb (2270 g) of baking soda, assuming 100% efficiency?
13. Freon-12, a valuable refrigerant and propellant gas, can be prepared by the vapor-phase reaction of carbon tetrachloride with hydrogen fluoride, using a special catalyst:

    $CCl_4(g) + 2HF(g) \longrightarrow CF_2Cl_2(g) + 2HCl(g)$

    How many grams of carbon tetrachloride are required for the production of 66 liters of Freon-12, measured at 500°C and 1000 torr, assuming 100% efficiency?
14. One of the ways of producing "dry ice" commercially utilizes the formation of the initially gaseous CO_2 required by the fermentation of glucose:

    $C_6H_{12}O_6(aq) \longrightarrow 2C_2H_5OH(aq) + 2CO_2(g)$

    What volume of CO_2, at 37°C and 720 torr, could be formed by the fermentation of 5.00 lb (2270 g) of glucose?
15. During the process of photosynthesis, a green plant utilizes water and carbon dioxide gas to form carbohydrates. Oxygen is a by-product, as illustrated by the simplified photosynthetic equation;

    $H_2O + CO_2 \longrightarrow \quad (CH_2O) \quad + O_2$
    ("carbohydrates")

    What mass of carbohydrate would correspond to the photosynthetic release of 25 liters of oxygen, measured at 37°C and 700 torr?
16. One of the waste products of operation of an ordinary internal combustion engine is nitric oxide, resulting from oxidation of some of the nitrogen in the air. The nitric oxide emitted from the engine exhaust is photochemically converted to NO_2:

    $2NO(g) + O_2(g) \longrightarrow 2NO_2(g)$

    What mass of NO_2 would result from the photochemical oxidation of 65 liters of nitric oxide, measured at 27°C and 700 torr?
17. The explosive decomposition of nitroglycerine may be formulated as

    $4C_3H_5(NO_3)_3(l) \longrightarrow 12CO_2(g) + 10H_2O(g) + 6N_2(g) + O_2(g)$

    What pressure would result from the detonation of 5.0 g of nitroglycerine in a sealed 1.00-liter container, assuming the resulting gases attain a temperature of 2457°C? (Ignore the effect of any air or vapors in the original container.)
18. An "effervescent antacid" tablet was analyzed by collecting

the CO_2 gas generated from reaction of a weighed sample with excess dilute hydrochloric acid. If a 0.550-g sample of the tablet generated 41.6 ml of dry carbon dioxide, measured at 23°C and 730 torr, what was the percentage (by weight) of HCO_3^- in the tablet?

Objective (d)

19. What mass of oxygen is represented by a volume of 64 liters of O_2 collected over water at 27°C and 720 torr? (Vapor pressure of water at 27°C is 30 torr.)
20. What mass of nitrogen is represented by a volume of 16 liters of N_2 collected over water at 37°C and 700 torr? (Vapor pressure of water at 37°C is 47 torr.)
21. What mass of nitrous oxide is represented by a volume of 22 liters of N_2O collected over water at 30°C and 690 torr? (Vapor pressure of water at 30°C is 32 torr.)
22. What mass of phosphine is represented by a volume of 25 liters of PH_3 collected over water at 27°C and 650 torr? (Vapor pressure of water at 27°C is 30 torr.)
23. What mass of propane is represented by a volume of 33 liters of C_3H_8 collected over water at 15°C and 275 torr? (Vapor pressure of water at 15°C and 275 torr? (Vapor pressure of water at 15°C is 14 torr.)
24. A 125-ml sample of oxygen was collected over water at 30°C

and a total pressure of 708 torr. Calculate
 a. the partial pressure of the oxygen
 b. the mole fraction of oxygen
 c. the mass of oxygen collected (in grams)

OPTIONAL

25. For each of the following systems, calculate the gas pressure predicted by the general gas law and compare that with the pressure calculated from van der Waals equation. Suggest explanations for the resulting comparisons.
 a. 4.0 g of helium in a 2.5-liter container at 300°K
 b. 71 g of chlorine in a 2.5-liter container at 300°K
 c. 4.0 g of helium in a 2.5-liter container at 50°K

BRAIN TINGLER

26. The gaseous effusion separation of two uranium isotopes uses uranium hexafluoride. The "faster" gas contains $^{235}_{92}U$ (isotopic mass 234.993 amu) and has an average molecular speed, under the conditions employed, of 1.004 times that of the "slower" UF_6. What isotope of uranium is incorporated in the "slower" UF_6?

Unit 10 Self-Test

(Consult Table 10.2 and Table 10.3 when necessary.)
 1. Fill in the blanks:
 a. By convention, for gases, *standard* temperature is _____ and *standard* pressure is _____ atm (_____ torr).
 b. At STP, 1 mole of a gas (assuming ideal behavior) occupies a volume of _____ liters.
 c. One form of the general gas law is expressed as $PV = $ _____ T.
 2. Calculate
 a. the volume occupied by 64.0 g of oxygen at STP
 b. the density (in grams per cubic centimeter) of oxygen at 27°C and 0.50 atm
 c. the volume resulting from heating 100 ml of oxygen from 0°C to 546°C and increasing the pressure from standard pressure to 1520 torr
 d. the pressure resulting from heating 100 ml of oxygen from 27°C to 127°C at constant volume if the original pressure was 600 torr

3. What is the molecular weight of a pure gaseous compound having a density of 3.46 g liter^{-1} at 87°C and 1080 torr?
4. A 50-ml sample of hydrogen was collected over water at 30°C and a total pressure of 630 torr. Calculate
 a. the mole fraction of water vapor
 b. the partial pressure of hydrogen
 c. the weight of hydrogen collected (in grams)
5. What volume (in liters) of dry hydrogen cyanide could be formed at 27°C and 710 torr from reaction of 98 g of sodium cyanide with excess sulfuric acid?

OPTIONAL

6. Use the van der Waals equation to calculate the pressure (in atmospheres) exerted by 51 g of ammonia in a 10.0-liter vessel at 227°C. Compare this with the pressure calculated from the general gas law.

Unit 11

Solids and liquids

Objectives

(a) After studying the appropriate definitions in Appendix E and the terms usage in this Unit, be able to demonstrate your familiarity with the following terms by matching them with appropriate definitions, descriptions, or specific examples:

allotropes	immiscible
amorphous	lattice point
critical temperature	liquid crystal
crystal habit	meniscus
crystal structure	miscible
crystal system	polymorphism
crystalline	surface tension
fluid	viscosity

(b) Be able to construct models and to draw three-dimensional representations for

    (1) body-centered cubic, face-centered cubic, simple cubic, and hexagonal closest-packed lattices

    (2) locations of tetrahedral and octahedral sites in face-centered cubic lattices

        (3) lattice structures for simple salts in the isometric system
        (4) symmetry axes, symmetry centers, and mirror planes for polyatomic or polyionic aggregates
    (c) On the basis of interparticle forces (Unit 9), be able to
        (1) classify solids by bond type
        (2) predict simple properties of solids, such as relative melting points or thermal and electrical conductivity
        (3) predict relative liquid viscosities
        (4) predict miscibility or immiscibility of liquid pairs
    (d) Using descriptions of simple crystal structures and the methods for assigning particles per unit cell, be able to determine the empirical formulas of simple salts (Section 11.7).
    (e) Given appropriate experimental data and equations, be able to calculate
        (1) densities of solids or liquids (Section 11.2)
        (2) liquid viscosity (equation 11.1)
        (3) surface tension of liquids (equation 11.2)

REVIEW

Interparticle forces (Unit 9)

A "liquid crystal" seems to imply some characteristics that are incompatible. Liquids, like gases, are fluid, and we think of "crystals" as hard, rigid solids. Crystals, as we have seen in Unit 9, are described as very orderly arrays of particles, whereas liquids are characterized by much more random orientations of particles. Nevertheless there are substances classified as liquid crystals, and they have many very interesting properties.

Certain liquid crystals change color dramatically with very small changes in temperature. An interesting demonstration of this phenomenon involves painting an area of the body with a particular liquid crystal system. The blood vessels appear as different colors against the background color of the painted skin because of the small temperature difference between the circulating blood and the surrounding tissue. The extreme sensitivity of some liquid crystal systems to tiny thermal gradients has led to their use in infrared sensing devices for use with highly sophisticated weapons systems.

Recent research has shown that liquid crystals play important roles in a number of biochemical events. The combination of fluidity with orderly arrangements of particles is ideally suited to the requirements of many aspects of molecular biology.

Liquid crystals offer an interesting study of the transitions from the fluid systems, gases and liquids, to the nonfluid solids. The reasons for the unique properties of liquid crystals are associated with aspects of molecular geometry (Unit 8), the motions of particles (Unit 9), and the nature of interparticle forces (Unit 9): Liquid crystals do have orderly arrangements of molecules—somewhat like those of normal crystals—and considerable freedom of translational motion—like normal fluids. The types of molecular substances forming liquid crystals are, in general, limited to fairly long, "cylindrical" molecules or fairly large, "flat" molecules. These requirements offer an important clue: Such molecules, if properly oriented, can "slide" fairly easily *without tumbling; that is, the fluidity is associated with directional translational motions and the order with a resistance to molecular rotation.*

As an analogy, consider what happens when an ordered stack of tiny bar magnets is shaken in a box: "Intermagnet forces" and shapes are quite unsymmetrical, so few magnets will rotate freely even when the clusters are shaken vigorously.

The molecular characteristics of typical materials for liquid crystals[1] are illustrated by two such substances:

p-azoxyanisole
(semicylindrical)

cholesterol
(semiflat)

In this unit we shall consider the properties of solids and liquids. We shall see that the true solids are not fluid at the molecular level, although collections of small particles of solids—powders, for instance—show bulk fluidity. We shall also see that some kinds of solids—the crystals—are characterized by highly ordered arrangements of unit particles, but that other types—the amorphous solids—may have a much greater molecular disorder than in liquid crystals. The disorder in liquids approaches that of the gaseous state. The properties of solids and liquids, like those of gases, can be explained by the atomic model of matter.

11.1 CHARACTERISTICS OF SOLIDS

The common physical states of pure substances are distinguished by differences in particle motions and the order of particle arrangements (Figure 11.1). The unit particles of a gas are considered to be in rapid random motion, traveling for relatively large distances between collisions, and involving rotations and vibrations in addition to translational motions (Unit 9). Gases assume the shapes of their containers and volumes that can vary tremendously with temperature or pressure (Unit 10). The unit particles in liquids also undergo rotational, vibrational, and translational motions with considerable freedom—although to a much smaller extent, in general, than in gases. Liquids undergo only small volume changes with variations in temperature or pressure, and their shapes are determined by the shapes of their containers, except for a surface in contact with a gas phase. Gas–liquid interfaces are essentially flat for liquids at rest when a large surface area is involved. The pronounced curvature associated with small-area interfaces is discussed in Section 11.9.

[1] For more details on liquid crystals, see "Liquid Crystals Get Improved Properties," in the September 27, 1971, issue of *Chemical and Engineering News* (pp. 42–43).

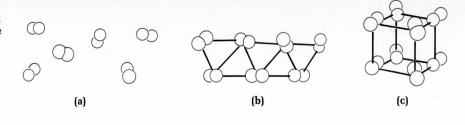

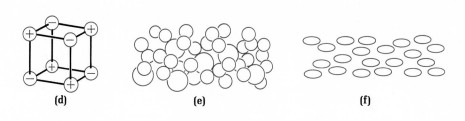

Figure 11.1 Order and motion. (a) Gas (little order, rapid motions). (b) Polar liquid (little order, slower motions). (c) Crystalline solid (highly ordered, little motion). (d) Ionic crystal. (e) Amorphous solid. (f) Liquid crystal.

"Groups" of solid particles, for example, sand, gravel, and powders, have certain characteristics of fluids, but the individual solid pieces are rigid.

In pure solids the particles have very little freedom of motion, and each unique solid "piece" has its own characteristic shape; that is, solids are not fluid. Although fine powders can be poured like fluids, they can be formed into stable patterns that do not settle spontaneously to the flattened surfaces characteristic of the stable condition of true liquids (Figure 11.2).

Figure 11.2 Powders and fluids. (a) Finely divided solid. (b) Liquid.

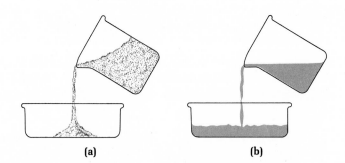

11.2 DENSITIES OF SOLIDS AND LIQUIDS

The densities of crystals are important data for the determination of crystal structures. The density (mass per unit volume) of a solid may be found in several different ways. The particular method selected depends on the characteristics of the sample and the accuracy required.

If the faces are not rectangular, the arithmetic is a bit more complicated. In such cases, edge angles must also be measured.

For a solid having a very regular shape (e.g., a cube), the volume can be calculated from direct measurement of the edge lengths, and the mass can be found by weighing the solid. If the solid is irregular, but fairly large, its volume can be determined by allowing it to sink in a liquid of lower density and measuring the volume of liquid displaced. Small solid objects, such as tiny crystals, are not amenable to either of these techniques (except when they are of uniform shape, in which case their dimensions can be measured by appropriate microtechniques). Estimates of the density of small solid objects, in particular small crystals, can be made with reasonable accuracy by a *flotation* method. By preparing a series of solutions of different concentrations, it is possible to find one whose density is the same as that of the crystal. When the crystal is dropped into an inert solution of lower density, the crystal will sink. It will float on a solution of higher density, but in a solution whose density is the same as that of the crystal, it will hang in suspension. Once the proper solution has been found, a measured volume can be removed and weighed.

The determination of the mass of a measured volume of a liquid is, of course, the normal way of finding the density of the liquid. Although volume changes as a function of temperature or pressure are very much smaller for liquids or solids than for gases (Unit 10), they are not necessarily insignificant. The expansion of mercury in a heated thermometer is a clear illustration of how liquid volume and density (since the mass of mercury in the thermometer doesn't change) may be affected by temperature change. Accurate tabulations of density must indicate the temperature at which the measurements were made. Since unit particles are quite closely packed in solids and liquids, densities of liquids and solids vary only slightly with pressure. Except under unusual conditions, pressure specification is rarely indicated.

Example 1

A crystal of potassium chlorate was found neither to float nor sink in a solution of bromoform (density 2.89 g cm^{-3}) and acetone (density 0.792 g cm^{-3}), at 25°C. A 0.250-cm^3 sample of the solution was removed and found to weigh 0.579 g. What was the density of the potassium chlorate crystal?

Solution

$$d_{\text{crystal}} = d_{\text{solution}}$$

$$d_{\text{solution}} = \frac{\text{mass}}{\text{volume}} = \frac{0.579 \text{ g}}{0.250 \text{ cm}^3}$$

$$d_{\text{crystal}} = 2.32 \text{ g cm}^{-3} \qquad (\text{at } 25°C)$$

11.3 CRYSTALS AND AMORPHOUS SOLIDS

If we compare small plastic cubes with cubic crystals of sodium chloride, several differences can be noted. The NaCl crystal can be cleaved along definite planes (e.g., parallel to the crystal faces), whereas attempted cleavage of the plastic material results in irregular fracture (Figure 11.3). The salt, when heated, increases in temperature until, within 1 or 2 degrees of 800°C, it suddenly liquefies. The plastic, on the other hand, softens gradually, with no sudden liquefaction, over a broad temperature range. When the liquid NaCl is slowly cooled, regular geometrical shapes appear as solid forms at 800° ± 2°C. These solids have planar surfaces and characteristic angles between adjacent faces. Under close examination they appear as more or less perfect copies of the original cubic crystal. As the liquid plastic cools, it gradually solidifies over a broad range of temperature into an irregular blob. Only if cooled in a proper mold will it assume a shape like that of the original cube.

Substances that have the general characteristics described for sodium chloride are said to be "crystalline," whereas those more similar to the plastic are called "amorphous" solids (or glasses).

If the atomic (particle) model of matter is satisfactory, it should help explain observed differences between crystalline and amorphous solids.[2] X-ray diffraction photographs from crystals (Excursion 1) show typically regular patterns, whereas those from amorphous solids do not. Evidence suggests that these patterns originate

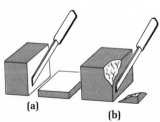

(a)   **(b)**

Figure 11.3 Crystalline and amorphous solids. (a) Cleavage along planar pattern in NaCl(s). (b) Irregular fracture of a solid plastic.

---

[2] For an interesting discussion of amorphous and crystalline solids, see the article by John Hicks in the January, 1974, issue of the *Journal of Chemical Education* (pp. 28–31).

Figure 11.4 Model for planar fracture of a crystalline salt. Such patterns of particle locations also explain why certain planes are observed in crystal fracture or crystal growth and others are not. Only the planes defined by unit particle locations are possible for fracture or growth surfaces. (Diamond cutters must be especially aware of the limitations on appropriate fracture patterns.)

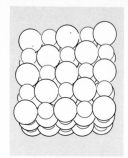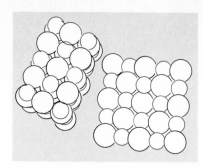

from particles (atoms or ions) in the crystal. The atomic model, then, postulates a highly ordered arrangement of unit particles in crystalline solids and a much less orderly arrangement in amorphous solids (Figure 11.1).

This model would predict that interparticle forces in pure crystalline substances would be essentially identical in all internal regions of the solid, whereas forces would vary within amorphous solids from one "microregion" to another. As a result, thermal energy sufficient to break particles loose from their normal positions in the solid (e.g., melting the solid) would be much more variable for amorphous than for crystalline materials—a situation consistent with the broad melting range of the former compared to the sharp melting range of the latter.

Orderly arrays of particles in crystals and disordered arrangements in glasses would also be consistent with the planar cleavages possible with crystalline substances (Figure 11.4).

11.4 THE LIQUID STATE

Liquids are intermediate between solids and gases. Like the solids, liquids have a certain degree of short-range order (Figure 11.5), short interparticle distances, and relatively strong interparticle forces. Like gases, they are fluid, permit more rapid translational motions than occur in solids, and have little (if any) long-range order.

Whether or not the unit particles of a substance will be arranged in a regular pattern depends on a balance between interparticle forces (Unit 9) and the random-

Figure 11.5 Solid, liquid, and gas. More ordered arrangements in fluids occur when particles are highly polar. In nonpolar liquids there are probably no significant regions of regular order.

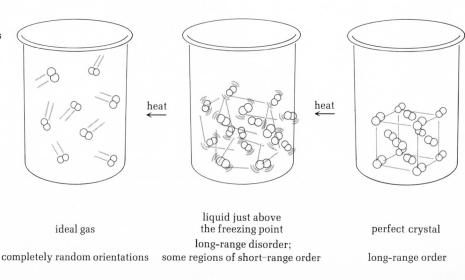

ideal gas

completely random orientations

liquid just above
the freezing point

long-range disorder;
some regions of short-range order

perfect crystal

long-range order

(a) **(b)** **(c)**

Figure 11.6 X-ray diffraction patterns (see Excursion 1) (a) A crystalline powder. (b) The same substance as the liquid just above the melting point. (c) The liquid just below its boiling point.

ness resulting from thermal motions (Unit 5). Experimental evidence from a variety of investigations[3] suggests that this balance results in only microregions of order in liquids, probably consisting of only a few molecules. As might be expected, nonpolar liquids are less ordered than polar liquids such as water, and the disorder increases with temperature.

The highest order in a liquid is usually found near its freezing point and resembles that of a powder that is microcrystalline, that is, that consists of tiny crystals randomly arranged with respect to each other (Figure 11.6).

Another factor that suggests a close relationship between the structures of solid and liquid is the sharp melting point of pure crystalline substances. One might suspect that some relatively minor change might account for this sudden transition from crystal to liquid. This need not suggest microcrystals that suddenly "fit together," but some relatively minor difference must exist.

One model has been suggested to account for these observations. If we view a solid as a simple orderly array of particles, then melting may be explained as a sudden dislocation of one (or a few) particle from their lattice positions, causing a complete disruption of the entire lattice (Figure 11.7).

The "microorder" observed in polar liquids near their freezing point begins to disappear with increasing temperature. Apparently thermal vibrations and collisions between clusters of liquid particles contribute to this change. Very little data are available for liquid structures near the boiling point, but it appears likely that random particle orientations are more common than orderly clusters, with the latter being rapidly changed by collisions and by vibrations of the clusters.

Two other observations support the idea that liquids at high temperatures more closely resemble gases than they do crystalline solids. The boiling point of a pure liquid, like the melting point of a pure crystal, represents a fairly sudden transition. Although *vaporization* occurs to some extent from the surfaces of solids or liquids at any temperature, boiling represents the rapid conversion of liquid to gas and frequently occurs "inside" liquids, at the boiling point, especially at points of contact between the liquid and irregular solids. Since all evidence supports the idea of maximum disorder in the gaseous state, it seems likely that a high degree of disorder exists in the liquid at a temperature where it is rapidly changing to gas.

Above certain temperatures (the *critical temperature* of the particular substance), only one phase can exist. Whether this is a highly compressed gas or a considerably expanded liquid is largely a question of how one wishes to define it. It is most commonly described simply as a one-phase *fluid* of very high particle density. In any event, as a liquid is heated in a sealed tube, the ratio of liquid to gas changes with temperature; this can be observed by the changing position of the meniscus (the curved boundary between liquid and gas phases). When the critical temperature is

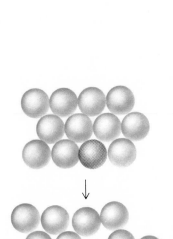

Figure 11.7 Model for melting of a simple crystal by dislocation of a single particle. To illustrate this model better, arrange a dozen or so marbles in a closest-packed two-dimensional pattern. Then push any one of the "interior" marbles to a new location not fitting the pattern. Observe the effect of this motion on the remaining marbles.

[3] For details, see the October, 1959, issue of *Scientific American* (pp. 133ff).

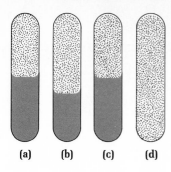

Figure 11.8 Liquid–vapor systems near the critical temperature. (a) Liquid in equilibrium with vapor at low temperature, t_a (rate of vaporization = rate of condensation; see also Unit 14.) (b) Temperature increased to t_b ($>t_a$) such that net volume decrease occurs because vaporization of some liquid exceeds volume increase by thermal expansion. (c) Temperature increased to t_c ($>t_b$) such that net volume increase occurs. At high temperatures, vapor pressure may be so large that volume lost by vaporization is less than liquid volume increase by thermal expansion. (d) Temperature increased to above *critical temperature*. Meniscus suddenly disappears and only one phase is observed.

reached, the meniscus suddenly disappears and only one phase is observed (Figure 11.8). Has the liquid expanded to fill the tube, or has it suddenly vaporized? Above this critical temperature, no pressure increase is sufficient to produce a two-phase system.

At high temperatures, then, liquids are quite similar to gases, although nonideal because of short interparticle distances and significant interparticle forces.

11.5 BONDING IN CRYSTALS

Unit particles in crystals may be atoms, ions, or neutral molecules (Figure 11.9). The forces between these unit particles may be any of the types discussed in Unit 9. One

Figure 11.9 Unit particles in crystals. (a) Neon atoms held in crystal lattice positions by weak London dispersion forces. Crystal melts at −249°C. (b) Carbon atoms—diamond—tetrahedrally bonded through shared pairs of electrons to adjacent carbon atoms. Melting range of diamond is >3500°C. (c) Monatomic ions of magnesium oxide with maximum electrostatic attractions of Mg^{2+} ions for O^{2-} ions. Melting point: 2640°C. (Mg^{2+}, dark; O^{2-}, light.) (d) Monatomic potassium ions arranged uniformly among pyramidal IO_3^- ions. Melting point: 560°C. (K^+, small dark; I, large light; O, small light.) Note that central I is bonded to only three O's of unit cell. Remaining O's are bonded to I's of surrounding unit cells (not shown). (e) Polar water molecules associated through hydrogen bonds with adjacent H_2O molecules so that the region around each oxygen is approximately tetrahedral in hydrogen atoms. Melting point: 0°C. (f) Nonpolar bromine molecules held in lattice postions by London forces. Melting point: −7°C.

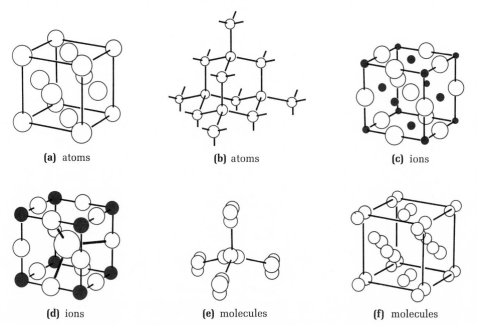

(a) atoms   **(b)** atoms   **(c)** ions

(d) ions   **(e)** molecules   **(f)** molecules

way of classifying crystals is on the basis of these interparticle forces, which may be directly related to bulk properties of the crystalline solids (Table 11.1).

Crystalline metals have already been described in terms of a model of positive ions in a "cloud" of mobile electrons (Unit 9). The atomic model for metals, like those for other crystalline solids, is consistent with observable properties.

Table 11.1 CRYSTAL CLASSIFICATION BY UNIT PARTICLES AND INTERPARTICLE FORCES

CLASSIFICATION	UNIT PARTICLES	INTERPARTICLE FORCES	GENERAL PROPERTIES	EXAMPLES
atomic	atoms	London dispersions (Unit 9)	soft, very low melting, poor electrical conductors, good thermal conductors	noble gases
molecular	molecules, polar or nonpolar	van der Waals forces (London dispersions, dipole-dipole, hydrogen bonds) (Unit 9)	fairly soft, low to moderately high melting, poor thermal and electrical conductors	"dry ice" [$CO_2(s)$], ice [$H_2O(s)$], "sugar," etc.
metallic	cations in electron cloud	metallic bond (Unit 9)	soft to very hard, low to very high melting, excellent thermal and electrical conductors, malleable (can be formed into thin sheets), ductile (can be drawn into wires)	all metallic elements
ionic	positive and negative ions	ionic bond (Unit 6)	hard, high melting, poor thermal and electrical conductors	typical salts, e.g., NaCl, $BaSO_4$, etc.
covalent (network)	atoms connected in covalent bond network	covalent bond (Unit 7)	very hard, very high melting, poor thermal and electrical conductors	diamond, quartz, silicon

11.6 CRYSTAL STRUCTURES

Well-formed crystals exhibit highly regular geometries—cubes, octahedra, and so on. The external appearance of such crystals is called the *crystal habit,* and this provides useful clues to the trained geologist or crystallographer to aid in the classification of crystals. Since we have described crystals as consisting of highly ordered and characteristic arrays of unit particles, it would be tempting to suggest that the crystal habit reflects the simplest basic pattern of these particles—*the unit cell.* Such is not necessarily the case: Consider, for example, the numerous rectangular shapes that could be made by stacking together sets of tiny cubic blocks. The important geometric patterns that define a *crystal system* are those related to the submicroscopic three-dimensional pattern of fundamental crystal units—*the crystal lattice.* Clues to some possibilities of lattice geometries may be gained from careful examination of the crystal habit, but the ultimate description of the lattice requires the determination of relative positions of unit particles. This determination can be made by techniques of x-ray diffraction, as described in Excursion 1.

Crystal lattices are represented by three-dimensional structures of lattice points (Figure 11.10), each of which represents the center of mass of a unit particle—atom, ion, or molecule. In crystals of metals or noble gases, the lattice points correspond to atomic locations.

The term *symmetry* has been used in connection with molecular geometry (Unit 8) as well as in the definition of crystal lattice points above. There are three simple *symmetry elements* of importance in crystal geometry, and these may be defined in terms of *symmetry operations.*

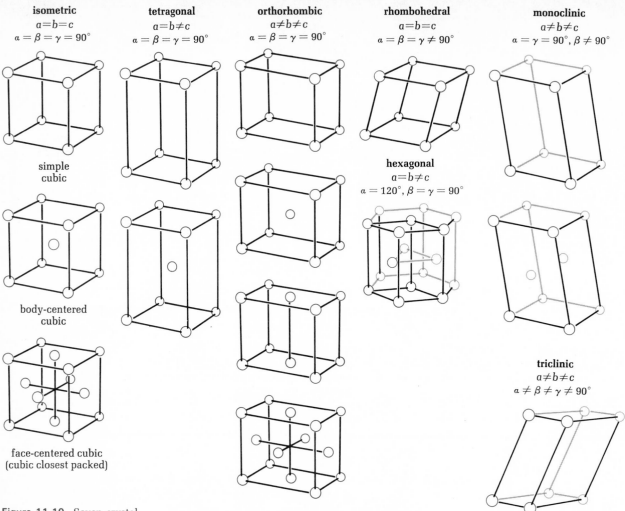

isometric
$a=b=c$
$\alpha = \beta = \gamma = 90°$

simple
cubic

body-centered
cubic

face-centered cubic
(cubic closest packed)

tetragonal
$a=b\neq c$
$\alpha = \beta = \gamma = 90°$

orthorhombic
$a\neq b\neq c$
$\alpha = \beta = \gamma = 90°$

rhombohedral
$a=b=c$
$\alpha = \beta = \gamma \neq 90°$

hexagonal
$a=b\neq c$
$\alpha = 120°, \beta = \gamma = 90°$

monoclinic
$a\neq b\neq c$
$\alpha = \gamma = 90°, \beta \neq 90°$

triclinic
$a\neq b\neq c$
$\alpha \neq \beta \neq \gamma \neq 90°$

Figure 11.10 Seven crystal systems and 14 Bravais lattices (edge lengths, a, b, c; angles between connecting faces, α, β, γ). Different crystallographers vary somewhat in the nomenclature of crystal systems. From the seven *crystal systems* come the fourteen *Bravais lattices* as determined by additional possibilities for lattice points.

Symmetry element	Symmetry operation
symmetry axis	rotation around the axis
mirror plane	reflection in a mirror plane
center of symmetry	projection through a point

If an object (e.g., a molecule or a unit cell lattice structure) can be rotated around an imaginary axis selected so that the equivalent image appears to the viewer two or more times during a rotation of 360°, the object is said to possess a symmetry axis. The number of times the equivalent image appears during a 360° rotation identifies this axis. For example, the trigonal planar BF_3 molecule has one *threefold* axis of symmetry and three *twofold* axes (Figure 11.11).

The use of an imaginary mirror plane can be illustrated with a real mirror. If you put your hands together, palm to palm with thumbs touching, there is an imaginary mirror plane between them. If you now separate your hands, holding the same orientations, and place your left hand behind a mirror and your right hand in front of the mirror, the reflection of the right hand will look like the left hand. The ability to

PLASTICS AND GLASS

The very high degree of order associated with typical crystals is difficult to achieve with very large molecules. The greater the symmetry of a unit particle, the easier it is to fit it into a simple lattice pattern. Metals and noble gases, for example, pack easily into simple arrays like stacks of cannonballs. Spherical particles require no special orientations. More complex particles are more difficult to pack in simple ways. Special orientations are required. These are favored among polar molecules by interparticle forces that help align the molecules, but nonpolar molecules have no such "help." Very large molecules require considerable energy to turn them for simple alignment and, since solidification typically occurs at temperatures low enough to reduce thermal motions below those typical of the liquid state, such molecules may find themselves rather randomly oriented in the solid.

Large complex molecules, then, frequently form amorphous solids rather than crystals. Many common plastics behave in this way. Polyethylene, for example, forms a highly disordered solid in which the long polymer molecules are intertwined much like a snarled bundle of threads. Most other plastics also form amorphous solids. Even the three-dimensional networks in plastics such as Bakelite are highly irregular as compared to the orderly arrays in typical crystals.

Ordinary glass is a mixture of silicate polymers and simple ions. "Soft" glass may be prepared by melting together a mixture of sodium carbonate, limestone, and sand. After the liquid melt has ceased to evolve bubbles (CO_2), it is poured into a mold, blown into vessels, or stamped with dies. As the liquid is cooled, a solid state is reached before the silicate polymers have been uniformly aligned. The rate of cooling is important. Submicroscopic disorder is desirable for the properties expected of typical glass, but large discontinuities produce undesirable stresses and these frequently occur when the cooling is done too rapidly. Within recent years it has become possible to effect techniques for producing crystalline "glass" whose strength and optical properties approach those of quartz.

position an imaginary planar mirror through a segment of a molecule or lattice so that the reflection is equivalent to what is behind the mirror is a necessary condition for a plane of symmetry (Figure 11.12).

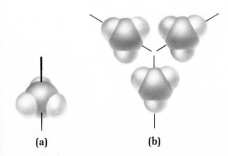

(a) **(b)**

Figure 11.11 Symmetry axes in BF_3. (a) Threefold axes. (b) Twofold axes. (*Suggestion:* Obtain four small styrofoam balls and some toothpicks. Assemble a model of BF_3 and demonstrate the nature of the symmetry axes by rotations.)

Figure 11.12 A plane of symmetry in methane. The mirror plane is perpendicular to the page along the heavy line.

279

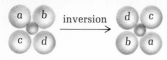

Application of more complex symmetry operations than the three described leads to the ultimate possibility of 230 different ways of forming repetitive three-dimensional patterns.

The operation required for location of a center of symmetry may be considered as requiring the imaginary projection of all elements of an object in straight lines passing through at a common point (the center of symmetry) such that the projection is an exact replica of the original object (Figure 11.13). Very few patterns are sufficiently symmetrical to possess a center of symmetry.

Each crystal system is characterized by certain minimum elements of symmetry. The isometric system, for example, is the only one that has four threefold symmetry axes. Symmetry elements revealed by x-ray crystallographic patterns (Excursion 1) are essential to the classification of a crystal in the proper system.

Symmetry elements are important in describing crystals. In simple crystals whose unit particles are atoms or monatomic ions, the crystal may be adequately described by use of one of the 14 Bravais lattices. However, when unit particles are molecules or polyatomic ions, these may introduce further symmetry considerations and therefore permit additional crystal geometries. It has been determined, however, that all natural crystals can be described in terms of one of 32 possible *crystal classes*. This limitation is an important factor in simplifying the determinations of crystal structure.

11.7 SIMPLE CRYSTALLINE ELEMENTS AND SALTS

The simplest types of unit cells are those containing only monatomic particles. Crystalline metals and noble gases, in which all particles are identical, are especially easy to characterize, although the very low melting points of the latter present serious experimental problems. Many of the metals crystallize in "closest-packed" arrays, either cubic or hexagonal (Figure 11.14). Others, such as the alkali metals, crystallize in the more open body-centered cubic structure.

To see how these arrays are formed, first consider two alternative ways of packing spheres in a single layer (Figure 11.16). The body-centered *tetragonal* structure results from stacking layers of square patterns of spheres, with alternating layers offset so that each sphere fits most snugly in the "hole" centered in a square of spheres in the next layer (Figure 11.17). A less efficient packing occurs in the primitive cubic lattice,

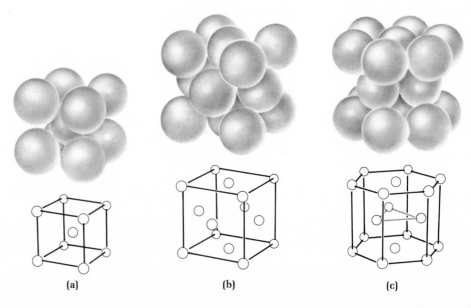

Figure 11.14 Some simple metallic crystal lattices. (a) Body-centered cubic; all group IA elements and barium. (b) Face-centered cubic (cubic closest packed); calcium and strontium. (c) Hexagonal closest packed; beryllium and magnesium. Other (less common) crystal forms (allotropes) are also known for many of these substances.

(a)   (b)   (c)

MACROMOLECULAR CRYSTALS

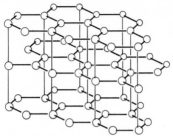

Figure 11.15 Crystal structure in graphite. In the crystal each planar network is an enormous "fused ring" molecule. Only a small segment is shown here.

The ease of "sliding" of network planes in graphite accounts for its slippery characteristics. Graphite is the material that rubs off when you write with a "lead" pencil. Graphite absorbs gases readily at the surface and between planes. This absorbed gas increases the "slipperiness" and undoubtedly contributes to the lubricant properties of powdered graphite.

Certain giant molecules (macromolecules) are sufficiently symmetrical to form crystalline aggregates. One of the most interesting of these is graphite (Figure 11.15). Layers of enormous molecules resembling hexagonal networks of chicken wire are stacked uniformly and held together by forces variously described as van der Waals attractions or weak end-to-end overlap of p orbitals.

The bonds within the plane of the hexagonal ring network appear to be like those in benzene (Unit 8). Adjacent carbon atoms are connected by σ bonds involving sp^2 hybrid orbitals, and the p orbitals perpendicular to this plane form a π-bond system that is continuous over the entire network. The bonding between the planes seems to involve some weak overlap of the p orbitals participating in the π-bond systems of consecutive layers. The weak interplanar forces allow the layers to slide over each other fairly easily. Unlike most other molecular crystals, graphite is a good electrical conductor—because of the mobile character of electrons in the delocalized π-bond system. At high temperatures and very high pressures, the crystal structure of graphite is converted to a three-dimensional tetrahedral network of carbon atoms—diamond. At high temperatures ($\sim 3500°C$) and low pressures, diamond reverts to the thermodynamically more stable graphite. The phenomenon in which a substance can form more than one type of crystal structure is known as *polymorphism*, and the different forms, such as graphite and diamond, are called *allotropes*.

Many interesting macromolecules, such as several types of proteins (Unit 32) and nucleic acids (Unit 34), form solids that are highly ordered but that do not have the rigidity of typical crystals. Such substances are known as *fibers*. Hair, for example, is a fiber consisting of ordered strands of very long protein molecules intertwined in a regular pattern like the tiny threads of a twisted rope.

The determination of the structure of fibers and macromolecular crystals has been made possible in recent years by very sophisticated applications of x-ray diffraction. These techniques and their applications to the description of crystal structure are discussed in Excursion 1.

Figure 11.16 Two packing patterns for spheres. The area occupied by the 12 spheres packed in square patterns (left) is $4d \times 3d = 12d^2$, where d is the diameter of the sphere. The area occupied by 12 spheres packed in hexagonal patterns (right) is calculated from simple trigonometric relationships to be $10.94d^2$. Thus the hexagonal pattern is a more efficient (closer-packing) way of arranging spheres than the square pattern.

which corresponds to square patterns stacked so that all spheres are aligned both horizontally and vertically. To form the body-centered cubic pattern, alternate layers must be "compressed" to provide larger "holes" to avoid the extended vertical dimension distinguishing the tetragonal from the cubic lattice. This makes for

(equivalent)

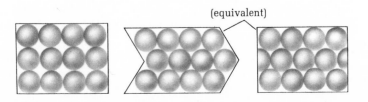

Figure 11.17 "Inefficient" packing of spheres. (a) Primitive cubic. (b) Body-centered tetragonal. (c) Body-centered cubic.

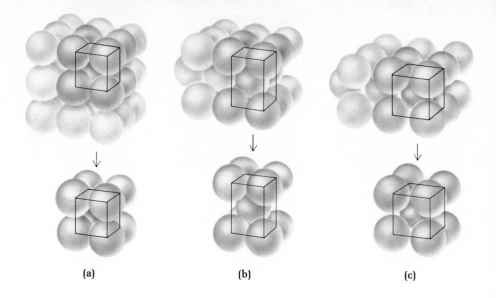

(a)　　　　(b)　　　　(c)

Each atom is said to have a co-ordination number of 12. This favors maximum London dispersion forces, as opposed, for example, to the body-centered cubic, in which each sphere only has a coordination number of 8.

inefficient packing, because atoms are no longer touching within layers—only along diagonals.

There are two simple ways to pack spheres more efficiently. Both start with a first layer of the hexagonal type. The differences in ways of adding successive layers lead to either hexagonal or cubic closest-packed structures (Figure 11.18). Both are very "efficient": 74 percent of the available volume is occupied by the spheres. In both, each sphere (in extended cell networks) has 12 nearest neighbors.

Even though the closest-packing arrays are more "efficient," there is not a large difference. In the body-centered cubic, for example, 68 percent of the available space is occupied, as compared to 74 percent in the face-centered cubic. Differences in the densities of metallic elements are much greater than can be accounted for solely by differences in atomic packing patterns and atomic weights. Interparticle forces, which are functions of coordination numbers and of individual atomic characteristics, contribute to differences in unit cell volumes. Stronger interparticle forces tend to decrease bond lengths and thus to increase densities. Tabulated values of metallic radii can be used for approximation of the volume of metallic lattices (Excursion 1), but care must be taken to select values corresponding to the particular crystal form under consideration. Allotropic modifications may have different effective particle radii.

The number of *particles* assigned to a unit cell is *not* the same as the number of *points* required by the Bravais lattice. Edge, corner, and face particles are shared among adjacent unit cells. For an atomic solid belonging to the isometric (cubic) system, for example, each corner atom is shared by eight unit cells (Figure 11.19), each face atom by two unit cells, and each edge atom by four. In assigning the number of particles per unit cell, only the segment of the particle *within* the imaginary "lattice box" is counted. The segments shared with additional unit cells are not. In the isometric system this means that we count

$\frac{1}{8}$ for each corner particle
$\frac{1}{4}$ for each edge particle
$\frac{1}{2}$ for each face particle
1 for each particle entirely enclosed by the "lattice box"

Figure 11.18 Closest-packing patterns. The first layer (viewed from above) is shown at (1). There are three possible positions for successive layers: (a) directly above the spheres in the first layer (dots); (b) above one triangular pattern of "holes" in the first layer (dark), and (c) above the inverted triangular pattern of holes in the first layer (white).

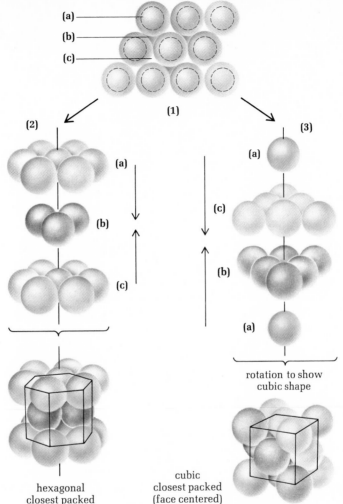

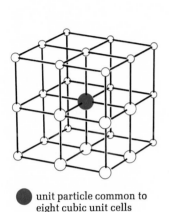

● unit particle common to eight cubic unit cells

○ other unit particles of the "basic" cubic cell

○ points defining additional cubic cells

Figure 11.19 Particle sharing among adjacent unit cells.

Orthogonal faces are mutually perpendicular.

Other crystal systems are treated in a similar fashion, but the fractions counted at nonorthogonal locations are different from those indicated above. The proper fractions can be determined by applications of the mathematics of solid geometry.

Example 2

A compound of iron with oxygen crystallizes in the isometric system. The unit cell may be described as a face-centered cubic array of oxide ions with an iron ion located at the center of the cube and additional iron ions at the centers of each edge of the cube. What is the empirical formula of this compound?

Solution

A cube has 8 corners, 12 edges, and 6 faces. Oxide ions are located at corners and in

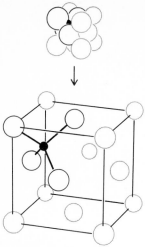

● tetrahedral site
(one under each corner)

Figure 11.20 Tetrahedral site in cubic closest-packed lattice.

Remember that the number of such ions assigned to the unit cell is 4, [($\frac{1}{8}$ × 8 corners) + ($\frac{1}{2}$ × 6 faces)].

It is, of course, possible to find salts of large cations with small anions. These follow similar patterns. For example, UO_2 and BaF_2 have crystal structures like those of Li_2O.

Still a third interstitial site exists, the *trigonal* site centered in each triangular pattern of spheres. Its radius ratio is only 0.155, so it is rarely occupied in closest-packed lattices.

the center of each face, so we count

$$\tfrac{1}{8} \times 8 + \tfrac{1}{2} \times 6 = 4 \text{ oxide ions}$$

Iron ions are located at the center of the cube and at the centers of each edge, so we count

$$1 + \tfrac{1}{4} \times 12 = 4 \text{ iron ions}$$

The unit cell is then assigned four oxide ions and four iron ions. The empirical formula (Unit 3) is then FeO.

In both the cubic and hexagonal closest-packed lattices there are different types of "holes," called *interstitial sites*. *Tetrahedral sites* occur as the small spaces left in each pattern of four spheres corresponding to the stacking pattern a, b (Figure 11.18). You may recall that the tetrahedral pattern is characteristic of certain simple molecules, such as methane (Unit 8), in which four atoms surround a fifth. In a crystal lattice, the tetrahedral *site* is the hole symmetrically surrounded by four lattice particles. One such site occurs just behind each corner particle in a cubic closest-packed lattice (Figure 11.20). Since the tetrahedral sites are contained *inside* the imaginary cube defining the unit cell, *eight* such sites (under the eight corners) are counted for a face-centered cubic unit cell.

The octahedral pattern consists of six particles symmetrically surrounding a seventh (Unit 8). Thus one octahedral site is located at the center of a face-centered cubic unit cell. The remaining octahedral positions are located at the centers of each edge of the unit cell (Figure 11.21) and since each edge is common to four unit cells, one-fourth of each such position is assigned to a single unit cell. The number of octahedral sites, then, assigned to a face-centered cubic unit cell is

$$(12 \text{ edges} \times \tfrac{1}{4}) + (1 \text{ center}) = 4$$

In cubic closest-packed lattices of large ions (e.g. monatomic anions) there are, then, two tetrahedral sites and one octahedral site per ion, which could be occupied by an ion of opposite charge if it were small enough to fit. Closest-packed arrays of anions are approximated in salts of the type M_2X (e.g., Li_2O, with Li^+ ions in all tetrahedral sites). If only half the available tetrahedral sites are occupied (e.g., ZnS, AgI) or if cations occupy octahedral sites (e.g., LiCl, AgF), then salts of the type MX are formed. The selection of an interstitial site depends to some extent on the size of the spherical "hole" available for occupancy. Simple geometry shows that for tetrahedral sites,

$$\frac{\text{hole radius}}{\text{sphere radius}} = 0.225$$

for octahedral sites,

$$\frac{\text{hole radius}}{\text{sphere radius}} = 0.414$$

In reality, the radius ratio may exceed these values without appreciable distortions, since ions are not rigid spheres and their electron clouds can overlap considerably. Thus anions may be closest packed with very small cations in tetrahedral sites or with somewhat larger cations in octahedral sites. If the cation-to-anion radius ratio exceeds about 0.5, the anions are forced apart somewhat from their closest-packed positions.

The type of bonding involved in a salt also determines the nature of site occupancy: Ions in tetrahedral holes have a coordination number of 4, whereas those in octahedral holes are surrounded by six nearest neighbors of opposite charge.

11.8 MASS AND ENERGY TRANSPORT

Evidence indicates that heat flow in mobile-electron systems, such as solid or liquid metals, is associated with electron movement, as well as with the normal motions of atomic-size particles.

Both thermal energy and particles can move through a condensed phase in either of two ways. The more rapid way is by bulk movement caused by "currents" in a liquid. Thermal currents, for example, occur when localized heating causes a greater increase in particle motion in one liquid region than in another. Mechanical currents result from any "stirring" effect in the liquid. Whatever the source, these bulk movements transport particles (e.g., liquid molecules, dissolved ions, etc.) from one place to another in the liquid. They also transport heat, as more rapidly moving particles are spread to regions where they can transfer energy to slower moving particles by collisions.

The alternative mechanism depends on individualized particle motion. Thermal conductivity involves a "jostling" motion transmitted between adjacent particles. In solids this effect is relatively efficient. The particles are close together and, in crystals, uniformly aligned. Thermal conductivity in gases is very inefficient since the distance particles travel between collisions (mean free path) is relatively large and the randomness characteristic of gases makes *directional* heat flow less likely. Liquids display a thermal conductivity more like solids than gases, primarily because of the short interparticle distances in liquids. Individualized particle motion occurs in *diffusion* processes, which are most rapid in gases and slowest in solids (Figure 11.22).

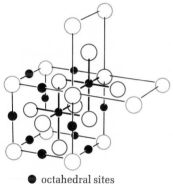

● octahedral sites

Figure 11.21 Octahedral sites in face-centered cubic lattice.

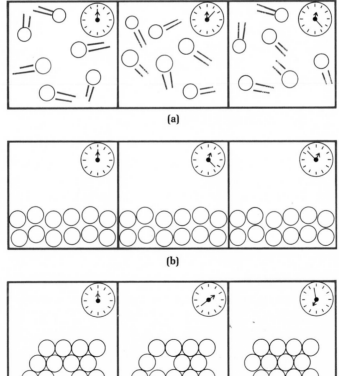

(a)

(b)

Figure 11.22 Relative diffusion rates (at the same temperature). (a) In gases a particle follows a zigzag path resulting from occasional collisions. (b) In liquids a particle slowly jostles in and out among liquid particles. (c) In solids a particle only rarely finds opportunity to change lattice location.

(c)

In diffusion, particles move into a region not previously occupied by such particles. Diffusion in gases is rapid because of the large empty spaces available. In liquids, empty spaces are much less frequently encountered, and diffusion is necessarily slower. Diffusion in solids is very slow, being favored by such factors as lattice vacancy defects.

11.9 VISCOSITY AND SURFACE TENSION

Viscosity is the opposite of fluidity; it is generally defined as a resistance to flow. We are all familiar with liquids of different viscosities: Water flows more easily than molasses, which in turn flows more easily than rubber cement. Motor oils are rated on a viscosity scale. Since viscosity normally increases with decreasing temperature, we replace our "summer weight" (higher viscosity) motor oil with one of lower viscosity for colder weather.

Like other bulk properties of matter, viscosity can be related to behavior at the atomic level. Remember that our model for liquids is that of a fairly random grouping of particles in continuous jostling motion. Any factor tending to retard bulk flow in a liquid will increase viscosity. Viscosity is increased by interparticle forces between flowing and nonflowing groups of molecules and by transfer of molecules back and forth between mobile and nonmobile groups. Small molecules are transferred more readily than larger ones at the same temperature. Intermolecular forces are, however, of greater effect in liquids, and the transfer from mobile to nonmobile groups is of principal interest in gases. The fact that water is more viscous than hexane at the same temperature, for example, can be explained by the stronger intermolecular forces in liquid water.

In liquids consisting of long-chain molecules, such as the hydrocarbons in oils and greases, viscosity is increased by molecular tangling.

Viscosity can be considered in terms of the quantity of matter flowing a particular distance in a given time. A viscosity of $1 \text{ g cm}^{-1} \text{ sec}^{-1}$ is designated as *poise*, in honor of J. L. M. Poiseville, who derived a simple equation for viscosities of liquids in 1844. Most common simple liquids have viscosities considerably less than 1; the unit millipoise (0.001 poise) is frequently convenient (Table 11.2).

Viscosities of liquids are commonly measured with an Ostwald viscometer or, for the more viscous liquids, with a falling-ball viscometer (Figure 11.23). In either case, the viscosity may be calculated from appropriate quantities that are functions of the liquid or of the apparatus and conditions employed. These calculations and the

This is an oversimplification. The coefficient of viscosity is actually defined as the force necessary to produce a unit velocity gradient per unit contact area. Appropriate units cancel to give the net dimensions of mass $\times$ length^{-1} $\times$ time^{-1}.

Table 11.2 VISCOSITIES OF SOME COMMON SUBSTANCES
(millipoise)

SUBSTANCE	0°C	20°C	50°C
water	17.92	10.02	5.49
acetone	3.99	3.27	~2.7
benzene	9.12	6.52	4.42
diethyl ether	2.84	2.33	1.81
ethanol	17.73	12.00	7.02
nitroglycerin	—	692	102
castor oil	>4000	986	~200
octane (C_8H_{18})	7.06	5.42	~4
sulfuric acid	484	254	88.2
glycerin	121,100	14,900	<400
air (1 atm)	0.171	0.184	0.196

Note that ordinary liquids decrease in viscosity (increase in fluidity) with increasing temperature, as would be expected since the increased thermal motions interfere with formation of stable particle clusters. Gases show the opposite effect. Since clustering is negligible, the principal effect in gases is transfer of particles between mobile and nonmobile groups.

Figure 11.23 Measurement of viscosity. (a) Ostwald viscometer. (b) Fallingball viscometer.

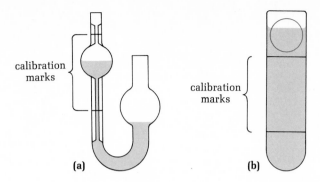

calibration marks

calibration marks

(a)   **(b)**

necessary measurements are tedious, and it is much more convenient to determine the viscosity of a liquid by comparison with that of a liquid whose viscosity is already known, using identical experimental conditions. With the Ostwald viscometer, such a determination requires values of the densities (d) of the two liquids and measurement of the times required for a given volume of the liquids to flow between the calibration marks on the instrument. The viscosity of the unknown liquid, then, can be related to that of the comparison liquid by

η = Greek eta.

$$\eta_{unk} = \eta_{comp} \times \frac{d_{unk} \times time_{unk}}{d_{comp} \times time_{comp}}$$ (11.1)

Example 3
___

A sample of turpentine (density 0.873 g cm^{-3}) was measured in an Ostwald viscometer, using ethanol (density 0.789 g cm^{-3}) as the comparison standard. The average running times of the two liquids at 20°C were

     ethanol: 93.4 sec
turpentine: 104.6 sec

What was the viscosity of the turpentine in millipoise?

Solution
From equation (11.1) and Table 11.2,

$$\eta_{turpentine} = 12.00 \left(\frac{0.873 \times 104.6}{0.789 \times 93.4} \right)$$

$$= 14.87 \text{ millipoise}$$
___

"Wetting" refers to a liquid film adhering to a surface.

A liquid–vapor interface appears flat when a large surface area is involved, but there is a pronounced curvature apparent for a small surface area. This curved surface is called the *meniscus* (Figure 11.24) and its shape is determined by the surface tension of the liquid and by the extent to which the liquid will "wet" the container walls.

Surface tension is related to the strength of interparticle forces (Unit 9). Within a liquid these forces are fairly uniform in all directions. Although individual polar molecules may exert stronger forces in one direction than in another, the random orientations of particles in liquids result in microregions whose net forces are

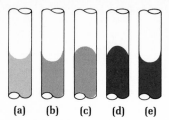

(a)   (b)   (c)   (d)   (e)

Figure 11.24 Liquid–gas interface showing meniscus. (a) Acetone (surface tension 23.7 dynes cm^{-1} at 20°C) in glass (1 dyne = 1 g cm sec^{-2}). Acetone wets glass. (b) Water (surface tension 73.0 dynes cm^{-1} at 20°C) in glass. Water wets glass. (c) Water in tube made from a hydrophobic plastic, not wetted by water. (d) Mercury (surface tension 490 dynes cm^{-1} at 20°C) in glass. Mercury does not wet glass. (e) Mercury in glass tube with translucent interior film of silver. Mercury wets silver.

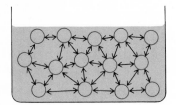

Figure 11.25 Surface tension in liquids. The circles represent particles (atoms or spherically symmetrical molecules or ions) or microregions of spherical symmetry.

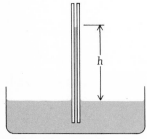

Figure 11.26 Capillary Technique for measuring surface tension.

Table 11.3 SURFACE TENSIONS OF SOME COMMON LIQUIDS

LIQUID	IN CONTACT WITH	SURFACE TENSION AT 20°C (dynes cm^{-1})
water	air	73
acetone	air	24
benzene	air	29
diethyl ether	vapor	17
ethanol	vapor	23
octane (C_8H_{18})	vapor	22
sulfuric acid	air	55
glycerin	air	63
mercury	air	490

spherically symmetrical. At the surface, however, there are no "upper" particles to participate in particle interactions (except for occasional gas particles), so the net attractions are distributed downwards and laterally (Figure 11.25). The result is a strong surface cohesion, which varies with the strength of the interparticle forces involved. The surface tension of water is large enough to support a thin sheet of steel if the surface is not broken.

A liquid–vapor interface also occurs in a bubble immersed in a liquid or in a gas. Almost every child has blown soap bubbles and noted the considerable strength of the thin liquid film of the bubble. One way of measuring surface tension is by determining the force needed to expand a bubble from one measured diameter to another.

An alternative method measures the force required to stretch a thin liquid film from one measured area to another.

For comparison purposes, some values of surface tensions of common substances are listed in Table 11.3.

The surface tension of a liquid may be estimated by measuring the height a liquid column will rise in a constricted tube (capillary) made from a material that is wetted by the liquid (Figure 11.26). The estimation is made from an equation whose derivation we shall omit:

$$\gamma = 490drh \tag{11.2}$$

in which γ (Greek *gamma*) is the surface tension, which will have the units of dynes per centimeter when the liquid's density d is expressed in grams per cubic centimeter; the capillary radius r is expressed in centimeters; and the height from the flat liquid surface to the center of the meniscus h is expressed in centimeters. The constant 490 is one-half the standard gravitational acceleration (980 cm sec^{-2}).

Example 4

How high should water rise in a wettable capillary of diameter 0.20 cm at 20°C?

Solution
From equation (13.2),

$$h = \frac{\gamma}{490dr}$$

From Table 13.2, $\gamma_{H_2O} = 73$ dynes cm^{-1} at 20°C. The density of water at 20°C is

NEGATIVE PRESSURES

The mangrove tree grows in swamplands near the ocean. Its roots are immersed in salt water but its sap, which rises to great heights, is nearly salt free. Osmotic effects (Unit 15) suggest that water would diffuse irreversibly from the roots to the surrounding salt water and that the trees would dehydrate if the pressure of the sap were greater than that of the surrounding water. The tall "columns" of sap in high trees should certainly exert a considerable pressure, but the trees do not dehydrate. It has been possible to measure the pressure of the sap in the roots of mangrove trees and, surprisingly, the results indicate *negative* pressures. The sap does not rise, as does the liquid mercury in a barometer, under a condition of positive pressure. Rather, it is pulled—not pushed—toward the top of the tree. Indeed, careful measurements have shown that sap rise under positive pressure is very rare—a fact that may surprise many biologists. Desert plants may have as much as 80 atm of negative pressure in their "pull" for moisture from the arid soil.

This is not some "magical" phenomenon of transfer of water from one living cell to another upward through the plant: Sap will rise easily for large distances through dead tissue. Apparently, the sap is pulled up from above. The exact mechanism of this action is not yet understood.[4]

The existence of negative pressures in liquids is an interesting phenomenon. There must be strong interparticle forces in liquids to account for cohesion in long columns under negative pressure. Liquids are more complex than either the simple gases or the crystalline solids. They are more difficult to investigate and to treat by simple theoretical models.

[4] The interested student may wish to read "Negative Pressure in Liquids: Can It Be Harnessed to Serve Man?," by Alan T. J. Hayward, in the *Scientific American* (Vol. 59, 1971, pp. 434ff).

approximately 1.0 g cm^{-3}, so

$$h = \frac{73 \text{ dynes cm}^{-1}}{490 \text{ cm sec}^{-2} \times 1.0 \text{ g cm}^{-3} \times 0.10 \text{ cm}} \times \frac{1 \text{ g cm sec}^{-2}}{1 \text{ dyne}}$$

$$= 1.5 \text{ cm}$$

11.10 LIQUID MISCIBILITY

"Negligible" is a rather subjective term and really depends on the particular circumstances involved.

All of the types of interparticle forces previously discussed (Unit 9) can occur in mixtures as well as in pure substances. When different substances have very similar interparticle forces, they will usually mix homogeneously (i.e., form *solutions*). In optimum cases, such substances may form solutions in any relative proportions. These substances are said to be *miscible*. Strictly speaking, some degree of mixing would be possible for any materials, but in practice substances with negligible mutual solubilities are termed *immiscible*.

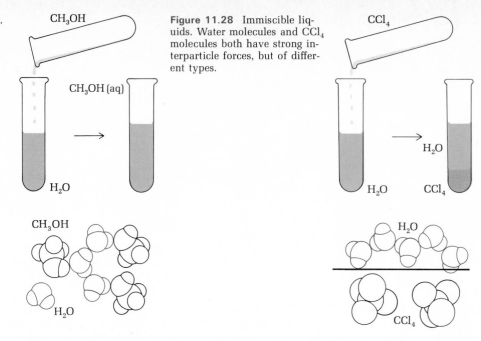

Figure 11.27 Miscible liquids. Mixing of methanol (CH_3OH) with water may occur in any relative amounts.

CH_3OH

CH_3OH (aq)

H_2O

CH_3OH

H_2O

Figure 11.28 Immiscible liquids. Water molecules and CCl_4 molecules both have strong interparticle forces, but of different types.

CCl_4

H_2O

H_2O CCl_4

H_2O

CCl_4

When two miscible liquids are mixed, a single liquid phase results (Figure 11.27). When two immiscible liquids are mixed (Figure 11.28), two liquid phases (or sometimes a colloidal dispersion—Unit 14) result. This may imply that each phase remains discrete because of equally strong, but different, types of interparticle forces. However, it is not uncommon for a liquid (e.g., water) of strong intermolecular attractions to "squeeze out" other molecules present, even though the second liquid consists of molecules with only weak attractions for each other.

Saturated solutions (Figure 11.29) occur when there is a limited solubility of one substance in another (i.e., they are not miscible). The extent to which any two substances will mix depends on a number of factors, including temperature and detailed structural features. Although quantitative prediction of solubility is difficult, reasonable qualitative guesses can be made by considerations of molecular structures and possible intermolecular attractions.

Figure 11.29 Saturated solution. 1-Butanol ($CH_3CH_2CH_2CH_2OH$) is only sparingly soluble in water.

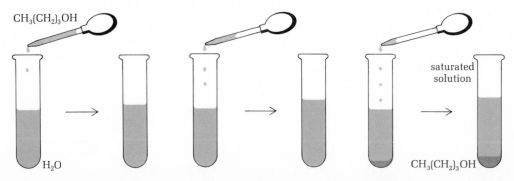

$CH_3(CH_2)_3OH$

H_2O

saturated solution

$CH_3(CH_2)_3OH$

SUGGESTED
READING

Bernal, J. D. 1960. "The Structure of Liquids," *Sci. Amer.,* August.
Claussen, W. F. 1967. "Surface Tension and Surface Structure of Water," *Science,* **156,** 1226.
Dreisbach, Dale. 1966. *Liquids and Solutions.* Boston: Houghton Mifflin.
Galway, A. K. 1967. *Chemistry of Solids.* New York: Barnes and Noble.
Moore, W. S. 1967. *Seven Solid States: An Introduction to the Chemistry and Physics of Solids.*
Menlo Park, Calif.: W. A. Benjamin.

EXERCISES

1. Select the term that is best associated with each of the following definitions, descriptions, or specific examples.
   TERMS: (A) allotropes (B) amorphous (C) crystal habits (D) crystal systems (E) fluid (F) immiscible (G) surface tension (H) viscosity
   a. the external appearance of well-formed crystals
   b. isometric and monoclinic
   c. resistance to flow
   d. can be measured by a "capillary rise" method
   e. not mutually soluble
   f. $H_2(g)$ and $H_2(l)$
   g. rhombic sulfur and monoclinic sulfur
   h. Pyrex glass and Lucite plastic

2. Draw a three-dimensional representation of a unit cell of a face-centered cubic lattice, then use toothpicks and styrofoam balls (or marshmallows, gumdrops, etc.) to construct a model of a simple cubic unit cell. Using this model, identify and draw representations of
   a. a threefold symmetry axis; **b.** a mirror plane; **c.** a center of symmetry

3. Classify each of the following crystalline solids by bond type and predict whether each might be expected to melt below 0°C and whether each is expected to be a good electrical conductor (as a solid).
   a. iodine; **b.** calcium; **c.** calcium chloride; **d.** graphite; **e.** diamond

4. For each of the following pairs, select which should be the more viscous. Explain your choices.
   a. pentane (C_5H_{12}) or ethylene glycol at 20°C
   b. a hydrocarbon motor oil at −10°C or at 40°C
   c. nitrogen gas (1 atm) at −10°C or at 40°C

5. Indicate which of the liquid pairs should be completely miscible. Explain your choices.
   a. water and acetic acid; **b.** benzene and acetic acid; **c.** pentane and octane
   [Check your prediction on pair (b) against some special information on this system given in Unit 9.]

6. The following salts are all derived from a cubic closest-packed arrangement of the larger ions. For each, draw a representation of the unit cell and calculate the empirical formula of the salt.
   a. an oxide of calcium with calcium ions in octahedral sites (the *rock salt* structure)
   b. an oxide of hafnium with oxide ions in all tetrahedral sites (the *fluorite* structure)

   c. an iodide of copper with copper ions in one-half the tetrahedral sites (the *zinc blende* structure)

7. A compound of potassium, iodine, and oxygen crystallizes in the isometric system. The unit cell is described as a primitive cubic lattice arrangement of potassium ions, with an iodine at the center of the cube and oxygens at the centers of all edges. What is the empirical formula and name of this compound?

8. Calculate the density of each of the following:
   a. a tetragonal solid of mass 18.4 g and edge lengths 2.0 cm, 2.0 cm, and 3.6 cm
   b. an irregular solid of mass 11.9 g that sinks to displace 5.20 ml of water
   c. a tiny solid that is suspended by a solution from which a 0.500-ml portion weighs 0.873 g

9. A liquid hydrocarbon was compared with benzene in an Ostwald viscometer at 20°C. The following data were obtained:

Compound	Density (g cm^{-3})	Running time (sec)
the hydrocarbon	0.779	78.4
benzene	0.879	86.3

What is the viscosity of this hydrocarbon at 20°C?

10. The same hydrocarbon was measured to rise 8.3 cm in a wettable capillary of diameter 0.20 mm at 20°C. What was the surface tension of the hydrocarbon?

PRACTICE FOR PROFICIENCY

Objective (a)

1. Select the term best associated with each definition, description, or specific example.
   TERMS: (A) critical temperature (B) crystal structure (C) crystalline (D) lattice point (E) liquid crystal (F) meniscus (G) miscible (H) polymorphism
   a. characterized by a narrow melting-point range and planar cleavage patterns
   b. complete description of a crystal lattice pattern
   c. location of center of mass of a unit particle in a crystal
   d. existence of rhombic and monoclinic forms of sulfur

e. above this point no pressure is great enough to produce a two-phase system

f. ethanol and water

g. characterized by fluidity, but higher degree of molecular order than in typical gases or liquids

h. curved liquid–gas interface

Objective (b)

2. Construct simple models and draw three-dimensional representations for unit cells classified as
   a. simple (primitive) cubic
   b. body-centered cubic
   c. hexagonal closest-packed

3. Draw representations clearly defining
   a. a tetrahedral site under a corner of a face-centered cubic unit cell
   b. an octahedral site on the edge of a face-centered cubic unit cell

4. Draw representations for unit cells of
   a. rock salt (chloride ions in a face-centered cubic array, with sodium ions in all octahedral sites)
   b. zinc blende (sulfide ions in a face-centered cubic array, with zinc ions in one-half the tetrahedral sites)

5. Draw representations of
   a. a two-fold symmetry axis in boron trifluoride
   b. a three-fold symmetry axis in boron trifluoride
   c. a mirror plane in ethylene
   d. a mirror plane for a hexagonal closest-packed unit cell
   e. a center of symmetry for a body-centered cubic unit cell

Objective (c)

6. A solid has a sharp melting point slightly above room temperature and is a poor thermal and electrical conductor. Its crystal classification by bond type is probably
   a. molecular; b. atomic; c. metallic; d. ionic; e. covalent (network)

7. A crystalline solid has a high melting point and is a poor thermal and electrical conductor. When the solid is melted, the resulting liquid is a good electrical conductor. Its crystal classification by bond type is probably
   a. covalent (network); b. molecular; c. metallic; d. atomic; e. ionic

8. For each of the following pairs, select which should be more viscous. Explain your choices.
   a. glycerin or benzene at 20°C
   b. benzene at 10°C or benzene at 70°C
   c. oxygen gas at −10°C or oxygen gas at 100°C

9. Indicate which of the liquid pairs should be completely miscible. Explain your choices.
   a. water and benzene
   b. water and ethylene glycol
   c. benzene and p-xylene $\left(CH_3-\bigcirc-CH_3\right)$

10. Which statement is incorrect?

a. Pentane ($CH_3(CH_2)_3CH_3$) and decane ($CH_3(CH_2)_8CH_3$) are not miscible with water.

b. Both pentane and decane are miscible with hexane (C_6H_{14}).

c. Decane is probably more viscous than pentane at 20°C.

d. Pentane vapor is more viscous at 120°C than at 220°C.

e. Liquid decane is more viscous at 20°C than at 80°C.

11. Which statement is incorrect?
   a. Ethylene diamine ($H_2NCH_2CH_2NH_2$) is probably less soluble in water than is benzene (C_6H_6).
   b. Ethylene diamine is probably more viscous than benzene at 10°C.
   c. Benzene is more viscous at 10°C than at 70°C.
   d. Benzene vapor is probably more viscous at 170°C than at 110°C.
   e. Ethylene diamine is probably somewhat soluble in ethanol.

Objective (d)

12. For each of the following crystalline salts, determine the empirical formula from the unit cell description:
   a. A compound of hafnium and oxygen crystallizes in a lattice pattern whose unit cell is described as cubic closest-packed hafnium ions with oxide ions in all tetrahedral sites.
   b. A compound of manganese and oxygen crystallizes in a lattice pattern whose unit cell is described as cubic closest-packed manganese ions with oxide ions in all octahedral sites.
   c. A compound of tin and sulfur crystallizes in a lattice pattern whose unit cell is described as cubic closest-packed tin ions with sulfide ions in half the tetrahedral sites.
   d. A compound of iron and sulfur crystallizes in a lattice pattern whose unit cell is described as cubic closest-packed sulfide ions with iron ions in all octahedral sites.
   e. A compound of copper and oxygen crystallizes in a lattice pattern whose unit cell is described as cubic closest-packed oxide ions with copper ions in half the tetrahedral sites.
   f. A compound of tin and chlorine crystallizes in a lattice pattern whose unit cell is described as cubic closest-packed tin ions with chloride ions in all tetrahedral sites.

13. A compound of barium and oxygen crystallizes in the isometric system. The unit cell may be described as a cubic closest-packed (face-centered cubic) array of barium ions with an oxygen at the center of the unit cell and additional oxygens at the centers of all edges of the unit cell. Is this compound barium oxide or barium peroxide (BaO_2)?

Objective (e)

14. Calculate the density of each of the following:
   a. a cubic solid of mass 5.8 g and edge length 1.2 cm
   b. an irregular solid of mass 9.7 g that sinks to displace 3.65 ml of water
   c. a tiny crystal that is suspended by a liquid from which a 0.500-ml sample weighs 0.708 g

15. An alcohol of density 0.835 g cm^{-3} had a running time of 116.8 sec in an Ostwald viscometer at 20°C. Water (density 1.00 g cm^{-3}, viscosity 10.02 millipoise) had a running time of 68.6 sec under the same conditions. What is the viscosity of the alcohol?

16. The alcohol in the preceding problem was measured to rise 12.4 cm in a wettable capillary of diameter 0.20 mm at 20°C. What is the surface tension of the alcohol?

17. An alcohol of density 0.784 g cm^{-3} had a running time in an Ostwald viscometer of exactly twice that of water ($d = 1.00$ g cm^{-3}) at the same temperature. The viscosity of water at that temperature is 10.2 millipoise. What was the viscosity of the alcohol, in millipoise?

BRAIN TINGLERS

18. Metallic potassium crystallizes in a body-centered cubic lattice. If the density of potassium at 25°C is 0.86 g cm^{-3}, what is the approximate edge length of the unit cell in crystalline potassium at 25°C (in angstroms)?

19. For an interesting problem describing why crystalline UO_2 has replaced uranium metal in fuel elements for nuclear reactors, see "Freshman-Level Chemistry Shapes the Nuclear Power Industry" in the August, 1975, issue of the *Journal of Chemical Education* (p. 523).

20. If you are interested in some novel applications of some of the ideas discussed in this unit, you will enjoy reading "Superhard Materials" (*Scientific American*, August, 1974) and "The Deformation of Metals at High Temperatures" (*Scientific American*, April, 1975).

Unit 11 Self-Test

1. Fill in the blanks, using the most appropriate term.
TERMS: (A) allotropes (B) amorphous (C) critical (D) crystal habit (E) crystal systems (F) liquid crystals (G) meniscus (H) poise (I) polymorphism (J) surface tension
  a. There are two common crystalline forms of sulfur, rhombic and monoclinic. These two forms are said to be _____.
  b. The curved surface at the liquid–vapor interface is called the _____.
  c. The basic unit of viscosity is the _____.
  d. A liquid–vapor two-phase system cannot exist above the _____ temperature.
  e. A solid that is characterized by a broad melting range and the formation of irregular fractures rather than planar cleavage patterns is said to be _____.
  f. The external appearance of a well-formed crystal is known as the _____.
  g. The phenomenon in which different solid forms of the same substance can exist is known as _____.
  h. Crystals can be classified as belonging to one of seven _____.

2. Draw representations for:
  a. a face-centered cubic lattice
  b. a threefold symmetry axis in ammonia

3. Which of the following molecules does not contain a mirror plane?

  a. H_2O; **b.** $CHCl_3$; **c.** $CHFClBr$; **d.** $HOCl$

4. Classify each of the following crystalline solids by bond type and identify the one expected to have the lowest melting point and the one expected to have the best electrical conductivity (as a solid).
  a. aluminum; **b.** barium chloride; **c.** carbon dioxide ("dry ice"); **d.** neon; **e.** silicon (diamondlike lattice)

5. Underline the member of each pair that probably has the higher viscosity:
  a. glycerin or ethanol at 20°C
  b. glycerin at 0°C or glycerin at 50°C
  c. air (1 atm) at 0°C or air (1 atm) at 50°C

6. Choose the liquid pairs expected to be completely miscible:
  a. glycerin and water
  b. benzene and toluene
  c. glycerin and benzene

7. A compound of tin and oxygen crystallizes in a lattice pattern whose unit cell is described as cubic closest-packed tin ions with oxide ions in all octahedral sites. What is the empirical formula of this salt?

8. An oxide of copper crystallizes in the isometric system. The unit cell may be described as a body-centered cubic lattice of oxide ions in which copper ions are arranged tetrahedrally around the central oxide ions so that each copper ion is midway between the central oxide ion and a corner oxide ion. What is the empirical formula of this compound?

9. A crystal of ammonium chloride is suspended by an inert solution from which a 0.500-ml aliquot (portion) weighs 0.769 g. What is the density of the crystal?

10. An alcohol of density 0.861 g cm^{-3} had a running time in an Ostwald viscometer at 20°C of 124.2 sec. Water (density 1.00 g cm^{-3}, viscosity 10.02 millipoise) had a running time of 73.4 sec in the same viscometer at 20°C

   a. What is the viscosity of the alcohol, given that

$$\eta_a = \eta_b \times \frac{d_a \times time_a}{d_b \times time_b}$$

The alcohol was measured to rise 11.2 cm in a wettable capillary of diameter 0.20 mm at 20°C.

   b. What is the surface tension of the alcohol? ($\gamma = 490drh$)

Change of state

The snowflake falls, yet lays not
   long
Its feath'ry grasp on Mother
   Earth
Ere Sun returns it to the vapors
Whence it came,
Or to waters tumbling down the
   rocky slope.
ROD O'CONNOR (1972)

Objectives

(a) After studying the appropriate definitions in Appendix E and the terms'
usage in this Unit, be able to demonstrate your familiarity with the following terms by matching them with corresponding definitions, descriptions, or
specific examples:

dynamic equilibrium	lyophilization
heat capacity	specific heat
heat of crystallization	steady state
heat of fusion	supercooling
heat of liquefaction	superheating
heat of vaporization	triple point
Le Chatelier's principle	

(b) Be able to interpret a simple phase diagram (Section 12.6).
(c) Be able to use a phase diagram applied to conditions for spontaneous
change of state (Section 12.7).
(d) Be able to calculate from appropriate data (Section 12.4).
   (1) atomic weights, using the law of Dulong and Petit

(2) enthalpy changes associated with physical processes
(e) Be able to apply Trouton's rule and the Clausius-Clapeyron relationship, in the form of equation (12.2), to estimate the variation of vapor pressure with temperature, or of boiling point with pressure, for "simple" liquids. (Section 12.8).

REVIEW

Units 5, 9, and 11

Instant coffee was originally prepared by evaporating the water from brewed coffee, which contained dissolved and suspended substances extracted from ground coffee beans, under reduced pressure at moderate temperatures. Unfortunately, many of the organic components of the coffee are fairly volatile and were lost during the evaporation process. When the dry residue was reconstituted with hot water, the aroma and flavor were noticeably different from that of freshly brewed coffee. Over the years, the food industries have improved this process considerably by varying evaporation conditions to minimize loss of volatile flavoring agents. The most recent innovation uses a process called lyophilization (freeze-drying). In this procedure the aqueous mixture is reduced to a very low temperature, forming a solid phase containing microcrystals of ice. The container of the solid is connected to a vacuum line, and the water is converted directly from the solid to the vapor phase, in which condition it is removed by the vacuum system. The differences in temperature dependence of the vapor pressures of water and the organic components result in a decreased loss of flavoring agents.

Freeze-drying has now been extended to a wide variety of foodstuffs. These need not be in solution or suspension. Natural moisture in solid foods may be removed readily by this process. In addition to the reduction in losses of the more volatile components, this process also preserves many of the natural components of the foods that would have been partially destroyed by hydrolysis or thermal degradation in processes requiring higher temperatures. Proteins in particular are sensitive to heat as are many of the vitamins, and these important ingredients may often be preserved by use of the techniques of freeze-drying.

Other types of change of state are also of importance in modern society. Fractional distillation is the basic operation in the production of hydrocarbon fuels, lubricating oils, and greases from natural petroleum. A cyclic process of vaporization and condensation of volatile liquids, usually Freons, is used in refrigeration systems. The pure crystals required for much of the modern solid state technology are prepared by controlled crystallization from the liquid state and in recent years zone melting has been employed in the production of ultrapure metals, often containing less than one part per billion impurities. In this technique an intense heat source is introduced as a narrow ring which can be moved along a metal rod. Melting occurs at the leading edge of the heated zone, and recrystallization takes place at the trailing edge of the zone. Slow movement of the hot zone permits crystallization of very pure metals, and impurities are thus moved along the rod in the liquid part of the zone. These are eventually concentrated near one end of the rod, which is then cut off to leave a long segment of very pure metal.

In this unit we shall review some of the events occurring at the molecular level during changes of state. We shall also see how thermodynamics (Unit 5) can be applied to these processes. The concept of dynamic equilibrium will be examined as it applies to physical changes.

12.1 CHANGE OF STATE

We have already described a number of thermal effects at the molecular level: Rotational, vibrational, and translational changes may occur in all states of matter as the temperature is varied. These changes are most pronounced in gases (Unit 10) and least evident in solids (Unit 11).

As heat is added to a solid, there is little opportunity for increased translational or rotational motions, so that the principal effect is one of increasing vibrations within the crystal lattice network. At the melting point of the solid, heat is absorbed without an increase in temperature; this heat is utilized primarily in destruction of the lattice, permitting particles to attain the freedom of rotation and translation characteristic of the liquid state. Continued addition of heat, resulting in a temperature increase, then serves to increase all types of possible motions within the liquid and to decrease the lifetime of microregions of ordered clusters. A second condition of heat absorption without temperature change occurs at the boiling point of the liquid, a temperature much more dependent on external pressure than is the melting point of the solid. Added heat at the boiling point acts to break down remaining particle clusters and to impart sufficient energy to permit particles to enter the gas phase. From this point on, additional heat simply increases the temperature of the gas by changing the average kinetic energy of the gas particles.

Some crystalline solids may undergo transitions from one allotropic form to another. Such transitions, like melting, occur at constant temperature.

Quantum mechanics is useful in a more detailed examination of the effects of heat at the molecular level. A rigorous application of these principles must, however, be reserved for more advanced texts.

Boiling occurs when the vapor pressure of the liquid just exceeds the external pressure.

12.2 BULK EFFECTS: HEAT CAPACITIES

The amount of heat required to raise the temperature of 1 g of any substance by one degree (Celsius or Kelvin), without change of state or allotropic form, is called the *heat capacity,* or the *specific heat.* Frequently a more useful quantity is the *molar heat capacity,* defined as the quantity of heat required to raise the temperature of 1 *mole* of a substance by 1 degree.

Heat capacities vary with temperature. Over a narrow temperature range, this variation can often be neglected without significant error, but it must be remembered that such an approximation will introduce some degree of error.

Since condensed phases (solids and liquids) undergo relatively minor volume changes with variations of temperature, as compared to those of gases, there are usually only minor differences between the heat capacity at constant volume (C_v) and the heat capacity at constant pressure (C_p) for liquids or solids. These differences can usually be ignored, except under special conditions requiring some rather complex corrections. With gases, however, there are many cases in which quite significant differences occur between C_p and C_v. At constant volume, none of the heat added to a gas can be used in pressure-volume work (gas expansion), but such is not the case at constant pressure. We shall not go into the details of these differences except to point out that care must be exercised in selecting appropriate values of heat capacities of gases, depending on whether a given situation involves constant-pressure or constant-volume conditions. It can be shown[1] that, for gases under conditions

Historically, the term *specific heat* was used for the ratio of the heat capacity (per gram) of a substance to that of liquid water at 15°C. The term is now commonly used as synonymous to heat capacity per gram.

The temperature range over which such approximations are justified depends on the degree of accuracy desired in the particular calculation and on the nature of the substance involved. Liquid water, for example, has a heat capacity varying from a maximum of 1.007 at 0° and 100°C to 0.998 cal g^{-1} deg^{-1} around 35°C. No serious error is involved in using 1.00 cal g^{-1} deg^{-1} for the entire range from 0° to 100°C.

---

[1] The interested student may wish to consult an elementary physical chemistry text for the derivation.

favoring ideal behavior,

$$C_p - C_v = R$$

that is, the two heat capacities differ by the value of R, the molar gas constant (2.0 cal mole^{-1} deg^{-1}) [as given in Unit 10].

12.3 HEATS OF TRANSITION, FUSION, AND VAPORIZATION

Enthalpy changes for the constant-temperature processes of transition between allotropic forms, melting (fusion), and boiling (vaporization) are useful in dealing with the thermal aspects of both physical change and chemical change.

The heat of transition (ΔH_{trans}), for example, can be used to calculate the standard heat of formation (Unit 5) of less stable allotropic forms, when coupled with appropriate corrections for differences between transition temperature and that of the defined standard state. Such calculations show, for example, that diamond has a standard heat of formation of 0.453 kcal mole^{-1}.

Heats of fusion (ΔH_{fusion}) and vaporization (ΔH_{vap}), when used with appropriate heat capacity terms, can also be used to calculate enthalpy changes associated with chemical reactions at other than standard state conditions. The reverse of fusion is crystallization (freezing), and the reverse of vaporization is condensation (liquefaction). Exactly the same enthalpy change occurs. (Remember that enthalpy is a *state function*.) Fusion and vaporization are endothermic processes, whereas crystallization and liquefaction are exothermic.

Thus

For state functions, reversing the direction of the process changes the algebraic sign of the thermodynamic quantity.

$$\Delta H_{cryst} = -\Delta H_{fusion}$$

and

$$\Delta H_{liq} = -\Delta H_{vap}$$

A number of useful calculations can be made from tabulated values of heats of fusion and vaporization (Table 12.1) and average values of heat capacities (Table 12.2). In situations in which heat capacities vary appreciably with temperature, precise calculations must use more complex mathematics involving empirical relationships for temperature dependence of heat capacities.

In general, heats of fusion reflect the energy differences between the condition of uniform interparticle forces in an orderly crystal lattice and the less uniform (and, on the average, weaker) interactions among the random, jostling particles in the liquid state. As we might expect, this difference is relatively large for ionic compounds,

Table 12.1 MOLAR HEATS OF FUSION AND VAPORIZATION (APPROXIMATE) FOR SOME COMMON SUBSTANCES

SUBSTANCE	MELTING POINT (°C)	ΔH_{fusion} (kcal mole^{-1})	BOILING POINT (1 atm)(°C)	$\Delta H_{vap.}$ (kcal mole^{-1})
aluminum	658	2.1	2330	68.0
argon	−190	0.27	−186	1.5
bromine	−7.3	2.6	63	6.4
chlorine	−104	1.6	−34	4.9
mercury	−39	5.6	357	14.1
potassium	62	0.55	757	18.5
potassium fluoride	860	6.3	1500	37.6
water	0	1.44	100	9.72
benzene	5.4	2.4	80	7.36
methane	−183	0.23	−159	2.2

**Table 12.2   AVERAGE VALUES OF MOLAR HEAT
CAPACITIES FOR SOME COMMON SUBSTANCES**

SUBSTANCE	C_p (avg) (cal mole^{-1} deg^{-1})	USEFUL OVER THE TEMPERATURE RANGE (°C)
Ag(s)	6.0	0–30
	6.1	30–100
	6.2	100–200
Al(s)	5.7	0–30
	5.9	30–100
	6.2	100–200
Ca(s)	6.0	0–30
	6.4	30–100
	6.8	100–200
Cu(s)	5.8	0–30
	5.9	30–100
	6.1	100–200
Fe(s)	6.0	0–30
	6.3	30–100
	6.9	100–200
Pb(s)	6.3	0–30
	6.6	30–100
	7.0	100–200
Pt(s)	6.3	0–30
	6.4	30–100
	6.5	100–200
C(s) (graphite)	2.0	0–30
	2.1	30–100
	3.0	100–200
C(s) (diamond)	1.3	0–30
	2.0	30–100
	2.9	100–200
Si(s)	4.8	0–30
	5.1	30–100
	5.8	100–200
AgBr(s)	13	0–30
	14	30–100
CaF$_2$(s)	16	0–30
	17	30–100
AlF$_3$(s)	19	0–30
	20	30–100
Hg(l)	6.7	0–30
	6.6	30–100
	6.5	100–200
H$_2$O(l)	18.0	0–100
C$_6$H$_6$(l)	31	6–15
	32	15–30
	34	30–50
	35	50–70
	37	70–80

compared to simple atomic species such as argon. Heats of vaporization are, in general, much larger than heats of fusion, since the former involve energy differences between closely packed (strongly interacting) particles in liquids and widely separated particles in the gaseous state, in which interactions are relatively slight. The relationship of ΔH_{fusion} and ΔH_{vap} (at constant temperature) with significant changes in particle interactions accounts for the much higher heat values associated with change of state compared to change of temperature (Figure 12.1). Heating a liquid to

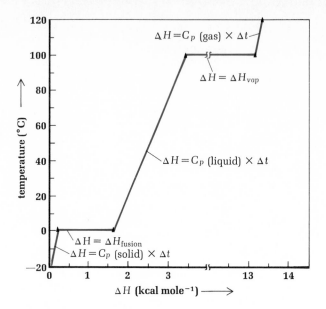

Figure 12.1 Change of state (at constant temperature) and change of temperature (for water).

simply change its temperature, for example, involves relatively minor changes in interactions among neighboring molecules. *Boiling* of the liquid, on the other hand, essentially "breaks" *all* the intermolecular forces. Obviously, interparticle forces (Unit 9) in condensed states of matter, even when only London dispersion forces are involved, are significant.

12.4 SOME CALCULATIONS INVOLVING ΔH AND C_p TERMS

Law of Dulong and Petit

The combining weight of an element is the weight that will combine with 8.0 g of oxygen ($\frac{1}{2}$ gram-atom) or with an equivalent amount of any other element (e.g., 35.45 g of chlorine).

In the early nineteenth century, there were no good methods for determining the atomic weight of an element. "Combining weights" could be measured, but these need not be the same as atomic weights. In 1819 Pierre Dulong and Alexis Petit discovered that a series of elements whose atomic weights were known with some degree of certainty had an interesting relationship. When their heat capacities *per gram* (quite different values) were multiplied by their atomic weights, the products were approximately constant. The *law of Dulong and Petit,* which could be applied to simple crystalline metallic elements over a temperature range of about 150°C, may be expressed as

$$c_p \times W = 6.2 \pm 0.4 \text{ cal mole}^{-1} \text{ deg}^{-1} \tag{12.1}$$

in which c_p is the heat capacity per gram and W is the atomic weight of the element.

This relationship is reasonably general (compare values in Table 12.2) and was used during the first part of the nineteenth century to estimate a number of atomic weights. The combining weights of metallic elements could be determined with considerable accuracy. Using approximate atomic weights based on heat capacity measurements, combining weights could then be converted to rather accurate values of atomic weights.

Example 1

A metal is found to have a heat capacity of approximately 0.10 cal g^{-1} deg^{-1} over the range 10°–50°C. When a 0.982-g sample of the metal was allowed to react with excess chlorine, 2.163 g of the crystalline chloride was obtained. What is the atomic weight of the element? What is the name of the element?

Solution

The *approximate* atomic weight can be estimated from the law of Dulong and Petit, using the average value of 6.2 cal mole^{-1} deg^{-1}:

$$W = \frac{6.2}{c_p} = \frac{6.2 \text{ cal mole}^{-1} \text{ deg}^{-1}}{0.10 \text{ cal g}^{-1} \text{ deg}^{-1}} = 62 \text{ g mole}^{-1}$$

The combining weight (cw) of the metal, defined in this case as the weight that will combine with 1 mole of atomic chlorine, can be calculated by the methods of simple stoichiometry:

$$\text{cw} = \frac{0.982 \text{ g (metal)}}{(2.163 - 0.982) \text{ g (Cl)}} \times \frac{35.45 \text{ g (Cl)}}{1 \text{ mole (Cl)}}$$

$$= 29.5 \text{ g (metal)/mole (Cl)}$$

Now, chlorine (or any element) must combine as whole numbers of atoms. Thus there must be some simple ratio between the atoms of metal and chlorine in the compound and, hence, between the *moles* of metal and chlorine. Since chlorine combines in a $-$ I oxidation state (Unit 3) with metals, this ratio must be a whole number equal to the oxidation state of the metal in the compound:

Remember that the Dulong and Petit relationship yields only an *approximate* atomic weight.

$$\frac{\sim 62 \text{ g (metal)}}{1 \text{ mole (metal)}} \times \frac{1 \text{ mole (Cl)}}{29.5 \text{ g (metal)}} = 2 \text{ moles (Cl)/mole (metal)}$$
$$\text{(nearest whole number)}$$

The accurate atomic weight, then, must be

$$\frac{29.5 \text{ g (metal)}}{1 \text{ mole (Cl)}} \times \frac{2 \text{ moles (Cl)}}{1 \text{ mole (metal)}} = 58.9 \text{ g mole}^{-1}$$

From the periodic table, the element having this atomic weight is *cobalt*.

Heat in change of state   Values of ΔH and C_p terms can be used to calculate heat effects in physical changes such as melting or boiling processes, simple change of temperature, or any combination of these events. Conversely, heat effects can be measured experimentally and used to determine ΔH or C_p quantities by techniques of *calorimetry* (Unit 5).

Example 2

What quantity of heat is required to convert 250 g of ice at $-10°$C to steam at 110°C (at 1.00 atm pressure)? The molar heat capacity of ice may be taken as 9.0 cal mole^{-1} deg^{-1} and that of steam as 8.6 cal mole^{-1} deg^{-1} over the temperature ranges involved.

Solution

It is, perhaps, easiest to treat this as a series of steps:

You may find it helpful to trace these "steps" against Figure 12.1.

STEP 1 *Ice from* $-10°$ *to* $0°$C

$$\Delta H = C_p \times \Delta t \times \text{number of moles}$$

$$= \frac{9.0 \text{ cal}}{\text{mole deg}} \times \frac{10°}{1} \times \frac{1 \text{ mole}}{18 \text{ g}} \times \frac{250 \text{ g}}{1} \times \frac{1 \text{ kcal}}{10^3 \text{ cal}}$$

$$= 1.2 \text{ kcal}   \text{ (to two significant figures)}$$

STEP 2 *Melting at constant temperature (0°C)*

$$\Delta H = \Delta H_{fusion} \times \text{number of moles}$$

From Table 12.1,

$$\Delta H_{fusion} = 1.44 \text{ kcal mole}^{-1}$$

$$\Delta H = \frac{1.44 \text{ kcal}}{\text{mole}} \times \frac{1 \text{ mole}}{18.0 \text{ g}} \times \frac{250 \text{ g}}{1}$$

$$= 20.0 \text{ kcal}$$

STEP 3 *Liquid water from 0° to 100°C*

$$\Delta H = C_p \times \Delta t \times \text{number of moles}$$

From Table 12.2,

$$C_p = 18.0 \text{ cal mole}^{-1} \text{ deg}^{-1}$$

$$\Delta H = \frac{18.0 \text{ cal}}{\text{mole deg}} \times \frac{100°}{1} \times \frac{1 \text{ mole}}{18.0 \text{ g}} \times \frac{250 \text{ g}}{1} \times \frac{1 \text{ kcal}}{10^3 \text{ cal}}$$

$$= 25.0 \text{ kcal}$$

STEP 4 *Boiling at constant temperature (100°C)*

$$\Delta H = \Delta H_{vap} \times \text{number of moles}$$

From Table 12.1,

$$\Delta H_{vap} = 9.72 \text{ kcal mole}^{-1}$$

$$\Delta H = \frac{9.72 \text{ kcal}}{\text{mole}} \times \frac{1 \text{ mole}}{18.0 \text{ g}} \times \frac{250 \text{ g}}{1}$$

$$= 135 \text{ kcal}$$

STEP 5 *Steam at 1.00 atm from 100° to 110°C*

$$\Delta H = C_p \times \Delta t \times \text{number of moles}$$

$$= \frac{8.6 \text{ cal}}{\text{mole deg}} \times \frac{10°}{1} \times \frac{1 \text{ mole}}{18 \text{ g}} \times \frac{250 \text{ g}}{1} \times \frac{1 \text{ kcal}}{10^3 \text{ cal}}$$

$$= 1.2 \text{ kcal} \quad \text{(to two significant figures)}$$

Then, for the overall process,

$$\Delta H = 1.2 + 20.0 + 25.0 + 135 + 1.2$$
$$= 182.4$$

The answer should be reported as $\Delta H = 182$ kcal, since the value calculated in the fourth step is limited to three significant figures.

Example 3

A piece of metal weighing 34.2 g was heated to 58.3°C and dropped into 100.0 g of water contained in an insulated vessel (calorimeter) equipped with a stirrer and thermometer. If the original temperature of the water was 23.7°C and the final temperature was measured to be 24.8°C, what is the approximate heat capacity of the metal in the temperature range involved? (Neglect any corrections for calorimeter factors.)

Accurate determinations require corrections for minor heat effects of stirring and for temperature change of the interior of the calorimeter. These corrections could be made in this type of situation by a second experiment using a metal whose specific heat is known.

Solution

heat lost by metal = heat gained by water
heat lost by metal = C_p (metal) $\times \Delta t$ (metal) $\times$ weight (metal)
heat gained by water = C_p (water) $\times \Delta t$ (water) $\times$ weight (water)

Thus

$$C_p \text{ (metal)} \times (58.3 - 24.8)° \times 34.2 \text{ g}$$
$$= (1.00 \text{ cal g}^{-1} \text{ deg}^{-1}) \times (24.8 - 23.7)° \times 100 \text{ g}$$

$$C_p \text{ (metal)} = \frac{(1.00 \text{ cal g}^{-1} \text{ deg}^{-1}) \times 1.1° \times 100 \text{ g}}{33.5° \times 34.2 \text{ g}}$$

$$= 0.096 \text{ cal g}^{-1} \text{ deg}^{-1}$$

Note that a sealed calorimeter represents a condition of constant volume, not constant pressure. The distinction may be ignored in simple cases of this type.

Corrections for nonstandard conditions

Note that the temperature for conventional *standard state* conditions (25°C, 298°K) is different from standard temperature (0°C, 273°K) for gases.

Enthalpy changes associated with chemical reactions can be calculated, under *standard state* conditions, from tabulated values such as standard heats of formation (Unit 5). However, few reactions actually occur under these particular conditions, and it is necessary to apply corrections for enthalpy terms for reactions taking place under conditions other than those of standard states. Since enthalpy is a state function, such corrections can be made by proper application of terms associated with simple temperature change (heat capacities) and change of state or form (ΔH_{fusion}, ΔH_{vap}, ΔH_{trans}). In cases involving temperature ranges over which heat capacities vary appreciably, these corrections require mathematical operations beyond the scope of this text.

Example 4

Initial reactants and all final products are at 1.00 atm.

A 24-g sample of methanol vapor was completely oxidized by oxygen gas under conditions of constant pressure (1.00 atm) and constant temperature (110°C). What was the approximate enthalpy change for this process? Use any of the following data and refer to Table 5.1, Unit 5, when necessary.

average C_p CH_3OH (liquid)	19.2 cal mole^{-1} deg^{-1}
average C_p CH_3OH (vapor)	13.8 cal mole^{-1} deg^{-1}
average C_p O_2 (gas)	7.0 cal mole^{-1} deg^{-1}
average C_p CO_2 (gas)	8.8 cal mole^{-1} deg^{-1}
average C_p H_2O (vapor)	8.6 cal mole^{-1} deg^{-1}
average C_p H_2O (liquid)	18.0 cal mole^{-1} deg^{-1}
ΔH_{vap} CH_3OH at boiling point	8.42 kcal mole^{-1} at 65°C
ΔH_{vap} H_2O at boiling point	9.72 kcal mole^{-1} at 100°C

Methanol boils at 65°C at 1.00 atm

Solution
The balanced equation for this process is

$$2CH_3OH + 3O_2 \longrightarrow 2CO_2 + 4H_2O$$

The 24 g of methanol is 0.75 mole (24/32), so

$$0.75 \text{ mole (CH}_3\text{OH)} \times \frac{2 \text{ moles (CO}_2)}{2 \text{ moles (CH}_3\text{OH)}} = 0.75 \text{ mole (CO}_2)$$

$$0.75 \text{ mole (CH}_3\text{OH)} \times \frac{4 \text{ moles (H}_2\text{O)}}{2 \text{ moles (CH}_3\text{OH)}} = 1.50 \text{ moles (H}_2\text{O)}$$

$$0.75 \text{ mole (CH}_3\text{OH)} \times \frac{3 \text{ moles (O}_2)}{2 \text{ moles (CH}_3\text{OH)}} = 1.125 \text{ moles (O}_2)$$

Then the question really asks "What is the enthalpy difference at 1.00 atm and 110°C between products (0.75 mole CO_2 and 1.50 moles H_2O) and reactants (0.75 mole CH_3OH and 1.125 moles O_2)?"

Since enthalpy is a state function, the actual process (e.g., the catalyst) is unimportant. Only initial and final states are considered, so we may select any imaginary sequence of events for convenience of calculation. Since the data in Table 5.1 (Unit 5) are for standard states, the following sequence may be employed:

STEP 1 Conversion of $CH_3OH(g)$ and $O_2(g)$ from 110°C to $CH_3OH(l)$ and $O_2(g)$ at 25°C at constant pressure of 1.00 atm.

STEP 2 Conversion of $CH_3OH(l)$ and $O_2(g)$ at 25°C to $CO_2(g)$ and $H_2O(l)$ at 25°C at constant pressure at 1.00 atm.

STEP 3 Conversion of $CO_2(g)$ and $H_2O(l)$ from 25°C to $CO_2(g)$ and $H_2O(g)$ at 110°C.

The *net* enthalpy change for these three steps must be that for the original process in question. These changes are then calculated as follows:

STEP 1 $\Delta H = C_p[CH_3OH(g)] \times \Delta t \times$ number of moles
$+ \Delta H_{cond}(CH_3OH) \times$ number of moles
$+ C_p[CH_3OH(l)] \times \Delta t \times$ number of moles
$+ C_p[O_2(g)] \times \Delta t \times$ number of moles

$$= \frac{13.8 \text{ cal}}{\text{mole deg}} \times \frac{(65° - 110°)}{1} \times \frac{0.75 \text{ mole}}{1}$$

$$+ \frac{-8.42 \text{ kcal}}{\text{mole}} \times \frac{0.75 \text{ mole}}{1} \times \frac{1000 \text{ cal}}{\text{kcal}}$$

$$+ \frac{19.2 \text{ cal}}{\text{mole deg}} \times \frac{(25° - 65°)}{1} \times \frac{0.75 \text{ mole}}{1}$$

$$+ \frac{7.0 \text{ cal}}{\text{mole deg}} \times \frac{(25° - 110°)}{1} \times \frac{1.125 \text{ moles}}{1}$$

$$= (-466) + (-6315) + (-576) + (-669)$$

ΔH (step 1) $= -8026 \text{ cal} = -8.026 \text{ kcal}$

STEP 2 $\Delta H = [\Delta H_f° (CO_2) \times$ number of moles $+ \Delta H_f° (H_2O) \times$ number of moles$]$
$- [\Delta H_f° (CH_3OH) \times$ number of moles $+ \Delta H_f° (O_2) \times$ number of moles$]$

$\Delta H_f°$ for O_2 (an element) is zero. The other values are found from Table 5.1 (Unit 5).

$$\Delta H = \left(\frac{-94.0 \text{ kcal}}{\text{mole}} \times \frac{0.75 \text{ mole}}{1} + \frac{-68.3 \text{ kcal}}{\text{mole}} \times \frac{1.50 \text{ moles}}{1} \right)$$

$$- \left(\frac{-57.0 \text{ kcal}}{\text{mole}} \times \frac{0.75 \text{ mole}}{1} + 0 \right)$$

$$= (-70.5) + (-102.4) - (-42.8)$$

ΔH (step 2) $= -130.1 \text{ kcal}$

STEP 3 $\Delta H = C_p\,[CO_2(g)] \times \Delta t \times$ number of moles
$\qquad + C_p\,[H_2O(l)] \times \Delta t \times$ number of moles
$\qquad + \Delta H_{vap}\,(H_2O) \times$ number of moles
$\qquad + C_p\,[H_2O(g)] \times \Delta t \times$ number of moles

$$= \frac{8.8\ \text{cal}}{\text{mole deg}} \times \frac{(110° - 25°)}{1} \times \frac{0.75\ \text{mole}}{1}$$

$$+ \frac{18.0\ \text{cal}}{\text{mole deg}} \times \frac{(100° - 25°)}{1} \times \frac{1.50\ \text{moles}}{1}$$

$$+ \frac{9.72\ \text{kcal}}{\text{mole}} \times \frac{1.50\ \text{moles}}{1} \times \frac{1000\ \text{cal}}{\text{kcal}}$$

$$+ \frac{8.6\ \text{cal}}{\text{mole deg}} \times \frac{(110° - 100°)}{1} \times \frac{1.50\ \text{moles}}{1}$$

$$= 561 + 2025 + 14{,}580 + 129$$

$\Delta H\ (\text{step 3}) = 17{,}295\ \text{cal} = 17.30\ \text{kcal}$

SUMMATION

$\Delta H_{net} = \Delta H\ (\text{step 1}) + \Delta H\ (\text{step 2}) + \Delta H\ (\text{step 3})$
$\qquad = (-8.026) + (-130.1) + (17.30)$
$\qquad = -121\ \text{kcal} \qquad (\text{to three significant figures})$

12.5 EQUILIBRIA AND STEADY STATES

If we place a quantity of liquid water in an open beaker above the flame of a laboratory burner, we observe that the water becomes hotter and eventually begins to boil. As boiling continues, the liquid volume decreases until all the liquid has boiled away. There are two ways of maintaining a constant liquid volume during boiling, and the distinction between these is of the utmost importance.

One method would involve the addition of liquid water at the same rate that liquid is converted to gas (Figure 12.2). This represents a condition of *steady state*. One part of the system is maintained constant by balancing the rates of opposing processes of continuous change. A considerable energy requirement is associated with such a process. The alternative procedure involves closing the system, such as by addition of a movable piston or some other constant-pressure device, and maintaining a constant temperature. In this method a minimum energy is required; that is, only enough heat must be added to maintain conditions of constant temperature. This second alternative, which is characterized by constant macroscopic properties, represents a condition of *equilibrium*.

Figure 12.2 Steady state and equilibrium. (a) Steady state: constant temperature, constant liquid volume. (b) Equilibrium: constant temperature, constant liquid volume, constant gas volume.

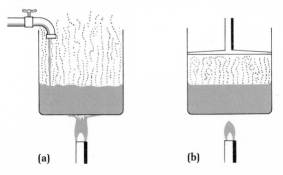

(a)   **(b)**

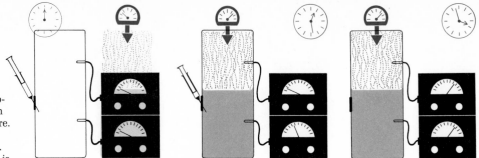

Figure 12.3 Dynamic equilibrium. The liquid-vapor system is held at constant temperature. Pressure and liquid-to-vapor volume ratio remain constant. A small amount of the liquid is withdrawn and replaced by an equal amount of the same liquid now containing a radioactive isotope. As time passes the pressure and volume ratio are unchanged, but radioactivity is detected in both the liquid and vapor phases. Since pressure is constant, some gas particles must return to liquid to replace those escaping into gas.

In the idealized case, no energy input is required for an equilibrium system that is perfectly insulated.

A child's teeter-totter can be arranged so that it is perfectly balanced; this is an example of *static* equilibrium. The equilibria involved in physical processes such as change of state or in various chemical systems appear to the casual observer to involve the same characteristics. The constant macroscopic properties of an equilibrium system, such as a liquid–vapor equilibrium, might result from a static condition. Liquid placed in a container begins to vaporize. When the container is closed and maintained at constant temperature, equilibrium is reached when no further change occurs in the amounts of liquid and vapor in the system. Has the process of vaporization simply stopped so that the system becomes static at equilibrium? Careful investigation (Figure 12.3) at the molecular level has shown that this is *not* the case. Rather, opposing processes are occurring at the same rate, so that the *net* change is zero. The principal difference between this *dynamic equilibrium* and the steady state condition is that the former occurs within a closed system and requires only the energy necessary to maintain constant temperature.

In equilibria at the atomic or molecular level, direct observation might be quite difficult, and in most cases impossible. The ideas of particle motions expressed as explanation of gas behavior and of change of state strongly suggest dynamic situations. It seems reasonable, for example, to explain the constant macroscopic properties of a liquid–vapor equilibrium (i.e., constant relative amounts of liquid and vapor in a closed system at constant temperature) on the basis that at equilibrium particles are escaping from the liquid surface at the same rate that other particles are entering the liquid phase.

An important rule of the game called "science" requires that any postulated explanation or prediction be subjected to experimental verification. It is not enough that it is *logical* that "molecular" equilibria are dynamic. Experiments must be devised to test this suggestion. Indeed, a number of experiments have been used to confirm the dynamic characteristics of these situations. One of the most straightforward is the addition of radioisotopes to equilibrium systems (Figure 12.3).

12.6 PHASE DIAGRAMS

Equilibria involving physical states or allotropic modifications, or both, may be displayed graphically as *phase diagrams* (Figure 12.4). The condition of equilibrium exists only at points along the plotted lines, which correspond to the temperature and pressure criteria for equilibrium. At any temperature–pressure combination not corresponding to equilibrium, spontaneous change should occur (Section 12.7) until the equilibrium condition is attained.

The vapor pressure of the liquid or solid may be determined at any temperature within the region of the phase diagram by finding the point on the liquid–gas (or

EQUILIBRIUM ECOLOGY?

The distinction between equilibrium and steady state is not generally understood. Much attention has been focused in recent years on the problems associated with the rapid diminution of our natural resources and the increased accumulation of liquid, gaseous, and solid wastes. Among the suggestions that have been offered is the idea of *recycling*. Some steps in this direction have already been taken, and a number of others are in various stages of research. Recycling of aluminum cans and various paper products has received considerable publicity. These ideas have produced suggestions, more philosophical than scientific, that we should plan toward an equilibrium ecology.

Unfortunately, this panacea is theoretically impossible. The equilibrium condition requires constant macroscopic properties within an isolated system and an energy input sufficient only to maintain constant temperature. These conditions cannot be attained by recycling processes. The most ideal circumstances would permit, at best, a steady state condition, and this automatically requires a greater energy input. Every recycling method requires considerable energy and thus continues the depletion of natural energy sources. Nevertheless, a steady state is the most favorable circumstance that can be hoped for, and every step that approaches this condition increases the survival time of society.

The equilibrium vapor pressure of a substance is the pressure exerted by the gas phase *of that substance* in a sealed vessel containing only the equilibrium system including the gas phase and the liquid or solid phase(s), or both.

Our use of such terms as "melting point," "boiling point," and "triple point" involves some degree of oversimplification, although this is relatively minor for pure one-component systems.[2]

solid–gas) equilibrium curve corresponding to the temperature given and projecting a horizontal line to the pressure axis. The data plotted in Figure 12.4 are adequate only for rough approximations, and more precise diagrams or tables must be consulted for accurate data.

Two particular phenomena may be considered as "temporary" exceptions to this statement. *Supercooling* occurs when a liquid is chilled to below its freezing point under a pressure at which solidification should be spontaneous without change of state, and *superheating* corresponds to the continued existence of liquid at conditions under which vaporization should be spontaneous. These conditions are said to be *metastable* and have only transient existence. Crystallization is spontaneous in a supercooled liquid if "seed" crystals are added, and boiling of a superheated liquid occurs rapidly whenever nucleation (formation of microbubbles) begins. Although the former process is often slow, laboratory accidents from the sudden boiling of superheated liquids are not uncommon.

Phase diagrams (Figure 12.4) typically reveal *triple points* corresponding to conditions at which three phases are in equilibrium. More than one triple point may be shown for substances having allotropic modifications in the solid state.

Critical temperatures have been discussed previously (Unit 11). *Above* the critical

[2]For more general and precise descriptions, see the article by R. C. Parker and D. S. Kristol in the October, 1974, issue of the *Journal of Chemical Education* (pp. 658–660).

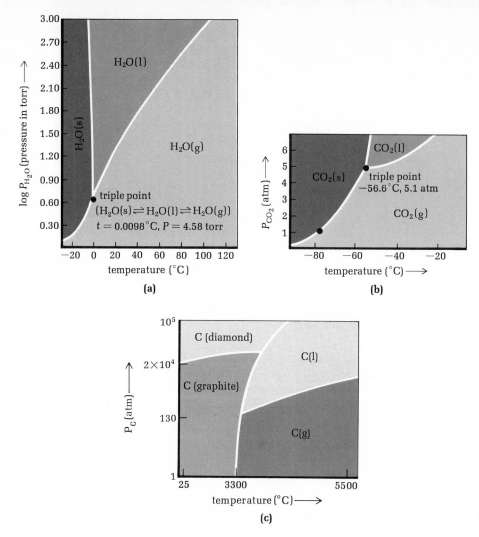

Figure 12.4 Phase diagrams (approximate). (Log P_{H_2O} has been plotted to compress the scale.) (The diagram for carbon also employs a nonlinear P/T scale.)

temperature the system consists of a single phase, no matter how large the pressure. The critical pressure is that required to produce a two-phase system *at* the critical temperature. For water, the critical pressure is 218 atm at the critical temperature of 374°C. Above this temperature water exists as only a single phase (highly compressed fluid).

Example 5

By reference to the phase diagrams in Figure 12.4, determine approximate values of

    a. the vapor pressure of liquid water at 50°C
    b. the vapor pressure of solid carbon dioxide ("dry ice") at −60°C
    c. the boiling point of water at a pressure of 400 torr (boiling occurs when the vapor pressure just exceeds the external pressure)
    d. the temperature at which graphite will sublime (spontaneously change from solid directly to gas) at 1 atm pressure

Solution

a. The vapor pressure of liquid water at 50°C is obtained by selecting a point on the liquid-gas equilibrium curve corresponding to a temperature coordinate of 50°C (by projecting a vertical line from 50° on the temperature axis to intersect the liquid–gas equilibrium curve), then determining the corresponding pressure coordinate. This is done by projecting a horizontal line to the pressure (log P) axis (intersecting at about 2.0) and finding the antilog. The antilog of 2.0 is 100, so the vapor pressure of water at 50°C is approximately 100 torr (reported experimental value, 92.51 torr).

b. A similar procedure is employed here, projecting vertically from −60°C on the temperature axis to intersect the solid-gas equilibrium curve, then projecting horizontally from the intersection point to the pressure axis. A direct pressure reading gives approximately 4 atm (reported experimental value, 4.043 atm).

c. To determine the approximate boiling point of water, we first convert 400 torr to the logarithmic equivalent (log 400 = 2.6). The value 2.6 is located on the log P axis and projected horizontally to the liquid-gas equilibrium curve. The corresponding temperature coordinate is located by vertical projection from the intersection point to the temperature axis, giving approximately 85°C (reported experimental value, 83°C).

d. The phase diagram for carbon indicates that the graphite-gas phase equilibrium exists at about 3300°C at 1 atm. A temperature slightly above that of the equilibrium system will result in spontaneous conversion of solid graphite to gaseous carbon.

12.7 EQUILIBRIA DISTURBED: LE CHATELIER'S PRINCIPLE

It is observed that an equilibrium can be disturbed by varying one or more of the factors that determine the equilibrium conditions. For example, lowering the temperature of a system involving liquid–vapor equilibrium in a constant volume will cause a decrease in pressure and an increase in liquid volume until, at a new temperature, equilibrium is reestablished. *Le Chatelier's principle*, first suggested by Henri Le Chatelier in 1888, is a generalization of such observations which applies to any type of equilibrium:

When any factor contributing to an equilibrium condition is varied, equilibrium will be destroyed and the system will return to a state of continuing change until a new equilibrium is attained whose properties balance the effect of the original variation.

A careful consideration of the competing dynamic processes in any given equilibrium will allow a qualitative prediction of the result of changes in many of the factors determining equilibrium. Thermodynamic considerations are necessary in some situations, and such cases as the influence of temperature changes on chemical reactions require a knowledge of experimentally determined values of ΔH and ΔS quantities (Units 5 and 17).

Example 6

By reference to the phase diagrams in Figure 12.4, suggest what changes, if any, should be observed in the following cases.

a. A liquid water-ice equilibrium system at 0°C is subjected to a continuing increase in pressure at constant temperature.

b. Carbon dioxide gas at 3 atm pressure is cooled from $-40°C$ to $-100°C$ at constant pressure.

c. A diamond is heated to $3000°C$ at a pressure of 1 atm.

d. Water vapor is cooled from $110°C$ to $50°C$ at a constant pressure of 50 torr.

Solution

a. From Figure 12.4(a), we note that the solid–liquid equilibrium curve slopes toward the pressure axis. This indicates that at a constant temperature of $0°C$, increasing pressure favors the transition from solid to liquid. We should observe the melting of ice as pressure is increased at $0°C$. (What does this suggest about possibilities for the surface of an ice skating rink under the pressure of a skate blade?)

b. At the point defined by $(-100°C, 3\ \text{atm})$ on the CO_2 phase diagram, the stable phase is solid CO_2. As a result, chilling of CO_2 gas at 3 atm pressure from $-40°$ to $-100°C$ should result in spontaneous solidification to "dry ice". Note that projection from the point $(-40°C, 3\ \text{atm})$ to the point $(-100°C, 3\ \text{atm})$ does not intersect either the gas–liquid or the liquid–solid equilibrium curves, so no intermediate liquid state is encountered in the process described.

c. The stable form of carbon at $3000°C$ and 1 atm is solid graphite [Figure 12.4(c)], so diamond should spontaneously change to graphite under the conditions indicated. [Note: We might not *observe* this change if it is very slow. Predictions of spontaneity based on interpretations of phase diagrams, like those made from thermodynamic calculations (Unit 5), do not carry any implications of *how rapidly* changes should occur.]

d. Both the points $(110°C, \log 50)$ and $(50°C, \log 50)$ lie in the gas region of the phase diagram for water $(\log 50 = 0.70)$. As a result, no change of state should occur during the indicated temperature change. However, at constant pressure, gas volume will decrease when the gas is cooled from $110°C$ to $50°C$ (Unit 10).

12.8 VAPOR PRESSURE AND BOILING

Particles in liquids, like those in gases, do not all have the same kinetic energy (Unit 9). More energetic particles may escape from the liquid surface into the gas phase. As the temperature is increased, the percentage of high-energy particles, and thus the tendency for *vaporization,* increases. In an open container, a liquid will evaporate at a rate determined by the temperature, the surface area, the external pressure, the rate of removal of vapor (e.g., by a moving stream of air passing over the liquid surface), and by the fundamental characteristics of the particular liquid. Since temperature is a measure of the average kinetic energy and it is the more energetic particles that escape most frequently, the temperature of the liquid is lowered by evaporation.

If the liquid is in a sealed container, evaporation may appear to stop while there is still liquid present. Careful experiments show that in such cases molecules continue to leave the liquid, but other molecules return from the vapor to the liquid phase, the condition corresponding to *dynamic equilibrium* (Section 12.5). The pressure exerted by a pure vapor in equilibrium with its liquid phase is called the "equilibrium vapor pressure" (or, commonly, just the vapor pressure) and is determined solely by the nature of the liquid and the temperature of the equilibrium system.

When the vapor pressure of a liquid just exceeds the external pressure on the system, the liquid will boil. Boiling is a condition in which liquid is converted spontaneously to gas at a constant temperature (the boiling point). This conversion occurs at the principal liquid–vapor interface and also at the liquid–vapor surface of any bubbles within the liquid. The reversal of vaporization is condensation. When

Figure 12.5 Simple distillation. (Boiling chips are small chips of a porous material such as silicon carbide which are used to provide an irregular surface on which bubbles can form, in order to prevent superheating.)

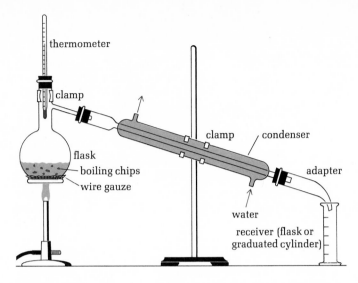

boiling and condensation are used successively, the net process is called "distillation"; this is one of the most useful ways of purifying liquids (Figure 12.5).

If we know the normal boiling point of a "simple" liquid, that is, the boiling point at 1 atm pressure, we can estimate the vapor pressure of the liquid at a given temperature from an empirical equation.

For purposes of applying equation (12.2), we consider "simple" liquids to be those in which intermolecular forces are weaker than typical hydrogen bonds (Unit 9).

$$\log P_a = \frac{4.6}{T_a} (T_a - T_b) \qquad \text{(12.2)}$$

in which P_a is the equilibrium vapor pressure (in atmospheres) at temperature T_a (in degrees Kelvin) and T_b is the boiling point of the liquid at 1 atm (in degrees Kelvin).

Equation (12.2) is a reliable approximation (within ±15%) for "simple" liquids. It cannot be used for accurate estimates of vapor pressure–temperature relationships for hydrogen-bonded liquids, such as water or alcohols, although a similar relationship can be employed by replacing the constant (4.6) with a value appropriate to the particular liquid–vapor system involved. Since the value of the "constant" is related to ΔH_{vap}, proportionally larger values are found for liquids having higher heats of vaporization.

Example 7

Estimate the vapor pressure of benzene at 25°C, using data from Table 12.1.

Solution
From Table 12.1, the boiling point of benzene at 1 atm is 80°C (353°K), so

$$\log P_{298^\circ} = \frac{4.6}{298} (298 - 353) = -0.85$$

To find the antilog (see Math Skills), we must write the logarithm in a form having a positive mantissa

$$-0.85 = (0.15 - 1)$$

Then

$$P_{298°} = 1.4 \times 10^{-1} = \mathbf{0.14}\ \text{atm}$$

(tabulated experimental value, 0.13 atm)

High boiling liquids are often distilled under conditions of reduced pressure (vacuum distillation) to take advantage of the lowered boiling point in minimizing thermal decomposition of the liquid compound. Equation (12.2) may also be used to estimate the boiling point of a "simple" liquid at a given pressure.

Example 8

Cinnamaldehyde, an organic compound having the aroma of cinnamon, boils at 253°C (at 1 atm) with significant thermal decomposition. It has been found that vacuum distillation at a pressure of 15 torr results in a low enough boiling point to reduce decomposition to an acceptably small percentage of the sample. What boiling point (in degrees centigrade) would be predicted for cinnamaldehyde at a pressure of 15 torr?

Solution

To utilize equation (12.2), pressure must be expressed in atmospheres,

$$15\ \text{torr} \times \frac{1\ \text{atm}}{760\ \text{torr}} = 0.020\ \text{atm}$$

and temperature in degrees Kelvin. Then

$$T_b = 253 + 273 = 526°\text{K}$$

$$\log(0.020) = \frac{4.6}{T_{15}}[T_{15} - 526]$$

$$-1.7 = \frac{4.6}{T_{15}}(T_{15} - 526)$$

$$-1.7T_{15} = 4.6T_{15} - 2420$$

$$T_{15} = \frac{2420}{6.3} = 384°\text{K}$$

The boiling point, then, is

$$384 - 273 = \mathbf{111}°\text{C}$$

(reported experimental value, 125°C)

SUGGESTED READING

Pimentel, George C., and Richard D. Spratley. 1969. *Understanding Chemical Thermodynamics.* San Francisco: Holden-Day.

Tabor, D. 1969. *Gases, Liquids, and Solids.* Baltimore, Md.: Penguin.

DERIVATION OF EQUATION (12.2)

Figure 12.6 Empirical basis for Trouton's rule.

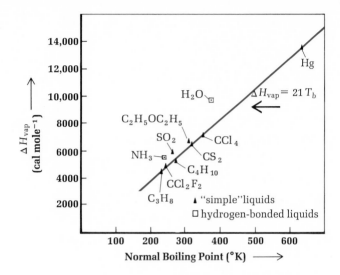

Equation (12.2) may be derived from two *empirical* relationships, that is, formulations based on correlations of experimental data without a quantitative theoretical model. The first of these relationships, called "Trouton's rule," was determined from observation of various ways of graphing boiling point and heat of vaporization data for a series of compounds (Figure 12.6). An essentially linear plot results, corresponding to the general equation $y = mx + b$, when the normal boiling point, in degrees Kelvin, is the y coordinate and the molar heat of vaporization is the x coordinate. The slope of the line is approximately 21 cal mole^{-1} deg^{-1} and the y intercept (extrapolated) is zero. As a result, Trouton's rule may be expressed as

$$\Delta H_{vap} = 21T_b$$

Since "simple" liquids show the best correlation with the linear plot, Trouton's rule is generally limited to approximations of ΔH_{vap} values from measured boiling points of liquids of this type.

   The other relationship, which may also be determined from graphical displays of experimental data, reveals that a linear plot is obtained for a particular liquid if the y coordinate is the logarithm of the equilibrium vapor pressure and the x coordinate is the *inverse* of the absolute temperature (i.e., $1/T$). The particularly interesting feature of such plots (Figure 12.7) is that the slope of the line for any specific liquid–vapor system is always found to be, within reasonable limits, a function of the molar heat of vaporization of the liquid:

$$\text{slope} = -\frac{\Delta H_{vap}}{4.6}$$

A theoretical model applicable to any equilibrium system accounts for this relationship. This model, which applies a thermodynamic treatment to the variation of equilibrium system composition as a function of temperature, is beyond the scope of

Figure 12.7 Variation of vapor pressure with temperature.

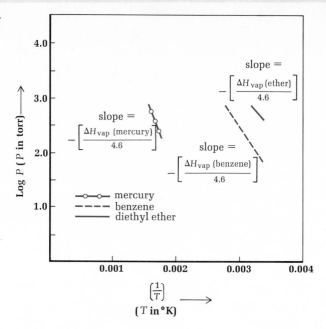

this text, but the interested student may wish to discuss this subject with a faculty member specializing in physical chemistry. For our purposes it is sufficient to note that the relationship may be established empirically from analysis of appropriate graphical displays of data.

For any one of the linear plots in Figure 12.7, we may express the equation for the line (in the form $y = mx + b$) as

$$\log P = -\left(\frac{\Delta H_{vap}}{4.6}\right)\left(\frac{1}{T}\right) + b$$

It is rarely feasible to determine the y intercept (b), since the linearity of the plot is actually limited to a fairly narrow temperature range. However, if we select two sets of temperature, pressure coordinates, we may write

$$\log P_a = -\left(\frac{\Delta H_{vap}}{4.6}\right)\left(\frac{1}{T_a}\right) + b$$

$$\log P_b = -\left(\frac{\Delta H_{vap}}{4.6}\right)\left(\frac{1}{T_b}\right) + b$$

Then, by subtracting equalities,

$$\log\left(P_a - P_b\right) = \left[-\left(\frac{\Delta H_{vap}}{4.6}\right)\left(\frac{1}{T_a}\right) + b\right] - \left[-\left(\frac{\Delta H_{vap}}{4.6}\right)\left(\frac{1}{T_b}\right) + b\right]$$

from which

$$\log\left(\frac{P_a}{P_b}\right) = \left(\frac{\Delta H_{vap}}{4.6}\right)\left(\frac{1}{T_b} - \frac{1}{T_a}\right)$$

or

$$\log\left(\frac{P_a}{P_b}\right) = \left(\frac{\Delta H_{vap}}{4.6}\right)\left(\frac{T_a - T_b}{T_a T_b}\right)$$

This form of the equation is referred to as the Clausius-Clapeyron relationship, in honor of two nineteenth-century physicists, Rudolph Clausius (German) and B. P. Clapeyron (French).

If we consider this mathematical formulation under the special condition in which T_b is the normal boiling point, so that P_b is 1.00 atm, and use Trouton's rule to approximate ΔH_{vap}, then for "simple" liquids,

$$\log\left(\frac{P_a}{1.00}\right) = \left(\frac{21T_b}{4.6}\right)\left(\frac{T_a - T_b}{T_a T_b}\right)$$

from which, with P_a in atmospheres, we have equation (12.2),

$$\log P_a = \frac{4.6}{T_a}(T_a - T_b)$$

EXERCISES

1. Select the term that best matches the definition, description, or specific example given.
   TERMS: (A) dynamic equilibrium (B) heat of vaporization (C) specific heat (D) supercooling (E) triple point
   a. the ratio $c_p(\text{substance})/c_p(\text{H}_2\text{O})$
   b. characterized by opposing processes at the same rate in a closed system
   c. temperature and pressure at which, for example,

   $$\text{CO}_2(\text{s}) \rightleftharpoons \text{CO}_2(\text{l}) \rightleftharpoons \text{CO}_2(\text{g})$$

   d. ΔH for $\text{H}_2\text{O}(\text{l}) \longrightarrow \text{H}_2\text{O}(\text{g})$
   e. $\text{H}_2\text{O}(\text{l})$ at $-10°\text{C}$, 760 torr
2. Referring to the phase diagrams in Figure 12.4 determine
   a. the approximate vapor pressure of water at $70°\text{C}$
   b. the approximate vapor pressure of dry ice (solid CO_2) at $-60°\text{C}$
   c. the approximate boiling point of water at 100 torr pressure
   d. what should happen to a piece of dry ice placed in an open container maintained at $-60°\text{C}$ at 760 torr
   e. what should happen to a sample of liquid water placed in a container at $0°\text{C}$ under a constant pressure of 2.0 torr, maintained by a vacuum pump
3. A metal is found to have a specific heat of 0.031 ± 0.001 cal g^{-1} deg^{-1} over the temperature range of 15–$75°\text{C}$. A 3.9393-g sample of the metal was converted to 6.0665 g of its chloride. What is the atomic weight of the metal? What is the name of the metal?
4. What quantity of heat is required to convert 39 g of crystalline benzene at $0°\text{C}$ to benzene vapor at $90°\text{C}$? For benzene vapor over the temperature range involved, C_p may be taken as 21.8 cal mole^{-1} deg^{-1} and C_p for crystalline benzene may be taken as 18 cal mole^{-1} deg^{-1} over the range involved.

5. Using equation (12.2), estimate
   a. the vapor pressure of carbon tetrachloride (normal boiling point $77°\text{C}$) at $30°\text{C}$
   b. the boiling point of cycloheptanone (normal boiling point $179°\text{C}$) at a reduced pressure of 38 torr

PRACTICE FOR PROFICIENCY

Objective (a)

1. Select the term best associated with each definition, description, or specific example.
   TERMS: (A) heat capacity (B) heat of crystallization (C) heat of fusion (D) heat of liquefaction (E) Le Chatelier's principle (F) lyophilization (G) steady state (H) supercooling (I) superheating
   a. ΔH for $\text{Hg}(\text{g}) \longrightarrow \text{Hg}(\text{l})$
   b. ΔH for $\text{Hg}(\text{s}) \longrightarrow \text{Hg}(\text{l})$
   c. ΔH for $\text{Hg}(\text{l}) \longrightarrow \text{Hg}(\text{s})$
   d. C_p (at constant pressure)
   e. $\text{H}_2\text{O}(\text{l})$ at $-5°\text{C}$, 760 torr
   f. $\text{H}_2\text{O}(\text{l})$ at $105°\text{C}$, 760 torr
   g. illustrated by the fact that increasing pressure on a liquid–vapor equilibrium system at constant temperature converts vapor to liquid until equilibrium is reestablished
   h. illustrated by maintaining constant air pressure in a punctured automobile tire by pumping in air at the same rate as air is lost through the puncture
   i. freeze-drying
2. Which statement is incorrect?
   a. The "heat of crystallization" is equal to the "heat of fusion," but of opposite algebraic sign.

315

b. The "heat of liquefaction" is equal to the "heat of fusion," but of opposite algebraic sign.

c. Three phases of a substance coexist in equilibrium at the "triple point" of the substance.

d. "Steady state" and "dynamic equilibrium" are *not* synonymous terms referring to the same condition.

e. An "equilibrium system" is characterized by zero *net* change.

3. Which statement is incorrect?

a. The "molar heat capacity" refers to an enthalpy change associated with a temperature change.

b. The "heat of transition" for conversion of diamond to graphite refers to an enthalpy change at constant temperature.

c. "Heat of fusion" refers to an enthalpy change at constant temperature.

d. The "vapor pressure" of a liquid is a function of temperature.

e. The composition of an "equilibrium system" never varies with temperature.

4. In a particular experiment a 100-ml sample of liquid water was introduced into an evacuated 250-ml container attached to a pressure gauge and maintained at a constant temperature of 50°C. It was observed that the liquid volume decreased to a constant value while the pressure reading increased to a constant value of 92 torr. At the point at which liquid volume, vapor pressure, and temperature were constant, the system was in a condition of

a. superheating; **b.** static equilibrium; **c.** dynamic equilibrium; **d.** steady state **e.** critical state

Objectives (b) and (c)

5. Referring to the phase diagrams in Figure 12.4 determine

a. the approximate vapor pressure of water at 35°C

b. the approximate temperature at which dry ice should sublime ($CO_2(s) \longrightarrow CO_2(g)$) at 1.00 atm pressure

c. the approximate boiling point of water at 380 torr

d. the approximate temperature and pressure for the graphite–liquid carbon–carbon vapor triple point

6. According to the phase diagram given, which description is correct?

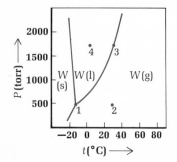

a. At the temperature and pressure of point 1,

equilibrium: $W(s) \rightleftharpoons W(l) \rightleftharpoons W(g)$

b. At the temperature and pressure of point 2,

spontaneously: $W(l) \longrightarrow W(s) \longrightarrow W(g)$

[starting with $W(l)$]

c. At the temperature and pressure of point 3,

equilibrium: $W(l) \rightleftharpoons W(s)$

d. At the temperature and pressure of point 4,

spontaneously: $W(l) \longrightarrow W(s)$

e. At 0°C and 1000 torr,

metastable: $W(l)$ (supercooled)

7. According to the phase diagram given, which of the following statements is wrong?

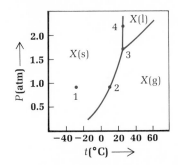

a. At the temperature and pressure of point 1, substance X exists as a three-phase equilibrium system.

b. At the temperature of point 2, a pressure of 1.5 atm is sufficient to solidify gaseous X.

c. If the $W(s) \rightleftharpoons W(l) \rightleftharpoons W(g)$ system is maintained at the temperature of point 3 while pressure is decreased, more X will vaporize.

d. If solid X is maintained at the pressure of point 4 while the temperature is increased to 40°C, the solid will melt.

e. The existence of liquid X at 40°C and 1 atm represents the metastable condition of "superheating."

8. According to the phase diagram given, which of the following statements is wrong?

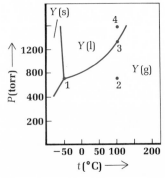

a. At the temperature and pressure of point 1, substance Y exists as a three-phase equilibrium system.

b. At the temperature and pressure of point 2, substance Y exists as a one-phase gaseous system.

c. If the $Y(l) \rightleftharpoons Y(g)$ system is maintained at the temperature of point 3 while pressure is increased, more Y will vaporize.

d. If liquid Y is maintained at the pressure of point 4 while the temperature is increased to 200°C, the liquid will vaporize.

e. The existence of liquid Y at 50°C and 400 torr represents the metastable condition of "superheating".

9. According to the phase diagram given, which description is correct?

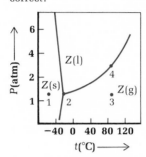

a. At the temperature and pressure of point 1,

equilibrium: $Z(s) \rightleftharpoons Z(l) \rightleftharpoons Z(g)$

b. At the temperature and pressure of point 3,

spontaneously: $Z(l) \longrightarrow Z(g)$

[starting with Z(l)]

c. At the temperature and pressure of point 4,

equilibrium: $Z(s) \rightleftharpoons Z(g)$

d. At the temperature and pressure of point 2,

spontaneously: $Z(l) \longrightarrow Z(s)$

[starting with Z(l)] [no Z(g) possible]

e. At 0°C and 4 atm,

metastable: Z(l)   (superheated)

Objective (d)

10. A metal was found to have a heat capacity of approximately 0.115 cal g⁻¹ deg⁻¹ over the temperature range 20°–80°C. A 0.8320-g sample of the metal combined with oxygen to form 1.216 g of the metal oxide. What was the metal?

11. A metal was found to have a heat capacity of approximately 0.056 cal g⁻¹ deg⁻¹ over the temperature range 20°–80°C. A 0.6744-g sample of the metal combined with oxygen to form 0.7704 g of the metal oxide. What was the metal?

12. Given: (for water) (melting point, 0°C; boiling point 100°C)

$$\Delta H_{fusion} = 1.44 \text{ kcal mole}^{-1} \quad \text{at } 0°C$$
$$\Delta H_{vap} = 9.77 \text{ kcal mole}^{-1} \quad \text{at } 100°C$$
$$C_p(l) = 18.0 \text{ cal mole}^{-1} \text{ deg}^{-1}$$

Calculate ΔH for conversion of 5.4 g of steam at 100°C to ice at 0°C (in kilocalories).

13. What quantity of heat is required to convert 36 g of ice at −5°C to steam at 105°C?

14. Given: (for Hg) (melting point, −39°C; boiling point, 357°C)

$$\Delta H_{fusion} = 5.6 \text{ kcal mole}^{-1} \quad \text{at } -39°C$$
$$\Delta H_{vap} = 14.1 \text{ kcal mole}^{-1} \quad \text{at } 357°C$$
$$C_p(l) = 6.6 \text{ cal mole}^{-1} \text{ deg}^{-1}$$
$$C_p(g) = 5.0 \text{ cal mole}^{-1} \text{ deg}^{-1}$$

Calculate ΔH for conversion of 80 g of crystalline mercury at −39°C to liquid mercury at 41°C (in kilocalories).

15. Calculate ΔH for conversion of 20.0 g of mercury vapor at 387°C to liquid mercury at 307°C (in kilocalories).

16. In the movie *Around the World in 80 Days,* a valet was directed to prepare a bath of an exact temperature. What bath temperature would he obtain by mixing 40 liters of 20°C water with 80 liters of 90°C water, assuming negligible heat transfer with the surroundings?

17. What would be the approximate equilibrium temperature of a system obtained by adding 25.0 g of ice at 0°C to 250 ml of "hot" (80°C) coffee, assuming the heat capacity of the coffee to be the same as that of water and that there was negligible heat transfer with the surroundings? (Heat of fusion of ice is 1.44 kcal mole⁻¹ at 0°C.)

Objective (e)

18. Nicotine is a very toxic substance found in tobacco. It has been employed as an insecticide. Although the normal boiling point of nicotine is about 250°C, caution must be used in working with the liquid even at lower temperatures to avoid inhalation of the vapor. Assuming nicotine to behave as a "simple" liquid, estimate its vapor pressure at

a. 20°C (temperature of a cool room)
b. 35°C (temperature of a storage room)
c. 45°C (temperature of a warehouse on a very hot day)

19. The compound 2,4,6-trinitrotoluene (TNT) may be purified by careful distillation. However, the normal boiling point (240°C) is too close to the explosive decomposition temperature, so vacuum distillation must be employed. If the temperature of the liquid drops below 82°C, the liquid crystallizes. Assuming liquid TNT to behave "simply", estimate

a. the boiling point of TNT at a reduced pressure of 500 torr
b. the boiling point at 25 torr, a pressure readily attained by a water aspirator vacuum system
c. the lowest pressure acceptable for vacuum distillation of TNT without cooling the liquid to its freezing point

BRAIN TINGLER

20. The simple sugar glucose is often referred to as a "biological fuel" because of its value as an energy source for a large number of living organisms. The *standard* heat of combustion of glucose is −673 kcal mole⁻¹. Within the living organism, of course, the conversion of glucose to CO_2 and water is a complex multistep process. However, remembering that ΔH is a state function, and assuming that the difference between solid glucose and aqueous glucose may be neglected, we can treat the process mathematically as though it occurred according to the equation

$$C_6H_{12}O_6(s) + 6O_2(g) \longrightarrow 6CO_2(g) + 6H_2O(l)$$

What should be the net enthalpy change for the complete combustion of 5.0 g of glucose at the normal temperature of a human body (37°C)? Assume: $C_P[O_2(g)] = 7.0$ cal mole^{-1} deg^{-1}, $C_P[CO_2(g)] = 8.8$ cal mole^{-1} deg^{-1}, $C_P[H_2O(l)] = 18.0$ cal mole^{-1} deg^{-1}, and $C_P[C_6H_{12}O_6(s)] = 73.0$ cal mole^{-1} deg^{-1}.

Unit 12 Self-Test

1. Fill in the blanks, using the most appropriate of the terms given.
   TERMS: (A) dynamic equilibrium (B) fusion (C) Le Chatelier's (D) lyophilization (E) specific heat (F) steady state (G) superheating (H) triple point (I) vaporization
   a. A metastable condition characterized by a liquid remaining under conditions of temperature and pressure at which the stable phase is vapor is called _____.
   b. The technical term for the process of freeze-drying is _____.
   c. The equilibrium condition involving three phases of a substance occurs at the _____.
   d. The quantity of heat required to melt 1 mole of solid at constant temperature is the molar heat of _____.
   e. _____ principle describes the behavior of an equilibrium system when it is subjected to conditions tending to alter the equilibrium.
   f. A system in a _____ condition is characterized by constancy of part, but not all of the components of the system.

2. According to the phase diagram given, which of the following statements is wrong?

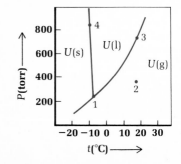

P(torr) →
t(°C) →

   a. At the temperature and pressure of point 1, substance U exists as a three-phase equilibrium system.
   b. At the temperature and pressure of point 2, substance U exists as a one-phase gaseous system.
   c. At the temperature and pressure of point 3, substance U exists as a two-phase gas–liquid equilibrium system.
   d. If the $U(s) \rightleftharpoons U(l)$ system is maintained at the temperature of point 4 while pressure is decreased, more U will freeze.
   e. There are no conditions of temperature and pressure under which solid U will vaporize without melting first.

3. A metal is found to have a specific heat of 0.032 ± 0.001 cal g^{-1} deg^{-1} over the temperature range 15–75°C. A 0.5853-g sample of the metal was converted to 1.0107 g of its chloride. Calculate the atomic weight of the metal and then use a periodic table to find the name of the metal.

$$(c_P \times W \approx 6.2)$$

4. What quantity of heat (in kilocalories) is required to convert 50 g of liquid mercury at 100°C to mercury vapor at 400°C at a constant pressure of 1.00 atm? (The boiling point of mercury is about 357°C at 1.00 atm, the heat of vaporization is 14.1 kcal mole^{-1} at 360°C, and the molar heat capacities of liquid and vapor over the temperature ranges involved may be taken, respectively, as 6.5 cal mole^{-1} deg^{-1} and 5.0 cal mole^{-1} deg^{-1}.)

5. The normal boiling point of toluene is 111°C. Estimate
   a. the vapor pressure of toluene at 37°C
   b. the boiling point of toluene at 15 torr

$$\left[\log P_a = \frac{4.6}{T_a} (T_a - T_b) \right]$$

Excursion 1

Crystal structure

In 1895 the German physicist Wilhelm Roentgen discovered a strange radiation formed when a beam of electrons struck a metal target in a vacuum tube. Because the nature of this radiation was unknown, Roentgen called it "x-radiation." It was so powerful that it could pass through an opaque shield around the vacuum tube and produce a glow on a fluorescent screen or a darkening of a photographic plate some distance away. Within a few weeks of Roentgen's discovery, physicians had begun to investigate these x-rays as a way of "seeing inside" the human body.

It was not until 1912 that the nature of x-rays was determined. They had the same properties as ordinary visible light except that their wavelengths were much shorter, thus accounting for their high energy. A year later, H. G. J. Moseley, an English physicist, showed that the wavelength of the x-rays formed was a function of the metal used as the electron target in the vacuum tube. He determined that the wavelength produced was a direct function of the atomic number of the element used as the target. This not only provided an experimental method for determining atomic number, but it also showed that the periodic table of the elements could best be based on atomic number rather than atomic weight, as first suggested by Mendeleev.

Visible light forms a diffraction pattern (Unit 1) when it is permitted to shine through a tiny hole in an opaque card. These patterns are formed only for holes of a limited diameter range, related to the wavelength of the light used. Similar diffraction occurs when the light is passed through a narrow slit. Attempts to use this technique with x-rays showed that their wavelengths were in the range of ∼1 Å, the order of magnitude of interatomic distances in crystals. The German physicist Max von Laue suggested that crystals might serve as diffraction gratings for x-rays and, indeed, the first experiment—using a copper(II) sulfate pentahydrate crystal—produced the expected diffraction patterns (Figure 1a).

The initial use of x-rays to determine the arrangement of atoms in a crystal was made by a student, W. Lawrence Bragg, at Cambridge University. He developed a technique for determining atomic locations from analysis of x-ray diffraction patterns and applied that to the first experimental description of a crystal structure, that of a cubic form of zinc sulfide. W. L. Bragg, working with his father, W. H. Bragg, then developed apparatus for experimental investigation of crystal structure. Their equipment was the model for modern-day x-ray investigation (Figure E1.1).

During the years since Bragg's investigation of the zinc sulfide crystal, x-ray diffraction has been applied to problems of increasing complexity. The vast amount of data generated by a complex crystal and the necessity for its interpretation by a repetitive trial-and-error method made x-ray studies very tedious and time consuming. With the advent of computers and new engineering techniques for automation of data collecting, x-ray diffraction has become a commonplace tool for studies of crystal and molecular structure.

In 1953, x-ray data were used to determine the structure of complex nucleic acids, the genetic material responsible for protein synthesis and the transmittance of hereditary traits. This work culminated in the Nobel Prize for the investigators,

Figure E1.1 X-ray diffraction. (a) von Laue experiment (1912). (b) Modern x-ray diffractometer.

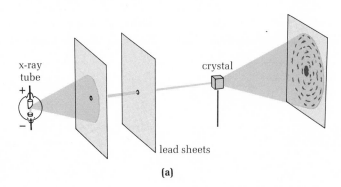

(a)

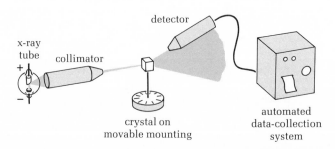

(b)

F. H. C. Crick and J. D. Watson.[1] *X-ray research has advanced a long way since the first applications to simple crystal structures.*

1 X-RAY DIFFRACTION

When water waves (Figure E1.2) strike a row of posts, new wave patterns are formed. At an appropriate distance it is noted that regions of maximum wave height are regularly interspersed with regions of calm water. Apparently troughs from some waves formed around the posts have coincided with crests from adjacent wave patterns to cancel each other (*destructive interference*). Likewise, some crests coinciding with other crests (or troughs with troughs) increase wave intensity (*constructive interference*).

It is observed that somewhat similar patterns are formed when a beam of x-rays is passed through a crystal. Since the atomic hypothesis views crystals as regular arrays of atoms, it is possible to suggest that x-rays are diffracted by sets of atoms in crystals in much the same way that water waves are diffracted by rows of posts. Such a proposal is, in fact, a considerable oversimplification. Evidence indicates that x-rays are actually absorbed by atoms or ions to cause electronic transitions (Unit 1). Diffraction patterns then result from mutual interactions of x-rays reemitted as electrons in atoms or ions return from higher excited states to lower-energy situations. Fortunately, this phenomenon of x-ray absorption and reemission is mathematically equivalent to the simpler situation of reflection from parallel planes (Figure E1.3).

The arithmetic of x-ray diffraction is treated as though the radiation were reflected from a series of parallel planes, spaced apart at a distance d. The laws dealing with reflection of visible light from a surface state that the angle of incidence equals the angle of reflection. When x-rays impinge on the crystal, a large fraction of the radiation passes straight through, but some is reflected. If the parallel rays strike the crystal planes at an angle θ (Greek, theta), reflection occurs at the same angle, θ, to the planar surface. Note that the crystal plane bisects the angle between the reflected and nonabsorbed beams, so that this angle must be 2θ. The intensity of a collection of reflected rays will be increased by constructive interference (maxima and minima correspond) or decreased by destructive interferences (waves out of phase). The spot appearing on the final x-ray film will correspond to a condition of

Figure E1.2 Diffraction patterns from water waves.

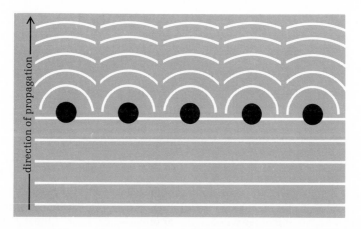

direction of propagation

[1] J. D. Watson, *The Double Helix* (New York: New American Library, 1968).

Figure E1.3 Reflections from parallel planes. (Waves are in phase, so that constructive rather than destructive interference occurs.)

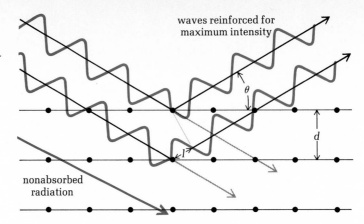

maximum constructive interference; that is, rays from successive planes must all be in phase. However, a ray penetrating the crystal to the second plane must travel farther than one reflected from the surface plane. The difference between the paths of the two rays will be $2l$ (Figure E1.3). In order for the rays to remain in phase, this extra distance must correspond to a whole number of wavelengths:

$$2l = n\lambda$$

where λ (Greek, lambda) is the symbol for wavelength. In order to relate this to the interplanar distance d, let us consider a simplified segment of Figure E1.3.

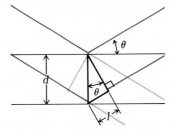

Then

$$\sin \theta = \frac{l}{d}$$

from which

$$l = d \sin \theta$$

The substitution of $d \sin \theta$ for l in the equation for constructive interference then gives a relationship between d, the interplanar spacing, and two measurable quantities: λ, the wavelength of the x-rays used, and θ, the angle of reflection. This relationship is known as the *Bragg equation*.

$$n\lambda = 2d \sin \theta \tag{1}$$

This equation applies both to the situation of simple reflection by parallel planes and to the analogous, but more complex, situation of X-ray diffraction. In the latter case n is some whole number of wavelengths, d is the distance between planes defined by positions of unit particles in the crystal lattice, and θ is one half the angle between

Figure E1.4 X-ray diffraction (simplified representation; in real situations, many different spots are formed).

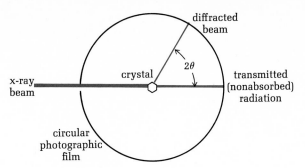

Table E1.1 SINES OF ANGLES (0–30°)

ANGLE (deg)	SIN
1	0.018
2	0.035
3	0.052
4	0.070
5	0.087
6	0.104
7	0.122
8	0.139
9	0.156
10	0.174
11	0.191
12	0.208
13	0.225
14	0.242
15	0.259
16	0.276
17	0.292
18	0.309
19	0.326
20	0.342
21	0.358
22	0.375
23	0.391
24	0.407
25	0.423
26	0.438
27	0.454
28	0.470
29	0.485
30	0.500

Monochromatic radiation consists of a single wavelength, as opposed to polychromatic (many wavelengths).

the nonabsorbed beam of x-rays and the reinforced beam of reemitted x-rays (Figure E1.4). The latter angle (2θ) is called the *diffraction angle*.

There are many more planes defined by lattice points than are indicated by the lattice faces alone. Figure E1.5, for example, shows some additional imaginary planes that correspond to a primitive lattice in the isometric system. More complex lattices have larger numbers of such planes.

It is important to note that calculation of *d* values from the Bragg equation will reveal a *series* of different interplanar spacings, not just the edge lengths of simple lattices. Lattice faces do, of course, correspond to parallel planes in the crystal, so that selections of the appropriate interplanar spacings can be used to determine edge lengths of crystal lattices.

Diffraction patterns can be obtained from powders as well as from single crystals. Since the powders consist of tiny crystals mixed in all possible orientations, the number of diffraction angles observed will represent the total number of possible diffracting planes in any of the microcrystals. These powder photos, called "Debye–Scherrer patterns," are useful in determining crystal lattice characteristics for substances in the isometric or tetragonal systems. They are less useful for the more complex systems; some crystals, for example, give several thousand different diffraction angles.

Measurements of diffraction angles (see Table E1.1 for sine functions) can be used for accurate determinations of interplanar distances in crystals (Example 1) when monochromatic x-rays of known wavelength are used. Conversely, such measurements can serve to determine the x-ray wavelength from diffraction patterns of a crystal whose lattice dimensions are known (Example 2). A value of $n = 1$ in the Bragg equation is said to indicate *first-order* diffraction. Situations in which the order of the diffraction are of consequence are beyond the scope of this text. The Suggested Reading at the end of the unit contains additional information for those interested in exploring this and other aspects of x-ray diffraction in greater detail.

Figure E1.5 Some imaginary planes in addition to faces of a primitive cubic cell. (a) Diagonal plane. (b) Oblique plane.

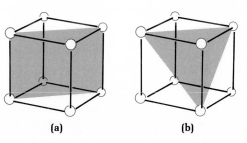

(a)   **(b)**

Example 1

An x-ray tube using a silver target emits x-rays of wavelength 0.558 Å. When these x-rays are directed at the proper angle against the face of a crystal of iron known to consist of body-centered cubic unit cells, a diffraction angle of 22.48° is observed to correspond to a diffracted beam of maximum intensity. What is the edge length of the unit cell, assuming that this diffraction is associated with the shortest distance between planes parallel to the face of the unit cell?

Solution

First, use the Bragg equation to find the interplanar spacing (assume first-order diffraction).

$$n\lambda = 2d \sin \theta$$
$$n = 1$$
$$\lambda = 0.558 \text{ Å}$$
$$\theta = \frac{22.48°}{2} = 11.24°$$

from Table 1,

$$\sin (11.24°) = 0.195$$

Then

$$d = \frac{n\lambda}{2 \sin \theta} = \frac{0.558 \text{ Å}}{2 \times 0.195} = 1.43 \text{ Å}$$

Construction of a model or drawing of a network of body-centered cubic unit cells will show that the closest planes are defined by a set of corner atoms and a set of center atoms. Since the latter are halfway between faces of the unit cell, the total distance between faces is 2×1.43 Å. Thus the edge length of the unit cell is 2.86 Å.

Example 2

A new x-ray tube is calibrated by using a crystal of aluminum. The metallic radius of aluminum is known to be 1.43 Å. The crystal structure of aluminum has been determined. It involves a face-centered cubic unit cell. The crystal is oriented in a way known to produce first-order diffraction from closest planes parallel to faces of the unit cell, and the resulting diffraction angle for maximum intensity is observed to be 40.0°. What is the wavelength of the x-rays used?

Solution

The interplanar spacing involved corresponds in this case to one half the edge length of the unit cell, which can be determined from a diagram of the face of the cell.

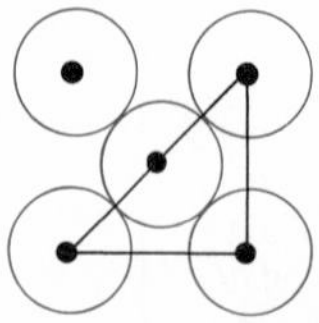

The right triangle drawn has a hypotenuse equal to four times the metallic radius. Each leg of the triangle is equal to the edge length of the unit cell, l. Then

$$(4 \times 1.43)^2 = l^2 + l^2$$

from which

$l = 4.04 \text{ Å}$

But the closest-spaced planes parallel to the face of the unit cell are defined by corner points and points at the centers of all faces; thus the interplanar spacing of concern is

$d = 2.02 \text{ Å}$

and

$n = 1$

$$\theta = \frac{40.0°}{2} = 20.0° \qquad (\sin 20.0° = 0.342)$$

so that

$$\lambda = \frac{2d \sin \theta}{n} = 2 \times 2.02 \text{ Å} \times 0.342$$

$$= 1.38 \text{ Å}$$

2 PARTIAL DESCRIPTION OF THE UNIT CELL

In addition to information on interplanar spacings, x-ray diffraction patterns can provide symmetry data (Unit 11) of critical importance in describing the crystal structure. Some symmetry information may be obtained from direct visual examination of diffraction patterns (Figure E1.6). In simple crystals the symmetry features and interplanar distances determinable from diffraction patterns may be sufficient for complete description of the crystal structure. Unfortunately, the more interesting substances require a much more complex analysis.

Lattice dimensions are not sufficient for the correct assignment of an appropriate lattice description (Unit 11). Symmetry elements must also be known and, in addition, it must be possible to determine the number of atoms of each type to be assigned to the unit cell. The latter can be calculated from a knowledge of unit cell dimensions, crystal system, and crystal density (Unit 11).

Figure E1.6 X-ray diffraction pattern showing two mirror planes.

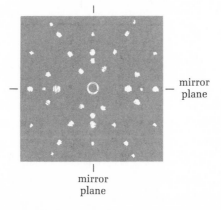

mirror plane

mirror plane

Example 3

A crystal of tungsten is found to have a density of 18.5 g cm^{-3}. Analysis of x-ray diffraction data reveals symmetry elements typical of the isometric system and an interplanar spacing parallel to lattice faces of 1.60 Å. What is the appropriate description for the unit cell in the tungsten crystal?

Solution

We shall assume that the unit cell must be primitive, body centered, or face centered (Unit 11). For a primitive cubic lattice, the interplanar spacing indicated corresponds to the edge length of the lattice. For either the body-centered or face-centered cubic systems, the edge length is twice the indicated interplanar spacing (Examples 1 and 2). Then the volume of the unit cell lattice (the edge length cubed) must be

$$V = (1.60x)^3$$

where x is either 1 or 2.

The volume of the unit cell lattice may also be estimated from the crystal density (and the atomic weight of tungsten, 183.8):

$$V = \frac{1 \text{ cm}^3}{18.5 \text{ g}} \times \frac{183.8 \text{ g}}{6.02 \times 10^{23} \text{ atoms}} \times \frac{10^{24} \text{ Å}^3}{1 \text{ cm}^3} \times \frac{n \text{ atoms}}{\text{unit cell}}$$

in which n is the number of atoms assigned to the unit cell (Unit 11).

We then have two equations:

$$V = 16.5n$$
$$V = 4.10x^3$$

Thus:

$$4.10x^3 = 16.5n$$

If x = 1, we have a primitive lattice, for which $n = 1$, but

$$4.10 \neq 16.5$$

Hence, x must be 2, from which

$$\frac{(4.10)(2)^3}{16.5} = n$$

Within the limits of experimental error, $n = 2$. The tungsten is then described in terms of a *body-centered cubic unit cell*.

---

3 INTENSITIES: KEY TO CRYSTAL STRUCTURE

Crystal structure implies the complete location of all atoms of the crystal. Since the crystal consists of identical unit cells in a repetitive three-dimensional array, this really means that the unit cell must be completely described. The selection of the correct Bravais lattice identifies only the centers of symmetry of the unit particles of the crystal, and unless these are monatomic particles this information does not tell us how the particles are oriented with respect to each other.

An x-ray diffraction photograph (Figure E1.6) reveals some symmetry features and spacings that can be related by the known geometry of the x-ray equipment to the diffraction angles. A close examination of Figure E1.6 shows a very important feature: The spots are not all of the same intensity.

The intensity of a particular spot is determined by a number of factors, including

intensity of the incident radiation
thickness of the crystal along the diffraction angle
sensitivity of the detection system used
experimental conditions, including exposure time, corrections for absorption within the apparatus, and so on

the diffraction angle

a structure factor (*F*), which is related to the characteristics of the electrons whose energy level transitions are involved in the diffraction phenomenon

For any particular experiment, all contributing parameters are constants that can be accounted for, except the diffraction angle (which can be measured directly) and the structure factor, *F*. This factor is of critical importance, because it is related to the possible energy states of the electrons involved (and therefore to the nature of the particular atoms with which they are associated) and to the relative locations of these atoms. The latter consideration arises from the fact that particular relative locations are required for the diffracted x-rays to be in phase. If all the structure factors can be determined, they will pinpoint the atomic positions so that the crystal structure and—of greater interest to the chemist—the geometry of the molecule or polyatomic ion in the crystal can be described. This is where the "fun and games" of x-ray research begins.[2]

X-ray studies may now be supplemented by the new technique of "position injection."[3]

There are now some very complex ways for direct calculation of *F* from intensity data—in limited situations. It is more common, and more generally applicable, to approach this problem by a series of successive approximations. The arithmetic of this procedure is far from simple, but basically the method starts with an educated guess of a possible structure. The more information available, such as the condensed structural formula (Unit 3), the better this first guess will be. The guess is then used to calculate a diffraction pattern corresponding to the proposed geometric features. Comparison of the calculated pattern with the experimental observations tells how good the guess was and, with luck, suggests some modifications that will improve the agreement between experimental and recalculated patterns. This process is continued until a structure is devised that fits the experimental data well.

In this way it is possible to determine the proper description for the unit cell and for the geometry of any polyatomic particle in the crystal.

One way of refining the data (improving the correlation between calculation and experiment) uses electron density maps (Figure E1.7). When a series of such drawings are properly combined, they can provide a three-dimensional picture.

A very interesting way of representing geometry has recently been devised. The data describing the refined structure are fed into a computer, which is programmed to draw a pair of stereo projections of the molecular or lattice structure. When properly viewed, these projections from a three-dimensional image[4] (Figure E1.8).

Figure E1.7 An electron density map. (Planar-NO_2 group in a crystalline organic compound.)

The assignment of correct *F* values from intensity data is known as *the phase problem,* and this represents the real challenge in interpreting x-ray data. Until recent years it was necessary to catalog each spot on a series of x-ray photographs taken

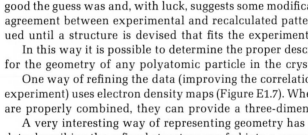

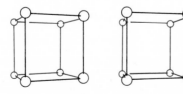

Figure E1.8 Stereo projections of a primitive cubic lattice. To obtain a three-dimensional image, look at the drawings from a distance of about 6 in. for a few seconds. Use the left eye for the figure on the left and the right eye for the other. Several seconds are necessary to adjust your vision for the proper effect.

[2] For a more detailed discussion of interpretation of x-ray data, see Jon Kapecki's article in the April, 1972, issue of the *Journal of Chemical Education* (pp. 231–236).

[3] Described by Werner Brandt in the July 1975 issue of *Scientific American* (pp. 34–42).

[4] The technique is described for a use of two-color stereo images by Edgar F. Meyer, Jr. in the *Journal of Chemical Documentation* (Vol. 10, 1970, pp. 85–86).

with different crystal orientations and to estimate the relative spot intensities by visual comparisons. With some types of modern equipment, the crystal is mounted in a device that can be revolved and rotated in unison with the movements of an x-ray detector system (Figure E1.1). The latter converts x-radiation into an electric signal, so that information on intensities and diffraction angles is automatically collected and converted into a form suitable for computer processing.

4 CRYSTAL DEFECTS

Silicon is the earth's second most abundant element (oxygen is more abundant), and compounds of silicon are estimated to constitute nearly 90 percent of the earth's crust. Most of these are complex silicates. Free silica (empirical formula SiO_2) occurs in several allotropic forms, including quartz, tridymite, cristobalite, and the amorphous "quartz glass." In all of the crystalline forms of silica, each silicon atom is surrounded by a tetrahedral pattern of oxygen atoms. In one of these forms the silicon atoms are arranged exactly like the carbon atoms in diamond (Unit 11). Ordinary quartz is a slight modification of this structure (Figure E1.9).

"Perfect" crystals of quartz are rare. Like other crystalline materials, quartz usually contains one or more types of crystal defects. Some of these are physical imperfections caused by irregularities in the lattice patterns. Others, such as the foreign ions responsible for the color of rose quartz, are chemical defects. A series of interesting silicates may be considered as representing departures from the perfect lattice structure of quartz.

The basic unit of the quartz structure is the tetrahedral SiO_4 aggregate. In quartz, each oxygen is shared by two silicon atoms so that the silicon-to-oxygen atomic ratio is $1:2$. Feldspar minerals contain substitutional impurities—some of the silicon atoms are replaced by aluminum atoms. Additional ions, such as those of simple metals, may then fit into the lattice structure. Although such substitutions usually occur in a stoichiometric ratio, they are not necessarily uniform throughout the crystal lattice; therefore a simple unit cell description of such minerals is less representative of the complete crystal lattice than would be the case with completely homogeneous substances.

In a second group of silicate minerals we find the three-dimensional SiO_4 network limited to fairly thin, planar regions. In talc, for example, SiO_4 tetrahedra are combined in sheets separated by rather nonuniform collections of cations. The forces between these sheets are much weaker than the covalent bonds within the sheets. As a result, silicate sheets in talc can slide over each other fairly easily—similar to the behavior of carbon sheets in graphite (Unit 11). Mica has a similar structure, except that aluminum has been substituted for about one-fourth of the silicon. This permits addition of more ions between the sheets, so that layers of mica crystals are held together more tightly than those in talc.

A third type of silicate is represented by fibrous minerals such as asbestos. Long chains of SiO_4 tetrahedra are surrounded by metal ions. This type of structure is a very significant departure from the highly ordered three-dimensional network of the pure quartz crystal.

Defects in the idealized quartz structure result in a host of important minerals. In addition to those described, there are the clays—similar in structure to mica, soft rocks such as bentonite (Ba^{2+} and Ti^{4+} cations associated with the $Si_6O_9^{6-}$ anion), and the polycrystalline materials such as granite—a mixture of tiny crystals of quartz, feldspar, and mica.

Imperfect crystals are more than just a "nuisance" to the theoretical crystallographer. In many cases the imperfections are responsible for very important properties (Figure E1.9). The art of growing crystals with *planned* defects has, in recent years, become one of the most challenging fields in solid state research. As we

Figure E1.9 Some important crystal defects (stylized representations). (a) "Perfect" lattice in pure quartz ("SiO_2"). (b) Mica (silicate sheets separated by mixtures of ions). (c) Asbestos (double-stranded silicate fibers surrounded by mixtures of ions). (d) Granite (microcrystalline aggregates). (e) "Perfect" lattice in pure silicon. (f) p-Type semiconductor (boron substitution). (g) n-Type semiconductor (arsenic substitution).

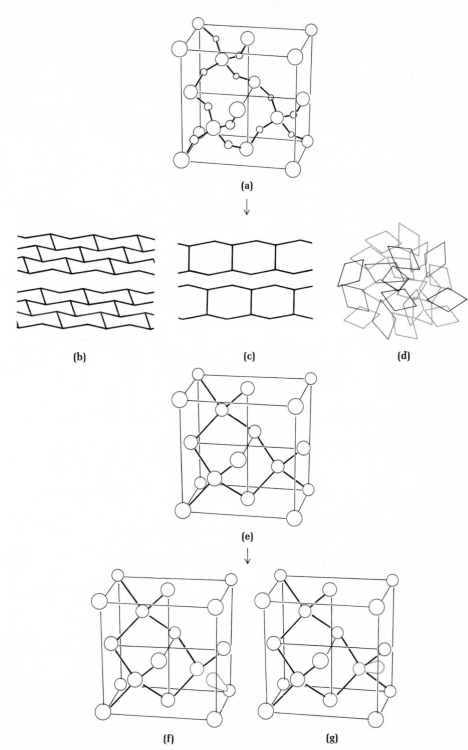

shall see, appropriate crystal defects are essential to the technology of space age science.

5 IMPERFECT CRYSTALS

If we think about the processes by which crystals form, we can see why perfect crystals are so rare in nature. Typically, particles enter the crystalline state from a melt (liquid) or solution, although a direct condensation from the gas phase is not unknown. Remember that particles in fluids are undergoing a variety of motions, including tumbling modes that lead to random orientations. Once a tiny collection of ordered particles has been formed (nucleation), the crystal begins to grow by addition of particles at the surface. Polyatomic particles not properly oriented may produce growth patterns that are not identical with those of the idealized lattice. Continuation of regularity is, of course, favored by *directional* interparticle forces, but even in the simple ionic crystals there is a strong probability that growth will occur at different rates at different locations on the surface. Additional lattice layers may build up before inner layers are completed. Unless the fluid is identical in composition with the crystal, foreign particles (impurities) may substitute for or fit between normal lattice components. These may, in turn, lead to further distortions in particle arrays.

The sublimation of iodine, formation of "dry ice", and crystallization of water vapor to snow are all examples of crystal formation directly from the gaseous state.

In very simple crystals, such as those of metals forming closest-packed lattices, "incompatible" packing patterns may be formed at different locations on a single face. As these grow together, defects appear (Figure E1.10).

We can see, then, that the description of a crystal as a uniform repetitive pattern of identical unit cells is likely to be an idealized picture. Most real crystals exhibit *short-range order,* that is microregions characterized by perfect lattices. *Long-range order* exists but is, in most cases, less regular than that of a perfect lattice. The microregions are called *crystallites,* and most large crystals consist of sets of such crystallites whose boundaries are defined by intersecting patterns of crystal growth.

Figure E1.10 Lattice defect in closest packing (Compare Figure 11.18, Unit 11).

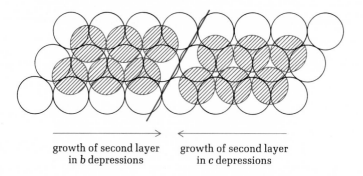

growth of second layer   growth of second layer
in *b* depressions   in *c* depressions

6 TYPES OF CRYSTAL DEFECTS

Crystal imperfections may occur in chemically pure substances as a result of irregular growth patterns. These are broadly classified as *physical defects*. Physical defects are also commonly found in impure crystals, but these also contain *chemical defects* in the form of substitutional or interstitial (or both) foreign particles. Since it is very difficult to produce crystals of even moderate size that are completely pure, most real crystalline substances contain both types of defects.

Physical defects

Physical lattice imperfections are very common in natural crystals. These are, of course, of a more fundamental nature than the scars resulting from fracturing or weathering of the crystals. Some defects, such as edge or screw dislocations (Figure E1.11) have long-range effects and may result in visible surface deformations, al-

Figure E1.11 Some lattice dislocations. (a) Edge dislocation (e.g., from irregular growth pattern resulting in a wedge plane part way into the lattice). (b) Screw dislocation (e.g., from warping during growth of competitive patterns).

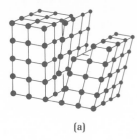

(a)

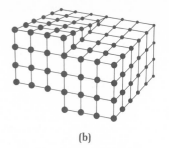

(b)

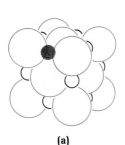

(a)

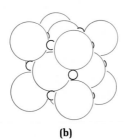

(b)

Figure E1.12 Some positional defects. (a) Lattice vacancy (e.g., FeS_2, shown here). (b) Interstitial displacement in AgI.

Nonstoichiometric compounds

A typical "correct" formulation is $Cu_{1.96}S$, rather than Cu_2S.

though some are too small for observation without the aid of tools such as the electron microscope. Surface irregularities apparently play an important role in many phenomena such as the catalytic activity of platinum and other metals.

Lattice vacancies or interstitial displacements (Figure E1.12) on the other hand, occur in isolated regions and may not be apparent even under very close examination of crystal exteriors. These may, however, have profound effects on crystal properties.

Many a tenderfoot miner, new to the West, has been excited by the discovery of metallic-appearing "fool's gold." This disappointing species is an iron pyrite, formulated as FeS_2. In this substance there are vacant cation sites in the lattice. Many of the iron ions are Fe^{3+} instead of the expected Fe^{2+}, so the electrical neutrality of the crystal is maintained. Electrons may "jump" rather easily from Fe^{2+} to Fe^{3+} particles, resulting in an electron mobility approaching that of metallic crystals. The luster of the pyrites, which causes them to glitter like gold in bright sunshine, is associated with this electron mobility.

Silver iodide is an interesting salt. The silver ion is quite small (radius = 1.26 Å) compared to the iodide ion (2.20 Å), and the electronegativity of silver is close enough to that of iodine to permit significant covalent bonding; that is, the Ag and I electron clouds show appreciable overlap. For these reasons, there is only a very small energy difference between tetrahedral and octahedral positions of Ag^+ ions in a closest-packed array of I^- ions. Under the influence of an applied voltage, silver ions migrate rather easily from one type of interstitial site to another. Silver iodide, unlike most simple salts, is an effective conductor of electricity, by movement of Ag^+ ions.

For many years a routine freshman chemistry experiment involved heating a weighed copper wire with excess sulfur, burning away any unreacted sulfur, and weighing the solid product to determine the formula of copper sulfide. The results were generally poor, but these were attributed to deficiencies in weighing equipment available or in student technique. Students were directed to round off their results to the nearest whole-number ratio in order to obtain a stoichiometric formula. We know now that copper sulfide prepared in this manner is *not* a stoichiometric compound. Very careful analyses show that the product of this reaction is not appropriately represented by a simple formula showing whole-number ratios. The lattice structure in the microcrystalline product is not a "perfect" arrangement of copper and sulfide ions. Some particles are missing. Some of the S^{2-} positions are occupied by atomic sulfur, and atomic copper, Cu^+, and Cu^{2+} are all found in the lattice. Many of the student results in these experiments were closer to the "truth" than their professors expected.

Nonstoichiometric crystals may involve mixed oxidation states of the lattice particles, as in the case of the copper sulfide. Alternatively, some particles may actually be missing from lattice positions without the compensation possibilities of variable oxidation states. When zinc oxide is heated, it begins to turn yellow. A tiny fraction of the oxide ions are converted to atomic oxygen and driven out of the lattice, eventually combining to form O_2 molecules. As a result, some of the zinc ions shift position slightly, and free electrons removed from the oxide ions are distributed within the lattice. The maximum deviation from the stoichiometric formula, ZnO, occurs at 800°C, and only 0.007 percent of the original oxygen positions are vacated. Obviously, such small deviations from stoichiometric ratios require very careful analyses to detect. Even these tiny defects, however, may have profound effects on crystal properties such as electrical conductivity.

Impurities in crystals

A gram of table salt contains more than 10^{22} ions each of sodium and chloride. The chances that a few particles of some other species will be incorporated in crystalline "salt" are very good. A purity that would easily meet government standards for human consumption would still permit a much higher percentage of foreign substance than would be desirable to the research chemist interested in the properties of perfect salt crystals. Much solid state research is aimed at improving techniques for obtaining ultrapure crystals.

Impurities in crystals may be adsorbed on the surfaces. The ratio of absorbed impurities to crystal mass is increased as crystal size is decreased, due to the larger ratios of surface area per unit mass of substance.

However, impurities may be—and usually are—found within the crystal lattice. These may involve fairly large regions, such as air bubbles trapped in ice or a small crystal imbedded in a larger one. From the chemist's point of view, the more interesting situations involve substitutional or interstitial impurities. The former occurs when normal lattice particles are replaced by foreign particles and the latter when foreign particles are trapped in vacant interstitial sites in the lattice. Both situations are important in the planned introduction of impurities to produce crystals having certain special properties. Two important examples are the metal alloys and the semiconductors.

7 ALLOYS

See Unit 14 for a discussion of colloids.

Alloys are metallic substances containing two or more elements, not all of which are necessarily metals. Three general types of alloys are commonly employed: compounds, solid solutions, and solid colloidal dispersions. The first two are characterized by microscopic homogeneity (Figure E1.13), the third by microscopic heterogeneity.

Alloys of the solid-solution type can be formed by addition of solute atoms to vacant interstitial sites in the "solvent" crystal or by substitution of solute for solvent atoms in the lattice. Both cases usually result in properties that are significantly different from those of the pure elements involved. The theories of bonding in alloys (or even in pure metals) are not sufficiently developed to permit consistently accurate predictions of alloy properties. Physical metallurgy is still largely an experimental science.

Figure E1.13 Common types of alloys. (a) Tantalum carbide, a compound (TaC). (b) Coinage gold, a microcrystalline mixture—each crystal a solid solution of copper atoms randomly distributed among gold atoms. (c) Lead shot, a colloidal dispersion of arsenic microcrystals in crystallites of lead.

(a)

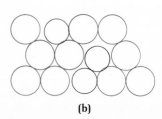
(b)

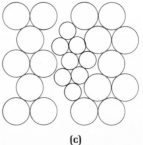
(c)

8 SEMICONDUCTORS   Typically, metals are good electrical conductors at low temperatures and decrease in conductivity when heated. Pure ionic solids are nonconductors but become excellent conductors when they melt. Some ionic solids show weak conductivity, in cases where lattice vacancies or interstitial defects allow some ion movements.

A completely different type of situation occurs in such substances as silicon or germanium, which act as insulators at low temperatures and conductors at high temperatures. These substances are called *metalloids* (Unit 2) and provide interesting examples of major changes in properties resulting from very minor crystal defects.

Perfect crystals of such elements as silicon are insulators at normal temperatures. Evidence suggests that the crystal lattice of silicon consists of silicon atoms covalently bonded together in tetrahedral arrays like the carbon atoms in diamond (Figure E1.9). Thus all valence electrons are tied up in bonding orbitals and are not free to move under the influence of an applied voltage. Only at temperatures high enough to disrupt covalent bonds (or, in photoconductivity, under exposure to light whose energy can break the bonds) will the material permit electron flow. Substitution of elements such as boron or arsenic (Figure E1.14) into crystalline silicon provides a situation in which some valence electrons are not in bonding orbitals and can move when a potential difference is applied to the crystal. Such defect crystals are called *semiconductors*.[5]

Semiconductor terminology refers to the movement, as a result of an applied voltage, of either *negative* electrons (n type) or *positive* "holes" (p type). In reality, of course, a "hole" cannot move: There is nothing there. Both types of semiconductors therefore involve electrons as the particles that actually move. However, in n-type semiconductors these are "extra" electrons, and their movement does not require any change in the covalent bond network of the crystal. In the p-type semiconductors

Figure E1.14   Semiconductors with substitutional defects. (a) n-Type semiconductor has substitutional atom with extra valence electrons. Conductivity is mainly movement of these "excess" electrons as replaced by those "pumped in." (b) p-Type semiconductor has substitutional atom with insufficient valence electrons to bond with all neighboring atoms. Conductivity mainly requires addition of electrons from external source.

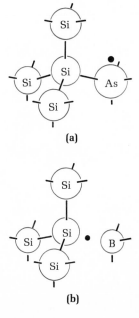

(a)

(b)

[5] For an excellent discussion of semiconductors as contrasted with metallic conductors and "superconductors", see the January, 1973, issue of *Scientific American* (pp. 88–98).

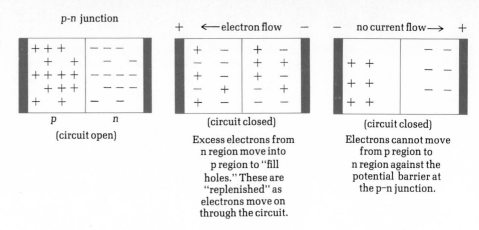

Figure E1.15 A p-n junction rectifier. Note that there are no net *charges* in the crystals. The + and − signs are only used symbolically to represent one-electron bonds and extra valence electrons, respectively.

p-n junction

p   n

(circuit open)

+   ←electron flow   −

(circuit closed)

Excess electrons from n region move into p region to "fill holes." These are "replenished" as electrons move on through the circuit.

−   no current flow→   +

(circuit closed)

Electrons cannot move from p region to n region against the potential barrier at the p-n junction.

there are no extra electrons; instead there are "holes," which may be thought of as one-electron bonds. Under the influence of an applied electric potential, an electron "jumps" from a normal covalent (two-electron) bond to the one-electron bond. Thus "holes move" in the opposite direction of electron flow.

Semiconductors need not involve metalloids. Certain ionic salts may be used. Zinc oxide, for example, is an n-type semiconductor, and "fool's gold" (pyrite) having lattice vacancies corresponding to missing iron atoms, is a p-type semiconductor.

In general, ionic semiconductors are stable over a wider temperature range than are those made from the metalloids, since the latter tend to become more metallic at temperatures high enough to promote electrons from their bonding orbitals.

Semiconductors are of immense importance in modern technology. As an example, they can be used to provide a "one-way street" for current flow (Figure E1.15) by formation of a *p-n junction*.

SUGGESTED READING

Bunn, Charles. 1964. *Crystals: Their Role in Nature and in Science.* New York: Academic Press.

Evans, R. C. 1964. *An Introduction to Crystal Chemistry.* New York: Cambridge University Press.

Greenwood, N. N. 1970. *Ionic Crystals, Lattice Defects, and Nonstoichiometry.* New York: Chemical Publishing Company. (Advanced, but highly recommended.)

Reiss, H. 1967. "Chemical Properties of Materials," *Sci. Amer.*, September.

Sands, D. E. 1969. *Introduction to Crystallography.* Menlo Park, Calif.: W. A. Benjamin.

Stout, G. H., and L. H. Jensen. 1968. *X-ray Structure Determination.* New York: Macmillan.

Overview Test

Section FOUR

1. Select the term best associated with each definition, description, or specific example.

   TERMS: (A) allotropes (B) amorphous (C) covalent radius (D) crystalline (E) diffusion (F) dipole–dipole forces (G) dipole-induced dipole forces (H) dynamic equilibrium (I) effusion (J) elastic collision (K) hydrogen bonds (L) London dispersion forces (M) metallic radius (N) meniscus (O) steady state (P) van der Waals radius

   a. distance indicated for crystalline iodine

   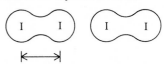

   b. distance indicated for crystalline iodine

   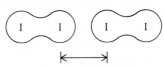

   c. distance indicated for crystalline sodium

   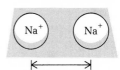

   d. characterized by sharp melting point and planar cleavage patterns
   e. characterized by broad melting range and irregular fracture patterns
   f. characterized by conservation of translational kinetic energy
   g. the curved surface at a liquid–gas interface
   h. constant volume of liquid water and constant water vapor pressure of 760 torr in a sealed $H_2O(l)$–$H_2O(g)$ system maintained at 100°C
   i. constant volume of liquid water and constant water vapor pressure in an open vessel containing boiling water

   at 100°C, with water added at a rate equal to the rate of liquid loss by vaporization
   j. movement of $HCl(g)$ through the air of a laboratory room
   k. escape of $HCl(g)$ through a porous barrier into an evacuated chamber
   l. "red" phosphorus and "white" phosphorus
   m. principal forces among neighboring molecules in liquid nitrogen
   n. contributing forces among neighboring molecules in liquid HCl
   o. contributing forces among neighboring molecules in liquid NH_3
   p. contributing forces between HCl and CCl_4 molecules in a solution of HCl in CCl_4

2. Carbon tetrachloride, CCl_4, is a tetrahedral molecule. Its various degrees of freedom are described as
   a. 3 translational, 2 rotational, 10 vibrational
   b. 3 translational, 10 rotational, 2 vibrational
   c. 3 translational, 3 rotational, 9 vibrational
   d. 3 translational, 9 rotational, 3 vibrational
   e. 5 translational, 5 rotational, 5 vibrational

3. Classify each of the following in terms of interparticle forces: a. liquid water; b. crystalline magnesium; c. crystalline magnesium bromide; d. liquid bromine; e. liquid hydrogen bromide

4. What is the maximum number of molecules of glycerol that could form hydrogen bonds with a single molecule of the compound?

$$\left(HOCH_2 - \overset{\overset{\displaystyle H}{|}}{\underset{\underset{\displaystyle OH}{|}}{C}} - CH_2OH \right)$$

   a. 1; b. 3; c. 5; d. 7; e. 9

5. Which probably has the lowest boiling point at 1.00 atm?
   a. CH_4; b. SiH_4; c. GeH_4; d. SnH_4; e. PbH_4

6. Which probably has the lowest boiling point at 1.00 atm?
   a. $H_2NCH_2CH_2NH_2$; b. $CH_3CH_2CH_2NH_2$; c. $(CH_3)_2CHNH_2$;

   d. $CH_3CH_2\overset{\underset{\displaystyle H}{|}}{N}CH_3$; e. $(CH_3)_3N$

7. Solution formation can be described by considering the three attractions listed below. Which of these will act to reduce the chances of solution formation?

   I. solvent–solute attractions

   II. solute–solute attractions

   III. solvent–solvent attractions

   a. I and II; **b.** II and III; **c.** I and III; **d.** I, II, and III; **e.** I

8. Which is probably least soluble in water?

   a. CH_3CO_2H;   **b.** HCO_2H;   **c.** HO_2CCO_2H;

   d. $CH_3(CH_2)_{10}CO_2H$;   **e.** All are infinitely soluble in water

9. Which statement is incorrect?

   a. Ethanolamine ($HOCH_2CH_2NH_2$) is probably more soluble in water than it is in hexane (C_6H_{14}).

   b. Hexane is probably immiscible with water.

   c. Hexane is probably less viscous than ethanolamine at 20°C.

   d. Ethanolamine is less viscous at 60°C than at 20°C.

   e. Hexane vapor is probably less viscous at 260°C than at 200°C.

10. According to the phase diagram given, which of the following statements is wrong?

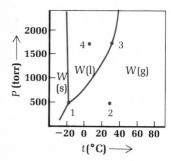

   a. At the temperature and pressure of point 1, substance W exists as a three-phase equilibrium system.

   b. At the temperature of point 2, a pressure of 500 torr is sufficient to liquify gaseous W.

   c. If the $W(l) \rightleftharpoons W(g)$ system is maintained at the temperature of point 3 while pressure is decreased, more W will vaporize.

   d. If liquid W is maintained at the pressure of point 4 while the temperature is increased to 80°C, the liquid will vaporize.

   e. The existence of liquid W at −40°C and 500 torr represents the metastable condition of "supercooling."

11. Which of the following molecules does not contain a mirror plane?

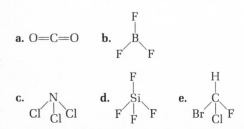

12. A compound of iron and sulfur crystallizes in a lattice pattern whose unit cell is described as cubic closest-packed sulfide ions with iron ions in all octahedral sites. The empirical formula of this salt is

   a. Fe_2S_3; **b.** Fe_2S; **c.** Fe_3S_4; **d.** FeS_2; **e.** FeS

13. What is the density of a tiny crystal that is suspended by a liquid from which a 0.500-ml sample weighs 0.624 g?

   a. 0.67 g ml^{-1}; **b.** 0.80 g ml^{-1}; **c.** 1.75 g ml^{-1}; **d.** 1.50 g ml^{-1}; **e.** 1.25 g ml^{-1}

14. A gas occupies a volume of 7.5 ml at 70°C and 900 torr. At 10°C and 300 torr, its volume would be approximately

   a. 3.2 ml; **b.** 30 ml; **c.** 27 ml; **d.** 24 ml; **e.** 19 ml

15. The density of phosgene ($COCl_2$) gas at 37°C and 700 torr, in grams per liter, is approximately

   a. 4.2; **b.** 4.6; **c.** 3.6; **d.** 2.8; **e.** 5.4

16. What is the molecular weight of a pure gaseous compound having a density of 4.36 g liter^{-1} at 57°C and 1020 torr?

   a. 48; **b.** 60; **c.** 72; **d.** 88; **e.** 96

17. What total gas volume (in liters) at 357°C and 600 torr would result from decomposition of 9.6 g of ammonium carbonate according to the equation

$$(NH_4)_2CO_3(s) \longrightarrow 2NH_3(g) + H_2O(g) + CO_2(g)$$

   a. 26; **b.** 20; **c.** 17; **d.** 15; **e.** 11

18. What mass of chlorine is represented by a volume of 75 liters of Cl_2 collected over water at 27°C and 610 torr? (Vapor pressure of water at 27°C is 30 torr.)

   a. 220 g; **b.** 185 g; **c.** 165 g; **d.** 146 g; **e.** 129 g

19. The Law of Dulong and Petit may be expressed mathematically as

$$c_p \times W = 6.2 \pm 0.4 \text{ cal mole}^{-1} \text{ deg}^{-1}$$

A metal was found to have a heat capacity of approximately 0.056 cal g^{-1} deg^{-1} over the temperature range 20°–80°C. A 0.6474-g sample of the metal combined with oxygen to form 0.6954 g of the metal oxide. What was the metal?

   a. silver; **b.** cadmium; **c.** tin; **d.** chromium; **e.** manganese

20. Given: (for Hg) (melting point, −39°C; boiling point, 357°C)

$$\Delta H_{\text{fusion}} = 5.6 \text{ kcal mole}^{-1} \quad \text{at } -39°C$$
$$\Delta H_{\text{vap}} = 14.1 \text{ kcal mole}^{-1} \quad \text{at } 357°C$$
$$C_p(\text{liq}) = 6.6 \text{ cal mole}^{-1} \text{ deg}^{-1}$$
$$C_p(g) = 5.0 \text{ cal mole}^{-1} \text{ deg}^{-1}$$

Calculate ΔH for conversion of 120 g of mercury vapor at 357°C to liquid mercury at 57°C (in kilocalories).

   a. −9.6; **b.** −9.1; **c.** −8.5; **d.** −6.3; **e.** −4.5

Section FIVE

SOLUTIONS AND COLLOIDS

Life on earth is absolutely dependent on aqueous solutions and colloids. Whether we consider the biochemical phenomena within the simplest living cell or the complex ecosystem of the oceans, we must deal with chemical reactions and transport systems involving water as the principal solvent or dispersing medium.

Even the air we breathe depends on aqueous systems for removal of CO_2 (and many "pollutants") and for replenishment of essential oxygen. Microscopic plants in the ocean contribute a major fraction to the CO_2-to-O_2 exchange process, and land plants require dissolved nutrients and aqueous transport media.

Man has changed, and is continuing to change, the composition of significant fractions of the natural aquatic systems. Oil spills, industrial wastes, agricultural runoff, and municipal sewage (sometimes "treated," sometimes "raw") are entering our rivers, lakes, and oceans in ever-increasing amounts. Natural waste-removal systems have become, in many cases, "overloaded" or are unable to cope with chemicals for which their mechanisms were not designed. In one *limited* two-month survey of *accidental* water pollution in the midwestern United States, reports were received of over 100,000 liters of concentrated fertilizer spilled into a small stream (with all aquatic life destroyed for more than a mile downstream); of nearly 4000 liters of concentrated hydrochloric acid spilled into the Ohio River; of more than 1,500,000 liters of paper-mill waste dumped into a Texas lake; and of six different spills of oil, totaling more than 12 million liters, into inland rivers and lakes. This is one small sample of *accidents*, within two months, and the total of industrial waste, municipal sewage, and other *planned* dumping into lakes and streams is sufficient to dwarf the accidental pollution to insignificance.

Other solutions and colloids are also important. In particular, pollutant gases "dissolved" in our air pose threats ranging from lung damage to destruction of the ozone layer, and colloidal smokes and smogs blanket urban and industrial areas.

To understand fully the threats of air and water pollution, of which we are all too painfully aware, and to evaluate potential ways of salvaging our environment, we must acquire the basic concepts of how solutions and colloidal dispersions are formed; of the factors that contribute to their stability; of the methods used in analyzing their compositions and concentrations; and of the ways that undesirable dissolved and colloidal material can be removed.

The units in this section are not "about our environment," although some environmental applications are mentioned. Rather, these units are designed to provide us with the basic understanding of solutions and colloids essential to a meaningful approach to some of our environmental problems, as well as to other important applications of the chemistry of solutions and colloids.

CONTENTS

Unit 13 **Aqueous Solutions**
Solution formation; ways of expressing solution concentration; stoichiometric problems involving aqueous solutions

Unit 14 **Colloids**
Classification of colloids; methods by which colloidal dispersions can be formed; factors determining colloid stability; some ways of destroying colloidal systems

Unit 15 **Colligative Properties**
Distinctions between constitutive and colligative properties; Raoult's law; boiling point elevation and freezing point depression; osmotic pressure; use of colligative properties in molecular weight determinations
Section Five Overview Test

Unit 13

Aqueous solutions

Objectives

(a) After studying the appropriate definitions in Appendix E and the terms'
   usage in this unit, be able to write useful working definitions of the concentration units (Table 13.2)

   formality   normality
   molality   parts per million (ppm)
   molarity   weight percent

(b) Be able to describe proper techniques for preparing solutions (Section
   13.4).
   (1) from pure solute and solvent
   (2) by dilution of concentrated solutions

(c) Given appropriate information, be able to calculate (Section 13.4)
   (1) solution concentration, in various units
   (2) weight of solute required for a particular solution
   (3) volume of concentrated solution required for a particular dilute solution

(d) Be able to determine the equivalent weight of a chemical substance and to
   use *normality* as a concentration unit (Section 13.5).

(e) Be able to apply the principles of stoichiometry (Unit 4) to reactions involving aqueous solutions (Section 13.6).

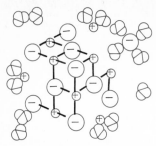

Figure 13.1 Formation of an aqueous solution. When salts dissolve in water, the process involves interactions between water molecules and ions at the crystal surfaces to overcome lattice forces and the formation of ion–water complexes in solution. The water molecules of these hydrates act as "shields" to hinder the return of ions to the crystal lattice.

Nearly three-fourths of the earth's surface is covered with water, and of this only about 3 percent is "fresh." This means that most of the earth is composed of enormous aqueous solutions—the oceans. It is said that we know less about these complex ocean systems than we do about outer space, yet the oceans hold keys to life on the earth. A few years ago it was hoped that the oceans might eventually provide vast resources of food, fuels, and vital minerals for the world's ever-expanding population. These hopes have been dimmed by the evidence of man's increasing pollution of his environment—pollution whose effects have been noted far beyond the shores of the major land masses.

Mercury from industrial waste and soluble agricultural fungicides threatens the global fishing industry; oil spills from tankers, refueling operations, and offshore wells have become serious hazards to aquatic life; significant concentrations of DDT and lead salts have been found in marine animals in the "isolated" Antarctic regions; the long-range effects of radioactive wastes, chemical warfare agents, and other "dumped" products of man's technology can only be guessed at present.

Ocean water contains, on the average, about 3.5 percent by weight of dissolved materials. Most of these are ionic species: Na^+, K^+, Mg^{2+}, Ca^{2+}, Cl^-, Br^-, HCO_3^-, and SO_4^{2-} are most common. "Average" figures on composition of solutions as complex as sea water should not be interpreted too literally. Factors such as distance from river junctions, depth, and temperature can have significant effects on various components of sea water.

In addition to the dissolved substances, the oceans contain many types of suspended materials as well as the fantastic variety of living systems found at different depths. These depend on dissolved gases (O_2 or CO_2) for their existence, except for the relatively few species that obtain oxygen or carbon dioxide directly from the air above the ocean surface.

The ecology of the oceans is of profound importance for all life on the earth. Apart from the food sources found in the sea, the sea plants now contribute the majority of the CO_2-to-O_2 conversion necessary for replenishment of atmospheric oxygen. Thus the effects of pollution on the ocean systems are of the utmost significance.

We live with complete dependence on two kinds of solutions: One is gaseous—the atmosphere—and the other is aqueous (Figure 13.1). These types of solutions are so common to us that we seldom think much about them, yet we are, at an alarmingly increasing rate, changing their compositions in ways we do not yet truly understand or whose long-term significance we have only begun to fear.

Three different aspects of solutions must be thoroughly understood and appreciated.

The *composition* of a solution is the sum total of all the ingredients of which it is composed, and the *concentration* of a solution is the relative amounts of these various components. Both these aspects must be known if a particular solution is to be accurately described and its properties are to be meaningfully estimated. The third aspect is the *quantity* of solution (mass or volume) involved in any particular process.

Given the choice between drinking a quart of "cyanide solution" or a quart of an "ethanol–water solution," the average person would quickly elect the latter. But a quart of "cyanide solution" containing less than 1.0 part per million cyanide would be quite harmless, whereas a quart of 95-percent ethanol would quickly prove fatal. Thus any meaningful description of the effects of a solution must consider quantity, composition, and concentration.

13.1 TYPES OF SOLUTIONS

Solutions have been defined (Unit 9) as homogeneous mixtures of two or more substances. Homogeneity, in this case, refers to readily measurable properties and not to the molecular-level characteristics, which will vary somewhat from one microregion to another.

It is possible to have solutions in any of the states of matter. All mixtures of true gases (Unit 10) are solutions. Solid solutions, such as certain alloys (Unit 11), are not uncommon. Most solutions, however, exist in the liquid state. These may involve gases dissolved in liquids (e.g., NH_3 in H_2O), liquids in liquids (e.g., ethanol in water), or solids in liquids (e.g., "saline solution"—NaCl in H_2O).

Many of the properties of true solutions are also exhibited by the closely related colloidal dispersions (Unit 14). The latter are, in fact, sometimes called "colloidal solutions." The distinctions between true solutions and colloids are more a matter of degree than of kind, and the classification is to a large extent arbitrary. In general, solutions are composed of intimate mixtures of particles of atomic, ionic, or molecular size, with the added limitation that the ions or molecules involved are relatively small. Colloids, on the other hand, contain larger molecules or ions or stable aggregates of smaller particles. Thus an operational classification may be used.

It should be noted that the "borderline" distinction between solutions and colloids is somewhat arbitrary. Some chemists prefer to think of dilute dispersions of large molecules such as proteins, for example, as solutions.

A solution is a homogeneous mixture whose component particles are smaller than $\sim 10^3 \text{ Å}^3$.

This definition may be applied to liquid or gaseous mixtures rather easily by passing the mixtures through filtration systems whose pore size is such as to permit the passage of only true solution components. Solid mixtures require more sophisticated investigation.

There are, of course, numerous "borderline" situations and cases for which other definitions are more appropriate. The above classification is, however, generally useful for aqueous systems.

13.2 SOLUTION TERMINOLOGY

The chemical and physical properties of solutions form the basis for most of our subsequent study of chemistry. A number of important terms are so frequently encountered that their meanings should be understood and remembered. These are summarized in Table 13.1.

Table 13.1   SOLUTION TERMINOLOGY

Aqueous (adj.): Water as solvent.
Concentrated (adj.): High solute-to-solvent ratio.
Concentration (n): Relative amounts of solute and solvent (or solute and total solution).
Concentrate (v): To increase the solute-to-solvent ratio.
Dilute (adj.): Low solute-to-solvent ratio.
Dilute (v): To decrease the solute-to-solvent ratio.
Dilution (n): The addition of more solvent to a solution.
Dissolve (v): To form a solution.
Hydration (n): Formation of clusters (hydrates) of water with other particles.
Insoluble (adj.): Dissolving to a negligible extent.
Saturated (adj.): Maximum solute-to-solvent ratio with solution in equilibrium with solute in another phase.
Solubility (n): The amount of solute that will dissolve in a given amount of solvent.
Soluble (adj.): Dissolving to an appreciable extent.
Solute (n): Usually refers to a minor component (or group of components) of the solution. When a liquid solution is formed, normally gaseous or solid components are considered solutes no matter what their relative amounts.
Solvation (n): Formation of discrete clusters of solute and solvent particles together.
Solvent (n): Usually refers to the major component of the solution or, in some cases, to the component that is normally a liquid when other components were originally gases or solids.

Table 13.1 (Continued)

Water is usually considered the solvent in aqueous solutions, even when it is a minor component.

Supersaturated (adj.): Metastable condition in which solution contains a higher solute-to-solvent ratio than possible at the equilibrium state (saturated). When disturbed sufficiently, a supersaturated solution loses solute until the equilibrium state is attained.

Tincture (n): Alcohol (usually ethanol) as solvent.

13.3 SOLUTION FORMATION AND COMPOSITION

It would be nice to have a set of simple theoretical procedures for estimating the solubility of a compound in water. Unfortunately, this is not possible. There are a number of competing factors that determine solubility, and it is not yet feasible to estimate the relative importance of these on any simple qualitative basis.

Consider as an example the competitions among interparticle forces (Figure 13.2) and among the thermodynamic factors (Figure 13.3) for the relatively simple case of sodium chloride in water.

Thermodynamic quantities have been tabulated for a number of systems involving changes associated with formation of aqueous solutions. From data such as lattice energies, heats of hydration, and so on, it is possible to calculate changes in entropy and enthalpy for a number of solution processes. Of perhaps greater importance, such quantities can be used to estimate the conditions necessary for equilibria involving saturated solutions (Unit 18). These data, unfortunately, are limited to a relatively small number of systems, and they often require calculations more demanding than are justified by the need for semiquantitative estimates. The common empirical solubility rules have been described previously (Unit 6 for ionic compounds and Unit 9 for covalent compounds). These rules should be reviewed as they apply to the formation of aqueous solutions.

Figure 13.2 Competing interparticle forces in the sodium chloride–water system. Solvent–solvent forces (hydrogen bonds) and solute–solute forces (ionic bonds) favor separate solid and liquid phases. Solvent–solute forces (polar molecules oriented around ions) favor solution formation.

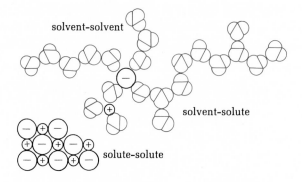

Figure 13.3 Competing thermodynamic factors in the sodium chloride–water system. The destruction of the regular crystal lattice and of water molecule clusters requires energy and leads to a decrease in order ($\Delta H > 0$, $\Delta S > 0$). Formation of more or less regular hydrates releases energy as polar molecules approach ions and leads to more orderly arrays than those of isolated ions and molecules ($\Delta H < 0$, $\Delta S < 0$).

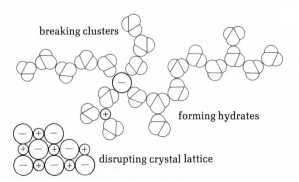

The composition of a solution is determined either by knowledge of the species used to prepare the solution or by *analysis* of a solution whose ingredients are not yet known. As an example of the former case, a solution prepared by adding a small amount of hydrogen chloride gas to water can be described as consisting of water, hydrated protons (H^+), and hydrated chloride ions. This description is based on previous knowledge that hydrogen chloride molecules ionize when dissolved in water (Unit 6). The analysis of solutions of unknown composition may be relatively simple or extremely complicated. Naturally occurring solutions often present a considerable challenge. The venom of the common honeybee, for example, has been under continuous investigation by a number of research groups for over 50 years, but its composition is still only partially described. Some of the more common methods for solution analysis are discussed in Excursion 2.

13.4 SOLUTION PREPARATION AND CONCENTRATION

The concentration of a component of a solution is usually described either in terms of a comparison to the amount of solvent present or to the total amount of solution involved. The difference between these two general types of concentration expressions is reflected in the different ways required for solution preparation or analysis. Solute–solvent concentration units require measurements of the amounts of solute and solvent separately, on some weight-to-weight, weight-to-volume, or volume-to-volume basis. On the other hand, solute–solution concentration units do not require any measurement of the amount of solvent involved. Rather, solute and total solution quantities are determined. In many cases the two different types of concentration expressions may be interconverted if appropriate densities are known. The most common concentration units are listed in Table 13.2.

In preparing solutions one must be aware of three factors:

limited solubilities;   *volume change on mixing;*   *heat of solution*

The first of these applies primarily to planning of solution preparation. It is not possible, for example, to prepare a stable solution whose concentration exceeds that of the saturated solution. Frequently persons preparing reagents heat a mixture to speed up the solution process and are then somewhat chagrined to note that the "solution" placed on the laboratory shelf is now cloudy or consists of two phases. Since solubility usually increases with temperature, they have probably prepared a hot solution of concentration greater than that of the saturated solution at room temperature. Tables of solubility data should always be consulted before unfamiliar solutions are prepared.

An interesting exception is calcium hydroxide, which is more than twice as soluble at 20°C than at 100°C. New stockroom personnel often prepare a solution of $Ca(OH)_2$ that is saturated at 100°C and are surprised to note that all the residual solid dissolves when the mixture is cooled to room temperature.

The fact that there is usually a volume change on mixing has two implications. Only for dilute solutions (and not even then if precise concentrations are important) can one measure the amount of solvent separately for solutions whose concentration is expressed in molarity or a similar solute–solution unit. Instead, the solution is prepared by adding the measured amount of solute to a portion of solvent (such that the initial mixture volume will definitely be less than the desired final solution volume) in a calibrated container (usually a volumetric flask, Figure 13.4). This is mixed thoroughly until it is homogeneous and, for precise work, is adjusted to the calibration temperature of the container. Then additional solvent is added until the bottom of the meniscus (curved liquid–air interface) is exactly level with the calibration mark on the container. This final mixture is shaken or stirred to assure homogeneity.

The second implication applies to solute–solvent concentration units involving mass. In this case (e.g., molality), solute and solvent amounts are measured separately

Table 13.2 COMMON UNITS OF SOLUTION CONCENTRATION

UNIT	WORKING DEFINITION	COMMENTS
RELATIVE SOLUTE-TO-SOLVENT QUANTITIES (solute and solvent amounts measured separately)		
solubility (g per 100 g solvent) (see also Relative Solute-to-Solution Quantities)	$sol = \dfrac{\text{no. g solute}}{\text{no. g solvent}} \times 100$	A common way of tabulating maximum solubility, at a specified temperature, e.g., $(NH_4)_2PtCl_6$ 0.61 (20°C), 1.10 (90°C)
weight percent	$\text{weight \%} = \dfrac{\text{no. g solute} \times 100\%}{\text{no. g (solute + solvent)}}$	10 percent NaOH, e.g., would be 10 g NaOH for each 90 g ($\sim$90 ml) of water
volume percent (ambiguous term, see also Relative Solute-to-Solution Quantities)	$\text{vol \%} = \dfrac{\text{vol solute} \times 100\%}{\text{vol solute + vol solvent}}$	Applies to volumes measured out to prepare solution. Volume change on mixing is common, e.g., 50 percent (vol.) ethanol is made by mixing equal volumes of ethanol and water.
mole fraction	$f_A = \dfrac{\text{no. moles A}}{\text{total no. moles of all components}}$	Used primarily in physical chemistry, e.g., in calculating vapor pressures of solutions. (Some chemists prefer x to symbolize mole fraction.)
molality	$m = \dfrac{\text{no. moles solute}}{\text{no. kg solvent}}$	(See also Unit 14) e.g., a solution containing 46 g of glycerine ($C_3H_8O_3$) and 500 g of water is 1.0 m in glycerine
RELATIVE SOLUTE-TO-SOLUTION QUANTITIES (solvent amount not measured separately)		
solubility (g per 100 g solution) (see also Relative Solute-to-Solvent Quantities)	$sol = \dfrac{\text{no. g solute} \times 100}{\text{no. g solution}}$	An alternative way of tabulating maximum solubility, e.g., NH_4Br 41.1 (15°C), 56.1 (100°F).
weight per volume	$\text{weight/vol} = \dfrac{\text{no. g solute}}{\text{no. vol units of final solution}}$	A solution, e.g., labeled NaCl 12 g/liter contains 12 g NaCl per liter of total solution (amount of water not indicated).
volume percent (ambiguous term, see also Relative Solute-to-Solvent Quantities)	$\text{vol \%} = \dfrac{\text{vol solute used} \times 100\%}{\text{total solution volume}}$	Volume of solute is measured, then sufficient solvent is added to produce final volume desired.
molarity[a] (formality)	$M \text{ (or } F) = \dfrac{\text{no. moles solute}}{\text{no. liters solution}}$	"Moles" refers to number of gram-formula weights of solute, e.g., 6.0 M H_2SO_4 contains 588 g sulfuric acid per liter of total solution.
normality	$N = \dfrac{\text{no. g-eq solute}}{\text{no. liters solution}}$	Depends on specific reaction type for definition of "equivalents." See Section 13.5 and Units 19 and 22.
parts per million	$ppm = \dfrac{\text{no. mg solute}}{\text{no. kg solution}}$	For dilute aqueous solutions (density $\approx$ 1.0 g ml^{-1}), $ppm = \dfrac{\text{no. mg solute}}{\text{no. liters solution}}$

[a] Some chemists use *formality* to refer to the number of gram formula weights (moles) of solute *used*, per liter, *in solution preparation*, reserving *molarity* for *actual concentrations* of species in the final solution. For example, if we dissolved 0.100 gfw of acetic acid (a *weak acid*, Unit 3) in sufficient water to form 1.00 liter of solution, we could label this solution 0.100 F CH_3CO_2H. Since acetic acid is partially ionized in solution, the final solution actually contains less than 0.100 mole per liter of *molecular* acetic acid. Analysis of the species present, in this particular case, would permit us to show "0.099 M CH_3CO_2H, 0.0013 M $CH_3CO_2^-$, 0.0013 M H^+." The distinction between *formality* and *molarity* is important only for certain situations (e.g., aqueous equilibria, Units 19–21). We shall not employ *formality* in our work. Rather, in cases requiring a distinction of this type, we shall use the term

Figure 13.4 Use of volumetric flask. (a) Solute is added to a portion of solvent, slowly, with swirling to mix. (b) More solvent is added with mixing, temperature is adjusted, and final solution is added dropwise. (c) When the bottom of the meniscus is even with the calibration mark, the flask is stoppered and shaken thoroughly (loosen stopper when finished).

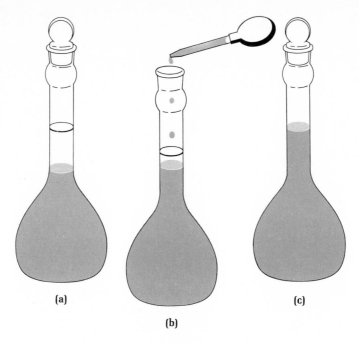

(a)

(b)

(c)

Calibration may involve placing an exactly measured liquid volume in the container at a controlled temperature and marking its meniscus level on the container (e.g., on the neck of a volumetric flask).

and then mixed carefully. The problem is one of preparing a definitely required final solution volume. Unless the amount of volume change for the given solution is known, it is necessary to measure out more solute and solvent than appears necessary, so that any shrinkage on mixing is allowed for.

Heat of solution is a consideration both for accuracy and for safety. When accurate solution volume measurement is required, the solution temperature must be the same as the temperature at which the container was calibrated. Heat is frequently evolved as a solution forms, and sometimes heat is absorbed (e.g., ammonium acetate + water). In such cases it is necessary that a constant temperature (the calibration temperature) be attained before addition of the final solvent increment. In some cases, such as H_2SO_4 or NaOH + water, heat evolution may be sufficient to cause water to boil or the container to break. Such cases require special precautions, especially a slow addition of solute to solvent with stirring to dissipate heat. When preparing unfamiliar reagents, it is always wise to mix a small amount of solute and solvent to determine heat problems.

A very common procedure used in preparing reagent solutions involves quantitative dilution of a concentrated solution. Table 13.3 lists concentrations of the more common commercial reagents. Since volume of concentrated *solution* is involved, *solvent* amount is unknown, but can be calculated if solution density and concentration are known. In the case of units such as molarity, of course, only solution volume is involved.

From the definition of molarity (Table 13.2), it follows that

$$M \text{ (in moles liter}^{-1}) \times V \text{ (in liters)} = \text{no. moles solute}$$

Adding water (dilution) will not change the number of moles of *solute;* that is,

$$\text{no. moles solute (originally)} = \text{no. moles solute (diluted)}$$

Figure 13.5 Transfer pipet (the type shown here is calibrated TD, leaving a small residue in the pipet tip after draining).

Step 1: A rubber suction bulb is used to draw a small portion of solution for transfer into the pipet. Note that the reagent to be used is transferred to a clean container to avoid possible contamination from insertion of a pipet into the stock reagent bottle. A fingertip is placed on top of the pipet to prevent draining, and the pipet is held horizontally and rotated to rinse the interior thoroughly. The rinse solution is discarded (Repeat as necessary.) Then solution is drawn into the pipet above the calibration mark, fingertip pressure is applied to the pipet top, the pipet is removed from solution, the tip is wiped dry with a clean Kimwipe, and the solution is allowed to drain back into beaker (using careful release of fingertip pressure) until the meniscus is level with the calibration mark on the pipet stem. If a drop remains on the tip, it is touched against the top of the beaker to remove the drop.

Step 2: The pipet is transferred to a dilution flask containing a portion of solvent to be used and inserted just inside the neck of the flask with the tip touching the flask interior. Fingertip pressure is removed and the solution is allowed to drain into the flask, with swirling to mix the solution and dissipate heat.

Step 3: A tip residue remains after drainage, usually taking 30 sec for the last residue to drain out. (Some pipets are calibrated for "blowout" of all solution.) Then dilution is completed as in Figure 13.4.

We can therefore write a very useful *dilution equation* as follows:

$$M_{conc} \times V_{conc} = M_{dil} \times V_{dil} \tag{13.1}$$

A moment's reflection will indicate that similar equations can be written for other solute–solution concentration terms, such as normality, formality, or parts per million (ppm).

To measure accurately the volume of concentrated solution to be used, one must employ a liquid transfer device calibrated TD (to deliver) rather than TC (to contain), equipment such as a graduated cylinder. Usually a transfer pipet (Figure 13.5) is used. If approximate concentrations are satisfactory, less accurate measurements may be used. In any case, the precautions described earlier must be taken to allow for volume change on mixing and heat of solution (in this case, heat of dilution).

Table 13.3 CONCENTRATIONS AND DENSITIES OF COMMON COMMERCIAL ("CONCENTRATED") REAGENTS

SOLUTION	MOLARITY	DENSITY AT 20°C (g ml^{-1})
acetic acid ("glacial")	17.4	1.05
ammonia, aqueous (sometimes labeled ammonium hydroxide)	14.8	0.90
hydrochloric acid (36%)	12.2	1.19
nitric acid (70%)	15.7	1.41
phosphoric acid (85%)	14.8	1.70
sulfuric acid (95%)	17.8	1.84

Example 1 (Solution preparation: Solute–solvent units)

A laboratory stock bottle, used to contain about 250 ml of 5 percent sulfuric acid (w/w), must be refilled. How would you prepare the reagent?

Solution

The problem is simplified by the facts that the concentration is expressed to only one significant figure and the exact solution volume is not critical ("about" 250 ml).

Since the solution is fairly dilute, no large error would be made in assuming that its density is about that of water (1.0 g ml^{-1}). Thus

$$\text{total solution weight} \approx \frac{1.0\ \text{g}}{\text{ml}} \times 250\ \text{ml} \approx 250\ \text{g}$$

Then

$$\text{weight } H_2SO_4 \text{ required} \approx 250 \text{ g} \times \frac{5}{100} \approx 12.5 \text{ g}$$

$$\text{weight } H_2O \text{ required} \approx 250 - 12.5 \approx 237.5 \text{ g}$$

The actual procedure would involve weighing about 12.5 g of concentrated sulfuric acid and adding it slowly to about 238 ml of water, with careful stirring. The mixing should be done in a Pyrex container, not the typical "soft-glass" bottle used for reagent storage. The solution would be transferred through a funnel to the reagent bottle when it was sufficiently cool to avoid any stress on the glass of the container.

Example 2 (Solution preparation: Solute–solvent units)

How would you prepare about 150 ml of a 0.500 m glucose solution? (Glucose may be represented as $C_6H_{12}O_6$.)

Solution
Again, the total solution volume is not critical, so the shrinkage expected on mixing can be tolerated. (If we really needed 150 ml and did not know the density of the final solution desired, we could calculate the amounts for preparation of an excess, e.g., 175 ml, and be reasonably sure we would have enough.) We shall base our calculation on using 150 ml of water and settle for whatever total solution volume actually results.
From Table 13.2,

$$m = \frac{\text{no. moles solute}}{\text{no. kg solvent}}$$

Thus,

no. moles solute $= m \times$ no. kg solvent   (the density of water is 1.00 g ml^{-1}).

$$\frac{\text{no. moles}}{\text{solute}} = \frac{0.500 \text{ mole}}{\text{kg } (H_2O)} \times \frac{150 \text{ ml } (H_2O)}{1} \times \frac{1.00 \text{ g } (H_2O)}{1.00 \text{ ml } (H_2O)} \times \frac{1 \text{ kg}}{10^3 \text{ g}} = 0.0750 \text{ mole}$$

Formula weight of glucose ($C_6H_{12}O_6$):

$$\begin{array}{ll} C_6\text{: } 6 \times 12.011 = & 72.066 \\ H_{12}\text{: } 12 \times 1.008 = & 12.096 \\ O_6\text{: } 6 \times 15.999 = & \underline{95.994} \\ & 180.156 \quad \text{(approx 180)} \end{array}$$

Then

$$\text{weight glucose required} = \frac{180 \text{ g}}{\text{mole}} \times 0.0750 \text{ mole} = 13.5 \text{ g}$$

The solution would then be prepared by mixing 13.5 g of glucose with 150 ml of water. Experience indicates that no special precautions are needed to avoid heat problems in this case. In the absence of such experience, every case should be treated as though heat problems might occur. Note that although we do not know the total volume of the solution very accurately (until we prepare it), we do know its concentration to three significant figures.

Example 3 (Solution preparation: Solute–solution units)

How would you prepare 500.0 ml of 0.150 M sodium oxalate?

Solution

(Note that the use of concentration units such as molarity permits accurate measurement of total solution volume, but not of solvent volume unless the solution density is known.)

From Table 13.2,

$$M = \frac{\text{no. moles solute}}{\text{no. liters solution}}$$

Then

$$0.150 = \frac{\text{no. moles solute}}{0.5000}$$

$$(500.0 \text{ ml} = 0.5000 \text{ liter})$$
$$\text{no. moles solute} = (0.150)(0.5000)$$
$$= 0.0750 \text{ mole}$$

Formula weight of sodium oxalate ($Na_2C_2O_4$):

$$
\begin{aligned}
Na_2\colon 2 \times 22.99 &= 45.98 \\
C_2\colon 2 \times 12.01 &= 24.02 \\
O_4\colon 4 \times 16.00 &= \underline{64.00} \\
&\ 134.00
\end{aligned}
$$

Then

$$\text{weight } Na_2C_2O_4 \text{ required} = \frac{134.0 \text{ g}}{\text{mole}} \times 0.0750 \text{ mole}$$

$$= 10.05 \text{ g}$$

The solution would then be prepared by adding 10.05 g of sodium oxalate, slowly with swirling, to about 400 ml of water in a 500-ml volumetric flask. After all solute had dissolved and the solution had reached the flask's calibration temperature, water would be added to bring the total volume to the 500.0-ml calibration mark. The final solution would be stoppered and shaken thoroughly to ensure homogeneity.

(In cases in which severe heat problems may occur, the initial mixing may be performed in a beaker. After this solution has cooled, it is poured through a funnel into the volumetric flask, being careful to rinse the beaker and funnel thoroughly, transferring all rinsings into the flask before final dilution.)

Example 4 (Solution preparation by dilution)

What volume of concentrated (85 percent) phosphoric acid is required to prepare 5.00 liters of 0.25 M H_3PO_4?

Solution

According to Table 13.3, 85 percent phosphoric acid is 14.8 M. Then, using equation (13.1),

$$M_{\text{conc}} \times V_{\text{conc}} = M_{\text{dil}} \times V_{\text{dil}}$$

$$14.8 \ M \times V_{conc} = 0.25 \ M \times 5.00 \ \text{liters}$$

$$V_{conc} = \frac{0.25 \ M \times 5.00 \ \text{liters}}{14.8 \ M}$$

$$= 0.0845 \ \text{liter} = 84.5 \ \text{ml}$$

Example 5 (Interconversion of concentration units)

A bottle is labeled "constant-boiling hydrochloric acid, 20.2 percent HCl by weight, density 1.096 g ml^{-1}." Calculate (a) the molality and (b) the molarity of the HCl.

Solution

a. Molality

$$m = \frac{\text{no. moles solute}}{\text{no. kg solvent}}$$

The formula weight of HCl = 1.008 + 35.453 = 36.461.

Alternatively, we could work with *millimoles* (mmol, Unit 4), in which case

$$M = \frac{\text{no. mmol solute}}{\text{no. ml solution}}$$

$$m = \frac{\text{no. mmol solute}}{\text{no. grams solvent}}$$

$$\text{HCl content} = \frac{1.096 \ \text{g}}{\text{ml}} \times \frac{20.2}{100} \times \frac{1 \ \text{mole}}{36.46 \ \text{g}}$$

$$= 6.07 \times 10^{-3} \ \text{mole ml}^{-1}$$

$$\text{H}_2\text{O content} = \frac{1.096 \ \text{g}}{\text{ml}} \times \frac{(100 - 20.2)}{100} \times \frac{1 \ \text{kg}}{1000 \ \text{g}}$$

$$= 8.75 \times 10^{-4} \ \text{kg ml}^{-1}$$

$$m = \frac{6.07 \times 10^{-3} \ \text{mole ml}^{-1}}{8.75 \times 10^{-4} \ \text{kg ml}^{-1}} = 6.93$$

b. Molarity:

$$M = \frac{\text{no. moles solute}}{\text{no. liters solution}}$$

$$M \ (\text{HCl}) = \frac{1.096 \ \text{g}}{\text{ml}} \times \frac{20.2}{100} \times \frac{1 \ \text{mole}}{36.46 \ \text{g}} \times \frac{1000 \ \text{ml}}{1 \ \text{liter}}$$

$$= 6.07$$

Example 6 (Parts per million)

Water "hardness" is often expressed as parts per million "ppm (Ca^{2+})." What mass of Ca^{2+} must be removed by an ion exchange "water softener" (Unit 6) from 2500 liters of water to reduce the "hardness" from 35 ppm (Ca^{2+}) to 0.10 ppm (Ca^{2+})?

Solution
From Table 13.2,

$$\text{ppm} = \frac{\text{no. mg solute}}{\text{no. kg solution}} \approx \frac{\text{no. mg solute}}{\text{no. liters solution}}$$

Thus the original amount of Ca^{2+} is given by

$$\text{no. mg (Ca}^{2+}) = \text{ppm} \times \text{no. liters solution}$$
$$= 35 \times 2500 = 87{,}500 \ \text{mg} \ (87.5 \ \text{g})$$

and for the "softened" water:

$$\text{no. mg } (Ca^{2+}) = \text{ppm} \times \text{no. liters solution}$$
$$= 0.10 \times 2500 = 250 \text{ mg } (0.25 \text{ g})$$

The mass of Ca^{2+} to be removed, then, is

$$87.5 \text{ g} - 0.25 \text{ g} = 87.25 \text{ g}$$

(87 g to two significant figures).

13.5 NORMALITY: CHEMICAL EQUIVALENTS

When dealing (Section 13.6) with the stoichiometry of reactions in which solution concentrations are expressed in molarity (or formality), it is necessary to have a balanced equation for the process or some other way of determining the mole ratios of species of interest (Unit 4). It would be convenient to simplify the arithmetic of such situations by having a concentration unit such that all ratios were 1:1. *Normality* is such a unit.

If we define the *equivalent weight* of a chemical substance as the mass of the substance, in atomic mass units, involved in the transfer of one unit charge, then it follows that an *equivalent weight* of one reactant will consume exactly an *equivalent weight* of the other reactant.

On a measurable scale, it is convenient to use the gram-equivalent weight (eq) or milligram-equivalent weight (meq) in the same way as the gram-formula weight (mole) or milligram-formula weight (millimole). We may relate these terms mathematically.

$$\text{equivalent weight} = \frac{\text{formula weight}}{n} \tag{13.2}$$

$$\text{no. eq} = \text{no. moles} \times n \tag{13.3}$$

in which n represents the number of unit charges transferred (or, in some cases, "transferable") per formula unit.

The equivalent weight concept is most commonly employed for either acid-base reactions (Unit 19) or oxidation-reduction reactions (Unit 22), although an equivalent weight may also be defined for a simple salt. The important factor in each case is the determination of n, the number of unit charges involved (Table 13.4).

Table 13.4 SUMMARY OF WAYS OF DETERMINING n

CHEMICAL SYSTEM	n DEFINED	EXAMPLES
oxidation-reduction (Unit 22)	number of electrons transferred per formula unit (of a reactant)	$\cdots CrO_4^{2-} + 3e^- \longrightarrow Cr^{3+} + \cdots$ (for CrO_4^{2-}, $n = 3$) $H_2C_2O_4 \longrightarrow 2H^+ + 2CO_2 + 2e^-$ (for $H_2C_2O_4$, $n = 2$)
acid-base (Units 6, 19)	number of protons *transferrable* per formula unit	H_2SO_4 [$n = 2$] H_3PO_4 [$n = 3$] H_2NNH_2 (capable of *accepting* 2 protons, Unit 6, so $n = 2$) $Ca(OH)_2$ (capable of furnishing $2OH^-$ as proton acceptors, Unit 6, so $n = 2$)
salt (not involved in electron transfer or proton transfer)	total *net* charge of either cations or anions of the salt (without regard to algebraic sign)	Na_3PO_4 ($3Na^+$ or PO_4^{3-} shows net charge of 3, so $n = 3$) $Al_2(CO_3)_3$ ($2Al^{3+}$ or $3CO_3^{2-}$ shows net charge of 6, so $n = 6$)

Table 13.5 SELECTED "HALF EQUATIONS" FROM APPENDIX D

$$ClO_4^- + 2H^+ + 2e^- \longrightarrow ClO_3^- + H_2O$$
$$ClO_3^- + 3H^+ + 2e^- \longrightarrow HClO_2 + H_2O$$
$$HClO_2 + 2H^+ + 2e^- \longrightarrow HOCl + H_2O$$
$$Cl_2 + 2e^- \longrightarrow 2Cl^-$$
$$H_2O_2 + 2H^+ + 2e^- \longrightarrow 2H_2O$$
$$MnO_4^- + e^- \longrightarrow MnO_4^{2-}$$
$$MnO_4^- + 2H_2O + 3e^- \longrightarrow MnO_2 + 4OH^-$$
$$MnO_4^- + 8H^+ + 5e^- \longrightarrow Mn^{2+} + 4H_2O$$
$$NO_3^- + H_2O + 2e^- \longrightarrow NO_2^- + 2OH^-$$
$$HNO_2 + H^+ + e^- \longrightarrow NO + H_2O$$

For oxidation-reduction reactions, n is determined by the number of electrons transferred per formula unit of the reactant substance. This is most readily seen from an equation for a "half reaction", for example,

$$MnO_4^- + 8H^+ + 5e^- \longrightarrow Mn^{2+} + 4H_2O$$

($n = 5$ for MnO_4^-).

We will learn how to develop such "half-reaction" equations in Unit 22. For now, we may rely on a set of tabulated equations given in Appendix D (Table 13.5), realizing that a more detailed understanding of oxidation-reduction processes must await subsequent developments. It is important to realize that n (and hence the equivalent weight) is determined by the particular *change* undergone by a chemical species and not (as in the case of formula weight) simply by the chemical formula. For example, permanganate ion may undergo a number of different electron-transfer processes, for each of which there is a different value of n (Table 13.5).

$$\underline{MnO_4^-} + e^- \longrightarrow MnO_4^{2-}$$

$$\left(n = 1, \text{ equivalent weight} = \frac{\text{formula weight}}{1} \right)$$

$$\underline{MnO_4^-} + 2H_2O + 3e^- \longrightarrow MnO_2 + 4OH^-$$

$$\left(n = 3, \text{ equivalent weight} = \frac{\text{formula weight}}{3} \right)$$

$$\underline{MnO_4^-} + 8H^+ + 5e^- \longrightarrow Mn^{2+} + 4H_2O$$

$$\left(n = 5, \text{ equivalent weight} = \frac{\text{formula weight}}{5} \right)$$

Caution must be used in "counting protons" in the formulas of acids. We know, for example, that only one hydrogen is "acidic" in acetic acid (CH_3CO_2H). Two special cases should also be noted: H_3PO_3 (two "acidic" hydrogens) and H_3PO_2 (one "acidic" hydrogen). In general, hydrogen bonded to oxygen, halogen, or sulfur is "acidic". In H_3PO_3 one hydrogen is bonded to phosphorus. Two hydrogens are bonded to phosphorus in H_3PO_2 (Unit 26).

For acid-base reactions, as considered by the Brønsted-Lowry definitions (Unit 6), n is determined by the number of *protons* transferred by the species. Strictly speaking, proton transfer, like electron transfer, should be analyzed in terms of the specific *change* involved. For example, with phosphoric acid (H_3PO_4), n should be 1 for a reaction removing only one of the protons. However, it is common practice in dealing with simple acids or bases to consider the *potential* for proton loss or gain. Thus for a polyprotic ("many-proton") acid, such as H_2SO_3, we simply count the number of protons represented by the formula ($n = 2$). With a base the practice is to consider the number of protons that could react with one formula unit, even when that "formula unit" is not strictly classified as a base in the Brønsted-Lowry system. For example, ethylenediamine ($H_2\overset{..}{N}CH_2CH_2\overset{..}{N}H_2$) has two "basic groups," so $n = 2$ and aluminum hydroxide (strictly classified as a salt containing the base OH^-) could react with three protons; so $n = 3$ for $Al(OH)_3$.

For simple salts not involved in either electron transfer or proton transfer, n is determined by the net positive charge of all cations in the salt or by the net negative charge of all anions (Table 13.4). This permits the use of *normality* for calculations involving precipitation reactions. However, this particular application is rarely used.

Example 7

Calculate the equivalent weight for
a. sulfuric acid for use in an oxidation-reduction reaction in which the appropriate "half equation" is

$$H_2SO_4 + 6H^+ + 6e^- \longrightarrow S + 4H_2O$$

b. sulfuric acid for use in an acid-base reaction
c. sodium sulfate, as a simple salt.

Solution
a. $n = 6$ (from the "half equation")

formula weight $H_2SO_4 = 98.076$

From equation (13.2),

$$\text{equivalent weight} = \frac{98.076}{6} = 16.346$$

b. For H_2SO_4 as an acid, $n = 2$ then

$$\text{equivalent weight} = \frac{98.076}{2} = 49.038$$

c. For Na_2SO_4, $n = 2$ (from $2Na^+$ or SO_4^{2-}), then

$$\text{equivalent weight} = \frac{142.04}{2} = 71.02$$

The concentration unit of normality is defined (Table 13.2) as the number of equivalents per liter (milliequivalents per milliliter). A comparison of the definitions of *molarity* and *normality,* along with the relationship expressed by equation (13.3) then permits us to write

$$N = M \times n \tag{13.4}$$

We can, then, work with normalities in the same way as with molarities. It is important to remember, however, that the *molarity* of a solution is determined only by how it was prepared[1] (i.e., the solute-to-solution ratio), whereas the *normality* of a solution depends also on how the solution is *used.*

Example 8

What mass of potassium permanganate is required for preparation of 250 ml of a 0.150 N solution to be used in an oxidation-reduction reaction in which the MnO_4^- is converted to Mn^{2+} (Table 13.5)? (The K^+ does not participate in the electron-transfer process, but it must be considered in the mass of the compound to be measured.)

---

[1] We are speaking here of the "label concentration," sometimes expressed as *formality* (Table 13.2).

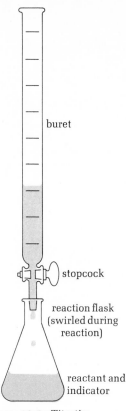

buret

stopcock

reaction flask
(swirled during
reaction)

reactant and
indicator

Figure 13.6 Titration.

Solution
From the definition of normality (Table 13.2),

no. g-eq solute $= N \times$ no. liters of solution
$$= 0.150 \text{ g-eq liter}^{-1} \times 0.250 \text{ liter}$$
$$= 0.0375 \text{ g-eq}$$

The formula weight of $KMnO_4$ is calculated to be 155 and n (from Table 13.5) is 5 for the conversion of MnO_4^- to Mn^{2+}. Thus the equivalent weight of $KMnO_4$ for this process [Equation (13.2)] is

$$\text{equivalent weight} = \frac{155}{5} = 31.0$$

From the definition of gram-equivalent weight, we may then calculate the required solute mass.

$$\frac{31.0 \text{ g}}{\text{g-eq}} \times 0.0375 \text{ g-eq} = \underline{1.16 \text{ g}}$$

Example 9

What volume of "concentrated" sulfuric acid (Table 13.3) is required for preparation of 500.0 ml of 0.60 N H_2SO_4 for use as an acid?

Solution
From Table 13.3, "concentrated" sulfuric acid is 17.8 M. For use as an acid, n will be 2, so [from equation (13.4)],

$$N_{conc} = 17.8 \text{ mole liter}^{-1} \times 2 \text{ g-eq mole}^{-1}$$
$$= 35.6 \text{ } N$$

and, by analogy to the "dilution formula" [Equation (13.1)],

$$N_{conc} \times V_{conc} = N_{dil} \times V_{dil}$$
$$35.6 \text{ } N \times V_{conc} = 0.60 \text{ } N \times 500.0 \text{ ml}$$
$$V_{conc} = \frac{0.60 \text{ } N \times 500.0 \text{ ml}}{35.6 \text{ } N} = \underline{8.4 \text{ ml}}$$

13.6 SOLUTION QUANTITY: TITRATION AND PRECIPITATION

Pure substances taking part in chemical reactions may be measured directly (mass, volume, etc.), but reactions involving solutions require a knowledge of both the amount of solution used, as such, and the *concentration* of the reacting species in the solution. Analytical reactions depending on the stoichiometry of solutions usually employ accurate measurement of solution volume by a delivery device called a buret (Figure 13.6). Solution concentration is determined either by accurate solution preparation or by *standardization* of the solution through reaction with a known amount of some other substance. An understanding of the working definitions of molarity (or formality) and normality will then permit calculation of the number of moles, or gram-equivalents, of solute reacting if the solution volume and concentration used are known.

moles = $M \times$ volume (liters)
millimoles = $M \times$ volume (ml)
(1 mmol = 10^{-3} mole)
gram-equivalents = $N \times$ volume (liters)
milligram-equivalents = $N \times$ volume (ml)
(1 meq = 10^{-3} eq)

Some method must be used to determine when reaction is complete, so that only the solution volume stoichiometrically required is added. The *equivalence point* (stage of complete reaction) for such *titration* processes may be determined by appropriate instruments or, more frequently, by the use of indicators, which are substances that change color when a very slight excess of one reactant is added. The stage of a titration at which the indicator changes color is called the *endpoint*. Obviously, it is desirable to select an indicator such that its endpoint will correspond as closely as possible to the equivalence point of the reaction. The subject of indicators is discussed in detail in Unit 19.

Example 10 (Titration, using molarities)

A 25.0-ml aliquot (portion) of barium hydroxide solution was exactly neutralized by 18.6 ml of 0.104 M HCl. What was the concentration of the barium hydroxide? [Neutralization refers to proton transfer (Unit 6).]

Solution
When molarity is employed as the concentration unit, the stoichiometric ratio involved must be determined (Unit 4). The use of normalities to avoid this necessity will be illustrated in Example 11.
  The complete equation for this reaction may be formulated as

$$Ba(OH)_2(aq) + 2HCl(aq) \longrightarrow BaCl_2(aq) + 2H_2O$$

from which the stoichiometric ratio of interest is

$$\frac{1 \text{ mole } Ba(OH)_2}{2 \text{ moles HCl}} \quad \text{or} \quad \frac{1 \text{ mmol } Ba(OH)_2}{2 \text{ mmol HCl}}$$

From equation (15.2),

no. mmol HCl = 0.104 $M \times$ 18.6 ml
= 1.93

Then

$$\text{no. mmol } Ba(OH)_2 = 1.93 \text{ mmol HCl} \times \frac{1 \text{ mmol } Ba(OH)_2}{2 \text{ mmol HCl}}$$

And

$$M[Ba(OH)_2] = \frac{0.965 \text{ mmol}}{25.0 \text{ ml}} = 3.86 \times 10^{-2}$$

The use of equivalents rather than moles allows us to perform stoichiometric calculations without establishing mole ratios. It is, however, necessary for us to know *what* chemical change is involved for a particular substance so that we can determine the correct value for n.

Example 11 (Titration, using normalities)

What volume of 0.187 N $KMnO_4$ is required for complete oxidation of 25.0 ml of 0.352 N Fe^{2+} solution. The $KMnO_4$ and Fe^{2+} solution concentrations were determined for the electron-transfer process involved in the titration reaction.

Solution
By the definition of chemical equivalents,

$$\text{no. eq } (MnO_4^-) = \text{no. eq } (Fe^{2+})$$

or, for a small-scale process,

$$\text{no. meq } (MnO_4^-) = \text{no. meq } (Fe^{2+})$$

From equations (13.5),

$$N_{MnO_4^-} \times V_{MnO_4^-} = N_{Fe^{2+}} \times V_{Fe^{2+}}$$
$$0.187 \text{ N} \times V_{MnO_4^-} = 0.352 \text{ N} \times 25.0 \text{ ml}$$
$$V_{MnO_4^-} = \frac{0.352 \text{ N} \times 25.0 \text{ ml}}{0.187 \text{ N}}$$
$$V_{MnO_4^-} = \underline{47.1 \text{ ml}}$$

Solution stoichiometry also finds applications in other types of reactions. As in the case of a titration, the relationship between solution concentration and volume provides the key to determining the quantity of chemical reactant used.

Example 12 (Precipitation)

What volume of 0.085 M sodium hydroxide (in milliliters) is required for quantitative removal of the Fe^{3+} from 750 ml of a solution that is 87 ppm in Fe^{3+}, assuming that complete precipitation of $Fe(OH)_3$ does not require excess OH^-?

Solution
From the formula of the precipitate, three OH^- ions are required per Fe^{3+} to be removed. We can write a series of relationships from which appropriate unity factors may be developed.
    To find: milliliters (NaOH)

from definition of M:	1 ml (OH^-) = 0.085 mmol (OH^-)
from definition of ppm:	1 liter (Fe^{3+}) = 87 mg (Fe^{3+})
from definition of mmol and atomic weight of Fe:	1 mmol Fe^{3+} = 56 mg (Fe^{3+})
from the formula $Fe(OH)_3$:	1 mmol Fe^{3+} = 3 mmol (OH^-)

Then

$$\frac{1 \text{ ml } (OH^-)}{0.085 \text{ mmol } (OH^-)} \times \frac{3 \text{ mmol } (OH^-)}{1 \text{ mmol } (Fe^{3+})} \times \frac{1 \text{ mmol } (Fe^{3+})}{56 \text{ mg } (Fe^{3+})} \times \frac{87 \text{ mg } (Fe^{3+})}{1 \text{ liter } (Fe^{3+})} \times \frac{0.750 \text{ liter } (Fe^{3+})}{1}$$

$$= \underline{41 \text{ ml } (OH^-)}$$

DRINKING WATER: A REAL SOLUTION PROBLEM

Within recent years serious concerns have been growing with respect to the quality of the water we drink. Even those living on remote farms and ranches, once proud of the excellence of their well or cistern water supplies, have been touched by the contamination seemingly inherent to our industrial society. Many rural wells show significant concentrations of nitrate, an ion of major concern as a health hazard, particularly to babies. High nitrate concentrations in drinking water have been linked directly with damage to the body's oxygen supply system and, in severe cases, with the incidence of "blue babies" (those suffering from major oxygen deficiency). Nitrates from commercial fertilizers may leach through the soil and accumulate in underground water supplies. Rainwater, once considered the ultimate in purity for storage in cisterns, may now contain airborne pollutants ranging from lead salts (primarily from automobile exhaust systems) to significant concentrations of acids (from reaction of atmospheric moisture with oxides of nitrogen and sulfur). Rainwater in the eastern half of the continental United States appears to be increasing in acidity. This is especially noted near heavily industrialized areas, where rainfall is often 100 to 1000 times as acidic as normal rainfall from relatively clean air.

Of even greater concern are the water supplies for major urban centers, where contaminated rivers and lakes may provide the only available large-scale water sources. Residues from sewage treatment plants, wastes from agricultural runoff, massive amounts of industrial chemicals, and even raw sewage contaminate most of our rivers and lakes, and this contamination—in spite of the efforts of environmentalists—is growing worse in many locations. In a recent analysis of the drinking water of a major city in the southern United States, 66 different organic chemicals were identified. Sixteen of these are very toxic, two are known carcinogens (cancer-causing), and eleven are suspected carcinogens. Fortunately, most of the dangerous pollutants in our municipal water supplies appear, at present, in very low concentrations.

About half of the present municipal water systems in this country offer no treatment of drinking water other than chlorination to kill bacteria. Only about 1 percent of the municipal water systems employ carbon filtration, the method now believed to be the most promising for large-scale water purification.

Our supplies of "good" drinking water are dwindling rapidly, yet an apathetic public continues to accept substandard waste treatment and water purification systems.

How much do *you* know about the water you drink. You might be surprised, and even concerned, if you looked into the situation.[2]

[2]For further information of interest, see "What is Happening to Our Drinking Water," by Eugenia Keller, in the February, 1975, issue of *Chemistry*.

SUGGESTED   Dreisbach, Dale. 1966. *Liquids and Solutions*. Boston: Houghton Mifflin.
READING   Moeller, Therald, and Rod O'Connor. 1971. *Ions in Aqueous Systems*. New York: McGraw-Hill.
             O'Connor, Rod, and Charles Mickey. 1974. *Solving Problems in Chemistry*. New York: Harper & Row.

EXERCISES

1. Fill in the missing items in the following:

   a. $F = \dfrac{\rule{3cm}{0.4pt}}{\text{no. liters solution}} = M$

   b. $m = \dfrac{\text{no. moles solute}}{\rule{3cm}{0.4pt}}$

   c. $N = \dfrac{\rule{3cm}{0.4pt}}{\text{no. liters solution}}$

   d. weight % = $\dfrac{\rule{3cm}{0.4pt}}{\rule{3cm}{0.4pt}}$ × 100%

   e. ppm = $\dfrac{\text{no. mg solute}}{\rule{3cm}{0.4pt}}$

2. How would you prepare

   a. 125 ml of 0.30 M nickel(II) acetate from the crystalline tetrahydrate (4H$_2$O) salt? Show calculations and explain in detail.

   b. 25.0 liters of 0.30 M sulfuric acid from the commercial acid (95 percent) (Table 13.3)? Show calculations and describe the procedure to be used.

   c. 100 ml of 0.35 N nitric acid from the commercial concentrated acid (Table 13.3) for use in an oxidation-reduction reaction involving conversion of nitrate to nitrite (Table 13.5)?

3. A bottle is labeled 70% nitric acid (by weight), density 1.41 g ml^{-1}. Calculate

   a. the molarity of the acid

   b. the molality of the acid

   c. the normality of the acid for use as an acid

4. Calculate the equivalent weights of

   a. perchloric acid for use as an acid

   b. magnesium perchlorate as a "simple salt"

   c. perchloric acid for use in an oxidation reaction involving conversion to HClO$_2$ (Table 13.5)

5. A 50.0-ml aliquot of a sodium hydroxide solution was exactly neutralized by 38.6 ml of 0.213 M sulfuric acid. What was the concentration (M) of the original sodium hydroxide?

6. What volume of a 0.30 M Na$_3$PO$_4$ solution is required to precipitate 0.755 g of silver phosphate from a solution of silver nitrate, assuming quantitative precipitation of Ag$_3$PO$_4$ without excess phosphate?

PRACTICE FOR PROFICIENCY

Objective (a)

1. Write operational definitions for
   a. formality; b. molality; c. normality; d. parts per million; e. weight percent

Objective (b)

2. Describe in detail how you would prepare each of the following solutions. Show all necessary calculations.

   a. 250 ml of 0.15 M sodium chromate, from the anhydrous salt

   b. 3.0 liters of 0.60 M hydrochloric acid from the commercial concentrated acid (Table 13.3)

   c. 125 ml of 0.20 N potassium permanganate from the anhydrous salt for use in an oxidation-reduction process involving its conversion to manganese dioxide (Table 13.5)

   d. 25 liters of a 35 ppm (Mg^{2+}) solution from anhydrous magnesium sulfate

   e. 1.00 liter of 0.86% sodium chloride (by weight) [Such a solution is referred to as "physiological saline" (Unit 15).]

Objectives (c, d)

3. A bottle is labeled 85% phosphoric acid (by weight), density 1.70 g ml^{-1}. Calculate

   a. the molarity of the acid

   b. the molality of the acid

   c. the normality of the acid, for use in preparing trisodium phosphate

4. Calculate the equivalent weights of

   a. nitric acid for use as an acid

   b. nitric acid for use in an oxidation-reduction reaction involving its conversion to nitric oxide (NO, Table 13.5)

   c. aluminum nitrate as a "simple salt"

   d. hypochlorous acid for use in an oxidation-reduction process involving its conversion to perchloric acid

   e. phosphoric acid for use as an acid

   f. phosphorous acid for use as an acid

   g. hydrazine (H$_2$NNH$_2$) for use as a base

5. For each of the following solutions, calculate the molarity, the molality, and the normality (for use in acid-base reactions).

   a. 40.0% HNO$_3$, by weight (density 1.25 g ml^{-1})

   b. 60% acetic acid, by weight (density 1.06 g ml^{-1})

   c. 25.0% sulfuric acid, by weight (density 1.18 g ml^{-1})

   d. 30.0% phosphoric acid, by weight (density 1.18 g ml^{-1})

   e. 20.0% aqueous ammonia, by weight (density 0.922 g ml^{-1})

6. Calculate the solute mass required (in grams) for preparation of

   a. 100 ml of 0.90 M solution from MgSO$_4 \cdot$ 7H$_2$O

   b. 300 ml of 0.50 M solution from Fe(NO$_3$)$_2 \cdot$ 6H$_2$O

   c. 75 ml of 0.60 M solution from CuSO$_4 \cdot$ 5H$_2$O

   d. 100 ml of 0.15 M solution from AlCl$_3 \cdot$ 6H$_2$O

   e. 50 ml of 0.94 M solution from KAl(SO$_4$)$_2 \cdot$ 12H$_2$O

   f. 250 ml of 0.20 N MnO$_4^-$ from KMnO$_4$ for use in an oxidation-reduction reaction involving conversion of MnO_4^- to Mn^{2+} (Table 13.5)

g. 5.0 liters of 0.030 N $Ba(OH)_2$ for use in an acid-base reaction

h. 5.0 liters of 0.30 N sodium oxalate solution for use in an oxidation-reduction reaction involving

$$C_2O_4{}^{2-} \longrightarrow 2CO_2 + 2e^-$$

i. 12 liters of 15 ppm $(NO_3{}^-)$ solution from potassium nitrate

j. 250 ml of 2.0 m glucose $(C_6H_{12}O_6)$ solution (density = 1.14 g ml^{-1})

7. Calculate the volume of the specified concentrated solution required for preparation of

a. 250 ml of 0.60 M H_2SO_4 from 18 M H_2SO_4

b. 3.0 liters of 0.36 M HCl from 12 M HCl

c. 750 ml of 0.78 N H_3PO_4 (for use as an acid) from 15 M H_3PO_4

d. 500 ml of 0.15 N H_2O_2 (for use in an oxidation-reduction reaction involving conversion of H_2O_2 to water [Table 13.5]) from 13 M H_2O_2

e. 12 liters of 15 ppm $(NO_3{}^-)$ from 0.30 M HNO_3

Objective (e)

8. What volume of 0.50 M phosphoric acid is required for complete conversion of 25.0 ml of 0.10 M barium hydroxide to barium phosphate?

9. Magnesium oxide, marketed as tablets or as an aqueous slurry called "milk of magnesia," is a common commercial antacid. What volume, in milliliters, of fresh gastric juice, corresponding in acidity to 0.17 M HCl, could be neutralized by 125 mg of magnesium oxide?

10. A 50.0-ml sample of a commercial "vinegar" (dilute acetic acid) was neutralized by titration with 41.6 ml of 0.814 M sodium hydroxide. How many grams of acetic acid were present in 2.50 liters of this "vinegar"?

11. "Baking soda," $NaHCO_3$, used to be a common remedy for "acid stomach." What weight of baking soda would be required to neutralize 85 ml of gastric juice, corresponding in acidity to 0.17 M HCl?

12. What volume of 0.185 N potassium permanganate would be needed for complete reaction with 25.0 ml of 0.212 N potassium nitrite? (The normalities given are appropriate for the oxidation-reduction reaction involved.)

13. What volume of 0.30 N nitric acid is required for complete oxidation of 16 g of copper? (The copper is involved as

$Cu \longrightarrow Cu^{2+} + 2e^-$. The normality of the acid is appropriate for the process.)

14. A commercially valuable pigment, "chrome yellow," can be prepared by the precipitation reaction,

$$Pb^{2+}(aq) + CrO_4{}^{2-}(aq) \longrightarrow PbCrO_4(s)$$

How many liters of 0.25 M sodium chromate solution would be required for preparation of 2.2 kg of "chrome yellow," assuming quantitative precipitation?

15. Nickel(II) hydroxide is used in the manufacture of certain flashlight cells. What volume of 0.30 M sodium hydroxide is needed for preparation of 175 g of $Ni(OH)_2$, assuming quantitative precipitation from a solution containing Ni^{2+}?

16. Barium sulfate is employed as an opaque material for x-ray studies. What volume of a solution labeled 0.36 N H_2SO_4 (for acid-base reaction) would be needed to prepare 30 g of $BaSO_4$, assuming quantitative precipitation from a solution of $Ba(OH)_2$?

17. What volume of 0.15 M sodium carbonate solution would be required for removal of essentially all the Ca^{2+} from 6.0 liters of 78 ppm (Ca^{2+}), assuming quantitative precipitation of $CaCO_3$?

BRAIN TINGLERS

18. When iron pyrite is treated with nitric acid, the reaction observed may be formulated as

$$FeS_2(s) + 18HNO_3(aq) \longrightarrow$$
$$Fe(NO_3)_3(aq) + 15NO_2(g) + 2H_2SO_4(aq) + 7H_2O(l)$$

What volume of concentrated nitric acid (17.4 M) would be required to consume 1.6 kg of the pyrite, assuming 100% efficiency?

19. The etching of glass by hydrofluoric acid may be represented, on a simplified basis, by the reaction of "silica" with HF:

$$SiO_2(s) + 6HF(aq) \longrightarrow H_2SiF_6(aq) + 2H_2O(l)$$

What mass of "silica" would be consumed by 250 ml of 6.5 M HF, assuming 100% efficiency?

20. Copper does not react with nonoxidizing acids, so the artistic etching of copper items is normally done with nitric acid:

$$3Cu(s) + 8HNO_3(aq) \rightarrow 3Cu(NO_3)_2(aq) + 2NO(g) + 4H_2O(l)$$

What mass of copper would be consumed by 125 ml of 8.5 M nitric acid, assuming 100% efficiency?

Unit 13 Self-Test

1. Fill in the missing items:

   a. $N = \dfrac{}{\text{no. liters solution}}$

   b. $m = \dfrac{\text{no. moles solute}}{}$

   c. $M =$

2. What is the molality of an aqueous nitric acid solution that is 70% HNO_3 by weight if its density is 1.41 g ml^{-1}?
3. What weight of sodium chromate is required to prepare 150 ml of 0.40 M solution?
4. What volume of glacial acetic acid (17.4 M) is required to prepare 8.7 liters of 0.60 M acetic acid?
5. What is the equivalent weight of
   a. $KClO_3$ (for use as $ClO_3^- + 5H^+ + 4e^- \longrightarrow HOCl + 2H_2O$)
   b. H_3PO_4 (as an acid)
   c. $Al_2(SO_4)_3$ (as a "simple salt")
6. A solution is labeled 44.0% H_2SO_4, by weight, (density 1.34 g ml^{-1}). What is
   a. the molarity of H_2SO_4
   b. the molality of H_2SO_4
   c. the normality of H_2SO_4 (as an acid)
7. A 25.0-ml aliquot of barium hydroxide is exactly neutralized by 21.2 ml of 0.186 M sulfuric acid. What was the molarity of the original barium hydroxide?
8. Aluminum hydroxide, sold under such names as Amphogel or Creamalin, is a valuable commercial antacid. What mass of aluminum hydroxide would neutralize 150 ml of fresh gastric juice, corresponding in acidity to 0.17 M HCl?

   $$Al(OH)_3 + 3HCl \longrightarrow AlCl_3 + 3H_2O$$

   a. 0.55 g; **b.** 0.44 g; **c.** 0.66 g; **d.** 0.33 g; **e.** 0.77 g

Unit 14

Colloids

Objectives

(a) Be able to relate common colloidal systems and the types of colloids as classified by continuous and discontinuous phases (Figure 14.1).

(b) Be able to give common examples of the various types of colloidal systems and to classify colloids by type (Figure 14.1).

(c) Be able to describe factors contributing to colloid stability and to suggest ways of preparing or destroying specific colloidal dispersions (Sections 14.4–14.6).

In Biblical times, Divine intervention was necessary for a really spectacular case of air pollution or ruination of the rivers. Thanks to modern technology, we can get along on our own now, and our "stinking rivers" and "smoking country" make these early records seem insignificant by comparison.

Much of the pollution of modern society has already been described in terms of solid wastes and solutions—both aqueous and gaseous. However, the enormous problems of water pollution by detergents and air pollution by particulate matter or smog belong in the area of colloid science.

In this unit we shall examine the types of colloidal dispersions (Figure 14.1), the ways they are formed, the factors contributing to their stability, and some methods by which they can be destroyed. These topics have considerable significance in dealing with various aspects of air and water pollution.

Figure 14.1 Common types of colloids.

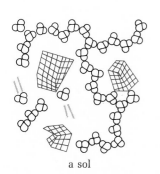

an aerosol smoke

a liquid foam

Common name	Colloidal particle (discontinuous phase)	Suspending medium (continuous phase)	Examples
aerosol (fog)	liquid	gas	insecticide sprays, paint sprays, fog
aerosol (smoke)	solid	gas	tobacco smoke, fine dust in air
foam (liquid)	gas	liquid	"head" on a beer, whipped cream
emulsion	liquid	liquid	"homogenized milk," mayonnaise
sol	solid	liquid	detergents in water, many paints
foam (solid)	gas	solid	marshmallows, Ivory soap
"colloidal solid"	solid	solid	many alloys, "smoky quartz"
gel	liquid–solid phases interdispersed (both phases continuous)		Jello, jellies

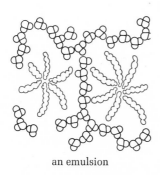

a sol

an emulsion

a gel

14.1 COLLOIDAL DISPERSIONS   The distinction between heterogeneous and homogeneous mixtures is an arbitrary one at best. On the one hand, there are the obviously heterogeneous mixtures such as sand and gravel together on a beach. Viewed closely, such a mixture readily reveals some regions of sand and other regions of large pebbles, possibly of distinctively different colors. From a modern jetliner at 30,000 feet, however, the beach might appear as a smooth homogeneous expanse. Like so many things, "it's all in how you look at it." An aqueous solution of copper(II) chloride appears homogeneous on

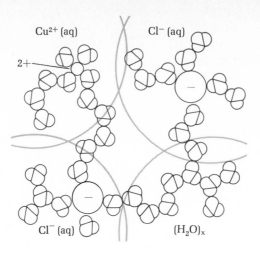

Figure 14.2 Submicroscopic heterogeneity in solution.

Cu^{2+} (aq)   Cl$^-$ (aq)

2+

Cl$^-$ (aq)   (H$_2$O)$_x$

examination under the most powerful microscope. Indeed, samples of the solution selected randomly from different regions can be shown to contain the same relative amounts of Cu^{2+}, Cl$^-$, and H$_2$O, within the limits of ordinary means of analysis. *Solutions* are, by definition, homogeneous mixtures, yet at the molecular level evidence suggests submicroscopic heterogeneity (Figure 14.2). Apparently, mixtures range over a continuum of heterogeneities, from obviously "rough" mixtures to pure solutions.

Somewhere between the rough heterogeneous mixture and the true solution is a class of mixtures called *colloidal dispersions*. Such mixtures are metastable (i.e., they will eventually revert to more stable systems, such as separate phases). The components of colloidal dispersions are not separable by most of the procedures used for separating rough mixtures. For example, aqueous colloids remain unchanged by simple filtration or centrifugation. Relatively recent techniques such as membrane filtration and ultracentrifugation (very high centrifugal forces) can frequently effect such separations, although even these techniques do not separate components of true solutions. This difference in separability is a reflection of differences in particle sizes. Particle size, then, is used as the principal criterion in defining colloids:

> Colloids are mixtures in which one or more species exists as particles of about 10^{-4}–10^{-7} cm in diameter.

Some colloids remain stable for very long periods. A colloidal gold dispersion prepared more than 100 years ago is still on display in a major European museum.

14.2 SOME TYPES OF COLLOIDS

Like solutions, colloidal dispersions may be considered as originating with mixtures of various phases, such as solid in solid, solid in gas, solid in liquid, liquid in solid, liquid in gas, liquid in liquid, gas in solid, gas in liquid. The particle size limitations for the gas phase suggest that mixtures consisting of true gases are best classed as solutions. It is, of course, possible to have even more complex systems by various mixtures of gas, liquid, and solid together. Figure 14.1 illustrates some of the more common types of simple colloids.

In addition to the classification based on types of phases present, colloids may also be distinguished as either *lyophobic* or *lyophilic*.

The term *lyo* refers to the suspending medium. The more familiar *phobic* (Gr., fearing) and *philic* (Gr., loving) serve to indicate whether or not the dispersed particles have a weak (lyophobic) or strong (lyophilic) affinity for the suspending medium. This distinction has an experimental basis. In general, lyophilic colloids are fairly easy to prepare, quite stable, and reasonably simple to reconstitute if the colloidal system has been broken up. The lyophobic colloids are generally less stable and are exceptionally difficult to reconstitute.

Soap dispersed in water is a common example of a lyophilic system. Oil suspended in water, perhaps by use of an ultrasonic dispersion technique, represents a typical lyophobic colloid.

Figure 14.3 Tyndall effect. First noted in 1857 by Michael Faraday and thoroughly investigated by an English physicist, John Tyndall, around 1869.

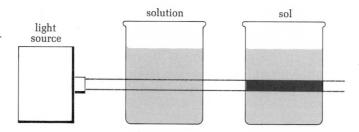

light source — solution — sol

14.3 SOME PROPERTIES OF COLLOIDS

Named in honor of Robert Brown, who first reported the "dancing" motion of pollen grain fragments suspended in water (1827).[1] Three Nobel prizes were awarded within two years for basic studies related to Brownian motion: Kerker, Milton, *J. Chem. Ed.,* **51,** 764 (1974).

First noted in 1857 by Michael Faraday and thoroughly investigated by an English physicist, John Tyndall, around 1869.

To demonstrate this, calculate the surface area of a cube having an edge length of 1.00 cm and compare this with the total surface area that would result if the original cube were cut into cubic fragments of edge length 1.00×10^{-4} cm.

Probably the most familiar property of colloids is the so-called *Tyndall effect,* in which a ray of light forms a visible diffuse beam when passed through an otherwise relatively transparent colloidal dispersion (Figure 14.3). We have all seen this effect—as headlight beams in fog or beams of sunlight through air in a dusty room. In an attempt to study this phenomenon more closely, Siedentopf and Zigmondy developed in 1903 an *ultramicroscope,* which directs a fine beam of bright light through a colloidal sample on a microscope stage. At high magnification, tiny points of light are observed bouncing in erratic random motion. This characteristic, observed with all types of transparent or translucent colloids having a fluid (gas or liquid) continuous phase, is called *Brownian motion.*

The bright spots observed with the ultramiscroscope were once thought to be the dispersed particles themselves. We know now that only the larger colloidal particles can be seen through an ordinary microscope. The bright spots in typical small-particle colloids result from the scattering of light by the particles, and the "spot" size is larger than that of the actual scattering particle.

Light scattering by colloids is a function of the particle size of the dispersed material and of the wavelength of light involved. It has been observed that larger particles scatter light of longer wavelength than smaller particles. A gold sol may be prepared which is ruby red (shorter wavelengths scattered so that principally red light is transmitted). If the particles are caused to aggregate, by heating or addition of a salt, the sol becomes blue. These larger dispersed particles scatter the longer wavelengths, transmitting blue light. The brilliant colors of a desert sunset are due to the scattering of short wavelengths by fine colloidal dust in the air. Similar sunsets near the ocean involve colloidal water droplets in the atmosphere.

The slow diffusion rates in colloids, as compared to those in true solutions, and the ability to separate dissolved species from colloids by ultrafiltration through membranes are also related to the sizes of the particles involved. The enormous increase in total surface area associated with the conversion of a macroscopic object to an equivalent mass of colloidal-size particles is also significant. The possibilities for surface adsorption are fantastically increased.

14.4 PREPARATION OF COLLOIDS

Since the principal distinguishing feature of the colloidal state is a defined range of particle sizes, the problem of preparing a colloidal dispersion becomes either (a) how

[1] See the article by David Layton in the *Journal of Chemical Education* (Vol. 42, p. 367, 1965).

to break up large particles into ones that are small enough to form a colloidal dispersion without going so far as to obtain a true solution or (b) how to increase the particle size of one or more components of a solution to colloidal dimensions without going so far as to obtain a rough heterogeneous mixture.

In some cases, such as the many types of colloidal dispersions occurring in living systems (e.g., the hemoglobin system of the blood), the inherent properties of the dispersed molecules and ions are such as to favor formation of stable dispersions. All that is necessary is the mechanism for manufacture of the proper species in the correct environment. Fortunately, biochemical processes do occur properly and quite efficiently, even though acceptable chemical explanations are available for only a few of these processes.

The preparation of colloidal dispersions in the laboratory or in industrial situations is, to a large extent, still a trial-and-error procedure, since many of the factors involved in the formation of stable colloids are still incompletely understood.

The most common method of preparing colloids involves a mechanical dispersion of two immiscible substances. Sometimes this is a very simple process.

For example, the manufacturers of a well-known soap have made an enormous profit from the simple process of whipping air bubbles into the liquefied soap so that when it solidifies a stable foam (gas dispersed in solid) results. The advantage, beyond the rather obvious one of finding the dropped bar more easily on the surface of the bath water, lies primarily in a simple advertising gimmick—a slogan associating purity with the "floating" of the soap. The implication to the average housewife (although carefully *implied,* rather than stated) is that any soap that does not float is not pure. Actually, of course, the purity of the soap has nothing to do with its density—the air bubbles take care of that. The purity alone could have made the soap worth buying, but the slogan was remarkably "catchy."

In other cases, such as the formation of stable oil-in-water dispersions, simple mechanical agitation is inadequate to break oil droplets into small enough particles. Ultrasonic generators (which produce very high-frequency sound waves) are now employed for these systems, with applications ranging from "ultrasonic cleaners" for research glassware to the manufacture of "one-phase" salad dressings.

The preparation of colloids by particle growth is, perhaps, more frequently seen in accidental, rather than planned, situations. Fog, smoke, smog, and water pollution by detergents and various industrial waste products are all examples of massive colloidal systems that pose major problems for modern society. Incomplete combustion, as in the internal combustion engine, forms smokes as carbon atoms combine to form colloidal particles and "fogs" as unburned or partially decomposed hydrocarbon fuel residues combine into fine droplets. Heavy industry belches forth tons of particulate (solid) waste into the air, along with various types of fog producers—including such substances as SO_2, which is strongly hydrated to form tiny water droplets of solvated sulfurous acid. The combination of different smokes and fogs (including *photochemical* fogs produced by light-catalyzed reactions of many chemical by-products) results in the smog now considered one of the most adverse side effects of modern technology.

Science has come a long way from the days when preparing a gold sol by an electric arc between two gold electrodes under water was a fascinating demonstration of growing particles limited to colloidal size. Now we have too many colloids in our environment, and attention must be centered on their removal and prevention.

14.5 STABILIZATION OF COLLOIDS

If, as has been suggested, colloids are *metastable* systems, how can the prolonged stability of so many dispersions be explained? What prevents these suspended particles from disolving or growing larger to form readily separable phases?

The inherent rigidity of nonfluid colloids, such as solid foams or colloidal alloys, is, of course, the principal factor determining their stability. One of the earliest observations on fluid colloidal systems was the appearance of the continuous jostling motion of the suspended particles, *Brownian motion,* now interpreted as the result of collisions of tiny particles of atomic or molecular dimensions with the larger colloidal aggregates. Although this irregular bouncing probably hinders the coagulation of colloidal particles, it is not believed to be a major factor. For all but the smallest colloids, thermal convection currents and slow rates of sedimentation are more important.

Probably the most common situation contributing to colloid stability is the accumulation of net charge of like sign on dispersed particles, so that further particle growth is inhibited by electrostatic repulsions. Since the suspended particles have the same composition, it is not surprising that they tend to accumulate the same kind of charge.

In some cases, large polyatomic ions such as protein "molecules" or soap anions (Figure 14.4) possess a characteristic net charge. Agglomerates of such particles in aqueous media may form *micelles,* in which the nonpolar segments of several "molecules" are squeezed into a common central region as surrounding water molecules form hydrogen-bonded clusters. Such *hydrophobic* ("water-fearing") regions may be surrounded by the charged (*hydrophilic,* i.e., "water-loving") groups of the "molecule-ions," orientated so that they may attain maximum degrees of hydration.

In other cases dispersed particles gain a net positive or negative charge electrostatically by loss or gain of electrons, respectively. Many industrial smokes, for example, contain particles that have become charged in much the same way that a comb picks up a charge when brushed through hair. Such charges are gradually lost, more rapidly in areas of high relative humidity. *Adsorbed* charges (charges distrib-

Figure 14.4 Soap micelle in water.

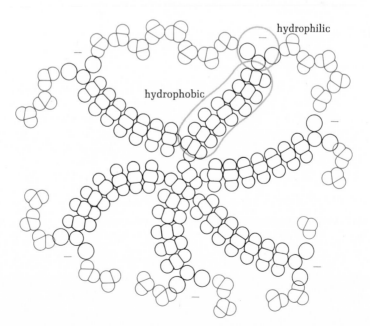

Figure 14.5 Stabilized sols. (a) Iron(III) oxide sol. These sols, often stabilized by adsorbed H+ ions, contribute to the faint orange color of many natural water supplies. (b) Arsenic(III) sulfide sol. Sulfide precipitations, especially in alkaline solutions, are frequently complicated by formation of sulfide sols stabilized by adsorbed OH- ions.

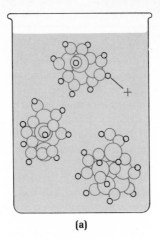

(a)

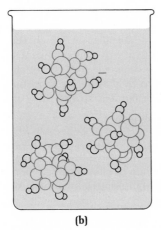

(b)

uted primarily on particle surfaces) are also common in aqueous colloids. Here, however, the charges are usually associated with simple ions that cling to the colloidal particles (Figure 14.5).

Still another type of net charge accumulation occurs in certain types of "protective" colloids, the most common examples being dispersions of oils and greases by aqueous detergents. Detergents are large ions or molecules having a significant nonpolar (hydrophobic) segment and one or more groups (hydrophilic) that can be strongly hydrated. The detergent particles act by mixing their hydrophobic segments with small clusters of the nonpolar substance (e.g., a hydrocarbon) in such a way that their hydrophilic groups project from the microglobule and are hydrated by water. The resulting micelles, like those in soap dispersions themselves (Figure 14.4), thus have a nonpolar region surrounded by hydrated groups. When the latter are ionic, the effect is similar to, but more stable than, the situations of adsorbed charges. The micelles may remain stable until precipitated by chemical action or destroyed by oxidative or bacterial processes. The massive accumulation of such dispersions in rivers, streams, and lakes has resulted in legislation requiring synthetic detergents to be "biodegradable," so that destruction by bacteria can reduce the concentration of these colloids. Even so, their addition to water resources is still more rapid than their degradation.

Since colloidal particles are very small, they would dissolve more rapidly than larger aggregates. Thus colloidal systems are stable for prolonged periods only when the dispersed substance is very insoluble in the dispersing medium.

14.6 THE DESTRUCTION OF COLLOIDS

In nonionic detergents the hydrophilic group is an uncharged region containing atomic groupings (usually OH) capable of forming hydrogen bonds with water molecules. Colloids "protected" by such detergents are, of course, electrically neutral unless ions are adsorbed from dispersing solutions.

Professor Robert Brasted, an old friend of the author, claims to have heard this same story from a PT-boat officer, relative to sabotage at a naval base in the Pacific. Evidently, saboteurs (or possibly yarn tellers) are widely distributed.

Much current research is directed at the prevention or destruction of colloidal systems responsible for environmental pollution. The problems are very complex and encompass areas of engineering, economics, and politics, as well as the chemistry required. Industries have frequently, although not always, resisted antipollution programs, and one of the largest difficulties is the development of processes that are not only reasonably inexpensive, but that may become profitable by reclamation of present "waste" for useful purposes. It should be understood that pollution is by no means limited to "capitalistic" societies. Indeed, much greater progress in appropriate research areas has been made by "free" industry than by government-controlled industry in most countries.

A number of ways of destroying colloidal dispersions have been developed. The selection and use of appropriate methods and the development of new ones will proceed at a rate determined largely by the allocation of research and development funds in government and industry. There is little doubt that this rate can be accelerated appreciably by pressure from interested and informed citizens.

One method of colloid destruction involves the conversion of large dispersed particles into ones small enough to form true solutions. This technique is quite limited, but a rather interesting example was reported in one of the stories that came out of World War II.

It seems that a local saboteur at a desert outpost had added syrup to the stored gasoline supply. Enough of this syrup was dispersed in the gasoline to render it useless as a fuel. The distillation of the enormous quantities involved with makeshift equipment was out of the question. The problem was solved by pouring water into the gasoline. The water effectively dissolved the dispersed syrup, forming a separate liquid phase which could be drained away, allowing most of the gasoline to be recovered.

A second, and more commonly employed, method involves heat coagulation of colloids. At normal temperatures suspended particles may not collide with sufficient energy to break loose surrounding clusters of the continuous phase. By heating the suspension the frequency of high-energy collisions is increased, and particle growth may occur as effective contact is made between colloidal particles. The sulfur sols often encountered in precipitations using hydrogen sulfide can usually be destroyed by heating. Protein suspensions are particularly sensitive to heat. These macromolecules have very specific geometries in their natural states, which favor formation of stable dispersions. At elevated temperatures portions of the molecules, through thermal vibrations and collisions with more rapidly moving particles in the immediate environment, begin to shift locations. Irreversible changes in geometry (*denaturation*) may occur, which so alter the molecular characteristics that the molecules precipitate (e.g., as in the cooking of egg white).

On a limited scale, ultrafiltration may remove colloidal particles. Certain types of air filtration systems have been designed, for example, to remove colloidal radioactive materials from air systems in radioisotope research laboratories or for fallout shelters. These filters are expensive and require frequent replacement as pores become clogged, so they are not feasible for large-scale industrial applications. Filter clogging is the principal problem in removal of colloidal pollutants from large volumes of air or water by filtration methods.

Still another method for colloid destruction utilizes the neutralization of adsorbed or inherent charges that, while present, prevent the coagulation of colloidal particles by electrostatic repulsion. Such techniques include:

Neutralization by added ions

Soap micelles, including those containing nonpolar "protected" molecules, may be precipitated by acidifying the solution (addition of proton):

$$CH_3(CH_2)_nCOO^- + H^+ \longrightarrow CH_3(CH_2)_nCOOH$$

"soluble"   insoluble

or by addition of ions such as Ca^{2+}, which, unlike the Na^+ normally present, form insoluble salts:

$$2CH_3(CH_2)_nCOO^- + Ca^{2+} \longrightarrow Ca(OOC(CH_2)_nCH_3)_2$$

"soluble"   insoluble

These reactions are undesirable when the soap is to be used for cleaning, but they do provide a method for colloid destruction. Most modern synthetic detergents contain ionic groups (sulfonates) that are not affected by H^+ or ions such as Ca^{2+}.

Dispersions stabilized by adsorbed ions can also be precipitated by addition of ions if those of opposite charge can also be adsorbed. Ammonium chloride, for instance, is often used to neutralize negatively charged colloids by adsorption of the positive NH_4^+ ions.

Electrical neutralization

Dispersed particles stabilized by a surplus or deficiency of electrons may be precipitated by passing the dispersion between plates of opposite charge. Positive particles gain electrons, and negative particles lose electrons. Applications of this process range from massive Cottrell precipitators (Figure 14.6), used to recover tons of

Figure 14.6 Cottrell precipitator.

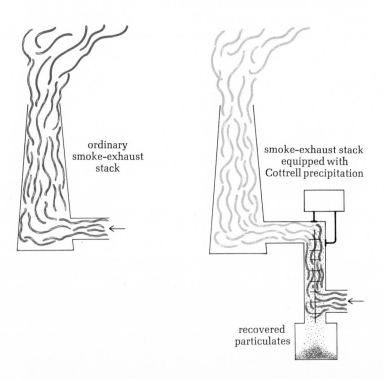

ordinary
smoke-exhaust
stack

smoke-exhaust stack
equipped with
Cottrell precipitation

recovered
particulates

chemicals from various industrial smokes, to the small electrostatic precipitators used in some household air systems to remove such airborne matter as pollens, dust, and certain viruses.

14.7 COLLOIDAL POLLUTANTS

The interested reader may find a frightening, but possibly prophetic, account of some future alternatives in Don Pendleton's *1989: Population Doomsday.*[2] Perhaps the most frightening aspect is the poetic license required to propose a suitably drastic solution to an otherwise realistic problem.

Environmental contamination involves an entire spectrum of particle sizes, from the rusting hulks of junked automobiles to the invisible molecules of carbon monoxide. The larger particles are, of course, the most visible forms of pollution, but they are usually the least hazardous. Belching smokestacks or massive oil spills are rather localized phenomena and relatively simple to control. Dissolved or colloidal substances, because of their unique stability, cause far more serious and widespread problems. Much of the technology necessary to avoid or eliminate the more common colloidal contaminants is well developed. The extent to which this technology will be applied depends largely on the strength of our commitment to an improved environment and our willingness to make the sacrifices necessary to achieve it.

SUGGESTED READING

Parfitt, G. D. 1966. *Surface and Colloid Chemistry.* Elmsford, N.Y.: Pergamon.
Vold, M. J., and R. D. Vold. 1964. *Colloid Chemistry.* New York: Van Nostrand Reinhold.

EXERCISES

1. Give two examples, other than those in the text, of each of the eight types of common colloids (Figure 14.1). Indicate whether you would expect the emulsions selected to represent lyophilic or lyophobic systems.
2. For each of the examples given for Exercise 1, suggest the principal factor (or factors) contributing to stability.
3. Select any four of the examples given for Exercise 1 and suggest how they might be prepared and how they might most easily be destroyed.

PRACTICE FOR PROFICIENCY

1. Classify the following colloidal systems:
   a. Ivory soap
   b. microscopic pollen grains in air
   c. homogenized oil and vinegar salad dressing
   d. an aqueous dispersion of a solid detergent
   e. whipped cream
2. Would system (c) above be classified as lyophilic or lyophobic?
3. Describe the principal factor(s) contributing to stability of the systems in question 1.
4. For any three of the systems in question 1 suggest
   a. a reasonable method of preparation
   b. a method for destroying the colloidal system
5. Read and discuss one of the following articles, or an alternate suggested by your instructor.
   a. "Brownian Movement and Molecular Reality," by Milton Kerker, in the December, 1974, issue of the *Journal of Chemical Education* (pp. 764–768).
   b. "Surface Chemistry in Industrial Processes," by J. Leja, in the March, 1972, issue of the *Journal of Chemical Education* (pp. 157–161).

[2] Don Pendleton, 1970. *1989: Population Doomsday.* New York; Bee-Line Books.

Unit 14 Self-Test

1. Matching: Select the example most likely to correspond to each type of colloid listed.
   COLLOIDS: (A) ammonium chloride in the laboratory atmosphere (B) gelatin (C) hemogenized salad dressing (D) household deodorizing spray (E) Jello in hot water (F) pumice (G) shaving lather
   a. aerosol fog; **b.** aerosol smoke; **c.** liquid foam; **d.** solid foam; **e.** emulsion; **f.** gel; **g.** sol

2. A dispersion of starch in water is an example of a lyo_____ system, whereas a salad oil–vinegar emulsion is probably lyo_____.

3. Indicate the principal factor (rigidity, surface charge, or "other factors") that probably contribute to the stability of each of the systems listed ("other factors" include Brownian motion, slow sedimentation rates, and thermal convection currents.):
   a. a marshmallow
   b. a "black diamond"
   c. an emulsion of oil in soapy water
   d. an antimony(III) sulfide sol prepared by precipitation in alkaline solution
   e. homogenized milk

4. Briefly describe
   a. two alternative ways of preparing a dilute aqueous sol of calcium carbonate
   b. the principle of the Cottrell precipitator

Unit 15

Until the classical studies of Raoult and van't Hoff about 1882 on colligative properties of solutions, no reliable method existed for the determination of molecular weights of non-volatile compounds.

Journal of Chemical Education 36 (1959)

GEORGE B. KAUFMANN

Colligative properties

Objectives

(a) Be able to distinguish between constitutive and colligative properties of solutions (Section 15.1).
(b) Given appropriate data on solution composition, be able to calculate:
   (1) the vapor pressure of a solution involving a nonvolatile solute (Section 15.2).
   (2) the initial boiling point (with nonvolatile solute) or freezing point of a solution for which the boiling and freezing points of the pure solvent are known (Sections 15.3 and 15.4).
   (3) the osmotic pressure of a solution (Section 15.5).
(c) Given the osmotic pressure of a solution or the melting point of a solid mixture, along with other appropriate data, be able to calculate the apparent formula weight of the solute component (Section 15.6).

REVIEW

Molarities and molalities (Unit 13)

Figure 15.1 Vapor pressure: a colligative property. (a) Solution. (b) Pure liquid. The vapor pressure of a solution containing nonvolatile solute is lower than that of the pure solvent. This effect depends, for a given solvent, primarily on the relative numbers of solute and solvent particles.

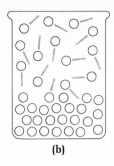

(a)   (b)

The blood is one of the most fascinating and complex of the aqueous systems in the human body. Biologists and biochemists have learned a great deal about the blood, yet many of its properties still offer mysteries for future research. The composition of the blood varies as it circulates through the body, but these changes are carefully controlled by the body chemistry, and seemingly minor variations beyond normal limits may have profound effects on the body.

Whole blood is a heterogeneous mixture of particles of varying sizes and characteristics. The larger particles are themselves complex living systems—cells such as the erythrocytes (red blood cells) or the leucocytes (white blood cells). In addition, there are many colloidal substances in blood—macromolecules such as the fibrinogen proteins (essential to the clotting mechanism) or the antibodies; molecular aggregates of certain hydrophilic lipids; and "protected" dispersions of hydrophobic lipids mixed with "biological soaps" (the bile salts). Finally, there are numerous types of dissolved substances—carbohydrates (mainly glucose), other organic compounds, and ions such as Na^+, K^+, H^+, and HCO_3^-. All these species have essential functions, and their concentrations are controlled by complex regulatory mechanisms.

Lipids is a broad term including steroids, such as cholesterol, "fats," and related compounds.

Although all human blood has certain common characteristics, every individual's blood has some unique features and, indeed, an individual's blood composition will vary somewhat almost from minute to minute. Exercise, diet, emotional state, and general health are some of the variables involved. Yet in spite of these variables, "normal" blood retains certain essential properties, and the mechanisms controlling these form one of the most fascinating studies in modern biochemistry.

The cellular components of blood are physically separated from the surrounding aqueous environment by complex membranes enclosing each cell. These membranes are highly selective in terms of the species they permit to enter or leave the cells. In general, they are quite permeable to water molecules—a factor that has considerable significance on limitations in concentrations of the dissolved or dispersed species in the cells' aqueous environment.

Blood plasma is obtained by removing the cellular components from the blood; a further separation of certain species, including fibrinogen, produces blood serum. Both plasma and serum have important medical uses, and both are still mixtures of dissolved and colloidal materials.

We have already discussed many aspects of solutions and colloids. In this unit we shall extend these studies to an examination, primarily, of certain colligative properties—properties that depend mainly on the relative numbers of the component particles in solutions or colloidal dispersions (Figure 15.1).

15.1 CONSTITUTIVE AND COLLIGATIVE PROPERTIES

The high-pressure air required for deep sea work (Unit 10) can be replaced by "air" of equal pressure in which helium is substituted for some or all of the nitrogen. The pressure is the same in either case, but other properties are quite different. Indeed, carbon monoxide could have been used instead of helium if pressure had been the sole consideration. Gas pressure, at constant temperature, is (under ideal gas conditions) a function only of the number of gas particles per unit volume in the container and not of the chemical nature of the particles. Such properties, dependent only on the *concentration* of particles, are called *colligative properties*. Characteristics that depend more on the *kind* of particles are said to be *constitutive* properties. These are reflected, in the gas mixtures described, by the quite different physiological properties of nitrogen, helium, and carbon monoxide.

The distinction between constitutive and colligative properties is more commonly reserved for discussions of liquid mixtures, either solutions or colloids. Like most attempts at categorization, these differences are to a considerable extent semantic, but they are useful for qualitative and semiquantitative work. Although most solution characteristics are *constitutive*—for example, the blue color of aqueous Cu^{2+} solutions or the sharp odor of dilute acetic acid—these also reflect relative particle numbers (concentration). Thus, although both a 1.0 M Cu^{2+} and a 0.010 M Cu^{2+} solution are blue and 20 percent acetic acid has the same "vinegar" odor as 3 percent acetic acid, the *intensities* of color or odor, respectively, depend on concentrations.

There are four physical properties of liquid solutions and colloids that are most appropriately considered as colligative properties:

Vapor pressure
Boiling point
Freezing point
Osmotic pressure

The terms "boiling *point*" and "freezing *point*" reflect common usage. Actual experimental measurements determine temperature *ranges* for boiling or freezing, and if only a single temperature is reported, it is usually the midpoint of the range measured. For very pure substances, careful measurement yields very narrow temperature ranges, essentially a constant freezing or boiling *point*.

15.2 VAPOR PRESSURE

The vapor pressure of a pure substance is determined only by the nature of the substance and the temperature (Unit 11). In solutions, however, the escaping tendency from the solution surface into the vapor phase is reduced from that of the pure component. We can suggest a simplified explanation for this observation on the basis of what we know about solutions and colloids.[1]

Consider, as an example, a solution of ethanol in water. Two factors will contribute to a difference in escaping tendency (and thus vapor pressure) for ethanol and water molecules in the solution as compared to the pure liquids. First, the *fraction* of water (or ethanol) molecules at the surface will be lower for the solution than for the pure liquids. In addition, the solution will contain water-ethanol molecular aggregates that "tie up" a certain fraction of the molecules which would otherwise have been available for escape into the vapor state.

Careful experimental studies of solutions and colloids have shown that the vapor pressures of such mixtures are colligative properties, depending on the relative numbers of solute and solvent particles. The relationship is expressed by *Raoult's law:*

François-Marie Raoult (1886).

$$P_A = f_A P_A^0 \tag{15.1}$$

in which P_A is the vapor pressure of solution component A, f_A is the mole fraction (Unit 13) of this component in the mixture, and P_A^0 is the vapor pressure of this component in its pure state.

Raoult's law can be used to estimate the initial boiling point of a solution at any

[1] For a more accurate and rigorous treatment, see K. J. Mysels, "The Mechanism of Vapor Pressure Lowering," in the *Journal of Chemical Education* (Vol. 32, p. 179, 1955), or Frank Rioux, "Colligative Properties," in the *Journal of Chemical Education* (Vol. 50, p. 490, 1973).

external pressure and to estimate the liquid and vapor compositions during distillation, as we shall see in later units. If the solute is nonvolatile, the vapor pressure of the solution should always be lower than that of the pure solvent by an amount that can be calculated from Raoult's law. It should be noted, however, that Raoult's law calculations are seldom very accurate for other than dilute solutions.

Example 1

Calculate the vapor pressure of an aqueous 1.00 m glucose solution at 100°C.

Solution
A 1.00 m solution contains 1.00 mole of solute per kilogram of solvent (Unit 13). Thus for an aqueous solution (molecular weight of H_2O = 18.0),

$$\text{no. moles } H_2O \text{ (per mole solute)} = \frac{1000 \text{ g}}{18.0 \text{ g/mole}}$$

$$= 55.6 \text{ moles}$$

and mole fraction of water,

$$f_{H_2O} = \frac{55.6}{1.00 + 55.6} = 0.982$$

The vapor pressure of pure water at 100°C is 760.0 torr (Unit 10), thus

$$P_{H_2O} = f_{H_2O} \times P^0_{H_2O}$$
$$P_{H_2O} = 0.982 \times 760.0 = 746 \text{ torr}$$

The vapor pressure of the 1.00 m glucose solution is lower than that required for boiling at normal atmospheric pressure (760 torr). As a result, the solution will not boil at 100°C unless the external pressure is decreased to 746 torr. Alternatively, the solution will boil at normal atmospheric pressure if it is heated above 100°C, to a temperature at which the vapor pressure becomes 760 torr.

It is important to note that calculations of this type are meaningful in terms of boiling-point effects only when the solute is nonvolatile. When volatile solutes are involved, the total vapor pressure of the solution is the sum of the vapor pressures of all volatile components as calculated from Raoult's law.

15.3 BOILING-POINT ELEVATION

If we had a phase diagram (Unit 12) for the particular solution involved, we could determine its initial boiling point. However, since the solute is nonvolatile, the composition of the solution changes as pure solvent boils away. As the solution becomes more concentrated by loss of solvent, the mole fraction of solvent decreases and the boiling point changes. Obviously, a simple phase diagram is not satisfactory for a solution of changing concentration.

It would be convenient to have some way of calculating the boiling point for any solution of known composition. Raoult's law suggests that the difference between the boiling point of a solution and that of a pure solvent should be a function of the mole fraction of the solvent. The exact function can be derived from thermodynamic considerations, but the derivation requires mathematics too advanced for our use. We can, however, note from Example 1 that the mole fraction of solvent can be related directly to the *molality* of the solution. We can reasonably suggest, then, that

[2] For examples, see M. L. McGlashan, "Deviations from Raoult's Law," in the *Journal of Chemical Education* (Vol. 40, p. 516, 1963).

the difference between the boiling points of solution and pure solvent is proportional to the molality of solute *particles,* that is,

$\Delta t_{b.p.}$ = initial b.p. (solution) − b.p. (pure solvent).

$$\Delta t_{b.p.} \propto \phi m$$

in which ϕ is the number of particles (e.g., molecules or ions) per formula unit of solute.

Another way of expressing this invokes a proportionality constant, which we shall call $k_{b.p.}$, the *molal boiling-point constant:*

$$\Delta t_{b.p.} = \phi m k_{b.p.} \qquad (15.2)$$

The value of this constant varies with the solvent (Table 15.1). To determine $k_{b.p.}$ experimentally, it is necessary only to measure the initial boiling points of a pure solvent and a solution of known molality. The constant, $k_{b.p.}$, has the units of °C kg (solvent) mole^{-1} (solute particles) and may be thought of as having the numerical value of Δt for a solution whose particle concentration is 1.00 m.

Table 15.1 MOLAL BOILING-POINT CONSTANTS FOR SOME COMMON SOLVENTS

SOLVENT	BOILING POINT (°C) AT 1 atm (pure solvent)	MOLAL BOILING-POINT CONSTANT ($k_{b.p.}$)
benzene	80.1	2.65
carbon tetrachloride	76.8	5.05
diethyl ether	34.6	2.11
ethanol	78.5	1.19
water	100.00	0.52

Further evidence that the boiling-point elevation is a colligative property is found in the behavior of solutions of electrolytes (Unit 6). If, in fact, it is the relative *number* of solute and solvent *particles* that determine this effect, then a mole of an ionic compound should have a greater influence on the boiling point of an aqueous solution than a mole of a covalent substance. Such is, indeed, the case. A mole of a salt such as NaCl or $CuSO_4$ has about twice the effect of a mole of glucose, and a mole of a more complex salt has a correspondingly larger effect.

That the effect is not *exactly* that predicted by the theoretical number of ions is evidence for ionic association in solution. The theoretical effect is approached in very dilute solutions.

Example 2

Calculate the initial boiling points of the following solutions:

a. 1.5 m glucose in ethanol
b. 1.5 m glucose in water
c. 1.5 m KNO_3 in water
d. 1.5 m K_2SO_4 in water

Solution
a. Glucose is a covalent compound, so $\phi = 1$. From Table 15.1, $k_{b.p.}$ for ethanol is 1.19.

$$\begin{aligned}
\Delta t_{b.p.} &= \phi m k_{b.p.} \\
&= 1 \times 1.5 \times 1.19 \\
&= 1.8°C
\end{aligned}$$

Then
$$\text{solution boiling point (initial)} = 78.5 + 1.8$$
$$= 80.3°C$$

b. $\Delta t_{b.p.} = 1 \times 1.5 \times 0.52 = 0.78°C.$
$$\text{solution boiling point (initial)} = 100.00 + 0.78$$
$$= 100.78°C$$

c. Potassium nitrate exists as potassium ions and nitrate ions, so $\phi = 2$.

$$\Delta t_{b.p.} = 2 \times 1.5 \times 0.52 = 1.6°C$$
$$\text{solution boiling point (initial)} = 100.0 + 1.6$$
$$= 101.6°C$$

d. $K_2SO_4(aq) = 2K^+ + SO_4^{2-}$, so $\phi = 3$.

$$\Delta t_{b.p.} = 3 \times 1.5 \times 0.52 = 2.3°C$$
$$\text{solution boiling point (initial)} = 100.0 + 2.3$$
$$= 102.3°C$$

15.4 FREEZING-POINT DEPRESSION

This statement is accurate for the usual case in which the solid initially formed is pure solvent. In those cases in which solid solutions form, the situation is more complex.

As in the case of boiling, the freezing out of pure solvent increases the concentration of the remaining solution. The temperature should, ideally, be measured at the time of appearance of the first solid formed on freezing, but it is more common to measure a freezing range. It should be noted that only approximate calculations are possible for other than *dilute* solutions (Section 15.2).

Salts such as rock salt (crude crystalline NaCl) or calcium chloride are frequently sprinkled on icy streets and sidewalks to help melt the ice. We add antifreeze to our automobile radiators in cold weather to prevent the cooling system from freezing. Both these familiar phenomena suggest that *solutions begin to freeze at lower temperatures than the pure solvent*. Like boiling-point elevation, the freezing-point depression of a solution is independent of the nature of the solute. Again, each particular solvent has its own unique molal *freezing-point constant* (Table 15.2).

We may think of this situation as somewhat analogous to boiling. In both cases, solvent escapes from solution to a separate pure solvent phase. The escaping tendency is a colligative property related to the mole fraction of solvent in the original solution and hence to the solution's molality.

Freezing-point effects can be calculated on a basis similar to that applied to boiling-point elevation, noting that the negative sign of the constant $k_{f.p.}$ (Table 15.2) results in a *lowered* temperature for freezing:

$$\Delta t_{f.p.} = \phi m k_{f.p.} \qquad (15.3)$$

Table 15.2 MOLAL FREEZING-POINT CONSTANTS FOR SOME COMMON SOLVENTS

SOLVENT	FREEZING POINT (°C) (pure solvent)	MOLAL FREEZING-POINT CONSTANT $(k_{f.p.})^a$
acetic acid	16.6	−3.9
benzene	5.6	−5.0
camphor	178.4	−37.7
water	0	−1.86

^{a} °C kg (solvent) mole^{-1} (solute particles).

Example 3

What relative proportions of ethylene glycol and water (by weight) should be mixed to form an antifreeze solution that will not start to freeze until the temperature reaches $-37°C$?

AUTOMOBILE ANTIFREEZE

One of the first compounds to be used as an additive in automobile radiator systems to protect the water from freezing during cold weather was methanol ("wood alcohol," CH_3OH). Methanol is relatively inexpensive and its low molecular weight (32) means that the number of moles *per gram* of particles available for freezing-point depression is relatively large. The principal disadvantage of methanol is its volatility (boiling point, 65°C). As a volatile solute it acts to *increase* the total vapor pressure of the solution (according to Raoult's law), thus resulting in increased loss of radiator coolant by evaporation.

Methanol is rarely employed today as an antifreeze for automobile radiator systems. It has been almost entirely replaced by ethylene glycol ($HOCH_2CH_2OH$), a *nonvolatile* liquid (boiling point, 197°C). The glycol, although more expensive than methanol, helps retard evaporation, in addition to acting as an antifreeze.

As we have mentioned, equations (15.1)–(15.3) apply accurately only to "ideal" solutions. Deviations from "ideal" behavior become increasingly significant at higher solution concentrations. Deviations also result from variations in solute–solvent interaction. The presence of two hydroxyl groups (OH) in the ethylene glycol molecule suggests that such a molecule might "tie up" more water molecules (by hydrogen bonding) than would, for example, the simpler CH_3OH molecule. We do, in fact, observe that ethylene glycol provides a lower freezing point for aqueous solutions (over a concentration range from 1 m to ~25 m) than would be calculated from equation (15.3). This deviation ranges from about 0.2°C at 1 m concentration to nearly 5°C at 25 m concentration.

Application of equation (15.3) might suggest that higher and higher concentrations of ethylene glycol could be employed to protect automobile radiator systems to very low temperatures. Such is not, in fact, the case. Pure ethylene glycol freezes at −16°C. Calculations from equation (15.3), based on water as the solvent, are reasonably reliable for concentrations of the ethylene glycol up to ~38 m. (This solution begins to freeze around −65°C.) Beyond this point, the system behaves as though the glycol were the solvent and water the solute, so that higher glycol-to-water ratios result in solutions freezing *above* −65°C (approaching the −16°C freezing point of the pure glycol.)

Most commercial antifreeze mixtures contain additives such as "rust inhibitors" (antioxidants). Ethylene glycol is still, however, the principal ingredient of most antifreeze preparations.

Solution
Ethylene glycol ($HOCH_2CH_2OH$) is a covalent compound, so $\phi = 1$.

$\Delta t_{f.p.}$ = freezing point (solution) − freezing point (pure solvent)
   $= -37°C - 0°C = -37°C$

$$= \phi m k_{f.p.}$$
$$-37 = 1 \times m \times -1.86$$
$$m = \frac{-37}{-1.86} = 20$$

Thus for each kilogram (1000 g) of solvent (H_2O), 20 moles of ethylene glycol must be added.

$$20 \text{ moles} \times \frac{62 \text{ g}}{\text{mole}} = 1240 \text{ g}$$

The proportions, by weight, may then be reported as

ethylene glycol : water = 1.24 : 1.00

15.5 OSMOTIC PRESSURE

If erythrocytes (red blood cells) are removed from their normal aqueous environment and suspended in distilled water, the cells gradually swell and eventually burst. Careful studies have shown that this is caused by the diffusion of water through the cell membrane into the cell. The opposite effect is observed if the erythrocytes are suspended in a concentrated saline solution. Water diffuses irreversibly out of the cells and shriveling (crenation) occurs. The cells can be maintained in their normal condition for a long time in a saline solution whose concentration approximates the total particle concentration of the blood.

Solutions or colloidal dispersions designed for injection into the bloodstream must have concentrations within the range that will avoid crenation or bursting (hemolysis) of the erythrocytes and other cellular components of the blood.

"Semipermeable" refers to the limited pore size of the membrane, which restricts the size of particles that can pass through.

When an aqueous solution enclosed by a *semipermeable membrane* reaches equilibrium with pure water outside the membrane, the resulting hydrostatic pressure (Figure 15.2) is called the *osmotic pressure* of the solution, in reference to the

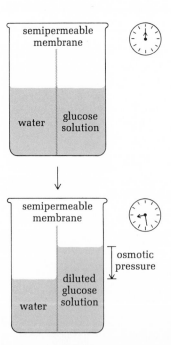

Figure 15.2 Osmosis and osmotic pressure.

Note that in the cases being considered this refers to concentration of *water*, not solute.

process of *osmosis*. In this process a solution component (e.g., water) passes through the pores of a membrane from a region of higher concentration of the component to one of lower concentration.

Equilibrium is reached when the osmotic pressure of the solution is just sufficient to prevent further net flow of solvent through the membrane into the solution. This process is another example of dynamic equilibrium, so it is important to note that solvent flow is continuous but no *net* change occurs at equilibrium.

Although osmotic pressure is caused by an accumulation of excess *solvent,* this accumulation occurs because of the *solute particles* that tie up the solvent in solute–solvent clusters. Thus the effect can be attributed to the concentration of solute particles. The *van't Hoff equation* bears a striking resemblance to an equation used to represent the ideal gas law [equation (10.2), Unit 10]:

J. H. van't Hoff (1885).

$$\Pi V = nRT$$

in which Π (capital Greek pi) is the osmotic pressure of a solution containing n moles of solute *particles* in a total solution volume V, R is the ideal gas constant (0.082 liter atm mole^{-1} deg^{-1}), and T is the absolute temperature.

It is frequently useful to express the van't Hoff equation in terms of solution concentration:

$$\left(\frac{n}{V} = \phi M \right)$$

$$\Pi = \phi MRT \tag{15.4}$$

Molarity is used because the relationship treats the solute particles as though they were gas particles, so total (solution) volume is involved. In very dilute aqueous solutions, there is no significant difference between molarity and molality units.

The concentration unit M, it should be noted, is *molarity,* rather than the *molality* used in Equations (15.2) and (15.3). The ϕ term has the same connotation as in the freezing-point and boiling-point expressions. The van't Hoff equation is reasonably accurate for dilute solutions or colloids.

Example 4
___

A 0.86% (by weight) solution of sodium chloride in water is often referred to as "physiological saline," since the osmotic pressure for this solution closely approximates that of normal blood cells. The density of "physiological saline" is 1.004 g ml^{-1} at 37°C (the normal temperature of the human body). What is the osmotic pressure of "physiological saline" at 37°C?

Solution

In order to employ the van't Hoff equation, we must express the concentration of NaCl (formula weight, 58.5) in moles per liter:

$$M_{(NaCl)} = \frac{0.86 \text{ g (NaCl)}}{100 \text{ g (solution)}} \times \frac{1.004 \text{ g (solution)}}{1 \text{ ml (solution)}} \times \frac{10^3 \text{ ml}}{1 \text{ liter}} \times \frac{1 \text{ mole (NaCl)}}{58.5 \text{ g (NaCl)}}$$

$$= 0.15 \, M$$

In water, sodium chloride provides two ions per formula unit (Na$^+$ and Cl$^-$), so $\phi = 2$.

Then, in equation (15.4),

[37°C = 310°K]
$\Pi = 2 \times 0.15$ (mole liter^{-1}) $\times 0.082$ (liter atm mole^{-1} deg^{-1}) $\times 310$ (deg)
$\Pi = 7.6$ atm

WATER PURIFICATION BY REVERSE OSMOSIS

If a solution separated from pure water by a semipermeable membrane is placed under external pressure, perhaps by a piston, then the osmotic process can be reversed and pure solvent can be forced out of the solution through the membrane. This technique, called *reverse osmosis,* is being investigated as a method of obtaining fresh water from sea water. Most cell membranes or larger vegetable or animal membranes are fairly fragile and are ruptured by osmotic pressures of more than a few atmospheres. A synthetic membrane capable of withstanding osmotic pressures of several hundred atmospheres can be prepared by precipitating gelatinous $Cu_2[Fe(CN)_6]$ on a support of unglazed porcelain. Recently relatively thin membranes of cellulose acetate have been developed for reverse osmosis in *desalinization* (salt removal) processes for producing pure water from seawater, or from water heavily contaminated by dissolved pollutants. Plants using such processes (Figure 15.3) are now operating at levels of several thousand gallons of water per day.

Reverse osmosis shows a considerable promise for water purification, although it will be some time before large enough plants are developed to supply the pure water requirements of major metropolitan areas or large agricultural irrigation projects.

Figure 15.3   Reverse osmosis.

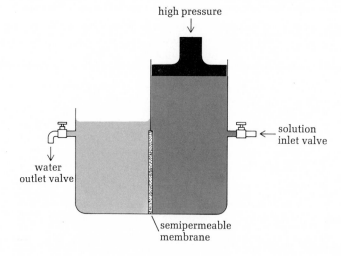

Example 5

What minimum pressure must be applied to a $0.20\,M$ aqueous sodium chloride solution enclosed by a semipermeable membrane at $27°C$ in order to initiate reverse osmosis? (This concentration approximates that of seawater.)

Solution

$$NaCl \longrightarrow Na^+ + Cl^-$$

so $\phi = 2$. Then

$$\Pi = \phi MRT$$
$$= (2 \times 0.20 \text{ mole liter}^{-1}) \times (0.082 \text{ liter atm mole}^{-1} \text{ deg}^{-1}(K)) \times (300°K)$$
$$= 9.8 \text{ atm}$$

Thus reverse osmosis would require an external pressure exceeding 9.8 atm.

15.6 FORMULA-WEIGHT APPROXIMATIONS

The values determined actually represent *average* formula weights of solute particles. Thus ionization (e.g., of acetic acid) or aggregation (e.g., in colloidal aggregates) will yield values different from "true" formula weights.

The same consideration should apply to any mixture of solid and solution as long as these have the same relative compositions.

Freezing-point depressions are generally easier to determine accurately than are boiling-point elevations. The latter are more subject to variations with fluctuations of atmospheric pressure, the techniques for small samples are less satisfactory, and the molal boiling-point constants for most solvents are smaller than their molal freezing-point constants. Chemists and biochemists find freezing-point depression a useful method for determining the apparent formula weights of a wide variety of compounds.

The temperature at which solid first appears from a freezing solution should be the same as that at which the last crystals of pure solvent melt from the corresponding frozen mixture (i.e., this phenomenon is a *state function*). Consequently, melting ranges of mixtures may be used rather than freezing ranges of solutions, and the average temperature (midpoint of the range) should be the same in either case, under ideal conditions. The choice of measuring melting or freezing range is usually a matter of convenience determined by the characteristics of the particular system being studied.

Example 6

A 21.6-mg sample of an organic compound was ground with 462 mg of camphor to form an intimate homogeneous mixture. A sample of the mixture was found to melt over the range 169.2–171.4°C. What is the apparent formula weight of the organic compound?

Solution

$$\begin{array}{r} 171.4 \\ -169.2 \\ \hline 2.2 \end{array} \qquad \tfrac{1}{2} \times 2.2 = 1.1$$

$$\text{midpoint of melting range} = 169.2 + 1.1 = 170.3°C$$
$$\Delta t_{f.p.} = 170.3 - 178.4 = -8.1°C \qquad \text{(Table 15.2)}$$
$$= \phi m k_{f.p.}$$

Since camphor is an unlikely solvent for ionic compounds, we may assume that $\phi = 1$ (an assumption implied by "apparent" formula weight).

$$-8.1 = 1 \times m \times -37.7 \qquad \text{(Table 15.2)}$$

$$m = \frac{-8.1}{-37.7} = 0.21$$

Then, from the definition of molality (Unit 13), if we let W represent the apparent formula weight,

$$0.21 \text{ mole (solute) kg}^{-1}\text{(solvent)} = \frac{(2.16 \times 10^{-2}) \text{ g(solute)}/W}{4.62 \times 10^{-4} \text{ kg(solvent)}}$$

$$W = \frac{(2.16 \times 10^{-2}) \text{ g(solute)}}{(4.62 \times 10^{-4}) \text{ kg(solvent)} \times 0.21 \text{ mole(solute) kg}^{-1}\text{(solvent)}}$$

$$= {\sim}220 \text{ g mole}^{-1}$$

Modern research laboratories use techniques of mass spectrometry for accurate measurements of molecular weights. However, the older methods are still commonly employed in many circumstances.

The apparent molecular weights of simple organic compounds are most frequently determined by measurements of freezing-point depression (Example 6). These methods are, however, relatively inaccurate for compounds of high molecular weight. Osmotic pressure measurements have proved most useful for the study of colloidal dispersions, especially those involving macromolecules such as proteins or synthetic polymers.

It should be noted that colloidal dispersions are significantly less than "ideal" as solutions. Although colloids do display colligative properties, these are not accurately represented, in many cases, by the simple equations, such as equations (15.1)–(15.3), for ideal solutions. Osmotic pressure measurements are more reliable for colloids, but even these must be taken as rather rough approximations.

Example 7

A protein isolated from a blood serum sample was found on careful analysis to contain 0.285 percent iron. A colloidal dispersion of 685 mg of the protein in sufficient water to form 10.0 ml of solution was found to have an osmotic pressure of 0.21 atm at 4°C. What is the apparent molecular weight of the protein, to three significant figures?

Solution
We assume the protein to act as a covalent compound, so $\phi = 1$.

$$\Pi = \phi M R T$$

from which

$$M = \frac{\Pi}{\phi R T}$$

$$= \frac{0.021 \text{ atm}}{(0.082 \text{ liter atm mole}^{-1} \text{ deg}^{-1}) \times 277°\text{K}}$$

$$= 9.2 \times 10^{-4} \text{ mole liter}^{-1}$$

If we represent the apparent molecular weight by W, then using the definition of molarity (Unit 13),

$$9.2 \times 10^{-4} \text{ mole liter}^{-1} = \frac{0.685 \text{ g}/W}{0.0100 \text{ liter}}$$

from which

$$W = \frac{0.685 \text{ g}}{(9.2 \times 10^{-4} \text{ mole liter}^{-1}) \times (0.0100 \text{ liter})}$$

$$= 7.4 \times 10^{4} \text{ g mole}^{-1}$$

Now, we can refine this value by introducing the analytical data on the iron content of the protein. If there are x atoms of iron in each protein molecule, then iron (atomic weight 55.85) contributes (55.85x) to the molecular weight of the protein. Then, to two significant figures,

$$\frac{0.285}{100} \times 7.4 \times 10^4 = 56x$$

$$x = \frac{0.285 \times 7.4 \times 10^4}{100 \times 56} = 3.8$$

Since x must be a whole number if all the protein molecules are identical, x must be 4 (i.e., four atoms of iron per protein molecule). Then, to *three* significant figures, letting W' represent the protein's molecular weight,

$$\frac{0.285}{100} \times W' = (55.85)(4)$$

From which

$$W' = \frac{(55.85)(4)(100)}{0.285} = 7.84 \times 10^4$$

The protein used for Example 7 was hemoglobin, so the calculation will give you some feel for the mass of this important molecule.

Note that the apparent molecular weight determined by osmotic pressure measurement is useful primarily in obtaining the correct order of magnitude and is far less accurate than that based on careful chemical analysis.

Note also that this problem illustrates an important consideration in dealing with significant figures. By convention, a calculated value is reported with the number of significant figures consistent with the least precisely known measured quantity. This is a matter of *precision* rather than *accuracy*.

The colligative properties of solutions and colloids are of significance in dealing with a wide variety of important phenomena, ranging from the calculation of appropriate antifreeze mixtures for automobiles to the preparation of solutions for use in the study of living cells.

SUGGESTED READING   Schaum, Daniel. 1966. *Theory and Problems of College Chemistry.* New York: McGraw-Hill.
Thain, J. E. 1967. *Principles of Osmotic Phenomena.* London: The Chemical Society.

EXERCISES

1. Which of the following observations illustrate *colligative* properties?
   a. Pure benzene boils 20° lower than pure water at 1.00 atm pressure.
   b. The blue color of 1.0 M $CuSO_4$ solution is more intense than that of 0.10 M $CuSO_4$.
   c. Saturated salt water boils at a higher temperature than pure water at 760 torr.
   d. The osmotic pressure of 1.0 M NaCl is nearly twice that of 1.0 M glucose, at the same temperature.
   e. The taste of 1.0 M NaCl is quite different from that of 1.0 M glucose.
2. From data in Table 10.2, Unit 10, calculate the vapor pressure at 30°C of a solution prepared from 25.0 g of glucose ($C_6H_{12}O_6$) and 100.0 g of water.
3. a. Estimate the initial boiling point at 1.00 atm of a saturated aqueous solution of sodium chloride, assuming saturation at 100°C. (Solubility of NaCl at 100°C is 39.1 g 100 ml^{-1}, $k_{b.p.}$ for water is 0.52.)
   b. Estimate the initial freezing point of a saturated aqueous solution of sodium chloride, assuming saturation at 0°C. (Solubility of NaCl at 0°C is 35.7 g 100 ml^{-1}, $k_{f.p.}$ for water is −1.86.)
   c. Estimate the osmotic pressure at 20°C of an aqueous solution containing 250 g of sodium chloride per liter. ($R = 0.082$ liter atm mole^{-1} deg^{-1}.)

4. A 16.8-mg sample of an organic compound was ground with 438 mg of camphor to form an intimate homogeneous mixture melting at 168.4–170.6°C. What is the apparent formula weight of the organic compound? (Pure camphor melts at 178.4°C; $k_{f.p.} = -37.7$.)

5. A colloidal dispersion of 12.3 g of a polystyrene plastic in 100 ml of toluene was found to have an osmotic pressure of 0.032 atm at 20°C. What is the apparent formula weight of the polystyrene? (Assume 100 ml total volume.)

PRACTICE FOR PROFICIENCY

Objective (a)

1. Indicate the observations illustrating colligative properties:
   a. A saturated aqueous solution of benzoic acid begins to freeze below 0°C.
   b. Pure liquid benzoic acid begins to freeze at 122°C, whereas pure water begins to freeze at 0°C.
   c. A 0.010 M sodium benzoate solution has an osmotic pressure nearly twice that of a 0.010 M benzoic acid solution.
   d. A saturated aqueous benzoic acid solution tastes sour, whereas a saturated sodium benzoate solution tastes somewhat "salty."

Objective (b)

2. Pure water has a vapor pressure of 233 torr at 70°C. What is the vapor pressure at this temperature of a solution containing 15.0 g of sucrose ($C_{12}H_{22}O_{11}$) in 100.0 ml of water?

3. What is the initial freezing point, at 1.00 atm, of a solution containing equal parts by weight of sucrose and water?

4. What is the osmotic pressure at 37°C of a 0.20 M sodium sulfate solution?

5. The initial boiling point (in °C) of a solution of 69 g of K_2CO_3 in 2.50 kg of water is approximately _____.

6. The initial freezing point (in °C) of a solution of 180 g of $MgSO_4$ in 2.5 kg of water is approximately _____.

7. A solution of 60.0 g of glucose ($C_6H_{12}O_6$) in 1.00 liter of water will have as osmotic pressure (in atmospheres) at 27°C of approximately _____.

8. The initial boiling point (in °C) of a solution of 238 g of KBr in 3.0 kg of water is approximately _____.

9. A solution of 39.6 g of $(NH_4)_2SO_4$ in 1.00 liter of water will have an osmotic pressure (in atmospheres) at 27°C of approximately _____.

10. An automobile cooling system was filled with a mixture of 8.0 liters of water and 12.0 liters of ethylene glycol. To what temperature (in °C) is the system protected against freezing? (That is, what is the initial freezing point of the mixture?) (The density of water is 1.00 g ml^{-1} and that of $HOCH_2CH_2OH$ is 1.11 g ml^{-1}.)

Objective (c)

11. A 13.6-mg sample of an organic compound was ground with 484 mg of camphor to form a mixture melting at 169.1°–170.9°C. What is the apparent formula weight of the compound? (Pure camphor melts at 178.4°C, $k_{f.p.} = -37.7$.)

12. A 23.2 mg sample of an organic compound was ground with 483 mg of camphor to form an intimate mixture melting at 169.6° ± 0.9°C. What is the apparent formula weight of the organic compound?

13. A 16.7-mg sample of an organic compound was ground with 421 mg of camphor to form an intimate mixture melting at 169.5° ± 0.9°C. What is the apparent formula weight of the organic compound?

14. A dispersion of 11.0 g of a polymer is 125.0 ml of toluene had an osmotic pressure of 0.055 atm at 27°C. What is the apparent formula weight of the polymer? (Assume a total solution volume of 125 ml.)

15. A dispersion of 16.4 g of a polymer in 150.0 ml of toluene had an osmotic pressure of 0.048 atm at 27°C. What is the apparent formula weight of the polymer? (Assume a total solution volume of 150 ml.)

BRAIN TINGLER

16. Raoult's law may be used to estimate the boiling point of a mixture of two or more volatile compounds and the composition of the initial distillate from such a mixture. The key to these predictions is an understanding of Raoult's law, Dalton's law of partial pressures, and the meaning of "mole fraction" ($f_a = n_a/n_{total}$). Given that pure aniline ($C_6H_5NH_2$) boils at 184°C at 760 torr and the vapor pressure of nitrobenzene ($C_6H_5NO_2$) at 184°C is 380 torr, and assuming ideal behavior, predict
   (a) the external pressure at which a mixture of aniline and nitrobenzene will boil at 184°C if the mole fraction of nitrobenzene is 0.15
   (b) the mole fraction of nitrobenzene in the first drop of liquid distilled

Unit 15 Self-Test

(Consult Table 15.1 or Table 15.2 when necessary.)

1. Pure water has a vapor pressure of 92.0 torr at 50°C. What is the vapor pressure (in torr) of a 2.0 m solution of sucrose at 50°C?

2. What is the initial boiling point (at 1.00 atm) of a solution containing equal parts, by weight, of ethanol and glycerin (nonvolatile)?

3. What is the initial freezing point of a solution containing 20.0 g of naphthalene ($C_{10}H_8$) in 100.0 g of benzene?

4. An aqueous solution containing 54.3 g of glucose per liter may be used for intravenous injection because it has the same osmotic pressure as blood. Calculate the osmotic pressure (in atmospheres) of this "isotonic" solution at 37°C.

5. A 23.2-mg sample of an organic compound was ground thoroughly with 485 mg of camphor. A sample of the mixture melted at 170.2°–172.6°C. What is the apparent formula weight of the organic compound?

Overview Test

Section FIVE

(Refer, as necessary, to Tables 15.1 and 15.2)

1. Write "working definitions" of
   a. molarity; **b.** formality; **c.** normality; **d.** molality; **e.** parts per million

2. The "$99\frac{44}{100}$% pure" Ivory soap is an example of a colloidal
   a. aerosol; **b.** emulsion; **c.** foam; **d.** gel; **e.** sol

3. A typical industrial "smog" is an example of a colloidal
   a. aerosol; **b.** emulsion; **c.** foam; **d.** gel; **e.** sol

4. A dispersion of liquid soap in water is an example of a colloidal
   a. aerosol; **b.** emulsion; **c.** foam; **d.** gel; **e.** sol

5. Submicroscopic soil particles dispersed in water is an example of a colloidal
   a. aerosol; **b.** emulsion; **c.** foam; **d.** gel; **e.** sol

6. A Cottrell precipitator is used for
   a. causing precipitation in supersaturated solutions
   b. inducing rain by cloud seeding
   c. producing river deltas by addition of salt water
   d. removing colloidal solids from certain smokes
   e. preparing aqueous gold sols by electrolytic reduction

7. Indicate whether each of the following illustrates a colligative property or a constitutive property.
   a. Water freezes at a different temperature than benzene.
   b. A 0.5 m aqueous NaCl solution freezes at a different temperature than a 0.5 m aqueous glucose solution.
   c. A 5% (by weight) aqueous $CuSO_4$ solution is blue, whereas a 5% (by weight) aqueous methanol solution is colorless.
   d. The vapor pressure of a 5% aqueous $CuSO_4$ solution is lower than that of a 5% aqueous methanol solution, at 75°C.
   e. A red blood cell suspended in distilled water at 37°C gradually swells until it bursts, but this phenomenon is not observed with red blood cells suspended in 0.86% aqueous NaCl at 37°C.

8. A solution is labeled "20.0% H_3PO_4, by weight (density 1.12 g ml^{-1})." What is
   a. the molality of the solution
   b. the molarity of the solution
   c. the normality of the solution, for use as an acid

9. The mass (in grams) of $FeSO_4 \cdot 7H_2O$ required for preparation of 60 ml of 0.25 M solution is
   a. 4.2; **b.** 7.3; **c.** 11; **d.** 15; **e.** 23

10. The volume (in milliliters) of 15 M ammonia required for preparation of 250 ml of 0.30 M solution is
    a. 2.25; **b.** 3.3; **c.** 4.5; **d.** 5.0; **e.** 6.6

11. The initial boiling point (in °C) of a solution of 32.8 g of Na_3PO_4 in 0.75 kg of water is approximately
    a. 101.0; **b.** 100.6; **c.** 100.3; **d.** 100.1; **e.** 98.4

12. A solution of 32.0 g of NH_4NO_3 in 1.00 liter of water will have an osmotic pressure (in atmospheres) at 27°C of approximately
    a. 5.0; **b.** 10; **c.** 15; **d.** 20; **e.** 90

13. An automobile cooling system was filled with a mixture of equal volumes of water and ethylene glycol. To what temperature (in °C) is the system protected against freezing? (That is, what is the initial freezing point of the mixture?) (The density of water is 1.00 g ml^{-1} and that of $HOCH_2CH_2OH$ is 1.11 g ml^{-1}.)
    a. -33°C; **b.** -40°C; **c.** -46°C; **d.** -53°C; **e.** -62°C

14. A 50.0-ml sample of a commercial "vinegar" (dilute acetic acid) was neutralized by titration with 31.6 ml of 0.814 M sodium hydroxide. How many grams of acetic acid were present in 2.50 liters of this "vinegar"?

    $$[CH_3CO_2H(aq) + OH^-(aq) \longrightarrow H_2O(l) + CH_3CO_2^-(aq)]$$

    a. 126; **b.** 102; **c.** 77.2; **d.** 53.5; **e.** 31.6

15. About half of the water used by the residents of San Francisco comes from the Sierra-Calaveras watershed and averages 2.4 ppm (mg liter^{-1}) Ca^{2+}. How many kilograms of sodium carbonate would be required to "soften" 8.00 million liters of this water, assuming stoichiometric removal of Ca^{2+}?

    $$[Na_2CO_3(aq) + Ca^{2+}(aq) \longrightarrow 2Na^+(aq) + CaCO_3(s)]$$

    a. 51; **b.** 43; **c.** 32; **d.** 24; **e.** 17

16. When iron pyrite is treated with nitric acid, the reaction observed may be formulated as

$FeS_2(s) + 18HNO_3(aq) \longrightarrow$
$$Fe(NO_3)_3(aq) + 15NO_2(g) + 2H_2SO_4(aq) + 7H_2O(l)$$

What volume of concentrated nitric acid (17.4 M) would be required to consume 1.6 kg of the pyrite, assuming 100% efficiency?

a. 14 liters; **b.** 10 liters; **c.** 8.7 liters; **d.** 5.5 liters; **e.** 3.2 liters

17. A 26.5-mg sample of an organic compound was ground with 423 mg of camphor to form an intimate mixture melting at 169.8° ± 0.9°C. What is the apparent formula weight of the organic compound?

a. 194; **b.** 230; **c.** 275; **.d.** 313; **e.** 364

18. A dispersion of 17.1 g of a polymer in 200.0 ml of toluene had an osmotic pressure of 0.036 atm at 27°C. What is the apparent formula weight of the polymer? (Assume a total solution volume of 200 ml.)

a. 40,000; **b.** 58,000; **c.** 80,000; **d.** 86,000; **e.** 94,000

Section SIX

KINETICS AND EQUILIBRIUM

Reaction kinetics is the study of factors that influence the speed of a chemical process. We are all familiar with many applications in this area: New automobiles are equipped with "catalytic converters" to speed up the oxidation of exhaust gases; dust explosions in grain elevators and coal mines represent safety hazards explained by the influence of particle size on reaction rate; "spontaneous combustion" is a common source of fire in hay barns, and illustrates the temperature dependence of reaction speeds; "No Smoking" signs in hospital areas in which oxygen is in use represent an awareness of concentration effects on the rate of combustion.

   Information obtained from studies of reaction kinetics can also provide clues to *mechanisms*, the steps by which chemical processes occur. Such mechanisms, in turn, can lead to ways of manipulating reaction conditions for improved product yield and purity.

Dynamic equilibrium is a topic we have touched on before in connection with the physical equilibrium involving different states of matter (Unit 12). We want now to extend the study of equilibrium to chemical processes, both homogeneous (Unit 17) and heterogeneous (Unit 18). An understanding of equilibrium is essential to the further exploration of many important chemical systems, such as solutions in contact with undissolved solute (Unit 18), the weak acids and bases (Unit 19), buffer systems (Unit 20), electrochemical processes (Units 23 and 24), and coordination compounds (Unit 27).

Within living organisms, many important biochemical systems—ranging from the buffer action of blood to the complex process of transport across cell membranes—may be studied by applications of equilibrium principles. Many, if not most, of these systems actually represent steady state phenomena (Unit 12), rather than true equilibria, but the mathematics of equilibrium analysis can still provide useful approximations for an understanding of numerous aspects of molecular biology.

CONTENTS

Reaction kinetics and mechanisms

... It made such a thunderclap
... that the stable-servant imag-
ined . . . either that someone
had shot at him through the
window or that the Devil himself
was active in the stable. (First investigation of mercury fulminate,
1716)
JOHANN KUNCKEL

Objectives

(a) After studying the appropriate definitions in Appendix E and the terms'
usage in this unit, be able to demonstrate your familiarity with the following
terms

autocatalysis	homogeneous catalysis
activation energy	inhibitor
chain reaction	intermediate
effective collision	molecularity
elementary process	transition state
flood	quench
free radical	reaction order
heterogeneous catalysis	

(b) Be able to make reasonable predictions and to offer logical explanations for
effects on reaction speeds of
(1) catalysis (Sections 16.5, 16.12).
(2) temperature variation (Sections 16.3, 16.13).
(3) collision geometry (Section 16.2).

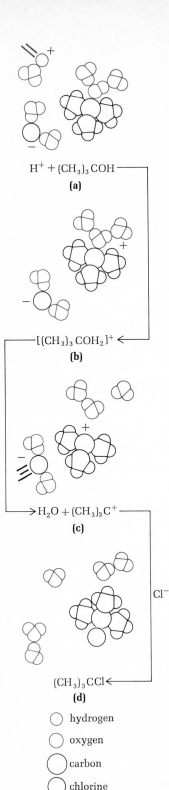

$H^+ + (CH_3)_3COH$

(a)

$[(CH_3)_3COH_2]^+$

(b)

$H_2O + (CH_3)_3C^+$

(c)

Cl^-

$(CH_3)_3CCl$

(d)

○ hydrogen

○ oxygen

○ carbon

○ chlorine

Chemical reactions occurring at very high temperatures are difficult to study. The importance of such processes in this era of space age research, however, has resulted in some very sophisticated experiments. The thermodynamic aspects of chemical reactions are relatively easy to describe because of their independence of specific reaction conditions. Thermodynamic treatments do, however, have limitations. It has proved more profitable, although considerably more difficult, to investigate the kinetic factors of reactions in rocket engines. Kinetics data have been found[1] to provide much better approximations of engine performance in rockets than the types of data based on thermodynamic and equilibrium considerations.

This same period has found heightened interest in a totally different aspect of reaction kinetics—the role of catalysts in influencing reaction speed. Research in the field of catalysis has ranged from the very practical aspects of development of more efficient and less expensive techniques for industrial processes to the highly theoretical studies of mechanisms of catalytic activity.[2] The important biological catalysts (enzymes) are gradually yielding their secrets to research scientists. The chemical industry is making increasing use of new catalytic methods for processes ranging from the manufacture of margarine to the production of complex plastics.

Catalysts are also playing significant roles in recent efforts to improve the environment. For many years tetraethyllead has been used as a gasoline additive to slow down the combustion rate for more uniform piston thrust. Lead compounds exhausted from automobile engines are major environmental contaminants. Similarly, carbon monoxide, hydrocarbon residues, and nitrogen oxides from automobile exhausts contribute to the air pollution problem. New research by petroleum chemists, automotive engineers, and environmental specialists promises major breakthroughs in the use of catalytic agents to decrease pollution from exhaust emissions and to produce fuels less dependent on lead additives for efficient engine performance.

In this unit we shall examine several factors that influence the speeds of chemical reactions. An understanding of these factors is a key to controlling chemical processes. We shall see how kinetics data can provide important clues to the steps by

[1] See Richard M. Lawrence and William H. Bowman, "High-Temperature Reactions Associated with Space Vehicles," in the *Journal of Chemical Education* (Vol. 48, pp. 458–460, 1971).

[2] See Heinz Heinemann, "Homogeneous and Heterogeneous Catalysis—Common Frontier or Common Territory?," in the May, 1971, issue of *Chemical Technology* (pp. 286–291).

Figure 16.1 *Mechanism of reaction of HCl with $(CH_3)_3COH$. It is suggested that reaction is initiated (a) by collision of H^+ (aq) with the oxygen of the hydroxyl group. Subsequently, (b) H_2O is eliminated to leave (c) a carbonium ion, $(CH_3)_3C^+$, which is then attacked (d) by Cl^- (aq).*

which reactions occur. An understanding of these steps can often permit the selection of reaction conditions for improvement of reaction yield, product purity, and process efficiency.

The substances we call "foods" consist primarily of three major classes of chemical compounds—carbohydrates (e.g., starches, sugars), proteins, and lipids (e.g., animal or vegetable fats and oils). These rather complex species are broken down in the body into simpler compounds by a wide variety of chemical reactions, most of which are catalyzed by specific enzyme systems. The reactions involved are known broadly as metabolism, and the steps in metabolic processes are probably the most thoroughly investigated reaction mechanisms in all of science.

Today we know a great deal about the mechanisms of metabolic reactions. We can outline, for example, the various steps through which starches are eventually transformed to carbon dioxide and water, and we can account rather well for the energy changes associated with these processes. Similar information is available for many of the stages of protein and lipid metabolism.

Biological systems are exceedingly complex, so much of our knowledge of mechanisms in biochemical pathways has been gained from chemical studies of simpler laboratory systems.

Reaction mechanisms, then, are not just the basis of intellectual games for theoretical chemists. Their elucidation is of importance in diverse fields ranging from agriculture to zoology. Modern mechanisms research is providing information of value to nutritionists, industrial chemists, physicians, ecologists concerned with formation and degradation of environmental contaminants, and the ever-growing groups involved in studies of drug action.

Recently, special "molecular beam" techniques have permitted the definitive description of some relatively simple one-step mechanisms, but these methods are not yet adaptable to complex systems.

In this unit we shall see how kinetics data can be interpreted to offer clues to reaction mechanisms (Figure 16.1). Rate laws alone, however, are never sufficient for the complete description of complex reactions. Other information, such as spectroscopic data or the use of special isotopes, may be needed.

Since we cannot see directly the ways that molecules behave, any mechanism suggested for a complex chemical reaction is only an educated guess. Descriptions of most reaction mechanisms, like most theories, must still be open to revision as new experimental evidence is obtained.

16.1 CHEMICAL KINETICS

Powdered glucose may sit in contact with air for years with no appreciable change, yet in certain biochemical systems glucose and oxygen are rapidly consumed, eventually liberating CO_2 and H_2O. The same net reaction occurs even more rapidly when glucose is dusted into a flame burning in contact with air—in fact, a proper mixture of very fine glucose "dust" and oxygen can be ignited by a spark to produce a violent explosion. Thus the speed of the combustion reaction of glucose must depend on a number of factors not indicated by the equation for the net process,

$$C_6H_{12}O_6 + 6O_2 \longrightarrow 6CO_2 + 6H_2O$$

It is important to recognize that thermodynamics is a useful "tool," but like any tool, its applications are limited. Both thermodynamic and kinetic considerations are important to an understanding of chemical processes.

Thermodynamic calculations (Unit 5) suggest that an equimolar mixture of $H_2(g)$ and $Cl_2(g)$ is "unstable" with respect to formation of $HCl(g)$ at room temperature. Reaction should occur spontaneously. Certainly, according to all concepts of gas behavior, hydrogen and chlorine molecules in such a mixture must collide frequently. Nonetheless, no reaction occurs at a measurable rate if the gas mixture is kept in the dark at room temperature. Exposure of the mixture to an electric spark or to light of appropriate wavelength produces an explosive reaction, with essentially quantitative formation of HCl. Apparently, predictions of spontaneity of a reaction do not provide information on the rate at which reaction occurs or the particular conditions necessary for a reasonable speed of product formation.

Chemical kinetics is the study of reaction rates as influenced by such factors as the "nature" of reacting species, the temperature of the system, the type of interparticle contact possible, the effect of catalysts (recoverable substances that change reaction speed), and the concentrations of chemical substances involved. Kinetics research has applications in such diverse fields as the elucidation of chemical changes in living organisms and the engineering requirements for maximum efficiency of industrial processes.

16.2 THE "NATURE" OF CHEMICAL REACTANTS

One of the reactions of considerable importance in organic synthesis is the conversion of alcohols to organic halides for use as intermediates in the preparation of many different types of compounds. Consider the following examples.

Molecular size

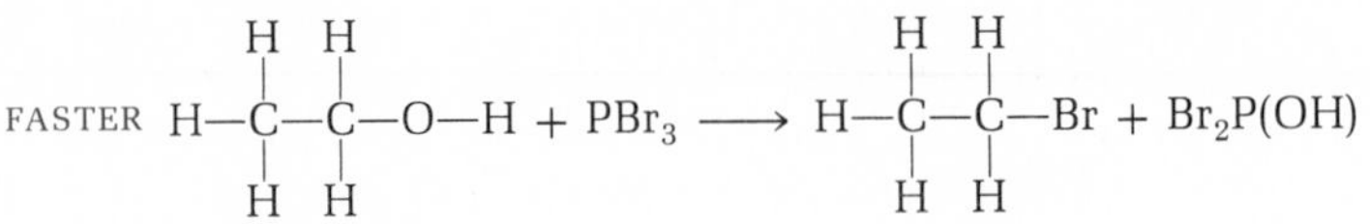

FASTER

$$\underset{\substack{\text{ethyl alcohol}\\\text{(ethanol)}}}{H-\overset{\overset{\displaystyle H}{|}}{\underset{\underset{\displaystyle H}{|}}{C}}-\overset{\overset{\displaystyle H}{|}}{\underset{\underset{\displaystyle H}{|}}{C}}-O-H} + PBr_3 \longrightarrow \underset{\substack{\text{ethyl bromide}\\\text{(bromoethane)}}}{H-\overset{\overset{\displaystyle H}{|}}{\underset{\underset{\displaystyle H}{|}}{C}}-\overset{\overset{\displaystyle H}{|}}{\underset{\underset{\displaystyle H}{|}}{C}}-Br} + Br_2P(OH)$$

SLOWER $\underset{\substack{\text{normal-octyl alcohol}\\\text{(1-octanol)}}}{CH_3(CH_2)_6CH_2OH}$ + PBr$_3$ $\longrightarrow$ $\underset{\substack{\text{n-octyl bromide}\\\text{(1-bromooctane)}}}{CH_3(CH_2)_6CH_2Br}$ + Br$_2$P(OH)

Note that this argument applies only to the length of time between collisions, since the energy of collision will be equivalent (average kinetic energy of colliding particles is determined by the temperature).

The influence of molecular size and geometry is commonly referred to as the "steric" effect.

Two reasons can be suggested for the observed difference in rate of reaction. First, the 1-octanol is a heavier molecule (130:46) than ethanol. At the same temperature both substances should (as a first approximation) have the same average molecular kinetic energy ($\frac{1}{2}ms^2$), so the heavier molecules will have a lower average speed. Since the reaction depends on collision of an alcohol molecule with a molecule of phosphorus tribromide with sufficient energy to break C—O and P—Br bonds, such effective collisions will occur less frequently with the heavier molecules.

The second, and much more important, factor concerns the probability that properly oriented PBr$_3$ molecules will collide *with the carbon bonded to the* —OH *(hydroxyl) group*, also properly oriented (Figure 16.2). The odds of a PBr$_3$ collision with some other part of the molecule are obviously considerably greater for the 1-octanol than for the ethanol. As a result, the rate of effective collision, and hence

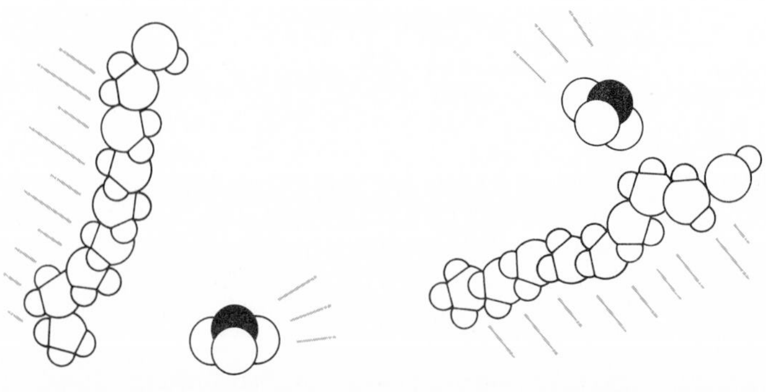

Figure 16.2 Collision efficiency (a) Ineffective (rebound—no reaction). (b) Effective (alcohol $\longrightarrow$ halide conversion). Note importance of "steric factor."

(a)   **(b)**

the rate of overall reaction, should be slower for the 1-octanol than for the ethanol.

Electrostatic effects

Isomeric compounds are those having the *same molecular formula* (in this case $C_4H_{10}O$) but *different structures* (i.e., different arrangements of the atoms within the molecule).

FAST

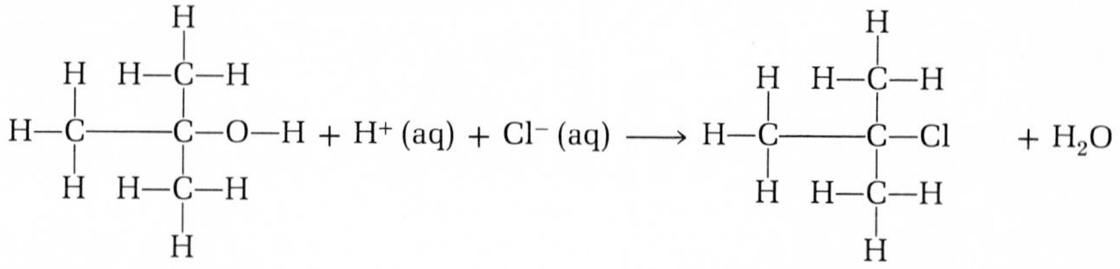

$$\text{tertiary-butyl alcohol} + H^+(aq) + Cl^-(aq) \longrightarrow \text{t-butyl chloride} + H_2O$$

tertiary-butyl alcohol
(2-methyl-2-propanol)

t-butyl chloride
(2-chloro-2-methylpropane)

SLOW

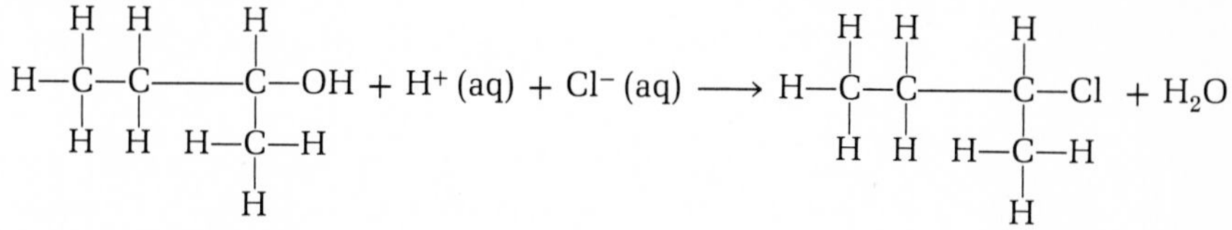

$$\text{secondary-butyl alcohol} + H^+(aq) + Cl^-(aq) \longrightarrow \text{secondary-butyl chloride} + H_2O$$

secondary-butyl alcohol
(2-butanol)

secondary-butyl chloride
(2-chlorobutane)

EXTREMELY SLOW

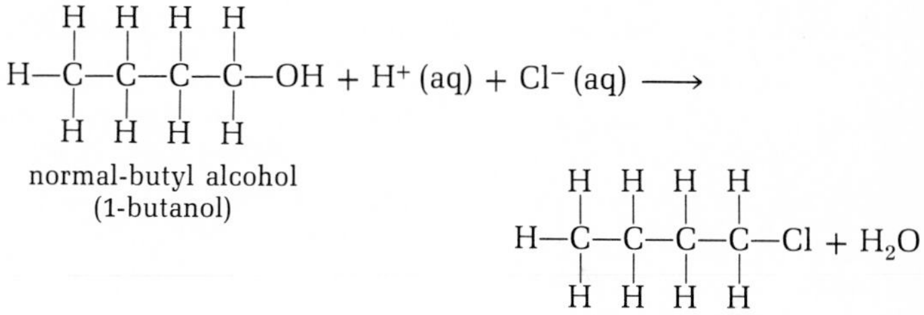

$$\text{normal-butyl alcohol} + H^+(aq) + Cl^-(aq) \longrightarrow$$

normal-butyl alcohol
(1-butanol)

$$\longrightarrow \text{n-butyl chloride} + H_2O$$

n-butyl chloride
(1-chlorobutane)

One can hardly invoke the explanations suggested for the 1-octanol-to-ethanol comparison for isomeric alcohols, since these all have the same mass. Admittedly, differences in molecular geometries (steric effects) among isomers should affect collision probabilities, but surely not to an extent commensurate with the large differences in observed reaction rates.

A better explanation in this case lies with the electrostatic effects of attached atoms on the electron density of the carbon atom bonded to oxygen. The mechanism by which an alcohol reacts with aqueous HCl involves the intermediate formation of a *carbonium* ion (Figure 16.1). The energy of such an ion is lowered by the presence of *alkyl* groups (e.g., methyl, CH_3) attached to the positive carbon atom, since such groups act to decrease the positive charge density to a greater extent than does a hydrogen atom alone. As we shall see, a lower "energy barrier" to overcome results in a more rapid reaction. These so-called electrostatic, or *inductive*, effects are discussed in more detail in later units.

16.3 TEMPERATURE EFFECTS

It is observed that most chemical reactions proceed more rapidly as the temperature is increased.

An oily dustcloth casually tossed into a box of excelsior in a storage closet under the stairs was the cause of a disastrous fire in a midwestern elementary school some years ago. *Spontaneous combustion* was the official report. The hydrocarbons in the oil slowly reacted with oxygen in the air, liberating a small amount of heat. Since the cloth was well insulated, the heat was only slightly dissipated. Most of the heat generated served to raise the temperature of the remaining oil. At the increased temperature oxidation was more rapid, producing heat more rapidly and again raising the temperature of the oil. This process was continuous, over a period of weeks, until the temperature of the residual oil finally reached the *combustion temperature* (the temperature at which flame is produced). The cloth and excelsior were ignited.

Biochemical reaction rates in living organisms are also temperature sensitive. Frozen or refrigerated foods resist spoilage much longer than foods at room temperature because of the temperature dependence of the chemical reactions in bacterial cells. Insects may be "anesthetized" by chilling them in a refrigerator. Metabolism rates decrease in hibernating animals during cold weather. Special chilling processes are even finding medical applications in situations in which it is desirable temporarily to reduce biochemical reaction speeds in humans.

Not all chemical reactions are influenced in the same way by temperature changes. Although a large number of simple reactions follow the rule of thumb that the reaction speed doubles for each 10°C increase in temperature, many reactions show a different rate dependence. Some are nearly independent of temperature, and a few proceed more rapidly at low temperatures than at high temperatures.

We shall see, in Section 16.13, how appropriate data may be employed to predict the temperature-dependence of reaction rates.

It is not possible to predict the rate of a reaction at any temperature without experimental data relating reaction kinetics and temperature.

A knowledge of the stoichiometry and thermodynamics of a reaction will not permit rate predictions. If the mechanism of the reaction is known, qualitative estimates of temperature effects may be possible, but usually only for relatively simple cases. Complex reactions require experimental determination of temperature dependence, as discussed in Section 16.13.

Since the temperature is associated with the average kinetic energy of all the particles in a reaction mixture, it is reasonable to suggest that average particle speed increases as the temperature is raised. It can be demonstrated, however, that the resulting increase in collision frequency is too small to account by itself for the rate increase in processes occurring by simple collision of two particles. The key to the problem of the large temperature dependence of such reactions lies with the *energy of collision*, which is a major factor to be considered with collision frequency (Figure 16.3).

It is important to realize that no chemical reaction can occur unless reactant particles collide with sufficient net energy to produce a chemical change. Even if the

Figure 16.3 Boltzmann energy distribution. The fraction of particles having energies high enough for *effective* collisions increases rapidly with increasing temperature.

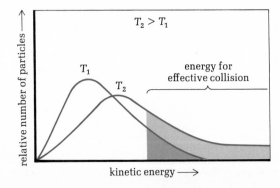

Figure 16.4 Particle size effects. (a) A cube of volume 1 cm³ containing pure X collides with an equal volume cube of pure Y. (X + Y ⟶ products). Maximum collision would involve face-to-face contact; that is, the maximum possible reactions on first collision of these cubes would be proportional to the surface areas, 1 cm² of X and 1 cm² of Y. Most of the particles would be unavailable for collision-produced reaction. (b) Each of the preceding cubes can be divided into 1000 equal cubes, each having single face areas of 0.01 cm². Maximum possible face-to-face collisions would now involve 10 cm² (1000 × 0.01) each of X and Y. Speed of product formation should be increased. (It is unlikely that this would be a simple tenfold increase. Why?) A larger fraction of the particles has become available for collision-produced reaction. (c) X and Y are dissolved. Maximum possible collisions have been greatly increased. All particles are "available" for reaction, and the rate should be vastly increased.

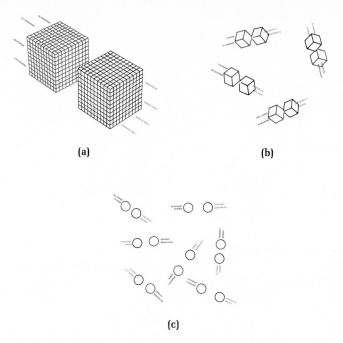

colliding particles strike each other with exactly optimum orientations (Figure 16.2), the collision will not be *effective* (result in reaction) unless it is sufficiently energetic. Otherwise the particles will simply "bounce off" each other. The *Boltzmann energy distribution* (Figure 16.3) indicates the influence of increasing the temperature of a chemical system. Not only does the average kinetic energy of the particles increase, but—more important—the *fraction of high energy particles* in the system increases with increasing temperature. As a result, the *frequency of effective collisions* (the reaction rate) is greater at higher temperatures.

In the few cases for which reaction rates do not increase with temperature, careful analysis reveals special complications. The explanations developed for these systems are still consistent with the Boltzmann distribution.

16.4 INTERPARTICLE CONTACT

When large crystals of potassium iodide are mixed with large crystals of mercury(II) nitrate, a red film of HgI_2 is formed slowly at the points of contact between crystals of the two reagents. If the crystals are ground together or if the substances are mixed as fine powders, the mercury(II) iodide forms more rapidly. If solutions of Hg^{2+} and I^- are mixed in stoichiometric proportions, an essentially instantaneous precipitation of $HgI_2(s)$ occurs.

Large lumps of coal are difficult to ignite and burn slowly once combustion begins. Fine coal dust is easily ignited by a spark, and many coal mining accidents have resulted from such dust explosions. Apparently particle size has a rather major effect on reaction speed. To understand this, consider the situations shown in Figure 16.4.

In addition to factors related directly to particle size, considerations of particle geometry are also important. Reactions occurring between ions of opposite charge and essentially spherical charge distribution (Figure 16.5) are usually rapid. Effective collisions can occur from any relative direction of approach. In other cases (e.g.,

Some ions that have spherical charge distribution when isolated (free from influence of other particles) are not spherically symmetrical when solvated.

Figure 16.5 Collision of "spherical" ions,

$Ag^+(aq) + Cl^-(aq) \longrightarrow AgCl(s)$

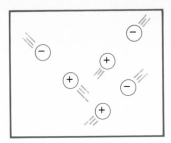

Figure 16.2) the relative particle geometry at the time of collision is of major importance in determining whether or not a reaction can result from the particular collision. Usually the more complex the particle geometry, the slower the reaction, because of the low probability of *effective* collision.

16.5 CATALYSIS

Activation energy is, in simple cases, the minimum net energy of particles that will result in effective collision.

The activation energy may be related to the difference between the energy states of reactants and the intermediate *transition state* (Section 16.12).

Charcoal heated to remove previously adsorbed gases is said to have been "activated."

Many reactions that proceed slowly under some specific conditions of temperature and concentration can be speeded up significantly by the addition of a small amount of a *catalyst*. These substances act to increase reaction rates, sometimes by providing a new mechanism of reaction of lower activation energy and sometimes by affording a surface on which reactants are adsorbed for more facile reaction. There are many unanswered questions about the roles of catalysts, and these offer interesting research areas in fields ranging from solid state physics to molecular biology. The only generalizations about catalytic behavior are that a definite increase in reaction rate can be observed, and that the catalyst can, theoretically, be completely recovered when reaction is complete.

In *heterogeneous* catalysis the reactants are in a separate phase (or phases) from the catalytic agent. Many of the reactions utilized by the chemical industry on a massive scale involve this type of catalysis, frequently using finely divided metals such as platinum or nickel as surface (contact) catalysts (Figure 16.6). It is postulated that there are two fundamentally different types of adsorption on surfaces of materials. *Physical adsorption,* such as probably occurs when activated charcoal "collects" various gases on its surface, is believed to involve van der Waals forces between the adsorbed molecules and surface particles of the solid material. *Chemisorption,* on the other hand, involves forces more closely approximating those of a covalent bond. In these situations, then, adsorbed molecules, ions, or atoms undergo chemical changes and if the chemisorption is part of a catalytic process, these changes render the adsorbed substances more reactive.

Some of the more important industrial processes involving heterogeneous catalysis include the following.

Figure 16.6 Postulated chemisorption of hydrogen on a platinum surface. The H—H bond is broken by formation of a Pt—H "bond."

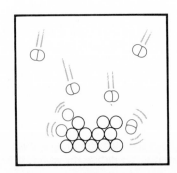

Note that catalysts do not *cause* reactions to proceed unless they are already spontaneous, nor do catalysts affect the relative amounts in an equilibrium mixture. They do decrease the time for equilibrium to be attained (Unit 17). The role of a catalyst in lowering the activation energy is discussed in Section 16.12.

It is estimated that industrial catalysis is involved in more than $100 billion in production annually.[3]

The Haber process

$$N_2(g) + 3H_2(g) \rightleftharpoons 2NH_3(g)$$

(catalyzed by iron containing, for better activity, a "promoter" such as traces of vanadium)

The contact process

$$2SO_2(g) + O_2(g) \rightleftharpoons 2SO_3(g)$$

(catalyzed by V_2O_5 or platinum; a step in conversion of SO_2 to H_2SO_4)

Hydrogenation of C—C double bonds

(catalyzed by platinum; used in conversion of vegetable oils to more viscous materials for margarine, etc.)

Homogeneous catalysis utilizes a catalytic agent in the same phase as one or more of the reactants. It is believed that this substance participates in the reaction to provide an alternative, and more rapid, reaction sequence, which eventually reforms the original catalyst. For example, the decomposition of aqueous hydrogen peroxide at room temperature is relatively slow:

$$2H_2O_2(aq) \longrightarrow 2H_2O(l) + O_2(g)$$

The reaction is catalyzed by $Fe^{3+}(aq)$, probably by the participation of Fe^{3+} and Fe^{2+} in two reactions that are more rapid than the simple decomposition of hydrogen peroxide alone.

FIRST $2Fe^{3+}(aq) + H_2O_2(aq) \longrightarrow 2Fe^{2+}(aq) + O_2(g) + 2H^+(aq)$

SECOND $2Fe^{2+}(aq) + H_2O_2(aq) + 2H^+(aq) \longrightarrow 2Fe^{3+}(aq) + 2H_2O(l)$

Note that the net effect of these two reactions is the same as the uncatalyzed reaction. The only difference is that the presence of a small amount of Fe^{3+} provides alternative routes that are more rapid. In homogeneous catalysis, the reaction rate is usually a function of the concentration of the catalyst. (See, for example, Practice for Proficiency, question 15.)

A special type of catalysis occurs when one of the products of a reaction catalyzes the reaction itself. An example of such an *autocatalytic* process is the reaction of permanganate ion with oxalic acid:

$$6H^+ + 2MnO_4^- + 5H_2C_2O_4 \longrightarrow 8H_2O + 2Mn^{2+} + 10CO_2$$

The reaction is catalyzed by Mn^{2+}.

Enzyme catalysis (Unit 34) is one of the aspects of biochemistry currently receiving a great deal of attention. Although many aspects of enzyme function are still largely speculative, it is apparent that these macromolecules are extremely efficient catalysts. The major portion of an enzyme molecule is protein (Unit 32). Usually participating with this protein during the catalytic function of the enzyme is a small organic molecule called a *coenzyme* or some metal ion such as Mg^{2+} or Fe^{3+}. The enzymes not only speed up reactions in living organisms to the rates required, but

[3] For an interesting discussion, see "Catalysis," by V. Haensel and R. L. Burwell, Jr., in the December, 1971, issue of *Scientific American* (pp. 46–58).

also select rather specifically the type of reaction they will catalyze. Thus the α-amylase enzyme system in human saliva catalyzes the reaction between starches and water to form glucose, but does not catalyze the reaction of water with celluloses to form glucose. Enzymes are usually *stereoselective* catalysts; that is, the molecular geometry of the reactant involved determines whether or not it can "fit" into the catalytic region (active site) of the enzyme.

Sometimes it is desirable to slow down the rate of reaction by addition of a chemical substance called an *inhibitor*. Inhibitors are not "negative catalysts" since, unlike true catalysts, the inhibitors generally act to "scavenge" chemical species that might otherwise participate in a reaction as a reactant or a catalyst or to consume reactive intermediates before these can take part in subsequent steps of a reaction sequence.

CATALYSIS IN THE BOMBARDIER BEETLE

The bombardier beetle utilizes a rather unique chemical defense system. When alarmed or angered, the beetle forcefully ejects a spray propelled by oxygen gas from the decomposition of hydrogen peroxide:

$$2H_2O_2(aq) \longrightarrow 2H_2O(l) + O_2(g)$$

Although hydrogen peroxide will decompose spontaneously to oxygen and water, the reaction is very slow unless a catalyst is present.

The bombardier beetle stores hydrogen peroxide (about 7 M) in one small reservoir and hydroquinone (about 1 M) in another. The reaction involves the mixing of these two solutions, resulting in both the decomposition of hydrogen peroxide and the conversion of hydroquinone to quinone:

(hydroquinone)   (quinone)

It has been demonstrated that quinone catalyzes both of the reactions, so that the "firing process" of the beetle is essentially autocatalytic. The quinone formed is not very soluble in water, so some of this product is expelled as a puff of smoke following the ejection of the spray. Some of the solid quinone apparently remains behind in the "mixing chamber" as a "primer" catalyst for the next time the beetle needs to mix its reagents.

16.6 CONCENTRATION EFFECTS: RATE LAWS

Iodine monobromide gas decomposes at elevated temperature to form bromine and iodine gases:

$$2IBr(g) \longrightarrow I_2(g) + Br_2(g)$$

The progress of the reaction can be monitored by the appearance of the violet color of iodine vapor. It is found experimentally that the rate of formation of I_2 is *four* times as great with an initial concentration of 2 moles liter^{-1} of IBr as it is when the concentration is originally 1 mole liter^{-1}, at constant temperature. We can express this effect in terms of what is called a *rate law*.

$$\text{rate of formation of } I_2 = k \times C_{IBr}^2$$

in which k is a *rate constant,* numerically equal to the speed of the reaction when $C_{IBr} = 1$ mole liter^{-1}. It should be noted that although the reaction *rate* (i.e., the change of reactant or product concentration with time) is a function of concentration, the *rate constant* is concentration independent at a particular temperature. The temperature variation of rate constants is discussed in Section 16.13.

Now, we may suggest a simple *mechanism* that is consistent with this experimental rate law. Suppose that the only "step" in this process involves the collision of two iodine monobromide molecules with sufficient energy and proper orientations to cause reaction to occur. Then how would a change in concentration affect this process? Remember that the average translational speed of the IBr molecules is determined only by the temperature and the characteristics of IBr molecules, so we may assume that the average speeds are independent of concentration. Now let us consider how the number of possible two-molecule collisions varies with concentration (Figure 16.7). It is apparent that the number of possible collisions more than doubles when the concentration is doubled, but it does not appear to be a function of the square of the concentration. However, such a function is approached as the concentration, in terms of particles per unit volume, gets larger.

C (particles per volume)	POSSIBLE COLLISIONS	$\frac{1}{2} C^2$ FACTOR[a]	PERCENT DIFFERENCE[b]
4	6		
8	28	32	12.5
16	120	128	6.2
32	496	512	3.1
64	2016	2048	1.6
128	8128	8192	0.8

[a] The $\frac{1}{2}$ factor accounts for the fact that collisions involves *pairs* of molecules, so the number of colliding pairs is only $\frac{1}{2}$ the number of molecules.
[b] Note that the difference between "real" and "theoretical" collisions decreases as the number of particles available increases.

The idea that the rate of a chemical reaction is proportional to the number of reactant collisions per unit time is the basis of the *collision theory* of reaction kinetics.

It can be shown that the number of possible collisions per unit time (i.e., the frequency of possible two-body collisions) *under conditions of the same average particle speeds* is proportional to a simple concentration function.

$$\text{two-body collision frequency} \propto (C)(C - 1)$$

In any realistic situation in a chemical process, $C - 1$ is insignificantly smaller than C. (Remember the magnitude of Avogadro's number.) As a result, for all practical purposes,

$$\text{two-body collision frequency} \propto C^2$$

Figure 16.7 Collision possibilities.

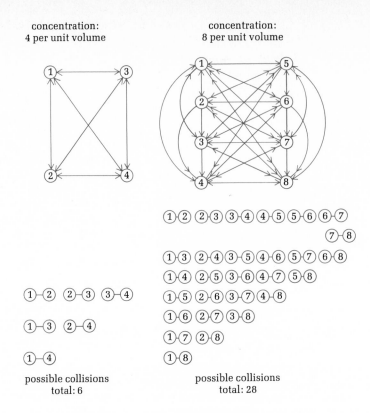

concentration: 4 per unit volume

concentration: 8 per unit volume

possible collisions total: 6

possible collisions total: 28

Now, we have already suggested that only a fraction of the total possible collisions will result in reaction. Collisions must be sufficiently energetic to supply the energy needed to break the iodine-bromine bond (Unit 5), but the fraction of high-energy particles is a function of temperature (Figure 16.3), not of concentration. In addition, molecules must collide with appropriate relative orientations if reaction is to occur. In any very large number of IBr molecules, the fraction oriented for proper collision will be determined by a statistical probability that is again independent of concentration.

These arguments suggest that the frequency of *effective* collisions (those resulting in a chemical reaction) should be proportional to the frequency of total two-body collisions. Thus for the reaction in question we may express this by use of a proportionality constant as

$$\text{frequency of effective collisions} = k \times C_{IBr}^2$$

We have intentionally omitted certain complications, such as the fact that some molecules may get in the way of the "possible" collisions described. This factor becomes increasingly important at high concentrations. Our simple treatment of reaction by collision must be recognized as an idealized, non-rigorous description.

This is identical with the experimental rate law, indicating that the simple mechanism proposed is consistent with the kinetic data.

Although reaction rates usually increase with increased initial concentrations of reactants, it is rare that such a simple situation occurs. Reactions involving combination of two ions of opposite charge may follow such a pattern, but the reaction speed in such cases is usually too rapid for simple study of concentration effects.

Most reactions involving molecules and most oxidation-reduction (electron-transfer) reactions proceed by a series of steps rather than by simple collisions.

The balanced equation for a reaction does not necessarily indicate anything about the concentration dependence of the reaction rate.

The rate law for any reaction must be determined experimentally by study of the reaction speed as influenced by variation in reactant concentrations. The techniques used for such studies vary widely, depending on the complications posed by different types of reactions. One procedure involves "flooding" the reaction with one reactant, that is, adding sufficient excess so that the change in concentration of that reactant is negligible, while studying the rate as influenced by changes in concentration of another reactant. An alternative approach uses sets of experiments in which the concentration of only one reactant is varied in each experiment, thus allowing the concentration-dependence of rate to be determined for each reactant by comparison of appropriate experimental data.

Rate data may be obtained in various formats. For reasons we shall explore later, it is particularly useful to employ the concentration-dependence of *initial rates* (during the period in which only a tiny fraction of reactants have been consumed). Examples 1 and 2 illustrate this technique.

Example 1

The rate of reaction of nitrogen dioxide with carbon monoxide,

$$NO_2(g) + CO(g) \longrightarrow CO_2(g) + NO(g)$$

can be determined by measuring the rate of disappearance of the brownish-orange NO_2, the only component of the reaction mixture that is colored. Experiments show that the following relative rates are observed at 300°C.

EXPERIMENT NUMBER	INITIAL C_{NO_2} (mole liter^{-1})	INITIAL C_{CO} (mole liter^{-1})	RELATIVE INITIAL RATES[a]
1	0.1	0.1	1
2	0.1	0.02	1
3	0.1	0.01	1
4	0.02	0.1	0.04
5	0.01	0.1	0.01

[a] Rate of disappearance of NO_2 color determined by time required for a 5-percent change in color intensity. *Relative* rates are compared to rate for Experiment 1.

Solution

Comparison of experiments 1, 2, and 3 indicate that variation in concentration of carbon monoxide has no effect on the reaction speed. Experiments 1, 4, and 5 reveal that NO_2 concentration *does* influence the rate. To determine the nature of this influence, we may compare two concentrations in terms of their corresponding relative rates.

Since, for simple cases, the reaction rate is proportional to some exponential function of concentration (rate $\propto C_{reactant}^x$), we may compare two such proportions when only the reactant involved is varied.

$$\left(\frac{C_a}{C_b}\right)^x = \frac{\text{rate}_a}{\text{rate}_b}$$

A value of $x = 1$ indicates that rate varies directly with concentration, whereas a value of $x = 2$ suggests that rate varies with the square of the concentration. Selecting, for example, experiments 4 and 5,

$$\left(\frac{0.02}{0.01}\right)^x = \frac{0.04}{0.01}$$

$$2^x = 4$$

$$x = 2$$

We may conclude, then, that rate $\propto C_{NO_2}^2$.

One way of expressing this relationship uses a mathematical symbolism:

$$-\frac{dC_{NO_2}}{dt} = k \times C_{NO_2}^2$$

This is called the *differential rate law* and represents a change of concentration with respect to time. The negative sign indicates concentration *decrease* with time. We could, in this case, have equally well selected

$$-\frac{dC_{CO}}{dt}, \frac{dC_{NO}}{dt}, \quad \text{or} \quad \frac{dC_{CO_2}}{dt}$$

to represent the *rate* of reaction. While these differential expressions will be familiar to the student of calculus, we will use the more generally familiar verbal terminology. In this case, for example, we would write:

rate of NO_2 consumption $= k \times C_{NO_2}^2$

The approach used in this example is limited to fairly simple cases, but we shall find it satisfactory for our purposes.

The concentration exponent for a species appearing in a rate-law expression is referred to as the *reaction order* in that species, and the sum of all rate-law exponents gives the *overall reaction order*. Concentration exponents may be whole numbers, fractions, or zeros, and the rate laws may include some or all of the reactant species, product species, or catalysts. We will limit our studies to the simple cases in which only reactants, or reactants and catalysts, appear in the rate laws with concentration exponents of two (second order), one (first order), or zero (zeroth order), and in which overall reaction orders are no greater than "third." In the case of "zeroth order," of course, the concentration term is not actually shown in the final rate law, since any quantity with an exponent of zero is equal to unity.

Although simple rate laws are easily determined from data on the variation of "initial rate" with concentration (Examples 1 and 2), these data are not always easy to obtain accurately. From a graphical display of kinetics data, a rate can be esti-

Figure 16.8 A concentration-versus-time plot for a simple second-order reaction. The tangent to the curve at the point (3.6 min, 0.65 mole liter^{-1}) indicates a rate of -6.3×10^{-2} mole liter^{-1} min^{-1}, from the slope (0.82/13). (Note that this is only an illustration. Rate estimation by this method would only be used when the rate law was not yet known.)

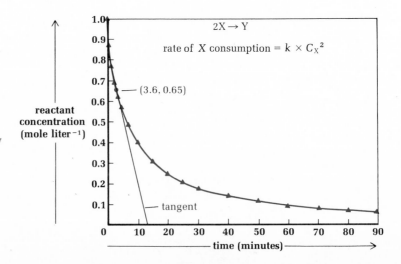

Concentration-versus-time curves are most nearly linear for first- or second-order processes near the high-concentration and low-concentration parts of the curve, so slopes are easiest to estimate in these regions. The high-concentration region is generally preferred since a minor estimation error represents a smaller *relative* uncertainty with respect to a large rate than to a small rate.

The limited solubility of bromine in water requires the use of dilute solutions for this particular study.

mated from the slope of the concentration-versus-time curve at a particular point (Figure 16.8). Such a graph clearly illustrates the concentration dependence of reaction rate and the value of using a rate estimation near the initial part of the curve where the slope is relatively large (i.e., large concentration change per unit time).

Example 2

The rate of reaction of hydrogen peroxide with hydrogen bromide in dilute aqueous solution can be measured in terms of the rate of formation of colored $Br_2(aq)$:

$$H_2O_2(aq) + 2H^+(aq) + 2Br^-(aq) \longrightarrow 2H_2O(l) + Br_2(aq)$$

Experiments show the following relationships between initial concentrations and initial rates:

EXPERIMENT NUMBER	INITIAL $C_{H_2O_2}$ (M)	INITIAL C_{H^+} (M)	INITIAL C_{Br^-} (M)	RELATIVE INITIAL RATES[a]
1	0.1	0.1	0.1	1
2	0.01	0.1	0.1	0.1
3	0.1	0.01	0.1	0.1
4	0.1	0.1	0.01	0.1

[a] Rate determined by appearance of color of Br_2 (aq) in terms of time required for formation of 5 percent of stoichiometric amount of Br_2. Relative rates are compared to rate for experiment 1.

Solution

Hydrogen peroxide (comparing experiments 1 and 2, in which only H_2O_2 is varied):

$$\left(\frac{0.1}{0.01}\right)^x = \frac{1}{0.1}$$

$$10^x = 10$$

$$x = 1$$

Hydrogen ion (comparing experiments 3 and 1, in which only H^+ is varied):

$$\left(\frac{0.01}{0.1}\right)^x = \frac{0.1}{1}$$

$$(0.1)^x = 0.1$$

$$x = 1$$

Bromide ion (comparing experiments 1 and 4, in which only Br^- is varied):

$$\left(\frac{0.1}{0.01}\right)^x = \frac{1}{0.1}$$

$$10^x = 10$$

$$x = 1$$

It is important to recognize that rate law exponents, unlike those used in equilibrium expressions (Unit 17), cannot be deduced from the coefficients of a balanced equation. The rate law exponents can *only* be determined from experimental kinetics data.

In this case, data show the reaction speed to be proportional to the concentrations of all three species. Hence the rate law may be expressed as

$$\text{rate of } Br_2 \text{ formation} = k \times C_{H_2O_2} \times C_{H^+} \times C_{Br^-}$$

Note that, in both Examples 1 and 2, the concentration exponents in the rate law are different from the numerical coefficients in the balanced net equation.

16.7 HALF-LIFE The use of kinetics data employing "relative initial rates" is a fairly simple way of determining reaction order by considering one reactant at a time (Example 2). An alternative method, often more accurate for interpreting "raw" kinetics data (i.e., measurements of reactant or product concentrations at different times during the course of a reaction), employs relationships obtained by the mathematical process of *integration* of the differential forms of simple rate laws (Table 16.1). This method, illustrated for the case of a single reactant in Example 3, may be applied to multiple-reactant processes (Example 4) in two ways. First, we may determine the order in each reactant by considering how the half-life is affected by varying the initial concentration of only one reactant at a time. Second, we may determine the rate constant for the process from a *half-life* (the time required for consumption of half of the original limiting reagent), measured under the *specific conditions* of initial reactant concentrations in the stoichiometric ratios given by the balanced equation. In this case the concentrations of all reactants may be converted to their *equivalents* in terms of the concentration of a single reactant (e.g., X in Table 16.1).

The derivation of the equations used in Table 16.1 requires the use of calculus. Interested students may wish to discuss these relationships with their instructor.

Students of calculus will note that the differential expressions can be integrated over a time interval t during which the concentration of X changes from some initial value, $C_{X(0)}$, to some smaller value at time t, $C_{X(t)}$. This time can be selected as any interval desired, but it is particularly useful to consider the special case in which the time is that required for conversion of X to one-half its original concentration. This interval is called the *half-life*, $t_{1/2}$. Appropriate integrations, then, yield two important relationships (Table 16.1).

These expressions offer two ways to find the reaction order in a single species. One method is to find the half-life of the reaction for several different initial concentrations.

If the half-life is constant, the reaction is first order in X. If the half-life varies inversely with concentration (e.g., half-life doubles when initial concentration is halved), the reaction is second order in X. If the half-life varies directly with concentration, then the reaction *rate* is independent of concentration of X (zero order). Determination of half-life time and reaction order also permits calculation of the rate constant k.

Note that for a zero order process the *speed* of the reaction is independent of the concentration of X, but the time required to consume half the quantity depends on the amount (concentration) initially present.

Table 16.1 ALTERNATIVE EXPRESSIONS FOR KINETICS DATA

REACTION ORDER IN X	DIFFERENTIAL RATE LAW[a]	INTEGRATED FORMULATION	HALF-LIFE
zero	$\dfrac{-dC_X}{dt} = k$	$C_{X(t)} = C_{X(0)} - kt$	$\dfrac{C_{X(0)}}{2k}$
first	$\dfrac{-dC_X}{dt} = kC_X$	$2.303 \log \dfrac{C_{X(t)}}{C_{X(0)}} = -kt$	$\dfrac{0.693}{k}$
second	$\dfrac{-dC_X}{dt} = kC_X^2$	$\dfrac{1}{C_{X(t)}} - \dfrac{1}{C_{X(0)}} = kt$	$\dfrac{1}{kC_{X(0)}}$
third	$\dfrac{-dC_X}{dt} = kC_X^2$	$\dfrac{1}{C_{X(t)}^2} - \dfrac{1}{C_{X(0)}^2} = 2kt$	$\dfrac{3}{2kC_{X(0)}^2}$

[a] Remember that $-dC_X/dt$ means "rate of X consumption."

Example 3

From the following data, determine the rate law and the rate constant at $400°C$ for the reaction,

$$2NO_2(g) \longrightarrow 2NO(g) + O_2(g)$$

The process may be monitored by the disappearance of the red-brown color of NO_2.

EXPERIMENT NUMBER	INITIAL C_{NO_2} (mole liter^{-1})	HALF-LIFE (seconds)	(temperature: 400°C)
1	0.25	0.83	
2	0.10	2.08	
3	0.05	4.17	

Solution

Since the half-life increases as the initial concentration of NO_2 is decreased, the reaction must be either second order or third order in NO_2 (Table 16.1). A quantitative comparison from any two experiments shows the variation of half-life to be inversely proportional to initial concentration (e.g., from experiments 2 and 3, the half-life is doubled when the initial concentration is halved). Thus the reaction is second order in NO_2 and the rate law is

rate of NO_2 consumption $= kC_{NO_2}^2$

The rate constant may then be calculated from any one of the experiments: (Table 16.1)

$$t_{1/2} = \frac{1}{kC_{X(0)}}$$

$$k = \frac{1}{t_{1/2}C_{X(0)}}$$

from experiment 1,

$$k = \frac{1}{0.83 \text{ sec} \times 0.25 \text{ mole liter}^{-1}}$$

$$k = 4.8 \text{ liter mole}^{-1} \text{ sec}^{-1}$$

Example 4

The reaction between nitric oxide and bromine to form nitrosyl bromide is rapid at 550°K:

$$2NO(g) + Br_2(g) \longrightarrow 2NOBr(g)$$

A spectrophotometric determination of bromine concentration as a function of time permits a relatively accurate determination of half-life. From the following data, find the rate law and the rate constant for this process at 550°K.

EXPERIMENT NUMBER	INITIAL C_{NO} (mole liter^{-1})	INITIAL C_{Br_2} (mole liter^{-1})	$t_{1/2}$ (hours)	(temperature: 550°K)
1	0.20	0.20	10.8	
2	0.20	0.10	10.8	
3	0.10	0.10	21.6	

Solution

Since we note that the half-life does not change when the initial concentration of bromine is varied, we may conclude (Table 16.1) that the reaction is first order in Br_2.

The inverse relationship between initial NO concentration and half-life (half-life doubled when concentration is halved, experiments 2 and 3) shows the reaction is second order in nitric oxide. The rate law, then, must be

$$\text{rate of NOBr formation} = k \times C_{NO}^2 \times C_{Br_2}$$

Since the overall reaction order is three, the rate constant may be determined from experiment 2, in which initial concentrations are in the stoichiometric ratio of the balanced equation ($2NO:1Br_2$). Thus all reactant concentrations are equivalent to 0.20 mole liter^{-1} (NO). From Table 16.1,

$$t_{1/2} = \frac{3}{2kC_{X(0)}^2}$$

$$k = \frac{3}{2t_{1/2}C_{X(0)}^2}$$

$$k = \frac{3}{2 \times 10.8 \text{ hours} \times (0.20 \text{ mole liter}^{-1})^2}$$

$$k = 3.5 \text{ liter}^2 \text{ mole}^{-2} \text{ hour}^{-1}$$

An alternative technique for handling kinetics data employs the graphical treatment of data to see which of the "integrated formulations" from Table 16.1 will yield a linear plot. For example,

For a zero-order reaction, the plot of C_X versus time is linear. The slope of the line is equal to $-k$, and its intercept at the concentration axis is $C_{X(0)}$.

For a first-order reaction, the plot of log C_X versus time is linear, with slope equal to $-k/2.303$ and intercept equal to log $C_{X(0)}$.

For a second-order reaction, the plot of $1/C_X$ versus time is linear, with slope equal to k and intercept equal to $1/C_{X(0)}$.

The choice of a method for interpreting kinetics data is largely determined by the relative convenience of obtaining the appropriate experimental information. For very slow reactions, for example, it may not be feasible to continue the experiments long enough for direct determination of the half-life and the method of "relative initial rates" may be most appropriate. With rapid reactions, "relative initial rate" data are often difficult to obtain accurately.

Each particular process investigated presents its own challenges. The experimental determination of reactant or product concentrations during the course of a reaction must be done in such a way as to avoid any effects on the process being studied, while at the same time guaranteeing that the concentration measured corresponds to the appropriate time interval of the process. One could not, for example, withdraw a portion of a reaction mixture from a rapid-process study and determine a concentration by the slow method of titration. Instrumental methods that can directly follow the concentration change of a particular species (e.g., by monitoring the color intensity due to the only colored species present) are particularly useful. In other cases it may be possible to withdraw aliquots (portions) of a reaction mixture and "quench" the process (stop the process by a sudden change of conditions) to permit a more "leisurely" analysis.

Once the appropriate data are available, interpretations to yield rate laws and rate constants can provide valuable clues to the ways in which reactions proceed.

**16.8 REACTION
MECHANISMS**

The conception of chemical reactions as proceeding by the simple collision of reactant particles to form products is, with the exception of a few gas-phase processes, vastly oversimplified. Even in such an "uncomplicated" case as the combination of $Ag^+(aq)$ with $Cl^-(aq)$ to form solid AgCl, it must be understood that ions in solution are surrounded by clusters of solvent molecules. Before ions of opposite charge can combine, they must "shoulder aside" these shielding molecules (Figure 16.9).

Some of the collisions between aqueous ions are ineffective, although the electrostatic forces do tend to accelerate oppositely charged ions toward each other so that most collisions can occur with sufficient impact to shake loose intervening solvent molecules. In some cases, such as crystallization of $CuSO_4 \cdot 5H_2O$, some water molecules are bound so tightly to ions that they accompany them into the crystal lattice (e.g., as $[Cu(H_2O)_4]^{2+}$ and $[SO_4(H_2O)]^{2-}$). A balanced equation for ion combination, such as

$$Ag^+(aq) + Cl^-(aq) \longrightarrow AgCl(s)$$

indicates only that on a macroscopic scale silver ions and chloride ions tend to combine in aqueous solution to form solid silver chloride. The *equation* is not descriptive of *actual processes* at the ionic level, but only of net changes.

Reactions other than simple ion combination are usually even more complex *and seldom occur in a single step*. For example, the gas-phase reaction between carbon monoxide and nitrogen dioxide, whose rate at 300°C was shown (Example 1) to be proportional to $C_{NO_2}^2$ and independent of C_{CO}, is believed to occur by two steps:

Figure 16.9 Ion collisions in solution. (a) Ineffective collision. (b) Effective collision.

SLOW $\quad 2NO_2 \longrightarrow NO + NO_3 \quad$ (rate determining)

RAPID $\quad NO_3 + CO \longrightarrow NO_2 + CO_2$

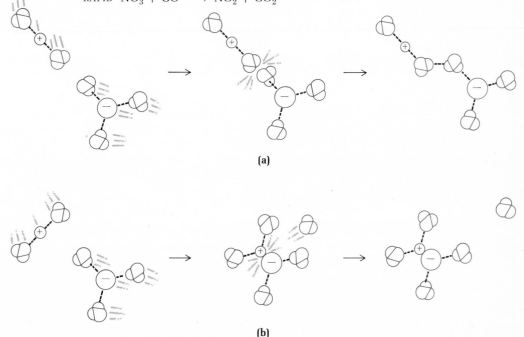

(a)

(b)

Since the second reaction is much faster than the first, *the overall rate depends only on the speed of the slow reaction.* The situation is not unlike that of registration for a freshman chemistry class at a large university. If such a process requires standing in line for 5 hours followed by filling out a form in 30 seconds, the overall rate of progress is determined, for all practical purposes, only by the speed of the first step (standing in line).

The description of a logical series of steps, the *reaction mechanism,* by which a chemical process may occur, is not an easy procedure. Chemical kinetics may determine the rate law for a reaction, but this provides information only about the slowest (rate-determining) step in the sequence. More rapid processes that precede or follow this step are more difficult to describe. Sometimes an intermediate substance can be isolated or instrumental techniques may permit detection of some transient species, but usually the rapid stages of a complex mechanism are described by a combination of various pieces of experimental data, a knowledge of the types of structures and reactions characteristic of possible species involved, and a certain chemical "intuition," based largely on experience. In any event, the equations proposed for various steps (so-called *elementary processes*) in the mechanism must add up to the overall equation for the total chemical process.

Even this is sometimes a bit misleading. Participation of one or more solvent molecules in the rate-determining step will not usually be apparent from the rate law, since solvent is usually present to such a large excess that a change in its concentration is not detectable.

In describing elementary processes, reference is frequently made to the *molecularity* of the reaction step. This term is used to refer to the number of unit particles involved as reactants in the elementary process. Thus for a single reactant particle we have a *unimolecular* process; for two particles, a *bimolecular* process; and for three, a *termolecular* process. The *collision theory* of reaction kinetics suggests that bimolecular processes would be most likely and, indeed, these are most often observed in postulated mechanisms. A termolecular process is quite unlikely for any system in which reactant particles are widely separated (e.g., in the gas phase or in dilute solution) since the *simultaneous* collision of three different particles is highly improbable. Unimolecular steps would at first glance appear to be inconsistent with the collision theory. However, we may reconcile this apparent anomaly by suggesting that the indicated reactant molecule may actually collide with some "nonreacting substance," such as a solvent molecule or the wall of a container, in order to obtain the energy necessary to "break apart," (assuming, of course, that the solvent molecule or container wall is of higher energy than the particular reactant molecule).

It should be noted that the molecularities of elementary processes are not necessarily apparent from the *reaction order.* However, for the simple mechanisms we shall consider, *the molecularity of the rate-determining step will be either the same as the reaction order* (Example 5) *or mathematically equivalent to the reaction order* (Example 6).

The description of a mechanism is not simplified by the fact that some reactions proceed by more than one mechanism. Sometimes one mechanism is favored at lower temperatures and a different mechanism is favored at higher temperatures. Sometimes reaction conditions, such as the particular solvent system involved or the acidity of the solution, favor one mechanism over another. Sometimes a catalyst, occasionally something as unexpected as the irregular surface of a broken piece of glass accidentally present in the reaction vessel, may determine which of a competitive set of mechanisms is followed. It is small wonder that the elucidation of reaction mechanisms is a research area in which the amateur investigation is seldom highly successful.

16.9 MECHANISM CLUES FROM RATE LAWS

Since the rate law for a reaction defines how the reaction speed is determined by the concentrations of particular species, and since the overall rate is determined by the

Additional clues to reaction mechanisms may be found from more complex rate laws, but we shall leave such cases for more advanced studies.

species involved in the slowest (rate-determining) elementary process, it follows that the rate law provides clues to the nature of the rate-determining step.

For the simple cases we are considering, the rate law can identify for us the *reactants* for the rate-determining step (Example 5) or for the combination of that step and a preceding "fast" process (Example 6). It is important to recognize that only the *reactants* are defined. For our purposes, the *reactants for the rate-determining step* (and, when appropriate, a preceding fast step) must be *only the species shown in the rate law* (or chemically equivalent thereto, when a preceding step is involved), with *coefficients equal to the corresponding rate-law exponents*. Reaction products, for elementary processes, must be selected to provide a balanced equation for the process, but they need not be the same as the final products for the overall reaction.

If any "intermediate" products are shown for an elementary process, subsequent processes must be suggested for consumption of these "intermediates" and for eventual formation of the final reaction products. Remember that all elementary processes must appear as balanced equations, that the total set of elementary processes must be equivalent to the overall equation (i.e., the summation of individual steps must give the overall equation), and that termolecular process are unlikely. Within the framework of these "rules," we can probably suggest a *number of plausible mechanisms* for any particular reaction. Other factors being equal, it is common practice to suggest the simplest mechanism (e.g., the one having the fewest steps) from among the alternatives devised. Let's consider first (Example 5) some second-order and first-order processes, since these do not require the suggestion of a "fast" step preceding the rate-determining step; that is, reactants for a plausible "slow" process are definable directly from the rate laws.

Example 5

Postulate plausible mechanisms for (a) the decomposition of nitrous oxide:

$$2N_2O(g) \longrightarrow 2N_2(g) + O_2(g)$$

rate of N_2O decomposition $= k \times C_{N_2O}$

and (b) the reaction between chlorine and carbon monoxide to produce phosgene:

$$Cl_2(g) + CO(g) \longrightarrow COCl_2(g)$$

rate of $COCl_2$ formation $= k \times C_{Cl_2} \times C_{CO}$

Solution

(a) $2N_2O(g) \longrightarrow 2N_2(g) + O_2(g)$

Since the rate law indicates that this is a first-order reaction, we postulate a unimolecular rate-determining step:

SLOW $N_2O \longrightarrow$ (?)

The product(s) for this step cannot be defined, but we know that molecular nitrogen is a final product, so we might suggest (to give a balanced equation)

SLOW $N_2O \longrightarrow N_2 + O$

Since atomic oxygen is not a *final* product (but a plausible intermediate that might be expected to react rapidly), we must suggest a second step in which atomic oxygen is consumed. The overall equation shows consumption of *two* N_2O species, so we might plausibly suggest that the second step involves a bimolecular reaction of N_2O with atomic oxygen. Molecular nitrogen and molecular oxygen are plausible products for this step and give a simple balanced equation:

$$\text{FAST} \quad N_2O + O \longrightarrow N_2 + O_2$$

These two elementary processes are sufficient, since they account for the consumption of two N_2O, the appearance and consumption of the intermediate, and the net formation of two N_2 and one O_2, and first order dependence on N_2O.

Proposed mechanism

$$\begin{array}{rl}
\text{SLOW} & N_2O \longrightarrow N_2 + O \\
\text{FAST} & N_2O + O \longrightarrow N_2 + O_2 \\
\hline
\text{SUMMATION} & 2N_2O \longrightarrow 2N_2 + O_2
\end{array}$$

Alternative mechanisms could be suggested that are equally consistent with the rate law. For example,

$$\text{SLOW} \quad N_2O \longrightarrow NO + N$$

$$\text{FAST} \quad NO + N_2O \longrightarrow N_2 + NO_2$$

$$\text{FAST} \quad NO_2 + N \longrightarrow N_2 + O_2$$

In the absence of other information (such as the identity of transient intermediates), we might "prefer" the original suggestion as the simpler mechanism. We must, of course, recognize this as only a *plausible* (not "definite") mechanism. (Can you suggest others?)

(b) $Cl_2 + CO \longrightarrow COCl_2$

Since the rate law indicates that this reaction is first order in each reactant, we might plausibly suggest a one-step mechanism:

$$Cl_2 + CO \longrightarrow COCl_2$$

Such a proposal is entirely feasible, and complies nicely with the model of the *collision theory*. [The *transition-state* theory (Section 16.12) suggests an alternative way of looking at such processes.]

Since we have indicated that termolecular processes are statistically unlikely, third-order reactions commonly invoke a preceding fast step before the rate-determining process. The most common suggestion involves a rapid bimolecular process establishing a *steady state* (Unit 12) in which some intermediate is consumed by a subsequent bimolecular rate-determining step as rapidly as it is generated by the initial step (thus maintaining a constant "steady state" concentration of the intermediate during most of the course of the reaction).

Let's consider a hypothetical case for a third order reaction

$$2A + B + C \longrightarrow E + F$$

for which the rate law is

$$\text{rate of E formation} = k \times C_A^2 \times C_B$$

and the proposed mechanism, involving a steady state species D, is

Since the steady state condition is mathematically analogous to equilibrium, we will use the double-arrow designation (Units 12 and 16).

$$\text{FAST} \quad 2A \rightleftharpoons D \quad \text{(steady state)}$$

$$\text{SLOW} \quad D + B \longrightarrow G \quad \text{(high energy intermediate)}$$

$$\text{FAST} \quad G + C \longrightarrow E + F$$

In order to maintain a constant concentration of the steady state species (D), the rate of its formation (in the initial step) must equal the rate of its consumption by the following bimolecular step. However, a *mathematically equivalent* situation would result if D were consumed instead by reversal of its formation, a unimolecular process

$$D \longrightarrow 2A$$

We can therefore state:

$$\underset{\text{(rate of } D \text{ formation)}}{k_1 \times C_A^2} = \underset{\text{("hypothetical" rate of } D \text{ consumption)}}{k_2 \times C_D}$$

from which

$$C_D = \left(\frac{k_1}{k_2}\right) C_A^2$$

Then the rate of the proposed bimolecular *slow* step is

$$k_3 \times C_D \times C_B = k_3 \left(\frac{k_1}{k_2}\right) \times C_A^2 \times C_B$$

and this is the same as the rate law for the overall process, with

$$k_3 \left(\frac{k_1}{k_2}\right) = k_{\text{overall}}$$

As a result, the combination of an initial steady state process with a subsequent bimolecular process involving the steady state intermediate is mathematically equivalent to the form of a third-order rate law.

Example 6

Suggest plausible mechanisms for (a) the reaction of nitric oxide with bromine (Example 4) and (b) the reaction between hydrogen peroxide and hydrobromic acid (Example 2).

Solution

(a) $2NO(g) + Br_2(g) \longrightarrow 2NOBr(g)$

From Example 4,

rate of NOBr formation $= k \times C_{NO}^2 \times C_{Br_2}$

Since a termolecular process is unlikely in a gas-phase reaction, we need to propose two elementary processes whose combination offers a plausible slow sequence consistent with the third-order rate law. In the absence of other information, we could suggest an initial fast step involving either of the following bimolecular processes,

$$2NO \rightleftharpoons N_2O_2 \qquad NO + Br_2 \rightleftharpoons NOBr_2$$

Thus two equally plausible mechanisms may be suggested.

FAST $2NO \rightleftharpoons N_2O_2$ and then	or	$NO + Br_2 \rightleftharpoons NOBr_2$ (steady state) and then
SLOW $N_2O_2 + Br_2 \longrightarrow 2NOBr$		$NOBr_2 + NO \longrightarrow 2NOBr$

(Can you suggest others?)

(b) $H_2O_2(aq) + 2H^+(aq) + 2Br^-(aq) \longrightarrow 2H_2O(l) + Br_2(aq)$

From Example 2,

$$\text{rate of } Br_2 \text{ formation} = k \times C_{H_2O_2} \times C_{H^+} \times C_{Br^-}$$

As indicated in Example 2, this reaction occurs in dilute aqueous solution. Again, a termolecular process is unlikely. With three different reactants involved in the rate law, it might seem that three alternative initial fast steps are plausible:

$$H_2O_2 + H^+ \rightleftharpoons H_3O_2^+$$
$$H_2O_2 + Br^- \rightleftharpoons H_2O_2Br^-$$
$$H^+ + Br^- \rightleftharpoons HBr$$

We should remember, however, that hydrobromic acid is a strong acid (Unit 3), so the third suggestion should be discarded. In the absence of other information, we may suggest either of the remaining steps. We will select the first proposal for this example and let you develop an alternative mechanism on your own from the initial bimolecular reaction of hydrogen peroxide with bromide ion.

The slow step must be selected so that its combination with the initial process is consistent with the rate law (the combination of both steps is equivalent to $H_2O_2 + H^+ + Br^-$);

FAST $H_2O_2 + H^+ \rightleftharpoons H_3O_2^+$ (steady state)

SLOW $H_3O_2^+ + Br^- \longrightarrow H_2O + HOBr$

We could have suggested a number of different products for the second step. Water was selected because it is one of the final products. (The earlier we can show final product formation, the simpler will be the total mechanism.) Since a balanced equation required us to account for one more H, O, and Br, and a zero net charge, we suggested HOBr. (Hypobromous acid does exist.)

Now, further unimolecular or bimolecular steps (or both) must be formulated to show consumption of the intermediate HOBr and additional H^+ and Br^- (from the original equation) and formation of Br_2 and a "second" water molecule. Several possibilities exist. For example,

FAST $HOBr + H^+ \longrightarrow H_2O + Br^+$ or $HOBr + Br^- \longrightarrow Br_2 + OH^-$

FAST $Br^+ + Br^- \longrightarrow Br_2$ or $OH^- + H^+ \longrightarrow H_2O$

Either sequence is "correct" according to our "rules" (although suitable experimental techniques might be devised to determine which, if either, "really" occurs). A complete, "plausible" mechanism, then, might be

FAST $H_2O_2 + H^+ \rightleftharpoons H_3O_2^+$ (steady state)

SLOW $H_3O_2^+ + Br^- \longrightarrow H_2O + HOBr$

FAST $HOBr + Br^- \longrightarrow Br_2 + OH^-$

FAST $OH^- + H^+ \longrightarrow H_2O$

SUMMATION $H_2O_2 + 2H^+ + 2Br^- \longrightarrow 2H_2O + Br_2$

16.10 MECHANISM CLUES FROM ISOTOPIC LABELING

All isotopes of an element have essentially identical chemical properties. Differences due to minor differences in mass are very slight, although these are sometimes used in sophisticated investigations in which such effects offer useful information. In most cases synthesis of a compound containing an "unusual" isotope in a particular location within the molecule can be expected to yield a substance sufficiently like the "normal" molecule for accurate investigations. However, the molecule now contains a "label" at a known location, and this "unusual" atom can be followed through reaction sequences. Sometimes radioactive isotopes are used because of their ease of detection. In other cases nonradioactive isotopes, either heavier or lighter than the "normal" atoms, are used and these may be found in isolated species by techniques of mass spectrometry.

In any event, the techniques of isotopic labeling often produce information about reaction mechanisms that could not have been obtained from rate laws alone. These techniques have proved particularly useful in studying the reactions of organic compounds (Unit 29).

For example, oxygen-18 ($^{18}_{8}O$) has been employed in the study of the reaction between organic carboxylic acids and alcohols (Unit 29) to demonstrate that it is the oxygen of the alcohol that appears in the organic product, whereas an oxygen from the acid "ends up" in water.

$$\underset{\text{(acetic acid)}}{CH_3\overset{\displaystyle O}{\overset{\|}{C}}O^*H} + \underset{\text{(methanol)}}{CH_3OH} \longrightarrow \underset{\text{(methyl acetate)}}{CH_3\overset{\displaystyle O}{\overset{\|}{C}}OCH_3} + H_2O^*$$

$$CH_3\overset{\displaystyle O}{\overset{\|}{C}}OH + CH_3O^*H \longrightarrow CH_3\overset{\displaystyle O}{\overset{\|}{C}}O^*CH_3 + H_2O$$

(The starred oxygen is $^{18}_{8}O$.)

16.11 MECHANISM CLUES FROM SPECIES ANALYSIS

If you are interested in details of techniques for "intermediate" analysis, you may wish to visit with a faculty member specializing in the area of chemical kinetics.

Within recent years, a number of sophisticated techniques have been developed that often permit the identification of transient chemical species, such as short-lived reaction intermediates, even at very low concentrations. When such analyses are possible, the identification of intermediates involved in one or more of the elementary processes of a reaction sequence can be used to eliminate some of the alternative mechanisms suggested by the rate law alone. Except in relatively simple cases, however, we do not yet have the ability to define the "true" mechanism of a reaction. We can approach this level of understanding by successively eliminating alternative possibilities through careful and innovative investigations.

An example of this approach is illustrated by the case of the gas-phase reaction of hydrogen with iodine:

$$H_2(g) + I_2(g) \longrightarrow 2HI(g)$$

Until 1967 this reaction was believed to occur by a simple bimolecular process.

SLOW $H_2 + I_2 \longrightarrow (H_2I_2)$ (an intermediate "high-energy" species)

FAST $(H_2I_2) \longrightarrow 2HI$

In fact, the well known CHEM Study film (which many of you have probably seen), *Introduction to Reaction Kinetics,* employs this mechanism in animated sequences to illustrate various aspects of kinetics studies. In 1967, careful investigations suggested that *atomic* iodine was probably involved as a reaction intermediate, rather than the earlier postulated (H_2I_2) (a metastable combination of a hydrogen molecule with an iodine molecule). Although the investigation could not distinguish between all possible alternatives, it did show that an initial fast step was probably involved.

FAST $I_2(g) \rightleftharpoons 2I(g)$ (steady state)

In 1974, further work on this system was reported suggesting that atomic iodine may not, after all, be involved. At present, all we are certain of is that this "simple" system is not really simple.[4]

Example 7

The rate law for the reaction

$$CO(g) + NO_2(g) \longrightarrow CO_2(g) + NO(g)$$

is different at different temperatures. At 300°C, for example,

rate of CO_2 formation $= k_{300°} \times C_{NO_2}^2$

while at 400°C,

rate of CO_2 formation $= k_{400} \times C_{NO_2} \times C_{CO}$

a. Suggest two plausible mechanisms consistent with the rate law at 300°C and two consistent with the rate law at 400°C.
b. On the basis of the following analytical information, show how some mechanisms suggested in (a) can be eliminated and propose mechanisms at 300°C and 400°C that are consistent with *both* rate law and analytical information.

Species analysis: at 300°C, NO_3 was detected as a transient intermediate; at 400°C, a transient species is indicated as having the molecular formula CNO_3.

Solution
a. At 300°C the rate-determining step must involve only the species shown in the rate law, with a coefficient equal to the exponent of the concentration term; that is,

SLOW STEP $2NO_2 \longrightarrow X$

(where X must be some substance *or substances* having a total content of two nitrogens and four oxygens). Such a substance might be N_2O_4, a compound known to be formed from NO_2 at low temperatures. Then, a postulated slow step is:

$$NO_2 + NO_2 \longrightarrow N_2O_4$$

The reaction also consumes carbon monoxide and forms carbon dioxide and nitric oxide, so a possible rapid step might be

$$N_2O_4 + CO \rightleftharpoons CO_2 + NO + NO_2$$

Then the entire postulated mechanism would be

SLOW $NO_2 + NO_2 \longrightarrow N_2O_4$
FAST $N_2O_4 + CO \longrightarrow CO_2 + NO + NO_2$
_____
SUMMATION $NO_2 + CO \longrightarrow CO_2 + NO$
(to equal overall reaction equation)

---

[4] If this aspect of reaction mechanism studies interests you, you may wish to read the report of the $H_2 + I_2$ investigation in the January 16, 1967, issue of *Chemical and Engineering News* (pp. 40–41), and in the December 11, 1974 issue of the Journal of the American Chemical Society (pp. 7621–7622).

Other mechanisms equally consistent with the rate law include (Can you suggest additional alternatives?):

Alternate 1

SLOW $2NO_2 \longrightarrow NO + NO_3$

FAST $NO_3 + CO \longrightarrow NO_2 + CO_2$

Alternate 2

SLOW $2NO_2 \longrightarrow N_2O + O_3$

FAST $O_3 + CO \longrightarrow CO_2 + O_2$

FAST $O_2 + N_2O \longrightarrow NO_2 + NO$

Alternate 3

SLOW $2NO_2 \longrightarrow N_2O_4$

FAST $N_2O_4 \longrightarrow NO + NO_3$

FAST $NO_3 + CO \longrightarrow NO_2 + CO_2$

At 400°C the rate law requires

SLOW $NO_2 + CO \longrightarrow X$

(where X is some substance or substances having a total content (mass balance) of one nitrogen, one carbon, and three oxygens). In this case, X may be the final reaction products, CO_2 and NO, or some transient intermediate(s). Thus one could write two (or more) plausible alternatives:

SLOW $NO_2 + CO \longrightarrow NO + CO_2$ (only step)

or

Alternate 1

SLOW $NO_2 + CO \longrightarrow NO_3 + C$

FAST $NO_3 + C \longrightarrow NO + CO_2$

Alternate 2

SLOW $NO_2 + CO \longrightarrow (CNO_3)$ ("high energy" intermediate)

FAST $(CNO_3) \longrightarrow CO_2 + NO$

b. The identification of NO_3 as an intermediate in the reaction at 300°C eliminates both the original proposal and alternate 2, since neither of these mechanisms indicates any role for NO_3. Both alternates 1 and 3 remain consistent with rate law and analytical information. The correct mechanism has now been shown as

SLOW $NO_2 + NO_2 \longrightarrow NO + NO_3$

RAPID $NO_3 + CO \longrightarrow NO_2 + CO_2$

At 400°C, the evidence for the transient appearance of a species characterized as CNO_3, suggests a mechanism such as that indicated by alternate 2.

Remember that in the absence of other information we would generally propose a "simpler" mechanism (such as the original suggestion or alternate 1). This does *not* mean, however, that more complex mechanisms can be disregarded in further studies. A *plausible* mechanism is satisfactory only as long as it remains consistent with all experimental information obtained.

Such *intermediates* may be energetically stable, but rapidly consumed by reaction with other species present. It is also possible for X to be a very high energy (unstable) aggregate of the atoms from the species colliding.

16.12 TRANSITION-STATE THEORY

The relatively simple model of the *collision theory* of reaction kinetics, suggesting that reaction rate is proportional to the frequency of effective collisions in the rate-determining step, is a useful way of looking at an elementary process involving the combination of initially separate reactants, but it offers no suggestion as to how interactions may occur. A modification of this simple model, proposed by Henry Eyring, is the *transition-state* theory. In this model, it is assumed that the combination of reactants is somewhat more complicated than a simple collision.[5] It is suggested that significant changes occur as reactants approach each other, well before actual collision occurs; that is, the electric fields of approaching reactant particles may interact to produce changes in bond strengths (and bond lengths). In its most sophisticated applications, transition-state theory applies a quantum mechanical treatment to these interactions.[6]

We can gain some useful insights to several aspects of reaction kinetics by considering the simpler ideas of transition-state theory. Let's consider, as an example, an analysis of the process

$$I_2(g) + Br_2(g) \longrightarrow 2IBr(g)$$

Transition-state theory suggests that as two properly oriented molecules of iodine and bromine approach each other, electronic interactions cause the I—I and Br—Br bonds to lengthen and weaken in each molecule. Since this condition is "less stable" than that of the normal molecules, the internal energy of the system increases. As the molecules continue to approach each other, I—Br distances decrease (with increasing "bonding"). At the highest internal energy of the system, some unique combination of the two I_2 and Br_2 molecules exists, so "unstable" that it should rapidly decompose. This *transition state* has certain characteristics of "normal" molecules, such as unique geometry and typical bond lengths, but its lifetime is so short it could not normally be identified by analytical techniques. The *transition state* is therefore different from a more stable *reaction intermediate*. The decomposition of the transition state may occur to form products or initial reactants. Only the former, of course, would result in progress of the reaction.

The internal energy difference between reactants and transition state is called the *activation energy* and it is this "energy barrier" that must be surmounted for the reaction to proceed. This corresponds in the *collision theory* to the minimum energy necessary for "effective collision" (Section 16.6). The internal energy difference between reactants and *products*, for a system in which no *work* is involved, is equivalent to the "ΔH of reaction" (Unit 5). An *internal energy diagram* (Figure 16.10) illustrates these ideas graphically.

Such graphical displays are also called *potential* energy diagrams.

Obviously, any two molecules might approach each other in a number of relative orientations, several of which might conceivably lead to eventual reaction. Various structures of alternative "high energy" intermediates might therefore be possible. Each different structure would correspond to a unique activation energy. It is observed in nature that processes generally follow the "path of least resistance," so reactions are presumed to occur via the transition states of lowest energy (Figure 16.11). Indeed, the Boltzmann distribution (Figure 16.3) suggests that only a small percentage of the molecules have "higher-than-average" energies at a particular temperature, so most molecules will have insufficient energy to react by high-energy paths. It is this assumption that permits some definition, in many cases, of plausible structures for transition state species.

[5] For a more detailed comparison of the *collision* and *transition-state* theories, with some special applications, see "Collision and Transition-State Theory Approaches to Acid-Base Catalysis," by H. B. Dunford, in the September, 1975, issue of the *Journal of Chemical Education* (pp. 578–580).

[6] For an example, see "Frontier Molecular Orbitals: A Link between Kinetics and Bonding Theory," by J. D. Bradley and G. C. Gerrans in the July, 1973, issue of the *Journal of Chemical Education* (pp. 463–466).

Figure 16.10 Internal energy diagram for the reaction

$$I_{2(g)} + Br_{2(g)} \longrightarrow 2IBr_{(g)}$$

The diagram indicates that formation of IBr is exothermic. The "reverse" process of IBr decomposition is endothermic and *both processes involve the same transition state.* (Note that $\Delta E_{Reaction} = \Delta H_{Reaction}$ when $w = 0$, Unit 5.)

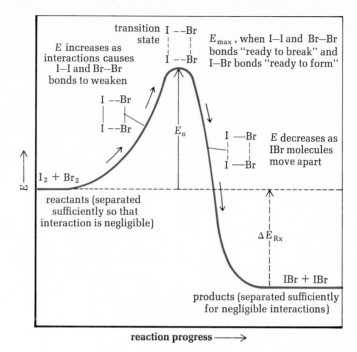

Figure 16.11 Alternative geometries for transition state species. "Effective collision" is most probable when optimum geometry results, and this corresponds to the lowest activation energy.

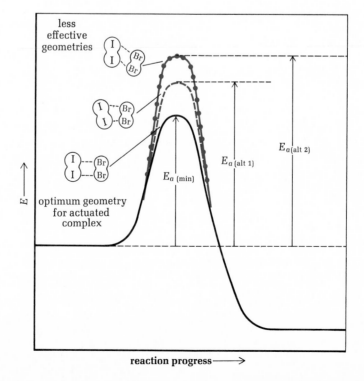

Figure 16.12 Effect of catalyst on activation energy.

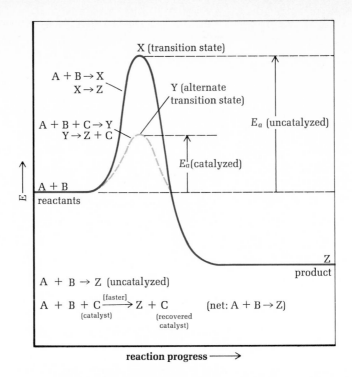

reaction progress ⟶

Evidence suggests that a catalyst (Section 16.5) functions by providing an alternative mechanism of reaction via a transition state of energy lower than that for the
uncatalyzed process (Figure 16.12). Again, since the reaction tends to proceed by the
path having the lowest "energy barrier" and since more molecules will have "average
energy" than "high energy," it follows that more reactants, per unit time, will
surmount the lowered energy barrier. As a result the rate of reaction is increased.

Since the activation energy is a thermodynamic quantity, it might appear that this
function would provide a way of relating the ΔH change for a reaction to the speed of
the process, a relationship we have indicated as "nonexistent" (Unit 5). The problem
lies with the extremely transient, and usually poorly defined, character of the
transition state. Whereas methods have been devised for estimating appropriate
thermodynamic properties of transition-state species, these have generally been
limited to relatively simple systems. It is much easier to determine activation energies
from experimental kinetics data, so we may still indicate that there is no simple,
general relationship between enthalpies of reaction and reaction speed.

In 1889, the Swedish chemist Svante Arrhenius suggested the concept of an
"energy barrier" (the activation energy) in an attempt to explain the influence of
temperature on reaction speed. On a purely empirical basis (i.e., by determining a
mathematical formulation properly correlating experimental data), Arrhenius developed the equation

$$k = Ae^{-E_a/RT} \tag{16.1}$$

in which k is the rate constant, E_a the activation energy, R the molar gas constant
(2.0 cal mole^{-1} deg^{-1}, Unit 10), and T the absolute temperature (degrees Kelvin). The A
term is a proportionality constant (sometimes referred to as the "Arrhenius factor")
and e is the base for natural logarithms.

In 1901, Arrhenius received a
Nobel prize, the third to be
awarded in chemistry.

The Arrhenius equation has now been theoretically related to the transition-state model for reaction kinetics. This relationship provides the basis for the experimental determination of activation energies and for the calculation of temperature effects on reaction rate. The mathematical formulation employed in the detailed transition-state approach is very complex, but it leads to results indicating that the simple empirical equation is adequate for most kinetic studies.

16.13 TEMPERATURE DEPENDENCE REVISITED

As mentioned earlier (Section 16.3), the speed of a chemical reaction varies more with temperature than would be expected simply on the basis of increased collision frequency. This variation is noted in a temperature dependence of the rate constant for the process. When competing mechanisms are involved, the different rate constants may vary sufficiently in their temperature dependences so that one mechanism predominates at one temperature and the other at a significantly different temperature (Example 7).

The key to this temperature effect is the *activation energy* for the process, the internal energy difference between reactants and the transition state.

The information required to determine activation energies is found by experimental investigation of the temperature dependence of reaction speed. Recall that the Boltzmann distribution curves (Figure 16.3) indicate an increasing percentage of high-energy collisions as the temperature is raised. The way in which this is manifested in increased reaction rate is most easily seen from a slight modification of the Arrhenius equation (16.1):

$$\log k = \log A - \frac{E_a}{2.303RT} \qquad \text{(16.2)}$$

Natural logarithms (base e), represented by ln, are related to common logarithms (base 10), in the following way:

$$e = 2.71828\ldots$$
$$e^{2.303} = 10$$
$$ln\ x = 2.303 \log x$$

(This equation is simply a logarithmic form of the Arrhenius equation, using base-10 logarithms.)

It is necessary only to recognize that with an activation energy greater than zero and the Arrhenius factor (A) a large positive number, a temperature increase is typically associated with an increased reaction rate. Since A is a constant and temperature appears as a denominator term in a negative fraction, the rate constant must increase with increasing temperature.

Equation (16.2) poses the difficulty of having three "unknown" terms (k, A, and E_a). We must find two of these before the third can be evaluated, or we must develop a way of determining two "unknowns" simultaneously. If we rearrange equation (16.2) slightly, it is obvious that it is an equation for a straight line, for a plot of $\log k$ versus $1/T$:

$$\log k = \frac{-E_a}{2.303R}\left(\frac{1}{T}\right) + \log A$$

$$y = mx + b \qquad \text{(general equation for a straight line)}$$

Then the experimental determination of rate constants at different temperatures permits us to plot a linear graph (Figure 16.13) from which the slope is $-E_a/2.303R$ and the intercept is ($\log A$).

An alternative approach, when two sets of sufficiently accurate data are available or as an approximation when rate constant-versus-temperature data are particularly difficult to determine over a wide range of temperatures, employs a relationship

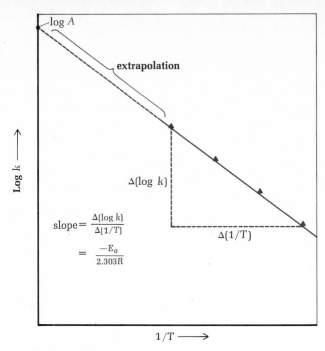

Figure 16.13 Graphical determination of activation energy and Arrhenius Factor.

log A

extrapolation

Log k →

$\Delta(\log\ k)$

$\text{slope} = \dfrac{\Delta(\log k)}{\Delta(1/T)}$

$\quad\quad = \dfrac{-E_a}{2.303R}$

$\Delta(1/T)$

1/T →

easily derived from equation (16.2):

$$\log\left(\frac{k_2}{k_1}\right) = \frac{E_a(T_2 - T_1)}{(4.6 \times 10^{-3})(T_1 \times T_2)}$$

(16.3)

(The constant, 4.6×10^{-3}, results from 2.303×2.0 cal mole^{-1} deg $\times 10^{-3}$ cal kcal^{-1}.)

A knowledge of the activation energy for a reaction permits calculations of reaction speeds (or rate constants) at different temperatures.

Example 8

For the reaction $2N_2O_5 \longrightarrow 2N_2O_4 + O_2$, the rate law is

rate of O_2 formation $= k \times C_{N_2O_5}$

The half-life for this process at 25°C is 1.69×10^4 sec. At 55°C the half-life is 330 sec.

a. Calculate the activation energy for this reaction.
b. Calculate the rate constant for the reaction at 35°C.

Solution
a. Finding E_a.
First, calculate the rate constants at 25°C (298°K) and 55°C (328°K) from the expression for first-order processes (Table 16.1).

$$t_{1/2} = \frac{0.693}{k}$$

$$k_{298°} = \frac{0.693}{1.69 \times 10^4 \text{ sec}} = 4.10 \times 10^{-5} \text{ sec}^{-1}$$

$$k_{328°} = \frac{0.693}{330 \text{ sec}} = 2.10 \times 10^{-3} \text{ sec}^{-1}$$

Second, use equation (16.3) to calculate the activation energy.

$$\log\frac{(2.10 \times 10^{-3})}{(4.10 \times 10^{-5})} = \frac{E_a(328 - 298)}{(4.6 \times 10^{-3})(298 \times 328)}$$

$$E_a = \frac{(1.710)(4.6 \times 10^{-3})(298)(328)}{30}$$

$$= 26 \text{ kcal mole}^{-1}$$

b. Finding $k_{308°}$.

Use equation (16.3), with the value of E_a as determined in part (a). Either $k_{298°}$ or $k_{328°}$ could be used. We shall use the former.

The assumption we have used that E_a and A are temperature independent is valid for most simple systems over temperature ranges of general chemical interest.

$$\log\frac{k_{308°}}{(4.10 \times 10^{-5})} = \frac{(26)(308 - 298)}{(4.6 \times 10^{-3})(298)(308)}$$

$$\log k_{308°} - \log(4.10 \times 10^{-5}) = 0.62$$

$$= 0.62 - 4.39 = -3.77$$

$$= 0.23 - 4$$

$$k_{308°} = 1.7 \times 10^{-4} \text{ sec}^{-1}$$

It should be pointed out that the use of the Arrhenius equation is limited to relatively simple cases. In systems for which different mechanisms are operable at different temperatures (e.g., Example 7) or for which highly complex sequences of elementary processes are involved, more than one of which contributes significantly to the net rate constant, linear relationships are not obtained for plots of k versus $(1/T)$. In fact, the failure of such data to fit a linear plot has often been an important clue to some unexpected complexities.

CHAIN REACTIONS

The types of reactions we have discussed so far are characterized by rates that decrease as the reaction progresses (i.e., as reactant concentrations decrease). Some reactions, however, exhibit a rapid *increase* in speed as the reaction progresses, sometimes achieving a rate corresponding to explosive violence.

Reactions of this type may occur when "fast" steps, following an initial "slow" process, form products that may subsequently react rapidly in additional steps to provide ways for the reaction to progress without continuation of the original slow step. When various steps form such rapidly reacting intermediates, a "chain process" occurs. The reaction between hydrogen and bromine gases is an example of such a *chain reaction*. The chain is originally *initiated* by the decomposition of a bromine molecule to form two free bromine atoms. Two subsequent steps serve to *propagate* the chain by forming, in addition to the final product, highly reactive atoms which, in turn, can attack original reactant species. The chain may be *inhibited* by reaction of the product (HBr) with a free atom (since loss of product is equivalent to a reduced rate of product formation) or *terminated* by combination of free atoms. Several possible steps might be suggested.

$$Br_2 \longrightarrow 2Br \quad \text{(chain initiating)}$$
$$\text{(free radicals)}$$

Chain propogating	Chain inhibiting	Chain terminating
$Br + H_2 \longrightarrow HBr + H$	$H + HBr \longrightarrow H_2 + Br$	$H + Br \longrightarrow HBr$
$H + Br_2 \longrightarrow HBr + Br$	$Br + HBr \longrightarrow Br_2 + H$	$H + H \longrightarrow H_2$
		$Br + Br \longrightarrow Br_2$

This particular reaction has been studied quite extensively. The rate law is quite complex (typical of chain reactions):

$$\text{rate of HBr formation} = k\left[\frac{C_{H_2} C_{Br_2}^{\frac{1}{2}}}{1 + C_{HBr}/k'C_{Br_2}}\right]$$

The mechanism proposed to account for this rate law does not incorporate all of the possible steps previously suggested.

$$Br_2 \longrightarrow 2Br \qquad \text{(chain initiating)}$$
$$\left.\begin{array}{l} Br + H_2 \longrightarrow HBr + H \\ H + Br_2 \longrightarrow HBr + Br \end{array}\right\} \quad \text{(chain propogating)}$$
$$H + HBr \longrightarrow H_2 + Br \qquad \text{(chain inhibiting)}$$
$$Br + Br \longrightarrow Br_2 \qquad \text{(chain terminating)}$$

Most chain reactions involve *free radicals,* which are atoms or polyatomic groups having one or more unpaired electrons. Free radicals may be generated from molecules by thermal decomposition, light absorption of appropriate wavelength, or high-energy radiation (Unit 35). Some of the effects of radiation on living systems are attributed to the generation of highly reactive free radicals. Molecular oxygen is already a free radical, having two unpaired electrons (Unit 7). This undoubtedly accounts, to some extent, for the reactive character of the O_2 molecule.

Chemical lasers were developed as "spin-off" of basic studies of reaction kinetics.

The detailed study of reaction mechanisms might appear to be of purely academic interest. In many significant cases, however, a knowledge of reaction mechanisms has been of great importance in the development of new synthetic routes to useful substances or in the planning of industrial processes for optimum efficiency. When reactions are rapid and quantitative, uncomplicated by formation of product mixtures or unwanted isomers, the mechanisms may appear to have no immediate applications. Even these cases, however, may provide useful analogies for more complex systems or develop techniques applicable to reactions of more obvious value. Certainly in those reactions—so typical of organic chemistry—characterized by molecular rearrangements, product mixtures, and less than quantitative yields, a knowledge of the mechanisms of reactions is the key to the selection of the best synthetic schemes for obtaining useful chemicals of maximum purity at minimum costs. Eventually it is hoped that techniques will become available for studying all of the reactions in living systems so that these mechanisms may be understood and, perhaps, beneficially controlled.

SUGGESTED READING

Abbott, D. 1968. *An Introduction to Reaction Kinetics.* Boston: Houghton Mifflin.
Gardiner, W. C., Jr. 1969. *Rates and Mechanisms of Chemical Reactions.* Menlo Park, Calif.: W. A. Benjamin.
Stevens, B. 1970. *Chemical Kinetics for General Students of Chemistry.* New York: Halsted Press.

EXERCISES

1. Select the term that best matches each definition, description, or specific example.

TERMS: (A) autocatalysis (B) effective collision (C) flood (D) heterogeneous catalysis (E) quench

a. a process in which the catalyst is in a separate phase from that of the reactants
b. a process in which one of the reaction products serves to increase the speed of the reaction
c. stopping a reaction suddenly by cooling or other means
d. reactant particles coming together with sufficient energy and appropriate orientations to permit chemical reaction
e. addition of a large excess of one reactant so that concentration effects of another reactant can be studied

2. For the following suggested mechanism, identify
a. the molecularity of the rate-determining step
b. the molecular formula of the transition state for the rate-determining step
c. two proposed reaction intermediates
d. the overall reaction order

FAST $2NO \rightleftharpoons N_2O_2$
SLOW $N_2O_2 + H_2 \longrightarrow N_2O + H_2O$
FAST $N_2O + H_2 \longrightarrow N_2 + H_2O$

SUMMATION $\overline{2NO + 2H_2 \longrightarrow N_2 + 2H_2O}$

3. Which of the reactions in each pair listed would you expect to be the more rapid? Give a brief explanation for your choices.
a. $Ag^+(aq) + Cl^-(aq) \longrightarrow AgCl(s)$ at 25°C (dark) or
$2Ag(s) + Cl_2(g) \longrightarrow 2AgCl(s)$ at 25°C (dark)
b. $N_2(g) + 3H_2(g) \longrightarrow 2NH_3(g)$ at 25°C (dark) or
$N_2(g) + 3H_2(g) \longrightarrow 2NH_3(g)$ at 250°C (dark)
c. $2H_2O_2(l) \longrightarrow 2H_2O(l) + O_2(g)$ at 25°C (dark) or

$2H_2O_2(l) \xrightarrow{Fe^{3+} \text{ (catalyst)}} 2H_2O(l) + O_2(g)$ at 25°C (dark)

4. Suggest explanations for
a. the increased speed of a simple bimolecular process when the temperature of the reaction mixture is increased

b. the action of a solid-phase catalyst to increase the rate of a unimolecular gas-phase decomposition reaction

5. Suggest a rate law for each of the following reactions:

a. $2N_2O_5(g) \longrightarrow 2N_2O_4(g) + O_2(g)$

EXPERIMENT NUMBER	INITIAL $C_{N_2O_5}$ (mole liter^{-1})	RELATIVE INITIAL RATES
1	0.10	1.0
2	0.050	0.50
3	0.25	2.5

b. $Cl_2(aq) + 2Fe^{2+}(aq) \longrightarrow 2Cl^-(aq) + 2Fe^{3+}(aq)$

EXPERIMENT NUMBER	INITIAL C_{Cl_2} (M)	INITIAL $C_{Fe^{2+}}$ (M)	RELATIVE INITIAL RATES
1	0.10	1.0	1.0
2	0.050	1.0	0.5
3	0.10	0.10	0.10
4	0.050	0.050	0.025

6. From the information given, and without looking at Table 16.1, determine the rate law for the reaction. Then use information from Table 16.1 to calculate the rate constant at 0°C.

for $2ClO_2(aq) + 2OH^-(aq) \rightarrow ClO_3^-(aq) + ClO_2^-(aq) + H_2O(l)$

EXPERIMENT NUMBER (at 0°C)	INITIAL C_{ClO_2} (mole liter^{-1})	INITIAL C_{OH^-} (mole liter^{-1})	$t_{1/2}$ (sec)
1	0.050	0.050	5.2
2	0.050	0.025	5.2
3	0.025	0.050	10.4
4	0.025	0.010	10.4

7. Suggest plausible mechanisms for the reactions given in Exercise 5.

8. Suggest two alternative mechanisms for the reaction given in Exercise 6.
9. It has been determined that the use of hydroxide ion containing oxygen-18 for reaction with ClO_2 (Exercise 6) results in the formation of ClO_2^- containing no $^{18}_{8}O$ and ClO_3^- containing one $^{18}_{8}O$. Suggest a mechanism for this reaction that is consistent with both the rate law and the "isotopic labeling" information.
10. For a particular gas-phase reaction, the rate constant at $250°$ is 3.1×10^{-7}. At $500°C$, $k = 3.8 \times 10^{-2}$. Calculate
    a. the activation energy for this reaction
    b. the rate constant at $375°C$

PRACTICE FOR PROFICIENCY

Objective (a)

1. Select the term that best matches each definition, description, or specific example.
   TERMS: (A) activation energy (B) chain reaction (C) elementary process (D) free radical (E) heterogeneous catalysis (F) homogeneous catalysis (G) intermediate (H) transition state

   a. $H:\overset{\displaystyle H}{\underset{\displaystyle H}{C}}\cdot$

   b. the internal energy difference between the reactants and the transition state
   c. characterized by an increase in reaction speed during the progress of a reaction
   d. a single step in an overall reaction sequence
   e. catalysis of the gas-phase decomposition of nitrous oxide by a small amount of chlorine gas
   f. catalysis of the gas-phase oxidation of carbon monoxide by metallic platinum
   g. NO_3 in the elementary processes

      SLOW $NO + O_2 \longrightarrow NO_3$ (detected)

      FAST $NO_3 + NO \longrightarrow 2NO_2$

      proposed as the mechanism for the overall reaction

      $2NO + O_2 \longrightarrow 2NO_2$

   h. a postulated (I_2Br_2) transient species suggested for the elementary processes

      SLOW $2IBr \longrightarrow (I_2Br_2)$ (proposed, not detected)

      FAST $(I_2Br_2) \longrightarrow I_2 + Br_2$

      to account for the overall reaction
      $2IBr \longrightarrow I_2 + Br_2$

Objective (b)

2. Which of the reactions in each pair listed would you expect to be more rapid? Explain your choices.
   a. $BaBr_2(aq) + 2AgNO_3(aq) \longrightarrow Ba(NO_3)_2(aq) + 2AgBr(s)$ at $25°C$, or

$BaBr_2(s) + 2AgNO_3(s) \longrightarrow Ba(NO_3)_2(s) + 2AgBr(s)$ at $25°C$
   b. $CO(g) + NO_2(g) \longrightarrow CO_2(g) + NO(g)$ at $25°C$, or
      $CO(g) + NO_2(g) \longrightarrow CO_2(g) + NO(g)$ at $250°C$
3. Suggest explanations for
   a. The rate of decomposition of hydrogen iodide is about 100,000 times greater at $500°C$ than at $250°C$.
   b. A solid-phase catalyst is employed in some automobile exhaust emission systems. One of the functions of this catalyst is to increase the rate of oxidation of carbon monoxide to CO_2.
   c. For the gas-phase process,

   it has been determined that only a fraction of the collisions occurring with net energy equal to E_a actually result in reaction.
   d. For the overall reaction,

   $$S_2O_8^{2-}(aq) + 3I^-(aq) \longrightarrow 2SO_4^{2-}(aq) + I_3^-(aq)$$

   it has been demonstrated that the initial rate of reaction is doubled if the initial concentration of either reactant is doubled.

Objective (c)

4. For the gas-phase decomposition of acetaldehyde at $530°C$,

   $$CH_3CHO \longrightarrow CH_4 + CO$$

   the initial rate is nine times as great for an initial concentration of 0.30 mole liter^{-1} as for an initial concentration of 0.10 mole liter^{-1}. What is the rate law for this decomposition?
5. Suggest a rate law for the reaction described in question 3d.
6. The half-life is 2.4 hours at all measured initial concentrations for the reaction at $30°C$,
   $2N_2O_5 \longrightarrow 4NO_2 + O_2$
   a. What is the reaction order?
   b. What is the value of the rate constant for this reaction at $30°C$?
   c. At what rate will the reaction proceed when the concentration of N_2O_5 is 0.10 mole liter^{-1}?
7. From the following information for the reaction:

   $$2H_2 + 2NO \longrightarrow N_2 + 2H_2O$$

   determine
   a. the rate law
   b. the overall reaction order
   c. the rate constant

EXPERIMENT NUMBER	INITIAL C_{NO} (mole liter^{-1})	INITIAL C_{H_2} (mole liter^{-1})	HALF-LIFE (hours)
1	0.50	0.50	4.3
2	0.50	0.25	4.3
3	0.25	0.50	8.6

8. It has been demonstrated that the reaction,

$$2ClO_2 + F_2 \longrightarrow 2FClO_2$$

follows a second-order rate law. The rate varies if the initial concentration of either reactant is varied. What is the rate-law expression for this reaction?

Objective (d)

9. Suggest plausible mechanisms for the reactions described in questions 4, 5, and 6.
10. Suggest two mechanisms for the reaction described in question 7, on the basis of kinetics data alone. Then reconsider these proposals in light of the information that N_2O_2 has been detected as a steady state intermediate in this reaction.
11. Propose a plausible mechanism for the reaction described in question 8. What is the molecular formula for the transition-state species?
12. The chain reaction between methane and chlorine gases to form hydrogen chloride and chloromethane (CH_3Cl) is initiated by the photochemical (light-induced) decomposition of chlorine molecules. A small amount of ethane is found in the final product mixture. Suggest a mechanism for this reaction that contains a chain-initiating step, two chain-propagating steps, and three chain-terminating steps. Identify all free radicals involved in the proposed mechanism.

Objective (e)

13. For the gas-phase decomposition of bromoethane (CH_3CH_2Br) to hydrogen bromide and ethylene (C_2H_4), the unimolecular rate constant at $300°C$ is 1.05×10^{-7} liter mole^{-1} sec^{-1}. At $450°C$,

$$k_{450°} = 1.36 \times 10^{-3} \text{ liter mole}^{-1} \text{ sec}^{-1}$$

Calculate
a. the activation energy for this reaction
b. the rate constant at $600°C$
c. the Arrhenius factor
14. For the unimolecular decomposition of N_2O_4, the activation energy is 13.0 kcal mole^{-1} and the Arrhenius factor is 10^4 sec^{-1}. Calculate
a. the rate constant for the reaction at $400°C$
b. the half-life of the reaction (in seconds) at $400°C$
c. the initial reaction rate at $400°C$ when the concentration of N_2O_4 is 0.15 mole liter^{-1}

BRAIN TINGLER

15. The decomposition of nitramide,

$$H_2NNO_2 \longrightarrow N_2O + H_2O$$

is slow in neutral aqueous solution, following the rate law,

$$\frac{-dC_{H_2NNO_2}}{dt} = k \times C_{H_2NNO_2}$$

In alkaline solutions, the reaction is much more rapid and the rate law is

$$\frac{-dC_{H_2NNO_2}}{dt} = k' \times C_{H_2NNO_2} \times C_{OH^-}$$

with $k' \gg k$. Note that OH^- does not appear in the overall equation. Discuss this information in terms of the *transition-state theory*, including appropriate mechanisms and description of the role of OH^- in the reaction.

Unit 16 Self-Test

1. Matching: Select the term most appropriate for each process.
   TERMS: (A) autocatalysis (B) heterogeneous catalysis (C) homogeneous catalysis (D) inhibition
   a. $2N_2O(g) \longrightarrow 2N_2(g) + O_2(g)$
      (catalyzed by nitric oxide gas)

   b. $2C_6H_{14}(g) + 19O_2(g) \longrightarrow 12CO_2(g) + 14H_2O(g)$
      (catalyzed by platinum metal)

   c. $NH_4^+(aq) + HNO_2(aq) \longrightarrow N_2(g) + 2H_2O(l) + H^+(aq)$
      (catalyzed by H^+)

d. $2C_2H_5OC_2H_5(l) + O_2(g) \longrightarrow 2C_2H_5OOC_2H_5(l)$

    (reaction slowed by presence of metallic iron)

2. Which is probably more rapid?

    a. $BaS(s) + CuSO_4(s) \longrightarrow BaSO_4(s) + CuS(s)$ at $25°C$

    or

    b. $BaS(aq) + CuSO_4(aq) \longrightarrow BaSO_4(s) + CuS(s)$ at $25°C$

3. Which is probably more rapid?

    a. $H_2C{=}CH_2(g) + HBr(g) \longrightarrow CH_3CH_2Br(g)$ at $250°C$

    or

    b.

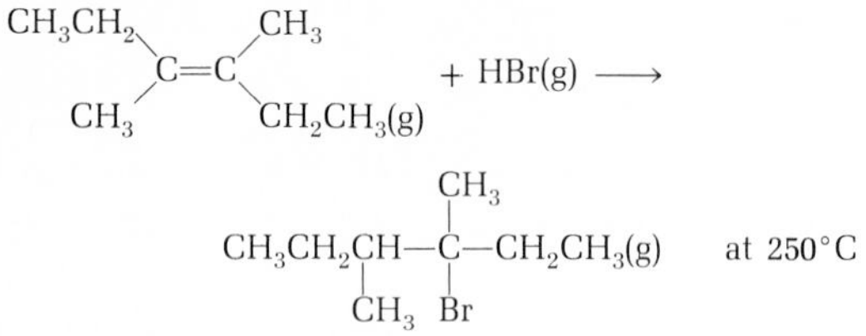

$$\begin{array}{c} CH_3CH_2 \quad\quad CH_3 \\ \diagdown\diagup \\ C{=}C \\ \diagup\diagdown \\ CH_3 \quad\quad CH_2CH_3(g) \end{array} + HBr(g) \longrightarrow$$

$$\begin{array}{c} CH_3 \\ | \\ CH_3CH_2CH{-}C{-}CH_2CH_3(g) \\ | \quad | \\ CH_3 \;\; Br \end{array} \quad \text{at } 250°C$$

4. a. Suggest a rate law for the reaction

$$2NO(g) + Cl_2(g) \longrightarrow 2NOCl(g)$$

EXPERIMENT NUMBER	C_{NO} (mole liter^{-1})	C_{Cl_2} (mole liter^{-1})	RELATIVE INITIAL RATES
1	0.10	0.10	1
2	0.20	0.20	8
3	0.10	0.20	2

b. Suggest a rate law for the reaction

$$2NO_2(g) + F_2(g) \longrightarrow 2NO_2F(g)$$

EXPERIMENT NUMBER	C_{NO_2} (mole liter^{-1})	C_{F_2} (mole liter^{-1})	RELATIVE INITIAL RATES
1	0.10	0.10	1
2	0.20	0.20	4
3	0.20	0.10	2

c. Suggest a rate law for the reaction

$$H_2(g) + 2NO(g) \longrightarrow H_2O(g) + N_2O(g)$$

EXPERIMENT NUMBER	C_{H_2} (mole liter^{-1})	C_{NO} (mole liter^{-1})	$t_{1/2}$ (seconds)
1	0.010	0.0010	15
2	0.010	0.0020	15
3	0.0010	0.010	720
4	0.0020	0.010	360

(*After* determining the rate law, refer to Table 16.1 to calculate the rate constant.)

5. Propose plausible mechanisms for the reactions in question 4.

6. For the decomposition of N_2O_5 to NO_2 and O_2, a first order process, the rate constant at $25°C$ is 3.5×10^{-5} liter mole^{-1} sec^{-1}. At $55°C$, $k = 1.5 \times 10^{-3}$ liter mole^{-1} sec^{-1}. Calculate (a) the activation energy for this reaction, (b) the rate constant at $37°C$, and (c) the Arrhenius factor.

Unit 17

One makes discoveries if he can
and not merely through the wish
to make them.
HENRI LE CHATELIER (1922)

Homogeneous
equilibria

Objectives

(a) Given the chemical equation for an equilibrium system, be able to write a conventional equilibrium constant expression (Example 1 and Section 17.3).

(b) Given the concentrations of all components of an equilibrium system, or appropriate information from which concentrations can be determined, be able to calculate the equilibrium constant for the system (Section 17.4).

(c) Given the equilibrium constant and other appropriate data, be able to calculate concentrations of specified equilibrium components (Section 17.5).

(d) Be able to estimate, on a qualitative basis, effects of changes in pressure, temperature, or concentration on an equilibrium system (Section 17.6).

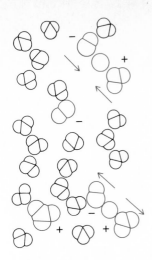

Figure 17.1 Equilibrium in pure water. Equilibrium results when the rate of ionization $(H_2O \longrightarrow H^+ + OH^-)$ equals the rate of ion combination $(H^+ + OH^- \longrightarrow H_2O)$. In pure water at $25°C$, the equilibrium composition is approximately 56 moles liter^{-1} H_2O, 10^{-7} mole liter^{-1} H^+ (aq) and 10^{-7} mole liter^{-1} OH^- (aq).

Ethyl acetate, $CH_3CO_2C_2H_5$, is a rather sweet-smelling organic compound, perhaps most familiar as nail polish remover. It is also a common solvent for certain types of model airplane cement. In recent years, ethyl acetate has frequently made news as one of the most hazardous agents involved in drug experimentation by young people. When inhaled in large quantities at high concentrations, this otherwise innocuous chemical can produce extremely serious effects.

Ethyl acetate may be synthesized from ethanol and acetic acid. The progress of the reaction can be monitored by withdrawing tiny samples from the homogeneous reaction mixture from time to time and analyzing these by any of a number of modern techniques. If equimolar amounts of the reactants are mixed and maintained at constant temperature, it is observed that these are slowly converted to ethyl acetate and water. After a very long time no further change occurs, even though the mixture still contains about one-third of the original reactants.

$$CH_3CO_2H + HOC_2H_5 \rightleftharpoons CH_3CO_2C_2H_5 + H_2O$$

The synthesis can be performed much more rapidly if a small amount of concentrated H_2SO_4 or anhydrous HCl is added to the reactant mixture. Now the reaction is completed in just a few hours, but the composition of the final equilibrium mixture is the same as that obtained by the much slower route, except for the presence of the added acid—which appears unchanged by the process.

Equilibrium may be reached by different paths at different speeds, but the composition of the equilibrium system (Figure 17.1) is independent of the pathway selected. In this unit we shall examine the subject of chemical equilibrium in some detail, and we shall see how a knowledge of reaction kinetics and mechanisms (Unit 16) can aid the chemist in selecting more efficient ways of controlling chemical processes, even those limited by equilibrium considerations.

17.1 REVERSIBLE AND IRREVERSIBLE PROCESSES

If we toss a piece of sodium into water, a violent reaction occurs, forming sodium hydroxide and hydrogen gas. Although we could calculate (Unit 24) the conditions under which this reaction could be reversed (i.e., reduction of aqueous NaOH by H_2 to sodium and water), it is not feasible to achieve these conditions experimentally. Reactions of this type are said to be *irreversible*.

Consider, on the other hand, the *reversible* reaction between metallic iron and steam. Again, hydrogen gas is formed (although much more slowly than in the case of sodium). The iron is converted to an iron oxide, Fe_3O_4. If the reaction is carried out in an open vessel, the hydrogen gas escapes and the reaction proceeds essentially to completion; that is, the iron is entirely converted to Fe_3O_4 if we supply enough steam. The conditions for quantitative reduction of the iron oxide to iron by hydrogen gas (the reverse reaction) can also be achieved rather easily in the laboratory. In either of these processes, in an open container, reaction is essentially complete because of removal of one of the products as a gas:

$$4H_2O(g) + 3Fe(s) \longrightarrow Fe_3O_4(s) + 4H_2(g)$$
steam   escapes

$$4H_2(g) + Fe_3O_4(s) \longrightarrow 3Fe(s) + 4H_2O(g)$$
swept out by
excess H_2 gas

If we combine the reactants for the oxidation of iron (first equation) in stoichiometric proportions in a *closed* container at constant temperature, the products gradually accumulate. As hydrogen and Fe_3O_4 are formed, occasional collisions of these species are effective and the reduction of Fe_3O_4 begins to occur. Eventually, a condition is reached at which there is no further net change. The *equilibrium* system contains iron, the iron oxide, hydrogen, and water in constant quantities.

Radioisotope studies show that, like earlier cases studied (Unit 12), this system is one of *dynamic* equilibrium.

Chemical equilibrium results whenever two exactly opposite chemical changes occur at the same rate within a closed system.

Theoretically, any chemical process should eventually reach equilibrium. The real difference, then, between reversible and irreversible reactions is whether or not it is feasible to achieve laboratory conditions at which *detectable* concentrations of all equilibrium components can coexist.

17.2 A SIMPLE MODEL FOR DYNAMIC EQUILIBRIUM

Let's consider a reversible process discussed in Unit 16, the formation and decomposition of hydrogen iodide. Although we now recognize the mechanism of reaction in this system to be fairly complex, let's *pretend* initially that we don't know about the information published since 1966 so that we can consider both reactions as though they involved simple bimolecular processes.

FORMATION $H_2(g) + I_2(g) \longrightarrow 2HI(g)$

rate of HI formation $= k_{formation} \times C_{H_2} \times C_{I_2}$

DECOMPOSITION $2HI(g) \longrightarrow H_2(g) + I_2(g)$

rate of HI decomposition $= k_{decomposition} \times C_{HI}^2$

Now, remember that rate constants, at a fixed temperature, are true "constants" (i.e., concentration independent), but reaction *rates* themselves are concentration dependent, as expressed by the rate laws. If we mixed hydrogen and iodine gases in a sealed container maintained at constant temperature, reaction would begin at a rate determined by the rate constant ($k_{formation}$) and the initial reactant concentrations. As reaction progressed, hydrogen and iodine concentrations would decrease and the rate of formation of hydrogen iodide (or rate of consumption of hydrogen) would get progressively slower. If the reaction were *irreversible,* this process would continue, at a continuously decreasing rate, until all of the hydrogen and iodine were converted to HI (assuming a stoichiometric ratio of reactants, Unit 4). However, the process *is* reversible. As hydrogen iodide is formed, some of these molecules decom-

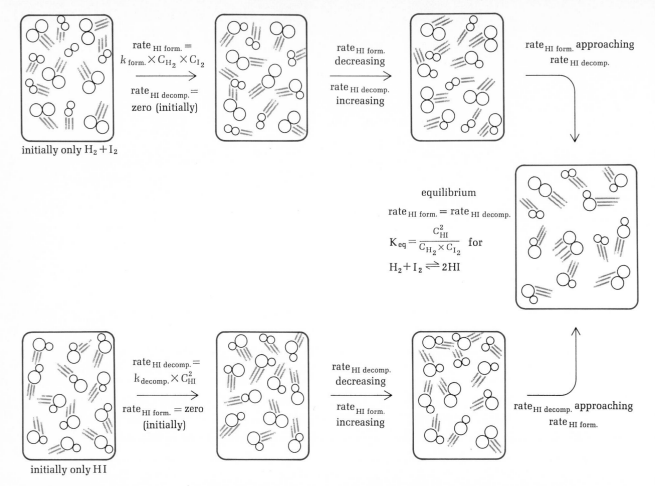

$$\text{rate}_{\text{HI form.}} = k_{\text{form.}} \times C_{\text{H}_2} \times C_{\text{I}_2}$$

$$\text{rate}_{\text{HI decomp.}} = \text{zero (initially)}$$

initially only $H_2 + I_2$

rate$_{\text{HI form.}}$ decreasing

rate$_{\text{HI decomp.}}$ increasing

rate$_{\text{HI form.}}$ approaching rate$_{\text{HI decomp.}}$

equilibrium

$$\text{rate}_{\text{HI form.}} = \text{rate}_{\text{HI decomp.}}$$

$$K_{\text{eq}} = \frac{C_{\text{HI}}^2}{C_{\text{H}_2} \times C_{\text{I}_2}} \quad \text{for}$$

$$H_2 + I_2 \rightleftharpoons 2HI$$

$$\text{rate}_{\text{HI decomp.}} = k_{\text{decomp.}} \times C_{\text{HI}}^2$$

$$\text{rate}_{\text{HI form.}} = \text{zero (initially)}$$

initially only HI

rate$_{\text{HI decomp.}}$ decreasing

rate$_{\text{HI form.}}$ increasing

rate$_{\text{HI decomp.}}$ approaching rate$_{\text{HI form.}}$

Figure 17.2 Dynamic equilibrium. Although the equilibrium state may be reached at different rates by different paths, the final composition of the equilibrium mixture is path independent. (The sequences illustrated depict hypothetical bimolecular processes. The "true" reaction mechanisms are probably more complex.)

pose to H_2 and I_2, at a rate determined by the HI concentration and the rate constant ($k_{\text{decomposition}}$). Thus two competing processes are taking place within the same closed system. As H_2 and I_2 are consumed, the rate of formation of HI decreases while the rate of decomposition of HI increases (with increasing concentration of HI). Eventually, these two rates must become equal and no further *net* change occurs in the composition of the system. The system has reached a state of *equilibrium* (Figure 17.2).

Chemical equilibrium is characterized by

a. constant system composition, resulting from
b. dynamic opposing processes occurring at equal rates within the same closed system.

A mathematical analysis of our example shows that, at equilibrium

$$H_2(g) + I_2(g) \underset{k_{\text{decomposition}}}{\overset{k_{\text{formation}}}{\rightleftharpoons}} 2HI(g)$$

rate of formation = rate of decomposition

$$k_{\text{formation}} \times C_{\text{H}_2} \times C_{\text{I}_2} = k_{\text{decomposition}} \times C_{\text{HI}}^2$$

from which

$$\frac{k_{\text{formation}}}{k_{\text{decomposition}}} = \frac{C_{\text{HI}}^2}{C_{\text{H}_2} \times C_{\text{I}_2}}$$

The ratio of two constants must also be a constant, in this case called the *equilibrium constant*, for the system characterized as

$$H_2 + I_2 \rightleftharpoons 2HI$$

$$K_{\text{eq}} = \frac{k_{\text{formation}}}{k_{\text{decomposition}}} = \frac{C_{\text{HI}}^2}{C_{\text{H}_2} \times C_{\text{I}_2}}$$

It is apparent that we could evaluate the equilibrium constant if we had *either* the experimentally determined rate constants or the experimentally determined equilibrium concentrations. It is the latter relationship that is particularly important to the very extensive utility of equilibrium calculations, for it permits us to work with equilibrium constants without any necessity of knowing the rate constants.

Note that we could have used the reciprocal relationships in developing the equilibrium constant, that is,

$$K'_{\text{eq}} = \frac{k_{\text{decomposition}}}{k_{\text{formation}}} = \frac{C_{\text{H}_2} \times C_{\text{I}_2}}{C_{\text{HI}}^2}$$

in which case $K'_{\text{eq}} = 1/K_{\text{eq}}$, as originally defined.

By convention, the equilibrium constant is written for the ratio of the concentration expression for the right-hand side of the equilibrium equation over that for the left-hand side.

As we shall see (in Section 17.3) some species shown in the chemical equations for equilibrium systems are excluded from the *conventional* equilibrium constant expressions.

Any chemical equilibrium may be described by an equilibrium constant. The numerical value of this constant is equal to the product of the concentrations of "right-hand" species divided by the product of the concentrations of "left-hand" species, with all concentration terms having exponents equal to the coefficients of the species in question as written in the balanced equilibrium equation.

EQUILIBRIUM CONSTANTS AS "STATE FUNCTIONS"

It would seem from our derivation of an equilibrium constant as a ratio of rate constants that only those species appearing in rate laws (Unit 16) for the "opposing processes" should be used in concentration terms for the equilibrium constant expressions. As we have seen, neither the rate-law concentration terms nor the exponents for these terms are necessarily related to the balanced equation for the overall reaction. This appears to be inconsistent with the development of equilibrium constant expressions directly from the chemical equations for equilibrium systems without regard to reaction mechanisms.

This is not, in fact, a real problem. A rigorous derivation of the equilibrium constant for any complex systems is, unfortunately, beyond the scope of this text, but such derivations show that equilibrium expressions are, indeed, independent of the mechanisms by which equilibrium states are reached. In this sense an equilibrium constant is a *state function* (Unit 5). Although we have not developed sufficient background in thermodynamics to permit us to present the rigorous and general derivations, we can illustrate the path independence of the equilibrium constant with a familiar example.[1]

As we have seen in Unit 16, the

$$H_2 + I_2 \rightleftharpoons 2HI$$

system may *not*, according to recent information, involve two simple opposing bimolecular processes. In fact, the *true* mechanisms for both the formation and decomposition of HI are still uncertain. Let's consider one of the sets of mechanisms proposed for this system to illustrate that we can still arrive at the same description of the equilibrium constant expression.

Postulated Mechanisms

Formation of HI	Decomposition of HI

FAST $I_2 \rightleftharpoons 2I$ (steady state)

SLOW $2I + H_2 \longrightarrow (H_2I_2)$ (transition state)

FAST $(H_2I_2) \longrightarrow 2HI$

rate of HI formation $= k'_{formation} \times C_I^2 \times C_{H_2}$

SLOW $2HI \longrightarrow (H_2I_2)$ (transition state)

FAST $(H_2I_2) \longrightarrow H_2 + I_2$

rate of HI decomposition $= k_{decomposition} \times C_{HI}^2$

We note that the rate law for decomposition is the same as that employed in our original derivation of the equilibrium process, but that for the formation reaction (involving a termolecular process) appears different.

From the definition of *steady state* (Units 12 and 16)

$$\frac{C_I^2}{C_{I_2}} = K_{steady\ state}$$

that is, this concentration-ratio is constant since iodine atoms are formed as rapidly as they are consumed by a subsequent process. Thus

$$C_I^2 = K_{steady\ state} \times C_{I_2}$$

and substitution of this relationship in the rate law suggested for the termolecular reaction yields

$$\text{rate of HI formation} = k'_{formation} \times K_{steady\ state} \times C_{I_2} \times C_{H_2}$$

The product of two constants is still a constant, in this case equal to the experimentally determined rate constant for the formation reaction,

$$\underbrace{k'_{formation} \times K_{steady\ state}}_{} = \underbrace{k_{formation}}_{}$$

(from proposed mechanism)   (experimental)

As a result, the ratio of the experimental rate constants still gives the correct value

---

[1]For an example of a more rigorous approach to equilibrium, see "Chemical Equilibrium as a State of Maximum Entropy," by Leonard Nash in the May, 1970, issue of *Journal of Chemical Education* (pp. 353–357).

for K_{eq} for the system

$$H_2 + I_2 \rightleftharpoons 2HI$$

When sufficient information is available, it can be demonstrated that alternative mechanisms can always be mathematically manipulated to yield the same equilibrium constant expression for any particular equilibrium system.

17.3 CONVENTIONAL EQUILIBRIUM CONSTANTS

We can determine the molar concentration of a solute in a solution by any of a number of *direct* measurements. If we know, for example, that a solute does not react with the solvent, we can determine the *molarity* (Unit 13) of the solution by finding the mass of solute (of known formula weight) present in a measured volume of the solution. In many cases we can determine the molar concentration of a solute by spectrophotometric measurements (Unit 30). Under appropriate conditions, the technique of titration (Unit 13) can provide an accurate analysis of the molarity of particular species.

Many *indirect* methods often suggest solute concentrations significantly different from those determined by direct measurement. For example, if we prepare a solution by dissolving 2.00 moles of sodium chloride in sufficient water to make 1.000 liter of solution, the concentration (from direct measurement) is 2.00 M. However, the molar concentration calculated from the experimentally determined osmotic pressure (Unit 15) is somewhat less than 2.00 M NaCl (2.00 M Na^+ and 2.00 M Cl^-). This may be explained by suggesting that measurements of colligative properties, such as osmotic pressure, reflect the *number* of solute particles, rather than the nature of the particles, so that closely associated ion pairs "behave" as "single particles." Many properties of solutions are determined by such "apparent molarities," referred to as *activities*.

The rigorous derivation of equilibrium constants requires the use of *activities,* rather than the molar concentrations we have employed (represented by C). Activities may be thought of as "effective" concentrations. By definition, the activity of a *pure* liquid or a *pure* solid is unity (one). As a result, *pure* liquids or solids are omitted from equilibrium constant expressions, since such terms (being defined as unity) would not affect the value of the equilibrium constant.

One of the best ways of determining solute *activities* employs electrochemical techniques and we shall return to this concept when we study electrochemical cells (Units 23 and 24). In dilute solutions, *activities* are closely approximated by molar concentrations, and such approximations are sufficiently accurate for our purposes even at moderately high concentrations.

We will therefore use equilibrium constants based on directly measured molar concentrations for aqueous solutions, remembering that this represents an approximation introducing some degree of uncertainty in calculated quantities. We will retain, however, the convention of excluding concentration terms for pure liquids or solids. When gases are involved, either concentrations (in moles liter^{-1}) or partial pressures (Unit 10) may be used. Care must be taken when gases are involved that pressure or concentration units employed are consistent with those used for the particular equilibrium constant selected, since different units are not generally *numerically* equivalent.

Because a chemical equilibrium is a *dynamic* process of opposing reactions, the terms "reactants" and "products" (as used for either irreversible processes or reactions *progressing toward* equilibrium) have no meaning. Rather, we may describe "right-hand" species (to the right of the double arrows) or "left-hand" species on the basis of *how we choose* to write the chemical equation. In a similar way we may describe the "forward reaction" as consumption of "left-hand" species or the "reverse reaction" as consumption of "right-hand" species. Since we will be excluding pure solids or liquids from our equilibrium constant expressions, it is important that we either remember the physical states of all species shown in the equation or that we indicate these states by (s), (l), (g), or (aq) (Unit 3).

We may summarize our conventions for equilibrium constant expressions as

In Unit 21 we shall investigate systems requiring more than a single chemical equation.

a. A balanced chemical equation is written to indicate the interrelationships of equilibrium species.

b. Physical states of all species in the chemical equation must be remembered or indicated.

c. Concentration terms, representing *equilibrium* concentrations (in moles liter^{-1}), for all "right-hand" species, *other than solids or pure liquids,* are written in the numerator, with exponents equal to the corresponding coefficients in the chemical equation. These concentration terms are multiplied together.

d. Similar concentration terms for "left-hand" species are written in the denominator.

e. If the only "right-hand" species are liquids or solids (or both), the numerator is written as 1 (unity). No denominator term is written when its value is one.

f. If "pressure constants" are employed,[2] partial pressures are used for gases, rather than concentrations in moles liter^{-1}.

Example 1

Write conventional equilibrium constant expressions for the equilibrium systems indicated.

a. $HCN(aq) \rightleftharpoons H^+(aq) + CN^-(aq)$
b. $NH_3(aq) + H_2O(l) \rightleftharpoons NH_4^+(aq) + OH^-(aq)$
c. $H_3PO_4(aq) \rightleftharpoons 3H^+(aq) + PO_4^{3-}(aq)$
d. $Ag_3PO_4(s) \rightleftharpoons 3Ag^+(aq) + PO_4^{3-}(aq)$
e. $2SO_2(g) + O_2(g) \rightleftharpoons 2SO_3(g)$ (for use with a "concentration-constant")
f. $N_2(g) + 3H_2(g) \rightleftharpoons 2NH_3(g)$ (for use with a "pressure-constant")
g. $[Al(OH)_4]^-(aq) + H^+(aq) \rightleftharpoons Al(OH)_3(s) + H_2O(l)$

Solution

a. $K_{eq} = \dfrac{C_{H^+} \times C_{CN^-}}{C_{HCN}}$

b. $K_{eq} = \dfrac{C_{NH_4^+} \times C_{OH^-}}{C_{NH_3}}$ (*liquid* H_2O excluded)

c. $K_{eq} = \dfrac{C_{H^+}^3 \times C_{PO_4^{3-}}}{C_{H_3PO_4}}$

d. $K_{eq} = C_{Ag^+}^3 \times C_{PO_4^{3-}}$ (*solid* Ag_3PO_4 excluded, so denominator = 1)

---

[2]The numerical value of a pressure constant will, in general, be different from that of the corresponding concentration constant. We shall use concentration constants unless otherwise specified.

e. $K_{eq} = \dfrac{C_{SO_3}^2}{C_{SO_2}^2 \times C_{O_2}}$

f. $K_{eq} = \dfrac{P_{NH_3}^2}{P_{N_2} \times P_{H_2}^3}$

g. $K_{eq} = \dfrac{1}{C_{[Al(OH)_4]^-} \times C_{H^+}}$ (*solid* $Al(OH)_3$ *and liquid* H_2O *excluded, so numerator* = 1)

Two other conventions may be mentioned at this point. When the equilibrium equation indicates the "left-hand" species as a single solid salt and the "right-hand" species as the aqueous ions derived from that solid, the conventional equilibrium constant is called the *solubility product,* symbolized by K_{sp} (Unit 18). The other specific convention is the use of K_w, called the *ion-product for water* (or the *autoprotolysis constant* for water), to represent $C_{H^+} \times C_{OH^-}$ from the equilibrium system

$$H_2O(l) \rightleftharpoons H^+(aq) + OH^-(aq)$$

Other special designations will be discussed later (e.g., K_a and K_b for acids and bases in Unit 19).

Except for practice in writing conventional equilibrium expressions, we shall restrict the remaining discussions in this unit to *homogeneous* equilibria, those in which all components are in the same phase (either gaseous or in solution). *Heterogeneous* equilibria, involving two or more phases, are described in Units 18 and 21.

17.4 EVALUATING EQUILIBRIUM CONSTANTS

We have already mentioned that an equilibrium constant could be calculated from the ratio of rate constants for the opposing processes. This is rarely practical. In most cases it is much easier to calculate equilibrium constants from experimentally determined concentrations (or *activities*) than it is to determine and use the appropriate rate constants.

It is important that we remember that concentration terms in equilibrium constant expressions represent *equilibrium* concentrations, so we must be certain that the system has actually reached equilibrium before concentration measurements are made, *and* that the measuring technique employed does not disturb the equilibrium in such a way as to affect the accuracy of measurements. Equilibrium constants, like rate constants, may vary significantly with temperature, so the temperature of the system must also be measured and held constant. The problem of ascertaining that equilibrium has been reached is often a significant one, especially for very slow processes. Since both the "forward reaction" and the "reverse reaction" involve the same transition state, it is often possible to use a catalyst (Unit 16) to reduce the time required for the system to reach equilibrium. Since the equilibrium constant is a state function, its value will not be affected by the use of a catalyst to provide an alternative reaction path.

There are many important systems for which the "true" equilibrium state is achieved very slowly. In such cases suitable approximations based on "pseudoequilibrium" considerations may be used.[3]

We can determine equilibrium concentrations in either of two ways. In some cases a separate analysis can be made for the concentration of each component of the equilibrium system (Example 2). Alternatively, information on *initial* reactant concentrations and the equilibrium concentration of one component permits us to

Remember (Unit 13) that *formality* (F) may be used to characterize a solution in terms of solute amount used in solution preparation. However, we shall continue to use *molarity,* as a matter of convenience.

[3]For a discussion, see "Practical Equilibrium Concentrations and Yields in Selective Catalyst Systems," by Bruce Hamilton and Martin Greenwald in the November, 1974, issue of the *Journal of Chemical Education* (pp. 732–734).

determine the equilibrium concentrations of the remaining components from stoichiometric relationships (Example 3).

Example 2

A mixture of oxygen, sulfur dioxide, and sulfur trioxide was observed to maintain constant pressure at 1000°K for a period of several days. On the assumption that the system was at equilibrium, a spectrophotometric analysis was made, revealing

$$C_{O_2} = 0.25 \text{ mole liter}^{-1}$$
$$C_{SO_2} = 0.31 \text{ mole liter}^{-1}$$
$$C_{SO_3} = 2.52 \text{ mole liter}^{-1}$$

If the assumption was valid, what is the equilibrium constant at 1000°K for the system described by

$$2SO_3(g) \rightleftharpoons 2SO_2(g) + O_2(g)$$

Solution
For the equation given, the conventional equilibrium constant is

$$K_{eq} = \frac{C_{SO_2}^2 \times C_{O_2}}{C_{SO_3}^2}$$

Using the measured concentrations,

$$K_{eq} = \frac{(0.31)^2 \times (0.25)}{(2.52)^2} = 3.8 \times 10^{-3}$$

Example 3

A solution was prepared by dissolving 286 mg of sodium bisulfate in sufficient water to make 75.0 ml of solution. The hydrogen-ion concentration was measured accurately by use of a pH meter (Units 19 and 24). Repeated measurements over a period of time showed that the initially determined value of

$$C_{H^+} = 1.5 \times 10^{-2} \, M$$

remained constant. On the assumption that the equilibrium system is satisfactorily represented by

$$HSO_4^-(aq) \rightleftharpoons H^+(aq) + SO_4^{2-}(aq)$$

what is the equilibrium constant?

Solution
The initial concentration of HSO_4^- is given by

$$\frac{286 \text{ mg (NaHSO}_4)}{75.0 \text{ ml}} \times \frac{1 \text{ mmol (NaHSO}_4)}{120 \text{ mg (NaHSO}_4)} \times \frac{1 \text{ mmol (HSO}_4^-)}{1 \text{ mmol (NaHSO}_4)} = 3.18 \times 10^{-2} \, M$$

From the stoichiometric ratios expressed by the balanced equation, it is seen that for each H^+ ion formed, one HSO_4^- ion is consumed and one SO_4^{2-} is produced. It follows, then, that for an equilibrium $C_{H^+} = 1.5 \times 10^{-2} \, M$,

$$C_{SO_4^{2-}} = C_{H^+} = 1.5 \times 10^{-2} \, M$$

and

$$C_{HSO_4^-} = \text{original } M \, (HSO_4^-) - C_{H^+}$$
$$= (3.18 \times 10^{-2}) - 1.5 \times 10^{-2} = 1.7 \times 10^{-2} \, M$$

The conventional equilibrium expression for the system is given by

$$K_{eq} = \frac{C_{H^+} \times C_{SO_4^{2-}}}{C_{HSO_4^-}}$$

and substitution of the experimental and calculated equilibrium concentrations gives

$$K_{eq} = \frac{(1.5 \times 10^{-2}) \times (1.5 \times 10^{-2})}{1.7 \times 10^{-2}} = 1.3 \times 10^{-2}$$

The choice of which method to use for a particular system depends on the nature of the data available. In most cases it is convenient to measure "initial" concentrations and the equilibrium concentration of an appropriate component of the system, from which the remaining equilibrium concentrations can be found by use of stoichiometric ratios. As we shall see, stoichiometric relationships are also useful in other applications in equilibrium analyses.

17.5 FINDING EQUILIBRIUM CONCENTRATIONS The equilibrium constants for a large number of chemical systems have been measured and extensive tabulations, over a considerable range of temperatures, are available in source books such as the Chemical Rubber Company's *Handbook of*

Table 17.1 SOME EXAMPLES OF HOMOGENEOUS EQUILIBRIA

EQUATION	EQUILIBRIUM CONSTANT
GAS PHASE	
$2SO_3(g) \rightleftharpoons 2SO_2(g) + O_2(g)$	$K_{eq} = \dfrac{C_{SO_2}^2 \times C_{O_2}}{C_{SO_3}^2} = 3.8 \times 10^{-3}$ (at 1000°K)
$N_2(g) + 3H_2(g) \rightleftharpoons 2NH_3(g)$	$K_{eq} = \dfrac{C_{NH_3}^2}{C_{N_2} \times C_{H_2}^3} = 2.4 \times 10^{-3}$ (at 1000°K)
$I_2(g) \rightleftharpoons 2I(g)$	$K_{eq} = \dfrac{C_I^2}{C_{I_2}} = 3.8 \times 10^{-5}$ (at 1000°K)
$2CO(g) + O_2(g) \rightleftharpoons 2CO_2(g)$	$K_{eq} = \dfrac{C_{CO_2}^2}{C_{CO}^2 \times C_{O_2}} = 2.2 \times 10^{22}$ (at 1000°K)
$2NO(g) + O_2(g) \rightleftharpoons 2NO_2(g)$	$K_{eq} = \dfrac{C_{NO_2}^2}{C_{NO}^2 \times C_{O_2}} = 6.4 \times 10^5$ (at 500°K)
AQUEOUS	
$H_2O(l) \rightleftharpoons H^+(aq) + OH^-(aq)$	$K_w = C_{H^+} \times C_{OH^-} = 1.0 \times 10^{-14}$ (at 298°K) (see Unit 19)
$NH_3(aq) + H_2O(l) \rightleftharpoons NH_4^+(aq) + OH^-(aq)$	$K_b = \dfrac{C_{NH_4^+} \times C_{OH^-}}{C_{NH_3}} = 1.8 \times 10^{-5}$ (at 298°K) (see Unit 19)
$CH_3CO_2H(aq) \rightleftharpoons H^+(aq) + CH_3CO_2^-(aq)$	$K_a = \dfrac{C_{H^+} \times C_{CH_3CO_2^-}}{C_{CH_3CO_2H}} = 1.8 \times 10^{-5}$ (at 298°K) (see Unit 19)
$[Ag(NH_3)_2]^+(aq) \rightleftharpoons Ag^+(aq) + 2NH_3(aq)$	$K_{inst} = \dfrac{C_{Ag^+} \times C_{NH_3}^2}{C_{[Ag(NH_3)_2]^+}} = 6.3 \times 10^{-8}$ (at 298°K) (see Unit 27)
$Sn^{2+}(aq) + 2Fe^{3+}(aq) \rightleftharpoons Sn^{4+}(aq) + 2Fe^{2+}(aq)$	$K_{eq} = \dfrac{C_{Sn^{4+}} \times C_{Fe^{2+}}^2}{C_{Sn^{2+}} \times C_{Fe^{3+}}^2} = 1.0 \times 10^{19}$ (at 298°K) (see Unit 24)

Chemistry and Physics. Some typical data are shown in Table 17.1 and K values for many aqueous systems at 20°–25°C, with which we shall be mainly concerned, are given in Appendix C.

When equilibrium constants are known, measurements of appropriate "initial" or equilibrium concentrations, together with the use of stoichiometric relationships, permit us to calculate the equilibrium concentrations of other components. The use of concentration constants, rather than those based on *activities*, is an approximation introducing some degree of uncertainty in our calculations. For most systems we will consider, this uncertainty is on the order of ±5 percent or less. Since we already have this inherent uncertainty, we will find it suitable to use certain mathematical approximations to simplify equilibrium calculations.

Useful mathematical simplifications may often be made in determining values to use for concentration terms in an equilibrium constant expression. Whenever a concentration is expressed by the sum or difference of two or more values, additive or subtractive factors contributing less than ~5 percent may usually be neglected. This depends, of course, on the overall accuracy desired. These simplifications become almost intuitive as familiarity with equilibrium problems increases. Let us consider two examples that illustrate this point.

Example 4

What is the approximate composition of the equilibrium system resulting from mixing 0.20 mole of carbon monoxide and 4.0 moles of oxygen in a 1.00-liter vessel maintained at 1000°K?

Solution
From Table 17.1

$$2CO + O_2 \rightleftharpoons 2CO_2$$

$$\frac{C_{CO_2}^2}{C_{CO}^2 \times C_{O_2}} = 2.2 \times 10^{22} \quad \text{(at } 1000°K\text{)}$$

The magnitude of the equilibrium constant and the initial use of a large excess of O_2 suggest that nearly all the carbon monoxide is converted to carbon dioxide at equilibrium.

If we let y represent the concentration of CO *remaining at equilibrium*, then the stoichiometry of the process suggests that, at equilibrium, if

$$C_{CO} = y \text{ mole liter}^{-1}$$

then

$$C_{CO_2} = (0.20 - y) \text{ mole liter}^{-1}$$
$$C_{O_2} = [4.0 - \tfrac{1}{2}(0.20 - y)] \text{ moles liter}^{-1}$$
$$\left(\frac{1 \text{ mole } O_2}{2 \text{ moles CO}}\right)$$

Since the equilibrium constant is so large, we suspect that y is very small compared to 0.20, so the concentrations may be closely approximated by neglecting y in subtractive terms:

$$C_{CO_2} = 0.20 \text{ mole liter}^{-1}$$
$$C_{O_2} = [4.0 - \tfrac{1}{2}(0.20)] = 3.9 \text{ moles liter}^{-1}$$
$$C_{CO} = y \text{ mole liter}^{-1}$$

Then, using these factors in the equilibrium constant expression,

$$2.2 \times 10^{22} = \frac{(0.20)^2}{y^2 \times 3.9}$$

$$y = \sqrt{\frac{(0.20)^2}{2.2 \times 10^{22} \times 3.9}} = 6.8 \times 10^{-13} \text{ mole liter}^{-1}$$

The equilibrium composition, then, is

$$C_{CO_2} \approx 0.20 \text{ mole liter}^{-1}$$
$$C_{CO} \approx 6.8 \times 10^{-13} \text{ mole liter}^{-1}$$
$$C_{O_2} = 3.9 \text{ moles liter}^{-1}$$

It is, then, obvious that no significant error was introduced by the simplification

$$(0.20 - y) \approx 0.20 \qquad [0.20 - (6.8 \times 10^{-13}) \approx 0.20]$$

Such approximations are not always justified. It is necessary to consider relative concentrations and magnitudes of equilibrium constants before deciding on appropriate simplifications.

Since the "simplified" solutions for equilibrium concentration problems are always easier than the "rigorous" solutions, the best general approach is to make the appropriate simplifying approximations, solve the resulting equations, and *then* test the validity of the approximations to see if a second ("rigorous") solution is necessary. This test of approximation validity should *always* be made.

Example 5

What is the composition of the equilibrium mixture resulting from introduction of 2.0×10^{-4} mole of iodine into a 1.00-liter vessel at $1000°K$?

Solution
From Table 17.1

$$I_2 \rightleftharpoons 2I$$

$$\frac{C_I^2}{C_{I_2}} = 3.8 \times 10^{-5}$$

If we represent the equilibrium concentration of atomic iodine by y, then at equilibrium

$$C_I = y$$
$$C_{I_2} = (2.0 \times 10^{-4} - 0.5y)$$

May we safely make the simplification that $(2.0 \times 10^{-4} - 0.5y) \approx 2.0 \times 10^{-4}$? Let us try it and see what happens.

$$3.8 \times 10^{-5} = \frac{y^2}{2.0 \times 10^{-4}}$$

$$y = \sqrt{7.6 \times 10^{-9}} = 8.7 \times 10^{-5}$$

Then

$$(2.0 \times 10^{-4} - 0.5y) = 1.6 \times 10^{-4}$$

so the simplification that

$$(2.0 \times 10^{-4} - 0.5y) = 2.0 \times 10^{-4}$$

was not justified if we desire less than a 20-percent error. If a more precise value is to be obtained, a more complex solution is required:

$$3.8 \times 10^{-5} = \frac{y^2}{(2.0 \times 10^{-4} - 0.5y)}$$

This leads to the expression

$$y^2 + (1.9 \times 10^{-5})\,y - (7.6 \times 10^{-9}) = 0$$

from which

$$y = 7.8 \times 10^{-5}$$

The equilibrium mixture then contains 7.8×10^{-5} mole liter^{-1} of atomic iodine and 1.6×10^{-4} mole liter^{-1} of molecular iodine at $1000°\mathrm{K}$.

For $ax^2 + bx + c = 0$,

$$x = \frac{-b \pm \sqrt{b^2 - 4ac}}{2a}$$

Note that of the two possible *mathematical* solutions, only one is a realistic *chemical* solution.

17.6 LE CHATELIER REVISITED

We may think of an equilibrium system as analogous to a *spring*. When you compress a spring, it "tries to *push* back." When you stretch a spring, it "tries to *pull* back." In a similar way, an equilibrium system "tries to undo" whatever you "do" to it.

The effects of changes in concentration, temperature, or, when gases are involved, pressure in an equilibrium system can be calculated with considerable accuracy. Temperature effects, however, require some thermodynamic considerations a bit too complex for our purposes.

We shall content ourselves at this point with qualitative applications of Le Chatelier's principle (Unit 12).

When any factor contributing to an equilibrium condition is varied, equilibrium is destroyed and the system returns to a state of continuing change until a new equilibrium is attained whose properties balance the effect of the original variation.

We can apply this principle to the situations in which concentration, temperature, or gas pressure is varied.

Concentration: If more of a particular component is added to an equilibrium system, reaction will occur to consume some of the new material. Conversely, removal of some of one component favors reaction to replace it.

Temperature: If an equilibrium system is heated to increase its temperature, that reaction that is endothermic will be favored in an attempt to use up added heat. Cooling an equilibrium system favors the exothermic process (tending to replace lost heat).

Pressure (when gases are involved): An increase of pressure favors the reaction that decreases the total number of moles of gas in the system in an attempt to counteract the pressure increase. Pressure reduction favors reaction *increasing* the total number of moles of gas. If neither the forward nor reverse reaction changes the number of moles of gas, equilibrium composition[4] is not affected by pressure change.

One additional factor should be considered. Substances called *catalysts* may be used to increase the speed of a chemical process (Unit 16). Catalysts, however, do not affect the *composition* of an equilibrium system, as we have seen. The only advantage of a catalyst for equilibrium systems is to reduce the time required to attain equilibrium.

Example 6

What is the effect ("increase," "decrease," or "no effect") on the equilibrium amount

---

[4]However, *concentrations* are changed as a result of change in gas volume (Unit 10). Note that we are talking here in terms of pressure change by compression or expansion (volume change). An increase in total gas pressure, at constant volume and temperature, by addition of some nonreactive gas does not affect the equilibrium, since concentrations (or *partial pressures*) of equilibrium components are not altered.

of carbon dioxide in the system

$$2CO(g) + O_2(g) \rightleftharpoons 2CO_2(g)$$

of each of the following:

a. increasing the pressure on the system at constant temperature by reducing the volume

b. increasing the temperature of the system (oxidation of carbon monoxide is exothermic)

c. introducing more oxygen gas into the system

Solution

a. Increased pressure can be counteracted by reducing the number of moles of gas in the system. This may occur by *increasing* CO_2 production:

$$\underbrace{2CO + O_2}_{\text{3 particles}} \longrightarrow \underbrace{2CO_2}_{\text{2 particles}}$$

b. Since formation of CO_2 is exothermic, the system should react to heating by *decreasing* the amount of CO_2:

$$\text{heat} + 2CO_2 \longrightarrow \underbrace{2CO + O_2}_{\text{endothermic}}$$

c. Addition of more oxygen favors an *increase* in CO_2, since this reaction consumes some of the added oxygen:

$$2CO + \underbrace{O_2 \longrightarrow 2CO_2}_{\text{consumed}}$$

Example 7

What is the effect on the equilibrium amount of I_2 in the system

$$H_2(g) + I_2(g) \rightleftharpoons 2HI(g)$$

of each of the following:

a. decreasing the pressure on the system at constant temperature by increasing the volume of the system

b. adding more hydrogen gas

c. adding a platinum catalyst to the equilibrium mixture

Solution

a. In this case, pressure change has *no effect* on the equilibrium composition:

$$\underbrace{H_2(g) + I_2(g)}_{\text{2 particles}} \rightleftharpoons \underbrace{2HI(g)}_{= \text{2 particles}}$$

b. Added hydrogen gas is consumed by the process that *decreases* the amount of iodine:

$$H_2 + I_2 \longrightarrow 2HI$$

c. A catalyst has *no effect* on equilibrium composition.

THE ECONOMIC GAME

When chemistry is the basis for an economic venture, there must be a profit involved. For any reaction in a closed system, the maximum yield of product is determined by the limitations of the equilibrium state. An understanding of equilibrium, including applications of Le Chatelier's principle, is essential to the planning of reaction conditions for industrial processes to permit an optimum combination of high yield and low cost.

An excellent example is found in the *Haber process*, the principal commercial method for manufacturing ammonia.

At equilibrium,

$$N_2(g) + 3H_2(g) \rightleftharpoons 2NH_3(g)$$

for which

$$K_{eq} = \frac{C_{NH_3}^2}{C_{N_2} \times C_{H_2}^3}$$

The maximum concentration of ammonia (for the equilibrium mixture) is then given by

$$C_{NH_3} = \sqrt{K_{eq} \times C_{N_2} \times C_{H_2}^3}$$

Le Chatelier's principle suggests that the yield of ammonia could be improved by increasing either the hydrogen or nitrogen concentrations by introducing more of one of these reactants into the equilibrium system. The mathematical formulation indicates that a greater effect would be seen with hydrogen than with nitrogen. If hydrogen were sufficiently inexpensive, then the yield of ammonia would be improved by using a reaction mixture in which the mole ratio of hydrogen to nitrogen was greater than 3:1.

All the components of the system are gases. If the reaction did not proceed at all, then on a molar scale we would have 4 moles of gas ($N_2 + 3H_2$)(89.6 liters at STP). If the reaction proceeded irreversibly, we would have only 2 moles of gas ($2NH_3$)(44.8 liters at STP). Higher pressures favor smaller volumes (at constant temperature), so the yield of ammonia in terms of the percent ammonia at equilibrium should be higher if we perform the reaction at increased pressure. It can be shown, for example, that at 500°C NH_3 is less than 1 percent of the equilibrium mixture at a total pressure of 1.00 atm, but it is almost 50 percent of the mixture when the pressure is 500 atm.

The conversion of nitrogen and hydrogen to ammonia is exothermic. Le Chatelier's principle suggests, then, that the yield of ammonia should decrease if additional heat is added.

On the basis of equilibrium considerations alone, the most favorable conditions for ammonia production should be

1. excess hydrogen in reaction mixture
2. high pressure
3. low temperature

However, these conclusions are necessary but not sufficient. *Remember that equilibrium is a state function and that equilibrium considerations provide no information as to the speed of the process.*

The reaction of hydrogen with nitrogen is very slow at low temperatures, but moderately rapid at high temperatures. Although the equilibrium considerations suggest a better yield at low temperatures, for industry "time is money," and it is better to reach equilibrium by a more rapid route even if the equilibrium yield is somewhat lower. *Reaction kinetics*, then, must also be a factor in the design of an industrial synthesis. Careful experimentation has led to the optimum conditions for the Haber process—a compromise among the various factors of kinetics, equilibrium, and economics:

high pressure (equilibrium favored)
high temperature (kinetics favored)
iron (+ additives) catalyst (kinetics favored)
stoichiometric reactant ratio (economics favored)

SUGGESTED READING   Hamm, Randall, and Carl Nyman. 1968. *Chemical Equilibrium*. Lexington, Mass.: Raytheon.
Moeller, Therald, and Rod O'Connor. 1971. *Ions in Aqueous Systems*. New York: McGraw-Hill.

EXERCISES

1. Write conventional equilibrium constant expressions for
   a. $N_2O_4(g) \rightleftharpoons 2NO_2(g)$
   b. $H_2S(aq) \rightleftharpoons 2H^+(aq) + S^{2-}(aq)$
   c. $CH_3NH_2(aq) + H_2O(l) \rightleftharpoons CH_3NH_3^+(aq) + OH^-(aq)$
   d. $B_4O_7{}^{2-}(aq) + 5H_2O(l) \rightleftharpoons 2H_2BO_3{}^-(aq) + 2H_3BO_3(aq)$
   e. $Co^{2+}(aq) + 6NH_3(aq) \rightleftharpoons [Co(NH_3)_6]^{2+}(aq)$
   f. $Ag_2S(s) \rightleftharpoons 2Ag^+(aq) + S^{2-}(aq)$

2. A 0.183-g sample of sodium cyanide was dissolved in sufficient water to form 25.0 ml of solution, and the hydroxide-ion concentration of the resulting equilibrium system was determined at 25°C, by use of a pH meter, to be 1.8×10^{-3} M. Assuming that the equilibrium system is adequately represented by

   $$CN^-(aq) + H_2O(l) \rightleftharpoons HCN(aq) + OH^-(aq)$$

   what is the value of the equilibrium constant, at 25°C?

3. What is the composition of the equilibrium system resulting from mixing 0.10 mole of nitric oxide and 4.0 moles of oxygen in a 5.0-liter reaction vessel maintained at 500°K? (See Table 17.1.)

4. (For $HCO_2H \rightleftharpoons H^+ + HCO_2^-$, $K_a = 2.0 \times 10^{-4}$.) What is the composition of the equilibrium system in a solution labeled 2.0 M formic acid?

5. What is the effect ("increase," "decrease," or "no change") on the equilibrium amount of SO_3 in the system

   $$2SO_2(g) + O_2(g) \rightleftharpoons 2SO_3(g)$$

   of each of the following?

   a. decreasing the total pressure on the gas mixture, at constant temperature, by increasing the volume of the system
   b. adding more oxygen
   c. raising the temperature (combustion of SO_2 is exothermic)
   d. adding a solid phase catalyst

PRACTICE FOR PROFICIENCY

Objective (a)

1. Write conventional equilibrium constant expressions for
   a. $2NO(g) + O_2(g) \rightleftharpoons 2NO_2(g)$ (as a concentration constant)
   b. $H_2SO_3(aq) \rightleftharpoons 2H^+(aq) + SO_3{}^{2-}(aq)$
   c. $F^-(aq) + H_2O(l) \rightleftharpoons HF(aq) + OH^-(aq)$
   d. $BaSO_4(s) \rightleftharpoons Ba^{2+}(aq) + SO_4{}^{2-}(aq)$
   e. $2CO(g) + O_2(g) \rightleftharpoons 2CO_2(g)$ (as a pressure constant)
   f. $Fe^{3+}(aq) + 6CN^-(aq) \rightleftharpoons [Fe(CN)_6]^{3-}(aq)$
   g. $CO_2(aq) + H_2O(l) \rightleftharpoons H^+(aq) + HCO_3^-(aq)$
   h. $Al^{3+}(aq) + 3OH^-(aq) \rightleftharpoons Al(OH)_3(s)$
   i. $MgF_2(s) \rightleftharpoons Mg^{2+}(aq) + 2F^-(aq)$
   j. $H_2O(l) \rightleftharpoons H^+(aq) + OH^-(aq)$

2. Identify the equilibrium constants written for question 1 that are conventionally represented by
   a. K_{sp}; b. K_w

Objective (b)

3. Analysis of an equilibrium mixture of phosgene, carbon monoxide, and chlorine at 100°C showed

$C_{COCl_2} = 0.186$ mole liter^{-1}
$C_{CO} = 1.3 \times 10^{-2}$ mole liter^{-1}
$C_{Cl_2} = 3.1 \times 10^{-9}$ mole liter^{-1}

Calculate the equilibrium constant at 100°C for the system represented by

$$COCl_2(g) \rightleftharpoons CO(g) + Cl_2(g)$$

4. The gas mixture described in question 3 was heated to 400°C and maintained at that temperature until the pressure was stabilized. The mixture was then analyzed and the composition, reported in partial pressures, was found to be

$P_{COCl_2} = 10.3$ atm
$P_{CO} = 0.71$ atm
$P_{Cl_2} = 0.66$ atm

What is the pressure constant for this system at 400°C, as described by the chemical equation given in question 3?

5. A 0.384-g sample of ammonium nitrate was dissolved in sufficient water to form 100.00 ml of solution. Careful measurements showed the equilibrium acidity of the resulting solution, at 22°C, to be represented by

$$C_{H^+} = 5.2 \times 10^{-6} \, M$$

Calculate the equilibrium constant for the system represented by

$$NH_4^+(aq) \rightleftharpoons NH_3(aq) + H^+(aq)$$

Objective (c)

6. What is the composition of the equilibrium system resulting from mixing 0.30 mole of carbon monoxide with 6.0 moles of oxygen in a 20.0-liter reaction vessel maintained at 1000°K? (See Table 17.1.)

7. What is the approximate equilibrium concentration (moles liter^{-1}) of $CH_3CO_2^-$ in a solution labeled 0.50 M CH_3CO_2H? ($K = 1.8 \times 10^{-5}$.)

$$CH_3CO_2H(aq) \rightleftharpoons CH_3CO_2^-(aq) + H^+(aq)$$

8. What is the approximate equilibrium concentration (moles liter^{-1}) of H^+ in a solution labeled 0.075 M HCN? ($K = 4.8 \times 10^{-10}$.)

$$HCN(aq) \rightleftharpoons H^+(aq) + CN^-(aq)$$

9. What is the approximate equilibrium concentration (moles liter^{-1}) of HCO_2^- in a solution labeled 0.32 M HCO_2H? ($K = 2.0 \times 10^{-4}$.)

$$HCO_2H(aq) \rightleftharpoons HCO_2^-(aq) + H^+(aq)$$

10. What is the approximate equilibrium concentration (moles liter^{-1}) of OCl^- in a solution labeled 0.20 M HOCl? ($K = 3.2 \times 10^{-8}$.)

$$HOCl(aq) \rightleftharpoons H^+(aq) + OCl^-(aq)$$

11. What is the approximate equilibrium concentration (moles liter^{-1}) of H^+ in a solution labeled 0.080 M HNO_2? ($K = 4.5 \times 10^{-4}$.)

$$HNO_2(aq) \rightleftharpoons H^+(aq) + NO_2^-(aq)$$

Objective (d)

12. What is the effect ("increase," "decrease," "no change") on the equilibrium amount of carbon monoxide in the system,

$$2CO(g) + O_2(g) \rightleftharpoons 2CO_2(g)$$

of each of the following?
a. increasing the total pressure on the gas mixture at constant temperature by reducing the volume of the system
b. adding solid sodium hydroxide, a CO_2 absorbent
c. adding a solid phase catalyst
d. cooling the gas mixture at constant pressure (Combustion of CO is exothermic.)

13. Consider the gas-phase equilibrium system represented by the equation,

$$NO_2 + O_2 \rightleftharpoons NO + O_3$$

Given that the conversion of "left-hand" species to "right-hand" species, as written, is endothermic, which of the following changes will decrease the equilibrium *amount* of NO?
a. adding more ozone
b. decreasing the pressure at constant temperature by increasing the volume
c. increasing the pressure at constant temperature by decreasing the volume
d. heating the system at constant pressure
e. none of these
Explain your choice.

14. Consider the gas-phase equilibrium system represented by the equation,

$$2H_2O \rightleftharpoons 2H_2 + O_2$$

Given that the conversion of "left-hand" species to "right-hand" species, as written, is endothermic, which of the following changes will decrease the equilibrium *amount* of H_2O?
a. adding more oxygen
b. adding a solid phase catalyst
c. increasing the pressure at constant temperature, by decreasing volume
d. increasing the temperature at constant pressure
e. none of these
Explain your choice.

15. Consider the gas-phase equilibrium system represented by the equation,

$$2HI \rightleftharpoons H_2 + I_2$$

Given that the conversion of "left-hand" species to "right-hand" species, as written, is endothermic, which of the following changes will decrease the equilibrium amount of HI?
a. adding more iodine
b. adding a catalyst
c. increasing pressure at constant temperature by decreasing the volume
d. decreasing pressure at constant temperature by increasing the volume
e. none of these
Explain your choice.

BRAIN TINGLER

16. It has been estimated that the world's oceans contain more than 70,000,000 tons of gold, most of which is present in an equilibrium system, represented for simplicity as

$$[AuCl_4]^-(aq) + H_2O(l) \rightleftharpoons [Au(OH)_4]^-(aq) + 4H^+(aq) + 4Cl^-(aq)$$

The equilibrium constant for this system, at 25°C, is 2.3×10^{-30}.

A seawater sample was analyzed and found to have

$$C_{H^+} = 7.9 \times 10^{-9} \ M$$
$$C_{Cl^-} = 0.55 \ M$$
$$C_{[AuCl_4]^-} = 1.6 \times 10^{-15} \ M$$

What volume of such seawater would have to be processed to recover 5.0 g of gold, assuming complete recovery of all gold from the equilibrium system described? Does this appeal to you as a potentially profitable method for "mining" gold?

Unit 17 Self-Test

1. Write conventional equilibrium constant expressions for
   a. $N_2(g) + O_2(g) \rightleftharpoons 2NO(g)$
   b. $Ag_2S(s) \rightleftharpoons 2Ag^+(aq) + S^{2-}(aq)$
   c. $CO_3^{2-}(aq) + H_2O(l) \rightleftharpoons HCO_3^-(aq) + OH^-(aq)$
   d. $H_2C_2O_4(aq) \rightleftharpoons 2H^+(aq) + C_2O_4^{2-}(aq)$
   e. $PO_4^{3-}(aq) + 2H_2O(l) \rightleftharpoons H_2PO_4^-(aq) + 2OH^-(aq)$

2. A solution prepared by dissolving 0.841 g of sodium acetate in sufficient water to form 25.0 ml of solution was found to have an equilibrium concentration of acetic acid of 1.5×10^{-5} mole liter^{-1} at 25°C. Calculate the equilibrium constant, at 25°C, for the system as represented by

   $$CH_3CO_2^-(aq) + H_2O(l) \rightleftharpoons CH_3CO_2H(aq) + OH^-(aq)$$

3. What is the approximate concentration (in moles per liter) of atomic iodine in the equilibrium system $I_2 \rightleftharpoons 2I$ for a

mixture prepared by introducing 0.10 mole of iodine into a 1.00-liter vessel maintained at 1000°K? ($K_{eq} = 3.8 \times 10^{-5}$.)

4. What is the approximate equilibrium concentration (moles liter^{-1}) of $C_6H_5CO_2^-$ in a solution labeled 0.66 M $C_6H_5CO_2H$? ($K = 6.6 \times 10^{-5}$.)

   $$C_6H_5CO_2H(aq) \rightleftharpoons C_6H_5CO_2^-(aq) + H^+(aq)$$

5. Indicate the effect of each of the following ("increase," "decrease," "no change") on the amount of I_2 in the equilibrium system $2HI(g) \rightleftharpoons H_2(g) + I_2(g)$.
   a. addition of more hydrogen
   b. increase in total gas pressure, resulting from reduced volume
   c. increase in temperature (formation of HI is endothermic)
   d. addition of a solid phase catalyst

. . . Equilibrium of the contend-
ing forces ends the operation,
and limits the effect.
CLAUDE BERTHOLLET (1803)

Solubility equilibria

Objectives

(a) Given the name or formula of a sparingly soluble salt, be able to write
(Section 18.2).
  (1) the equilibrium equation for its saturated solution
  (2) the solubility product expression for the salt
(b) Be able to calculate solubility products from experimentally determined
solubilities (Example 2).
(c) Be able to use tabulated values of solubility products.
  (1) to calculate maximum solubilities (Example 4).
  (2) to predict whether or not precipitation should occur when particular
  solutions are mixed (Section 18.3).
  (3) to determine common-ion effects in the absence of complex formation
  (Section 18.4).
(d) Given the solubilities of a compound in two immiscible solvents, be able to
estimate the distribution coefficient for the solute in the two-phase solvent
mixture, and be able to use these distribution coefficients in calculations
involving "extraction efficiencies" for liquid–liquid extractions (Section 18.5).

In 1884 a gang of robbers secreted their loot in a little-known cavern in the Rincon Mountains southeast of Tucson, Arizona. Three of the bandits later "bit the dust" in a gunfight in Willcox, Arizona, and the fourth was imprisoned. In 1902 he returned to the area, presumably to recover the treasure. Whether or not he ever found the money is still a matter of speculation, and the story of the hidden wealth is a feature of any trip through the caverns—now a tourist site known as Colossal Cave.

Many famous caverns, like Colossal Cave, have been formed by an interesting process involving solubility equilibria. Most such caves have been formed by the action of natural waters on underground deposits consisting principally of calcite, calcium carbonate.

Calcium carbonate is quite insoluble in pure water. Its saturated solution may be represented by the equilibrium equation

$$CaCO_3(s) \rightleftharpoons Ca^{2+}(aq) + CO_3^{2-}(aq)$$

The double arrows are used to indicate an equilibrium condition, and their unequal lengths are a qualitative indication of the relative extents to which the opposing processes are favored. Calcite, then, should not dissolve appreciably in pure water, but natural rain water or, in fact, any water exposed to the atmosphere, is not pure H_2O. Atmospheric gases dissolve in the water and, if contact with the atmosphere continues long enough, the water becomes saturated with oxygen, nitrogen, carbon dioxide, and other (minor) components of the air. Carbon dioxide constitutes only about 0.05 percent by weight of dry air, but it is quite soluble in water. Water saturated with air at 1.00 atm and 20°C contains about 1.7 g of CO_2 per liter, compared to ~0.03 g of N_2 and ~0.06 g of O_2. The dissolved carbon dioxide reacts with the water, and eventually an equilibrium is reached whose principal composition may be represented as

$$CO_2(aq) + H_2O \rightleftharpoons H^+(aq) + HCO_3^-(aq)$$

Thus water saturated with air is slightly acidic because of the excess H^+ released by reaction between dissolved CO_2 and water. Calcite will dissolve in this aqueous system:

$$CaCO_3(s) + H^+(aq) \rightleftharpoons Ca^{2+}(aq) + HCO_3^-(aq)$$

It is this process that slowly dissolves calcite deposits to form the huge underground caverns and, incidentally, to form ground water whose "hardness" (Unit 6) results from the presence of Ca^{2+}.

Often large lakes are formed within the caverns. Within the water, bicarbonate and carbonate are in equilibrium:

$$HCO_3^-(aq) \rightleftharpoons H^+(aq) + CO_3^{2-}(aq)$$

Under the water there may still be significant deposits of calcite, so that the lakes represent enormous saturated solutions of calcite. The dynamic equilibria involved may result in the growth of large crystals of calcite at various locations. When the lakes recede, perhaps through earthquake cracks or holes leached in the lake beds, new caverns are formed which are decorated by clusters of these beautiful crystal formations.

As calcite solutions seep through small holes from aqueous regions into lower empty caverns, the dripping solutions may slowly lose water by evaporation. As the droplets evaporate, calcite formations gradually accumulate:

$$Ca^{2+}(aq) + 2HCO_3^-(aq) \rightleftharpoons CaCO_3(s) + H_2O(g) + CO_2(g)$$

Air has the approximate composition 75.5 percent N_2, 23.2 percent O_2, 1.3 percent Ar, 0.05 percent CO_2, 0.001 percent Ne (by weight), with trace amounts of He, CH_4, Kr, H_2, and Xe. Man has added a few other ingredients, most less innocuous than these natural components.

Figure 18.1 *Saturated solution.*

1. *Ion-water interactions weaken crystal lattice.*
2. *Solution formation occurs as ion leaves crystal lattice in association with water molecules.*
3. *Hydrated ion in solution.*
4. *Water molecules occasionally break free from molecular clusters.*
5. *Small ion sets may form as part of precipitation process.*
6. *Ions return to crystal lattice in competition with solution process.*

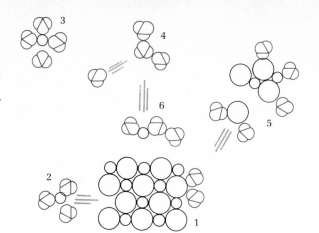

This process results in stalactites hanging from cave ceilings and stalagmites "growing" up from the cavern floor. Large columns may form as two opposite segments "grow" together. These formations are produced at varying rates, estimated to range from about 0.1 in. per year to 0.005 in. per year, depending on conditions such as temperature, relative humidity, air circulation pattern, and water flow rate.

In this unit we shall examine the competing processes of solution formation and precipitation and the nature of equilibria associated with saturated solutions (Figure 18.1).

18.1 SATURATED SOLUTIONS

Many substances are completely miscible with water. Ethanol, acetone, and glycerin, for example, form solutions with water in any proportions. Other substances, such as diamond or quartz, are for all practical purposes insoluble in water. Between these extremes lie thousands of chemicals of varying solubilities for which the compositions of saturated solutions can be meaningfully described.

It is important to distinguish between *solubility* and *rate of solution formation*. Ordinary granular table salt, for example, dissolves more rapidly in water at room temperature than does a sugar cube (sucrose), but more slowly than confectioner's sugar (powdered glucose). However, the *solubilities* of these compounds (Table 18.1) are in a quite different order. The rate at which a substance dissolves, at a given temperature, is quite dependent on the particle size of the sample and the stirring efficiency involved. Since *solubility* is measured for a saturated solution, the mechanism (e.g., stirring efficiency) by which the solution was formed is not a factor, and variation in particle size of solute has a relatively small effect.

A *saturated* solution has been defined as one that contains the maximum possible concentration of solute in equilibrium with the pure solute in a separate phase. In many cases, especially gas–liquid or liquid–liquid solutions, we are really dealing with two *solutions* in different phases. When water is saturated with air in a closed container, for example, the air is also saturated with water vapor; when ether and water are allowed to reach equilibrium, each liquid phase is saturated with the other. These considerations are important, as we shall see later, in describing phenomena such as liquid–liquid extraction and various types of chromatography (Excursion 4). Since diffusion is slow in solids, compared to gases and liquids, this phenomenon is less important in dealing with solid–liquid mixtures. However, the physical characteristics of solid phases in prolonged contact with their saturated solutions may be appreciably altered by incorporation of solvent molecules into the solid. Common laboratory reagents such as saturated $Ca(OH)_2$, observed over a period of months,

Table 18.1 WATER SOLUBILITIES OF TABLE SALT, TABLE SUGAR, AND CONFECTIONER'S SUGAR

COMPOUND	APPROXIMATE SOLUBILITY AT 20°C	
	GRAMS PER 100 ml	MOLES PER LITER
sodium chloride	~37	~6
sucrose	~250	~7.5
glucose	~84	~5

This phenomenon is not due, in most cases, to diffusion of water into solid. It is associated with the dynamic processes of the solubility equilibrium and incorporation of water of hydration with ions returning from solution to the crystals.

show this effect as the initial microcrystalline solid assumes the more gelatinous appearance of the fully hydrated compound.

Some observations of solute behavior in saturated solutions suggest useful clues as to the nature of the equilibria involved. A saturated solution will vary in concentration somewhat depending on the particle size of the solute used. Large particles not only dissolve more slowly than small ones, they *are also slightly less soluble than small ones.* This suggests that large crystals are more stable than small ones with respect to their solutions. Careful thermodynamic treatment reveals that the difference in stabilities in such cases can be related directly to the energy required to break the larger crystals into pieces the size of the smaller crystals. When a saturated solution is left in contact with residual solute of varying particle size, it is noted, over a very long time period, that the original large particles grow slightly larger, small particles decrease in size, and very tiny particles dissolve—even though the solution concentration remains nearly constant. Very careful observations of crystalline solutes of uniform particle size show that the particles often change in shape. Both of these observations suggest a dynamic process.

The use of radioactive isotopes in studies of saturated solutions has confirmed that these systems, like phase equilibria in pure substances (Unit 12), represent a condition of dynamic equilibrium.

Chemists, as you may have noted, do not fetter themselves with their own definitions. It is quite simple to separate physically a saturated solution from the residual solute phase without altering the composition of the solution. Thus it is not uncommon to find only clear solutions in reagent bottles labeled "saturated $Ca(OH)_2$" or when the solution has been diluted with an equal volume of water, "half-saturated $Ca(OH)_2$." Although such terms given an indication of the solution composition, the *equilibrium* condition is found only when residual undissolved solute is present.

18.2 THE SOLUBILITY PRODUCT

Such simple techniques are, of course, limited to situations in which solvent evaporation leaves solute of the same composition as that originally used. Formation of stable hydrates or decomposition of the solute must be considered as possible complications.

An electrochemical method for determining solubility products is described in Unit 24.

The concentration of a solution can be indicated in a number of different ways (Unit 13). A saturated solution may be described by specifying the solubility of the particular solute involved. Solubilities on the order of 10 mg or more per 100 ml can be measured fairly accurately by simple methods, such as evaporation of the solvent from measured volumes of saturated solutions, after removal of the residual solute phase. As a result, it is quite common to describe the composition of saturated solutions of the relatively soluble compounds in terms of such solubility units (e.g., see Table 18.1, or the extensive tabulations in references such as the Chemical Rubber Company's *Handbook of Chemistry and Physics*).

Much more sophisticated techniques are required to determine the concentration of saturated solutions of the sparingly soluble compounds. [For instance, CuS has a solubility in water at 20°C of approximately 1.9×10^{-18} g $(100 \text{ ml})^{-1}$.] Saturated solutions of such low concentrations are most commonly described by constants called *solubility products,* partly because these constants can be quite easily related to certain experimental data from techniques suitable to measurement of very low concentrations and partly because of the utility of such constants in dealing with equilibria in saturated solutions. The solubility product of any salt (M_aX_b) is expressed by

$$K_{sp} = C_M^a \times C_X^b \tag{18.1}$$

in which K_{sp} is called the "solubility product," C represents a "saturation" concentration in moles per liter, and a and b are the numerical values for the subscripts for cation (M) and anion (X), respectively, in the formula of the salt. This follows directly

A noteworthy exception will be any salt of the mercury(I) ion, since this ion exists as the diatomic Hg_2^{2+} (Unit 3). For example,

$$Hg_2Cl_2(s) \rightleftharpoons$$
$$Hg_2^{2+}(aq) + 2Cl^-(aq)$$
$$K_{sp} = C_{Hg_2^{2+}} \times C_{Cl^-}^2$$

from the general conventions described in Unit 17 for equilibrium constants, for any salt simply dissociating in solution to the ions represented by the formula of the salt. Obviously, we should be able to write the chemical equation for any simple saturated solution of a salt, even if the salt contains more than two ions, thus extending equation (18.1) to more complex cases. For example,

$$Ca_2F(PO_4)(s) \rightleftharpoons 2Ca^{2+}(aq) + F^-(aq) + PO_4^{3-}(aq)$$
$$K_{sp} = C_{Ca^{2+}}^2 \times C_{F^-} \times C_{PO_4^{3-}}$$

Note that, in accord with the K_{sp} convention described in Unit 17, the undissolved (solid) salt is written as the "left-hand" species and its component ions as the "right-hand" species in the equilibrium equation.

Example 1

Write the solubility product expressions for the saturated solutions of each of the following salts, on the assumption that the formulas of the salts supply all the necessary information. Then show that the K_{sp} expressions written are consistent with the conventions described in Unit 17 for general equilibrium systems by writing chemical equations depicting the solubility equilibria.

    a. AgCl   **b.** CuS
    c. MgF_2 **d.** Ag_2S
    e. $Fe(OH)_3$ **f.** $Ca_3(PO_4)_2$

Solution
From the chemical formulas [on the basis of equation (18.1)],

    a. for AgCl, $K_{sp} = C_{Ag^+} \times C_{Cl^-}$
    b. for CuS, $K_{sp} = C_{Cu^{2+}} \times C_{S^{2-}}$
    c. for MgF_2, $K_{sp} = C_{Mg^{2+}} \times C_{F^-}^2$
    d. for Ag_2S, $K_{sp} = C_{Ag^+}^2 \times C_{S^{2-}}$
    e. for $Fe(OH)_3$, $K_{sp} = C_{Fe^{3+}} \times C_{OH^-}^3$
    f. for $Ca_3(PO_4)_2$, $K_{sp} = C_{Ca^{2+}}^3 \times C_{PO_4^{3-}}^2$

These are consistent with the general equilibrium conventions developed in Unit 17, since the equilibrium systems are conventionally represented by

    a. $AgCl(s) \rightleftharpoons Ag^+(aq) + Cl^-(aq)$
    b. $CuS(s) \rightleftharpoons Cu^{2+}(aq) + S^{2-}(aq)$
    c. $MgF_2(s) \rightleftharpoons Mg^{2+}(aq) + 2F^-(aq)$
    d. $Ag_2S(s) \rightleftharpoons 2Ag^+(aq) + S^{2-}(aq)$
    e. $Fe(OH)_3(s) \rightleftharpoons Fe^{3+}(aq) + 3OH^-(aq)$
    f. $Ca_3(PO_4)_2(s) \rightleftharpoons 3Ca^{2+}(aq) + 2PO_4^{3-}(aq)$

Expressions in the *form of* equilibrium constant expressions, representing *ion products*, may be written for any aqueous solution, whether or not an equilibrium state exists. *Ion products*, however, are not constants except for the specific condition of ions at the equilibrium concentrations.

The ion product for any salt that dissolves completely in water is determined from the equation for the solution process by using the coefficients of the aqueous ionic products as exponents on the corresponding concentration terms; for example, for

$$MgF_2(s) \longrightarrow Mg^{2+}(aq) + \boxed{2}\,F^-(aq)$$
$$\text{ion product} = C_{Mg^{2+}} \times C_{F^-}^{\boxed{2}}$$

When the solution is saturated, the equation is written with a double arrow to

The K_{sp} value itself is "constant" only for a particular temperature and varies somewhat even then as a function of particle size of the residual solute. It is, however, quite useful for semiquantitative work.

indicate a state of equilibrium, and the ion product is then—*and only then*—equal to the solubility product. *At equilibrium*

$$MgF_2(s) \rightleftharpoons Mg^{2+}(aq) + 2F^-(aq)$$
$$K_{sp} = C_{Mg^{2+}} \times C_{F^-}^2$$

The ion product may not exceed the solubility product (i.e., ion concentrations may not be greater than those of the saturated solution) except for the *metastable* condition of *supersaturation* (Unit 13). Supersaturated solutions will eventually revert to the stable equilibrium system by precipitation of solid until the ion product equals K_{sp} for the salt. The restriction of an ion product to a number $\leqslant K_{sp}$ is a useful criteria for "predicting precipitation," as we shall see in Section 18.3.

When a pure salt dissolves in water, the concentrations of the resulting ions may be determined from the quantity of salt dissolved (in moles), the volume of solution formed (at low concentrations, the volume of water used), and the stoichiometric ratios indicated by the equation for the solution process.

Example 2

The solubility of magnesium fluoride in water at 18°C is tabulated as 0.0076 g (100 ml)$^{-1}$. Calculate (a) the solubility product for magnesium fluoride at 18°C and (b) the ion product for magnesium fluoride in a half-saturated solution at 18°C.

Solution

a. Find K_{sp}.

First, convert the solubility from grams per 100 ml to moles per liter.

Formula weight of MgF_2

$$Mg: \quad 1 \times 24.3 = 24.3$$
$$F_2: \quad 2 \times 19.0 = \underline{38.0}$$
$$MgF_2: \quad 62.3$$

$$\frac{0.0076 \text{ g}}{100 \text{ ml}} \times \frac{1000 \text{ ml}}{1 \text{ liter}} \times \frac{1 \text{ mole}}{62.3 \text{ g}} = 1.22 \times 10^{-3} \text{ M}$$

Second, write the equation for the solution process and use the stoichiometric ratios to find ion concentrations.

$$MgF_2(s) \longrightarrow Mg^{2+}(aq) + 2F^-(aq)$$
$$C_{Mg^{2+}} = \text{concentration of } MgF_2 \text{ dissolved}$$
$$C_{F^-} = 2 \times (\text{concentration of } MgF_2 \text{ dissolved})$$
$$C_{Mg^{2+}} = 1.22 \times 10^{-3} \text{ M}$$
$$C_{F^-} = 2 \times 1.22 \times 10^{-3} = 2.44 \times 10^{-3} \text{ M}$$

Third, write the solubility product expression and solve for K_{sp}.

$$[MgF_2(s) \rightleftharpoons Mg^{2+}(aq) \times 2F^-(aq)]$$
$$K_{sp} = C_{Mg^{2+}} \times C_{F^-}^2$$
$$= (1.22 \times 10^{-3}) \times (2.44 \times 10^{-3})^2$$

Rounded off to the original two significant figures,

$$K_{sp} = 7.3 \times 10^{-9}$$

b. Find the ion product at half-saturation.

First, find the concentration of MgF_2 at half-saturation from part (a).

$$\text{concentration} = \tfrac{1}{2} \text{ concentration of saturated solution}$$
$$= \tfrac{1}{2} \times 1.22 \times 10^{-3} \, M$$
$$= 6.1 \times 10^{-4} \, M$$

Second, convert to ion concentrations.

$$[MgF_2(s) \longrightarrow Mg^{2+}(aq) + 2F^-(aq)]$$
$$C_{Mg^{2+}} = 6.1 \times 10^{-4} \, M$$
$$C_{F^-} = 2 \times 6.1 \times 10^{-4} = 1.22 \times 10^{-3} \, M$$

Third, calculate the ion product.

$$[MgF_2 \longrightarrow Mg^{2+} + 2F^-]$$
$$\text{ion product} = C_{Mg^{2+}} \times C_{F^-}^2$$
$$= (6.1 \times 10^{-4}) \times (1.22 \times 10^{-3})^2$$
$$= 9.1 \times 10^{-10}$$

18.3 PRECIPITATION PROCESSES AND SOLUBILITY

So far we have only considered the formation of a saturated solution by dissolving a salt. The reverse of this process is the precipitation of the salt by mixing solutions of soluble salts containing reactant ions, for example,

$$MgCl_2(aq) + 2NaF(aq) \longrightarrow 2NaCl(aq) + MgF_2(s)$$

Reactions of this type may be generalized by *net equations* (Unit 3). For example,

$$Mg^{2+}(aq) + 2F^-(aq) \longrightarrow MgF_2(s)$$

Since the ion product may not exceed the solubility product for the stable equilibrium condition, precipitation reactions, like solution processes, eventually reach a state of equilibrium for which appropriate equations may be written. It is important to realize that equilibrium is characterized by no *net* change, so that the equations for precipitation and solution formation are equivalent *at equilibrium*.

Solution formation before equilibrium

$$MgF_2(s) \longrightarrow Mg^{2+}(aq) + 2F^-(aq)$$

At equilibrium

$$MgF_2(s) \rightleftharpoons Mg^{2+}(aq) + 2F^-(aq)$$

Precipitation before equilibrium

$$Mg^{2+}(aq) + 2F^-(aq) \longrightarrow MgF_2(s)$$

At equilibrium

$$Mg^{2+}(aq) + 2F^-(aq) \rightleftharpoons MgF_2(s)$$

By convention, a continuously changing process is represented by an equation with a single arrow[1] showing the direction of net change. An equilibrium state is indicated by double arrows, and the specification of reactants and products is purely a matter of convenience since these terms are truly meaningful only for continuing processes. It is most common to express solubility equilibria by equations showing the solid phase substance at the left of the arrows and the dissolved species at the right, whether equilibrium was reached via precipitation or via solution formation. The

[1] As mentioned earlier, we will also use the "single arrow" representation as a matter of convenience for systems not requiring equilibrium considerations.

Table 18.2   SELECTED SOLUBILITY PRODUCTS

SUBSTANCE	K_{sp} AT 20°C
ACETATES	
$Ag(CH_3CO_2)$	4.1×10^{-3}
$Hg_2(CH_3CO_2)_2$	3.6×10^{-10}
BROMATES	
$AgBrO_3$	6.0×10^{-5}
$Ba(BrO_3)_2$	5.6×10^{-6}
BROMIDES	
$AgBr$	5.0×10^{-13}
Hg_2Br_2	5.0×10^{-23}
$PbBr_2$	1.0×10^{-6}
CARBONATES	
Ag_2CO_3	8.2×10^{-12}
$BaCO_3$	1.6×10^{-9}
$CaCO_3$	4.7×10^{-9}
CHLORIDES	
$AgCl$	2.4×10^{-10}
Hg_2Cl_2	1.2×10^{-18}
$PbCl_2$	1.6×10^{-5}
FLUORIDES	
BaF_2	1.0×10^{-7}
CaF_2	1.0×10^{-10}
MgF_2	6.8×10^{-9}
HYDROXIDES	
$Ca(OH)_2$	4.0×10^{-6}
$Cu(OH)_2$	1.8×10^{-19}
$Fe(OH)_3$	6.0×10^{-38}
$Mg(OH)_2$	1.2×10^{-11}
SULFATES	
$BaSO_4$	7.9×10^{-11}
$CaSO_4$	2.5×10^{-5}
$PbSO_4$	1.3×10^{-8}

solubility product, then, is a *state function* (Unit 17) independent of the path of the process.

It is interesting to note that precipitation is typically much more rapid than the dissolving of a salt.

Equilibrium may be reached at different speeds depending on the routes selected, but the final state is independent of the route used.

Precipitation or solution formation will continue spontaneously until equilibrium is reached, at which time the two opposing processes continue at equal rates so that no net change occurs. This is an alternative way of stating that *the ion product may not exceed the solubility product.* Some solubility products for common substances are given in Table 18.2. For a more complete listing, see Appendix C.

Example 3 (Predicting precipitation)
___

Use solubility products from Table 18.2 to predict which of the following should result in precipitation:

a. 50 ml of 10^{-3} M $AgNO_3$ + 50 ml of 10^{-3} M $NaBrO_3$
b. 50 ml of 10^{-3} M $Pb(NO_3)_2$ + 50 ml of 10^{-3} M Na_2SO_4
c. 50 ml of 10^{-3} M $AgNO_3$ + 50 ml of 10^{-2} M $NaCl$
d. 50 ml of 10^{-5} M $FeCl_3$ + 50 ml of 10^{-5} M $NaOH$

Solution
(Note that the mixing of equal solution volumes reduces concentrations to one-half their original values.)

a. $NaNO_3$ is soluble (Unit 6), so the only possible precipitation reaction is

$$Ag^+(aq) + BrO_3^-(aq) \longrightarrow AgBrO_3(s)$$

for which the ion product, after mixing, is

$$C_{Ag^+} \times C_{BrO_3^-} = (5 \times 10^{-4}) \times (5 \times 10^{-4}) = 2.5 \times 10^{-7}$$

From Table 18.2, K_{sp} for $AgBrO_3$ is 6.0×10^{-5}. Since

ion product $<$ K_{sp}
(2.5×10^{-7}) (6.0×10^{-5})

precipitation should not occur.

b. The only possible precipitation reaction is

$$Pb^{2+}(aq) + SO_4^{2-}(aq) \longrightarrow PbSO_4(s)$$

for which the ion product, after mixing, is

$$C_{Pb^{2+}} \times C_{SO_4^{2-}} = (5 \times 10^{-4}) \times (5 \times 10^{-4}) = 2.5 \times 10^{-7}$$

From Table 18.2, K_{sp} for $PbSO_4$ is 1.3×10^{-8}. Since

ion product $>$ K_{sp}
(2.5×10^{-7}) (1.3×10^{-8})

net precipitation should occur until equilibrium is reached.

c. The only possible precipitation reaction is

$$Ag^+(aq) + Cl^-(aq) \longrightarrow AgCl(s)$$

for which the ion product, after mixing, is

$$C_{Ag^+} \times C_{Cl^-} = (5 \times 10^{-4}) \times (5 \times 10^{-3}) = 2.5 \times 10^{-6}$$

Note that ion concentrations from the *dissolving* of a pure salt may be determined from stoichiometric ratios, but those involved in precipitation processes are determined only by reactant concentrations used.

Since ion product $> K_{sp}$ (2.4 $\times 10^{-10}$, Table 18.2), net precipitation should occur until equilibrium is reached.

d. NaCl is soluble (Unit 6), so the only possible precipitation reaction is

$$Fe^{3+}(aq) + 3OH^-(aq) \longrightarrow Fe(OH)_3(s)$$

for which the ion product, after mixing, is

$$C_{Fe^{3+}} \times C_{OH^-}^3 = (5 \times 10^{-6}) \times (5 \times 10^{-6})^3 = 6.2 \times 10^{-22}$$

Since ion product $> K_{sp}$ (6.0 $\times 10^{-38}$, Table 18.2), net precipitation should occur until equilibrium is reached.

Example 4 (Estimating solubility)

What is the approximate solubility of barium bromate, in grams per 100 milliliter, as based on the K_{sp} given in Table 18.2?

Solution

The dissolving of barium bromate should continue until the ion product equals the K_{sp}, as formulated by

$$Ba(BrO_3)_2(s) \longrightarrow Ba^{2+}(aq) + 2BrO_3^-(aq)$$

for which the maximum ion product (equal to K_{sp}) is given by

$$C_{Ba^{2+}} \times C_{BrO_3^-}^2 = 5.6 \times 10^{-6}$$

If we represent the solubility of barium bromate, in moles per liter, by s, then the stoichiometric ratio indicates that in a saturated solution

$$C_{Ba^{2+}} = s$$
$$C_{BrO_3^-} = 2s$$

Substitution in the ion product equation then gives

$$(s) \times (2s)^2 = 5.6 \times 10^{-6}$$

from which

$$4s^3 = 5.6 \times 10^{-6}$$
$$s^3 = 1.4 \times 10^{-6}$$
$$s = \sqrt[3]{1.4 \times 10^{-6}} = 1.1 \times 10^{-2} \ M$$

The problem then resolves to one of converting moles per liter to grams per 100 ml.

Formula weight of $Ba(BrO_3)_2$

Ba: $1 \times 137 = 137$
Br_2: $2 \times 80 = 160$
O_6: $6 \times 16 = \underline{\ 96}$
   $Ba(BrO_3)_2$: 393

$$\text{solubility} = \frac{1.1 \times 10^{-2} \text{ mole}}{\text{liter}} \times \frac{0.1 \text{ liter}}{(100 \text{ ml})} \times \frac{393 \text{ g}}{\text{mole}}$$

Note that 100 ml is used here as an intact unit so that the 100 is not part of the mathematical operation.

$$= \frac{1.1 \times 10^{-2} \times 10^{-1} \times 3.93 \times 10^2}{(100 \text{ ml})}$$

$$= {\sim}0.43 \text{ g}(100 \text{ ml})^{-1}$$

18.4 LE CHATELIER'S PRINCIPLE AND SOLUBILITY

The randomness and mobility of ions in solution can be considered as being similar in many respects to their characteristics in liquid salts. Thus solubility equilibria are not unlike the solid–liquid phase equilibria studied previously (Unit 12). If free-energy changes (ΔG, Units 5 and 24) are determined for the competing processes of precipitation and dissolving, then the requirement that $\Delta G_{net} = 0$ at equilibrium can be used to predict equilibrium conditions. Similarly, if the temperature dependence of ΔG for the opposing processes can be estimated, this information can be used to predict the effects of temperature variation on solubility. In many cases, over a moderate temperature range, entropy effects are negligible, so that the more easily measurable ΔH values can be used to predict solubility changes as the temperature is increased or decreased. These effects are in accord with Le Chatelier's principle. It is most common to find that solubility increases with increasing temperature, although there are notable exceptions.

A second application of Le Chatelier's principle is noted in the effects of ions introduced into a solubility equilibrium system. One of the most important cases, from a practical standpoint, is the so-called *common-ion effect*. Consider, for example, the equilibrium system described by

$$AgCl(s) \rightleftharpoons Ag^+(aq) + Cl^-(aq)$$

and characterized by constant concentrations of Ag^+ and Cl^-. If additional Cl^- is added in such a way as to produce an increase in Cl^- concentration, the equilibrium is disturbed. According to Le Chatelier's principle, the system will undergo change until a new equilibrium is established whose properties balance the effect of the added Cl^-. Since the ion product cannot exceed the value of K_{sp}, some Cl^- must combine with Ag^+ to form solid silver chloride. Thus the concentration of Ag^+ in the new equilibrium system will be lower than that of the original saturated solution. The extent of this common-ion effect can be estimated from calculations based on solubility product expressions. In general, the solubility of a salt is decreased by addition of another substance furnishing ions common to the original salt.

This effect may be offset if the added ion results in formation of a soluble complex. For example, lead chloride dissolves in concentrated sodium chloride solutions because of the reaction

$$PbCl_2(s) + Cl^-(aq) \rightleftharpoons [PbCl_3]^-(aq)$$

Problems of complex formation are discussed in Unit 27 and the effects of ions other than those in the original salt are described in Unit 21.

Example 5

Compare the concentration of Ca^{2+} in a saturated aqueous solution of $Ca(OH)_2$ with that of a saturated solution of this compound to which additional sodium hydroxide has been added to the extent that at equilibrium the concentration of hydroxide ion is 2.0 M.

Solution

In both cases the concentration of Ca^{2+} is expressed by

$$C_{Ca^{2+}} = \frac{K_{sp}}{C_{OH^-}^2} \qquad (\text{because } C_{Ca^{2+}} \times C_{OH^-}^2 = K_{sp})$$

The stoichiometric ratio shown by the formula of the compound indicates that for the saturated aqueous solution, in which dissolved $Ca(OH)_2$ is the only significant source of Ca^{2+} and OH^- ions (ignoring the negligible C_{OH^-} in pure water, $\sim 10^{-7} M$),

$$C_{OH^-} = 2 \times C_{Ca^{2+}}$$

Thus

$$C_{Ca^{2+}} = \frac{K_{sp}}{(2 \times C_{Ca^{2+}})^2}$$

from which

$$C_{Ca^{2+}} = \sqrt[3]{\frac{K_{sp}}{4}}$$

and, using Table 18.2, $K_{sp} = 4.0 \times 10^{-6}$, so that

$$C_{Ca^{2+}} = \sqrt[3]{\frac{4.0 \times 10^{-6}}{4}} = 10^{-2}\ M$$

In the second equation, $C_{OH^-} = 2.0\ M$, so that

$$C_{Ca^{2+}} = \frac{K_{sp}}{(2.0)^2} = \frac{4.0 \times 10^{-6}}{4.0} = 10^{-6}\ M$$

Thus calcium hydroxide is 10,000 times more soluble in water than in 2.0 M sodium hydroxide solution.

Example 6

In which of the following solutions will silver chromate have the smallest *molar* solubility:

a. pure water
b. 0.10 M silver nitrate
c. 0.10 M sodium nitrate
d. 0.10 M sodium chromate?

Solution
The solubility equilibrium for silver chromate is

$$Ag_2CrO_4(s) \rightleftharpoons 2Ag^+(aq) + CrO_4^{2-}(aq)$$

for which

$$K_{sp} = C_{Ag^+}^2 \times C_{CrO_4^{2-}} = 1.9 \times 10^{-12} \qquad \text{(from Appendix C)}$$

a. If we express the molar solubility in water as s_{H_2O} [in moles (Ag_2CrO_4) liter^{-1}], then, from the chemical equation, each mole of Ag_2CrO_4 dissolved yields 2 moles of Ag^+ and one of CrO_4^{2-}; so

$$\begin{aligned} C_{Ag^+} &= 2s_{H_2O} \\ C_{CrO_4^{2-}} &= s_{H_2O} \\ K_{sp} &= (2s_{H_2O})^2(s_{H_2O}) = 4s_{H_2O}^3 \end{aligned}$$

from which

$$s_{H_2O} = \sqrt[3]{K_{sp/4}} = \sqrt[3]{\frac{1.9 \times 10^{-12}}{4}} = 7.8 \times 10^{-5}\ M$$

b. When the solid Ag_2CrO_4 is added to silver nitrate, a "common-ion effect" is noted, due to the presence of Ag^+ in addition to that resulting from the Ag_2CrO_4. Le Chatelier's principle suggests that some of the additional Ag^+ will react with CrO_4^{2-} to establish a new equilibrium. Since the K_{sp} value for Ag_2CrO_4 is very small, we may safely conclude that the C_{Ag^+} from the dissolved salt is $\lll 0.10\ M$ and the reduction in CrO_4^{2-} consumes a negligible amount of Ag^+ from the silver nitrate. Then, representing the molar solubility of Ag_2CrO_4 in the silver nitrate by s_{AgNO_3} and noting that each mole of Ag_2CrO_4 dissolved yields 1 mole of CrO_4^{2-}

we have at equilibrium

$$C_{Ag^+} \approx 0.10 \ M \qquad \text{(essentially that of the } AgNO_3 \text{ solution)}$$
$$C_{CrO_4{}^{2-}} = s_{AgNO_3}$$
$$K_{sp} = (0.10)^2(s_{AgNO_3})$$

from which

$$s_{AgNO_3} = \frac{K_{sp}}{(0.10)^2} = \frac{1.9 \times 10^{-12}}{(0.10)^2} = 1.9 \times 10^{-10} \ M$$

c. Since both $AgNO_3$ and Na_2CrO_4 are soluble to the extent of at least 0.10 M [from information on the solutions used for parts (b) and (d)], the Na^+ and NO_3^- ions do not affect the solubility of silver chromate and it should have the same solubility in 0.10 M $NaNO_3$ as in pure water, 7.8×10^{-5} M [part (a)].

d. In a manner similar to that used in part (b), and representing the molar solubility of Ag_2CrO_4 in sodium chromate by $s_{Na_2CrO_4}$,

$$C_{Ag^+} = 2s_{Na_2CrO_4}$$
$$C_{CrO_4{}^{2-}} \approx 0.10 \ M$$
$$K_{sp} = (2s_{Na_2CrO_4})^2(0.10)$$

from which

$$s_{Na_2CrO_4} = \sqrt{\frac{K_{sp}}{(2)^2(0.10)}} = \sqrt{\frac{1.9 \times 10^{-12}}{0.40}} = 2.2 \times 10^{-6}$$

Comparison of the calculated molar solubilities then shows that the molar solubility of Ag_2CrO_4 is smallest in 0.10 M $AgNO_3$, of the four systems compared.

18.5 DISTRIBUTION EQUILIBRIA: LIQUID–LIQUID EXTRACTION

We have previously discussed the factors contributing to the solubility of covalent compounds in various liquids (Units 9 and 11). A compound can be extracted from its solution in one liquid by a second, immiscible, liquid in which the compound is more soluble. This is the principle on which the useful technique of liquid–liquid extraction is based (Figure 18.2).

Liquid–liquid extraction is used to separate components of mixtures on the basis of differing solubilities of the various components in two immiscible liquids. For example, caffeine may be separated from most of the other components of an aqueous solution of coffee by extraction with chloroform. Caffeine is more soluble in chloroform than in water, whereas most of the other components are more soluble in water.

Organic compounds (Unit 28) are compounds of carbon.

Liquid–liquid extraction, involving an organic solvent and an aqueous phase, utilizes the equilibrium

$$\text{substance (aq)} \rightleftharpoons \text{substance (org)}$$

for which the equilibrium constant, called the *distribution coefficient,* is

$$K_D = \frac{C_{\text{substance (org)}}}{C_{\text{substance (aq)}}} \tag{18.2}$$

The distribution coefficient may be estimated from the relative solubilities of the substance in the two liquids used, although this is sometimes a poor approximation since the actual extraction involves water saturated with the organic solvent and the

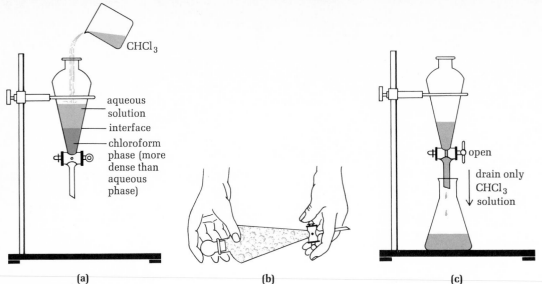

Figure 18.2 Simple liquid–liquid extraction. (a) Chloroform is added to an aqueous solution in a *separatory funnel* to extract a solute more soluble in $CHCl_3$ than in water (original solvent). (b) The funnel is stoppered and shaken to establish distribution equilibrium. (In laboratory practice, the funnel must be vented periodically during extraction to release excess vapor pressure.) (c) The chloroform solution is removed. (The stopcock is closed before aqueous solution can drain too.)

Labels in figure (a): CHCl₃; aqueous solution; interface; chloroform phase (more dense than aqueous phase)

Labels in figure (c): open; drain only CHCl₃ solution

latter saturated with water. The solvent characteristics are altered for both phases by this mutual saturation effect.

In selecting a solvent for liquid–liquid extraction, the prime consideration is again the desire to isolate selectively the desired compound from accompanying "impurities." Trial-and-error approaches may be necessary, but some knowledge of the characteristics of the desired product and of the impurities present will help in selecting a procedure.

It can be demonstrated that a single extraction with a large volume of solvent is less efficient than successive extractions with smaller volumes of solvent. This is particularly important when the distribution coefficient is small, since the equilibrium distribution will then leave a significant fraction of the solute in the aqueous phase.

EXTRACTION EFFICIENCY

Consider the general case for the extraction of a solute from an aqueous solution. The number of moles of solute is n_s. From the definition of molar concentration,

$$n_{s\ original} = C(aq)_{original} \times V(aq)$$

After extraction, we have the same number of moles of solute, but now distributed at equilibrium between the two immiscible solvents so that, at equilibrium

$$n_{\text{s original}} = [C(\text{org})_{\text{equil}} \times V(\text{org})] + [C(\text{aq})_{\text{equil}} \times V(\text{aq})]$$

The number of moles remaining in the aqueous solution after extraction by the organic solvent is given by

$$n_{\text{s remaining}} = C(\text{aq})_{\text{equil}} \times V(\text{aq})$$

and the *fraction* of solute remaining in the aqueous solution at equilibrium is

$$f_{\text{remaining}} = \frac{n_{\text{s remaining}}}{n_{\text{s original}}} = \frac{C(\text{aq})_{\text{equil}} \times V(\text{aq})}{[C(\text{org})_{\text{equil}} \times V(\text{org})] + [C(\text{aq})_{\text{equil}} \times V(\text{aq})]}$$

If we now divide both the numerator and denominator of the right-hand expression by $[(C(\text{aq})_{\text{equil}} \times V(\text{aq}))]$,

$$f_{\text{remaining}} = \frac{\dfrac{C(\text{aq})_{\text{equil}} \times V(\text{aq})}{C(\text{aq})_{\text{equil}} \times V(\text{aq})}}{\dfrac{C(\text{org})_{\text{equil}} \times V(\text{org})}{C(\text{aq})_{\text{equil}} \times V(\text{aq})} + \dfrac{C(\text{aq})_{\text{equil}} \times V(\text{aq})}{C(\text{aq})_{\text{equil}} \times V(\text{aq})}}$$

Since

$$K_D = \frac{C(\text{org})_{\text{equil}}}{C(\text{aq})_{\text{equil}}}$$

this rather cumbersome fraction resolves to

$$f_{\text{remaining}} = \frac{1}{K_D[V(\text{org})/V(\text{aq})] + 1}$$

The *fraction* remaining unextracted is then seen to be independent of the original concentration and determined only by the distribution coefficient and the relative volumes of solvents.

For a second extraction of the original aqueous solution, using the same ratio of volumes, the fraction of the *original* solute remaining in the aqueous phase would be

$$\underset{\substack{\text{(of original)}}}{f_{\text{remaining}}} = \underset{\substack{\text{(after second extraction)}}}{f_{\text{remaining}(2)}} \times \underset{\substack{\text{(after original extraction)}}}{f_{\text{remaining}(1)}}$$

that is,

$$\underset{\substack{\text{(of original)}}}{f_{\text{remaining}}} = \left[\frac{1}{K_D[V(\text{org})/V(\text{aq})] + 1}\right] \times \left[\frac{1}{K_D[V(\text{org})/V(\text{aq})] + 1}\right]$$

$$= \left[\frac{1}{K_D[V(\text{org})/V(\text{aq})] + 1}\right]^2$$

Obviously, this could be extended to any number of extractions, n, such that

$$\underset{\substack{\text{(of original, after}\\ n \text{ extractions)}}}{f_{\text{remaining}}} = \left[\frac{1}{K_D[V(\text{org})/V(\text{aq})] + 1}\right]^n \qquad \text{and, since}$$

$$\underset{\substack{\text{(of original)}}}{f_{\text{remaining}}} + \underset{\substack{\text{(of original)}}}{f_{\text{extracted}}} = 1$$

the *extraction efficiency*, in terms of fraction of original solute extracted is

$$f_{\text{extracted}} = 1 - \left[\frac{1}{K_D[V(\text{org})/V(\text{aq})] + 1}\right]^n \tag{18.3}$$

A number of applications of liquid–liquid extraction are discussed in Excursion 4. For now we are interested only in illustrating this system as another example of solubility equilibria.

<hr>

Example 7

The solubility of m-hydroxybenzoic acid at 25°C is listed as 1.04 g (100 ml)$^{-1}$ in water and 9.8 g (100 ml)$^{-1}$ in ether.
  a. Estimate the distribution coefficient for this acid in a water–ether system.
  b. Estimate the weight of the acid extracted from 100 ml of its saturated aqueous solution by a single extraction with 100 ml of ether.
  c. Estimate the weight of the acid extracted from 100 ml of its saturated aqueous solution by two successive extractions, each using 50 ml of ether.
  d. Calculate the minimum number of successive extractions, using equal volumes of ether and the aqueous solution, required for removal of 99 percent of the acid from its aqueous solution.

Solution

  a. Since any conversion factors will cancel, concentration units (approximated by solubilities) can be used as grams per 100 ml rather than molarities.

$$K_D = \frac{9.8 \text{ g } (100 \text{ ml})^{-1)}}{1.04 \text{ g } (100 \text{ ml})^{-1}} \qquad \text{[equation (18.2)]}$$

$$= 9.4$$

  b.

$$f = 1 - \left(\frac{1}{9.4 + 1}\right)^1 \qquad \text{[equation (18.3)] (when } V(\text{aq}) = V(\text{org)})$$

$$= 1 - \frac{1}{10.4} = 1 - 0.096 = 0.904$$

$$\text{weight extracted} = 0.904 \times \frac{1.04 \text{ g}}{100 \text{ ml}} \times 100 \text{ ml} = (0.94 \text{ g})$$

  c.

$$f = 1 - \left[\frac{1}{9.4(50/100) + 1}\right]^2 \qquad \text{[equation (18.3)]}$$

$$= 1 - \left(\frac{1}{5.7}\right)^2 = 1 - 0.031 = 0.969$$

$$\text{weight extracted} = 0.969 \times \frac{1.04 \text{ g}}{100 \text{ ml}} \times 100 \text{ ml} = 1.01 \text{ g}$$

  d.

$$0.99 = 1 - \left(\frac{1}{9.4 + 1}\right)^n \qquad \text{[equation (18.3)] (when } V(\text{aq}) = V(\text{org)})$$

$$\left(\frac{1}{10.4}\right)^n = 0.01 = \frac{1}{100}$$

Without resorting to a "mathematical" solution, it is apparent that n must be 2.

<hr>

GAS SOLUBILITY

When some gases, such as CO_2 or NH_3, dissolve in water, chemical reactions occur, resulting in relatively complex equilibrium systems. Such *interactive equilibria* are discussed in Unit 21.

For gases not interacting with the solvent, the situation is much simpler. We need only recognize that for gaseous solutes the solute–solute forces are extremely weak. Gas solubility then is principally determined by the nature and strength of solute–solvent forces (favoring solubility) and solvent–solvent forces (opposing solubility). Thus gases, such as formaldehyde, having molecules that can "accept" hydrogen bonds (Unit 9) are expected to be quite soluble in water, whereas gases having only "London force" interactions with the solvent should be relatively insoluble in water. For example,

Gas	Water solubility	Conditions
H_2CO (formaldehyde)	very soluble (40% aqueous solution is the biological preservative, *formalin*)	solubility decreases as temperature increases
N_2	2.3 ml $(100 \text{ ml})^{-1}$ 1.4 ml $(100 \text{ ml})^{-1}$	at 5°C, 1 atm at 37°C, 1 atm
O_2	4.9 ml $(100 \text{ ml})^{-1}$ 3.0 ml $(100 \text{ ml})^{-1}$ 2.3 ml $(100 \text{ ml})^{-1}$	at 5°C, 1 atm at 37°C, 1 atm at 100°C, 1 atm
H_2	2.1 ml $(100 \text{ ml})^{-1}$ 1.9 ml $(100 \text{ ml})^{-1}$	at 5°C, 1 atm at 37°C, 1 atm

As the preceding data suggest, the solubility of gases generally decreases as temperature increases. This is not surprising, since the increased kinetic energy at higher temperatures should permit more gaseous molecules to escape from the relatively weak attractions of solvent molecules. This effect has been noted in a number of cases of *thermal pollution* of natural waters in which water has been pumped from a lake or stream for use as a heat-exchange agent in various industrial applications. The warmer water returned is a less effective solvent for oxygen, thus decreasing the dissolved oxygen required by the aquatic ecosystem.

The effect of pressure on gas solubility can be understood in terms of Le Chatelier's principle. For the equilibrium system,

$$X(aq) \rightleftharpoons X(g)$$

an increase in external pressure may be counteracted by the dissolving of more gaseous particles, thus reducing the total number of moles in the gas phase. As in other equilibria, this system may be represented mathematically by

$$K_g = \frac{P_X}{C_X}$$

in which K_g is referred to as Henry's law constant (in honor of William Henry, who discovered this relationship in 1803); P_X represents the partial pressure of $X(g)$ and C_X represents the molar concentration of the dissolved gas.

We have already mentioned the higher solubilities of gases at high pressures in Unit 10 in connection with the problems of breathing mixtures for deep-sea diving. Unfortunately, Henry's law is accurate only for slightly soluble gases and over only a moderate pressure range, so we cannot use this relationship to calculate gas solubility at pressures much greater than 1 atm. The solubility of nitrogen, for example, increases much more at high pressures than does that of helium or hydrogen. It is this factor that has suggested the use of He or H_2 for "deep-sea atmospheres."

The decreased gas solubility when pressure is reduced accounts for the familiar "fizz" when we open a bottle or can of a "carbonated" beverage.

Solubility equilibria are of utmost importance in processes for a wide variety of analytical and preparative methods. Many of these, along with additional applications of Le Chatelier's principle, will be described in later units.

SUGGESTED READING

Hamm, Randall, and Carl Nyman. 1968. *Chemical Equilibrium.* Lexington, Mass.: Raytheon.

Moeller, Therald, and Rod O'Connor. 1971. *Ions in Aqueous Systems.* New York: McGraw-Hill.

O'Connor, Rod, and Charles Mickey. 1974. *Solving Problems in Chemistry.* New York: Harper & Row.

EXERCISES

1. Write equilibrium equations representing the saturated solutions of
   a. magnesium carbonate; **b.** $Al(OH)_3$; **c.** copper(I) sulfide
2. Write solubility product expressions and calculate K_{sp} values for
   a. barium chromate, solubility 0.00038 g (100 ml)$^{-1}$
   b. silver oxalate, solubility 0.0034 g (100 ml)$^{-1}$
   c. lead phosphate, solubility 0.000014 g (100 ml)$^{-1}$
3. Use ion products and K_{sp} values (Table 18.2) to calculate which of the following should result in precipitation:
   a. 10 ml of 10^{-3} M $BaCl_2$ + 10 ml of 10^{-3} M Na_2CO_3
   b. 10 ml of 0.010 M $AgNO_3$ + 10 ml of 0.10 M $K(CH_3CO_2)$
   c. 10 ml of 2.0×10^{-2} M $BaCl_2$ + 10 ml of 1.0×10^{-3} M NH_4F
4. Use K_{sp} values (Table 18.2) to estimate maximum solubilities (in grams per 100 ml) of
   a. silver bromide
   b. lead chloride
   c. mercury(I) acetate (*Hint:* Remember the Hg_2^{2+} ion?)
5. Calculate the solubility (in grams per 100 ml) of silver carbonate, assuming no competitive reactions, in
   a. pure water; **b.** 2.0 M sodium carbonate; **c.** 2.0 M silver nitrate
6. Oxalic acid has a solubility at 25°C of 9.8 g (100 ml)$^{-1}$ in water and 16.9 g (100 ml)$^{-1}$ in ether. Calculate
   a. the weight of oxalic acid that could be extracted from 50 ml of its saturated aqueous solution by two successive extractions, each using 25 ml of ether.

   b. the minimum number of extractions, using equal volumes of ether and the aqueous solution, required for extraction of 95 percent of the oxalic acid from its aqueous solution.

PRACTICE FOR PROFICIENCY

Objective (a)

1. Write equilibrium equations representing the saturated solutions of
   a. barium sulfate; **b.** iron(III) hydroxide; **c.** silver chromate;
   d. lead chloride; **e.** lithium carbonate; **f.** calcium phosphate
2. Write K_{sp} expressions for the salts given in question 1.
3. Write K_{sp} expressions for
   a. silver acetate; **b** lead bromide; **c.** silver phosphate;
   d. chromium(III) hydroxide; **e.** lead arsenate, $Pb_3(AsO_4)_2$

Objective (b)

4. Write solubility product expressions and calculate K_{sp} values for
   a. magnesium fluoride, solubility 0.0074 g (100 ml)$^{-1}$
   b. barium phosphate, solubility 0.00066 g (100 ml)$^{-1}$
5. For each of the following salts, the molar solubility indicated was measured on the solution prepared by addition of excess solid salt to pure water. Calculate the solubility products for the salts.

a. silver bromide, $C_{Ag^+} = 7.1 \times 10^{-7} M$
b. lead bromide, $C_{Pb^{2+}} = 6.3 \times 10^{-3} M$
c. silver phosphate, $C_{PO_4^{3-}} = 1.56 \times 10^{-5} M$
d. magnesium hydroxide, $C_{OH^-} = 2.9 \times 10^{-4} M$
e. strontium phosphate, $C_{PO_4^{3-}} = 1.3 \times 10^{-6} M$

Objective (c)

6. Use ion products and K_{sp} values (Appendix C) to determine which of the following should result in precipitation:
   a. 10 ml of $10^{-4} M$ $Pb(NO_3)_2$ + 15 ml of $10^{-4} M$ H_2SO_4
   b. 10 ml of $10^{-4} M$ $CaCl_2$ + 15 ml of $10^{-2} M$ NaF

7. Use K_{sp} values (Appendix C) to estimate maximum solubilities in g $(100 ml)^{-1}$ of
   a. calcium carbonate; **b.** iron(III) hydroxide

8. Calculate the solubility, in g $(100 ml)^{-1}$, of iron(III) hydroxide, assuming no competitive reactions, in
   a. 0.20 M iron(III) nitrate; **b.** 0.20 M sodium hydroxide

9. The K_{sp} for each of the following salts is 1.0×10^{-6}. Which salt has the greatest *molar solubility*?
   a. MX; **b.** MX_2; **c.** MX_3; **d.** M_2X; **e.** All are equally soluble.

10. The molar solubility of silver carbonate in a saturated aqueous solution of Ag_2CO_3 is given by
    a. $\sqrt{K_{sp}}$; **b.** $\sqrt{K_{sp}/2}$; **c.** $\sqrt[3]{K_{sp}/4}$; **d.** $\sqrt[3]{K_{sp}/27}$; **e.** $\sqrt[6]{K_{sp}/64}$

11. The relationship between the molar concentration of Ag^+ in a saturated aqueous solution of silver phosphate and the solubility product for Ag_3PO_4 may be expressed as
    a. $K_{sp} = 9C_{Ag^+}^3$; **b.** $K_{sp} = 27C_{Ag^+}^4$; **c.** $K_{sp} = \sqrt[3]{C_{Ag^+}}$; **d.** $K_{sp} = \frac{1}{8}C_{Ag^+}^3$; **e.** $K_{sp} = \frac{1}{3}C_{Ag^+}^4$

12. The molar solubility of calcium phosphate in a saturated aqueous solution of $Ca_3(PO_4)_2$ is given by
    a. $\sqrt[6]{K_{sp}/125}$; **b.** $\sqrt[5]{K_{sp}/108}$; **c.** $\sqrt[4]{K_{sp}/27}$; **d.** $\sqrt[3]{K_{sp}/4}$; **e.** $\sqrt{K_{sp}}$

13. An aqueous slurry of barium sulfate is used in radiographic studies of the gastrointestinal tract. How many milligrams of dissolved Ba^{2+} are present in 2.00 liters of saturated aqueous $BaSO_4$?
    a. 5.0; **b.** 3.7; **c.** 3.0; **d.** 2.4; **e.** 1.9

14. A famous limestone cavern contains a clear lake through which beautiful calcite formations on the bottom appear as "rock flowers." What is the concentration of Ca^{2+} in the lake water, in ppm (mg liter^{-1}), assuming that the lake water is essentially a saturated aqueous solution of calcite ($CaCO_3$)?
    a. 7.5; **b.** 6.9; **c.** 5.0; **d.** 3.8; **e.** 2.7

15. A concentration of approximately 1 ppm (1 mg liter^{-1}) of Cu^{2+} is useful in preventing the growth of algae in tropical fish tanks. To avoid having to reconstitute the solution each time the tank water is changed, a lump of an appropriate copper salt can be placed in the bottom of the tank so that its saturated solution provides an appropriate concentration of Cu^{2+}. Assuming distilled water is used, which salt would provide the saturated solution closest to the desired Cu^{2+} concentration?
    a. $CuSO_4$; **b.** CuS; **c.** $Cu(OH)_2$; **d.** $CuCO_3$; **e.** $Cu(NO_3)_2$

16. When equal volumes of the solutions indicated are mixed, precipitation should occur only for
    a. $2 \times 10^{-5} M$ Ag^+ + $2 \times 10^{-5} M$ CO_3^{2-}

b. $2 \times 10^{-5} M$ Ca^{2+} + $2 \times 10^{-5} M$ CO_3^{2-}
c. $2 \times 10^{-5} M$ Ca^{2+} + $2 \times 10^{-2} M$ F^-
d. $2 \times 10^{-5} M$ Mg^{2+} + $2 \times 10^{-6} M$ F^-
e. $2 \times 10^{-3} M$ Ba^{2+} + $2 \times 10^{-3} M$ F^-

17. When equal volumes of the solutions indicated are mixed, precipitation should occur only for
    a. $2 \times 10^{-3} M$ Ag^+ + $2 \times 10^{-6} M$ SO_4^{2-}
    b. $2 \times 10^{-6} M$ Ba^{2+} + $2 \times 10^{-6} M$ SO_4^{2-}
    c. $2 \times 10^{-5} M$ Mg^{2+} + $2 \times 10^{-5} M$ OH^-
    d. $2 \times 10^{-2} M$ Mg^{2+} + $2 \times 10^{-5} M$ OH^-
    e. $2 \times 10^{-10} M$ Cu^{2+} + $2 \times 10^{-3} M$ OH^-

18. When equal volumes of the solutions indicated are mixed, precipitation should occur only for
    a. $2 \times 10^{-8} M$ Mg^{2+} + $2 \times 10^{-4} M$ OH^-
    b. $2 \times 10^{-2} M$ Mg^{2+} + $2 \times 10^{-5} M$ OH^-
    c. $2 \times 10^{-5} M$ Cu^{2+} + $2 \times 10^{-10} M$ OH^-
    d. $2 \times 10^{-8} M$ Cu^{2+} + $2 \times 10^{-5} M$ OH^-
    e. $2 \times 10^{-3} M$ Pb^{2+} + $2 \times 10^{-6} M$ SO_4^{2-}

19. Which will dissolve the *least* amount of barium fluoride?
    a. 50 ml of water
    b. 50 ml of 2.0 M HNO_3
    c. 50 ml of 2.0 M $Ba(NO_3)_2$
    d. 50 ml of 2.0 M NaF
    e. All will dissolve the same amount.

20. Which of the following should dissolve the smallest amount of silver sulfide per liter, assuming no complex ion formation?
    a. 0.10 M HNO_3; **b.** 0.10 M Na_2S; **c.** 0.10 M $AgNO_3$; **d.** 0.10 M $NaNO_3$; **e.** pure water

21. An aqueous slurry of barium sulfate is used in radiographic studies of the gastrointestinal tract. If you wished to reduce the concentration of dissolved Ba^{2+} in the mixture for use with a patient unusually sensitive to Ba^{2+}, a useful procedure would be
    a. addition of dilute aqueous Na_2SO_4
    b. addition of dilute aqueous HCl
    c. addition of dilute aqueous $BaCl_2$
    d. addition of more solid $BaSO_4$
    e. heating the slurry (dissolving of $BaSO_4$ is endothermic)

22. A particular soil is rich in $Fe(OH)_3$. It is planned to establish a shrubbery planting using certain evergreens that respond unfavorably to an excess of dissolved Fe^{3+}. Of the following soil additives, which would best reduce the concentration of dissolved Fe^{3+} in groundwater in this soil?
    a. limewater, $Ca(OH)_2(aq)$
    b. saltpeter, $KNO_3(s)$
    c. ferric chloride, $FeCl_3(aq)$
    d. ammonium nitrate, $NH_4NO_3(s)$
    e. ammonium sulfate, $(NH_4)_2SO_4(s)$

23. To a *saturated* solution of $BaCO_3$ is added just enough Na_2SO_4 so that no $BaSO_4$ precipitates. What is the molarity of SO_4^{2-} in the solution?
    a. 2.8×10^{-5}; **b.** 6.7×10^{-6}; **c.** 2.0×10^{-6}; **d.** 4.0×10^{-5}; **e.** 8.9×10^{-5}

Objective (d)

24. The solubility of 2,4-dinitrophenol at 25°C is 0.68 g $(100 \text{ ml})^{-1}$ in water and 3.27 g $(100 \text{ ml})^{-1}$ in ether. Calculate
   a. the distribution coefficient of 2,4-dinitrophenol in an ether–water system.
   b. the weight of 2,4-dinitrophenol which could be extracted from 1.00 liter of its saturated aqueous solution by two successive extractions, each using 250 ml of ether.

BRAIN TINGLER

25. The acidity of the water in swimming pools must be controlled to maintain a chlorine concentration sufficient to prevent bacterial growth. When chlorine is dissolved in water, an equilibrium system is established that may be represented as

$$Cl_2(aq) + H_2O(l) \rightleftharpoons H^+(aq) + Cl^-(aq) + HOCl(aq)$$

As suggested by Le Chatelier's principle, an increase in H^+ concentration would result in the production of more chlorine and if this occurred to a large enough extent, the limited solubility of chlorine could be exceeded so that chlorine escaped from solution.

Since the pool water is exposed to the atmosphere, equilibrium is eventually established between gases in the atmosphere and the same gases in the pool water. Among these gases is carbon dioxide, for which the following equilibria are established:

$$CO_2(g) \rightleftharpoons CO_2(aq)$$

$$CO_2(aq) + H_2O(l) \rightleftharpoons H^+(aq) + HCO_3^-(aq)$$

$$HCO_3^-(aq) \rightleftharpoons H^+(aq) + CO_3^{2-}(aq)$$

When the H^+ concentration of the pool water is reduced to minimize loss of chlorine, the CO_2 equilibria are also affected, with the ultimate result that CO_3^{2-} concentration is increased as H^+ concentration is decreased.

If the pool water contains Ca^{2+} (typically present in natural "hard" waters) at a sufficient concentration, solid calcium carbonate is formed, often as sharp crystals of calcite growing on the pool ladder or in other inconvenient locations.

What is the maximum water hardness (in ppm Ca^{2+}) that can be permitted in a swimming pool if it is desired to avoid calcite formation and the pool water is found to have a carbonate ion concentration of 0.057 M? *(1 ppm = 1 mg/liter). You will need to consult a solubility product constant table.*

Unit 18 Self-Test

(Refer to Appendix C when necessary.)

1. What is the molarity of a solution labeled "half-saturated calcium hydroxide"?
2. Calculate the solubility product for copper(II) iodate, solubility 0.12 g $(100 \text{ ml})^{-1}$.
3. Will any precipitate form if equal volumes of 10^{-4} M $MgCl_2$ and 10^{-4} M NaOH are mixed?
4. Calculate the solubility (in grams per 100 ml) of copper(II) hydroxide.
5. Calculate the solubility (in grams per 100 ml) of copper(II) hydroxide in 6.0 M NaOH.
6. Salicylaldehyde has a solubility at 25°C of 1.42 g $(100 \text{ ml})^{-1}$ in water and 66 g $(100 \text{ ml})^{-1}$ in benzene. Estimate the distribution coefficient of this compound in a benzene–water system.
7. The solubility of 2,4-dinitrophenol at 25°C is 0.68 g $(100 \text{ ml})^{-1}$ in water and 3.27 g $(100 \text{ ml})^{-1}$ in ether. What is the minimum number of extractions required, using equal volumes of ether and aqueous solution, to remove 95 percent of the compound from its aqueous solution?

(Consult Appendix C, as necessary.)

1. Which of the following correctly complete the statement: An autocatalyst . . .

   (I) is always one of the products of the reaction
   (II) is always a heterogeneous catalyst.
   (III) always lowers the activation energy of the reaction.

   a. I, II, and III; **b.** only I and II; **c.** only II; **d.** only III; **e.** only I and III

2. In the Haber process for the manufacture of ammonia from hydrogen and nitrogen gases, a solid phase catalyst is employed to
   a. increase the product–reactant ratio at equilibrium
   b. increase the pressure of the reactant mixture
   c. increase the rate of ammonia formation at constant temperature
   d. increase the magnitude of the equilibrium constant
   e. increase the value for the standard enthalpy of reaction

3. A piece of rusty iron, when suspended in the blue solution of metallic potassium in liquid ammonia, speeds up the formation of potassium amide (KNH_2):

$$2K + 2NH_3 \longrightarrow 2KNH_2 + H_2(g)$$
$$\text{(in solution)}$$

The rusty iron is recovered unchanged. This set of information suggests that
   a. Rusty iron raises the activation energy for this reaction.
   b. Rusty iron lowers $\Delta H°$ for this reaction.
   c. Rusty iron is a heterogeneous catalyst for this reaction.
   d. Rusty iron changes the equilibrium constant for this system.
   e. The reaction of potassium with ammonia is autocatalytic.

4. Most reactions are more rapid at high temperatures than at low temperatures. This is consistent with

   I. an increase in the rate constant with increasing temperature
   II. an increase in the activation energy with increasing temperature

   III. an increase in the percentage of "high-energy" collisions with increasing temperature.

   a. only I; **b.** only II; **c.** III; **d.** only I and III; **e.** I, II, and III

5. The reaction $2NO(g) + 2H_2(g) \longrightarrow N_2(g) + 2H_2O(g)$ follows the rate law,

$$\text{rate} = k \times C_{NO}^2 \times C_{H_2}$$

From this information, we can predict that the initial rate of reaction will double if
   a. the initial C_{NO} is doubled, while initial C_{H_2} remains constant
   b. the initial C_{H_2} is doubled, while initial C_{NO} remains constant
   c. the initial C_{NO} and C_{H_2} are both doubled
   d. the temperature is doubled, without changing initial concentrations
   e. the initial reactants are mixed in the ratio of

     1 mole NO:1 mole H_2

6. A postulated mechanism for the reaction between nitric oxide and hydrogen (question 5) is

   FAST (steady state) $NO + H_2 \rightleftharpoons H_2NO$
   SLOW $H_2NO + NO \longrightarrow H_2O + N_2O$
   FAST $N_2O + H_2 \longrightarrow N_2 + H_2O$

On the basis of this suggested mechanism:
   a. What is the molecularity of the rate-determining step?
   b. What is the overall reaction order?
   c. What are the formulas of two proposed "intermediates"?
   d. What is the molecular formula of the transition state species for the slow step?
   e. Is the mechanism consistent with the rate law given in question 5?

7. For the reaction,
$$O_2(g) + 2NO(g) \longrightarrow 2NO_2(g)$$
the postulated mechanism, consistent with rate data, is:

   FAST (steady state) $2NO(g) \rightleftharpoons N_2O_2(g)$
   SLOW $N_2O_2(g) + O_2(g) \longrightarrow 2NO_2(g)$

The reaction *rate*, then, must equal:

a. $k \times C_{NO}^2$;　**b.** $k \times C_{O_2}$;　**c.** $k \times C_{NO}$;　**d.** $k \times C_{O_2} \times C_{NO}^2$;
e. $k \times C_{O_2} \times C_{NO}$

8. Select the mechanism consistent with the kinetics data given for the reaction,

$$(CH_3)_3CBr + OH^- \longrightarrow (CH_3)_3COH + Br^-$$

EXPERIMENT NUMBER	INITIAL CONCENTRATION (moles liter^{-1})		RELATIVE INITIAL RATES
	$(CH_3)_3CBr$	OH^-	
1	0.1	0.1	1.0
2	0.1	0.01	1.0
3	0.01	0.01	0.1

a. SLOW $(CH_3)_3CBr \longrightarrow (CH_3)_3C^+ + Br^-$
　　FAST $(CH_3)_3C^+ + OH^- \longrightarrow (CH_3)_3COH$
b. SLOW $(CH_3)_3CBr + OH^- \longrightarrow (CH_3)_3CBrOH^-$
　　FAST $(CH_3)_3CBrOH^- \longrightarrow (CH_3)_3COH + Br^-$
c. SLOW $2(CH_3)_3CBr \longrightarrow (CH_3)_3C=C(CH_3)_3 + 2Br^-$
　　FAST $(CH_3)_3C=C(CH_3)_3 + OH^- \longrightarrow (CH_3)_3COH$
　　　　　　　　　　　　　　　　　　$+ (CH_3)_3C$
d. SLOW $(CH_3)_3CBr + OH^- \longrightarrow (CH_3)_3CO^- + HBr$
　　FAST $(CH_3)_3CO^- + HBr \longrightarrow (CH_3)_3COH + Br^-$
e. none of these

9. For the reaction between ethoxide ion and iodomethane,

$$C_2H_5O^- + CH_3I \longrightarrow C_2H_5OCH_3 + I^-$$

the rate constant at $10°C$ is 2.2×10^{-4}. At $25°C$, $k = 1.1 \times 10^{-3}$. What is the activation energy for this reaction, in kcal mole^{-1}?

$$\log \frac{k_2}{k_1} = \frac{E_a(T_2 - T_1)}{(4.6 \times 10^{-3})(T_1 \times T_2)}$$

a. 5; **b.** 11; **c.** 18; **d.** 37; **e.** 52

10. $S^{2-}(aq) + 2H_2O(l) \rightleftharpoons H_2S(aq) + 2OH^-(aq)$

The conventional equilibrium constant expression for the system formulated above is

a. $\dfrac{C_{S^{2-}} \times C_{H_2O}^2}{C_{H_2S} \times C_{OH^-}^2}$　**b.** $\dfrac{C_{H_2S} \times C_{OH^-}^2}{C_{S^{2-}} \times C_{H_2O}^2}$

c. $\dfrac{C_{H_2S} \times C_{OH^-}^2}{C_{S^{2-}}}$　**d.** $\dfrac{C_{H_2S} \times 2C_{OH^-}}{C_{S^{2-}}}$

e. none of these

11. A solution was prepared by dissolving 0.863 g of sodium propionate in sufficient water to form 50.0 ml of solution. The equilibrium concentration of hydroxide ion in the resulting system was found to be 1.13×10^{-5} M. What is the approximate value of the equilibrium constant for the system represented by

$$CH_3CH_2CO_2^-(aq) + H_2O(l) \rightleftharpoons CH_3CH_2CO_2H(aq)$$
$$\text{(propionate ion)} \qquad\qquad + OH^-(aq)$$

a. 5.4×10^{-8}; **b.** 6.0×10^{-9}; **c.** 7.1×10^{-10}; **d.** 8.3×10^{-11}; **e.** 9.5×10^{-12}

12. The conventional equilibrium constant at $25°C$ is 2×10^{-5}

for the aqueous cyanide ion equilibrium system:

$$CN^-(aq) + H_2O(l) \rightleftharpoons HCN(aq) + OH^-(aq)$$

What is the approximate equilibrium concentration of hydrocyanic acid in this system when the cyanide ion is 2.45 M?

a. 0.007 M; **b.** 0.0045 M; **c.** 0.003 M; **d.** 0.02 M; **e.** 0.45 M

13. What is the approximate equilibrium concentration (moles liter^{-1}) of CN^- in a solution labeled 0.048 M HCN? ($K = 4.8 \times 10^{-10}$)

a. 1×10^{-6};　**b.** 4.8×10^{-6};　**c.** 1×10^{-5};　**d.** 4.8×10^{-5}
e. 6.9×10^{-5}

14. Consider the gas-phase equilibrium system represented by the equation,

$$2SO_3 \rightleftharpoons 2SO_2 + O_2$$

Given that the conversion of "left-hand" species to "right-hand" species, as written, is endothermic, which of the following changes will decrease the equilibrium *amount* of SO_2?

a. adding a solid phase catalyst
b. adding more SO_3
c. increasing the pressure at constant temperature, by reducing volume
d. increasing the temperature at constant pressure
e. none of these

15. At $25°C$, $K_{eq} = 5.0$ for

$$A(g) + B(g) \rightleftharpoons C(g) + D(g)$$

Consumption of A and B is endothermic. If we wish to *increase* the value of K_{eq}, we must

a. increase pressure at constant temperature, by decreasing volume
b. decrease pressure at constant temperature, by increasing volume
c. add a catalyst
d. increase the temperature
e. decrease the temperature

16. The solubility product expression for magnesium hydroxide, using x to represent the molar concentration of magnesium and y to represent the molar concentration of hydroxide is formulated as

a. xy; **b.** x^2y; **c.** xy^2 **d.** x^3y; **e.** xy^3

17. For a compound M_3X_2 whose water solubility can be represented by s(mole liter^{-1}), the solubility product (K_{sp}) is

a. $4s^3$; **b.** $4s^4$; **c.** $108s^5$; **d.** $27s^5$; **e.** $27s^4$

18. The relationship between the molar concentration of silver ion in a saturated aqueous solution of silver sulfide and the solubility product for Ag_2S may be expressed as

a. $K_{sp} = 8C_{Ag^+}^3$
b. $K_{sp} = \frac{1}{3}C_{Ag^+}^2$
c. $K_{sp} = 4C_{Ag^+}^3$
d. $K_{sp} = \frac{1}{2}C_{Ag^+}^3$
e. $K_{sp} = \sqrt[3]{C_{Ag^+}}$

19. What is the minimum mass required (in grams) of $PbBr_2$ (formula weight 367) for formation of 1.00 liter of its saturated solution?

a. 1.8; **b.** 2.3; **c.** 3.1; **d.** 3.7; **e.** 4.4

20. An aqueous slurry of barium sulfate is used in radiographic studies of the gastrointestinal tract. How many milligrams of dissolved Ba^{2+} are present in 3.00 liters of saturated aqueous $BaSO_4$?

    a. 5.0; **b.** 3.7; **c.** 3.0; **d.** 2.4; **e.** 1.9

21. When equal volumes of the solutions indicated are mixed, precipitation should occur only for

    a. 2×10^{-3} M Ca^{2+} + 2×10^{-2} M OH^-

    b. 2×10^{-2} M Mg^{2+} + 2×10^{-5} M OH^-

    c. 2×10^{-5} M Ca^{2+} + 2×10^{-5} M CO_3^{2-}

    d. 2×10^{-3} M Ag^+ + 2×10^{-5} M CO_3^{2-}

    e. 2×10^{-1} M Ba^{2+} + 2×10^{-4} M BrO_3^-

22. Which of the following should dissolve the smallest amount of magnesium hydroxide per liter, assuming no complex ion formation?

    a. 0.10 M $NaNO_3$; **b.** 0.10 M HNO_3; **c.** 0.10 M $Mg(NO_3)_2$;

d. 0.10 M NaOH; **e.** pure water

23. Salicylic acid has a water solubility at 20°C of 0.18 g (100 ml)$^{-1}$. At 20°C, 52 g of the acid will dissolve per 100 ml of ether. The approximate distribution coefficient for salicylic acid in a water–ether system is

    a. 290; **b.** 52; **c.** 9.4; **d.** 0.18; **e.** 3.5×10^{-3}

24. On the basis of the K_D value from question 23, what percentage of the salicylic acid would be extracted from 50.0 ml of its saturated aqueous solution by a single extraction with 10 ml of ether?

    a. 100%; **b.** 98%; **c.** 74%; **d.** 26%; **e.** 2%

25. What is the minimum number of extractions, each by 20 ml of ether, necessary to remove 1.7 g of salicylic acid from 1.00 liter of its saturated aqueous solution (on the basis of the K_D value from question 23)?

    a. 1; **b.** 2; **c.** 3; **d.** 4; **e.** 5

Section
SEVEN

ACID-BASE
CHEMISTRY

As we shall see, there are several different ways of defining acids and bases. At first, this might seem an unnecessary complication, but the versatility provided by alternative ways of describing acid-base behavior is important in a number of applications. Organic chemists, for example, find it useful to employ the Lewis model of "competition for an electron pair" in discussing the catalytic activity of such "Lewis acids" as $AlCl_3$ or BF_3 for certain reactions in nonaqueous solvents. In aqueous systems, with which we shall be mainly concerned, the Brønsted-Lowry model of acid-base systems as "competition for protons" is particularly convenient. As we develop the general concepts of acid-base chemistry (Unit 19), the benefits of alternative ways of looking at acid-base behavior will become more obvious.

An important concept to be considered is that of acid-base *strength*. We use this term in a special way to apply to the *percentage ionization* of an acid in solution, as has been mentioned in Unit 3. Some confusion often arises from the common idea that an acid is strong if it "burns your skin" or, as popularized in a television commercial, household ammonia is "strong if you can really *smell* it." We prefer other terms, such as *corrosive* or *concentrated*, to describe such phenomena. In connection with *strength* and *concentration*, we shall apply the principles of equilibrium (Unit 17) to the analysis of aqueous solutions of *weak* acids or bases and we shall develop, in Unit 19, the idea of *pH* as a useful way of working with hydrogen-ion concentrations in the general range of $1\,M - 10^{-14}\,M$.

For many laboratory procedures, and in numerous biochemical systems, some method is needed to maintain the hydrogen-ion concentration within a particular range. As we shall see in Unit 20, *buffer systems* provide a useful way of controlling hydrogen-ion concentration. An understanding of the characteristics of *buffers* offers important clues to processes ranging from the preparation of "carbonated" beverages to the complex physiology of the blood.

In most natural waters, we find measurable concentrations of both Ca^{2+} and CO_3^{2-}. In some cases these concentrations are sufficient to result in the formation of crystalline calcium carbonate, as described in Unit 18. In other cases, precipitation is avoided by a control of carbonate concentration through a series of *interactive equilibria*:

$$H_2O(l) \rightleftharpoons H^+(aq) + OH^-(aq)$$
$$CO_3^{2-}(aq) + H^+(aq) \rightleftharpoons HCO_3^-$$
$$HCO_3^-(aq) + H^+(aq) \rightleftharpoons H_2O(l) + CO_2(aq)$$
$$CO_2(aq) \rightleftharpoons CO_2(g)$$

It might appear that such complicated systems would be quite difficult to analyze by the mathematics of chemical equilibria. However, as we shall discover in Unit 21, the careful use of suitable approximations offers a generally appropriate way of dealing with interactive equilibria.

Although increasing numbers of chemical analyses are done by instrumental methods, *wet chemical* techniques are still important. Excursion 2 offers some examples of analytical procedures involving applications of acid-base chemistry, buffer systems, and interactive equilibria.

CONTENTS

equilibria in aqueous solutions of weak acids or bases; pH and pOH; hydrolysis of salts; acid-base titrations

Unit 20 Buffer systems
Types and characteristics of buffers; buffer pH; pH change and buffer capacity

Unit 21 Interactive equilibria
Description of complicated equilibrium systems; approximations in the analysis of interactive equilibria; pH-controlled precipitations and separations.

Section Seven Overview Test

Excursion 2 Wet chemical analysis
Some common analytical procedures illustrating applications of acid-base chemistry, buffer systems, and interactive equilibria.

Unit 19

In a very real sense, we can make an acid be anything we wish—the differences between the various acid-base concepts are not concerned with which is "right" but which is *most convenient to use in a particular situation.*

Inorganic Chemistry: Principles of Structure and Reactivity (1972)
JAMES E. HUHEEY

Acids and bases

Objectives

(a) Given the name or formula of a compound or ion and, when necessary, an indication of its behavior in a particular reaction, be able to classify the species as an acid, a base, or neither in the system proposed by
   (1) Arrhenius
   (2) Brønsted and Lowry
   (3) Lewis
(b) Be able to identify the members of a conjugate pair, as defined by Brønsted and Lowry (Section 19.1).
(c) Be able to calculate the pH of an aqueous solution:
   (1) of a strong acid or base, given the name or formula of the solute and its molar concentration (Example 4).
   (2) of a weak acid or base, given the name or formula of the compound, the "label" concentration of the solution, and the ionization constant for the acid or base involved (Section 19.5).
   (3) of a salt of a weak acid or base, given the name or formula of the salt, indication of the solution concentration, and the ionization constant for the acid or base from which the salt was formed (Section 19.6).

(d) Be able to make qualitative estimates of relative acid strengths of oxyacids and of successive ionization constants for polyprotic acids (Sections 19.2 and 19.3).

(e) Be able to calculate the pH at the equivalence point for an acid-base titration and to select an appropriate indicator for the titration (Section 19.7).

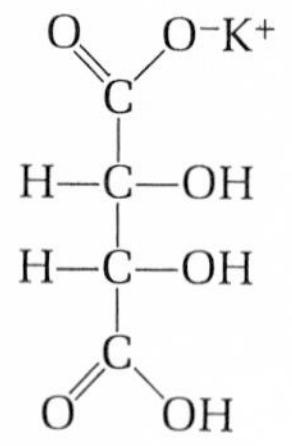
potassium hydrogen tartrate

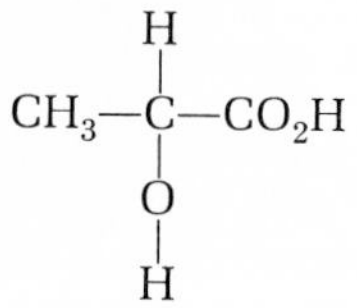
lactic acid

The weak acid itself may also react with bicarbonate, e.g.,

$$H_2PO_4^- + HCO_3^- \longrightarrow H_2O + HPO_4^{2-} + CO_2$$

In this modern age of packaged mixes and canned biscuits, it is rare to find a cook who still uses "baking powder" or "baking soda." Not too many years ago, however, these were found in nearly every kitchen. Baking powder is a mixture of sodium bicarbonate (baking soda) with some salt containing an acidic ion. The most common of these salts are cream of tartar (potassium hydrogen tartrate, $KHC_4H_4O_6$*), calcium dihydrogen phosphate* $[Ca(H_2PO_4)_2]$*, sodium dihydrogen phosphate, and sodium aluminum sulfate* $[NaAl(SO_4)_2]$*. Generally, baking powder also contains some starch, which adsorbs atmospheric moisture to prevent decomposition of the other ingredients. Baking soda can be used in the preparation of dough when some other acid is added, typically cream of tartar or sour milk (which contains lactic acid). In any of these cases, the reaction of importance involves generation of carbon dioxide gas from decomposition of the sodium bicarbonate in the mixture. The gas forms bubbles in the dough and these expand, causing the dough to "rise" as more* CO_2 *is released and as the temperature is increased while the dough is baked in an oven. The general reaction may be formulated as*

$$H^+(aq) + HCO_3^-(aq) \longrightarrow H_2O(l) + CO_2(g)$$

In the case of the sour milk, cream of tartar, or dihydrogen phosphate salts, the aqueous H^+ *originates from a weak acid, such as*

$$H_2PO_4^- \rightleftharpoons H^+ + HPO_4^{2-}$$

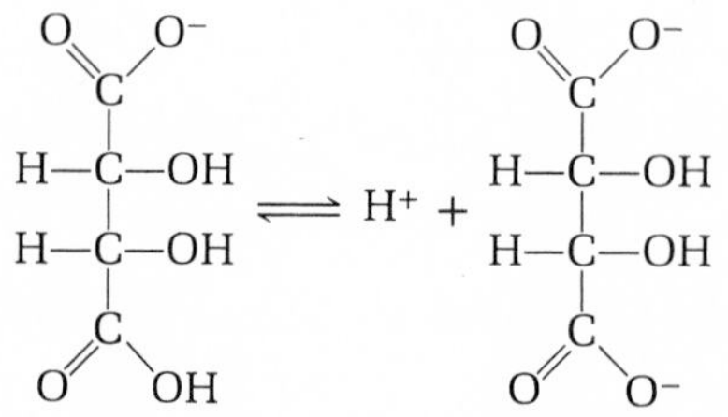

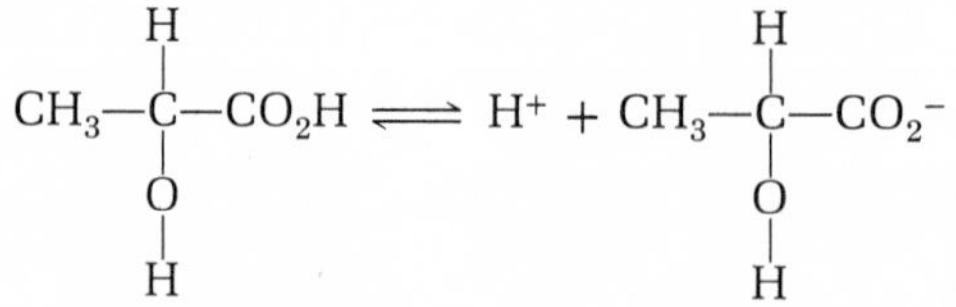

The acidity of moistened sodium aluminum sulfate results from the presence of the hydrated aluminum ion, for instance,

$$[Al(H_2O)_6]^{3+} \rightleftharpoons H^+ + [Al(H_2O)_5(OH)]^{2+}$$

The reaction between strong acids and bases is typically quite exothermic. Attempts to neutralize spilled acids must be cautious ones. Contact with any of the common strong acids or bases in the home, on the job, or in the laboratory should never be treated by an attempt at neutralization. Almost invariably, this results in a severe heat burn in addition to the chemical burn from the reagent. Proper first aid for acid or base contact requires immediate and copious washing with water, not addition of a neutralizing reagent.

This unit will consider several aspects of acid-base chemistry (Table 19.1). Some of these topics have been introduced before, and these should be reviewed (Units 3, 6, and 17). In particular, the names of common acidic, basic, and amphoteric species should be restudied carefully (Units 3 and 6).

19.1 ACID-BASE SYSTEMS

Historically, the earliest definitions of acids and bases still in common usage are attributed to Svante Arrhenius (1887). These were based on a series of observations of properties common to certain types of compounds (Table 19.2). It was suggested that the properties of *acids* could be explained by defining them as *substances that furnish hydrogen ions (protons, H^+) in aqueous solution*. Bases were defined as *substances that furnish hydroxide (OH^-) ions in solution*. A neutralization reaction in water could then be written generally as, for example,

$$\text{acid} + \text{base} \longrightarrow \text{a salt} + \text{water}$$
$$HCl + KOH \longrightarrow KCl + H_2O$$

Some examples of compounds classified as acids or bases by these definitions are given in Table 19.3. Although later descriptions are more useful in understanding the actual behavior of compounds involved in acid–base reactions, chemists still use the classifications of Arrhenius in many qualitative situations.

Many reactions, such as the gas-phase reaction of ammonia with hydrogen chloride,

$$NH_3(g) + HCl(g) \longrightarrow NH_4Cl(s)$$

have similar characteristics to the *neutralization* processes described by Arrhenius in the sense that they involve a proton transfer. However, OH^- and H_2O are not present.

Table 19.1 ACID-BASE DEFINITIONS

ARRHENIUS

acid: H^+ source in water
base: OH^- source in water
e.g., neutralization,

$$HBr(aq) + NaOH(aq) \longrightarrow$$
$$\underset{\text{acid}}{} \qquad \underset{\text{base}}{}$$
$$\underset{\text{salt}}{NaBr(aq)} + \underset{\text{water}}{H_2O}$$

BRØNSTED-LOWRY

acid: H^+ donor
base: H^+ acceptor
e.g., competition for proton,

$$\underset{\text{acid}_1}{NH_4^+} + \underset{\text{base}_2}{H_2O} \rightleftharpoons \underset{\text{base}_1}{NH_3} + \underset{\text{acid}_2}{H_3O^+}$$

LEWIS

acid: electron-pair acceptor
base: electron-pair donor
e.g., competition for electron pair,

$$Al\overset{\cdot\cdot}{\underset{\cdot\cdot}{Cl}}: + :\overset{\cdot\cdot}{\underset{\cdot\cdot}{Cl}}:\overset{\cdot\cdot}{\underset{\cdot\cdot}{Cl}}: \rightleftharpoons$$

$$:\overset{\cdot\cdot}{\underset{\cdot\cdot}{Cl}}:Al:\overset{\cdot\cdot}{\underset{\cdot\cdot}{Cl}}:^- + \overset{\cdot\cdot}{\underset{\cdot\cdot}{Cl}}:^+$$

Table 19.2 PROPERTIES OF ARRHENIUS ACIDS AND BASES

ACIDS	BASES
turn blue litmus red	turn red litmus blue
neutralize bases	neutralize acids
taste sour (e.g., vinegar)	taste bitter [e.g., $Al(OH)_3$]
liberate H_2 by reaction with active metals (e.g., Mg, Zn)	feel soapy when moist
furnish excess H^+ in water	*furnish excess OH^- in water*

Table 19.3 EXAMPLES OF ARRHENIUS ACIDS AND BASES

FORMULA	NAME
HCN(aq)	hydrocyanic acid
HNO_3	nitric acid
C_6H_5COOH	benzoic acid
HOOCCOOH	oxalic acid
H_2SO_4	sulfuric acid
H_3PO_4	phosphoric acid
LiOH	lithium hydroxide
NaOH	sodium hydroxide
KOH	potassium hydroxide
$Mg(OH)_2$	magnesium hydroxide
$Al(OH)_3$	aluminum hydroxide
$NH_4OH(aq)$[a]	aqueous ammonia (sometimes called "ammonium hydroxide")

[a] Evidence is now definite that so-called ammonium hydroxide solutions contain mostly $NH_3(aq)$, some NH_4^+ and OH^- ions, and only traces of NH_4OH. The solution is therefore more properly called *aqueous ammonia*.

Alternative formulations are possible, depending on the particular types of acids and bases involved, for example,

$$HB_1^+ \; + \; B_2^- \; \rightleftharpoons$$

cationic   anionic
acid   base
(e.g., NH_4^+) (e.g., OH^-)

$$HB_2 \; + \; B_1$$

molecular molecular
acid   base
(e.g., H_2O) (e.g., NH_3)

Note that Arrhenius thought of aqueous HCl as containing the acid *HCl*, while the Brønsted-Lowry concept considers the main species in such a solution (HCl is essentially 100 percent ionized in water) to be H_3O^+ or H^+(aq). Similar distinctions apply to other strong acids (Unit 3). Recent studies indicate that the hydrated proton is probably most commonly in the form $H(H_2O)_4^+$, rather than the H_3O^+ frequently used.

Table 19.4 TYPICAL BRØNSTED-LOWRY BASES[a]

FORMULA	NAME
ANIONS	
OH^-	hydroxide
SH^-	hydrosulfide[b]
CH_3O^-	methoxide
$CH_3CH_2O^-$	ethoxide
CN^-	cyanide
NH_2^-	amide
HCO_3^-	bicarbonate[b]
$H_2PO_4^-$	dihydrogen phosphate[b]
HPO_4^{2-}	monohydrogen phosphate[b]
S^{2-}	sulfide
NEUTRAL MOLECULES	
H_2O	water[b]
NH_3	ammonia[b]
CH_3NH_2	methylamine[b]
CATIONS (RARE)	
$H_2N(CH_2)_4NH_3^+$	

[a] See also Unit 3.
[b] May also function as acids in the presence of appropriate proton acceptors.

Liquid ammonia has many of the solvent properties of water, and reactions in liquid ammonia may involve proton transfer in a fashion quite analogous to the reaction in aqueous solution, such as in the following examples.

General neutralization in water

$$H^+(aq) + OH^-(aq) \longrightarrow H_2O(l)$$
"H_3O^+"

General neutralization in liquid ammonia

$$H^+(NH_3) + NH_2^-(NH_3) \longrightarrow NH_3(l)$$
"NH_4^+"   amide ion

In order to expand the definitions for acid–base systems, J. N. Brønsted and J. M. Lowry in 1923 proposed the following, more generally applicable descriptions:

Acids: substances that *act* as proton donors.
Bases: substances that *act* as proton acceptors (by furnishing an unshared pair of valence electrons for bonding with H^+).
Acid-base reaction: a competition for formation of a covalent bond to H^+

conjugate pair

$$HB_1 + B_2 \rightleftharpoons HB_2^+ + B_1^-$$

conjugate pair

Conjugate pair: a pair of substances differing only by one H^+.

Brønsted–Lowry acids would include most of the species commonly thought of as Arrhenius acids and, in addition, such species as HSO_4^-, HCO_3^-, NH_4^+, and so on (Unit 3). The most important difference, however, is that the newer definitions would classify OH^- as a base, not LiOH, NaOH, and so on. Other bases would include such species as those shown in Table 19.4. *Compounds* of hydroxide ion are considered to be *salts* that *contain* the base OH^-.

Note that the key to the Brønsted–Lowry system is how a species *acts*. Table 19.4 then, simply indicates species that often act as proton acceptors, but correct classi-

Table 19.5 SOME TYPICAL LEWIS ACIDS

SIMPLE CATIONS

(acid strength increases with availability of a bonding orbital region forming stable covalent bonds and with positive charge density)

e.g., $Fe^{3+} > Fe^{2+}$
$B^{3+} > Be^{2+} > Li^+$
$Li^+ > Na^+ > K^+$

MOLECULES CONTAINING AN ELECTRON-PAIR DEFICIENT ATOM

e.g., $AlCl_3$, BF_3, $FeBr_3$, SO_3, $ZnCl_2$

MOLECULES CONTAINING AN ATOM CAPABLE OF VALENCE EXPANSION[a]

e.g., $SiF_4 + 2F^- \longrightarrow SiF_6^{2-}$

[a] Valence expansion refers to the ability of a "central atom" (Unit 8) to form stable species by accommodating more than the "normal" four pairs of electrons in the "valence shell" (Unit 7). This is associated with atoms for which unoccupied *d* orbitals have energies close to those of the next lower *s* and *p* orbitals. This topic is considered in Unit 27.

fication depends on the particular reaction in which the species participates. For example,

$$HF(aq) + H_2O(l) \rightleftharpoons H_3O^+(aq) + F^-(aq)$$
(water acts as a base)

$$NH_3(aq) + H_2O(l) \rightleftharpoons NH_4^+(aq) + OH^-(aq)$$
(water acts as an acid)

Note that three definitions would describe the acidity of aqueous HCl differently:

Arrhenius acid: HCl
Brønsted–Lowry acid: H_3O^+
  [more accurately, $H(H_2O)_4^+$]
Lewis acid: H^+

One of the more recent generalizations is that an *acid* is any electrophilic ("negative charge loving") species that reacts to accept an electron pair at a speed determined by diffusion rate, and that a *base* is any nucleophilic ("positive charge loving") species that reacts to furnish an electron pair at a speed determined by diffusion rate.

Still another set of generalizations for acid–base behavior was proposed in 1923, this time by G. N. Lewis. Lewis recognized that the factor common to all "neutralization" reactions was the formation of a covalent bond by the donation of an unshared valence electron pair to an electron-deficient species. Lewis suggested that the most general description would simply define an *acid* as *some species "seeking" an electron pair* and a *base* as *some species furnishing an electron pair.* Thus the exchange of a proton would be just a special case of a much more general phenomenon. *Lewis bases* would include any substance having one or more unshared pairs of valence electrons—the same requirement as the Brønsted–Lowry system. Lewis acids (Table 19.5), however, would include many substances in addition to the proton.

The fact that there are different ways of looking at acid–base systems poses no serious problem. In fact, additional suggestions appear periodically. Chemists simply use the convention most appropriate to a particular situation. *In general, we shall use the Brønsted–Lowry system,* since it is a convenient way of looking at many characteristics of aqueous solutions.

19.2 ACID-BASE STRENGTH

Under the Brønsted–Lowry system, any acid–base reaction involves a competition for protons. *Strong acids* are those that, in aqueous solution, donate protons very readily to water. The reaction is essentially irreversible[1]:

$$HX(aq) + H_2O(l) \longrightarrow H_3O^+(aq) + X^-(aq)$$

Strong bases tend to accept protons to such an extent that, in aqueous solution, they have an affinity for protons almost equal to (or greater than) that of OH^-.

The strength of an acid or base can be expressed by an equilibrium constant (Unit 17), which indicates the extent of the competition with the solvent for protons. For *aqueous solutions*

Similar expressions may be written for *ionic* acids and bases, except that the charges will be different for the acid-base conjugate pair species.

Acids

$$HX(aq) + H_2O(l) \rightleftharpoons H_3O^+(aq) + X^-(aq)$$

$$K_a = \frac{C_{H_3O^+} \times C_{X^-}}{C_{HX}} \quad (acid \text{ dissociation constant})$$

Bases

$$B(aq) + H_2O(l) \rightleftharpoons BH^+(aq) + OH^-(aq)$$

$$K_b = \frac{C_{BH^+} \times C_{OH^-}}{C_B} \quad (base \text{ constant})$$

Note that C_{H_2O} is omitted, since water appears as a liquid (Unit 17).

---

[1] Careful measurements suggest that many strong acids appear to be less than 100 percent ionized in aqueous solution, but this is believed to result, in most cases, from closely associated hydrated ion pairs rather than residual molecular species. For our purposes, however, we will assume complete ionization of strong acids (Unit 3).

Table 19.6 ACID AND BASE STRENGTHS[a]

	ACID	K_a	BASE	K_b	
strong	HClO$_4$	$\sim 10^{10}$			strong
	HI	$\sim 10^9$	OH$^-$	(reference std.)	
	HBr	$\sim 10^9$	S^{2-}	$\sim 10^{-1}$	
	HCl	$\sim 10^6$	NH$_3$	1.8×10^{-5}	
	HNO$_3$	~ 30	SO$_3^{2-}$	7.7×10^{-8}	
weak	HCO$_2$H	2.1×10^{-4}	CH$_3$CO$_2^-$	5.6×10^{-10}	weak
	HF	6.9×10^{-4}	F$^-$	1.4×10^{-11}	
	CH$_3$CO$_2$H	1.8×10^{-5}	HCO$_2^-$	4.7×10^{-13}	
	HSO$_3^-$	1.3×10^{-7}	NO$_3^-$	$\sim 10^{-16}$	
	HOCl	3.2×10^{-8}	Cl$^-$	$\sim 10^{-20}$	
	NH$_4^+$	5.6×10^{-10}			
	HS$^-$	1.3×10^{-13}			

[a] Shaded areas include species for which the ionization "constants" are quite concentration dependent. For simple calculations these species are treated as being completely ionized in aqueous solution, i.e., K values are not used.

For aqueous strong acids, the K_a value is not particularly useful. It represents only an "apparent" constant and will vary considerably with concentration. Such acids may be more meaningfully compared in solvents other than water. It is important to realize that the *strength* of an acid is different from the *concentration* of the acid. For example, a $10^{-6}\,M$ solution of the strong acid HClO$_4$ is only slightly acidic ($C_{H^+} \approx 10^{-6}\,M$).

Some comparisons of acids and bases are given in Table 19.6. A more complete listing will be found in Appendix C. The common strong acids should be remembered. Sulfuric acid is strong with respect to loss of one proton, but HSO$_4^-$ is weak ($K_a = 1.3 \times 10^{-2}$).

There are a number of influences that may contribute to the strength of an acid (or base). To understand these, we need to consider (in the Brønsted–Lowry system) (a) factors tending to affect the "ease of proton escape" from an acid by weakening the bond to hydrogen, and (b) factors tending to affect the "ease of proton capture" by the conjugate base.

Unfortunately, there are many cases in which these two factors have opposing influences, so that no qualitative predictions are possible. In such cases we must rely on experimentally determined values of K_a or K_b (although more advanced theoretical models often permit a quantitative evaluation of competing influences).

Remember (Unit 6) that *charge density* is "charge per unit volume."

Let's consider first the "ease of proton capture" as a measure of base strength. We can relate this, for anionic bases, to a concept already developed: *charge density* (Unit 6). Electrostatic considerations suggest *that the attraction for a proton (base strength) should decrease with decreasing negative charge density*. Thus for a comparison of ions of roughly the same volume, base strength decreases with decreasing negative charge (e.g., ClO$_4^-$ < SO$_4^{2-}$ < PO$_4^{3-}$). For ions of the same net charge, we must consider two factors: ionic volume and the influence of delocalized π bonds. When no π-bonding considerations are necessary, we can predict relative strengths on the basis of relative ionic volumes (Unit 2) (e.g., I$^-$ < Br$^-$ < Cl$^-$). When delocalized π bonding (Unit 8) can occur, its effect in reducing charge density on *a specific atom* is reflected in lowered base strength, for example,

To decide if π bonding considerations are important, construct an electron-dot or line-bond formulation for the species and analyze the bonding description.

$$\begin{array}{ccc} \overset{\displaystyle O^-}{\underset{\displaystyle }{\text{CH}_3\text{C}}}\!\!-\!\!\text{O} & < & \text{CH}_3\text{CH}_2\text{O}^- \\ \text{acetate ion} & & \text{ethoxide ion} \\ \text{(charge delocalized} & & \text{(high charge density} \\ \text{by } \pi\text{-bond system,} & & \text{on oxygen)} \\ \text{Unit 8)} & & \end{array}$$

With electrically neutral molecular bases, we cannot consider *charge* density, but a somewhat analogous relationship is noted in terms of decreasing base strength with increasing atomic volume (e.g., $AsH_3 < PH_3 < NH_3$).

In terms of factors relating acid strength to "ease of proton escape," some useful insight can be gained from the comparison of electronegativities (Unit 7). For example, the more highly polarized bond in HCl should permit a more facile proton escape than should the less polar bond in HBr. Electronegativity influences may extend through more than a single bond:

$$Cl—O—H \rightleftharpoons Cl—O^- + H^+$$

The highly electronegative chlorine helps "pull electrons" away from hydrogen, facilitating H^+ escape.

$$Br—O—H \rightleftharpoons Br—O^- + H^+$$

Since bromine is less electronegative than chlorine, the effect on the O—H bond is less. (BrOH < ClOH in acid strength.)

An entire polyatomic group may be thought of as having a single influence on the "ease of proton escape." This is most readily assessed on the basis of *formal charge* on the atom bonded to the OH grouping.

It should be noted that *formal charge*, like *oxidation number* (Unit 3), is a "bookeeping assignment" and has no physical reality.

The *formal charge* of an atom is determined by writing an "octet formula" of the compound (Unit 7) and assigning half the bond-pair electrons to each atom in the bond. Then the *formal charge* is defined as the difference between the number of valence electrons of the *neutral* atom and the number of valence electrons *assigned to the atom, that is,*

formal charge = neutral atom valence electrons − assigned valence electrons

We may think of *formal charge* as representing a positive field attraction for the bond pair electrons of the O—H group.

The higher the *formal charge* on an atom bonded to an OH group, the greater will be the influence in weakening the O—H bond. For example, in comparing chloric acid with perchloric acid, we predict the latter to be the strong acid because of the higher *formal charge* on chlorine:

assigned 5 valence electrons
:Cl· neutral chlorine atom (Group VIIA), 7 valence electrons
(formal charge = 7 − 5).

assigned 4 valence electrons
(formal charge = 7 − 4).

Formal charge is useful in comparing acid strengths only for different oxyacids of the same element. A general "rule of thumb" has been developed for oxyacids that combines both electronegativity and *formal charge* considerations (Table 19.7).

For an acid to be strong, its conjugate base must be weak; so it is easy to predict acid strength when all contributing factors favor "ease of proton loss" and oppose "ease of proton capture." For example, the high electronegativity of chlorine weakens the H—Cl bond (favoring proton loss) and the relatively large ionic volume of Cl^- (with a resulting low charge density) does not particularly favor proton capture by Cl^-. The problem in predicting acid strength occurs when various factors are opposing. Thus although the electronegativity of fluorine suggests that H—F would be a strong acid, the high charge density of F^- suggests that proton capture by F^- is also

Table 19.7 PREDICTION OF RELATIVE STRENGTHS OF OXYACIDS

FOR AN OXYACID OF FORMULA $XO_m(OH)_n$
1. If $m = 0$, the acid is *very weak;* for example, for ClOH, $K_a = 3.2 \times 10^{-8}$.
2. If $m = 1$, the acid is *moderately weak;* for example, for NO(OH) [HNO_2], $K_a = 4.5 \times 10^{-4}$.
3. If $m = 2$ or 3, the acid is *strong;* for example, for $NO_2(OH)$ [HNO_3], $K_a = \sim 30$.

A special consideration applies to the case of HF. Because of strong hydrogen bonding among HF molecules (Unit 8), the proton is effectively "trapped" in a hydrogen-bond situation from which "escape is difficult":

H—F:---H—F:

favored. Experimentally, we find that HF is a *weak* acid, but our simple considerations would not have permitted that prediction. We shall have to be content with the limitations of acid strength prediction on the basis of the qualitative factors described, leaving more quantitative prediction for advanced studies.

Example 1

On the basis of charge density considerations, predict which member of each pair should be the stronger base:

a. chloride ion or iodide ion
b. hydrosulfide ion (HS^-) or hydroxide ion
c. formate ion 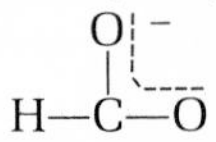 or methoxide ion (CH_3O^-)

Solution

a. Since both chloride and iodide ions have the same charge, the one with the smaller ionic volume should have the higher charge density. Thus we predict Cl^- to be the stronger base.
b. Again, the difference in ionic volumes (corresponding to $S > O$) suggests a higher charge density for hydroxide ion, so we predict OH^- to be stronger.
c. The delocalized π bonding in formate ion, by reducing charge density on oxygen, suggests that methoxide ion is a stronger base.

Example 2

On the basis of electronegativity effects and *formal charge,* as summarized in Table 19.7, predict which member of each pair is the stronger acid:
a. nitric acid or nitrous acid
b. nitrous acid or perchloric acid
c. sulfuric acid or phosphoric acid, in terms of loss of a single proton

Solution
Using the "summary predictions" from Table 19.7, with formulations as $XO_m(OH)_n$:

a. $HNO_3 \equiv NO_2(OH)$ $(m = 2)$
$HNO_2 \equiv NO(OH)$ $(m = 1)$
$HNO_3 > HNO_2$

b. $HNO_2 \equiv NO(OH)$ $(m = 1)$
$HClO_4 \equiv ClO_3(OH)$ $(m = 3)$
$HClO_4 > HNO_2$

c. $H_2SO_4 \equiv SO_2(OH)$ $(m = 2)$
$H_3PO_4 \equiv PO(OH)_3$ $(m = 1)$
$H_2SO_4 > H_3PO_4$

Example 3

On the basis of Table 19.7, it would appear that phosphoric acid (H_3PO_4) should be significantly stronger than phosphorous acid (H_3PO_3). However, the K_a values for loss of a single proton are

$$H_3PO_4, K_{a_1} = 7.5 \times 10^{-3}$$

and

$$H_3PO_3, K_{a_1} = 1.6 \times 10^{-2}$$

Given that phosphoric acid contains four P—O bonds and phosphorous acid contains three P—O bonds and one P—H bond, show that the *formal charge* on phosphorous is consistent with the similar strengths of these two acids.

Solution

H_3PO_4

H_3PO_3

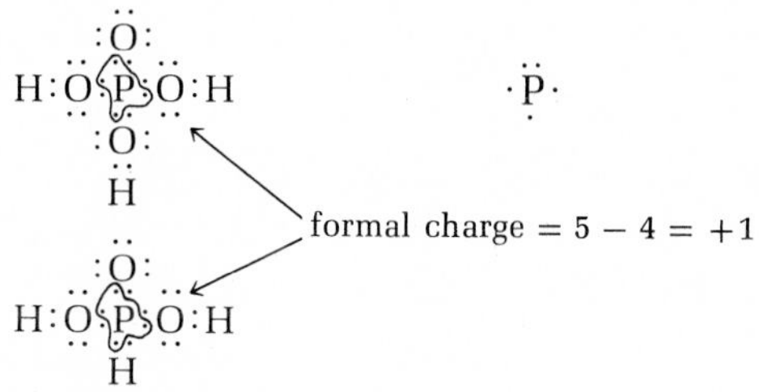

formal charge $= 5 - 4 = +1$

19.3 POLYPROTIC SYSTEMS

Experiments have shown that two diprotic acids, carbonic and sulfurous, do not exist as the molecular acids in any significant concentrations. Rather, what is called "carbonic acid, H_2CO_3" is principally $CO_2(aq)$ and what is called "sulfurous acid, H_2SO_3" is principally $SO_2(aq)$. As a matter of convenience, however, we may still discuss these systems in terms of H_2CO_3 or H_2SO_3, respectively, while recognizing that the molecular acids are not major components of the aqueous equilibrium systems.

Salts that furnish more than 1 mole of hydroxide per mole of salt (e.g., $Ba(OH)_2$, Example 4) represent simple situations, since the complete ionization of the salt makes it easy to calculate the hydroxide concentration. Acids that may furnish more than one proton (e.g., H_2SO_4, H_3PO_4) are not so simple. Polyprotic acids may react completely with strong bases to lose all available protons. The extent of reaction depends on the relative strengths of the acids and bases involved and on the relative amounts of reactants mixed. In aqueous solutions of these acids alone, however, the base is relatively weak (H_2O), and the equilibrium mixtures are complex. For example,

In aqueous sulfuric acid

$$H_2SO_4(aq) + H_2O(l) \longrightarrow H_3O^+(aq) + HSO_4^-(aq)$$

(essentially 100%)

$$HSO_4^-(aq) + H_2O(l) \rightleftharpoons H_3O^+(aq) + SO_4^{2-}(aq)$$
$$(K_{a_2} = 1.3 \times 10^{-2})$$

In aqueous phosphoric acid

$$H_3PO_4(aq) + H_2O(l) \rightleftharpoons H_3O^+(aq) + H_2PO_4^-(aq)$$
$$(K_{a_1} = 7.5 \times 10^{-3})$$

$$H_2PO_4^-(aq) + H_2O(l) \rightleftharpoons H_3O^+(aq) + HPO_4^{2-}(aq)$$
$$(K_{a_2} = 6.2 \times 10^{-8})$$

$$HPO_4^{2-}(aq) + H_2O(l) \rightleftharpoons H_3O^+(aq) + PO_4^{3-}(aq)$$
$$(K_{a_3} = 2.0 \times 10^{-13})$$

Note that subscript numbers are used to distinguish between successive ionization constants.

The successive ionization constants for a polyprotic acid are approximately in the ratios, as illustrated by phosphoric acid,

$$K_{a_1} : K_{a_2} : K_{a_3} = 1 : 10^{-5} : 10^{-10}$$

In the equilibrium system referred to as aqueous phosphoric acid, then, there are various species (H_3PO_4, $H_2PO_4^-$, HPO_4^{2-}, PO_4^{3-}, H^+) whose relative concentrations depend on the magnitudes of different ionization constants. Although we may write several equations for this system, it is important to realize that the total equilibrium mixture contains specific characteristic concentrations of each species. We shall consider these systems in more detail in Unit 21.

19.4 pH, pOH, AND pK NOTATIONS

The relationship between cell voltage (E) and concentration quotient (Q) is given by the Nerst equation (Unit 24),

$$E = E^0 - \frac{0.0592}{n} \log Q$$

pX notation has the added convenience of avoiding very small numbers, for example, pH = 9.7 rather than $C_{H^+} = 2.0 \times 10^{-10}$ M.

As we shall see later (Unit 24), electrochemical cells may be used to measure ion concentrations. These concentrations are related to cell voltage by a logarithmic factor. Measurements of this type are commonly made on acidic or alkaline (basic) solutions by an instrument called a "pH meter" (Figure 19.1). Because of the logarithmic relationship between concentration and cell voltage, it is convenient to use a "pX" notation:

$$pX = -\log X \tag{19.1}$$

When dealing with H^+ concentration, we use pH ($-\log C_{H^+}$). Similarly, hydroxide concentrations or equilibrium constants may be expressed as pOH ($-\log C_{OH^-}$) or pK ($-\log K$), respectively. As was discussed in Unit 17, the ion product for water at room temperature is given by

$$K_w = 1.0 \times 10^{-14} = C_{H^+} \times C_{OH^-}$$

In pure water, $C_{H^+} = C_{OH^-}$ so that $C_{H^+} = \sqrt{1.0 \times 10^{-14}} = 10^{-7}$ M. Since the concentration of H^+ in pure water is 10^{-7} M, we define a neutral aqueous solution as one of pH 7. Since negative logarithms are used, the pH decreases as the H^+ concentration increases (Table 19.8).

Table 19.8 ACIDITY SCALE AND pH

SPECIES	pH	$C_{H^+}(M)$	$C_{OH^-}(M)$	pOH
acidic	−1	10	10^{-15}	15
	0	1	10^{-14}	14
	1	10^{-1}	10^{-13}	13
	2	10^{-2}	10^{-12}	12
	3	10^{-3}	10^{-11}	11
	4	10^{-4}	10^{-10}	10
	5	10^{-5}	10^{-9}	9
	6	10^{-6}	10^{-8}	8
neutral	7	10^{-7}	10^{-7}	7
alkaline	8	10^{-8}	10^{-6}	6
	9	10^{-9}	10^{-5}	5
	10	10^{-10}	10^{-4}	4
	11	10^{-11}	10^{-3}	3
	12	10^{-12}	10^{-2}	2
	13	10^{-13}	10^{-1}	1
	14	10^{-14}	1	0
	15	10^{-15}	10	−1

Since $C_{H^+} \times C_{OH^-} = 10^{-14}$ (K_w, Unit 17), pH + pOH = 14

Figure 19.1 A pH meter.

Since strong acids and salts of hydroxide ion are essentially 100 percent ionized in water, the pH of such solutions can be determined directly from molar concentrations.

It is worth noting that the pH and pOH of an aqueous solution are related by the ion product for water (Unit 17). Since $C_{H^+} \times C_{OH^-} = 10^{-14}$, appropriate use of logarithms permits us to write

$$pH = 14.00 - pOH \qquad (19.2)$$

Example 4

Calculate the pH of:
  a. 0.10 M $HClO_4$
  b. 0.020 M HClO4
  c. 0.050 M NaOH
  d. 0.0020 M $Ba(OH)_2$

Solution
  a. $HClO_4$ is a strong acid, so $C_{H^+} = 0.10$ M.

$$pH = -\log(0.10) = -\log(1.0 \times 10^{-1})$$
$$= 1.00$$

  b. $C_{H^+} = 0.020$ $M = 2.0 \times 10^{-2}$ M.
$$pH = -\log(2.0 \times 10^{-2}) = -(\log 2.0 + \log 10^{-2})$$
$$= -0.30 - (-2) = 1.70$$

  c. The salt NaOH (containing the base OH^-) is completely ionized in water.

$$NaOH \longrightarrow Na^+ + OH^-$$

So

$$C_{OH^-} = 0.050\ M = 5.0 \times 10^{-2}\ M.$$
$$pOH = -\log(5.0 \times 10^{-2}) = -(\log 5.0 + \log 10^{-2})$$
$$= -0.70 - (-2) = 1.30$$
$$pH = 14.00 - pOH = 14.00 - 1.30 = 12.70$$

d. For $Ba(OH)_2$,

$$Ba(OH)_2 \longrightarrow Ba^{2+} + 2OH^-$$

So

$$C_{OH^-} = 2 \times 0.0020 = 4.0 \times 10^{-3} \; M.$$
$$pOH = -\log(4.0 \times 10^{-3}) = -(\log 4.0 + \log 10^{-3})$$
$$= -0.60 - (-3) = 2.40$$
$$pH = 14.00 - pOH = 14.00 - 2.40 = 11.60$$

When the C_{H^+} (or C_{OH^-}) from a solute approaches $10^{-7}\,M$, the contribution of water ionization becomes significant.

For weak acids and bases, the pH must be determined from considerations of the equilibria involved (Unit 17). These considerations may involve the detailed analysis of mass balance and electroneutrality (Unit 21), including allowance for the ionization of water.

19.5 EQUILIBRIA INVOLVING WEAK ACIDS AND BASES

For *monoprotic* acids or bases in aqueous solutions, we must consider the total equilibrium system (Table 19.9).

A rigorous treatment of acid–base equilibria requires a careful assessment of all equilibrium species and their interrelationships. In a dilute solution of acetic acid, for example, we should note that both the acid and the water contribute to the total hydrogen ion concentration. The equilibrium concentration of undissociated acetic acid molecules in solution will be less than the "label" concentration by the amount ionized to acetate and hydrogen ions. In many cases, detailed analysis of the equilibrium system is quite complex, as illustrated in Unit 21.

The "label" concentration is defined as the concentration the solute would have if it underwent no change when dissolved in water, that is, the concentration typically printed on the reagent bottle label. Some texts express this concentration by *formality* (Unit 13).

Fortunately, most of the solutions of interest to us can be treated by some simplifying approximations with satisfactory accuracy. It is, however, important that we recognize the limitations of these approximations.

In practice, as we have suggested in Unit 17, it is best to make an initial calculation based on the simplifying assumptions. After checking on the validity of these assumptions, a more rigorous calculation can be made, if necessary.

If K_a (or K_b) is in the range $\sim 10^{-4}$ to $\sim 10^{-9}$ and the label concentration of the solution is in the range from $\sim 1\,M$ to $\sim 10^{-3}\,M$, then no serious error is made if

1. the *equilibrium* concentration of the "parent" acid or base is approximated by its *label* concentration, and
2. the contribution of water ionization to the total H^+ concentration of the acid solution (or to the OH^- concentration of a base solution) is neglected.

Table 19.9   EQUILIBRIA IN AQUEOUS SOLUTIONS OF SIMPLE WEAK ACIDS AND BASES

ACIDS

$$HX(aq) + H_2O(l) \rightleftharpoons H_3O^+(aq) + X^-(aq)$$
$$2H_2O(l) \rightleftharpoons H_3O^+(aq) + OH^-(aq)$$

or, in simpler terms,

$$HX(aq) \rightleftharpoons H^+(aq) + X^-(aq)$$
$$H_2O(l) \rightleftharpoons H^+(aq) + OH^-(aq)$$

BASES

Molecular
$$B(aq) + H_2O(l) \rightleftharpoons BH^+(aq) + OH^-(aq)$$
$$H_2O(l) \rightleftharpoons H^+(aq) + OH^-(aq)$$

Anionic (-1)

$$X^-(aq) + H_2O(l) \rightleftharpoons HX(aq) + OH^-(aq)$$
$$H_2O(l) \rightleftharpoons H^+(aq) + OH^-(aq)$$

Example 5

What is the pH of a solution labeled 0.020 M hypochlorous acid?

Solution

From Table 19.6 (or Appendix C), K_a for HOCl is 3.2×10^{-8}, so we may neglect the ionization of water as a source of H^+ in the ~0.02 M solution and represent the equilibrium of principal importance by

$$HOCl(aq) \rightleftharpoons H^+(aq) + OCl^-(aq)$$

This equation tells us that H^+ and OCl^- ions are produced in equal numbers, so

$$C_{OCl^-} = C_{H^+}$$

and, under our simplifying approximations, the equilibrium concentration of molecular HOCl may be taken as 0.020 M (the label concentration).

The equilibrium constant expression is

$$K_a = 3.2 \times 10^{-8} = \frac{C_{H^+} \times C_{OCl^-}}{C_{HOCl}}$$

Substituting 0.020 for C_{HOCl} and C_{H^+} for its equality, C_{OCl^-}, gives

$$3.2 \times 10^{-8} = \frac{C_{H^+} \times C_{H^+}}{0.020}$$

from which

$$C_{H^+} = \sqrt{0.020 \times 3.2 \times 10^{-8}} = 2.5 \times 10^{-5} \, M$$

Then

$$pH = -\log C_{H^+} = -\log(2.5 \times 10^{-5})$$
$$= 4.60$$

Were our approximations justified? At equilibrium,

$$C_{HOCl} = C_{HOCl} \qquad\qquad - \quad C_{H^+}$$

(equil) (label) (equal to HOCl loss by ionization)

$$0.020 = \qquad\qquad \underbrace{0.020 - 0.000025}$$

(our approximation) (result of our calculation)

and

$$C_{H^+} = \quad C_{H^+} \quad + \quad C_{H^+}$$

(equil) (from HOCl) (from H_2O)

$$2.5 \times 10^{-5} = \underbrace{2.5 \times 10^{-5} + \sim 10^{-7}}$$

(result of our approximation)

Obviously, the simplifications used introduced no significant error.

Any weak acid or weak base may be treated in this way, as long as the equilibrium constant and label concentration are within the limits described. No "formula" approach is needed if the principles of equilibrium are applied, together with appropriate approximations. When repeated calculations of this type are necessary, however, it is convenient to have a simple equation.

$$K_a = \frac{C_{H^+} \times C_{X^-}}{C_{HX}} = \frac{C_{H^+}^2}{C_{HX}}$$

$$C_{H^+} = \sqrt{K_a C_{HX}}$$

So that

$$pH = -\log \sqrt{(K_a C_{HX})} = -\tfrac{1}{2}\log(K_a C_{HX})$$

Then, for cases within our limits for approximation,

$$pH = -\tfrac{1}{2}\log(K_a C_{acid}) \tag{19.3}$$

A similar treatment for weak bases (Table 19.6), remembering that $pH = 14.00 - pOH$ [equation (19.2)] gives

$$pH = 14.00 + \tfrac{1}{2}\log(K_b C_{base}) \tag{19.4}$$

In equations (19.3) and (19.4), C_{acid} and C_{base} refer to label concentrations.

Example 6

What is the pH of a solution labeled 0.50 M ammonia?

Solution

From Table 19.6 (or Appendix C), K_b for NH_3 is 1.8×10^{-5}. Using Equation (19.4),

$$pH = 14.00 + \tfrac{1}{2}\log(1.8 \times 10^{-5} \times 0.50)$$
$$= 14.00 + \tfrac{1}{2}\log(9.0 \times 10^{-6})$$
$$= 14.00 + \tfrac{1}{2}(0.95 - 6) = 14.00 + \tfrac{1}{2}(-5.05)$$
$$= 11.48$$

The use of "formulas" should be minimized except for repetitive situations or when the quantitative application of basic principles is relatively complex. Simple equilibria can easily be handled by

1. describing the system by an appropriate chemical equation,
2. identifying the equilibrium concentrations of all species in terms of the single component of interest, and
3. writing and using the proper equilibrium constant expression.

Alternative Solution to Example 6

a. For 0.50 M NH_3 ($K_b = 1.8 \times 10^{-5}$), we may safely ignore the ionization of water and write

$$NH_3(aq) + H_2O(l) \rightleftharpoons NH_4^+(aq) + OH^-(aq)$$

b. Satisfactory approximations suggest that

$$C_{NH_4^+} = C_{OH^-}$$

and

$$C_{NH_3} = \sim 0.50 \ M$$

(equil)

c. Then, from

$$K_b = 1.8 \times 10^{-5} = \frac{C_{NH_4^+} \times C_{OH^-}}{C_{NH_3}} = \frac{C_{OH^-} \times C_{OH^-}}{C_{NH_3}}$$

we may write

$$1.8 \times 10^{-5} = \frac{C_{OH^-}^2}{0.50}$$

$$C_{OH^-} = \sqrt{1.8 \times 10^{-5} \times 0.50} = 3.0 \times 10^{-3}\ M$$

$$pOH = -\log(3.0 \times 10^{-3}) = 2.52$$

and

$$pH = 14.00 - 2.52 = 11.48$$

To illustrate that simplifying approximations *are* limited, consider what would happen if 0.50 M ammonia were diluted, by a series of successive dilutions, until the label concentration was $5.0 \times 10^{-12}\ M$. Application of Equation (19.4) would then give

$$pH = 14 + \tfrac{1}{2}\log(1.8 \times 10^{-5} \times 5 \times 10^{-12})$$
$$pH = \sim 6$$

But a solution of pH 6 is *acidic* (Table 19.8), and reason tells us that no solution containing neutral water and a base should be acidic. In this case OH⁻ from ionization of water is significant.

Other situations not amenable to simplified calculation are discussed in Unit 21.

19.6 HYDROLYSIS Strictly speaking, any reaction in which water is a reactant may be classified as *hydrolysis*. Historically, this term has been associated with the reaction of a salt of a weak acid or a salt of a weak base with water; for example, when sodium acetate is dissolved in water,

$$CH_3CO_2^- + H_2O \rightleftharpoons CH_3CO_2H + OH^-$$

or when ammonium chloride is dissolved in water,

$$NH_4^+ + H_2O \rightleftharpoons NH_3 + H_3O^+$$

Certain cations also react with water to form acidic hydrates. For these, the hydrolysis reactions may be formulated, for example, as

Complex ions, such as $[Al(H_2O)_5(OH)]^{2+}$ are discussed briefly in Unit 21 and in detail in Unit 27.

$$Al^{3+} + 6H_2O \rightleftharpoons [Al(H_2O)_5(OH)]^{2+} + H^+$$

or

$$[Al(H_2O)_6]^{3+} + H_2O \rightleftharpoons H_3O^+ + [Al(H_2O)_5(OH)]^{2+}$$

Within the Brønsted–Lowry system, no special treatment of hydrolysis is necessary, since any species identifiable as a proton donor or proton acceptor can be treated appropriately. A hydrolysis constant (K_h), then, is just an alternative designation for the more familiar acid or base constant (K_a or K_b). If such a constant is not readily available for a reaction of interest, it may be calculated easily if the constant for the conjugate acid or base is known.

Consider, for example, the situation in an aqueous solution of sodium hypochlorite. This salt is completely ionized in water, and the sodium ion (like most ions of group IA or IIA metals) undergoes no significant reaction with water. The hypochlorite ion, however, is formed from the weak acid HOCl ($K_a = 3.2 \times 10^{-8}$, Table 19.6), so in aqueous solution,

$$OCl^- + H_2O \rightleftharpoons HOCl + OH^-$$

Note that hypochlorite ion, in accepting a proton from water, acts as a *base*.

Suppose we wish to know the hydrolysis constant (i.e., $K_{h(b)}$) for this reaction:

$$K_{h(b)} = \frac{C_{HOCl} \times C_{OH^-}}{C_{OCl^-}}$$

The $K_{h(b)}$ value can be determined from known constants for two other equilibria:

$$HOCl \rightleftharpoons H^+ + OCl^- \qquad K_a = \frac{C_{H^+} \times C_{OCl^-}}{C_{HOCl}}$$

and

$$H_2O \rightleftharpoons H^+ + OH^- \quad K_w = C_{H^+} \times C_{OH^-}$$

The equilibrium expression for hydrolysis of hypochlorite ion is obtained by dividing the ion product for water by the ionization expression for hypochlorous acid:

$$\frac{C_{H^+} \times C_{OH^-}}{(C_{H^+} \times C_{OCl^-})/C_{HOCl}} = \frac{C_{H^+} \times C_{OH^-}}{1} \times \frac{C_{HOCl}}{C_{H^+} \times C_{OCl^-}}$$

$$= \frac{C_{HOCl} \times C_{OH^-}}{C_{OCl^-}}$$

It follows, then, that the equilibrium constant for hydrolysis of hypochlorite ion must equal K_w/K_a:

The value of K_w at 20°–25°C (1.0×10^{-14}) should be remembered.

$$K_{h(b)} \text{ (for } OCl^-) = \frac{K_w}{K_{a(HOCl)}} = \frac{1.0 \times 10^{-14}}{3.2 \times 10^{-8}} = 3.1 \times 10^{-7}$$

This is a general approach, and it can be shown that for any conjugate-pair system in water:

$$K_{h(a)} = \frac{K_w}{K_{b(conj)}} \tag{19.5}$$

$$K_{h(b)} = \frac{K_w}{K_{a(conj)}} \tag{19.6}$$

In deciding how to write the chemical equation for an equilibrium system involving hydrolysis, we need to recognize that *the conjugate base of a weak acid will tend to recapture a proton from water*. Then, to express a hydrolysis equilibrium involving the salt of a weak acid, we should write water and the anion from the weak acid as "left-hand" species, with the conjugate weak acid and hydroxide (resulting from proton loss by water) as "right-hand" species. A similar approach is used for hydrolysis of the salt of a weak base, although water may be omitted from the equation as a simplification, for example, for an ammonium salt,

$$NH_4^+ + H_2O \rightleftharpoons NH_3 + H_3O^+$$

or (simplified)

$$NH_4^+ \rightleftharpoons NH_3 + H^+$$

An important exception is Be^{2+} (Unit 25), which does react with water. To a slight extent Mg^{2+} reacts with water, but this reaction is not significant in the cases we will consider.

We should remember that cations of Groups IA and IIA do not undergo any significant hydrolysis, nor do anions from strong acids (Table 19.6).

Example 7

For each of the following salts, write the chemical equations descriptive of the hydrolysis equilibria involved and calculate the hydrolysis constants from K_a and K_b values given in Appendix C:

a. potassium cyanide
b. sodium sulfate
c. ammonium nitrate

Solution

a. For potassium cyanide, only CN^- (from the weak acid HCN) undergoes hydrolysis:

$$CN^-(aq) + H_2O(l) \rightleftharpoons HCN(aq) + OH^-(aq)$$

for which, from equation (19.6),

$$K_{h(b)} = \frac{C_{HCN} \times C_{OH^-}}{C_{CN^-}} = \frac{K_w}{K_{a(HCN)}}$$
(for CN^-)

From Appendix C, K_a for HCN = 4.8×10^{-10}, so

$$K_{h(b)} = \frac{1.0 \times 10^{-14}}{4.8 \times 10^{-10}} = 0.21 \times 10^{-4} = 2.1 \times 10^{-5}$$
(for CN^-)

b. For Na_2SO_4, only SO_4^{2-} (from the weak acid HSO_4^-) undergoes hydrolysis:

$$SO_4^{2-}(aq) + H_2O(l) \rightleftharpoons HSO_4^-(aq) + OH^-(aq)$$

(Note that bisulfate ion, from the *strong* acid H_2SO_4, does not undergo hydrolysis.)
Then

$$K_{h(b)} = \frac{C_{HSO_4^-} \times C_{OH^-}}{C_{SO_4^{2-}}} = \frac{K_w}{K_{a(HSO_4^-)}}$$
(for SO_4^{2-})

From Appendix C, K_a for HSO_4^- = 1.3×10^{-2}, so

$$K_{h(b)} = \frac{1.0 \times 10^{-14}}{1.3 \times 10^{-2}} = 0.77 \times 10^{-12} = 7.7 \times 10^{-13}$$
(for SO_4^{2-})

c. For ammonium nitrate, only NH_4^+ undergoes significant hydrolysis (since NO_3^- is the conjugate base of the *strong* acid HNO_3):

$$NH_4^+(aq) + H_2O(l) \rightleftharpoons NH_3(aq) + H_3O^+(aq)$$

or

$$NH_4^+(aq) \rightleftharpoons NH_3(aq) + H^+(aq)$$

then

$$K_{h(a)} = \frac{C_{NH_3} \times C_{H^+}}{C_{NH_4^+}} = \frac{K_w}{K_{b(NH_3)}}$$
(for NH_4^+)

From Appendix C, K_b for $NH_3 = 1.8 \times 10^{-5}$, so

$$K_{h(a)} = \frac{1.0 \times 10^{-14}}{1.8 \times 10^{-5}} = 0.56 \times 10^{-9} = 5.6 \times 10^{-10}$$
$$\text{(for } NH_4^+\text{)}$$

Calculations involving hydrolysis equilibria are made in exactly the same way as those for any aqueous solution of a weak acid or weak base. The only added step is the determination of the hydrolysis constant from the tabulated K_a or K_b value for the appropriate conjugate acid or base.

Example 8

What is the approximate pH of a solution labeled 0.50 M sodium fluoride?

Solution

Since only the F^- ion (from the weak acid HF) should affect the pH,

$$F^-(aq) + H_2O(l) \rightleftharpoons HF(aq) + OH^-(aq)$$

for which

$$K_{h(b)} = \frac{K_w}{K_{a(HF)}} = \frac{C_{HF} \times C_{OH^-}}{C_{F^-}} = \frac{1.0 \times 10^{-14}}{6.9 \times 10^{-4}} = 1.4 \times 10^{-11}$$
$$\text{(for } F^-\text{)} \qquad\qquad\qquad\qquad \text{(from Appendix C)}$$

On the basis of our "simplifying approximations"

$$C_{HF} \simeq C_{OH^-}$$
$$C_{F^-} \simeq 0.50 \text{ M}$$

then

$$1.4 \times 10^{-11} = \frac{C^2_{OH^-}}{0.50}$$

from which

$$C_{OH^-} = \sqrt{0.50 \times 1.4 \times 10^{-11}} = 2.6 \times 10^{-6} \text{ M}$$

As a test on our approximations,

$$C_{F^-} = C_{F^-} - C_{OH^-}$$
$$\text{(equil)} \quad \text{(label)(equal to loss of } F^-\text{)}$$

$$0.50 = 0.50 - 2.6 \times 10^{-6} \qquad \text{(close enough)}$$

and

$$C_{OH^-} = C_{OH^-} + C_{OH^-}$$
$$\text{(equil)} \quad \text{(calculated)} \quad \text{(from water)}$$

$$2.6 \times 10^{-6} = 2.6 \times 10^{-6} + {\sim}10^{-7}$$

(within acceptable limits, Unit 17).

then

$$pOH = -\log(2.6 \times 10^{-6}) = 5.58$$

and

$$pH = 14.00 - pOH = 14.00 - 5.58 = 8.42$$

Table 19.10 ACID-BASE REACTIONS IN WATER (MONOPROTIC ACIDS AND BASES)

ACID REACTANT	BASE REACTANT	PRODUCTS	pH AT EQUIVALENCE POINT
strong (H_3O^+) [e.g., HCl(aq)]	strong (OH^-) [e.g., NaOH(aq)]	water + "neutral" ions (e.g., Na^+, Cl^-)	~ 7
strong (H_3O^+) [e.g., HCl(aq)]	weak (e.g., NH_3 or $CH_3CO_2^-$)	weak acid (e.g., NH_4^+ or CH_3CO_2H) + "neutral" ions	< 7
weak (e.g., CH_3CO_2H or NH_4^+)	strong (e.g., NaOH)	water + "neutral" ions + weak base (e.g., NH_3 or $CH_3CO_2^-$)	> 7
weak (e.g., CH_3CO_2H or NH_4^+)	weak (e.g., NH_3 or $CH_3CO_2^-$)	conjugate weak base + conjugate weak acid	pH < 7 if $K_a > K_b$ pH $= 7$ if $K_a = K_b$ pH > 7 if $K_a < K_b$ (requires detailed analysis)

19.7 ACID-BASE REACTIONS

That is, the quantity of base is exactly sufficient to accept all protons available from the quantity of acid present, assuming complete proton transfer. When these quantities have reacted, the process is said to have reached the *equivalence point*.

Indicators are typically used with *titrations* (Unit 13) to establish the equivalence point of the reaction.

If the solution is not boiled to expel CO_2, the pH will be < 7, because of the equilibrium system

$$CO_2(aq) + H_2O(l) \rightleftharpoons H^+(aq) + HCO_3(aq)$$

The generalization that "an acid reacts with a base to form a salt and water" is a holdover from the early work of Arrhenius. It is perhaps more useful now to consider what happens when equivalent amounts of acids and bases react in terms of the relative "strengths" of the reactants. The several possibilities are summarized in Table 19.10.

A knowledge of the pH at the equivalence point is necessary if we are to be able to determine when stoichiometric combination has occurred. Reactions of this type are common in many analytical situations, and colored compounds called "pH indicators" are frequently employed to show when reaction is complete. These indicators are themselves weak acids or weak bases. They are typically organic molecules that change color (Figure 19.2) when the pH changes by an appropriate amount.

When a strong acid is titrated with a strong base, the pH at the equivalence point will be 7.0. However, the equivalence point will not be at pH 7.0 if either the acid or the base (or both, if $K_a \neq K_b$) is weak. A knowledge of acid–base equilibria is essential if an indicator is to be properly selected for a titration so that the endpoint (where the indicator changes color, Table 19.11) is to correspond to the pH of the equivalence point (see Table 19.10). An indicator is typically selected such that the midpoint of the indicator pH range is approximately the same as the pH at the equivalence point. We can estimate the pH at the equivalence point of a titration involving a weak acid and a strong base (or a weak base and a strong acid) by

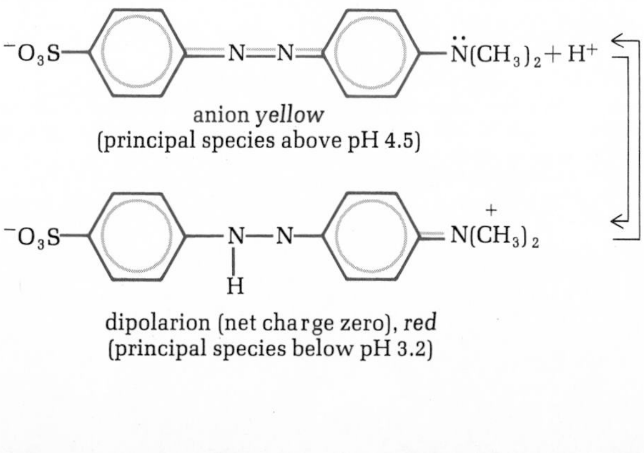

anion *yellow*
(principal species above pH 4.5)

dipolarion (net charge zero), *red*
(principal species below pH 3.2)

Figure 19.2 Methyl orange, a pH indicator. (For explanation of dotted lines and hexagon representations, see Unit 8.)

Table 19.11 SOME COMMON pH INDICATORS

INDICATOR	COLOR AND pH
thymol blue (acidic)	red below pH 1.2 orange pH 1.2–2.8 yellow above pH 2.8
methyl orange	red below pH 3.1 orange pH 3.1–4.4 yellow-orange above pH 4.4
methyl red	red below pH 4.4 orange pH 4.4–6.2 yellow above pH 6.2
litmus	red below pH 4.5 purple pH 4.5–8.3 blue above pH 8.3
bromthymol blue	yellow below pH 6.0 green pH 6.0–7.6 blue above pH 7.6
phenol red	yellow below pH 6.4 orange pH 6.4–8.2 red above pH 8.2
thymol blue (basic)	yellow pH 2.8–8.0 green pH 8.0–9.6 blue above pH 9.6
phenolphthalein	colorless below pH 8.3 light pink pH 8.3–10.0 red above pH 10.0
alizarin yellow	yellow below pH 10.0 tan pH 10.0–11.0 violet above pH 11.0

potassium salt of
  benzaldehyde
  p-nitrophenylhydrazone[a]
	yellow below pH 11.0
	tan pH 11.0–12.0
	purple above pH 12.0

trinitrobenzene	colorless below pH 12.0 yellow pH 12.0–14.0 orange above pH 14.0

[a] See O'Connor, R., W. Rosenbrook, and G. Anderson, "A New Indicator for pH 11–12," *Anal. Chem.*, **33**, 1282 (1961)

recognizing that the equivalence point corresponds to a solution of a salt involved in a hydrolysis equilibrium (Section 19.6).

Example 9

a. What is the pH at the equivalence point for the titration of 0.30 M acetic acid with 0.15 M sodium hydroxide?

b. Which indicator from Table 19.11 would be most appropriate for this titration?

Solution

a. We may represent the reaction during the course of the titration by

$$CH_3CO_2H(aq) + NaOH(aq) \longrightarrow CH_3CO_2Na(aq) + H_2O(l)$$

or, in *net* form (Unit 3) as

$$CH_3CO_2H(aq) + OH^-(aq) \longrightarrow CH_3CO_2^-(aq) + H_2O(l)$$

At the equivalence point, equimolar amounts of acetic acid and hydroxide have reacted. The resulting solution of acetate ion is represented by the hydrolysis equilibrium,

$$CH_3CO_2^-(aq) + H_2O(l) \rightleftharpoons CH_3CO_2H(aq) + OH^-(aq)$$

for which

$$\underset{\text{(for } CH_3CO_2^-)}{K_{h(b)}} = \frac{C_{CH_3CO_2H} \times C_{OH^-}}{C_{CH_3CO_2^-}} = \frac{K_w}{K_{a(CH_3CO_2H)}} = \frac{1.0 \times 10^{-14}}{1.8 \times 10^{-5}} = 5.6 \times 10^{-10}$$

(from Appendix C)

The concentration of acetate ion in the equilibrium system should be closely approximated by the number of moles of sodium acetate formed divided by the *total volume* of the final reaction mixture. Since all reactants and products appear in equimolar amounts, in this particular reaction

no. moles $CH_3CO_2^-$ formed = no. moles CH_3CO_2H consumed
= no. moles OH^- consumed

and, from the definition of *molarity* (Unit 13), letting V_{acid} represent the volume of acetic acid used

no. moles $CH_3CO_2^- = M_{CH_3CO_2H} \times V_{acid} = (0.30)V_{acid}$

and

An alternative approach, that you may find simpler, is to arbitrarily select some volume of one reactant (e.g., 25 ml) and calculate the other reactant volume from

$$M_{acid} \times V_{acid} = M_{base} \times V_{base}$$

This approach, limited of course to equimolar reactant ratios, avoids the algebraic work with an "unknown" volume, V.

$$V_{base} = \frac{M_{CH_3CO_2H} \times V_{acid}}{M_{OH^-}} = \frac{(0.30)V_{acid}}{(0.15)} = 2V_{acid}$$

The total volume of the final reaction mixture is

$$V_{acid} + V_{base} = V_{acid} + 2V_{acid} = 3V_{acid}$$

and the molar concentration of acetate ion is then

$$M_{CH_3CO_2^-} = \frac{\text{no. moles } CH_3CO_2^-}{\text{no. liters solution}} = \frac{(0.30)V_{acid}}{3V_{acid}} = 0.10 \text{ M}$$

We can now estimate the pH at the equivalence point from considerations of the hydrolysis equilibrium (using our "simplifying approximations"):

$$5.6 \times 10^{-10} = \frac{C_{CH_3CO_2H} \times C_{OH^-}}{C_{CH_3CO_2^-}} = \frac{C_{OH^-}^2}{0.10}$$

$$C_{OH^-} = \sqrt{0.10 \times 5.6 \times 10^{-10}} = 7.5 \times 10^{-6} \text{ M}$$

$$pOH = -\log(7.5 \times 10^{-6}) = 5.12$$

$$pH = 14.00 - 5.12 = 8.88$$

b. From Table 19.11, thymol blue (basic) would appear to be the most appropriate indicator,[2] since the midpoint of its pH range is ~8.8. A slight excess of base should produce a readily observable color change (from green to blue).

---

[2] In most simple laboratory titrations of acetic acid with sodium hydroxide, phenolphthalein is used as the indicator. Although the midpoint of its pH range is slightly above the pH of the equivalence point, the change from colorless to pink is easier to see, particularly for the relatively inexperienced investigator.

Typical pH indicators change color over a range of one or more full pH units. In the titrations of very weak acids or bases, the pH change near the equivalence point[3] is so gradual that a large volume of titrant may be required to cause a pH change of one unit. This will, of course, seriously affect the accuracy of the titration (which should be within ±0.1 ml of titrant). To check on this possibility, it is necessary to calculate the pH changes caused by addition of 0.1-ml aliquots of titrant to the solution at the equivalence point. If this calculation shows that an indicator will not permit sufficient accuracy, then the titration must use a pH meter so that smaller pH changes can be observed. Titration curves, graphs of pH versus titrant volume, are particularly useful in noting the "sharpness" of the pH change at the equivalence point and the pH range for indicator selection.

The concept of equivalence is applied to the definition of *normality* for acids and bases (Unit 13):

$$\left(\begin{matrix}\text{equivalent}\\\text{weight}\end{matrix}\right) = \frac{\text{formula weight}}{\text{no. of protons exchanged per formula unit}}$$

$$N = \frac{\text{no. gram-equivalent weights}}{\text{no. liters solution}}$$

Normality is particularly useful for calculations pertaining to titration data for aqueous acids and bases, since at the equivalence point we may write

$$N_{acid} \times V_{acid} = N_{base} \times V_{base} \qquad (19.7)$$
$$\text{no. equivalents acid} = \text{no. equivalents base}$$

Although many textbooks, particularly those in analytical chemistry, make extensive use of *normality* as a concentration unit, we shall continue to work primarily with *molarities* for solution stoichiometry.

---

[3] For a detailed discussion of the problem of pH change near the equivalence point, see "Acid-Base Reaction Parameters," by Henry Freiser, in the December, 1970, issue of the *Journal of Chemical Education* (pp. 809–811).

THE CHANGING pH OF RAIN

Within recent years scientists have noted significant variation in the pH of rainwater and environmentalists are concerned that this may be an indication of serious long-range problems.

Normally, rainwater saturated with carbon dioxide should have a pH of about 5.7. In the Scandinavian countries, the acidity of rainwater has been studied carefully since around 1900. Acidity increases have been particularly noted within the last 20 years. Although rainfall pH there now averages around 4.7, values as low as 2.8 have been recorded. Investigations suggest the increased acidity to be the result of heavy atmospheric pollutants in air masses moving from England and central Europe, among which are significant concentrations of nitrogen and sulfur oxides (which form acids by reaction with atmospheric moisture). The increasingly acidic rains have been linked directly to damages to aquatic ecosystems, including fish depopulation.

The problem is not limited to Scandinavia. In Hubbard Brook, New Hampshire, rain of pH 2.1 has been detected, and acid rains have been reported in such remote areas as the Amazon Valley, western Alaska, and portions of the South Pacific. Rainfall in the alkaline range has also been observed, with rainfall as alkaline as pH 8.2 in some areas of Michigan.

To what extent this pH variation poses a global threat to aquatic environments cannot yet be assessed, nor is there yet sufficient evidence to attribute the problem entirely to man-made air pollution. That a potential threat exists and that more extensive investigations are indicated were evidence by the convening of the first International Symposium on Acid Precipitation in the spring of 1975.[4]

[4] *Chemical and Engineering News*, June 9, 1975, pp. 19–20.

SUGGESTED READING

Drago, R. S., and N. A. Matwiyoff. 1968. *Acids and Bases.* Lexington, Mass.: Raytheon.

O'Connor, Rod, and Charles Mickey. 1975. *Solving Problems in Chemistry.* New York: Harper & Row.

van der Werf, C. A. 1961. *Acids, Bases, and the Chemistry of the Covalent Bond.* New York: Van Nostrand Reinhold.

EXERCISES

1. a. Give an example of a chemical compound that Arrhenius would have defined as a base, and the Lewis system would define as containing both an acid and a base.

b. Write a net equation for the equilibrium system resulting from dissolving sodium fluoride in water. Indicate the conjugate pairs, and in each pair label the acid and the base.

2. Calculate the pH of
a. 0.0010 M hydrochloric acid
b. 0.050 M barium hydroxide
c. 0.10 M acetic acid ($K_a = 1.8 \times 10^{-5}$)
d. 0.050 M ammonia ($K_b = 1.8 \times 10^{-5}$)

3. Calculate the pH of:
a. 0.20 M sodium benzoate (K_a for $C_6H_5CO_2H$ is 6.6×10^{-5})
b. 0.20 M methylammonium chloride (K_b for CH_3NH_2 is 5.0×10^{-4})
c. The solution resulting from mixing equal volumes of 0.10 M NaOH and 0.10 M CH_3CO_2H.

4. a. Estimate the relative strengths of each of the following acids:

boric acid, H_3BO_3 [$B(OH)_3$]
chloric acid, $HClO_3$ [$ClO_2(OH)$]
sulfurous acid, H_2SO_3[$SO(OH)_2$]

b. The first ionization constant for arsenic acid, H_3AsO_4, is $\sim 10^{-2}$. Estimate K_{a_2} and K_{a_3} for this acid.

5. Estimate the pH at the equivalence point for the titration of 0.25 M propionic acid ($CH_3CH_2CO_2H$, $K_a = 1.4 \times 10^{-5}$) with 0.050 M sodium hydroxide. Suggest an appropriate indicator from Table 19.11 for this titration.

PRACTICE FOR PROFICIENCY (Consult Appendix C, as necessary)

Objectives (a, b)

1. Arrhenius would have classified aluminum hydroxide as a base. How would G. N. Lewis have classified aluminum hydroxide in terms of the reaction,

$$Al(OH)_3(s) + OH^-(aq) \rightleftharpoons [Al(OH)_4]^-$$

Would the Brønsted–Lowry system have applied the same classification as Lewis to this reaction?

2. Write a net equation for the equilibrium system resulting from dissolving sodium acetate in water. Indicate the conjugate pairs and label the acid and base in each pair.

3. Which of the following would *not* have been classified as a base by Arrhenius?

   a. CH_3OH; **b.** KOH; **c.** LiOH; **d.** $Mg(OH)_2$; **e.** $Cr(OH)_3$

4. In the Brønsted–Lowry system a base is defined as

   a. an electron-pair acceptor; **b.** a hydroxide donor; **c.** a proton donor; **d.** a proton acceptor; **e.** none of these

5. In the Lewis system, a base is defined as

   a. an electron-pair acceptor; **b.** an electron-pair donor; **c.** a proton donor; **d.** a proton acceptor; **e.** a hydroxide donor

6. $CN^- + H_2O \rightleftharpoons HCN + OH^-$. In this system, the *conjugate pairs* are

   a. CN^-/OH^- and H_2O/HCN; **b.** CN^-/H_2O and HCN/OH^-; **c.** CN^-/HCN and H_2O/OH^-; **d.** only CN^- and OH^-; **e.** only H_2O and HCN

7. $HF + H_3O^+ \rightleftharpoons H_2F^+ + H_2O$. For the system formulated here, Brønsted would classify the *basic* species as

   a. H_3O^+ and H_2O; **b.** H_2F^+ and H_2O; **c.** HF and H_3O^+; **d.** HF and H_2O; **e.** H_3O^+ and H_2F^+

Objective (c)

8. Calculate the pH of

   a. 0.0020 M nitric acid

   b. 0.0020 M calcium hydroxide

   c. 0.20 M nitrous acid ($K_a = 4.5 \times 10^{-4}$)

   d. 0.20 M dimethylamine ($K_b = 5.1 \times 10^{-4}$)

9. Calculate the pH of

   a. 0.20 M sodium nitrite

   b. 0.20 M dimethylammonium chloride

10. Calculate the approximate pH of each of the solutions labeled as

   a. 8×10^{-4} M hydrobromic acid

   b. 0.0004 M sodium hydroxide

   c. 1.3×10^{-4} M nitric acid

   d. 3.0×10^{-4} M barium hydroxide

   e. 2.0×10^{-3} M sulfuric acid (Note: This is a bit "tricky.")

11. Calculate the approximate pH of each of the solutions labeled as

   a. 0.050 M ammonia

   b. 0.30 M propionic acid

   c. 0.30 M trimethylamine [$(CH_3)_3N$, $K_b = 5.3 \times 10^{-5}$]

   d. 0.20 M hypochlorous acid

   e. 0.50 M sodium bisulfate

12. Which of the following, in 0.2 M solution, has the highest pH?

   a. boric acid; **b.** formic acid; **c.** benzoic acid; **d.** propionic acid; **e.** nitric acid

13. Which of the following solutions has the lowest pH at 25°C?

   a. 0.5 M hydrofluoric acid; **b.** 0.4 M formic acid; **c.** 0.3 M hydrocyanic acid; **d.** 0.2 M boric acid; **e.** 0.1 M nitric acid

14. Calculate the approximate pH of each of the solutions labeled as

   a. 0.40 M ammonium bromide

   b. 0.20 M sodium formate

   c. 0.20 M ammonium nitrate

   d. 0.50 M potassium cyanide

   e. 0.10 M calcium hypochlorite

Objective (d)

15. Arrange the following in order of acid strength (strongest first):

   boric acid, $H_3BO_3[B(OH)_3]$

   chloric acid, $HClO_3[ClO_2(OH)]$

   nitrous acid, $NHO_2[NO(OH)]$

16. Which of these acids is probably the weakest?

   a. HIO_4; **b.** HIO_3; **c.** HIO_2; **d.** HIO; **e.** HI

17. Which of these acids is probably the strongest?

   a. HSO_4^-; **b.** H_2SO_4; **c.** H_2SO_3; **d.** HSO_3^-; **e.** H_2S

18. Which of these acids is probably the weakest?

   a. HCl; **b.** HCN; **c.** HNO_3; **d.** H_2SO_4; **e.** $HClO_4$

19. Which of these acids is probably the strongest?

   a. H_3AsO_4; **b.** H_3AsO_3; **c.** $H_2AsO_4^-$; **d.** $H_2AsO_3^-$; **e.** $HAsO_4^{2-}$

20. Use *formal charge* calculations to estimate the relative acid strengths of

   a. nitrous acid and nitric acid

   b. sulfuric acid and sulfurous acid

   c. phosphorous acid [$HPO(OH)_2$] and hypophosphorous acid [$H_2PO(OH)$]

21. The first ionization constant of phosphoric acid is 7.5×10^{-3}. Estimate K_{a_2} and K_{a_3} for phosphoric acid. How do your estimates compare with values tabulated in Appendix C?

Objective (e)

22. What is the pH at the equivalence point for reaction between 0.20 M sodium hydroxide and 0.20 M nitrous acid?

23. What is the approximate pH at the equivalence point for the titration of 0.050 M methylamine with 0.15 M hydrochloric acid?

24. What is the approximate pH at the equivalence point for the titration of 0.030 M hydrocyanic acid with 0.15 M sodium hydroxide?

25. Suggest an appropriate indicator from Table 19.11 for each of the titrations in questions 22, 23, and 24.

BRAIN TINGLER

26. One of the most important analyses in many fields of applied chemistry is the Kjeldahl determination for nitrogen compounds. This procedure is applied to such diverse assays as the nitrogen content of food proteins and the concentration of nitrates in waters heavily polluted by agricultural nitrate fertilizers. The procedure depends eventually on the titration of a sulfuric acid solution that has been partially neutralized by ammonia formed from the nitrogen compound being analyzed.[5]

   Two alternative procedures are involved in the initial stages of the analysis. If the nitrogen compounds being investigated contain nitrogen in the -III oxidation state, the

[5] One modification often used employs a 4-percent solution of boric acid. Ammonia absorbed by this solution is titrated by standardized hydrochloric acid.

sample is digested in concentrated sulfuric acid. This process converts the nitrogen (e.g., from protein) to ammonium sulfate.[6] If the nitrogen of interest is in a higher oxidation state, it must first be reduced. For example, analysis for nitrate involves initial treatment of the sample with a mixture of sulfuric and salicylic acids. The nitrated salicylic acid is reduced with sodium thiosulfate, and the mixture is then subjected to the sulfuric acid digestion.

After digestion is complete, excess sodium hydroxide solution is added carefully[7] and the ammonia liberated[8] is distilled into a measured volume of dilute sulfuric acid of known concentration.

The ammonia acts to neutralize some of the sulfuric acid. The remaining excess acid is titrated by standardized sodium hydroxide ($\sim$0.1 N), using methyl red as indicator.[9] The difference between the original concentration of the sulfuric acid and the concentration determined by the titration is a measure of the amount of ammonia absorbed.

In a particular analysis of a protein mixture, a 1.1841-g sample was digested in 25 ml of concentrated sulfuric acid to which a small crystal of $CuSO_4 \cdot 5H_2O$ had been added. When digestion was complete, as indicated by the disappearance of the dark carbon initially formed, the mixture was cooled and carefully diluted with 250 ml of cold water. A 100-ml portion of 40 percent sodium hydroxide was carefully poured down the wall of the digestion flask so as to form two liquid phases. A few zinc pellets were added to keep the mixture agitated by slow formation of hydrogen, and the flask was connected to a distillation apparatus. The condenser outlet was arranged to dip below the surface of 50.00 ml of 0.1897 N sulfuric acid in the receiving flask. The digestion flask was heated until the solution was maintained at a gentle boil. Boiling was continued until about half the contents of the digestion flask had distilled. The heat source was removed, and the receiving flask was immediately lowered to avoid "suck back" of liquid as the system cooled. The solution in the receiving flask was titrated with 0.1089 N sodium hydroxide to a methyl red endpoint. The volume of base required was 38.72 ml.

Calculate the percent nitrogen in the original sample.

---

[6] Reaction is usually slow, so a catalyst such as a crystal of copper sulfate is generally employed.

[7] The reaction between sodium hydroxide and sulfuric acid is very exothermic.

[8] $NH_4^+(aq) + OH^-(aq) \longrightarrow NH_3(aq)$

[9] Neither sodium hydroxide nor sulfuric acid can be used as a primary standard (i.e., neither can be obtained in sufficient purity for routine use as an accurate standard). The sodium hydroxide is standardized against a carefully weighed sample of some solid acid, such as potassium hydrogen phthalate. The standardized sodium hydroxide is then used to determine the concentration of the sulfuric acid.

Methyl red changes color in the pH range corresponding to $(NH_4)_2SO_4$ at the concentrations employed.

Unit 19 Self-Test

1. Write a net equation for the equilibrium system resulting from dissolving ammonium iodide in water. Indicate the conjugate pairs, and in each pair label the acid and the base.

2. Briefly describe the differences in definition of *base* by the Arrhenius, the Brønsted–Lowry, and the Lewis systems.

3. Calculate the pH of
    a. 0.0020 M perchloric acid
    b. 0.050 M potassium hydroxide
    c. 0.50 M hydrofluoric acid ($K_a = 6.9 \times 10^{-4}$)
    d. 0.074 M ammonia ($K_b = 1.8 \times 10^{-5}$)

4. Calculate the pH of:
    a. 0.20 M sodium cyanide (K_a for HCN is 4.8×10^{-10})
    b. 0.20 M dimethylammonium chloride (K_b for $(CH_3)_2NH$ is 5.1×10^{-4})

5. Arrange the following acids in order of acid strength (for first ionization): sulfuric $[SO_2(OH)_2]$, silicic $[Si(OH)_4]$, nitrous $[NO(OH)]$, perchloric $[ClO_3(OH)]$.

6. What is the approximate pH at the equivalence point for the titration of 0.30 M ammonia by 0.10 M hydrochloric acid? Suggest an appropriate indicator from Table 19.11 for this titration.

Unit 20

Buffer systems

Objectives

(a) Given the composition of a conjugate-pair buffer system and access to the appropriate equilibrium constant, be able to calculate the pH of the buffer solution (Section 20.2).

(b) Be able to calculate the pH of a solution of a simple amphoteric species (Examples 3 and 4).

(c) Given the composition of a conjugate-pair buffer system and the amount of strong acid or strong base to be added to a particular volume of the buffer solution, be able to calculate the pH change of the buffer system (Section 20.3).

(d) Be able to calculate buffer capacity and pH change for buffer depletion (Section 20.4).

The enzymes that catalyze most biochemical processes are generally active only within a narrow pH range. It is, then, essential that living organisms have some system for controlling the pH of the aqueous mixtures in which enzymes function. These "pH controls" must avoid significant general fluctuations of pH when localized acidity changes occur.

The blood, for example, is maintained at about pH 7.4—rarely varying from this value by more than 0.1 pH unit. A variation of ±0.4 or more pH units generally results in death. The composition of biological pH control systems is usually complex, but the basic principles are relatively simple. Solutions that resist pH change when diluted with water or when treated with acid or base are said to be buffered. The action of buffering agents can be understood in terms of acid–base equilibria and Le Chatelier's principle.

Buffer action is not unfamiliar. Television advertisements have stressed the values of buffered aspirin preparations and antacids that do not add excessive alkali to the stomach. Soft drinks are buffered, and tropical fish enthusiasts are familiar with buffering agents used to control aquarium pH. These agents, like the

Amphoteric species are capable of both donating and accepting protons (Unit 6).

common biological buffers, are generally amphoteric species or mixtures of weak acids and weak bases.

In this unit we shall examine some common buffer systems to see how they function, and in Unit 21 we shall see how buffers may be used to control the concentrations of species other than hydrogen ion.

20.1 BUFFER SYSTEMS

When acid or base added to a solution produces a smaller pH change than would have occurred in water, the solution is said to be *buffered* (Figure 20.1). A number of different types of substances can produce this effect, as illustrated in Table 20.1.

Although acids and bases themselves have some buffering activity (Table 20.1), it

Figure 20.1 Buffer action. The graphs show pH change as 0.01 M NaOH is added to 100 ml of water or solution.

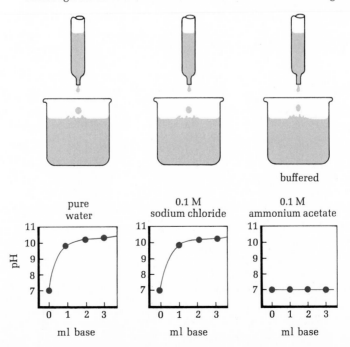

pure water

0.1 M sodium chloride

0.1 M ammonium acetate

buffered

Table 20.1 pH CHANGES IN VARIOUS SOLUTIONS

TYPE	SOLUTION	ORIGINAL	pH AFTER ACID[a]	AFTER BASE[a]
——	Pure water	7.00	2.00	12.00
Salt containing "neutral" ions	0.10 M NaCl	7.00	2.00	12.00
weak acid	0.10 M CH_3CO_2H	2.87	2.00	3.79
weak base	0.10 M NH_3	11.13	10.21	12.00
strong acid	0.10 M HCl	1.00	0.95	1.05
strong base	0.10 M NaOH	13.00	12.95	13.05
weak acid with weak base $(K_a \approx K_b)$	0.10 M $NH_4(CH_3CO_2)$	7.00	5.70	8.30
weak acid with weak base $(K_a \neq K_b)$	0.10 M CH_3CO_2H with 0.10 M $CH_3CO_2^-$	4.74	4.66	4.83
amphoteric species	0.10 M HCO_3^-	8.35	7.33	9.36

[a] After addition of 0.010 mole H^+ (or OH^-) per liter, without significant volume change.

is more common to consider buffers as substances or mixtures that can *react* with added acid *and* with added base. Two types of buffers illustrate this situation.

A second criterion for classification as a buffer requires that pH be concentration independent over a reasonable range of concentrations. Acids or bases alone do not demonstrate this characteristic unless they are amphoteric species (Table 20.2).

AMPHOTERIC SPECIES (See also Unit 6)

Bicarbonate ion

$$HCO_3^- + H^+ \rightleftharpoons H_2CO_3 \rightleftharpoons H_2O + CO_2$$
$$HCO_3^- + OH^- \rightleftharpoons H_2O + CO_3^{2-}$$

Glycine

$$\overset{+}{H_3}NCH_2CO_2^- + H^+ \rightleftharpoons \overset{+}{H_3}NCH_2CO_2H$$

$$\overset{+}{H_3}NCH_2CO_2^- + OH^- \rightleftharpoons H_2NCH_2CO_2^- + H_2O$$

MIXTURES OF WEAK ACIDS WITH WEAK BASES

Ammonium acetate

$$CH_3CO_2^- + H^+ \rightleftharpoons CH_3CO_2H$$
$$NH_4^+ + OH^- \rightleftharpoons NH_3 + H_2O$$

Acetic acid with sodium acetate

$$CH_3CO_2^- + H^+ \rightleftharpoons CH_3CO_2H$$
$$CH_3CO_2H + OH^- \rightleftharpoons CH_3CO_2^- + H_2O$$

Buffers of either of these types have a special advantage over acids or bases alone in that their pH remains fairly constant over a wide range of dilution (Table 20.2). This is easily demonstrated for the acetate–acetic acid buffer, an example of the very common *conjugate-pair* buffer systems.

Table 20.2 pH AND DILUTION

SOLUTION	ORIGINAL pH	pH AFTER DILUTION BY		
		1:10	1:100	1:1000
1.0 M HCl	0.00	1.00	2.00	3.00
1.0 M CH_3CO_2H	2.37	2.87	3.37	3.87
1.0 M CH_3CO_2H with 1.0 M $CH_3CO_2^-$	4.74	4.74	4.74	4.74
1.0 M HCO_3^-	8.35	8.35	8.35	8.35

That is, when C_{H^+} or C_{OH^-} from water is negligible and when $C_{CH_3CO_2H}$ and $C_{CH_3CO_2^-}$ are equally affected by dilution.

Note that this buffer system may be described by the same expressions we used in Unit 19 for the ionization of a weak acid.

Under simplest conditions, the equilibrium in this buffer is closely approximated by

$$CH_3CO_2H(aq) \rightleftharpoons H^+(aq) + CH_3CO_2^-(aq)$$

for which

$$K_a = \frac{C_{H^+} \times C_{CH_3CO_2^-}}{C_{CH_3CO_2H}}$$

Then

$$C_{H^+} = K_a \frac{C_{CH_3CO_2H}}{C_{CH_3CO_2^-}}$$

Since dilution affects both acetate and acetic acid to the same extent, the pH is determined only by the K_a value and the *ratio* of the concentrations of the conjugate pair.

It is apparent from Table 20.1 that considerations of "pH stability" only for added H^+ or OH^- would suggest use of a moderately concentrated strong acid, strong base, or conjugate-pair system. On the other hand Table 20.2 indicates that "pH stability" only for dilution would suggest a conjugate-pair system, an amphoteric species (e.g., HCO_3^-), or a weak acid–weak base mixture [e.g., $NH_4(CH_3CO_2)$]. When *both* factors are considered, it is obvious that the conjugate-pair buffer system best meets both criteria for effective buffer action.

Many important buffer systems, such as those in the blood or those used in commercial "carbonated" beverages have several components. Analysis of such systems requires detailed statements of mass balance and electroneutrality expressions (Unit 21) and careful consideration of possible simplifications. We shall restrict our study in this unit to the simple amphoteric species and two-component buffers, leaving more complex cases for later study.

We have alluded to three aspects of buffer behavior: buffer pH, change of pH on addition of acid or base, and effect of dilution. These factors are important to any understanding of buffers. Let us examine them in more detail, using simple buffer systems.

In the subsequent discussions we shall use two assumptions.

Autoprotolysis refers to the "self-ionization" of a species, e.g.,

$$H_2O \rightleftharpoons H^+ + OH^-$$

1. Buffer concentrations are large enough to permit omission of autoprotolysis effects of water (Unit 17) on pH.
2. Equilibrium concentrations of buffer components are sufficiently approximated by label concentrations and simple dilution factors.

These assumptions provide the limits under which the treatments given are acceptable. These limits must be recognized.

20.2 BUFFER pH

We suggested earlier that the hydrogen ion concentration of an acetic acid–acetate buffer depends only on the K_a value for acetic acid and the *ratio* of acetic acid and acetate concentrations. Since both components are present in the same solution, the concentration *ratio* will not be affected by dilution. This is one of the attributes of a good buffer system.

Example 1

Calculate the pH of a solution in which the acetic acid concentration and the acetate concentration are each 1.0 M. Then calculate the pH for 1:10, 1:100, and 1:1000 dilutions of this buffer.

Solution
The acetic acid–acetate system in aqueous solution may be represented by

$$CH_3CO_2H(aq) \rightleftharpoons H^+(aq) + CH_3CO_2^-(aq)$$

for which (from Appendix C)

$$K_a = 1.8 \times 10^{-5} = \frac{C_{H^+} \times C_{CH_3CO_2^-}}{C_{CH_3CO_2H}}$$

Note that we cannot find the pH of this solution by the method used for a simple weak acid (Unit 19) since

$$C_{H^+} \neq C_{CH_3CO_2^-}$$

(In the buffer, the acetate is primarily from an added salt, such as sodium acetate, rather than from ionization of the acetic acid.)

Solving the equilibrium expression for C_{H^+} gives

$$C_{H^+} = 1.8 \times 10^{-5} \times \frac{C_{CH_3CO_2H}}{C_{CH_3CO_2^-}}$$

Then, for the original buffer,

$$C_{H^+} = 1.8 \times 10^{-5} \times \frac{1.0}{1.0} = 1.8 \times 10^{-5}$$

and

$$pH = -\log(1.8 \times 10^{-5}) = 4.74$$

For 1:10 dilution,

$$C_{CH_3CO_2H} = 1.0 \ M \times \tfrac{1}{10} = 0.10 \ M$$
$$C_{CH_3CO_2^-} = 1.0 \ M \times \tfrac{1}{10} = 0.10 \ M$$
$$C_{H^+} = 1.8 \times 10^{-5} \times \frac{0.10}{0.10} = 1.8 \times 10^{-5}$$
$$pH = 4.74$$

Continuation of this procedure for 1:100 and 1:1000 dilutions would show that the pH remains 4.74 (Table 20.2).

The general case of a system capable of reacting with both acid and base can be represented as an HA^a/B^b buffer, in which HA^a represents the acid component (of charge a) and B^b represents the base component (of charge b). Note that HA^a and B^b may be completely different species (e.g., NH_4^+ and $CH_3CO_2^-$ ions), a conjugate pair (e.g., NH_4^+ and NH_3), different groups within the same molecule (e.g., H_3N^+— and —CO_2^- in glycine), or the same substance (e.g., bicarbonate ion). Within the limits of our two assumptions, all of these may be considered on the basis of *buffer equilibria*.

$$HA^a \rightleftharpoons H^+ + A^{a-1}$$
$$B^b + H_2O \rightleftharpoons HB^{b+1} + OH^-$$

for which

$$K_a = \frac{C_{H^+} \times C_{A^{a-1}}}{C_{HA^a}}$$

and

$$K_b = \frac{C_{HB^{b+1}} \times C_{OH^-}}{C_{B^b}}$$

$$K_w = C_{H^+} \times C_{OH^-}$$

From the autoprotolysis constant for water, C_{OH^-} in the K_b expression may be replaced by K_w/C_{H^+}:

$$K_b = \frac{C_{HB^{b+1}} \times K_w}{C_{B^b} \times C_{H^+}}$$

Then

$$\frac{K_a}{K_b} = \frac{C_{H^+}^2 \times C_{B^b} \times C_{A^{a-1}}}{C_{HA^a} \times C_{HB^{b+1}} \times K_w}$$

from which

$$C_{H^+} = \sqrt{\frac{K_a K_w}{K_b} \times \frac{C_{HA^a} \times C_{HB^{b+1}}}{C_{B^b} \times C_{A^{a-1}}}} \tag{20.1}$$

In specific cases this may be reduced to a simpler formulation. For example, in a conjugate-pair buffer,

$$HA^a \equiv HB^{b+1}$$

$$B^b \equiv A^{a-1}$$

$$K_a = \frac{K_w}{K_b} \qquad \text{(Unit 19)}$$

Thus a general formula for conjugate-pair buffers, *under restrictions of our two assumptions,* is

$$C_{H^+} = K_a \frac{C_{HA^a}}{C_{A^{a-1}}} \tag{20.2}$$

Note that pH will depend only on the concentration ratio.

Example 2

What is the pH of an ammonia–ammonium chloride buffer in which $C_{NH_3} = 0.20 \ M$ and $C_{NH_4^+} = 0.30 \ M$?

Solution

If we wish to use equation (20.2), we must consider the buffer system as

$$NH_4^+ \rightleftharpoons H^+ + NH_3$$

for which (Unit 19)

$$K_a = \frac{K_w}{K_{b(NH_3)}} = \frac{1.0 \times 10^{-14}}{1.8 \times 10^{-5}} = 5.6 \times 10^{-10}$$

$$\text{(for } NH_4^+\text{)} \qquad\qquad \text{(from Appendix C)}$$

Then

$$C_{H^+} = 5.6 \times 10^{-10} \left(\frac{0.30}{0.20}\right) = 8.4 \times 10^{-10}$$

$$pH = -\log(8.4 \times 10^{-10}) = 9.08$$

Note that we could have worked equally well with the system formulated as

$$NH_3 + H_2O \rightleftharpoons NH_4^+ + OH^-$$

for which, as a buffer,

$$C_{OH^-} = K_b \frac{C_{NH_3}}{C_{NH_4^+}}$$

and

$$pH = 14.00 - pOH$$

For an amphoteric substance, such as bicarbonate ion or glycine,

$$HA^a \equiv B^b$$

(but A^{a-1} and HB^{b+1} are different species) and, in the absence of other acids or bases,

$$HA^a + B^b \rightleftharpoons HB^{b+1} + A^{a-1}$$

so that

$$C_{HB^{b+1}} = C_{A^{a-1}}$$

A general formula for such species, then, *remembering stated limitations*, is

$$C_{H^+} = \sqrt{\frac{K_a K_w}{K_b}} \tag{20.3}$$

Note that pH is independent of concentration, within the limits of our assumptions.

Example 3

The general formula for a simple bifunctional amino acid is

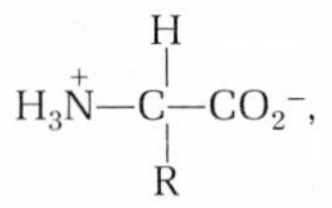

$$H \atop H_3\overset{+}{N}-\underset{R}{\overset{|}{C}}-CO_2^-,$$

in which R represents some neutral grouping such as —H or —CH_3. Amino acids are discussed in Unit 32.

The isoelectric point of an amino acid is defined as the pH at which the net charge of the principal species is zero. What is the pH of the isoelectric point for glycine ($K_a = 1.6 \times 10^{-10}$, $K_b = 2.5 \times 10^{-12}$)?

Solution
Using equation (20.3),

$$C_{H^+} = \sqrt{\frac{1.60 \times 10^{-10} \times 10^{-14}}{2.5 \times 10^{-12}}} = \sqrt{6.4 \times 10^{-13}}$$

$$pH = -0.5 \log(6.4 \times 10^{-13}) = 6.10$$

Example 4

The principal species is actually aqueous CO_2 (Unit 19), but H_2CO_3 is often used as a convenient formulation.

Show that the pH recorded in Table 20.2 (pH 8.35) for 1.0 M HCO_3^- is correct. (K_{a_1} for $H_2CO_3 = 4.2 \times 10^{-7}$, $K_{a_2} = 4.8 \times 10^{-11}$.)

Solution

For bicarbonate acting as an acid,

$$HCO_3^- \rightleftharpoons H^+ + CO_3^{2-} \qquad K_a = 4.8 \times 10^{-11}$$

(i.e., this represents the second ionization of carbonic acid.)

For bicarbonate acting as a base,

$$HCO_3^- + H_2O \rightleftharpoons H_2CO_3 + OH^-$$

$$K_b = \frac{K_w}{K_{a(conj)}} \qquad \text{(Unit 19)}$$

$$= \frac{1.0 \times 10^{-14}}{4.2 \times 10^{-7}} \overset{\curvearrowleft}{\underset{(K_{a_1} \text{ for } H_2CO_3)}{}}$$

Then, in equation (20.3),

$$C_{H^+} = \sqrt{\frac{4.8 \times 10^{-11} \times 10^{-14}}{10^{-14}/4.2 \times 10^{-7}}}$$

$$= \sqrt{4.8 \times 10^{-11} \times 4.2 \times 10^{-7}} = 4.5 \times 10^{-9}$$

$$pH = -\log(4.5 \times 10^{-9}) = 8.35$$

In a weak acid–weak base buffer in which the principal components are not members of a conjugate pair (e.g., ammonium acetate or a mixture of unequal amounts of ammonium chloride and sodium acetate), the pH may be estimated from equation (20.1), as will be illustrated in Unit 21.

Note that a more detailed analysis is required for cases in which conditions are such that the assumptions listed earlier are not applicable.

Even for simple cases, it is neither necessary nor particularly desirable to rely on specific formulas, except when they are convenient for repetitive calculations. It is preferable to treat each situation by the general methods applicable to chemical equilibria.

20.3 pH CHANGE When acid is added to buffer, the *relative* amounts of buffer components will change as added H^+ is consumed. A similar change in composition occurs when base is added to the system. Typical equations for these processes are shown in Section 20.1. As long as appreciable concentrations of buffer components remain, the pH of the system may be calculated by the appropriate equations for the particular type of buffer involved. For simple cases, pH changes can be estimated with sufficient accuracy by assuming stoichiometric reaction with the added H^+ or OH^- and including any dilution that may occur. Cases in which a buffer component is essentially all consumed are discussed in Section 20.4.

Example 5

Compare the pH change when 10.0 ml of 0.10 M NaOH is added to 90.0 ml of distilled water with that resulting from addition of 10.0 ml of 0.10 M NaOH to 90.0 ml of a 1.0 M NH_3/1.0 M NH_4^+ buffer. (K_a for $NH_4^+ = 5.6 \times 10^{-10}$.)

Solution

Original pH:

Pure water | The buffer [using Equation (20.2)]

$$C_{H^+} = 1.0 \times 10^{-7} M \qquad C_{H^+} = 5.6 \times 10^{-10}\left(\frac{1.0}{1.0}\right)$$

pH = 7.0 | pH = 9.25

After addition of NaOH (assuming no volume change on mixing),
To pure water

$$C_{OH^-} = 0.10\, M\left(\frac{10.0\ ml}{10.0\ ml + 90.0\ ml}\right) = 1.0 \times 10^{-2}$$

$$C_{H^+} = \frac{K_w}{C_{OH^-}} = \frac{1.0 \times 10^{-14}}{1.0 \times 10^{-2}}$$

$$pH = 12.0$$

pH change = 12.0 − 7.0 = 5.0 pH units

To the buffer
Assuming stoichiometric reaction,

$$NH_4^+ + OH^- \longrightarrow NH_3 + H_2O$$

(no. mmol = no. ml × M)(Unit 13)

no. mmol NH_3 formed = 10.0 ml × 0.10 M

= 1.0

no. mmol NH_4^+ consumed = 1.0

no. mmol NH_3 originally = 90.0 ml × 1.0 M = 90

no. mmol NH_4^+ originally = 90

no. mmol NH_4^+ after NaOH addition = 90 − 1 = 89

no. mmol NH_3 after NaOH addition = 90 + 1 = 91

At equilibrium,

$$C_{NH_4^+} = \frac{89\ mmol}{(10.0 + 90.0)\ ml} = 0.89\ M$$

$$C_{NH_3} = \frac{91\ mmol}{(10.0 + 90.0)\ ml} = 0.91\ M$$

Then, using equation (20.2),

$$C_{H^+} = 5.6 \times 10^{-10}\left(\frac{0.89}{0.91}\right) = 5.5 \times 10^{-10}$$

$$pH = 9.26$$

pH change = 9.26 − 9.25 = 0.01 pH unit

Since both components of a conjugate-pair buffer are present in the same solution, the solution volume "cancels" and we may use either the mole ratio or the millimole ratio of buffer components in lieu of the *concentration* ratio employed in equation (20.2), that is,

$$C_{H^+} = K_a\frac{moles\ HA^a}{moles\ A^{a-1}} = K_a\frac{mmols\ HA^a}{mmols\ A^{a-1}} \qquad (20.2a)$$

Alternate Solution (to "buffer pH change" from Example 5)

no. mmol NH_3 formed = 1.0

$$\text{no. mmol NH}_4^+ \text{ consumed} = 1.0$$
$$\text{no. mmol NH}_3 \text{ originally} = 90$$
$$\text{no. mmol NH}_4^+ \text{ originally} = 90$$
$$\text{no. mmol NH}_3 \text{ after NaOH addition} = 90 + 1 = 91$$
$$\text{no. mmol NH}_4^+ \text{ after NaOH addition} = 90 - 1 = 89$$

from equation (20.2a),

$$C_{H^+} = (5.6 \times 10^{-10})\left(\frac{89}{91}\right) = 5.5 \times 10^{-10}$$

$$\text{pH} = 9.26$$

20.4 BUFFER CAPACITY AND BUFFER DEPLETION

Although the pH of a typical buffer is independent of actual concentrations of buffer components (Section 20.2), *the quantity of acid or base that can be absorbed without a large pH change* is concentration dependent. Buffer capacity is a measure of the effectiveness of a buffer system in maintaining a nearly constant pH.

Buffer capacity can, in fact, be defined mathematically in various ways, depending on the circumstances in which a particular buffer is used. The most common definition employed by analytical chemists is

Buffer capacity is the number of moles H+ (or OH−) that will change the pH of 1.00 liter of buffer by 1.0 pH unit. It should be noted that, except for an *equimolar* solution, the buffer capacity of a conjugate-pair buffer may be significantly different for added H+ than for added OH−.

When sufficient strong acid or base has been added to a conjugate-pair buffer to reduce the concentration of one of the components below that necessary for significant buffer action, the buffer is said to be *depleted*, and the pH must then be calculated on a different basis than that used for a buffer [equation (20.2)].

Example 6

Calculate the buffer capacity of
a. 0.10 M CH_3CO_2H/0.30 M $CH_3CO_2^-$ buffer, for added H+
b. 0.10 M CH_3CO_2H/0.30 M $CH_3CO_2^-$ buffer, for added OH−
c. 0.010 M CH_3CO_2H/0.030 M $CH_3CO_2^-$ buffer, for added OH−

Solution
First, let's note that the original pH of all three buffers is [from equation (20.2)]

$$C_{H^+} = K_a\left(\frac{C_{CH_3CO_2H}}{C_{CH_3CO_2^-}}\right) = (1.8 \times 10^{-5})\left(\frac{1}{3}\right) = 6.0 \times 10^{-6}\ M$$
$$\text{(Appendix C)} \qquad \text{(components ratio)}$$

Then, from the definition of *buffer capacity*, for a pH change of 1.0 pH unit per liter of buffer, the C_{H^+} must change by a factor of 10.
a. For added H+, C_{H^+} will change from $6.0 \times 10^{-6}\ M$ to $6.0 \times 10^{-5}\ M$ on addition of y moles of H+ (the buffer capacity) per liter. After addition of the H+ (assuming negligible volume change),

$$C_{CH_3CO_2H} = 0.10\ M + y$$
$$C_{CH_3CO_2^-} = 0.30\ M - y$$

since

$$H^+ + CH_3CO_2^- \longrightarrow CH_3CO_2H$$

and

$$6.0 \times 10^{-5} = (1.8 \times 10^{-5})\frac{(0.10 + y)}{(0.30 - y)}$$

from which

$$(6.0 \times 10^{-5})(0.30 - y) = (1.8 \times 10^{-5})(0.10 + y)$$
$$(1.8 \times 10^{-5}) - (6.0 \times 10^{-5})y = (1.8 \times 10^{-6}) + (1.8 \times 10^{-5})y$$
$$y = \frac{1.62 \times 10^{-5}}{7.8 \times 10^{-5}} = 0.20$$

That is, this buffer can consume 0.20 mole H^+ per liter of buffer with a pH change of only 1.0 pH unit.

b. For added OH^-, C_{H^+} will change from 6.0×10^{-6} M to 6.0×10^{-7} M, and for y moles of OH^- per liter of buffer,

$$C_{CH_3CO_2H} = 0.10\ M - y$$
$$C_{CH_3CO_2^-} = 0.30\ M + y$$

since

$$OH^- + CH_3CO_2H \longrightarrow CH_3CO_2^- + H_2O$$

and

$$6.0 \times 10^{-7} = (1.8 \times 10^{-5})\frac{(0.10 - y)}{(0.30 + y)}$$

from which

$$y = 0.087$$

Note that the smaller buffer capacity for OH^- with this buffer reflects the ratio of $C_{CH_3CO_2H}/C_{CH_3CO_2^-}$ as less than 1.

c. In a manner similar to that for (b),

$$C_{CH_3CO_2H} = 0.010\ M - y$$
$$C_{CH_3CO_2^-} = 0.030\ M + y$$
$$6.0 \times 10^{-7} = (1.8 \times 10^{-5})\frac{(0.010 - y)}{(0.030 + y)}$$

from which

$$y = 0.0087$$

In comparing buffer capacities for (b) and (c), we see that the buffer capacity, unlike the buffer pH, *is* highly dependent on buffer concentration.

Example 7

Compare the pH values resulting from addition of 10.0 ml of 0.10 M HCl to 100.0 ml of
  a. 0.10 M NH_3/0.10 M NH_4^+ buffer
  b. 0.010 M NH_3/0.010 M NH_4^+ buffer
  c. 0.0010 M NH_3/0.0010 M NH_4^+ buffer
Note that all three buffer solutions originally have the same pH (9.25).

Solution

Assume that added H^+ consumes an equivalent quantity of NH_3 until the latter is all consumed.

$$H^+ + NH_3 \longrightarrow NH_4^+$$

$$\text{(no. mmol = no. ml} \times M)$$
$$\text{no. mmol } H^+ \text{ added} = 10.0 \text{ ml} \times 0.10 \ M = 1.0 \text{ mmol}$$

a. no. mmol NH_3 available $= 100.0 \text{ ml} \times 0.10 \ M$
$$= 10 \text{ mmol}$$
no. mmol NH_4^+ originally $= 100.0 \text{ ml} \times 0.10 \ M$
$$= 10 \text{ mmol}$$
no. mmol NH_3 consumed $= 1.0 \text{ mmol}$
no. mmol extra NH_4^+ formed $= 1.0 \text{ mmol}$

At equilibrium

$$C_{NH_4^+} = \frac{(10 + 1) \text{ mmol}}{(10.0 + 100.0) \text{ ml}} = 0.10 \ M$$

$$C_{NH_3} = \frac{(10 - 1) \text{ mmol}}{(10.0 + 100.0) \text{ ml}} = 0.082 \ M$$

Using equation (20.2),

$$C_{H^+} = 5.6 \times 10^{-10}\left(\frac{0.10}{0.082}\right) = 6.9 \times 10^{-10}$$

$$pH = 9.16 \qquad \text{(change of 0.09 pH unit)}$$

b. no. mmol NH_3 available $= 100.0 \text{ ml} \times 0.010 \ M$
$$= 1.0 \text{ mmol}$$
no. mmol NH_4^+ originally $= 100.0 \text{ ml} \times 0.010 \ M$
$$= 1.0 \text{ mmol}$$
no. mmol NH_3 consumed $= 1.0 \text{ mmol}$
no. mmol extra NH_4^+ formed $= 1.0 \text{ mmol}$

Since most of the NH_3 component has been consumed, this buffer is *depleted* in terms of further reaction with added H^+.

The solution, then, is essentially

$$C_{NH_4^+} = \frac{(1.0 + 1.0) \text{ mmol}}{(10.0 + 100.0) \text{ ml}} = 1.8 \times 10^{-2} \ M$$

Then, since the solution is that of an ammonium salt,

$$NH_4^+ \rightleftharpoons NH_3 + H^+$$

$$C_{NH_3} = C_{H^+}$$

$$K_a = \frac{C_{NH_3} \times C_{H^+}}{C_{NH_4^+}}$$

$$5.6 \times 10^{-10} = \frac{C_{H^+}^2}{1.8 \times 10^{-2}}$$

$$C_{H^+} = \sqrt{1.0 \times 10^{-11}}$$

$$pH = 0.5 \log(1.0 \times 10^{-11})$$

$$= 5.50 \qquad \text{(change of 3.75 pH units)}$$

c. no. mmol NH_3 available $= 100.0$ ml $\times 0.0010\ M = 0.10$ mmol

no. mmol NH_4^+ originally $= 100.0$ ml $\times 0.0010\ M = 0.10$ mmol

no. mmol NH_3 consumed $= 0.10$ (maximum available)
no. mmol extra NH_4^+ formed $= 0.10$ mmol
no. mmol excess $H^+ = 1.0 - 0.10 = 0.90$ mmol

Since NH_4^+ is a very weak acid, the H^+ furnished by 0.20 mmol of NH_4^+ (per 110 ml) is negligible compared to the excess of 0.90 mmol of H^+. Therefore

$$C_{H^+} = \frac{\sim 0.90\ \text{mmol}}{(10.0 + 100.0)\ \text{ml}} = \sim 8.2 \times 10^{-3}$$

pH $= 2.09$ (change of 7.16 pH units)

20.5 HOW BUFFERS ARE SELECTED

When pH control by buffers is desirable, several factors must be considered. These include

1. The pH required. (Different buffer components produce different pH values. See Table 20.1 for examples. Component *ratios* also affect pH.)
2. The buffer capacity needed. (The quantity of acid or base to be consumed and the permissible limits of pH variation will determine the necessary concentrations of buffer components.)
3. Possible "extra" reactions of buffer components. (For example, an ammonia–ammonium chloride buffer would not be suitable for a reaction involving silver ion because of possible precipitation of silver chloride.)

In Unit 21 we shall see examples of the use of buffers in chemical analysis to control concentrations of other ions.

BLOOD BUFFERS

Your blood, if you are an "average" individual, comprises almost 10% of your body weight. The blood is a complex aqueous system containing circulating cells, colloidal solids, and a variety of dissolved species.

The pH of circulating blood is maintained around 7.4 by the action of several buffer systems. Blood proteins and amino acids have a relatively minor role in control of pH. Of somewhat greater significance is the $H_2PO_4^-/HPO_4^{2-}$ conjugate-pair buffer, but the principal buffer action is due to the CO_2/HCO_3^- system,

$$CO_2(aq) + H_2O(l) \rightleftharpoons HCO_3^-(aq) + H^+(aq)$$

About 90% of the carbon dioxide dissolved in the blood is present as bicarbonate ion and the regulation of the $CO_2 \rightleftharpoons HCO_3^-$ system is critical to the proper functioning of the whole body.

Carbon dioxide enters the bloodstream through cell membranes, having been produced within the cell as a product of various metabolic processes. Transported through the circulatory system principally as bicarbonate, the CO_2 is eventually exhaled from the lungs. Although the CO_2 content of the blood does vary somewhat as a function of body activity, its concentration is maintained within close limits by the *steady state* system of CO_2 (production $\rightleftharpoons$ exhalation).

The regulation of blood pH is essential to the proper functioning of a number of vital enzyme systems (Unit 34) for which the catalytic activity is quite pH dependent. When the $CO_2 \rightleftharpoons HCO_3^-$ system is disturbed, as by *hyperventilation* (excessive exhalative loss of CO_2) or by insufficient oxygen supply, the blood pH may drop significantly (a condition referred to as *acidosis*). Localized acidosis can result in significant muscle pain and "cramping."

One of the effects of the disease *cholera* is a disruption of the $CO_2 \rightleftharpoons HCO_3^-$ control mechanism. If the condition is not corrected to restore proper blood buffering action, death can result.

SUGGESTED READING

Drago, Russell S., and Nicholas A. Matwiyoff. 1968. *Acids and Bases*. Lexington, Mass.: Raytheon.

Moeller, Therald, and Rod O'Connor. 1972. *Ions in Aqueous Systems*. New York: McGraw-Hill.

O'Connor, Rod, and Charles Mickey. 1975. *Solving Problems in Chemistry*. New York: Harper & Row.

EXERCISES

1. Calculate the pH of each of the following solutions:
   a. 0.10 M benzoic acid/0.10 M sodium benzoate (K_a for $C_6H_5CO_2H$ is 6.6×10^{-5})
   b. 0.10 M sodium bisulfite (K_{a_1} for sulfurous acid is 1.6×10^{-2}; K_{a_2} is 1.3×10^{-7})
   c. 0.10 M ammonium benzoate (K_b for ammonia is 1.8×10^{-5})
2. Calculate the pH change when 10.0 ml of 0.10 M NaOH is added to
   a. 90.0 ml of distilled water
   b. 90.0 ml of the buffer in Exercise 1a
3. Calculate the pH change when 10.0 ml of 0.10 M NaOH is added to 100.0 ml of
   a. 0.10 M acetic acid/0.10 M acetate
   b. 0.010 M acetic acid/0.010 M acetate
   c. 0.0010 M acetic acid/0.0010 M acetate
4. Calculate the buffer capacity of
   a. 0.25 M NH_3/0.15 M NH_4^+ buffer, for added H^+
   b. 0.25 M NH_3/0.15 M NH_4^+ buffer, for added OH^-
   c. 0.50 M NH_3/0.30 M NH_4^+ buffer, for added H^+

OPTIONAL

5. Estimate the isoelectric point of a simple bifunctional amino acid for which $K_a = 3.4 \times 10^{-10}$ and $K_b = 2.1 \times 10^{-11}$.

PRACTICE FOR PROFICIENCY (Consult Appendix C, as necessary)

Objectives (a, b)

1. Calculate the pH of the following solutions:
   a. 2.0 M acetic acid/1.0 M sodium acetate
   b. 0.50 M acetic acid/0.25 M sodium acetate (K_a for acetic acid is 1.8×10^{-5}).
2. What is the pH of 0.50 M sodium bisulfite? (*Hint*: for H_2SO_3, $K_{a_1} = 1.6 \times 10^{-2}$ and $K_{a_2} = 1.3 \times 10^{-7}$.)
3. What is the approximate pH of a solution in which the concentration of formic acid is 0.40 M and the concentration of sodium formate is 0.40 M? (K_a for formic acid is 2.0×10^{-4}.)
4. What is the approximate pH of a solution in which the concentration of ammonia is 0.10 M and the concentration of ammonium nitrate is 0.90 M? (K_b for ammonia is 1.8×10^{-5}.)
5. What is the approximate pH of a solution in which the concentration of ammonia is 0.90 M and the concentration of ammonium nitrate is 0.10 M?
6. What is the approximate pH of a solution prepared by mixing equal volumes of 0.50 M ammonia and 0.40 M hydrochloric acid?
7. A lab technician thought he had prepared a buffer by mixing 0.100 mole of acetic acid with 300 ml of water. To the total amount of this "buffer" he added 10.0 ml of 7.00 M sodium

hydroxide and diluted the resulting solution to a total volume of 1.00 liter. What was the pH of this solution?

8. What is the pH of a solution prepared by mixing 25 ml of 0.60 M HCl and 1.64 g of sodium acetate with sufficient water to form 100 ml of solution? (K_a for acetic acid is 1.8×10^{-5}.)

9. What is the approximate pH of the isoelectric point for a simple bifunctional amino acid for which $K_a = 4.0 \times 10^{-11}$ and $K_b = 2.5 \times 10^{-10}$?

Objectives (c, d)

10. Calculate the pH *change* when 10 ml of 0.20 M nitric acid is added to
    a. 90.0 ml of distilled water
    b. 90.0 ml of the buffer in question 1a.

11. A buffer (pH 4.74) was prepared by mixing 1.00 mole of acetic acid and 1.00 mole of sodium acetate to form a 1.0-liter aqueous solution. To 100 ml of this solution, 10.0 ml of 2.00 M NaOH was added.
    The new pH is ——.

12. A buffer (pH 9.56) was prepared by mixing 2.00 moles of ammonia and 1.00 mole of ammonium chloride to form an aqueous solution with a total volume of 1.00 liter. To 200 ml of this solution, was added 10.0 ml of 10.0 M sodium hydroxide.
    The new pH is ——.

13. A student prepared 200 ml of a buffer (pH 3.16) by mixing 100 ml of 0.100 M HF and 100 ml of 0.100 M NaF. To this solution he added 100 ml of 0.400 M NaOH. What is the pH of the resulting 300 ml of solution?

14. A buffer (pH 4.88) was prepared by mixing 0.10 mole of benzoic acid 0.5 mole of sodium benzoate to form a 1.0-liter aqueous solution. To 70.0 ml of this solution, 2.00 ml of 2.00 M HI was added.
    The new pH is ——.

15. Calculate the buffer capacity for
    a. 0.50 M benzoic acid/0.10 M sodium benzoate buffer, for added H$^+$

    b. 0.50 M benzoic acid/0.50 M sodium benzoate, both for added H$^+$ and for added OH$^-$
    c. 0.10 M benzoic acid/0.020 M sodium benzoate buffer, for added H$^+$

16. Calculate the pH change when 1.0 ml of 1.0 M NaOH is added to 49 ml of each of the buffer solutions in question 15.

17. Calculate the pH change when 1.0 ml of 1.0 M HNO$_3$ is added to 49 ml of each of the buffer solutions in question 15.

18. Calculate the pH change when 10 ml of 0.20 M nitric acid is added to 90 ml of
    a. 0.50 M ammonia/0.50 M ammonium nitrate
    b. 0.050 M ammonia/0.050 M ammonium nitrate
    c. 0.0050 M ammonia/0.0050 M ammonium nitrate (K_b for ammonia is 1.8×10^{-5}).

19. A buffer was prepared by adding 2.4 g of ammonium nitrate to 100.0 ml of 0.30 M ammonia. Calculate the pH change that would result from
    a. adding another 2.4 g of ammonium nitrate to this solution
    b. adding another 100.0 ml of 0.30 M ammonia to the original buffer
    c. adding 10.0 ml of 0.30 M sodium hydroxide to the original buffer

BRAIN TINGLER

20. Biochemists have found it useful to express the pH of a conjugate-pair buffer solution in terms of a relationship known as the *Henderson-Hasselbach equation*:

$$pH = pK_a + \log \frac{C_{\text{base component}}}{C_{\text{acid component}}}$$

Using information presented in Units 19 and 20, derive the *Henderson-Hasselbach equation*.

Unit 20 Self-Test

1. What is the pH of a solution that is 0.40 M in hydrofluoric acid and 0.40 M in sodium fluoride? (K_a for hydrofluoric acid is 6.9×10^{-4}.)

2. What is the pH of 0.50 M NaHCO$_3$? [K_{a_1} for H$_2$CO$_3$(H$_2$O + CO$_2$) is 4.2×10^{-7}; K_{a_2} is 4.8×10^{-11}.]

3. What is the pH of a solution that was originally 0.20 M in propionic acid and 0.30 M in sodium propionate after addition of 1.0 ml of 0.10 M HCl to 10.0 ml of the solution? (K_a for

CH$_3$CH$_2$CO$_2$H is 1.4×10^{-5}.)

4. What is the pH after addition of 3.0 ml of 1.0 M HCl to 10.0 ml of 0.20 M CH$_3$CH$_2$CO$_2$H/0.30 M CH$_3$CH$_2$CO$_2^-$?

5. What is the buffer capacity of the buffer in question 3?

6. What would be the pH of a solution prepared by mixing equal volumes of 0.30 M HCl and the buffer solution described in question 3?

Unit 21

Interactive equilibria

Objectives

(a) Be able to use a systematic approach to mathematical analysis of simultaneous equilibria, involving:
(1) charge-balance equations (principle of electroneutrality)
(2) mass-balance equations
(3) stoichiometric considerations
(b) Be able to determine the suitability of possible simplifying approximations in
simultaneous equilibria.
(c) When suitable approximations can be used, be able to analyze the composition of aqueous solutions involving polyprotic acid systems
(Section 21.3).
(d) When suitable approximations can be used, be able to calculate
appropriate conditions for pH-controlled precipitation separations (Sections
21.4 and 21.5).
(e) When suitable approximations can be used, be able to analyze the composition of aqueous solutions of complex ions (Section 21.8).

515

In real situations it is rare to encounter a simple equilibrium system involving only three or four different species. It has already been noted that aqueous solutions frequently represent situations of simultaneous equilibria in which the same species participates in two equilibrium systems; for example,

$$Ba(OH)_2 (s) \rightleftharpoons Ba^{2+} (aq) + 2 \boxed{OH^- (aq)}$$

$$H_2O (l) \rightleftharpoons \boxed{OH^- (aq)} + H^+ (aq)$$

or

$$HCN (aq) \rightleftharpoons CN^- (aq) + \boxed{H^+ (aq)}$$

$$H_2O (l) \rightleftharpoons \boxed{H^+ (aq)} + OH^- (aq)$$

It should be noted that in situations in which some species is a component of two or more simultaneous equilibria, the concentration of that species is the same in all the equilibrium systems in which it is involved.

Many seemingly simple mixtures may actually contain a very large number of equilibrium species. Consider, for example, an aqueous solution of copper(II) sulfate and aluminum sulfate to which is added sufficient aqueous ammonia to produce a precipitate of aluminum hydroxide and to convert most of the copper(II) ions to the deep blue tetraaminecopper(II), $[Cu(NH_3)_4]^{2+}$. If the mixture is sealed in a bottle and maintained at constant temperature, leaving some air space above the solution surface, the following equilibria will be established:

Can you think of others?

equilibria involving only water and its ions

$$H_2O(l) \rightleftharpoons H_2O(g)$$
$$H_2O(l) \rightleftharpoons H^+(aq) + OH^-(aq)$$

One could include other minor components, such as helium, or, in most areas today, a variety of "pollutants" such as SO_2.

as gaseous components of the air form saturated aqueous solutions

$$N_2(g) \rightleftharpoons N_2(aq)$$
$$O_2(g) \rightleftharpoons O_2(aq)$$
$$CO_2(g) \rightleftharpoons CO_2(aq)$$

equilibria involving $CO_2(aq)$ and associated ions

$$CO_2(aq) + H_2O(l) \rightleftharpoons H_2CO_3(aq)$$
$$H_2CO_3(aq) \rightleftharpoons H^+(aq) + HCO_3^-(aq)$$
$$HCO_3^-(aq) \rightleftharpoons H^+(aq) + CO_3^{2-}(aq)$$

We shall employ the convention of square brackets, [ ], to designate a complex ion or coordination compound (Unit 27).

equilibria involving solution components

$$NH_3(aq) \rightleftharpoons NH_3(g)$$
$$NH_3(aq) + H_2O(l) \rightleftharpoons NH_4^+(aq) + OH^-(aq)$$
$$SO_4^{2-}(aq) + H_2O(l) \rightleftharpoons HSO_4^-(aq) + OH^-(aq)$$
$$[Cu(NH_3)_4]^{2+}(aq) \rightleftharpoons [Cu(NH_3)_3]^{2+}(aq) + NH_3(aq)$$
$$[Cu(NH_3)_3]^{2+}(aq) \rightleftharpoons [Cu(NH_3)_2]^{2+} + NH_3(aq)$$
$$[Cu(NH_3)_2]^{2+}(aq) \rightleftharpoons [Cu(NH_3)]^{2+}(aq) + NH_3(aq)$$
$$[Cu(NH_3)]^{2+}(aq) \rightleftharpoons Cu^{2+}(aq) + NH_3(aq)$$
$$Al(OH)_3(s) + OH^- \rightleftharpoons [Al(OH)_4]^-(aq)$$
$$Al(OH)_3(s) \rightleftharpoons [Al(OH)_2]^+(aq) + OH^-(aq)$$
$$[Al(OH)_2]^+(aq) \rightleftharpoons [Al(OH)]^{2+}(aq) + OH^-(aq)$$
$$[Al(OH)]^{2+}(aq) \rightleftharpoons Al^{3+}(aq) + OH^-(aq)$$

K_b and K_w reflect "room temperature" ($\sim$20-25°C).

A large fraction of the equilibria shown contain $NH_3(aq)$, $OH^-(aq)$, and $H^+(aq)$. The concentrations of these species are related by the expressions

$$NH_3 + H_2O \rightleftharpoons NH_4^+ + OH^-$$

$$K_b = 1.8 \times 10^{-5} = \frac{C_{NH_4^+} \times C_{OH^-}}{C_{NH_3}}$$

and

$$H_2O \rightleftharpoons H^+ + OH^-$$

$$K_w = 1.0 \times 10^{-14} = C_{H^+} \times C_{OH^-}$$

Addition of sodium hydroxide or perchloric acid would change the pH of the solution without introducing any new equilibria (neither Na^+ nor ClO_4^- react with any of the components of the mixture in dilute aqueous solution at $20°–25°C$). Changing the pH would, however, affect many of the equilibria in ways that can be estimated qualitatively from the chemical equations. Quantitative estimates of the changes could be made if the equilibrium constants were known for all equilibria involved.

A description of certain of the principal components of the system given is relatively simple. The mixture could be filtered and the precipitate analyzed to show that it consists of aluminum hydroxide. Gravimetric techniques could determine the quantity of the $Al(OH)_3$ quite accurately. The pH of the solution could be measured accurately by a pH meter to determine C_{H^+} and, from $K_w = 1.0 \times 10^{-14}$, C_{OH^-}. The concentration of the deep blue $Cu(NH_3)_4^{2+}$ could be estimated rather closely by colorimetric methods, although this would be complicated by the presence of other colored $Cu(II)$ species. The "free" ammonia, although a major component, is more difficult to determine. A titration with acid would not be satisfactory, since it would disturb all equilibria in which H^+, OH^-, or NH_3 participate. Analyses satisfactory for accurate estimation of C_{NH_3} and for concentrations of many other species in the equilibrium mixture are very complex and require a more advanced treatment than is feasible at this stage. Indeed, in many cases acceptable methods for detection of minor components in complex mixtures in terms of their equilibrium concentrations have yet to be developed.

It should be apparent that even without a detailed knowledge of all equilibria involved, chemists can control such processes as the formation or dissolution of precipitates or the concentrations of certain complexes by adjusting the concentrations of species participating in the equilibria, such as H^+. Applications of such controls in the area of wet chemical analysis are described in Excursion 2.

In even the simplest living organisms, the number of different chemical reactions

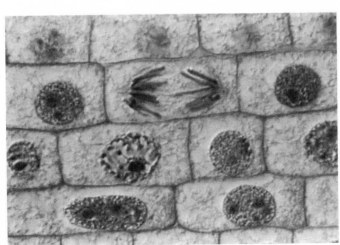

Figure 21.1 Living systems: complex chemistry. The equilibrium and nonequilibrium processes in even the simplest living organism are exceedingly complex. (Photograph by R. D. Allen, State University of New York.)

occurring at any given time is very large. Many of these involve reversible processes that, at times, reach conditions of equilibrium. As determined by the requirements of the organism, such equilibria may be disturbed by pH changes or changes in the concentrations of various participating species. The total system is so complex (Figure 21.1) that it is not surprising that our present knowledge of the chemistry of the "life process" is still quite limited.

21.1 MANY-COMPONENT SYSTEMS

It is an unfortunate fact of life that most really interesting systems are not as simple as we would like them to be. Part of the fascination of chemistry is the challenge this presents, coupled with the hope that ways can be found of systematically gaining some understanding of even the complex systems. In many cases the complexities are essentially mathematical or experimental. We believe the fundamental laws governing chemical behavior to be rather straightforward. With increasingly sophisticated experimental methods and with the mathematical tools of modern technology—high-speed computers—it seems feasible to tackle some problems that are not really simple.

Many-component equilibria are now within the limits of our capabilities, at least for cases whose components are reasonably small molecules and ions. Indeed, many of the complicated biochemical systems are rapidly becoming meaningful areas for analysis.

In this unit we shall consider only three types of many-component systems: aqueous acids, pH-controlled precipitations, and complex ions. These are examples of *interactive* systems; that is, the equilibria involve species common to more than one reaction. Noninteractive systems can be treated by considering each equilibrium separately—a reasonably good approximation in dilute solutions. We shall see that, within certain limits, simultaneous (interactive) equilibria are amenable to relatively straightforward analysis, employing appropriate simplifications. Beyond these limits the mathematical treatments will require methods that are not within the scope of our considerations. One of our goals will be to try to decide whether or not a particular system is simple enough for our limited methods to handle.[1]

21.2 MASS BALANCE AND ELECTRONEUTRALITY

All chemical processes are governed by the laws of conservation of mass and energy. In addition, every stable system is electrically neutral. These natural laws provide two useful relationships for treating equilibrium mixtures.

The *principle of mass balance* states that *the sum of the quantities of all related equilibrium species equals the sum of the quantities of original species from which they were formed. If all substances are in solution, then "quantities" may be replaced by "concentrations."*

The *principle of electroneutrality* requires that *the total positive charge in an equilibrium system must equal the total negative charge.*

Let us see how these principles apply to a particular example.

Example 1

What is the composition of an equilibrium mixture resulting from addition of 0.41 g

---

[1]For some examples of approaches useful for more complicated systems, see "The Equilibrium Between a Solid Solution and an Aqueous Solution of Its Ions," by A. F. Berndt and R. I. Stearns in the *Journal of Chemical Education*, June, 1973, (pp. 415–417) and "Molar Solubility Calculations and the Control Equilibrium," by S. H. H. Chaston in the April, 1975, issue of the *Journal of Chemical Education* (pp. 206–209).

of sodium acetate and 1.00 ml of glacial acetic acid (17.4 M, Unit 13) to sufficient water to form 100.0 ml of solution?

Solution

First, find the label concentrations of acetic acid and sodium acetate.

$$C_{NaCH_3CO_2} = \frac{0.41 \text{ g}}{0.1000 \text{ liter}} \times \frac{1 \text{ mole}}{82 \text{ g}} = 5.0 \times 10^{-2} \text{ M}$$

$$C_{CH_3CO_2H} = 17.4 \text{ M} \times \frac{1.00 \text{ ml}}{100.0 \text{ ml}} = 1.74 \times 10^{-1} \text{ M}$$

Second, consider all possible contributions to the final equilibrium system.

a. Ionization of acetic acid (involves acetate ion)

$$CH_3CO_2H \rightleftharpoons H^+ + CH_3CO_2^-$$

From Appendix C,

$$K_a = \frac{C_{H^+} \times C_{CH_3CO_2^-}}{C_{CH_3CO_2H}} = 1.8 \times 10^{-5}$$

b. Ionization of water

$$H_2O \rightleftharpoons H^+ + OH^-$$

From Unit 17,

$$K_w = C_{H^+} \times C_{OH^-} = 1.0 \times 10^{-14}$$

c. Effect of sodium ion

Sodium acetate is a salt that is completely ionized in water. The sodium ion undergoes no reaction with water, so

$$C_{Na^+} = C_{Na^+} = 5.0 \times 10^{-2} \text{ M}$$

(at equil)   (label)

Third, describe the equilibrium concentrations in terms of the principles of mass balance and electroneutrality.

a. Mass balance

$$C_{CH_3CO_2H} + C_{CH_3CO_2^-} = 1.74 \times 10^{-1} + 5.0 \times 10^{-2}$$

(at equilibrium)   (original label concentrations)

b. Electroneutrality

$$C_{H^+} + C_{Na^+} = C_{OH^-} + C_{CH_3CO_2^-}$$

that is,

$$C_{H^+} + (5.0 \times 10^{-2}) = \left(\frac{10^{-14}}{C_{H^+}}\right) + C_{CH_3CO_2^-}$$

Fourth, consider the complete equilibrium system and how concentrations are related.

a. For the equilibrium

$$CH_3CO_2H \rightleftharpoons H^+ + CH_3CO_2^-$$

Le Chatelier's principle (Unit 17) suggests that the additional acetate ion in the mixture will reduce the concentration of H$^+$ and increase the concentration of CH_3CO_2H correspondingly, as compared to a 0.174 M "acetic acid" solution alone.

b. For the equilibrium

$$H_2O \rightleftharpoons H^+ + OH^-$$

the fact that the complete equilibrium mixture contains more acid than base $(C_{CH_3CO_2H} > C_{CH_3CO_2^-})$ means that the concentration of H$^+$ will be higher than in pure water and the concentration of OH$^-$ (K_w/C_{H^+}) correspondingly lower.

The "exact" concentration terms can then be related through the equilibrium expressions and the principles of mass balance and electroneutrality.

From electroneutrality

$$C_{CH_3CO_2^-} = (5.0 \times 10^{-2}) + C_{H^+} - \left(\frac{10^{-14}}{C_{H^+}}\right)$$

From mass balance

$$\begin{aligned}
C_{CH_3CO_2H} &= (2.24 \times 10^{-1}) - C_{CH_3CO_2^-} \\
&= (2.24 \times 10^{-1}) - \left[(5.0 \times 10^{-2}) + C_{H^+} - \left(\frac{10^{-14}}{C_{H^+}}\right)\right] \\
&= (1.74 \times 10^{-1}) - C_{H^+} + \left(\frac{10^{-14}}{C_{H^+}}\right)
\end{aligned}$$

Then, in the K_a expression for acetic acid,

$$1.8 \times 10^{-5} = \frac{(C_{H^+})[(5.0 \times 10^{-2}) + C_{H^+} - (10^{-14}/C_{H^+})]}{[(1.74 \times 10^{-1}) - C_{H^+} + (10^{-14}/C_{H^+})]}$$

Fifth, introduce appropriate simplifications and solve. Considerations of the magnitudes of K_a and K_w and of the effect of the added acetate (Le Chatelier's principle) suggest that

$$C_{H^+} \ll 5.0 \times 10^{-2}$$

and

$$\frac{10^{-14}}{C_{H^+}} \ll 5.0 \times 10^{-2}$$

Thus

$$1.8 \times 10^{-5} \approx \frac{(C_{H^+})(5.0 \times 10^{-2})}{(1.74 \times 10^{-1})}$$

from which

$$C_{H^+} \approx 6.3 \times 10^{-5}$$

The remaining concentrations can then be determined from previously established relationships.

Total equilibrium system

$$\begin{aligned}
C_{CH_3CO_2H} &= 1.74 \times 10^{-1}\ M \\
C_{CH_3CO_2^-} &= 5.0 \times 10^{-2}\ M \\
C_{Na^+} &= 5.0 \times 10^{-2}\ M \\
C_{H^+} &= 6.3 \times 10^{-5}\ M \\
C_{OH^-} &= 1.6 \times 10^{-10}\ M
\end{aligned}$$

In most simple cases of homogeneous equilibria, approximate solutions are obtained much more rapidly than this lengthy example might suggest. There are many cases, however, in which applications of the principles of mass balance and electroneutrality are crucial to any reasonably accurate description of the equilibrium system.

In Unit 19 we considered situations in which simplifying approximations were satisfactory for aqueous weak acids and bases. One of the limitations imposed on this treatment was the requirement that K_a (or K_b) be $\sim 10^{-4}$ or smaller. Let us consider a case in which $K_a > 10^{-4}$ to see the principles of mass balance and electroneutrality can be applied to situations not amenable to a simpler treatment.

Example 2

The formula for dichloroacetic acid is Cl_2CHCO_2H.

What is the pH of a solution labeled 0.020 M dichloroacetic acid ($K_a = 5 \times 10^{-2}$)?

Solution
$$Cl_2CHCO_2H(aq) + H_2O(l) \rightleftharpoons H_3O^+(aq) + Cl_2CHCO_2^-(aq)$$
At equilibrium

$$C_{Cl_2CHCO_2^-} = C_{H_3O^+} \quad \text{(electroneutrality)}$$

(assuming C_{OH^-} from water is negligible)

$$\underset{\text{equilibrium}}{C_{Cl_2CHCO_2H}} + C_{Cl_2CHCO_2^-} = \underset{\text{label}}{0.020\ M} \quad \text{(mass balance)}$$

From combination of electroneutrality and mass balance relationships,

$$C_{Cl_2CHCO_2H} = 0.020 + C_{H_3O^+}$$
(equil)

For an equation in the form:
$ax^2 + bx + c = 0$,

$$x = \frac{-b \pm \sqrt{b^2 - 4ac}}{2a}$$

$$K_a = 5 \times 10^{-2} = \frac{C_{H_3O^+} \times C_{Cl_2CHCO_2^-}}{C_{Cl_2CHCO_2H}} = \frac{C^2_{H_3O^+}}{0.020 - C_{H_3O^+}}$$

Solution of this equation gives

$$C_{H^+} = 1.5 \times 10^{-2}$$

from which

$$pH = 1.82$$

If we had treated this case by the method applied for simpler situations,

$$C_{H^+} = \sqrt{K_a C_{acid}} \quad \text{(Unit 19)}$$
(label)

we would have calculated a hydrogen ion concentration of $3.2 \times 10^{-2}\ M$. This is obviously impossible for a solution that contained no more than 2.0×10^{-2} mole liter^{-1} of monoprotic acid.

21.3 POLYPROTIC ACIDS

Sulfuric acid is described as a diprotic acid. When treated with sodium hydroxide, for example, proton transfer is essentially a quantitative reaction:

$$H_2SO_4 + 2NaOH \longrightarrow Na_2SO_4 + 2H_2O$$

When the base is water, however, the proton transfer is less than quantitative:

$$H_2SO_4(aq) + H_2O(aq) \longrightarrow H_3O^+(aq) + HSO_4^-(aq)$$
$$\text{(essentially 100\%)}$$

$$HSO_4^-(aq) + H_2O(aq) \rightleftharpoons H_3O^+(aq) + SO_4^{2-}$$
$$\text{(measurably} < 100\%)$$

Although we could write a "combined" equation,

$$H_2SO_4 \rightleftharpoons 2H^+ + SO_4^{2-}$$

the equation might appear misleading in terms of the way we have been treating equilibrium systems. The H^+ concentration, for example, is *not* twice the SO_4^{2-} concentration, and there is a measurable equilibrium component, HSO_4^-, not indicated.

The situation is even more complicated with polyprotic acids weaker than H_2SO_4 or with those having even more equilibrium components. With phosphoric acid, for example, the equilibrium is best represented by three equations

$$\begin{array}{lll} H_3PO_4 \rightleftharpoons H^+ + H_2PO_4^- & K_{a_1} = 7.5 \times 10^{-3} \\ H_2PO_4^- \rightleftharpoons H^+ + HPO_4^{2-} & K_{a_2} = 6.2 \times 10^{-8} \\ HPO_4^{2-} \rightleftharpoons H^+ + PO_4^{3-} & K_{a_3} = 2.0 \times 10^{-13} \end{array}$$

Equilibria involving polyprotic acids may get a bit tricky. If we know the pH of the solution, a factor rather easy to measure, we can often decide on the principal species of importance and simplify the mathematics by omitting species of negligible concentration. In aqueous phosphoric acid, at moderate concentrations, we would be justified in neglecting the concentration of PO_4^{3-} as a first approximation, since it is a very minor component of the system at low pH, as shown by the very small K_{a_3}. Simplifying approximations become almost intuitive with practice. The best procedure, in general, is to work with the simplest reasonable mathematical situation, see how the results affect the validity of any approximations made, and then go on to more complicated analysis only when necessary. The approach to solving problems dealing with polyprotic acids may be summarized as follows:

1. Write equations and equilibrium constant expressions for all components of the interactive equilibrium system.
2. Write electroneutrality and mass-balance statements.
3. a. Decide on what seems to be a set of appropriate approximations in terms of species whose concentrations appear to be negligible.
   b. Express concentrations in terms of a single variable, typically C_{H^+}.
4. Set up equilibrium constant expressions to solve for total C_{H^+}, omitting terms that appear to be negligible. Use the simplest relationship to find a first approximation for C_{H^+}. Use this to check the suitability of any simplifications made. Correct by using more detailed analysis as necessary.
5. Use C_{H^+} to solve for concentrations of other components.

If the situation cannot be reduced to a relatively simple mathematical form, identify the system as "too complex." Interested students may wish to discuss such cases with faculty specializing in analytical chemistry to see how these systems may be handled.

Compare the system $Ca(IO_3)_2(s) \rightleftharpoons Ca^{2+}(aq) + 2IO_3^-(aq)$, in which $C_{IO_3^-} = 2 \times C_{Ca^{2+}}$.

Example 3

What is the composition of the equilibrium mixture in a solution labeled 0.010 M sulfuric acid? ($K_{a_2} = 1.3 \times 10^{-2}$.)

Solution

1. Equilibria

$$H_2SO_4 \longrightarrow H^+ + HSO_4^- \qquad \text{quantitative}$$

$$HSO_4^- \rightleftharpoons H^+ + SO_4^{2-} \qquad K_{a_2} = 1.3 \times 10^{-2} = \frac{C_{H^+} \times C_{SO_2^{2-}}}{C_{HSO_4^-}}$$

$$H_2O \rightleftharpoons H^+ + OH^- \qquad K_w = 1.0 \times 10^{-14} = C_{H^+} \times C_{OH^-}$$

2. a. Electroneutrality statement

$$C_{H^+} = C_{HSO_4^-} + C_{OH^-} + (2 \times C_{SO_4^{2-}})$$

Note that electroneutrality requires that $C_{SO_4^{2-}}$ be multiplied by 2, since each $2-$ ion must always be associated with two $+1$ ions.

b. Mass balance statement

$$0.010\ M = C_{HSO_4^-} + C_{SO_4^{2-}}$$

($C_{H_2SO_4}$ is too small to consider.)

3. Simplifying approximations and reduction of variables

a. C_{OH^-} is negligible compared to other concentrations, so

$$C_{H^+} \approx C_{HSO_4^-} + 2C_{SO_4^{2-}}$$

b. Then, from mass balance,

$$(C_{SO_4^{2-}} = 0.010 - C_{HSO_4^-})$$
$$C_{H^+} \approx C_{HSO_4^-} + 2(0.010 - C_{HSO_4^-})$$
$$\approx 0.020 - C_{HSO_4^-}$$
$$C_{HSO_4^-} \approx 0.020 - C_{H^+}$$

and, from mass balance,

$$C_{SO_4^{2-}} = 0.010 - (0.020 - C_{H^+})$$
$$= C_{H^+} - 0.010$$

4. Substituting in the K_{a_2} expression

$$\left(K_{a_2} = 1.3 \times 10^{-2} = \frac{C_{H^+} \times C_{SO_4^{2-}}}{C_{HSO_4^-}} \right)$$

$$1.3 \times 10^{-2} = \frac{C_{H^+} \times (C_{H^+} - 0.010)}{(0.020 - C_{H^+})}$$

from which

$$C_{H^+}^2 + (3 \times 10^{-3})C_{H^+} - (2.6 \times 10^{-4}) = 0$$

Solution of this equation gives

$$C_{H^+} = 1.47 \times 10^{-2}\ M$$

Then

$$C_{HSO_4^-} = 5.3 \times 10^{-3}\ M$$

$$C_{SO_4^{2-}} = 4.7 \times 10^{-3}\ M$$

and

$$C_{OH^-} = K_w/C_{H^+} = 6.8 \times 10^{-13}$$

Example 4

What is the pH of a solution labeled $0.10\ M$ phosphoric acid?

Solution

1. Equilibria

$$H_3PO_4 \rightleftharpoons H^+ + H_2PO_4^-$$

$$K_{a_1} = 7.5 \times 10^{-3} = \frac{C_{H^+} \times C_{H_2PO_4^-}}{C_{H_3PO_4}}$$

$$H_2PO_4^- \rightleftharpoons H^+ + HPO_4^{2-}$$

$$K_{a_2} = 6.2 \times 10^{-8} = \frac{C_{H^+} \times C_{HPO_4^{2-}}}{C_{H_2PO_4^-}}$$

$$HPO_4^{2-} \rightleftharpoons H^+ + PO_4^{3-}$$

$$K_{a_3} = 2.0 \times 10^{-13} = \frac{C_{H^+} \times C_{PO_4^{3-}}}{C_{HPO_4^{2-}}}$$

$$H_2O \rightleftharpoons H^+ + OH^-$$

$$K_w = 1.0 \times 10^{-14} = C_{H^+} \times C_{OH^-}$$

2. a. Electroneutrality statement

$$C_{H^+} = C_{H_2PO_4^-} + (2 \times C_{HPO_4^{2-}}) + (3 \times C_{PO_4^{3-}}) + C_{OH^-}$$

b. Mass balance statement

$$0.10\ M = C_{H_3PO_4} + C_{H_2PO_4^-} + C_{HPO_4^{2-}} + C_{PO_4^{3-}}$$

3. Simplifying approximations and reduction of variables
  a. The simplest assumption would be that, at the label concentration given, $C_{PO_4^{3-}}$, $C_{HPO_4^{2-}}$, and C_{OH^-} are negligible, so (from mass balance)
  b. $C_{H_3PO_4} = 0.10 - C_{H^+}$
  (i.e., $C_{H_2PO_4^-} \approx C_{H^+}$)
4. Substituting in the K_{a_1} expression

$$7.5 \times 10^{-3} = \frac{C_{H^+}^2}{0.10 - C_{H^+}}$$

Solution of the resulting quadratic equation gives

$$C_{H^+} = {\sim}2.4 \times 10^{-2}\ M$$
$$pH = 1.62$$

Then, as a check on the validity of our assumption that

$$C_{H_2PO_4^-} \approx C_{H^+}$$

substitution of 2.4×10^{-2} for C_{H^+} and $C_{H_2PO_4^-}$ in the equations for K_{a_2}, K_{a_3}, and K_w shows that the concentrations of HPO_4^{2-}, PO_4^{3-}, and OH^- were, indeed, negligible compared to H^+, $H_2PO_4^-$, and H_3PO_4 for this solution.

(*Suggestion:* Trace through the same analysis for $10^{-3}\ M$ phosphoric acid and $10^{-6}\ M$ phosphoric acid to see if either (or both) of these solutions would result in equations too complex for our methods to solve.)

21.4 pH-CONTROLLED PRECIPITATIONS

Many common precipitations, such as those involving sulfides or carbonates, involve anions that are conjugate bases of weak acids. As a result, solubility equilibria are often complicated by hydrolysis effects (Unit 19). These equilibria are not as simple as the ion-product treatment (Unit 18) seems to suggest, although such an approach is quite often a satisfactory approximation.

The use of anions from weak acids as precipitating agents has a particular utility in analytical chemistry. By controlling the pH of a solution, usually by an appropriate buffer system, the concentration of precipitating anion can be maintained at a level frequently useful in obtaining *selective* precipitation from cation mixtures. We shall consider three examples of such methods in this unit, and others are discussed in Excursion 2.

Hydroxide precipitations

Examples of the further complications occurring with amphoteric hydroxides are given in Section 21.9. Sparingly soluble acidic hydroxides are discussed in Unit 25.

When cations form insoluble basic hydroxides, two equilibria are important. The solubility equilibrium assumes that the hydroxides are completely ionized in solution:

$$M(OH)_x(s) \rightleftharpoons M^{x+}(aq) + X(OH^-)(aq) \qquad K_{sp} = C_{M^{x+}} \times C_{OH^-}^x$$
$$H_2O \rightleftharpoons H^+ + OH^- \qquad K_w = C_{H^+} \times C_{OH^-}$$

Since precipitation will occur whenever the ion product exceeds the solubility product, these precipitations may be controlled by controlling the pH of the solution.

Example 5

What ratio of $C_{NH_4^+}$ to C_{NH_3} will provide a buffer whose pH is just below the minimum for precipitation of iron(II) hydroxide from a solution of 0.10 M $Fe(NO_3)_2$? (K_{sp} for $Fe(OH)_2$ is 1.8×10^{-15}.)

Solution
The equilibria involved are

$$NH_4^+ \rightleftharpoons NH_3 + H^+ \qquad K_a = 5.6 \times 10^{-10} = \frac{C_{NH_3} \times C_{H^+}}{C_{NH_4^+}}$$
$$\text{(Unit 20)}$$

If precipitation occurs,

$$Fe(OH)_2(s) \rightleftharpoons Fe^{2+}(aq) + 2OH^-(aq) \qquad K_{sp} = 1.8 \times 10^{-15} = C_{Fe^{2+}} \times C_{OH^-}^2$$
$$H_2O \rightleftharpoons H^+ + OH^- \qquad K_w = 1.0 \times 10^{-14} = C_{H^+} \times C_{OH^-}$$

From the solubility product, at saturation,

$$C_{OH^-} = \sqrt{\frac{K_{sp}}{C_{Fe^{2+}}}} = \sqrt{\frac{1.8 \times 10^{-15}}{0.10}} = 1.34 \times 10^{-7}$$

If C_{OH^-} equals or exceeds this value, precipitation will occur. To avoid precipitation, then, C_{H^+} must be greater than

$$\frac{1.0 \times 10^{-14}}{1.34 \times 10^{-7}} \qquad \text{(i.e., } C_{H^+} > 7.5 \times 10^{-8}\text{)}$$

Then, since

$$C_{H^+} = K_a(C_{NH_4^+}/C_{NH_3}) \qquad \text{(Unit 20)}$$
$$\frac{C_{NH_4^+}}{C_{NH_3}} = \frac{C_{H^+}}{K_a} = \frac{7.5 \times 10^{-8}}{5.6 \times 10^{-10}} = 134$$

That is, the $C_{NH_4^+}$ must exceed $134 \times C_{NH_3}$ to provide a pH low enough to avoid precipitation.

Example 6

A 10.0-ml aliquot of saturated ammonium acetate is added to 10.0 ml of a solution that is 0.10 M in Mg^{2+} and 0.10 M in Zn^{2+}.

a. Will any precipitate form?
b. What is the composition of the equilibrium system in terms of metal ion concentrations?

(K_{sp} for $Mg(OH)_2$ is 1.2×10^{-11}; K_{sp} for $Zn(OH)_2$ is 4.5×10^{-17}.)

Solution
Ammonium acetate solution is effectively buffered at pH 7.0 (Unit 20). Thus C_{OH^-} is 1.0×10^{-7} M, and the ion products for the two hydroxides would be, respectively,
For $Mg(OH)_2$
(0.10 M Mg^{2+}, diluted $\frac{10}{20}$)

$$\text{ion product} = C_{Mg^{2+}} \times C_{OH^-}^2$$
$$= (5.0 \times 10^{-2}) \times (1.0 \times 10^{-7})^2 = 5.0 \times 10^{-16}$$

Since this value is smaller than the solubility product, magnesium hydroxide will not precipitate, and

$$C_{Mg^{2+}} = 5.0 \times 10^{-2}\, M \text{ at equilibrium.}$$

For $Zn(OH)_2$
(0.10 M Zn^{2+}, diluted $\frac{10}{20}$)

The "dilution factor" may be found from the relationship (Unit 13)

$$M_{conc} \times V_{conc} = M_{dil} \times V_{dil}$$

$$\text{ion product} = (5.0 \times 10^{-2}) \times (1.0 \times 10^{-7})^2 = 5.0 \times 10^{-16}$$

Since this value is larger than the solubility product, zinc hydroxide will precipitate, and the equilibrium concentration of Zn^{2+} can be estimated.

$$C_{Zn^{2+}} = \frac{K_{sp}}{C_{OH^-}^2} = \frac{4.5 \times 10^{-17}}{(1.0 \times 10^{-7})^2} = 4.5 \times 10^{-3}\, M$$

[Note: This solution assumes that the buffering action of saturated ammonium acetate is sufficient to maintain a pH of 7.0 even after removal of some OH^- by precipitation. It also assumes no complex formation by Mg^{2+} or Zn^{2+} (Sections 23.8 and 23.9).]

Carbonate precipitations

Carbon dioxide is quite soluble in water. Careful analysis of aqueous solutions of CO_2 indicates that the principal species at low pH is best represented as $CO_2(aq)$ rather than as carbonic acid, H_2CO_3. The equilibrium system may be formulated as

$$CO_2(g) \rightleftharpoons CO_2(aq) \qquad K_{eq} = \frac{C_{CO_2(aq)}}{P_{CO_2(g)}} = 3 \times 10^{-2} \text{ mole liter}^{-1}\text{ atm}^{-1}$$

It is more common to find the *net* equilibrium equation

$CO_2(aq) + H_2O(l) \rightleftharpoons$
$\qquad H^+(aq) + HCO_3^-(aq)$

for which $K_{a_1} = 4.2 \times 10^{-7}$.

$$CO_2(aq) + H_2O(l) \rightleftharpoons H_2CO_3(aq) \qquad K_{eq} = \frac{C_{H_2CO_3}}{C_{CO_2}} = 3.2 \times 10^{-3}$$

$$H_2CO_3(aq) \rightleftharpoons H^+(aq) + HCO_3^-(aq) \qquad K_{a_1} = 1.3 \times 10^{-4}$$
$$HCO_3^-(aq) \rightleftharpoons H^+(aq) + CO_3^{2-}(aq) \qquad K_{a_2} = 4.8 \times 10^{-11}$$
$$H_2O \rightleftharpoons H^+ + OH^- \qquad K_w = 1.0 \times 10^{-14}$$

Le Chatelier's principle suggests that CO_2 should dissolve more readily in alkaline solutions than in acidic solutions. (Note the effect of OH^- in successive equilibria.) This is, in fact, a significant problem in working with aqueous bases. Absorption of CO_2 when these solutions are exposed to the air may produce significant changes in concentrations.

Again, control of solution pH will determine concentrations of species such as CO_3^{2-} or HCO_3^- for possible precipitations. A common commercial method, the *Solvay Process,* for production of baking soda ($NaHCO_3$) uses concentrated aqueous ammonia saturated with sodium chloride. When this solution is saturated with carbon dioxide, the sparingly soluble $NaHCO_3$ is precipitated.

Swimming pools must be maintained at an alkaline pH to avoid loss of chlorine, the antibacterial additive. If the water is fairly hard (Unit 6), CO_2 absorbed from the air may result in growth of calcite crystals ($CaCO_3$) on metal or concrete surfaces. These sharp crystals are an unpleasant experience to swimmers.

Example 7

A swimming pool is buffered at pH 8.7 and is found to contain a total inorganic carbon concentration equivalent to 0.040 M CO_2. What is the maximum hardness in parts per million (ppm) of Ca^{2+} that can be permitted if it is desired to avoid formation of calcite deposits? (K_{sp} for $CaCO_3$ is 4.7×10^{-9}.)

Solution
Mass balance:

$$0.040\ M = C_{CO_2} + C_{H_2CO_3} + C_{HCO_3^-} + C_{CO_3^{2-}}$$

We cannot write an electroneutrality statement, since we do not know the complete composition of the solution (e.g., no details were given on the buffer system involved).

Since pH = 8.7,

$$C_{H^+} = 2.0 \times 10^{-9}\ M$$

Then (from K_{a_2}),

$$\frac{C_{CO_3^{2-}}}{C_{HCO_3^-}} = \frac{K_{a_2}}{C_{H^+}} = \frac{4.8 \times 10^{-11}}{2.0 \times 10^{-9}} = 2.4 \times 10^{-2}$$

This suggests that less than 5 percent error would be introduced by neglecting $C_{CO_3^{2-}}$ in the initial approximation. Since we know that $C_{H_2CO_3}$ is very small, the principal species are $CO_2(aq)$ and $HCO_3^-(aq)$. Then

$$4.2 \times 10^{-7} = \frac{C_{H^+} \times C_{HCO_3^-}}{C_{CO_2}}$$

and, from approximations in the mass-balance statement,

$$C_{CO_2} = 0.040 - C_{HCO_3^-}$$

so that

$$4.2 \times 10^{-7} = \frac{2.0 \times 10^{-9} \times C_{HCO_3^-}}{(0.040 - C_{HCO_3^-})}$$

from which

$$C_{HCO_3^-} = 3.98 \times 10^{-2} \approx 4.0 \times 10^{-2}$$

(i.e., at pH 8.7 nearly all the "inorganic carbon" is present as HCO_3^-).
Then (from K_{a_2}),

$$C_{CO_3^{2-}} = \frac{(4.8 \times 10^{-11})(4.0 \times 10^{-2})}{(2.0 \times 10^{-9})} = 9.6 \times 10^{-4} \, M$$

The maximum $C_{Ca^{2+}}$, then, is

$$C_{Ca^{2+}} = \frac{K_{sp}}{C_{CO_3^{2-}}} = \frac{4.7 \times 10^{-9}}{9.6 \times 10^{-4}} = 4.9 \times 10^{-6} \, M$$

In parts per million (for dilute aqueous solutions parts per million $\approx$ milligrams per liter),

$$\frac{4.9 \times 10^{-6} \, \text{mole}}{\text{liter}} \times \frac{4.0 \times 10^4 \, \text{mg}}{\text{mole}} = \sim 0.2 \, \text{ppm}$$

Concentrations of Ca^{2+} exceeding *20* ppm are not unusual in many municipal water systems.

Thus calcite formation would be expected if the water exceeded 0.2 ppm Ca^{2+}.

Sulfide precipitations

Hydrogen sulfide was for many years the most generally useful reagent for separation and analysis of a very large number of common cations. The value of the reagent was based partly on the variety of insoluble sulfides of differing solubility products (Appendix C) and partly on the wide range of S^{2-} concentration that can be obtained by simple pH control. Equilibria in aqueous hydrogen sulfide, saturated with H_2S gas, may be formulated as

$$H_2S(g) \rightleftharpoons H_2S(aq) \qquad K = \frac{C_{H_2S(aq)}}{P_{H_2S(g)}} = 1.0 \times 10^{-1} \, \text{mole liter}^{-1} \, \text{atm}^{-1}$$

$$H_2S(aq) \rightleftharpoons H^+(aq) + HS^-(aq) \qquad K_{a_1} = 1.0 \times 10^{-7}$$

$$HS^-(aq) \rightleftharpoons H^+(aq) + S^{2-}(aq) \qquad K_{a_2} = 1.3 \times 10^{-13}$$

$$H_2O(l) \rightleftharpoons H^+(aq) + OH^-(aq) \qquad K_w = 1.0 \times 10^{-14}$$

When the author was working his way through college, one of the jobs involved filling the Chemistry Building's low-pressure H_2S tank from a high-pressure gas cylinder. While the students and faculty were attending an assembly one day, the building custodian decided to disassemble the H_2S gas taps in the fume hoods to clean them. On returning to the building, the group discovered one unconscious custodian and a "redecorated" building. All of the nice white walls had been turned streaky brown as the H_2S combined with the lead ions in the paint.

Hydrogen sulfide gas has some serious disadvantages. It is unpleasant to smell (odor of "rotten eggs") and highly toxic (more toxic than HCN). Its reaction with metal ions causes deterioration of many types of paint, so that its use often resulted in very untidy laboratory areas.

By the late 1950s, hydrogen sulfide gas had been largely replaced by solutions of thioacetamide, $CH_3\overset{\overset{\displaystyle S}{\|}}{C}NH_2$. When these solutions are heated, the thioacetamide is decomposed, forming an aqueous solution of H_2S without the dangerous and unpleasant disadvantages of the older H_2S gas systems:

$$CH_3\overset{\overset{\displaystyle S}{\|}}{C}NH_2 + H_2O + H^+ \longrightarrow CH_3CO_2H + NH_4^+ + H_2S$$

Sulfide precipitations using pH control of S^{2-} ion concentrations follow the same general patterns as the carbonate precipitations described earlier. However, there are many more precipitations and a much wider range of anion concentrations possible with sulfides than with carbonates.

Today it is rare to find an analytical laboratory routinely using precipitation analyses. Instrumental methods have largely supplanted the older wet chemical techniques. There are, however, still many situations in which precipitation analyses are important, and the principles involved are basic to many aspects of modern chemistry.

21.5 EQUILIBRIA AND STOICHIOMETRY

It is easy to fall into the trap of concentrating on one aspect of a problem to the extent that other important features are neglected. This is true not only in mathematical problems but—even more embarrassing—in actual laboratory situations. It is not uncommon to consider simultaneous equilibria on a simplified basis without thinking back to the principles of stoichiometry. It is important to remember stoichiometric relationships, including the idea of the "limiting reagent" (Unit 4).

Every problem dealing with the establishment of an equilibrium system should consider, as a check, the principles of stoichiometry. This will avoid "impossible" answers to mathematical problems and errors in laboratory results from failure to obtain "complete" reaction.

Example 8

A precipitant solution was prepared by dissolving 0.10 mole of H_2S in a liter of concentrated buffer of pH 3.0. What is the concentration of Cu^{2+} in the equilibrium system resulting from mixing 10.0 ml of this solution with 10.0 ml of 0.30 M $Cu(NO_3)_2$? (K_{sp} for CuS is 8.0×10^{-37}.)

Solution
Analysis of the preequilibrium mixture gives

$$C_{Cu^{2+}} = 0.30 \ M \times \frac{10.0 \ ml}{20.0 \ ml} = 0.15 \ M$$

$$C_{H_2S} \approx 0.10 \ M \times \frac{10.0 \ ml}{20.0 \ ml} = 0.050 \ M$$

(H_2S is the principal form at pH 3.0.)

$$C_{S^{2-}} \approx \frac{K_{a_1} \times K_{a_2} \times C_{H_2S}}{C_{H^+}^2} \approx 6.5 \times 10^{-16} \ M$$

Then

ion product $= 0.15 \times 6.5 \times 10^{-16} = 9.8 \times 10^{-17}$

and, since ion product $> K_{sp}$, CuS will precipitate.

However, not all the Cu^{2+} can be precipitated even if all the H_2S is consumed. Complete consumption of all available H_2S (the limiting reagent in this case) could remove only one-third of the Cu^{2+}.

Since CuS is very insoluble and since Cu^{2+} is present in excess, it should be safe to assume that precipitation is essentially quantitative until the H_2S is consumed. The concentration of Cu^{2+} remaining at equilibrium, then, will be closely approximated by

$\frac{2}{3} \times 0.15 \ M = 0.10 \ M$

A description of $C_{S_2^-}$, C_{HS^-}, and C_{H_2S} would require a rather complex analysis. The interested student may wish to attempt this and check his work with the instructor.

There is no real difference between a coordinate covalent
and a "simple" covalent bond.
The former designation merely
indicates the origin of the electron pair, but there is nothing
unique about the bond itself.
For example, one of the bonds
in NH_4^+ is "coordinate covalent," but all four bonds are of
equal length and strength.

Covalent bonds have been described (Unit 7) as a pair of electrons shared between
the positive fields of two (or more) nuclei. A rather special case is sometimes made
for the situation in which one atom furnishes both electrons for the bond pair; this is
called a *coordinate* covalent bond. Species that can furnish one or more valence
electron pairs are considered *Lewis bases,* and species that accept a pair of electrons
to form a coordinate covalent bond are *Lewis acids* (Unit 19).

A number of interesting chemical species result from the formation of coordinate
covalent bonds. Species of this type having a net charge are called *complex ions.*
Neutral species, either molecular compounds or compounds containing a complex
ion associated with oppositely charged ions of the same net charge, are called
complex compounds or *coordination compounds.*

Examples

COORDINATION OF MOLECULES WITH ATOMS. The *Mond process* is used on a large
scale to separate nickel from cobalt, an element of very similar properties. The
mixture of nickel and cobalt is treated with carbon monoxide at 1 atm and 25°C.
Nickel atoms coordinate readily with four carbon monoxide molecules, whereas
cobalt reacts with the carbon monoxide very slowly. The products, $Ni(CO)_4$ and some
$Co_2(CO)_8$, are separated by distillation of the low-boiling nickel complex. When the
latter is heated to a high temperature, the complex is decomposed to metallic nickel
and carbon monoxide, which can be recycled through the process.

COORDINATION OF MOLECULES WITH IONS. When ionic substances dissolve in water,
the ions form more or less regular hydrates (Unit 13). The hydration of cations by
water results in complex ions, for example,

$$Cu^{2+} + 6H_2O \rightleftharpoons [Cu(H_2O)_6]^{2+}$$
$$\text{hexaaquocopper(II)}$$

This ion probably exists in
aqueous solutions principally
as diaquotetraaminecopper(II),
$[Cu(H_2O)_2(NH_3)_4]^{2+}$.

Ammonia, through the unshared electron pair on nitrogen, forms similar complexes.
The deep blue tetraaminecopper(II) is a familiar example.

COORDINATION OF ANIONS WITH CATIONS. A number of colored complexes result
from combination of cations and anions. The $[Fe(NCS)]^{2+}$ species, for example, is a
deep red complex often used as qualitative evidence for the presence of Fe^{3+}.
Occasionally, the formation of cation–anion complexes poses a problem in analytical
procedures relying on precipitations. Thus the concentration of Cl^- must be controlled for precipitation of lead(II) chloride.

$$Pb^{2+}(aq) + 2Cl^-(aq) \rightleftharpoons PbCl_2(s)$$

because, in excess Cl^-, precipitate is dissolved by the reaction

$$PbCl_2(s) + Cl^-(aq) \rightleftharpoons [PbCl_3]^-(aq)$$

A few complex ions have common ("trivial") names, which have been in use for so
long that they are essentially "permanent fixtures" in the chemical literature. In
particular, *ferrocyanide,* $[Fe(CN)_6]^{4-}$, and *ferricyanide,* $[Fe(CN)_6]^{3-}$, should be remembered. Modern terminology employs a systematic approach, which is described
in Appendix F.

No attempt need be made at this time to learn the details of the IUPAC nomenclature system. We shall consider it in detail in Unit 27. It is usually simple to deduce
the formula of the complex from its name, providing you are a bit familiar with the

prefix terms, ligand names, and some of the more common Latin names for metals as described in Appendix F.

21.8 EQUILIBRIA INVOLVING COMPLEX IONS

The species furnishing the electron pair for formation of the coordinate covalent bond are called *ligands*. For a more detailed discussion, see Unit 27.

See, for example, the equilibria described for the Cu^{2+}/NH_3 system in the introduction to this unit.

The formation of complex ions from simple cations and appropriate *ligands* is an example of Lewis acid–base chemistry, and the replacement of one ligand by another is a competitive process depending on such factors as the availability of the electron pair and the type (e.g., sigma or pi) of bond formed. It is unlikely that ligand exchange involves simultaneous replacement at all coordinated positions, as might be inferred from commonly used *net* equilibrium equations, for example,

$$[Cu(H_2O)_6]^{2+} + 4NH_3 \rightleftharpoons [Cu(H_2O)_2(NH_3)_4]^{2+} + 4H_2O$$

Rather, stepwise exchange is a more reasonable process, and equilibrium mixtures contain a number of intermediate species whose concentrations may or may not be negligible. In most cases in which the free ligand is present in large excess, a single complex ion may be treated as the principal species containing the coordinated metal ion. Many other situations are not so simple, and major errors may result from neglecting intermediate complexes. The equilibria involving complex ions can be expressed in terms of *formation constants,* which consider the complex as the "product," or in terms of *instability constants,* with the complex as "reactant." Appendix C lists some values of stepwise formation constants. The product of the stepwise constants is the overall (net) formation constant for the complex as normally written.

$$K_{net} = K_1 \times K_2 \times K_3 \cdots$$

(21.1)

Example 9

The stepwise equilibrium constants for formation of diamminesilver(I) are

$$\begin{aligned} Ag^+ + NH_3 &\rightleftharpoons [Ag(NH_3)]^+ & K_1 &= 2.0 \times 10^3 \\ [Ag(NH_3)]^+ + NH_3 &\rightleftharpoons [Ag(NH_3)_2]^+ & K_2 &= 8.0 \times 10^3 \end{aligned}$$

Calculate (a) the net formation constant, and (b) the net instability constant for diamminesilver(I).

Solution

$$K_1 = \frac{C_{[Ag(NH_3)]^+}}{C_{Ag^+} \times C_{NH_3}}$$

$$K_2 = \frac{C_{[Ag(NH_3)_2]^+}}{C_{[Ag(NH_3)]^+} \times C_{NH_3}}$$

The *net* equilibrium equation for formation is

$$Ag^+ + 2NH_3 \rightleftharpoons [Ag(NH_3)_2]^+$$

for which

$$K_{net} = \frac{C_{[Ag(NH_3)_2]^+}}{C_{Ag^+} \times C_{NH_3}^2}$$

The right-hand expression may be obtained from multiplication of the expressions for K_1 and K_2; that is,

$$\frac{C_{[Ag(NH_3)_2]^+}}{C_{Ag^+} \times C_{NH_3}^2} = \underbrace{\frac{C_{[Ag(NH_3)]^+}}{C_{Ag^+} \times C_{NH_3}}}_{(K_1)} \times \underbrace{\frac{C_{[Ag(NH_3)_2]^+}}{C_{[Ag(NH_3)]^+} \times C_{NH_3}}}_{(K_2)}$$

hence [equation (21.1)]:

$$K_{net} = K_1 \times K_2 = 2.0 \times 10^3 \times 8.0 \times 10^3$$
$$= 1.6 \times 10^7 \quad (formation \text{ constant})$$

The conventional *net* expression for *instability* is

$$[Ag(NH_3)_2]^+ \rightleftharpoons Ag^+ + 2NH_3$$

for which

$$K_{inst} = \frac{C_{Ag^+} \times C_{NH_3}^2}{C_{[Ag(NH_3)_2]^+}}$$

This is the inverse of the expression for the net *formation* constant, so

$$K_{inst} = \frac{1}{K_{formation}} \tag{21.2}$$

$$K_{inst} = \frac{1}{1.6 \times 10^7} = \sim 6.2 \times 10^{-8}$$

It should be apparent that systems involving complex ions are examples of simultaneous equilibria in which several interacting species must be considered. The detailed calculations associated with such systems require very careful use of the principles of mass balance and electroneutrality. Situations in which the ligand species is present in large excess are relatively simple, since "intermediate" complexes may usually be neglected on the assumption that the principal complex is that involving the maximum number of ligands. If the ligand is *not* in excess, significant errors may be introduced by neglecting intermediate species. We shall content ourselves with identifying such cases, since the mathematics required for accurate analysis will usually be quite cumbersome.

Example 10

What is the approximate concentration of free Ag^+ in the equilibrium mixture resulting from mixing 10.0 ml of 0.20 M silver nitrate with 15.0 ml of 3.0 M ammonia?

Solution
The complete equilibrium system is described by the equations

$$Ag^+(aq) + NH_3(aq) \rightleftharpoons [Ag(NH_3)]^+(aq)$$
$$[Ag(NH_3)]^+(aq) + NH_3(aq) \rightleftharpoons [Ag(NH_3)_2]^+(aq)$$
$$NH_3(aq) + H_2O(l) \rightleftharpoons NH_4^+(aq) + OH^-(aq)$$
$$H_2O(l) \rightleftharpoons H^+(aq) + OH^-(aq)$$

Since the final system will have a large excess of ammonia, we may consider the principal species of interest to be NH_3 and $[Ag(NH_3)_2]^+$. For these, we can work (as a reasonable approximation) with the net equation

$$[Ag(NH_3)_2]^+ \rightleftharpoons 2NH_3 + Ag^+$$

for which K_{inst} (Example 9) is 6.2×10^{-8}.

Then, before mixing

$$C_{Ag^+} = 0.20 \ M$$
$$C_{NH_3} = 3.0 \ M$$

After mixing, before reaction

$$C_{Ag^+} = 0.20 \ M \times \frac{10 \ ml}{25 \ ml} = 0.080 \ M$$

$$C_{NH_3} = 3.0 \ M \times \frac{15 \ ml}{25 \ ml} = 1.8 \ M$$

And at equilibrium

$$C_{[Ag(NH_3)_2]^+} = \sim 0.080 \ M$$

(nearly all Ag^+ converted to complex in excess NH_3) and

$$C_{NH_3} = 1.8 - 2(0.08) = 1.64 \ M$$

(two NH_3 consumed per complex formed). Then

$$K_{inst} = \frac{C_{NH_3}^2 \times C_{Ag^+}}{C_{[Ag(NH_3)_2]^+}}$$

$$6.2 \times 10^{-8} = \frac{(1.64)^2 \times C_{Ag^+}}{0.080}$$

from which

$$C_{Ag^+} = \frac{6.2 \times 10^{-8} \times 8.0 \times 10^{-2}}{2.69}$$

$$= \sim 1.8 \times 10^{-9} \ M$$

To check the validity of neglecting $[Ag(NH_3)]^+$, from K_1 (Example 9)

$$C_{[Ag(NH_3)]^+} = 2.0 \times 10^3 \times 1.8 \times 10^{-9} \times 1.64 = 5.9 \times 10^{-6}$$

$$(K_1 \times C_{Ag^+} \times C_{NH_3})$$

Thus, $C_{[Ag(NH_3)]^+}$ *was* negligible compared to $C_{[Ag(NH_3)_2]^+}$.

Example 11

Show that $C_{[Ag(NH_3)]^+}$ cannot be neglected without significant error in calculating the C_{Ag^+} in the equilibrium mixture resulting from mixing equal volumes of 0.40 M $AgNO_3$ and 0.80 M NH_3.

Solution
Before mixing

$$C_{Ag^+} = 0.40 \ M$$
$$C_{NH_3} = 0.80 \ M$$

After mixing, before reaction

$$C_{Ag^+} = 0.40 \ M \times \tfrac{1}{2} = 0.20 \ M$$
$$C_{NH_3} = 0.80 \ M \times \tfrac{1}{2} = 0.40 \ M$$

If we assume stoichiometric combination to form 0.20 M $[Ag(NH_3)_2]^+$ and neglect $[Ag(NH_3)]^+$, then

$$[Ag(NH_3)_2]^+ \rightleftharpoons Ag^+ + 2NH_3 \qquad K_{inst} = 6.2 \times 10^{-8}$$

and (remember that $[Ag(NH_3)]^+$ is neglected)

$$C_{NH_3} = 2 \times C_{Ag^+}$$

Then

$$6.2 \times 10^{-8} = \frac{C_{Ag^+} \times (2C_{Ag^+})^2}{0.20}$$

$$C_{Ag^+} = \sqrt[3]{\frac{6.2 \times 10^{-8} \times 0.20}{4}} = 1.46 \times 10^{-3} \, M$$

and

$$C_{NH_3} = 2 \times 1.46 \times 10^{-3} = 2.92 \times 10^{-3} \, M$$

Now, as a rough check on our simplifications, let us see what the $C_{[Ag(NH_3)]^+}$ would have been if the $C_{[Ag(NH_3)_2]^+}$ were really 0.20 M and the C_{NH_3} were really 2.9×10^{-3}. From K_2,

$$8 \times 10^3 = \frac{C_{[Ag(NH_3)_2]^+}}{C_{[Ag(NH_3)]^+} \times C_{NH_3}}$$

from which

$$C_{[Ag(NH_3)]^+} = \frac{0.20}{8 \times 10^3 \times 2.9 \times 10^{-3}} = 8.7 \times 10^{-3}$$

But this suggests that more Ag^+ is tied up in the $[Ag(NH_3)]^+$ complex than was available in our original estimate of free Ag^+ ($1.46 \times 10^{-3} \, M$). As a result we must conclude that neglecting $[Ag(NH_3)]^+$ in this particular case introduces major errors.

A complete analysis of the equilibrium system is quite complicated. It shows, however, that the concentration of free Ag^+ is less than one-half that calculated by the "simplified" method.

It is important to realize that use of any net equilibrium expression must be considered as a rough approximation at best. The accuracy of the approximation will depend on the relative concentrations of species involved and on the magnitudes of the stepwise equilibrium constants.[2]

21.9 AMPHOTERIC HYDROXIDES   The hydroxides of most metals are basic. The oxygen–metal bond in the typical metal hydroxides is essentially ionic, so that these salts act to furnish the basic hydroxide ion; for example,

$$NaOH(s) \longrightarrow Na^+(aq) + OH^-(aq)$$
$$Ba(OH)_2(s) \longrightarrow Ba^{2+}(aq) + 2OH^-(aq)$$

Some hydroxides are acidic. In particular, $B(OH)_3$ and $Si(OH)_4$ (Unit 25) are weak acids. $B(OH)_3$, boric acid, acts as an acid in the Lewis sense, but the net result in aqueous solution is an increase in the hydrogen ion concentration:

$$B(OH)_3(aq) + H_2O(l) \rightleftharpoons [B(OH)_4]^-(aq) + H^+(aq)$$

[2] For further discussion of this point, see J. E. Banks, "The Equilibria of Complex Formation," in the *Journal of Chemical Education* (Vol. 38, p. 391, 1961).

Both boron and silicon form strong covalent bonds to oxygen.

Intermediate between the basic and acidic hydroxides is a group in which the metal–oxygen bond may be thought of as polar covalent (i.e., having significant ionic character). These *amphoteric* (Unit 3) hydroxides are characterized by central metal atoms whose ions would have a high charge density. In addition, the valence orbitals of the metals involved are such as to permit formation of stable bonds to oxygen. Examples of common amphoteric hydroxides include

Beryllium hydroxide
   as a base:

$$Be(OH)_2(aq) + 2H^+(aq) \rightleftharpoons Be^{2+}(aq) + 2H_2O(l)$$

   as an acid:

$$Be(OH)_2(aq) + 2OH^-(aq) \rightleftharpoons [Be(OH)_4]^{2-}(aq)$$

Aluminum hydroxide
   as a base:

$$Al(OH)_3(s) + 3H^+(aq) \rightleftharpoons Al^{3+}(aq) + 3H_2O(l)$$

   as an acid:

$$Al(OH)_3(s) + OH^-(aq) \rightleftharpoons [Al(OH)_4]^-(aq)$$

Chromium(III) hydroxide
   as a base:

$$Cr(OH)_3(s) + 3H^+(aq) \rightleftharpoons Cr^{3+}(aq) + 3H_2O$$

   as an acid:

$$Cr(OH)_3(s) + OH^-(aq) \rightleftharpoons [Cr(OH)_4]^-(aq)$$

Tin(IV) hydroxide
   as a base:

$$Sn(OH)_4(s) + 4H^+(aq) \rightleftharpoons Sn^{4+}(aq) + 4H_2O(l)$$

   as an acid:

$$Sn(OH)_4(s) + 2OH^-(aq) \rightleftharpoons [Sn(OH)_6]^{2-}(aq)$$

As in other cases involving complex ions, the *net* equations for amphoteric behavior are oversimplifications, and intermediate species may be quite significant. Because of the very high charge density of +4 ions, for example, the "free" hydrated Sn^{4+} has no measurable concentration in aqueous solution. Additional details on amphoteric hydroxides, including equilibrium constants, are given in Appendix D.

Further examples of important complexes will be encountered in later units.

SUGGESTED READING   **Moeller, Therald, and Rod O'Connor.** 1972. *Ions in Aqueous Systems.* New York: McGraw-Hill.
O'Connor, Rod, and Charles Mickey. 1975. *Solving Problems in Chemistry.* New York: Harper & Row.
Quagliano, J. V., and L. J. Vallarino. 1969. *Coordination Chemistry.* Indianapolis, Ind.: Heath-Raytheon.

EXERCISES

1. Describe the composition of the equilibrium system labeled 0.20 M phosphorous acid. ($K_{a_1} = 1.6 \times 10^{-2}$; $K_{a_2} = 6.9 \times 10^{-7}$; the third hydrogen does not dissociate.)

2. What is the minimum pH at which manganese(II) hydroxide will precipitate from a solution 0.30 M in Mn^{2+}? (K_{sp} for $Mn(OH)_2$ is 2.0×10^{-13}.)

3. Within what pH range must a solution be buffered in order selectively to precipitate barium carbonate from a solution that is 0.10 M in Ba^{2+} and 0.10 M in Mg^{2+} by addition of an equal volume of 0.50 M sodium bicarbonate solution? Assume buffer capacity is sufficient to maintain the pH constant during the entire process. (K_{sp} for $MgCO_3$ is 4.0×10^{-5}; K_{sp} for $BaCO_3$ is 1.6×10^{-9}.)

4. What is the minimum pH for precipitation of MnS from a 0.30 M solution of Mn^{2+} if the solution is saturated with H_2S gas at 1.0 atm? (K_{sp} for MnS is 7.0×10^{-16}.) Will MnS be "completely" precipitated (i.e., > 99.9 percent) under these conditions? Explain.

5. a. What is the approximate concentration of free Hg^{2+} in the equilibrium system resulting from mixing 10.0 ml of 0.060 M $Hg(NO_3)_2$ with 20.0 ml of 6.0 M HCl? The net instability constant for $[HgCl_4]^{2-}$ is 8.5×10^{-16}.

   b. Explain why the equilibrium concentration of free Hg^{2+} in a solution prepared by mixing equal volumes of 0.20 M $Hg(NO_3)_2$ and 0.80 M HCl cannot be calculated accurately from the expression for the net instability constant of tetrachloromercurate(II).

PRACTICE FOR PROFICIENCY

Objectives (a), (b), and (c)

1. Describe the composition of the equilibrium system labeled 0.50 M sulfurous acid. ($K_{a_1} = 1.6 \times 10^{-2}$, $K_{a_2} = 1.3 \times 10^{-7}$.)

2. Use *mass balance* and *electroneutrality* considerations to verify the pH information given for ammonium acetate and for bicarbonate ion in Table 20.1 (Unit 20).

3. For each of the following solutions, for which *label* concentrations are given, make a careful analysis of the various interactive equilibria. Identify those cases for which simplifying approximations are inadequate. For cases in which appropriate approximations *can* be made, calculate the equilibrium concentrations of all components.
   a. 0.30 M H_2SO_4; b. 0.30 M Na_2SO_4; c. 0.15 M H_3PO_4; d. 0.010 M KH_2PO_4; e. 0.010 M Na_3PO_4

Objectives (a), (b), and (d)

4. What is the minimum pH at which cobalt(II) hydroxide will precipitate from a solution 0.020 M in Co^{2+}? (K_{sp} for $Co(OH)_2$ is 2.0×10^{-16}.)

5. What is the minimum pH for precipitation of FeS from a 0.20 M solution of Fe^{2+} if the solution is saturated with H_2S gas at 1.0 atm? (K_{sp} for FeS is 4.0×10^{-19}.) Will FeS be "completely" precipitated (i.e., > 99.9 percent) under these conditions? Explain.

6. What is the maximum pH at which a solution may be buffered to avoid precipitation of manganese(II) hydroxide from a 0.25 M solution of Mn^{2+}?

7. To a concentrated buffer of pH 7.0 was added an equal volume of a solution which was 0.20 M in each of the ions Mn^{2+}, Cu^{2+}, and Fe^{3+}. The precipitate expected should contain
   a. only $Mn(OH)_2$ b. only $Cu(OH)_2$
   c. only $Fe(OH)_3$ d. only $Cu(OH)_2$ and $Fe(OH)_3$
   e. $Mn(OH)_2$, $Cu(OH)_2$, and $Fe(OH)_3$
   (K_{sp} for $Mn(OH)_2$ is 2.0×10^{-13}; K_{sp} for $Cu(OH)_2$ is 1.8×10^{-19}; and K_{sp} for $Fe(OH)_3$ is 6.0×10^{-38})

8. Which ratio of $C_{NH_4^+}/C_{NH_3}$ would provide a buffer of pH low enough to avoid precipitation of $Co(OH)_2$ from a 0.0002 M Co^{2+} solution? (K_b for NH_3 is 1.8×10^{-5}; K_{sp} for $Co(OH)_2$ is 2.0×10^{-16}.)
   a. 1/20; b. 1/4; c. 1/1; d. 4/1; e. 20/1

9. What is the maximum pH at which a solution may be buffered to avoid precipitation of iron(III) hydroxide from a 7.5×10^{-3} M solution of Fe^{3+}?

10. Which ratio of $C_{NH_4^+}/C_{NH_3}$ would provide a buffer of pH low enough to avoid precipitation of $Cd(OH)_2$ from a 0.50 M Cd^{2+} solution? (K_b for NH_3 is 1.8×10^{-5}; K_{sp} for $Cd(OH)_2$ is 2.0×10^{-14}.)
   a. 1/50; b. 1/100; c. 1/1; d. 100/1; e. 50/1

11. To a concentrated buffer of pH 9.0 was added an equal volume of a solution which was 0.20 M in Ca^{2+}, Cd^{2+}, and Cu^{2+}. The precipitate expected should contain
    a. only $Ca(OH)_2$ b. only $Cd(OH)_2$
    c. only $Cu(OH)_2$ d. only $Cd(OH)_2$ and $Cu(OH)_2$
    e. $Ca(OH)_2$, $Cd(OH)_2$, and $Cu(OH)_2$
    (K_{sp} for $Ca(OH)_2$ is 4.0×10^{-6}; K_{sp} for $Cd(OH)_2$ is 2.0×10^{-14}; K_{sp} for $Cu(OH)_2$ is 1.8×10^{-19}.)

12. Four waste reagents were dumped into the sink so that their contents were mixed in equal volumes, along with an equal volume of distilled water, in the sink trap. The reagents were 0.05 M NaOH, 0.05 M $CuSO_4$, 0.05 M $CaCl_2$, and 0.05 M $Mg(NO_3)_2$. A precipitate formed in the sink trap, probably consisting of
    a. only $Ca(OH)_2$ b. only $MgSO_4$
    c. only $NaNO_3$ d. only $MgCl_2$
    e. only $CaSO_4$, $Cu(OH)_2$, and $Mg(OH)_2$

13. A swimming pool was sufficiently alkaline so that CO_2 absorbed from the air produced in the pool a solution that was 2×10^{-4} M in CO_3^{2-}. If the pool water was originally 4×10^{-3} M in Mg^{2+}, 6×10^{-4} M in Ca^{2+}, and 8×10^{-7} M in Fe^{2+}, then a precipitate should form of
    a. only $MgCO_3$ b. only $CaCO_3$
    c. only $FeCO_3$ d. only $CaCO_3$ and $FeCO_3$
    e. $MgCO_3$, $CaCO_3$, $FeCO_3$

14. The water distributed by a municipal water company was chlorinated and fluoridated. Analysis showed it to be 3×10^{-3} M in Cl^-, 5×10^{-4} M in F^-, and 7×10^{-5} M in SO_4^{2-}. If the original water had a "hardness" of 2×10^{-2} M Ca^{2+}, then
    a. A precipitate of $CaCl_2$ should have formed.

b. A precipitate of CaF_2 should have formed.

c. A precipitate of $CaSO_4$ should have formed.

d. A precipitate containing both CaF_2 and $CaSO_4$ should have formed.

e. No precipitate of $CaCl_2$, CaF_2, or $CaSO_4$ should have formed.

15. The water in a municipal sewage plant was analyzed as 2×10^{-6} M OH^-, 2×10^{-5} M CO_3^{2-}, 2×10^{-25} M S^{2-}, and 2×10^{-8} M SO_4^{2-}. To the water was added copper sulfate as an algaecide, in the amount of 4×10^{-8} mole per liter of water. Which of the following should precipitate?

a. only $Cu(OH)_2$ and CuS; **b.** only $CuCO_3$ and $CuSO_4$; **c.** only $CuSO_4$; **d.** only $Cu(OH)_2$; **e.** only CuS

Objectives (a), (b), and (e)

16. What is the approximate concentration of free Zn^{2+} in a solution prepared by mixing equal volumes of 0.10 M $Zn(NO_3)_2$ and 5.0 M ammonia? (The net instability constant for tetraamminezinc(II) is 3.6×10^{-10}.)

17.

$Cd^{2+} + Cl^- \rightleftharpoons [CdCl]^+$	$K_1 = 100$
$[CdCl]^+ + Cl^- \rightleftharpoons [CdCl_2]$	$K_2 = 5$

$[CdCl_2] + Cl^- \rightleftharpoons [CdCl_3]^-$	$K_3 = 0.4$
$[CdCl_3]^- + Cl^- \rightleftharpoons [CdCl_4]^{2-}$	$K_4 = 0.2$

For tetrachlorocadmate(II):

a. In a solution prepared by mixing equimolar amounts of Cd^{2+} and Cl^-, the concentration of $[CdCl_4]^{2-}$ is zero.

b. The net instability constant is approximately 2.5×10^{-2}.

c. The net formation constant is approximately 106.

d. In a solution prepared by mixing equal volumes of 0.1 M Cd^{2+} and 0.4 M Cl^-, the concentration of $[CdCl]^+$ is zero.

e. None of the above is correct.

18. What is the approximate concentration (in mole liter^{-1}) of free Ag^+ ion in a solution prepared by mixing equal volumes of 0.050 M Ag^+ and 10.0 M NH_3? (The net formation constant for $[Ag(NH_3)_2]^+$ is 1.6×10^7.)

19. What is the approximate concentration (in mole liter^{-1}) of free Hg^{2+} ion in a solution prepared by mixing equal volumes of 0.02 M Hg^{2+} and 8.0 M Cl^-? (The net formation constant for $[HgCl_4]^{2-}$ is 1.2×10^{15}.)

20. What is the approximate concentration (in mole liter^{-1}) of free Fe^{3+} ion in a solution prepared by mixing equal volumes of 0.06 M Fe^{3+} and 4.0 M F^-? (The net formation constant for $[FeF_5(H_2O)]^{2-}$ is 2.0×10^{15}.)

Unit 21 Self-Test

1. What is the pH of a solution labeled 0.50 M sulfurous acid? ($K_{a_1} = 1.6 \times 10^{-2}$. $K_{a_2} = 1.3 \times 10^{-7}$.)

2. What is the minimum pH for precipitation of cadmium hydroxide from a 0.30 M solution of Cd^{2+}? (K_{sp} for $Cd(OH)_2$ is 2.0×10^{-14}.)

3. What is the composition of the precipitate formed when 10.0 ml of 0.10 M $NaHCO_3$ is added to 10.0 ml of a solution of 0.010 M Mg^{2+} and 0.010 M Sr^{2+} in saturated ammonium acetate (pH 7.0)? (K_{sp} for $MgCO_3$ is 4.0×10^{-5}; K_{sp} for $SrCO_3$ is 7.0×10^{-10}; K_{a_2} for carbonic acid is 4.8×10^{-11}.)

4. What is the composition of the precipitate formed when H_2S is bubbled through a solution of 0.010 M Zn^{2+}, 0.010 M Pb^{2+}, and 0.010 M Mn^{2+} buffered at pH 2.0 until 0.10 mole liter^{-1} of H_2S has been added? (K_{sp} for ZnS is 1.6×10^{-23}; K_{sp} for PbS is 7.0×10^{-29}; K_{sp} for MnS is 7.0×10^{-16}; K_{a_1} for H_2S is 1.0×10^{-7}; K_{a_2} is 1.3×10^{-13}.)

5. What is the approximate concentration of free Co^{2+} in a solution prepared by mixing 20.0 ml of 0.010 M $Co(NO_3)_2$ with 30.0 ml of 5.0 M NH_4NCS? (The net instability constant for $[Co(NCS)_4]^{2-}$ is 3.8×10^{-3}.)

Overview Test Section SEVEN

(Consult Appendix C, as necessary.)

1. In the Lewis system, a base is defined as
   a. an electron-pair acceptor; **b.** a proton acceptor; **c.** a proton donor; **d.** a hydroxide donor; **e.** an electron-pair donor

2. In the Brønsted–Lowry system, a base is defined as
   a. a proton donor; **b.** a hydroxide donor; **c.** an electron-pair acceptor; **d.** a water former; **e.** none of these

3. $NH_4^+ + H_2O \rightleftharpoons NH_3 + H_3O^+$
   In the above system, the *conjugate pairs* are
   a. NH_4^+/NH_3 and H_2O/H_3O^+; **b.** NH_4^+/H_3O^+ and H_2O/NH_3; **c.** NH_4^+/H_2O and NH_3/H_3O^+; **d.** only NH_4^+ and H_3O^+; **e.** only H_2O and NH_3

4. What is the approximate pH of a solution labeled 3.8×10^{-5} M HNO_3?
   a. 3.0; **b.** 4.4; **c.** 5.0; **d.** 6.1; **e.** 8.3

5. What is the approximate pH of a solution labeled 0.0003 M $Ca(OH)_2$?
   a. 11.8; **b.** 10.8; **c.** 9.4; **d.** 8.6; **e.** 7.8

6. What is the approximate pH of a solution labeled 0.30 M $CH_3CH_2CO_2H$? ($K_a = 1.4 \times 10^{-5}$)
   a. 2.7; **b.** 3.2; **c.** 3.8; **d.** 4.1; **e.** 5.3

7. What is the approximate pH of a solution labeled 0.050 M NH_3? ($K_b = 1.8 \times 10^{-5}$)
   a. 8.1; **b.** 9.1; **c.** 11.5; **d.** 11.3; **e.** 11.0

8. What is the approximate pH of a solution labeled 0.20 M NH_4NO_3?
   a. 3.9; **b.** 4.3; **c.** 4.7; **d.** 5.0; **e.** 5.3

9. What is the approximate pH of a solution labeled 0.50 M potassium cyanide? (K_a for HCN is 4.8×10^{-10}.)
   a. 13.2; **b.** 12.7; **c.** 10.0; **d.** 10.3; **e.** 11.5

10. Which of these acids is probably the weakest?
    a. HI; **b.** HIO; **c.** HIO_2; **d.** HIO_3; **e.** HIO_4

11. Which of these acids is probably the strongest?
    a. $H_2PO_2^-$; **b.** $H_2PO_3^-$; **c.** H_3BO_3; **d.** H_3PO_4; **e.** $H_2PO_4^-$

12. Which of these acids is probably the weakest?
    a. HF; **b.** HCl; **c.** HBr; **d.** HI; **e.** HNO_3

13. What is the approximate pH at the equivalence point for the titration of 0.25 M methylamine with 0.50 M hydrochloric acid?

    a. 5.2; **b.** 5.7; **c.** 6.1; **d.** 6.5; **e.** 7.0
    What indicator (from Table 19.11) would be most appropriate for this titration? _____

14. What is the approximate pH of a solution in which the concentration of ammonia is 0.20 M and the concentration of ammonium nitrate is 1.0 M? (K_b for ammonia is 1.8×10^{-5}.)
    a. 7.6; **b.** 8.1; **c.** 10.0; **d.** 10.8; **e.** 8.6

15. What is the approximate pH of a 0.25 M solution of sodium bicarbonate?
    a. 6.57; **b.** 7.00; **c.** 8.35; **d.** 9.65; **e.** 10.0

16. What pH *change* would result from the addition of 5.0 ml of 0.10 M HCl to 50.0 ml of a 0.10 M NH_3/0.10 M NH_4^+ buffer?
    a. 0.01; **b.** 0.03; **c.** 0.09; **d.** 0.20; **e.** 0.60

17. What is the buffer capacity of a 0.10 M NH_3/0.30 M NH_4^+ buffer, for added H^+?
    a. 0.09; **b.** 0.17; **c.** 0.25; **d.** 0.30; **e.** 0.66

18. What pH *change* would result from the addition of 10.0 ml of 1.0 M NaOH to 50.0 ml of 0.10 M NH_3/0.20 M NH_4^+ buffer?
    a. 0.16; **b.** 0.84; **c.** 1.3; **d.** 1.6; **e.** 2.4

19. What is the approximate pH of a solution labeled 0.036 M H_2SO_4?
    a. 1.15; **b.** 1.36; **c.** 1.45; **d.** 2.55; **e.** 2.85

20. What is the maximum pH at which a solution may be buffered to avoid precipitation of cadmium hydroxide from a 0.50 M solution of Cd^{2+}? (K_{sp} for $Cd(OH)_2$ is 2.0×10^{-14}.)
    a. 5.7; **b.** 6.8; **c.** 7.0; **d.** 7.3; **e.** 8.5

21. To a concentrated buffer of pH 3.0 was added an equal volume of a solution that was 0.20 M in each of the ions Mn^{2+}, Cu^{2+}, and Fe^{3+}. The precipitate expected should contain
    a. only $Mn(OH)_2$ **b.** only $Cu(OH)_2$
    c. only $Fe(OH)_3$ **d.** only $Cu(OH)_2$ and $Fe(OH)_3$
    e. $Mn(OH)_2$, $Cu(OH)_2$, and $Fe(OH)_3$

22. What is the approximate concentration (in mole liter^{-1}) of free Cd^{2+} ion in a solution prepared by mixing equal volumes of 0.06 M Cd^{2+} and 4.0 M CN^-? (The net formation constant for $[Cd(CN)_4]^{2-}$ is 5.9×10^{18}.)
    a. $\sim 10^{-15}$;   **b.** $\sim 10^{-17}$ **c.** $\sim 10^{-20}$ **d.** $\sim 10^{-22}$ **e.** $\sim 10^{-25}$

Excursion 2

Wet chemical analyses

The chemical laboratory today bears little resemblance to the dark, bottle-strewn room of the analytical chemist a few years ago. Modern instruments in neat air-conditioned areas have largely replaced the collections of filters, centrifuges, and reagent racks used for routine analytical tests. The ink-smeared notebooks have given way to punchcards and computer tape. Ion-exchange columns and other new techniques have provided rapid quantitative separation methods to supplant the tedious sulfide precipitations, and the "heady aroma" of hydrogen sulfide has almost disappeared from the halls. "Wet chemical" analyses, however, still play a vital role in modern research, analysis, and quality-control laboratories.

Anions are rarely suitable for analysis by instrumental techniques, and these species are now the most common challenge to the analyst. The chemistry of some common anions is discussed in Unit 26. Metals and metal ions, on the other hand, are now usually assayed by spectroscopic methods. In complex mixtures it is sometimes expeditious to effect partial separations of major components or species of particular interest by "traditional" precipitations to simplify data obtained from subsequent instrumental procedures. Ore samples, for example, are frequently subjected to such preliminary treatment.

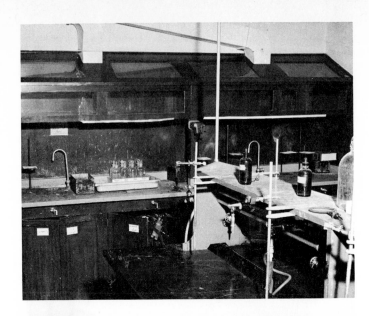

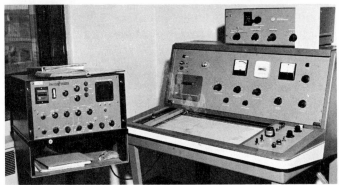

1 COMPLEX IONS IN QUALITATIVE ANALYSIS

Qualitative analysis of inorganic materials involves a determination of the cations and anions present in the sample. The formation and stability of complexes are important in many of the tests used for common ions.

Examples

1. The diaquotetraaminecopper(II) ion has a deep blue color, which is so intense that the formation of the complex can be used to detect Cu^{2+} in very dilute solutions:

$$[Cu(H_2O)_6]^{2+} + 4NH_3 \rightleftharpoons [Cu(NH_3)_4(H_2O)_2]^{2+} + 4H_2O$$
light blue   dark blue
(intense)

Although aquo complexes are shown in these examples, it is more common to formulate complexes in aqueous solution without indicating the associated water molecules, for example,

$$[Cu(NH_3)_4(H_2O)_2]^{2+}$$
$$\equiv [Cu(NH_3)_4]^{2+}$$
$$[Fe(NCS)(H_2O)_5]^{2+}$$
$$\equiv [Fe(NCS)]^{2+}$$

2. Iron(III) ion forms a deep wine red complex with thiocyanate ion:

$$[Fe(H_2O)_6]^{3+} + NCS^- \rightleftharpoons [Fe(NCS)(H_2O)_5]^{2+} + H_2O$$

3. A common test for the presence of cyanide ion involves a series of reactions eventually producing the salt of a complex anion:

a. Separation of cyanide from mixture

$$CN^-(aq) + H^+(aq) \longrightarrow HCN(g)$$

b. Recovery of cyanide

$$HCN(g) + OH^-(aq) \longrightarrow H_2O(l) + CN^-(aq)$$

c. Conversion to complex anion

$$6CN^-(aq) + Fe^{2+}(aq) \longrightarrow [Fe(CN)_6]^{4-}(aq)$$

d. Precipitation for detection

$$Na^+(aq) + Fe^{3+}(aq) + [Fe(CN)_6]^{4-}(aq) \longrightarrow NaFe[Fe(CN)_6](s)$$
$$\text{blue-green}$$

4, The competition among different ligands for particular metal ions is often useful in tests for ions in mixtures. For example, both Fe^{3+} and Co^{2+} form colored thiocyanato complexes (deep red and pale blue, respectively). The intense color of $[Fe(NCS)]^{2+}$ obscures the much less intense blue of the $[Co(NCS)_4]^{2-}$. Fortunately, F^- has a greater affinity for Fe^{3+} than does NCS^-, but the latter has a greater affinity than F^- for Co^{2+}. Cobalt(II) and iron(III) may, then, both be detected in the same solution by treating separate samples with thiocyanate alone or a thiocyanate-fluoride mixture.

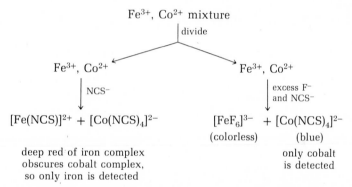

deep red of iron complex obscures cobalt complex, so only iron is detected	only cobalt is detected

2 STOICHIOMETRY IN ANALYSIS

Many of the more important analytical methods depend on reactions performed under essentially irreversible conditions. Gravimetric analysis, for example, involves precipitation of some sparingly soluble salt—usually in the presence of excess precipitating ion—and isolation of the precipitate. Titrations of strong acids or bases, typically to an endpoint near pH 7, depend on the reaction $H^+ + OH^- \longrightarrow H_2O$, and the reverse reaction (autoprotolysis of water) is of negligible magnitude. Some analyses measure loss of weight when a gas is evolved, and these are performed under conditions such that reaction is quantitative. Inorganic carbon, for example, present as CO_3^{2-} or HCO_3^- in a sample, may be converted by excess acid to $CO_2(g)$ and removed irreversibly by heating. Silicon analysis by evolution of $SiF_4(g)$ produced when the sample is treated with excess hydrofluoric acid is another common example.

Although the end results of such procedures are not measured under equilibrium conditions, the principles of equilibrium are, as we shall see later, important in devising proper experimental methods for quantitative analyses.

3 GRAVIMETRIC ANALYSES

Gravimetric methods depend on weight data. Both the original sample and the final product must be weighed accurately. For precise work, multiple analyses are performed, and samples are usually "heated to constant weight." That is, they are heated in an oven or by some other heat source, cooled and weighed, reheated, recooled, reweighed, and so on, until successive weights are in satisfactory agreement.

Modern analytical balances permit very accurate mass measurements to be made, and *microbalances* have been developed to extend this accuracy to very small samples.

Example 1: Analysis by precipitation

Gravimetric methods are rarely used for analysis of iron or steel samples because of the greater convenience of volumetric procedures (titration) or instrumental methods. Iron ores, however, typically require more complicated treatments, one stage of which involves precipitation of hydrated iron(III) hydroxide.[1] Since this precipitate is available as a result of the separation techniques used, it is convenient to convert it to a form suitable for weighing. The original precipitate is gelatinous and varies considerably in water content. When heated to a high temperature, however, this is converted to a product of definite composition:

$$2Fe(OH)_3 \cdot nH_2O(s) \xrightarrow{\text{heat}} Fe_2O_3(s) + (n + 3)H_2O(g)$$

In a particular analysis, a 1.2806-g sample of an iron ore (heated to constant weight) was treated with 5.0 ml of concentrated nitric acid[a] and boiled to expel oxides of nitrogen. Insoluble residues (not containing iron) were removed by filtration, and the filtrate was diluted to 100 ml. This solution was placed on a hotplate and boiled gently, with stirring, while 6 M ammonia was slowly added until precipitation appeared complete. The solution was removed from the hotplate, and the gelatinous precipitate was allowed to settle. A few drops of ammonia were added to the supernatant liquid[b] to test for complete precipitation. The mixture was covered and kept warm for about an hour to permit digestion[c] of the precipitate. The precipitate was isolated by filtration through a "low-ash" paper[d] and washed with hot ammonium nitrate solution[e]. The paper containing the precipitate was transferred to a crucible which had previously been heated to constant weight. The crucible and contents were heated over a low flame until the paper was completely charred, then the heat was increased until the crucible bottom glowed a bright red for 15 to 20 minutes. This ignition step was repeated until a constant weight of the cooled[f] product was obtained.

The final product, Fe_2O_3, weighed 0.6204 g.

What was the percent iron in the ore sample, based on this determination?

Solution
Each mole of Fe_2O_3 contains 2 moles of iron. The weight of iron in the sample, then, was

$$0.6204 \text{ g} \times \frac{2 \times \text{atomic weight Fe}}{\text{formula weight Fe}_2\text{O}_3}$$

wt. Fe in sample $= 0.6204 \text{ g} \times \dfrac{111.69}{159.69} = 0.4359 \text{ g}$

The percent iron was

$$\frac{\text{weight Fe}}{\text{weight sample}} \times 100\%$$

[1] Portions of the procedure described in this example are shown in the film *Gravimetric Techniques* (New York: Harper & Row, 1968).

Notes:

[a] Nitric acid is used to ensure that all iron is in the +3 oxidation state, for example, $3Fe^{2+} + NO_3^- + 4H^+ \longrightarrow 3Fe^{3+} + NO + 2H_2O$.

[b] Solution in contact with the precipitate.

[c] Digestion of a precipitate allows time for included impurities to dissolve and for larger particles to grow at the expense of smaller ones (Unit 18). Thus both purity and filterability of the precipitate are improved by digestion. Heating the mixture also helps coagulate any $Fe(OH)_3$ remaining in colloidal suspension.

[d] Special filter papers are used which leave ignition residues of negligible mass. Papers of various pore sizes are available, and a knowledge of the properties of the precipitate permits selection of the appropriate filter.

[e] Ammonium nitrate is used in the wash solution to avoid *peptization* (formation of a sol by adsorption of OH^- ions on colloidal-size particles). The NH_4NO_3 is completely destroyed during ignition. If impurities had been washed out by water alone, some $Fe(OH)_3$ would probably have been lost as colloidal particles small enough to pass through the pores of the paper.

[f] Cooling is always done in a desiccator containing a moisture absorbent. The weighed crucible is never touched directly with the fingers to avoid weight changes from oils from the skin.

$$\text{percent iron in ore} = \frac{0.4359 \text{ g}}{1.2806 \text{ g}} \times 100\% = 34.04\%$$

Example 2: Analysis by weight loss

"Limestones" consist mainly of calcium carbonate, smaller amounts of magnesium carbonate, and traces of other metals and anions such as phosphate, sulfate, and silicates. On ignition, the carbonates are converted to oxides with evolution of CO_2 gas. One of the routine analyses performed on limestone samples is "loss on ignition." This assay provides a reasonably good approximation of the carbonate content of the original limestone, although ignition also expels trace moisture not removed by routine drying. Any hydroxides present in the sample are converted to oxides (see Example 1), and other minor changes in composition may occur, so the analysis cannot be interpreted as revealing the exact carbonate content of the sample. Usually, however, weight changes other than those from loss of CO_2 are relatively small, and the results are generally satisfactory.

Magnesium carbonate decomposes rapidly, so initial heating must be done carefully to avoid sudden gas evolution, which might blow some of the powdered sample out of the crucible.

A sample of a particular limestone was ground to a find powder and dried overnight at $\sim 130°C$. The sample was then cooled to room temperature in a desiccator, and a portion was transferred to a platinum crucible which had been previously heated to constant weight. The crucible was covered and heated slowly to about 600°C, and this temperature was maintained for 15 minutes. Heat was then increased to $\sim 900°C$, and the crucible and contents were ignited at this temperature to constant weight. After cooling in a desiccator, the crucible was reweighed. The following data were recorded:

weight of platinum crucible + cover: 38.8763 g
weight of crucible + cover + sample
before ignition: 39.7195 g
after ignition: 39.4394 g

What was the loss on ignition, expressed as percent CO_2, from the original sample?

Solution

crucible + cover + sample: 39.7195 g
crucible + cover: 38.8763 g
sample wt: 0.8432 g
before ignition: 39.7195 g
after ignition: 39.4394 g
loss (assumed CO_2): 0.2801 g

$$\text{percent } CO_2 = \frac{0.2801 \text{ g}}{0.8432 \text{ g}} \times 100\% = 33.22\%$$

---

4 VOLUMETRIC ANALYSES

Titrations are usually much more rapid and convenient than gravimetric analyses, but many assays are not suitable for volumetric methods. The basic ideas of titrations have been discussed earlier (Units 13 and 19) using the covenient millimole (mmol) unit.

An alternative method for calculations based on titration data uses normality (Unit 13). Normality is defined as the number of gram-equivalent weights of solute per liter of solution.

The gram-milliequivalent (or milligram-equivalent) weight (meq) is 1/1000 gram-equivalent weight, just as the millimole (mmol) is 1/1000 mole. Since titrations most commonly involve volumes conveniently expressed in milliliters, the terms millimole and milliequivalent are quite useful.

$$N = \frac{\text{no. g-eq weights (solute)}}{\text{no. liters of solution}} = \frac{\text{no. meq (solute)}}{\text{no. ml solution}}$$

For acids:

$$\text{g-eq weight} = \frac{\text{g-formula weight}}{\text{no. available protons}}$$

For bases:

$$\text{g-eq weight} = \frac{\text{g-formula weight}}{\text{no. proton acceptors}}$$

Since acid–base reactions involve proton transfer, 1 gram-equivalent weight of any acid will always furnish the number of protons required for 1 gram-equivalent weight of any base, that is,

$$\text{no. eq acid} = \text{no. eq base}$$

from which

$$N_{\text{acid}} \times V_{\text{acid}} = N_{\text{base}} \times V_{\text{base}}$$

It should be noted that the concept of normality is not essential to the stoichiometry of volumetric analysis, since use of the balanced equation for the process permits easy calculations on the basis of molarity (Unit 13). Normality is simply a convenience to avoid the necessity of determining stoichiometric ratios.

Example 3 was used as "Practice for Proficiency" question 26 (Unit 19), with *normalities*.

Notes:

[a] One modification often used employs a 4-percent solution of boric acid. Ammonia absorbed by this solution is titrated by standardized hydrochloric acid.

[b] Reaction is usually slow, so a catalyst such as a crystal of copper sulfate is generally employed.

[c] The reaction between sodium hydroxide and sulfuric acid is very exothermic.

[d] $NH_4^+(aq) + OH^-(aq) \longrightarrow$ $NH_3(aq)$

[e] Neither sodium hydroxide nor sulfuric acid can be used as a primary standard (i.e., neither can be obtained in sufficient purity for routine use as an accurate standard). The sodium hydroxide is standardized against a carefully weighed sample of some solid acid, such as potassium hydrogen phthalate. The standardized sodium hydroxide is then used to determine the concentration of the sulfuric acid.

[f] Methyl red changes color in the pH range corresponding to $(NH_4)_2SO_4$ at the concentrations employed.

Example 3

One of the most important analyses in many fields of applied chemistry is the Kjeldahl determination for nitrogen compounds. This procedure is applied to such diverse assays as the nitrogen content of food proteins and the concentration of nitrates in waters heavily polluted by agricultural nitrate fertilizers. The procedure depends eventually on the titration of a sulfuric acid[a] solution that has been partially neutralized by ammonia formed from the nitrogen compound being analyzed.

Two alternative procedures are involved in the initial stages of the analysis. If the nitrogen compounds being investigated contain nitrogen in the -III oxidation state, the sample is digested in concentrated sulfuric acid.[b] This process converts the nitrogen (e.g., from protein) to ammonium sulfate. If the nitrogen of interest is in a higher oxidation state, it must first be reduced. For example, analysis for nitrate involves initial treatment of the sample with a mixture of sulfuric and salicylic acids. The nitrated salicylic acid is reduced with sodium thiosulfate, and the mixture is then subjected to the sulfuric acid digestion.

After digestion is complete, excess sodium hydroxide solution is added carefully,[c] and the ammonia liberated[d] is distilled into a measured volume of dilute sulfuric acid of known concentration.

The ammonia acts to neutralize some of the sulfuric acid. The remaining excess acid is titrated by standardized[e] sodium hydroxide ($\sim$0.1 M), using methyl red[f] as indicator. The difference between the original concentration of the sulfuric acid and the concentration determined by the titration is a measure of the amount of ammonia absorbed.

In a particular analysis of a protein mixture, a 1.1841-g sample was digested in 25 ml of concentrated sulfuric acid to which a small crystal of $CuSO_4 \cdot 5H_2O$ had been added. When digestion was complete, as indicated by the disappearance of the dark carbon initially formed, the mixture was cooled and carefully diluted with

250 ml of cold water. A 100-ml portion of 40 percent sodium hydroxide was carefully poured down the wall of the digestion flask so as to form two liquid phases. A few zinc pellets were added to keep the mixture agitated by slow formation of hydrogen, and the flask was connected to a distillation apparatus. The condenser outlet was arranged to dip below the surface of 50.00 ml of 0.09458 M sulfuric acid in the receiving flask. The digestion flask was heated until the solution was maintained at a gentle boil. Boiling was continued until about half the contents of the digestion flask had distilled. The heat source was removed, and the receiving flask was immediately lowered to avoid "suck back" of liquid as the system cooled. The solution in the receiving flask was titrated with 0.1089 M sodium hydroxide to a methyl red endpoint. The volume of base required was 38.72 ml.

Calculate the percent nitrogen in the original sample.

Solution

We may represent the reactions with sulfuric acid as

With the distilled ammonia

$$2NH_3(aq) + H_2SO_4(aq) \longrightarrow (NH_4)_2SO_4(aq)$$

Remaining H_2SO_4, with NaOH

$$2NaOH(aq) + H_2SO_4(aq) \longrightarrow Na_2SO_4(aq) + 2H_2O(l)$$

The "remaining H_2SO_4" is given by

$$\frac{0.1089 \text{ mmols (NaOH)}}{\text{ml (NaOH)}} \times \frac{38.72 \text{ ml (NaOH)}}{1} \times \frac{1 \text{ mmol (H}_2\text{SO}_4)}{2 \text{ mmol (NaOH)}}$$
$$= 2.108 \text{ mmols (H}_2\text{SO}_4)$$

The amount of H_2SO_4 originally in the receiver was

$$50.00 \text{ ml (H}_2\text{SO}_4) \times \frac{0.09485 \text{ mmols (H}_2\text{SO}_4)}{\text{ml (H}_2\text{SO}_4)} = 4.742 \text{ mmols (H}_2\text{SO}_4)$$

Then, the amount of H_2SO_4 consumed by the distilled ammonia was

$$4.742 - 2.108 = 2.634 \text{ mmols}$$

from which, the amount of ammonia distilled must have been

$$2.634 \text{ mmols (H}_2\text{SO}_4) \times \frac{2 \text{ mmols (NH}_3)}{1 \text{ mmol (H}_2\text{SO}_4)} = 5.268 \text{ mmols (NH}_3)$$

and thus the percent of N in the sample was

$$\frac{5.268 \text{ mmols (NH}_3)}{1.1841 \text{ g (sample)}} \times \frac{1 \text{ mmol (N)}}{1 \text{ mmol (NH}_3)} \times \frac{14.01 \times 10^{-3} \text{ g (N)}}{1 \text{ mmol (N)}} \times 100\% = 6.233\%$$

5 EQUILIBRIUM CONSIDERATIONS

We have already discussed, in Units 19 and 20, some of the equilibrium considerations necessary for the planning of titration analysis. The calculations described in Unit 19 of pH at the equivalence point were necessary for the selection of methyl red as the indicator for the titration described in Example 3.

Gravimetric analyses may also involve significant use of equilibrium concepts. When the precipitate is very insoluble and no possibility of complex ion formation exists, equilibrium situations may be omitted from consideration. For more soluble species, or when complex formation is possible, equilibrium factors are very important.

Normally, an excess of precipitating ion is used to ensure complete precipitation. In some cases, however, a significant excess of precipitating ion will result in loss of product by formation of a soluble complex. Amphoteric hydroxides are notable in this respect. Aluminum may be determined by a method similar to that used for iron (Example 1). However, aluminum hydroxide is amphoteric and if an excess of concentrated ammonia is added, some precipitate is lost by the reactions

$$NH_3(aq) + H_2O(l) \rightleftharpoons NH_4^+(aq) + OH^-(aq)$$
$$Al(OH)_3(s) + OH^-(aq) \rightleftharpoons [Al(OH)_4]^-(aq)$$

Precipitates are typically washed to remove impurities. The volume of wash solution that may be safely used depends on the solubility of the precipitate. The more soluble the salt, the greater will be the loss on washing.

Example 4

What is the maximum percent loss to be expected from a lead(II) chloride precipitate of approximately 2.0 g if it is washed with 100 ml of water? (K_{sp} for $PbCl_2$ is 1.6×10^{-5}.)

Solution

Note that washing a precipitate on a filter would reduce loss, since the wash liquid would not usually be in contact with the precipitate long enough for saturation to occur.

The maximum loss would occur if the wash liquid were saturated with $PbCl_2$.

$$K_{sp} = 1.6 \times 10^{-5} = C_{Pb^{2+}} \times C_{Cl^-}^2$$

Let the solubility of $PbCl_2$ (in moles per liter) be y; then

$$C_{Pb^{2+}} = y$$
$$C_{Cl^-} = 2y$$
$$1.6 \times 10^{-5} = y \times (2y)^2$$

from which

$$y = \sqrt[3]{4.0 \times 10^{-6}} = {\sim}1.6 \times 10^{-2}$$

Then (for 100 ml wash)

$$\text{weight loss} = \frac{1.6 \times 10^{-2}\ \text{mole}}{\text{liter}} \times \frac{0.100\ \text{liter}}{1} \times \frac{278\ \text{g}}{\text{mole}} = 0.44\ \text{g}$$

and

$$\text{percent loss} = \frac{0.44\ \text{g}}{2.0\ \text{g}} \times 100\% = 22\%$$

There are many types of analysis in addition to the relatively simple titrations and precipitations described in this unit. Every analysis, no matter how "simple," must be carefully planned and properly executed if meaningful results are to be expected. A knowledge of stoichiometry, equilibria, and chemical properties is essential to any understanding of analytical chemistry.

SUGGESTED READING   Skoog, Douglas, and Donald West. 1969. *Fundamentals of Analytical Chemistry* (2nd ed.). New York: Holt, Rinehart and Winston.

EXERCISES

1. An ore was assayed for titanium by the method described for iron in ores (Example 1). If a 1.6903-g sample of the ore yielded 0.3819 g of titanium(IV) oxide, what was the percent titanium in the ore?

2. A limestone sample was dried to constant weight, then heated to 900°C to convert all carbonate to $CO_2(g)$. From the following data, what was the percent CO_3^{2-} in the limestone sample

weight of crucible + cover:	37.5711 g
weight of crucible + cover + sample	
before ignition at 900°C:	38.8302 g
after ignition at 900°C:	38.5028 g

3. From the data given below for Kjeldahl analysis of a blood protein sample, calculate the percent nitrogen in the sample.

sample wt.: 1.0274 g
volume of H_2SO_4 in receiver: 50.00 ml
initial concentration of H_2SO_4 in receiver: 0.09235 M
concentration of NaOH titrant: 0.1080 M
volume titrant used: 41.34 ml

4. What is the maximum percent loss to be expected from a silver chromate precipitate of approximately 0.80 g if it is washed with 100 ml of water? (K_{sp} for silver chromate is 1.9×10^{-12}.)

Section EIGHT

ELECTRON TRANSFER PROCESSES

One of the most important concepts in chemistry and biochemistry is that of chemical reactions involving electron transfer. These *oxidation-reduction* processes have applications in such diverse areas as the electroplating industries, the use of batteries and fuel cells as "energy sources," the analytical determination of ions in solutions by voltage measurements, and the elucidation of many of the biochemical processes within living organisms.

In Unit 22 we shall study the characteristics of electron-transfer processes and learn a systematic method for balancing the equations descriptive of oxidation-reduction reactions. By examining a number of examples of important electron-transfer processes, we shall demonstrate the utility of the balanced equation for process description and for applications to stoichiometric analysis of oxidation-reduction reactions.

Electron transfer may occur either by direct collision between reactant species or by *indirect* transfer of electrons through an external circuit. Significant differences are observed for these alternatives in terms of energy utilization, as illustrated by the increased efficiency of fuel utilization by *fuel cells* as compared to internal combustion engines. Indirect electron transfer is also the basis for such important industrial activities as metal plating and "electrorefining." In Unit 23 we shall see how electric current and voltage may be related to the chemical process of oxidation and reduction, and we shall extend our considerations of stoichiometry to include the relationship between chemical change and charge transfer.

In Unit 24 we shall study a key relationship in electrochemistry, the *Nernst equation*. It is this equation which permits us to determine the voltage of an electrochemical cell as a function of the concentrations of the chemical components of the cell. This has applications both in designing electrochemical cells for particular voltage requirements and in using voltage measurement for concentration determination, as with the pH meter.

An understanding of oxidation-reduction processes will provide basic knowledge for many fields of chemistry, engineering, and the biomedical sciences.

CONTENTS

. . . Sulfur in burning, far from
losing weight, on the contrary
gains it. . . .
(First step toward modern understanding of oxidation, 1772)
ANTOINE LAVOISIER

Oxidation-reduction reactions

Objectives

(a) Be able to identify oxidizing and reducing agents in electron-transfer processes (Section 22.1).

(b) Be able to balance equations for oxidation-reduction reactions, using a systematic method (Section 22.3).

(c) Be able to apply the principles of stoichiometry to oxidation-reduction
processes, including the use of normality for solution reactions (Section
22.4).

Up to this point we have concentrated primarily on metathetical reactions, such as those involving precipitations, proton exchange, or formation of complex ions. Such reactions are important and have numerous practical applications. Oxidation–reduction reactions (electron-transfer processes), however, are even more important. Modern society is energy dependent, and most of this energy is produced by oxidation–reduction reactions. The combustion of fuels for heating, cooking, or production of electrical or mechanical power most commonly involves direct electron transfer—the reactant species, usually molecules, exchange electrons during actual collisions. Indirect electron transfer, on the other hand, is the basis for electrochemical processes in batteries, fuel cells, and electrolytic operations.

Combustion processes may be as simple as the burning of charcoal for a steak fry or as complex as the timed ignition of an air-fuel mixture in a finely tuned racing engine. Direct electron transfer, such as that involved in the burning of fossil fuels, is generally an inefficient way of obtaining energy and usually results in the formation of unwanted by-products—pollution. Indirect processes are usually cleaner and more efficient, but the speed of energy release is much slower than in the simpler combustion processes. The distinction is an important one. If we want cleaner and more efficient methods of power production, we must be willing to reduce our demands for large blocks of power per unit time.

As an example, it is possible to manufacture an automobile run by clean fuel cells that give excellent mileage. Such automobiles, however, cannot be the massive, high-speed, long-distance, luxury cars now available with the internal-combustion engine.

In this unit we shall consider some basic aspects of electron-transfer processes, and in Units 23 and 24, we shall explore factors related to voltage and current flow in electrochemical systems.

22.1 ELECTRON-TRANSFER PROCESSES

Every chemical change (Unit 3) involves some redistribution of electrons relative to the positive fields of participating nuclei. Consider the following examples.

1. $H^+(g) + F^-(g) \longrightarrow HF(g)$

   The isolated proton (H^+) has an electron density of zero. When it combines with fluoride ion, the electron density around hydrogen is greatly increased.

2. $H_2(g) + F_2(g) \longrightarrow 2HF(g)$

   The electron density of each hydrogen atom is decreased slightly because of the higher electronegativity of fluorine (Unit 7).

3. $HF(g) \longrightarrow H^+(g) + F^-(g)$

   The electron density of hydrogen is drastically decreased by formation of the free proton.

Chemists do not always agree on the "best" ways of defining terms.[1]

The most commonly used definitions describe oxidation as a loss of electrons and reduction as a gain of electrons. Under these definitions, which—if any—of the preceding examples would be classified as an oxidation–reduction process? Surprisingly enough, only the second example—the one involving the *smallest* change in electron density—is considered to be an oxidation–reduction reaction.

Although we can write simple equations indicating an electron transfer, the real processes are not as simple as these equations might suggest. We can, for example,

[1] See, for example, "The Lion Roars: Rebuttals to 'What Is Oxidation?,'" by J. D. Herron in the September, 1975, issue of the *Journal of Chemical Education* (pp. 602–603).

prepare an electrochemical cell (Units 23 and 24) for which we can write the electrode reactions

$$Ag^+(aq) + e^- \longrightarrow Ag(s)$$
$$Cu(s) \longrightarrow Cu^{2+}(aq) + 2e^-$$

The hydrated (aqueous) cations, however, do not have a charge *on the metal ion* of a "full" +1 or +2, respectively. The positive charge density of the metal ion is reduced by formation of coordinate covalent bonds with electrons pairs of the water molecules, so the charge is actually delocalized throughout the entire hydrate cluster. Thus although the silver ion has "gained an electron" in becoming a silver atom and the copper atom has "lost two electrons" in becoming a copper(II) ion, the change in actual electron density of the unit particles is less than this might suggest.

Once again, simple definitions are convenient, but we must recognize their limitations relative to real systems.

Experimentally, we can determine whether or not electron transfer is involved by detecting a current flow—providing the system is arranged for indirect electron transfer and providing the instruments available are sufficiently sensitive for the particular reaction system involved. It is obviously not feasible to subject every reaction to individual laboratory investigation, so some convenient classification method is desirable. Since "complete" electron transfer is really an idealized situation, we must use a purely arbitrary system that represents a matter of bookkeeping convenience rather than a description of any real entities. Our classification scheme will use the concept of an oxidation number (or oxidation state) as previously described (Unit 3). Then, *on the basis of changes in assigned oxidation numbers,*

Oxidation is electron loss (algebraic increase in oxidation number); for example,

$$Cu \longrightarrow Cu^{2+} + 2e^- \qquad Cu(0) \longrightarrow Cu(II)$$

Reduction is electron gain (algebraic decrease in oxidation number); for example,

$$e^- + Ag^+ \longrightarrow Ag \qquad Ag(I) \longrightarrow Ag(0)$$

The *oxidizing agent is the complete reactant species* acting to accept electrons. The *reducing agent is the complete reactant species* acting to donate electrons; for example,

$$\underset{\substack{\text{reducing} \\ \text{agent}}}{3CH_3OH} + \underset{\substack{\text{oxidizing} \\ \text{agent}}}{4MnO_4^-} \longrightarrow 3HCO_2^- + 4MnO_2 + 4H_2O + OH^-$$

accepting electrons: $\overset{+VII}{MnO_4^-} + 3e^- \longrightarrow \overset{+IV}{MnO_2} + \cdots$

donating electrons: $\overset{-II}{C}H_3\overset{}{OH} \longrightarrow \overset{+II}{H}CO_2^- + 2e^- + \cdots$

In order to classify a reaction as *oxidation–reduction* (sometimes referred to as "redox"), as opposed to *metathetical,* it is necessary only to locate any one element symbol that has changed oxidation number as a result of the reaction involved. The method for assigning oxidation numbers should be reviewed (Unit 3) as well as the common oxidation numbers used (Table 3.5, Unit 3). An abbreviated list is given in Table 22.1.

It should be noted, of course, that oxidation-reduction reactions require that *two* species change oxidation states, one being oxidized and the other being reduced.

Table 22.1 SOME TYPICAL OXIDATION STATES[a]

SPECIES	OXIDATION STATE
all free elements	0
hydrogen combined with nonmetals	+I
group IA elements (combined)	+I
group IIA elements (combined)	+II
oxygen in most compounds	−II

[a] See also Table 3.5, Unit 3.

Example 1

Classify each of the following equations as metathetical or oxidation–reduction (redox). For each of the latter, determine the oxidizing agent and the reducing agent.

a. $2H^+ + S^{2-} \longrightarrow H_2S$
b. $H_2 + S \longrightarrow H_2S$
c. $2MnO_4^- + 5C_2O_4^{2-} + 16H^+ \longrightarrow 2Mn^{2+} + 10CO_2 + 8H_2O$
d. $3MnO_4^{2-} + 4H^+ \longrightarrow 2MnO_4^- + MnO_2 + 2H_2O$
e. $2HCO_3^- \longrightarrow CO_3^{2-} + CO_2 + H_2O$

Solution

Remember, if any element symbol changes oxidation number assignment, the reaction is classified as *redox*. Thus if the first element symbol checked (the one appearing simplest) changes oxidation number, we can classify the reaction immediately. However, if this is not the case, we must check other elements until a definite decision is possible. We must be certain that *no* change in oxidation state occurs before classifying a reaction as metathetical.

a. Oxidation numbers assigned by algebraic method (Unit 3):

$$\overset{+\text{I}}{2H^+} + \overset{-\text{II}}{S^{2-}} \longrightarrow \overset{+\text{I} \;-\text{II}}{H_2S}$$

No change in any oxidation number occurs, so the reaction is *metathetical.*

b. Oxidation numbers assigned by algebraic method:

$$\overset{0}{H_2} + \overset{0}{S} \longrightarrow \overset{+\text{I} \;-\text{II}}{H_2S}$$

Oxidation states change, so the reaction is *redox.*
　　Sulfur is the *oxidizing agent* (0 in reactant species, −II in product species).
　　Hydrogen (H_2) is the *reducing agent* (0 in reactant species, +I in product species).

c. The easiest species to check is H:

$$\overset{+\text{I}}{H^+} \longrightarrow \overset{+\text{I}}{H_2O}$$

No change occurs, so check some other species, such as

$$\overset{+\text{VII}\,-\text{II}}{MnO_4^-} \longrightarrow \overset{+\text{II}}{Mn^{2+}}$$

The reaction is of the *redox* type since Mn changes oxidation states.
　　Permanganate ion (*not* Mn) is the *oxidizing agent* (Mn is +VII in reactant species and +II in corresponding product species).
　　Oxalate ion ($C_2O_4^{2-}$) must be the *reducing agent,* since the only other reactant (H^+) does not change oxidation state.

Note that the *complete species* containing the element symbol involved is considered the oxidizing agent (or, in the case of $C_2O_4^{2-}$, the reducing agent).

d. Hydrogen does not change oxidation state (+I as H^+ or H_2O), so we might be tempted to classify this as a metathetical reaction. (Remember that both oxidation and reduction must occur, so *two* species must change oxidation state.) However, we note that manganese occurs in three different oxidation states (+VI in MnO_4^{2-}, +VII in MnO_4^-, and +IV in MnO_2). This *is* an oxidation–reduction reaction. It is an example of *disproportionation,* a process in which the same species is both oxidizing agent and reducing agent. We could indicate this as

$$2MnO_4^{2-} + MnO_4^{2-} + 4H^+ \longrightarrow 2MnO_4^- + MnO_2 + 2H_2O$$

reducing   oxidizing
agent   agent

e. At first glance this reaction might appear to be another example of disproportionation. However, assignment of oxidation numbers (by the algebraic method) shows that it is a metathetical reaction, in this case an illustration of the amphoteric properties of bicarbonate ion.

$$\overset{+I+IV-II}{2HCO_3^-} \longrightarrow \overset{+IV-II}{CO_3^{2-}} + \overset{+IV-II}{CO_2} + \overset{+I-II}{H_2O}$$

22.2 EQUATIONS FOR HALF-REACTIONS

Processes occurring at *single electrodes* in electrolytic or Galvanic cells (Units 23 and 24) will be either oxidation (electron loss) or reduction (electron gain). As such, they may be represented by equations in which electrons (e^-) are shown as reactants or products. It is important to realize that balanced oxidation–reduction equations will *not* contain electron symbols, since the balancing process will ensure that electron loss and gain are equal. Balanced equations for oxidation or reduction *half-reactions* will contain electron symbols as required to indicate charge balance.

For example, in the electroplating of aluminum, one may symbolize the reduction process as

$$Al^{3+} + 3e^- \longrightarrow Al$$

and in the electrolysis of molten sodium chloride the oxidation process may be represented by

$$2Cl^- \longrightarrow Cl_2 + 2e^-$$

Equations for half-reactions are most appropriate for describing electrochemical processes, in which the indirect electron transfer via an external circuit requires half-reactions at the two electrodes. Any oxidation–reduction reaction, whether electron transfer is direct or indirect, may be represented by a pair of equations for net half-reactions—one representing oxidation and the other reduction. Such equations should not be interpreted as indicating the *mechanism* (Unit 16) of the overall process. They are simply another bookkeeping convenience.

22.3 BALANCING OXIDATION-REDUCTION EQUATIONS

Redox processes are frequently more complex than the metathetical reactions previously studied. Equations for oxidation-reduction reactions may sometimes be balanced by "simple inspection" (Unit 3) or by considerations of changes in oxidation state, noting that total electron loss must equal total electron gain. However, neither of these approaches is guaranteed to work without frequent trial-and-error manipulations, and both are limited to fairly simple species.

There is a systematic method that may be used to balance any oxidation–reduction equation, no matter how complex. It eliminates trial-and-error fumblings and the necessity for determining oxidation numbers. At first glance the steps may appear a bit complicated, but careful study and practice will make them relatively simple. The method is so universally useful that it should be committed to memory (Table

22.2). After sufficient practice it will be possible to combine various steps, but this should not be attempted until considerable experience has been accumulated.

Balancing redox equations by the method of half-reactions

If the original equation is not in *net* form (Unit 3), first convert it to a net equation. After this is balanced, restore all species to the formulations used in the original equation, being careful to preserve mass and charge balance.

Table 22.2 ABBREVIATED RULES FOR BALANCING REDOX EQUATIONS

1. Convert net equation to two half-reaction equations.
2. Balance all symbols except H and O.
3. Use H_2O to balance O.
4. Use H^+ to balance H.
5. Use e^- to balance charge.
6. Multiply to equalize e^- numbers.
7. Add equations, eliminating duplications.
8. Check
9. *For alkaline solutions,* convert H^+ to H_2O, add that number OH^- to opposite side, eliminate duplications of H_2O.

STEP 1 From the net overall equation, select the two equations representing oxidation and reduction half-reactions. (Omit H^+, OH^-, and H_2O unless these species contain elements changing oxidation state.)

STEP 2 Use appropriate coefficients to balance all symbols *except* H and O in each half-reaction equation.

STEP 3 Add sufficient H_2O units to the oxygen-deficient side to balance O symbols in each equation.

STEP 4 Add sufficient H^+ units to the hydrogen-deficient side to balance H symbols in each equation.

STEP 5 Add sufficient e^- units to the side having the larger net positive charge (or smaller net negative charge) to balance net reactant and product charges in each equation.

STEP 6 Multiply all coefficients in each half-reaction equation by the smallest whole number that will equalize electrons in the two equations.

STEP 7 Add the half-reaction equations resulting from step 6, subtracting or combining numbers of duplicated species so that no species appears in more than one place in the final equation.

STEP 8 Check charge and mass balance. Be sure all coefficients are the smallest whole numbers consistent with balance, and that no electron symbols have been retained.

STEP 9 (*For reactions in alkaline solution*): Be sure all species are formulated as they would exist in an alkaline medium. Change all H^+ units to H_2O units and add that number of OH^- to the other side of the equation. Subtract or combine "duplicated" water molecules.

Example 2

Balance (Note: H^+, OH^-, and H_2O not shown)

$$HgS(s) + NO_3^-(aq) + Cl^-(aq) \longrightarrow [HgCl_4]^{2-}(aq) + NO_2(g) + S(s) \qquad \text{at pH} < 7$$

This is the equation for the decomposition of mercury(II) sulfide by *aqua regia,* a mixture of one part concentrated nitric acid and three parts concentrated hydrochloric acid.

Solution
(Refer to "rules")

STEP 1

$$HgS + Cl^- \longrightarrow [HgCl_4]^{2-} + S \qquad\qquad NO_3^- \longrightarrow NO_2$$

STEP 2

$$HgS + 4Cl^- \longrightarrow [HgCl_4]^{2-} + S \qquad\qquad \text{same}$$

STEP 3

$$\text{same} \qquad\qquad NO_3^- \longrightarrow NO_2 + H_2O$$

STEP 4

$$\text{same} \qquad\qquad NO_3^- + 2H^+ \longrightarrow NO_2 + H_2O$$

STEP 5

$$HgS + 4Cl^- \longrightarrow [HgCl_4]^{2-} + S + 2e^- \qquad NO_3^- + 2H^+ + e^- \longrightarrow NO_2 + H_2O$$

STEP 6

same
$$2NO_3^- + 4H^+ + 2e^- \longrightarrow 2NO_2 + 2H_2O$$

STEP 7

$$HgS + 4Cl^- + 2NO_3^- + 4H^+ + \cancel{2e^-} \longrightarrow [HgCl_4]^{2-} + S + \cancel{2e^-} + 2NO_2 + 2H_2O$$

STEP 8 (Check)

Mass balance (checking that element symbols balance)

$$Hg + S + 4(Cl) + 2(N) + 6(O) + 4(H) = Hg + 4(Cl) + S + 2(N) + 6(O) + 4(H)$$

Charge balance

$$(4-) + (2-) + (4+) = 2-$$

Coefficients are smallest possible. Electrons have been eliminated.

Answer

$$HgS + 4Cl^- + 2NO_3^- + 4H^+ \longrightarrow [HgCl_4]^{2-} + S + 2NO_2 + 2H_2O$$

Example 3

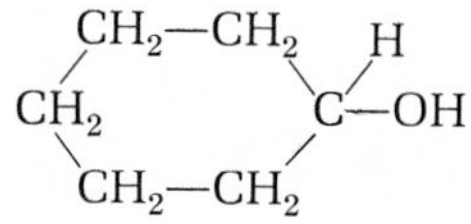

$$\begin{matrix} CH_2-CH_2 \quad H \\ CH_2 \qquad\qquad C-OH + NO_3^- \longrightarrow \\ CH_2-CH_2 \end{matrix}$$

$$HO_2CCH_2CH_2CH_2CH_2CO_2H + NO_2 \qquad \text{at pH} < 7$$

Cyclohexanol may be converted to adipic acid, one of the starting materials for preparation of nylon, by oxidation with nitric acid.

Solution

To simplify balancing, use molecular formulas (Unit 7) until the final answer.

$$C_6H_{12}O + NO_3^- \longrightarrow C_6H_{10}O_4 + NO_2$$

STEP 1

$$C_6H_{12}O \longrightarrow C_6H_{10}O_4 \qquad\qquad\qquad NO_3^- \longrightarrow NO_2$$

STEP 2

same $\qquad\qquad\qquad\qquad\qquad\qquad\qquad\qquad$ same

STEP 3

$$C_6H_{12}O + 3H_2O \longrightarrow C_6H_{10}O_4 \qquad\qquad NO_3^- \longrightarrow NO_2 + H_2O$$

STEP 4

$$C_6H_{12}O + 3H_2O \longrightarrow C_6H_{10}O_4 + 8H^+ \qquad NO_3^- + 2H^+ \longrightarrow NO_2 + H_2O$$

STEP 5

$$C_6H_{12}O + 3H_2O \longrightarrow C_6H_{10}O_4 + 8H^+ + 8e^- \quad NO_3^- + 2H^+ + e^- \longrightarrow NO_2 + H_2O$$

STEP 6

same
$$8NO_3^- + 16H^+ + 8e^- \longrightarrow 8NO_2 + 8H_2O$$

STEP 7

$$C_6H_{12}O + \overset{}{3}H_2O + 8NO_3^- + \underset{8}{16}H^+ + 8e^- \longrightarrow$$
$$C_6H_{10}O_4 + 8H^+ + 8e^- + 8NO_2 + \underset{5}{8}H_2O$$

STEP 8 (Check)

Mass balance (checking that element symbols balance)

$$6(C) + 20(H) + 8(N) + 25(O) = 6(C) + 20(H) + 8(N) + 25(O)$$

Charge balance

$$(8-) + (8+) = 0$$

Coefficients are smallest possible.
Electrons have been eliminated.

Answer (reverting to original structural formulas)

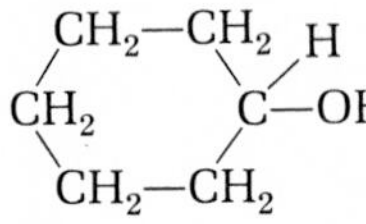

$$\begin{array}{c} CH_2-CH_2 \ H \\ CH_2 \qquad C-OH + 3H_2O + 8NO_3^- + 8H^+ \longrightarrow \\ CH_2-CH_2 \end{array}$$

$$HO_2CCH_2CH_2CH_2CH_2CO_2H + 8NO_2 + 5H_2O$$

Example 4

$$ClO^-(aq) \longrightarrow Cl^-(aq) + ClO_3^-(aq) \qquad (\text{at } pH > 7)$$

This disproportionation is responsible for the decomposition of hypochlorite bleaching solutions when they are not stored in cool areas.

Solution

STEP 1

$$ClO^- \longrightarrow Cl^- \qquad\qquad\qquad ClO^- \longrightarrow ClO_3^-$$

STEP 2

same   same

STEP 3

$$ClO^- \longrightarrow Cl^- + H_2O \qquad\qquad ClO^- + 2H_2O \longrightarrow ClO_3^-$$

STEP 4

$$ClO^- + 2H^+ \longrightarrow Cl^- + H_2O \qquad ClO^- + 2H_2O \longrightarrow ClO_3^- + 4H^+$$

STEP 5

$$ClO^- + 2H^+ + 2e^- \longrightarrow Cl^- + H_2O \qquad ClO^- + 2H_2O \longrightarrow ClO_3^- + 4H^+ + 4e^-$$

STEP 6

$$2ClO^- + 4H^+ + 4e^- \longrightarrow 2Cl^- + 2H_2O \qquad \text{same}$$

STEP 7

$$3ClO^- + \cancel{4H^+} + \cancel{4e^-} + \cancel{2H_2O} \longrightarrow 2Cl^- + ClO_3^- + \cancel{2H_2O} + \cancel{4H^+} + \cancel{4e^-}$$

STEP 8 (Check)

Mass balance (checking that element symbols balance)

$$3(Cl) + 3(O) = 3(Cl) + 3(O)$$

Charge balance

$$3- = 3-$$

Coefficients are smallest possible.
Electrons have been eliminated.

Answer

$$3ClO^- \longrightarrow 2Cl^- + ClO_3^-$$

Example 5

$$H_2C{=}CH_2(g) + KMnO_4(aq) \longrightarrow KOH(aq) + HOCH_2CH_2OH(aq) + MnO_2(s)$$
$$(\text{at pH} > 7)$$

Ethylene glycol, for use in permanent antifreeze solutions, may be prepared from ethylene gas and aqueous potassium permanganate. The product must be purified by removal of the inorganic by-products.

Solution

Write as the net equation, omitting OH^-. (K^+ ions are "spectator" ions, so they are omitted from the net equation.) Simplify by using molecular formulas.

$$C_2H_4 + MnO_4^- \longrightarrow MnO_2 + C_2H_6O_2$$

STEP 1

$$C_2H_4 \longrightarrow C_2H_6O_2 \qquad\qquad\qquad MnO_4^- \longrightarrow MnO_2$$

STEP 2

same   same

STEP 3

$$C_2H_4 + 2H_2O \longrightarrow C_2H_6O_2 \qquad\qquad MnO_4^- \longrightarrow MnO_2 + 2H_2O$$

STEP 4

$$C_2H_4 + 2H_2O \longrightarrow C_2H_6O_2 + 2H^+ \qquad MnO_4^- + 4H^+ \longrightarrow MnO_2 + 2H_2O$$

STEP 5

$$C_2H_4 + 2H_2O \longrightarrow C_2H_6O_2 + 2H^+ + 2e^- \qquad MnO_4^- + 4H^+ + 3e^- \longrightarrow$$
$$MnO_2 + 2H_2O$$

STEP 6

$$3C_2H_4 + 6H_2O \longrightarrow 3C_2H_6O_2 + 6H^+ + 6e^- \qquad 2MnO_4^- + 8H^+ + 6e^- \longrightarrow$$
$$2MnO_2 + 4H_2O$$

STEP 7

$$3C_2H_4 + \cancel{6}H_2O + 2MnO_4^- + \cancel{8}H^+ + \cancel{6e^-} \longrightarrow 3C_2H_6O_2 + 2MnO_2 + \cancel{6H^+} + \cancel{4}H_2O + \cancel{6e^-}$$
$$\quad\quad\quad\quad 2 \quad\quad\quad\quad\quad\quad\quad 2$$

STEP 8 (Check)

Mass balance (*checking that element symbols balance*)

$$6(C) + 18(H) + 10(O) + 2(Mn) = 6(C) + 18(H) + 10(O) + 2(Mn)$$

Charge balance

$$(2-) + (2+) = 0$$

Coefficients are smallest possible.
Electrons have been eliminated.
Balanced net equation (in acidic solution).

$$3C_2H_4 + 2H_2O + 2MnO_4^- + \boxed{2H^+} \longrightarrow 3C_2H_6O_2 + 2MnO_2$$

STEP 9 (Add 2OH$^-$ to each side.)

$$(H^+ + OH^- \longrightarrow H_2O) \cdots + 2H_2O \longrightarrow \cdots + 2OH^-$$

Balanced net equation (*in alkaline solution*)

$$3C_2H_4 + 4H_2O + 2MnO_4^- \longrightarrow 3C_2H_6O_2 + 2MnO_2 + 2OH^-$$

Answer (*reverting to original formulations*):

$$3H_2C{=}CH_2 + 4H_2O + 2KMnO_4 \longrightarrow 3HOCH_2CH_2OH + 2MnO_2 + 2KOH$$

METHANOL POISONING

It was not until 1856 that methanol (CH_3OH, "wood alcohol") was recognized as a toxic chemical. In spite of this discovery, the use of methanol in numerous "medicinal" preparations was continued well into the twentieth century.

Evidence suggests that the biological damage results from the activities of one or more *oxidation products*, rather than from methanol itself. Methanol may be oxidized by various chemical reagents through three different stages, each involving carbon in successively higher oxidation states:

$$\overset{-\text{II}}{CH_3OH} \longrightarrow \overset{0}{H_2CO} \longrightarrow \overset{+\text{II}}{HCO_2H} \longrightarrow \overset{+\text{IV}}{CO_2} + H_2O$$
$$\text{(methanol)} \quad\quad \text{(formaldehyde)} \quad\quad \text{(formic acid)}$$

Although the "final" oxidation products are innocuous, both of the "intermediate stages" (formaldehyde and formic acid) may cause significant biological damage.

Methanol poisoning may result in *acidosis* (Unit 20), partial or complete loss of vision, or (in severe poisoning) death. Formaldehyde has been associated with eye damage in cases of methanol poisoning. Although formaldehyde has a short biological half-life, the retinal cells contain the enzyme *alcohol dehydrogenase* (which catalyzes the conversion of methanol to formaldehyde), so that formaldehyde is generated from methanol directly within the retinal cells. Apparently formaldehyde interferes with cell metabolism and its generation may thus be self-destructive to the retinal cells.

Acidosis is known to cause problems with the body's oxygen-transport system. Since retinal cells are particularly sensitive to oxygen depletion, formic acid (also an enzyme-catalyzed oxidation product of methanol) may contribute to eye damage through its effect on the pH of biological fluids. If sufficient formic acid accumulates, the blood pH may drop sufficiently to result in coma, convulsions, or death.

Although methanol is no longer permitted in pharmaceutical preparations, cases of methanol poisoning still occur. Contaminated "moonshine" ethanol has been reported as the cause of a number of cases of methanol poisoning. (You may remember such a case from the motion picture *Walking Tall*.) From our point of view, it is important to recognize that methanol *does not have to be swallowed* to cause poisoning. It is absorbed through the skin, and cases of poisoning from repeated or prolonged contact with methanol in laboratories are occasionally reported. Inhalation of methanol vapor is also hazardous.

Since it is the oxidation of methanol in the body that results in biological damage, treatment for methanol poisoning must either minimize the oxidation or counteract the oxidation products. One method of treatment, when the poisoning is quickly recognized, takes advantage of the difference in biological oxidation *rates* of methanol and ethanol. The latter is much more rapidly oxidized, so that administration of ethanol successfully competes for the catalytic enzymes and permits the unreacted methanol to be excreted. The oxidation products of ethanol do not appear to be harmful to retinal cells at the concentration levels involved. When the poisoning is not discovered soon enough, *acidosis* may be treated by appropriate administration of bicarbonate to restore proper pH of body fluids.

22.4 APPLICATIONS OF REDOX PROCESSES

In nearly every important use of electron-transfer processes it is necessary, at some stage, to establish a stoichiometric reaction ratio. Usually this requires a balanced equation or at least determination of the mole ratio of oxidizing and reducing agents. Sometimes, particularly in routine repetitive analyses, it is convenient to work with equivalents (or milliequivalents). For oxidizing or reducing agents,

$$\text{equivalent weight} = \frac{\text{formula weight}}{\text{change in oxidation state}} \tag{22.1}$$

$$\overset{+\text{VII}}{\text{KMnO}_4} \longrightarrow \overset{+\text{IV}}{\text{MnO}_2} + \cdots$$

$$\text{eq weight} = \frac{(39.1 + 54.9 + 64.0)}{(7 - 4)} = 52.7$$

and for

$$\overset{+\text{VII}}{\text{KMnO}_4} \longrightarrow \overset{+\text{II}}{\text{Mn}^{2+}} + \cdots$$

$$\text{eq weight} = \frac{(39.1 + 54.9 + 64.0)}{(7 - 2)} = 31.6$$

The use of equivalents or milliequivalents is especially convenient for treating data for solution reactions such as analytical titrations. Because of the requirement that electron loss equal electron gain, the number of equivalents (or milliequivalents) of oxidizing and reducing agents for stoichiometric reaction will be the same. Then, from the definition of normality (Unit 13),

$$N_{\text{ox}} \times V_{\text{ox}} = N_{\text{red}} \times V_{\text{red}} \tag{22.2}$$

This equation greatly simplifies calculations involving titration data.

Example 6

What is the normality of a potassium permanganate solution from which a 25.0-ml aliquot is completely reduced to Mn^{2+} by 37.4 ml of 0.308 N Fe^{2+} solution?

Solution [using equation (22.2)]

$$N_{\text{KMnO}_4} \times 25.0 \text{ ml} = 0.308 \ N \times 37.4 \text{ ml}$$

$$N_{\text{KMnO}_4} = \frac{0.308 \ N \times 37.4 \text{ ml}}{25.0 \text{ ml}} = 0.461 \ N$$

One difficulty in using *normality* as a concentration unit is the possible confusion that can result from using a solution for a reaction involving an oxidation state change different from that for which the *normality* was established. Similar confusion is possible between "*normality* as an acid (or base)" and "*normality* as an oxidizing (or reducing) agent." Although the *molarity* (or *formality*) of a solution is unambiguous, because of its relationship to *how the solution was prepared*, *normality* is an ambiguous term because of its dependence on how the solution is *used*. To avoid confusion, bottles labeled with *normality* should include on the label the basis on which the normality was determined.

It is convenient to be able to relate *normality* to *molarity*. For this purpose, the relationship between equivalent weight and formula weight [equation (22.1)] and the respective definitions of *normality* and *molarity* (Unit 13) permit us to write

$$N = x M \tag{22.3}$$

in which x represents the number of protons involved (for an acid or base) or the change in oxidation state (for an oxidizing agent or reducing agent).

Example 7

A solution was labeled "0.150 N nitric acid (for use as an acid)." How should the solution be relabeled for *normality* of nitric acid for use as an oxidizing agent

a. in a reaction involving reduction of nitric acid to nitric oxide (NO)
b. in a reaction involving reduction of nitric acid to ammonium ion

Solution
Since the *molarity* of a solution is independent of how the solution is to be used, we

may first determine

$$M = \frac{N}{x} \quad \text{[from equation (22.3)]}$$

and for HNO_3 "to be used as an acid,"

$$x = 1,$$

so

$$M = N = 0.150$$

a. For the oxidation state change represented by

$$\overset{+V}{HNO_3} \longrightarrow \overset{+II}{NO}$$
$$x \quad = \quad 3$$

so

$$N = 3 \times M = 3 \times 0.150 = 0.450$$

and, the label should read

"0.450 N HNO_3 (for oxidation state change = 3)."

b. For the oxidation state change represented by

$$\overset{+V}{HNO_3} \longrightarrow \overset{-III}{NH_4{}^+}$$
$$x \quad = \quad 8$$

so,

$$N = 8 \times M = 8 \times 0.150 = 1.20$$

and, the label should read

"1.20 N HNO_3 (for oxidation state change = 8)."

Although the use of *normalities* has some advantages, particularly in simplifying calculations for repetitive determinations of the same general type, the unambiguous unit of *molarity* is more generally useful. This is well illustrated by a reaction in which a particular reagent "plays more than one role." For example, when copper reacts with nitric acid, some of the acid is involved as an oxidizing agent, but some simply "furnishes nitrate ions" without undergoing a change in oxidation state:

$$3Cu + 8HNO_3 \longrightarrow 3Cu(NO_3)_2 + 2NO + 4H_2O$$

Example 8

What volume of 17.4 M nitric acid, in milliliters, is required to convert 25.0 g of copper to copper(II) nitrate?

Solution

According to the balanced equation for this reaction, as previously given, 8 moles of nitric acid are required for each 3 moles of copper.

$$\frac{25.0 \text{ g (Cu)}}{1} \times \frac{1 \text{ mole (Cu)}}{63.5 \text{ g (Cu)}} \times \frac{8 \text{ moles (HNO}_3)}{3 \text{ moles (Cu)}} \times \frac{1 \text{ liter (HNO}_3)}{17.4 \text{ moles (HNO}_3)} \times \frac{10^3 \text{ ml}}{1 \text{ liter}}$$

$$= 60.3 \text{ ml}$$

We shall see additional applications for oxidation-reduction processes in Units 23 and 24.

SUGGESTED READING Moeller, Therald, and Rod O'Connor. 1972. *Ions in Aqueous Systems*. New York: McGraw-Hill.
O'Connor, Rod, and Charles Mickey. 1975. *Solving Problems in Chemistry*. New York: Harper & Row.

EXERCISES

1. Assign oxidation numbers for all element symbols in
   a. nitrous acid; **b.** bisulfite ion; **c.** acetone; **d.** acetic acid
2. Balance (H_2O, H^+, and OH^- not shown), then underline each oxidizing agent.
   a. $Cu + NO_3^- \longrightarrow Cu^{2+} + NO_2$ (in acidic solution)
   b. $Al + Cr_2O_7^{2-} \longrightarrow Al^{3+} + Cr^{2+}$ (in acidic solution)
   c. $(CH_3)_2CHOH + MnO_4^- \longrightarrow CH_3\overset{\overset{O}{\|}}{C}CH_3 + MnO_2$ (in alkaline solution)
   d. $Mn^{2+} + MnO_4^- \longrightarrow MnO_4^{2-}$ (in alkaline solution)
   e. $KOCl + KCl + HNO_3 \longrightarrow KNO_3 + Cl_2 + H_2O$
3. Calculate the percent yield in each of the following processes (refer to Examples 3 and 5):
   a. A 52-g sample of cyclohexanol was oxidized by excess nitric acid to 54 g of adipic acid.
   b. A 74-g sample of ethylene was oxidized by excess potassium permanganate to 62 g of ethylene glycol.
4. A solution labeled 0.350 N potassium permanganate had been standardized by titration with oxalic acid, a reaction in which the permanganate is reduced to Mn^{2+}. This solution is now to be used for a reaction in which the permanganate is reduced to manganese dioxide. How should the label be changed to indicate the normality of the solution for this latter process?

PRACTICE FOR PROFICIENCY

Objective (a)

1. Assign oxidation numbers for all element symbols in
   a. nitrate ion;   **b.** $Ni(HCO_3)_2$;   **c.** methanol;   **d.** K_2MnO_4;
   e. copper(II) arsenate; **f.** $K_2Cr_2O_7$; **g.** benzene; **h.** $KAl(SO_4)_2$;
   i. sodium oxalate; **j.** $H_4P_2O_7$
2. In the reaction between chlorine and iodide ion, which species would be referred to as "being reduced"?

   $Cl_2 + 2I^- \longrightarrow 2Cl^- + I_2$

3. In the reaction between iron and moist air, what is the oxidizing agent?

   $2Fe + 2H_2O + O_2 \longrightarrow 2Fe(OH)_2$

4. In the reaction between oxalic acid and permanganate, what is the reducing agent?

   $5H_2C_2O_4 + 2MnO_4^- + 6H^+ \longrightarrow 10CO_2 + 2Mn^{2+} + 8H_2O$

5. In the reaction between mercury(II) sulfide and aqua regia, the species actually oxidized would be _____.

   $3HgS + 8H^+ + 2NO_3^- + 12Cl^- \longrightarrow$
   $\qquad\qquad 3[HgCl_4]^{2-} + 4H_2O + 3S + 2NO$

Objectives (a, b)

6. Balance (H_2O, H^+, and OH^- not shown), then underline each oxidizing agent.
   a. $Ag + NO_3^- \longrightarrow Ag^+ + NO_2$ (in acidic solution)
   b. $MnO_4^- + HSO_3^- \longrightarrow Mn^{2+} + SO_4^{2-}$ (in acidic solution)
   c. $C_6H_5CH_3 + MnO_4^- \longrightarrow C_6H_5CO_2^- + MnO_2$ (in alkaline solution)
   d. $Cr(OH)_4^- + H_2O_2 \longrightarrow CrO_4^{2-} + H_2O$ (in alkaline solution)
   e. $Al + NaOH \longrightarrow NaAl(OH)_4 + H_2$ (in alkaline solution)
7. For the reactions given in question 6,
   a. what is the reducing agent in each case?
   b. what might be an oxidizing agent in addition to (or instead of) NaOH in the reaction between aluminum and *aqueous* sodium hydroxide?
8. Complete and balance each of the following equations, then draw one line under each oxidizing agent and two lines under each reducing agent. (If any of the equations represent disproportionation reactions, draw three lines under the species that is disproportionated.)
   a. $I^- + Cl_2 \longrightarrow IO_4^- + Cl^-$ (in alkaline solution)
   b. $Co^{2+} + HNO_2 \longrightarrow Co^{3+} + NO$ (in acidic solution)
   c. $Zn + Cr^{3+} \longrightarrow Zn^{2+} + Cr^{2+}$ (in acidic solution)
   d. $I_2 + H_2O \longrightarrow I^- + IO_3^- + H^+$
   e. $I_2 + S_2O_3^{2-} \longrightarrow I^- + S_4O_6^{2-}$
   f. $Cu^{2+} + CN^- \longrightarrow Cu^+ + CNO^-$ (in alkaline solution)
   g. $MnO_2(s) + NaBiO_3(aq) + HNO_3(aq) \longrightarrow$
   $\qquad\qquad NaMnO_4(aq) + Bi(NO_3)_3(aq)$

h. $MnO_4^{2-} \longrightarrow MnO_2 + MnO_4^-$ (in acidic solution)

i. $H_3AsO_3 + BrO_3^- \longrightarrow H_3AsO_4 + Br^-$ (in acidic solution)

j. $H_3AsO_4(aq) + H_2SnCl_4(aq) + HCl(aq) \longrightarrow$
$$As(s) + H_2SnCl_6(aq)$$

Objective (c)

9. Calculate the percentage yield in each of the following processes:

   a. A 25-g sample of toluene ($C_6H_5CH_3$) was converted to 31 g of potassium benzoate ($KC_6H_5CO_2$) by oxidation with excess hot potassium permanganate (see question 6c).

   b. A 50-g sample of nitrobenzene ($C_6H_5NO_2$) was reduced by excess hydrogen, using a platinum catalyst, to 28 g of aniline ($C_6H_5NH_2$).

10. What volume of 0.174 N potassium permanganate (standardized for $MnO_4^- \longrightarrow Mn^{2+}$) is required for complete conversion of 5.6 g of ethylene to ethylene glycol by the process shown in Example 5?

11. What volume of chlorine gas, measured at 20°C and 700 torr, would be required for the complete oxidation of 15.0 g of potassium iodide to potassium periodate, according to the reaction in question 8a?

12. What volume of 0.30 M nitrous acid would be consumed by complete reaction with 25.0 ml of 0.28 M Co^{2+}, according to the reaction in question 8b?

13. Calculate the mass of "red lead" (Pb_3O_4) that would be consumed by reaction with 50.0 ml of 6.2 M nitric acid, according to the equation,

$$Pb_3O_4 + 4HNO_3 \longrightarrow 2Pb(NO_3)_2 + PbO_2 + 2H_2O$$

14. What is the molarity of a thiosulfate solution from which a 23.7-ml aliquot is required for the complete reduction of a 12.7-g sample of iodine, according to the reaction in question 8e?

15. Calculate the volume of 12.0 M sulfuric acid that would be required for the complete conversion of 16.3 g of copper to copper(II) sulfate. (The other products are sulfur dioxide and water.)

16. What volume of a solution labeled "0.100 N nitric acid (standardized for $NO_3^- \longrightarrow NO_2$)" would be required for oxidation of 0.36 g of silver according to the equation given? (Assume that the nitric acid calculated is required only as a source of oxidizing agent.)

$$3Ag + 4H^+ + NO_3^- \longrightarrow 3Ag^+ + NO + 2H_2O$$

a. 11 ml; **b.** 24 ml; **c.** 38 ml; **d.** 47 ml; **e.** 55 ml

17. What volume of a solution labeled "1.00 N sulfuric acid (standardized for $HSO_4^- \longrightarrow S$)" would be required for oxidation of 0.10 g of copper, according to the equation given? (Assume that the sulfuric acid calculated is required only as a source of oxidizing agent.)

$$Cu + 3H^+ + HSO_4^- \longrightarrow Cu^{2+} + SO_2 + 2H_2O$$

a. 9.4 ml; **b.** 14 ml; **c.** 19 ml; **d.** 22 ml; **e.** 28 ml

18. What volume of a solution labeled "1.00 N nitric acid (standardized for $NO_3^- \longrightarrow NO$)" would be required for oxidation of 0.25 g of zinc, according to the equation given? (Assume that the nitric acid calculated is required only as a source of oxidizing agent.)

$$Zn + 4H^+ + 2NO_3^- \longrightarrow Zn^{2+} + 2NO_2 + 2H_2O$$

a. 13 ml; **b.** 18 ml; **c.** 23 ml; **d.** 29 ml; **e.** 33 ml

19. What volume of 0.100 N nitrous acid (standardized for $HNO_2 \longrightarrow NO_3^-$) is required for conversion of 13.0 g of $CoCl_2$ to $CoCl_3$? (Under acidic conditions in dilute HCl, the nitrous acid is reduced to NO.)

a. 1.50 liters; **b.** 2.00 liters; **c.** 2.25 liters; **d.** 3.00 liters; **e.** 4.50 liters

BRAIN TINGLER

20. In a case of oxalic acid poisoning, it was planned to administer orally a solution of potassium permanganate as an emergency antidote, to be followed immediately by stomach pumping. The amount of oxalic acid ingested was believed to be about 250 mg. Any significant excess of permanganate had to be avoided because it is toxic and might also oxidize chloride ion in the gastric juices to chlorine. What molar concentration of potassium permanganate should have been employed so that 200 ml of the solution would be just sufficient to convert the suspected amount of oxalic acid to carbon dioxide? During the reaction, at the pH of the gastric juices, permanganate is reduced to Mn^{2+}. What assumption must have been made in the planned procedure concerning the relative rates of oxidation of chloride and oxalic acid? Can you suggest an alternative procedure that might reduce the risk of using so much permanganate?

1. Label each of the following equations as metathetical or redox. In each redox equation, draw a circle around the oxidizing agent and a rectangle around the reducing agent.
   a. $S(s) + SO_3^{2-}(aq) \longrightarrow S_2O_3^{2-}(aq)$
   b. $SO_3^{2-}(aq) + 2H^+(aq) \longrightarrow H_2O(l) + SO_2(g)$
   c. $2ClO_2(g) + 2OH^-(aq) \longrightarrow ClO_3^-(aq) + ClO_2^-(aq) + H_2O(l)$
   d. $2HClO_4(l) \longrightarrow Cl_2O_7(l) + H_2O(l)$
   e. $XeO_3(aq) + 4OH^-(aq) + O_3(g) \longrightarrow$
   $$XeO_6^{4-}(aq) + 2H_2O(l) + O_2(g)$$

2. Assign oxidation numbers to each element symbol in
   a. Na_2O_2; b. $Ca(OCl)_2$

3. Balance (H_2O, H^+, and OH^- not shown)

   a. $Al + NO_3^- \longrightarrow Al^{3+} + NO_2$ (in acidic solution)
   b. $CH_3OH + MnO_4^- \longrightarrow HCO_2^- + MnO_2$ (in alkaline solution)
   c. $C_6H_5CH_3 + K_2Cr_2O_7 + H_2SO_4 \longrightarrow$
   $$C_6H_5CO_2H + K_2SO_4 + Cr_2(SO_4)_3$$

4. What is the percent yield for a reaction in which 46 g of toluene ($C_6H_5CH_3$) is oxidized by excess potassium dichromate to 48 g of benzoic acid? (See question 3c.)

5. What volume of 0.585 N potassium permanganate (standardized for $MnO_4^- \longrightarrow Mn^{2+}$) is required for complete conversion of 0.800 g of methanol to formate ion by the process shown in question 3b?

Unit 23

Faraday's laws are exact; present views admit no other possibility.

Introduction to Electrochemistry (1970)
ERNEST H. LYONS, JR.

Electro-chemistry

Objectives

(a) Given the chemical system for a Galvanic or electrolytic cell, be able to draw a simple representation for the cell, observing standard conventions for labeling electrodes (e.g., Figure 23.2).

(b) Be able to apply the principles of stoichiometry to electrochemical processes, using balanced equations for half-reactions and the value of the Faraday constant (Section 23.2 and Example 8).

(c) Be able to calculate standard cell potentials using the data tabulated in Appendix D (Sections 23.4 and 23.5).

(d) Be able to calculate standard free-energy and entropy changes for electrochemical processes, given ΔH^0 data and standard cell or electrode potentials (Section 23.3).

567

Silver ions may be reduced to metallic silver in several different ways. A mirror may be prepared by treating one side of a glass sheet with an alkaline solution of diamminesilver(I) to which has been added a reducing agent such as glucose or formaldehyde. With the latter, the reaction may be formulated as

$$[Ag(NH_3)_2]^+ \rightleftharpoons Ag^+ + 2NH_3$$

$$3NH_3 + H_2O + 2Ag^+ + H_2C{=}O \longrightarrow 2Ag + HCO_2^- + 3NH_4^+$$

mirror
coating

This reaction is an example of direct electron transfer between reactant species.

An alternative process for reducing silver ion is used in the silverplating industry. The object to be plated is connected to the negative terminal of some source of direct current. This electrode is then immersed in a solution containing dicyano-argentate(I). The positive terminal is connected to a silver electrode. When current is passed through the circuit, the following reactions and half-reactions occur:
at the silver electrode

$$Ag \longrightarrow Ag^+$$
$$Ag^+ + 2CN^- \rightleftharpoons [Ag(CN)_2]^-$$

at the electrode being plated

$$[Ag(CN)_2]^- \rightleftharpoons Ag^+ + 2CN^-$$
$$Ag^+ + e^- \longrightarrow Ag$$

In this indirect electron transfer the electric circuit is maintained by a flow of electrons through the external wiring and a migration of ions within the solution. Although the process could use some simpler salt such as silver nitrate as the electrolyte, it has been found that this produces an irregular plating. The dicyano complex permits a high electrolyte concentration, which reduces the resistance of the cell, while maintaining a low uniform concentration of free Ag^+ for a smooth, firm deposition of silver.

Still a third type of process is worth noting. If a silver electrode is placed in a solution containing silver ions, this forms a half-cell. When another half-cell containing an appropriate metal, such as copper, in a solution of its ions (e.g., Cu^{2+}) is added, a galvanic cell is formed. Current flow will occur when the electrodes are connected by a wire and the electrolyte solutions are connected by a salt bridge or some other device that maintains electroneutrality without mixing of the different metal ions. Note that there is no external source of electrons. When the concentration ratio of Ag^+ and Cu^{2+} ions is within certain limits (Unit 24), the electrode processes may be formulated as
at the silver electrode:

$$Ag^+ + e^- \longrightarrow Ag$$

at the copper electrode:

$$Cu \longrightarrow Cu^{2+} + 2e^-$$

Again, electron transfer is indirect. Electrons flow spontaneously through the external circuit from the copper electrode to the silver electrode. Electroneutrality is maintained by migration of ions (other than Ag^+ and Cu^{2+}) between the two elec-

The invention of the electro-chemical cell is usually credited to the Italian physicist, Alessandro Volta, in 1800. It has recently been suggested, however, that a "magician" may have developed such cells around 200 B.C.[1]

---

[1]For an interesting glimpse of this story, see "Electric Cells from Mesopotamia?" by Paul Lauren, in the December, 1974, issue of Chemistry magazine (pp. 6–9).

trolyte solutions. Galvanic cells, such as the silver–copper cell, are useful when a relatively small current is required at a fairly low voltage.

In this unit we shall examine the significant features of direct versus indirect electron transfer. We shall then concentrate on two different characteristics of electrochemical processes: current (the speed of electron transfer) and voltage (the "electrochemical potential"). In Unit 24 we shall see how cell voltage varies with concentration of participating chemical species. This relationship, expressed by the Nernst equation, is the key to many important techniques of electrochemistry and to the study of equilibria in oxidation–reduction systems.

23.1 DIRECT AND INDIRECT ELECTRON TRANSFER

Oxidation–reduction reactions that occur as a result of the mixing of oxidizing and reducing agents involve electron transfer directly between colliding species. When copper is placed in a silver nitrate solution (Figure 23.1), for example, silver ions may collide with the metallic copper. Electrons are held more tightly in the silver atom than in the copper atom. As a result, the silver ions are able to pull electrons away from the copper atoms and metallic silver is formed—two silver atoms for each Cu^{2+} ion released into solution. This process is exothermic and most of the energy released is wasted as heat, which is difficult to harness for useful work.

Inefficient conversion of energy to work is typical of direct electron-transfer processes, as we have discussed earlier in connection with the combustion of fossil fuels (Unit 5).

Indirect electron transfer, involving the use of an external conductor to move electrons from the region in which oxidation occurs to that in which reduction occurs, still requires collisions. In the silver–copper electrochemical cell (Figure 23.2), for example, the *net* change ($Cu + 2Ag^+ \longrightarrow Cu^{2+} + 2Ag$) is the same as that which would take place if a copper wire were placed in a solution of Ag^+ of appropriate concentration. In the cell, however, the moving electrons can easily be used for work—lighting a bulb, running a motor, or producing some other chemical change in a different type of cell.

In order for indirect electron transfer to occur, it is necessary to prevent direct mixing of oxidizing and reducing agents. These must be confined to separate regions, or their mobilities must be restricted by mixing them in solid phases in which diffusion is quite slow. Electrochemical cells may be prepared by either of these methods. When solutions are employed, the electrolytes of the two half-cells are kept in separated containers. However, no more than a brief, almost instantaneous electron flow can be obtained unless there is some way of maintaining electroneutrality in the cell solutions. To effect this, some ions must be able to migrate between the separated half-cells. This is made possible either by using a *salt bridge,* containing a solution of an electrolyte not involved in the electron transfer, or by using a porous partition through which hydrated ions smaller than those of the oxidizing and reducing agents can migrate. In either case it is not possible to avoid some mixing of half-cell reactants, but this can be minimized.

Some electrochemical cells, such as the familiar "dry cell," employ a reactant mixture in the form of a moistened paste. Diffusion is slow enough to avoid significant electron transfer over fairly long time periods. It is not uncommon, however, for such cells to deteriorate because of direct oxidation–reduction processes, and dry cell flashlight batteries, for example, that are stored in a warm place may eventually "leak" as the metal container is oxidized.

Every electrochemical process must involve both oxidation (electron loss) and reduction (electron gain). The electrode at which oxidation occurs is called the *anode*

Figure 23.1 Direct electron transfer: Chemical change and "wasted" energy. When copper wire is placed in a silver nitrate solution, beautiful silver crystals begin to grow on the wire and the solution begins to turn blue as Cu^{2+} ions are formed. The energy of the direct electron transfer

$$(2Ag^+ + Cu \longrightarrow Cu^{2+} + 2Ag)$$

appears mostly as heat, and only a small fraction of this can be utilized for work. Direct electron-transfer processes are generally inefficient energy sources.

Additional confusion sometimes results from the electrical convention that "current" flows from the positive to the negative terminal. The actual particles that "flow" are, of course, electrons, and these move from the − terminal to that labeled + in the external circuit. Persons concerned with the external circuit use a different terminology, referring to the negative terminal as the *external cathode* and the positive terminal as the external anode.

Chemists, being interested primarily in cell reactions, usually employ the terms *cathode* and *anode* in their *internal* sense, whereas those more concerned with the electrical processes use the terms in reference to the *external* terminals. We shall use the terms *anode* and *cathode* exclusively for *internal* electrodes, specifying any reference to the outside terminals of the cell by addition of the word "external." It is necessary, then, only to remember that *oxidation is an anode process.*

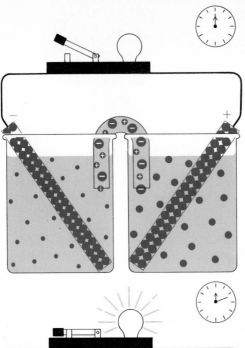

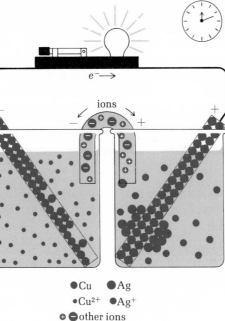

●Cu ●Ag
•Cu²⁺ ●Ag⁺
⊕ ⊖ other ions

Figure 23.2 Indirect electron transfer: electric energy from chemical change.

$$Cu(s) \longrightarrow Cu^{2+}(aq) + 2e^-$$
$$2Ag^+(aq) + 2e^- \longrightarrow 2Ag(s)$$

Indirect electron-transfer processes, such as those in the copper-silver electrochemical cell, are much more efficient energy sources than the corresponding direct electron-transfer reactions.

and since electrons exit through this electrode, the external terminal of this electrode is labeled −. Reduction, then, takes place at the *cathode,* the electrode labeled + on the outside of a cell. Electrons traveling through the external circuit enter the cell at this electrode.

**23.2
ELECTROCHEMICAL
STOICHIOMETRY**

The basis for the stoichiometry of electrochemical processes was first enunciated by Michael Faraday, an English scientist, in 1834. In modern terminology:

1. The quantity of chemical change associated with an electric current is proportional to the number of electrons transferred.

2. The masses of different substances changed by an electrochemical process involving a fixed electron transfer are proportional to the equivalent weights of the substances.

We recognize now the relationship responsible for Faraday's laws. Every chemical change is governed by the laws of conservation of mass, charge, and energy. For electrochemical processes, that is, those that require indirect electron transfer, we can write *half-reaction equations*. When these are balanced, they represent the stoichiometric ratios of moles of chemical substances *and moles of electrons*. For example,

$$Ag^+ + e^- \longrightarrow Ag$$

(for each mole of silver formed, 1 mole of electrons is required), and

$$Cu \longrightarrow Cu^{2+} + 2e^-$$

(for each mole of copper consumed, 2 moles of electrons are released).

Chemical quantities can be measured in several ways. We can weigh solids or liquids, measure the volumes of liquids or gases, and determine the volumes and concentrations of solutions. None of these methods can be applied to the electrons in an oxidation or reduction process. We can, however, measure electric charge or—for a *current*—the speed of charge movement. The most common unit for expressing a current flow rate is the *ampere* (amp) defined as the number of coulombs of charge moving past a reference point per unit time.

$$\text{amperes} = \frac{\text{coulombs}}{\text{seconds}} \tag{23.1}$$

Now, we know the charge of a single electron to be 1.6021×10^{-19} C (from Millikan's oil-drop experiments). We also know the number of unit particles per mole to be 6.0225×10^{23} (Avogadro's number, Unit 4). From these data we can calculate a useful quantity called the *Faraday constant*:

$$\frac{1.6021 \times 10^{-19} \text{ C}}{1 \text{ electron}} \times \frac{6.0225 \times 10^{23} \text{ electrons}}{\text{mole } (e^-)} = 9.65 \times 10^4 \text{ C mole}^{-1} (e^-)$$

For measuring small current flow rates, meters may be used that are calibrated in milli-amperes (milliamp). Special instruments called galvanometers are employed for very sensitive current measurements.

The quantity of charge associated with an electrochemical process can then be measured by inserting an ammeter in the electric circuit and determining the current flow rate for a measured time interval. Since current may fluctuate, it is common to report the *average* amperage measured.

We may now write the two sets of unity factors that provide the basis for calculations in electrochemical stoichiometry:

1. from the Faraday constant,

$$\frac{9.65 \times 10^4 \text{ C}}{1 \text{ mole } (e^-)} \quad \text{or} \quad \frac{1 \text{ mole } (e^-)}{9.65 \times 10^4 \text{ C}}$$

2. from equation (23.1),

$$\frac{1 \text{ C}}{1 \text{ amp sec}} \quad \text{or} \quad \frac{1 \text{ amp sec}}{1 \text{ C}}$$

Two different types of electrochemical processes are of great importance. Cells used to *produce* an electric current are called *galvanic* cells. These utilize some oxidation–reduction reaction that will take place spontaneously when the electric circuit is completed (e.g., Figure 23.2). The other type, called an *electrolytic* cell, requires an external source of electric energy to produce an oxidation–reduction reaction that would not occur spontaneously. The electrolysis of water, the production of "anodized" aluminum, and silverplating are examples of processes using such electrolytic cells. Electrolytic cells usually do not require a separation of electrode compartments, since electrolyte mixing is seldom a problem.

The relationship between coulombs and moles of electrons is analogous to that between mass and moles of a chemical substance. As a result, these relationships may be conveniently employed in working with the stoichiometry of electrochemical reactions in a manner similar to those used in weight-to-weight problems (Unit 4) or weight-to-volume problems (Unit 10). The principal difference is that electrochemical stoichiometry requires a knowledge of the electron-to-chemical species ratio, and this is most readily determined from balanced equations for *half-reactions*.

Example 1

We shall use the common convention of "current" to mean *rate* of electron transfer.

In the older type of flashlight cell, the electrolyte is contained in a zinc can that also serves as one of the electrodes. What mass of zinc is oxidized to Zn^{2+} during discharge of a cell of this type for a 30.0-min period involving a current of 125 milliamp?

Solution
The equation for the half-reaction concerned is

$$Zn \longrightarrow Zn^{2+} + 2e^-$$

From equation (23.1)

$$\text{no. coulombs transferred} = \text{amperes} \times \text{seconds}$$

$$= \frac{125 \text{ milliamp}}{1} \times \frac{1 \text{ amp}}{10^3 \text{ milliamp}} \times \frac{30.0 \text{ min}}{1} \times \frac{60 \text{ sec}}{1 \text{ min}} \times \frac{1 \text{ C}}{\text{amp sec}}$$

$$= 225 \text{ C}$$

Then, using the method outlined for conventional weight-to-weight problems (Unit 4),

$$1 \text{ mole (Zn)} = 2 \text{ moles } (e^-) \quad \text{(from the chemical equation)}$$
$$1 \text{ mole (Zn)} = 65.4 \text{ g (Zn)} \quad \text{(from the periodic table)}$$
$$1 \text{ mole } (e^-) = 9.65 \times 10^4 \text{ C} \quad \text{(Faraday constant)}$$

$$\frac{225 \text{ C}}{1} \times \frac{1 \text{ mole } (e^-)}{9.65 \times 10^4 \text{ C}} \times \frac{1 \text{ mole (Zn)}}{2 \text{ moles } (e^-)} \times \frac{65.4 \text{ g (Zn)}}{1 \text{ mole (Zn)}} = 7.62 \times 10^{-2} \text{ g (Zn)}$$

$$\left(\begin{array}{c} \text{C} \\ \downarrow \\ \text{mole } (e^-) \end{array} \right) \quad \left(\begin{array}{c} \text{mole } (e^-) \\ \downarrow \\ \text{mole (Zn)} \end{array} \right) \quad \left(\begin{array}{c} \text{mole (Zn)} \\ \downarrow \\ \text{g (Zn)} \end{array} \right)$$

Example 2

What average current must be maintained for an industrial process designed to produce 250 kg of "anodized" aluminum per hour by electrolysis of a molten aluminum salt?

Solution

The half-reaction may be formulated as

$$Al^{3+} + 3e^- \longrightarrow Al$$

Equivalents for unity factors required:

1 mole (Al)	= 3 moles (e⁻)	(from the chemical equation)
1 mole (Al)	= 27.0 g (Al)	(from the periodic table)
1 mole (e⁻)	= 9.65 × 10⁴ C	(Faraday constant)
1 C	= 1 amp sec	(Equation 23.1)
1 hr	= 3600 sec	
1 kg	= 10³ g	

$$\frac{250 \text{ kg (Al)}}{1 \text{ hr}} \times \frac{10^3 \text{ g}}{1 \text{ kg}} \times \frac{1 \text{ mole (Al)}}{27.0 \text{ g (Al)}} \times \frac{3 \text{ moles (e}^-)}{1 \text{ mole (Al)}} \times \frac{9.65 \times 10^4 \text{ C}}{1 \text{ mole (e}^-)} \times \frac{1 \text{ amp sec}}{1 \text{ C}} \times \frac{1 \text{ hr}}{3600 \text{ sec}}$$

$$= 7.4 \times 10^5 \text{ amp}$$

watts = amps × volts

That is a lot of current. For comparison, a 250-watt (110–115 volt) light bulb "pulls" about 2 amp. Although electroplating industries are "cleaner" than many other types of industry, they may still represent significant environmental problems if their current must come from a power plant using "environmentally dirty" fuel.

Example 3

How long would it take (in hours) to product 1.00 liter of hydrogen gas (measured at STP) by electrolysis of water using an apparatus designed to maintain an average current of 300 milliamps (0.300 amp)?

Solution

The half-reaction may be represented as

$$2H_3O^+ + 2e^- \longrightarrow H_2 + 2H_2O$$

(Electrolysis of "water" is typically performed with a dilute acid solution, since pure water is a very poor electrical conductor. Alkaline solutions are avoided because of the action of alkalis on the glass of the equipment used.)

Then we may define appropriate equivalents for use as unity factors:

1 mole (H₂) = 22.4 liters (at STP)
1 mole (H₂) = 2 moles (e⁻)
1 mole (e⁻) = 9.65 × 10⁴ C
1 C = 1 amp sec
1 hr = 3600 sec

$$\frac{1.00 \text{ liter}}{1} \times \frac{1 \text{ mole (H}_2)}{22.4 \text{ liters}} \times \frac{2 \text{ mole (e}^-)}{1 \text{ mole (H}_2)} \times \frac{9.65 \times 10^4 \text{ C}}{1 \text{ mole (e}^-)} \times \frac{1 \text{ amp sec}}{1 \text{ C}}$$

$$\underbrace{}_{\text{at STP}}$$

$$\times \frac{1}{0.300 \text{ amp}} \times \frac{1 \text{ hr}}{3600 \text{ sec}} = 7.98 \text{ hr}$$

The preceding examples assume 100 percent efficiency. Real electrochemical processes are less than 100 percent efficient, so calculations of these types yield only the theoretical values. Some reasons for this will be discussed later.

23.3 VOLTAGE: THE ELECTROCHEMICAL POTENTIAL

There is an energy difference between different oxidation states of an element. We have already seen how this difference may be expressed for monatomic species by ionization potentials and electron affinities (Unit 2). We have also seen how enthalpy differences may be used (Unit 5).

One important aspect of electrochemistry can be used to determine another type of energy difference—ΔG, the Gibbs free-energy change. This determination is particularly useful because of the relationship between ΔG and reaction spontaneity and equilibrium.

FREE ENERGY REVISITED

In Unit 5 we defined the *Gibbs free-energy* change for a process in terms of equation (5.5):

$$\Delta G = \Delta H - T\,\Delta S \qquad \text{(at constant pressure and temperature)}$$

Let's examine the origin of this equation and see what insight it can provide in terms of reaction spontaneity and equilibrium, with particular application to electrochemistry.

The *second law of thermodynamics* (Unit 5) tells us that *any spontaneous change is associated with an increase in entropy* ("randomness") *of the universe*, that is,

$$\Delta S_{\text{universe}} > 0 \qquad \text{(for a spontaneous process)}$$

We recognize that this statement must be applied to the "universe," that is, to a thermodynamic system *and* its total surroundings. We know that many "localized" processes associated with "decreased entropy" are spontaneous, such as the crystallization of liquid water to ice at temperatures of $0°C$ or lower. In order for such processes to conform to the second law of thermodynamics, the entropy increase of the *surroundings* must exceed the entropy decrease of the *system*. In the case of the freezing of water, for example, the heat transferred from the water to the surroundings serves to increase the "randomness" of the surroundings. Thus we should consider entropy changes of both our *system* and its *surroundings*:

$$(\Delta S_{\text{universe}} = \Delta S_{\text{system}} + \Delta S_{\text{surroundings}}) > 0$$

The entropy change of the *surroundings* is a function of both the amount of heat transferred to the surroundings and the temperature:

$$\Delta S_{\text{surroundings}} = \frac{q_{\text{transferred}}}{T}$$

Although the mathematical basis for the preceding equation requires a more extensive background in thermodynamics and in calculus, we can see the conceptual basis for the relationship. At high temperatures molecules move more rapidly than at low temperatures; that is, "randomness" is already greater at higher temperatures. Thus a particular amount of heat will make less *change* in the randomness of a system at high temperatures than at low temperatures, hence the inverse relationship of ΔS to T.

Now, at the particular condition of constant pressure and temperature when no work other than $P \Delta V$ is done on or by the system (Unit 5),

$$q_{transferred} = -\Delta H_{system}$$

that is, for the system to maintain constant temperature, all heat produced must be transferred to the surroundings (or for an endothermic process, all heat required must be transferred *from* the surroundings.) Thus

$$\Delta S_{universe} = \Delta S_{system} + \frac{(-\Delta H_{system})}{T}$$

or

$$T \Delta S_{universe} = T \Delta S_{system} - \Delta H_{system} = -(\Delta H_{system} - T \Delta S_{system})$$

Since the entropy change of the *universe* would be difficult to measure, it is convenient to note that this can now be related to *system* properties. The *Gibbs free energy* (named for the American theoretician J. Willard Gibbs) is now *defined* in terms of these system properties:

$$\Delta G_{system} = \Delta H_{system} - T \Delta S_{system}$$

and since $T \Delta S_{universe}$ then equals $-\Delta G_{system}$, the statement of process spontaneity in terms of *system* properties follows from

$$\Delta S_{universe} > 0$$

but

$$\Delta G_{system} = -T \Delta S_{universe}$$

and, at any temperature above *absolute zero*, $T > 0$; so

$$\Delta G_{system} < 0 \qquad \text{for any spontaneous process}$$

Note that because of the way this relationship was derived, this statement is true *only* at constant temperature and pressure. These constraints impose some limitations, but there are many chemical systems for which such conditions can be considered, and essentially all the chemical reactions within living organisms occur at constant temperature and pressure.

Since the condition of chemical equilibrium corresponds to a system of constant properties, that is, of *zero net change*, it follows that at equilibrium $\Delta S_{universe} = 0$, so that

$$\Delta G_{system} = 0 \qquad \text{at equilibrium}$$

It is not necessary that *all* the enthalpy change for a process result in an entropy change of the surroundings. We know from experience that some heat can be converted to useful work. Of the total heat produced by a process, only that part "lost" (*transferred*) to the surroundings is unavailable for useful work; that is,

$$q_{useful} = q_{total} - q_{transferred}$$

Now, at constant temperature and pressure,

$$q_{total} = -\Delta H_{system}$$

and

$$q_{transferred} = T \, \Delta S_{surroundings}$$

(Note that $q_{transferred}$ is *not* $-\Delta H_{system}$ when we are considering work *other* than $P \, \Delta V$.) Since

$$T \, \Delta S_{surroundings} = T \, \Delta S_{universe} - T \, \Delta S_{system}$$

then

$$q_{useful} = -\Delta H_{system} - (T \, \Delta S_{universe} - T \, \Delta S_{system})$$

that is, by definition of ΔG

$$q_{useful} = -\Delta G_{system} - T \, \Delta S_{universe}$$

The *maximum* conversion of heat to work would occur if *none* of the heat resulted in an entropy change of the universe (i.e., as $\Delta S_{universe} \longrightarrow 0$), thus

the $-\Delta G_{system}$ represents the maximum useful work, other than $P \, \Delta V$, that could be obtained from any spontaneous process at constant temperature and pressure.

Now, what kind of work other than $P \, \Delta V$ (gas expansion) could we consider? The answer to this is *electrical work* and the key to an understanding of electrical work is the definition of the *electrochemical potential,* or *voltage,* as the maximum electrical work available from a system per unit charge to be transferred; that is,

$$\mathcal{E} = \frac{-\Delta G_{system}}{nF} \tag{23.2}$$

in which n is the number of moles of electrons theoretically available for transfer, F is the Faraday constant [$9.65 \times 10^4 \, C \, mole^{-1} \, (e^-)$], and $\mathcal{E}$ is the electrochemical potential (voltage) of the cell.

From this definition of voltage it is apparent that we could determine the maximum electrical work (i.e., the theoretical limit) available for any electron-transfer process from an *experimentally measurable* quantity (voltage) and a knowledge of the chemical limitations on charge transfer.

Electrical work can be expressed as the product of the quantity of charge transferred and the potential difference (voltage) through which the charge is moved. For an electrochemical cell, then, the maximum work theoretically obtainable will be the total number of coulombs available for complete consumption of the limiting reagent of the cell components multiplied by the cell voltage; that is,

As we shall see in Unit 24, the voltage of a galvanic cell varies with concentrations, so the voltage does not remain constant during discharge.

$$\Delta G = -nF\mathcal{E} \tag{23.3}$$

A conversion factor is necessary to convert the units *volt coulombs* to the more familiar energy units of calories. This factor is

$$1.000 \text{ cal} = 4.186 \text{ VC}$$

Although a volt coulomb sounds like a strange unit of energy, remember that electric energy is sold in units of kilowatt hours. Since wattage is the product of current (amperes) and voltage, the kilowatt hour can easily be shown to be reducible to volt coulombs.[2]

Cell voltage, then, is a very useful quantity to know. It can be measured easily (with a voltmeter) and accurately. The use of cell voltage to calculate ΔG, when combined with an independent measurement of ΔH, permits the calculation of entropy changes for chemical processes.

Example 4

The electrochemical cell whose net reaction is $2H_2(g) + O_2(g) \longrightarrow 2H_2O(l)$ has a potential under standard state conditions of 1.23 V. The standard heat of formation of liquid water is -68.3 kcal mole^{-1}. Calculate the standard entropy of formation of water.

Solution

For conversion of two oxygens from 0 to $-$II oxidation state (or for four hydrogens from 0 to $+$I) a total of four electrons must be transferred. Thus, *per mole of water formed, n = 2*. Then

$\mathcal{E}^0$ represents the *standard cell potential,* that is, the potential of an electrochemical cell under conditions of standard states (Unit 5). (See also Section 23.4.)

$$\Delta G^0 = -nF\mathcal{E}^0$$

$$= -\left(\frac{2 \text{ moles } e^-}{1 \text{ mole } H_2O} \times \frac{96,500 \text{ C}}{1 \text{ mole } e^-} \times \frac{1.23 \text{ V}}{1} \times \frac{1 \text{ cal}}{4.186 \text{ VC}} \times \frac{1 \text{ kcal}}{10^3 \text{ cal}} \right)$$

$$= -56.7 \text{ kcal mole}^{-1}$$

$$\Delta G^0 = \Delta H^0 - T\,\Delta S^0$$

Under standard state conditions,

$$T = 298°\text{K}$$

This applies only to the standard states we have employed thus far. As we shall see in Unit 24, 298°K is not a requirement of the *generalized* "standard state" concept.

so

$$\Delta S^0 = \frac{\Delta H^0 - \Delta G^0}{298°}$$

$$= \frac{(-68.3) - (-56.7)}{298} = -3.89 \times 10^{-2} \text{ kcal mole}^{-1} \text{ deg}^{-1}$$

or, in more common terms,

$$\Delta S_f{}^0(\text{water}) = -38.9 \text{ cal mole}^{-1} \text{ deg}^{-1}$$

23.4 SINGLE ELECTRODE POTENTIALS

Voltage is measured as the *difference* between two electric potentials. Thus voltage measurements can be made only on the basis of comparison between two half-cells. There is no way of measuring the "absolute voltage" of a single half-cell. Nevertheless it is desirable to be able to assign potentials to half-cells, since this will enable us to calculate ΔG for half-reactions and thus to use free-energy changes associated with oxidation or reduction half-reactions. Since ΔG is a state function, single electrode potentials would also be state functions. This would permit us to make a variety of

[2] For an interesting application, showing that the average person's daily energy expenditure is about that of an ordinary light bulb, see "Electrochemistry in Organisms," by Thomas Chirpick, in the February, 1975, issue of the *Journal of Chemical Education* (pp. 99–100).

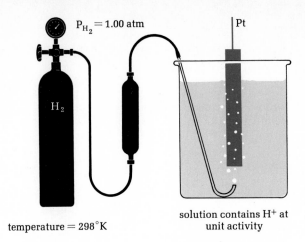

$P_{H_2} = 1.00$ atm

Pt

H₂

temperature = 298°K

solution contains H⁺ at unit activity

useful calculations, including the voltage to be expected from a cell made by combining any two half-reactions for which the potentials are known.

Since only the potential *difference* can be measured, we must select some half-cell as a reference standard against which other electrode potentials may be compared. The universally accepted reference is the *standard hydrogen electrode* (Figure 23.3), which is *assigned* a potential of exactly 0 V.

Theoretically, an electrochemical cell could be formed by connecting any other half-cell (whose reactive components are under standard state conditions) to the reference hydrogen electrode. If a voltmeter were used that had a zero center reading with possibility for both positive and negative deflections, then the voltage registered (cell potential) would represent the single electrode potential of the half-cell being compared (Figure 23.4). To maintain the correct sign convention, the hydrogen electrode would always be connected to the positive terminal of the meter. Standard electrode potentials ($\mathcal{E}^0$) have been tabulated for a wide variety of chemical systems. For a sample listing and a description of their use, see Appendix D. It will be noted that the most widely used convention lists $\mathcal{E}^0$ values with reference to *reduction*

Systems involving conditions other than those of the standard state are discussed in Unit 24.

Figure 23.4 *Hypothetical* cell for determination of single electrode potential.

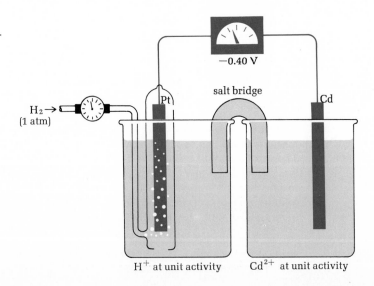

−0.40 V

salt bridge

Pt

Cd

H₂ →
(1 atm)

H⁺ at unit activity

Cd²⁺ at unit activity

half-reactions. To use *these* potentials for a calculation of cell voltage, it is first necessary to identify the internal cathode and anode processes (Section 23.1). Then

$$\mathcal{E}^0_{\text{cell}} = \mathcal{E}^0_{\text{cathode}} - \mathcal{E}^0_{\text{anode}} \qquad (23.4)$$

Since ΔG^0 must be less than zero for a spontaneous process, it follows that $\mathcal{E}^0_{\text{cell}}$ must be positive for a spontaneous cell reaction. A negative voltage reading, such as that obtained for the hydrogen–cadmium cell in Figure 23.4, simply means that the voltmeter terminals have been connected to the electrodes of *opposite* sign for spontaneous cell function.[3] The true cell potential is still [equation (23.4)].

$$\mathcal{E}^0_{\text{cell}} = \mathcal{E}^0_{\text{H}_2} - \mathcal{E}^0_{\text{Cd}} = 0 - (-0.40) = 0.40 \text{ V}$$

It is important to note that the calculation of voltage for spontaneous processes is applicable to electrolytic as well as galvanic cells. For example, the electrolysis of water under standard state conditions requires a minimum voltage of 1.23 V—the same as the *maximum* voltage obtainable from a galvanic cell in which water is formed from its elements (Example 4).

23.5 SAMPLE CALCULATIONS INVOLVING STANDARD ELECTRODE POTENTIALS

We have referred to *hypothetical* cells and the *hypothetical* standard hydrogen electrode. This might seem to suggest that standard electrode potentials are not "real" values. Such is not the case. The problem is an experimental one, coupled with the "proper" definition of *standard state*[4] for an electrolyte solution. The standard state of an electrolyte solution is *correctly* defined as

that of a hypothetical solution of mean ionic molality of unity (1) for which the reference state is so chosen that the mean ionic activity coefficient approaches unity as the concentration of the solute approaches zero.

The *activity coefficient* is the ratio of the thermodynamic activity to the molality of the solute.

Without taking the time to go into the complete analysis of this definition, we may point out some important implications. First, the requirements of the proper definition of standard state means *that it is not possible actually to prepare a standard hydrogen electrode.* However, other reference electrodes based on the defined value of zero for $\mathcal{E}^0$ of the standard hydrogen electrode *can* be prepared and these can, indeed, be used for experimental determinations. Second, the use of $\mathcal{E}^0$ values implies that dissolved electrolytes are expressed in *activities,* and *activities* are only approximated by *molalities,* and even less accurately so by *molarities.* In very dilute solutions these approximations are actually quite close, but the differences between *activities, molalities,* and *molarities* increase with increasing solute concentration (Unit 13).

Because *molarity* is such a useful concentration term in a large number of chemical applications, we will continue to use *molar* concentrations in our study of electrochemical cells. This will, of course, impose some limitations on the accuracy of our calculations. In general, the use of *molarities* will represent an approximation

[3]For an interesting application of single electrode potentials as measures of oxidizing-reducing *strength,* see "Conjugate Acid–Base and Redox Theory," by Richard Pacer in the March, 1973, issue of the *Journal of Chemical Education* (pp. 178–180).

[4]For detailed explanations of the problems imposed by *standard state* limitations, see "The Standard Hydrogen Electrode—A misrepresented concept," by T. Biegler and R. Woods in the September, 1973, issue of the *Journal of Chemical Education* (pp. 604–605) and "The Proper Definition of the Standard Electromotive Force," by Omer Robbins, Jr. in the November, 1971, issue of the *Journal of Chemical Education* (pp. 737–740).

ranging from about ±10 percent for moderately concentrated solutions to less than ±1 percent for very dilute solutions.

The following examples, and others found in Unit 24, employ the approximation of *molar* concentrations.

Before analyzing the following examples, study Appendix D carefully.

Example 5

---

    a. What is the standard cell potential for a galvanic cell in which one electrode is copper immersed in 1.0 M Cu^{2+} solution and the other is magnesium immersed in 1.0 M Mg^{2+} solution?

    b. Which electrode is labeled − at it's external terminal?

    c. What is the net equation for the spontaneous cell process?

Solution

    a. From Appendix D,

$$\mathcal{E}^0_{cell} = \mathcal{E}^0_{cathode} - \mathcal{E}^0_{anode} \qquad \text{[equation (23.4)]}$$

For $Mg^{2+} + 2e^- \longrightarrow Mg$,

$$\mathcal{E}^0 = -2.37 \text{ V}$$

For $Cu^{2+} + 2e^- \longrightarrow Cu$,

$$\mathcal{E}^0 = +0.34 \text{ V}$$

Since $\mathcal{E}^0_{cell}$ must be >0 for spontaneity, $\mathcal{E}^0$ for magnesium must be subtracted from that for copper:

$$\mathcal{E}^0_{cell} = +0.34 - (-2.37) = 2.71 \text{ V}$$

    b. The negative terminal corresponds to the *internal* anode. Since the preceding calculation required $\mathcal{E}^0$ for magnesium to be used as the anode term, the magnesium electrode must be labeled negative; that is, for the spontaneous process, electrons leave the cell through the magnesium electrode.

    c. From the definitions of oxidation and reduction, it follows that the spontaneous half-reactions are formulated by

$$\text{oxidation} \quad Mg \longrightarrow Mg^{2+} + e^-$$
$$\text{reduction} \quad Cu^{2+} + 2e^- \longrightarrow Cu$$

The net cell process, then, is

$$Mg + Cu^{2+} \longrightarrow Mg^{2+} + Cu$$

Example 6

Calculate the standard potential for the silver–copper electrochemical cell (Figure 23.2).

Solution

From Appendix D,

$$Ag^+ + e^- \longrightarrow Ag \qquad \mathcal{E}^0 = +0.80 \text{ V}$$
$$Cu^{2+} + 2e^- \longrightarrow Cu \qquad \mathcal{E}^0 = +0.34 \text{ V}$$

To make $\mathcal{E}^0_{cell} > 0$,

$$\mathcal{E}^0_{cell} = \mathcal{E}^0_{Ag} - \mathcal{E}^0_{Cu} = +0.80 - (+0.34) = +0.46 \text{ V}$$

(i.e., silver must be the cathode).

Note that in Example 6 the equation for the cathode process shows a single electron, whereas that for the anode reaction shows two electrons. This difference has no bearing on the calculation of the *cell* potential, since only the number of electrons transferred for the *complete* process ($2e^-$ for the *balanced* equation) is involved [equation (23.3)]. However, the combination of two electrode potentials to find a third *electrode* potential *does* require the use of the numbers of electrons in each half-reaction (Appendix D).

It may not always be obvious as to when the number of electrons transferred must be considered. If in doubt, you should see how the net free energy change for the particular process is related to ΔG values for individual electrode processes involved, in terms of equation (23.3), $\Delta G = -nF\mathcal{E}$.

Example 7

Given:

$$Ag^{2+} + e^- \longrightarrow Ag^+ \qquad \mathcal{E}^0 = +1.98 \text{ V}$$
$$Ag^+ + e^- \longrightarrow Ag \qquad \mathcal{E}^0 = +0.80 \text{ V}$$

Calculate $\mathcal{E}^0$ for $Ag^{2+} + 2e^- \longrightarrow Ag$.

Solution
From Appendix D,

$$(n_a + n_b)\mathcal{E}^0_{\text{net}} = n_a\mathcal{E}_a^{\,0} + n_b\mathcal{E}_b^{\,0}$$

Then

$$(1 + 1)\mathcal{E}^0_{\text{net}} = (1)(1.98) + (1)(0.80)$$

$$\mathcal{E}^0_{\text{net}} = \frac{2.78}{2} = +1.39 \text{ V}$$

23.6 LIMITATIONS ON CURRENT AND VOLTAGE

Remember that maintenance of electroneutrality requires a migration of nonreactant ions (e.g., through a salt bridge or porous barrier) at the rate determined by changes in reactant ion concentrations. Thus an additional limitation on current flow is imposed by this requirement.

The *apparent* diffusion rate of H^+ is a notable exception. In water, however, this does not involve a true diffusion process, since the *same* proton need not move very far. Rather, "new" protons are generated by O—H bond rupture in water molecules.

Galvanic cells are limited in terms of both maximum voltage and maximum current-flow rate. The latter limitation arises from the dependence of current flow on the migration of reactant ions through the electrolyte. Similar problems occur in electrolytic cells. Diffusion rates in aqueous solutions are relatively slow, and they are appreciably slower in the moist pastes used as electrolytes in many commercial cells. Galvanic cells cannot therefore be used when high currents are required.

Cells may be connected in series to increase voltage. The familiar automobile battery, for example, consists of three cells (for 6.0 V) or six cells (for 12.0 V) arranged in series. It might seem that this technique would provide a way for an unlimited voltage increase. Such is not the case, however, because voltage drop from *resistance* effects also increases as additional cells are connected.

As galvanic cells are discharged, the concentrations of necessary reactants are decreased. Voltage then begins to drop, reaching zero when the cell is "dead." Maximum current-flow rates also decrease as discharge continues, since at lower concentrations fewer reactant ions can reach the electrode per unit time. In many cases it is possible to reverse the cell reactions by application of an external source of current and recharge the cells.

The variations of electrochemical potential with reactant concentration form the subject of the next unit.

There are other "practical" limitations on electrochemical processes. The electrical *resistance* of the conducting medium (e.g., a wire or an aqueous electrolyte solution) results in the conversion of some of the electric energy to heat. Competing redox processes, such as the electrolysis of water during an electroplating operation, also reduce the efficiency of a particular electrochemical process.

Example 8

How long, in hours, would be required for the electroplating of 39 g of chromium from a chromium(III) solution, using an average current of 30 amp at an 80% electrode efficiency?

Solution

A direct stoichiometric calculation permits the calculation of *theoretical* quantities (i.e., at 100% efficiency). For

$$Cr^{3+} + 3e^- \longrightarrow Cr$$

$$\frac{\text{theoretical}}{\text{time}} = \frac{1 \text{ amp sec}}{1 \text{ coulomb}} \times \frac{1 \text{ hr}}{3600 \text{ sec}} \times \frac{96,500 \text{ C}}{1 \text{ mole } e^-} \times \frac{3 \text{ mole } e^-}{1 \text{ mole Cr}} \times \frac{1 \text{ mole Cr}}{52 \text{ g Cr}} \times \frac{39 \text{ g Cr}}{30 \text{ amps}}$$

$$= 2.0 \text{ hours}$$

But since the process is less than 100% efficient, more time will be required,

theoretical time = 0.80 (actual time)

from which

$$\text{actual time} = \frac{\text{theoretical time}}{0.80} = \frac{2.0 \text{ hr}}{0.80} = 2.5 \text{ hr}$$

SUGGESTED READING

Crow, D. R. 1974. *Principles and Applications of Electrochemistry.* New York: Halsted Press.

Lyons, Ernest H., Jr. 1967. *Introduction to Electrochemistry.* Lexington, Mass.: Raytheon.

O'Connor, Rod, and Charles Mickey. 1975. *Solving Problems in Chemistry.* New York: Harper & Row.

Robbins, J. 1972. *Ions in Solution: An Introduction to Electrochemistry.* New York: Oxford University Press.

EXERCISES

1. Draw a representation for a galvanic cell using Al, Al^{3+} and Cu, Cu^{2+} half-cells. Label correct signs for external electrodes. Assume ion concentrations to be approximately 1.0 *M*.

2. The net reaction during discharge of the lead storage battery may be formulated as

$$Pb + PbO_2 + 2H^+ + 2HSO_4^- \longrightarrow 2PbSO_4 + 2H_2O$$

What mass of lead would be consumed during a 3.0-hr discharge involving an average current of 3.2 amp?

3. Metallic sodium may be prepared by electrolysis of molten sodium chloride.

   a. How long must a current of 6.1 amp be maintained to produce 1.00 kg of sodium?

   b. The other product of this electrolysis is chlorine gas. How long would the current of 6.1 amp need to be maintained to form 5.0 liters of chlorine gas, measured at 37°C and 620 torr?

4. a. For each of the following standard cells, calculate $\mathcal{E}^0$ and indicate which electrode is the anode.

   (1) Ag, Ag^+ versus Cd, Cd^{2+}

   (2) H_2, H^+ versus Cl^-, Cl_2

   (3) Pb, $PbSO_4$ versus $PbSO_4$, PbO_2

   (4) Cs, Cs^+ versus F^-, F_2

   b. Given:

   $$Co^{2+} + 2e^- \longrightarrow Co \qquad \mathcal{E}^0 = -0.28$$
   $$Co^{3+} + e^- \longrightarrow Co^{2+} \qquad \mathcal{E}^0 = +1.82$$

   Calculate $\mathcal{E}^0$ for $Co^{3+} + 3e^- \longrightarrow Co$.

5. Given:

   $$\Delta H_f^0(AgCl) = -30.4 \text{ kcal mole}^{-1}$$

   $$AgCl + e^- \longrightarrow Ag + Cl^- \qquad \mathcal{E}^0 = +0.25$$
   $$Cl_2 + 2e^- \longrightarrow 2Cl^- \qquad \mathcal{E}^0 = +1.36$$

   Calculate ΔG_f^0 and ΔS_f^0 for AgCl.

PRACTICE FOR PROFICIENCY

Objective (a)

1. Draw a representation for a galvanic cell using Ni, Ni^{2+} (1.0 M) and Cd, Cd^{2+} (1.0 M) half-cells. Label correct signs for external electrodes.

2. Draw representations for galvanic cells, with each having one electrode involving hydrogen gas (1 atm) bubbled through a solution of ~1 M H^+ in contact with a platinum electrode, and the other electrode as indicated. Label correct signs for external electrodes such that the cell potential will be > 0, in each case.
   a. Al, Al^{3+} (~1 M)
   b. Ag, Ag^+ (~1 M)
   c. Zn, Zn^{2+} (~1 M)
   d. Cu, Cu^{2+} (~1 M)
   e. $Cl_2(g)$ (1 atm), Cl^- (~1 M), in contact with a platinum electrode.

3. Draw a representation of an electrolytic cell, using $[-|\,|\,|\,|\,|\,|+]$ to represent an external source of current, for the production of metallic copper and gaseous chlorine by the electrolysis of aqueous $CuCl_2$. Use copper and (inert) platinum electrodes, respectively.

4. Which would correctly refer to the *anode* of a galvanic cell?
   I. Electrons are furnished to the external circuit at this electrode.
   II. Within the cell, reduction is occurring at this electrode.
   III. This electrode must be a metal having an electrode potential greater than zero.
   a. I, II, and III; b. only I and III; c. only III; d. only II; e. only I

5. Which would correctly refer to the *cathode* of a galvanic cell?
   I. Electrons are furnished to the external circuit at this electrode.
   II. Within the cell, reduction is occurring at this electrode.
   III. This electrode must be a metal having an electrode potential greater than zero.
   a. I, II, and III; b. only I and III; c. only III; d. only II; e. only I

6. For a galvanic cell using Fe, Fe^{2+} (1.0 M) and Pb, Pb^{2+} (1.0 M) half-cells, which of the following statements is correct?
   a. The mass of the iron electrode will increase during discharge.
   b. Electrons leave the lead electrode to pass through the external circuit during discharge.
   c. The iron electrode is the anode.
   d. The concentration of Pb^{2+} increases during discharge.
   e. When the cell has "completely discharged" (to zero voltage), the concentration of Pb^{2+} will be zero.

7. For a galvanic cell using Cu, Cu^{2+} (1.0 M) and Zn, Zn^{2+} (1.0 M) half-cells, which of the following statements is incorrect?
   a. The copper electrode is the cathode.

b. Electrons will flow through the external circuit from the zinc electrode to the copper electrode.
   c. Reduction occurs at the copper electrode during discharge.
   d. The mass of the zinc electrode will decrease during discharge.
   e. The concentration of Cu^{2+} will increase during discharge.

Objective (b)

8. What mass of lead sulfate would be formed during a 12.0-hr discharge of a lead storage battery at an average current of 0.60 amp?

$$Pb + PbO_2 + 2H^+ + 2HSO_4^- \longrightarrow 2PbSO_4 + 2H_2O$$

9. Potassium metal may be produced by electrolysis of molten potassium chloride.
   a. How long must a current of 10.0 amp be maintained to produce 780 g of potassium?
   b. What volume of chlorine gas, measured at 27°C and 708 torr, would be produced by the current required for part (a) above?

10. What mass (in grams) of copper could be electroplated from a solution of a copper(II) salt by a current of 0.50 amp flowing for 12 hrs?

11. What volume (in liters) of dry hydrogen gas (STP) could be generated by the electrolysis of water over a 12.0-hr period with an average current of 25 amp?

12. How long (in hours) must a current of 0.60 amp be maintained to electroplate 66 g of gold from a solution of gold(III) chloride?

13. What average current (in amperes) must be maintained to electroplate 2.9 g of nickel per hour from a solution of a nickel(II) salt?

14. What mass of iron per hour is converted to Fe(III) by the electrolytic corrosion of a metal drainpipe at a rate equivalent to a current of 0.30 amp?

15. How long (in hours) must a current of 0.40 amp be maintained to electroplate 66 g of gold from a solution of gold(III) chloride?

16. What average current (in amperes) must be maintained to electroplate 11 g of manganese per hour from a solution of a manganese(II) salt?

17. The anode reaction for an acetylene-oxygen fuel cell may be formulated as

$$H-C\equiv C-H(g) + 14OH^-(aq) \longrightarrow$$
$$2CO_3^{2-}(aq) + 8H_2O(l) + 10e^-$$

How many milliliters of acetylene (measured at STP) would be consumed during the operation of such a cell to produce a current of 2.5 amp for 15 min, assuming 100% efficiency?

18. If a current of 20 amp flowing for 1.3 hr plated out 39 g of platinum from a solution of $[PtCl_6]^{2-}$, what was the approximate electrode efficiency?

19. A special cobalt electrorefining process uses a "trickle cur-

rent" of 0.20 amp at 80% efficiency to reduce a cobalt(II) complex to metallic cobalt. At what rate (in grams per hour) is the metal formed?

a. 0.10; **b.** 0.17; **c.** 0.27; **d.** 0.41; **e.** 0.54

20. What volume of dry oxygen gas (measured at 37°C and 650 torr) would be released by electrolysis of water over a 30.0-min period by a current of 0.200 amp at an electrode efficiency of 85%? (Hydrogen gas is the other product.)

21. How long, in hours, would be required to electroplate 39 g of platinum from a solution of $[PtCl_6]^{2-}$, using a current of 20 amp at an 80% electrode efficiency?

22. What average current (in amperes must be maintained for a tinplating operation forming 1.2 kg of tin per hour from a solution of a tin(IV) complex, at an electrode efficiency of 90%?

23. The anode reaction for a propane-oxygen fuel cell may be formulated as

$$CH_3CH_2CH_3(g) + 26OH^-(aq) \longrightarrow$$
$$3CO_3{}^{2-}(aq) + 17H_2O(l) + 20e^-$$

How many milliliters of propane (measured at STP) would be required for such a fuel cell to produce an average current of 3.0 amp for 45 min, assuming 100% efficiency?

Objective (c)

24. For the standard cell Cu, Cu^{2+} versus Fe, Fe^{2+}, the anode and $\mathcal{E}^0$ for the cell are, respectively,

a. Fe, 0.78 V; **b.** Cu, 0.10 V; **c.** Cu, 0.34 V; **d.** Fe, 0.10 V; **e.** Cu, 0.78 V

25. For the standard cell Cu, Cu^{2+} versus Au, Au^{3+}, the cathode and $\mathcal{E}^0$ for the cell are, respectively,

a. Au, 1.16 V; **b.** Cu, 1.16 V; **c.** Cu, 1.84 V; **d.** Au, 1.50 V; **e.** Au, 1.84 V

26. For the standard cell Fe, Fe^{2+} versus Au, Au^{3+}, the anode and $\mathcal{E}^0$ for the cell are, respectively,

a. Fe, −0.44 V; **b.** Au, 1.94 V; **c.** Fe, 1.06 V; **d.** Au, 1.06 V; **e.** Fe, 1.94 V

27. For the standard cell Al, Al^{3+} versus Fe, Fe^{2+}, the cathode and $\mathcal{E}^0$ for the cell are, respectively,

a. Al, 2.10 V; **b.** Fe, 1.22 V; **c.** Al, 1.22 V; **d.** Fe, −0.44 V; **e.** Fe, 2.10 V

28. For the standard cell Ag, Ag^+ versus Cd, Cd^{2+}, the anode and $\mathcal{E}^0$ for the cell are, respectively.

a. Cd, 1.20 V; **b.** Ag, 1.20 V; **c.** Ag, 0.40 V; **d.** Ag, 0.80 V; **e.** Cd, 0.40 V

29. For the standard cell Zn, Zn^{2+} versus Cu, Cu^{2+}, the cathode and $\mathcal{E}^0$ for the cell are, respectively,

a. Zn, 1.10 V; **b.** Cu, 0.42 V; **c.** Cu, 0.34 V; **d.** Cu, 1.10 V; **e.** Zn, 0.42 V

30. **a.** For each of the following standard cells, calculate $\mathcal{E}^0$ and indicate which electrode is the anode.

(1) Al, Al^{3+} versus Cu, Cu^{2+}
(2) Al, Al^{3+} versus Mg, Mg^{2+}
(3) Cu, Cu^{2+} versus Mg, Mg^{2+}

b. Given:

$$Cu^{2+} + 2e^- \rightleftharpoons Cu \qquad \mathcal{E}^0 = +0.34$$
$$Cu^+ + e^- \rightleftharpoons Cu \qquad \mathcal{E}^0 = +0.52$$

Calculate $\mathcal{E}^0$ for $Cu^{2+} + e^- \longrightarrow Cu^+$

Objective (d)

31. The standard heat of formation for aqueous hydrogen peroxide is −45.7 kcal mole⁻¹. For $O_2(g) + 2H^+(aq) + 2e^- \longrightarrow H_2O_2(aq)$, $\mathcal{E}^0$ is +0.68. Calculate ΔG_f^0 and ΔS_f^0 for aqueous H_2O_2.

32. From information in Appendix D, calculate ΔG_f^0 for
a. $CuF_2(aq)$; **b.** $CrBr_3(aq)$; **c.** $MgCl_2(aq)$; **d.** $AlF_3(aq)$; **e.** $ZnCl_2(aq)$

BRAIN TINGLER

33. A typical industrial electricity rate is 11.2¢ per kilowatt hour. Assume that a small electroplating company uses an average current of 12.0 amp, delivered at 220 V, with an 85% electrode efficiency. Calculate the "electric bill" for production of 1.00 metric ton (1000 kg) each of

a. aluminum, from Al^{3+}
b. copper, from Cu^{2+}
c. silver, from Ag^+

Suggest *two* reasons for the differences in electricity costs. Do the relative market prices of these metals follow the same trends as the "electric bills"? Explain.

Unit 23 Self-Test

(Refer to Appendix D when necessary.)

1. a. Draw a representation for a galvanic cell using Ag, Ag^+ and Co, Co^{2+} half-cells. Label correct signs for external electrodes (assume standard states).

   b. Draw a representation for an electrolytic cell to be used for collecting H_2 and O_2 from electrolysis of water. Label electrodes. (*Hint:* Use $-||||||+$ to represent an external source of current and use inert platinum electrodes.)

2. What mass of copper could be electroplated from a Cu^{2+} solution by an average current of 255 milliamp flowing for 15 min?

3. How long could the cell drawn for question 1a maintain an average current of 150 milliamp before 25 mg of the cobalt electrode would be consumed?

4. a. Calculate $\mathcal{E}^0_{cell}$ for each of the following. Underline the reactants employed for the anode half-cell.
   (1) Cl_2, Cl^- versus Mg, Mg^{2+}
   (2) Cl_2, Cl^- versus F_2, F^-
   (3) Ag, Ag^+ versus Al, Al^{3+}

   b. Given:

$$Hg_2^{2+} + 2e^- \longrightarrow 2Hg \qquad \mathcal{E}^0 = +0.79$$
$$Hg^{2+} + 2e^- \longrightarrow Hg \qquad \mathcal{E}^0 = 0.85$$

Calculate $\mathcal{E}^0$ for $2Hg^{2+} + 2e^- \longrightarrow Hg_2^{2+}$.

5. The standard heat of formation of aqueous Cl^- is -40.0 kcal mole^{-1}. Calculate ΔG_f^0 and ΔS_f^0 for aqueous HCl.

The Nernst equation

Living organisms are composed mainly of biochemical fuel cells. Though their functioning is not understood, they oxidize fuel to produce electric currents with which the organism carries out its activities.

Introduction to Electrochemistry
ERNEST H. LYONS, JR.

Objectives

(a) Be able to write standard shorthand cell notations (Section 24.3).
(b) Be able to use the Nernst equation to calculate electrode and cell potentials (Section 24.4).
(c) Be able to calculate equilibrium constants for redox systems (Section 24.6).
(d) Be able to use the Nernst equation to calculate concentrations of cell components (Section 24.7).

Most of the familiar electrochemical processes do not involve standard state conditions. Thus the voltages expected from common galvanic cells and the voltages required for most electrolytic cells cannot be calculated from standard electrode potentials alone. The Nernst equation,

$$\mathcal{E} = \mathcal{E}^0 - \frac{0.0592}{n} \log Q$$

is one of the most important relationships in electrochemistry, since it permits a wide variety of useful calculations for cells at other than standard state conditions.

Perhaps the most familiar electrochemical cell is that used in constructing the lead storage battery (Figure 24.1). During discharge the half-cell reactions may be represented as

anode

$$Pb(s) + HSO_4^-(aq) \longrightarrow PbSO_4(s) + H^+(aq) + 2e^-$$

cathode

$$PbO_2(s) + HSO_4^-(aq) + 3H^+(aq) + 2e^- \longrightarrow PbSO_4(s) + 2H_2O(l)$$

The lead sulfate formed coats the electrodes. The battery consists of a set of cells connected in series, each cell designed for a potential of about 2.0 V. Since both oxidizing and reducing agents are solids, the cell potential is maintained fairly constant. If discharge continues too long, however, several problems may be encountered. The electrolyte (dilute sulfuric acid) is consumed during discharge, and the cell resistance is thereby increased. If the concentration of sulfuric acid becomes too low, the cell resistance may increase to the point at which the battery can no longer generate current at the rate required. In addition, prolonged discharge may lead to irreversible deterioration of the electrodes or flaking off of the lead sulfate coatings (or both). Automobile batteries are designed for reversible use. After short discharge periods, the battery is recharged by current from the automobile generator. This recharge simply reverses the half-reactions previously listed. If regular periodic recharging does not occur, as in the case of generator malfunction, the battery may be permanently damaged by electrode deterioration or loss of some of the $PbSO_4$ needed for the recharge process.

Before 1960 the most common flashlight "battery" (cell) was of the "dry cell" type, similar to the larger "dry cell" shown in Figure 24.1, except that the bottom of the can is used as the external cathode in the flashlight cell. These cells are not truly "dry" in that the electrolyte is a moist paste. The cell reactions are relatively complex, but when current rate is small they are probably best formulated as

anode

$$Zn(s) \longrightarrow \underset{\text{in paste}}{Zn^{2+}} + 2e^-$$

cathode

$$\underset{\text{in paste}}{MnO_2(s)} + H_2O(l) + e^- \longrightarrow \underset{\text{in paste}}{MnO(OH)(s)} + OH^-(aq)$$

In recent years the "dry cell" has been largely replaced by "mercury" or "nickel–cadmium" cells. The latter can be easily recharged, a process generally inefficient for the other cells because of the poor adhesion of their discharge products to electrode surfaces.

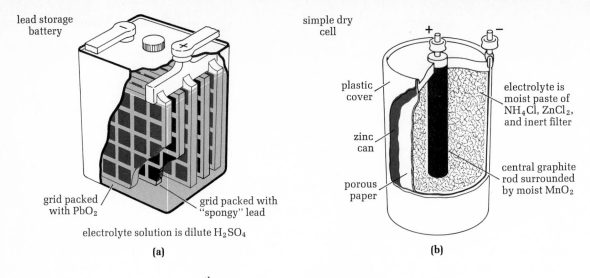

lead storage battery

grid packed with PbO$_2$ grid packed with "spongy" lead

electrolyte solution is dilute H$_2$SO$_4$

(a)

simple dry cell

plastic cover

zinc can

porous paper

electrolyte is moist paste of NH$_4$Cl, ZnCl$_2$, and inert filter

central graphite rod surrounded by moist MnO$_2$

(b)

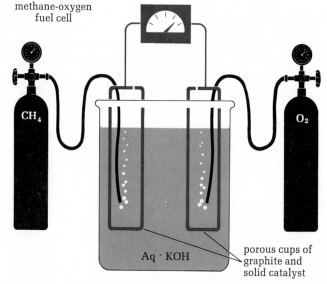

methane-oxygen fuel cell

CH$_4$

O$_2$

Aq · KOH

porous cups of graphite and solid catalyst

electrolyte solution is dilute KOH

(c)

Figure 24.1   Some electrochemical cells. For more details, see "Electrochemical Reactions in Batteries," by Akiya Kozawa and R. A. Powers in the September, 1972, issue of the Journal of Chemical Education (pp. 587–591) and "Electrochemical Principles Involved in a Fuel Cell," by Ashok K. Vijh, in the October, 1970, issue of the Journal of Chemical Education (pp. 680–682).

DISCHARGE PROCESSES ("ALKALINE" CELLS)

THE "MERCURY" CELL

anode $Zn(s) + 2OH^- \longrightarrow Zn(OH)_2(s) + 2e^-$
                   in paste

cathode $HgO(s) + H_2O + 2e^- \longrightarrow Hg(l) + 2OH^-$
                     in paste   in paste

THE "NICKEL–CADMIUM" CELL

anode $Cd(s) + 2OH^-(aq) \longrightarrow Cd(OH)_2(s) + 2e^-$
cathode $NiO_2(s) + 2H_2O(l) + 2e^- \longrightarrow Ni(OH)_2(s) + 2OH^-(aq)$

The fuel cell is a fairly recent development, having graduated from the class of scientific curiosities to become an important feature of many aspects of space age technology. The methane–oxygen fuel cell (Figure 24.1) is a cleaner and more efficient energy source than the direct combustion of methane (Unit 5). This particular cell is not reversible, but the electrolyte solution can be drained periodically and replaced conveniently for continued operation. The half-reactions during discharge of this cell may be formulated as

anode

$$CH_4(g) + 10OH^-(aq) \longrightarrow CO_3^{2-}(aq) + 7H_2O(l) + 8e^-$$

cathode

$$O_2(g) + 2H_2O(l) + 4e^- \longrightarrow 4OH^-(aq)$$

One electrochemical process that has undesirable consequences is electrolytic corrosion.[1] When metals are in contact with moisture, each different metal will be oxidized to a slightly different extent. Usually the initial oxidation is almost infinitesimal, but the differences will result in an electrochemical potential. When the metallic regions are connected, current can flow and the moisture provides an electrolyte medium to complete the circuit. As a result, corrosion rates increase and the metals deteriorate. Electrolytic corrosion may occur where two different metals are in contact, as in a joint between copper pipe and "galvanized" iron pipe. It is possible, however, for this process to take place between different regions of the same metal, since inhomogeneity of the metal or even minor differences in crystallite structures may be sufficient to develop the necessary potential.

In this unit we shall examine factors that influence electrode and cell potentials, and we shall see how the measurement of cell voltage can be used for important analytical procedures.

24.1 FACTORS AFFECTING CELL POTENTIAL

Let us think for a moment about some of our common experiences with electrochemical cells in light of what we know now about their chemistry.

A car is harder to start on a cold morning than on a warm day, especially when the battery is weak.

This is not due to any single factor, but to a combination of effects. The lubricants are more viscous at lower temperatures. Current flow from the battery is reduced somewhat, because diffusion rates in the electrolyte solution are lower when the battery is cold. If we checked the battery with a voltmeter, we would find that the voltage also varies with temperature.

Cell potential may vary with temperature (although not always), but the variation is a characteristic of each particular type of cell.

Strictly speaking, a *battery* is a set of connected cells. Thus it is more accurate to speak of a flashlight *cell,* although the term battery is common usage.

Flashlight batteries grow weaker with use. A "battery tester" is actually a voltmeter and shows that the voltage of the cell decreases with use.

It we cut open flashlight cells that had been discharged for varying lengths of time and analyzed the contents, we would find that reactants are consumed during discharge in a way that can be mathematically related to the change in voltage.

Cell potential is a function of reactant concentration.

---

[1] For an interesting discussion of several aspects of electrolytic corrosion, see "Corrosion," by Wendell Slabaugh, in the April, 1974, issue of the *Journal of Chemical Education* (pp. 218–220).

A tiny penlight battery, a regular flashlight battery, and a large "dry" cell all have the same voltage (1.48 V). However, the use time increases with the size of the battery.

The total amount of charge that can be transferred by a battery (i.e., current × time) depends on the *quantity* of reactant materials available.

The potential of a cell is not determined by the quantity of reactant materials, but only by the nature of the reactants and their concentrations.

24.2 THE NERNST EQUATION

In 1889, Walther Nernst developed an expression from thermodynamic considerations that relates cell (or single electrode) potential, temperature, and component concentrations. We can arrive at the most useful form of this *Nernst equation* by considering the Gibbs free energy terms for appropriate electron-transfer processes.

What we ultimately need is a way of relating the potential of any cell to the *standard* potential for a cell involving the same components, since the *standard* potential can be calculated directly from tabulated electrode potentials (Appendix D). To develop the desired relationship, let's consider a hypothetical situation, depicted in Figure 24.2, in which a cell of potential $\mathcal{E}$ is produced simply by diluting both electrode solutions of a standard cell. For our present purposes, we shall add the constraint that the dilution does not change the signs of the electrodes. This is not a general requirement, but it does simplify our derivation a bit.

One key relationship was described in Unit 23:

$$\Delta G = -nF\mathcal{E} \qquad [\text{equation (23.3)}]$$

The other relationship of importance concerns the free energy change associated with dilution. Thermodynamicists have defined this for us, for an ideal solution, as

$$\Delta G_{\text{dilution}} = RT \ln \frac{a_{\text{final}}}{a_{\text{initial}}}$$

in which a_{final} represents the solute activity in the diluted solution and a_{initial} the solute activity in the original solution.

Now let us consider the two cells depicted in Figure 24.2, for which the respective discharge reactions are

"*initial*" *cell:* $2A(s) + B^{2+}(aq) \longrightarrow 2A^+(aq) + B(s)$
$$[a = 1] \qquad\qquad [a = 1]$$

"*final*" *cell:* $2A(s) + B^{2+}(aq) \longrightarrow \quad 2A^+(aq) \quad + B(s)$
$$[a = a_{B_f}(<1)] \qquad [a = a_{A_f}(<1)]$$

We can relate the free energy terms for the respective cells by a stepwise summation process similar to that used with enthalpy calculations (Hess's law, Unit 5), remembering that all terms must be in the same units before summation.

		ΔG (in kcal mole^{-1} of B^{2+})
STEP 1 ("initial" cell)	$2A(s) + B^{2+}(aq) \longrightarrow 2A^+(aq) + B(s)$ $(a = 1) \qquad\qquad (a = 1)$	ΔG^0_{cell}
STEP 2 (ΔG for concentration difference in cathode compartment)	$B^{2+}(aq) \longrightarrow B^{2+}(aq)$ $(a = a_{B_f}) \qquad (a = 1)$	$RT \ln \left[\dfrac{1}{a_{B_f}}\right]$

Figure 24.2 Concentration and cell potential. Spontaneous discharge reaction:
$$2A^+(aq) + B(s) \longrightarrow$$
$$2A(s) + B^2(aq)$$

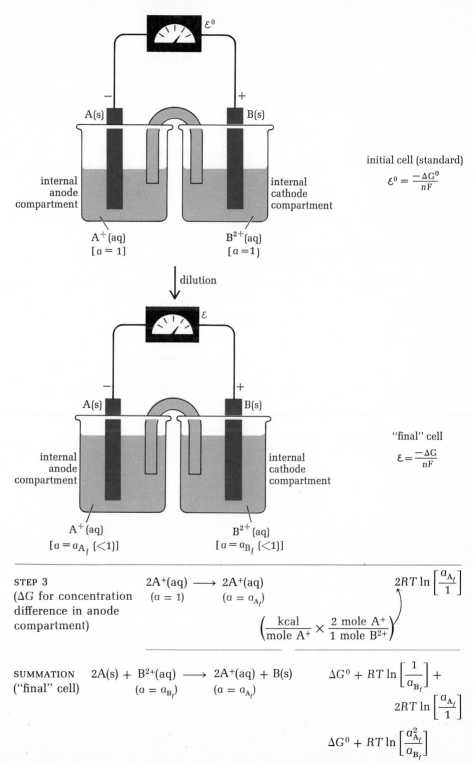

initial cell (standard)

$$\mathcal{E}^0 = \frac{-\Delta G^0}{nF}$$

dilution

"final" cell

$$\mathcal{E} = \frac{-\Delta G}{nF}$$

STEP 3 (ΔG for concentration difference in anode compartment)	$2A^+(aq)$ $(a=1)$	$\longrightarrow$	$2A^+(aq)$ $(a=a_{A_f})$	$2RT \ln\left[\dfrac{a_{A_f}}{1}\right]$

$$\left(\frac{kcal}{mole\ A^+} \times \frac{2\ mole\ A^+}{1\ mole\ B^{2+}}\right)$$

SUMMATION ("final" cell)	$2A(s) + B^{2+}(aq)$ $(a=a_{B_f})$	$\longrightarrow$	$2A^+(aq) + B(s)$ $(a=a_{A_f})$	$\Delta G^0 + RT \ln\left[\dfrac{1}{a_{B_f}}\right] +$

$$2RT \ln\left[\frac{a_{A_f}}{1}\right]$$

$$\Delta G^0 + RT \ln\left[\frac{a_{A_f}^2}{a_{B_f}}\right]$$

Since [equation (23.3)] $\Delta G = -nF\mathcal{E}$, we can write

$$\Delta G = \Delta G^0 + RT \ln \left[\frac{a_{A_f}^2}{a_{B_f}} \right]$$

$$-nF\mathcal{E} = -nF\mathcal{E}^0 + RT \ln \left[\frac{a_{A_f}^2}{a_{B_f}} \right]$$

from which

$$\mathcal{E} = \mathcal{E}^0 - \frac{RT}{nF} \ln \left[\frac{a_{A_f}^2}{a_{B_f}} \right]$$

and this equation gives us the desired relationship between $\mathcal{E}$ and $\mathcal{E}^0$ in terms of the activities of electrode compartment solutes.

By extending our considerations to more complicated cells with multicomponent electrode systems, we could arrive at the generalization that for a cell for which the spontaneous discharge reaction is represented by,

$$mM + nN + \cdots \longrightarrow wW + xX + \cdots$$

the logarithmic term is

$$\ln \frac{a_W^w \times a_X^x \times \cdots}{a_M^m \times a_N^n \times \cdots}$$

Note that this expression, called the "activity quotient" (Q), is represented by "right-hand" species in the numerator and "left-hand" species in the denominator, as determined by the cell equation; that is, the "activity quotient" has the same *form* as a conventional equilibrium constant (Unit 17). For pure solids or liquids, the activity is defined as unity. For an ideal gas, the activity is replaced by the partial pressure of the gas, expressed in *atmospheres*.

It should be apparent that the preceding derivation of *cell potential* could equally well apply to a *single electrode potential,* if we chose to use the hypothetical standard hydrogen electrode as the appropriate half-cell in both the "initial" and "final" cells, with dilution of only the other half-cell components.

In its most general form, then, the *Nernst equation* may be expressed as

$$\mathcal{E} = \mathcal{E}^0 - \frac{RT}{nF} \ln Q \tag{24.1}$$

As in earlier cases, we shall use molar concentrations rather than activities. This approximation introduces no serious errors when dealing with simple dilute solutions. It is worth noting, however, that the Nernst equation does provide a way of determining *activities* from cell potential measurements.

where $\mathcal{E}$ is the actual cell or electrode potential, $\mathcal{E}^0$ is the standard potential, R is the familiar molar gas constant, F is the Faraday constant (Unit 23), n is the number of electrons required for the balanced equation for the cell process, and Q is the "activity quotient" in the *form* of an equilibrium constant expression (Unit 17).

At first glance it would appear that the Nernst equation could be used to calculate voltage as a function of temperature from $\mathcal{E}^0$ values such as those in Appendix D. Such is not the case. We have consistently used standard states at $298°K$, but this temperature is not a necessary constant—that is, standard states may be *defined* at any temperature as long as that temperature is specified. The Nernst equation *can* be used to find $\mathcal{E}^0$ values at temperatures other than $298°K$ from experimental measurements of potential ($\mathcal{E}$) and activities.

Since most routine electrochemical work is performed around 25°C, it is convenient to reduce the Nernst equation to a simpler form suitable for easy calculations. If we set $T = 298°K$, insert the values of R and F (with appropriate conversion factors for their units), and apply the factor 2.303 to convert from ln (base e) to log (base 10), we obtain the most useful form of the Nernst equation:

$$\mathcal{E} = \mathcal{E}^0 - \frac{0.0592}{n} \log Q \qquad \text{at } 298°K$$

(24.2)

This is the form we shall use for our study, and it is a useful equation to remember.

24.3 CELL NOTATION

In subsequent discussions we shall employ a convenient shorthand notation for describing an electrochemical cell:

anode|anode compartment| |cathode compartment|cathode

This notation is represented symbolically as

$$A \,|\, A^{a+}\,(M)\|C^{c+}\,(M)\,|\,C$$

or

$$I_A \,|\, A_{red}\,(M),\ A_{ox}\,(M)\|C_{ox}\,(M),\ C_{red}\,(M)\,|\,I_C$$

The single vertical lines represent phase boundaries at the half-cell electrodes. The double line represents the half-cell boundary (e.g., a salt bridge or porous partition). When there is no boundary between half-cell compartments, as in the lead storage battery, the double line is replaced by a single dotted line.

Note: Selection of the correct half-cell can be made from tabulated $\mathcal{E}^0$ values only for standard state conditions. Otherwise, $\mathcal{E}$ values from the Nernst equation must be used, as we shall see later.

By convention the half-cell represented at the left is that in which oxidation occurs (the *internal* anode). The letter A represents species in the oxidation half-reaction and C represents those in the reduction half-reaction (cathode). Concentrations (molarities) must be specified for all dissolved materials—the (M) term in the notation. This is replaced by pressure for gaseous species. The I represents an inert electrode (e.g., Pt).

The simpler first notation is used when the electrodes themselves are involved in electron transfer (e.g., as in a copper–zinc cell). The other notation is used when inert electrodes (e.g., platinum) are employed. The subscripts *red* and *ox* refer to "reduced" or "oxidized" forms of the species concerned. Certain cells may be represented by a combination of half of each notation form. In every case the species are in the same order as in the equation for the spontaneous cell process.

Example 1

Write the shorthand notation for each of the following galvanic cells under standard state conditions (each cell includes a salt bridge).

a. Ag, Ag^+ versus Cu, Cu^{2+}
b. H_2, H^+ versus Cl^-, Cl_2 (Pt electrodes)
c. H_2, H^+ versus Cu, Cu^{2+} (Pt electrode for H_2)

Solution
Use $\mathcal{E}^0$ values (Appendix D) to decide which half-reaction will occur spontaneously as oxidation and which as reduction. [The half-reaction associated with the more negative (or less positive) $\mathcal{E}^0$ value will occur as oxidation.]

$$Ag^+ + e^- \longrightarrow Ag \qquad \mathcal{E}^0 = +0.80$$

a. *Oxidation* $\qquad Cu^{2+} + 2e^- \longrightarrow Cu \qquad \mathcal{E}^0 = +0.34$

Thus

$$Cu \,|\, Cu^{2+}\,(1.0\ M)\|Ag^+\,(1.0\ M)\,|\,Ag$$

b. *Oxidation* $2H^+ + 2e^- \longrightarrow H_2$ $\mathcal{E}^0 = 0$
$Cl_2 + 2e^- \longrightarrow 2Cl^-$ $\mathcal{E}^0 = +1.36$

Thus

$Pt \,|\, H_2 \,(1.0\ atm),\ H^+ \,(1.0\ M) \,\|\, Cl_2 \,(1.0\ atm),\ Cl^- \,(1.0\ M) \,|\, Pt$

c. *Oxidation* $2H^+ + 2e^- \longrightarrow H_2$ $\mathcal{E}^0 = 0$
$Cu^{2+} + 2e^- \longrightarrow Cu$ $\mathcal{E}^0 = +0.34$

Thus

$Pt \,|\, H_2 \,(1.0\ atm),\ H^+ \,(1.0\ M) \,\|\, Cu^{2+} \,(1.0\ M) \,|\, Cu$

24.4 CALCULATING CELL AND ELECTRODE POTENTIALS

Substitution of appropriate values for the $\mathcal{E}^0$, n, and Q terms in equation (24.2) permits calculation of any cell or electrode potential at $298°K$. It is important to note that

a cell reaction is spontaneous only under conditions in which the cell potential is positive (i.e., $\Delta G < 0$).

This means that we cannot select the half-reactions to be specified as cathode and anode processes until the individual electrode potentials have been calculated. Electrode potentials vary with concentration and may even change sign. Thus we have two choices for calculation of the cell potential:

1. Calculate both single electrode potentials separately by application of the Nernst equation. Then designate the larger (algebraic) $\mathcal{E}$ value as that for the internal cathode.

If the calculated $\mathcal{E}_{cell}$ value is zero, this represents a condition of equilibrium corresponding to the completely discharged cell (Section 24.6).

2. Arbitrarily write the balanced net equation for the cell reaction as it would proceed spontaneously under standard state conditions. Calculate $\mathcal{E}_{cell}$ from the Nernst equation. If the calculated $\mathcal{E}_{cell}$ value is negative, then the actual spontaneous reaction must be represented by the *reverse* of the equation originally written.

Example 2

Calculate single electrode potentials for
  a. copper metal in contact with a 0.10 M Cu^{2+} solution.
  b. silver metal in contact with a saturated solution of silver sulfide ($K_{sp} = 5.5 \times 10^{-51}$).

Solution
Remember that, by the IUPAC conventions, single electrode processes are always indicated by equations for *reduction* half-reactions.
  a. $Cu^{2+}(0.10\ M) + 2e^- \longrightarrow Cu$.

$$\mathcal{E} = \mathcal{E}^0 - \frac{0.0592}{n} \log Q \qquad \text{[equation (24.2)]}$$

$n = 2 \qquad (2e^-)$

$\mathcal{E}^0 = +0.34 \qquad \text{(Appendix D)}$

$$Q = \frac{1}{C_{Cu^{2+}}} = \frac{1}{0.10} = 10$$

Note that Q, like the equilibrium constant expressions used earlier (Unit 17), contains concentration terms for "right-hand" species in the numerator and "left-hand" species in the denominator. Terms for solids or pure liquids are not included (since their activity values are unity, by definition).

Then

$$\mathcal{E} = +0.34 - \frac{0.0592}{2} \log 10$$

$$= +0.34 - 0.0296$$

$$= +0.31 \text{ V}$$

b. We must first calculate the Ag^+ concentration in a saturated silver sulfide solution. We shall assume that hydrolysis effects are negligible (Units 18 and 21).

For $Ag_2S(s) \rightleftharpoons 2Ag^+ + S^{2-}$,

$$K_{sp} = 5.5 \times 10^{-51} = C_{Ag^+}^2 \times C_{S^{2-}}$$

$$C_{S^{2-}} = \tfrac{1}{2}C_{Ag^+}$$

so

$$5.5 \times 10^{-51} = C_{Ag^+}^2 \times (0.5)C_{Ag^+}$$

$$C_{Ag^+} = \sqrt[3]{11 \times 10^{-51}} = 2.2 \times 10^{-17} \text{ M}$$

Then the half-cell is represented by

$$Ag^+ \, (2.2 \times 10^{-17} \text{ M}) + e^- \longrightarrow Ag$$

for which

$$n = 1 \quad (e^-)$$

$$\mathcal{E}^0 = +0.80 \quad \text{(Appendix D)}$$

$$Q = \frac{1}{C_{Ag^+}} = \frac{1}{2.2 \times 10^{-17}} = 4.5 \times 10^{16}$$

then

$$\mathcal{E} = +0.80 - \frac{0.0592}{1} \log(4.5 \times 10^{16})$$

$$\mathcal{E} = +0.80 - (0.0592)(16.65) = +0.80 - 0.99$$

$$\mathcal{E} = -0.19 \text{ V}$$

Note that the electrode potential has changed sign.

Example 3

---

a. Calculate the potential for a silver–copper cell in which

$$C_{Cu^{2+}} = 0.10 \text{ M} \quad \text{and} \quad C_{Ag^+} = 2.2 \times 10^{-17} \text{ M}$$

b. Write the conventional shorthand notation to represent this cell.

Solution

a. We could calculate the cell potential by either of the methods given.

By method 1 (from Example 2) for Cu,

$$Cu^{2+} \, (0.10 \text{ M}), \, \mathcal{E} = +0.31 \text{ V}$$

for Ag,

$$Ag^+ \ (2.2 \times 10^{-17} \ M), \qquad \mathcal{E} = -0.19 \ V$$

Then

$$\begin{aligned}
\mathcal{E}_{cell} &= \mathcal{E}_{cathode} - \mathcal{E}_{anode} \\
&= +0.31 - (-0.19) \\
&= +0.50 \ V
\end{aligned}$$

To convince yourself that both methods are equivalent, derive the relationship

$$\mathcal{E}_{cell} = \mathcal{E}^0_{cell} - \frac{0.0592}{2} \log \frac{C_{Cu^{2+}}}{C^2_{Ag^+}}$$

from the two Nernst equations for the copper and silver half-cells.

By Method 2 (Note that this method requires fewer steps and is therefore preferred for cases in which the single electrode potentials are not yet known)

$$Cu^{2+} + 2e^- \longrightarrow Cu \qquad \mathcal{E}^0 = +0.34$$
$$Ag^+ + e^- \longrightarrow Ag \qquad \mathcal{E}^0 = +0.80$$

Then the spontaneous reaction under standard state conditions is represented as

$$Cu + 2Ag^+ \longrightarrow Cu^{2+} + 2Ag$$

for which

$$\mathcal{E}^0 = +0.80 - (+0.34) = +0.46 \ V$$

$$n = 2 \qquad (2e^- \text{ transfer for balanced equation})$$

$$Q = \frac{C_{Cu^{2+}}}{C^2_{Ag^+}} = \frac{0.10}{(2.2 \times 10^{-17})^2}$$

$$= 2.1 \times 10^{32}$$

Then

$$\mathcal{E} = \mathcal{E}^0 - \frac{0.0592}{n} \log Q$$

$$= +0.46 - \frac{0.0592}{2} \log(2.1 \times 10^{32})$$

$$= +0.46 - (0.0296)(32.32) = +0.46 - 0.96$$

$$= -0.50$$

However (remember that $\Delta G = -nF\mathcal{E}$), $\mathcal{E}$ must be positive for a spontaneous cell reaction, so $\mathcal{E}_{cell} = +0.50$ V for the spontaneous process formulated as

$$Cu^{2+} + 2Ag \longrightarrow Cu + 2Ag^+$$

b. Now that we know the spontaneous reaction, we can write the proper cell notation (Section 24.3) as

$$Ag \,|\, Ag^+ \ (2.2 \times 10^{-17} \ M) \,\|\, Cu^{2+} \ (0.10 \ M) \,|\, Cu$$

24.5 CONCENTRATION CELLS

Since electrode potential is a function of concentration, it is possible to use the same chemical species for both components of an electrochemical cell. It is necessary only that there be a difference in concentration of some species participating in electron transfer in the two half-cells. A silver–silver ion concentration cell is illustrated in Figure 24.3.

A metal pipe laid through soil of varying pH regions may develop different metal ion concentrations at various distances along the pipe. When the soil becomes wet, the necessary electrolyte solution may form, and electrolytic corrosion may occur as a result of what is basically a concentration cell.

Figure 24.3 A concentration cell. For Ag, Ag^+ $(10^{-1}\,M)$, $\varepsilon = +0.80 - (0.0592/1)\log(1/10^{-1})$ $= +0.74\,V$. For Ag, Ag^+ $(10^{-5}\,M)$, $\varepsilon = +0.80 - (0.0592/1)\log(1/10^{-5})$ $= 0.50\,V$. Thus $\varepsilon = +0.24\,V$ for the cell $Ag|Ag^+\,(10^{-5}\,M)\|Ag^+\,(10^{-1}\,M)|Ag$.

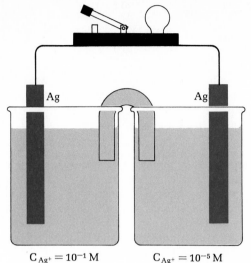

$C_{Ag^+} = 10^{-1}\,M$ $C_{Ag^+} = 10^{-5}\,M$

24.6 REDOX EQUILIBRIA: $\varepsilon = 0$

The thermodynamic criterion for equilibrium is $\Delta G = 0$ (Unit 23). Since $\Delta G = -nF\varepsilon$, it follows that $\varepsilon = 0$ is a condition for equilibrium in an electrochemical cell. That this equilibrium is dynamic (Unit 17) can easily be demonstrated by use of radioactive isotopes. Chemical change, then, has not *stopped* when a cell has been completely discharged. A zero voltage simply indicates that no further *net* change in cell composition will occur under existing conditions.

The Nernst equation allows us to use the condition of $\varepsilon = 0$ to calculate equilibrium constants for redox systems. When $\varepsilon = 0$, the Q term represents the equilibrium constant for the system.

In mathematical terms, when $\varepsilon = 0$, $Q = K_{eq}$, so

$$0 = \varepsilon^0 - \frac{0.0592}{n}\log K_{eq}$$

from which

$$\log K_{eq} = \frac{n\varepsilon^0}{0.0592} \qquad \textbf{(24.3)}$$

Example 4

Calculate the equilibrium constant for the system formulated as

$$Cu^{2+}(aq) + 2Ag(s) \rightleftharpoons Cu(s) + 2Ag^+(aq)$$

Solution

Note that it is quite legitimate to have a *calculated* value of ε^0 that is negative. The chemical species do not particularly care how we write the equation.

$$\begin{aligned} Cu^{2+} + 2e^- &\longrightarrow Cu & \varepsilon^0 &= +0.34 \\ Ag^+ + e^- &\longrightarrow Ag & \varepsilon^0 &= +0.80 \end{aligned}$$

For the equation given, then,

$$\varepsilon^0 = +0.34 - (+0.80) = -0.46$$
$$n = 2 \quad \text{(2e}^- \text{ for the balanced equation)}$$

We must have a positive mantissa before we can find the antilog

$$-15.5 = (+0.5 - 16)$$
$$\text{antilog } 0.5 = 3$$
$$\text{antilog}(-16) = 10^{-16}$$

and, using equation (24.3),

$$\log K_{eq} = \frac{(2)(-0.46)}{0.0592} = -15.5$$

from which

$$K_{eq} = 3 \times 10^{-16}$$

Since $\mathcal{E}^0$ values are measured quantities we should, to be strictly correct, have reported $\log K_{eq}$ as -16 (rounded to two significant figures), from which $K_{eq} = \sim 10^{-16}$. This bit of "poetic license" was used to review the procedure for manipulating negative logarithms.

24.7 CONCENTRATIONS (ACTIVITIES) FROM CELL POTENTIAL MEASUREMENTS

The Q term in the Nernst equation actually involves *activities* (Unit 17). The relationship between activity (a) and concentration (C) is expressed mathematically as $\gamma = a/C$, in which γ (Gr. gamma) is called the *activity coefficient*. This value approaches unity at very low concentrations. Activity coefficients may be determined by comparing *activities* from electrochemical data with *concentrations* calculated from other types of data. For example, the concentration might be calculated simply on the basis of how the solution was prepared.

One of the simplest ways of determining activities involves use of a concentration cell in which one half-cell has a solution sufficiently dilute so that the activity coefficient of the species involved is essentially unity. For our purposes, we shall continue to work with molar concentrations, even in terms of values calculated from the Nernst equation. We must realize, however, that equating "molarity" and "activity" is an approximation of convenience whose reliability decreases with increasing concentration and solution complexity. Remember that activities are most commonly related to *molalities*, but

Probably the most important practical application of the Nernst equation is in the calculation of concentrations from measured values of cell potentials. For an approximate determination it is necessary only to measure the potential of a cell in which a single reactant concentration is unknown. This method may be used, for example, to determine approximate ion concentrations in very dilute solutions such as those in the saturated solutions of sparingly soluble salts.

Example 5

An electrochemical cell is prepared by using a Cu, Cu^{2+} (0.10 M) half-cell connected to a half-cell consisting of a pure silver electrode immersed in a saturated silver chloride solution. It is determined that the copper electrode is the internal anode and the cell potential at 25°C is measured to be 0.20 V. What is the approximate solubility product for AgCl based on this determination?

Solution
Since the copper electrode is found to be the internal anode, the spontaneous cell reaction must involve

$$Cu + 2Ag^+ \longrightarrow Cu^{2+} + 2Ag$$

For which

$$\left(\mathcal{E} = \mathcal{E}^0 - \frac{0.0592}{n} \log Q \right)$$

$$\mathcal{E} = +0.20 \text{ V}$$
$$\mathcal{E}^0 = +0.46 \text{ V} \qquad (\text{Example 4})$$
$$n = 2$$
$$Q = \frac{C_{Cu^{2+}}}{C^2_{Ag^+}} = \frac{1.0 \times 10^{-1}}{C^2_{Ag^+}}$$

so that

$$+0.20 = 0.46 - \frac{0.0592}{2} \log \left(\frac{1.0 \times 10^{-1}}{C^2_{Ag^+}} \right)$$

Then, solving for $\log(C^2_{Ag^+})$, having noted that for AgCl(s) $\rightleftharpoons$ $Ag^+ + Cl^-$, since $C_{Ag^+} = C_{Cl^-}$, $K_{sp} = C^2_{Ag^+}$,

$$\log \left(\frac{1.0 \times 10^{-1}}{C^2_{Ag^+}} \right) = \frac{(2)(0.46 - 0.20)}{0.0592} = 8.8$$

molality and molarity are essentially equivalent in *dilute* aqueous solution.

$$\log(1.0 \times 10^{-1}) - \log(C_{Ag^+}^2) = 8.8$$
$$\log(C_{Ag^+}^2) = \log(1.0 \times 10^{-1}) - 8.8$$
$$= -1.0 - 8.8$$
$$\log(C_{Ag^+}^2) = -9.8 = 0.2 - 10$$

from which

$$C_{Ag^+}^2 = K_{sp} = \sim 1.6 \times 10^{-10}$$

Highly accurate determinations require special techniques.[2]

24.8 THE pH METER

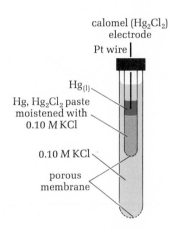

calomel (Hg$_2$Cl$_2$) electrode

Pt wire

Hg$_{(l)}$

Hg, Hg$_2$Cl$_2$ paste moistened with 0.10 M KCl

0.10 M KCl

porous membrane

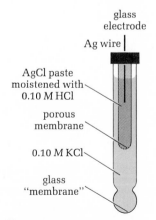

glass electrode

Ag wire

AgCl paste moistened with 0.10 M HCl

porous membrane

0.10 M KCl

glass "membrane"

Figure 24.4 Common pH meter electrodes (two-electrode system). The potential of the calomel electrode is pH independent, but that of the glass electrode varies with pH.

One of the most familiar applications of the Nernst equation is found in the pH meter (Unit 19). This instrument is basically a very sensitive electronic voltmeter (electrometer) which can be used to determine the potential for a wide variety of cells. Typically, a pH meter scale is calibrated in both pH and millivolt units. Measurement of pH actually involves a voltage measurement related logarithmically to hydrogen ion concentration, hence the choice of the definition pH $= -\log C_{H^+}$. As predicted by the Nernst equation, the voltage varies by 0.0592 V for a change of 1.0 pH unit, so one scale is simply marked in pH units to correspond to each 59.2 increment on the millivolt scale.

Although a variety of electrodes are usually available for use with pH meters in different applications, the most common system for pH determination itself uses the glass electrode together with the calomel electrode (Figure 24.4). The more modern pH meters combine these in a single two-segment tube, thus permitting measurements on solution volumes too small for insertion of two separate electrodes.

The key to the function of the glass electrode is the thin glass membrane, which was once thought to be selectively permeable to H$^+$, apparently by transfer of H$^+$ through water contained in the pores of the membrane. The transfer of H$^+$ through the membrane in the direction of the solution of higher pH was believed to account for the development of the electrochemical potential. However, recent experiments with radioactive tritium ions ($_1^3$H$^+$) have shown that H$^+$ transport through the glass membrane does *not* occur. The more recent interpretation of the functioning of the glass electrode involves three factors:

1. There are very tiny pores in the glass membrane. When the membrane is soaked in water (or an aqueous solution), hydrated glass layers form at the glass–solution interfaces within the pores.

2. Cation exchange occurs at all glass–solution interfaces, involving H$^+$ from the solution and Na$^+$ from the glass (a sodium silicate):

$$Na^+(glass) + H^+(aq) \rightleftharpoons H^+(glass) + Na^+(aq)$$

The quantity of H$^+$ exchanged is proportional to the activity of H$^+$ in the solution. If the activity of H$^+$ in the solution *outside* the electrode membrane differs from that in the solution *inside* the glass electrode, a phase boundary potential is established. The magnitude of this potential is a function of the pH of the solution outside the membrane.

3. Electrical conductivity of the glass itself depends on the migration of Na$^+$ ions among interstitial holes in the glass itself.

---

[2] For a description of some of these, see *Electroanalytical Principles*, by R. E. Murray and C. N. Reilley (New York: Wiley, 1964).

It should be mentioned that this explanation has not yet been fully accepted. Additional experiments will be needed before the mechanism of action of the glass electrode is firmly established. We do know that the glass electrode is pH dependent and that it will not function properly if allowed to "dry out" so that "pore water" is lost.

ION-SELECTIVE ELECTRONS

Although the glass electrode, utilized in most pH meters (Figure 24.4), has been in use for over 60 years, other ion-selective electrodes have only recently been developed. The design of special electrodes to permit the selective determination of ions other than H^+ has many obvious applications. An extensive discussion of such electrodes was held at a symposium at the National Bureau of Standards in January of 1969 (NBS Special Publication 314), by which time more than 20 different ion-selective electrodes had been developed.

Three general types of ion-selective electrodes are now in common use: glass electrodes for ions other than H^+, solid state electrodes, and liquid ion-exchange electrodes. Other types have been developed for various specific applications.

The first type utilizes specially formulated glass membranes designed to favor the exchange of ions other than H^+ by a mechanism presumed to be similar to that of the glass electrode described for the pH meter in Section 24.8. For example, a sodium-ion glass electrode has been made by using glass in which a fairly high percentage of the silicon has been replaced by aluminum. One important application of the sodium-ion electrode is in the automatic monitoring of certain ion-exchange "water softeners" (Unit 6), since a sudden drop in Na^+ concentration in the effluent solution may indicate a need to recharge the ion-exchange resin.

Solid state electrodes are now commercially available in various forms for the determination of a number of different cations and anions. The most common solid state electrodes utilize either pure single-crystal membranes or silicone rubber membranes containing imbedded salts. Obviously, for measurements in aqueous solutions the solid salts employed must be very insoluble in water. Most solid state electrodes are used for the determination of univalent $(+1, -1)$ ions.

The liquid ion-exchange electrodes are more useful for determination of multi-valent ions $(+2, +3, -2)$, since the mobility in a liquid phase better permits such ions to "locate" binding sites of appropriate charge density. Typically, the liquid ion-exchange layer is separated from the solution to be measured by a porous membrane of glass or plastic.[3]

---

[3] If you are interested in the subject of ion-specific electrodes, you may wish to read Robert Fischer's "Ion-Selective Electrodes" in the June, 1974, issue of the *Journal of Chemical Education* (pp. 387–390).

SUGGESTED
READING

Lyons, Ernest H., Jr. 1967. *Introduction to Electrochemistry.* Lexington, Mass.: Heath.
Murray, R. E., and C. N. Reilley. 1964. *Electroanalytical Principles.* New York: Wiley.
O'Connor, Rod, and Charles Mickey. 1974. *Solving Problems in Chemistry.* New York: Harper & Row.

EXERCISES

[Refer to Appendix D, as necessary.]

1. Calculate the electrode potentials for
   a. zinc in contact with 1.0×10^{-3} M Zn^{2+}.
   b. hydrogen gas at 0.80 atm in contact with a platinum electrode immersed in 0.50 M HCl.

2. a. Calculate the potential for a cell utilizing the two half-cells in Exercise 1.
   b. Calculate the potential for a cell utilizing the half-cells Cd, Cd^{2+} (0.30 M) and Al, Al^{3+} (2.0×10^{-5} M).
   c. Write standard shorthand notations for the cells in parts (a) and (b).

3. Calculate the equilibrium constants for
   a. $Cu + 2H^+ \rightleftharpoons H_2 + Cu^{2+}$
   b. $Zn + 2H^+ \rightleftharpoons Zn^{2+} + H_2$
   c. $2Fe + 6H^+ \rightleftharpoons 2Fe^{3+} + 3H_2$
   d. $2Fe + 3Zn^{2+} \rightleftharpoons 2Fe^{3+} + 3Zn$

4. Calculate the solubility product for lead sulfate, given that the potential is 1.00 V for a cell whose components are silver in contact with 0.20 M Ag^+ and lead in contact with saturated lead sulfate solution. The lead electrode is found to be the anode.

PRACTICE FOR PROFICIENCY

(Refer to Appendix D, as necessary.)

Objective (a)

1. Write standard shorthand cell notations for each of the following cells, for which the *internal* electrodes are described.
   a. anode: copper in contact with 0.10 M Cu^{2+}
      cathode: silver in contact with 0.10 M Ag^+
   b. anode: lead in contact with 0.05 M Pb^{2+}
      cathode: copper in contact with 0.05 M Cu^{2+}
   c. anode: hydrogen gas (1 atm) bubbling over a platinum electrode in contact with 0.0010 M H^+
      cathode: liquid mercury in contact with 0.10 M Hg^{2+}
   d. anode: hydrogen gas (0.8 atm) bubbling over a platinum electrode in contact with 0.005 M H^+
      cathode: chlorine gas (0.8 atm) bubbling over a platinum electrode in contact with 0.30 M Cl^-

Objectives (a) and (b)

2. For each of the following cells, for which half-cell descriptions are given, calculate the cell potential and determine the appropriate equation for spontaneous cell discharge. Then write the standard shorthand cell notation for each.
   a. Ni, Ni^{2+} (0.10 M) and Cu, Cu^+ (1.0×10^{-6} M)
   b. Cu, Cu^{2+} (0.10 M) and Ag, Ag^+ (1.0×10^{-10} M)

c. Pt(s), H_2(g) (1.0 atm), H^+ (1.0×10^{-3} M) and Zn, Zn^{2+} (0.10 M)
d. Pt(s), H_2(g) (1.0 atm), H^+ (1.0×10^{-10} M) and Zn, Zn^{2+} (0.10 M)
e. Al Al^{3+} (0.050 M) and Pt(s), H_2(g) (1.0 atm), H^+ (0.010 M)

3. Calculate the potential for each of the galvanic cells indicated:
   a. $Al|Al^{3+}$ (10^{-5} M)$\|Sn^{2+}$ (10^{-3} M)$|Sn$
   b. $Mg|Mg^{2+}$ (10^{-4} M)$\|Al^{3+}$ (10^{-7} M)$|Al$
   c. $Mn|Mn^{2+}$ (10^{-2} M)$\|Cr^{3+}$ (10^{-5} M)$|Cr$
   d. $Cr|Cr^{3+}$ (10^{-1} M)$\|Cd^{2+}$ (10^{-6} M)$|Cd$
   e. $Ag|Ag^+$ (10^{-5} M)$\|Au^{3+}$ (10^{-1} M)$|Au$

Objective (c)

4. Calculate the approximate equilibrium constant at 25°C for each of the systems indicated:
   a. $2Al(s) + 3Cu^{2+}(aq) \rightleftharpoons 2Al^{3+}(aq) + 3Cu(s)$
   b. $3Mn^{2+}(aq) + 2Cr(s) \rightleftharpoons 3Mn(s) + 2Cr^{3+}(aq)$
   c. $3Ag(s) + Au^{3+}(aq) \rightleftharpoons 3Ag^+(aq) + Au(s)$
   d. $4P(s) + 6H_2O(l) \rightleftharpoons 4PH_3(g) + 3O_2(g)$
   e. $4AsH_3(g) + 3O_2(g) \rightleftharpoons 4As(s) + 6H_2O(l)$

5. From the information given, and other data from Appendix D, calculate the net formation constant (Unit 21) for each complex ion:
   a. $[Cd(NH_3)_4]^{2+}(aq) + 2e^- \rightleftharpoons Cd(s) + 4NH_3(aq),$
   $$\mathcal{E}^0 = -0.61 \text{ V}$$
   b. $[Cd(CN)_4]^{2-}(aq) + 2e^- \rightleftharpoons Cd(s) + 4CN^-(aq),$
   $$\mathcal{E}^0 = -0.96 \text{ V}$$
   c. $[Ag(NH_3)_2]^+(aq) + e^- \rightleftharpoons Ag(s) + 2NH_3(aq),$
   $$\mathcal{E}^0 = +0.39 \text{ V}$$
   d. $[Fe(CN)_6]^{3-}(aq) + 3e^- \rightleftharpoons Fe(s) + 6CN^-(aq),$
   $$\mathcal{E}^0 = -0.651 \text{ V}$$
   e. $[FeF_5(H_2O)]^{2-}(aq) + 3e^- \rightleftharpoons Fe(s) + 5F^-(aq) + H_2O(l),$
   $$\mathcal{E}^0 = -0.333 \text{ V}$$

Objective (d)

6. Calculate the solubility product for silver chromate, given that the potential is 0.25 volt for a cell whose components are copper in contact with 0.30 M Cu^{2+} and silver in contact with saturated silver chromate solution.

7. Calculate the approximate pH of a solution of hydrochloric acid in an electrode compartment through which hydrogen gas is bubbled at a pressure of 1.00 atm if this electrode system, when combined with a nickel/0.10 M Ni^{2+} electrode system, forms a 0.15 V cell. The spontaneous reaction for discharge of this cell may be formulated as

$$Ni(s) + 2H^+(aq) \longrightarrow Ni^{2+}(aq) + H_2(g)$$

8. Calculate the approximate ionization constant for butyric acid, given that the potential is 0.62 V for a cell whose components are zinc in contact with 0.10 M Zn^{2+} and hydrogen gas at 1.00 atm bubbling through a 0.10 M solution of $C_3H_7CO_2H$ in contact with a platinum electrode. The spontaneous reaction for discharge of this cell may be formulated as

$$Zn(s) + 2H^+(aq) \longrightarrow Zn^{2+}(aq) + H_2(g)$$

9. Calculate the approximate solubility product for copper(I) cyanide, given that the potential is 1.87 V for a cell whose components are aluminum in contact with 0.10 M Al^{3+} and copper in contact with saturated CuCN solution. The spontaneous reaction for discharge of this cell may be formulated as

$$Al(s) + 3Cu^+(aq) \longrightarrow Al^{3+}(aq) + 3Cu(s)$$

10. Calculate the approximate pH of a solution of hydrochloric acid in an electrode compartment through which hydrogen gas is bubbled at a pressure of 1.00 atm if this electrode system, when combined with a zinc/0.10 M Zn^{2+} electrode system, forms a 0.65-V cell. The spontaneous reaction for discharge of this cell may be formulated as

$$Zn(s) + 2H^+(aq) \longrightarrow Zn^{2+}(aq) + H_2(g)$$

11. Calculate the approximate ionization constant for p-chlorobenzoic acid given that the potential is 2.25 V for a cell whose components are magnesium in contact with 0.10 M Mg^{2+} and hydrogen gas at 1.00 atm bubbling through a 0.10 M solution of $ClC_6H_4CO_2H$ in contact with a platinum electrode. The spontaneous reaction for discharge of this cell may be formulated as

$$Mg(s) + 2H^+(aq) \longrightarrow Mg^{2+}(aq) + H_2(g)$$

12. Calculate the approximate solubility product for aluminum arsenate, given that the potential is 2.56 V for a cell whose components are silver in contact with 0.10 M Ag^+ and aluminum in contact with saturated $AlAsO_4$ solution. The spontaneous reaction for discharge of this cell may be formulated as

$$Al(s) + 3Ag^+(aq) \longrightarrow Al^{3+}(aq) + 3Ag(s)$$

13. Calculate the approximate pH of a solution of hydrochloric acid in an electrode compartment through which hydrogen gas is bubbled at a pressure of 1.00 atm if this electrode system, when combined with an aluminum/0.10 M Al^{3+} electrode system, forms a 1.46-V cell. The spontaneous reaction for discharge of this cell may be formulated as

$$2Al(s) + 6H^+(aq) \longrightarrow 2Al^{3+}(aq) + 3H_2(g)$$

BRAIN TINGLERS

(In answering the following questions, consider the relationships between free energy and spontaniety and free energy and cell potential.)

14. Which of the following reactions will proceed spontaneously as formulated, at 25°C, with the *initial* concentrations indicated?

I. $Sn(s) + Fe^{2+}(aq) \longrightarrow Sn^{2+}(aq) + Fe(s)$
$$(C_{Fe^{2+}} = 0.10 \text{ M}, C_{Sn^{2+}} = 0.10M)$$
II. $2H^+(aq) + Cu(s) \longrightarrow H_2(g) + Cu^{2+}(aq)$
$$(C_{Cu^{2+}} = 10^{-6} \text{ M}, pH = 3.0, P_{H_2} = 1.0 \text{ atm})$$
III. $Au^{3+}(aq) + 3Ag(s) \longrightarrow Au(s) + 3Ag^+(aq)$
$$(C_{Au^{3+}} = 10^{-6} \text{ M}, C_{Ag^+} = 10^{-2} \text{ M})$$

a. only I **b.** only II
c. only III **d.** only I and II
e. I, II, and III

15. Which of the following reactions will proceed spontaneously as formulated, at 25°C, with the *initial* concentrations indicated?

I. $Mn(s) + Cd^{2+}(aq) \longrightarrow Mn^{2+}(aq) + Cd(s)$
$$(C_{Cd^{2+}} = 0.10 \text{ M}, C_{Mn^{2+}} = 0.10 \text{ M})$$
II. $2H^+(aq) + Fe(s) \longrightarrow H_2(g) + Fe^{2+}(aq)$
$$(C_{Fe^{2+}} = 10^{-4} \text{ M}, pH = 2.0, P_{H_2} = 1.0 \text{ atm})$$
III. $Al^{3+}(aq) + 3Ag(s) \longrightarrow Al(s) + 3Ag^+(aq)$
$$(C_{Al^{3+}} = 10^{-9} \text{ M}, C_{Ag^+} = 10^{-3} \text{ M})$$

a. only I **b.** only II
c. only III **d.** only I and II
e. I, II, and III

Unit 24 Self-Test

(Refer to Appendix D when necessary.)

1. Calculate the electrode potential for aluminum in contact with 5×10^{-8} M Al^{3+}.
2. Calculate the potential for a cell utilizing the half-cells Mg, Mg^{2+} (1.0×10^{-3} M) and Ag, Ag^+ (2.0×10^{-6} M).
3. Write the standard shorthand notation for the cell in question 2.

4. Calculate the equilibrium constant for
$$Ni + Cu^{2+} \rightleftharpoons Ni^{2+} + Cu$$

5. Calculate the solubility product for copper(I) bromide, given that the potential is 0.017 V for a cell whose components are copper in contact with 0.10 M Cu^{2+} and copper in contact with saturated CuBr solution. The electrode immersed in the saturated solution is the anode.

Overview Test
Section EIGHT

(Consult Appendix D, as necessary.)

1. In the reaction between hypochlorite ion and hydrochloric acid, what is the oxidizing agent?

$$OCl^- + 2H^+ + Cl^- \longrightarrow Cl_2 + H_2O$$

a. H_2O; **b.** Cl^-; **c.** H^+; **d.** OCl^-; **e.** Cl_2

2. What is the *sum* of all coefficients when the following equation is balanced? (*Hint:* Remember "implied" coefficients of unity.)

$$Co^{2+} + HNO_2 \longrightarrow Co^{3+} + NO \quad \text{(in acidic solution)}$$

a. 4; **b.** 6; **c.** 7; **d.** 13; **e.** 15

3. What is the *sum* of all coefficients when the following equation is balanced? (*Hint:* Remember "implied" coefficients of unity.)

$$I^- + Cl_2 \longrightarrow IO_4^- + Cl^- \quad \text{(in alkaline solution)}$$

a. 7; **b.** 12; **c.** 19; **d.** 26; **e.** 29

4. What volume of a solution labeled "0.100 N nitric acid (standardized for $NO_3^- \longrightarrow NO_2$)" would be required for oxidation of 0.36 g of zinc, according to the equation given? (Assume that the nitric acid calculated is required only as a source of oxidizing agent.)

$$3Zn + 8H^+ + 2NO_3^- \longrightarrow 3Zn^{2+} + 2NO + 4H_2O$$

a. 11 ml; **b.** 24 ml **c.** 38 ml **d.** 47 ml; **e.** 55 ml

5. What average current, in amperes, must be maintained to electroplate 5.9 g of nickel per hour from a solution of a nickel(II) salt?

a. 2.6; **b.** 2.9; **c.** 3.7; **d.** 5.4; **e.** 5.9

6. What volume (in liters) of dry oxygen gas (at STP) could be generated by the electrolysis of water over a 6.0-hr period with a current of 1.5 amp?

a. 1.9; **b.** 2.1; **c.** 2.5; **d.** 3.2; **e.** 5.0

7. What mass of iron per hour is converted to Fe(III) by the electrolytic corrosion of a metal drainpipe at a rate equivalent to a current of 0.20 amp?

a. 0.45 g hr^{-1}; **b.** 0.30 g hr^{-1}; **c.** 0.21 g hr^{-1}; **d.** 0.14 g hr^{-1}; **e.** 0.06 g hr^{-1}

8. How long, in hours, would be required for electroplating 196 g of nickel from a solution of nickel(II) ion, using an average current of 20 amp with 80% electrode efficiency?

a. 27; **b.** 18; **c.** 21; **d.** 11; **e.** 4

9. For a galvanic cell using Cu, Cu^{2+} (1.0 M) and Zn, Zn^{2+} (1.0 M) half-cells, which of the following statements is incorrect?

a. The copper electrode is the cathode.
b. Electrons will flow through the external circuit from the zinc electrode to the copper electrode.
c. Reduction occurs at the copper electrode during discharge.
d. The mass of the zinc electrode will increase during discharge.
e. The concentration of Cu^{2+} will decrease during discharge.

10. For the standard cell Zn, Zn^{2+} versus Fe, Fe^{2+}, the cathode and $\mathcal{E}^0$ for the cell are, respectively,

a. Fe, 0.32 V; **b.** Zn, 0.32 V; **c.** Zn, 1.20 V; **d.** Zn, −0.76 V; **e.** Fe, 1.20 V

11. Given the following information, calculate ΔG_f^0 for aqueous CuF_2, in kilocalories per mole (1.0 cal = 4.2 VC).

$$Cu^{2+}(aq) + 2e^- \rightleftharpoons Cu(s) \qquad \mathcal{E}^0 = +0.34$$
$$F_2(g) + 2e^- \rightleftharpoons 2F^-(aq) \qquad \mathcal{E}^0 = +2.87$$

a. −34; **b.** −65; **c.** −116; **d.** −133; **e.** −148

12. Calculate the potential for the galvanic cell indicated, in volts.

$$Mg|Mg^{2+}\ (10^{-3}\ M)\|Ag^+\ (10^{-6}\ M)|Ag$$

$$Ag^+ + e^- \rightleftharpoons Ag \qquad \mathcal{E}^0 = +0.80$$
$$Mg^{2+} + 2e^- \rightleftharpoons Mg \qquad \mathcal{E}^0 = -2.37$$

a. 3.44; **b.** 3.23; **c.** 2.90; **d.** 3.11; **e.** 3.17

13. Estimate the equilibrium constant for the system indicated (at 25°C).

$$2Al + 3Cu^{2+} \rightleftharpoons 2Al^{3+} + 3Cu$$

a. 5×10^{202}; **b.** 3.7×10^{67}; **c.** 6.1×10^{33}; **d.** 2.7×10^{-68}; **e.** 2×10^{-203}

14. Calculate the approximate pH of a solution of hydrochloric acid in an electrode compartment through which hydrogen gas is bubbled at a pressure of 1.00 atm if this electrode system, when combined with a nickel/0.10 M Ni^{2+} electrode system, forms a 0.10 V cell. The spontaneous reaction for discharge of this cell may be formulated as

$$Ni(s) + 2H^+(aq) \longrightarrow Ni^{2+}(aq) + H_2(g)$$

a. 2.2; **b.** 4.7; **c.** 6.1; **d.** 3.0; **e.** 1.3

15. Calculate the approximate solubility product for copper(I) cyanide, given that the potential is 0.98 V for a cell whose components are zinc in contact with 0.10 M Zn^{2+} and copper in contact with saturated CuCN solution. The spontaneous reaction for discharge of this cell may be formulated as

$$Zn(s) + 2Cu^+(aq) \longrightarrow Zn^{2+}(aq) + 2Cu(s)$$

a. 6.4×10^{-8}; **b.** 5.0×10^{-9}; **c.** 7.3×10^{-12}; **d.** 1.4×10^{-16}; **e.** 3.3×10^{-23}

Section NINE

SOME ASPECTS OF INORGANIC CHEMISTRY

There are more compounds of carbon than of any other element, since carbon is unique in the nature of its chemical bonding. The ability of carbon atoms to form stable bonds with each other or with a large variety of other atoms accounts for the vast number of covalent compounds of carbon. The study of these compounds is the subject of *organic chemistry,* an area we will investigate in some detail in Units 28–34. The study of the other elements and their compounds is considered as the field of *inorganic chemistry.*

Inorganic chemistry, then, encompasses the investigation of more than a hundred different elements, with all of their compounds. Obviously, we can only touch on some aspects of this large area of chemical science.

In Unit 25 we will consider some of the more common *representa-*

tive metals and metalloids, in particular those of groups IA, IIA, IIIA, and IVA. The more common nonmetals of groups VIA and VIIA are discussed in Unit 26. Many of these elements form useful and important compounds, ranging from ordinary "table salt" to the lead sulfate and lead dioxide of the automobile battery.

The simple metals exhibit only one, or at most two, oxidation states in their compounds. Many of the *d* transition metals (Unit 27), on the other hand, exhibit a wide range of oxidation states in their compounds. In addition, the *d* transition metal ions form a number of interesting complexes, from the familiar deep blue tetraamminecopper(II) to the biologically important hemoglobin. Unlike the simpler elements, for which valence electrons in *s* or *p* orbitals are of principal importance, the *d* transition metals utilize valence-shell *d* orbitals. We shall have to extend our considerations of electron configuration and bonding orbitals to include the 3*d* and 4*d* orbitals to permit us to understand the chemistry of the *d* transition elements.

Inorganic chemistry constitutes a large fraction of the work of the chemical industry. Industrial chemists and chemical engineers are involved in continuing efforts of research and development for new and improved processes ranging from the isolation of crude metals from their ores to the production of ultrapure chemicals for specialty uses. Within recent years much attention has been directed toward recycling and waste-reclamation projects, many of which have provided economical routes to important inorganic chemical processes. In Excursion 3 we shall examine some of these industrial processes.

CONTENTS

Section Nine Overview Test

Excursion 3 Some industrial processes
Metal production; inorganic compound preparation; recycling and waste reclamation

You are the salt of the earth. . . .
MATTHEW 5:13

Representative metals and metalloids

Objectives

(a) Given access to a simple periodic table, be able to
  (1) identify the more common representative metals and metalloids (Section 25.1)
  (2) indicate general trends in physical and chemical properties of the representative metals and metalloids
  (3) recognize "diagonal relationships" in some chemical and physical properties
(b) Be able to identify the hydroxides of representative metals and metalloids as "basic," "amphoteric," or "acidic" (Section 25.7).
(c) Be able to predict the products and to write balanced equations for reactions of representative metals and metalloids with halogens, hydrogen, nitrogen, oxygen, or water

REVIEW

(d) Be able to apply the principles of stoichiometry, equilibrium, and electrochemistry to the representative metals and metalloids, or their compounds.

Salt—ordinary sodium chloride, a cheap, common chemical without which life could not exist. On the freedom to make it hung the fate of an empire.

India in 1930 was ripe for rebellion. Many leaders of the Congress Party urged open violence and revolution, but one of the great men of all time, Mahatma Gandhi, chose salt as the focal point for an act of nonviolent protest. At the age of 61, Gandhi led a march in civil disobedience 200 miles across India to the sea. There he and his followers made salt in protest of the British requirement of controlled purchase of government-taxed salt. Many of the marchers were beaten by government troops; some were killed; thousands were imprisoned. The news that a mighty empire could act in this way spread rapidly throughout the world. The "Great Salt March" led to further acts of nonviolent protest, eventually gaining independence for India, and serving as an example for peaceful men seeking social reform, even within our own "enlightened" society.

Salt—a standard of exchange in primitive societies, a "pillar" from Lot's wife, a term indicating great importance ("the salt of the earth"), a problem in arid lands needing water from the sea, and an essential ingredient of living organisms. In spite of its apparent simplicity, there is much to be learned about it, especially in the field of biochemistry.

"Table salt" is a compound of sodium, a member of one of the simplest group of elements (IA). All the IA and IIA elements are metallic and, except for the lighter members of these groups (lithium and beryllium), the general chemical and physical properties of the IA and IIA metals are quite similar.

Differences in "group properties" become more pronounced in the elements of group IIIA and there are more differences than similarities in the group IVA elements, ranging from the nonmetallic character of diamond to the typically metallic nature of lead. Interestingly enough, the group IVA elements are all found in the modern automobile, with some rather peculiar relationships:

> Carbon, which forms the largest variety of natural compounds, occurs in the largest number of different sources in the automobile, for instance, in pigments used in dyeing fabrics, in the various plastic parts, in the rubber of the tires and electrical insulation, in the fuel and lubricating oils, in the "antifreeze," and in structural steel alloys.
>
> Silicon, which occurs in nature almost always in combination with oxygen, occurs in the automobile mainly with oxygen as silicates in window glass, mirror glass, headlight glass, and so on.
>
> Germanium, the rarest of the group IVA elements in nature, is also the rarest of these elements in the automobile, occurring in tiny transistors in the automobile radio.
>
> Tin, naturally more abundant than germanium but less abundant than carbon, silicon, or lead, has a similar relative abundance in the automobile. It is used

as a minor component of certain metal alloys and, to a larger extent, in solder for electric connections.

The most common additive is tetraethyllead, $Pb(C_2H_5)_4$, a substance of wide publicity as a source of environmental pollution.

Lead, the most naturally abundant metal of group IVA, is the most abundant metal of this group in the automobile. It is found, for example, in the battery, in many of the metal alloys, and frequently as an additive to the gasoline fuel.

With this unit we begin the study of the elements and some of their more important compounds. We shall use this study to apply many of the principles described in earlier units.

25.1 METALS AND METALLOIDS

Most of the chemical elements exhibit the typical metallic characteristics (Unit 2) of surface luster, ductility, malleability, high thermal conductivity, and electrical conductivity that decreases with increasing temperature. Of the remaining elements, boron, silicon, germanium, arsenic, antimony, tellurium, and polonium are usually classified as the *metalloids* (Figure 25.1). A major distinction between metals and

Figure 25.1 Metals and metalloids.

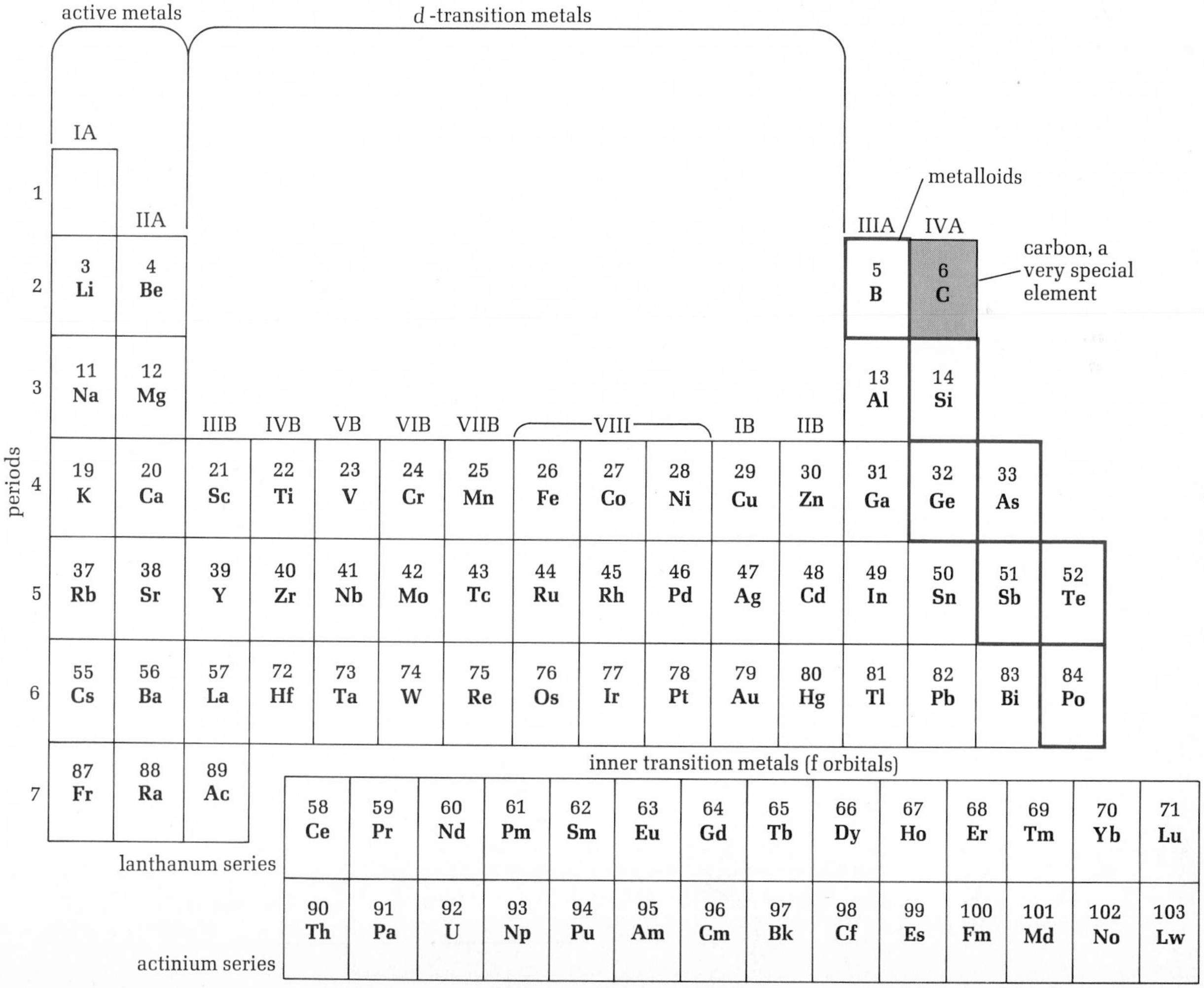

The decrease in electrical conductivity of metals with increasing temperature is explained by the interference of random thermal motions of metal ions in the crystal lattice with the directional flow of mobile electrons (Unit 11).

metalloids is generally made on the basis of electrical conductivity. The metalloids, unlike typical metals, exhibit *increasing* electrical conductivity in the solid state with increasing temperature. This characteristic reflects the differences in interatomic forces within the crystalline metals and metalloids. In the latter, the valence electrons are "immobilized" by participation in covalent bonding among neighboring atoms. Until sufficient thermal energy is available to rupture some of these bonds, the electron mobility required for electric current flow cannot be attained.

Carbon is a very special element in many ways. It is neither a typical metal nor a metalloid. In the allotropic form of diamond (Unit 2), carbon is an electrical nonconductor, but graphite conducts an electric current readily. In many respects carbon could be considered a nonmetal. Since such classifications are quite arbitrary, we shall simply consider carbon as a rather unique case. Hydrogen (Unit 26) is another "unique" element and many of the other elements are difficult to classify because of the major differences among allotropic modifications (Unit 26).

25.2 REACTIVE METALS

All elements classified as metals have a number of common properties. Electrical conductivity, relatively easy oxidation, and general tendency to form stable positive ions are attributed to loosely held valence electrons in metal atoms.

Still, metals vary widely in a number of properties. Some, like potassium, are very soft, whereas others are quite hard—and mercury is a liquid at room temperature. Most have a shiny, "silvery" surface when pure, but gold and copper have very characteristic colors. Cesium, one of the most reactive elements, bursts into flame when exposed to moist air, whereas gold or platinum is inert even to strong acids.

The most reactive metals, in the sense of easy electron loss, are found in groups IA and IIA of the periodic table. These elements are unique in having their valence electrons only in s orbitals. The nondirectional character of s orbitals accounts for the fact that these metals form relatively simple crystal lattices (Figure 25.2) of the types associated with easy ways of packing spheres together. Since loss of only one (group IA) or two (group IIA) electrons forms an ion isoelectronic with an inert gas atom, these elements most commonly form simple positive ions.

Group IA elements are called "alkali metals" because they all form water-soluble hydroxides that are strongly basic ("alkaline"). Group IIA elements are called "alkaline earth" metals because alchemists applied the term "earth" to any substance that would neither dissolve in water nor be changed by heating, and the particular "earths" MgO, CaO, and so on were alkaline in that they reacted rapidly with dilute acids. Note that Li_2O, Na_2O, and so on, were not "earths," because they dissolved readily in water to form the corresponding hydroxides.

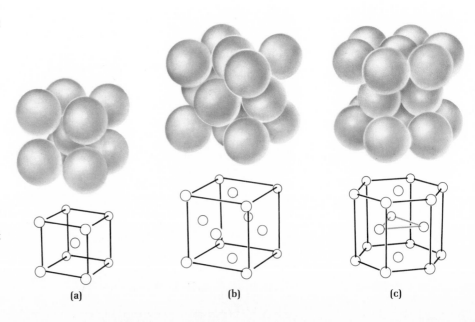

Figure 25.2 Crystal structures of group IA and IIA metals. (a) Body-centered cubic; all group IA elements and barium. (b) Face-centered cubic (cubic closest packed); calcium and strontium. (c) Hexagonal closest packed; beryllium and magnesium. Other (less common) crystal forms (allotropes) are also known for many of these elements.

(a)   (b)   (c)

Figure 25.3 Relative sizes and ionization potentials (eV atom^{-1}). (1.00 eV atom^{-1} = 23 kcal mole^{-1}.)

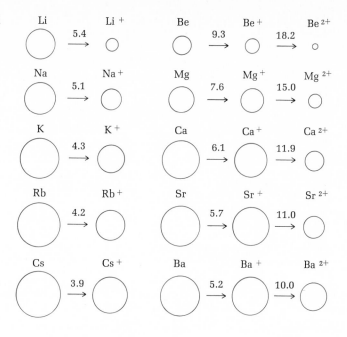

Table 25.1 ELECTRON CONFIGURATIONS (GROUP IA AND IIA ELEMENTS)

Li: (He), 2s↑	Be: (He), 2s↑↓
Na: (Ne), 3s↑	Mg: (Ne), 3s↑↓
K: (Ar), 4s↑	Ca: (Ar), 4s↑↓
Rb: (Kr), 5s↑	Sr: (Kr), 5s↑↓
Cs: (Xe), 6s↑	Ba: (Xe), 6s↑↓
Fr: (Rn), 7s↑	Ra: (Rn), 7s↑↓

The electron configurations (Table 25.1) for these elements correlate with a number of observable properties. For example, relative atomic and ionic sizes and ionization potentials (Figure 25.3) are consistent with the idea that each successive s orbital (e.g., 1s, 2s, 3s, . . .) extends farther from the nucleus.

A number of physical properties (Figure 25.4) can be related to factors such as electron configuration, particle size, and ionization potential. The bonding between atoms in metallic crystals, for example, is believed to involve delocalization of valence electrons throughout the crystal lattice. Thus these metals are viewed as orderly arrays of positive ions in a "sea" of mobile electrons (Unit 11). Two factors, then, influence the strengths of interparticle forces in the crystals. The higher the charge density on the ion (charge per unit volume), the more strongly it is held in the electron "sea." This favors small ions over large ones and 2+ ions over 1+ ions. In addition, forces of the van der Waals type will increase with increasing atomic

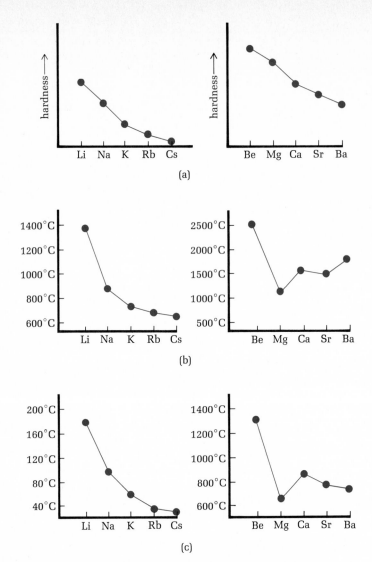

Figure 25.4 Some physical properties of group IA and IIA elements. (a) "Hardness." (b) Boiling point (°C) at 1 atm. (c) Melting point (°C).

number, and valence electrons of large atoms are more readily lost, increasing the degree of electron delocalization. These factors favor larger atoms. Since interparticle forces are reflected in such properties as hardness, boiling point, and melting point, these properties show the trends tabulated in Figure 25.4. Note that trends for group IIA elements are less regular than those for group I. To some extent this reflects differences in crystal lattice patterns, with resulting differences in interparticle forces.

Since atomic mass increases more rapidly than effective atomic diameter going down a given column in the periodic table, densities tend to show a general increase. The least dense element in these groups is lithium (0.53 g cm^{-3}) and the most dense is radium ($\sim$5 g cm^{-3}).

In terms of chemical reactions, there are many similarities among these elements. All are good reducing agents and, as such, react rather readily with substances acting as electron acceptors. Several of the more common reactions of these metals are

Table 25.2 GENERAL REACTIONS COMMON TO GROUP IA AND IIA ELEMENTS

REACTANT	WITH GROUP IA	WITH GROUP IIA
any halogen e.g., Cl_2	$2M + X_2 \longrightarrow 2MX$ e.g., $2Na + Cl_2 \longrightarrow 2NaCl$	$M + X_2 \longrightarrow MX_2$ e.g., $Mg + Cl_2 \longrightarrow MgCl_2$
sulfur, selenium, or tellurium	$2M + X \longrightarrow M_2X$ e.g., $2K + S \longrightarrow K_2S$	$M + X \longrightarrow MX$ e.g., $Ca + Se \longrightarrow CaSe$
hydrogen	$2M + H_2 \longrightarrow 2MH$ e.g., $2Li + H_2 \longrightarrow 2LiH$	$M + H_2 \longrightarrow MH_2$ e.g., $Ba + H_2 \longrightarrow BaH_2$
water	$2M + 2H_2O \longrightarrow 2MOH + H_2$ e.g., $2Cs + 2H_2O \longrightarrow 2CsOH + H_2$	$M + 2H_2O \longrightarrow M(OH)_2 + H_2$ (Be unreactive) $Mg + H_2O \longrightarrow MgO + H_2$

listed in Table 25.2. Although the reactions shown are similar in character, they may differ considerably in speed. For example, all group IA elements and calcium, strontium, and barium react vigorously with cold water. Magnesium reacts vigorously only with hot water or steam, and beryllium reacts very slowly with superheated steam at temperatures below the melting point of the metal.

Reactions involving these metals with oxygen or nitrogen are quite variable (Table 25.3). Note the similarities between lithium and magnesium.

Table 25.3 REACTIONS OF OXYGEN AND NITROGEN WITH SOME GROUP IA AND IIA METALS

$$4Li + O_2 \longrightarrow 2Li_2O$$
$$\text{oxide}$$

$$2Mg + O_2 \longrightarrow 2MgO$$
$$\text{oxide}$$

$$2Na + O_2 \longrightarrow Na_2O_2$$
$$\text{peroxide}$$

$$2Ba + O_2 \longrightarrow 2BaO$$
$$\text{oxide}$$

$$\text{and}$$

$$M + O_2 \longrightarrow MO_2$$
$$\text{superoxide}$$

$$Ba + O_2 \longrightarrow BaO_2$$
$$\text{peroxide}$$

$$(M = K, Rb, Cs)$$

$$6Li + N_2 \longrightarrow 2Li_3N$$
$$\text{nitride}$$

$$3M + N_2 \xrightarrow{\text{heat}} M_3N_2$$
$$\text{nitride}$$

$$(Na, K, Rb, Cs \text{ no reaction})$$

$$\text{e.g., } 3Mg + N_2 \longrightarrow Mg_3N_2$$

25.3 "DIAGONAL" RELATIONSHIPS

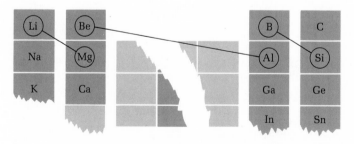

Although there are definite similarities among elements in a given vertical column in the periodic table, the first member of each group always differs in some respects from the other elements in the group. Frequently there are striking resemblances between the first element of a group and the second element of the next group to the right (diagonal relationship). These observations may perhaps be explained on the basis of similar charge densities (Figure 25.5), for the ions or, in the neutral atoms, for the kernel regions (Unit 7).

Figure 25.5 Approximate charge densities for "isolated" ions (C cm^{-3}).

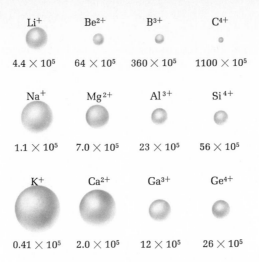

Li$^+$	Be^{2+}	B^{3+}	C^{4+}
4.4×10^5	64×10^5	360×10^5	1100×10^5

Na$^+$	Mg^{2+}	Al^{3+}	Si^{4+}
1.1×10^5	7.0×10^5	23×10^5	56×10^5

K$^+$	Ca^{2+}	Ga^{3+}	Ge^{4+}
0.41×10^5	2.0×10^5	12×10^5	26×10^5

As an example of diagonal relationships, some of the ways in which lithium resembles magnesium include the following:

1. Similar reactions with O_2 and N_2 (Table 25.3).
2. Similar solubilities of fluorides, carbonates, and phosphates (Table 25.4).
3. Formation of similar "organometallic" compounds involving covalent bonds, for example, CH_3Li and $(CH_3)_2Mg$. Other group IA and IIA elements do not form such compounds.

Organomagnesium compounds are very important in a variety of organic reactions. Such compounds are known as "Grignard reagents."

25.4 IONS OF GROUPS IA AND IIA

The alkali and alkaline earth metal ions form water-soluble combinations with more different anions than do most other groups of cations (Table 25.4). Because of their relatively high charge density, these cations are strongly hydrated, although such hydrates do not have the stable geometric patterns associated with hydrates of the transition metal ions (Unit 27).

The group IIA salts are generally less soluble than those of group IA, partly

Table 25.4 WATER SOLUBILITY OF SOME ION COMBINATIONS[a]

CATION	ANION						
	F$^-$	Cl$^-$	OH$^-$	S^{2-}	SO$_4^{2-}$	CO$_3^{2-}$	PO$_4^{3-}$
Li$^+$	X					X	X
Na$^+$							
K$^+$							
Rb$^+$							
Cs$^+$							
Be^{2+}						X	
Mg^{2+}	X		X			X	X
Ca^{2+}	X		X		X	X	X
Sr^{2+}	X				X	X	X
Ba^{2+}	X				X	X	X
Al^{3+}	X		X				X
Ag$^+$		X	X (oxide)	X	X	X	X
Pb^{2+}	X	X	X	X	X	X	X
Sn^{2+}			X	X	X		
Cu$^+$		X	X (oxide)	X	X	X	X
Cu^{2+}			X	X		X	X
Zn^{2+}			X	X		X	X

[a] X = solubility less than 0.1 mole liter^{-1} at 20°C.

Table 25.5 SOME FLAME COLORS

ION	FLAME COLOR
Na^+	yellow (intense)
K^+	violet (faint)
Ca^{2+}	brick red (fleeting)
Sr^{2+}	carmine (intense)
Ba^{2+}	green (weak)

because of the higher charge on their ions, leading to stronger forces within crystal lattices.

Analyses for group IA and IIA ions, like those for most cations, are most commonly carried out using spectroscopic instruments. In complex mixtures, ions of most other groups may be removed by precipitation as chlorides, sulfides, or hydroxides before analysis for the group IA and IIA ions is attempted. If only one of the more common ions (except Mg^{2+}, which gives no flame color) is present, it may be identified by the color imparted when the salt (preferably the "volatile" chloride) is placed in a flame (Table 25.5). If more than one of these ions is present, it may obscure the flame color of another. In particular, sodium obscures most of the others quite effectively. Instruments are available that can analyze spectra of mixtures with fair accuracy. Sometimes, however, older methods of wet chemical analysis are employed, particularly when detailed quantitative analysis of mixtures is required. A systematic method using sequential precipitations is outlined in Figure 25.6. A schematic outline of this type is called a *flow chart*.

Figure 25.6 Separation and detection of common ions in groups IA and IIA (after removal of insoluble chlorides, sulfides, and hydroxides).

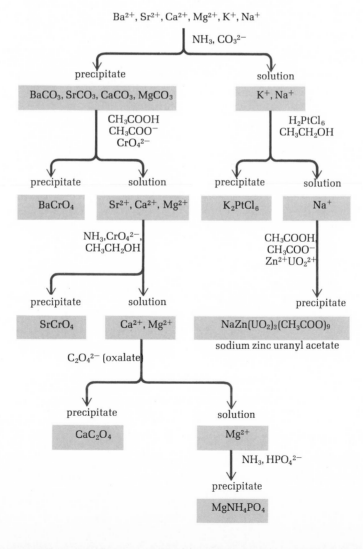

Chlorides of all the group IA and IIA ions are quite water soluble. The most common one, of course, is NaCl, ordinary "table salt." A number of large deposits of almost pure "salt" have been found, and many of these are mined commercially. "Salt" is also obtained by evaporation of seawater, but the product is generally contaminated by traces of $CaCl_2$ and $MgCl_2$. Because of the high charge density of Mg^{2+} and Ca^{2+} these ions hydrate much more readily than Na^+, and "salt" containing too much $MgCl_2$ or $CaCl_2$ absorbs moisture from the air, making it difficult to pour or shake from salt shakers. Very pure NaCl can be obtained by saturating seawater concentrates with HCl gas. A well-known brand of table salt uses trace additives of dextrose and sodium silicoaluminate to keep its salt crystals dry so that the salt will pour "when it rains." "Iodized salt" contains about 0.01 percent KI to provide iodine necessary for proper thyroid function. A number of "salt substitutes" have been devised for persons needing to reduce their Na^+ intake, but none have been found that taste exactly like "salt."

Potassium chloride is used to make "soluble potash" (K_2O) for use in fertilizers. It is estimated that more than half of the yearly requirement of KCl in this country could be obtained from "smog" from iron smelters and cement kilns should we be so fortunate as to convince industry to recover chemicals now "dumped" into the air.

Calcium chloride is commonly used to sprinkle snow-laden streets because of its high solubility and the resulting low freezing range of its solutions. An inexpensive by-product from a number of industrial processes, it permits large-scale use.

The *sulfates* of the alkali metals, beryllium, and magnesium are water soluble; $MgSO_4 \cdot 7H_2O$ (Epsom salts) is used for such diverse purposes as a fire retardant for fabrics and a stimulant for bile secretion; K_2SO_4 is something used in fertilizer mixtures, particularly in tobacco production.

The sulfates of the remaining group IIA elements are quite insoluble in water; $CaSO_4 \cdot 2H_2O$ occurs as gypsum and alabaster. When the former is carefully heated, it is converted to $(CaSO_4)_2 \cdot H_2O$, "plaster of paris." When mixed with water, this material is converted back to gypsum as the plaster "sets."

Barium sulfate is used as a suspension for x-rays of the gastrointestinal tract. The dense Ba^{2+} scatters x-rays very effectively. More soluble salts cannot be used because of the toxicity of Ba^{2+}.

Hydroxides and oxides are strongly alkaline, although $Be(OH)_2$ is amphoteric because of the very high charge density of Be^{2+}, resulting in considerable covalent character of the Be—O bond.

NaOH ("lye") is used extensively in textile, paper, and detergent industries. Soap can be made from animal fats or vegetable oils by a process called *saponification*:

The *n* varies depending on the source of the fat or oil, but is typically a number around 16 or so; the carbon chain may contain one or more C—C double bonds.

$$
\begin{array}{c}
\text{H} \quad\quad \text{O} \\
| \quad\quad\ \ \| \\
\text{H}-\text{C}-\text{O}-\text{C}(\text{CH}_2)_n\text{CH}_3 \\
|\quad\quad \text{O} \\
\quad\quad\ \ \| \\
\text{H}-\text{C}-\text{O}-\text{C}(\text{CH}_2)_n\text{CH}_3 \ +\ 3\text{NaOH(aq)} \longrightarrow \\
|\quad\quad \text{O} \\
\quad\quad\ \ \| \\
\text{H}-\text{C}-\text{O}-\text{C}(\text{CH}_2)_n\text{CH}_3 \\
| \\
\text{H}
\end{array}
\quad
3\left[\text{Na}^+ \ \begin{array}{c} {}^-\text{O} \\ \diagdown \\ \diagup \\ \text{O} \end{array} \text{C}(\text{CH}_2)_n\text{CH}_3 \right]
\ +\
\begin{array}{c}
\text{H} \\
| \\
\text{H}-\text{C}-\text{O}-\text{H} \\
| \\
\text{H}-\text{C}-\text{O}-\text{H} \\
| \\
\text{H}-\text{C}-\text{O}-\text{H} \\
| \\
\text{H}
\end{array}
$$

triglyceride   "soap"   glycerine

Potassium hydroxide (more expensive than NaOH) may be used to make corresponding potassium compounds, called "soft soaps" because of their higher water solubility.

The oxides of group IIA elements are very thermally stable, remaining stable at temperatures as high as 3000°C. The oxides, except for BeO, hydrate readily. The heat released when CaO (lime) is mixed with small ratios of water may be sufficient to ignite paper or wood. Magnesium oxide (milk of magnesia as the aqueous suspension) fortunately hydrates slowly without appreciable heat evolution.

The *carbonates* of group IA are water soluble, with the exception of Li_2CO_3, whereas group IIA carbonates are insoluble. Calcium carbonate, for example, is a principal component of chalk, marble, pearl, and the stalactites and stalagmites found in many natural caverns (Unit 18).

Na_2CO_3, sometimes called "soda ash," is used as the decahydrate, $Na_2CO_3 \cdot 10H_2O$, as "washing soda" both because of its alkaline properties and because it precipitates ions such as Ca^{2+} and Fe^{3+}, which would otherwise precipitate added soap.

Sodium bicarbonate is used as "baking soda" or when mixed with a weak acid such as $Ca(H_2PO_4)_2$, as "baking powder." When the latter mixture is moistened or when $NaHCO_3$ is acidified, such as by "sour" milk, carbon dioxide is liberated:

$$HCO_3^- + H^+ \longrightarrow H_2O + CO_2$$

In baking dough the bubbles of $CO_2(g)$ are trapped and expand as the dough is heated, resulting (with a little skill and luck) in a "light," fluffy pastry.

25.6 GROUP IIIA ELEMENTS

Aluminum, gallium, indium, and thallium are typical metallic elements, similar in many respects to the elements of groups IA and IIA, although considerably less reactive. Boron is the "maverick" of group IIIA, having many properties more like silicon than like aluminum.

Whereas aluminum, gallium, indium, and thallium crystallize in typical metallic lattices and show typical metallic conductivity, boron forms a rather unique crystal pattern (Figure 25.7) and, like silicon, is classified as a metalloid. The valence electrons of boron, unlike those of the other group IIIA elements, are highly localized into what are essentially covalent bonds similar to those in silicon. At high temperatures electrons are broken free from bonding positions, giving rise to the increasing conductivity with increasing temperature that is characteristic of simple metalloids.

As a consequence of their crystal structures and delocalized valence electrons, the elements aluminum, gallium, indium, and thallium show such additional metallic properties as malleability (formation into thin sheets or foils) and ductility (formation into wires). These metals are, however, not quite as soft as the group IA and IIA elements.

Although aluminum and boron typically form trivalent compounds, both ele-

Figure 25.7 Crystal lattices. (a) Boron icosahedron (B_{12} unit). (b) Silicon crystallizes in a pattern of repetitive tetrahedral units like those in diamond (Unit 11).

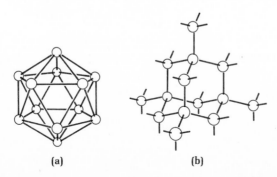

(a)   (b)

ments exhibit quadravalent character in complexes such as $[AlCl_4]^-$ or $[BF_4]^-$; and dimeric Al_2X_6-type compounds, in which each aluminum is surrounded by a distorted tetrahedron of X groups, are well known.[1]

No $+I$ compounds of either boron or aluminum are known, but the $+I$ oxidation state becomes more stable for the larger members of this group. The Ga^+ and In^+ ions are known in a few cases (e.g., Ga_2O, In_2O), but are not stable in aqueous solution. On the other hand, Tl^+ is more stable in water than Tl^{3+} and has many properties similar to Ag^+; for example,

$$Tl^+(aq) + Cl^-(aq) \rightleftharpoons TlCl(s)$$

Compare: $\quad Ag^+(aq) + Cl^-(aq) \rightleftharpoons AgCl(s)$

$$Tl_2O(s) + H_2O(l) \rightleftharpoons 2Tl^+(aq) + 2OH^-(aq)$$

Compare: $\quad Ag_2O(s) + H_2O(l) \rightleftharpoons 2Ag^+(aq) + 2OH^-(aq)$

25.7 BEHAVIOR OF HYDROXIDES

In line with the "diagonal relationships" so frequently observed in the periodic table, there are strong similarities between beryllium and aluminum and between boron and silicon. These similarities are pronounced for the hydroxides of these elements (Table 25.6).

Table 25.6 SOME HYDROXIDES COMPARED

ACIDIC

$$H_2O(l) + B(OH)_3(aq) \rightleftharpoons H^+(aq) + [B(OH)_4]^-(aq) \qquad K_a = 6 \times 10^{-10}$$
$$Si(OH)_4(aq) \rightleftharpoons H^+(aq) + [SiO(OH)_3]^-(aq) \qquad K_a = 2 \times 10^{-10}$$

AMPHOTERIC

$[Be(OH)]^+(aq) \xleftarrow{H^+(aq)} Be(OH)_2(s) \xrightarrow{OH^-(aq)} [Be(OH)_3]^-(aq)$
$[Be(H_2O)_3(OH)]^+ \qquad\qquad\qquad\qquad\qquad [Be(H_2O)(OH)_3]^-$

$[M(OH)_2]^+(aq) \xleftarrow{H^+(aq)} M(OH)_3(s) \xrightarrow{OH^-(aq)} [M(OH)_4]^-(aq)$
$[M(H_2O)_2(OH)_2]^+ \qquad\qquad\qquad\qquad (M = Al, Ga)$

BASIC

$$NaOH(s) \rightleftharpoons Na^+ + OH^- \qquad K_{sp} = {\sim}2 \times 10^2$$
$$Ca(OH)_2(s) \rightleftharpoons Ca^{2+} + 2OH^- \qquad K_{sp} = 4 \times 10^{-6}$$
$$Tl(OH)_3(s) \rightleftharpoons Tl^{3+} + 3OH^- \qquad K_{sp} = 1 \times 10^{-44}$$

The very high charge densities of the (hypothetical) B^{3+} and Si^{4+} ions (Figure 25.5) would be expected to produce a considerable polarization of the oxygen in OH^-. Apparently this effect is strong enough to produce B—O and Si—O covalent bonding and consequent weakening of O—H bonds to the extent that $B(OH)_3$ and $Si(OH)_4$ are acidic. A similar, but weaker, bonding is suggested for the amphoteric $Be(OH)_2$, $Al(OH)_3$, and $Ga(OH)_3$. The larger In^{3+} and Tl^{3+} ions, with correspondingly reduced charge densities, are more similar to the ions of group IIA and their hydroxides are typical strong bases, although not very water soluble.

25.8 THE GROUP IVA ELEMENTS

Groups IA and IIA contain only simple metals. Although the lighter elements show some anomalous properties, all have the typical metallic characteristics associated with loosely held valence electrons. Boron, on the other hand, illustrates a case in which properties vary more widely within a group, although the remaining group IIIA elements are fairly similar to each other.

---

[1]For a discussion of compounds of the type $\begin{smallmatrix} X & & X & & X \\ & \diagdown & & \diagup & \\ & & Al & & Al \\ & \diagup & & \diagdown & \\ X & & X & & X \end{smallmatrix}$ see "Bond Length-Bond Energy Term Correlations for Bonds in Al_2X_6 Compounds," by K. Wade, in the July, 1972, issue of the *Journal of Chemical Education* (pp. 502–504).

BORON: A "UNIQUE" ELEMENT

In the early history of the Southwest, prospectors wandered the deserts and mountains searching for gold or silver. Many even braved the dangers of Death Valley, often leaving their sun-bleached skeletons as warnings to others seeking an "easy" way to wealth. Little did they realize that the many acres of white barrenness often encountered were to prove more valuable than all the gold and silver found in that region.

By the time that sizable deposits of *colemanite* $(Ca_2B_6O_{11} \cdot 5H_2O)$ had been discovered in Death Valley, borax was already recognized as a valuable commercial product. Useful as a cleansing agent, a fire retardant, an ingredient in many types of glass, and an aid in certain large-scale welding operations, borax could be readily prepared by treatment of *colemanite* with sodium carbonate:

(a) $2Ca_2B_6O_{11}(H_2O)_5 + 4CO_3{}^{2-}(aq) \longrightarrow$

$$4H_2O(l) + 3B_4O_7{}^{2-}(aq) + 2OH^-(aq) + 4CaCO_3(s)$$

(b) $2Na^+(aq) + B_4O_7{}^{2-}(aq) + 10H_2O(l) \xrightarrow{\text{evaporation}} Na_2B_4O_7 \cdot 10H_2O(s)$
$$\text{borax}$$

Not too many years ago, boron chemistry was considered fairly pedestrian. A few companies marketed borax and similar products as cleansing agents, and pharmaceutical firms made a modest profit from boric acid preparations for eyewashes. Except for the fairly routine chemistry of these operations, however, there was little interest in boron outside a few academic research centers. With the advent of government-sponsored rocket research following World War II and, particularly, the "space race" between the United States and the U.S.S.R., interest in several aspects of boron chemistry began a rapid period of growth. Much of this interest was centered on the boron hydrides because of their promise as rocket fuels. These compounds combine high heats of combustion with low molecular weights, promising rich sources of energy per pound of fuel required.

COMPARISON OF A CONVENTIONAL HYDROCARBON FUEL WITH A BORON HYDRIDE

COMPOUND	HEAT OF COMBUSTION (kcal mole^{-1})	MOLECULAR WEIGHT	FUEL VALUE (kcal lb^{-1})
ethane (C_2H_6)	$\sim$380	30.0	$\sim$6000
diborane (B_2H_6)	$\sim$480	27.6	$\sim$8000

Free boron does not occur naturally, mainly because of its easy oxidation. The element is relatively rare, constituting about 0.001 percent of the earth's crust. The chief sources of boron in the United States are deposits of borax $(Na_2B_4O_7 \cdot 10H_2O)$, kernite $(Na_2B_4O_7 \cdot 4H_2O)$, and colemanite. Many of the more complex rocks contain some boron (e.g., tourmaline), and boric acid is found in several "mineral springs."

Unlike aluminum, boron cannot be satisfactorily prepared by electrolytic methods. On a small scale the metal can be produced in very high purity by decompositon of halides at high temperatures in the presence of hydrogen:

$$2BCl_3(g) + 3H_2(g) \xrightarrow{1500°C} 2B(s) + 6HCl(g)$$

Pure crystalline boron, a glossy black solid, may be produced in this way with less than 0.0001 percent impurity. Crystalline boron is almost as hard as diamond. On a larger scale, brown amorphous boron, containing traces of impurities such as magnesium boride (Mg_3B_2), can be prepared by heating a mixture of the oxide or boric acid with magnesium:

$$B_2O_3 + 3Mg \longrightarrow 3MgO + 2B$$
$$\text{amorphous}$$

Boron is an example of an element exhibiting significant "diagonal" similarities. Like silicon, boron is a metalloid and very high melting (2300°C).

The chemical properties of boron and many of its compounds are also more like those of silicon than of aluminum. For example, boron is resistant to nonoxidizing acids:

$$2Al + 6HCl \longrightarrow 2AlCl_3 + 3H_2$$
$$B + HCl \longrightarrow \text{no reaction}$$
$$Si + HCl \longrightarrow \text{no reaction}$$

Boron is oxidized very slowly by air at moderate temperatures, but the element will burn in air, oxygen, or the gaseous halogens. When burned in air, boron emits a distinctive green flame.

Although trivalent boron is stable in the boron halides (and in many other compounds), the trivalent hydride (BH_3) is unstable except as a transient product of the decomposition of other hydrides. The simplest stable compound of boron with hydrogen is diborane.

Aside from their more "mundane" commercial interests, the boron hydrides have a considerable appeal to theoretical chemists because of their most unusual bonding. The simplest possible hydride, BH_3, is not stable, although the simple tetrahedral borohydride ion (BH_4^-) is a valuable reducing agent, commercially available as $LiBH_4$, $NaBH_4$, and so on.

Boron forms a number of more complex hydrides, such as B_2H_6, B_5H_9, B_4H_{10}, $B_{10}H_{16}$, and others. One of the most thoroughly studied is diborane, B_2H_6. Remembering that boron has only three valence electrons, one finds it a bit tricky to write a conventional electron-dot formula for diborane:

$$
\begin{array}{cc}
\text{H H} & \text{H H} \\
\text{H:C:C:H} & \text{H:B\ \ B:H} \\
\text{H H} & \text{H H} \\
\text{ethane} & \text{conventional electron pairs}
\end{array}
$$

The formula suggested is unlike any previously encountered:

$$
\begin{array}{c}
\text{H\ \ H\ \ H} \\
\text{B\ \ B} \\
\text{H\ \ H\ \ H}
\end{array}
$$

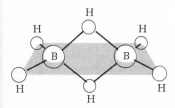

Diborane: the three-center ("banana") bond.

The so-called "three-center" or "banana" bond, involving a hydrogen "bridge" between two boron atoms, is consistent with data concerning relative bond lengths and nonequivalent hydrogens.

Diborane can be prepared from boron trifluoride and lithium hydride:

$$8BF_3 + 6LiH \longrightarrow B_2H_6 + 6LiBF_4$$

The diborane must be kept from air or moisture to avoid combustion or hydrolysis to boric acid, respectively. Boric oxide itself may be considered the anhydride of boric acid:

An anhydride is a species resulting from the *reversible* loss of water.

$$2B(OH)_3(s) \xrightarrow{\text{heat}} B_2O_3(s) + 3H_2O(g)$$

$$B_2O_3(s) + 3H_2O(l) \longrightarrow 2B(OH)_3(s)$$

Polymeric oxides are used in the production of borosilicate glass, probably most familiar under the trade name of Pyrex. Such glass is much higher melting and has a much lower thermal expansion range than ordinary "soft" glass.

Boric acid is often formulated as H_3BO_3, but it is important to realize that the bonding is better represented by $B(OH)_3$; that is, there are no boron–hydrogen bonds. The more specific name for this compound is orthoboric acid, although the simpler name is most commonly employed. The dehydration of this acid occurs in stages, leading to the formation of other acids of boron:

$$B(OH)_3 \longrightarrow H_2O + \underset{\substack{\text{metaboric} \\ \text{acid (HBO}_2\text{)}}}{BO(OH)}$$

tetraboric acid

$$4BO(OH) \longrightarrow H_2O + \underset{\substack{\text{tetraboric} \\ \text{acid}}}{H_2B_4O_7}$$

$$H_2B_4O_7 \longrightarrow H_2O + 2B_2O_3$$

The reactions of boric acid and borates in water are by no means simple. Three successive ionization constants have been reported, but the acid is generally considered monoprotic when water is the base.

This reaction may, perhaps, be formulated more correctly as showing the $B(OH)_3$ acting as a *Lewis* acid:
$$B(OH)_3 + H_2O \rightleftharpoons$$
$$[B(OH)_4]^- + H^+$$

$$B(OH)_3 \rightleftharpoons [BO(OH)_2]^- + H^+ \qquad K_{a_1} = 6.0 \times 10^{-10}$$

$$[BO(OH)_2]^- \rightleftharpoons [BO_2(OH)]^{2-} + H^+ \qquad K_{a_2} = 2 \times 10^{-13}$$

$$[BO_2(OH)]^{2-} \rightleftharpoons BO_3^{2-} + H^+ \qquad K_{a_3} = 2 \times 10^{-14}$$

In more concentrated solutions (e.g., $>0.5\ M$), the boric acid solutions are complicated by the formation of more complex species such as $HB_4O_7^-$.

Borax is a weak base due to the presence of the tetraborate ion ($B_4O_7^{2-}$). This is by far the most valuable compound of boron from a commercial standpoint. It is used as a cleaning agent, as a fire retardant for fabrics, and as a "flux" in soldering and welding operations. In this respect it acts to dissolve metal oxide films to provide a clean surface for metal-to-metal bonding.

This ester, $B(OCH_3)_3$, does not contain the double-bonded oxygen typical of the organic esters (Unit 29). The reaction is not limited to methanol, and a large number of borate esters are known.

Boric acid reacts with methanol in the presence of catalytic amounts of sulfuric acid to form a volatile ester, methyl borate. When aspirated into a flame, this ester burns to impart a bright green color to the flame. Measurement of the intensity of this green light is used in a technique called "flame emission spectroscopy" to estimate the amount of boron in a sample after its conversion to methyl borate.

Peroxyborates, most commonly prepared by electrolysis of borate solutions, are commercially important as bleaching and antiseptic agents. Sodium peroxyborate, $NaBO_3$, is probably the most common of these.

Figure 25.8 Some crystal lattices of group IVA elements. (a) "Diamond" lattice. Found in less stable allotropes of carbon (diamond) and tin ("gray" tin) and as a stable form of silicon and germanium. (Involves sp^3 hybridization.) (b) Graphite lattice. Strong covalent bonds within polycyclic sheets, involving sp^2 hybridization. The bonding between sheets involves a highly delocalized system of overlapping p orbitals, with electron mobility approaching that of metallic bonding. More stable allotrope of carbon. (c) Tetragonal lattice. More stable form of tin ("white" tin). Metallic. (d) Face-centered cubic lattice. Only form for crystalline lead. Simple metallic lattice.

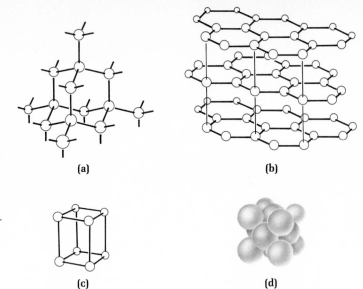

(a)　　　　　　**(b)**

(c)　　　　　　**(d)**

In group IVA we note a pronounced gradation of properties from the nonmetallic carbon and silicon to the metals tin and lead (Figure 25.8). Carbon, in the form of diamond, is a nonconductor. Silicon and germanium are typical metalloids, and tin and lead show metallic conductivity. Carbon[2] (in any of its allotropic forms) is quite inert toward simple acids and alkalis, although it will react with superheated steam at about 600°C:

$$C(s) + H_2O(g) \xrightarrow{\ (600°C)\ } CO(g) + H_2(g)$$

Silicon and germanium are also resistant to simple acids, but liberate hydrogen when treated with aqueous OH^-. Both tin and lead react slowly with acids to evolve hydrogen.

The elements in this group have four valence electrons. In order to achieve rare gas configuration, then, the elements must lose four electrons, gain four electrons, or participate in electron sharing (covalent bonds). The $4-$ ions are unknown, as might be expected from considerations of electrostatic repulsions in such ions. Compounds in which these elements are assigned a $+IV$ oxidation number are very common, but this should not be confused with the existence of a tetrapositive ion. Since $+4$ ions would have a very high charge density, the chances are slim that such ions could exist as such, except in rather unusual cases. Indeed, it appears likely that most $+4$ "ions" actually involve some covalent bonding, so that "$+4$" is more appropriately represented as an assigned oxidation number ($+IV$) rather than an actual charge. It is conceivable, perhaps, that the larger ions Sn^{4+} and Pb^{4+} do exist as such in certain crystals (e.g., SnO_2 or PbO_2), but it must be remembered that the designation of ionic bond versus covalent bond is a rather arbitrary one at best, since all bonds involve some degree of electron sharing (orbital overlap).

There are a number of cases in which simple molecules or polyatomic ions of

[2]For some interesting information concerning carbon as *graphite* or as *coal*, see "The Chemistry and Manufacturing of the Lead Pencil," by Frederick L. Enke, in the August, 1970, issue of the *Journal of Chemical Education* (pp. 575–576) and "The Chemistry of Coal and its Constituents," by Maurice E. Bailey, in the July, 1974, issue of *Journal of Chemical Education* (pp. 446–448).

group IVA elements are found to have the tetrahedral geometry generally ascribed to sp^3 orbitals (Figure 25.9). This type of bonding is most common in compounds of carbon and is occasionally encountered even in the large elements of the group.

25.9 SOME COMPOUNDS OF CARBON

Organic compounds are those that contain covalently bonded carbon. Units 28–34 are concerned with some aspects of organic chemistry.

It is of interest to note that less than 150 years ago it was universally believed that organic compounds could be synthesized only *within* a living organism. In 1828, Frederick Wöhler was attempting to prepare and isolate a simple "inorganic" salt, ammonium cyanate (NH_4^+, CNO^-). The white crystalline product obtained did not have the properties expected of an ionic substance. Careful study proved it to be *urea*, a covalent organic compound:

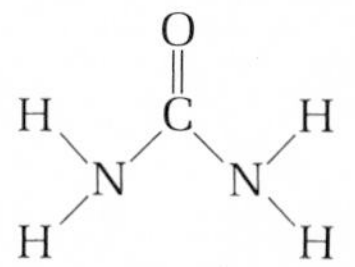

bond angles 109.5°

methane (CH_4)   silane (SiH_4)

germane (GeH_4)   stannane (SnH_4)

plumbane (PbH_4)

Figure 25.9 *sp³* Hybrids and tetrahedral molecules. Carbon forms thousands of different compounds containing C—C and C—H bonds. The simplest set of these (homologous series) has the general formula C_nH_{2n+2}. Silicon and germanium form a few analogous hydrides such as Si_2H_6 and Ge_4H_{10}, but only the simpler ones have been prepared. Tin and lead form only the hydrides SnH_4 and PbH_4, neither of which is stable except at relatively low temperatures.

Quite by accident, *an organic compound had been made synthetically.* Scientists were amazed and, though often skeptical, began hundreds of synthetic attempts. Leading theologians were aghast. Man was infringing on areas reserved for God alone. Really serious disputes often resulted in blanket condemnations of scientists as "Godless men, in league with Satan to shatter the faith of our youth by false claims. . . ."

Today, such a furor seems a bit overdramatic. Synthetic fabrics, dyed with synthetic pigments, cover plastic pews in churches cooled by systems using synthetic refrigerants. Do not feel sorry, however, that you missed the great battles. Human nature being what it is, innovative dogmatists in both religion and science will surely find new areas for lively conflict.

There are a number of simple molecules and ions containing carbon. Carbon monoxide (CO), for example, is a product of the incomplete combustion of hydrocarbon fuels. It has the ability to form *carboxyhemoglobin* by complexing with the Fe(II) of the hemoglobin in the blood. This complex is stronger than the oxygen complex, so competition favors the carbon monoxide. Concentrations as low as 0.2 percent in air may lead to unconsciousness within 30 minutes and death within three hours. The carboxyhemoglobin is only slowly converted back to normal oxygenated hemoglobin. Areas high in "smog" from such sources as automobile or airplane engines may accumulate sufficient carbon monoxide to disrupt seriously the normal oxygen distribution systems in persons continuously exposed to the polluted atmosphere.

Carbon dioxide (CO_2) is a product of normal animal respiration and a reactant in photosynthetic processes in plants. A normal balance between respiration and photosynthesis is believed to maintain about a 0.04 percent CO_2 concentration in the air. Under high pressures, CO_2 may be solidified to dry ice (sublimes at $-79°C$ at 1 atm pressure). When CO_2 dissolves in water, the resulting solution is weakly acidic:

$$CO_2(g) + H_2O(l) \rightleftharpoons CO_2(aq)$$
$$CO_2(aq) + H_2O(l) \rightleftharpoons H^+(aq) + HCO_3^-(aq) \qquad K_{a_1} = 4.2 \times 10^{-7}$$
$$HCO_3^-(aq) + H_2O(l) \rightleftharpoons H^+(aq) + CO_3^{2-}(aq) \qquad K_{a_2} = 4.8 \times 10^{-11}$$

Carbides (C_2^{2-}) are formed by treatment of elemental carbon with group IA and group IIA metals (except beryllium). These carbides, most important of which is CaC_2, hydrolyze to produce acetylene:

$$CaC_2(s) + 2H_2O(l) \longrightarrow Ca(OH)_2(s) + H—C\equiv C—H(g)$$

Cyanide (CN^-), the anion of HCN (sometimes called "prussic acid"), is an exceedingly poisonous ion. Part of the reason for its toxicity to humans is the fact that cyanide complexes *very* strongly with Fe(II) in hemoglobin. Hydrogen cyanide is liberated as certain plant products decompose in soils, and Dr. Gary Strobel of Montana State University has recently demonstrated the existence of soil fungi and bacteria capable of metabolizing cyanide. This is somewhat surprising in view of the very general toxicity of cyanide, and the biochemistry of its metabolism should prove most interesting.

25.10 SOME COMPOUNDS OF SILICON

Silicon has never been found in nature as the free element, although its compounds constitute nearly 90 percent of the earth's crust (about one-fourth of this being silicon and the rest mostly oxygen and metal ions).

Silicon–oxygen compounds vary in complexity, but none are simple. Silicon dioxide (silica) occurs as three-dimensional covalent substances that can be considered as giant molecules. Quartz is the most common example of crystalline SiO_2, and each piece of quartz may be considered as a molecule containing tetrahedral units:

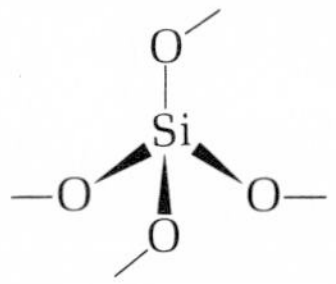

Sand, flint, and agate are also primarily three-dimensional SiO_2. In mica, on the other hand, sheets of alternating Si—O units are present. Between these covalent sheets are ions electrostatically held to negative oxygens in certain positions in the Si—O sheets. As a result, thin sections of mica crystals can be peeled away. Asbestos contains linear Si—O strands held together by ionic bonds as in the case of mica. These asbestos fibers are especially valuable because of the unusual thermal stability of the Si—O bond.

Approximate compositions of some silicate minerals are given in Table 25.7.

Silicates are used in making glass and as mineral sources of a number of metals. In the latter case, the chief problem is in getting rid of the silica.

An inexpensive "soft glass" can be made by fusing a mixture of 75 percent SiO_2 (sand), 15 percent sodium carbonate, 8 percent calcium carbonate, and 2 percent Al_2O_3 (alumina). Carbon dioxide is evolved as the mixture melts. The relative amounts of the ingredients for any particular type of glass are determined largely by trial-and-error processes to obtain the specific properties desired.

Fluorine is the only element that bonds more strongly to silicon than does oxygen.

Table 25.7 SOME COMMON SILICATES

GENERAL NAME	EXAMPLE
beryl	$Be_3Al_2(SiO_3)_6$
garnet	$Fe_3Al_2(SiO_4)_3$
kaolinite (a "clay")	$Al_2(Si_2O_3)(OH)_4$
talc	$Mg_3(Si_4O_{10})(OH)_2$

Table 25.8   SYNTHESIS OF A LINEAR SILICONE POLYMER

$$SiO_2 + C \xrightarrow[\text{Cu catalyst}]{\text{heat}} Si + CO_2$$

$$Si + \underset{\text{(methyl chloride)}}{2CH_3Cl} \xrightarrow{\text{heat}} \underset{\text{(about 40\% yield)}}{(CH_3)_2SiCl_2} + \cdots$$

$$(CH_3)_2SiCl_2 + 2H_2O \longrightarrow (CH_3)_2Si(OH)_2 + 2HCl$$

$$(CH_3)_2Si(OH)_2 \xrightarrow{H^+} H-O-\underset{\underset{CH_3}{|}}{\overset{\overset{CH_3}{|}}{Si}}-O\left(\underset{\underset{CH_3}{|}}{\overset{\overset{CH_3}{|}}{Si}}-O\right)_n \underset{\underset{CH_3}{|}}{\overset{\overset{CH_3}{|}}{Si}}-O-H$$

As a result silicates react with fluorides or HF to form either the volatile SiF_4 or the water-soluble SiF_6^{2-}, depending on the reaction conditions. These reactions are used in etching glass and in quantitative determinations for silicon in complex silicates.

Within recent years, some interesting organosilicon polymers, called *silicones,* have been prepared. Both linear and cross-linked polymers are known. Because of the high stability of the Si—O bond, these polymers show greater resistance to thermal decomposition than do most similar organic polymers. An example of a reaction sequence useful in preparing certain types of silicones is given in Table 25.8.

25.11 GERMANIUM, TIN, AND LEAD

As atomic diameter increases from germanium to lead, the valence electrons become successively easier to remove (Figure 25.10). This is consistent with the trends toward metallic character discussed earlier.

Unlike carbon and silicon, which use all four valence electrons in their stable compounds, germanium, tin, and lead form both divalent and tetravalent compounds. Such compounds as GeS and $GeCl_2$, however, readily disproportionate:

Simultaneous oxidation and reduction of a single species (Unit 22).

$$2GeCl_2 \longrightarrow GeCl_4 + Ge$$

Figure 25.10 Approximate ionization potentials (eV per electron lost).

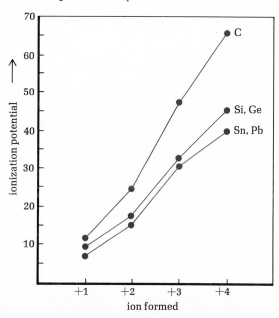

Since Pb^{4+} would have a lower charge density than Sn^{4+}, one might expect the Pb^{4+} to be more stable. In fact, Sn^{4+} is more stable than Sn^{2+}, but Pb^{4+} is *less* stable than Pb^{2+}. It is important to remember that these ions do not occur "free." Charge dissipation by hydration or partial covalent bonding increases stability, and the smaller Sn^{4+} would have the greater tendency for such charge dissipation.

Both tin and lead form complex ions; for example,

$$SnS_2(s) + S^{2-}(aq) \rightleftharpoons SnS_3^{2-}(aq)$$

$$PbCl_2(s) + 2Cl^-(aq) \rightleftharpoons PbCl_4^{2-}(aq)$$

Recently, data have been released that "pencil chewing may be hazardous to the health." The paint used on many pencils has a much higher lead content than federal standards allow for house paint.[3]

A number of lead salts are used as commercial pigments, for example, $PbCrO_4$ ("chrome yellow"). Lead ion is extremely toxic, and cases of lead poisoning of small children from eating paint chips are not uncommon. It has been suggested that the use of newly developed lead cooking vessels by the wealthy upper class in the latter days of the Roman Empire may have contributed to the "decline" of the Empire.

The use of lead and lead compounds in the automobile storage battery has been discussed in Unit 24.

25.12 ANALYSIS FOR THE GROUP IVA ELEMENTS

Inorganic carbon is generally considered as that present in the form of carbonate or bicarbonate. Quantitative analysis may involve either a determination of loss of weight (as CO_2) on acidification or heating (Excursion 2) or an acid-base titration. In any case, other ions that would interfere must be absent.

Organic compounds usually burn to evolve carbon dioxide. Thus tests for organic carbon generally involve oxidation to CO_2, followed by detection of this gas:

$$CO_2(g) + \underbrace{Ca^{2+}(aq) + 2OH^-(aq)}_{\substack{\text{as saturated} \\ Ca(OH)_2 \text{ solution}}} \rightleftharpoons CaCO_3(s) + H_2O$$

Carbon tetrachloride is a noteworthy exception. It is sometimes used in fire extinguishers because of its high vapor density, which results in displacement of the air needed for combustion. However, at elevated temperatures, CCl_4 is oxidized to the very toxic gas phosgene, $Cl_2C{=}O$.

Quantitative assay is commonly performed by "trapping" the CO_2 as an insoluble carbonate and determining the weight increase of the "trap." Within recent years, instrumental techniques have found increasing use in analysis for CO_2.

Silicon, usually in the form of some silicate, may be detected by conversion to volatile silicon tetrafluoride, followed by hydrolysis to the insoluble silicon dioxide. On a small scale the test is performed by treating the sample with hydrofluoric acid. The vapors of the SiF_4 formed are allowed to contact a tiny drop of water held above the original solution. The appearance of a cloudiness in the droplet indicates formation of hydrated silica:

Silicon dioxide (silica) is formulated as SiO_2. Unlike CO_2, this substance is a complex polymer. When formed in water, silica is hydrated, and the product is known as "silica gel."

$$(SiO_2)_x(s) + 4xHF(aq) \longrightarrow 2xH_2O(l) + xSiF_4(g)$$

$$ySiF_4(g) + (2y + z)H_2O \longrightarrow 4yHF + (SiO_2)_y \cdot zH_2O$$
$$\text{hydrated silica}$$

Formation of SiF_4 may also be used for quantitative assay of silica in various materials such as the many silicate minerals. The dried sample is weighed, treated with liquid HF in an inert platinum container, and heated to expel the SiF_4, water, and excess HF. The sample is then cooled and reweighed, the loss of weight being attributed to removal of the silica. The procedure is not highly accurate, since other species present may also react with HF.

Germanium, tin, and lead are routinely determined in industry by spectroscopic methods. Simple qualitative tests usually depend on some appropriate combination, as determined by other ions present, of the following:

[3] For an excellent survey of the problems of lead poisoning, see "Lead—A Case Study," by Richard F. Jones, in the March, 1975, issue of *Chemistry Magazine* (pp. 12–14, 21).

Germanium

Formation of insoluble $Ge(OH)_4$ and its amphoteric properties.
Formation of the insoluble white GeS_2 and its solubility in aqueous LiOH.

Tin

Formation of the insoluble brown SnS or yellow SnS_2, both of which dissolve in aqueous Na_2S or LiOH.
Action of tin(II) as a reducing agent to convert soluble $HgCl_2$ to the insoluble white Hg_2Cl_2 (usually mixed with free mercury).

Lead

Precipitation as the insoluble $PbCl_2$ (white), $PbSO_4$ (white), or $PbCrO_4$ (yellow).

SUGGESTED READING

Brotherton, R. J., and H. Steinberg. 1970. *Progress in Boron Chemistry*. Elmsford, N.Y.: Pergamon.
Laidler, Keith J., and M. H. Ford-Smith. 1970. *The Chemical Elements*. Tarrytown-on-Hudson, N.Y.: Bogden & Quigley.
Rochow, Eugene. 1968. *The Metalloids*. Lexington, Mass.: Raytheon.
Steele, David. 1966. *The Chemistry of the Metallic Elements*. Elmsford, N.Y.: Pergamon.

EXERCISES

1. Circle the symbol(s) or formula(s) that fit(s) each description given (refer to a periodic table):
   a. largest atomic radius   (Li, Na, K)
   b. smallest first ionization potential   (Li, Na, K)
   c. isoelectronic with neon (Na^+, Mg^+, Al^{2+})
   d. forms amphoteric hydroxide   (K, Ba, Al)
   e. least soluble in water $(LiF, NaCl, BaBr_2)$
   f. a typical metalloid   (Cs, Sr, Si)
   g. an acidic hydroxide $[Ba(OH)_2, B(OH)_3, KOH]$
   h. highest charge density (K^+, Al^{3+}, Si^{4+})
   i. most similar in chemical properties $[Be(OH)_2$ and $Al(OH)_3$, $Si(OH)_4$ and $Sn(OH)_4$, NaOH and $B(OH)_3]$
   j. exhibits *only* a +II oxidation state in its compounds (Ca, C, Sn)

2. Predict the normal products that are expected from the following reactions and write balanced equations for each process:
   a. $Li(s) + H_2(g) \longrightarrow$
   b. $Na(s) + H_2O(l) \longrightarrow$
   c. $Ca(s) + Cl_2(g) \longrightarrow$
   d. $Mg(s) + O_2(g) \longrightarrow$
   e. $Na(s) + O_2(g) \longrightarrow$
   f. $K(s) + O_2(g) \longrightarrow$
   g. $Sr(s) + N_2(g) \longrightarrow$
   h. $Li(s) + N_2(g) \longrightarrow$
   i. $Sr(s) + H_2O(l) \longrightarrow$
   j. $Mg(s) + F_2(g) \longrightarrow$

3. a. What weight of Epsom salts $(MgSO_4 \cdot 7H_2O)$ could be prepared from 5.0 kg of magnesium?
   b. A laboratory-scale version of the Solvay process $(NaCl + NH_3 + H_2O + CO_2 \longrightarrow NH_4Cl + NaHCO_3)$ consumed 50 ml of 15 M ammonia, 5.8 g of sodium chloride, and 10 g of "dry ice." A total of 4.7 g of sodium bicarbonate was isolated from the reaction mixture. What was the percent yield for the process, based on the limiting reagent used?
   c. Potassium superoxide is used in air rebreathing systems. It acts to absorb carbon dioxide, forming potassium carbonate and releasing oxygen gas. What weight of potassium superoxide is required for each 1.00 liter of carbon dioxide absorbed (measured at STP)?

4. The free metals of the group IA and IIA elements are normally prepared by electrolysis of their salts. If an electrolysis operation were devised to use a constant current of 15.0 amp, what weight per hour could be obtained (assuming 100 percent efficiency) of
   a. sodium;
   b. magnesium;
   c. aluminum

5. In order to determine the solubility product of thallium(I) bromide, a concentration cell was used in which thallium electrodes were immersed in 0.10 M thallium(I) nitrate and saturated thallium(I) bromide solution, respectively. The cell potential was 0.06 V. If $\mathcal{E}^0$ for $Tl^+ + e^- \longrightarrow Tl$ is -0.34 V, what is the approximate solubility product of TlBr?

6. What weight of borax could be prepared from 25 kg of colemanite, assuming a 95-percent conversion, using the sodium carbonate method?

7. What is the approximate pH of an eyewash solution prepared by adding 1.5 g of boric acid to sufficient water to form 100 ml of solution?

8. The compound 1,1,1-trichloro-2,2-bis-(*p*-chlorophenyl) ethane (DDT) is manufactured commercially by a process formulated as

$$2\,\text{C}_6\text{H}_5{-}\text{Cl} + \text{Cl}_3\text{CC}{=}\text{O} \xrightarrow[\text{catalyst}]{\text{H}_2\text{SO}_4}$$

chlorobenzene   trichloro-
acetaldehyde

DDT

If 5.0 kg of chlorobenzene were used with excess trichloro-acetaldehyde to produce 6.1 kg of DDT, what was the percent yield for the process?

9. What weight of lead sulfate is formed during discharge of a lead storage battery over a 3.0-hr period at an average current of 0.50 amp? (*Hint:* Check both electrode processes, Unit 24.)

10. What potential would be expected for a galvanic cell using Mg, Mg^{2+} (0.10 M) and Al, Al^{3+} (0.050 M) half-cells?

PRACTICE FOR PROFICIENCY

Objective (a)

1. Circle the symbol(s) or formula(s) that fit(s) each description given. (Use a periodic table.)
   a. smallest atomic radius   (Ca, K, Sr)
   b. largest first ionization potential   (Ca, Mg, Sr)
   c. softest metal   (Li, Mg, Na)
   d. forms basic hydroxide   (Ba, Be, B)
   e. least soluble in water   (K_2SO_4, $MgSO_4$, $CaSO_4$)
   f. exhibit typical "diagonal relationships"   (K and Mg, Be and Al, Li and Si)
   g. typical metalloid   (Al, Ge, Cs)
   h. reacts vigorously with cold water   (K, Be, C)
   i. forms the largest number of chemical compounds (Na, Fr, C)
   j. forms a superoxide on reaction with molecular oxygen (K, Mg, Na)

Objective (b)

2. Indicate the character (acidic, basic, or amphoteric) of each of the following by writing balanced net equations for reactions of the compounds with aqueous H^+ or OH^- (or both).
   a. $Al(OH)_3(s)$; b. $B(OH)_3(s)$; c. $Ba(OH)_2(s)$; d. $LiOH(s)$;
   e. $Be(OH)_2(s)$; f. $Si(OH)_4(s)$; g. $Mg(OH)_2(s)$; h. $Ga(OH)_3(s)$;
   i. $Ca(OH)_2(s)$

Objective (c)

3. Predict the normal products expected from the following reactions and write balanced equations for each process:
   a. $Ca(s) + Cl_2(g) \longrightarrow$ _____
   b. $Na(s) + O_2(g) \longrightarrow$ _____
   c. $Li(s) + H_2(g) \longrightarrow$ _____

d. $Cs(s) + H_2O(l) \longrightarrow$ _____
e. $Mg(s) + Br_2(l) \longrightarrow$ _____
f. $Ca(s) + O_2(g) \longrightarrow$ _____
g. $Cs(s) + O_2(g) \longrightarrow$ _____
h. $Mg(s) + N_2(g) \longrightarrow$ _____
i. $Ba(s) + O_2(g) \longrightarrow$ _____
j. $Sr(s) + H_2O(l) \longrightarrow$ _____

Objective (d)

4. a. What weight of calcium carbonate could be produced from 25.0 grams of calcium oxide and excess carbon dioxide?
   b. What volume of carbon dioxide, at 37°C and 650 torr, could be consumed by reaction with 5.0 g of potassium superoxide? (K_2CO_3 and O_2 are produced.)
   c. What volume of oxygen would be released, at 37°C and 650 torr, by the process in part (b) above?

5. What weight of colemanite ($Ca_2B_6O_{11} \cdot 5H_2O$) is required for production of 50 metric tons of borax, assuming a 90-percent conversion, using the sodium carbonate process?

6. What is the approximate pH of a 1.0-percent (by weight) boric acid solution?

7. How long must a current of 15.0 amp be maintained through a solution of borax to produce 60 g of sodium peroxyborate? (*Hint:* Boron is in the +III oxidation state in both compounds.)

8. The insecticide parachlor can be prepared by a process formulated as

If 25.0 metric tons of benzene were used with excess chlorine (in the presence of $FeCl_3$ catalyst) to form 41.2 metric tons of parachlor, what was the percentage yield for the process?

9. How long can a lead storage battery maintain an average current of 0.15 amp before 114 g of lead are consumed?

BRAIN TINGLER

10. By reference to Table 25.5 and Figure 25.6, interpret the following analytical information:
    a. An unknown salt was water soluble and its aqueous solution was unaffected by addition of HCl, H_2S, NaOH, or Na_2CO_3. When a sample of the original salt solution was aspirated into a burner flame, a faint violet color was observed. What was probably the cation of the salt?
    b. An unknown salt was water soluble and its aqueous solution was unaffected by addition of HCl, H_2S, or NaOH. When a solution of the salt was treated with Na_2CO_3, a white precipitate was formed. When a portion of this precipitate was dissolved in dilute HCl and aspirated into a flame, a fleeting green color was observed. What was probably the cation of the salt?

Unit 25 Self-Test

1. Circle the symbol or formula that fits each description given:
   a. smallest atomic radius (Na, Mg, Ba)
   b. largest first ionization potential (Li, Na, K)
   c. softest metal (Li, Na, K)
   d. forms acidic hydroxide (Ba, Be, B)
   e. least soluble in water (LiBr, KCl, MgF_2)
2. Complete and balance:

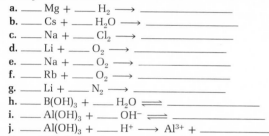

   a. ____ Mg + ____ H_2 $\longrightarrow$ _____
   b. ____ Cs + ____ H_2O $\longrightarrow$ _____
   c. ____ Na + ____ Cl_2 $\longrightarrow$ _____
   d. ____ Li + ____ O_2 $\longrightarrow$ _____
   e. ____ Na + ____ O_2 $\longrightarrow$ _____
   f. ____ Rb + ____ O_2 $\longrightarrow$ _____
   g. ____ Li + ____ N_2 $\longrightarrow$ _____
   h. ____ $B(OH)_3$ + ____ H_2O $\rightleftharpoons$ _____
   i. ____ $Al(OH)_3$ + ____ OH^- $\rightleftharpoons$ _____
   j. ____ $Al(OH)_3$ + ____ H^+ $\longrightarrow$ Al^{3+} + _____
3. Freon-12 is the most common substance now used in refrigeration systems such as freezers and air-conditioning systems. This compound, CCl_2F_2, is prepared commercially by combination of carbon tetrachloride and antimony(III) fluoride. The latter is converted to antimony(III) chloride. If reaction of 25 kg of antimony(III) fluoride with a stoichiometric amount of carbon tetrachloride produced 18 kg of Freon-12, what was the percent yield for the process?
4. What weight of borax could be prepared from 75 metric tons of colemanite ($Ca_2B_6O_{11} \cdot 5H_2O$), assuming an 85 percent efficiency, using the sodium carbonate process?
5. What is the approximate pH of a 3.0-percent (by weight) boric acid solution? (K_{a_1} for boric acid is 6.0×10^{-10}.)
6. What average current (in amperes) is required for production of 6.0 g of calcium per hour by electrolysis of a molten calcium salt?
7. What potential (in volts) would be expected for a galvanic cell using a magnesium, 0.20 M magnesium ion half-cell and a thallium, 0.010 M thallium(I) ion half-cell?

$$Mg^{2+} + 2e^- \longrightarrow Mg \qquad \mathcal{E}^0 = -2.37 \text{ V}$$
$$Tl^+ + e^- \longrightarrow Tl \qquad \mathcal{E}^0 = -0.34 \text{ V}$$

Unit 26

Chemistry is a terrible science, costs a lot of money, and stinks horribly.

KARL KAPPELER
(d. 1888; Director of what is now the Technical University of Zurich)

Representative nonmetals

Objectives

(a) Learn (or review), for recall and use, the names and formulas of
  (1) the simpler compounds of the nonmetals with hydrogen
  (2) the more common nonmetal oxides and halides
  (3) the more common oxyacids and oxyanions
(b) Be able to write electron-dot formulas for compounds of the nonmetals and to use these in predicting the geometries of molecules and polyatomic ions (review Units 7 and 8).
(c) Be able to relate structural characteristics and properties of allotropic modifications of some common nonmetals.
(d) Be able to write balanced equations for
  (1) reactions useful in preparing the more common nonmetals
  (2) typical reactions of the more common nonmetals and their compounds

REVIEW

(e) Be able to apply the principles of stoichiometry, equilibrium, and electrochemistry to the reactions of nonmetals and their compounds.

Several ways are known of making sulfuric acid commercially. The most interesting of these is applicable to large-scale recovery of a dangerous component of smog from smelters using sulfide ores or industrial processes powered from sulfur-containing fuels. When sulfide ores are roasted in air (a typical part of the processes for isolating metals from the ores), sulfur dioxide is formed. A large number of smelting operations now dump this into the atmosphere, where it may react with water or with other compounds. Smog containing high concentrations of SO_2 can be very dangerous, and a number of deaths have been attributed to prolonged inhalation of exceptionally concentrated "SO_2 smog." In smaller concentrations it is less dangerous, but its odor is irritating and it must be considered a public nuisance when present in obvious amounts. The SO_2 can instead be used for preparation of sulfuric acid, a procedure that not only reduces smog problems but provides a valuable by-product for smelter operations. In fairness to industry, it should be noted that the initial investment is appreciable, and markets must be found for excess H_2SO_4 produced before such processes become feasible.

One of the oldest procedures involves catalytic combination of SO_2 gas with oxygen from the air (at temperatures above $400°C$, using V_2O_5 or platinum catalysts) to form SO_3 gas. The SO_3 gas will react with water to form H_2SO_4, but the process is quite exothermic, and it is difficult to control product concentration. Instead, the contact process bubbles the SO_3 through concentrated sulfuric acid to produce the mixture of polymeric acids called "fuming sulfuric acid" or "oleum." This can then be mixed (carefully) with controlled amounts of water to produce sulfuric acid of any desired concentration. Unless contaminants have been removed from the SO_2 gas used, the sulfuric acid may be relatively impure. The crude acid still has many commercial uses, but the pure acid can usually be obtained from the crude mixtures only by fairly expensive processes. The overall equations for conversion of SO_2 to H_2SO_4 may be written as

$$2SO_2(g) + O_2(g) \xrightarrow[\text{(catalyst)}]{\text{heat}} 2SO_3(g)$$

$$SO_3(g) + H_2SO_4(l) \longrightarrow \underset{\text{polymer mixture (l)}}{H_2S_2O_7, H_2S_3O_{10}, \text{etc.}}$$

$$\underset{\text{polymer mixture (l)}}{H_2S_2O_7, \text{etc.}} + nH_2O(l) \longrightarrow \underset{\text{"concentrated"}}{mH_2SO_4(l)}$$

In this unit we shall survey some important aspects of the chemistry of the typical nonmetals and of the rather "unique" elements hydrogen and the "noble" gases.

26.1 GROUP VA ELEMENTS

We have seen that all group IA elements have the same simple crystal structure and typically metallic properties. In group IIA we noted some relatively minor differences in crystal lattice patterns and some rather "unusual" properties for the lightest group IIA element, beryllium. Still greater variation was observed in group IIIA, including a distinct difference between boron and the other group IIIA elements. In group IVA there were more differences than similarities. The group IVA elements display a number of allotropic modifications, so properties vary from those of an effective nonconductor such as diamond to the characteristic metallic substances, even for a single element (e.g., "gray" tin versus "white" tin).

The behavior of group IA–IVA elements can be explained, at least on a qualitative basis, by two main factors: atomic size and valence electron configuration. The former helps to explain why metallic properties are more pronounced for larger atoms (i.e., valence electrons are farther from the positive nucleus and are shielded from the nuclear charge by more "kernel" electrons). Valence electron configuration can be related to the types of bonding found in compounds of the elements and, to some extent, to the types of crystal lattices formed. For example, the nondirectional character of the s orbital electrons of group IA and IIA metals allows their atoms to pack into crystal lattices in simple patterns associated with easy ways of packing spheres. In elements such as carbon or silicon, sp^3 hybrids for the four valence electrons allow tetrahedral bonding in simple compounds and in crystal lattices such as diamond.

The group VA elements also show large variations in properties, many of which depend on differences among allotropic forms. Nitrogen is a nonmetal. As the diatomic gas N_2 it constitutes about 78 percent of dry air by volume (75 percent by weight), and its compounds are found in the oceans and the earth's crust. It is relatively inert chemically, although it will "burn" at high temperatures.

Phosphorus exists as P_4 tetrahedra (Figure 26.1) in the gas and liquid states. Phosphorus exhibits a very large number of allotropic modifications (Table 26.1), which vary from relatively stable "red" phosphorus to "white" phosphorus, which ignites spontaneously when exposed to air.

Arsenic and antimony are similar to phosphorus in many respects. Both have amorphous nonmetallic, and crystalline metallic forms as solids. Both form tetrahedral molecules, As_4 and Sb_4, in the vapor and liquid states.

Bismuth is another example of the trend toward metallic character as the sizes of the atoms increase. Only one solid form is known, a crystalline metallic lattice of unusually open structure. The metal is light gray and is very brittle.

Unlike nitrogen, the other group VA elements are fairly rare. Phosphorus constitutes about 0.1 percent of the earth's crust, mainly as various orthophosphate minerals [e.g., $Ca_3(PO_4)_2$ in bone]. Arsenic, antimony, and bismuth together account for less than 0.001 percent of the earth's crust. They are not found in the free (elemental) state. Their principal ores contain the sulfides or oxides.

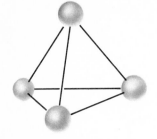

Figure 26.1 Tetrahedral P_4 unit.

26.2 PREPARATION AND CHEMICAL PROPERTIES OF THE GROUP VA ELEMENTS

Nitrogen is prepared commercially by the distillation of liquid air. The air is chilled at high pressure to around $-200°C$ (below the critical temperature) and the resulting liquid is slowly warmed to around $-196°C$, at which temperature the nitrogen begins to distill, leaving the higher-boiling oxygen (boiling point $-183°C$) as a liquid. Nitrogen prepared in this way contains small amounts of oxygen and argon. If

Table 26.1 SOME ALLOTROPIC FORMS OF PHOSPHORUS

NAME	STRUCTURE	SOME CHARACTERISTICS
white (β white)	P_4 clusters in hexagonal closest-packed lattice	stable only below $-80°C$
yellow (α white)	P_4 clusters in cubic closest-packed lattice	stored under water; ignites when exposed to air; very toxic; soluble in CS_2
red	amorphous polymers of P_4 units	does not ignite below 240°C; only sparingly soluble in CS_2; not very poisonous
scarlet	another amorphous polymeric form	similar to red
violet	amorphous polymers	most stable form; prepared pure by precipitation from solution in molten lead
black	corrugated layers similar to graphite lattice	semimetallic; electrical properties similar to graphite

oxygen-free nitrogen is required, the impure gas can be passed through a reducing agent to remove the residual O_2. When ultrapure N_2 is needed, it can be produced on a small scale by heating a solution of freshly prepared ammonium nitrite. Since this salt is quite unstable, it is made just before it is needed by mixing solutions of sodium nitrite and ammonium chloride. The decomposition reaction is formulated as

$$NH_4^+(aq) + NO_2^-(aq) \longrightarrow 2H_2O(l) + N_2(g)$$

Nitrogen forms a wide variety of chemical compounds, exhibiting every oxidation state from $-III$ to $+VI$. Although molecular nitrogen is relatively inert, it does react with certain metals to form ionic nitrides (Unit 25), and with some metalloids and nonmetals to form covalent nitrides, for example,

$$3Si(s) + 2N_2(g) \longrightarrow Si_3N_4(s)$$
$$2B(s) + N_2(g) \longrightarrow BN(s)$$

(Boron nitride (BN) is isoelectronic with carbon and forms two allotropes, one like graphite and the other like diamond.)

$$3P_4(g) + 10N_2(g) \longrightarrow 4P_3N_5(s)$$

Nitrogen also reacts with oxygen to form a number of different oxides (Table 26.2). The combustion of nitrogen during the burning of fuels in air is a major source of nitrogen oxide pollutants, for example,

$$N_2(g) + O_2(g) \longrightarrow 2NO(g)$$

Phosphorus was first prepared in 1669 by the alchemist Hennig Brand by heating a mixture of sand and dried urine in an attempt to obtain a substance that would

Table 26.2 SOME OXYGEN COMPOUNDS OF GROUP VA ELEMENTS

NITROGEN

known oxides: N_2O, NO, N_2O_3, NO_2, N_2O_4, N_2O_5, NO_3, N_2O_6
known oxyacids: $H_2N_2O_2$, $H_2N_2O_3$, H_2NO_2, HNO_2, HNO_3,[a] HNO_4 (peroxynitric acid)

PHOSPHORUS

known oxides: P_4O_6, P_4O_{10},[b] $(PO_2)_n$ (polymeric crystal)
known oxyacids: phosphorous type: H_2PO_2, HPO_2, $H_4P_2O_5$, H_3PO_3
        phosphoric type: $H_4P_2O_6$, HPO_3, $H_5P_3O_{10}$, $H_4P_2O_7$, H_3PO_4
        peroxyphosphoric type: H_3PO_5, $H_4P_2O_8$

ARSENIC

known oxides: As_4O_6, As_2O_5
known oxyacids: $HAsO_2$, H_3AsO_3, $HAsO_3$, H_3AsO_4, $H_4As_2O_7$

ANTIMONY

known oxides: Sb_4O_6, Sb_2O_4, Sb_2O_5
known oxyacids: $HSbO_2$, H_3SbO_3, H_7SbO_6
oxycation: SbO^+

BISMUTH

known oxides: BiO, Bi_4O_6, Bi_2O_4, Bi_2O_5
hydroxide: $Bi(OH)_3$ (basic)
oxycation: BiO^+

[a] HNO_3 is the only one of these that can be isolated in anhydrous form. The others decompose readily.
[b] P_4O_{10} is often called phosphorous pentoxide because its *empirical* formula is P_2O_5.

turn "baser" metals into gold. He found instead a residue that glowed in the dark and burned when exposed to air. We now know this substance as *white* phosphorus.

White phosphorus is commercially prepared by the high-temperature decomposition of a mixture of sand, coke, and phosphate rock. The gaseous P_2 generated is cooled and the white phosphorus is collected and stored under water, to prevent its oxidation by air:

$$3SiO_2(s) + 5C(s) + Ca_3(PO_4)_2(s) \longrightarrow 3CaSiO_3(s) + 5CO(g) + P_2(g)$$

$$2P_2(g) \xrightarrow{\text{cooling}} P_4(g) \xrightarrow{\text{bubbling through water}} P_4(s)$$

The more stable *red* phosphorus, which can be stored in contact with air, is made by heating white phosphorus to $\sim 400°C$ in an oxygen-free atmosphere.

Phosphorus normally forms compounds in which $+III$ or $+V$ oxidation states are exhibited. The common oxides (Table 26.2) react with water to form oxyacids. A number of molecular and ionic phosphorus halides are known, the geometries of which are consistent with the predictions of *Gillespie's rules* (Unit 8).

Example 1

When excess phosphorus is heated with chlorine gas, the major product is phosphorus trichloride. If chlorine is used in excess, phosphorus pentachloride is formed. Although the latter exists as a simple trigonal bipyramidal molecule in the gas phase, the crystalline compound has been shown to exist as an ionic mixture of PCl_4^+ and PCl_6^-. Draw representations for the geometries of PCl_3, PCl_5, PCl_4^+, and PCl_6^-.

Solution (based on *Gillespie's rules,* Unit 8)

Phosphorus (a VA element) has five valence electrons and chlorine (VIIA) has seven. The PCl_4^+ cation has one electron less than the neutral component atoms, and the PCl_6^- anion has one "extra" electron. Then

SPECIES	DOT FORMULA	NO. GROUPS AROUND "CENTRAL" ATOM	PREDICTED GEOMETRY
PCl_3	:Cl : P : Cl : : Cl :	4 (3 atoms, 1 lone pair)	 (tetrahedral *orbitals* pyramidal *atomic* pattern)
PCl_5	:Cl : P : Cl : : Cl :	5 (all atoms)	 (trigonal bipyramidal)

SPECIES	DOT FORMULA	PREDICTED GEOMETRY
PCl_4^+	$\overset{..}{:}\overset{..}{Cl}\overset{..}{:}\,^+$ $:\overset{..}{Cl}:P:\overset{..}{Cl}:$ $:\overset{..}{Cl}:$	(tetrahedral)
PCl_6^-	$\overset{..}{:}Cl\overset{..}{:}\,^-$ $Cl\;\overset{..}{P}\;Cl$ $Cl\;\;\;Cl$ $\overset{..}{:}Cl\overset{..}{:}$	(octahedral)

Arsenic, antimony, and bismuth are produced commercially by roasting their oxides with coke, for example,

$$As_4O_6(s) + 6C(s) \longrightarrow As_4(g) + 6CO(g)$$

Interestingly enough, environmentalists have been partly responsible for the exploitation of a valuable source of the oxides. Certain types of copper and lead smelters produce significant amounts of these oxides, particularly As_4O_6, as by-products. Many of these smelters formerly allowed the oxide dusts to escape into the atmosphere through waste stacks. Air quality control standards have now induced most smelters to remove the oxide dusts from effluent gases. The recovered oxides have proved to be valuable raw materials for the production of arsenic, antimony, and bismuth.

26.3 HYDRIDES OF THE GROUP VA ELEMENTS

Ammonia is a weak base ($K_b = 1.8 \times 10^{-5}$), although appreciably stronger than water. To some extent, NH_3 may be considered amphoteric. In liquid ammonia, for example,

$$NH_3 + NH_3 \rightleftharpoons NH_4^+ + NH_2^- \qquad K = \sim 10^{-30} \text{ at } -35°C$$

Appreciable concentrations of amide ion (NH_2^-) may be obtained by iron-catalyzed reaction of sodium with liquid ammonia, but the amide ion is such a strong base that no measurable concentration exists in aqueous solution, that is, the reaction of amide ion with water ($NH_2^- + H_2O \longrightarrow NH_3 + OH^-$) is for all practical purposes irreversible.

The ability of ammonia to accept a proton is ascribed to an unshared electron pair in an sp^3-type orbital on nitrogen (Figure 26.2). The ground-state electron configuration for nitrogen suggests that such hybrids are feasible:

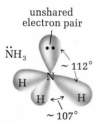

unshared electron pair

$\ddot{N}H_3$

~112°

H

H H

~107°

Figure 26.2 Ammonia molecule. Modern theory explains N—H bond angles of 107° by

H

suggesting that repulsion between orbitals increases in the order bond:bond < lone pair:bond < lone pair:lone pair (Unit 8).

hybridization

(sp^3)

(ground state) N; $1s\uparrow\downarrow, 2s\uparrow\downarrow, 2p_x\uparrow, 2p_y\uparrow, 2p_z\uparrow$

Factors that increase electron density in the lone-pair orbital would tend to increase the base strength of the molecule, whereas factors that decrease electron density should have the reverse effect. Thus replacement of a hydrogen on ammonia by an *alkyl group* ($-C_nH_{2n+1}$) generally increases basicity:

$$(K_b > K_{b(NH_3)})$$

methylamine, $K_b = 5.0 \times 10^{-4}$

Systems that reduce electron density on nitrogen by influence of a more electronegative atom or by involvement of a π-bond system tending to delocalize the electron pair will decrease basicity:

$$(K_b < K_{b(NH_3)})$$

hydroxylamine, $K_b = 6.6 \times 10^{-9}$

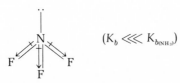

delocalization

$$(K_b \ll K_{b(NH_3)})$$

acetamide, $K_b = \sim 10^{-15}$

$$(K_b \lll K_{b(NH_3)})$$

nitrogen trifluoride,
neutral—i.e., K_b
too small to measure

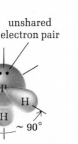

unshared
electron pair

PH$_3$

H   H
H
$\sim 90°$

Figure 26.3 Phosphine molecule.

Phosphine is a very poisonous gas, as are the hydrides AsH$_3$, SbH$_3$, and BiH$_3$.

Phosphorus forms a simple hydride, phosphine, which is analogous to ammonia but much less basic. Evidence indicates that bond angles in PH$_3$ (Figure 26.3) are close to 90 deg, suggesting that P—H bonds involve the $3p$ orbitals on phosphorus, leaving the unshared pair in a nondirectional $3s$ orbital. The low base strength of PH$_3$ is consistent with such a description, since electron density in a given direction convenient for bonding with an approaching proton is less for an electron pair in a spherical $3s$ orbital than for a pair in a directional sp^3 hybrid orbital.

Arsine, AsH$_3$, is very similar to phosphine. The corresponding hydrides of antimony and bismuth (stibine, SbH$_3$, and bismuthine, BiH$_3$) are unstable, but it is known that stibine is weakly acidic, not basic.

26.4 OXIDES AND HYDROXIDES OF GROUP VA

All of the group VA elements form strong bonds with oxygen, and a fantastic variety of oxides and hydroxides are known (Table 26.2). A detailed treatment of these is unwarranted here, but a few are of such interest as to justify further discussion.

Nitrous oxide (N_2O) is an odorless, tasteless, nontoxic gas which is quite soluble in oils and fairly soluble in water. It was once useful only as "laughing gas," a mild anesthetic often employed in dentistry. Today it is extensively employed as the compressed gas in a variety of aerosols and foams, such as "instant whipped cream." It can be made by a *very carefully controlled* decomposition of ammonium nitrate at about 200°C:

So called because of the mild hysteria frequently resulting from inhalation of small amounts of the gas.

$$NH_4NO_3(s) \xrightarrow{200°C} N_2O(g) + 2H_2O(g)$$

Precautions must be rigorously observed to avoid the *explosive* decomposition of ammonium nitrate, which may occur at higher temperatures:

$$2NH_4NO_3(s) \xrightarrow{\text{explosion}} 2N_2(g) + O_2(g) + 4H_2O(g)$$

Nitric oxide (NO), unlike N_2O, is a poisonous gas, although fairly high concentrations or prolonged inhalation are required to produce serious toxic effects. The nitric oxide is quite similar to carbon monoxide and can, apparently, form strong complexes with Fe(II) in hemoglobin. Nitric oxide is formed when certain metals (e.g., copper or silver) dissolve in nitric acid. It is also formed (in low yields) when an electric discharge is passed through a mixture of nitrogen and oxygen. This reaction, occurring during lightning flashes in the atmosphere, is an important step in the "nitrogen fixation" processes of plants. Nitric oxide reacts rapidly with oxygen to form *nitrogen dioxide:*

Nitric oxide is formed from N_2 and O_2 in the air during operation of internal combustion engines, as has been mentioned earlier. Nitric oxide ($\cdot \ddot{N} : \ddot{O}$) is a highly reactive *free radical* species (Unit 16).

$$2NO(g) + O_2(g) \rightleftharpoons 2NO_2(g)$$
colorless   red-brown

Reaction of nitrogen dioxide with water produces *nitric acid:*

$$3NO_2(g) + H_2O(l) \longrightarrow 2HNO_3(l) + NO(g)$$

In thunderstorms the conditions for conversion of atmospheric nitrogen and oxygen to nitric acid are ideal. This is one of the ways in which nitrogen is supplied for life processes in plants.

Other ways include direct conversion of N_2 to nitrogen compounds by certain bacteria, formation of ammonia by decomposition of proteins and other nitrogen compounds, and direct addition of nitrogen compounds to soils by decay of organic matter containing nitrogen compounds or, artificially, by addition of soil additives such as NH_4NO_3.[1]

Nitric acid is a strong acid (essentially 100 percent ionized in dilute solution; $K_a \approx 20$ in 3 M solution) and a powerful oxidizing agent. The nitrate ion, NO_3^-, is a planar ion with highly delocalized charge (Unit 8). Nitrates are among the most water-soluble salts, so nitrate ion is detected by a color-producing reaction rather than by precipitation. Quantitative estimation most commonly involves reduction to ammonia (e.g., by aluminum and aqueous sodium hydroxide) and distillation of the ammonia for titration (Excursion 2).

Nitrates are serious water pollutants, often in the form of nitrate fertilizers. In lakes, high nitrate concentrations result in rapid growth of algae and certain amoeba. Pollution of farm water supplies may also result from excessive use of nitrate fertilizers, and this poses a health hazard—especially to small children.

The 1975 U.S. Public Health Service limits on NO_3^- in drinking water were 45 ppm for adults and 10 ppm for infants.

Unlike nitric acid, nitrous acid (HNO_2), is a weak acid ($K_a = 4.5 \times 10^{-4}$) and unstable except in cold dilute solutions. It is of interest to organic chemists because

[1] For an excellent survey, see "Nitrogen Fixation," by David R. Safrany, in the October, 1974, issue of *Scientific American* (pp. 64–80).

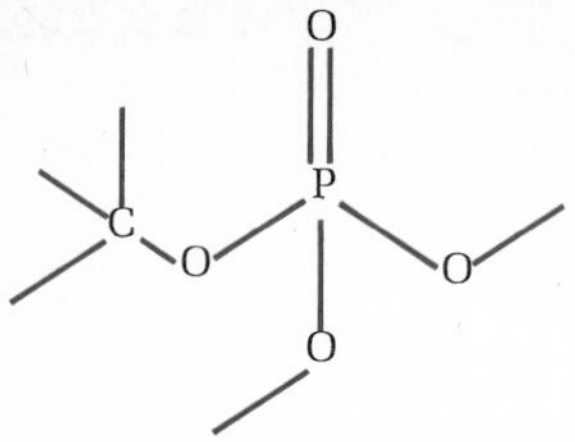

Figure 26.4   Phosphate ester.

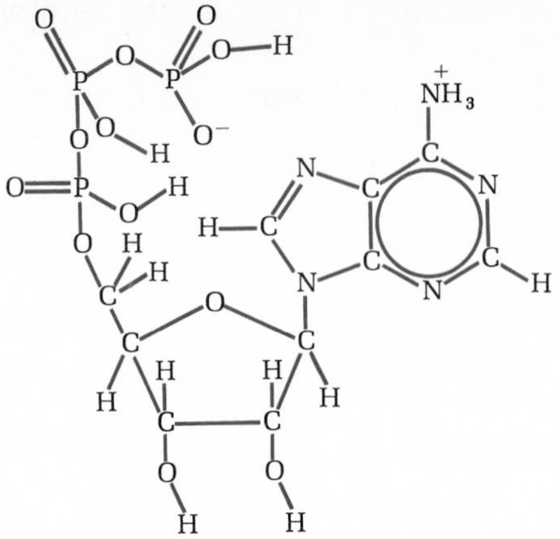

Figure 26.5   Adenosine triphosphate (ATP). The free energy change for the hydrolysis of a single P—O—P bond of ATP, under physiological conditions, is estimated as -10 to $-12\,\text{kcal mole}^{-1}$. Thus the P—O—P bonds of ATP represent significant "energy storehouses." (For comparison, ΔG^0 for the combustion of hydrogen is $-57\,\text{kcal mole}^{-1}$.)

of its characteristic reactions with *amines* (compounds similar to NH_3, but having one or more C—N bonds).

The most common ion containing phosphorus is $PO_4{}^{3-}$, phosphate (orthophosphate). Calcium phosphate is the principal inorganic component of bones and teeth. A number of organic compounds essential to living organisms contain phosphate esters (Figure 26.4). Among these are adenosine triphosphate, ATP (Figure 26.5), an important compound involved in biological energy conversion processes, and nucleic acids (DNA and RNA) responsible for control of protein biosynthesis and transmission of genetic information. These are discussed in Unit 34.

Phosphate is one of the ions obtained from *orthophosphoric acid* (often called simply phosphoric acid). This acid, containing three O—H bonds, is "triprotic":

$$H_3PO_4 \rightleftharpoons H^+ + H_2PO_4{}^- \qquad K_{a_1} = 7.5 \times 10^{-3}$$
$$H_2PO_4{}^- \rightleftharpoons H^+ + HPO_4{}^{2-} \qquad K_{a_2} = 6.2 \times 10^{-8}$$
$$HPO_4{}^{2-} \rightleftharpoons H^+ + PO_4{}^{3-} \qquad K_{a_3} = 2.0 \times 10^{-13}$$

Some of the other acids of phosphorus contain P—H bonds, which are not easily broken. Orthophosphorus acid (phosphorous acid) is diprotic,

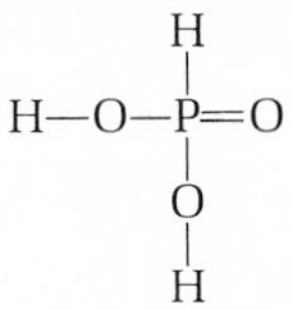

$$K_{a_1} = 1.6 \times 10^{-2}$$
$$K_{a_2} = 6.9 \times 10^{-7}$$

and hypophosphorous acid is monoprotic,

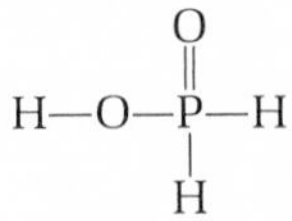

$$K_a = 8 \times 10^{-2}$$

A number of compounds containing P—O—P bonds are known. The simplest of these is *pyrophosphoric acid*:

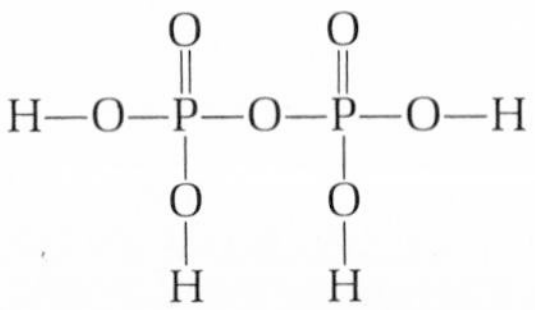

Esters are products of the reaction of an alcohol with an acid (Units 25 and 29).

Although these bonds (P—O—P) are only slowly hydrolyzed, a considerable amount of energy is released. This is the type of bond referred to as the "high-energy bond" in molecules such as ATP (Figure 26.5)

26.5 GROUP VIA ELEMENTS

O$_3$ (ozone) is an allotropic modification of oxygen produced by electric discharge or by light of wavelength <1860 Å. A rare O$_4$ molecule is also found as trace amounts in natural oxygen.

Oxygen is the most abundant element in the earth's crust (~49 percent by weight, ~54 percent of atoms), constitutes 21 percent by volume of the atmosphere, and is nearly 90 percent of all water on the earth. It occurs principally as oxygen gas (O$_2$), in liquid water, and in solid silicates, carbonates, phosphates, and oxides. Evidence indicates that the oxygen molecule contains two unpaired electrons,

$$\left(\cdot \ddot{O} \! : \! \ddot{O} \cdot \right)$$

a factor believed responsible for the brilliant blue color of liquid oxygen, its rather unusual magnetic properties, and the high reactivity of the oxygen molecule.

Sulfur[2] is one of the few elements that occur free in nature despite its chemical reactivity. If one includes sulfur in all natural forms (mainly sulfides and sulfates), then this element represents about 0.05 percent of the earth's crust. Three allotropic forms of the element are known (Table 26.3).

Conductivity resulting from the action of light to free valence electrons for current flow is referred to as *photoconductivity*.

Selenium and tellurium are relatively rare in nature. Both occur in several allotropic forms (Table 26.3), some of which exhibit photoconductivity. Tellurium has the distinction of being the only element found naturally in chemical combination with gold.

Polonium is a rare, radioactive element (Unit 35) that has no stable isotopes. Its natural abundance is fairly constant, because it is formed by radioactive decay of other elements about as rapidly as it decomposes:

$$^{210}_{84}\text{Po} \longrightarrow \, ^{206}_{82}\text{Pb} \, + \, ^{4}_{2}\text{He}$$
$$\quad\quad\quad\quad\text{stable}\quad\quad\text{alpha}$$
$$\quad\quad\quad\quad\quad\quad\quad\text{particle}$$

Table 26.3 SOME ALLOTROPIC FORMS OF SULFUR, SELENIUM, AND TELLURIUM

SULFUR

Rhombic: crystalline aggregates of S$_8$ rings; stable at room temperature
Monoclinic: probably a crystalline modification of the rhombic form; stable above 96°C
Rhombohedral: probably S$_6$ rings; rather unstable
Amorphous: helical chains with eight sulfur atoms per turn of the helix; slowly converts
    spontaneously to rhombic form

SELENIUM

Metallic: hexagonal crystal lattice; typical metallic conductor *except* that conductivity increases
    on exposure to light (photoconductivity)
α-Monoclinic: Se$_8$ rings as in monoclinic sulfur
Amorphous: several modifications known

TELLURIUM

Metallic: similar to Se; most stable form; weaker photoconductor than Se
Others: various amorphous forms and at least one other crystalline modification are known, but
    all are relatively unstable

---

[2] For some interesting details, see "Sulfur," by Chistopher Pratt, in the May 1970, issue of *Scientific American* (pp. 63–72).

26.6 PREPARATION AND CHEMICAL PROPERTIES OF THE GROUP VIA ELEMENTS

Approximately 97 percent of the oxygen commercially produced is obtained by the distillation of liquid air after removal of the lower-boiling nitrogen. The only other significant commercial source is the electrolysis of water, and this method is economically feasible only because hydrogen is the other product. On a small scale, oxygen can be prepared by the cautious heating of certain salts containing oxyanions. Potassium chlorate is the most common example:

$$2KClO_3(s) \longrightarrow 2KCl(s) + 3O_2(g)$$

Molecular oxygen is highly reactive, combining readily with all but the most inert elements to form oxides (O^{2-}), peroxides (O_2^{2-}), or superoxides (O_2^-), as described in Unit 25 for some reactive metals.

Sulfur was once produced primarily from underground deposits of the free element. In the *Frasch process,* hot water is pumped at high pressures (170°C, 7 atm) into the sulfur. The molten sulfur (melting point, 113°C) is then forced to the surface by compressed air and cooled. Most sulfur mines obtain the element directly with less than 0.5 percent impurity by the *Frasch* method. Within recent years, attempts to reduce sulfur dioxide pollution have led to the development of sulfur recovery processes and these, predominately from natural gas and oil supplies, now constitute more than half of the world's source of elemental sulfur. Unfortunately, sulfur recovery processes are not always profitable because of problems in the transportation of recovered sulfur to appropriate markets. It is anticipated that sulfur production may experience significant economic problems for several years.[3]

Although sulfur is relatively inert at low temperatures, at higher temperatures it will react with any element except tellurium, gold, platinum, palladium, nitrogen, iodine, or the "noble" gases.

26.7 SIMPLE HYDRIDES OF GROUP VIA

The use of p orbitals is undoubtedly a serious oversimplification. The formation of hybrids involving d orbitals (Unit 27) is more nearly correct. We shall, however, reserve such descriptions for more advanced study.

The hydrides of group VIA elements show the same sudden changes in properties exhibited by those of group VA (Figure 26.6). Again, these may be attributed to a change from sp³ hybridization for the lightest element to the use of simple p orbitals by the others (Figure 26.7). The unshared electron pairs on oxygen are, then, readily available for formation of hydrogen bonds or for accepting protons from acids.

Water owes its unique properties to its electronic geometry. Because of its bent shape and the large difference in electronegativity between oxygen and hydrogen, water is a very polar molecule, capable of furnishing hydrogen (usually as hydrogen bond) to any region of high electron density. Thus water solvates anions and neutral atoms having one or more unshared electron pairs (Figure 26.8). In addition, the

Figure 26.6 Boiling-point comparisons. Noble gases indicate the expected trend when principal interparticle forces are simple London dispersions.

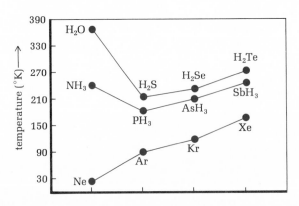

[3]See *Chemical and Engineering News,* June 9, 1975.

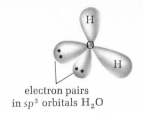

electron pairs
in sp^3 orbitals H_2O

electron pairs in
H_2S, H_2Se, H_2Te

Figure 26.7 Group VI hydrides.

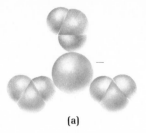

(a)

(b)

Figure 26.8 Water as an electrophilic agent (attracted to negative region). (a) Hydration of a negative ion. (b) Hydrogen bond from water to ammonia molecule ($H—O—H$) $\cdots$:NH_3).

readily available electron pairs on the oxygen in water can coordinate readily with any Lewis acid, or can form hydrogen bonds with hydrogen attached to an electronegative element (usually nitrogen, oxygen, or fluorine). As a result, water solvates cations readily and forms strong hydrogen bonds with such species as NH_3, HF, or other water molecules (Figure 26.9).

In H_2S, H_2Se, and H_2Te, the relative electronegativities (Unit 7) are such that the S—H, Se—H, and Te—H bonds are successively less polar than the O—H bond.

Since oxygen is the most electronegative element in this group, one might expect the O—H bond to have the highest ionic character and water, therefore, to be the most acidic of the hydrides. It is always dangerous to consider only a single factor in making such predictions, and it is always easier (although we hate to admit it) to determine an answer experimentally and then "explain it" than to predict it in advance. In this case, experimental data show that water actually is the weakest acid (Table 26.4).

The explanation suggested for this observed trend is twofold. Hydrogen bonding in water tends to trap protons in a stable electrostatic field between two water molecules. In addition, the high charge density on OH^- would favor accepting a proton to reform H_2O. These factors probably account for the exceptionally small K_{a_1} for H_2O. In the cases of the other three hydrides, hydrogen bonding is unlikely, so the principal factor is probably the difference in charge density (caused by difference in particle size) from SH^- to SeH^- to TeH^- ion. The larger the particle, the smaller the charge density and, correspondingly, the weaker the electrostatic attraction for H^+.

The hydrides of sulfur, selenium, and tellurium are notably toxic. For example, H_2S is more poisonous than hydrogen cyanide. Its odor generally provides ample warning before serious dosages are inhaled, but there are cases on record of persons recovered from near-fatal doses who reported noticing that the odor "disappeared" shortly before they became unconscious.

Table 26.4 APPROXIMATE VALUES OF K_{a_1} FOR
$$H_2X \rightleftharpoons H^+ + HX^-$$

H_2X	$K_{a_1}{}^a$
H_2O	$\sim 10^{-14}$ (K_w)
H_2S	$\sim 10^{-7}$
H_2Se	$\sim 10^{-4}$
H_2Te	$\sim 10^{-3}$

$${}^a K_{a_1} = \frac{C_{H^+} \times C_{HX^-}}{C_{H_2X}}$$

Note that K_w for water is $C_{H^+} \times C_{OH^-}$.

26.8 SULFURIC ACID

Sulfur forms a number of oxyacids (Table 26.5), the most important of which is H_2SO_4. Sulfuric acid is one of the most profitable industrial chemicals. In a crude form it is used as a drain-cleaning agent and in manufacturing a variety of detergents.

Figure 26.9 Water as a nucleophilic agent (attracted to positive region). (a) Hydration of a positive ion. (b) Hydrogen bond from HF to H_2O. (c) Hydrogen bond between two water molecules.

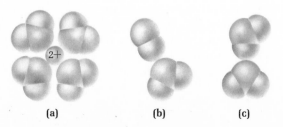

(a)   (b)   (c)

Tabe 26.5   SOME OXYACIDS OF SULFUR

FORMULA	NAME	CHARACTERISTICS
H_2SO_3	sulfurous acid	$K_{a_1} = 1.6 \times 10^{-2}$ $K_{a_2} = 1.3 \times 10^{-7}$ (cannot be isolated in anhydrous form)
H_2SO_4	sulfuric acid	first proton $\sim$100% ionized. $K_{a_2} = 1.3 \times 10^{-2}$ (powerful dehydrating agent; moderate oxidizing agent)
$H_2S_2O_7$, $H_2S_3O_{10}$, $H_2S_4O_{13}$, etc.	polysulfuric acids (components of fuming sulfuric acid or "oleum")	addition of SO_3(g) to H_2SO_4 produces a mixture of polymers
H_2SO_5, $H_2S_2O_8$, etc.	peroxysulfuric acids	contain —O—O— bonds; powerful oxidizing agents; hydrolyze in dilute acid to form H_2O_2
$H_2S_2O_3$	thiosulfuric acid	reducing agent or complexing agent; used as sodium salt for photographer's hypo: $Ag^+ + 2S_2O_3^{2-} \longrightarrow [Ag(S_2O_3)_2]^{3-}$
$H_2S_3O_6$, $H_2S_4O_6$, $H_2S_5O_6$, $H_2S_6O_6$	polythionic acids	contain two —SO_3H groups connected through one or more sulfur atoms

Figure 26.10 Sulfuric acid (H_2SO_4).

The purified acid is used as the electrolyte in automobile batteries and in a large number of synthetic processes.

Sulfuric acid is a tetrahedral-base molecule (Figure 26.10) in which all the S—O bonds are somewhat shorter than would normally be expected for simple σ bonds. It has been suggested that two of the $3d$ orbitals are used by sulfur to overlap somewhat with oxygen orbitals containing unshared electron pairs. This overlap would contribute some double-bond character and would be consistent with the bond distances found.

26.9 GROUP VIIA ELEMENTS: THE HALOGENS

The halogens were named from a Greek word meaning "salt formers."

Astatine is a radioactive element whose properties have been only briefly studied. It is not included in this discussion of the chemistry of the halogens.

The group VIIA elements are among the most reactive elements known. In addition, all of them—with the exception of the radioactive astatine—are familiar from the names of such common substances as "fluorides" in toothpaste, chlorine for swimming pools, "bromides" as tranquilizers, or "tincture of iodine" as an antiseptic agent.

There are a number of similarities among the group VIIA elements—many more than were observed in groups IVA, VA, or VIA. The halogens exist "free" as diatomic molecules, and all are effective oxidizing agents, with fluorine being the most powerful oxidizing agent of all the chemical elements. In addition, all the halogens are toxic to most life forms in high concentrations. Dilute solutions of chlorine or iodine (the latter usually as a "tincture," i.e., alcohol solution) are used to kill microorganisms. Tincture of iodine is legally classified as a poison in most states and usually cannot be purchased by minors. Chlorine has been used as a war gas, although it has largely been supplanted by more "exotic" chemical warfare agents. The halogens are also hazardous in case of skin or eye contact because of the severe and painful chemical burns produced.

Variations in properties do occur, generally in ways expected from simple periodic trends (Table 26.6).

Table 26.6 SOME PROPERTIES OF THE HALOGENS

ELEMENT	COLOR[a]	MELTING POINT (°C)	BOILING POINT (°C) (1 atm)	IONIZATION POTENTIAL (eV)	ELECTRON AFFINITY (eV)	ELECTRO-NEGATIVITY
F_2	pale yellow	-218	-188	17	3.6	4.0
Cl_2	yellow-green	-100	-35	13	3.8	3.0
Br_2	red-orange	-8	59	12	3.5	2.8
I_2	violet	112	183	10	3.2	2.5

[a] As vapor or solution where "true" color can be best observed. Solid iodine appears blue-black.

26.10 PREPARATION AND CHEMICAL PROPERTIES OF THE HALOGENS

Because of their high chemical reactivity, none of the halogens occurs free in nature. Seawater and natural salt deposits, containing Cl^-, Br^- and I^-, are the principal commercial sources of Cl_2, Br_2 and I_2. Fluorine is produced mainly from fluorspar (CaF_2) or cryolite (Na_3AlF_6), an important aluminum ore. Since fluorine is such a powerful oxidizing agent, it cannot be prepared by *chemical* oxidation of fluoride, so an electrolytic process is employed:

$$2F^- \longrightarrow F_2 + 2e^-$$

Typically, a solution of potassium fluoride in liquid HF is used as the electrolyte. Liquid HF alone is too weakly ionized (Unit 19) to provide a good electrolyte.

Chlorine is also prepared by electrolysis. To increase the economic efficiency of chlorine production, it is generally coupled with the manufacture of sodium hydroxide or one of the IA or IIA metals, for example,

$$\underbrace{2Na^+(aq) + 2Cl^-(aq)}_{\text{(from seawater)}} + 2H_2O(l) \xrightarrow{\text{electrolysis}} \underbrace{2Na^+(aq) + 2OH^-(aq)}_{\text{(aq NaOH)}} + H_2(g) + Cl_2(g)$$

$$\underset{\text{(molten salt)}}{MgCl_2(l)} \xrightarrow{\text{electrolysis}} Mg(s) + Cl_2(g)$$

In addition, some chlorine is now manufactured from waste-product HCl by the *Kel-Chlor process*. A series of reactions is involved, but the overall procedure may be represented by

$$4HCl(g) + 2NOCl(g) + O_2(g) \longrightarrow 3Cl_2(g) + 2NO(g) + 2H_2O(g)$$

The differences in reduction potentials for the halogens (Appendix D) permit chlorine to oxidize both bromide and iodide ions (and Br_2 to oxidize I^-). The oxidation by Cl_2 is a commercial method for preparing bromine and iodine from impure "salt," from seawater or salt mines:

$$2Br^-(aq) + Cl_2(g) \longrightarrow Br_2(l) + 2Cl^-(aq)$$
$$2I^-(aq) + Cl_2(g) \longrightarrow I_2(s) + 2Cl^-(aq)$$

Iodine is also manufactured from sodium iodate, an impurity in natural deposits of *Chile saltpeter* ($NaNO_3$).

$$2IO_3^-(aq) + 5HSO_3^-(aq) \longrightarrow I_2(s) + 5SO_4^{2-}(aq) + 3H^+(aq) + H_2O(l)$$

The halogens are sufficiently insoluble in water to permit their recovery from processes carried out in aqueous systems.

Fluorine forms compounds in which only the $-I$ oxidation state is exhibited, since fluorine is the most electronegative element (Unit 7). The other halogens, however, exhibit oxidation states ranging from $-I$ to $+VII$. The halogens are powerful oxidizing agents; fluorine is so reactive that even asbestos burns in fluorine gas.

26.11 HALIDE IONS

There is evidence for transient $+1$ ions, increasing in stability from Cl^+ to I^+, but these exist only under very special circumstances and they are quite reactive. The F^+ ion is unknown in chemical systems although, like many unusual ions, it may be formed in high-energy situations such as electric discharges.

The only stable monatomic ions are the -1 halide ions: F^-, Cl^-, Br^-, and I^-. The exceptionally high charge density (Figure 26.11) on F^- causes it to form covalent bonds readily (e.g., HF) and to be strongly hydrated in aqueous solutions. This, plus the tendency for formation of hydrogen bonds, accounts for the low ionization constant (Table 26.7) of HF compared to the other hydrogen halides.

As in so many other groups, the lightest member is the "unusual" one. It is not too surprising then, that F^- is unique in many respects (Table 26.8).

All the halide ions act as Lewis bases, furnishing electron pairs for formation of many different complex ions; for example,

$$Fe^{3+}(aq) + 6F^-(aq) \rightleftharpoons [FeF_6]^{3-}(aq)$$
$$PbCl_2(s) + Cl^-(aq) + H_2O(l) \rightleftharpoons [PbCl_3(H_2O)]^-(aq)$$

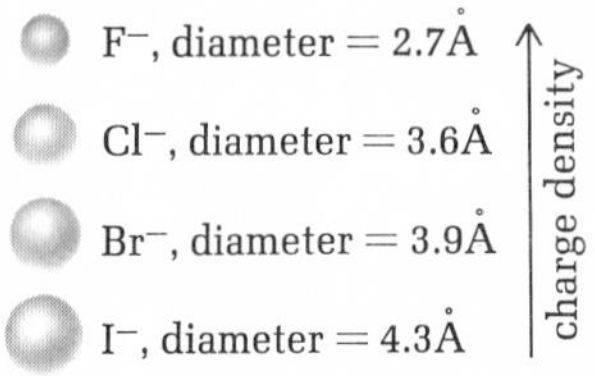

F$^-$, diameter $= 2.7$Å

Cl$^-$, diameter $= 3.6$Å

Br$^-$, diameter $= 3.9$Å

I$^-$, diameter $= 4.3$Å

charge density

Figure 26.11 Relative charge densities of halide ions.

Table 26.7 THEORETICAL K_a FOR HX[a]

HX	K_a
HF	6.7×10^{-4}
HCl	2×10^6
HBr	5×10^8
HI	2×10^9

[a] $K_a = \dfrac{a_{H^+} \times a_{X^-}}{a_{HX}}$, that is, K_a calculated using activities rather than molarities.

Table 26.8 WATER SOLUBILITY OF SOME HALIDES (in grams of solute per 100 ml of H_2O at 20°C)

ANION	CATION						
	Li^+	Na^+	Ag^+	Hg_2^{2+}	Pb^{2+}	Mg^{2+}	Ca^{2+}
F^-	0.30	4.5	36	very soluble	0.06	9×10^{-3}	2×10^{-3}
Cl^-	30	26	1.5×10^{-4}	2.9×10^{-4}	~ 1	16	25
Br^-	35	38	8.4×10^{-6}	3.7×10^{-5}	~ 1	33	59
I^-	43	51	2.1×10^{-7}	1.9×10^{-8}	~ 0.1	40	40

26.12 OXYGEN COMPOUNDS OF THE HALOGENS

There are a number of different kinds of halogen–oxygen compounds and ions. Oxygen difluoride (OF_2), for example, is one in which oxygen is assigned an oxidation number of $+II$—a very unusual situation. It is prepared by bubbling fluorine gas through a dilute alkaline solution and is a stable gas that is a very potent oxidizing agent. The compounds Cl_2O and Br_2O would appear to be similar to OF_2, but they are actually quite different since, because oxygen is more electronegative than any of the halogens except fluorine, Cl_2O and Br_2O contain "$-II$" oxygen. Thus Cl_2O is an explosive compound that reacts with water to form hypochlorous acid, whereas OF_2 is not explosive and is inert to water. Other oxides, such as F_2O_2, ClO_2, and I_2O_5 have been prepared.

There is a series of oxyacids and oxyanions for each of the halogens *except fluorine*[4] (Table 26.9). Probably the best known is hypochlorite, since it is the active ingredient in most laundry bleaches. Evidence obtained by the use of oxygen-18

[4] There is some evidence for the *transient* existence of HOF in aqueous solution (See *Journal of The American Chemical Society*, Vol. 93, p. 2350, 1971).

Table 26.9 OXYACIDS AND OXYANIONS OF THE HALOGENS

HALOGEN	OXIDATION NUMBER OF HALOGEN			
	+I	+III	+V	+VII
Cl	HOCl hypochlorous acid	HOClO chlorous acid	HOClO$_2$ chloric acid	HOClO$_3$[a] perchloric acid
	OCl$^-$ hypochlorite ion	ClO$_2^-$ chlorite ion	ClO$_3^-$ chlorate ion	ClO$_4^-$ perchlorate ion
Br	HOBr hypobromous acid	HOBrO bromous acid	HOBrO$_2$ bromic acid	only recently prepared[b]
	OBr$^-$ hypobromite ion	BrO$_2^-$ bromite ion	BrO$_3^-$ bromate ion	only recently prepared[b]
I	HOI hypoiodous acid	not well characterized[c]	HOIO$_2$ iodic acid	HOIO$_3$ periodic acid
	OI$^-$ hypoiodite ion	not well characterized[c]	IO$_3^-$ iodate ion	IO$_4^-$ periodate ion

[a] HOClO$_3$ (or HClO$_4$) is the strongest acid. Perchlorate ion, because of the high degree of electron delocalization, is a very poor Lewis base and is often used when a "noncomplexing" anion is required. Chloric, bromic, and iodic acids are also "strong," although weaker than perchloric acid (see Unit 19).
[b] Perbromates were unknown until 1968. They have since been prepared by the reaction

$$BrO_3^- + XeF_2 + H_2O \longrightarrow Xe + 2HF + BrO_4^-$$

Perbromates are, in general, much less stable than perchlorates and periodates.
[c] Iodite has been reported as a transient intermediate in some oxidation-reduction processes. It is very difficult to study because of its rapid disproportionation.

isotope indicates that hypochlorite may be one of the few oxidizing agents that acts by direct oxygen transfer:

$$Cl^{18}O^- + NO_2^- \longrightarrow Cl^- + O_2N^{18}O^-$$

All of the oxyacids and oxyanions are effective oxidizing agents. Laboratory explosions resulting from accidental contact of chlorates or perchlorates with organic matter in a situation involving heat are very common.

26.13 INTERHALOGEN COMPOUNDS

Diatomic molecules of two different halogens have been prepared for ClF, BrF, BrCl, ICl, and IBr. All are polar and can decompose to furnish the positive ion of the larger halogen and the negative ion of the smaller. Thus electrolysis of liquid ICl produces I$_2$ at the internal cathode and Cl$_2$ at the internal anode.

More complex interhalogen compounds, such as ClF$_3$, BrF$_5$, IF$_7$, ICl$_3$, and so on, have also been reported. At present they are of principal interest as examples of compounds for which simple Lewis "octet" structures cannot be written. It is believed that the heavier halogens (Cl, Br, I) utilize hybrid orbitals having some d character in these compounds. Orbitals of this type are discussed in Unit 27.

26.14 HYDROGEN: A "UNIQUE" ELEMENT

Although hydrogen is shown in many periodic tables at the top of both columns IA and VIIA, it is neither a metal nor a typical nonmetal. The "hydrogen ion," or proton (H$^+$), is the smallest possible ion and has the highest charge density. The *hydride* ion (H$^-$) is formed by reaction of various metals with hydrogen gas (Unit 25), but cannot exist in aqueous solution:

$$H^- + H_2O \longrightarrow H_2 + OH^-$$

Free molecular hydrogen has been detected in volcanic gases and in trace amounts in some "natural gas" supplies, but most of the hydrogen in nature occurs as compounds (e.g., water, metal hydroxides, hydrocarbons). Most of the hydrogen

(Continued on p. 650.)

ORGANIC HALOGEN COMPOUNDS

Compounds of carbon with various halogens are among the most important commercial products of modern chemical industry. Many of these substances have been developed to the stage of real economic value within the past 20 years. Table 26.10 lists some common examples of organic halides. It is interesting to note that many organic fluorides are nontoxic, whereas a high percentage of the organic chlorides are dangerous as poisons that accumulate in fatty tissues and slowly produce irreversible damage to a number of vital organs. This is in sharp contrast to the high toxicity of F^- for living systems that require Cl^- for survival. The C—F bond is very stable, while the C—Cl bond may undergo either *heterolytic* cleavage,

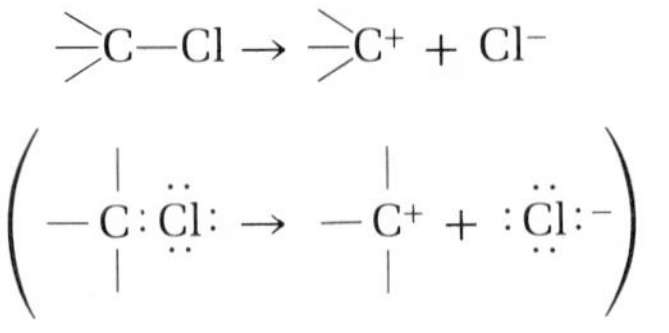

$$\geq\!\!C\!\!-\!\!Cl \rightarrow\ \geq\!\!C^+ + Cl^-$$

$$\left(-\underset{|}{\overset{|}{C}}:\overset{..}{\underset{..}{Cl}}: \rightarrow\ -\underset{|}{\overset{|}{C}}{}^+ + :\overset{..}{\underset{..}{Cl}}:^- \right)$$

(harmless)

or *homolytic* cleavage,

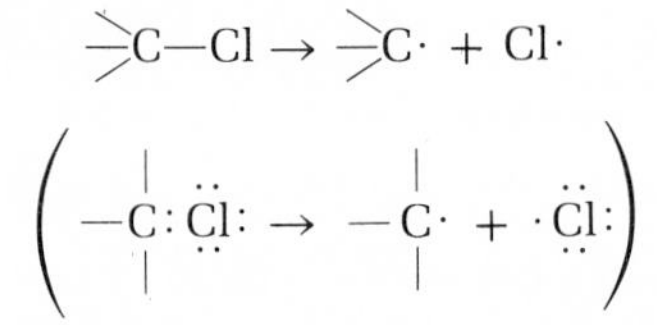

$$\geq\!\!C\!\!-\!\!Cl \rightarrow\ \geq\!\!C\cdot + Cl\cdot$$

$$\left(-\underset{|}{\overset{|}{C}}:\overset{..}{\underset{..}{Cl}}: \rightarrow\ -\underset{|}{\overset{|}{C}}\cdot + \cdot\overset{..}{\underset{..}{Cl}}: \right)$$

(very reactive free radicals, Unit 16)

Table 26.10 SOME COMMON ORGANIC HALIDES

NAME	STRUCTURE	COMMENTS
Freon-12 (difluorodichloromethane)	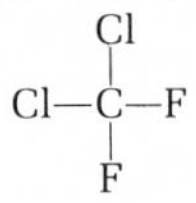 Cl—C(Cl)(F)—F	a refrigerant gas and propellant gas for aerosol "bombs"; nontoxic, odorless, nonflammable; of environmental concern in connection with O_3 destruction.
Teflon (polytetrafluoroethylene)	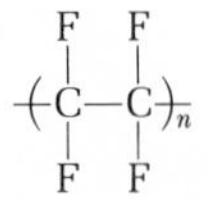 $+\!(C(F)(F)\!-\!C(F)(F))\!\!+_n$	a "nonstick" plastic very resistant to heat and chemical agents; nontoxic.
sodium fluoroacetate	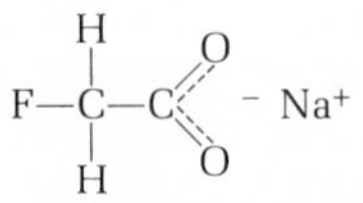 F—C(H)(H)—C(=O)(O$^-$) Na$^+$	very toxic rodenticide (note that the *trifluoro-acetate* is nontoxic)
chloroform	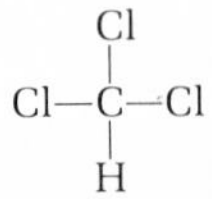 Cl—C(Cl)(H)—Cl	formerly used as an anesthetic, chloroform has been generally abandoned for human use because of evidence that it may cause serious damage to the heart

Table 26.10 (Continued)

NAME	STRUCTURE	COMMENTS
triclene (trichloroethylene)	H, Cl, Cl, Cl on C=C	a common dry-cleaning agent; prolonged inhalation may be hazardous
carbon tet (carbon tetrachloride)	Cl—C—Cl with Cl above and Cl below	another dry-cleaning agent, but much more toxic than triclene; used to extinguish fires, but *very dangerous* since it oxidizes to phosgene
phosgene	Cl, Cl on C=O	very toxic "war" gas
chloral hydrate	Cl—C—C—O—H (with Cl, Cl on first C and O—H, H on second C)	a potent "sleeping drug" available by prescription only; at one time this was an ingredient of the "Mickey Finn"—a drink used for a very real "knockout"
chlordane	(polycyclic chlorinated structure)	a potent insecticide, now banned from most uses because of possible link to cancers in test animals
hexachlorophene	(two chlorinated phenol rings joined by a CH₂ bridge)	bacteriostatic and bacteriocidal agent used in "mouthwashes," disinfectant soaps, etc.; possible hazards to humans first reported in 1971
tyrian purple (6,6′-dibromoindigo)	(dibromoindigo structure)	the famous "royal purple" once used exclusively for dyeing robes, etc. of "important personages"; originally obtained by pressing a type of sea snail to obtain a fluid that turned purple on exposure to sunlight and air; now made synthetically, but largely replaced by more brilliant and less expensive dyes
iodoform	I—C—I with I above and H below	a common antiseptic agent

Table 26.10 (Continued)

NAME	STRUCTURE	COMMENTS
thyroxine		principal hormone of the thyroid gland

produced commercially is a by-product of electrolysis processes (Sections 26.6 and 26.10) or of petroleum "cracking" (Unit 28). Although hydrogen is relatively expensive, it is an "environmentally clean" fuel,

$$2H_2(g) + O_2(g) \longrightarrow 2H_2O(g)$$

and the impetus of the "energy crisis" has led to a number of investigations of methods for producing hydrogen fuels. Among these is a "resurrection" of the *water-gas* process used for manufacture of a gas mixture for heating and lighting prior to the large-scale conversion to "natural gas" (methane):

$$\underset{\text{(coke)}}{C(s)} + H_2O(g) \longrightarrow \underset{(water\ gas)}{\underbrace{CO(g) + H_2(g)}}$$

On a small scale, hydrogen can be prepared by the treatment of certain metals with aqueous acids or by the reaction of some metals or metalloids with aqueous hydroxides, for example,

$$Mg(s) + 2H^+(aq) \longrightarrow Mg^{2+}(aq) + H_2(g)$$
$$Zn(s) + 2H^+(aq) \longrightarrow Zn^{2+}(aq) + H_2(g)$$
$$2Al(s) + 6H^+(aq) \longrightarrow 2Al^{3+}(aq) + 3H_2(g)$$
$$Zn(s) + 2OH^-(aq) + 2H_2O(l) \longrightarrow [Zn(OH)_4]^{2-}(aq) + H_2(g)$$
$$2Al(s) + 2OH^-(aq) + 6H_2O(l) \longrightarrow 2[Al(OH)_4]^-(aq) + 3H_2(g)$$
$$Si(s) + 2OH^-(aq) + H_2O(l) \longrightarrow SiO_3^{2-}(aq) + 2H_2(g)$$

Hydrogen reacts readily with most nonmetals. The *Haber process* is commercially important for the manufacture of ammonia,

$$3H_2(g) + N_2(g) \rightleftharpoons 2NH_3(g)$$

and the *Bergius process* for the catalyzed reaction of hydrogen with powdered carbon offers a way of supplementing natural supplies of methane and other hydrocarbons:

$$xH_2(g) + C(s) \longrightarrow CH_4(g) + \text{other hydrocarbons}$$

Two "complex" hydrides, lithium aluminum hydride and sodium borohydride, are very useful reducing agents for a number of types of organic compounds (Unit 29). Since hydrides react vigorously with water, $LiAlH_4$ and $NaBH_4$ are typically prepared in anhydrous organic solvents:

$$4LiH + AlCl_3 \longrightarrow LiAlH_4 + 3LiCl$$
$$4NaH + B(OCH_3)_3 \longrightarrow NaBH_4 + 3NaOCH_3$$
$$\text{[a boron ester,}$$
$$\text{Unit 25]}$$

26.15 "NOBLE" GASES

By 1871 the periodic table was quite similar to the one we see today. There were a few gaps corresponding to elements such as gallium, which had been predicted but not yet discovered. The principal difference was the complete absence of the helium group from the early periodic table.

As early as 1784, evidence was beginning to accumulate that suggested the existence of elements (or compounds) unlike any previously discovered or predicted. Henry Cavendish was studying the conversion of nitrogen (a relatively unreactive gas) to its oxides by electric spark through oxygen–nitrogen mixtures. He discovered that the resulting oxides were soluble in aqueous alkalis, but such gaseous mixtures prepared from air always left a small bubble of "air" unabsorbed. This unusual gas, constituting about 1 percent of the original air, was found in 1894 to be more inert than nitrogen and to have its own unique emission spectrum. Careful study eventually confirmed that this gas was a new chemical element. Because of its inert response to chemical reagents, it was called *argon* (Gr., "the lazy one").

The existence of another new element was reported in 1868. The unusual factor in this report was that the element, called *helium* (Gr. *helios,* the sun), was first found on the sun rather than on the earth. Careful study of the solar spectrum revealed certain missing wavelengths not corresponding to the absorption spectra of any known atoms or ions. It was not until 27 years later that helium was found on the earth.

By 1900, the entire group of "inert" gases had been characterized, the last being the radioactive gas, *radon,* found in gases resulting from nuclear decay of radium. The group is sometimes referred to as the "rare" gases because most of them occur in only trace amounts (Table 26.11). We shall continue to call these elements "noble" gases since, like certain of the human "nobility," they only rarely "associate" with the more common elements.

It had long been noted by chemists that certain elements formed only one "stable" monatomic ion. Thus lithium, sodium, potassium, rubidium, and cesium formed only $+1$ ions; fluorine, chlorine, bromine, and iodine formed only -1 monatomic ions that remained "stable." These observations suggested some "magic numbers" of electrons associated with unusual resistance to further electron loss or gain. In every simple case, these numbers corresponded to the atomic numbers of the nearest noble gas elements (Table 26.12).

In addition, prior to 1962 no reports of stable compounds of the noble gases had been accepted. The unique stability of ions isoelectronic with the noble gases and the apparent inability of these gases to form chemical compounds came to be one of the

Table 26.11 PERCENT "NOBLE" GASES IN AIR

GAS	PERCENT (BY VOLUME)
He (helium)	$\sim5 \times 10^{-4}$
Ne (neon)	$\sim2 \times 10^{-3}$
Ar (argon)	~1
Kr (krypton)	$\sim1 \times 10^{-4}$
Xe (xenon)	$\sim9 \times 10^{-6}$
Rn (radon)	$\sim6 \times 10^{-18}$

Some species had been reported, but these were "novelties" that existed only under very special conditions. For example, spectral data from gas discharge tubes containing mixtures of hydrogen and helium appeared to correspond to the high-energy ions HeH^+ and $HeH_2{}^+$. Some complexes of argon with BF_3 were reported at temperatures below $-130°C$; and "hydrates" of the type $Kr \cdot 5H_2O$ appeared to exist at low temperatures and very high pressures.

Table 26.12 SOME ISOELECTRONIC SPECIES

NUMBER OF ELECTRONS	SOME STABLE SPECIES
2	He, H^-, Li^+, Be^{2+}
10	Ne, F^-, O^{2-}, Na^+, Mg^{2+}, Al^{3+}
18	Ar, Cl^-, S^{2-}, K^+, Ca^{2+}, Sc^{3+}
36	Kr, Br^-, Se^{2-}, Rb^+, Sr^{2+}, Y^{3+}
54	Xe, I^-, Te^{2-}, Cs^+, Ba^{2+}, La^{3+}

most cherished beliefs in the chemical community. The group of the elements helium through radon (He–Rn) was called the *inert gases,* and every introductory treatment of chemical bonding emphasized the fact that *only these elements formed* no compounds, so that inert gas electronic structures were certainly stable. The existence of thousands of compounds in which electronic arrangements unlike those of the helium group appeared quite stable was conveniently ignored.

If there was any single fact that chemists universally accepted, the inertness of this group of elements must have been the one.

26.16 STABLE COMPOUNDS OF "INERT" ELEMENTS

In 1962, Professor Neil Bartlett at the University of British Columbia found that oxygen gas reacted with PtF_6 to form an ionic substance described as $(O_2^+)(PtF_6^-)$. Since the ionization potential for xenon (12.1 eV) is slightly less than that for oxygen (13.6 eV), it seemed reasonable that an ionic compound might result from the reaction of xenon with PtF_6. The reaction was attempted, and Bartlett's report of a stable crystalline solid identified as $(Xe^+)(PtF_6^-)$ was something of a shock to chemists. Had the research stopped here, it might have been dismissed as one of those novel cases that occasionally receive short-term publicity but soon lapse into the obscurity of unimportant footnotes in textbooks.

Selig, Malm, and Claassen.

However, a group working at Argonne National Laboratory had been studying PtF_6 and other fluoridating agents and decided to try direct reaction of xenon with molecular fluorine. By simply heating xenon and fluorine together, they were able to prepare various compounds, among them *xenon tetrafluoride,* the first stable covalent compound of an inert gas to be characterized.

Within less than five years, a number of compounds had been prepared (Table 26.13). It was evident that these elements, although much less reactive than most, had no magical resistance to chemical change.

There are difficulties in describing some of the noble gas compounds by current bonding theories. This is at least partly caused by experimental uncertainties in bond lengths and bond angles in several cases, so that descriptions of orbitals involved must await better data. In some cases, current theory is simply inadequate to account

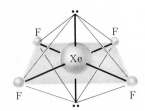

Figure 26.12 Xenon tetrafluoride, a chemical surprise.

Table 26.13 SOME COMPOUNDS OF "INERT" GASES (PREPARED 1962–1967)

COMPOUND	CHARACTERISTICS
$XePtF_6$	yellow solid, probably ionic
XeF_4	colorless crystalline solid, m.p. 100°C, contains XeF_4 molecules
XeF_2, XeF_6	solids, fairly stable in dry air, reactive with H_2O
XeO_3, XeO_4	solids, explosive except at low temperatures
KrF_2, KrF_4, RnF_4	fairly stable compounds

for observed properties. Some noble gas compounds have been amenable to relatively simple bonding descriptions. It is possible, with XeF_4 for example, to suggest that sp^3d^2 hybrids (Unit 27) would be consistent with the experimentally determined structure (Figure 26.12).

SUGGESTED READING

Claassen, Howard H. 1966. *The Noble Gases*. Lexington, Mass.: Raytheon.
Jolly, William L. 1966. *The Chemistry of the Non-Metals*. Englewood Cliffs, N.J.: Prentice-Hall.
Sherwin, Ernest, and Gordon J. Weston. 1966. *The Chemistry of the Non-Metallic Elements*. Elmsford, N.Y.: Pergamon.

EXERCISES

1. Fill in the missing names or formulas:
   a. KNO_2; **b.** orthophosphoric acid; **c.** $Na_2S_2O_3$; **d.** arsine; **e.** O_3; **f.** XeF_4; **g.** chloric acid; **h.** KIO_4; **i.** sodium hypochlorite; **j.** bromous acid
2. Write electron-dot formulas for each of the following molecules and polyatomic ions, then predict both *orbital* and *atomic* geometries for each.
   a. carbon disulfide; **b.** nitrogen trichloride; **c.** phosphorus pentachloride; **d.** nitrite ion; **e.** chlorate ion; **f.** nitric acid; **g.** xenon hexafluoride
3. Offer reasonable explanations for the following experimental observations:
   a. White phosphorus is quite soluble in carbon disulfide, but red phosphorus is not.
   b. A solution of 3.2 g of sulfur in 100 ml of carbon disulfide has a considerably higher freezing point than a solution of 3.1 g of phosphorus in 100 ml of carbon disulfide.
4. Give balanced equations illustrating:
   a. a reaction suitable for the small-scale generation of very pure nitrogen gas
   b. a commercial method for preparing white phosphorus
   c. a procedure for recovering arsenic from oxide wastes
   d. the laboratory preparation of oxygen from potassium chlorate
   e. the production of bromine from "sea salt" by chlorine oxidation
   f. the reaction of aluminum with aqueous sodium hydroxide to generate hydrogen
5. Complete and balance:

   a. $NO(g) + O_2(g) \longrightarrow$ _____

   b. $H_3PO_3(aq) + OH^-(aq) \longrightarrow$ _____
      excess

   c. $CH_3NH_2(aq) + H_2O(l) \rightleftharpoons$ _____

   d. $H_2S(aq) + OH^-(aq) \longrightarrow$ _____
      excess

   e. $P_4(g) + Cl_2(g) \longrightarrow$ _____
      excess

   f. $P_4(g) + Cl_2(g) \longrightarrow$ _____
      excess

   g. $IO_3^-(aq) + HSO_3^-(aq) \longrightarrow$ _____

   h. $N_2(g) + H_2(g) \rightleftharpoons$ _____

   i. $CaH_2(s) + H_2O(l) \longrightarrow$ _____

   j. $LiH + AlCl_3 \longrightarrow$ _____
   (in anhydrous diethyl ether)

6. Calculate the pH of
   a. 0.3 M nitric acid
   b. 0.3 M hydroxylamine
   c. 0.3 M sodium hypophosphite
   d. a solution 0.10 M in methylamine and 0.30 M in methylammonium chloride
7. What volume of nitrous oxide (measured at 200°C and 1.00 atm) could be prepared by controlled decomposition of 28 g of ammonium nitrate?
8. What volume of 17 M nitric acid is required to dissolve 15 g of copper? [The products are aqueous copper(II) nitrate, nitric oxide gas, and water.]
9. An insecticide particularly useful against the cotton boll weevil, γ-hexachlorocyclohexane, is prepared by photochlorination of benzene. This process involves a free-radical reaction, and a mixture of isomeric products is formed. When the reaction is performed in acetic anhydride at about 5°C, a 30-percent yield of the desired isomer (γ) is obtained:

γ-hexachlorocyclohexane

An excess of 10 percent of the stoichiometric amount of chlorine is used. What weights of benzene and of chlorine are required to produce 1.00 kg of the desired product under the conditions described?
10. A bleaching solution can be prepared by bubbling chlorine gas through dilute sodium hydroxide:

$$Cl_2 + 2OH^- \longrightarrow Cl^- + OCl^- + H_2O$$

The chlorine required may be generated by electrolysis of molten sodium chloride. What volume of 0.30 M hypochlorite solution could be prepared from the chlorine generated by electrolysis using a current of 0.30 amp for 25 min?

PRACTICE FOR PROFICIENCY

Objective (a)

1. Fill in the missing names or formulas:
   a. HNO_2; **b.** hypophosphorous acid; **c.** $KHSO_3$; **d.** sodium arsenate; **e.** Na_2Se; **f.** $HOCl$; **g.** potassium perchlorate; **h.** $Mg(ClO_3)_2$; **i.** chlorous acid; **j.** $LiAlH_4$; **k.** pyrophosphoric acid; **l.** $AgNO_3$; **m.** antimony trichloride; **n.** phosphine; **o.** O_3; **p.** sodium borohydride

Objective (b)

2. Write electron-dot formulas for each of the following molecules and polyatomic ions, then predict both *orbital* and *atomic* geometries for each.
   a. PF_3; **b.** $AsCl_5$; **c.** SO_2; **d.** SO_3; **e.** PO_4^{3-}; **f.** PCl_6^-; **g.** XeF_4; **h.** NO_3^-; **i.** SF_6; **j.** ClO_2^-

Objective (c)

3. Offer reasonable explanations for the following experimental observations:
   a. White phosphorus has a much higher vapor pressure than red phosphorus, at the same temperature.
   b. Both *rhombic* and *monoclinic* sulfur are much more soluble in carbon disulfide than is amorphous ("plastic") sulfur.
   c. "Ordinary" oxygen (O_2) is strongly attracted to a magnet, but ozone is not. (*Hint:* Consider the electronic requirements for paramagnetism.)
   d. One allotropic form of selenium exhibits metallic conductivity, but the α-monoclinic form is a nonconductor.
   e. So-called *black* phosphorus exhibits electrical conductivity similar to that of graphite, but white phosphorus is a nonconductor.

Objective (d)

4. Complete and balance:

   a. $NH_4NO_3(s) \xrightarrow{\text{explosive decomposition}}$ _____

   b. $H_3PO_4(aq) + \underset{\text{excess}}{NaOH(aq)} \longrightarrow$ _____

   c. $H_2NOH(l) + HCl(g) \longrightarrow$ _____

   d. $Ca(OH)_2(aq) + H_2SO_3(aq) \longrightarrow$ _____

   e. $NO_2(g) + H_2O(l) \longrightarrow$ _____

   f. $SO_3(g) + H_2O(l) \longrightarrow$ _____

   g. $NH_4^+(aq) + NO_2^-(aq) \longrightarrow$ _____

   h. $KClO_3 \xrightarrow{\text{heat}}$ _____

   i. $H_3PO_2(aq) + NaOH(aq) \longrightarrow$ _____

   j. $NaH + B(OCH_3)_3 \longrightarrow$ _____

5. Give balanced equations illustrating
   a. the preparation of phosphorus from calcium phosphate
   b. the production of antimony from Sb_4O_6
   c. the electrolysis of water
   d. the electrolysis of aqueous sodium chloride
   e. the preparation of iodine from sodium iodate

Objective (e)

6. Calculate the pH of
   a. 0.50 M methylamine; **b.** 0.050 M sulfuric acid; **c.** 0.50 M sodium cyanide

7. What volume of 17 M nitric acid is required for conversion of 9.0 g of aluminum to aluminum nitrate? (Some nitric acid is reduced to nitric oxide.)

8. Iodoform may be prepared by treating acetone with iodine and aqueous sodium hydroxide. Sodium acetate is the other organic product. What weight of acetone is required for production of 250 g of iodoform, assuming the reaction yield is 82 percent? (Acetone is $CH_3\overset{\displaystyle O}{\overset{\|}{C}}CH_3$.)

9. What is the approximate pH of 0.50 M sodium hypochlorite?

10. What mass of lithium nitride could be formed from 104 g of lithium and excess nitrogen gas?

11. What volume (in liters) of dry hydrogen gas (at STP) could be generated by the electrolysis of water over a 12-hr period with an average current of 50.0 amp?

BRAIN TINGLER

12. Many organic, and some inorganic, wastes are degraded in natural waters by oxidation processes. In some cases these involve direct reaction of chemical wastes with molecular oxygen dissolved in the water. More commonly, aerobic bacteria in the water degrade the chemicals while utilizing dissolved oxygen. In either case, oxygen is consumed and if waste degradation demands abnormally high amounts of O_2, fish and other aquatic life cannot survive in the oxygen-depleted waters.

   The biological oxygen demand (BOD) is defined as

   $$BOD = \frac{\Delta(\text{no. mg } O_2)}{\text{no. liters water}} \quad (\text{in ppm } O_2)$$

   and is determined by mixing a measured sample of polluted water with a known volume of oxygen-saturated pure water. The total O_2 concentration in the mixture is measured just after mixing and again after five days incubation at 20°C. It is the *change* in O_2 concentration (Δ) over this period that is used to calculate the BOD.

   Calculate
   a. the volume of oxygen (measured at 20°C and 150 torr) required over a five-day period by a sewage-treatment plant processing 25,000 liters per day with a BOD of 152 ppm
   b. the mass of carbon dioxide formed if 38% of the utilized oxygen is converted to CO_2

Unit 26 Self-Test

1. Supply mixing names or formulas:
   a. periodic acid; **b.** Ca(ClO)$_2$; **c.** potassium chlorate;
   d. magnesium bromite; **e.** sodium thiosulfate; **f.** AsH$_3$;
   g. pyrophosphoric acid; **h.** Bi$_2$O$_5$; **i.** nitrous oxide; **j.** H$_2$Se;
   k. MgSO$_3$

2. Write electron-dot formulas and predict *orbital* and *atomic* geometries for
   a. nitrous acid; **b.** ClO$_3^-$; **c.** PCl$_5$; **d.** SO$_2$; **e.** XeF$_6$

3. Monoclinic sulfur appears to consist of covalent S$_8$ rings arranged at the points of a regular crystal lattice. Which of the following properties would be expected for this substance?
   a. moderate solubility in CS$_2$ **b.** broad melting range
   c. good electrical conductivity   **d.** very high melting point
   e. zero vapor pressure

4. "α-White" phosphorus is not an effective electrical conductor. It has a sharp melting point, a moderately high vapor pressure, and is quite soluble in carbon disulfide. Its vapor density (corrected to STP) is 5.5 g liter^{-1}. The pure crystalline solid is probably described by
   a. a three-dimensional covalent lattice of P atoms
   b. covalent P$_4$ units held by van der Waals forces
   c. long chains of covalent P polymers
   d. P^{4+} ions in a sea of mobile electrons
   e. free P atoms connected by ionic bonds

5. Complete and balance:
   a. _____ NH$_4$NO$_3$(s) $\xrightarrow{200°C}$ _____

   b. _____ NO$_2$(g) + _____ H$_2$O(l) $\longrightarrow$ _____

   c. _____ H$_3$PO$_2$(aq) + _____ OH$^-$(aq) $\longrightarrow$ _____

   d. __ Ag(s) + __ H$^+$(aq) + __ NO$_3^-$(aq) $\longrightarrow$ _____

   e. _____ NH$_4^+$(aq) + NO$_2^-$(aq) $\longrightarrow$ _____

   f. _____ Cl$_2$(g) + _____ I$^-$(aq) $\longrightarrow$ _____

6. Calculate the pH of
   a. 0.50 M nitrous acid; **b.** 0.50 M sodium nitrite; **c.** 0.50 M sulfuric acid

7. What *total* gas volume (in liters at 600°C and 1.00 atm) would be formed by the explosive decomposition of 16 g of ammonium nitrate?

8. What volume (in liters) of phosgene gas (at 257°C and 710 torr) could be formed from reaction of 7.7 g of carbon tetrachloride with a stoichiometric amount of oxygen? (Phosgene is COCl$_2$.)

9. What is the approximate pH of 0.30 M sodium hypochlorite solution? (K_a for hypochlorous acid is 3.2 × 10^{-8}.)

10. What volume (in liters) of dry hydrogen gas (at STP) could be generated by the electrolysis of water over a 6.0-hr period with an average current of 25 amp?
    a. 63; **b.** 91; **c.** 120; **d.** 210; **e.** 250

d-Transition elements

From 1874 to 1877, Jacob Walzer deposited $254,000 in gold from a secret mine in the Superstition Mountains. He died in 1891 without revealing the location of his supply. Prospectors still search for this "Lost Dutchman Mine."

(Narrative of the Old West)

(a) Be able to recall the most common oxidation states exhibited by the first-row *d* transition elements and by silver, cadmium, gold, and mercury.

(b) Be able to write electron configurations for the first-row *d* transition elements and their stable monatomic ions and to predict paramagnetism of hypothetical "isolated" atoms and cations (Section 27.2).

(c) Be able to use the IUPAC naming rules for complex ions and coordination compounds (Appendix F).

(d) Be able to predict the geometric and magnetic characteristics of complex ions and to suggest orbital descriptions consistent with these characteristics (Sections 27.11 and 27.12).

(e) Be able to use an analytical flow chart to interpret test results from analyses for ions of the more common *d* transition elements (Section 27.8).

REVIEW

(f) Be able to apply the principles of stoichiometry, equilibrium, and electrochemistry to the chemistry of the *d* transition elements.

In 1848, John Sutter hired a carpenter named Jim Marshall to help build a sawmill on some of Sutter's holdings on the American River in the Sacramento Valley. On January 24 of that year Marshall found gold in the river bed. The news spread like wildfire, and the California Gold Rush was on. They came by horseback, on foot, in wagon trains or sailing ships—all lured by fantastic tales of easy wealth. Many of the "forty-niners" found their wildest dreams come true: More than $600,000,000 in gold was taken out of California by 1860. Others found an unmarked grave, often after an exchange of their few nuggets for a lead slug from someone who would "rather shoot than dig."

For some reason gold has always held a unique fascination for man. No other metal has been sought with such zeal, guarded with such care, or desired with such single-minded passion. Maybe this reflects some innate desire to own something whose value is secure—gold will not rust or decay; it will not break down or need constant repair; its market value is usually safe. Maybe, for the prospector, it represents that rare chance for the individual in a complex, interdependent society. Gold may be found in the free state, and one man alone can, with luck, find it and sell it—he does not have to depend on massive mining and milling operations, marketing experts, advertising firms, executives, stockholders, and all the hectic pressures of the modern business world. Only if he finds too much does he have a problem.

Silver is another metal of the Old West. All over the West there are ghost towns—silent testimonies of the epic days when gold and silver lured the adventuresome from the security of the Eastern cities. Many of these towns burst into existence with the discovery of rich silver ores and died a few years later as the high-grade ores disappeared and the price of silver dropped on the world markets.

Gold and silver are two of the "coinage metals" (group IB in the periodic table). The other is copper. Can you see John Wayne, six-gun smoking, calling out to the bad guys, "All right, you varmints! Drop that bag of copper!"?

Copper somehow lacks the excitement of gold or silver, yet it is a much more useful metal. If all the gold and silver in the world were to change suddenly to lead, modern technology would be little affected. But copper is essential to modern society. Copper carries the electric power on which society "runs," and no other metal has such desirable properties for this and many other necessary jobs.

Arizona is among the leaders in production of the coinage metals, currently contributing more than half the world's supply of copper. If the total value of gold, silver, and copper taken from Arizona alone during the past 75 years were distributed equally among that state's residents, every man, woman, and child would get about $190 in gold, $170 in silver, and $5500 in copper. Maybe the old prospectors were looking for the wrong metal.

In this unit we shall examine some of the properties of the transition elements, in particular those commonly referred to as the d transition elements. These include not only the coinage metals, but also such important metals as iron, nickel, mercury, platinum, and tungsten.

27.1 THE TRANSITION METALS

The simplest periodic trends are observed among the first 18 elements. From lithium to neon and from sodium to argon, the regular change from metallic through non-metallic to "inert" properties can usually be explained rather well in terms of the simple s and p orbitals and the types of hybrids formed from these orbitals. For larger elements, the d and f orbitals are involved, and these are of particular importance in understanding the chemistry of three sets of elements, the *transition metals*. Begin-

Sc	Ti	V	Cr*	Mn	Fe	Co	Ni	Cu*	Zn
(Ar)	(Ar)	(Ar)	(Ar)	(Ar)	(Ar)	(Ar)	(Ar)	(Ar)	(Ar)
$4s^2 3d^1$	$4s^2 3d^2$	$4s^2 3d^3$	$4s^1 3d^5$	$4s^2 3d^5$	$4s^2 3d^6$	$4s^2 3d^7$	$4s^2 3d^8$	$4s^1 3d^{10}$	$4s^2 3d^{10}$
Y	Zr	Nb*	Mo*	Tc*	Ru*	Rh*	Pd*	Ag*	Cd
(Kr)	(Kr)	(Kr)	(Kr)	(Kr)	(Kr)	(Kr)	(Kr)	(Kr)	(Kr)
$5s^2 4d^1$	$5s^2 4d^2$	$5s^1 4d^4$	$5s^1 4d^5$	$5s^1 4d^6$	$5s^1 4d^7$	$5s^1 4d^8$	$5s^0 4d^{10}$	$5s^1 4d^{10}$	$5s^2 4d^{10}$
La	Hf	Ta	W	Re	Os	Ir	Pt*	Au*	Hg
(Xe)	(Xe, $4f^{14}$)	(Xe, $4f^{14}$)	(Xe, $4f^{14}$)	(Xe, $4f^{14}$)	(Xe, $4f^{14}$)	(Xe, $4f^{14}$)	(Xe, $4f^{14}$)	(Xe, $4f^{14}$)	(Xe, $4f^{14}$)
$6s^2 5d^1$	$6s^2 5d^2$	$6s^2 5d^3$	$6s^2 5d^4$	$6s^2 5d^5$	$6s^2 5d^6$	$6s^2 5d^7$	$6s^1 5d^9$	$6s^1 5d^{10}$	$6s^2 5d^{10}$
Ac									
(Rn)									
$7s^2 6d^1$									

*Note "anomalous" configurations

Inner Transition Metals

lanthanum series

Ce	Pr	Nd	Pm	Sm	Eu	Gd*	Tb	Dy	Ho	Er	Tm	Yb	Lu
(Xe, $6s^2$)	(Xe, $6s^2$)	(Xe, $6s^2$)	(Xe, $6s^2$)	(Xe, $6s^2$)	(Xe, $6s^2$)	(Xe, $6s^2$)	(Xe, $6s^2$)	(Xe, $6s^2$)	(Xe, $6s^2$)	(Xe, $6s^2$)	(Xe, $6s^2$)	(Xe, $6s^2$)	(Xe, $6s^2$)
$5d^0 4f^2$	$5d^0 4f^3$	$5d^0 4f^4$	$5d^0 4f^5$	$5d^0 4f^6$	$5d^0 4f^7$	$5d^1 4f^7$	$5d^0 4f^9$	$5d^0 4f^{10}$	$5d^0 4f^{11}$	$5d^0 4f^{12}$	$5d^0 4f^{13}$	$5d^0 4f^{14}$	$5d^1 4f^{14}$

actinium series

Th	Pa*	U*	Np*	Pu	Am	Cm*	Bk	Cf	Es	Fm	Md	No	Lr
(Rn, $7s^2$)	(Rn, $7s^2$)	(Rn, $7s^2$)	(Rn, $7s^2$)	(Rn, $7s^2$)	(Rn, $7s^2$)	(Rn, $7s^2$)	(Rn, $7s^2$)	(Rn, $7s^2$)	(Rn, $7s^2$)	(Rn, $7s^2$)	(Rn, $7s^2$)	(Rn, $7s^2$)	(Rn, $7s^2$)
$6d^2 5f^0$	$6d^1 5f^2$	$6d^1 5f^3$	$6d^1 5f^4$	$6d^0 5f^6$	$6d^0 5f^7$	$6d^1 5f^7$	$6d^0 5f^9$	$6d^0 5f^{10}$	$6d^0 5f^{11}$	$6d^0 5f^{12}$	$6d^0 5f^{13}$	$6d^0 5f^{14}$	$6d^1 5f^{14}$

man-made elements

Figure 27.1 Electron configurations for transition in elements (as determined from spectroscopic data).

ning with the third horizontal row of the periodic table, we find a series of elements interposed between groups IIA and IIIA. These "B group" elements are commonly referred to as the d transition elements. In the ordinary periodic table there are two horizontal rows set below the rest of the table. These so-called "inner transition elements" are classified as two separate sets, the lanthanum series (elements 58–71) and the actinium series (elements 90–103).

It is no accident that there are 10 horizontal rows for the d transition elements and 14 for the inner transition elements. According to modern theory (Unit 1) there are five d orbitals, each capable of holding a maximum of two electrons. This accounts for the 10 columns of the group B elements. There are seven f orbitals, and these are involved in the inner transition elements (Figure 27.1). The inner transition elements, often referred to as the "rare earth elements," have very similar properties (within each series) and are quite difficult to separate by conventional chemical methods. This is ascribed to the fact that their valence electrons are all in the same orbitals, the differences being in the f orbitals, which generally act as kernel electrons.

We shall leave these rather exotic species for more advanced texts[1] and concentrate our attention now on the more common d transition elements.

[1] An excellent treatment of the simpler aspects of rare earth chemistry is found in *The Chemistry of the Lanthanides,* by Therald Moeller (New York: Van Nostrand Reinhold, 1964).

All of the transition elements are metals. Most exhibit more than one stable oxidation state in compounds, and many of their compounds are paramagnetic because of unpaired electrons in the transition element species. Notable exceptions that should be remembered are silver (+I), zinc and cadmium (+II), and scandium (+III). Perhaps the most notable property of the transition elements is their ability to form a wide variety of coordination compounds with *d* orbitals involved in bonding.

27.2 ELECTRON CONFIGURATION AND PARAMAGNETISM

The magnetic properties of materials are related to the magnetic and electrical characteristics of electrons and nuclear particles. The electron contribution is by far the more important, although the magnetic properties of certain nuclei are now used in a valuable analytical technique, nuclear magnetic resonance (Unit 30). The principal factor in determining magnetic characteristics of matter is concerned with electron spin (Unit 1). Most chemical compounds and many elements are *diamagnetic* (repelled by a magnetic field). This is because such compounds have no unpaired electrons, and the opposing spins of electron pairs sharing the same orbital result in a zero magnetic moment. Species having unpaired electrons are attracted by a magnetic field and are said to be *paramagnetic*. Paramagnetism can be determined experimentally. On a simple basis, for example, it can easily be demonstrated that a sample of a paramagnetic substance weighs more in a magnetic field than in the absence of the field. More sophisticated techniques are used for quantitative studies.

Quantum mechanics can be used to calculate the magnitude of the magnetic moment of any species as a function of the properties of its electrons. As a rough approximation, the magnetic moment may be estimated from

$$\mu = \sqrt{n(n + 2)} \tag{27.1}$$

Table 27.1   MAGNETIC MOMENTS OF SELECTED IONS

ION	UNPAIRED ELECTRONS	μ CALCULATED FROM EQUATION (27.1)	μ (EXPERIMENTAL)[a]
K^+	0	0	0
Ca^{2+}	0	0	0
Sc^{3+}	0	0	0
Ti^{3+}	1	1.73	1.77 ± 0.07
V^{2+}	3	3.88	3.85 ± 0.05
V^{3+}	2	2.83	2.75 ± 0.15
V^{4+}	1	1.73	1.75 ± 0.05
Cr^{2+}	4	4.90	4.8 ± 0.1
Cr^{3+}	3	3.88	3.8 ± 0.1
Mn^{2+}	5	5.92	5.85 ± 0.25
Mn^{3+}	4	4.90	4.95 ± 0.05
Mn^{4+}	3	3.88	3.9 ± 0.1
Fe^{2+}	4	4.90	5.1 ± 0.3
Fe^{3+}	5	5.92	5.85 ± 0.15
Co^{2+}	3	3.88	3.75 ± 0.45
Co^{3+}	4	4.90	5.1 ± 0.3
Ni^{2+}	2	2.83	3.4 ± 0.6
Cu^+	0	0	0
Cu^{2+}	1	1.73	1.95 ± 0.25
Zn^{2+}	0	0	0
Ga^{3+}	0	0	0

[a] Based on observations of a number of compounds.

in which μ is the magnetic moment (expressed in units of Bohr magnetons) and n is the number of unpaired electrons in the species involved.

Although this equation is an oversimplification of the quantum mechanical treatment, agreement with experimental values (Table 27.1) is sufficiently accurate in most cases to permit assignment of the proper number of unpaired electrons.

For the simpler elements, a straightforward approach could be used for assignment of electron configurations (Unit 1). Such an approach is *not* possible for the transition elements or their ions. The relative energies of the various orbitals depend on the nuclear charge, the nuclear shielding of inner electrons, and electron–electron repulsions. In the larger elements, the energy difference between s, p, d, and f orbitals in the valence electron region becomes fairly small. As a result there is considerable variation in valence electron configuration in the transition elements. Although there is apparently some extra stability associated with filled or half-filled sublevels (e.g., Cu is $4s^1 3d^{10}$ rather than $4s^2 3d^9$ and Cr is $4s^1 3d^5$ rather than $4s^2 3d^4$), this is not a universal rule (e.g., Tc is $5s^1 4d^6$ rather than $5s^2 4d^5$). The problem is even more complex with the monatomic ions, since their interactions with neighboring ions in salts or complexes may result in slight energy-level differences unlike those guessed for the theoretical "isolated" ions.

A set of rules (Hund's rules) can be used for predicting electron configurations of transition elements and ions with generally good success. These require, however, a theoretical treatment beyond the scope of this text. We shall have to be content with using experimentally determined configurations, although the following empirical rules are generally helpful in dealing with the *first-row* transition elements (Sc–Zn) and their monatomic ions:

Elementary texts, particularly many of those designed for high schools, often give the impression that a simple chart of energy levels can be applied in assigning electron configurations for all the elements and monatomic ions. This is simply not true, as can be seen from the "anomalous" cases in Figure 27.1. Such charts are, however, useful guidelines for most simple species.

1. The $4s$ is generally of lower energy than the $3d$ in the neutral atom,[2] *but the reverse is true in the cations.*
2. Within the $3d$, the maximum number of orbitals are used (i.e., unpaired electrons are favored).
3. Filled or half-filled d orbitals are unusually stable.

Example 1

Write electron configurations, showing the noble gas kernel and valence electron spin by arrows (Unit 1) for Co, Cu^{2+}, Ni, Cu^+, and Zn^{2+}. Omit orbital orientation notations (e.g., d_{z^2}).

Solution (According to "first-row rules")
Co (27 electrons):

 Ar, $4s$↑↓, $3d$↑↓, ↑↓, ↑, ↑, ↑

Cu^{2+} (27 electrons):

 Ar, $3d$↑↓, ↑↓, ↑↓, ↑↓, ↑

Ni (28 electrons):

 Ar, $4s$↑↓, $3d$↑↓, ↑↓, ↑↓, ↑, ↑

---

[2]The relative energies of 4s and 3d orbitals vary considerably, so that this "rule" is only a useful simplification for the particular elements involved. See, for example, *Inorganic Chemistry: Principles of Structure and Reactivity*, by James E. Huheey (New York: Harper & Row, 1972), or "The Energies of the Electronic Configurations of Transition Metals," by Robin Hochstrasser, in the March, 1965, issue of the *Journal of Chemical Education* (pp. 154–156).

Cu^+ (28 electrons):

Ar, $3d$↑↓, ↑↓, ↑↓, ↑↓, ↑↓

Zn^{2+} (28 electrons):

Ar, $3d$↑↓, ↑↓, ↑↓, ↑↓, ↑↓

Note that the number of *unpaired* electrons represented according to these "first-row rules" is consistent with observed paramagnetic properties (Table 27.1).

It must be noted that the approach described here to electron configuration and paramagnetism is limited to the hypothetical case of "isolated" atoms and cations. This situation is, of course, closely approximated by gaseous atoms or ions, at low gas densities. In crystalline metals the high degree of valence electron mobility represents a situation quite different from that of the "isolated" metal atom and, as we shall see later in this unit, the electronic characteristics of transition metal atoms or cations may be significantly altered by neighboring ions or molecules.

27.3 THE GROUP IB ELEMENTS: THE COINAGE METALS

For $Cu^+ + e^- \longrightarrow Cu$,
$\quad \mathcal{E}^0 = +0.52$ V

For $Ag^+ + e^- \longrightarrow Ag$,
$\quad \mathcal{E}^0 = +0.80$ V

For $Au^+ + e^- \longrightarrow Au$,
$\quad \mathcal{E}^0 = +1.7$ V

Copper, silver, and gold are all found in nature both in the free state and combined with other elements as alloys and salts. The elements are relatively inert to mild oxidizing agents, acids, and alkalis, with reactivity decreasing from copper to gold. This trend in reactivity is illustrated by the common ores of the elements (Table 27.2). The +I oxidation state (d^{10} configuration) is most stable for compounds of all three elements, although copper and silver form +II and +III compounds and gold forms

Table 27.2 COMMON ORES OF THE COINAGE METALS

ORE	CONTAINS
COPPER	
chalcocite	Cu_2S
chalcopyrite	$CuFeS_2$
cuprite	Cu_2O
malachite	$CuCO_3 \cdot Cu(OH)_2$
azurite	$(CuCO_3)_2 \cdot Cu(OH)_2$
chrysocolla	$CuSiO_3 \cdot 2H_2O$
SILVER	
argentite (silver glance)	Ag_2S
chlorargyrite (horn silver)	$AgCl$
stephanite	Ag_5SbS_4
GOLD	
gold telluride (only naturally occurring gold salt)	$AuTe_4$

Table 27.3 SALTS OF THE COINAGE METALS

SALT	USES
COPPER	
$CuSO_4 \cdot 5H_2O$ (blue vitriol)	used in the electrotyping process and as an additive to swimming pools to kill algae
$[CuSO_4]_2 \cdot [Cu(OH)_2]_3 \cdot 2H_2O$ (bordeaux mixture)	fungicide
$[Cu(CH_3CO_2)_2]_2 \cdot [Cu(OH)_2]$ (verdigris)	a green pigment
$[Cu(CH_3CO_2)_2] \cdot [Cu_3(AsO_3)_2]$ (Paris green)	insecticide
SILVER	
$AgNO_3$ (silver nitrate)	used in dilute solutions as a mild antiseptic and in concentrated solutions as a caustic
$AgBr$ (silver bromide)	used in photographic emulsions[a]
GOLD	
$NaAuCl_4 \cdot 2H_2O$	used in photographic processes for "toning" prints

[a] In black-and-white photography, the "emulsion" contains a colloidal dispersion of tiny silver bromide crystals in gelatin. On exposure to light, silver bromide is "activated" (electrons are raised to excited states) so that it reacts with mild reducing agents (developers) that do not affect the normal silver bromide unexposed to light. After developing, black metallic silver is formed in concentrations depending on the intensity of light exposure. The photograph is then "fixed" by treatment with sodium thiosulfate solution (photographer's hypo) to remove residual silver bromide:

$$AgBr(s) + 2S_2O_3^{2-}(aq) \longrightarrow [Ag(S_2O_3)_2]^{3-}(aq) + Br^-(aq)$$

The negative is then used as the image for preparation of a "positive" print which is produced by the same chemical processes as those used for the original negative.

The techniques of modern-day color and Polaroid photography, of course, involve more complex chemistry.

$+$III and $+$IV compounds. Of these, the only common simple ions stable *in aqueous solution* are Cu^{2+} and Au^{3+}. (All the ions Ag^+, Cu^+, and Au^+ are stable).

The coinage metals are relatively soft, so they are usually alloyed with other metals. Bronzes, for example, are alloys of copper with tin as the principal additive and are used for casting metals. Brasses of various types, used where softer metals are desired, are copper alloyed with zinc and various other metals as determined by the specific properties desired. Silver alloys often contain copper additives, particularly for coins and inexpensive jewelry. Gold, the most malleable and ductile of all the metals, is frequently alloyed with silver ("white gold" alloy) or with both copper and silver. The fraction of gold in an alloy is often expressed on the "carat" scale as parts of pure gold per 24 parts of total metal.

Some of the more common salts of the coinage metals and their uses are listed in Table 27.3.

27.4 THE GROUP IIB ELEMENTS: THE ZINC FAMILY

Zinc and cadmium are very similar to magnesium, although considerably less reactive. These elements are not found free in nature. Yellow CdS is occasionally found in a fairly pure form (greenockite ore), but cadmium is most commonly obtained as a by-product from zinc ores, in which it usually amounts to about 0.5 percent of the zinc content. Mercury is obtained principally from cinnabar, which contains HgS, but free mercury is often found as tiny droplets dispersed in these ore samples.

The principal zinc ores are zinc blende (containing ZnS) and smithsonite (containing $ZnCO_3$). Cadmium may be prepared by distillation of the metal from the crude zinc recovered from its ores.

Recent evidence suggests that the monatomic Hg^+ ion or, perhaps, its monomeric chloride (HgCl) may exist at high temperatures in liquid salts.

Cesium melts around 29°C, gallium at 30°C, and rubidium around 39°C.

Both zinc and cadmium form compounds in which their stable oxidation state is $+II$. Mercury forms both $+II$ and $+I$ compounds. In the latter, however, is found the unusual *diatomic ion* Hg_2^{2+}.

Mercury is the only common metal that is a liquid at room temperature. Although it has a fairly low vapor pressure (Unit 10), the repeated inhalation of mercury vapor from mercury spills in laboratory areas may be quite hazardous. Salts containing mercury(II) are very toxic, and evidence suggests that mercury(II) may be a cumulative poison. Considerable publicity has centered on mercury(II) contamination of natural waters and on the consequences of its accumulation in fish. Interestingly enough, mercury(I) compounds are generally nontoxic, calomel (Hg_2Cl_2) being used medicinally as a stimulant to bile secretion.

Zinc is used as a coating to protect iron from rusting (galvanized iron), as the container cathode for various electrochemical cells (Unit 23), and in many important alloys. Although the metal oxidizes fairly easily, the thin surface oxide coating formed protects the inner metal from further oxidation. Cadmium is also used in this way for metals on which a shiny surface is important.

Mercury forms unique types of alloys called *amalgams* with all other metals except iron and platinum. Many of these are of considerable importance. The particular silver amalgam melting around 100°C is the most common type of material for dental fillings. Amalgamation is a common technique used in recovering certain metals from their ores. The amalgam isolated is decomposed by heating to liberate the free metal and mercury vapor, which is condensed and recycled through the process.

Like mercury(II) compounds, soluble salts containing Zn^{2+} or Cd^{2+} are poisonous. Sparingly soluble compounds of these ions are often employed as bactericidal agents, since the small concentration of metal ion in the saturated solution is sufficient to destroy many microorganisms. Zinc oxide, for example, is employed in many medicinal ointments and salves. Other compounds of interest are listed in Table 27.4.

Table 27.4 COMPOUNDS OF THE GROUP IIB ELEMENTS

COMPOUND	USES
ZINC	
$ZnS \cdot BaSO_4$ (lithopone)	white pigment for paints
$Zn_3(PO_4)_2$ (zinc phosphate)	dental cement
$ZnCl_2$ (zinc chloride)	flux in "soft" soldering (to dissolve metal oxides)
CADMIUM	
CdS (cadmium sulfide)	yellow pigment
$CdSO_4$ (cadmium sulfate)	electrolyte in "cadmium" cells
MERCURY	
HgS (vermilion)	red pigment
K_2HgI_4 (Nessler's reagent)	test reagent for small concentrations of ammonia
$Hg(CNO)_2$ (fulminate of mercury)	extremely shock-sensitive explosive used in blasting caps

Most of these elements are fairly rare in nature and have rather limited and specialized uses. A discussion of their chemistry would be largely of academic interest at this point and will be reserved for more advanced texts dealing with inorganic chemistry.

In contrast to most of these elements, titanium and zirconium are quite abundant. Titanium is the fourth most abundant metal, and zirconium is more abundant than both copper and lead. Both are commercially important components of light-weight structural metal alloys and the oxides (TiO_2 and ZrO_2) are valuable white pigments, commonly used in enamels.

Titanium tetrachloride has a rather interesting use. It is a liquid (boiling around 136°C) and reacts with water to form the white dioxide and hydrogen chloride. When liquid $TiCl_4$ is sprayed into moist air, the hydrolysis results in a dense white smoke of TiO_2; $TiCl_4$ sprays are used in skywriting and in producing military smoke screens.

Chromium and manganese are the most important of these elements from the chemical point of view, although tungsten (wolfram) is certainly a close third. It is probably most familiar as the filament material in incandescent light bulbs. Molybdenum is quite rare in nature, and is chiefly important as an additive to steel to increase its strength and chemical resistance for high-temperature and high-friction uses. Technetium is a man-made element. It is radioactive, and no stable isotope is known, a very unusual circumstance for such a light element. Rhenium is one of the least abundant elements.

The elements of groups VIB and VIIB form compounds in which a very wide range of oxidation states occur (Table 27.5). We shall concentrate our attention on compounds of chromium and manganese, since these are very common in a variety of oxidation–reduction systems.

"Filament life" can be extended by including a trace of I_2 vapor in the light bulb. Tungsten vaporized from the filament reacts with $I_2(g)$. When the resulting $WI_4(g)$ contacts the hot filament, the compound decomposes, redepositing tungsten atoms on the filament.

Chromium compounds

Most compounds of chromium are colored. (The element is named for the Greek word *chroma,* meaning "color.") The hydrated Cr^{2+} ion is blue, and chromium(II) compounds can be prepared by reduction of dichromates by zinc; for example,

$$4Zn(s) + Cr_2O_7{}^{2-}(aq) + 14H^+(aq) \longrightarrow 4Zn^{2+} + 2Cr^{2+} + 7H_2O$$

Chromium(II) is easily oxidized; for example,

$$2Cr(OH)_2(s) \xrightarrow{\text{heat}} Cr_2O_3(s) + H_2O(g) + H_2(g)$$

Although chromium(III) oxide is green, the hydrated Cr^{3+} ion is purple. The Cr_2O_3 is amphoteric, like Al_2O_3, and its reactions with acids and bases may be formulated as

$$Cr_2O_3(s) + 6H^+(aq) \rightleftharpoons 2Cr^{3+}(aq) + 3H_2O(l)$$
$$Cr_2O_3(s) + 2OH^-(aq) + 7H_2O(l) \rightleftharpoons 2[Cr(OH)_4(H_2O)_2]^-(aq)$$

or

$$Cr_2O_3(s) + 2OH^-(aq) \rightleftharpoons \underset{\text{chromite}}{2CrO_2{}^-(aq)} + H_2O(l)$$

The deep green chromite [or tetrahydroxodiaquochromate(III)] ion may be oxidized in alkaline solution to the yellow chromate. This is a common analytical test for chromium:

$$2\underset{\text{green}}{CrO_2{}^-} + 3H_2O_2 + 2OH^- \longrightarrow 2\underset{\text{yellow}}{CrO_4{}^{2-}} + 4H_2O$$

In the presence of other colored ions, the chromate is detected by conversion to an

Table 27.5 **OXIDATION STATES OF THE ELEMENTS IN GROUPS VIB AND VIIB**

ELEMENT	OXIDATION STATE	EXAMPLE
chromium	+II	$Cr(OH)_2$ (unstable to heat: $2Cr(OH)_2 \longrightarrow Cr_2O_3 + H_2O + H_2$)
	+III[a]	Cr_2O_3 (chrome oxide green)
	+IV	CrF_4 (disproportionates in solution)
	+V	$Ba_3(CrO_4)_2$ (disproportionates in solution)
	+VI[a]	$PbCrO_4$ (chrome yellow)
molybdenum	+II	$MoCl_2$ (disproportionates in solution)
	+III	K_3MoCl_6
	+IV[a]	MoS_2 (molybdenite, most common ore)
	+V	Mo_2Cl_{10}
	+VI[a]	$(NH_4)_2MoO_4$ (ammonium molybdate, valuable test reagent for phosphate analysis; yields yellow ammonium phosphomolybdate, $(NH_4)_3PO_4 \cdot 12MoO_3$)
tungsten (wolfram)	+ II–+V	few compounds, mostly of theoretical interest (e.g. WI_4)
	+VI[a]	$CaWO_4$ (in sheelite, a tungsten ore)
manganese	+II[a]	MnS (a *pink* salt)
	+III	MnF_3 (disproportionates in solution to Mn^{2+} and MnO_2)
	+IV[a]	MnO_2 (in pyrolusite, the most important ore)
	+V	Na_3MnO_4 (disproportionates in solution)
	+VI	K_2MnO_4 (potassium *manganate,* stable in alkaline solution, disproportionates to MnO_2 and MnO_4^- in acidic solution)
	+VII[a]	$KMnO_4$ (potassium *permanganate,* important oxidizing agent)
technetium	similar to manganese, +VII most stable (chemically)	*radioactive*
rhenium	similar to manganese	perrhenates (ReO_4^-) are powerful oxidizing agents

[a]Most stable.

Solutions of chromates and dichromates are complicated by the presence of other species. For example, the orange color of "dichromate" solutions is partially due to the presence of $HCrO_4^-$. At low concentrations this is the principal species, but $Cr_2O_7^{2-}$ predominates at high concentrations.

insoluble salt; for example,

$$Ba^{2+}(aq) + CrO_4^{2-}(aq) \longrightarrow BaCrO_4(s)$$
$$\text{yellow}$$

Chromate ion is the principal +VI species in alkaline solution, but on acidification it is converted to dichromate:

$$2CrO_4^{2-} + 2H^+ \rightleftharpoons Cr_2O_7^{2-} + H_2O$$

Both chromate and dichromate are useful oxidizing agents, most frequently being reduced to chromium(III) species.

Manganese compounds

The hydrated Mn^{2+} ion is pink, as are many of the manganese(II) salts. Although the +II oxidation state is relatively stable, conversion to higher oxidation states is not difficult.

Dilute solutions of Mn^{2+} appear colorless because of the very pale nature of the pink color of the hydrated ion.

The most valuable manganese compound from a commercial standpoint is potassium permanganate, used as a disinfectant (particularly in veterinary medicine) and an oxidizing agent for a variety of reactions. Its industrial production illustrates an important point:

Whereas analytical laboratory tests are usually selected for simplicity of reactions, commercial syntheses must consider the most economical route for large-scale production.

Conventional laboratory test for manganese

$$2Mn^{2+}(aq) + \underset{\text{sodium bismuthate}}{5NaBiO_3(s)} + 14H^+ \longrightarrow 2MnO_4^- + 5Na^+ + 5Bi^{3+} + 7H_2O$$

Industrial synthesis of potassium permanganate

1. Fused salt oxidation:

The bismuthate ion is usually represented as BiO_3^-, but it is actually a more complex species, probably a mixture of different oxyanions.

$$2MnO_2(s) + 4KOH(s) + O_2(g) \xrightarrow{\text{fuse}} 2K_2MnO_4(s) + 2H_2O(g)$$

$\underset{\substack{\text{pyrolusite}\\\text{ore}}}{\text{from}}$ $\qquad$ $\underset{\text{air}}{\text{from}}$

2. Electrolytic oxidation (aq K_2MnO_4):

$$\underset{\text{(green)}}{MnO_4^{2-}(aq)} \longrightarrow \underset{\text{(purple)}}{MnO_4^-(aq)} + e^-$$

3. Concentration and recovery:

$$K^+(aq) + MnO_4^-(aq) \xrightarrow[\substack{\text{by}\\\text{evaporation}\\\text{of}\\\text{water}}]{\substack{\text{solution}\\\text{volume}\\\text{reduced}}} KMnO_4(s)$$

Permanganates are among the most important oxidizing agents. In general, MnO_4^- is reduced to Mn^{2+} in acidic solution, but only to MnO_2 in alkaline solution. This difference may be accounted for by the very low solubility of MnO_2 in alkaline solutions. As MnO_4^- is reduced, various intermediate oxidation states of manganese are formed. In acidic solutions, the +IV oxidation state occurs in a *soluble* species, possibly MnO^{2+}, but in alkaline solutions the insoluble MnO_2 precipitates so that it is unavailable for further reduction.

A rather unusual oxide, manganese heptoxide (Mn_2O_7), is formed when permanganates are treated with concentrated sulfuric acid. The heptoxide is a powerful explosive and very shock sensitive.

Examples of the use of permanganates for oxidations have already been discussed (Unit 22).

27.7 THE ELEMENTS OF GROUP VIIIB

$H_2C{=}CH_2 + H_2$

$\xrightarrow{\text{Pt catalyst}} CH_3CH_3$

These elements are commonly classified as the "light platinum triad" (Ru, Rh, Pd) and the "heavy platinum triad" (Os, Ir, Pt).

Chemical similarities are most commonly observed for elements in the same *vertical* columns of the periodic table. With the group VIIIB elements, however, a more striking similarity occurs for elements in the same *horizontal* rows. Iron, nickel, and cobalt (the "iron triad"), for example, have many more similar characteristics than iron, ruthenium, and osmium. The most striking characteristics common to all the group VIIIB elements is their affinity for hydrogen. Finely divided nickel, palladium, and platinum in particular are employed as catalysts (Unit 16) for *hydrogenation* reactions.

The elements of the iron triad are much more reactive than those of the "platinum metals." Most of the latter are sufficiently inert to be found free in nature. Platinum is very inert to acids (except aqua regia) but is readily attacked by hot alkalis. The

Table 27.6 SOME COMPOUNDS OF THE GROUP VIIB ELEMENTS

ELEMENT	COMMON ORES	OXIDATION STATES	EXAMPLE
iron	hematite [Fe_2O_3] magnetite [Fe_3O_4]	$+I$-$+VI$ (only $+II$ and $+III$ are common)	FeS_2 (fool's gold), $FeSO_4 \cdot 7H_2O$ (green vitriol, used in manufacture of inks and dyes), Fe_2O_3 (rust, rouge)
cobalt	cobalt glance [CoAsS] smaltite [$CoAs_2$]	$+I$-$+IV$ (only $+II$ and $+III$ are common)	$Co(OH)_2$ (used as a blue pigment for ceramics), CoF_3 (capable of oxidizing water to oxygen)
nickel	pentlandite [$NiS \cdot (FeS)_2$] garnierite (complex silicate)	$+I$-$+IV$ (only $+II$ is common)	$NiSO_4$ (used to form the ammine complex for nickel plating processes)
ruthenium	found in nature principally as the trimetallic alloy	Ru $+II$-$+VIII^a$	compounds mostly of theoretical interest: OsO_4 is a very unusual metal oxide, melting at $40°C$ and boiling at $100°C$—a useful catalyst for certain organic reactions, but dangerous to work with because it is very corrosive.
osmium		Os $+II$-$+VIII^a$	
iridium		Ir $+I$-$+VI^a$	
rhodium	occurs naturally alloyed with platinum	$+I$-$+IV$ and $+VI$ ($+III$ most common)	$KRh(SO_4)_2$ (rhodium alum, a deep pink salt)
palladium	occurs naturally alloyed with platinum	$+II$-$+IV$ ($+II$ most common)	$PdCl_2$ (used in preparing hydrogenation catalysts)
platinum	found as fairly pure nuggets or, more often, alloyed with other "platinum metals"	$+I$-$+IV$ and $+VI$ ($+II$ and $+IV$ are most common)	H_2PtCl_6 (chloroplatinic acid, can be used to precipitate K^+ or NH_4^+)

^{a}Not $+V$ or $+VII$.

reaction of platinum with aqua regia involves oxidation by NO_3^- and complexing of platinum cations with Cl^-. Both aspects are essential to the overall reaction.

The oxidation states of these elements, together with some of their common compounds and ores, are listed in Table 27.6.

27.8 ANALYTICAL TESTS   The transition metals are most commonly detected by spectroscopic methods, although some wet chemical separations are often used to reduce the complexity of samples prior to spectroscopic assay.

In dealing with complex mixtures, such as polymetallic alloys and certain ores, it is convenient to employ a systematic scheme of separation and identification of the metal ions. An example of such a scheme, designed for analysis of dilute aqueous solutions, is given in Figure 27.2.

Example 2

An ore sample was "field tested" by a survey geologist for evidence of Ag^+, Cd^{2+}, Co^{2+}, Cr^{3+}, Cu^{2+}, Fe^{3+}, Hg^{2+}, Mn^{2+}, Ni^{2+}, and Zn^{2+}. After dissolving a portion of the sample in hot concentrated nitric acid, the geologist diluted the solution, noting first that the concentrated solution was yellow in color. To the diluted solution was added a few drops of 0.3 M HCl. A white precipitate was formed and removed by filtration. This precipitate was insoluble in hot distilled water, thus proving that it was not $PbCl_2$, and was dissolved by 3 M ammonia, which proved that no Hg_2Cl_2 was present. After removal of the original chloride precipitate, an equal volume of 0.1 M thioacetamide, buffered at pH 5, was added to the remaining solution. When this solution was warmed to generate H_2S, a yellow precipitate was formed. This precipitate was removed by filtration and dissolved in dilute nitric acid. An equal volume of 0.1 M thioacetamide in 3 M ammonia was added to the resulting solution. No change was observed until the solution was warmed to generate H_2S, after which a yellow precipitate was formed. An excess of 0.5 M NaOH was added to the original solution

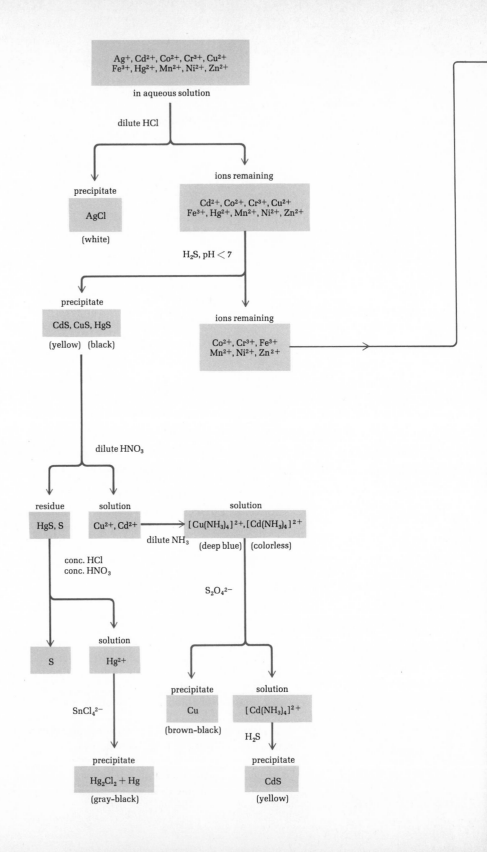

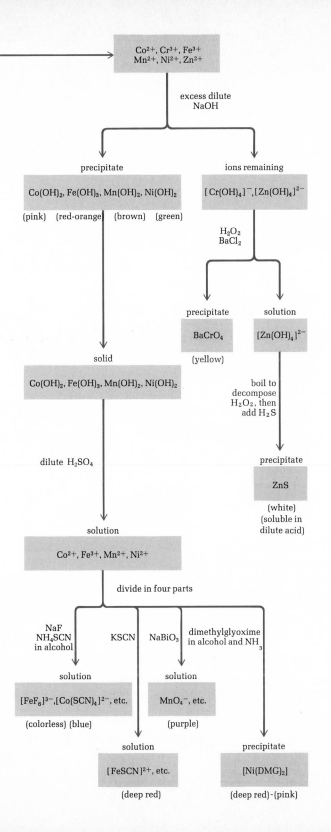

Figure 27.2 Separation and identification of some transition metal ions.

from which the white chloride and yellow sulfide precipitates had been removed. A red-orange precipitate was formed and removed by filtration. The remaining solution showed no change on addition of hydrogen peroxide and barium chloride, but addition of thioacetamide (after removal of residual H_2O_2) and warming resulted in formation of a white, acid-soluble precipitate. When the previously isolated red-orange precipitate was dissolved in 0.5 M H_2SO_4 and the resulting solution was divided into four portions, the following test results were observed:

PORTION NUMBER	TESTED WITH	OBSERVATION
1	NaF, NH_4SCN (in alcohol)	no change
2	KSCN(aq)	deep red solution
3	$NaBiO_3$(s)	no change
4	dimethylglyoxime in alcohol and NH_3	no change

By referring to Figure 27.2, determine which cations were detected in the ore sample.

Solution

1. The description of the white chloride precipitate and the information ruling out $PbCl_2$ and Hg_2Cl_2 (the only other insoluble chlorides, Appendix C) confirm the presence of Ag^+.

2. The description of the yellow sulfide precipitated confirms the presence of Cd^{2+}. The complete solubility of this sulfide in dilute HNO_3 shows that no detectable amount of Hg^{2+} was present, and the failure to observe any blue color on addition of aqueous ammonia rules out Cu^{2+}.

3. The red-orange precipitate formed on addition of sodium hydroxide suggests Fe^{3+} and this is confirmed by the test with KSCN. Specific tests for Co^{2+}, Mn^{2+}, and Ni^{2+} were negative.

4. The precipitation of a white, acid-soluble sulfide confirmed Zn^{2+}.

We conclude, then, that the only ions detected in the ore sample were Ag^+, Cd^{2+}, Fe^{3+}, and Zn^{2+}. (The color of the original sample solution offered some clues, as can be noted from Table 27.8.)

27.9 COORDINATION COMPOUNDS

In Unit 15 we studied the colligative properties of solutions, properties that depend on the *concentration* of solute particles and not on the characteristics of these particles themselves. Osmotic pressure is one such property that can be measured fairly accurately. Let us see how the osmotic pressures of three simple solutions compare, as shown in Table 27.7.

Experimental measurements like these, although crude, can indicate that solute species are combining in solution to form new chemical entities. We know, of course, that ammonia is a base and that as such it would be expected to combine with a proton through formation of a coordinate covalent bond:

Table 27.7   COLLIGATIVE PROPERTIES COMPARED

SOLUTION PREPARED BY MIXING EQUAL VOLUMES OF:	"PARTICLE" CONCENTRATION AFTER MIXING, IF NO REACTION OCCURRED	"THEORETICAL" (AT 27°C) OSMOTIC PRESSURE, IF NO REACTION ($\Pi = M_{particles} \times RT$)	EXPERIMENTAL VALUE OF OSMOTIC PRESSURE AT 27°C	EXPLANATION
0.80 M NH_3 + 0.20 M NaCl	$C_{NH_3} = 0.40\ M$ $C_{Na^+} = 0.10\ M$ $C_{Cl^-} = 0.10\ M$ ——— $0.60\ M$	$\Pi = 0.60 \times 0.082 \times 300$ $\Pi = 15\ atm$	~15 atm	no reaction $C_{NH_3} = 0.40\ M$ $C_{Na^+} = 0.10\ M$ $C_{Cl^-} = 0.10\ M$
0.80 M NH_3 + 0.20 M HCl	$C_{NH_3} = 0.40\ M$ $C_{H^+} = 0.10\ M$ $C_{Cl^-} = 0.10\ M$ ——— $0.60\ M$	$\Pi = 0.60 \times 0.082 \times 300$ $\Pi = 15\ atm$	~12 atm	$NH_3 + H^+ \rightleftharpoons NH_4^+$ $C_{NH_3} = {\sim}0.30\ M$ $C_{NH_4^+} = {\sim}0.10\ M$ $C_{Cl^-} = \underline{\quad 0.10\ M}$ ${\sim}0.50\ M$
0.80 M NH_3 + 0.20 M $CuSO_4$	$C_{NH_3} = 0.40\ M$ $C_{Cu^{2+}} = 0.10\ M$ $C_{SO_4{}^{2-}} = 0.10\ M$ ——— $0.60\ M$	$\Pi = 0.60 \times 0.082 \times 300$ $\Pi = 15\ atm$	~5 atm	$4NH_3 + Cu^{2+} \rightleftharpoons [Cu(NH_3)_4]^{2+}$ $C_{[Cu(NH_3)_4]^{2+}} = {\sim}0.10\ M$ $C_{SO_4{}^{2-}} = \underline{\quad 0.10\ M}$ ${\sim}0.20\ M$

Apparently the copper(II) ion, like H^+, can act as a Lewis acid in seeking the electron pairs of ammonia molecules:

$$4\ H{:}\underset{\cdot\cdot}{\overset{\cdot\cdot}{N}}{:}H + Cu^{2+} \rightleftharpoons \left[\begin{array}{c} H \\ H{:}\underset{\cdot\cdot}{\overset{\cdot\cdot}{N}}{:}H \\ H \qquad H \\ H{:}\underset{\cdot\cdot}{N}{:}\ Cu\ {:}\underset{\cdot\cdot}{N}{:}H \\ H \qquad H \\ H{:}\underset{\cdot\cdot}{\overset{\cdot\cdot}{N}}{:}H \\ H \end{array} \right]^{2+} \left(\begin{array}{c} H \\ H{-}N{-}H \\ H \downarrow H \\ H{-}N{\rightarrow} Cu {\leftarrow}N{-}H \\ H \uparrow H \\ H{-}N{-}H \\ H \end{array} \right)^{2+}$$

This is a simplified representation for the reaction between aqueous Cu^{2+} and aqueous NH_3. Careful studies have shown that there are intermediate species in the equilibrium mixture and that water molecules also coordinate with the Cu^{2+} (Unit 21).

We shall employ the IUPAC nomenclature for complex ions and coordination compounds. You should study this naming system carefully (Appendix F).

The simplest cation, H^+, forms many coordination compounds such as NH_4^+ or H_3O^+. All cations in aqueous solution may be thought of as involving some degree of coordination with lone-pair electrons of oxygen in water molecules.

Complexes such as the tetraamminecopper(II) ion can be characterized, like simple molecules, by particular bond strengths, bond lengths, and bond angles. They differ from most of the simple molecules in the ease with which the metal-ligand bonds form and break. In solution, complex ions should be thought of as dynamic species in which ligands are always leaving and reentering the complex. Thus we are dealing with *average* compositions and *average* geometries. It is still useful to describe these species, however, in the same basic ways used for the more "permanent" molecules such as methane.

Why does the Cu^{2+} ion form a strong discrete complex with ammonia whereas the Na^+ maintains an essentially independent existence? Are complexes limited to the transition elements? The answers to these questions require some understanding of the nature of the bonding in coordination species. As we shall see, coordination compounds are not peculiar to the transition elements, but many of the most interesting complexes do involve these elements. Complexes of the transition elements are often paramagnetic and are frequently colored (Table 27.8). Whereas many complexes of the simpler elements can be explained reasonably well on the basis of simple electrostatic interactions of positive ions with lone-pair electrons of ligands,

Table 27.8 COLORS OF SOME COMPLEXES IN AQUEOUS SOLUTION

ION	COLOR
$[Co(H_2O)_6]^{2+}$	pink
$[Co(NH_3)_6]^{2+}$	rose red
$[Co(H_2O)_2(NCS)_4]^{2-}$	violet
$[Co(NO_2)_6]^{3-}$	yellow
$[Cr(H_2O)_6]^{3+}$	violet
$[Cu(H_2O)_6]^{2+}$	light blue
$[Cu(H_2O)_2(NH_3)_4]^{2+}$	dark blue
$[Fe(H_2O)_6]^{2+}$	pale green
$[Fe(H_2O)_5Cl]^{2+}$	pale yellow
$[Fe(H_2O)_5NO]^{2+}$	brown
$[Fe(H_2O)_5(NCS)]^{2+}$	deep red
$[Fe(CN)_6]^{4-}$	yellow
$[Fe(CN)_6]^{3-}$	red
$[Mn(H_2O)_6]^{2+}$	pink[a]
$[Ni(H_2O)_6]^{2+}$	green
$[Ni(NH_3)_6]^{2+}$	blue

[a] No color is observable in *dilute* solutions.

Figure 27.3 Observed and absorbed colors.

	Absorbed	Observed
7500	infrared	(none)
6500	red	blue-green
5900	orange	blue
5600	yellow	indigo
5400	yellow-green	violet
5100	green	purple
4900	blue-green	red
4600	blue	orange
4200	indigo	yellow
4000	violet	yellow-green
	ultraviolet	(none)

Wavelength (Å)

the transition element complexes require more sophisticated theories of bonding to explain such phenomena as color. These theories are only partially satisfactory at present, but they indicate that the explanations are probably tied to the involvement of d orbitals of the metals.

27.10 COLOR IN COMPLEX IONS

"White" light consists of a continuum of energies, each associated with a particular wavelength of electromagnetic radiation through the equation $E = hc/\lambda$, where E is the energy of the radiation, h is Planck's constant, c is the velocity of light $(3 \times 10^{10} \text{ cm sec}^{-1})$, and λ (lambda) is the wavelength. The visible region in the spectrum constitutes only a small fraction of the range of electromagnetic radiation. Many aqueous solutions of complex ions appear colored because of absorption by the ions of visible light in certain specific wavelength regions. The color *observed* is a combination of those not absorbed, that is, in general, the complementary color of the absorbed color (Figure 27.3).

The question remains as to why some complexes are colored and others are not. The key lies with an analysis of the electronic energy states of the species. Absorption of visible light requires an electronic transition between two energy states separated by an energy equal to that of the light involved. If available energy states are too widely separated, the higher-energy ultraviolet light will be absorbed. If they are too closely spaced, infrared light (lower energy) will be absorbed. Only in the case of energy-level spacings corresponding to the rather narrow spectral region of visible light will color be seen.

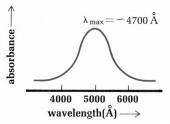

Figure 27.4 Visible absorption spectrum of a complex ion.

Apparently the energy differences involved in electronic energy states of many transition metal complexes are appropriate for absorption of visible light.

The electronic environment of the central ion in a complex may be such that relatively small energy differences can exist between possible states for certain electrons (e.g., alternative ligand–ion orbitals or alternative energy levels for loosely held electrons of the metal ion). The energy associated with a specific wavelength of light may, then, induce an electronic transition from a lower to a higher energy condition. On spontaneous return to the lower energy state, an equivalent amount of energy will be lost, some as light of the original wavelength (but scattered in all directions so that only a small percentage continues in the direction of the incident light beam) and some as other forms of energy, such as vibrational energy of the complex.

Since complex ions in solution exist as families of structures of continually varying distortions from the idealized complex geometry, the electronic environments around central ions will also have minor variations. Thus light energy appropriate for electronic transitions will be a more or less narrow band of wavelengths rather than a single wavelength. Vibrational and rotational excitations, in addition to the electronic transitions, also contribute to the breadth of the absorption band. Different wavelength components of the absorption band are absorbed in different amounts (intensities), so that the wavelength at which maximum absorption occurs (λ_{max}) may be reported as a characteristic of the substance (Figure 27.4).

For a more detailed discussion of absorption spectroscopy, see Unit 30.

27.11 GEOMETRIES OF COMPLEXES

Complex ions and coordination compounds are characterized by specific shapes, although these are subject to distortions by various factors such as thermal vibrations or collisions with other species. Some of the typical structures most frequently encountered are reminiscent of the simple molecules studied earlier (Unit 8), such as the linear *dicyanoargentate*(I),

$$[N\equiv C-Ag-C\equiv N]^-$$

or the tetrahedral *tetrahydroxozincate*(II).

$$\left[\begin{array}{c} O-H \\ \backslash \\ Zn \\ \end{array}\right]^{2-}$$

H—O
H—O
O—H

Bonds are represented by solid lines. Dashed lines are used here to define geometric features, such as coplanar regions.

Others have shapes not often found in simple molecules. The octahedral geometry and the closely related tetragonal arrangement are particularly common to complexes of *d* transition elements, such as the *octahedral* hexanitrocobaltate(III),

$$\left[\begin{array}{c} NO_2 \\ O_2N \cdots NO_2 \\ Co \\ O_2N \cdots NO_2 \\ NO_2 \end{array}\right]^{3-}$$

(all Co—N bonds equal)

or the *tetragonal* diaquotetraaminecopper(II),

$$
\left[
\begin{array}{c}
\text{H} \quad \text{H} \\
\diagdown \; \diagup \\
\text{O} \\
| \\
\text{H}_3\text{N} \text{------}|\text{------} \text{NH}_3 \\
\text{Cu} \\
\text{H}_3\text{N} \text{------}|\text{------} \text{NH}_3 \\
\text{O} \\
\diagup \quad \diagdown \\
\text{H} \quad \quad \text{H}
\end{array}
\right]^{2+}
$$

(Cu—O bond longer than Cu—N bond)

Others are less common, such as the *square planar* tetrachloroplatinate(II),

$$
\left[
\begin{array}{c}
\text{Cl} \text{-----} \text{Cl} \\
| \quad \diagdown \quad \diagup \quad | \\
| \quad \text{Pt} \quad | \\
| \quad \diagup \quad \diagdown \quad | \\
\text{Cl} \text{-----} \text{Cl}
\end{array}
\right]^{2-}
$$

or the *trigonal bipyramidal* pentacarbonyliron(0),

$$
\begin{array}{c}
\text{CO} \\
| \quad \text{CO} \\
| \diagup \\
\text{OC} \text{===} \text{Fe} \; | \\
| \diagdown \\
| \quad \text{CO} \\
\text{CO}
\end{array}
$$

The geometries of complexes could be predicted on the basis of *Gillespie's rules* (Unit 8) *if* appropriate electron-dot or line-bond formulas could be drawn. Therein lies the problem. With the simpler elements (A groups), we could determine the number of valence electrons of the neutral atoms from their group numbers (e.g., nitrogen, a VA element, has five valence electrons). In complex ions and coordination compounds of the transition elements, group numbers generally provide no useful information. The bonding in complexes usually involves electron *pairs* provided by the ligands, rather than valence electrons of the metal atom or ion. However, in some cases (particularly with *square planar* complexes) valence electron lone pairs must be considered as "groups" under the *Gillespie rules*. The electron-dot formulas so useful for compounds and polyatomic ions of the simpler elements are inappropriate for many of the complexes of the *d* transition elements because of the complications of *d*-orbital electrons. We *can* use a modification of *Gillespie's rules* in some cases by disregarding valence electrons of the metal atom or ion and considering ligands as the only "groups." This is satisfactory for two cases: six ligands (octahedral pattern) and five ligands (trigonal bipyramidal pattern). It is, however, limited to these two cases and *there is no simple method for predicting other cases* (e.g., whether four ligands are located around the "central atom" in a tetrahedral or a square planar pattern). Advanced quantum mechanics has been able to make correct predictions in some cases, but we shall have to be content with checking that geometric descriptions are *consistent* with bonding descriptions and experimental information on paramagnetism, as described in Section 27.12.

The *tetragonal* geometry is also referred to as *distorted octahedral*.

27.12 BONDING IN COMPLEXES

We have indicated that the bonding between the central metal atom or ion and the ligand is generally of the coordinate covalent type; that is, the bond pair has been donated by some atom of the ligand. However, once a "bond" has been formed, the origin of the electron pair is less important than some description of the bond itself. There are various ways of describing the bonds in complexes. Ideally, a bond description should be consistent with experimental evidence of coordination number (Table 27.9), bond strengths, and bond angles (geometries of complexes). In addition, other properties of the complex, such as color or paramagnetism, should be explained by the bond theory employed.

Let us consider four theories of bonding that have been applied to complexes.

Electrostatic theory

When the ligands are anions (e.g., F^-, OH^-, CN^-) it is reasonable to suggest that they will be attracted by the opposite charge of the cation and repelled by each other. Then some optimum stability might be expected when a cluster of anions has formed around a central cation in such a way as to minimize repulsive forces (i.e., maximum distance between ligands) while "saturating" the positive field of the cation with the opposite charge of the anions. This model for complexes (Figure 27.5) would, then, suggest a colinear arrangement for complexes containing two ligands (e.g., $[Ag(CN)_2]^-$), a trigonal coplanar geometry with three ligands (e.g., $B(OH)_3$, if we think of this as B^{3+} coordinated with three OH^- ions), a tetrahedral geometry (e.g., $[PbCl_4]^{2-}$) when there are four ligands, or an octahedral shape (e.g., $[Fe(CN)_6]^{2+}$) with six ligands.

The simple electrostatic theory does not consider a "bond" in the sense of the molecular orbital model commonly employed for covalent species. Rather, the strength of the metal ion–ligand attraction is dependent only on the magnitude of the electrostatic interaction.

Although the electrostatic model has the virtue of simplicity, it offers no clues to the problem of color in complexes or to the formation of the many complexes involving neutral species (e.g., $[Cu(H_2O)_6]^{2+}$ or $[Fe(CO)_5]$). In addition, there are cases in which the geometry predicted by electrostatic theory is inconsistent with experimental observation. The tetrachloroplatinate(II) ion, for example, is found to be a square planar complex rather than tetrahedral.

Perhaps the strongest argument against the electrostatic theory is the observation, made very early in the study of coordination compounds, that many complexes of "cations" with "anions" behaved as though the species were connected by a covalent

Table 27.9 USUAL COORDINATION NUMBERS OF SOME COMMON METAL IONS (NUMBER OF METAL-LIGAND BONDS IN STABLE COMPLEXES)

ION	COORDINATION NO.
Ag^+	2
Au^+	2, 4
Cu^+	2, 4
Au^{3+}	4
Cd^{2+}	4
Pt^{2+}	4
Zn^{2+}	4
Al^{3+}	4, 6
Co^{2+}	4, 6
Cu^{2+}	4, 6
Ni^{2+}	4, 6
Ca^{2+}	6
Co^{3+}	6
Cr^{3+}	6
Fe^{2+}	6
Fe^{3+}	6
Pt^{4+}	6

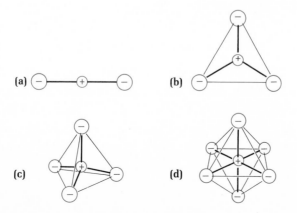

Figure 27.5 Stable "electrostatic" geometries. (a) Linear. (b) Trigonal coplanar. (c) Tetrahedral. (d) Octahedral.

Alfred Werner (1893), the "father of coordination chemistry," was the first inorganic chemist to be awarded the Nobel Prize (1913).

bond. For example, two of the "chlorides" of the species formulated as $CoCl_3 \cdot 4NH_3$ could not be removed by $AgNO_3$, as would have been expected for simple electrostatically bound Cl^-.

$$CoCl_3 \cdot 4NH_3 + \underset{\text{excess}}{Ag^+} \longrightarrow 1AgCl + \cdots$$

The simple electrostatic theory has little to offer. In many cases its prediction of the geometry of complexes is not in agreement with experimental evidence. In addition, it provides no explanation for the colors of complexes and no reliable correlation of magnetic properties. Its only virtue is its simplicity in making "first guesses" of complex geometry.

Valence bond theory (localized molecular orbitals)

Since the model of a covalent bond in terms of overlapping atomic orbitals has proved useful in simple molecules (Unit 8), it seems appropriate that some similar treatment be attempted for bonding in complexes. In particular, the concept of *hybrid* orbitals was especially useful in describing molecular geometry. More than 40 years ago, Linus Pauling suggested that a hybridization involving some *d*-orbital contribution could account for the octahedral and tetragonal shapes of many of the complexes of *d* transition elements. An important feature of this suggestion was the idea that the orbitals to be used were *vacant* orbitals of the metal atom or ion. Electrons were to be supplied by the ligands.

Figure 27.6 Six d^2sp^3 hybrid orbitals (consistent with octahedral geometry). The alternative sp^3d^2 octahedral set would utilize $4s$, $4p_x$, $4p_y$, $4p_z$, $4d_{z^2}$, and $4d_{x^2-y^2}$ orbitals.

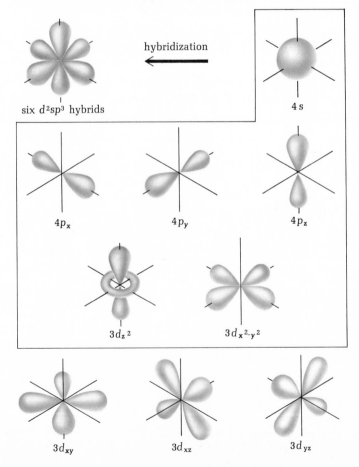

hybridization

six d^2sp^3 hybrids

$4s$

$4p_x$ $4p_y$ $4p_z$

$3d_{z^2}$ $3d_{x^2-y^2}$

$3d_{xy}$ $3d_{xz}$ $3d_{yz}$

(a)

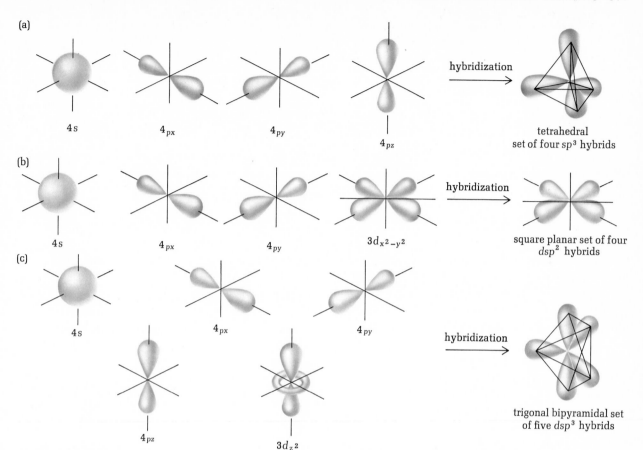

4 s   4 px   4 py   tetrahedral
                                    4 pz   set of four sp³ hybrids

(b)

4 s   4 px   4 py $3d_{x^2-y^2}$ square planar set of four
                                                                         dsp^2 hybrids

(c)

4 s   4 px   4 py

4 pz $3d_{z^2}$ trigonal bipyramidal set
                                                                     of five dsp^3 hybrids

Figure 27.7 Hybridizations yielding. (a) Tetrahedral sp^3. (b) Square planar dsp^2. (c) Trigonal bipyramidal dsp (alternative sp^3d, using 4s, 4p, and $4d_{z^2}$ orbitals).

The other *d* orbitals are oriented *between* the x, y, z axes, so they will not contribute to the necessary shape.

In the elements of the second period (Li–Ne), there is a large energy difference between the valence-electron ground states and the next principal energy level (3). In larger elements the higher energy levels are more closely spaced; there are, for example, relatively small differences among the energies of the 3s, 3p, 3d, 4s, 4p, and 4d orbitals. If we are to "prepare" six hybrid orbitals pointing to the corners of an octahedron, we must select six simple atomic orbitals whose directional character-istics will contribute properly to the desired geometry. These are the s, p_x, p_y, p_z, d_{z^2}, and $d_{x^2-y^2}$ orbitals. Hybridization, then, results in six orbitals all mutually perpen-dicular (Figure 27.6). Such hybrids are called d^2sp^3 if the d orbitals used have a lower principal quantum number than the s and p orbitals used (e.g., $3d_{z^2}$, $3d_{x^2-y^2}$, 4s, $4p_x$, $4p_y$, $4p_z$) or sp^3d^2 when all the orbitals used have the same principal quantum number.

The valence bond theory suggests that, as a first approximation, the metal–ligand bond is described as a molecular orbital resulting from overlap of an electron-pair orbital from the ligand with a vacant hybrid orbital of the metal. It is generally agreed today that this theory was a valuable first step toward bond descriptions for complexes. Although we have emphasized the octahedral case, valence bond theory is equally applicable to other geometries (e.g., tetrahedral sp^3, square planar dsp^2, or trigonal bipyramidal dsp^3 or sp^3d, Figure 27.7). Although 4d orbitals can be used for descriptions of octahedral hybrids (sp^3d^2) or trigonal bipyramidal hybrids (sp^3d) when no *vacant* 3d orbitals are available, it should be noted that square planar orbitals always utilize a vacant d orbital of principal quantum number *lower* than that of the s and p orbitals used. Since the d orbitals within a principal quantum level are always of higher energy than the s and p orbitals in the same level, sp^3 hybrids are energetically favored over the sp^2d alternative. Thus zinc(II) ion, having no vacant 3d orbitals, is expected to form only tetrahedral complexes when four ligands are involved.

These hybrid orbital descriptions are not limited to complex ions of the d transition metals. The trigonal bipyramidal PCl_5 and octahedral PCl_6^- (Unit 26), for example, are described by dsp^3 and d^2sp^3 hybridization, respectively.

One problem of the valence bond theory arises from the fact that both d^2sp^3 and sp^3d^2 hybrids are consistent with octahedral geometry and both dsp^3 and sp^3d sets are trigonal bipyramidal. How do we know which model to select for a particular complex? In many cases we have two equally plausible choices, and some experimental information is needed before the proper choice can be made. In either case, it is important to remember that the atomic orbitals selected for hybridization must be vacant when *coordinate* covalent bond descriptions are involved.

Example 3

Both hexaamminecobalt(III) and hexafluorocobaltate(III) are octahedral complexes. The former is diamagnetic, and the latter is paramagnetic. Suggest appropriate valence bond descriptions for the hybrid orbitals of cobalt(III) in these complexes.

Solution

Cobalt atom is described as $4s^23d^7$ and the cobalt(III) ion as $4s^03d^6$. For d^2sp^3 hybrids, we would require vacant $3d_{z^2}$, $3d_{x^2-y^2}$, $4s$, $4p_x$, $4p_y$, and $4p_z$ orbitals. For this to occur, the six valence electrons of cobalt(III) ion would have to exist as three pairs ($3d_{xy}\uparrow\downarrow$, $3d_{yz}\uparrow\downarrow$, $3d_{xz}\uparrow\downarrow$). This would be consistent with the diamagnetic properties of hexaamminecobalt(III), so we suggest d^2sp^3 hybridization in this complex.

The paramagnetism of hexafluorocobaltate(III) requires some unpaired electrons. Since this is not possible with six 3d electrons if the necessary two orbitals are left vacant, we assume that vacant $4d_{z^2}$ and $4d_{x^2-y^2}$ orbitals are used instead. This complex is then described as involving sp^3d^2 hybrids.

Note that experimental information (in this case, paramagnetism) is usually necessary when the valence bond theory permits two possible hybridization descriptions.

Example 4

Tetracarbonylnickel(0) has a low boiling point (43°C), suggesting a highly symmetrical structure. Remembering that electron configurations of "isolated" atoms may be altered by interactions with ligands, suggest the probable geometry of tetracarbonylnickel(0). Is this compound likely to be paramagnetic? [Carbon monoxide is represented as ($:C:::O:$).]

Solution

The valence electron configuration of the "isolated" nickel atom is $4s^23d^8$. If this configuration were retained by nickel in the coordination compound, neither of our bonding descriptions for tetracoordinated species (sp^3 or dsp^2) could be employed since the necessary 4s orbital would not be vacant. We may assume, however, that coordination of nickel with carbon monoxide could supply sufficient energy to provide the higher energy $4s^0 3d^{10}$ configuration. This still leaves no vacant $3d$ orbital, so we propose the sp^3 hybridization rather than dsp^2. As a result, we predict a *tetrahedral* geometry and since all valence electrons are paired, the compound is not expected to be paramagnetic. (Both predictions are consistent with experimental observations.)

Example 5

Suggest two alternative hybrid orbital descriptions for pentacarbonyliron(0). What experimental evidence might be used to distinguish between these alternatives?

Solution

Since we expect a pentacoordinated species to exhibit trigonal bipyramidal geometry, either dsp^3 or sp^3d orbitals could be involved as consistent with spatial arrangements. Both descriptions require a vacant $4s$ orbital on iron, leaving it as a $4s^0 3d^8$ species (rather than the $4s^2 3d^6$ of the "isolated" atom). If the eight d electrons exist as four pairs, the resulting dsp^3 description would be feasible and the compound should be diamagnetic. If, on the other hand, the eight d electrons are distributed among all five d orbitals, we would have sp^3d hybridization (using the vacant $4d$ orbital) and a paramagnetic compound ($\uparrow\downarrow, \uparrow\downarrow, \uparrow\downarrow, \uparrow, \uparrow$). Thus experimental evidence of paramagnetic or diamagnetic character could distinguish between the alternative bonding descriptions. (It has been determined that this compound is diamagnetic.)

For many years the valence bond theory was the principal description applied to coordination chemistry. Its virtue was simplicity in explaining geometric and magnetic properties of complexes, although it was of doubtful value in *predicting* these properties. In order to do more than "fit" experimental observations, the valence bond approach required expansion to the extent that its simplicity was lost. Since the basic theory failed to include descriptions for excited states, it did not offer any explanation for the color of complexes. There is no question, however, of the value of this theory as a useful step toward a broader theoretical treatment.

Crystal field theory

H. Bethe (1939) and J. H. Van Fleck (1932) were primarily responsible for introducing the crystal field model.

At the same time that Pauling and other chemists were developing the valence bond model (beginning around 1930), a group of physicists were exploring an alternative theoretical approach. The *crystal field theory,* as originally proposed, considered the interaction of metal ion and ligands as purely electrostatic; that is, the bonding description did not concern itself with covalency in the sense of orbital *overlap.* However, unlike the simple electrostatic theory, which considered the metal ion as a spherically symmetrical positive charge region, the crystal field theory included the directional characteristics of d orbitals on the metal ion and the ways these might be affected by the charge field of a group of surrounding ligands.

In an isolated ion (or in an ion surrounded by a spherically symmetrical field of opposite charge), the d orbitals are degenerate (of equivalent energy). If, however, the metal ion were placed in an *unsymmetrical* field of opposite charge, the d orbitals would no longer have the same energy. Negative ions or the negative region of an electron pair (such as the lone pair in an ammonia molecule) would exert an electrostatic repulsion on metal-ion electrons in the d orbitals, and this repulsion would be greater for orbitals directed toward the ligand than for orbitals not

Figure 27.8 Energy-level splitting by ligands in an octahedral complex. (a) Complete set (doughnut-shaped segment in xy plane of d_{z^2} orbital omitted from drawing for simplicity) of d orbitals in an octahedral array of ligands, showing that d_{z^2} and $d_{x^2-y^2}$ orbitals are directed toward ligands (thus representing "unstable" locations for the metal's electrons). The remaining d orbitals are directed *between* ligands, so electrons are more "stable" (of lower energy) in these locations. (b) Energy-level diagram showing splitting in an octahedral field.

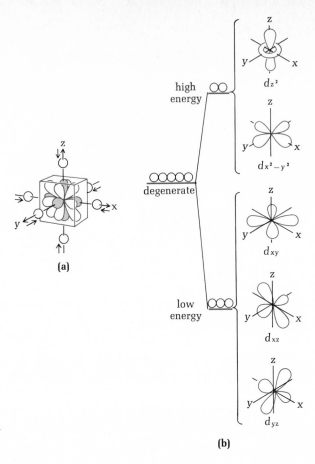

(a)

(b)

Unfortunately, the interaction between the electrostatic field of the ligand and the metal atom or ion is a function of *both* species. As a result, a ligand having a *strong* influence on d-orbital electrons of one metal ion may have only a *weak* influence for a different metal ion.

"pointed at" the ligand. We can then consider that a splitting of energy levels occurs, with some d orbitals of higher energy than others. This is fairly easy to see for the case of an octahedral complex (Figure 27.8). The magnitude of this energy difference would depend on the field strength of the ligand.

Crystal field theory addresses two important questions not answered by the valence bond model. (1) By providing two different energy levels for d-orbital electrons, crystal field theory offers a ground-state and an excited-state description appropriate to the explanation of color in complexes. By absorption of light of energy equal to the difference between energy levels, electrons may be excited to higher energy states. (2) The tendency for electrons to be forced into lower energy orbitals by approaching ligands favors maximum electron pairing (low spin). However, electrostatic repulsion of electrons favors minimum pairing (high spin). In complexes for which both possibilities exist, we may now *predict* that only ligands of high field strength can overcome the electron–electron repulsions to produce a low-spin complex.

Based on experimental determinations, a "ranking" of ligands in terms of their apparent electrostatic fields has been made (Table 27.10). Crystal field theory, then, would have correctly predicted that $(CoF_6)^{3-}$ is more likely to be a high-spin complex than $[Co(NH_3)_6]^{3+}$.

The crystal field theory, then, considers the bonds to be essentially electrostatic and the complex geometry to be that permitting maximum distances between ligands. In addition, it offers a theoretical explanation for color in complexes and a model suitable for prediction of magnetic properties.

Ligand field theory Neither the valence bond nor the crystal field theory is entirely satisfactory. Both encounter difficulties in making quantitative correlations with experimental results, and neither is completely reliable in predicting geometries of complexes. A modification of crystal field theory to introduce some factors for covalent bonding (ligand field theory) has proved quite useful in improving agreement between observations and theory.

Ligand field theory is an extension of the basic ideas of crystal field theory to incorporate contributions of covalent bonding in terms of molecular orbital concepts. In its present state, ligand field theory is still a semiempirical treatment of bonding. The details of the quantum mechanics are quite complicated; nevertheless this model is fundamentally similar to the more elegant molecular orbital treatments now used for other types of molecules. While retaining the crystal field ideas of energy-level splitting, ligand field theory adds the concepts of molecular orbital formation between metal and ligand. Both σ and π bonding may be invoked, the latter involving formation of a π bond (Figure 27.9) between a d orbital of the metal and a p or d orbital of the ligand. However, the most accurate and reliable treatment promises to come from a detailed quantum mechanical approach, utilizing wave equations descriptive of both covalent and ionic contributions to ligand-metal interaction. The description of this technique is beyond the scope of this text, and its implementation requires the use of computer technology for solution of the very complex wave functions involved.

We shall find the valence bond and crystal field theories satisfactory for qualitative explanations of geometric, magnetic, and optical properties of complex ions, recognizing their limitations to fairly simple situations.

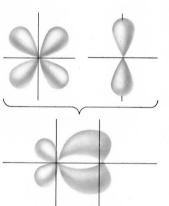

Figure 27.9 *p-d* π Bond.

Example 6

Experiments show that both hexafluoroferrate(II) and hexacyanoferrate(II) are octahedral complexes, but only one is paramagnetic. Use crystal field theory to predict which of these is the paramagnetic complex, then write valence bond descriptions for the hybrid orbitals for iron in both complexes.

Solution

Iron(II) is designated as $4s^0 3d^6$. Two alternative arrangements are possible for the d orbital electrons in a complex (Figure 27.8):

Low spin:

$3d_{xy}\uparrow\downarrow$, $3d_{yz}\uparrow\downarrow$, $3d_{xz}\uparrow\downarrow$

High spin:

$3d_{xy}\uparrow\downarrow$, $3d_{yz}\uparrow$, $3d_{xz}\uparrow$, $3d_{z^2}\uparrow$, $3d_{x^2-y^2}\uparrow$

Since CN^- is a high-field ligand and F^- is a low-field ligand (Table 27.10), the former is more likely to have sufficient field strength to crowd the d electrons into the low-spin pattern. Hence hexafluoroferrate(II) is probably the paramagnetic complex.

With the above information, we can now select appropriate hybridizations for valence bond descriptions of the complex, remembering that *vacant d* orbitals are required since these are to be filled by lone-pair electrons from the ligands. For the high-spin complex, $(FeF_6)^{4-}$, none of the $3d$ orbitals are vacant, so $4d$ orbitals are used, giving sp^3d^2 hybrids. Two of the $3d$ orbitals are vacant in the low-spin complex, $[Fe(CN)_6]^{4-}$, so it is described by d^2sp^3 hybrids.

Example 7

Aqueous solutions of titanium(III) salts appear violet because of the absorption of light in the yellow-green region of the spectrum, with principal absorption (λ_{max}) around 4930 Å. Use the crystal field theory to suggest electron configurations for the ground state and the excited state for titanium(III), in the octahedral $[Ti(H_2O)_6]^{3+}$, then (Unit 1) calculate the energy difference between these two states, assuming that this corresponds to the energy of light absorbed by the hydrated ion, hexa-aquotitanium(III).

$$E = \frac{hc}{\lambda} \qquad \text{(Unit 1)}$$

(Appendix A gives appropriate conversion factors.)

Note that in assigning electron configurations, any of a set of degenerate orbitals may be specified. More advanced discussions use special designations for such situations.

Solution

Titanium(III) is characterized as $4s^03d^1$, so we can assign the ground-state configuration as

Equivalent choices would be $(Ar)3d_{yz}\uparrow$ or $(Ar)3d_{xz}\uparrow$.

$$(Ar)3d_{xy}\uparrow$$

For the splitting induced by an octahedral array of ligands, an excited state may be represented as

Equivalent alternative would be $(Ar)3d_{x^2-y^2}\uparrow$.

$$(Ar)3d_{z^2}\uparrow$$

The energy difference corresponding to excitation by light of wavelength 4930 Å, as calculated *per atom* (Unit 1) is

$$E = \frac{hc}{\lambda} = \frac{1.58 \times 10^{-34}\ \text{cal sec} \times 3.0 \times 10^{10}\ \text{cm sec}^{-1}}{4.93 \times 10^{-5}\ \text{cm}}$$

This corresponds to 58 kcal mole^{-1} (Appendix A).

$$= 9.6 \times 10^{-20}\ \text{cal}$$

27.13 LIGAND CHARACTERISTICS

Electron-pair donors may be simple anions (e.g., Cl^-, CN^-), neutral molecules (e.g., H_2O, NH_3, $H_2NCH_2CH_2OH$), or molecular-type substances containing ionic groups (e.g., EDTA, Figure 27.10, or macromolecules such as proteins). The term *ligands* has now been generally accepted to describe substances acting as electron-pair donors, either before or after formation of the coordinate bond. Lewis bases are substances

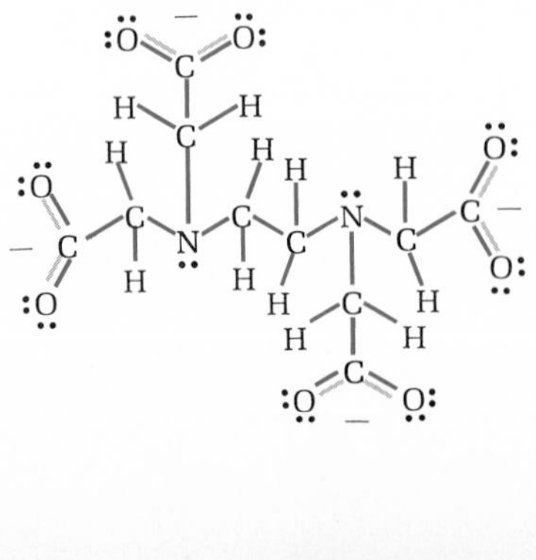

Figure 27.10 EDTA (ethylene-diaminetetraacetate).

Figure 27.11   Chelation.

$$2H_2O: \ + \ 2H_2\overset{..}{N}CH_2CH_2\overset{..}{N}H_2 + Cu^{2+}$$

monodentate   ethylenediamine—
ligand   a chelate ligand
                   (bidentate)

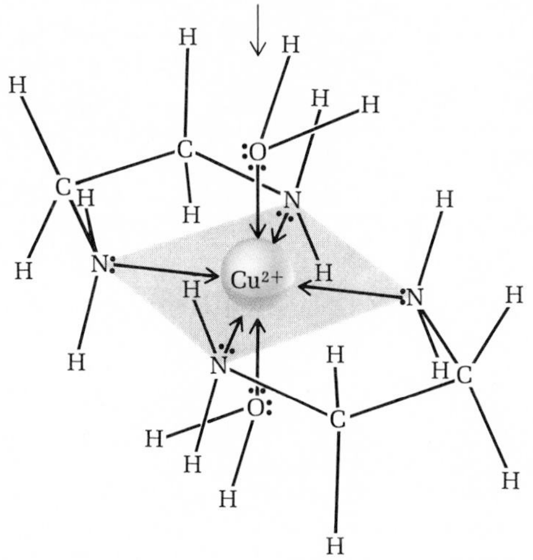

that have available electron pairs, and this term would not, strictly speaking, apply to the substance once its electrons are involved in a coordinate bond.

Ligands are frequently classified in terms of the number of electron pairs the ligand donates to a specific cation (or atom). *Monodentates,* then, donate a single electron pair; *bidentates,* two electron pairs; *tridentates,* three pairs; and so on. *Multidentate* is a general term for ligands donating more than one electron pair.

The term *dentate* offers the picture of the ligand "sinking its teeth" into the cation. (Compare "dentures," "dentist," etc.)

The ability of multidentate ligands to "clasp" around a central ion or atom has led to use of the term *chelation* (from the Greek *chele,* "claw") to describe such complex formation (Figure 27.11).

Affinity relates to the strength of the ligand–metal bond and does not necessarily correlate with the ligand field strength (Table 27.10).

The affinities of various ligands for any given metal ion depend on a number of factors, many of which are difficult to assess from simple considerations.[3] Such factors include

Electrical fields of the ligand and the metal: This is related, but not in a simple way, to charge density and electron density.

Type of bond (e.g., σ or π) possible between the ligand group and the metal ion or atom: Available orbitals on both the ligand group and the cation must be considered.

Steric factors: Ligands must be able to approach the central ion closely enough for effective coordination. If more than one ligand group is involved (as is usually the case), then the geometric arrangement of other ligands, including solvent molecules, must be considered in determining whether the new group can fit properly into a suitable region for bonding.

Thermodynamic factors: Changes in order (related to entropy) associated with complex formation, as well as changes in other thermodynamic quantities, are useful considerations, although often difficult to evaluate quantitatively.

[3]For an advanced discussion of some ligand characteristics, see "Ligand Group Orbitals and Normal Molecular Vibrations," by Chao-Yang Hsu and Milton Orchin, in the November, 1974, issue of the *Journal of Chemical Education* (pp. 725–729).

Some generalizations are available for qualitative discussions of complex formation, but there are numerous exceptions that require detailed study at a more advanced level. Simple electrostatic forces between ions and ligand electron-pair centers appear to account fairly well for complexes of alkali, alkaline earth, some first-row transition metal, lanthanide series, and actinide series ions. For ions of the more electronegative metals (e.g., Pb, Au, Ag, Pt, Hg) and such transition metal ions as Cu^{2+}, Ni^{2+}, Co^{2+}, covalent-type bond interactions complicate simple electrostatic considerations. These ions, unlike the former groups, form strong complexes with such species as I^-, S^{2-}, CN^-, CO, NO_2^-, in which rather complicated bond descriptions are applied. Values of formation constants for complexes (Appendix C) give the most reliable information on ligand–ion interactions from an experimental point of view. Such constants were discussed in Unit 21.

HEMOGLOBIN: A BIOCHEMICAL COMPLEX

If you watch television very much, you have surely seen commercials warning of "iron-poor tired blood." With the proper iron supplement, the formerly drab, droopy housewife suddenly meets her husband at the door with a rose in her teeth or wins the Mt. Lemmon uphill bicycle race. Such results may be a bit overdramatic, but there is no question of the critical importance of iron for the human body.

If all the iron in the average man were isolated, there would only be enough to make a couple of nails, but this iron is essential to the entire life process. Iron is involved in a number of important enzyme systems, in the hemoglobin which serves as the oxygen carrier in the blood, in the myoglobin which acts to store oxygen in muscle tissue, and in the cytochromes which function in the complex scheme of oxidation of organic compounds in the body.

Atomic iron and the simple ions Fe^{2+} and Fe^{3+} do not have the characteristics needed for these critical functions. In every case, the iron species involved is bound by coordinate covalent bonds to a complex organic compound, and it is the resulting *coordination compound* that possesses the necessary biochemical activity.

Probably the most interesting and important of all the coordination compounds are those in which iron(II) is bound in the center of an organic compound called a *porphyrin*. The iron–porphryin complex, called a *heme* group, is attached by covalent bonds to large protein molecules in hemoglobin, myoglobin, and the cytochromes. The type of attachment varies in these three classes of compounds, but in every case the heme group is present as a flat (coplanar) region with the central iron coordinated to four nitrogens of the porphyrin ring system. Iron(II) is most stable when coordinated with six ligand groups in something approximating an octahedral geometry. The remaining two ligands may be furnished by amino acids, which are part of the complex chain of the protein involved. In cytochrome *c*, for example, the sulfur of a methionine segment furnishes the electron pair for one coordinate bond and a ring nitrogen of a histidine segment furnishes the electron pair for the remaining bond. Alternatively, one of the ligands may be molecular oxygen, and this is the

A heme group, an important type of coordination compound. Iron(II) is coordinated with the $2-$ ion of a porphyrin to yield a coordination compound of zero net charge. Note the fully delocalized electron system of the coplanar porphyrin ring system.

Hemoglobin is the principal *reactive* component of erythrocytes. These cells contain about 60 percent water.

important feature of the *heme* group—its ability to bind an oxygen molecule to the central iron.

In the very simplest organisms, the diffusion of dissolved oxygen may be sufficient to distribute the oxygen throughout the organism. This is far too inefficient a process for the higher organisms. In man, oxygen is transported through the bloodstream as O_2 molecules coordinated to the iron in hemoglobin. This complex protein (molecular weight $\sim$68,000) constitutes about 35 percent of the red blood cells (erythrocytes). Each hemoglobin molecule contains four heme groups and is thus capable of transporting four O_2 molecules. The bound oxygen is transferred to the heme groups of myoglobin proteins in muscle tissue, where it is stored until needed for oxidation of organic "fuel." The actual transfer mechanisms and oxidation steps, the latter eventually involving other heme groups in the cytochrome systems, are still not completely understood.

In addition to O_2, the heme group will bind strongly to molecules such as carbon monoxide or ions such as cyanide. These complexes are more stable than that with molecular oxygen, which helps to explain the toxicity of CO and CN^- to humans.

Iron is not the only metal involved in biochemical coordination compounds. Some lower organisms utilize copper complexes in their oxygen-transport systems, magnesium is the key metal in the chlorophyll system of plants, and a variety of metals are employed in other complex systems such as the catalytic enzymes.

The subject of coordination chemistry is not simple. We have been able to touch on only a few of its more important aspects. For those interested in exploring the subject in greater depth, the following references will provide a useful start.

SUGGESTED READING

Basolo, Fred, and Ronald Johnson. 1964. *Coordination Chemistry*. Menlo Park, Calif.: W. A. Benjamin.

Earnshaw, A. 1974. *The Chemistry of the Transition Elements*. New York: Oxford.

Huheey, James E. 1972. *Inorganic Chemistry: Principles of Structure and Reactivity*. New York: Harper & Row. (Advanced.)

Quagliano, J. V., and L. J. Vallarino. 1969. *Coordination Chemistry*. Indianapolis, Ind.: Heath-Raytheon.

EXERCISES

1. Give the most common oxidation states for titanium; manganese; cobalt; copper; zinc

2. The electron configuration for titanium may be represented as (Ar), $4s\uparrow\downarrow$, $3d\uparrow$, $\uparrow$.
   a. Write electron configurations in this form for Cr, Cr^{2+}, Mn^{2+}, Fe, and Ni^{2+}.
   b. For each of the preceding species, calculate the magnetic moment [equation (27.1)].

3. Draw representations for
   a. dichloroaurate(I) (linear); b. tetrachloronickelate(II) (tetrahedral); c. tetracyanonickelate(II) (square planar); d. hexaamminecobalt(III) (octahedral).

4. Predict which member of each pair is more likely to be paramagnetic. Explain the reasons for each choice.
   a. $[Zn(NH_3)_4]^{2+}$ or $[Cu(NH_3)_4(H_2O)_2]^{2+}$; b. $[Fe(CN)_6]^{4-}$ or $[FeF_6]^{4-}$

5. Suggest a valence bond description (i.e., specify probable hybrid orbitals of the metal) for each of the following complexes:
   a. $[Zn(NH_3)_4]^{2+}$ (tetrahedral); b. $[PtCl_4]^{2-}$ (square planar); c. $[Fe(CN)_6]^{4-}$ (octahedral) (See Exercise 4b); d. $[FeF_6]^{4-}$ (octahedral) (See Exercise 4b.)

6. A solution may contain any or all of the ions listed in Figure 27.2. From the following observations on the results of a systematic analysis, which ions were present in the original solution? Write balanced equations for each reaction observed.

PROCEDURE	OBSERVATIONS
1. add aq HCl	1. no apparent reaction
2. add H_2S at pH < 7	2. black precipitate
3. dissolve ppt from (2) in HNO_3, then add excess NH_3	3. deep blue solution
4. add $S_2O_4^{2-}$ to solution from (3)	4. brown-black precipitate
5. add H_2S to solution from (4)	5. yellow precipitate
6. add excess NaOH to solution from (2)	6. no apparent reaction
7. add H_2O_2 and $BaCl_2$ to solution from (6)	7. no apparent reaction
8. add H_2S to solution from (7); after H_2O_2 removed	8. white precipitate

7. Zinc oxide, a useful fungicide, can be prepared from zinc blende ore by roasting the ore in air and then purifying the crude oxide. What weight (in kilograms) of zinc oxide could be produced from 5.00 kg of a zinc blende ore that analyzed at 48.6% zinc sulfide, if the net conversion is 66.7% efficient? ($ZnS + 2O_2 \longrightarrow ZnO + SO_3$)

8. A 25.0-ml aliquot of a solution containing Fe^{2+}, prepared during an assay of a pyrite ore, was titrated to a faint purple endpoint by 38.2 ml of 0.020 M $KMnO_4$. What was the molarity of the Fe^{2+} solution? (The MnO_4^- was reduced to Mn^{2+}.)

9. The net formation constant for tetraamminezinc(II) is 3×10^9. What would be the approximate potential of the electrochemical cell symbolized by

   $Zn\,|\,[Zn(NH_3)_4]^{2+}\,(1.0\,M)\,\|\,Zn^{2+}\,(1.0\,M)\,|\,Zn$

   (The anode compartment contains aqueous ammonia. Assume $C_{NH_3} = 1.0\,M$ and that intermediate complexes are negligible.)

10. What average current (in amperes) must be maintained to electroplate 33 g of manganese per hour from a solution of manganese(II) salt?

PRACTICE FOR PROFICIENCY

Objective (a)

1. Give the most common oxidation states for scandium; chromium; nickel; gold; silver; vanadium; iron

Objective (b)

2. The electron configuration for titanium may be represented as (Ar) $4s\uparrow\downarrow$, $3d\uparrow$, $\uparrow$.
   a. Write electron configurations in this form for Fe; Fe^{2+}; Fe^{3+}; Sc; Sc^{3+}; V; V^{3+}; Cr^{3+}; Mn; Co; Co^{2+}; Co^{3+}; Ni; Cu; Cu^+; Cu^{2+}; Zn; Zn^{2+}
   b. For each of the preceding species, calculate the magnetic moment [equation (27.1)].

Objective (c)

3. Supply the missing formulas and IUPAC names for
   a. diamminesilver(I) _____
   b. $[Cu(CN)_4(H_2O)_2]^{2-}$ _____
   c. tetrachlorodiaquoaurate(III) _____

d. $[Co(NH_3)_6]^{2+}$ _____

e. hexacyanoferrate(II) _____

f. $K_3[Co(NO_2)_6]$ _____

g. tetraamminecopper(II) sulfate _____

h. $Na[Ag(CN)_2]$ _____

i. hexaaquoscandium(III) nitrate _____

j. $Cu_3[Fe(CN)_6]_2$ _____

Objective (d)

4. Draw representations for
   a. dicyanoargentate(I) (linear); **b.** tetrachloroplumbate(II) (tetrahedral); **c.** tetrachloroplatinate(II) (square planar); **d.** hexaaquocopper(II) (octahedral); **e.** pentacarbonyliron(0)

5. Select the most plausible and consistent combination of hybrid orbital notation and magnetic-geometric description for the hexachlorocobaltate(III) ion.
   a. dsp^2, diamagnetic-tetrahedral
   b. d^2sp^3, paramagnetic-octahedral
   c. sp^3d^2, diamagnetic-hexagonal
   d. sp^3, paramagnetic-square planar
   e. d^2sp^3, diamagnetic-octahedral

6. Select the most plausible and consistent combination of hybrid orbital notation and magnetic-geometric description for the tetraamminezinc(II) ion.
   a. dsp^2, diamagnetic-square planar
   b. sp^2d, paramagnetic-tetrahedral
   c. sp^3, diamagnetic-tetrahedral
   d. sp^3, paramagnetic-square planar
   e. sp^3d^2, diamagnetic-square planar

7. Select the most plausible and consistent combination of hybrid notation and magnetic-geometric description for the hexacyanoferrate(III) ion.
   a. dsp^2, paramagnetic-tetrahedral
   b. sp^3d^2, paramagnetic-hexagonal
   c. d^2sp^3, diamagnetic-octahedral
   d. d^2sp^3, paramagnetic-octahedral
   e. sp^3, paramagnetic-square planar

8. Select the most plausible and consistent combination of hybrid orbital notation and magnetic-geometric description for the hexacyanoferrate(II) ion.
   a. dsp^2, diamagnetic-tetrahedral
   b. sp^2d, paramagnetic-tetrahedral
   c. sp^3d^2, diamagnetic-hexagonal
   d. d^2sp^3, paramagnetic-octahedral
   e. d^2sp^3, diamagnetic-octahedral

9. Select the most plausible and consistent combination of hybrid orbital notation and magnetic-geometric description for the hexaaquoiron(II) ion.
   a. d^2sp^3, paramagnetic-octahedral
   b. d^2sp^3, diamagnetic-hexagonal
   c. sp^3d^2, paramagnetic-octahedral
   d. dsp^2, paramagnetic-tetrahedral

   e. sp^3d^2, diamagnetic-square planar

10. Select the most plausible and consistent combination of hybrid orbital notation and magnetic-geometric description for the tetracyanonickelate(II) ion.
    a. sp^3, diamagnetic-tetrahedral
    b. dsp^2, paramagnetic-square planar
    c. d^2sp^3, paramagnetic-octahedral
    d. dsp^2, diamagnetic-square planar
    e. sp^3d^2, paramagnetic-tetrahedral

Objective (e)

11. A solution may contain any or all of the ions listed in Figure 27.2. From the following observations, which ions were definitely present in the solution? Which were definitely absent?

PROCEDURE	OBSERVATIONS
1. add aq HCl	1. white ppt insoluble in hot water, soluble in aq NH_3
2. remove insoluble chloride, add H_2S at pH $<$ 7 to solution	2. no apparent reaction
3. add excess NaOH to solution from (2)	3. red-brown ppt
4. add H_2O_2, $BaCl_2$ to supernatant solution from (3)	4. no apparent reaction
5. remove H_2O_2, add H_2S to solution from (4)	5. no apparent reaction
6. dissolve ppt from (3) in dil H_2SO_4, divide in 4 parts	6. pale yellow solution
7. to one part from (6), add NaF, NH_4SCN, ethanol	7. no apparent reaction
8. to second part from (6), add KSCN	8. deep red solution
9. to third part from (6), add $NaBiO_3$	9. no apparent reaction
10. to fourth part from (6), add dimethylglyoxime	10. deep red ppt

12. A solution may contain any or all of the ions listed in Figure 27.2. When 0.1 M HCl was added to the solution, no change was observed. Addition of H_2S to the acidic solution produced a black precipitate, which was removed by filtration. When the filtered solution was treated with excess 0.5 M sodium hydroxide, no change was observed. Which cations *may* have been present in the original solution? Which were *not* present in detectable concentrations?

Objective (f)

13. **a.** In the industrial synthesis of potassium permanganate, what weight of manganese dioxide is required for production of 25 kg of potassium permanganate, assuming the complete process is 78 percent efficient?
    b. How long must a current of 15.0 amp be maintained for

production of the 25 kg of $KMnO_4$, assuming complete efficiency of the electrolytic oxidation?

14. What volume of sulfur dioxide gas (at 27°C and 740 torr) would be formed by roasting 12.0 metric tons of zinc blende ore (84 percent ZnS) in air?[4] (Assume no other sulfide components of the ore.)

$$(2ZnS + 3O_2 \longrightarrow 2ZnO + 2SO_2)$$

15. What volume of 18 M sulfuric acid could be prepared from the SO_2 produced from the smelting of 5.00 metric tons of Cu_2S, assuming that only 50 percent of the SO_2 produced can be converted to the acid?

16. What mass (in grams) of copper could be electroplated from a solution of a copper(II) salt by a current of 0.30 amp flowing for 24 hr?

a. 7.1; **b.** 7.9; **c.** 8.5; **d.** 13; **e.** 19

17. What average current must be maintained to electroplate 6.5 g of chromium per hour from a solution of a chromium(III) salt, in amperes?

a. 6.5; **b.** 10; **c.** 33; **d.** 24; **e.** 16

18. How long, in hours, would be required to electroplate 325 g of zinc from a solution of a zinc(II) salt, using a current of 50 amp at an 80% electrode efficiency?

[4](1 metric ton = 1000 kg)

a. 24.0; **b.** 12.5; **c.** 8.0; **d.** 7.5; **e.** 6.7

19. The net formation constant for hexacyanoferrate(II) is 1×10^{24}. Calculate the approximate standard electrode potential for

$$[Fe(CN)_6]^{4-}(aq) + 2e^- \rightleftharpoons Fe(s) + 6CN^-(aq)$$

(for $Fe^{2+} + 2e^- \rightleftharpoons Fe$, $\mathcal{E}^0 = -0.44$)

a. +0.77 V; **b.** +0.36 V; **c.** −0.44 V; **d.** −1.15 V; **e.** −1.27 V

20. **a.** Estimate the net formation constant for diamminesilver(I) from the following data:

$$Ag^+ + e^- \rightleftharpoons Ag \qquad \mathcal{E}^0 = +0.80$$
$$[Ag(NH_3)_2]^+ + e^- \rightleftharpoons Ag + 2NH_3 \qquad \mathcal{E}^0 = +0.37$$

b. Estimate the net formation constant for tetrahydroxoaluminate ion from the following data:

$$Al^{3+} + 3e^- \rightleftharpoons Al \qquad \mathcal{E}^0 = -1.66$$
$$[Al(OH)_4]^- + 3e^- \rightleftharpoons Al + 4OH^- \qquad \mathcal{E}^0 = -2.35$$

BRAIN TINGLER

21. EDTA forms a chelate complex with Ni^{2+} that is given the shorthand notation $[Ni(EDTA)]^{2-}$. The complex (in terms of coordination centers) is approximately octahedral. Draw a plausible structural formula for this species.

Unit 27 Self-Test

1. Give the most common oxidation states for
   a. Sc; **b.** Cr; **c.** Mn; **d.** Fe; **e.** Zn
2. The electron configuration for nickel may be represented as (Ar), 4s↑↓, 3d↑↓, ↑↓, ↑↓, ↑, ↑. Write electron configurations in this form for each of the following species and circle each species expected to be paramagnetic.
   Ti^{2+}; Zn^{2+}; Mn^{2+}; Cu^+; Cu^{2+}
3. Supply the missing IUPAC names or formulas:
   a. $[Cu(NH_3)(H_2O)_5]^{2+}$_____
   b. $[CuCl_3(H_2O)]^-$_____
   c. $K_3[Co(NO_2)_6]$_____
   d. ammonium hexachloroplatinate(IV)
   e. ethylenediaminetetraacetatoferrate(III)
4. Select the member of each pair that is more likely to be paramagnetic:

   a. $[CuCl_4(H_2O)_2]^{2-}$ or $[CdCl_4]^{2-}$
   b. $[Co(H_2O)_6]^{3+}$ or $[CoCl_6]^{3-}$
5. What hybridization would probably be selected for the metal ion in a valence bond description of
   a. $[Ni(NH_3)_4]^{2+}$ (square planar)
   b. $[Ag(CN)_2]^-$ (linear)
   c. $[PbCl_4]^{2-}$ (tetrahedral)
   d. $[Co(H_2O_6]^{3+}$ (octahedral) }(see 4b)
   e. $[CoCl_6]^{3-}$ (octahedral)
6. (Refer to Figure 27.2.) A solution contains only ions listed in Figure 27.2. The solution is colorless. Addition of HCl produced a white precipitate. When the filtrate from this step was treated with H_2S at pH > 7, no reaction was observed.
   a. What ions were definitely present?
   b. Which were definitely absent?

7. What volume of carbon monoxide (at 1200°C and 720 torr) is released when 5.0 kg of ore containing 45% zinc oxide is smelted?

$$(ZnO(s) + C(s) \longrightarrow CO(g) + Zn(g))$$

a. 1.2×10^2 liters; **b.** 1.5×10^3 liters; **c.** 3.5×10^3 liters; **d.** 7.8×10^3 liters; **e.** 7.0×10^3 liters

8. A 25.0-ml aliquot of a solution containing Fe^{2+}, prepared during an assay of a pyrite ore, was titrated to a faint purple endpoint by 28.2 ml of 0.020 M $KMnO_4$. What was the molarity of the Fe^{2+} solution? (The MnO_4^- was reduced to Mn^{2+}.)

a. 0.11; **b.** 0.23; **c.** 0.30; **d.** 0.15; **e.** 0.060

9. How long, in hours, would be required for electroplating 587 g of nickel from a solution of nickel(II) ion, using an average current of 20 amp with 80% electrode efficiency?

a. 34; **b.** 21; **c.** 16; **d.** 9; **e.** 5

10. Estimate the net formation constant for tetraamminezinc(II) from the following data:

$$Zn^{2+} + 2e^- \rightleftharpoons Zn \qquad \mathcal{E}^0 = -0.76$$
$$[Zn(NH_3)_4]^{2+} + 2e^- \rightleftharpoons Zn + 4NH_3 \qquad \mathcal{E}^0 = -1.03$$

11. Draw representations for the complexes in question 5.

Overview Test

Section NINE

1. Which statement is incorrect?
   a. The atomic radius of calcium is smaller than that of strontium.
   b. The first ionization potential of calcium is smaller than that of strontium.
   c. Strontium is a softer metal than calcium.
   d. Strontium hydroxide is classified as "basic."
   e. Strontium hydroxide is more soluble in water than is calcium hydroxide.

2. Which of the following trends would be correct for the property indicated?
   I. first ionization potential: $Na < Mg$
   II. atomic radius: $B < Be < Li$
   III. basic character: $B(OH)_3 < Be(OH)_2 < LiOH$
   a. only II; b. only III; c. only I and II; d. only II and III; e. I, II, and III

3. In terms of the formation of nitrides and oxides, lithium is most similar to
   a. sodium; b. potassium; c. carbon; d. magnesium; e. barium

4. In terms of the acidic-basic character of its hydroxide, beryllium is most like
   a. aluminum; b. boron; c. lithium; d. magnesium; e. sodium

5. Which is most likely to exhibit typical metallic properties?
   a. boron; b. thallium; c. arsenic; d. fluorine; e. radon

6. Which element is most likely to form a stable oxide in which the element is in the $+IV$ oxidation state?
   a. thallium; b. gallium; c. tellurium; d. lead; e. antimony

7. Which is probably the most effective electrical conductor in its most common solid form?
   a. aluminum; b. silicon; c. phosphorus; d. arsenic; e. sulfur

8. Diamond is an allotropic form of carbon in which each atom is connected by covalent bonds to its nearest neighbors to form a repetitive tetrahedral pattern. Which is *not* an expected property of diamond?
   a. low solubility; b. low vapor pressure; c. hardness d. good electrical conductivity; e. high melting point

9. Which is much more soluble in 0.10 M NaOH than in water or 0.10 M HCl?
   a. $Be(OH)_2$; b. $B(OH)_3$; c. $Ca(OH)_2$; d. CsOH; e. none of these

10. Which name-formula combination is wrong?
    a. sulfurous acid/H_2SO_3
    b. hypophosphorous acid/H_3PO_4
    c. ammonium fluoride/NH_4F
    d. sodium bromite/$NaBrO_2$
    e. nitrogen(I) oxide/N_2O

11. Which name-formula combination is wrong?
    a. arsine/AsH_3
    b. thiosulfate/$S_2O_3^{2-}$
    c. phosphorous acid/H_3PO_3
    d. sodium nitrite/$NaNO_2$
    e. hydrogen cyanide/HCNO

12. Which is probably the most effective ("strongest") oxidizing agent?
    a. H_2; b. F_2; c. Cl_2; d. Br_2; e. I_2

13. "Red" phosphorus is insoluble in carbon disulfide. It has a broad melting range and a very low vapor pressure. It is not an effective electrical conductor. Evidence suggests that this substance is probably
    a. crystalline and metallic
    b. a simple crystal lattice of P_4 molecules
    c. amorphous and polymeric
    d. an irregular arrangement of unbonded P atoms
    e. P^{4+} ions in a sea of mobile electrons

14. Monoclinic sulfur appears to consist of covalent S_8 rings arranged at the points of a regular crystal lattice. Which of the following properties would be expected for this substance?
    a. good electrical conductivity
    b. moderate solubility in CS_2
    c. broad melting range
    d. very high melting point
    e. zero vapor pressure

15. Which is the correct electron-dot formulation for nitrous acid?

    a. H:N:O:O: b. H:O:O:N: c. H:O:N:O:

    d. H:O:N::O e. :O:N:O:
                       H

16. In which molecule are all atoms coplanar?
a. CH_4; **b.** $SiCl_4$; **c.** PF_3; **d.** NH_3; **e.** BF_3

17. Which is probably the strongest base?
a. NH_3; **b.** H_2NOH; **c.** PH_3; **d.** CH_3NH_2; **e.** NF_3

18. Which is the most probable ground-state configuration for iron(II) (beyond Ar)?
a. $4s^2 3d^4$; **b.** $4s^1 3d^5$; **c.** $4s^0 3d^4 4p^3$; **d.** $4s^0 4p^6$; **e.** $4s^0 3d^6$

19. Which is the most probable ground-state configuration for atomic titanium (beyond Ar)?
a. $4s^0 3d^4$; **b.** $4s^1 3d^3$; **c.** $4s^2 3d^2$; **d.** $4s^1 4p^3$; **e.** $4s^1 3d^2 4p^1$

20. Select the correct IUPAC name for $[Co(NH_3)_6]^{2+}$
a. hexammoniacobaltate(II)   **b.** hexaamminecobaltate(II)
c. hexammoniacobalt(II)   **d.** hexaamminecobalt(II)
e. none of these

21. Select the correct IUPAC name for $K_3[Co(NO_2)_6]$
a. tripotassiumhexanitrocobalt(0)
b. potassium hexanitrocobalt(III)
c. potassium hexanitrocobaltate(III)
d. tripotassiumhexanitrocobalt(III)
e. none of these

22. Select the correct IUPAC name for $[AuCl_4(H_2O)_2]^-$
a. tetrachlorodiaquoaurate(III)
b. tetrachlorodiaquoaurate(I)
c. tetrachlorodiaquogold(III)
d. tetrachlorodiaquogold(I)
e. none of these

23. Select the most plausible and consistent combination of hybrid orbital notation and magnetic-geometric description for the tetrachlorozincate(II) ion.
a. dsp^2, diamagnetic-octahedral
b. sp^3, paramagnetic-tetrahedral
c. dsp^2, paramagnetic-square planar
d. sp^3, diamagnetic-tetrahedral
e. d^2sp^3, diamagnetic-square planar

24. Select the most plausible and consistent combination of hybrid orbital notation and magnetic-geometric description for the hexaaquoiron(II) ion.
a. d^2sp^3, paramagnetic-octahedral
b. sp^3d^2, diamagnetic-hexagonal
c. dsp^2, diamagnetic-square planar
d. dsp^2, paramagnetic-tetrahedral
e. sp^3d^2, paramagnetic-octahedral

25. What mass of potassium superoxide could be prepared from 15.6 g of potassium metal and excess oxygen?
a. 35 g; **b.** 28 g; **c.** 24 g; **d.** 21 g; **e.** 16 g

26. What volume of oxygen gas (at STP) would be required for preparation of 39 g of sodium peroxide from sufficient sodium metal?
a. 11.2 liters; **b.** 9.6 liters; **c.** 7.4 liters; **d.** 5.6 liters; **e.** 2.2 liters

27. What is the approximate pH of a solution labeled 0.020 M potassium hypochlorite? (K_a for HOCl is 3.2×10^{-8}.)
a. 9.9; **b.** 7.4; **c.** 6.2; **d.** 8.6; **e.** 10.2

28. What average current (in amperes) must be maintained for a tinplating operation forming 2.4 kg of tin per hour from a solution of a tin(IV) complex, at an electrode efficiency of 90%?
a. 750; **b.** 1000; **c.** 1600; **d.** 2400; **e.** 3000

29. How long (in hours) must a current of 5.0 amp be maintained to electroplate 60 g of calcium from molten $CaCl_2$?
a. 27; **b.** 16; **c.** 11; **d.** 8.3; **e.** 5.9

30. The net formation constant for tetrachloromercurate(II) is 1×10^{15}. Calculate the approximate standard electrode potential for

$$[HgCl_4]^{2-}(aq) + 2e^- \rightleftharpoons Hg(l) + 4Cl^-(aq)$$

(for $Hg^{2+} + 2e^- \rightleftharpoons Hg$, $\mathcal{E}^0 = +0.85$)
a. -0.39 V; **b.** -0.63 V; **c.** $+0.79$ V; **d.** $+0.55$ V; **e.** $+0.41$ V

Excursion 3

Some industrial processes

1 METALLURGY AND THE ENVIRONMENT

Modern society cannot survive without the continued production of metals. That is obvious. It should now be equally obvious that modern society cannot *long* survive the present technology for producing metals or the present philosophy that production must steadily be increased. The following examples indicate two of the most common methods for procuring metals from natural ores. All of these procedures irreversibly consume energy and mineral resources and all contribute, in one way or another, to the deterioration of the environment. There are better ways—often more expensive. The ultimate solution must involve changes in technology and, more important, basic changes in philosophy. We cannot continue to measure basic human values in terms of "gross national product."

The basic key to the smelting of sulfide ores (major sources of copper, silver, zinc, cadmium, mercury, vanadium, and molybdenum) has traditionally been the fact that the sulfides, when roasted in air, evolve SO_2, leaving behind either the crude metals or their oxides.

Some of the new processes under development do not lead to formation of SO_2.

Before the roasting stage, the ores must go through some preliminary operations. The exact steps, of course, vary with the particular ore used. Generally, the preliminary work-up includes various physical operations such as crushing and washing. Crude separations may also be made to isolate the particular salt desired in a more "concentrated" form.

During the roasting stages, oxygen is consumed and sulfur dioxide is evolved. This gas, along with particulate matter, is the principal source of smelter pollution from operations using sulfide ores. Much attention has been focused on ways of reducing smelter pollution, and progress is being made. The result of the initial roasting may be either the metal or its oxide. The remaining steps vary considerably, depending on the metal involved and on the procedures deemed most economical by the particular smelter.

Example 1 (Copper from a low-grade copper ore)

STEP 1 Concentration by the flotation method

The crushed ore is treated with an oil–water emulsion. The sulfides are wetted by the oil and the "soils" by the water. Detergent is added, and the mixture is agitated vigorously. The oily sulfides are carried off by the foam, and the "soils" settle out.

STEP 2 Roasting in air

The sulfide concentrates (mainly Cu_2S and FeS, with traces of other sulfides) are roasted in a blast of air; SO_2 is evolved, along with colloidal smokes of various volatile oxides, such as As_2O_5. The product contains mainly copper(II) oxide and oxides and sulfides of iron. (FeS is much more resistant to oxygen than is Cu_2S.)

STEP 3 Smelting to form "blister copper"

The roast products are mixed with sand and coke and heated in a blast of hot air. Under these conditions the iron is converted to a molten silicate ("slag"), which is drained away. The coke is oxidized to carbon monoxide, and the CuO is reduced to Cu_2S (obtaining the sulfur from the iron sulfide). After removal of the slag, the Cu_2S is partially converted by hot air to Cu_2O. The air is then shut off, and the Cu_2O is permitted to react with residual Cu_2S, forming SO_2 and molten copper:

$$Cu_2S + 2Cu_2O \longrightarrow 6Cu + SO_2$$

The SO_2 is moderately soluble in the liquid copper. As the metal cools, SO_2 escapes, leaving the metal with a blistered appearance.

STEP 4 Electrolytic refining

The cast plates of crude blister copper are made the anodes in an electrolytic cell. Thin plates of pure copper are used as cathodes, and the electrolyte contains a copper(II) salt. Electrolytic refining typically produces copper containing less than 0.1 percent impurity.

Blister copper is typically contaminated by silver and gold from the original ore. These metals are recovered as valuable by-products from most copper smelters. Some lead, iron, nickel, and zinc are usually present as well.

Example 2 (Zinc from high-grade zinc blende)

STEP 1 Roasting in air

$$2ZnS(s) + 3O_2(g) \longrightarrow 2ZnO(s) + 2SO_2(g)$$
$\quad$ crude $\qquad\qquad\qquad\qquad$ crude

STEP 2 Smelting with coke (at 1200°C; zinc boils at 910°C)

$$ZnO(s) + C(s) \longrightarrow \text{solid residues} + CO(g) + Zn(g)$$
$\quad$ crude

$$Zn(g) \xrightarrow{\text{cool}} Zn(s)$$
$\qquad\qquad$ (as zinc dust)

(May be purified electrolytically.)

Example 3 (Mercury from high-grade cinnabar)

$$HgS(s) + O_2(g) \xrightarrow{\text{heat}} \text{solid residues} + SO_2(g) + Hg(g)$$
(crude)

$$Hg(g) \xrightarrow{\text{cool}} Hg(l)$$
(Purified by redistillation)

The treatment of oxide ores (major sources of titanium, zirconium, hafnium, chromium, manganese, and iron) is usually a multistep process. At first glance it would seem that the processing of oxide ores should be much "cleaner" than operations involving sulfides. In terms of what you *smell*, this is probably true, although some SO_2 is a by-product due to sulfides in the crude ores or sulfur in the coke used. However, many of the more dangerous pollutants are invisible and odorless. Oxide ores are typically treated by procedures involving carbon monoxide as a product or a reactant. Although provision is made to minimize carbon monoxide emission into the atmosphere, some invariably escapes. This gas is far more dangerous than SO_2. Fortunately, it is usually emitted in low concentrations. The chief pollution problem from oxide processing is the particulate matter, colloidal smokes containing various volatile metal oxides and carbon. Cottrell precipitators (Unit 14) have proved useful in reducing smoke emissions, but many plants still belch oxide smokes into the atmosphere and will continue to do so until stringent pollution laws are enforced.

Example 4 ("Pig iron" from hematite ore (the blast furnace process))

Iron ore, coke, and limestone ($CaCO_3$) are mixed and added at the top of the blast furnace and the mixture, heated initially to about $250°C$, is allowed to sift down through the furnace against a rising blast of hot air ($\sim1500°C$). The following reactions occur as the mixture falls through the hot blast:

$$CaCO_3(s) \longrightarrow CaO(s) + CO_2(g)$$
$$C(s) + CO_2(g) \longrightarrow 2CO(g)$$
coke
$$3Fe_2O_3(s) + CO(g) \longrightarrow 2Fe_3O_4(s) + CO_2(g)$$
$$Fe_3O_4(s) + CO(g) \longrightarrow 3FeO(s) + CO_2(g)$$
$$FeO(s) + CO(g) \longrightarrow Fe(l) + CO_2(g)$$
$$[C(s) + CO_2(g) \longrightarrow 2CO(g)]$$
$$xCaO(s) + (SiO_2)_x(s) \longrightarrow xCaSiO_3(l)$$
silicate
impurities

The calcium silicate (slag) is less dense than the molten iron. The two liquid phases are separated near the bottom of the furnace by first draining off the slag when the phase boundary reaches the upper drain outlet and then draining off the liquid iron from the lower furnace outlet. The iron obtained from the blast furnace (about 90 percent purity) is called "pig iron" because the molds originally used to collect the molten iron for cooling were shaped like fat pigs.

Various purification and alloying steps are used to convert this crude iron into different types of steel.

Example 5 (Chromium from chromite ore)

Crude chromium, for eventual use in chrome plating by electrodeposition, is prepared by heating a mixture of chromite ($FeCr_2O_4$) and coke in an electric furnace. Carbon monoxide is a by-product.

When high-purity chromium is desired, the following sequence is employed:

STEP 1 Alkali fusion

$$4FeCr_2O_4 + 16NaOH + 7O_2 \longrightarrow 2Fe_2O_3(s) + 8Na_2CrO_4(s) + 8H_2O(g)$$
$$\text{molten mixture}$$

STEP 2 Leaching and acidification

$$(Fe_2O_3, Na_2CrO_4) \xrightarrow{\ H_2O\ } Fe_2O_3(s) + Na_2CrO_4(aq)$$
$$\text{solid mixture}$$

$$2Na_2CrO_4(aq) + 2H^+(aq) \longrightarrow Na_2Cr_2O_7(aq) + H_2O(l) + 2Na^+(aq)$$
$$Na_2Cr_2O_7(aq) \xrightarrow{\ \text{evaporation}\ } Na_2Cr_2O_7(s)$$

STEP 3 Reduction by coke

$$Na_2Cr_2O_7(s) + 2C(s) \xrightarrow{\ \text{heat}\ } Na_2CO_3(s) + CO(g) + Cr_2O_3(s)$$

$$(Na_2CO_3, Cr_2O_3) \xrightarrow{\ H_2O\ } Na_2CO_3(aq) + Cr_2O_3(s)$$
$$\text{solid mixture}$$

STEP 4 Thermite process

$$Cr_2O_3 + 2Al \xrightarrow{\ \text{ignite}\ } Al_2O_3(s) + \quad 2Cr(l)$$
$$\text{(drained off)}$$

2 INORGANIC CHEMICAL PRODUCTION

A major fraction of the chemical industry is devoted to the preparation of *organic* compounds for pharmaceuticals, insecticides, pigments, polymers, explosives, and a host of other uses. Inorganic compounds, although a small portion of the total chemical production, still constitute a multimillion dollar enterprise. Ranging from such simple processes as the isolation of sodium chloride from seawater to multistep syntheses, such as the preparation of potassium permanganate (Unit 27), the methods of inorganic compound production represent a broad spectrum of techniques for synthesis, isolation, and purification. The following examples illustrate the diversity of this area of industrial chemistry.

Example 6 (Sodium tripolyphosphate)

"Phosphate rock" is a readily available and inexpensive natural source for a number of chemical processes. Its use in preparing elemental phosphorus was described in Unit 26. Since phosphorus (as phosphate) is an essential plant nutrient, much of the world industrial consumption of phosphate rock (around 20-million metric tons annually) is involved in the production of agricultural fertilizers, such as ammonium phosphate or "triple superphosphate," $Ca(H_2PO_4)_2$. A sizable fraction of the phosphate industry, however, is devoted to the production of detergent additives. The future of "phosphate detergents" is not entirely secure, since water pollution by phosphates has become a source of serious environmental concern. Some states have banned such detergents and the U.S. government has considered a number of alternatives to reduce phosphate pollution of natural waters. This pollution is partly the result of "run-off" of agricultural phosphates and partly from "phosphate detergents," mainly using sodium tripolyphosphate. This compound, constituting 35 to 75 percent of the "phosphate detergents," is used to form water-soluble complexes with Ca^{2+} and other ions found in "hard water" (Unit 6). Alternative complexing agents that have been tried as substitutes for phosphate additives have so far been unsatisfactory. Some have proved to be health hazards and others, such as the tetraborates, have been found to be more damaging to the environment than phosphate. Until such time as suitable alternatives are developed, the production of sodium tripolyphosphate will continue to be a commercially important process.

STEP 1 Phosphoric acid from phosphate rock

$$Ca_3(PO_4)(s) + 3H_2SO_4(l) \longrightarrow 2H_3PO_4(l) + 3CaSO_4(s)$$

$$2H_3PO_4(l),\ CaSO_4(s) \xrightarrow{\text{filtration}} \begin{array}{l} CaSO_4(s) \\ \text{(for "gypsum")} \\ H_3PO_4(l) \end{array}$$

STEP 2 Preparation of sodium salts

$$H_3PO_4(l) + 2NaOH(aq) \longrightarrow Na_2HPO_4(aq) + 2H_2O(l)$$
$$\xrightarrow{\text{evaporation}}$$
$$Na_2HPO_4(s) + H_2O(g)$$

$$H_3PO_4(l) + NaOH(aq) \longrightarrow NaH_2PO_4(aq) + H_2O(l)$$
$$\xrightarrow{\text{evaporation}}$$
$$NaH_2PO_4(s) + H_2O(g)$$

STEP 3 Thermal polymerization

$$2Na_2HPO_4(s) + NaH_2PO_4(s) \xrightarrow{\text{heat}} Na_5P_3O_{10}(s) + 2H_2O(g)$$
$$\text{(sodium tripolyphosphate)}$$

$$\left(\text{The tripolyphosphate anion is formulated as } \begin{bmatrix} & O & & O & & O & \\ & \| & & \| & & \| & \\ O-&P&-O-&P&-O-&P&-O \\ & | & & | & & | & \\ & O & & O & & O & \end{bmatrix}^{5-} \right)$$

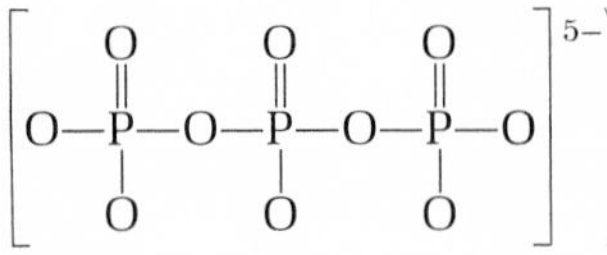

Example 7 (Calcium carbide and cyanamide)

A chemical process requiring only inexpensive and readily available raw materials can be the basis for a profitable enterprise when a large product market is available. An example is the case of calcium carbide, readily prepared in an electric furnace from lime and coke, both relatively cheap starting materials. The product can be used to generate acetylene, for welding operations or as a starting material for preparation of a number of organic compounds. Alternatively, some of the calcium carbide can be converted to the cyanamide, useful as a fertilizer or for ammonia production.

STEP 1 Formation of calcium carbide

$$CaO(s) + 3C(s) \xrightarrow{\sim 3000°C} CO(g) + CaC_2(s)$$

    (lime)   (coke)   (calcium carbide)

STEP 2 Generation of acetylene

$$CaC_2(s) + 2H_2O(l) \longrightarrow Ca(OH)_2(s) + H-C\equiv C-H(g)$$

Alternative Step 2 Conversion to the cyanamide

$$CaC_2(s) + N_2(g) \xrightarrow{\sim 1000°C} C(s) + CaCN_2(s)$$

                                     (calcium cyanamide)

The cyanamide can be used directly as a fertilizer or treated with steam to produce ammonia. In either case, the reaction with water is

$$CaCN_2 + 3H_2O \longrightarrow CaCO_3 + 2NH_3$$

3 RECYCLING AND WASTE PRODUCT RECOVERY

Many of the metals so essential to modern society were once produced exclusively from natural ores. With the increasing problems of depletion of high-grade ore supplies, attention has been focused on methods for recycling "scrap" metals. It is apparent that recycling processes, possibly coupled with the development of alternatives to some of our present uses of metals, are essential. It is estimated, for example, that *under the most favorable circumstances,* we may anticipate severe shortages within the next 10 to 30 years of such metals as gold, silver, copper, mercury, tin, lead, and zinc.

Significant progress has been made within the last few years in metal-recycling processes. Nearly half of our iron and steel now comes from "scrap" metal, rather than from natural iron ores. Almost all of the lead used in automobile batteries (about one-third of the total lead consumption) is recycled, and it is estimated that we will soon recycle more than one-fourth of all aluminum cans.

Recently a number of processes have been developed for the recovery of industrial chemical waste products by methods found to be economically profitable. An example of the use of waste oxides for the production of arsenic, antimony, and bismuth has been described in Unit 26, and the recovery of valuable organic compounds from coal tar is discussed in Unit 28.

We still have a long way to go. In spite of increased attention to problems of environmental pollution and natural resource conservation, we are still in the phase of continuing deterioration of the environment and depletion of valuable resources. The scientific and technical "know-how" *is* available to solve many of these problems, but the solutions to the associated economic and political problems have not yet been developed. As a people we must find the courage to reduce our demands for material things. If we do not, the legacy we leave for our children will not be a pleasant one.

SUGGESTED READING

Garrels, Robert M., Fred T. MacKenzie, and Cynthia Hunt. 1975. *Chemical Cycles and the Global Environment.* Los Altos, Calif.: William Kaufmann.

Hamilton, Carole L., *et. al.* 1973. Readings from *Scientific American: Chemistry in the Environment.* San Francisco: Freeman.

Section TEN

INTRODUCTION TO ORGANIC CHEMISTRY

More than 99 percent of the atoms in your body are hydrogen, oxygen, nitrogen, and carbon. Inorganic species other than water, though many are essential to life, are essentially trace components. The vast majority of the chemical compounds of your body—and, indeed, of all living systems—are molecules composed principally of carbon: *organic compounds*. Within every living organism a host of organic compounds are being synthesized, while others are being consumed or chemically modified. A living organism is a marvelous "chemical factory," often utilizing processes whose efficiency and selectivity far surpass the capabilities of our most advanced laboratories.

With this section we begin our study of organic chemistry, with the ultimate goal of gaining a measure of understanding of some important biochemical processes. To do this, we must first consider the simplest organic compounds, the hydrocarbons. In Unit 28 we shall see how these hydrocarbons, compounds containing only carbon and hydrogen, illustrate the unique characteristics of carbon that account for the existence of millions of organic compounds. We shall also learn a systematic method for naming organic compounds and a useful way of classifying compounds by *functional groups*, special atomic arrangements that contribute characteristic chemical properties to molecules.

Unit 29 offers a more extensive examination of the functional groups and their typical properties, with emphasis on the functional groups most common in biologically important compounds.

Not many years ago, the determination of the chemical identity or structure of an unknown organic compound was a laborious task, sometimes requiring months or years of careful research. Now, with modern instrumentation, structural determinations can often be made within a few weeks and, for simple compounds, in less than a day. Unit 30 provides an introduction to the important instrumental methods of spectroscopic investigation of organic compounds.

Since living systems contain hundreds of organic compounds, the task of isolating and purifying a single compound from a plant or animal source is far from simple. Even laboratory syntheses frequently result in complex mixtures, for which special separation techniques are required. In Excursion 4 we will examine some of the more common methods for the isolation and purification of an organic compound from a mixture of substances.

CONTENTS

Organic compounds

In fact, no one has been able to prepare uric acid . . . , sugar. . . . Here chemistry has been powerless and, if my guess is right, will always be so.

(In opposition to the "synthetic" chemistry of carbon compounds, 1842).

C. F. GERHARDT

a change of attitude in 12 years

Natural products and the artificial products of our laboratories are links of the same chain. . . .

C. F. GERHARDT (1854)

Objectives

(a) Be able to classify organic compounds on the basis of characteristic functional groups (Table 28.2).

(b) Be able to draw structural formulas that distinguish between
   (1) positional isomers (Sections 28.6 and 28.7)
   (2) conformational isomers (Section 28.8)
   (3) geometric isomers (Section 28.9)
   (4) optical isomers (Section 28.10)
   (5) functional group isomers (Section 28.11 and Table 28.2)

(c) Be able to use the IUPAC nomenclature rules for (Tables 28.4, 28.7, 28.8, 28.9 and Appendix G):
   (1) cycloalkanes
   (2) simple aromatic compounds
   (3) alkenes
   (4) alkanes

(d) Be able to predict simple variations in physical properties associated with isomeric differences in molecular structure (Section 28.11 and Unit 9).

Before man had a written history, he had learned to extract useful substances from sources provided by nature. Herbs provided flavors for foods or medicines for treating the sick. Bark extracts helped him in tanning animal hides, and volatile chemicals in wood smoke helped in preserving his food. He could protect himself with arrows tipped in concentrated poisons from natural sources, celebrate his victories with alcoholic beverages, or seek temporary escape from the realities of his life with a variety of drugs. Often the arts of extraction and concentration of physiologically active compounds were carefully guarded secrets and provided their owner with wealth, prestige, and power in his society. He could choose to use his knowledge for the benefit of others or for purposes of destruction and personal gain. The latter were the more common choices.

Today man has progressed far beyond this primitive state. He need no longer rely primarily on natural sources for useful chemicals. He can isolate an active substance, purify it carefully, determine its complete structure, and, frequently, make it synthetically for wide-scale use much more economically than can nature. In many cases he can prepare synthetic substitutes better than the original compound. He can choose to use his present knowledge for the benefit of others or for purposes of destruction and personal gain. Yes, man has progressed quite far. Synthetic antibiotics promise new eras of health; synthetic fabrics, synthetic pesticides, and synthetic fertilizers promise improved standards of living; and the major world powers have manufactured enough chemical warfare agents to kill every person on earth 30 times over. Man has progressed a long way. Or has he?

The arts of the chemist are no longer secret. A general understanding of the concepts involved and the techniques employed can be gained rather easily. Such knowledge should help in the wise determination of the uses for which these arts will be employed. In a free society the choices can no longer be made by the tribe's "wizard." In this day of public funding of research, the credit for developing new substances of potential benefit to man is shared to some extent by every citizen, and each must share in the blame for those new products that threaten the environment or that become new weapons in the arsenals of mass destruction.

In an attempt to produce more effective drugs with fewer adverse side effects, pharmaceutical chemists have synthesized thousands of compounds with structural variations designed to reveal those molecular characteristics essential for optimum biological activity. As an example, quinine has been largely replaced by synthetic antimalarial drugs developed over years of research into synthetic, structural, and biochemical areas. One of these new antimalarial drugs has the structure

$$\underset{}{\text{benzene ring with } CH_3 \text{ substituent}} - CH_2CH_2N(C_2H_5)_2$$

It is quite common to observe that seemingly minor variations in structure may have profound influences on biological activity. Neither of the following compounds is effective as an antimalarial drug, although they differ from the preceding substance only in the relative positions of the groups attached to the benzene ring:

$$CH_3-\hspace{-0.3em}\bigcirc\hspace{-0.3em}-CH_2CH_2N(C_2H_5)_2 \qquad \overset{CH_3}{\underset{}{\bigcirc}}-CH_2CH_2N(C_2H_5)_2$$

inactive

I have proposed to call substances of the same composition but differing properties isomeric, from the Greek ισομερης (composed of equal parts).
J. J. BERZELIUS (1832)

Isomeric compounds may vary widely in physical and chemical properties. Some, however, are essentially identical in these respects but show significant variations in their reactions within living organisms. The chemist must always be aware of isomeric differences and the problems they may pose in various aspects of his work. In no field are these problems of such importance as in the area of chemicals associated with living systems.

Organic chemistry is primarily the study of covalent compounds of carbon. These compounds include a fantastic variety of substances ranging from the simple hydrocarbons used as fuels and lubricants to the complex nucleic acids of the genetic material. Organic chemistry is, fortunately, a relatively systematic science—at least in its simpler aspects. When we learn some of the characteristic properties of alcohols, for example, we shall find that these give valuable clues to many attributes of carbohydrates and other complex hydroxyl compounds.

This unit begins a series of discussions of the introductory aspects of organic chemistry. We shall have to learn to recognize classes of compounds, to use the systematic IUPAC naming rules (as well as many "trivial" names), and to remember a host of reaction types. Organic chemistry requires steady, systematic study, but it is an interesting and rewarding field.

28.1 ORGANIC COMPOUNDS

Carbon forms more different chemical compounds than any other element. This is attributed to the fact that carbon–carbon σ or π bonds are exceptionally stable, permitting the formation of long chains or rings of carbon atoms for a wide variety of compounds. In addition, carbon forms stable covalent bonds with hydrogen and with many other elements, the most important being oxygen, nitrogen, and the halogens.

More than 2 million organic compounds are now known. This is at least 10 times the number of all other known compounds not containing carbon.

The simplest organic compounds are the hydrocarbons (Table 28.1). We shall concentrate on these in this unit, reserving the more complex classes of compounds (Table 28.2) for later study. Organic compounds may be categorized according to certain arrangements of atoms, the *functional groups* (Table 28.3), which contribute characteristic properties to the molecules in which they occur. Such a classification is important in systematizing the study of the enormous number of organic compounds. As we shall see (Unit 29), there are certain typical properties associated with each functional group, but individual molecular variations influence these properties to some extent. For example, all alcohols react with metals such as sodium to liberate hydrogen gas, but the speed of this reaction varies greatly with the nature of the particular alcohol involved.

Since carbon normally forms four covalent bonds, there are many different ways in which the same atoms can be "connected" (Figure 28.1). The existence of *isomers*, molecules having the same molecular formulas but different structures, is another reason for the immense number of different organic compounds.

Table 28.1 TYPES OF HYDROCARBONS [$C_nH_{2(n-y+1)}$][a]

CLASSIFICATION	EXAMPLE	MOLECULAR FORMULA	n	y
alkane ("straight chain")	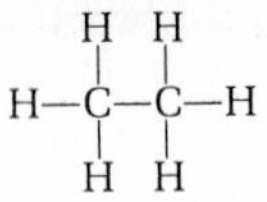 $\underset{\text{ethane}}{H-\overset{\overset{\displaystyle H}{\vert}}{\underset{\underset{\displaystyle H}{\vert}}{C}}-\overset{\overset{\displaystyle H}{\vert}}{\underset{\underset{\displaystyle H}{\vert}}{C}}-H}$	C_2H_6	2	0
cycloalkane	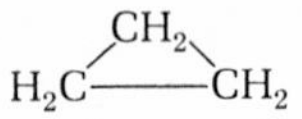 cyclopropane*	C_3H_6	3	1
alkene	$\underset{\text{ethylene}}{H_2C{=}CH_2}$	C_2H_4	2	1
cycloalkene	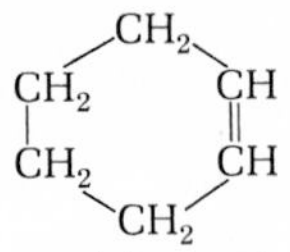 cyclohexene†	C_6H_{10}	6	2
polyene	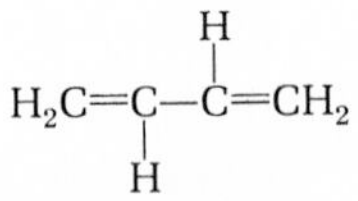 $\underset{\text{1,3-butadiene}}{H_2C{=}C{-}C{=}CH_2}$	C_4H_6	4	2
alkyne	$\underset{\text{acetylene}}{H-C{\equiv}C-H}$	C_2H_2	2	2
aromatic	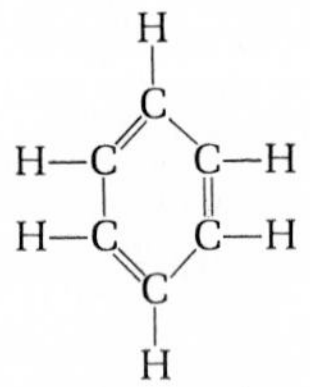 benzene‡	C_6H_6	6	4

[a] n = number of carbon atoms in the molecule, y = number of rings and/or pi bonds (counted from localized pi-bond representation)

Common notations:

* cyclopropane,

† cyclohexene,

‡ benzene,   or   or   ←→

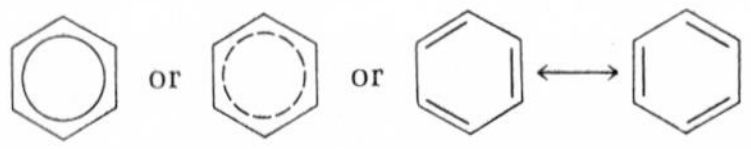

resonance hybrids (Unit 8)

Table 28.2 CLASSIFICATION BY FUNCTIONAL GROUP[a]

CLASS	FUNCTIONAL GROUP	EXAMPLE
alkane	none (only C—C and C—H bonds)	propane, $CH_3CH_2CH_3$
cycloalkane	none (only C—C and C—H bonds)	cyclopentane, 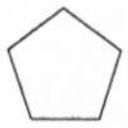
alkene	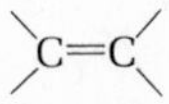 $\diagup C{=}C \diagdown$	propene, $H_2C{=}CHCH_3$

Table 28.2 (Continued)

CLASS	FUNCTIONAL GROUP	EXAMPLE
cycloalkene	$\diagdown C = C \diagup$	cyclopentene,
alkyne	$-C \equiv C-$	acetylene, $HC \equiv CH$
aromatic hydrocarbon	ring(s) with continuous pi-bond system	benzene,
halide	$-\overset{\|}{\underset{\|}{C}}-X$ (X = F, Cl, Br, I)	chloroform, $CHCl_3$
ether	$-\overset{\|}{\underset{\|}{C}}-O-\overset{\|}{\underset{\|}{C}}-$	dimethyl ether, CH_3OCH_3
alcohol	$-\overset{\|}{\underset{\|}{C}}-OH$	methanol, CH_3OH
phenol	OH on benzene ring	o-cresol,
aldehyde	$-\overset{H}{\underset{\|}{C}}=O$	acetaldehyde, $CH_3\overset{H}{\underset{\|}{C}}=O$
ketone	$-\overset{\|}{\underset{\|}{C}}-\overset{O}{\overset{\|}{C}}-\overset{\|}{\underset{\|}{C}}-$	acetone, $CH_3\overset{O}{\overset{\|}{C}}CH_3$
carboxylic acid	$-\overset{O}{\overset{\|}{C}}-OH$	acetic acid, $CH_3\overset{O}{\overset{\|}{C}}OH$
amine	$-\overset{\|}{\underset{\|}{C}}-\ddot{N}:$	methylamine, $CH_3\ddot{N}H_2$
carboxylate salt	$\left[-\overset{O}{\overset{\|}{C}}{\overset{\cdot\cdot}{\cdot}}O\right]^- M^+$ (M$^+$ = cation)	sodium acetate, $Na(CH_3CO_2)$
ammonium-type salt	$\left(-\overset{\|}{\underset{\|}{C}}-\overset{\|}{\underset{\|}{N}}{\overset{+}{-}}H\right)X^-$ (X$^-$ = anion)	methylammonium chloride, CH_3NH_3Cl
ester	$-\overset{O}{\overset{\|}{C}}-O-\overset{\|}{\underset{\|}{C}}-$	ethyl acetate, $CH_3\overset{O}{\overset{\|}{C}}-OCH_2CH_3$
amide	$-\overset{O}{\overset{\|}{C}}-\overset{\|}{N}-$	acetamide, $CH_3\overset{O}{\overset{\|}{C}}NH_2$
acid chloride	$-\overset{O}{\overset{\|}{C}}-Cl$	acetyl chloride, $CH_3\overset{O}{\overset{\|}{C}}Cl$
nitro compound	$-\overset{\|}{\underset{\|}{C}}-NO_2$	nitrobenzene,

[a] A few other, less common, functional groups will be introduced in later units.

Table 28.3 SOME EXAMPLES OF CHEMICAL PROPERTIES CHARACTERISTIC OF PARTICULAR FUNCTIONAL GROUPS

CLASS	FUNCTIONAL GROUP	REACTS READILY WITH				
		COLD(aq) H^+	COLD(aq) OH^-	$Br_2(aq)$ (IN THE DARK)	*MILD* OXIDIZING AGENTS	Na(s)
alkane	none (only C—C and C—H bonds)	—	—	—	—	—
alkene	C=C	—	—	X	—	—
alcohol	—C—O—H	—	—	—	—	X
carboxylic acid	—C(=O)—O—H	—	X	—	—	X
amine	C—N	X	—	—	—	only if N—H
aldehyde	—C(=O)—H	—	—	—	X	—
phenol	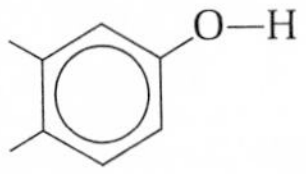 (aromatic)—O—H	—	X	some phenols do	some phenols do	X

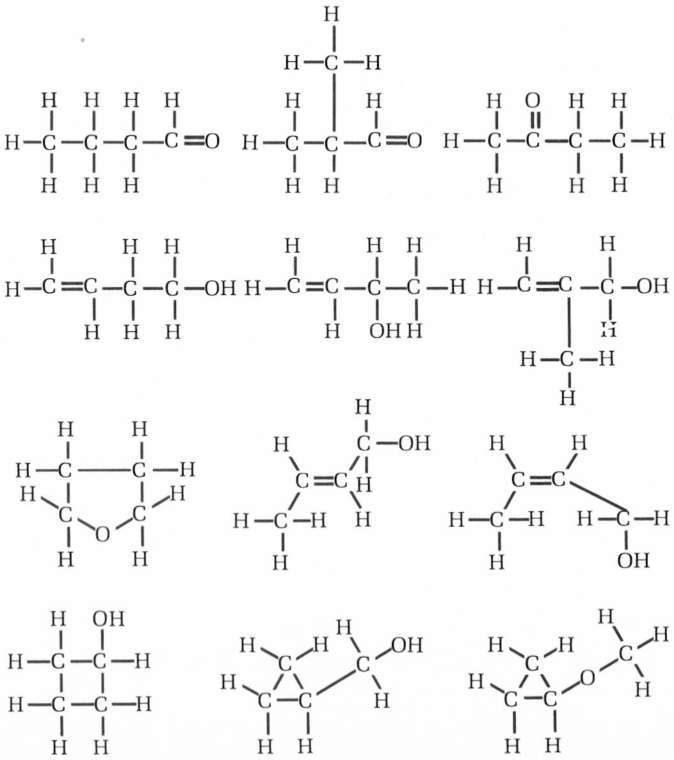

Figure 28.1 Some isomers having the formula C_4H_8O. Not all possible isomers have been shown. Can you suggest others?

28.2 MOLECULAR FORMULAS

In referring to inorganic compounds, simple formulas are generally adequate. Thus, NaCl can be used to refer to an isolated sodium chloride molecule (ion pair) in the gas phase, the rather random mixture of sodium and chloride ions in the molten salt or in a solution, or the orderly pattern of the ions in the crystalline salt. The formula tells us simply that the Na$^+$ and Cl$^-$ ions exist in 1:1 ratios under all these conditions. In a few cases, such as Hg_2Cl_2, an attempt is made to indicate some structural feature in addition to the stoichiometric ratio [e.g., that mercury(I) ion exists as Hg_2^{2+}].

Organic compounds, on the other hand, are so numerous that simple empirical formulas do not distinguish adequately between different compounds. For example, CH_2 would be the empirical formula for such widely different substances as cyclohexane, cyclopropane, and ethylene. Even molecular formulas, which show the actual numbers of each component element, provide only limited descriptions of the compounds. The determination of such molecular formulas is, however, a necessary step in the eventual determination of an organic structure.

To establish a molecular formula for a hydrocarbon most commonly involves complete combustion of the hydrocarbon to CO_2 and H_2O with accurate measurement of the amount of each produced from a weighed hydrocarbon sample. Traditionally, CO_2 and H_2O were determined by the increase in mass of appropriate absorbants [e.g., $Mg(ClO_4)_2$ for water and soda lime for carbon dioxide, Figure 28.2]. Today it is possible to determine CO_2 and H_2O both rapidly and accurately by vapor-phase chromatography using the combustion products from very small hydrocarbon samples. In any event, the percentage composition may then be calculated from simple considerations of stoichiometry:

> For a *hydrocarbon* either the percent carbon or percent hydrogen is sufficient, since the other may be found by difference. Measurement of both CO_2 and H_2O, however, provides a convenient check.

$$\text{percent C} = \frac{\text{mass of } CO_2}{\text{mass of sample}} \times \frac{12.01 \text{ g (C)}}{44.01 \text{ g } (CO_2)} \times 100\%$$

$$\text{percent H} = \frac{\text{mass of } H_2O}{\text{mass of sample}} \times \frac{2.016 \text{ g (H)}}{18.02 \text{ g } (H_2O)} \times 100\%$$

The *empirical formula*, C_xH_y (where x and y are the smallest possible integers), may then be determined by appropriate mathematical analysis:

percent C = g (C) per 100 g (compound)
percent H = g (H) per 100 g (compound)

$$\frac{\text{g (C)}}{100 \text{ g (cpd)}} \times \frac{1 \text{ mole (C)}}{12.01 \text{ g (C)}} = \frac{\text{mole (C)}}{100 \text{ g (cpd)}}\left(\frac{\%C}{12.01}\right)$$

$$\frac{\text{g (H)}}{100 \text{ g (cpd)}} \times \frac{1 \text{ mole (H)}}{1.008 \text{ g (H)}} = \frac{\text{mole (H)}}{100 \text{ g (cpd)}}\left(\frac{\%H}{1.008}\right)$$

Figure 28.2 Combustion train for determination of C and H.

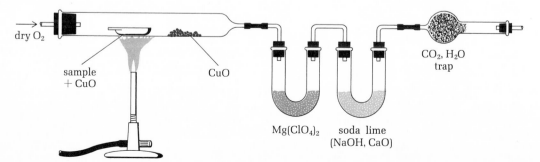

dry O$_2$

sample + CuO

CuO

$Mg(ClO_4)_2$

soda lime (NaOH, CaO)

CO_2, H_2O trap

Then the atomic ratio (y/x) is given by

$$\frac{y}{x} = \frac{\text{mole (H)}}{\cancel{100\text{ g (cpd)}}} \times \frac{\cancel{100\text{ g (cpd)}}}{\text{mole (C)}} \left(\frac{\%\text{H} \times 12.01}{\%\text{C} \times 1.008}\right)$$

from which

$$y = (x)\left(\frac{\%\text{H}}{\%\text{C}}\right)\left(\frac{12.01}{1.008}\right) \qquad (28.1)$$

where x is selected to be the smallest whole number that will make y also a whole number.

The next step in finding the molecular formula consists of a molecular weight determination. Generally, approximate values of molecular weights are sufficient and for volatile compounds these may be found from application of ideal gas equations to measured vapor densities (Unit 10).

Molecular weights of nonvolatile samples may be determined from vapor-pressure measurements on solutions of the substances or from cryoscopic data determined from freezing-point (or melting-point) depression of a mixture of the sample and a substance such as camphor. These techniques were described in Unit 15.

Once the molecular weight and empirical formula are known, the molecular formula may be expressed as $C_{xz}H_{yz}$, where x and y are the numbers in the empirical formula and

$$z = \frac{\text{molecular weight}}{\text{empirical formula weight}} \qquad (28.2)$$

Example 1

Calculate the molecular formula for a hydrocarbon for which the following data were determined:

Combustion analysis
---

sample mass:	0.2404 g
mass of CO_2:	0.7922 g
mass of H_2O:	0.2162 g

Vapor density measurement
---

sample mass:	0.3761 g
vapor volume:	0.1830 liter
temperature:	160°C
pressure:	0.91 atm

Solution

$$\text{percent C} = \frac{0.7922\text{ g (CO}_2)}{0.2404\text{ g (cpd)}} \times \frac{12.01\text{ g (C)}}{44.01\text{ g (CO}_2)} \times 100\% = 89.93\%$$

$$\text{percent H} = \frac{0.2162\text{ g (H}_2\text{O)}}{0.2404\text{ g (cpd)}} \times \frac{2.016\text{ g (H)}}{18.02\text{ g (H}_2\text{O)}} \times 100\% = 10.06\%$$

For C_xH_y,

$$y = x\left(\frac{10.06 \times 12.01}{89.93 \times 1.008}\right)$$

$$= (x)(1.333)$$

for which the smallest whole number for x that will make y a whole number is 3. Thus y = 4. The empirical formula, then, is C_3H_4.

Then, to find the molecular weight (Unit 10):

$$\underbrace{\frac{0.3761\ g}{0.1830\ liter} \times \frac{433°C}{273°C} \times \frac{1.0\ atm}{0.91\ atm}}_{\substack{\text{density corrected} \\ \text{to STP}}} \times \frac{22.4\ liters}{1.00\ mole} = 80\ g\ mole^{-1}$$

at STP

The empirical formula weight ($3 \times 12 + 4 \times 1$) is half the molecular weight, so the empirical formula must be only half the molecular formula: C_6H_8.

It might seem that the determination of the empirical formula is an unnecessary step, since the percent composition and molecular weight should be sufficient. However, vapor density or cryoscopic data are usually less accurate than percent composition data, and are rarely good enough for determination of exact numbers of hydrogens.

For compounds containing elements in addition to carbon and hydrogen, more extensive analyses are required to determine the percentage composition.[1] A method for nitrogen determination has been described in Excursion 2. Accurate procedures are available for determination of all the elements found in organic compounds (the most common of which, in addition to carbon, hydrogen, and nitrogen, are the halogens, phosphorus, and sulfur). Direct analyses for oxygen are rarely possible, so the oxygen content is generally found by difference.

28.3 BONDING IN HYDROCARBONS

The molecular formula tells us very little about the structure of the molecule except in the unambiguous cases of very simple molecules such as CH_4, C_2H_6, C_2H_4, or C_2H_2. In more complex molecules, many isomers are often possible and further analysis is required to determine a unique molecular description. Once we know the structural formula, that is, the way in which the atoms are "connected," we can approximate the molecular geometry fairly well on the basis of *Gillespie's rules* (Unit 8), by considering each atom in the molecule separately. Atoms, such as hydrogen, having only one bond contribute no special geometric features.

There are special cases, such as cyclopropane, in which "normal" bond angles are not possible because of other geometric restrictions. (Carbon–carbon bond angles in cyclopropane are 120 degrees, rather than the 109.5 degrees typical of carbon bonded to "four groups," because of the triangular pattern of carbon atoms in the cyclopropane ring). For most molecules, however, *Gillespie's rules* permit satisfactory prediction of geometry.

Example 2

Draw a representation of propene, $H_2C{=}\overset{\displaystyle H}{\underset{\displaystyle |}{C}}{-}CH_3$, indicating approximate bond angles and geometrical features of regions within the molecule.

---

[1]For a discussion of the calculations of empirical formulas for compounds other than hydrocarbons, see *Solving Problems in Chemistry*, by Rod O'Connor and Charles Mickey (Harper & Row, 1974).

Solution

Only the carbon atoms in propene are bonded to more than one group. An expanded line-bond representation (Unit 8) permits us to "count groups" on each carbon:

$$\underset{\textcircled{1}}{H-C}=\underset{\textcircled{2}\,\textcircled{3}}{C-C}-H$$

Carbons ① and ② are each connected to three groups, suggesting a trigonal planar geometry, whereas carbon ③ is connected to four groups in a tetrahedral pattern. (Remember that *Gillespie's rules* apply to the *number* of attached groups, irrespective of *how* they are attached; that is, we do not care whether single, double, or triple bonds are involved.)

Figure 28.3 Sigma (σ) bonds in methane.

The bond angles and geometric features (Unit 8) may then be indicated as

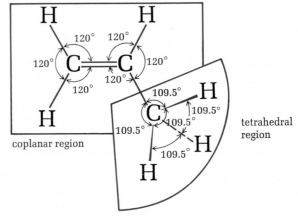

Figure 28.4 Pi (π) bond in ethylene. In viewing the drawing, consider the π bond to lie in the plane of the page and the "atomic skeleton" to lie in a plane perpendicular to the page.

Once the geometric features of a molecule are known, simple molecular orbital descriptions can be applied. Remember (Unit 8) that π bonding on carbon is described in terms of overlap of simple p orbitals from the bonded atoms, while σ-bond descriptions employ sp^3 (tetrahedral), sp^2 (trigonal planar), or sp (linear) hybridization.

We can illustrate the use of simple orbital descriptions for hydrocarbons by the following cases.

Case 1: Methane

Lewis structure

$$\begin{array}{c} H \\ H:\overset{..}{C}:H \\ H \end{array}$$

For CH_4, we must have four σ bonds on carbon. Since sp^3 hybrid orbitals form a tetrahedral pattern, these are selected as consistent with group geometry. The C—H bond is approximated by the overlap of a one-electron s orbital of hydrogen and a one-electron sp^3 orbital of carbon (Figure 28.3).

Case 2: Ethylene

Lewis structure $\begin{array}{c} H:C::C:H \\ H\ \ H \end{array}$

From the dot formula for C_2H_4, we conclude that each carbon must form only three σ bonds in ethylene. The fourth bond must be of the π type. Using sp^2 hybrid orbitals for the σ bonds, we predict a coplanar structure for the "atomic skeleton":

$$\begin{array}{cc} H & H \\ \diagdown & \diagup \\ & C—C \\ \diagup & \diagdown \\ H & H \end{array}$$

The two p orbitals used in forming sp^2 hybrids are coplanar, so the remaining p orbital is perpendicular to this plane. The π bond, formed from overlap of parallel p orbitals on the adjacent carbon atoms, must then have one segment above and one segment below the plane of the "atomic skeleton" (Figure 28.4).

Case 3: Benzene

Lewis structure, as a resonance hybrid (Unit 8), shows three σ bonds per carbon:

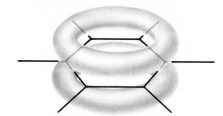

In benzene, C_6H_6, the σ bonding is described, as in ethylene, as involving sp^2 hybridization on carbon. As a result, the bonding predicted suggests a coplanar geometry that includes the carbon and hydrogen skeleton:

Since each carbon atom retains a one-electron p orbital, these overlap completely around the ring to provide a delocalized π-bond system[2] (Figure 28.5). Such delocalization is characteristic of molecules classified as *aromatic,* a term whose historical origin is traced to the pungent aroma of benzene and similar compounds.

Figure 28.5 Pi (π) bonding in benzene.

This same general procedure can be applied to each carbon atom in a more complex molecule to obtain an approximation of the total molecular geometry. Basically, the atoms bonded to a carbon atom having four σ bonds will be arranged in a tetrahedral pattern; the atoms around a carbon atom having three σ bonds will lie in the same plane with bond angles around carbon of about 120 deg; and the atoms bonded to a carbon atom having only two σ bonds will be collinear. A template set that can be used in constructing inexpensive models is described at the end of this unit.

It should be remembered that any static description of a molecule is, at best, a model for a "frozen instant in time," since vibrational and rotational motions (Unit 9) are continually altering the molecular shape. It has been suggested that a department store's stylized mannikin probably bears a closer resemblance to a real person than a molecular model does to a real molecule. Nevertheless, such models do serve a useful purpose.

Distortions from the idealized predictions will occur as a result of electrostatic effects (Unit 8), steric effects (restrictions on the space available to bulky atomic groupings), and bond-angle changes necessitated by the angular requirements of certain ring structures. The use of simple hybrid orbital descriptions is still a valuable basic approximation for most organic molecules.

28.4 NATURALLY OCCURRING HYDROCARBONS

These terms are used here in a completely different sense than when they are applied to solutions, although we might think of "saturated" in a general sense as "being as full as possible" (of solute, for a solution, or of sigma bonds, for a molecule).

Hydrocarbons are subclassified in several different ways (Table 28.1). Aromatic hydrocarbons, so designated because of their rather characteristic "medicinal" odors, are those that contain ring systems having delocalized π bonding, such as in benzene. Compounds not containing such ring systems are referred to as *aliphatic,* and these may be cyclic ("alicyclic") or "straight-chain" (acyclic) structures. Indeed, many hydrocarbons contain both ring and chain formations. Further description is provided by the terms *saturated* (containing only σ bonds) and *unsaturated* (containing both π and σ bonds). These latter terms will probably be familiar from modern

[2]For an excellent discussion of π-bond delocalization, see "The Nature of Aromatic Molecules," by Ronald Breslow, in the August, 1972, issue of *Scientific American* (pp. 32–40).

Table 28.4 THE FIRST TEN ALKANES

NUMBER OF CARBON ATOMS	MOLECULAR FORMULA	NAME	APPROXIMATE MELTING POINT (°C)	APPROXIMATE BOILING POINT °(C) AT 1.00 atm	DENSITY (g cm^{-3}) AT 20°C
1	CH_4	methane	−180	−160	(gas)
2	C_2H_6	ethane	−170	−90	(gas)
3	C_3H_8	propane	−190	−40	(gas)
4	C_4H_{10}	butane	−140	0	(gas)
5	C_5H_{12}	pentane	−130	40	0.63
6	C_6H_{14}	hexane	−95	70	0.66
7	C_7H_{16}	heptane	−90	100	0.68
8	C_8H_{18}	octane	−60	125	0.70
9	C_9H_{20}	nonane	−50	150	0.72
10	$C_{10}H_{22}$	decane	−30	170	0.73

advertisements for cooking oils and margarines ("polyunsaturated" refers to the presence of several π bonds in a molecule). Unsaturated hydrocarbons are also sometimes referred to as *olefins* ("oil forming"), a term based on the observation that ethylene and similar alkenes react with chlorine to form "oily" liquids.

A very large number of hydrocarbons occur naturally. Methane, originally called "marsh gas," was first identified in bubbles escaping from decaying vegetable matter in swampy areas. Petroleum is the chief source of most hydrocarbons, although the smaller molecules, such as methane, ethane, propane, and butane, occur in large underground pockets of "natural gas."

Since hydrogen and carbon differ only slightly in electronegativity (Unit 7) and the hydrocarbon molecules are fairly symmetrical in shape, hydrocarbons are essentially nonpolar. As a result, they have rather weak interparticle forces, primarily of the London dispersion type (Unit 9), and hydrocarbon mixtures can usually be separated by careful fractional distillation (Table 28.4). This technique is one of the main processes in petroleum refining. Low-boiling fractions (principally C_5–C_9 hydrocarbons) are used primarily for fuels such as lighter fluid, gasoline, and kerosene. Higher-boiling components (e.g., C_{10}–C_{16} hydrocarbons) are used in lubricating oils and greases, and those components (C_{17} and larger) that are solids at room temperature are used in paraffin and other "wax" products.

Most of the aliphatic hydrocarbons are obtained commercially from petroleum and natural gas deposits. These sources contain only small amounts of aromatic compounds, although many can be made synthetically from aliphatic compounds. The principal natural source of the aromatic hydrocarbons is *coal tar*, a liquid mixture distilled from coal heated in the absence of air. Only about 6 percent of soft coal is converted to coal tar by the "coking process," but coke is in such large demand for industrial smelting operations (Excursion 3) that this process provides an enormous supply of aromatic compounds. More than 200 different compounds have been isolated from coal tar. Most are aromatic, but not all are hydrocarbons (Table 28.5).

The production of benzene, for example, from coal tar is in excess of 2-million gallons per year.

The production of coal tar chemicals is an excellent example of the industrial recovery of waste products from one process, coke formation, for use in other processes. What might otherwise have been just another source of industrial pollution was instead turned into an exceedingly profitable venture.

28.5 THE NOMENCLATURE OF ORGANIC CHEMISTRY

The study of carbon compounds is one of the oldest fields of chemistry. As a result there are many "trivial" names of organic compounds that have been accepted for so long that they have become permanent parts of the language of chemistry. Some of these can be seen to be related to certain atomic groupings (Table 28.6), but others, such as phenol or aniline, seem to reflect mainly the whimsy of some poetic chemist.

Table 28.5   SOME AROMATIC COMPOUNDS FROM COAL TAR (APPROXIMATE YIELDS EXPRESSED AS GRAMS OF COMPOUND PER KILOGRAM OF COAL)

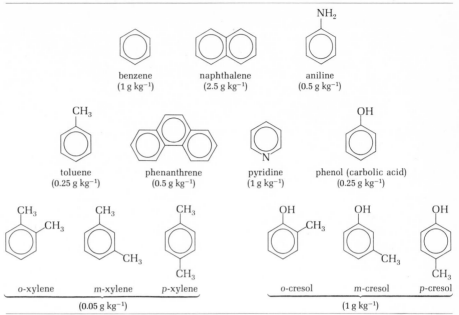

benzene
(1 g kg⁻¹)

naphthalene
(2.5 g kg⁻¹)

aniline
(0.5 g kg⁻¹)

toluene
(0.25 g kg⁻¹)

phenanthrene
(0.5 g kg⁻¹)

pyridine
(1 g kg⁻¹)

phenol (carbolic acid)
(0.25 g kg⁻¹)

o-xylene   *m*-xylene   *p*-xylene
(0.05 g kg⁻¹)

o-cresol   *m*-cresol   *p*-cresol
(1 g kg⁻¹)

Table 28.6   TRIVIAL NAMES REFLECTING THE GROUPING
CH_3C— (*acet*-)
$\overset{\|}{O}$

$CH_3\overset{O}{\overset{\|}{C}}OH$ *acetic acid*

$CH_3\overset{O}{\overset{\|}{C}}$—O⁻   *acetate ion*

$CH_3\overset{O}{\overset{\|}{C}}NH_2$ *acetamide*

$CH_3\overset{O}{\overset{\|}{C}}Cl$ *acetyl chloride*

$CH_3\overset{O}{\overset{\|}{C}}O\overset{O}{\overset{\|}{C}}CH_3$ *acetic anhydride*

$CH_3\overset{O}{\overset{\|}{C}}H$ *acetaldehyde*

$CH_3\overset{O}{\overset{\|}{C}}OCH_3$ *methyl acetate*

$CH_3\overset{O}{\overset{\|}{C}}CH_3$ *acetone*

$CH_3\overset{O}{\overset{\|}{C}}$—⟨ ⟩   *acetophenone*

*acetylsalicylic acid
(aspirin)*

Many of these trivial names will be encountered during the further study of organic chemistry, and the more common ones should be remembered.

As the number and complexity of characterized organic compounds increased, it became obvious that some systematic approach to naming must be developed. Today, such a system is universally recognized, although many of the trivial names are still used either through habit or as a convenience when the systematic name is unduly complex. The key to this IUPAC nomenclature system is a set of rules based on the alkanes (Table 28.7). These should be remembered. Additional rules for other classes of compounds will be introduced later. (See also Appendix G.)

Table 28.7   IUPAC[a] NOMENCLATURE FOR ALKANES (SIMPLIFIED RULES)

1. The base name is that of the alkane (Table 28.4) having the same carbon number as that of the longest *continuous carbon chain* in the compound being named.
2. Each carbon in the basic chain is numbered sequentially, starting at the end that will result in the smallest sum[b] of numbers for those carbon atoms attached to *branch groups.*
3. Hydrocarbon branch groups are named as *alkyl* groups, using the name of the corresponding alkane with the *ane* ending changed to *yl* (e.g., —CH_3 is methyl, —CH_2CH_3 is ethyl).
4. *Each* alkyl group is given the number of the carbon atom to which it is attached in the basic chain.
5. If there is more than one branch group of the same type, this is indicated by use of a prefix on the alkyl group name. (di = 2, tri = 3, tetra = 4, penta = 5, hexa = 6, hepta = 7, octa = 8, nona = 9, deca = 10.) *The location numbers of all identical groups are retained.*
6. The combined branch group names (either alphabetically or in order of increasing size) and location numbers serve as a prefix to the *base name.* Separate numbers from numbers by commas, numbers from words by hyphens, and do not separate words from other words.

[a]IUPAC stands for International Union of Pure and Applied Chemistry.
[b]This is a convenient oversimplification. For the detailed rule, see Appendix G.

Example 3

Give the IUPAC name for

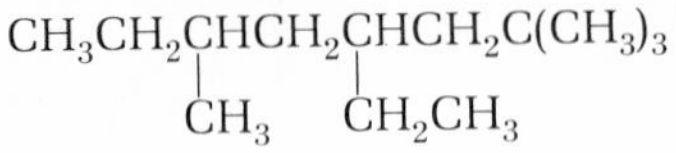

$$CH_3CH_2CHCH_2CHCH_2C(CH_3)_3$$
$$\quad\quad\quad | \quad\quad\quad |$$
$$\quad\quad\quad CH_3 \quad\quad CH_2CH_3$$

Note that these two —C*H$_3$ groups are part of the basic chain.

Solution

First, redraw the structural formula in expanded form, keeping only branch groups "compressed" (until some experience has been gained, it may be useful to mark the basic chain).

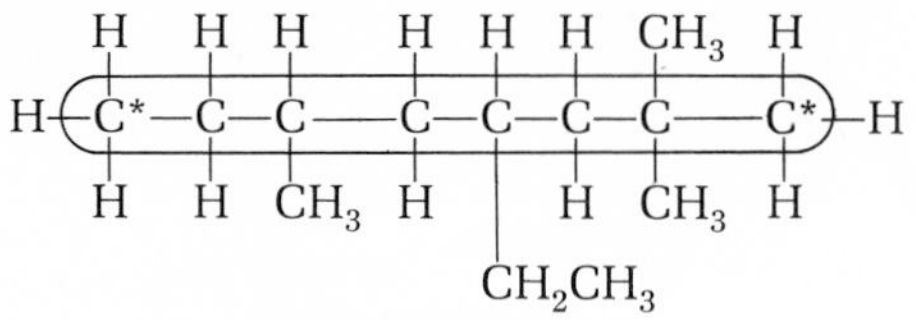

Now apply naming rules, in order.

1. There are eight carbons in the longest continuous chain, so the base name (Table 28.4) is *octane*.
2. We have two choices for numbering. If we start at the left end, as drawn, there are groups attached to carbons 3, 5, 7, 7. From the right, groups are numbered 2, 2, 4, 6. Since the latter yields the smaller sum of numbers ($3 + 5 + 7 + 7 = 22$, $2 + 2 + 4 + 6 = 14$), this is the sequence selected:

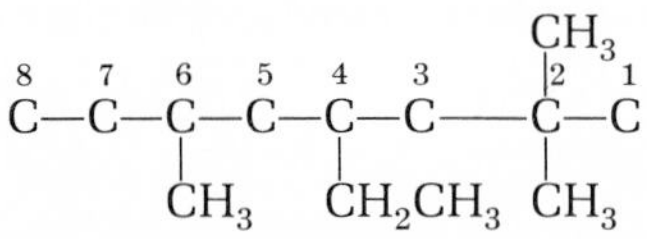

3. There are three methyl groups and one ethyl group attached to the basic chain.
4. The methyl group location numbers are 2, 2, 6, and that for the ethyl group is 4.
5. The three methyl groups are indicated as *2,2,6-trimethyl*.
6. The combined name, then, is written as either 4-ethyl-2,2,6-trimethyloctane (alphabetical) or 2,2,6-trimethyl-4-ethyloctane (groups in order of size).

Example 4

Give the IUPAC name for

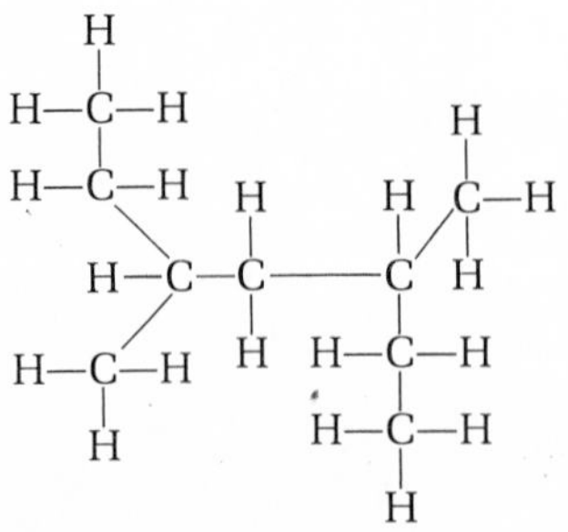

Solution

Since the structural formula may be drawn in a way that represents the longest chain

GASOLINE

The most familiar hydrocarbon fuel is probably the mixture of compounds known as gasoline. So-called "straight-run" gasoline, the hydrocarbon mixture obtained directly from petroleum distillation, is a poor fuel for modern internal combustion engines. In general, straight-chain hydrocarbons do not have desirable explosion characteristics. In high-compression engines a "smooth" explosion of the fuel–air mixture is necessary for proper operation, and straight-chain hydrocarbons explode too rapidly under normal engine conditions. This very fast explosion is known as "knocking" and reduces engine power.

Fuels may be rated in terms of their tendency for smooth explosion (i.e., explosion at a more suitable rate). The "octane" rating of fuels increases with their antiknock properties. An arbitrary scale has been established, using experimental comparisons of engine power. On this scale, heptane is rated as zero (very bad "knocking" properties) and 2,2,4-trimethylpentane (isooctane) is rated as 100.

It is found that higher octane ratings are favored by chain branching, unsaturation, or addition of special ingredients such as benzene or tetraethyllead. Theoretical explanations for these factors are still largely speculative, and the production of fuels for particular octane ratings is still mainly an empirical science.

Pyrolysis is a general term used to refer to any process of purely thermal decomposition. Hydrogen is a by-product of the fraction of the alkanes converted to alkenes. This is a valuable commercial source of H_2 (Unit 26).

When heated to high temperatures in the absence of air, hydrocarbons undergo a process called *cracking*. Changes in molecular structure occur, and straight-chain alkanes are converted to mixtures of alkenes and branched-chain alkanes. In this process, typically performed around 500°C over metal oxide catalysts, naturally occurring hydrocarbons are converted to more efficient fuels for modern high-compression engines. Synthetic hydrocarbons, usually prepared from low molecular weight alkenes, are also produced for modern gasolines.

In recent years, much attention has been devoted to environmental problems resulting from gasoline additives such as tetraethyllead. Research is still in progress to find efficient gasolines that do not require such additives. Aromatic hydrocarbons, such as benzene, are alternative antiknock agents, but many of these are considered health hazards potentially more dangerous than tetraethyllead.

The gasoline industries maintain some of the finest chemical research programs in the country, and there is real hope that new fuels will be developed that, in conjunction with improved engine designs, will significantly reduce pollution from automobiles.

Remember that carbon is bonded in a tetrahedral geometry in alkanes. Formulas may be written many different ways, and "linear" formulas are simply a matter of convenience, not representative of actual molecular geometry.

as other than linear, we shall mark the longest chain to avoid ambiguities:

Then

1. The base name (Table 28.4) is *heptane* (seven carbons).
2. Numbering from either end of the chain gives the same group location numbers (3 and 5):

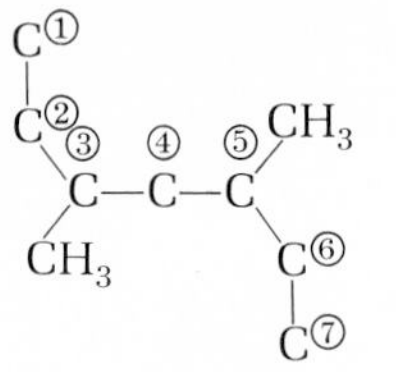

3, 4, 5. Groups are named and numbered as *3,5-dimethyl*.
6. The complete name, then, is 3,5-dimethylheptane.

28.6 ISOMERISM

An organic compound with the molecular formula C_6H_8 (Example 1) might have several different structures, including, among others,

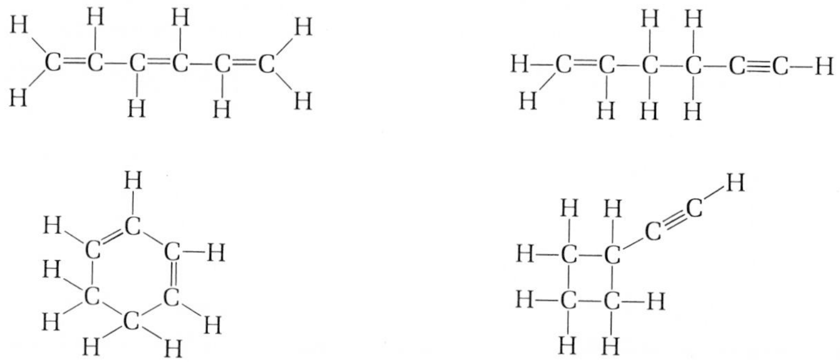

Substances like the preceding ones that have the same molecular formula but different structural formulas are called *isomers*. Carbon's ability to form stable σ or π bonds to other carbon atoms (as well as to atoms of many other elements) and to extend such bonding through long chains or rings of atoms accounts for the large number of isomeric compounds encountered. The determination of a complete (three-dimensional) structural formula for a compound is therefore a very complex operation, especially when there are many isomeric structures possible. Some of the modern methods employed for structural elucidation are described in later units.

There are several types of isomerism. *Positional isomerism* is, perhaps, the simplest to describe by two-dimensional formulas. The isomeric structures shown below differ only in the position of the methyl (CH_3) group attached to the five-carbon chain. It should be remembered that the most stable geometry around a carbon having four σ bonds is the tetrahedral pattern, corresponding to the sp^3

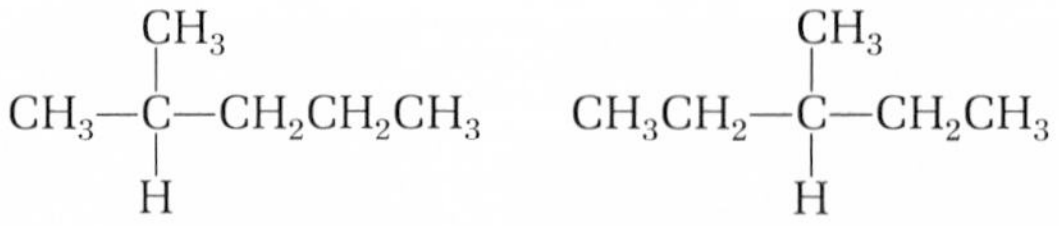

orbital description. Hence, structures such as those shown are only simplified two-dimensional representations. It will prove useful in your study of organic

Two trivial designations are commonly encountered for alkanes. These are *n-* (meaning "normal") for unbranched alkanes and *iso* for alkanes having only a single methyl group attached at the *second* carbon of the chain. For example, $CH_3CH_2CH_2CH_2CH_3$ may be called *n*-pentane, and $(CH_3)_2CHCH_2CH_3$ may be called isopentane. Note that the latter terminology counts *all* carbons in the molecule to determine the base name. [One exception to this use of the *iso* prefix is the particular compound called isooctane (see

Gasoline).] Similar trivial designations are often employed for alkyl groups or for other classes of compounds that may be considered as derived from alkanes. Again, the *n*- prefix indicates an unbranched chain (e.g., $CH_3CH_2CH_2$— is an *n*-propyl group, and $CH_3CH_2CH_2CH_2OH$ may be called *n*-butyl alcohol). The *iso* prefix now refers to the case of a methyl group (as the only "branch") at the *next to last* carbon, counting the carbon attached to some other functional group as the "first" [e.g., $(CH_3)_2CHCH_2OH$ may be called isobutyl alcohol].

structures if you will purchase about 20 small styrofoam spheres from an art store and assemble these with toothpicks to make three-dimensional models, as described at the end of this unit. When you recall that free rotation is possible around σ bonds, it will be obvious that a substance such as 2-methylpentane could be described by several formulas like those below, but that all are equivalent.

2-methylpentane

28.7 POSITIONAL ISOMERS OF RING COMPOUNDS

Positional isomers of acyclic alkanes may be distinguished either by appropriate structural formulas or by their IUPAC names. In simple cycloalkanes, a somewhat analogous nomenclature is used. The base name is determined by the number of carbon atoms in the ring, and attached groups are identified by location numbers assigned sequentially to ring carbons in the order that gives the smallest sum of location numbers. If only one ring carbon has attached groups, no location numbers are used, since all ring positions in cycloalkanes are equivalent.

Examples

1,2-dimethylcyclohexane

1,3-dimethylcyclopentane
(note: *not* 1,4-)

1,1-dimethyl-3-ethylcycloheptane

methylcyclobutane
(note: no location
number needed)

Although the IUPAC system is "universally accepted," many current papers and most of the older literature uses "trivial" naming for many compounds.

Location *numbers* are used for substituted benzenes when there are three or more groups attached to the ring. If there is only one group on the ring, no number is added. To distinguish disubstituted benzenes, letters rather than numbers are usually employed in deference to tradition. The IUPAC numbers are rarely used for such compounds. Groups on adjacent ring carbons are indicated by *o* (*ortho*); groups directly across the ring from each other are specified by *p* (*para*); and groups in what would correspond to 1,3 positions on cyclohexane are designated as *m* (*meta*). Although *benzene* is used as the base name on most of its substituted derivatives, many such compounds (Table 28.8) have common names that are almost invariably used instead of the more systematic nomenclature.

Table 28.8 SOME COMMON NAMES FOR SUBSTITUTED BENZENES[a]

FORMULA	NAME
CH_3 benzene ring	toluene
CH_3, CH_3 benzene ring	o-xylene (p- and m- isomers also possible)
OH benzene ring	phenol
OH, CH_3 benzene ring	o-cresol (p- and m- isomers also possible)
OH, OH benzene ring	hydroquinone (p-dihydroxybenzene)
NH_2 benzene ring	aniline
NH_2, CH_3 benzene ring	o-toluidine (p- and m- isomers also possible; sometimes called o-methylaniline)
$\overset{O}{\overset{\|}{C}}-CH_3$ benzene ring	acetophenone
$\overset{H}{\underset{}{C}}=O$ benzene ring	benzaldehyde
CO_2H benzene ring	benzoic acid

Table 28.8 (Continued)

FORMULA	NAME
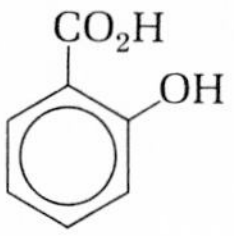	salicylic acid
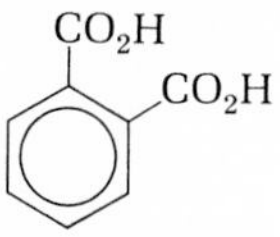	phthalic acid

[a] Names shown are often used as base names for compounds derived from these structures.

Examples

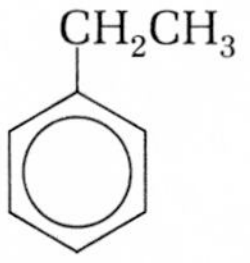

ethylbenzene

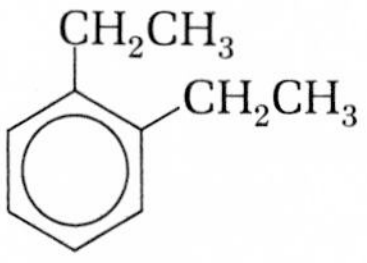

o-diethylbenzene
(less commonly:
1,2-diethylbenzene)

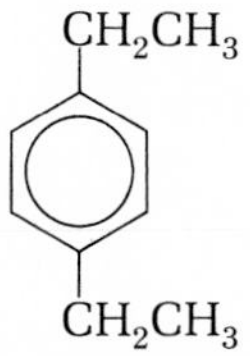

p-diethylbenzene
(less commonly:
1,4-diethylbenzene)

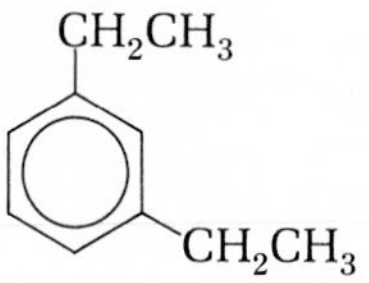

m-diethylbenzene
(less commonly:
1,3-diethylbenzene)

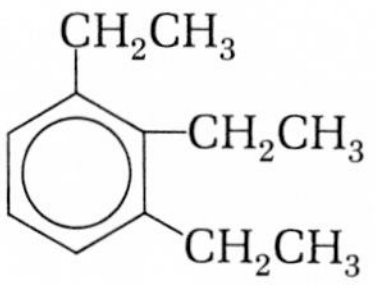

1,2,3-triethylbenzene

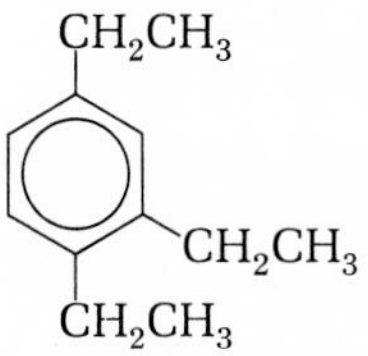

1,2,4-triethylbenzene

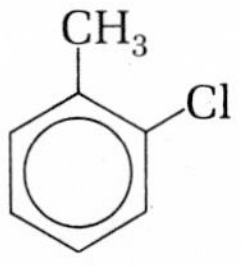

o-chlorotoluene
(less commonly:
2-chlorotoluene)

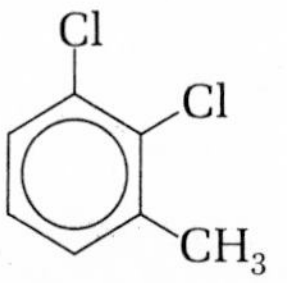

2,3-dichlorotoluene

When a benzene ring is attached to some other structure that determines the compound's base name, the benzene ring is named as a *phenyl* group.

phenyl = C_6H_5- ≡

≡

Examples
_____

cyclohexane

phenylcyclohexane

CH_3CO_2H

acetic acid

$-CH_2CO_2H$

phenylacetic acid

$$CH_3-\underset{\underset{NH_3{}^+}{|}}{\overset{\overset{H}{|}}{C}}-CO_2{}^-$$

alanine

$-CH_2-\underset{\underset{NH_3{}^+}{|}}{\overset{\overset{H}{|}}{C}}-CO_2{}^-$

phenylalanine

28.8 CONFORMATIONAL ISOMERISM

We have stressed several times that internal rotations may occur within molecules around σ bonds. As such rotations proceed in ethane molecules, for example, the relative positions of the hydrogens in one methyl group change with respect to those in the other group (Figure 28.6). The different structures involved are known as *conformational* isomers. These are the only isomers that can be interconverted without breaking covalent bonds.[3] Unlike the positional isomers, the different conformations of a simple molecule do not, in general, represent unique structures that can be separated and isolated as such. Even at low temperatures, the normal thermal energies of most molecules are such as to cause a continuous interconversion of conformational isomers. When we speak of "free" rotation, we do not mean that there is *no* energy barrier to rotation, but simply that the energy differences among the possible conformations are quite small relative to their normal thermal energies. Of all the possible conformations, we can distinguish two general types—the staggered conformations (more stable since like atoms are at greater distances from each other) and the *eclipsed* conformations (least stable since like atoms are at minimum separations). The *energy barrier* to rotation is simply the difference between the energies of the extreme conformations. As the size of attached atoms increases, the energy barrier increases since the eclipsed conformation becomes less stable as "crowding" increases (Figure 28.7). In some cases, particularly with certain cyclic compounds (Figure 28.8), energy differences among possible conformations are

[3] Conformational isomers are not generally considered as "true isomers" since conformational differences are usually quite small, in terms of energy, compared to those of "true" isomers. For a more detailed treatment of isomerism, including conformational isomerism, see "The Shapes of Organic Molecules," by Joseph Lambert, in the January, 1970, issue of *Scientific American* (pp. 58–70).

Figure 28.6 Some conformational isomers of ethane. (a) Staggered ("anti"; most stable). (b) Staggered ("skew"). (c) Eclipsed (least stable). (d) Staggered ("skew").

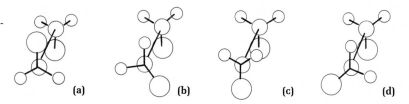

(a)   **(b)**

(d)   **(c)**

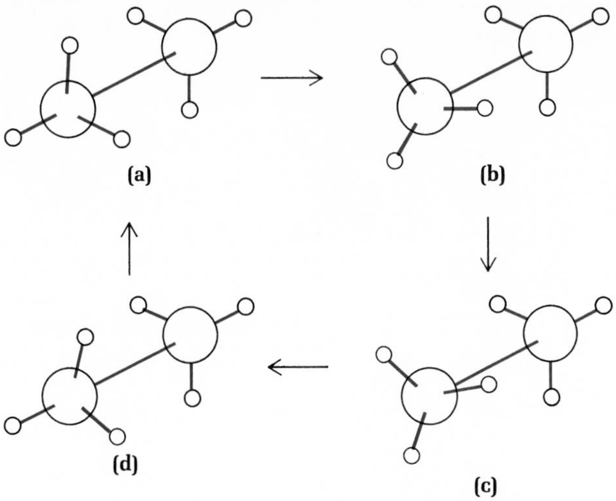

Figure 28.7 Some conformational isomers of 1,2-dibromoethane. (a) Staggered ("anti"; most stable). (b) Staggered ("skew"). (c) Eclipsed (least stable). (d) Staggered ("skew").

(a)   **(b)**   **(c)**   **(d)**

Figure 28.8 Some conformational isomers of 1,4-dibromocyclohexane. Axial positions are those approximately perpendicular to the ring's central plane. Equatorial positions lie roughly within or parallel to this plane. Note that the simplified representations (Table 28.1) for cycloalkanes are not geometrically accurate; that is, the cyclohexane ring, unlike the benzene ring, is not "flat."

● = bromine

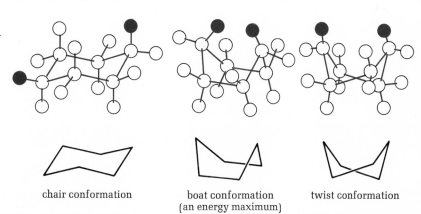

chair conformation   boat conformation (an energy maximum)   twist conformation

sufficiently large to permit the actual separation of stable conformational isomers.

Suggestion: Construct models of ethane, 1,2-dibromoethane, and 1,4-bromocyclohexane. By physical manipulation of the models, see which conformational isomers appear to have the largest energy differences. (Use large spheres for bromine atoms.)

Figure 28.9 Some geometric isomers

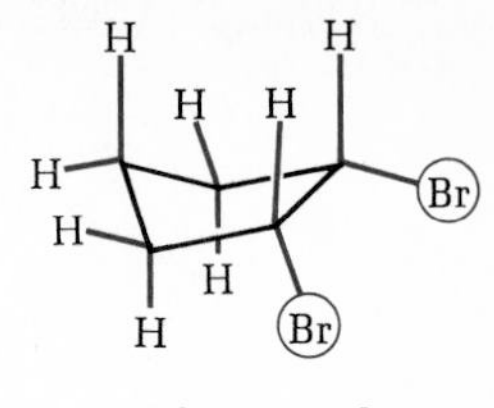

cis-1, 2-dibromocyclopentane

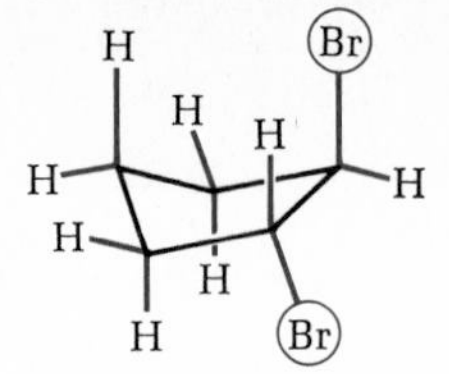

trans-1, 2-dibromocyclopentane

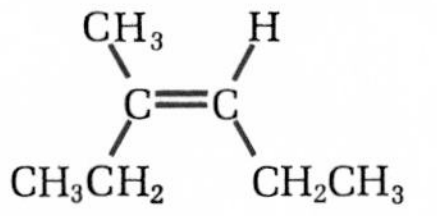

3-methyl-cis-3-hexene

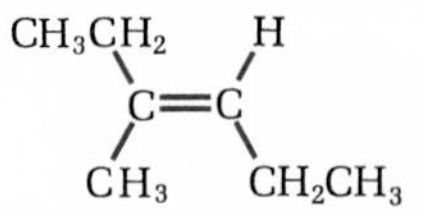

3-methyl-trans-3-hexene

28.9 GEOMETRIC ISOMERS

Certain structures that might appear at first glance to be simply different conformations actually represent two different chemical compounds. Geometric isomers are species that differ, as in the case of conformational isomers, only in the relative spatial arrangement of atoms of otherwise equivalent groups. Geometric isomers, however, *cannot be interconverted without breaking one or more covalent bonds.* Such isomers may occur when some structural feature restricts the rotations necessary for isomeric interconversion. Two such features are rings and double bonds (Figure 28.9). In the former case, σ bonds must be broken before isomer conversion is possible, and the latter requires rupture of the π-bond system. In either situation, considerable energy is needed and, as a result, geometric isomers are normally distinct, separable compounds.

A functional group determining the name ending is always given the smallest possible "location number."

To distinguish between geometric isomers, if their structures are known, one can either draw appropriate formulations or use an accepted naming notation. To indicate that two groups are in closest allowed geometric proximity, they are specified as *cis*. For groups at maximum possible separation, the term *trans* is used. In cyclic compounds, the groups being described are specifically named. In alkenes, the two groups are taken as the *largest groups* within *the carbon chain containing the double bond, so they are indicated as part of the base name* (Table 28.9).

Table 28.9 IUPAC NOMENCLATURE FOR SIMPLE ALKENES

1. The base name is formed from that of the *alkane* having the number of carbons found in the longest carbon chain in the alkene *that contains the double bond*. The name ending for alkenes is *ene* (rather than *ane* as for alkanes).
2. The carbon atoms of the basic chain are numbered sequentially, *beginning at the end nearest[a] the double bond*. The location number of the first atom of the double bond is inserted just before the base name. (For a cycloalkene, the double bond is assumed to be between carbons 1 and 2, so no double-bond location number is needed for the name.)
3. If the compound can exist as a pair of geometric isomers, indicate the specific isomer (if known) by -*cis*- or -*trans*- just preceding the location number of the double bond.

[a] If the double bond is in the center of the chain, number from the end giving the smallest sum of group location numbers.

Example 1

Give correct IUPAC names for

1.

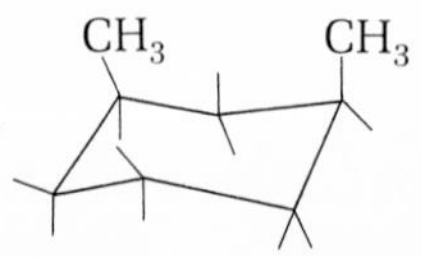

2.

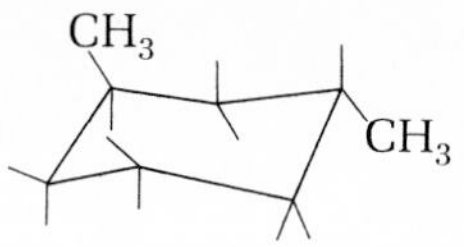

3.

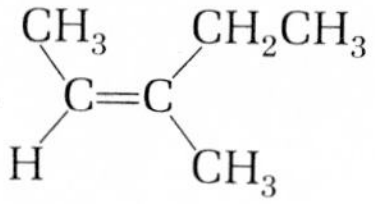

4.

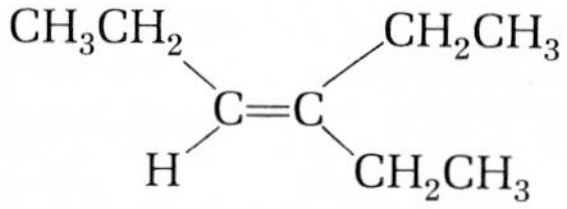

5.

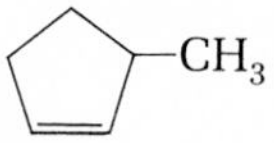

6.

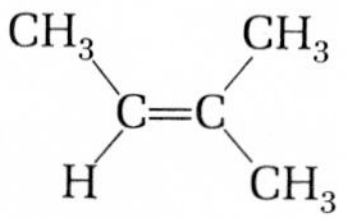

Solution

1. The compound would be named as a 1,3-dimethylcyclohexane. Since methyl groups are at closest allowed positions, they are specified as *cis*, hence
cis-1,3-dimethylcyclohexane

2. *trans*-1,3-dimethylcyclohexane (methyl groups have maximum separation)

Note: Name determined by relative positions of segments of longest carbon chain, *not* of "similar groups".

3.

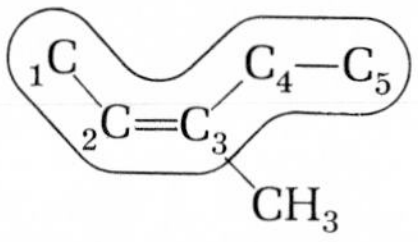

3-methyl-*cis*-2-pentene

4. This compound has no geometric isomers, since 180-degree rotation around the double bond results in an identical structure. The double bond is in the center of the longest chain, so we begin numbering with the chain end nearest the attached group.

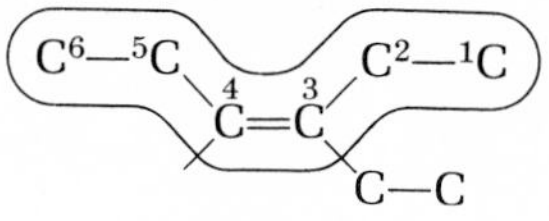

3-ethyl-3-hexene

5. The double-bond carbons must be numbered 1 and 2. A counterclockwise sequence is selected in this case so as to yield the smaller location number for the methyl group.

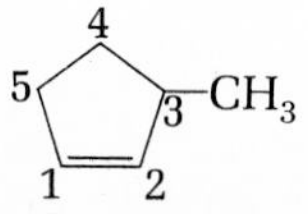

3-methylcyclopentene

6. Note that geometric isomers are not possible when there are two identical groups attached to one of the double-bond carbons (rotation would give the same structure). Thus the compound is called simply 2-methyl-2-butene.

28.10 OPTICAL ISOMERS

Ordinary light is described as electromagnetic radiation, and we know that such radiation is capable of interacting with the electric field of electrons in atoms and molecules (e.g., as in absorption spectroscopy). The electric field of ordinary light is considered as being symmetrical about the line representing the light path, but asymmetry is introduced when such light is reflected from a flat surface or passed through certain materials such as Polaroid. This *plane-polarized light* is now represented as having its electric field described by vectors confined to a single plane perpendicular to the light path.

If we view a light beam through a pair of identical Polaroid lenses from sunglasses and begin to turn one of the lenses, we notice that the light intensity begins to drop, reaching a minimum when the two lenses are in a position such that their longest dimensions are perpendicular (crossed). The plane-polarized light from the first lens is absorbed by the second lens unless the lenses are arranged in a parallel fashion. If we now insert a sample tube between the two lenses, we shall have prepared a simple instrument called a *polarimeter* (Figure 28.10) with which we can observe the effects of plane-polarized light on a liquid or solution.

If we place water, methanol, or chloroform, for example, in the sample tube, the maximum light intensity is found with the same relative orientations of polarizer and analyzer as with the sample tube empty—even though the molecules of these compounds are unsymmetrical. However, if a saturated aqueous solution of table sugar (sucrose) is placed in the sample tube, the maximum light intensity is not observed until the analyzer has been turned to a new angle relative to the alignment of the polarizer. Apparently the sucrose molecules interact in some way with the plane-polarized light to rotate the plane of polarization.

Substances capable of rotating the polarization plane of plane-polarized light are said to be optically active.

It has been found that optical activity is associated only with substances whose molecules (or polyatomic ions) are not superimposable on their mirror images.

Optical isomers differ with respect to their interactions with plane-polarized light. The simplest optical isomers occur in organic molecules containing an asymmetric carbon (Figure 28.11). Two optical isomers that are mirror images are called *enantiomers*. There are other types of optical isomers, but further discussion is beyond the

An asymmetric carbon is one that is bonded to four *different* groups.

Figure 28.10 A simple polarimeter. Commercial polarimeters are typically of the double-beam type. Half the incident light passes through the sample, and the remainder is reflected around the sample. The two light beams are displayed on separate halves of a circular eyepiece, and the analyzer is rotated until the light intensities are matched. Normally, monochromatic light (often from a sodium vapor lamp) is employed.

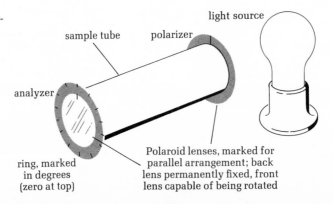

light source

sample tube   polarizer

analyzer

ring, marked in degrees (zero at top)

Polaroid lenses, marked for parallel arrangement; back lens permanently fixed, front lens capable of being rotated

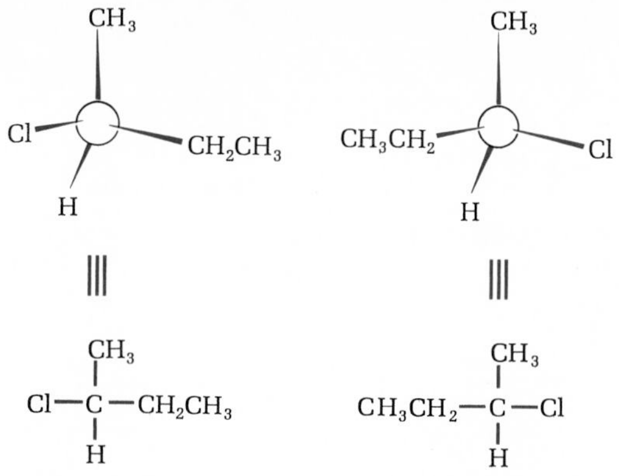

Figure 28.11 Simple optical isomers (enantiomers, nonsuperimposable mirror images). Conventional structural formulas may be used to represent optical isomers. Side groups are assumed to project up from the plane of the paper, and top and bottom groups project back behind the plane of the paper.

scope of this text. A special naming system is used to distinguish among optical isomers. For details, see the Suggested Reading at the end of this unit.

28.11 ISOMERS AND VARIATION OF PROPERTIES

The molecular formula C_2H_6O may refer to either of two compounds:

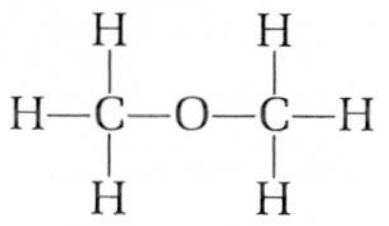

dimethyl ether

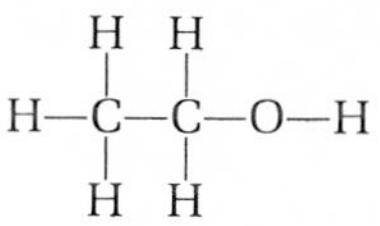

ethyl alcohol
(ethanol)

The former is a gas at room temperature, sparingly soluble in water. The latter is a liquid (boiling point, 78.5°C) and completely miscible with water. Even greater differences are observed in the chemical properties of these compounds. Ethanol reacts with carboxylic acids to form esters, with PCl_3 to form ethyl chloride, and may easily be "dehydrated" to form ethylene. Dimethyl ether undergoes none of these reactions and, in fact, is inert to a variety of chemicals that would easily react with ethanol.

Special types of *structural* isomers.

Such *functional group isomers* display the greatest differences in chemical and physical properties of all isomeric species. This subject is discussed in Unit 29.

Other types of isomers may also differ to some extent in their physical and chemical properties. These differences are usually much less than those of functional group isomers (Table 28.10). To some extent the differences in properties may be

(continued on p. 728.)

Table 28.10 ISOMERIC VARIATIONS IN BOILING POINTS

TYPE OF ISOMERISM	EXAMPLES	BOILING POINT (°C) AT 1 atm	COMMENTS
positional	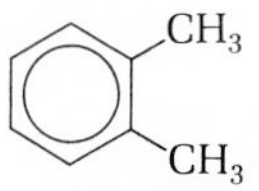	144	b.p.'s reflect polarity differences
	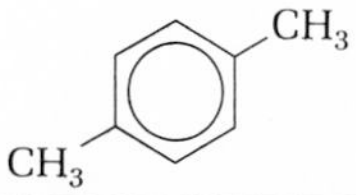	138	

Table 28.10 (Continued)

TYPE OF ISOMERISM	EXAMPLES	BOILING POINT (°C) AT 1 atm	COMMENTS
		240	low b.p. of *o*-isomer reflects *internal* hydrogen bonding
		286	
		276	
geometric		4	b.p.'s reflect polarity differences
		1	
		60	
		48	
optical	$HOCH_2-\overset{\overset{\displaystyle C_2H_5}{\vert}}{\underset{\underset{\displaystyle CH_3}{\vert}}{C}}-H$	129	other identical properties include density, m.p., solubility, and chemical reactivity in solution; the first isomer rotates plane-polarized light in a clockwise direction, the other gives counterclockwise rotation
	$H-\overset{\overset{\displaystyle C_2H_5}{\vert}}{\underset{\underset{\displaystyle CH_3}{\vert}}{C}}-CH_2OH$	129	
functional group	CH_3OCH_3	−24	water solubility at 25°C = ∼1.5 mole liter^{-1};
	CH_3CH_2OH	78	completely miscible with water
	$CH_3CH_2\overset{\overset{\displaystyle H}{\vert}}{C}=O$	49	water solubility at 25°C = ∼4 mole liter^{-1}
	$CH_3\overset{\overset{\displaystyle O}{\Vert}}{C}CH_3$	56	completely miscible with water

DIRECTIONS FOR MAKING STYROFOAM MODELS

Use "expanded styrofoam" (soft) spheres for ease of handling.

1. Measure the diameter of the spheres to be used and note this value as d.
2. Using the d value obtained, draw templates as shown in Figure 28.12 on stiff paper (heavy construction paper or manila folder). Cut, punch, and fold the templates as indicated.
3. If desired, paint spheres different colors to distinguish between elements (e.g., brown for C, red for H, blue for O, green for N, yellow for Cl, etc.). Most art supply stores have special paints for styrofoam.
4. Mark spheres for tetrahedral bonding by enclosing a sphere in the folded tetrahedron and marking a spot with a fine-tip felt pen through each punched hole. Insert a toothpick into the sphere, about halfway, through the hole in the tetrahedron base. Remove the sphere and use the alignment guide to position toothpicks at the three remaining spots.
5. For trigonal or collinear bonding, insert the sphere into the hole in the appro-

The store from which you purchase the spheres may know their sizes. If not, a simple measurement can be made by either measuring the "jaw width" of a wrench set so the sphere will just fit within the jaws or measuring the diameter of the hole formed when the sphere is pushed directly down into a pan of moist sand.

For the "tetrahedral" template, construct the large equilateral triangle first. Then draw the center triangle by connecting the midpoints of the three sides of the large triangle. Locate the center of each small triangle by finding the intersection point of any two lines drawn to bisect two angles.

template for tetrahedral positions

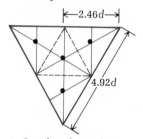

1. Cut along heavy lines.
2. Punch holes size of pen point at solid circles.
3. Fold along dashed lines.

template for trigonal coplanar bonds

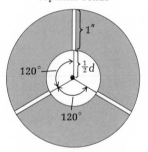

1. Mark double lines.
2. Cut out ring, discard center circle.

alignment guide for tetrahedral bonds

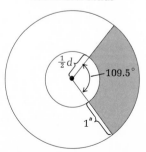

Cut out shaded segment for spacer

template for collinear bonds

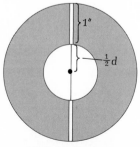

1. Mark double lines.
2. Cut out ring, discard center circle.

Figure 28.12 Templates for molecular models (using styrofoam spheres of diameter d).

priate template until the template bisects the sphere. Insert toothpicks halfway into the sphere along the marked double lines on the template ring.

6. For double bonds, use two parallel toothpicks, each spaced $\frac{1}{4}$ in. from the template line. For triple bonds, use two toothpicks spaced as for double bonds and a third directly between them.
7. Connect spheres together, using one toothpick per bond, so that sphere centers are collinear.

(These models are inexpensive, but commercial model kits are sturdier and you may prefer to obtain a commercial "stick-and-ball" kit.)

predicted on the basis of structural features. For example, variation in boiling points may often be estimated qualitatively on the basis of molecular arrangements related to polarity differences (Unit 8).

Variations in physical properties among isomeric compounds can often be explained and predicted in terms of relatively simple considerations of intermolecular forces. It might prove worthwhile to review the discussion of these forces in Unit 9.

Mirror-image optical isomers (enantiomers) are identical in many respects. Their melting points, boiling points, solubilities, and most reactions in solution are the same. Such isomers differ in two important respects: They rotate the polarization plane of plane-polarized light in opposite directions (but by the same amount), and their *biochemical* properties are usually *very* different because of the stereoselective properties of the catalytic enzymes involved (Unit 34) in the chemical functions of living systems.

It has recently been demonstrated[4] that optical isomers can have significantly different *odors*. This evidence supports the theory that olfactory receptor sites are quite *stereospecific*, that is, sensitive to unique geometrical features of adsorbed molecules.[5]

SUGGESTED READING

Benfey, O. T. 1966. *The Names and Structures of Organic Compounds*. New York: Wiley.
Goldwhite, Harold. 1972. *An Introduction to Organic Chemistry*. Columbus, Ohio: Merrill.
Stille, John K. 1968. *Industrial Organic Chemistry*. Englewood Cliffs, N.J.: Prentice-Hall.

EXERCISES

1. Make flash cards[6] for study of
   a. names of the first 10 alkanes (Table 28.4)
   b. classification by functional groups (Table 28.2)
   c. IUPAC nomenclature for alkanes (Table 28.7)
   d. common aromatic compounds (Table 28.8)
   e. IUPAC nomenclature for alkenes (Table 28.9)

[4] *Science*, Vol. 172, pp. 1043–1044, 1971.
[5] The "Stereochemical Theory of Odor" is described by Amoore, Johnston, and Rubin in the February, 1967, issue of *Scientific American*.
[6] For this and subsequent "memory work," flash cards are recommended. For functional group flash cards, draw the general group structure on one side of a small card and write the "class" name on the other side. Study carefully, then review by shuffling the cards and looking at the "structure" sides while writing the class names. If you miss one, insert that card near the bottom of the deck and try again. Then shuffle the cards again and look at the "name" sides, writing down the structures.

2. Identify each of the following compounds by functional group "class":

a. CHI_3

b. $H_2C{=}O$

c.

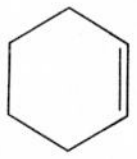

d.

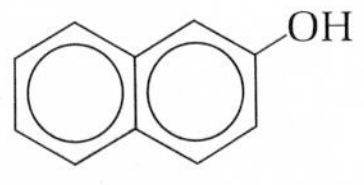

e.

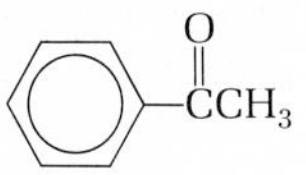

f. $(CH_3)_2CHOH$

g.

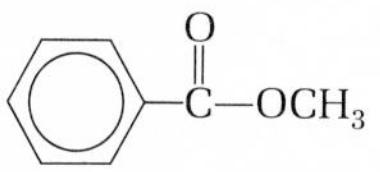

h. $CH_3\overset{O}{\overset{\|}{C}}N(CH_3)_2$

i.

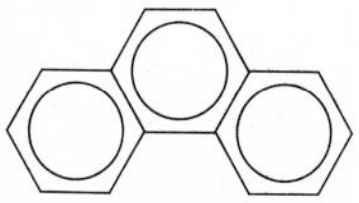

j.

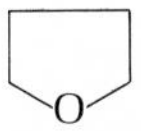

3. Supply the missing names or compressed structural formulas for

a. $(CH_3)_2CHCH_2\underset{\underset{\displaystyle CH_2CH_3}{|}}{C}HCH_2CH_3$

b. 2-methyl-4,4-diethyldecane

c. $CH_3CH_2\underset{\underset{\displaystyle CH_2CH_2CH_3}{|}}{\overset{\overset{\displaystyle CH_3}{|}}{C}}CH_2CH_2CH_3$

d. dimethylpropane

e. $CH_3CH_2CH_2CH_2\underset{\underset{\displaystyle CH_2CH_3}{|}}{\overset{\overset{\displaystyle CH_3}{|}}{C}}{-}CH_2\overset{\overset{\displaystyle CH_3}{|}}{C}HCH_2CH_2CH_3$

4. Draw all possible structures as indicated and give the IUPAC name for each:
a. all positional isomers having the molecular formula C_6H_{14}
b. all positional and geometric isomers of *acyclic* ("non-ring") compounds having the formula C_6H_{12}

5. Give accepted names for

a.

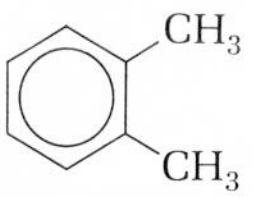

b.

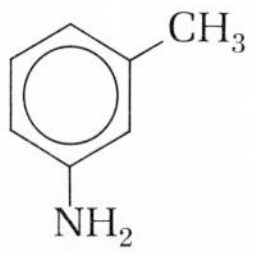

c.

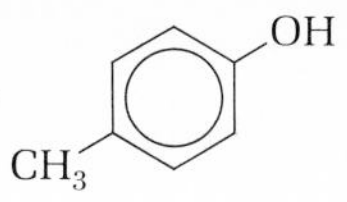

d.

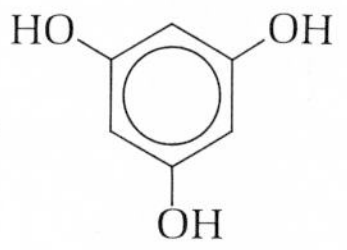

e.

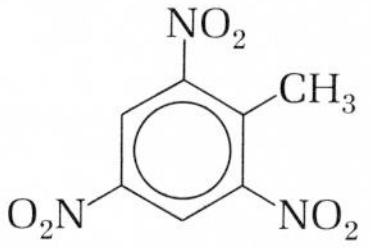

TNT

(*Hint:* Name as nitro derivative of toluene.)

6. Draw representations for
a. staggered and eclipsed conformations of butane (representing methyl groups by large circles)
b. boat and chair conformations of cyclohexane
c. optical isomers of 3-methylhexane

7. Predict which isomer in each pair is probably higher boiling. Explain the reasons for each choice.
a. *cis*-3-hexene or *trans*-3-hexene
b. 1,3,5-trimethylbenzene or 1,2,3-trimethylbenzene
c. *p*-cresol or *p*-xylene

OPTIONAL

8. Using styrofoam spheres, toothpicks, and the appropriate templates,[7] construct models of

$CH_3CH_2CH_2CH_3$
$CH_3C{\equiv}CCH_3$

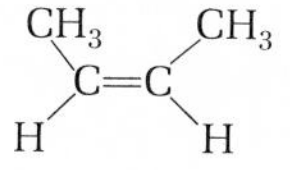

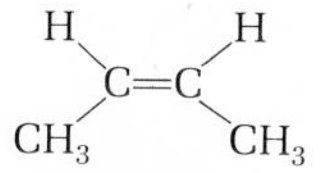

On the basis of your models, answer the following questions:
a. Can atomic groupings be rotated around single bonds without breaking the bonds?
b. Can atomic groupings be rotated around double or triple bonds without breaking the bonds?

[7] If a commercial model set is available, use it for this exercise.

c. Which of the following pairs represents true isomers?

$$\underset{\underset{H}{|}}{\overset{\overset{CH_3}{|}}{H-C}}-\underset{\underset{H}{|}}{\overset{\overset{CH_3}{|}}{C}}-H \quad \text{and} \quad \underset{\underset{H}{|}}{\overset{\overset{CH_3}{|}}{H-C}}-\underset{\underset{CH_3}{|}}{\overset{\overset{H}{|}}{C}}-H$$

or

$$\underset{H}{\overset{CH_3}{\diagdown}}C=C\underset{H}{\overset{CH_3}{\diagup}} \quad \text{and} \quad \underset{H}{\overset{CH_3}{\diagdown}}C=C\underset{CH_3}{\overset{H}{\diagup}}$$

d. Which of the four compounds given is probably the most polar? Explain.

PRACTICE FOR PROFICIENCY
Objective (a)

1. Give the name of the "class" to which each of the following belongs:

a. (cyclohexanone structure)

b. (tetrahydropyran structure)

c. $(CH_3)_2CHC{=}O$ with H below

d. $C_6H_5CO_2H$

f. H_2N- (benzene ring)

e. $H-\overset{\overset{O}{\|}}{C}-O-CH_2CH_3$

h. (benzene ring with CH_2OH)

g. $CH_3C{\equiv}CCH_3$

i. $H\overset{\overset{O}{\|}}{C}NH_2$

j. (benzene ring with three CH_3 groups)

2. Indicate the "class" for each of the following compounds, for which some familiar applications are given.

a. (nicotine structure)   *Nicotine*, a component of tobacco and a deadly poison, sometimes used as an insecticide

b. $\underset{CH_2\!-\!-\!-\!CH_2}{\overset{CH_2}{}}$ a common anesthetic

c. $H_2C{=}CH_2$ starting material for *polyethylene* plastic

d. (benzpyrene structure)   *3,4-benzpyrene*, a carcinogenic ("cancer-causing") agent found in cigarette smoke

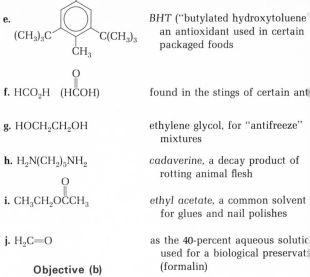

e. (BHT structure with OH, $(CH_3)_3C$, $C(CH_3)_3$, CH_3)   *BHT* ("butylated hydroxytoluene, an antioxidant used in certain packaged foods

f. HCO_2H $(H\overset{\overset{O}{\|}}{C}OH)$ found in the stings of certain ant

g. $HOCH_2CH_2OH$ ethylene glycol, for "antifreeze" mixtures

h. $H_2N(CH_2)_5NH_2$ *cadaverine*, a decay product of rotting animal flesh

i. $CH_3CH_2O\overset{\overset{O}{\|}}{C}CH_3$ *ethyl acetate*, a common solvent for glues and nail polishes

j. $H_2C{=}O$ as the 40-percent aqueous solution used for a biological preservati (formalin)

Objective (b)

3. Draw structural formulas for
   a. all isomers having the molecular formula C_4H_8
   b. all isomers having the molecular formula C_3H_6O, and containing no carbon–carbon double bonds
   c. all isomers containing a six-member aromatic ring and having the molecular formula C_9H_{12}
   d. all isomers containing a six-member aromatic ring and having the molecular formula C_7H_8O
   e. all isomers classified as ethers and having the molecular formula $C_4H_{10}O$
   f. four *functional group* isomers having the formula $C_5H_{10}O$
   g. geometric isomers of 2-hexene
   h. geometric isomers of 3-methyl-2-pentene
   i. optical isomers of 2-butanol ($CH_3CH_2\underset{\underset{OH}{|}}{C}HCH_3$)
   j. optical isomers of 3-methylhexane
4. Draw representations for
   a. staggered and eclipsed conformations of ethylene glycol
   b. optical isomers of 1-bromo-2-butanol ($BrCH_2\underset{\underset{OH}{|}}{C}HCH_2CH_3$)

Objective (c)
5. Give IUPAC names for all structural formulas drawn for question 3a, c, g, and h.
6. Give IUPAC names for

   a. $CH_3(CH_2)_8CH_3$
   b. $(CH_3)_2CHCH_2CH_3$
   c. $CH_3CH_2\underset{\underset{CH_3}{|}}{C}HCH_2\underset{\underset{CH_3}{|}}{C}HCH_3$
   d. (triangle/cyclopropane)

   e. (square with CH_3)
   f. (cyclopentene pentagon)
   g. (cyclohexane with CH_2CH_3)

h. 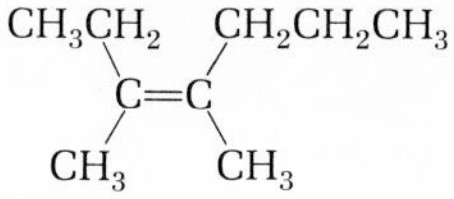 CH$_3$CH$_2$ CH$_2$CH$_2$CH$_3$
$$\underset{\underset{CH_3}{|}}{C}=\underset{\underset{CH_3}{|}}{C}$$

i. 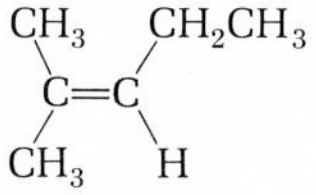 CH$_3$ CH$_2$CH$_3$
$$\underset{\underset{CH_3}{|}}{C}=\underset{\underset{H}{|}}{C}$$

j. 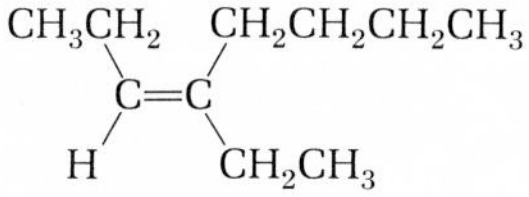 CH$_3$CH$_2$ CH$_2$CH$_2$CH$_2$CH$_3$
$$\underset{\underset{H}{|}}{C}=\underset{\underset{CH_2CH_3}{|}}{C}$$

7. Give accepted names for each of the following:

a.

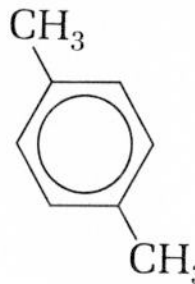

b.

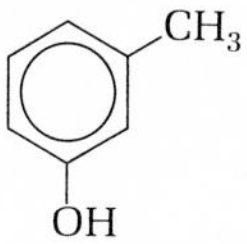

c.

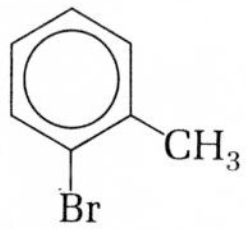

d.

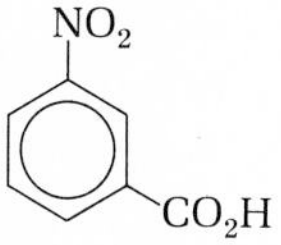

e.

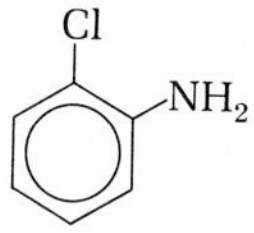

8. Supply the missing names or structural formulas for
a. CH$_3$(CH$_2$)$_5$CH$_3$
b. pentane
c. CH$_3$CH$_2$CH$_2$CH$_3$
d. decane

e. 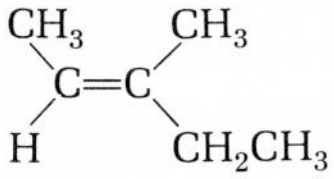 CH$_3$ CH$_3$
$$\underset{\underset{H}{|}}{C}=\underset{\underset{CH_2CH_3}{|}}{C}$$

9. Supply the missing names or structural formulas for
a. (CH$_3$)$_3$CCH$_2$CH$_3$
b. 3,5-dimethylnonane

c. 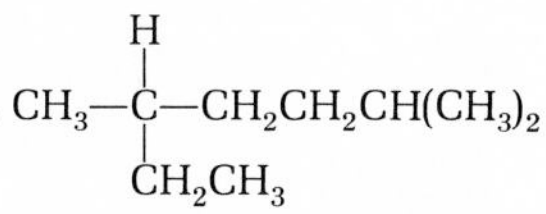 CH$_3$—$\overset{\overset{H}{|}}{\underset{\underset{CH_2CH_3}{|}}{C}}$—CH$_2CH_2$CH(CH$_3$)$_2$

d.

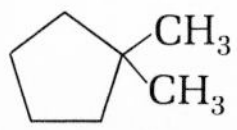

e.

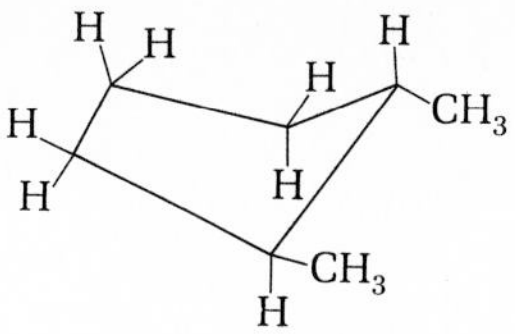

f.

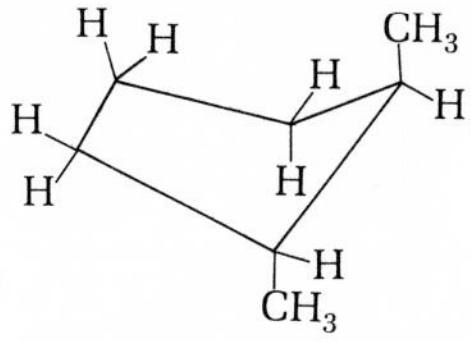

g. *cis*-2-pentene
h. 3-methyl-*trans*-3-octene
i. *o*-xylene
j. salicylic acid

Objective (d)

10. Predict which isomer in each pair is probably higher boiling. Explain each choice.

a. 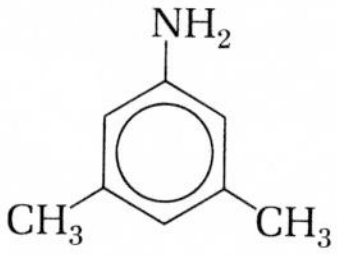   or

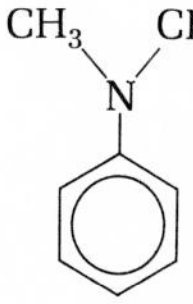

b. *p*-dichlorobenzene or *o*-dichlorobenzene

c. 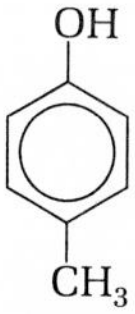   or

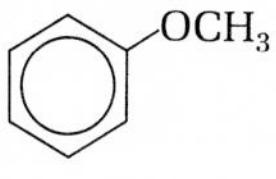

d. *cis*-2-butene or *trans*-2-butene

e. 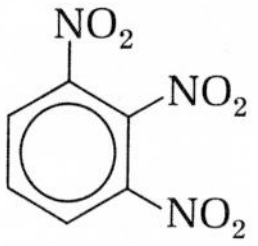   or

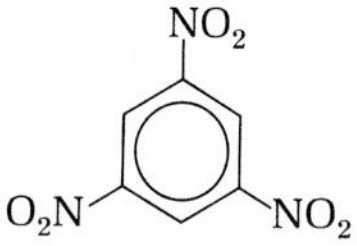

11. Which of the following pairs of isomers would be expected to have *identical* boiling points? Explain.

a. 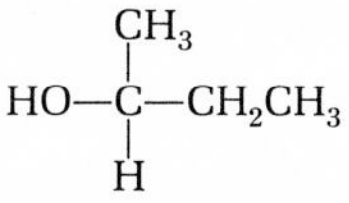 HO—$\overset{\overset{CH_3}{|}}{\underset{\underset{H}{|}}{C}}$—CH$_2CH_3$ and   CH$_3$CH$_2$—$\overset{\overset{CH_3}{|}}{\underset{\underset{H}{|}}{C}}$—OH

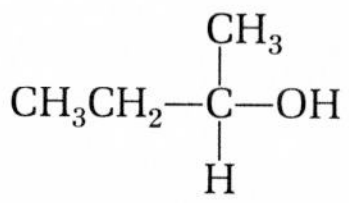

b. 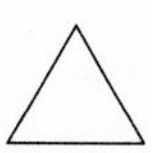   and   H$_2$C=CHCH$_3$

c.

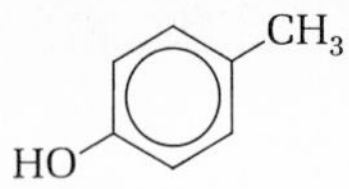

[structure: para-cresol type — benzene ring with CH₃ and HO] and [benzene ring with CH₃ and OH ortho]

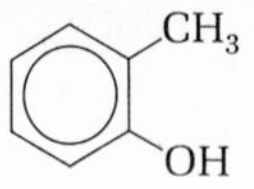

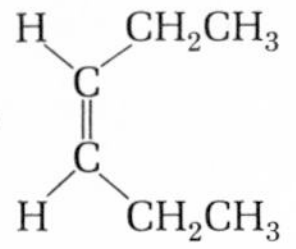

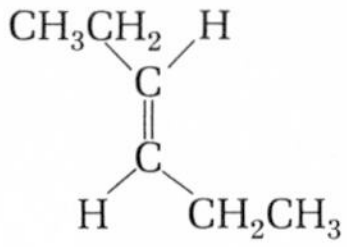

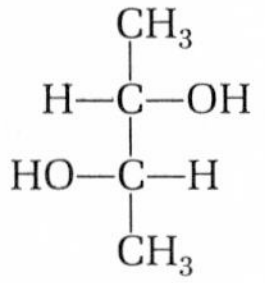

d.

$$\underset{H}{\overset{H}{}}\text{C}=\text{C}\ \text{with } CH_2CH_3$$ and [corresponding isomer]

H, CH₂CH₃ on one carbon, H, CH₂CH₃ — and — CH₃CH₂, H / H, CH₂CH₃

e.

$$\begin{array}{c} CH_3 \\ H-C-OH \\ HO-C-H \\ CH_3 \end{array}$$ and $$\begin{array}{c} CH_3 \\ HO-C-H \\ H-C-OH \\ CH_3 \end{array}$$

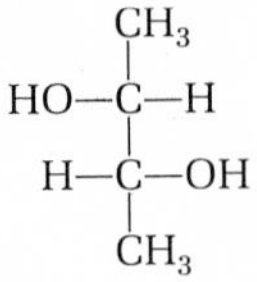

BRAIN TINGLER

12. A 0.2834-g sample of a hydrocarbon was completely oxidized, and the oxidation products were trapped by previously weighed reagents. The increase in weight of the soda lime trap was 0.8680 g, and the increase in weight of the magnesium perchlorate trap was 0.4147 g. A second sample of the hydrocarbon was vaporized in a 1.000-liter bulb until all air was expelled and the bulb contained only the hydrocarbon vapor. The bulb was sealed at 227°C at a pressure of 700 torr. The vapor mass was determined to be 1.935 g.
  a. Calculate the molecular formula of the hydrocarbon.
  b. Draw expanded structural formulas for all possible isomers of the hydrocarbon. (*Hint:* If you suspect that two or more formulas are equivalent, name them to see if they are the same or different.)

Unit 28 Self-Test

1. Give the name of the "class" to which each of the following belongs:

a. HC≡CH

b. 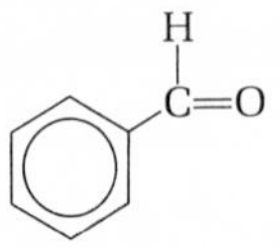 [benzene ring with —C(=O)H substituent]

c. [cyclohexane ring with —O—CH₃]

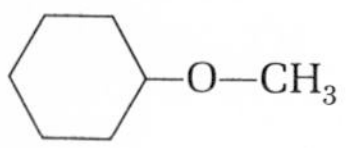

d. 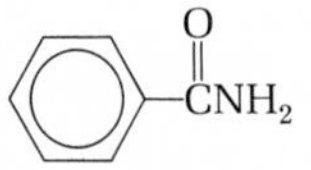 [benzene ring with —C(=O)NH₂]

e. 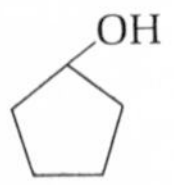 [cyclopentane ring with OH]

f. 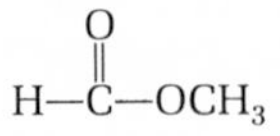 $$H-\overset{\overset{\textstyle O}{\|}}{C}-OCH_3$$

g. 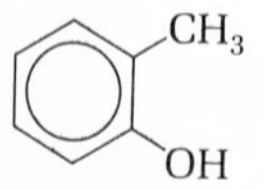 [benzene ring with CH₃ and OH]

h. [pyrrolidine ring with N—H]

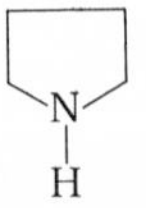

i. [cyclopentane ring with =O]

j. [three fused benzene rings — anthracene]

2. Give correct IUPAC names for

a. 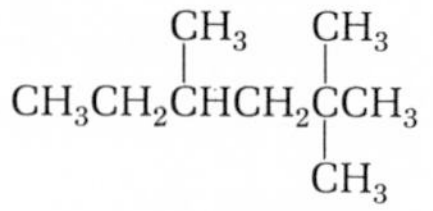 $$CH_3CH_2\overset{\overset{\textstyle CH_3}{|}}{C}HCH_2\overset{\overset{\textstyle CH_3}{|}}{\underset{\underset{\textstyle CH_3}{|}}{C}}CH_3$$

b. 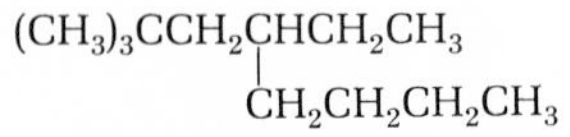 $$(CH_3)_3CCH_2\overset{\overset{\textstyle}{}}{\underset{\underset{\textstyle CH_2CH_2CH_2CH_3}{|}}{C}}HCH_2CH_3$$

3. Draw structural formulas for
  a. 2,2-dimethyl-5-ethylnonane
  b. 3-ethylheptane

4. Draw all positional and geometric isomers of *acylic* compounds having the formula C_5H_{10}. Identify each by a letter (a, b, c, . . .).

5. Give IUPAC names for all the compounds in question 4. Identify by the letters used for the formulas.

6. a. Give accepted names for

1. 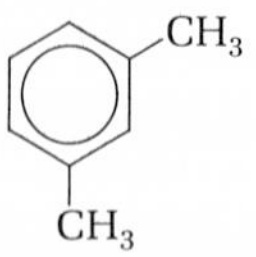 [benzene ring with CH₃ (meta) and CH₃]

2. 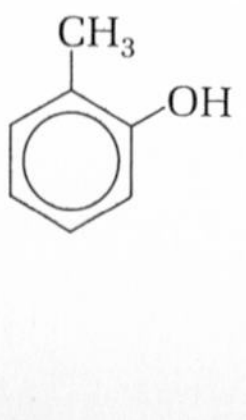 [benzene ring with CH₃ and OH]

3.

CH_3

NH_2

4.

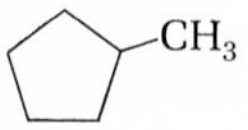

$-CH_3$

5.

CH_3CH_2 $CH_2CH_2CH_3$

$C=C$

H CH_2CH_3

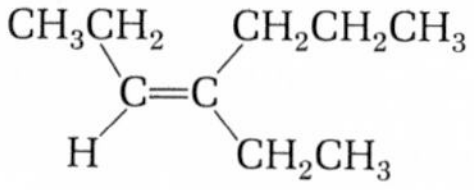

b. Draw formulas representing
1. p-chlorotoluene
2. hydroquinone
3. 2,4,6-trinitrophenol (picric acid)
4. the chair conformation of cyclohexane
5. 3-methyl-trans-2-hexene
7. Draw representations for
   a. staggered and eclipsed conformations of ethane
   b. optical isomers of 2-bromobutane
8. Select the member of each isomeric pair that probably has the higher boiling point:
   a. trans-2-butene or cis-2-butene
   b. o-xylene or p-xylene

   c. m-cresol or anisole ($-OCH_3$)

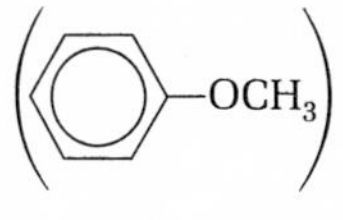

... organic chemistry just now
is enough to drive one mad."
FRIEDRICH WÖHLER (1835)

Functional groups

Objectives

(a) Be able to use the IUPAC nomenclature rules (Table 29.3 and Appendix G) for assigning correct names to monofunctional compounds, given their structural formulas, or correct formulas, given their names.

(b) Be able to use the trivial nomenclature for common carboxylic acids and related compounds (Table 29.4).

(c) Be able to relate structural features of organic molecules, in a qualitative way, to
   (1) solubility in water (Section 29.3)
   (2) acidic or basic properties (Section 29.4)

(d) Be able to apply information on the typical chemical behavior of selected classes of organic compounds to:
   (1) interpretation of analytical data
   (2) synthetic preparation of monofunctional compounds

REVIEW

(e) Be able to make calculations involving pH and K_a values for organic acids.

One of the most staggering problems facing modern society is the control of narcotics. It is impossible to estimate the human misery associated with the traffic in addictive drugs. Of the many substances involved, none has received such widespread attention as heroin, a compound prepared by a simple chemical treatment of the morphine obtained from Papaver somniferum, *the opium poppy.*

It is, perhaps, ironic that much of the present "drug problem" has been blamed on Communist China, since it was the British and American merchants who introduced opium into China on a massive scale in the seventeenth century. Although the opium trade could not have flourished without the cooperation of certain influential Chinese, it is beyond question that it depended primarily on the British and American interests in the highly profitable exchange of opium from British India for the tea of China.

Once the opium poppy has fully matured, it is basically harmless—the drug harvest is found in the unripened seed pod. When cut, the pod exudes a sticky gray fluid. This is collected and dried in the open air to leave a gray-brown solid. Pressed into bricks, this crude opium is the first step in the long chain from grower to addict. At this stage, the "opium farmer" can sell his product for about $25 per pound. At the other end of the market chain, the final product, usually heroin, will sell for $15,000 to $20,000 for the amount prepared from the original pound of crude opium.

Alkaloids are complex amines (hence, "alkaline") of plant origin.

Opium contains about two dozen different alkaloids. *The most familiar is morphine (Figure 29.1), constituting 10 to 15 percent of the crude opium. This compound was the first alkaloid ever isolated. Although morphine was recovered pure from*

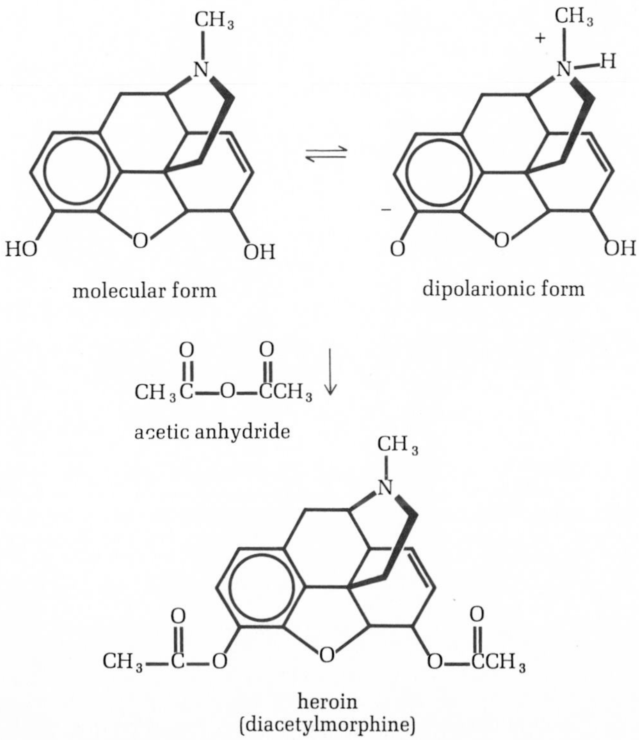
CH3
N
HO
O
OH
molecular form
CH3
+
N—H
−
O
O
OH
dipolarionic form
O O
CH3C—O—CCH3
acetic anhydride
CH3
N
O
CH3—C—O
O
O
O—CCH3
heroin
(diacetylmorphine)

Figure 29.1 *Morphine, a compound containing many functional groups (the ring system is not coplanar).*

opium by the German pharmacist Serturner in the early nineteenth century, its structure was not elucidated until 1925.

Morphine is a complex polyfunctional compound. Although its biochemical role in the body is only hazily understood, its chemical properties are essentially those associated with the simple functional groups. Like other amines, morphine is basic. Its phenol group is weakly acidic, so morphine is properly classified as an amphoteric compound (Unit 20). Since both phenols and alcohols form esters, morphine can be esterified at two positions. It is this simple reaction that leads to the production of heroin (Figure 29.1). Indeed, "heroin factories" may be detected[1] by the "vinegar" odor associated with this process.

For years chemists have sought ways of modifying morphine's structure or of preparing synthetic substitutes in attempts to find compounds whose analgesic properties are as good as those of morphine but that are nonaddictive. So far this research has met with only limited success.

The key to structural and synthetic chemistry of complex organic molecules is a knowledge of the characteristic behavior of functional groups. In this unit we shall study simple monofunctional compounds. Although polyfunctional molecules may pose special complications, their reactions quite often reflect those of the simple compounds containing similar functional groups.

To examine all the different functional groups, even in an introductory manner, would require at least a semester's course in organic chemistry. We shall limit most of our discussion in this unit to only a few of the properties characteristic of some of the functional groups found in biologically important molecules (Table 29.1). Some additional groups will be described in Units 31–33.

Table 29.1 SOME FUNCTIONAL GROUPS FOUND IN BIOLOGICALLY
IMPORTANT MOLECULES

CLASS	FUNCTIONAL GROUP	EXAMPLES
alcohol	$-C-O-H$	ethanol $[C_2H_5OH]$, a fermentation product of carbohydrates glycerol $[HOCH_2CH(OH)CH_2OH]$, a hydrolysis product of animal fats and vegetable oils *carbohydrates* [Unit 31].
aldehyde	$-C=O$ $\mid$ H	acetaldehyde $[CH_3CHO]$, a product of ethanol metabolism glucose and other *aldoses* [Unit 31]
amide	$-\overset{O}{\overset{\|}{C}}-N\langle$	urea $\left[H_2N\overset{O}{\overset{\|}{C}}NH_2\right]$, normal end product of nitrogen metabolism in humans *lidocaine* a common local anesthetic *peptides and proteins* [Unit 32]
amine	$-\overset{..}{N}\langle$	putrescine $[H_2NCH_2CH_2CH_2CH_2NH_2]$, a product of animal decay histamine $H-\overset{..}{N}$ ⎯ $CH_2CH_2\overset{..}{N}H_2$, the chemical responsible

---

[1] See The Spoilers, by Desmond Bagley (New York: Pyramid, 1971).

Table 29.1 (Continued)

CLASS	FUNCTIONAL GROUP	EXAMPLES
		for many of the effects of allergenic reactions amino acids [Unit 32]
carboxylic acid		acetic acid [CH_3CO_2H], a product of ethanol metabolism lactic acid [$CH_3CH(OH)CO_2H$], the acid in soured milk long-chain "fatty acids," hydrolysis products of fats, oil, and waxes amino acids [Unit 32]
ester		octyl acetate [$CH_3CO_2CH_2(CH_2)_6CH_3$], a flavoring agent in oranges esters of glycerol, principal components of animal fats
ketone		acetone $\left[CH_3\overset{O}{\overset{\|}{C}}CH_3 \right]$, formed in significant amounts during carbohydrate metabolism in diabetics vanillin $\left[HO{-}\hexagon{-}\overset{H}{\underset{}{C}}{=}O \right]$, vanilla flavoring agent fructose and other *ketoses* [Unit 31]
phenol		methyl salicylate $\left[\hexagon \overset{-CO_2CH_3}{\underset{-OH}{}} \right]$, "oil of wintergreen" flavoring agent and the "external aspirin" of preparations such as *Ben-Gay* tyrosine $\left[HO{-}\hexagon{-}CH_2{-}\overset{H}{\underset{+NH_3}{C}}{-}CO_2^- \right]$, an amino acid [Unit 32]

29.1 FUNCTIONAL GROUP ISOMERISM

Just as we may have positional and geometric isomers because of different arrangements of bonds in organic molecules, it is also possible for such different arrangements to result in functional group isomers (Unit 28). Functional groups are particular arrangements of atoms and bonds in molecules that produce characteristic properties and reactions. Thus formic acid (HCOOH), acetic acid (CH_3COOH), and butyric acid ($CH_3(CH_2)_2COOH$) all have certain properties, such as acidity, that may be attributed to the *carboxyl* (—COOH) group. Compounds may have the same molecular formula but different arrangements of bonds such that they have widely different properties. Both ethyl alcohol (ethanol) and dimethyl ether have the molecular formula C_2H_6O, but their properties are quite different. The use of functional groups in classifying compounds has already been introduced (Unit 28) and should be reviewed.

Before studying the reactions of various classes of compounds, we need to extend our nomenclature systems to permit us to discuss such substances by their names.

29.2 NOMENCLATURE OF MONOFUNCTIONAL COMPOUNDS

Some functional groups do not affect the name endings used in the IUPAC nomenclature system. These groups are treated in the same way as alkyl groups attached to the basic carbon chain (Unit 18), as illustrated in Table 29.2.

In other cases, such as the alkenes (Unit 28), the functional group is reflected in the name ending. *In such cases, the basic carbon chain is always numbered beginning at the end nearest the functional group that determines the name ending.*

The procedures used are illustrated in Table 29.3 and, in greater detail, in Appendix G.

Table 29.2 SOME COMMON FUNCTIONAL GROUPS NOT AFFECTING NAME ENDINGS[a]

CLASS	FUNCTIONAL GROUP		EXAMPLE
amine	$-\ddot{N}\big\langle$ (amino)	$H_2N(CH_2)_4NH_2$	1,4-diaminobutane
		$CH_3-\underset{NH_2}{\overset{H}{C}}-CH_2CH_2\underset{OH}{\overset{H}{C}}-CH_2CH_3$	6-amino-3-heptanol
		$CH_3CH_2\underset{N(CH_3)_2}{CHCH_3}$	2-(N,N-dimethylamino)butane
halide	—F (fluoro)	◯—F	fluorobenzene
	—Cl (chloro)	$CHCl_3$	trichloromethane (chloroform)
	—Br (bromo)	$CH_3CH_2CH_2Br$	1-bromopropane
	—I (iodo)	$(CH_3)_2CHCH_2CH_2\underset{I}{CHCH_2CH_3}$	5-iodo-2-methylheptane
ether	$-O-C\big\langle$ (alkoxy)	$\underset{OCH_3}{\overset{CH_3}{◯}}$	*p*-methoxytoluene
		$(CH_3)_3CCH_2CH_2OCH_2CH_3$	2,2-dimethyl-4-ethoxybutane

[a] For additional details, see Appendix G.

Table 29.3 COMMON FUNCTIONAL GROUPS DETERMINING NAME ENDINGS[a]

CLASS	FUNCTIONAL GROUP	NAME ENDING	EXAMPLE
alkene (Unit 28)	$\big\rangle C=C\big\langle$	. . . ene	$CH_3-\underset{CH_3}{\overset{H}{C}}-CH_2CH_2\underset{H}{\overset{}{C}}=CH_2$ 5-methyl-1-hexene
alcohol	—OH (hydroxyl group)	. . . ol	$CH_3-\underset{CH_3}{\overset{H}{C}}-CH_2CH_2\underset{OH}{\overset{H}{C}}-CH_3$ 5-methyl-2-hexanol

Table 29.3 (Continued)

CLASS	FUNCTIONAL GROUP	NAME ENDING	EXAMPLE
aldehyde	$-C=O$ / H / ($-C=O$ is called / a carbonyl group)	. . . *al*	$CH_3-\overset{\underset{\mid}{CH_3}}{\underset{\mid}{C}}-CH_2CH_2CH_2\overset{H}{C}=O$ / 5-methylhexanal[b]
ketone	$>C-C=O$ / (carbonyl group)	. . . *one*	$CH_3-\overset{H}{\underset{\mid}{\underset{CH_3}{C}}}-CH_2\overset{O}{C}CH_2CH_3$ / 5-methyl-3-hexanone
carboxylic acid	$\overset{O}{\overset{\|}{-C}}-OH$ / (carboxyl group)	. . . *oic acid*	$CH_3-\overset{H}{\underset{\mid}{\underset{CH_3}{C}}}CH_2CH_2CH_2CO_2H$ / 5-methylhexanoic[b] acid
salt of carboxylic acid	$\overset{O}{\overset{\|}{-C}}-O^-$ / (carboxylate group)	metal *(space)* / . . . *oate*	$CH_3-\overset{H}{\underset{\mid}{\underset{CH_3}{C}}}CH_2CH_2CH_2CO_2^-\,Na^+$ / sodium 5-methylhexanoate[b]
ester	$\overset{O}{\overset{\|}{-C}}-O-C<$	alkyl *(space)* / . . . *oate*	$CH_3-\overset{H}{\underset{\mid}{\underset{CH_3}{C}}}CH_2CH_2CH_2\overset{O}{C}OC_2H_5$ / ethyl 5-methylhexanoate[b]
amide	$\overset{O}{\overset{\|}{-C}}-N-$ / (note: N- substituted amides are indicated as in the cases of N- substituted amines, Table 29.2)	. . . *amide*	$CH_3-\overset{H}{\underset{\mid}{\underset{CH_3}{C}}}CH_2CH_2CH_2\overset{O}{C}NH_2$ / 5-methylhexanamide[b] $CH_3\overset{H}{\underset{\mid}{\underset{CH_3}{C}}}CH_2CH_2CH_2\overset{O}{C}N(CH_3)_2$ / N,N-dimethyl-5-methylhexanamide[b]

[a] For further details of nomenclature, see Appendix G.
[b] Note that for functional groups in which the group carbon is, of necessity, the end carbon in the basic chain, no group location number is used. This carbon is always the number 1 carbon, so no ambiguity is involved.

Although the IUPAC nomenclature is now universally accepted, a trivial name system for aldehydes, carboxylic acids, and derivatives of the acids has been used so long that such names are still quite commonly employed. This system is outlined in Table 29.4, and many similarities in name endings may be noticed on comparison with the IUPAC system (Appendix G).

Table 29.4 TRIVIAL NAMING FOR ALDEHYDES, ACIDS, AND RELATED COMPOUNDS

I. MONOCARBOXYLIC ACID NAMES

acid name	formic	acetic	propionic	butyric	valeric	caproic
acid formula	HCO_2H	CH_3CO_2H	$C_2H_5CO_2H$	$C_3H_7CO_2H$	$C_4H_9CO_2H$	$C_5H_{11}CO_2H$

II. NAME ENDINGS

CLASS	NAME ENDING	EXAMPLE	
aldehyde	. . . aldehyde	$H_2C{=}O$	formaldehyde
acid	. . . ic acid	CH_3CO_2H	acetic acid
salt	*metal* (space) . . . ate	$Na(CH_3CH_2CO_2)$	sodium propionate
ester	*alkyl* (space) . . . ate	$CH_3CH_2CH_2\overset{\displaystyle O}{\overset{\|}{C}}OCH_3$	methyl butyrate
amide	. . . amide	$CH_3CH_2CH_2CH_2\overset{\displaystyle O}{\overset{\|}{C}}NH_2$	valeramide
acid chloride	. . . yl (space) chloride	$CH_3\overset{\displaystyle O}{\overset{\|}{C}}Cl$	acetyl chloride
anhydride	. . . ic (space) anhydride	$CH_3\overset{\displaystyle O}{\overset{\|}{C}}O\overset{\displaystyle O}{\overset{\|}{C}}CH_3$	acetic anhydride

III. SUBSTITUTED ACIDS

Attached groups are identified by Greek location letters ($\alpha, \beta, \gamma, \ldots$), beginning with carbon adjacent to C=O. Thus, α in the trivial system corresponds to 2 in the IUPAC system; e.g., $CH_3CH_2CH_2\underset{\underset{\displaystyle Br}{\|}}{C}HCO_2H$ is α-bromovaleric acid.

IV. DICARBOXYLIC ACIDS

[In the IUPAC system, such acids are given the base name determined by the number of carbons in the chain, followed by the ending . . . dioic acid (e.g., $HO_2CCH_2CO_2H$ is propanedioic acid).]

acid name	oxalic	malonic	succinic	glutaric	adipic
acid formula	$HO_2C{-}CO_2H$	$HO_2CCH_2CO_2H$	$HO_2C(CH_2)_2CO_2H$	$HO_2C(CH_2)_3CO_2H$	$HO_2C(CH_2)_4CO_2H$

29.3 WATER SOLUBILITY

You may wish to review the discussions of water solubility given in Units 9, 11, and 13.

Functional groups that can be involved in hydrogen bonding tend to be readily solvated by water molecules. Such groups may act either as hydrogen donors or as hydrogen acceptors

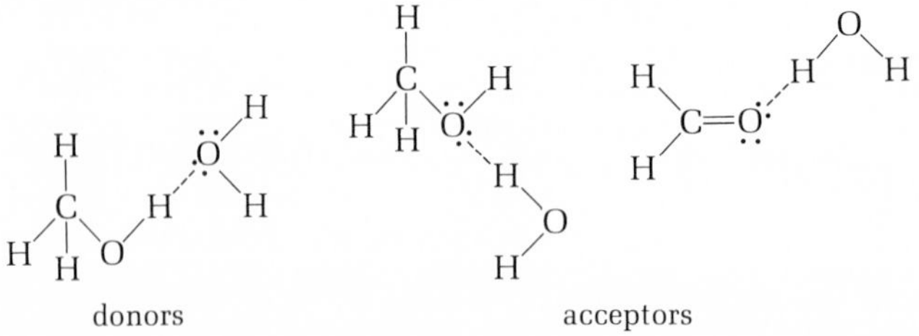

donors   acceptors

Regions within organic molecules where hydrogen bonding can occur will therefore be centers for clustering of water molecules. Examples of functional groups that can form hydrogen bonds include hydroxyl, carbonyl, carboxyl, amino, and amide groupings.

The presence of these functional groups does not, however, assure water solubility. Since interparticle forces between water molecules themselves are quite strong, nonpolar (hydrophobic) portions of organic molecules are "squeezed out" by formation of hydrogen-bonded water clusters. For alcohols a nonpolar region equivalent to an unbranched four-carbon chain is sufficient to reduce the water solubility quite significantly. Thus methanol and ethanol are completely miscible with water, 1-butanol is only slightly soluble, and long-chain alcohols such as 1-decanol are so insoluble that they may form "monolayers" on the surface of water in which the hydroxyl group of the alcohol is in the water and the nonpolar region protrudes almost entirely above the water surface. The effect of a nonpolar region on water solubility is related primarily to the "linear bulk" of the region. Long chain hydrocarbon regions reduce water solubility more than "compact" branched chains. For example, 1-butanol is much less water soluble than the isomeric 2-methyl-2-propanol.

Charged groups enhance water solubility considerably. Thus salts of acids or amines normally form aqueous solutions (or stable colloids) quite readily.

29.4 ACIDITY AND BASICITY

Both carboxylic acids and phenols are more acidic than water. In both cases the anion formed by transfer of a proton to a water cluster is stabilized by *delocalization of charge*. For carboxylic acids, the negative charge on the carboxylate ion is spread over three atoms, thus reducing the charge density per atom and contributing to increased stability.

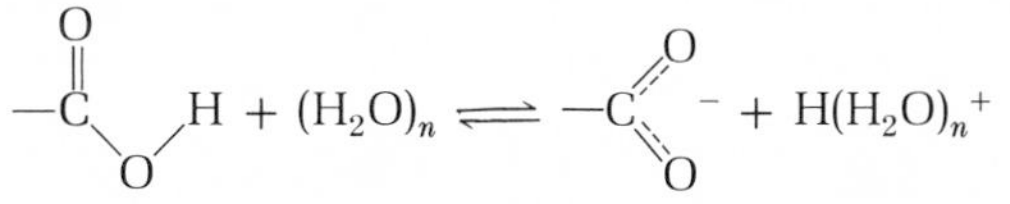

$$-\overset{\overset{\displaystyle O}{\|}}{C}\diagup^{\displaystyle O}_{\diagdown O}H + (H_2O)_n \rightleftharpoons -\overset{\overset{\displaystyle O}{}}{C}\diagup^{\displaystyle O}_{\diagdown O}{}^- + H(H_2O)_n{}^+$$

Phenols are also stabilized by charge delocalization, dissipating the surplus electron density through the π-bond system of the aromatic ring (Unit 28).

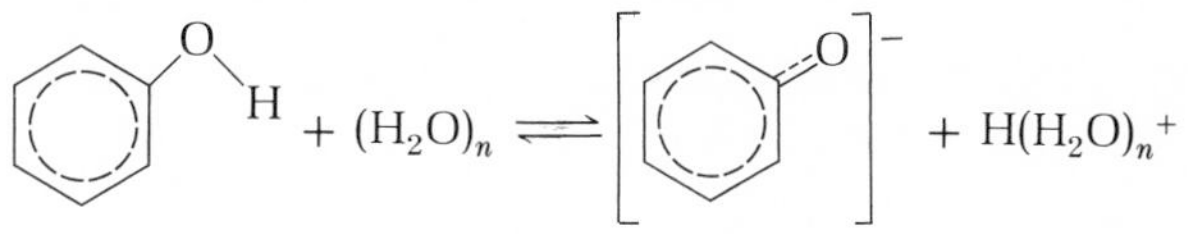

$$\langle\!\!\!\bigcirc\!\!\!\rangle\!-\!O\!\diagdown\!H + (H_2O)_n \rightleftharpoons \left[\langle\!\!\!\bigcirc\!\!\!\rangle\!-\!O\right]^- + H(H_2O)_n{}^+$$

Amine groups, in which the nitrogen atom has an unshared electron pair, are the most common basic functional groups, having K_b values similar to that of ammonia. Aromatic amines are generally less basic than aliphatic (nonaromatic) amines because of an apparent partial overlap of the lone-pair orbital on nitrogen with the π-bond system in the ring, a factor contributing to decreased availability of this unshared pair for coordination with a proton or other Lewis acid.

Alcohols, on the other hand, are less acidic than water. The *alkoxide ion* $(R\!-\!O^-)$ is not stabilized by charge delocalization, and the C—O bond polarity increases the electron density on the oxygen. Alkoxides are very strong bases.

When oxygen is involved only in σ bonding, the orbitals used are approximated as sp^3 hybrids (Unit 8). Formation of a π bond requires a p orbital on oxygen, so the remaining lone-pair and bond-pair electrons are then described by sp^2 hybridization. The energy difference between these two states is relatively small, and when charge delocalization is possible by π-bond formation, the resulting stability more than offsets the difference. As a result, oxyanions stabilized by π bonding on oxygen are of much lower energy than those in which charge density is localized on a single atom. Consider the following examples:

The influence of other functional groups and of alkyl groups on the strengths of organic acids and bases is discussed in Unit 31.

$$CH_3OH \quad \rightleftharpoons \quad CH_3O^- + H^+$$

$$K_a = <10^{-15} \qquad \text{localized charge}$$

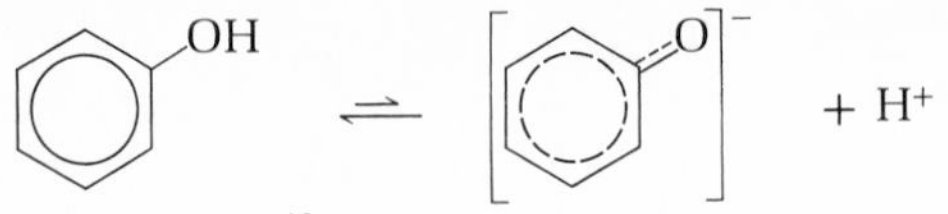

$$K_a = 1.1 \times 10^{-10} \qquad \text{delocalized charge}$$

$$CH_3CO_2H \quad \rightleftharpoons \qquad + H^+$$

$$K_a = 1.8 \times 10^{-5}$$

delocalized charge, two
oxygens participating

The ionization constants for monoprotic organic acids may be determined experimentally by measuring the pH of solutions in which the acid concentration is known. When very pure acids are available in a form suitable for use as primary standards, the "label concentrations" may be determined simply by accurate preparation of the solution. In other cases, concentrations are found by titrations using some standardized base.

You may wish to review such calculations in Unit 19.

29.5 GENERAL TYPES OF REACTIONS

Addition reactions are those in which π bonds are converted to σ bonds to newly added atoms. Both polar and nonpolar multiple bonds undergo a variety of such reactions.

Examples

1. Catalytic hydrogenation of carbon–carbon double bonds (used industrially to convert *unsaturated* vegetable oils to more viscous products such as margarine or shortening)

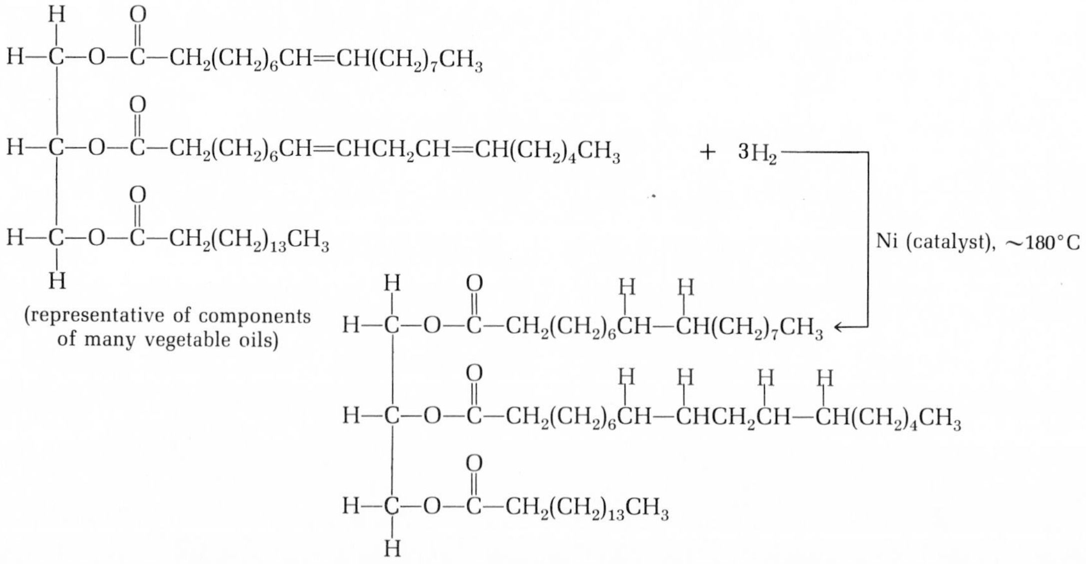

(representative of components
of many vegetable oils)

Ni (catalyst), $\sim$180°C

(representative of a "hydrogenated vegetable oil")

2. Addition of bromine to a carbon–carbon double bond (used as an analytical test for alkenes by observation of disappearance of the red-orange color of molecular bromine)

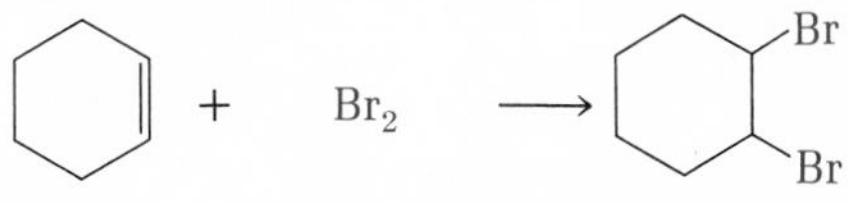

(colorless)   (red-orange)   (colorless)

Elimination reactions involve the removal of σ-bonded atoms to form π bonds. Examples include dehalogenation (loss of halogen) and dehydration (loss of water).

Examples

1. Debromination to recover an alkene as one step in isolation of an alkene from a mixture of similar compounds (used in the laboratory recovery of cholesterol from gallstones)

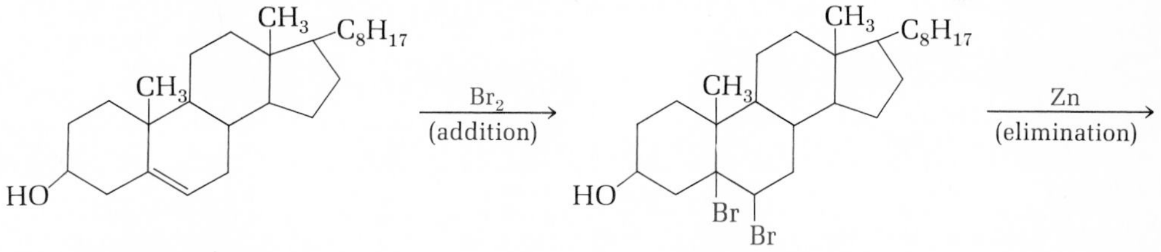

cholesterol

(a major component of gallstones, but difficult to isolate from the complex mixture)

dibromocholesterol

(easy to isolate by fractional crystallization)

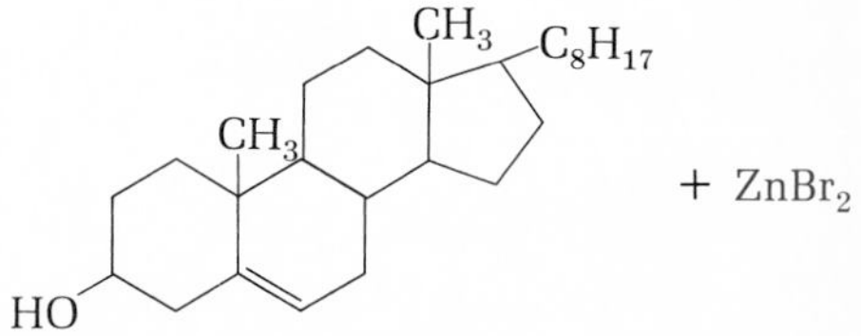

$+ \; ZnBr_2$

2. Dehydration of an alcohol to form an alkene (a "side reaction" in the production of diethyl ether from ethanol)

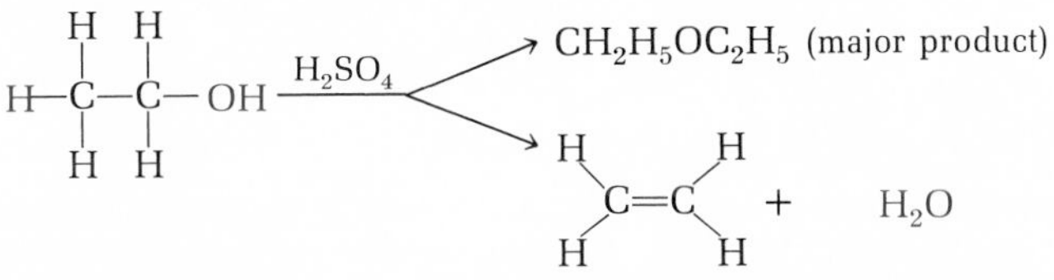

$CH_2H_5OC_2H_5$ (major product)

(by-product)   (absorbed by the H_2SO_4)

$+ \quad H_2O$

Substitution reactions occur when one atom or group replaces another in an organic molecule. Such reactions as replacement of halogen by hydroxyl or alkoxyl are typical substitutions.

1. Conversion of a halide to an alcohol (an older method of preparing ethylene glycol for "antifreeze")

$$H_2C{=}CH_2(g) + HOCl(aq) \longrightarrow HOCH_2CH_2Cl(aq) + OH^-(aq) \longrightarrow HOCH_2CH_2OH + Cl^-$$
$$\text{(ethylene glycol)}$$

2. The *Williamson ether synthesis* (used to convert morphine to codeine, a somewhat less addictive drug than morphine, used in some prescription "pain killers" and "cough syrups")

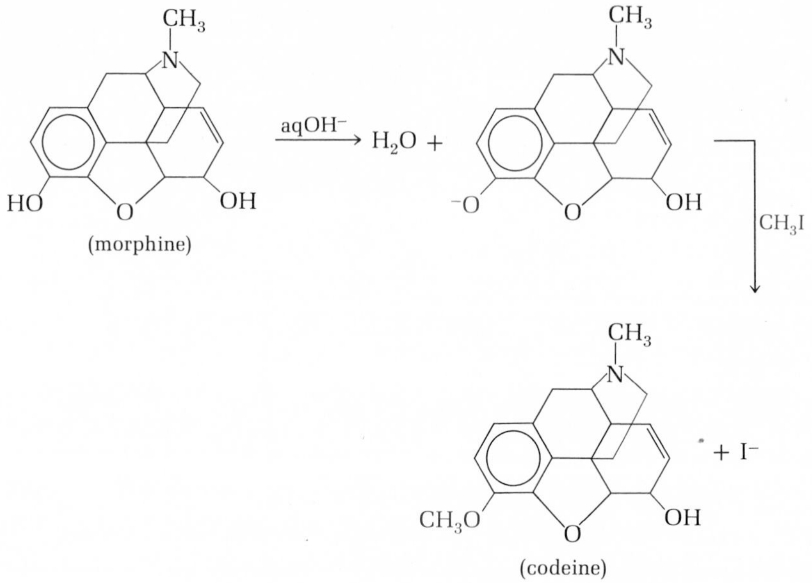

(morphine)

(codeine)

Oxidation–reduction reactions of many organic compounds are also of considerable importance in synthetic or analytical procedures, as well as in many biochemical processes.

Examples

1. Reduction of carbonyl groups by lithium aluminum hydride (used to convert aldehydes, ketones, carboxylic acids, or esters to alcohols via intermediate metal alkoxides subsequently decomposed by hydrolysis)

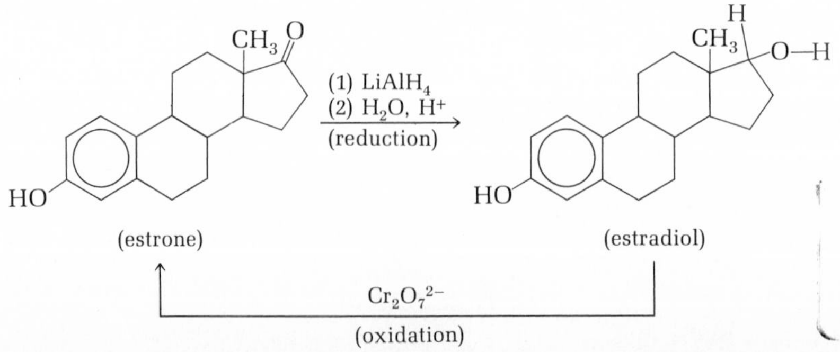

(estrone)

(1) LiAlH$_4$
(2) H$_2$O, H$^+$
(reduction)

(estradiol)

Cr$_2$O$_7^{2-}$
(oxidation)

2. Oxidation of alcohols to aldehydes or ketones. (The oxidation–reduction transformations between estrone and extradiol occur in the body via an enzymatic process that may be coupled to an important step in carbohydrate metabolism.)

Table 29.5 contains a summary of reactions typical of many of the more common functional groups. The listing is far from complete, but this should be useful as a reference for understanding some of the characteristic reactions of simple organic compounds. We shall consider some of these in more detail.

Table 29.5 SOME TYPICAL REACTIONS OF MONOFUNCTIONAL COMPOUNDS

CLASS	REACTION TYPE	EXAMPLES AND COMMENTS
alkenes	addition	hydrogenation: $\text{C=C} + H_2 \xrightarrow{\text{(Pt)}} \text{C-C}$ (with H H)
		e.g., $CH_3CH=CH_2 + H_2 \xrightarrow{\text{(Pt)}} CH_3CH_2CH_3$
		halogenation: $\text{C=C} + Br_2 \longrightarrow \text{C-C}$ (with Br, Br)
		e.g., cyclohexene $+ Br_2 \longrightarrow$ 1,2-dibromocyclohexane
alcohols	redox	O—H bond cleavage by reactive metals:
		e.g., $2CH_3OH + 2Na \longrightarrow H_2 + 2NaOCH_3$
		conversion to carbonyl:
		(Primary alcohols, —CH_2OH, yield aldehydes or, by further oxidation, carboxylic acids; secondary alcohols, >CHOH, yield ketones.)
		e.g., $CH_3CH_2OH + Cr_2O_7^{2-} \longrightarrow CH_3CHO + Cr^{3+}$
		(not balanced)
		(Unless aldehyde is removed as formed, oxidation continues to yield CH_3CO_2H.)
		$(CH_3)_2CHOH + Cr_2O_7^{2-} \longrightarrow CH_3\overset{\text{O}}{\underset{\|}{C}}CH_3 + Cr^{3+}$
		(not balanced)
	substitution	C—O bond cleavage:
		(Alcohols may be converted to halides by reagents such as PBr_3.)
		e.g., $(CH_3)_2CHOH + PBr_3 \longrightarrow (CH_3)_2CHBr + H_3PO_3$
		(not balanced)
		O—H bond cleavage:
		(*Esterification* may employ a carboxylic acid, an inorganic acid, an acid chloride, or an acid anhydride.)
		e.g.,
		$CH_3CH_2OH + CH_3\overset{\text{O}}{\underset{\|}{C}}OH \underset{\text{(H}^+)}{\rightleftharpoons} CH_3CH_2-O-\overset{\text{O}}{\underset{\|}{C}}CH_3 + H_2O$
		$\begin{array}{l}CH_2OH\\ \mid\\ CHOH\\ \mid\\ CH_2OH\end{array} + 3HNO_3 \longrightarrow \begin{array}{l}CH_2ONO_2\\ \mid\\ CHONO_2\\ \mid\\ CH_2ONO_2\end{array} + 3H_2O$
		glycerol (glycerine) glyceryl trinitrate (nitroglycerine)

Table 29.5 (Continued)

CLASS	REACTION TYPE	EXAMPLES AND COMMENTS

$$CH_3OH + CH_3\overset{O}{\underset{\|}{C}}Cl \longrightarrow CH_3O\overset{O}{\underset{\|}{C}}CH_3 + HCl$$

phenols — acid–base

(K_a values for simple phenols are around 10^{-10}; electronegative groups such as $-NO_2$ or $-Cl$ increase acidity, Unit 31)
e.g.,

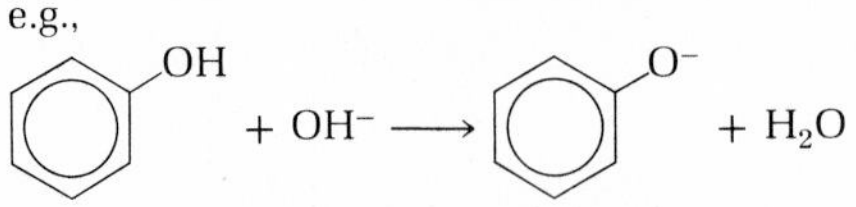

$$\text{(phenol)} + OH^- \longrightarrow \text{(phenoxide)} + H_2O$$

substitution

(Phenols do not react with PBr_3, etc., to form halides; the C—O bond is quite stable. Phenols may be esterified, usually by acid chlorides or anhydrides.)
e.g.,

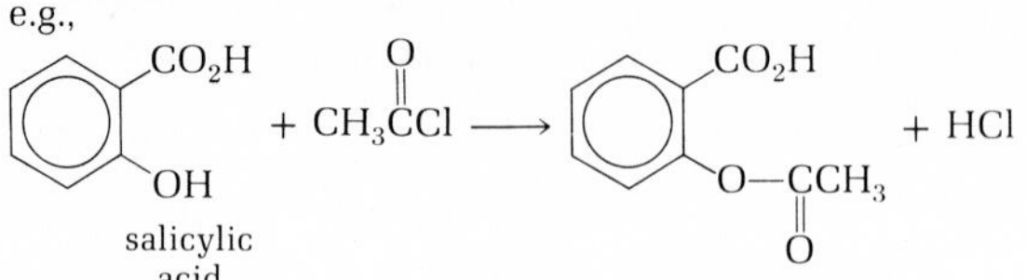

salicylic acid $+ CH_3\overset{O}{\underset{\|}{C}}Cl \longrightarrow$ acetylsalicylic acid (aspirin) $+ HCl$

aldehydes — redox

oxidation:
(Aldehydes are easily oxidized, even by air at room temperature; simple oxidations by metal ions are used in analytical tests.)
e.g., Tollen's test (most aldehydes, either aliphatic or aromatic)

$$CH_3-\overset{H}{\underset{\|}{C}}=O + 2[Ag(NH_3)_2]^+ + 3OH^- \longrightarrow 4NH_3 + 2Ag + CH_3CO_2^- + 2H_2O$$
$$\text{(ppt.)}$$

reduction:
(Aldehydes may be reduced to alcohols by a variety of methods, one of the most common using lithium aluminum hydride.)

$$4CH_3CHO + LiAlH_4 \longrightarrow (CH_3CH_2O)_3Al + CH_3CH_2OLi \xrightarrow{4H_2O}$$
$$LiOH + Al(OH)_3 + 4CH_3CH_2OH$$

ketones — redox

(Ketones are resistant to mild oxidizing agents, such as air, Benedict's reagent, and Tollen's reagent. Ketones may be reduced to secondary alcohols by reagents similar to those used for aldehydes.)
e.g.,

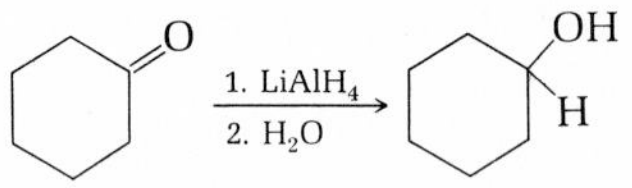

carboxylic acids — acid-base

(K_a values for simple acids are around 10^{-5}) [Units 19 and 31].
e.g., $CH_3CO_2H + NH_3 \longrightarrow CH_3CO_2^- + NH_4^+$

substitution

(see *alcohols*, esterification)
formation of acid chlorides:
e.g.,

$$CH_3\overset{O}{\underset{\|}{C}}OH + PCl_3 \longrightarrow CH_3\overset{O}{\underset{\|}{C}}Cl$$
$$\text{(not balanced)}$$

redox

(Carboxylic acids may be reduced to primary alcohols by $LiAlH_4$.)
e.g.,

$$CH_3CO_2H \xrightarrow[\text{2. } H_2O]{\text{1. } LiAlH_4} CH_3CH_2OH$$

Table 29.5 (Continued)

CLASS	REACTION TYPE	EXAMPLES AND COMMENTS
amines	acid-base	(K_b values of simple amines are around 10^{-4}. Aromatic amines are appreciably less basic than aliphatic amines) [Units 19 and 31]. e.g., $CH_3\ddot{N}H_2 + HCl \longrightarrow CH_3\overset{+}{N}H_3 + Cl^-$
	substitution	amide formation: (Primary and secondary amines react with acid chlorides or anhydrides to form amides.) e.g., 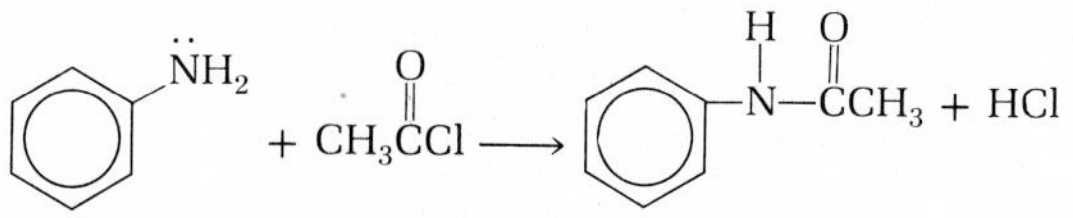 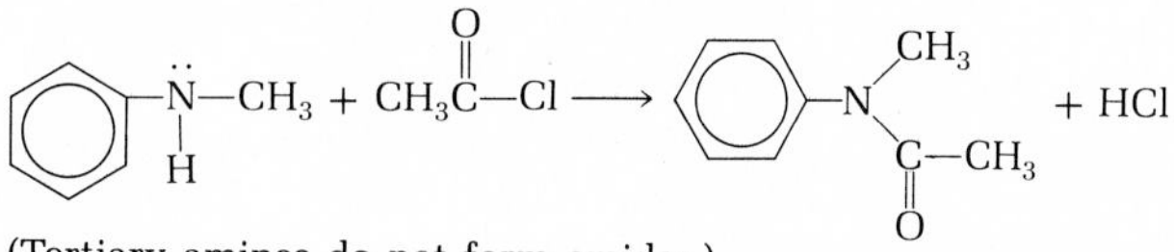 (Tertiary amines do not form amides.)
ammonium salts	acid-base	(K_a values (aliphatic) are around 10^{-11}) [Units 19 and 31]. e.g., $CH_3\overset{+}{N}H_3 + OH^- \longrightarrow H_2O + CH_3NH_2$
carboxylate salts	acid-base	(K_b values are around 10^{-10}) [Units 19 and 31]. e.g., $CH_3CO_2^- + HCl \longrightarrow CH_3CO_2H + Cl^-$
amides and esters	substitution	hydrolysis: *in aqueous acid* [Amides yield carboxylic acids and ammonium salts; esters yield carboxylic acids and alcohols (or phenols).] e.g., $CH_3CH_2\overset{O}{\overset{\Vert}{C}}\underset{H}{\underset{\vert}{N}}CH_3 + H_2O + H^+ \longrightarrow CH_3CH_2CO_2H + CH_3\overset{+}{N}H_3$ 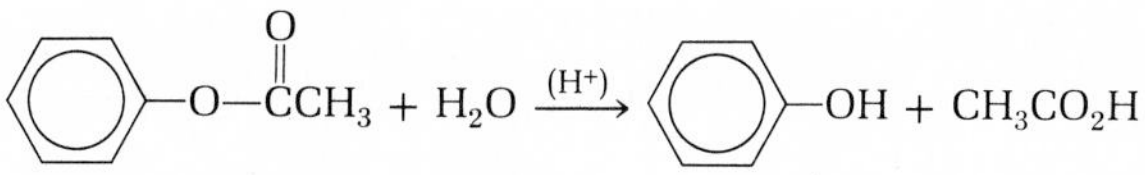 *in aqueous base* [Amides yield carboxylate salts and amines (or NH_3); esters yield carboxylate salts and alcohols (or phenoxide ions).] e.g., $CH_3\overset{O}{\overset{\Vert}{C}}NH_2 + OH^- \longrightarrow CH_3CO_2^- + NH_3$ $CH_3\overset{O}{\overset{\Vert}{C}}OCH_3 + OH^- \longrightarrow CH_3CO_2^- + CH_3OH$ [Alkaline hydrolysis of esters is called saponification.]

Table 29.5 (Continued)

CLASS	REACTION TYPE	EXAMPLES AND COMMENTS
halides	substitution	[Halogen may be replaced by any one of a number of Lewis bases.] e.g.,

$$CH_3Br + CH_3CH_2O^- \longrightarrow CH_3OCH_2CH_3 + Br^-$$
$$CH_3Br + OH^- \longrightarrow CH_3OH + Br^-$$
$$CH_3Br + \overset{..}{N}H_3 \longrightarrow CH_3NH_3^+ + Br^-$$

29.6 "VISUALIZING" REACTIONS

A knowledge of reaction mechanisms (Unit 16) is particularly useful to the understanding of organic chemistry. Unfortunately, a development of such mechanisms, that are often quite complex, requires more in-depth study and more background than we can justify at this point. We can, however, utilize some simple ways of "visualizing" many reactions. Although these visualizations should not be construed as actual *mechanisms*, they are useful in an initial study of organic compounds.

Many oxidation–reduction reactions of organic compounds may be visualized as hydrogen addition (for reduction) and as either hydrogen removal or oxygen insertion (for oxidation). In this respect a reducing agent acts as a "hydrogen source" and an oxidizing agent acts as an "oxygen source." The oxygen source may furnish oxygen either to combine with "removed" hydrogen or to be "inserted."

Examples

1. Reduction of a ketone to an alcohol (hydrogen addition)

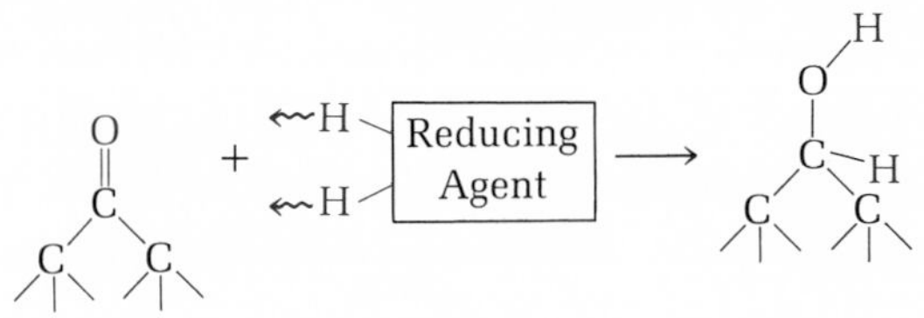

2. Oxidation of an alcohol to a ketone (hydrogen removal)

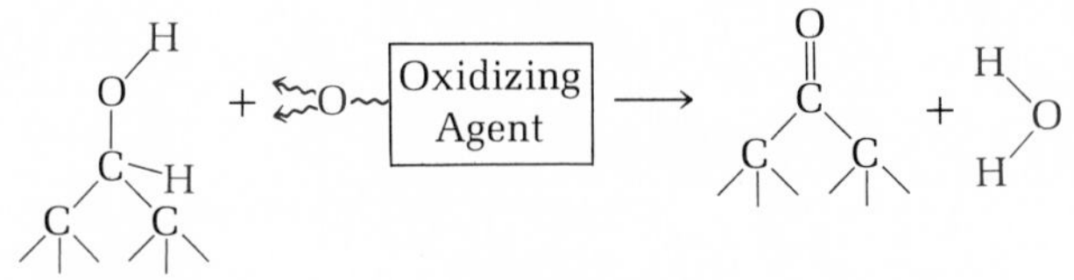

3. Oxidation of an aldehyde to a carboxylic acid (oxygen insertion)

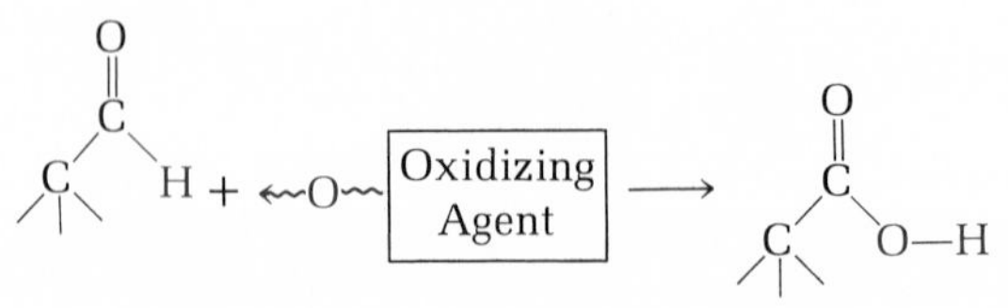

Acid chlorides, such as acetyl chloride, may react with primary or secondary amines (that is, with amines containing an N—H bond) to form amides and with alcohols or phenols to form esters. We may visualize these reactions in terms of the "splitting out" of HCl.

Examples

1. Amide formation, involving N—H of an amine

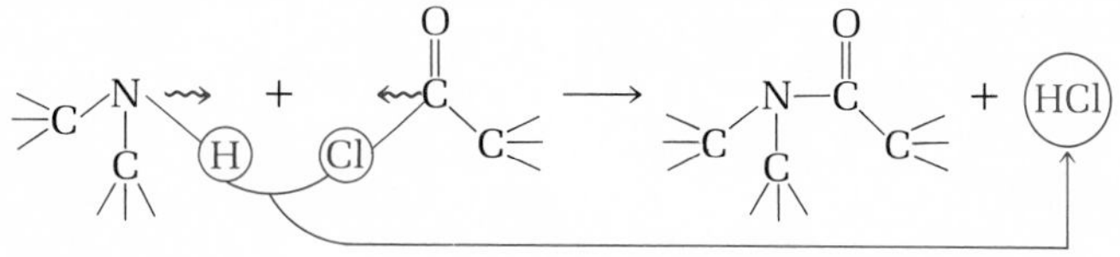

2. Ester formation, involving O—H of an alcohol or phenol

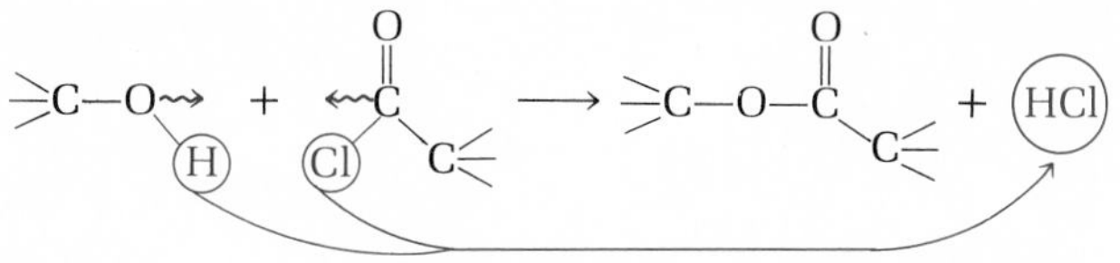

Esters may also be formed by the direct reaction of a carboxylic acid with an alcohol (or, in some special cases, with a phenol). This reaction may be visualized as the "splitting out" of H_2O by combination of the "H" from the alcohol with the "OH" from the acid.[2] Unlike the case of an acid chloride reaction, the direct esterification is reversible. Esters react with water (hydrolyze) to form carboxylic acids and alcohols (or phenols, from phenol esters).

Amides also undergo hydrolysis (Table 29.5). In the hydrolysis reactions of amides and esters, the "OH" of the carboxyl group is replaced from the water involved. Hydrolysis reactions are typically slow unless excess H^+ or OH^- catalyst is present. The formulation of the products depends on the pH of the reaction mixture (Table 29.5).

Examples

1. Direct esterification (reversible)

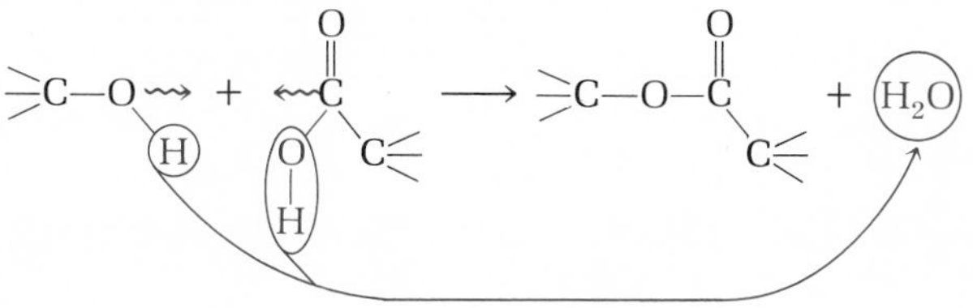

2. Ester hydrolysis

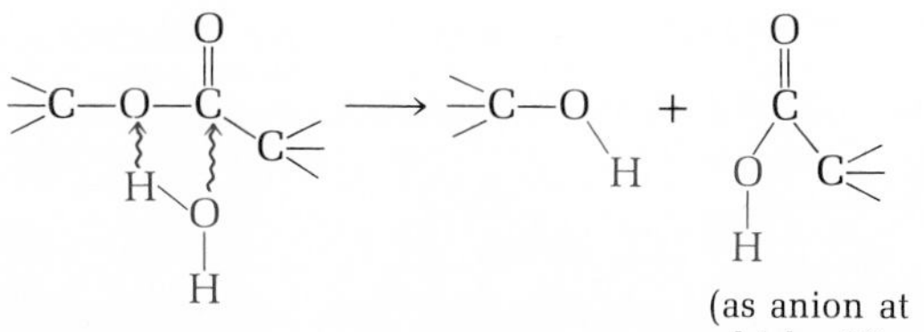

(as anion at high pH)

---

[2] We might expect the acid to furnish the "H", rather than the alcohol, but studies with oxygen-18 have shown this does not occur (Unit 16).

3. Amide hydrolysis

| (as ammonium-type cation at low pH) | (as anion at high pH) |

These simplified "visualizations" should not be taken too literally. At best, they indicate which bonds are broken and which new bonds are formed during some organic reactions. If you construct models of the compounds involved and use the examples given as guides to bond rupture and bond formation, you can develop a reasonably good (but simplified) conceptual grasp of some important reactions. With a little practice you can extend this approach to other reactions, such as those summarized in Tables 29.5 and 29.7.

29.7 FUNCTIONAL GROUP ANALYSIS

Modern instrumental techniques (Unit 30) have largely replaced the more traditional wet chemical tests for functional groups except for compounds whose spectra may be sufficiently complex to necessitate additional chemical testing. Although spectroscopic analyses are rapid, clean, and generally unambiguous, simple chemical tests are often less expensive and are still quite useful to the organic chemist.

The odor of a compound often yields valuable clues to its classification. Analysis of monofunctional compounds typically involves a physical examination (odor, color, solubility, melting or boiling range), simple chemical tests for functional groups, and some additional reactions to identify the specific compound. A general procedure for functional group analysis is summarized in Table 29.6.

Table 29.6 SIMPLE FUNCTIONAL GROUP ANALYSIS (SELECTED COMMON CLASSES)[a]

CLASS	SOLUBILITY	ODOR	QUICK CHEMICAL TEST[b]
alkanes	insol. in H_2O, NaOH, HCl	faint (e.g., lighter fluid)	negative results with all other group tests
alkenes	insol. in H_2O, NaOH, HCl	similar to alkanes	decolorize Br_2 solutions
aromatic hydrocarbons	insol. in H_2O, NaOH, HCl	pungent (e.g., mothballs)	burn with very sooty flame
halides	insol. in H_2O, NaOH, HCl	faint acrid to sweet (e.g., chloroform or DDT)	copper wire dipped in halide and held in burner flame produces bright green flame
alcohols	small molecules sol. in H_2O (pH ~7); larger molecules insol. in H_2O	sweet (e.g., rubbing alcohol)	liberate H_2 when treated with sodium metal
phenols	sparingly sol. in H_2O (pH < 7); sol. NaOH	antiseptic (e.g., Lysol)	give colored products when treated with $FeCl_3$ solution
aldehydes	small molecules sol. in H_2O (pH ~7); larger molecules insol. in H_2O	aliphatic: pungent; aromatic: pleasant (e.g., formalin; cinnamon)	form orange ppt with 2,4-DNP[c]; positive to Tollens' reagent
ketones	same as aldehydes	sweet (e.g., acetone)	form orange ppt with 2,4-DNP[c]; *negative* tests with Tollens' reagent

Table 29.6 (Continued)

CLASS	SOLUBILITY	ODOR	QUICK CHEMICAL TEST[b]
carboxylic acids	small molecules sol. in H_2O (pH $<$ 7); larger molecules insol. in H_2O, sol. in NaOH	irritating to rancid (e.g., small molecules have "vinegar" odor, larger molecules have odor of sweaty feet)	acidic, but negative to $FeCl_3$ (phenol test)
esters	most are insol. in H_2O, cold NaOH, or cold HCl	sweet, "fruity" (e.g., banana oil)	when boiled with NaOH, hydrolyze to form carboxylate salts and alcohols; products may be separated and identified
amines	small molecules sol. in H_2O (pH $>$ 7); larger molecules insol. in H_2O, sol. in HCl	ammonia odor to "fishy" (e.g., fish oils)	test alkaline (pH $>$ 7) in solution (usually in aqueous ethanol)
amides	most are insol. in H_2O, cold NaOH, or cold HCl. Note: most amides are solids	odorless	when boiled with NaOH, hydrolyze to form carboxylate salts and ammonia or amines (detect by odor)

[a] For reactions involved, see Table 29.5.

[b] No single test is universally satisfactory. Those listed are generally accurate when coupled with other observations.

[c] 2,4-DNP represents 2,4-dinitrophenylhydrazine, which reacts with the carbonyl group of an aldehyde or ketone as

Example 1

In cleaning out a chemical laboratory, a technician found three unlabeled bottles on a shelf and three labels lying on the floor. The labels read "acetone," "ethanol," and "benzene," respectively. Since the technician had a cold that day, the differences in odors of the compounds could not be detected. The technician placed a few drops of each liquid on separate pieces of aluminum foil, folded to make little "dishes," and ignited each liquid with a match. The liquids from the first two bottles burned with clean, almost colorless flames, but that from the third bottle burned with a yellow, sooty flame. The technician then placed a few drops of each liquid in three separate clean, dry test tubes and added a tiny piece of freshly cut sodium to each tube. Bubbles were observed only in the tube containing liquid from the first bottle. How should the bottles be relabeled, assuming that the labels found on the floor actually belonged to these bottles?

Solution

Burning with a sooty flame is typical of aromatic compounds (Table 29.6), so the benzene label belongs with the third bottle.

Alcohols react with sodium to generate hydrogen gas (Table 29.6), so the bubbles observed when sodium reacted with the liquid from the first bottle would suggest that the ethanol label belongs with this bottle.

Since the liquid from the second bottle burned cleanly and failed to react with sodium, the acetone label belongs with the second bottle. (Note: A good technician would take this evidence as "preliminary" only. Bottles should never be labeled by "probable" contents. There are, for example, several cases on record of serious laboratory explosions resulting from undergraduate organic experiments on preparation of nitrobenzene in which bottles labeled "benzene" actually contained acetone. Mixtures of acetone and nitric acid often decompose explosively.)

29.8 SYNTHESES   The production of synthetic organic compounds is one of the major functions of the chemical industry. In planning a synthesis procedure, research chemists must consider the kinds of reactions available, the starting materials that can be obtained, the reaction conditions required, and the numerous complications that may be anticipated. A knowledge of reaction mechanisms often permits the selection of synthetic routes for best yields and highest product purity. Attempts must be made to minimize undesirable side reactions or formation of isomeric mixtures requiring tedious separation procedures. In many cases, particularly for commercial processes, synthesis must be selected for most economical production on a large scale.

Complex compounds, including natural products such as morphine, present a real challenge to the synthetic chemist. The basic approach to any synthesis, no matter how complicated, begins with a "drawing-board" planning that works backward from the desired product to available starting materials. Once several alternative syntheses have been outlined in this way, comparisons can be made of possible complications, and the most promising approach can be selected for laboratory testing.

Examination of the reactions outlined in Table 29.5 reveals a number of potentially useful synthetic processes. For reference, we may conveniently select some of these and tabulate them according to the *class* of compound to be synthesized (Table 29.7). These reactions, together with information from Table 29.5, may prove useful in devising synthetic procedures.

Table 29.7 SOME SYNTHETIC PROCEDURES

CLASS TO BE SYNTHESIZED	POSSIBLE SYNTHETIC STEP	EXAMPLE
alcohol	halide + aq OH⁻	
	aldehyde, ketone, ester, or carboxylic acid +LiAlH₄ (followed by hydrolysis of intermediates)	$CH_3CHO \xrightarrow[\text{2. }H_2O,\ H^+]{\text{1. LiAlH}_4} CH_3CH_2OH$ (primary)
		(secondary)
		(+CH_3OH)
		(Note: carbon–carbon double bond unaffected)

Table 29.7 (Continued)

CLASS TO BE SYNTHESIZED	POSSIBLE SYNTHETIC STEP	EXAMPLE
aldehyde	primary alcohol + $Cr_2O_7{}^{2-}$ (with care to avoid complete oxidation to carboxylic acid)	$CH_3CH_2OH \xrightarrow{Cr_2O_7{}^{2-}} CH_3CHO$
amide	amine + acid chloride (primary or secondary)	$(CH_3)_2NH + CH_3\overset{O}{\overset{\|}{C}}Cl \longrightarrow CH_3\overset{O}{\overset{\|}{C}}N(CH_3)_2$
amine	*primary*	
	halide + NH_3 (aliphatic)	$CH_3Br + NH_3 \longrightarrow CH_3NH_2$
carboxylic acid	primary alcohol + $Cr_2O_7{}^{2-}$ (or aldehyde) (no attempt to "save" aldehyde product)	$CH_3CH_2OH \xrightarrow{Cr_2O_7{}^{2-}} CH_3CO_2H$ $CH_3\overset{H}{\underset{}{C}}=O \xrightarrow{O_2} CH_3CO_2H$
ester	alcohol or phenol + acid chloride	
	alcohol + carboxylic acid (an equilibrium process)	$CH_3CH_2OH + CH_3\overset{O}{\overset{\|}{C}}OH \underset{}{\overset{(H^+)}{\rightleftharpoons}} CH_3CH_2O{-}\overset{O}{\overset{\|}{C}}CH_3 + H_2O$
ketone	secondary alcohol + $Cr_2O_7{}^{2-}$	$(CH_3)_2CHOH \xrightarrow{Cr_2O_7{}^{2-}} (CH_3)_2C=O$

The general approach to any synthetic problem may be summarized:

1. Identify the *class* of the compound (or group within a polyfunctional compound) to be prepared. If the compound is indicated by name, draw the structural formula.
2. Consider possible reactions (e.g., Tables 29.5 and 29.7) that might be used to produce the desired compound or group.
3. Consider available "starting materials":
   a. If all reactants needed for a synthesis are available, write equations using the necessary reactants for the required synthesis.
   b. If no synthesis is feasible using the *available* "starting materials," consider necessary, but "unavailable" reactants, as products to be made, starting with step (1).

Example 2

Outline plausible syntheses of

    a. ethyl acetate, starting only with ethanol and any desired inorganic reagents

    b. ethyl acetate, starting only with acetaldehyde (ethanal) and inorganic reagents

    c. N-ethylbenzamide, starting with benzoic acid, any two-carbon compound, and inorganic reagents

    d. isopropyl acetate, starting with any aldehyde or ketone (or both), and inorganic reagents

    e. acetylsalicylic acid (*aspirin*), starting with salicylic acid, acetic acid, and inorganic reagents

Solution (Following the suggested "steps")

    a. *Ethyl acetate*

       STEP 1 Ethyl acetate, $C_2H_5OCCH_3$, is classified as an *ester*.

       STEP 2 Esters may be made (Table 29.7) from the appropriate alcohol and either a carboxylic acid or an acid chloride.

       STEP 3 One of the necessary "starting materials" (ethanol) is available. The other must be prepared. We might, for example, decide to use the direct reaction of an alcohol with an acid. The particular acid needed to make ethyl *acetate* is *acetic* acid and this can be prepared by oxidation (Table 29.7) of some of the available ethanol.

A plausible synthesis, then is (without necessarily balancing equations)

$$C_2H_5OH + Cr_2O_7{}^{2-} \longrightarrow CH_3CO_2H$$

$$CH_3CO_2H + C_2H_5OH \xrightarrow{\text{(H}^+\text{)}} CH_3\overset{\overset{\displaystyle O}{\|}}{C}OC_2H_5 + H_2O$$

This may be represented by a "sequential equation" (ignoring equilibrium considerations) as

$$C_2H_5OH \xrightarrow{Cr_2O_7{}^{2-}} CH_3CO_2H \xrightarrow{C_2H_5OH,\ \text{(H}^+\text{)}} CH_3\overset{\overset{\displaystyle O}{\|}}{C}OC_2H_5$$

    b. *Ethyl acetate*

       STEP 1 Same as for step 1 in (a).

       STEP 2 Same as for step 2 in (a).

       STEP 3 In this case, neither of the necessary "starting materials" are available, but both can be prepared from the available aldehyde (Table 29.7):

$$CH_3\overset{\overset{\displaystyle H}{|}}{C}{=}O + LiAlH_4 \longrightarrow \underset{\text{(alkoxides)}}{\text{intermediates}} \xrightarrow{H_2O,\ H^+} CH_3CH_2OH$$

$$CH_3\overset{\overset{\displaystyle H}{|}}{C}{=}O + O_2 \longrightarrow CH_3CO_2H$$

$$CH_3CH_2OH + CH_3CO_2H \xrightarrow{\text{(H}^+\text{)}} CH_3CH_2O\overset{\overset{\displaystyle O}{\|}}{C}CH_3$$

or

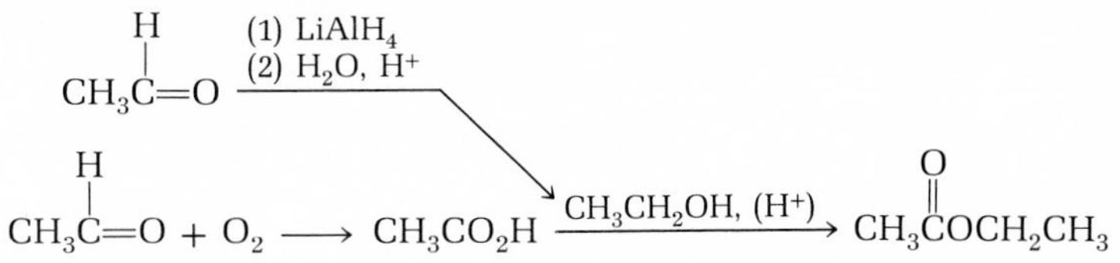

$$\underset{\underset{\text{H}}{|}}{\text{CH}_3\text{C}}{=}\text{O} \xrightarrow[\text{(2) H}_2\text{O, H}^+]{\text{(1) LiAlH}_4}$$

$$\underset{\underset{\text{H}}{|}}{\text{CH}_3\text{C}}{=}\text{O} + \text{O}_2 \longrightarrow \text{CH}_3\text{CO}_2\text{H} \xrightarrow{\text{CH}_3\text{CH}_2\text{OH, (H}^+)} \text{CH}_3\overset{\overset{\text{O}}{\|}}{\text{C}}\text{OCH}_2\text{CH}_3$$

c. *N-ethylbenzamide*

STEP 1 The desired product is an *amide*, specifically

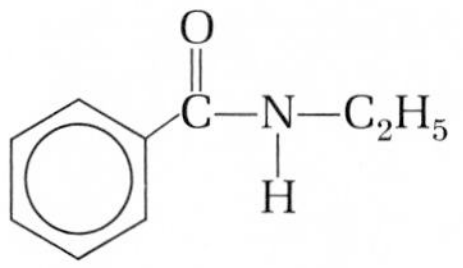

STEP 2 Amides are most readily prepared from amines and acid chlorides (Table 29.7).

STEP 3 Since "any two-carbon" compound is available, the specific amine required ($\text{CH}_3\text{CH}_2\text{NH}_2$) is available. The necessary acid chloride (benzoyl chloride) is not available, but it may be prepared from benzoic acid and PCl_3 (Table 29.5).

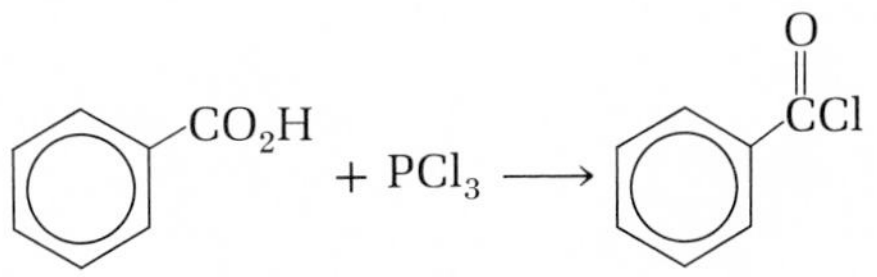

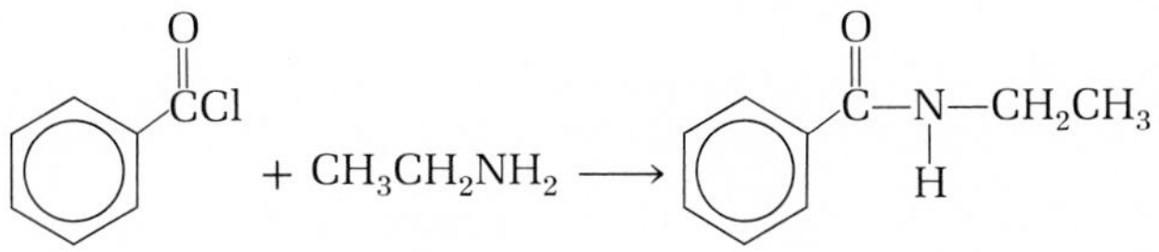

or

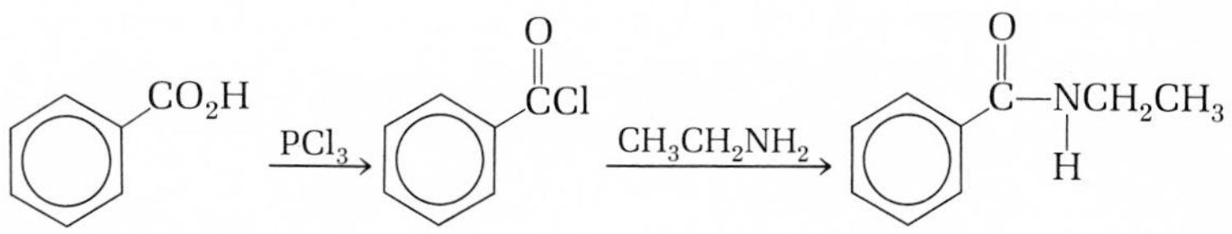

d. *Isopropyl acetate*

STEP 1 The desired product is an *ester*, specifically,

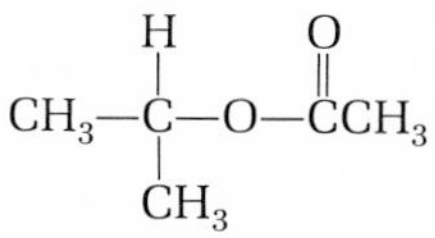

$$\text{CH}_3{-}\underset{\underset{\text{CH}_3}{|}}{\overset{\overset{\text{H}}{|}}{\text{C}}}{-}\text{O}{-}\overset{\overset{\text{O}}{\|}}{\text{C}}\text{CH}_3$$

STEP 2 Esters may be prepared from alcohols and either carboxylic acids or acid chlorides (Table 29.7).

STEP 3 Neither type of necessary "starting material" is available, but the alcohol needed (2-propanol) can be made from an "available ketone" (in this case, acetone) by reduction (Table 29.7), and the acid needed (acetic) can be made from an "available aldehyde" (ethanal) by oxidation (Table 29.7):

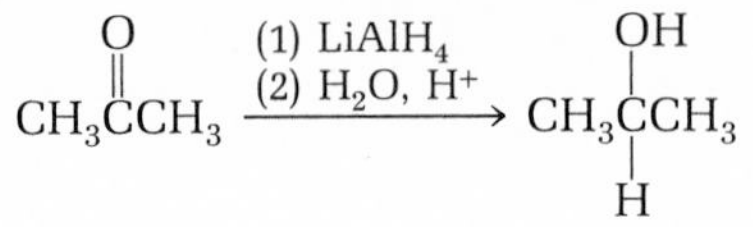

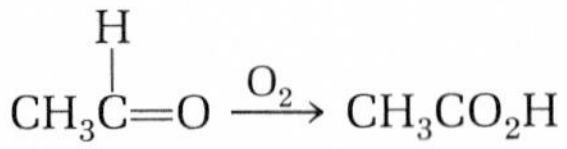

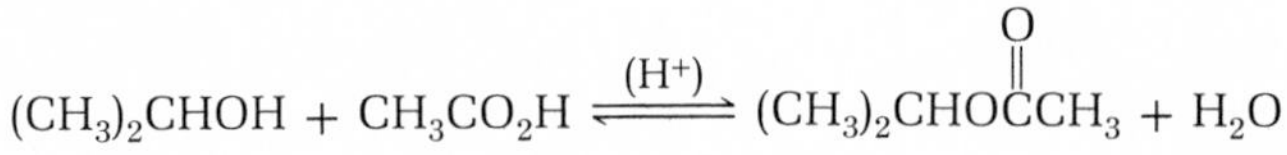

$$\underset{\text{CH}_3\overset{\displaystyle O}{\overset{\|}{C}}\text{CH}_3}{} \xrightarrow[\text{(2) H}_2\text{O, H}^+]{\text{(1) LiAlH}_4} \underset{\text{CH}_3\overset{\displaystyle OH}{\underset{\displaystyle H}{\overset{|}{C}}}\text{CH}_3}{}$$

$$\underset{\text{CH}_3\overset{\displaystyle H}{\overset{|}{C}}=O}{} \xrightarrow{O_2} \text{CH}_3\text{CO}_2\text{H}$$

$$(\text{CH}_3)_2\text{CHOH} + \text{CH}_3\text{CO}_2\text{H} \underset{}{\overset{(\text{H}^+)}{\rightleftharpoons}} (\text{CH}_3)_2\text{CHO}\overset{\displaystyle O}{\overset{\|}{C}}\text{CH}_3 + \text{H}_2\text{O}$$

or

$$\underset{\text{CH}_3\overset{\displaystyle O}{\overset{\|}{C}}\text{CH}_3}{} \xrightarrow[\text{(2) H}_2\text{O, H}^+]{\text{(1) LiAlH}_4}$$

$$\underset{\text{CH}_3\overset{\displaystyle H}{\overset{|}{C}}=O}{} \xrightarrow{O_2} \text{CH}_3\text{CO}_2\text{H} \xrightarrow{(\text{CH}_3)_2\text{CHOH, (H}^+)} (\text{CH}_3)_2\text{CHO}\overset{\displaystyle O}{\overset{\|}{C}}\text{CH}_3$$

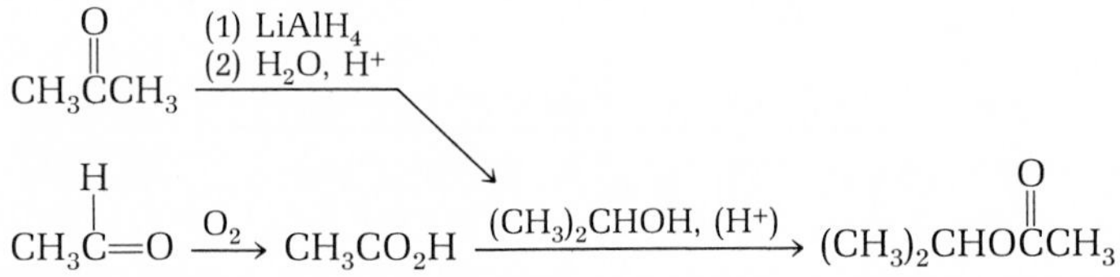

e. *Acetylsalicylic acid*

STEP 1 Acetylsalicylic acid is a *difunctional* compound (i.e., containing two functional groups). In drawing its structure, we can focus attention on the specific group to be prepared, that is, the group not present in the salicylic acid "starting material." This new group is an *ester*.

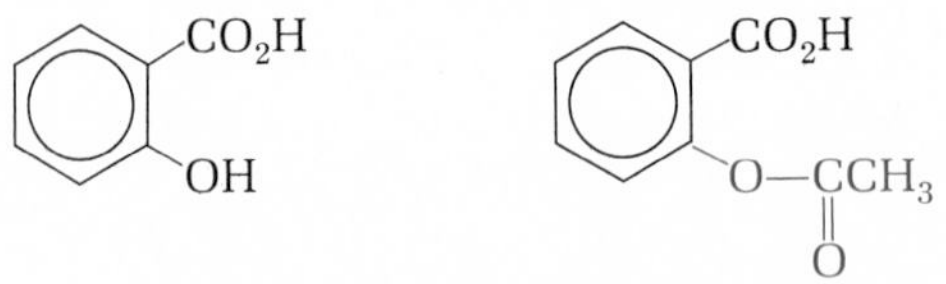

salicylic acid (Unit 28)   acetylsalicylic acid (Unit 28),
                                     showing the *ester* group

STEP 2 Esters of *phenols* are best prepared from the phenol and an acid chloride (Table 29.5).

STEP 3 The necessary phenol group is available in the salicylic acid, but the acid chloride needed (acetyl chloride) must be made from the available acetic acid.

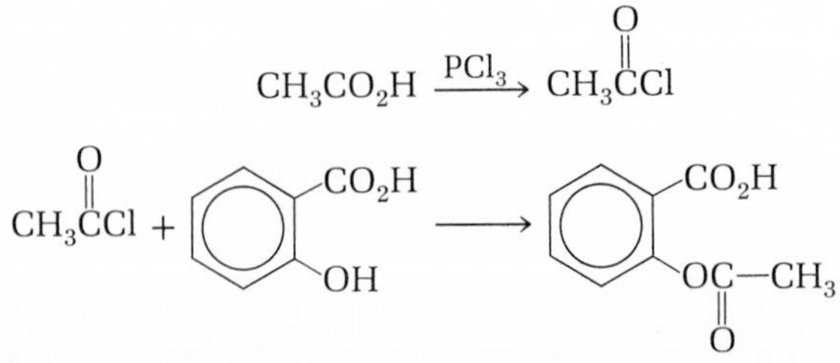

$$\text{CH}_3\text{CO}_2\text{H} \xrightarrow{\text{PCl}_3} \text{CH}_3\overset{\displaystyle O}{\overset{\|}{C}}\text{Cl}$$

or

$$CH_3CO_2H \xrightarrow{PCl_3} CH_3\overset{\overset{\displaystyle O}{\|}}{C}Cl \longrightarrow$$

We have touched only briefly on a few aspects of the reactions of organic compounds to illustrate how some functional groups may be involved in synthetic or analytical organic chemistry. If you pursue further study in this field, you will find better explanations and mechanisms for the reactions we have presented on a largely empirical basis.

Many of the more interesting organic compounds, especially those of biological importance, contain more than one functional group. To understand the chemistry of such species we must focus our attention on one functional group at a time. In this way, even our simplified treatment of functional group properties can provide a useful insight into the behavior of complex molecules.

SUGGESTED READING

Allinger, N. L., and J. Allinger. 1965. *Structures of Organic Molecules.* Englewood Cliffs, N.J.: Prentice-Hall.

Gutsche, C. D. 1967. *The Chemistry of Carbonyl Compounds.* Englewood Cliffs, N.J.: Prentice-Hall.

Jackson, R. A. 1973. *Mechanism: An Introduction to the Study of Organic Reactions.* New York: Oxford.

Stille, John K. 1968. *Industrial Organic Chemistry.* Englewood Cliffs, N.J.: Prentice-Hall.

Stock, L. 1968. *Aromatic Substitution Reactions.* Englewood Cliffs, N.J.: Prentice-Hall.

Trahanovsky, Walter S. 1971. *Functional Groups in Organic Chemistry.* Englewood Cliffs, N.J.: Prentice-Hall.

EXERCISES

1. Give correct IUPAC names for

a. $(CH_3)_2CHCH_2CH_2CHCH_2CH_3$
$\qquad\qquad\qquad\qquad\quad |$
$\qquad\qquad\qquad\qquad\ \ Br$

b.

$\qquad$ OH
$\qquad$ —OCH$_3$

c. $(CH_3)_2CHCH_2CH_2CHCH_2CH_3$
$\qquad\qquad\qquad\qquad\quad |$
$\qquad\qquad\qquad\qquad\ \ OH$

d. $(CH_3)_2CHCH_2CH_2CHCH_2\overset{\overset{\displaystyle H}{|}}{C}{=}O$
$\qquad\qquad\qquad\qquad\quad |$
$\qquad\qquad\qquad\qquad\ \ CH_3$

e. $(CH_3)_2CHCH_2CH_2\overset{\overset{\displaystyle O}{\|}}{C}CH_2CH_3$

f. $(CH_3)_2CHCH_2CH_2CHCH_2CO_2H$
$\qquad\qquad\qquad\qquad\quad |$
$\qquad\qquad\qquad\qquad\ \ CH_3$

g. $(CH_3)_2CHCH_2CH_2CHCH_2CH_3$
$\qquad\qquad\qquad\qquad\quad |$
$\qquad\qquad\qquad\qquad\ \ NH_2$

h. $(CH_3)_2CHCH_2CH_2\overset{\overset{\displaystyle O}{\|}}{C}HCOCH_2CH_3$
$\qquad\qquad\qquad\qquad\quad |$
$\qquad\qquad\qquad\qquad\ \ CH_3$

i. $(CH_3)_2CHCH_2CH_2CHCH_2\overset{\overset{\displaystyle O}{\|}}{C}NH_2$
$\qquad\qquad\qquad\qquad\quad |$
$\qquad\qquad\qquad\qquad\ \ CH_3$

2. Give correct *trivial* names for
a. HCO_2H
b. $K(CH_3CO_2)$

c. $CH_3CH_2\overset{\underset{\displaystyle H}{|}}{C}=O$

3. Which member of each pair would be more soluble in water? Explain.
a. 2-propanol or 2-chloropropane
b. 2-propanol or 2-hexanol
4. Which exhibit acidic properties? Explain.

a. [benzene ring with —CO$_2$H] **b.** [benzene ring with —CH$_2$OH] [benzene ring with —OH and CH$_3$]

5. A water-soluble neutral organic compound had the molecular formula $C_2H_6O_2$ and could be oxidized to oxalic acid.
a. What was the structure of the compound?
b. Suggest a plausible synthesis of the compound, starting with any desired halide and inorganic reagents.
6. Using any alcohol and any desired inorganic reagents, outline plausible syntheses for
a. ethyl acetate; **b.** acetophenone; **c.** ethyl benzoate
7. A solution of an organic acid in pure water had a pH of 2.30. A 25.0-ml aliquot of the solution required 20.4 ml of 0.128 M sodium hydroxide for neutralization. Calculate the approximate ionization constant for the acid.

PRACTICE FOR PROFICIENCY

Objectives (a, b)

1. Give correct IUPAC names for

a. $CH_3CH_2\overset{\underset{\displaystyle CH_3}{|}}{C}HCH_2CH_2\overset{\underset{\displaystyle OH}{|}}{C}HCH_2CH_2CH_3$

b. $CH_3CH_2\overset{\underset{\displaystyle CH_3}{|}}{C}HCH_2CH_2\overset{\underset{\displaystyle NH_2}{|}}{C}HCH_2CH_2CH_3$

c. $CH_3CH_2\overset{\underset{\displaystyle CH_3}{|}}{\overset{\displaystyle CH_3}{C}}CH_2CH_2CH_2CH_2CO_2H$

d. $CH_3CH_2\overset{\underset{\displaystyle Br}{|}}{C}HCH_2CH_2\overset{\displaystyle O}{\overset{\|}{C}}OCH_2CH_3$

e. [benzene ring with CH$_3$ and OCH$_3$]

2. Draw correct structures for
a. butyraldehyde; **b.** sodium formate; **c.** methyl acetate
3. Select the best name for
$(CH_3)_2CH\overset{\underset{\displaystyle OH}{|}}{C}HCH_2C(CH_3)_3$
a. 1,1,1,4,4-pentamethylbutanol
b. 1,1-dimethylisopentanol

c. 2,5,5-trimethyl-3-hexanol
d. 2,2,5-trimethyl-4-hexanol
e. none of these
4. Select the best name for

$CH_3\overset{\underset{\displaystyle OH}{|}}{C}HCH_2CH_2CH_2NH_2$

a. 5-amino-2-pentanol
b. 1-amino-4-pentanol
c. 2-hydroxy-5-pentamine
d. 4-hydroxy-1-pentamine
e. none of these
5. Select the best name for

$CH_3CH_2CH_2CH_2\overset{\underset{\displaystyle CH_3}{|}}{C}H\overset{\displaystyle O}{\overset{\|}{C}}OH$

a. 1-hydroxy-2-methylhexanone
b. 6-hydroxy-5-methylhexanal
c. 5-methylhexanoic acid
d. 2-methylhexanoic acid
e. none of these
6. Select the best name for

[benzene ring with $\overset{\displaystyle O}{\overset{\|}{C}}OCH_2CH_3$ and NO_2]

a. methyl o-nitrobenzylate
b. ethyl m-nitrobenzoate
c. m-nitroacetophenone
d. o-nitroacetophenone
e. none of these
7. Select the best name for

$CH_3\overset{\underset{\displaystyle Cl}{|}}{C}HCH_2CH_2\overset{\displaystyle O}{\overset{\|}{C}}CH_2CH_3$

a. 6-chloro-3-heptanone
b. 2-chloro-5-heptanone
c. ethyl 4-chloropentanoate
d. 1-acetyl-3-chlorobutane
e. none of these
8. Select the best name for

[benzene ring with $\overset{\displaystyle O}{\overset{\|}{C}}OCH_3$ and Cl]

a. o-chloroacetophenone
b. o-acetylchlorobenzene

c. m-methylchlorobenzoate
d. methyl p-chlorobenzoate
e. none of these

9. Select the best name for

$$(CH_3)_2CHCH_2CH_2CH_2\overset{\overset{\displaystyle O}{\|}}{C}H$$

a. 5-methyl-1-hexanone
b. 1,1-dimethyl-5-pentanone
c. 5-methylhexanal
d. 5,5-dimethylpentanal
e. none of these

10. Select the best name for

$$CH_3O\overset{\overset{\displaystyle O}{\|}}{C}CH_2\underset{\underset{\displaystyle Cl}{|}}{C}HCH_3$$

a. methyl 3-chlorobutanoate
b. 4-chloro-2-pentanone
c. 1-methoxy-3-chlorobutanal
d. 2-butanoylchloride
e. none of these

Objective (c)

11. Which would be more soluble in water, acetone or 2-hexanone? Explain.
12. Which is least soluble in water?
  a. CH_3I; **b.** CH_3NH_2; **c.** $H_2C{=}O$; **d.** CH_3OH;
  e. All are very soluble in water.
13. Which is least soluble in water?
  a. CH_3CO_2H; **b.** HCO_2H; **c.** HO_2CCO_2H;
  d. $CH_3(CH_2)_{10}CO_2H$; **e.** All are infinitely soluble in water.
14. Which is least soluble in water?
  a. $H_2NCH_2CH_2OH$; **b.** $H_2NCH_2CH_2NH_2$; **c.** $ClCH_2CH_2Cl$;
  d. $HOCH_2CH_2OH$;
  e. None are very soluble in water.
15. Which is least soluble in water?
  a. $CH_3CH_2CH_2CH_2OH$; **b.** $H_2NCH_2CH_2OH$;
  c. $HOCH_2CH_2OH$; **d.** $NaOH$; **e.** CH_3CO_2H
16. Which will not react with aqueous sodium hydroxide?

  a. CH_3CO_2H

  b. CH_3—⬡—OH

  c. ⬡—OH

  d. $H{-}\overset{\overset{\displaystyle O}{\|}}{C}{-}OH$

  e. $CH_3NH_3{}^+$

17. Which will not react with aqueous HCl?

  a. CH_3NH_2 **b.** ⬡—NH_2

c. $(CH_3)_3N$ **d.** ⬡—$CO_2{}^-$

e. ⬡—$\overset{\overset{\displaystyle O}{\|}}{C}CH_3$

18. How many moles of sodium hydroxide could react with 1 mole of salicylic acid (o-hydroxybenzoic acid)?
19. How many moles of hydrochloric acid could react with 1 mole of sodium p-aminobenzoate?

Objective (d)

20. Indicate the class to which each of the following compounds probably belongs:
  a. a colorless liquid of pronounced "rancid" odor, insoluble in water, but soluble in dilute sodium bicarbonate solution with evolution of CO_2
  b. a colorless liquid having a pungent odor, containing only hydrogen and carbon, and burning with a very sooty flame
21. Select the correct structural formula for compound N.

$$\text{butanal} \xrightarrow{\text{LiAlH}_4} \text{Cpd. I} \xrightarrow{\overset{\overset{\displaystyle O}{\|}}{CH_3CCl}} \boxed{\text{Cpd. N}}$$

a. $CH_3CH_2\overset{\overset{\displaystyle O}{\|}}{C}\underset{\underset{\displaystyle Cl}{|}}{C}HCH_2CH_3$ **b.** $CH_3\overset{\overset{\displaystyle O}{\|}}{C}CH_2CH_2CH_2CH_3$

c. $CH_3\overset{\overset{\displaystyle O}{\|}}{C}OCH_2CH_2CH_2CH_3$ **d.** $HO{-}\underset{\underset{\displaystyle CH_3}{|}}{\overset{\overset{\displaystyle CH_3}{|}}{C}}{-}\overset{\overset{\displaystyle O}{\|}}{C}CH_3$

e.
$$\begin{array}{c} \overset{\overset{\displaystyle O}{\|}}{CH_2CCH_3} \\ | \\ CH_2 \\ | \\ CH_2 \\ | \\ H{-}C{-}OH \\ | \\ Cl \end{array}$$

22. Select the correct structural formula for compound L.

$$\text{bromocyclopentane} \xrightarrow{OH^-} Br^- + \text{Cpd. I} \xrightarrow{Cr_2O_7{}^{2-}} \boxed{\text{Cpd. L}}$$

a. (structure: cyclopentane with OH and Br)

b. (structure: cyclopentane with CO₂H)

c. (structure: cyclopentane with CHO)

d. (structure: cyclopentane with Br and OH)

e. (structure: cyclopentanone)

23. Select the correct structural formula for compound U.

$$\text{1-propanol} \xrightarrow{Cr_2O_7{}^{2-}} \text{Cpd. I} \xrightleftharpoons{CH_3OH} H_2O + \boxed{\text{Cpd. U}}$$

a. $CH_3OCCH_2CH_3$ (with O double bond) **b.** $CH_3CCH_2CH_2CH_3$ (with O double bond)

c. $CH_3CCH_2CH_2OH$ (with O double bond) **d.** $CH_3OCH_2CH_2CH_2OH$

e. $CH_3OCH_2CHCH_3$ (with OH)

24. Which of the following proposed reaction sequences would be plausible for the conversion of salicylic acid (o-hydroxybenzoic acid) to "aspirin" (acetylsalicylic acid)?

a. salicylic acid + CH_3OH $\xrightarrow{(H_2SO_4)}$ "aspirin"

b. salicylic acid + NaOH $\longrightarrow$
  sodium salicylate $\xrightarrow{CH_3I}$ "aspirin"

c. salicylic acid + CH_3CCl (with O double bond) $\longrightarrow$ "aspirin"

d. salicylic acid + $LiAlH_4$ $\longrightarrow$
  o-hydroxybenzyl alcohol $\xrightarrow{CH_3CCH_3}$ "aspirin"

e. none of these

25. Which of the following proposed reaction sequences would be plausible for the conversion of salicylic acid (o-hydroxybenzoic acid) to "oil of wintergreen" (methyl salicylate)?

a. salicylic acid + CH_3CCl (with O double bond) $\longrightarrow$ "oil of wintergreen"

b. salicylic acid + CH_3OH $\xrightarrow{H_2SO_4}$ "oil of wintergreen"

c. salicyclic acid + $LiAlH_4$ $\longrightarrow$
  o-hydroxybenzyl alcohol $\xrightarrow{MnO_4{}^-}$ "oil of wintergreen"

d. salicylic acid + NaOH $\longrightarrow$
  sodium salicylate $\xrightarrow{CH_3CCH_3}$ "oil of wintergreen"

e. none of these

26. Starting with any alcohol and any desired inorganic reagents, outline plausible syntheses for
a. acetone; **b.** benzaldehyde; **c.** benzoic acid; **d.** methyl benzoate; **e.** benzamide; **f.** benzyl acetate; **g.** N-propylbutanamide

Objective (e)

27. A 0.25 M solution of an organic monoprotic acid had a pH of 4.0. What is the approximate ionization constant of the acid?

28. A solution of an organic monoprotic acid was found to have a pH of 3.30. A 25.0-ml aliquot of the solution was exactly neutralized by 18.5 ml of 0.135 M NaOH. What is the approximate ionization constant of the acid?

29. When a 0.25-g sample of an organic acid of molecular weight 124 was dissolved in 250 ml of pure water, the pH of the solution was found to be 3.30. What is the approximate ionization constant of the acid?

30. A solution of an organic monoprotic acid was found to have a pH of 4.70. A 25.0-ml aliquot of the solution was exactly neutralized by 22.5 ml of 0.111 M NaOH. What is the approximate ionization constant of the acid?

31. When a 0.30-g sample of an organic acid of molecular weight 225 was dissolved in 150 ml of pure water, the pH of the solution was found to be 4.30. What is the approximate ionization constant of the acid?

BRAIN TINGLER

32. *Papaverine* is an antispasmodic drug isolated from the opium poppy. The drug can be prepared synthetically from vanillin, a natural flavoring agent from the vanilla bean. Deduce the structures of the lettered compounds prepared as intermediates in the multistep synthesis of *papaverine*. (If you are interested in the complete synthesis, see Excursion 5.)

STEP 1 CH_3O — (ring with HO substituent, C=O with H) $\xrightarrow{OH^-}$ Ⓐ + H_2O
  (vanillin)

STEP 2 Ⓐ $\xrightarrow{CH_3I}$ Ⓑ + I^-

STEP 3 Ⓑ $\xrightarrow{LiAlH_4}$ Ⓒ + inorganic salts

STEP 4 (C) $\xrightarrow{\text{PBr}_3}$ (D) + inorganic species

ADDITIONAL STEPS $\rightsquigarrow$ CH$_3$O

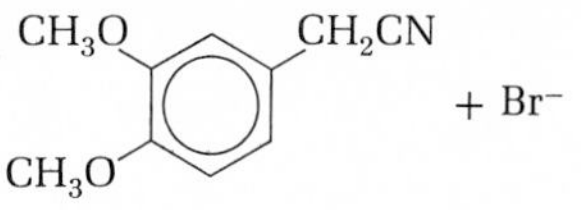

STEP 5 (D) $\xrightarrow{\text{CN}^-}$ CH$_3$O — CH$_2$CN + Br$^-$

CH$_3$O

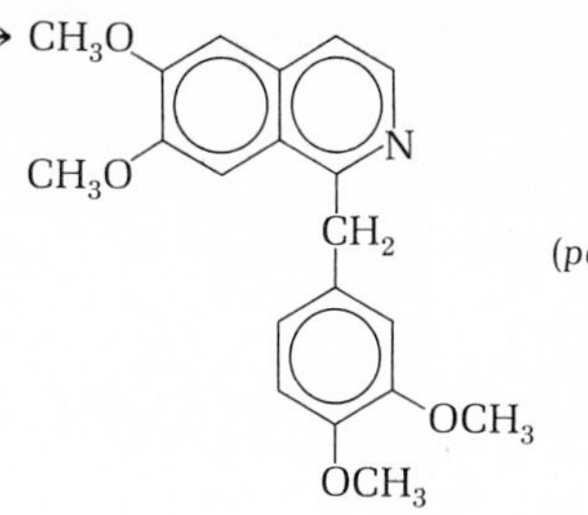

(papaverine)

OCH$_3$

OCH$_3$

Unit 29 Self-Test

1. Give correct IUPAC names for
  a. (CH$_3$)$_3$CCH$_2$CH$_2$Cl
  b. (CH$_3$)$_3$CCH$_2$CH$_2$OH
  c. (CH$_3$)$_3$CCH$_2$CO$_2$H
  d. (CH$_3$)$_3$CCH$_2$CO$_2$CH$_3$

  e.  CH$_3$O— —NH$_2$

2. Give correct *trivial* names for

  a. CH$_3$(CH$_2$)$_2$CO$_2$H

  b. CH$_3$(CH$_2$)$_3$$\overset{\text{O}}{\overset{\|}{\text{C}}}OCH_2CH_3$

  c. H$_2$C=O

  d. CH$_3$$\overset{\text{O}}{\overset{\|}{\text{C}}}$Cl

3. a. Select the formula of the most water-soluble compound:

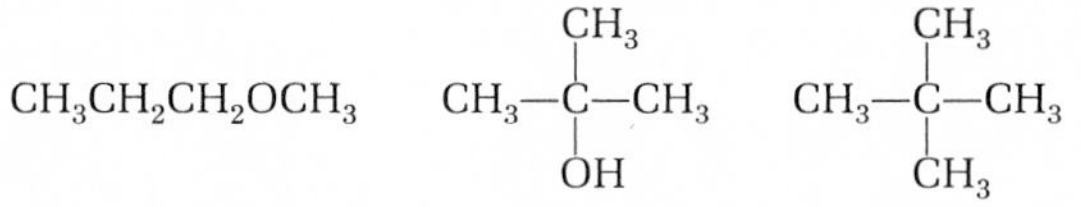

  CH$_3$CH$_2$CH$_2$OCH$_3$ CH$_3$—$\overset{\text{CH}_3}{\underset{\text{OH}}{\overset{|}{\underset{|}{\text{C}}}}}$—CH$_3$ CH$_3$—$\overset{\text{CH}_3}{\underset{\text{CH}_3}{\overset{|}{\underset{|}{\text{C}}}}}$—CH$_3$

b. Select the formula of the acidic compound:

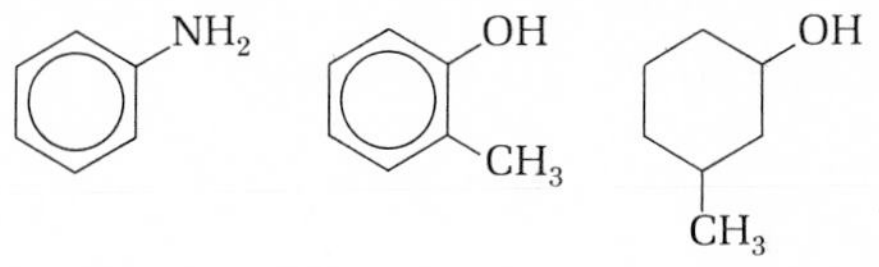

  NH$_2$ OH   OH
  CH$_3$ CH$_3$

4. Indicate the class (e.g., "ester") to which each of the following probably belongs:
  a. an oily liquid of pronounced "fishy" odor, insoluble in water, but soluble in dilute hydrochloric acid
  b. a colorless, nearly odorless liquid that burns with a clean flame and decolorizes bromine solution
  c. a white crystalline solid, soluble in 50 percent aqueous ethanol to give a neutral solution, which liberates a product smelling like ammonia when the solid is boiled with aqueous NaOH

5. Starting with any aldehyde and inorganic reagents, outline plausible synthesis for
  a. 1-butanol; **b.** acetic acid; **c.** benzyl alcohol; **d.** 1-chloropentane; **e.** hexanamide

6. A 0.236-g sample of *o*-nitrobenzoic acid was dissolved in sufficient water to form 100.0 ml of solution. The solution had a pH of 2.16. Calculate the ionization constant for this acid.

Unit 30

... an indispensable tool for the determination of structural information ...

Infrared Spectroscopy (1966)
ROBERT CONLEY

Spectroscopic investigations

Objectives

(a) Be able to make calculations employing Beer's law relationships (Section 30.5).

(b) Be able to relate ultraviolet and visible absorption spectra to the presence of certain types of *chromophore* groups in molecules (Sections 30.4 and 30.5).

(c) Be able to interpret infrared spectra of simple molecules in terms of probable structural features (Sections 30.6 and 30.7).

(d) Be able to deduce structural features of simple molecules from nuclear magnetic resonance spectra (Sections 30.8–30.12).

Until the advent of spectroscopic instruments, determination of the structure of an organic compound required extensive chemical investigations ranging from tests for functional groups to complex degradations of the molecule into simpler, identifiable

Figure 30.1   Electromagnetic spectrum.

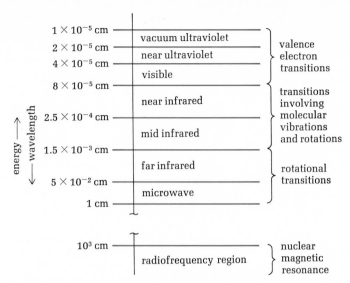

substances. The fragments of information obtained were fitted together like a jigsaw puzzle, with the added complication of several alternative ways of assembling the pieces. Ultimately, a synthesis following well-established routes would produce a substance identical with the original unknown compound. Structural determinations of this type were time consuming and required a large quantity of the pure compound. It was not unusual for such an analysis to require many years of research by many different chemists.

Technical advances in the early years of this century began to introduce nondestructive methods of analysis to supplement and, in some cases, to replace wet chemical investigations. Yet as late as 1960, few organic chemists would accept a structural determination based exclusively on spectroscopic data. Today it is not uncommon for a compound to be completely described by instrumental methods, and the suggestion that any structure may be determined from instrumental data, with computer-assisted interpretation, now seems an attainable goal rather than a chapter from science fiction.

The modern structural research laboratory, then, bears little resemblance to the classical picture of a collection of beakers, flasks, and odd-shaped glassware. Although the chemist still uses many of the standard procedures, he must now be able to use and interpret the data from modern electronic equipment ranging from the spectrophotometer to the high-speed computer.

The wide range of energies in the electromagnetic spectrum (Figure 30.1) offers a number of ways for obtaining structural information on organic molecules. In this unit we shall examine three of the more common types of spectroscopic techniques useful in the study of organic compounds.

30.1 COLOR IN ORGANIC MOLECULES

White light is actually a mixture of light of many different colors, as can be readily demonstrated with a prism or diffraction grating. Each of the so-called primary colors is itself a continuum of many wavelengths, each of which has a specific energy described by the equation.

$$E = \frac{hc}{\lambda}$$

(30.1)

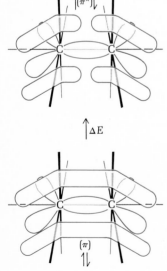

Figure 30.2 n-π* transition in azobenzene (one-electron excitation).

Figure 30.3 Delocalized π-π* transition (one-electron excitation).

where E is the energy of the light (in *calories* when Planck's constant, h, is expressed as 1.59×10^{-34} cal sec), c is the speed of light (3.00×10^{10} cm sec^{-1}), and λ is the wavelength of the radiation (in centimeters).

In most of the colored inorganic substances, light of a specific wavelength is absorbed in a process described as electronic transition involving d orbitals (or f orbitals, or hybrid orbitals with some d or f character). Few organic compounds, however, contain atoms having d-orbital electrons, and the more common colored organic molecules contain only carbon and hydrogen or, in some cases, nitrogen, oxygen, sulfur, or the halogens. In most cases, then, electron transitions associated with color of organic compounds must be described in terms of s, p, sp, sp^2, sp^3 orbitals or the σ or π bonds formed from such orbitals.

There are two common types of colored organic molecules, and both types are found to contain large, delocalized π-bond systems. In addition, one type involves unshared electron pairs (so-called n electrons). A rigorous treatment of the interaction of light with such electron systems is beyond the scope of this text, but the brief qualitative discussion in the following examples is useful in visualizing the general considerations involved.

Before considering these, however, we need to introduce the concept of *antibonding* orbitals. In our discussions of atomic and molecular orbitals, we have employed a much simplified pictorial approach. Any more rigorous treatment requires the use of algebraic signs ($+$ or $-$) for the wave functions describing the orbitals. Combination of atomic orbitals (or orbital lobes) of *like sign* results in the bonding (molecular) orbital of the type we have employed in our bond descriptions. If, on the other hand, combinations of *unlike signs* are considered, the result is a higher-energy system called an *antibonding* orbital. For our purposes, we may consider such cases to represent *excited states* in which the bond pair electrons have zero probability of being located exactly between the nuclei involved. We shall continue our use of pictorial representations without sign designation for the examples in this unit, leaving the more elegant treatments for advanced study.

Examples

1. Azobenzene is an orange crystalline solid that can exist in either the cis or trans form, the latter being the more stable around room temperature.

cis-azobenzene   *trans*-azobenzene

Light energy of about 10^{-19} cal ($\lambda = \sim 4400$ Å) is believed to cause one of the unshared electrons (n electrons) on nitrogen to enter the delocalized π system of the molecule in a delocalized antibonding orbital (Figure 30.2).

2. Azulene is a beautiful deep blue solid having the structure

Azulene has a strong absorption in the red region of the visible spectrum, so its observed color is blue (complementary to red). Again, a relatively low-energy transition occurs as a result of the absorption of light, this time described as a delocalized $\pi-\pi^*$ transition (Figure 30.3).

30.2 WHY ABSORPTION IS NOTICED

If light energy causes electrons to be promoted from a lower to a higher energy state and these electrons spontaneously return to their more stable condition, why isn't the light reemitted so that no net absorption occurs?

Remember that covalent bonds in molecules are always undergoing motions described as bending, stretching, and rotation. Some of the reemitted energy may appear as energy causing changes in these molecular motions; that is, some of the light energy may be converted to kinetic energy. Indeed, covalent bonds may actually be broken in rare cases. However, only some of the light energy is lost by such processes in most cases, and some other explanation must be invoked to account for additional absorption.

The other reason for observed absorption is the scattering of the reemitted radiation. The original light beam is directional (toward the observer). When absorption occurs, followed by reemission, the light given off by the sample is scattered in all directions, so that only a small fraction continues on in the direction of the observer (Figure 30.4).

Figure 30.4 Light absorption.

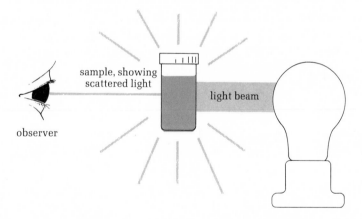

sample, showing scattered light

light beam

observer

30.3 THE SPECTROPHOTOMETER

A recording double-beam spectrophotometer is most commonly used for ultraviolet and visible spectra. Such an instrument is depicted schematically in Figure 30.5. The light originates from either a tungsten lamp (visible) or a hydrogen lamp (ultraviolet) and enters the monochromator area as a narrow beam. Here it passes through an optical system containing a movable prism or grating so that one wavelength at a time can be focused on the sample area. The beam is eventually split so that equal intensities are passed through the sample (usually in solution) and the reference (usually the pure solvent). Detectors (photomultiplier tubes similar in principle to a light meter) convert the light signals into electric energy, and the difference in intensities between the sample and reference beams is recorded on a moving chart. In this way a graphical plot of wavelength versus absorption can be made.

30.4 ULTRAVIOLET SPECTRA

Very few organic molecules are colored. Hence information about bonds that can be involved in $n-\pi^*$ or $\pi-\pi^*$ transitions in highly delocalized systems is quite limited and,

Figure 30.5 Optical layout of a double-beam spectrophotometer.

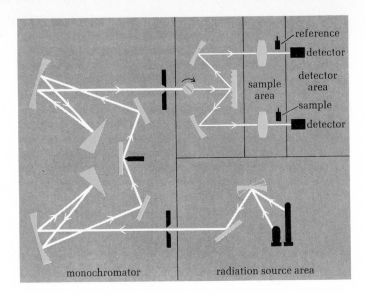

reference
detector

detector area

sample

detector

sample area

detector area

monochromator

radiation source area

indeed, need not involve spectroscopic instruments since color can be observed directly.

A very large number of compounds, however, contain electronic systems that can be excited by the higher energy light in the ultraviolet region (Figure 30.1). In particular, molecules containing π bonds or n-type electrons (or both) will frequently have characteristic ultraviolet absorptions that can be used to assign structural features. Substances such as the aromatic amines or phenols may undergo π–π^* or n–π^* transitions involving delocalized bonds. In most cases such transitions require energy greater than that needed for similar situations in the more highly delocalized

Figure 30.6 Idealized transitions in an isolated carbonyl group.

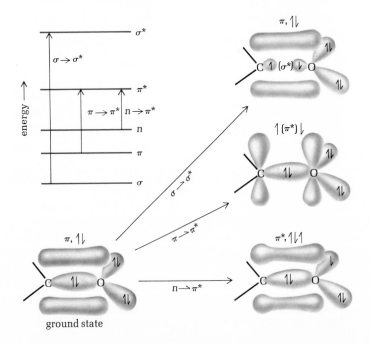

ground state

Figure 30.7 Idealized ultra-violet spectrum of a simple aliphatic ketone.

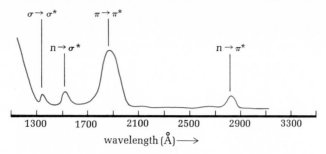

$$\sigma \rightarrow \sigma^*$$ $$\pi \rightarrow \pi^*$$

$$n \rightarrow \sigma^*$$ $$n \rightarrow \pi^*$$

| 1300 | 1700 | 2100 | 2500 | 2900 | 3300 |

wavelength (Å) $\longrightarrow$

bonds of compounds such as azulene or azobenzene, so that ultraviolet, rather than visible, spectra are obtained.

In other cases, excitations may occur in isolated rather than delocalized bonds. Figures 30.6 and 30.7 show an idealized series of transitions possible for a simple aliphatic ketone.

30.5 USES OF ULTRAVIOLET AND VISIBLE SPECTRA

Chromophore groups are electronic systems involved in reversible energy changes produced by absorbed light.

The visible and ultraviolet spectral regions usually supply only limited structural information. The appearance of intense ultraviolet absorptions does indicate a π-bond system in the molecule and, by comparison with spectra of known compounds, may give additional structural clues. Chromophore groups may be suggested by absorptions at characteristic wavelengths, but additional information is always needed before definitive structures can be drawn.

Quantitative uses of ultraviolet spectra are of more practical importance than are the qualitative uses for structural elucidation. Beer's law, equations (30.2) and (30.3), is an idealized relationship that can be used to determine a characteristic ε (Greek epsilon, molar absorptivity) value for a given compound, or when ε is known, the concentration of the compound in a solution. Table 30.1 gives some typical data.

$$A = \log \frac{I_0}{I} = kcb \qquad \text{Beer's law} \tag{30.2}$$

where A is the absorbance, I_0 is the intensity of the reference light beam, I is the intensity of light transmitted through the sample, k is the constant characteristic of the compound, c is concentration, and b is the path length of light through the sample.

Table 30.1 ULTRAVIOLET ABSORPTIONS FOR SOME COMMON SUBSTANCES

CLASS	EXAMPLE	λ_{max} (Å)[a]	ε_{max}
aliphatic aldehyde	CH_3CHO	2900	16
	CH_3CH_2CHO	2920	21
aliphatic ketone	CH_3COCH_3	2790	13
	⬡=O	2850	14
aliphatic ester	$CH_3CH_2COOCH_3$	2080	65
aliphatic amide	CH_3CONH_2	<2100	weak
aliphatic halide	$CH_3CH_2CH_2Br$	2080	300
	CH_3CH_2I	2580	400
alkene	$CH_3CH{=}CHCH_3$	<2000	—

Table 30.1 (Continued)

CLASS	EXAMPLE	λ_{max} (Å)[a]	ε_{max}
aromatic aldehyde	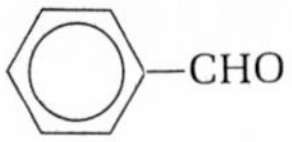 ⬡—CHO	3280; 2800; 2440	20; 1500; 15,000
aromatic ketone	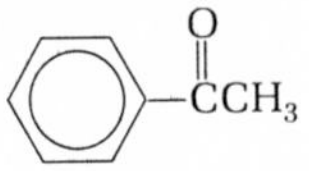 ⬡—CCH$_3$ (C=O)	3190; 2780; 2400	50; 1100; 13,000
	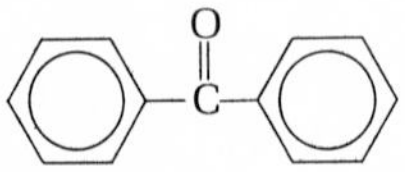 ⬡—C—⬡ (C=O)	3250; 2520	180; 20,000
aromatic hydrocarbon	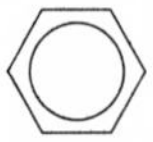 ⬡	2550	215
	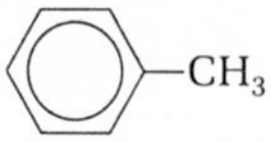 ⬡—CH$_3$	2600	300
	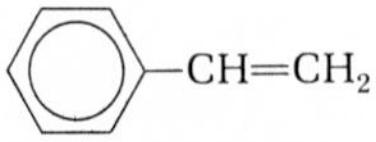 ⬡—CH=CH$_2$	2820; 2440	450; 12,000
carboxylic acid	CH_3COOH	2040	40
	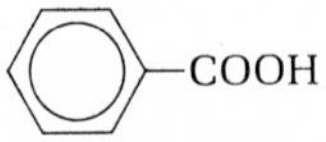 ⬡—COOH	2700; 2300	800; 10,000
phenol	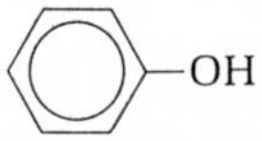 ⬡—OH	2700; 2100	1400; 6200

[a] Vacuum ultraviolet ($\lambda < 2000$ Å) may reveal useful data but requires special, very expensive equipment. Only the strongest absorptions in the near ultraviolet (2000–4000 Å) are recorded. Complete spectra may be more complex. Wavelengths are often reported in units of nanometers (nm; 1 nm = 10 Å). For some purposes *wave numbers* [1/λ (cm)] are preferred.

Beer's law is most commonly used, however, in the form

$$A = \varepsilon b C \tag{30.3}$$

where ε is the molar absorptivity, C is the concentration in moles per liter, and b is the path length of light in centimeters.

Example 1

A solution of pure phenol in ethanol had an absorbance of 0.83 at 270 nm, using a 1.00-cm cell. What was the concentration of phenol?

Solution

$A = \varepsilon b C$ [equation (30.3)]

From Table 38.1, ε at 270 nm (2700 Å) is 1400, so

$0.83 = 1400 \times C \times 1.00$

$$C = \frac{0.83}{1400} = 5.9 \times 10^{-4}\ M$$

Example 2

For a synthesis of acetophenone, a 39-g sample of benzene was treated with a slight excess of acetyl chloride, using aluminum chloride catalyst. After three hours, the reaction was stopped by the addition of water to destroy the residual acetyl chloride. The acetophenone and unreacted benzene were extracted by ether, and the aceto-phenone was recovered by fractional distillation. After purification by repeated distillation, a 41-g sample of pure acetophenone was recovered.

a. What was the percent yield for the overall process?

To determine the efficiency of the separation methods used, the reaction was repeated and after three hours the entire reaction mixture was diluted with sufficient 95 percent ethanol to form 500.0 ml of solution. A 0.100-ml aliquot of this homogene-ous solution was withdrawn and diluted to 100.0 ml. The absorbance of an aliquot of this solution in a 1.00-cm cell was measured to be 0.94 at 278 nm, a wavelength at which only the acetophenone contributes appreciably to the absorption.

b. What was the percent yield for the reaction as determined by this method?

c. What was the percent loss of acetophenone during the separation and purifica-tion procedures employed for the first process, assuming actual reaction yields were identical in both cases?

Solution

a. 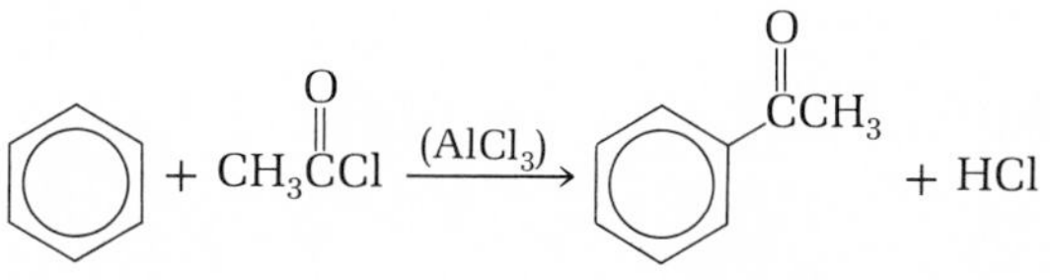
$$\text{C}_6\text{H}_6 + \text{CH}_3\overset{\text{O}}{\underset{\|}{\text{C}}}\text{Cl} \xrightarrow{\text{(AlCl}_3)} \text{(acetophenone)} + \text{HCl}$$

(C_6H_6) $\qquad\qquad\qquad\qquad (\text{C}_8\text{H}_8\text{O})$

$$\begin{array}{ccc} 39\text{ g} & & t \\ \text{C}_6\text{H}_6 + \cdots & \longrightarrow & \text{C}_8\text{H}_8\text{O} + \cdots \\ 78 & & 120 \end{array}$$

$$t = \frac{(120)(39\text{ g})}{78} = 60\text{ g}$$

$$\text{percent yield} = \frac{41\text{ g}}{60\text{ g}} \times 100\% = 68\%$$

b. $C_{\text{acetophenone}} = \dfrac{A}{\varepsilon b}$

From Table 30.1, $\varepsilon = 1100$ at 278 nm (2780 Å).

$$C_{\text{acetophenone}} = \frac{0.94}{1100 \times 1} = 8.5 \times 10^{-4}\text{ M}$$

$$\text{weight acetophenone formed} = \frac{8.5 \times 10^{-4}\text{ mole}}{\text{liter}} \times \underbrace{\frac{100\text{ ml}}{0.100\text{ ml}} \times \frac{0.500\text{ liter}}{1}}_{\text{(dilution factor)}} \times \frac{120\text{ g}}{1\text{ mole}}$$

$$= 51\text{ g}$$

$$\text{percent yield} = \frac{51\text{ g}}{60\text{ g}} \times 100\% = 85\%$$

c. weight acetophenone formed = 51 g
weight acetophenone recovered pure = 41 g
weight loss during work-up = 10 g

$$\text{percent loss} = \frac{10\text{ g}}{51\text{ g}} \times 100\% = {\sim}20\%$$

The principal use of ultraviolet spectroscopy is in quantitative studies. In terms of structural information, ultraviolet spectra may offer clues to the presence of certain chromophore groups, but infrared and nuclear magnetic resonance spectroscopy are much more powerful tools for structure elucidation.

30.6 PRINCIPLES OF INFRARED SPECTROSCOPY

Wave numbers 12,500 cm^{-1} (8000 Å) to 200 cm^{-1} (500,000 Å).

Even at absolute zero, molecules possess *vibrational* energy.

Infrared radiation, commonly thought of as "heat," covers the wavelength range from about 8000 to 500,000 Å. Most useful information concerning structure of organic molecules is obtained in the region 25,000 to 150,000 Å. Radiation in this region is of insufficient energy to induce electronic transitions in most organic molecules. What types of excitations, then, occur by absorption of infrared light?

Molecules are in constant motion above absolute zero. These motions may be classified in three general categories (Figure 30.8). Of these, translational motion is, for all practical purposes (Unit 9), a continuous function of the temperature of the system (remember $PV = nRT$?), but the other types are found to be characterized by energy discontinuities; that is, they are *quantized*. All covalent bonds are normally vibrating with characteristic frequencies dependent on the masses of the atoms in the bond and the force constant ("stretchiness") of the bond. Changes in vibrational (or rotational) motions may be caused by interparticle collisions or by absorption of light of frequency equal to the vibrational (or rotational) frequency involved. Such excitations do not change the frequency of the motion but may increase the amplitude (i.e., atoms may move farther apart).

Figure 30.8 Types of molecular motion. (a) Translation. (b) Rotation. (c) Vibration: left, symmetric stretching vibration in H$_2$O; center, asymmetric stretching vibration in H$_2$O; right, bending vibration in H$_2$O.

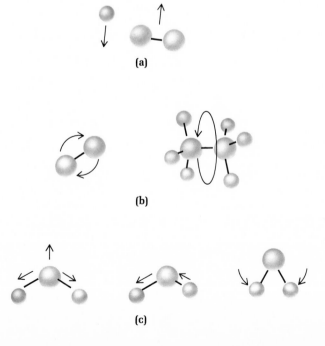

Figure 30.9 Vibrating dipole. (a) Frequency of the vibration is determined only by the masses of the two particles and the "force constant" characteristic of the "springlike" properties of the bond. (b) When the frequency of an applied alternating electrical field matches the vibrational frequency, the *amplitude* (stretching distance) is increased.

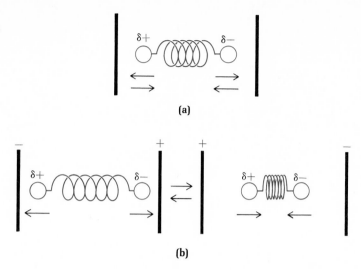

Only amplitude changes associated with changes in dipole moment are observed as infrared absorptions, since it appears that excitations result from the interaction of the oscillating electromagnetic field of the radiation with the alternating electric field of a changing dipole (Figure 30.9).

Since the frequency of vibration is a function of each kind of bond (atomic masses and force constant of bond), the frequencies of radiation absorbed by a compound will provide valuable clues to the types of bonds in the molecule. Infrared spectra are therefore very useful in the determination of structural formulas. Such spectra are usually quite complex (Figure 30.10) and generally can be used to determine a unique structure only when a spectrum is superimposable on the spectrum of a known comparison compound, or when sufficient additional data are available. Functional groups and types of C—H or C—C bonds may often be determined from infrared absorptions, but for complex compounds unambiguous interpretations are difficult.

Because both glass and quartz absorb infrared radiation, sample containers and optics of the infrared spectrophotometer differ from those of the ultraviolet and visible instrument. Sodium chloride, potassium bromide, cesium bromide, or silver iodide crystals are transparent in the infrared region, so such materials may be employed for prisms, other optical components, and sample containers, rather than glass or quartz.

Figure 30.10 Infrared spectrum of methyl butanoate. Note that absorptions are recorded either in wavelength (microns (μ) = 10^{-4} cm) or in wavenumbers [frequency = 1/wavelength(cm)]. The absorbance goes from top to bottom in typical infrared spectra, rather than from bottom to top as in most visible and ultraviolet spectra.

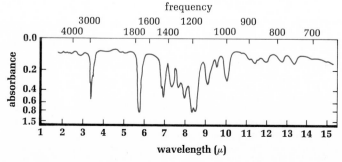

Table 30.2 INFRARED SPECTRA BAND ASSIGNMENTS (SELECTED REGIONS)

BAND (cm⁻¹)	ASSIGNMENTS
3650–3000	O—$\boxed{\text{H}}$ or N—$\boxed{\text{H}}$ (usually broad band)
3100–3000	aromatic C—$\boxed{\text{H}}$ (usually sharp band)
3040–3010	$\displaystyle \mathop{\text{C}}\limits={\text{C}}$ with $\boxed{\text{H}}$ (usually weak)
2970–2840	aliphatic C—$\boxed{\text{H}}$
2900–2700	C=O (usually two bands, 1 near 2720 cm⁻¹) with $\boxed{\text{H}}$

BAND (cm⁻¹)	ASSIGNMENTS	
1750–1735	aliphatic ester C=O	
1740–1720	aliphatic aldehyde C=O	
1730–1715	aromatic ester C=O	
1725–1705	aliphatic ketone C=O	A *strong* absorption in the region 1660–1750 should be interpreted as evidence for a *carbonyl* group (C=O).
1725–1700	aliphatic acid C=O	
1715–1695	aromatic aldehyde C=O	
1700–1680	aromatic acid C=O	
1700–1680	"mixed" ketone C=O (RĈAr)[a]	
1680–1620	alkene C=C (typically weak absorption)	
1670–1660	aromatic ketone (ArĈAr)[a]	
1600, 1500, 1450	aromatic C=C (usually three bands)	

[a] R = alkyl group
 Ar = aromatic ring

30.7 INTERPRETATION OF INFRARED SPECTRA

Spectra are interpreted by scanning portions of the recording of absorption versus wavelength (Table 30.2) to determine possible structural features. The following sequence is a generally useful procedure.

1. Look at the region 3600 to 2500 cm⁻¹ for bands characteristic of stretching of C—H, O—H, and N—H bonds of various types. List possibilities.
2. Look at the region 1750 to 1500 cm⁻¹ for bands characteristic of vibrations of carbon-carbon and carbon-oxygen double bonds of various types. Correct possibilities list from step 1.
3. (*Optional*) Look at region 1450 to 650 cm⁻¹ for bands confirming possibilities from step 2 (e.g., C—O, C—N, stretch, etc.). If spectra of known compounds are available, compare for identities. Step 3 is usually omitted except when comparison spectra are available because of the large variety of different bonds absorbing radiation in this region.

Absorptions in the region 1500 to 650 cm⁻¹ are not listed in this text. If you are interested in this region, consult the Suggested Reading at the end of the unit.

In the absence of other data or comparison spectra, an infrared spectrum can only provide some general structural clues. When combined with other information, such as molecular formulas or chemical analyses, the infrared spectrum may offer the definitive results needed for structural elucidation. The "fingerprint region" of the infrared spectrum (~1500–650 cm⁻¹) may be used for an exact identification when suitable comparison spectra are available. In general, infrared spectra vary somewhat with experimental conditions, so comparison spectra should be run under conditions as nearly identical as possible to those used for the unknown compound.

We shall be concerned primarily with the use of infrared spectroscopy to provide evidence on molecular structure. The Suggested Reading section at the end of this unit offers more details of the theory of infrared spectroscopy for those interested in pursuing this topic further.

The best way to learn to interpret infrared spectra is to follow some examples and then to practice with a series of spectra. The following section consists of a series of spectra of monofunctional compounds for study. They were prepared by Mr. James Weber of Tucson, Arizona.

Example 3

From the infrared spectrum shown, suggest a proper classification for the monofunctional compound.

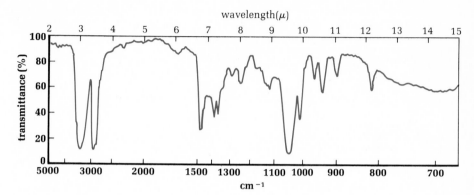

Solution

STEP 1

The broad absorption band around 3300 cm^{-1} suggests O—H or N—H, whereas that just below 3000 cm^{-1} suggests aliphatic C—H. Initial possibilities, then, include alcohols, amides, amines, carboxylic acids, or phenols (containing aliphatic C—H groups).

STEP 2

The *absence* of absorption in the region 1800 to 1450 cm^{-1} eliminates carbonyl groups and (probably) aromatic rings.

The compound, then, is probably an aliphatic alcohol or an aliphatic amine (primary or secondary, since tertiary amines have no N—H).

(The spectrum shown is actually that of 2-methyl-1-butanol.)

Example 4

The compound whose infrared spectrum is shown has the molecular formula C_7H_8O. Suggest plausible structural formulas.

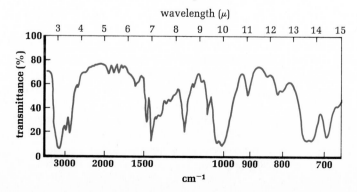

Solution

1. Since the compound contains no nitrogen, the broad band near 3300 cm^{-1} must be O—H absorption.
2. An aromatic ring is suggested by the C—H band at 3000 cm^{-1} and bands near 1600, 1500, and 1450 cm^{-1}.
3. Aliphatic C—H is indicated by the band near 2800 cm^{-1}.
4. Absence of strong absorption around 1700 cm^{-1} eliminates carbonyl groups. The only structures fitting the spectrum and molecular formula given are

(The spectrum is actually that of ⟨O⟩—CH$_2$OH, benzyl alcohol.)

30.8 PRINCIPLES OF NUCLEAR MAGNETIC RESONANCE (PROTON)

Other nuclei also produce nuclear magnetic resonance spectra, but not generally in a region to be confused with that of proton spectra. Information on organic molecules is usually obtained from proton spectra, although considerable recent attention has been focused on $^{13}_{6}$C spectra.

The nuclei of hydrogen atoms can be considered as spinning positive charges that produce magnetic fields. If samples containing such nuclei are placed in an external magnetic field, we can consider two possible orientations of the *nuclear* magnetic fields—either aligned with the external field or aligned opposing the magnetic field. The former condition should be more stable than the latter. Again we have quantized energy states, this time differing in energy by an amount like that associated with very long wavelength electromagnetic radiation—in the radiofrequency region. Given proper conditions, we should be able to place a sample in an external magnetic field and pass radiofrequency radiation through it until a wavelength is located that is absorbed by the sample, causing excitation of nuclei aligned with the external field to the "excited state" of opposing alignment.

In actual practice, it is more convenient to irradiate the sample with a specific radiofrequency while varying the applied magnetic field until a condition of "resonance" (energy difference between ground state and excited state equals energy of the radiofrequency employed) is attained.

A nuclear magnetic resonance (nmr) spectrometer is a sophisticated electronic instrument that uses radiofrequency radiation to induce and detect transitions between the nuclear spin orientations in a strong magnetic field. Basically, an nmr spectrometer consists of a transmitter that can apply radiofrequency radiation to a sample placed in a powerful magnetic field. The changes in the spin orientations produced by the applied radiation are detected by means of a radiofrequency receiver, amplified, and registered by means of a recording device. Inhomogeneities in the magnetic field tend to broaden the nmr signal. In addition to electronic correction, these effects can be reduced by rapidly spinning the sample (Figure 30.11).

30.9 CHEMICAL SHIFT

Each hydrogen nucleus in an organic compound is located in an environment of moving electrons (the covalent bond) that produce a magnetic field of their own. Thus each hydrogen nucleus (proton) is affected by the external magnetic field and by the magnetic field of the bond electrons. Since hydrogens in different types of chemical bonds will be in different electronic fields, dependent on such factors as the polarity of the bond or types of orbitals involved, the net magnetic influence on the proton will vary with the type of chemical environment in which it is localized. In order for a condition of resonance to be attained, the external magnetic field must be varied slightly for each different type of hydrogen so that the proper effective field

Table 30.3 APPROXIMATE[a] CHEMICAL SHIFTS FOR PROTON MAGNETIC RESONANCE
(R = alkyl, Ar = aromatic ring, X = halogen)

δ RANGE (RELATIVE TO TMS = 0)	TYPE OF PROTON
0.8–1.6	R—CH_3
1.2–1.8	R—CH_2—R
1.4–2.0	R_3CH
1.6–2.2	X—$\overset{\mid}{\underset{\mid}{C}}$—$CH_3$
1.9–2.5	Ar—CH_3
2.3–2.8	Ar—CH_2—R
1.8–3.0	$\overset{O}{\overset{\|}{-C}}$—$\overset{\mid}{C}$—H
2.6–3.2	Ar—$\overset{R}{\underset{R}{\overset{\mid}{\underset{\mid}{C}}}}$—H
3.0–3.6	X—CH_2—R
3.3–4.3	—O—$\overset{\mid}{\underset{\mid}{C}}$—H
3.6–4.6	X—$\overset{\mid}{\underset{\mid}{C}}$—H
4.5–6.0	$\overset{\diagup}{\diagdown}$C=C$\overset{\diagup}{\underset{H}{\diagdown}}$
5.8–8.6	Ar—H
1–5	—$\overset{\cdot\cdot}{N}$—H
1–6	R—O—H
4–12	Ar—O—H
9–10	$\overset{O}{\overset{\|}{-C}}$—H
10–12	$\overset{O}{\overset{\|}{-C}}$—O—H

[a] Nuclear magnetic resonance (nmr) absorption regions are *much* more variable than those associated with infrared absorption. As a result, chemical shift data are, except to the experienced operator, the least useful feature of the nmr spectrum. The *number* of different absorption bands, splitting patterns, and relative peak areas are more useful clues for the novice.

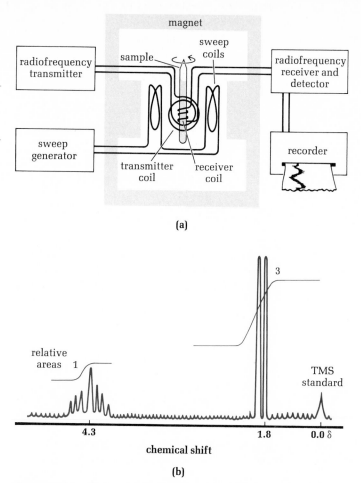

(a)

(b)

Figure 30.11 Nuclear magnetic resonance. (a) Instrument schematic based on drawing by Varian Associates, Palo Alto, California. (b) Nuclear magnetic resonance (nmr) spectrum for 2-bromo-propane, idealized.

(external ± electronic) is reached and transitions from absorbed radiofrequency energy become possible.

This variation of external field is referred to as *chemical shift* relative to the field needed for some specific reference-type hydrogen, usually the methyl hydrogens of tetramethylsilane (TMS), $(CH_3)_4Si$, (Table 30.3).

Information concerning the number of different types of hydrogens in a molecule can be obtained by finding the number of different external fields that will produce the resonance condition. The actual type of each hydrogen, that is, the type of chemical bond involved, can often be determined from the magnitude of the chemical shift associated with each resonance signal. The units in which chemical shift data are reported vary somewhat, but the most commonly used convention employs the so-called delta (δ) scale.

On the δ scale, the frequency (in cycles per second, *cps*) corresponding to energy absorption for excitation of the methyl hydrogen nuclei of tetramethylsilane is assigned a value of zero, as the reference standard. The *difference* between that frequency and the excitation frequency for some other hydrogen nucleus is defined as *chemical shift*. So that numerical values of chemical shift data will be characteristic of "hydrogen types" (that is, independent of the radiofrequency employed by a specific spectrometer), the δ value is obtained as the *ratio* of the chemical shift (in cps) to the frequency of the spectrometer used. Typical chemical shifts are small ($<$1000 cps) compared to the radiofrequency region employed (in the *megacycle*, 10^6 cps, range), so the δ values are reported in units of parts per million (ppm).

$$\delta = \frac{\text{observed chemical shift (cps)}}{\text{radiofrequency used (cps)}} \times \frac{10^6 \text{ parts}}{\text{million parts}}$$
$$\text{(conversion to ppm)}$$

For example, for a particular nmr signal observed at a frequency 90 cps different from that of the reference TMS signal, using a 60 megacycle (60×10^6 cps) spectrometer:

$$\delta = \frac{90 \text{ cps}}{60 \times 10^6 \text{ cps}} \times \frac{10^6 \text{ parts}}{\text{million parts}} = 1.5 \text{ ppm}$$

The interpretation of numerical chemical shift data requires a rather extensive background and experience in nmr spectrometry. For our purposes, it is generally sufficient to note the number of different principal nmr absorption regions as characteristic of the number of "different hydrogen types" (different chemical environments of hydrogen nuclei) in a molecule. In simple nmr spectra these absorption regions are clearly separated, but the spectra of more complex molecules may contain overlapping absorption regions difficult to interpret.

30.10 SPIN–SPIN SPLITTING Although the major influences on a proton are the applied external field and the electronic environment (chemical bond), the magnetic effects of protons bonded with nearby atoms in the molecule may also have an influence on the proton's nmr signal. The effect is transmitted through bonding electrons. This is, perhaps, best illustrated by a specific example.

Consider the case of ethyl chloride:

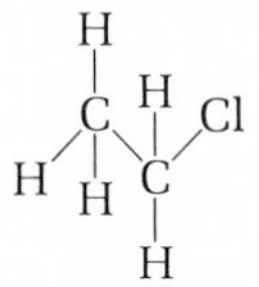

The methyl hydrogens are in an environment influenced by a carbon–carbon bond, while the methylene hydrogens are in an environment influenced by a carbon–carbon bond *and* a carbon–chlorine bond. We expect, therefore, two nmr signals, one corresponding to the methyl hydrogens and the other to the methylene hydrogens. Such is actually observed (CH_3 protons at δ 1.5 and CH_2 protons at δ 3.4).

These signals are observed to be closely spaced groups rather than single signals. These patterns can be explained by considering the effects of methylene protons on the methyl proton signal and methyl protons on the methylene proton signal.

Each methylene proton can have its magnetic field aligned either with the external field or opposed to it. The net magnetic effect of the two proton fields can then be of three types, both proton fields with the applied external field (increasing the effective field for methyl protons), both proton fields opposing the applied external field (decreasing the effective field for methyl protons), or one proton field with the external field and one opposed (no effect on effective field for methyl protons). Since the magnetic fields of these protons should be quite small compared to the large external magnetic field, only a small change should be apparent. The signal from the methyl protons appears as a "triplet," as indicated below:

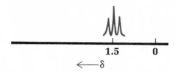

In a similar fashion, we can suggest that methyl protons could produce four different net effects on neighboring methylene protons, that is, three methyl proton fields with the external field, three opposed, two with and one opposed, or one with and two opposed. The nmr signal from the methylene protons then appears as a "quartet," as shown below:

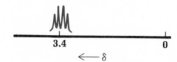

The spin–spin splitting patterns are thus useful in deducing the types of neighboring protons. Such splitting is generally observed only between nonequivalent neighboring protons (i.e., protons having different chemical shifts).

It is worth noting that the relative intensities of multiplet components can be correlated with the statistical probabilities of the proton fields affecting the signal. For the triplet produced by the methylene protons, for example, the central peak has twice the intensity of the "side" peaks since the "probabilities" are

(1 chance)   one side: both proton fields opposed to external
(1 chance)   other side: both proton fields with external
(2 chances)   central: proton A opposed and proton B *with* or proton B opposed and proton A with

The detailed interpretation of spin–spin splitting patterns can provide the experienced investigator with valuable structural information.

30.11 RELATIVE PROTON NUMBERS

The relative areas under signal peaks in an nmr spectrum are *proportional to* the relative numbers of protons responsible for the signals. This information, when used with the chemical shift and spin–spin splitting data, may reveal the relative numbers of hydrogens in the different chemical environments of the molecule. These areas are determined, with modern instrumentation, by an integration device.

Example 5

A compound of molecular formula C_2H_6O could be either ethyl alcohol, CH_3CH_2OH, or dimethyl ether, CH_3OCH_3. Deduce the structure of the compound from the following nmr spectrum.

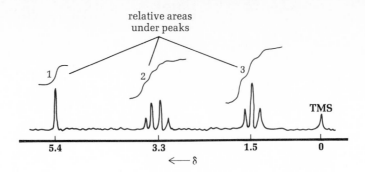

relative areas
under peaks

5.4   3.3   1.5   0

$\leftarrow \delta$

TMS

Solution

The alcohol should have three different types of protons: OH, CH_2, and CH_3. The ether should have only CH_3 protons. The chemical shift of 1.5 is characteristic for CH_3, that of 3.3 for CH_2 and that of 5.4 for OH.[1] (Note that the absence of splitting in the OH peak is caused by external chemical effects.) The splitting of the peak at δ 3.3 into a quartet is typical of CH_2 affected by adjacent CH_3 protons, while the triplet at δ 1.5 is typical of CH_3 affected by adjacent CH_2 protons. The relative peak areas of 1:2:3 are consistent with the relative numbers of hydrogens in ethyl alcohol—one OH proton, two CH_2 protons, and three CH_3 protons.

30.12 APPLICATIONS OF NUCLEAR MAGNETIC RESONANCE

The nmr spectrum of a compound provides several important structural clues:

1. The number of different types of hydrogen (from the number of different nmr signals).
2. The nature of the chemical environment for each type of hydrogen (from the magnitude of the chemical shift).
3. The nature of hydrogens on adjacent atoms (from spin–spin splitting patterns).
4. The relative numbers of each type of hydrogen in the molecule (from relative areas of signal peaks).

It is often possible to assign a structural formula for a simple molecule solely on the basis of an nmr spectrum. The spectra of more complex molecules are, of course, more difficult to interpret, and additional information is usually required.

The discussion in this unit is necessarily limited, but some practice with the spectra given in the following examples and exercises should be sufficient to give a general idea of how simple nmr spectra are interpreted.

Example 6

a. Draw all possible structural formulas for compounds having the molecular formula $C_4H_{10}O$.

b. Select the unique formula from the preceding set that would correspond to the following nmr spectrum:

---

[1] The position of an OH or NH nmr signal is quite dependent on the degree of hydrogen bonding in the sample. This is determined by such factors as temperature, the nature of the solvent used, and the concentration of the solution. These factors may cause as much as 4 ppm variation in the signal location.

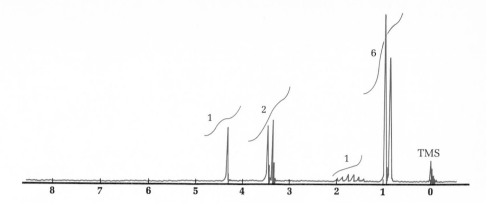

Solution

a. $CH_3CH_2CH_2CH_2OH$ $CH_3CH_2\underset{\overset{|}{OH}}{CH}CH_3$ $CH_3\underset{\overset{|}{CH_3}}{CH}CH_2OH$

(1)   (2)   (3)

$(CH_3)_3COH$ $CH_3OCH_2CH_2CH_3$ $CH_3OCH(CH_3)_2$
(4)   (5)   (6)

$CH_3CH_2OCH_2CH_3$
(7)

b. As a first step when alternative structures are being compared, simply count the number of absorption bands and the relative areas to eliminate structures that do not fit this information.

For more complex compounds, absorption bands may overlap, so the number of absorption peaks will indicate only the *minimum* number of different "types" of hydrogen.

The spectrum given has four bands, so there are four different "types" of hydrogen.

The relative peak areas are in the same ratio as that of the types of hydrogens, 1:2:1:6.

Only structures (2) and (3) fit this information. [Actually, the two methyl groups in structure (2) are nonidentical.]

Next, examine the spin–spin splitting patterns for further clues. The six-proton peak (two CH_3 groups) appears as a doublet, so both methyl groups are probably identical and attached to an atom having a single hydrogen. This information fits only structure (3).

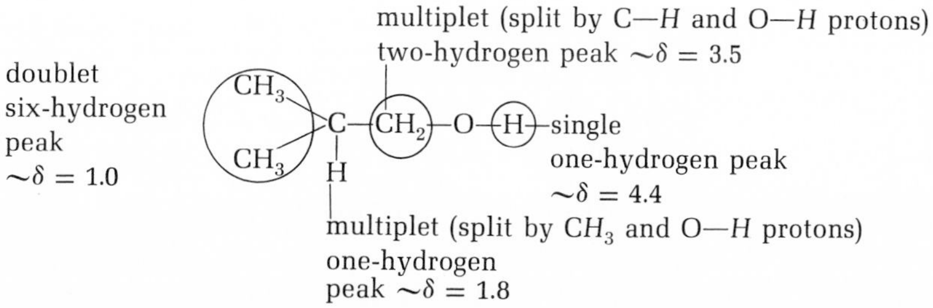

doublet
six-hydrogen
peak
$\sim\delta = 1.0$

multiplet (split by C—*H* and O—*H* protons)
two-hydrogen peak $\sim\delta = 3.5$

$\overset{CH_3}{\underset{CH_3}{\diagup}}C—(CH_2)—O—(H)$—single
$\overset{}{\underset{H}{|}}$ one-hydrogen peak
                                        $\sim\delta = 4.4$

multiplet (split by CH_3 and O—*H* protons)
one-hydrogen
peak $\sim\delta = 1.8$

Example 7

A compound having the molecular formula $C_7H_{12}O$ has an infrared spectrum suggesting that the compound is an aliphatic ketone. Assign a plausible structure consistent with the nmr spectrum given below:

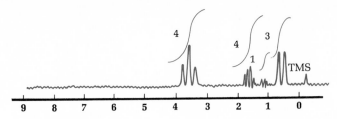

Solution

The molecular formula indicates two π bonds, two rings, or one π bond and one ring.

The infrared spectrum, indicating a carbonyl group, accounts for one π bond, thereby ruling out the second possibility above. The three-hydrogen doublet at $\delta = 0.9$ suggests

$$
\begin{array}{c}
| \\
-\!\!\overset{|}{\underset{|}{C}}\!\!-CH_3 \\
| \\
H
\end{array}
$$

and the four-hydrogen triplet at $\delta = 3.8$ suggests two identical methylene groups of the type

$$-CH_2-CH_2-$$

The multiplet four-hydrogen peak around $\delta = 1.8$ suggests two additional methylene groups.

When all the information is put together and considered along with possible structures, the only one found consistent with all data is

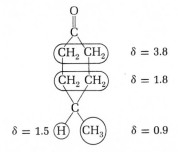

The compound, then, must be 4-methylcyclohexanone.

SUGGESTED READING

Akitt, J. W. 1973. *NMR and Chemistry*. Reading, Mass.: Addison-Wesley.

McLafferty, F. W. 1973. *Interpretation of Mass Spectra*. Reading, Mass.: Addison-Wesley.

Parikh, V. M. 1974. *Absorption Spectroscopy of Organic Molecules*. Reading, Mass.: Addison-Wesley.

Williams, D. H., and I. Fleming. 1973. *Spectroscopic Methods in Organic Chemistry* (2nd ed.). New York: McGraw-Hill.

(The best introductory discussion of spectroscopy is given by a series of articles by Jeff Davis in *Chemistry* magazine, beginning with the October, 1974, issue.)

MASS SPECTRA

The mass spectrometer is one of the most recent additions to the collection of research instruments used in structural determinations. Although J. J. Thomson used the prototype of the modern instrument as early as 1910, it was not until the late 1950s that organic chemists really began to appreciate the potential of this device for structural studies.

Charged particles in motion produce a magnetic field. Interaction of this field with an external magnetic field will result in a change in the motion of the particles. The radius of curvature of moving charged particles under the influence of an accelerating voltage and an external magnetic field is given by

$$r = \sqrt{\frac{2\,Vm}{H^2 q}}$$

where V is the accelerating voltage, H is the field strength of the external magnet, m is the mass of the ion, and q is the ionic charge.

Figure 30.12 Mass spectrometer.

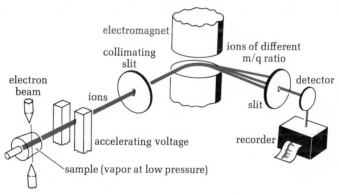

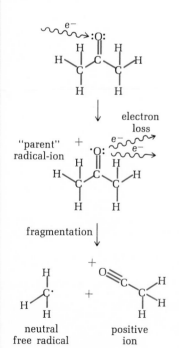

"parent" radical-ion

fragmentation

neutral free radical   positive ion

Figure 30.13 Formation of ions in the mass spectrometer. The examples shown are only a few of the possible ions that might be formed.

From this equation, one can see that a particular ion of characteristic mass and charge could be "focused" by adjustment of the accelerating voltage or the magnetic field strength (or both) in an instrument such as the mass spectrometer shown schematically in Figure 30.12.

An organic compound is injected into the sample region of the spectrometer as a vapor at low pressure. A beam of electrons is passed through the sample. When one of these electrons strikes an atom in a sample molecule, ionization may result by loss of one (usually) or more electrons. The resulting radical ion will have a positive charge and an unpaired electron. These ions are accelerated by a negative potential and can be directed toward a detector by the influence of a magnetic field. The ions are detected by the current produced in a sensitive galvanometer system. The parent ion might, indeed, fragment to form other ions, or neutral particles, or both (Figure 30.13). Additional positive ions may be directed toward the detector region by varying the accelerating voltage or the magnetic field (or both). The ions may be characterized by their masses (assuming q, V, and H are known, as well as the radius of curvature of the ion beam), and the intensity of an ion signal will be indicative of the stability of the ion and the probability of its formation.

Although mass spectral data are often extremely useful to the experienced investigator, they are not readily interpretable without a considerable background in the field and access to comparison data. For those interested in exploring this area, some of the books listed as Suggested Reading may prove a useful starting point.

In the period since 1960, mass spectral data have been accumulated on thousands of compounds. The suggestion that fragmentation patterns may be programmed for computerized interpretation, eventually leading to complete structural elucidation of any complex molecule, now seems a real possibility. At present, however, the accumulation of structural information from ultraviolet, infrared, nuclear magnetic resonance, and mass spectra are all useful pieces of the puzzle of molecular structure.

EXERCISES

1. Which of the following molecules would be expected to absorb light at the longest wavelength? Explain.
   acetophenone; benzophenone; ethylbenzene

2. What concentration of styrene (Table 30.1) would be required for a solution to be used in a 1.00-cm cell so that the observed absorbance is
   a. 0.75 at 244 nm; **b.** 0.75 at 282 nm

3. A 0.25-g sample of radioactive benzaldehyde (containing carbon-14) was treated with a slight excess of lithium aluminum hydride. After a two-hour reaction period, the mixture was poured carefully into 95 percent ethanol and filtered to remove insoluble aluminum salts. The filtrate was diluted to 500.0 ml with 95 percent ethanol and an aliquot was placed in a 1.00-cm spectrophotometer cell. The absorbance at 280 nm was measured to be 0.91. What was the percent yield for the reduction, assuming no competitive reactions and contribution to absorption at 280 nm by no species other than unreacted benzaldehyde?

4. **a.** Suggest appropriate classification possibilities for the compound with the infrared spectrum below.

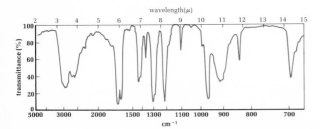

b. The compound with the infrared spectrum below has the molecular formula C_9H_{12}. Give IUPAC names for all possible compounds consistent with this formula and spectrum.

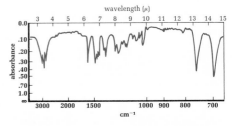

5. a. A compound has the molecular formula $C_4H_{10}O$ and the nmr spectrum below. What is the only possible structure for this compound?

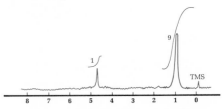

b. A compound having the molecular formula $C_5H_{10}O$ and the nmr spectrum below contains a carbonyl group. What is the IUPAC name of the compound?

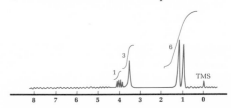

c. *Pheromones* are chemical communicating agents. The common honeybee uses several such compounds, one of which has been shown to be an alarm pheromone having the molecular formula $C_7H_{14}O_2$. The infrared spectrum of this compound shows absorption bands characteristic of aliphatic C—H and ester carbonyl. The nmr spectrum is shown below. Suggest a plausible structure for this pheromone and outline any additional investigations you feel necessary for complete structural proof.

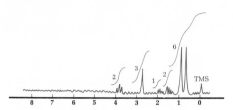

PRACTICE FOR PROFICIENCY

Objective (a)

1. What concentration of acetophenone (Table 30.1) would be required for a solution to be used in a 1.00-cm cell so that the observed absorbance is
   a. 0.80 at 319 nm
   b. 0.80 at 240 nm
2. What concentration of benzoic acid would be required for a solution to be used in a 1.00-cm cell so that the observed absorbance is 0.72 at 230 nm? [For benzoic acid: 270 nm ($\varepsilon = 800$), 230 nm ($\varepsilon = 10{,}000$)]
3. What concentration of phenol would be required for a solution to be used in a 1.00-cm cell so that the observed absorbance is 0.84 at 270 nm? [For phenol: 210 nm ($\varepsilon = 6200$), 270 nm ($\varepsilon = 1400$)]
4. What concentration of benzophenone would be required for a solution to be used in a 1.00-cm cell so that the observed absorbance is 0.54 at 252 nm? [For benzophenone: 325 nm ($\varepsilon = 180$), 252 nm ($\varepsilon = 20{,}000$).]
5. What concentration of benzophenone would be required for a solution to be used in a 1.00 cm cell so that the observed absorbance is 0.54 at 325 nm? [For benzophenone: 325 nm ($\varepsilon = 180$), 252 nm ($\varepsilon = 20{,}000$).]
6. In a study of phenylketonuria therapy, the serum tyrosine level was measured by determination of the absorbance of tyrosine solutions at 280 mμ. If a particular solution in a 1.00-cm cell had an absorbance of 0.95 at 280 mμ, what was the tyrosine concentration of the solution, in milligrams per liter?

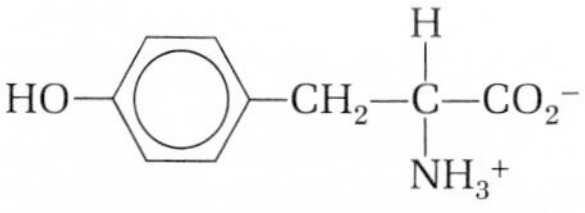

HO—⟨⟩—CH_2—$\overset{\overset{\text{H}}{|}}{\underset{\underset{NH_3^+}{|}}{C}}$—$CO_2^-$

(tyrosine, $\varepsilon = 12{,}000$ at 280 mμ)

7. A 53-g sample of benzaldehyde was dissolved in 250 ml of tetrahydrofuran and treated with lithium aluminum hydride. At the end of one hour a 1.00-ml aliquot was withdrawn through a filter and diluted to 50 ml with 95% ethanol. To determine the percentage reduced, the ethanol solution was used to fill a 1.00-cm cell and the absorbance was measured at 328 mμ. If the absorbance was 0.24, what percentage of original benzaldehyde had been reduced. (For benzaldehyde, $\varepsilon = 20$ at 328 mμ. The reduction product, benzyl alcohol, has no significant absorption at 328 mμ.)

8. A 23-g sample of toluene was treated with an excess of hot alkaline potassium permanganate. After a three-hour mixing period, the reaction mixture was acidified and extracted with 100 ml of ether to remove the benzoic acid formed. To check on the efficiency of extraction, a 0.10-ml sample of the ether solution was diluted to 50.0 ml with 95 percent ethanol. If the resulting solution, used to fill a 1.00-cm cell, had an absorbance of 2.00 at 270 mμ, what was the percentage efficiency of the process? (At 270 mμ, ε for benzoic acid is 800. Toluene has no significant absorption at this wavelength.)

Objective (b)

9. Which of the following molecules would be expected to absorb light at the longest wavelength? Explain.
   p-nitrobenzaldehyde; 4-nitrocyclohexanone; p-xylene
10. Only one of the following compounds is colored (orange). Identify the colored compound and explain your choice.

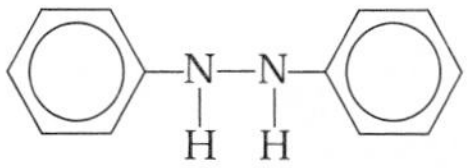

⟨⟩—$\overset{\overset{}{|}}{\underset{\underset{H}{|}}{N}}$—$\overset{\overset{}{|}}{\underset{\underset{H}{|}}{N}}$—⟨⟩   ⟨⟩—N=N—⟨⟩

(hydrazobenzene) 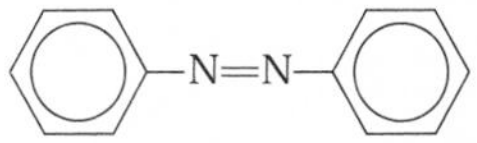   (azobenzene)

CH_3—N=N—CH_3

(azomethane)

11. Could ultraviolet spectroscopy probably be used to determine the percentage of
   a. toluene in a mixture of methylcyclohexane and toluene?
   b. o-xylene in a mixture of o-xylene and p-xylene?
   c. benzaldehyde in a mixture of benzaldehyde and acetaldehyde?
   Explain your conclusions.

Objective (c)

12. Select the compound most consistent with the infrared spectrum given.

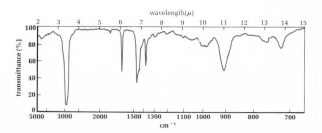

a. $CH_3\overset{OH}{\underset{OH}{CH}}CHCH_3$ b. $(CH_3)_2CHCO_2H$

c. $(CH_3)_2CH\overset{O}{C}OCH_3$ d. alkene structure e. CH_3 ... $C=O$

13. Select the compound most consistent with the infrared spectrum given.

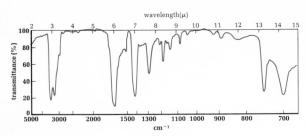

a. $CH_3O\overset{O}{C}CH(CH_3)_2$ b. $(CH_3)_3CHCH_2CO_2H$

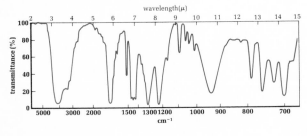

c. ; d. ; e.

14. Select the compound most consistent with the infrared spectrum given.

a. $\text{—}CH_2CH_2OH$ b. $\text{—}OCH_2CH_3$

c. OH / OH d. $\text{—}CH_2CO_2H$

e.

15. Select the compound most consistent with the infrared spectrum given.

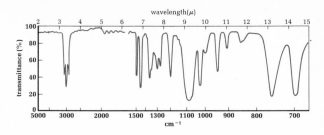

a. OH / CH₃ b. $\overset{O}{C}CH_2CH_3$

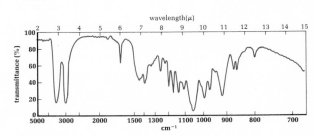

c. CH_2OCH_3 d. CH_3 / CH_3 e. CO_2H

16. Which compound is most consistent with the infrared spectrum?

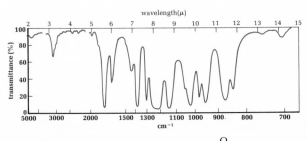

a. OCH_3 b. OH

c. ketone d. $N(CH_3)_2$ e. CO_2H

17. Which compound is most consistent with the infrared spectrum?

a. $H_2C=CHCH_2OH$ b. $H_2C=CHCH_2O\overset{O}{C}CH_3$

c. $H_2C=CH_2CH_3$ **d.** $HOCH_2CH_2O\overset{\displaystyle O}{\overset{\displaystyle \|}{C}}CH_3$

e. 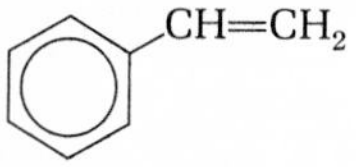 $\overset{\displaystyle CH=CH_2}{}$

a. $(CH_3)_2CH\overset{\displaystyle O}{\overset{\displaystyle \|}{C}}NH_2$

b. $(CH_3)_2CHCH_2OCH_2CH(CH_3)_2$

c. 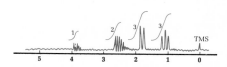—$CH_2\overset{\displaystyle O}{\overset{\displaystyle \|}{C}}OH$

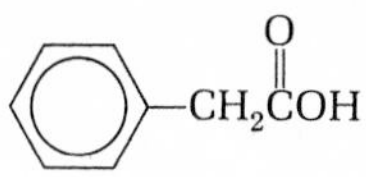

d. $(CH_3)_2CHCH_2\overset{\displaystyle O}{\overset{\displaystyle \|}{C}}CH_2CH(CH_3)_2$

e. (structure: benzene ring with two CH_3 groups and Cl)

Objectives (c) and (d)

18. Which structure is consistent with both the infrared (upper) and nmr (lower) spectra given?

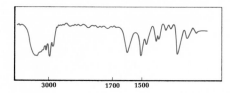

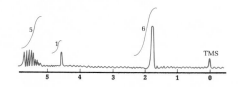

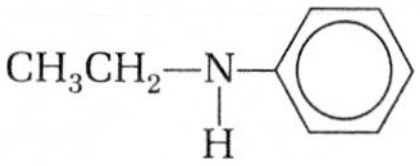

a. 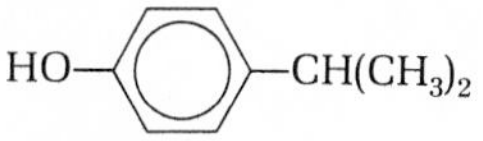 HO—(ring)—$CH(CH_3)_2$ **b.** 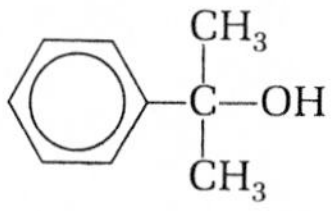 (ring)—$\overset{\displaystyle CH_3}{\underset{\displaystyle CH_3}{C}}$—$OH$

c. CH_3CH_2—$\overset{\displaystyle N}{\underset{\displaystyle H}{}}$—(ring)   **d.** 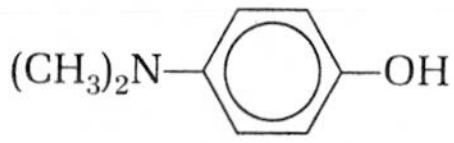 $(CH_3)_2N$—(ring)—OH

e. 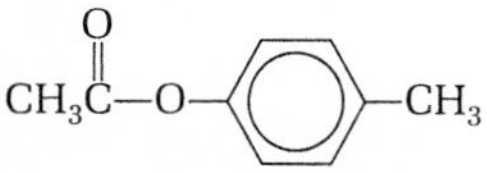 $CH_3\overset{\displaystyle O}{\overset{\displaystyle \|}{C}}$—$O$—(ring)—$CH_3$

19. Which structure is consistent with both the infrared (upper) and nmr (lower) spectra given?

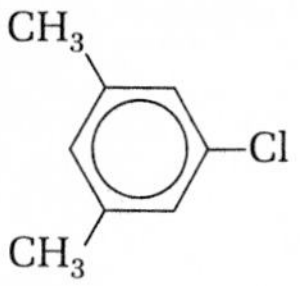

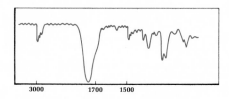

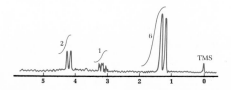

20. Which structure is consistent with both the infrared (upper) and nmr (lower) spectra given?

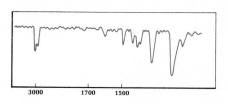

a. $CH_3CH_2\overset{\displaystyle H}{\underset{\displaystyle CH_3}{C}}OH$

b. $CH_3CH_2\overset{\displaystyle H}{\underset{\displaystyle CH_3}{C}}$—$O$—$\overset{\displaystyle H}{\underset{\displaystyle CH_3}{C}}CH_2CH_3$

c. $(CH_3)_2CHCH_2Cl$

d. $CH_3CH_2\overset{\displaystyle O}{\overset{\displaystyle \|}{C}}\underset{\displaystyle H}{N}CH_3$

e. 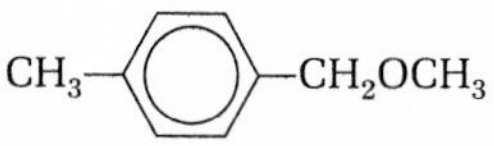 CH_3—(ring)—CH_2OCH_3

21. Suggest a plausible structure for an organic compound, $C_4H_8O_2$, for which the following spectral data are reported.
Ultraviolet: no detectable absorption above 250 nm
Principal infrared absorptions (3600 − 1600 cm^{-1}): 2960, 2900, 1745 cm^{-1}

Nuclear magnetic resonance (nmr) spectrum:

CHEMICAL SHIFT (δ, ppm)	PATTERN	RELATIVE AREA
1.1	triplet	3
2.3	quartet	2
3.4	singlet	3

Unit 30 Self-Test

1. Which of the following molecules would be expected to absorb light at the longest wavelength?
nitrobenzene; nitromethane; toluene

2. What concentration of benzoic acid (λ_{max} 270 nm, ε_{max} 800) is required for a solution to be used in a 1.00-cm cell so that the observed absorbance is 0.80 at 270 nm?

3. A 2.50-g sample of iodoethane (λ_{max} 258 nm, ε_{max} 400) containing carbon-14 was boiled for two hours with excess sodium ethoxide in pure ethanol in an attempted synthesis of "labeled" diethyl ether. At the end of that time the reaction was stopped by adding to the mixture a 5-percent solution of hydrogen chloride in ethanol until all ethoxide was neutralized. The resulting solution was diluted with ethanol to 500 ml, and an aliquot was placed in a 1.00-cm cell for analysis. The absorbance observed at 258 nm was 0.88. Assuming that iodoethane was the only component with significant absorption near this wavelength, what was the maximum yield of diethyl ether that could be expected? ($C_2H_5I + C_2H_5O^- \longrightarrow C_2H_5OC_2H_5 + I^-$)

4. a. Suggest possible classifications for the compound corresponding to the infrared spectrum below.

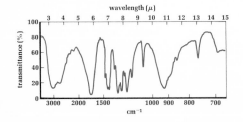

b. The compound whose infrared spectrum is shown below is monofunctional and has the molecular formula $C_8H_8O_2$. Draw four possible structures consistent with this formula and spectrum.

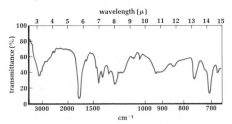

5. a. The compound whose nmr spectrum is shown below has the molecular formula $C_6H_{14}O$. What is its structure?

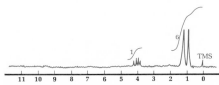

b. The compound whose nmr spectrum is shown below has the molecular formula $C_5H_{10}O_2$. A dilute solution of the compound in water has a pH less than 7. What is the IUPAC name of the compound?

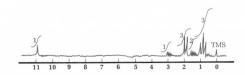

1. Select the best classification for

[structure]

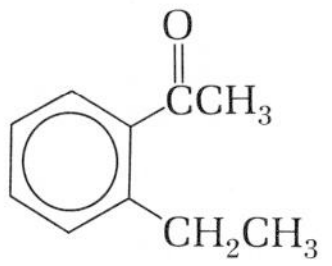

   a. aldehyde; **b.** carboxylic acid; **c.** ester; **d.** phenol; **e.** ketone

2. Select the best classification for

[structure]

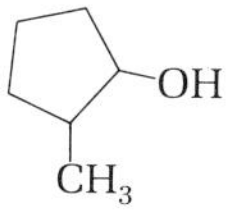

   a. phenol;   **b.** aldehyde;   **c.** alcohol;   **d.** carboxylic   acid;
   e. hemi-acetal

3. Select the best classification for

[structure]

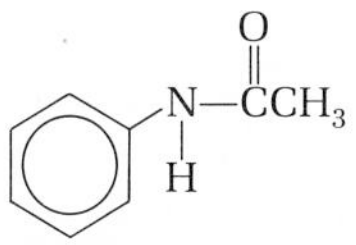

   a. amide; **b.** amine; **c.** ester; **d.** ketone; **e.** aniline

4. According to the stereochemical theory of odor, 1,4-dimethoxy-*cis*-2-butene should have a "pepperminty" odor. The structure of this compound may be represented as:

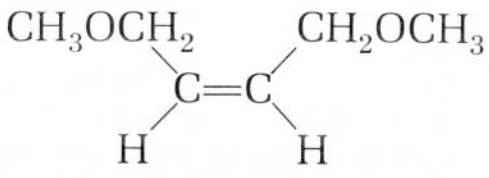

$$CH_3OCH_2 \quad CH_2OCH_3$$
$$\diagdown C=C \diagup$$
$$H \qquad H$$

Which of the following represents a geometric isomer of 1,4-dimethoxy-*cis*-2-butene?

a.

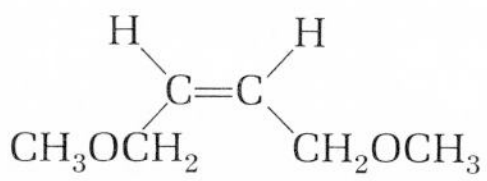

$$H \qquad H$$
$$\diagdown C=C \diagup$$
$$CH_3OCH_2 \quad CH_2OCH_3$$

b.

$$H \qquad CH_2OCH_3$$
$$\diagdown C=C \diagup$$
$$H \qquad CH_2OCH_3$$

c.

$$CH_3OCH_2 \quad H$$
$$\diagdown C=C \diagup$$
$$H \qquad CH_2OCH_3$$

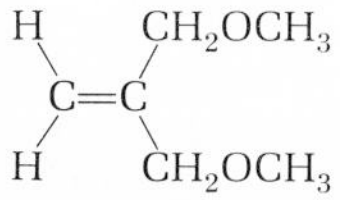

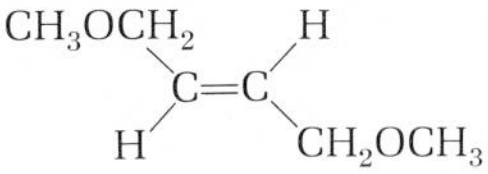

d.

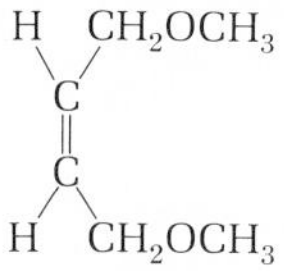

$$H \quad CH_2OCH_3$$
$$\diagdown C \diagup$$
$$\diagup C \diagdown$$
$$H \quad CH_2OCH_3$$

e.

$$HOCH_2CH_2 \quad H$$
$$\diagdown C \diagup$$
$$\diagup C \diagdown$$
$$H \quad CH_2CH_2OH$$

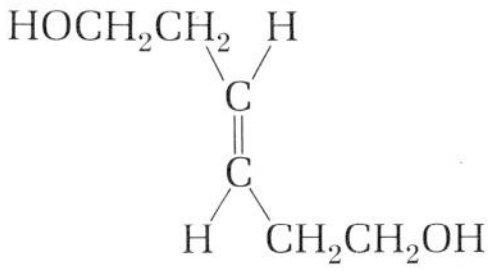

5. The compound 2-heptanone, $CH_3(CH_2)_4CCH_3$, has been identified as an ant alarm pheromone. Which of the following is a ketone *isomeric* with 2-heptanone?

   a. $CH_3C(CH_2)_4CH_3$ **b.** $CH_3(CH_2)_4COCH_3$

   c. 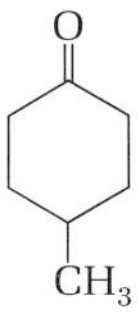 [structure with CH_3]   **d.** $CH_3(CH_2)_2C(CH_2)_2CH_3$

   e. $CH_3(CH_2)_5CH$

6. Select the correct IUPAC name for

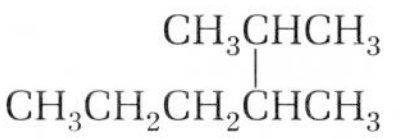

$$CH_3CHCH_3$$
$$\mid$$
$$CH_3CH_2CH_2CHCH_3$$

   a. 2-isopropylpentane;   **b.** 2,3-dimethylhexane;   **c.** 4,5-dimethylhexane; **d.** 1,1,2,4-tetramethylbutane; **e.** none of these

7. How many different *alkenes* have the molecular formula C_4H_8?
   a. 2; **b.** 3; **c.** 4; **d.** 5; **e.** more than 5

8. How many different aldehydes have the molecular formula $C_5H_{10}O$?

a. 2; b. 3; c. 4; d. 5; e. more than 5

9. How many different halides containing a benzene ring have the molecular formula C_7H_7Cl?
a. 2; b. 3; c. 4; d. 5; e. more than 5

10. Select the best name for

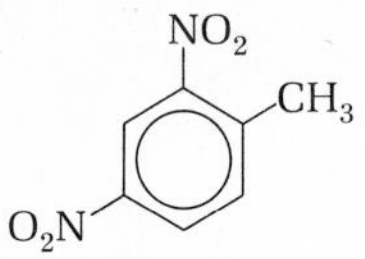

a. 2,4-dinitrotoluene; b. *m*-dinitrotoluene; c. *o*-nitroxylene; d. *p*-nitrotoluidine; e. none of these

11. Select the best name for

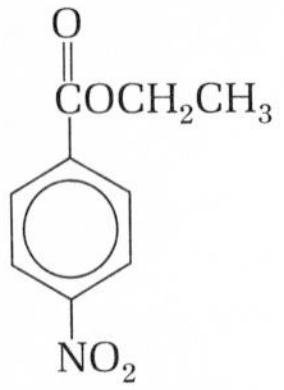

a. *p*-nitroacetophenone; b. *o*-ethyltoluidine; c. methyl *m*-nitrobenzylate; d. ethyl *p*-aminobenzoate; e. none of these

12. Select the best name for

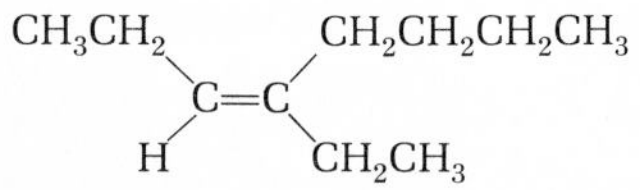

a. 4-ethyl-*cis*-3-octene; b. 4-ethyl-*trans*-3-octene; c. 4-butyl-*cis*-3-hexene; d. 5-ethyl-*trans*-5-octene; e. none of these

13. Select the best name for

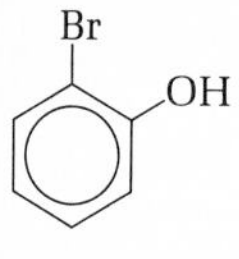

a. *o*-bromophenol; b. *o*-bromocresol; c. *m*-bromoaniline; d. *p*-hydroxybromobenzene; e. none of these

14. Which probably has the lowest boiling point at 1.00 atm pressure?

a. 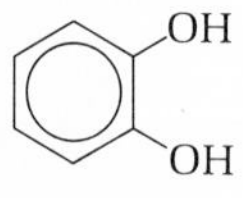

b. 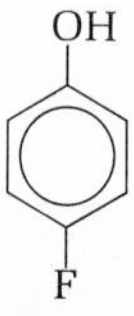 c. 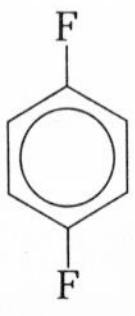 d. 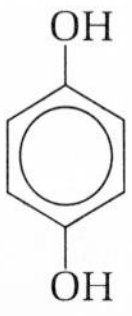 e. 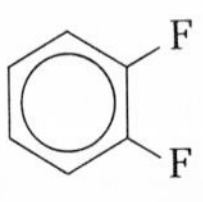

15. Which probably has the lowest boiling point at 1.00 atm pressure?

a. 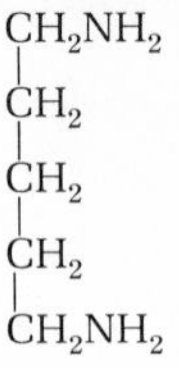 b. 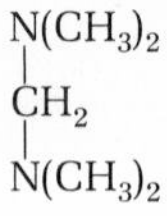

c. 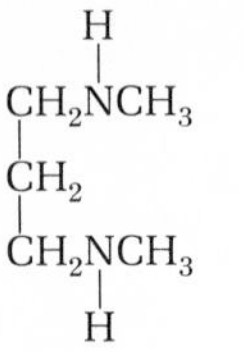 d. 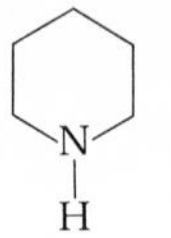 e. 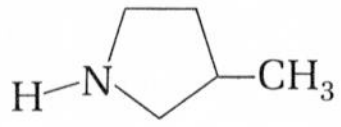

16. Which of the following is a pair of optical isomers?

a. 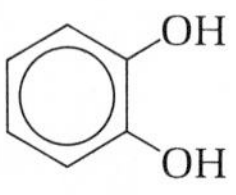 and 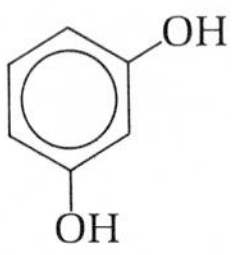

b. 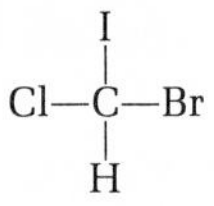 and 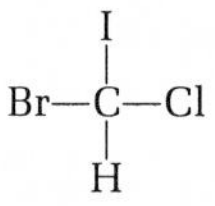

c. 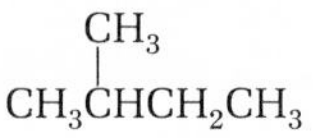 and 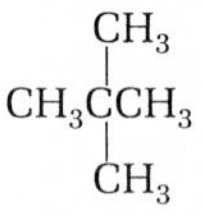

d. 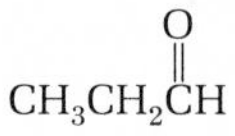 and 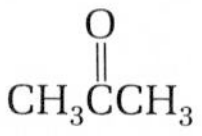

e. 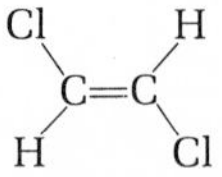 and 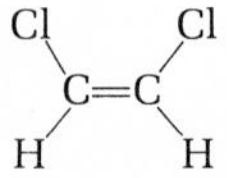

17. Which of the following could exist as a pair of *optical* isomers?
a. 1-butanol; b. 2-butanol; c. 2-methyl-2-propanol; d. 2-methyl-1-propanol; e. none of these

18. Which compound is least likely to react with sodium hydroxide?

a. 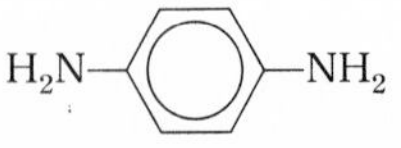 b. 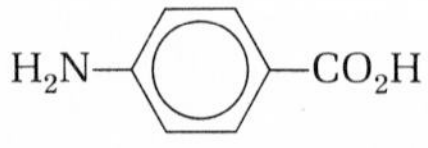

c. 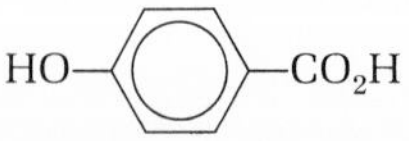 d. 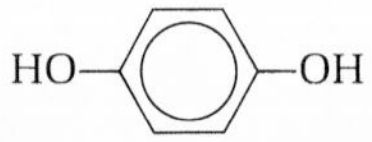

e. 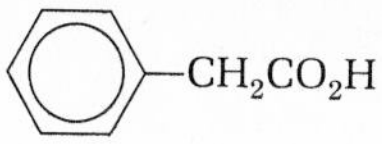 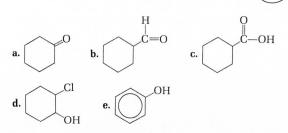 —CH₂CO₂H

19. Which compound is least likely to react with acetyl chloride?

a. 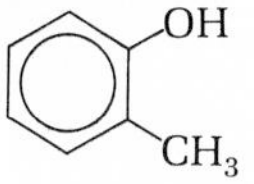 (benzene ring with OH and CH₃)   b. 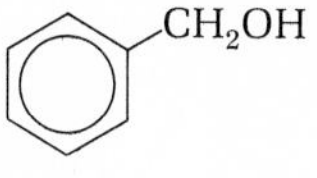 (benzene ring with CH₂OH)

c. 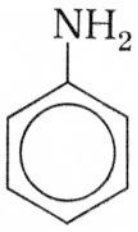 (benzene ring with NH₂)   d.  (benzene ring with OCH₃)   e. 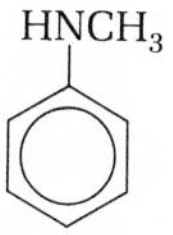 (benzene ring with HNCH₃)

20. Which compound is least likely to react with acetic acid?

a. CH₃CH₂—N—CH₂CH₃
          |
          CH₂CH₃

b. CH₃CH₂OH

c. CH₃CH₂OCH₂CH₃

d. 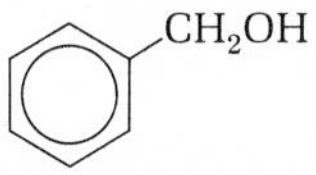 (benzene ring with CH₂OH)

e. 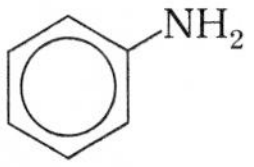 (benzene ring with NH₂)

21. Select the correct structural formula for compound V.

propanal $\xrightarrow{Cr_2O_7{}^{2-}}$ Cpd. I $\xrightleftharpoons{CH_3OH}$ H₂O + (Cpd. V)

a. CH₃OCH₂CH₂CH₂OH

b. CH₃OCH₂CHCH₃
                    |
                    OH

c. CH₃OCCH₂CH₂OH
        ‖
        O

d. CH₃CH₂COCH₃
              ‖
              O

e. CH₃CH₂CCH₃
            ‖
            O

22. Select the correct structural formula for compound K.

chlorocyclohexane $\xrightarrow{OH^-}$ Cl⁻ + Cpd. I $\xrightarrow{Cr_2O_7{}^{2-}}$ (Cpd. K)

a. (cyclohexane with =O)

b. (cyclohexane with CH=O)

c. (cyclohexane with C—OH, ‖O)

d. (cyclohexane with Cl and OH)

e. (benzene ring with OH)

23. Which compound is most consistent with the infrared spectrum?

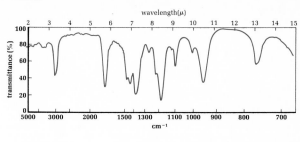

wavelength(μ)

transmittance (%)

cm⁻¹

a. (cyclohexanone)   b. (phenol)

c. (tetrahydropyran)   d. CH₃CNH₂ (with ‖O)   e. (toluene, CH₃)

24. Which compound is most consistent with the infrared spectrum?

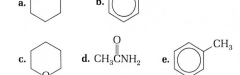

wavelength(μ)

transmittance (%)

cm⁻¹

a. (benzaldehyde, CH=O)   b. (benzoic acid, CO₂H)

c. (benzyl alcohol, CH₂OH)   d. (cyclohexanone, =O)   e. (N-methylpiperidine, N—CH₃)

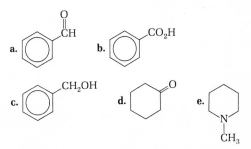

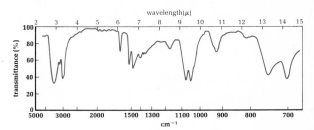

25. Which compound is consistent with both the infrared (upper) and nmr (lower) spectra given?

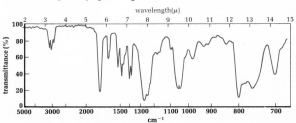

wavelength(μ)

transmittance (%)

cm⁻¹

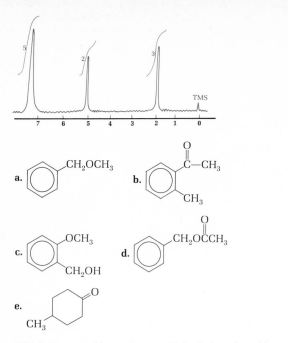

a. benzyl methyl ether CH_2OCH_3

b. o-methylacetophenone (ring with $\overset{O}{\underset{\|}{C}}-CH_3$ and CH_3)

c. (ring with OCH_3 and CH_2OH)

d. (ring with $CH_2O\overset{O}{\underset{\|}{C}}CH_3$)

e. (cyclohexanone ring with CH_3 and $=O$)

26. Which compound is consistent with both the infrared (upper) and nmr (lower) spectra given?

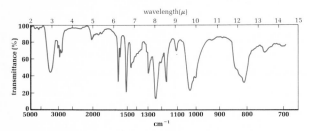

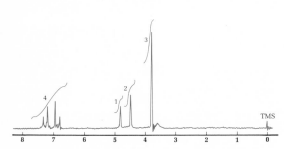

a. (ring with CH_2CO_2H and CH_3)

b. (ring with $\overset{O}{\underset{\|}{}}CH_2\overset{O}{\underset{}{C}}H$ and CH_3)

c. (ring with CH_2OH and OCH_3)

d. (ring with CO_2H and $OCCH_3$, with $=O$)

e. (ring with CH_2CH_2OH)

27. An aqueous solution of an organic acid had a pH of 3.60. A 25.0-ml aliquot of the solution was neutralized by 19.6 ml of 0.125 M sodium hydroxide. The ionization constant of this acid is approximately
a. 1.9×10^{-8};   **b.** 3.2×10^{-9};   **c.** 6.4×10^{-7};   **d.** 4.7×10^{-6};
e. 2.6×10^{-4}

28. What concentration of phenol would be required for a solution to be used in a 1.00-cm cell so that the observed absorbance is 0.84 at 210 nm? [For phenol: 210 nm ($\varepsilon = 6200$), 270 nm ($\varepsilon = 1400$)]
a. 6.0×10^{-4} M;   **b.** 4.0×10^{-3} M;   **c.** 1.4×10^{-4} M;
d. 3.1×10^{-3} M; **e.** 2.1×10^{-5} M

29. A 23-g sample of toluene was treated with an excess of hot alkaline potassium permanganate. After a three-hour mixing period, the reaction mixture was acidified and extracted with 100 ml of ether to remove the benzoic acid formed. To check on the efficiency of extraction, a 0.10-ml sample of the ether solution was diluted to 50.0 ml with 95 percent ethanol. If the resulting solution, used to fill a 1.00-cm cell, has an absorbance of 0.60 at 270 mμ, what was the percentage efficiency of the process? (At 270 mμ, ε for benzoic acid is 800. Toluene has no significant absorption at this wavelength.)
a. 67%; **b.** 45%; **c.** 33%; **d.** 22%; **e.** 15%

30. A commercial insecticide has the empirical formula C_3H_2Cl. Its vapor density (corrected to STP) is approximately 6.7 g liter^{-1}. What is the molecular formula of the insecticide?

Excursion 4

Isolation and purification

The folklore of primitive societies is rich with ideas of plant sources for the natural product chemist. From the deadly poisons of pygmy darts to the hallucinatory drugs of secret religious rites, chemists have sought for the intricate molecules of the alkaloids responsible for the biological effects of these exotic materials (Figure E4.1).

The chemist seeks to isolate the biologically active compounds from interesting natural sources, to determine their structures, to make them synthetically—to free man from dependence on the natural sources, and, often, to modify their structures in order to improve their beneficial properties.

It is rare that a plant will produce a single active chemical substance. More often "families" of similar compounds are formed, as in the opium alkaloids. The chemist's first task is to devise some way of isolating the particular compound of interest in a pure form. A few years ago this was an exceedingly tedious and laborious task. It is still not simple, but modern techniques of chromatography have made it much easier, especially when only small amounts of the compound are available.

Isolation and purification procedures can be planned much more easily if some aspects of the compound's structure are known. Alkaloids, for example, are complex organic amines, so the chemist can usually count on their ability to form salts with acids as one possible step in a purification method. In many cases trial-and-error approaches are required to perfect most efficient separation schemes.

Once the active compound is available in pure form, various methods can be used to determine its structure. Many of these are discussed in later sections. The ultimate step is an unambiguous synthesis (Excursion 5), not only to confirm the

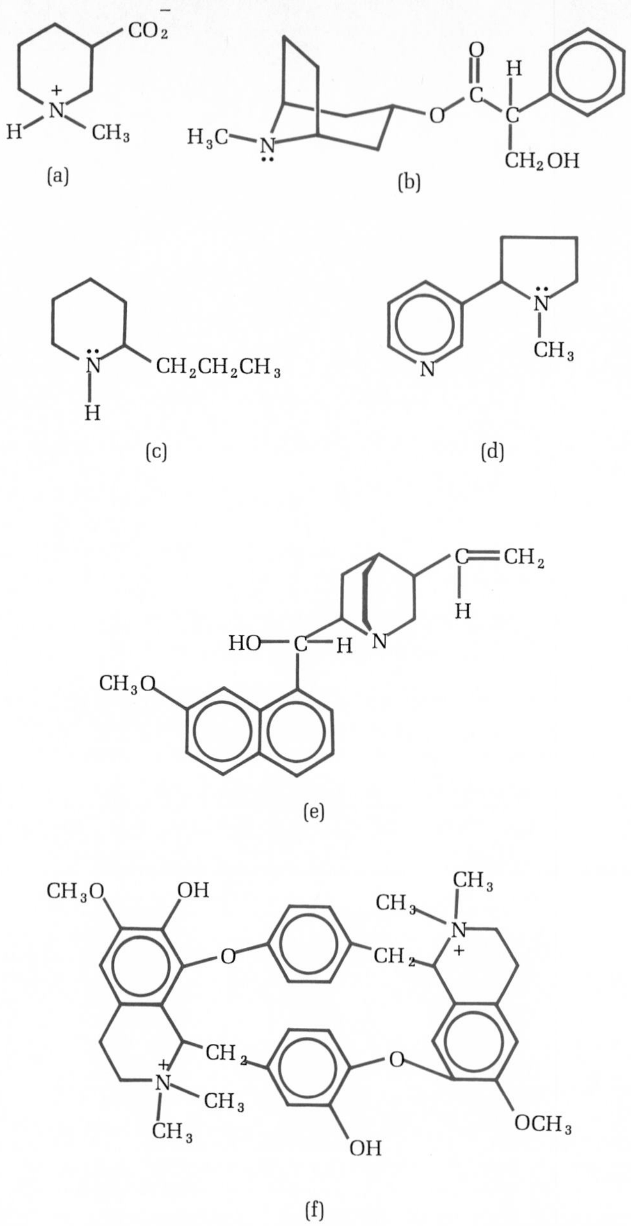

Figure E4.1 Some active compounds from plant sources. (a) Arecaidine, from betel nut (familiar from the song "Bloody Mary" in the Rodgers and Hammerstein musical, South Pacific); chewed as a mild stimulant in areas near the Indian Ocean. (b) Atropine, from Atropa belladonna ("deadly nightshade"); used to dilate the pupil of the eye. (c) Coniine, from hemlock; the poison used to kill Socrates. (d) Nicotine, from tobacco; a deadly poison sometimes used in insecticides. (e) Quinine, from the cinchona tree; an anti-malarial drug. (f) Tubocurarine, from the shrub Chondodendron tomentosum, the active principle of curare, a nerve poison used by the Jivaro Indians of the Amazon Valley.

postulated structure, but—hopefully—to provide an alternative to the natural supply of the compound.

Isolation and purification techniques are important in many other aspects of chemistry in addition to those dealing with natural products. Unlike the simple reactions of most ionic species, covalent compounds frequently undergo many competing reactions, so quantitative yields of pure products are rare in organic syntheses. Reaction mixtures, then, also require work-ups involving one or more isolation and purification steps.

In this excursion we shall consider briefly some of the more common ways of securing pure chemical compounds from natural sources or from reaction product mixtures.

1 ISOLATION OF SOLID SUBSTANCES

A large percentage of the most interesting compounds isolated from natural sources have been crystalline solids when purified. These compounds seldom constitute more than a small fraction of the original material, so their isolation in reasonable quantity may involve a very large amount of the raw product. If the isolation process is to be relatively inexpensive, any extraction involved should employ either a very cheap solvent or a continuous process requiring only a small amount of solvent. The *Soxhlet extraction* apparatus (Figure E4.2), or some modification thereof, is the most common way of performing continuous extraction from solid materials. In this procedure the solvent is boiled so that the vapors are condensed back to liquid above the sample. This liquid then seeps through the sample, extracting the desired compound and other soluble materials. The resulting solution accumulates in the boiling pot. Nonvolatile substances are thus concentrated as the solvent is continuously volatilized and recondensed. Such extracts are generally very crude mixtures and require further separation before the desired product can be obtained in reasonable purity.

If any extraction is to be feasible, a solvent must be used that dissolves a minimum number of additional components of the original mixture. Such requirements are difficult to meet because, although macromolecules such as celluloses in plants are insoluble in most solvents, a wide variety of compounds of similar solubilities occur in most natural products. At best, a crude mixture of extracted materials will be obtained. If the desired compound is polar, a polar solvent is used; and if it is nonpolar, a nonpolar solvent is selected (Table E4.1). A trial-and-error process is usually required to determine optimum extraction conditions.

Figure E4.2 Soxhlet extraction apparatus. The Soxhlet cup is made of a heavy, porous membrane inert to typical organic solvents.

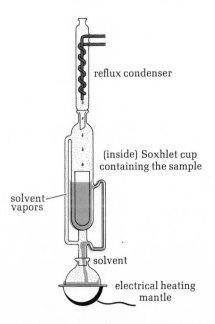

reflux condenser

(inside) Soxhlet cup containing the sample

solvent vapors

solvent

electrical heating mantle

Table E4.1 SOME COMMON SOLVENTS

polarity	water	
	ethanol	
	acetone	
	chloroform	
	diethyl ether	
nonpolar	benzene	
	n-hexane	
	carbon tetrachloride	

2 LIQUID–LIQUID EXTRACTION

Once a crude mixture has been obtained from a natural source (or from a synthetic procedure yielding a product mixture), additional separations must be devised. One possibility involves the partitioning of materials between two immiscible liquids, one of which is usually water or some aqueous solution. Since most organic compounds are less soluble in water than in other liquids, this procedure can seldom be expected to separate complex mixtures completely. Water-insoluble neutral compounds can, however, be separated from substances more soluble in water. Organic acids (e.g., phenols or carboxylic acids) or bases (amines) can be readily extracted by aqueous solutions whose pH has been adjusted to convert the organic compounds into their more soluble ionic forms. For example, a benzene solution of phenol and naphthalene could be extracted with aqueous sodium hydroxide to remove the phenol as the water-soluble ion.

Solubility of a substance in any solvent is governed primarily by three considerations.

1. *Solvent–solute* forces determine the strength of interactions between solvent molecules and dissolved particles. *Protic* solvents (those having highly polarized bonds involving hydrogen) can form strong hydrogen bonds with unshared electron pairs on oxygen or nitrogen in organic molecules, as discussed in Unit 29. Other types of polar solvents will form relatively tight solvate clusters with polar solutes, and nonpolar solvents such as carbon tetrachloride will attract solute particles only by London dispersion forces, whose strength depends roughly on the number of electrons in the particles and in the solvent molecules.

2. *Solute–solute* forces are those between particles of the substance being dissolved.

3. *Solvent–solvent* forces exist within clusters of solvent molecules.

In general, substances will dissolve to the extent that solvent–solute forces equal or exceed solute–solute and solvent–solvent forces. Examples of the types of interactions mentioned are shown in Table E4.2.

Molecules of organic compounds may contain both polar and nonpolar regions. As a rough approximation, nonpolar regions equivalent to or greater than an unbranched four-carbon chain are sufficient to reduce drastically the solubility of substances such as alcohols in solvents such as water. In such cases, solvent–solvent forces cause clustering of solvent molecules, which effectively "squeeze out" the nonpolar regions of the potential solute.

To use liquid–liquid extraction for the separation of components of a mixture, we must employ two immiscible liquids selected so that solutes differ significantly in their solubilities in the two liquids employed. Typically, one liquid is water or an aqueous solution having a pH favorable to differential solubilities of solute components. A knowledge of the structural features affecting solubility and of acid-base properties (Unit 29) is the key to planning a liquid–liquid extraction.

Table E4.2 TYPES OF INTERPARTICLE FORCES

TYPE OF INTERACTION	EXAMPLES		
	SOLUTE-SOLUTE	SOLVENT-SOLVENT	SOLUTE-SOLVENT
hydrogen bonding			
dipole–dipole			
London dispersions			

Example 1

Outline a scheme, using liquid–liquid extraction, to separate a mixture of benzalde-hyde (water solubility 0.33 g 100 ml^{-1}, very soluble in ether) and N-methylaniline (slightly soluble in water, very soluble in ether).

Solution

Since both compounds have similar solubilities, extraction of the neutral species using an ether-water system is not feasible. If one can be converted to a salt, the salt will be much more soluble in water than in ether. The N-methylaniline is basic, and the benzaldehyde is neutral. The separation scheme, then, is outlined as follows:

Figure E4.3 Extraction using a separatory funnel.

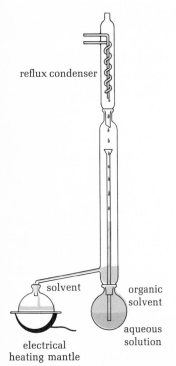

reflux condenser

solvent

organic solvent

aqueous solution

electrical heating mantle

Figure E4.4 Continuous liquid-liquid extraction (solvent less dense than water). Equipment is also available for solvents more dense than water.

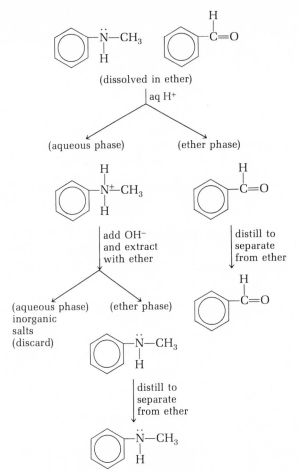

(dissolved in ether)

aq H$^+$

(aqueous phase)   (ether phase)

add OH$^-$ and extract with ether

distill to separate from ether

(aqueous phase) inorganic salts (discard)   (ether phase)

distill to separate from ether

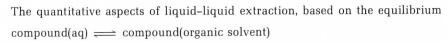

The quantitative aspects of liquid–liquid extraction, based on the equilibrium

compound(aq) $\rightleftharpoons$ compound(organic solvent)

have been discussed in Unit 18. When the distribution coefficient (Unit 18) is large, a simple extraction in a separatory funnel (Figure E4.3) is usually adequate. If the distribution coefficient is so small that a large series of extractions would be required, a continuous process may be employed (Figure E4.4).

3 RECRYSTALLIZATION

Temperature dependence refers to the variation in solubility with solvent temperature and may be defined as:

temperature dependence =
(solubility at t_2)/(solubility at t_1)

Solid products obtained by evaporation of extraction solvent are seldom very pure. If there is an appreciable difference in the solubilities of the desired product and of the impurities present, considering temperature-dependence effects, it may be possible to purify the product by recrystallization, assuming that the product will crystallize.

A typical procedure would involve addition of solvent at or near its boiling point to the crude mixture until the solubility of the desired compound is just exceeded. The resulting solution would be filtered while hot, if insoluble residue remained, and then allowed to cool slowly for formation of crystals of the product. Chilling would then be employed to increase the yield of crystalline material, and the final solid would be filtered rapidly by vacuum. Repeated recrystallizations will often effect satisfactory purification. There is, perhaps, more art than science to this procedure, and the novice may find it rather frustrating and time consuming, especially if both yield and purity of product are important.

4 SUBLIMATION

A substance that has a reasonably high vapor pressure in the solid state may be separated from less volatile impurities by sublimation. Typically, the solid mixture is heated to below its melting range (with applied vacuum when the product's vapor pressure is fairly low) until all of the desired product has vaporized and recondensed to solid on a chilled surface (Figure E4.5). Materials such as iodine and naphthalene are purified on a commercial scale by this method, which is applicable to crystalline compounds having relatively weak interparticle forces.

5 CHROMATOGRAPHY

Adsorption chromatography (Figure E4.6) is a separation process depending on the equilibrium between a substance adsorbed on the surface of some solid and the substance dissolved in a solvent in contact with the solid. Alumina (Al_2O_3) is a

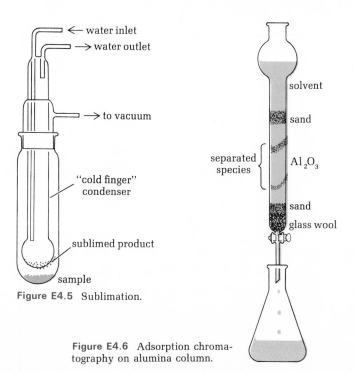

Figure E4.5 Sublimation.

Figure E4.6 Adsorption chromatography on alumina column.

Table E4.3 TYPICAL SOLVENT SEQUENCE FOR ADSORPTION CHROMATOGRAPHY

polarity

1. hexane
2. 9:1 hexane/benzene
3. 4:1 hexane/benzene
4. 3:2 hexane/benzene
5. benzene
6. 9:1 benzene/chloroform
7. 4:1 benzene/chloroform
8. 3:2 benzene/chloroform
9. chloroform
10. 9:1 chloroform/methanol

etc.

glacial acetic acid

common adsorbent for chromatographic separation of organic compounds. If a mixture of compounds is added to a column of alumina, the more polar components of the mixture are bound more tightly to the alumina particles than are the less polar components. Species, such as phenols, that can hydrogen bond to the oxygens of the alumina are adsorbed very strongly. When the column is eluted (washed) with a nonpolar solvent, the more soluble, least strongly adsorbed compounds are eluted first. By gradually increasing the solvent polarity (Table E4.3), sequential elution of the components of the original mixture may be possible.

Example 2

Predict the elution sequence of a mixture of p-cresol, chlorobenzene, and naphthalene chromatographed on an alumina column, using a solvent system of gradually increasing polarity.

Solution

Since the solvent system is initially nonpolar and nonpolar compounds are adsorbed only weakly on alumina, we expect the least polar component to be eluted first. Components capable of forming hydrogen bonds to oxygen in the Al_2O_3 should be eluted last. The structures of the compounds in question (Unit 28) then allow a prediction of elution sequence:

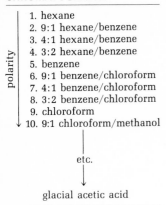

naphthalene
(nonpolar)
[eluted first]

p-cresol
(capable of hydrogen bonding
to Al_2O_3)
[eluted last]

chlorobenzene
(polar)
[eluted second]

Partition chromatography uses as a solid support either paper or some powder such as cellulose or silica gel on a glass or plastic backing. In true partition chromatography a mixed solvent is employed, one component of which forms solvate clusters on functional groups in the solid support substance (e.g., water could be used to form hydrate clusters around hydroxyl groups in cellulose molecules). These solvate clusters are called the *stationary phase,* and the unbound components of the solvent mixture, which move through the support by capillary action, constitute the *mobile phase.* By a process similar to liquid–liquid extraction, components of a mixture are separated by a continuous partitioning between the stationary and mobile phases. The components least soluble in water and most soluble in the mobile solvent move more rapidly. Rapid and efficient separations on a small scale are usually possible if the correct solvent system is employed.

It should be noted that some adsorption occurs during partition chromatography and, when mixed solvents are used, some partitioning occurs during adsorption chromatography.

When colorless compounds are involved, they must be detected by a color-producing reagent, fluorescence on exposure to ultraviolet light, or elution of portions of the chromatogram for other means of detection.

Partition chromatography is discussed in more detail in connection with its use in amino acid analysis (Unit 32).

Figure E4.7 Basic components of a simple vapor-phase chromatograph.

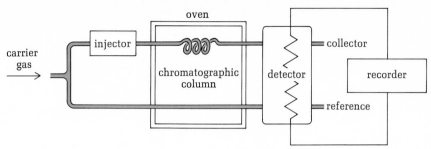

In *vapor-phase chromatography* (Figure E4.7) a mixture of volatile components is injected into a stream of a nonreactive gas (typically helium or nitrogen) and the gaseous mixture is passed through a column of adsorbent material. The column temperature is carefully regulated and separation is achieved on the basis of differences in vapor pressure and adsorption characteristics of the components. This technique is one of the most effective separation methods yet devised and the *retention time* of a compound (based on the time required for the compound to pass through the column under carefully controlled conditions) can be as characteristic for a volatile compound as the melting point is for a pure crystalline substance. *Vapor-phase chromatography* can be employed with very small samples or, with appropriate equipment, for the separation of components from a large amount of a mixture. Although only volatile compounds can be separated by this technique, it is often possible to convert nonvolatile substances into volatile derivatives. This method has been employed, for example, in the separation of amino acids (Unit 32).

The relatively new technique of *high-pressure liquid chromatography* (Figure E4.8) shows great promise as a versatile separation method. By using solvent mixtures passed through columns of absorbent material under high pressure, this technique combines the advantages of rapid separation characteristic of vapor-phase chromatography with the added capability of working with both volatile and nonvolatile components. Separations involve both adsorption and partition chromatography, since mixed solvents are typically employed.

Figure E4.8 Components of a *high-pressure liquid chromatography* system.

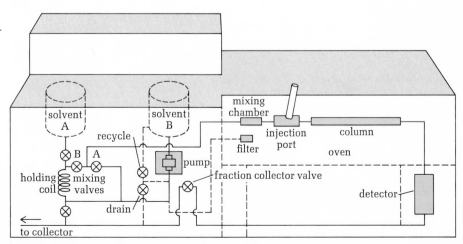

6 PURIFICATION OF LIQUIDS: DISTILLATION

Some of the techniques, such as liquid–liquid extraction, described for purification of crystalline solids can be employed with liquid compounds as well. The most common procedure, however, involves heating the liquid until it boils and condensing the vapor back to the liquid state for collection in a separate container. This process is called *distillation*.

Substances whose boiling ranges differ by about 30°C or more may be separated by simple distillation (Figure E4.9). Sometimes repetition of this procedure one or two times will effect an adequate purification, although components of a mixture whose boiling ranges differ by a smaller increment usually require a much more sophisticated technique. Special distilling columns are available for the separation of liquids differing by only a few degrees in boiling ranges (Figure E4.10). For high-boiling liquids, a vacuum pump can be used to reduce the pressure in the system so that boiling points are lowered, thus minimizing thermal decompositions.

A liquid will boil when its vapor pressure equals the external pressure on the system (Unit 12). If only one component of a mixture is volatile, then the boiling point will depend only on the vapor pressure of the volatile component. This is, of course, affected by dissolved materials (Unit 15). For a mixture of two volatile liquids, however, the total vapor pressure will be a function of the vapor pressures of each

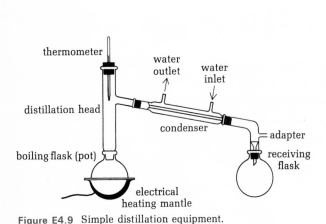

Figure E4.9 Simple distillation equipment.

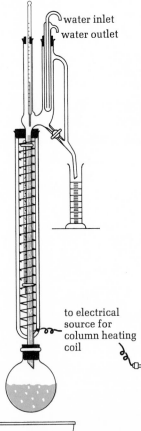

Figure E4.10 Distilling column for separation of liquids having only small boiling-range differences.

liquid and the mole fractions of the components. This relationship is expressed by *Raoult's law* as

$$P_T = f_A P_A + f_B P_B \tag{1}$$

in which P_T is the total vapor pressure of the mixture at a given temperature, P_A is the vapor pressure of pure A at that temperature, P_B is the vapor pressure of pure B at that temperature, and f_A and f_B are the mole fraction of A and mole fraction of B, respectively, in the mixture.

From Raoult's law it is possible to calculate the composition of the distillate from a mixture of two volatile liquids. This composition is given by

Azeotropic mixtures, for example, 95.5 percent ethanol and 4.5 percent water, boil at a constant temperature, as though the mixture were a pure liquid. Such azeotropes are deviations from "ideal" behavior and must be excluded from calculations of this type.

$$\frac{n_A}{n_B} = \frac{f_A P_A}{f_B P_B} \tag{2}$$

in which n_A is the number of moles of A in the distillate, n_B is the number of moles of B in the distillate, P_A and P_B are the respective vapor pressures of the pure liquids at the temperature of distillation, and f_A and f_B are the mole fractions of A and B, respectively, in the boiling mixture at the point during the distillation for which the calculation is being made.

As distillation continues, the distillate and the "pot residue" will continuously change in composition. Initial fractions collected will be richer in the lower-boiling components than will later fractions. By proper selection of fractions for successive redistillations, an eventual separation of the two components may be feasible.

Example 3
_____

A mixture of n-hexane (boiling point 69°C) and n-heptane (boiling point 98°C) begins to boil at 86°C at 760 torr when the mole fraction of hexane in the mixture is 0.20. The vapor pressure of pure n-heptane at 86°C is 520 torr. What is the mole fraction of n-hexane in the first drop of distillate?

Solution
At 86°C
[equation (1)]

$$760 \text{ torr} = (0.80)(520 \text{ torr}) + (0.20)(P_{\text{hexane}})$$
$$f_A P_A \text{ (heptane)} = 0.80(520) = 416 \text{ torr}$$
$$f_B P_B \text{ (hexane)} = 760 - 416 = 344 \text{ torr}$$

Then, [equation (2)]

$$\frac{n_{\text{heptane}}}{n_{\text{hexane}}} = \frac{416}{344} = 1.21$$

$$n_{\text{heptane}} = 1.21\, n_{\text{hexane}}$$

$$\underset{\text{(in vapor)}}{f_{\text{hexane}}} = \frac{n_{\text{hexane}}}{n_{\text{hexane}} + n_{\text{heptane}}}$$

$$= \frac{n_{\text{hexane}}}{(1 + 1.21)n_{\text{hexane}}}$$

$$f_{\text{hexane}} = \frac{1}{2.21} = 0.45$$

Thus initial distillation increased the mole fraction of the lower-boiling compound from 0.20 to 0.45.
_____

If a substance of known structure is being isolated, its properties may be compared with those of a pure sample of the compound. In the case of an unknown sample, however, such a comparison is not possible.

The purity sought will depend on the use designed for the substance. If structural determination is planned, and particularly in cases of physiologically active compounds, a very high state of purity is essential.

A substance is assumed to be pure when repeated attempts at further separations by a variety of techniques fail to alter the properties of the substance. Modern chromatographic procedures are especially valuable for demonstration of purity and, indeed, reagents labeled "chromatographically pure" are now considered the ultimate grade of sample commercially available.

**SUGGESTED
READING**
Bates, Robert B. and John P. Schaefer. 1971. *Research Techniques in Organic Chemistry.* Englewood Cliffs, N.J.: Prentice-Hall.
Cason, James, and Henry Rapoport. 1962. *Laboratory Text in Organic Chemistry.* Englewood Cliffs, N.J.: Prentice-Hall.

EXERCISES

1. A reaction mixture consisting of cyclohexanol, cyclohexene, and cyclohexanone was chromatographed on an alumina column, using a sequential solvent system of gradually increasing polarity. Predict the elution sequence and explain your prediction.

2. A reaction mixture consisting of p-toluidine, N,N-dimethylaniline, and p-xylene was chromatographed on an alumina column, using a sequential solvent system of gradually increasing polarity. Predict the elution sequence and explain your reasoning.

3. A reaction mixture consisting of p-chloroaniline, p-chloronitrobenzene, and p-dichlorobenzene was chromatographed on an alumina column, using a sequential solvent system of gradually increasing polarity. Predict the elution sequence to be expected.

4. Pure chlorobenzene boils at 132°C at 1.00 atm. The vapor pressure of bromobenzene is 400 torr at 132°C.
   a. At what external pressure will a mixture of chlorobenzene and bromobenzene begin to distill at 132°C if the mole fraction of bromobenzene is 0.25?
   b. What will be the mole fraction of chlorobenzene in the first drop of distillate collected?

5. Pure aniline boils at 184°C at 760 torr. The vapor pressure of nitrobenzene at 184°C is 380 torr.
   a. At what external pressure will a mixture of aniline and nitrobenzene begin to distill at 184°C if the mole fraction of nitrobenzene is 0.15?
   b. What will be the mole fraction of aniline in the first drop of distillate collected?

6. Pure toluene boils at 108°C at 1.00 atm. The vapor pressure of ethylbenzene at 108°C is 312 torr.
   a. At what external pressure will a mixture of toluene and ethylbenzene begin to distill at 108°C if the mole fraction of ethylbenzene is 0.25?
   b. What will be the mole fraction of toluene in the first drop of distillate collected?

7. Outline a plausible sequence for separation, using liquid–liquid extractions, of an equimolar mixture of 2,4-dinitrophenol [water solubility 0.68 g (100 ml)$^{-1}$, ether solubility 3.27 g (100 ml)$^{-1}$] and benzaldehyde [water solubility 0.33 g (100 ml)$^{-1}$, very soluble in ether].

8. Outline a plausible sequence for separation, using liquid–liquid extractions, of an equimolar mixture composed of N,N-diethylaniline [water solubility 1.6 g (100 ml)$^{-1}$, very soluble in ether], acetophenone (insoluble in water, soluble in ether) and 2,4,6-trichlorophenol [water solubility 0.08 g (100 ml)$^{-1}$, very soluble in ether].

9. Outline a liquid–liquid extraction scheme for separation and isolation of the components of an equimolar mixture of salicylic acid [water solubility 0.18 g (100 ml)$^{-1}$, ether solubility 52 g (100 ml)$^{-1}$] and methyl o-methoxybenzoate (insoluble in water, very soluble in ether).

Section
ELEVEN

POLY-
FUNCTIONAL
COMPOUNDS

Most biologically important compounds contain more than one functional group. In many cases these functional groups exhibit independent characteristics; that is, their chemical properties are essentially those of the corresponding monofunctional compounds. In such cases we can consider each functional group in the molecule separately in terms of the properties characteristic of the corresponding class of compound (Units 28 and 29). However, there are two possible complications with polyfunctional compounds. Electronic interactions may occur among groups located in close proximity or separated only by delocalized π-bond systems. We have already seen examples of such interactions in our considerations of factors influencing the strengths of organic acids (Unit 29). It is also possible that different functional groups within the same molecule may react with each other. As we shall see in Unit 31, these *intramolecular* reactions are typical of certain hydroxyacids (forming cyclic esters called *lactones*) and of hydroxyaldehydes and hydroxyketones, such as the simple sugars.

In Unit 32, we shall consider some of the most important of natural compounds: the amino acids, peptides, and proteins. Ranging from the simple difunctional amino acid glycine to the enormous and complex structural proteins, these compounds play important roles in a variety of biochemical processes.

Synthetic polymers are so much a part of our everyday lives that we would have to make some drastic changes to get by without them. Most of our modern fabrics are partly or entirely composed of synthetic fibers such as *nylon* or *Dacron*. We use plastic food wraps to cover leftovers from foods purchased in plastic containers, now tossed into plastic garbage bags. Synthetic polymers are a major product of the modern chemical industry but, like so many other aspects of science, these must be considered as "mixed blessings." Many types of polymers are now made from petroleum products and thus constitute a competition for needed energy resources. In addition, a number of our current types of polymers pose significant environmental problems. Some are "waste-accumulation" problems because they are degraded very slowly by natural waste-removal systems, and others may release potentially hazardous degradation products into our air or natural waters. In Unit 33 we shall consider some aspects of the chemistry of synthetic polymers.

Excursion 5 offers a brief survey of the methods employed by the natural product chemist for the structure determination and synthetic production of polyfunctional compounds. We will consider a specific example, *papaverine*, an antispasmodic drug from the opium poppy.

CONTENTS

Unit 31

Difunctional molecules

Objectives

(a) Be able to identify common functional groups (Unit 28) in complex molecules and to apply a knowledge of typical group properties (Unit 29) to polyfunctional compounds in which the groups are essentially ''independent'' (Section 31.1).
(b) Be able to predict inductive effects on the strengths of simple organic acids and bases (Section 31.2 and Units 19, 29).
(c) Be able to identify possible intramolecular reactions involving hydroxyl groups and leading to cyclizations (Section 31.3).
(d) Be able to write equations and structural formulas illustrating some characteristics of simple carbohydrates (Section 31.5).

Ethanol may react with sulfuric acid in several different ways. When heated with concentrated H_2SO_4 to about 170°C, ethanol is "dehydrated" to ethylene:

$$CH_3CH_2OH \xrightarrow[170°C]{H_2SO_4} H_2C{=}CH_2$$

With cold concentrated H_2SO_4, the principal reaction is esterification:

$$CH_3CH_2OH \xrightarrow[\text{cold}]{H_2SO_4} CH_3CH_2O-\overset{\displaystyle O}{\underset{\displaystyle OH}{\overset{\|}{S}}}=O$$

ethyl hydrogen sulfate

At intermediate temperatures, using excess ethanol, ether formation occurs:

$$\underset{\text{(excess)}}{CH_3CH_2OH} \xrightarrow[130°C]{H_2SO_4} CH_3CH_2OCH_2CH_3$$

Chemists know enough about the mechanisms of these competing reactions to have selected the optimum conditions for formation of any one of the products desired. However, in every case a product mixture is obtained; that is, ethanol and sulfuric acid always react to yield a mixture of ethylene, ethyl hydrogen sulfate, and diethyl ether (along with a few other minor by-products). Conditions of the reaction may be varied to favor one product, but a quantitative yield of a single product is never obtained.

Ethanol is a simple monofunctional compound, and reaction conditions for its treatment with sulfuric acid can be controlled fairly easily. But the chemical changes occurring are not simple ones. If polyfunctional compounds are involved in a reaction, the possible complications are increased enormously.

Living systems contain an immense number of complex molecules participating in a variety of chemical reactions, many of which are interdependent. If an ethanol–H_2SO_4 system is difficult to understand and control, it is not surprising that the chemistry of life processes is still largely a mystery. Although we know many of the chemical changes involved in living organisms, a complete description of the biochemistry of even the simplest cell is still a long way in the future.

In this unit and those that follow we shall consider some common types of polyfunctional compounds. In some cases we shall see that the functional groups are essentially independent of each other. In other cases the properties of one group are altered appreciably by the presence of another. Sometimes a molecule contains groups that can react with each other and, when the molecular geometry is suitable, intramolecular reactions may occur.

Most of the more important biologically active compounds are polyfunctional (Figure 31.1), and an understanding of interactive functional groups is essential to a description of their behavior in living organisms.

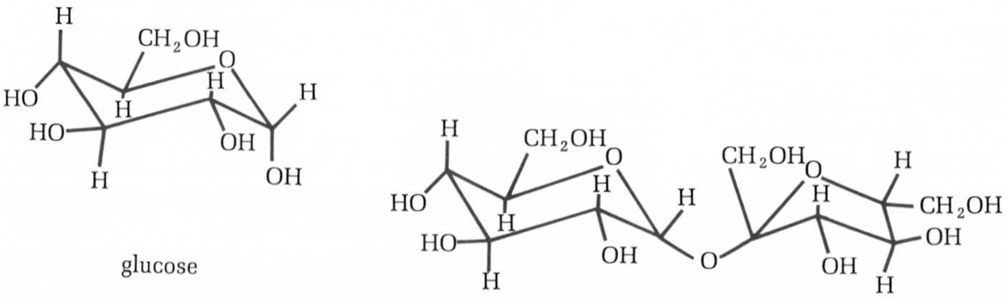

glucose

sucrose

Figure 31.1 Some polyfunctional compounds of biological importance. (a) Carbohydrates. (b) Lipids. (c) Amino acids and peptides.

(a)

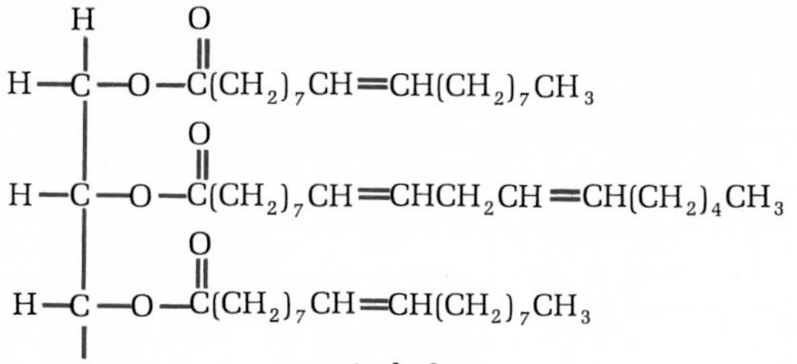

a typical "fat"

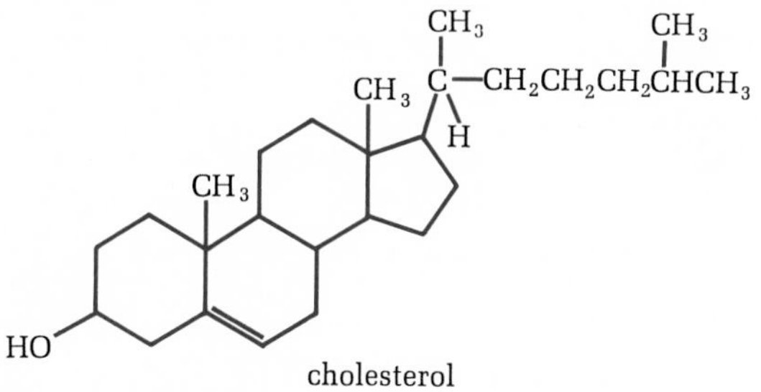

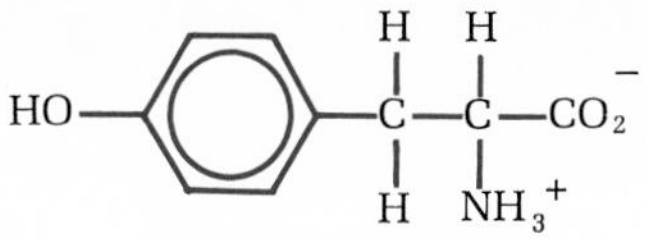

tyrosine

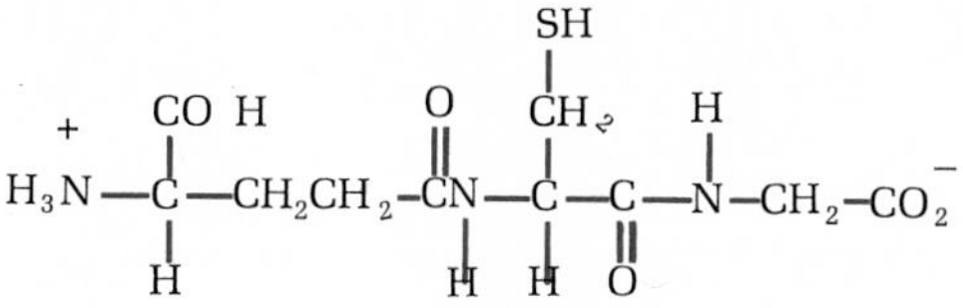

cholesterol

glutathione, a peptide

(b)

(c)

Figure 31.1 (Continued)

31.1 INDEPENDENT FUNCTIONAL GROUPS

Group interactions may be transmitted through delocalized π-bond systems over large portions of molecules, but only over small regions when σ bonds are involved.

Molecules that contain two or more functional groups that do not react with each other behave as though they were equimolar mixtures of simple monofunctional compounds, provided such groups are sufficiently far apart (usually four or more

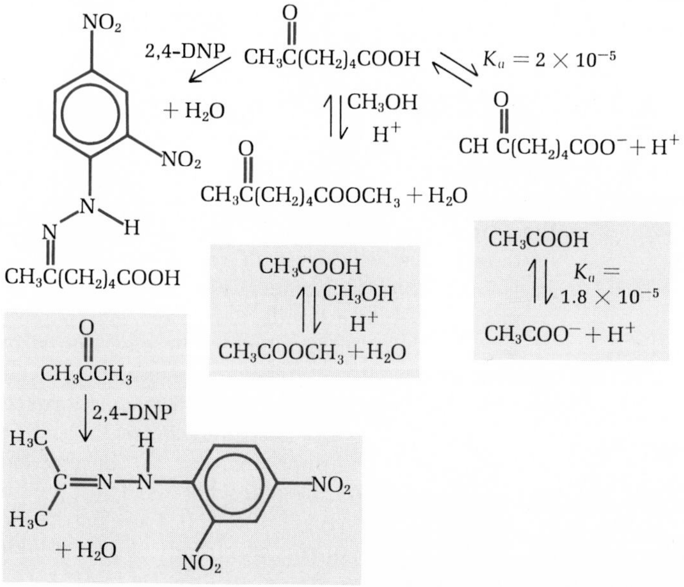

Figure 31.2 Some reactions of 6-oxoheptanoic acid. Note that this molecule contains two functional groups normally indicated in the name ending. In this case, the ketone carbonyl is named as an "oxo" group. Other polyfunctional compounds may also deviate from the nomenclature used for monofunctional substances (Appendix G).

carbons) to avoid influencing each other's properties. In such cases, the chemistry of the molecule is essentially that of the separate classes represented by the functional groups (Unit 29). For example, 6-oxoheptanoic acid (Figure 31.2) shows both the typical reactions of a simple aliphatic ketone and those of a carboxylic acid.

When groups are close enough together within a molecule, one may influence the properties of another by changing the polarity of various bonds. Usually such influences do not change the *type* of reaction of a functional group, but they may alter the speed of a reaction or the relative composition of an equilibrium reaction mixture. These effects, called *inductive effects,* may be transmitted through short σ-bond systems, longer π-bond systems, or through space between such groups. The magnitude of the effect will depend on the distance between the groups and on the relative electronegativities involved (Unit 7).

As can be seen from the data in Table 31.1, electron-withdrawing groups (electronegativity of atoms greater than that of carbon) tend to increase the acidity of carboxylic acids, and electron-releasing groups tend to decrease the acidity. Two factors can be considered as contributing to these effects. In the undissociated acid the presence of an electronegative group will affect the polarity of the O—H bond in the carboxyl group, pulling the electron pair of the bond farther away from the hydrogen and making it easier for H$^+$ to escape.

weakened

Since the net electron density of the entire carboxylate group is decreased, there will also be a weaker electrostatic attraction for the returning proton.

In the ion formed by proton loss, an electronegative atom will add stability by reducing the electron density on the carbon of the carboxylate ion.

more stable

Electropositive groups will have the opposite effect in each case.

Inductive effects drop off rapidly as the distance between interacting groups increases (Table 31.2), generally becoming negligible when the groups are separated by more than four σ bonds.

Since any factor tending to increase the strength of an acid will make the conjugate base (Unit 19) weaker, we expect these effects to be noted in such cases as the hydrolysis of salts (Unit 19). Sodium trichloroacetate, for example, is only slightly hydrolyzed; that is, aqueous solutions of the salt are only slightly alkaline. Electronegative atoms or groups decrease the base strengths of amines by reducing the electron density on nitrogen, whereas electropositive groups (such as alkyl groups) have the opposite effect (Table 31.3). The electron pair on a nitrogen atom attached to an aromatic ring is, of course, already "less available" because of involvement in the delocalized electron system of the ring (Unit 29).

Table 31.1 IONIZATION CONSTANTS FOR SOME CARBOXYLIC ACIDS[a] $(RCOOH \rightleftharpoons RCOO^- + H^+)$

NAME	STRUCTURE	K_a
formic acid (methanoic)	H ⟷ COOH	1.8×10^{-4}
Reference: acetic acid (ethanoic)	$H \leftrightarrow \overset{\displaystyle H}{\underset{\displaystyle H}{C}} \leftrightarrow COOH$	1.8×10^{-5}
isobutyric acid (2-methylpropanoic)	$H \leftrightarrow \overset{\displaystyle H}{\underset{\displaystyle H}{C}} \leftrightarrow \overset{\displaystyle H}{\underset{\displaystyle \underset{\displaystyle H \leftrightarrow \overset{\displaystyle }{\underset{\displaystyle H}{C}} \leftrightarrow H}{}}{C}} \leftrightarrow COOH$	1.4×10^{-5}
chloroacetic acid (chloroethanoic)	$Cl \leftrightarrow \overset{\displaystyle H}{\underset{\displaystyle H}{C}} \leftrightarrow COOH$	1.4×10^{-3}
dichloroacetic acid (dichloroethanoic)	$Cl \leftrightarrow \overset{\displaystyle H}{\underset{\displaystyle Cl}{C}} \leftrightarrow COOH$	5×10^{-2}
trichloroacetic acid (trichloroethanoic)	$Cl \leftrightarrow \overset{\displaystyle Cl}{\underset{\displaystyle Cl}{C}} \leftrightarrow COOH$	1×10^{-1}

[a] See also Units 19 and 29.

Table 31.2 INDUCTIVE EFFECT VERSUS DISTANCE BETWEEN GROUPS

NAME	STRUCTURE	K_a
2-chloropentanoic acid	$CH_3CH_2CH_2\overset{\displaystyle H}{\underset{\displaystyle Cl}{C}}-COOH$	1.4×10^{-3}
3-chloropentanoic acid	$CH_3CH_2\overset{\displaystyle H}{\underset{\displaystyle Cl}{C}}CH_2COOH$	8.9×10^{-5}
4-chloropentanoic acid	$CH_3\overset{\displaystyle H}{\underset{\displaystyle Cl}{C}}CH_2CH_2COOH$	3×10^{-5}
5-chloropentanoic acid	$Cl-CH_2CH_2CH_2CH_2COOH$	2×10^{-5}
Reference: acetic acid (ethanoic)	CH_3COOH	1.8×10^{-5}

Table 31.3 INDUCTIVE EFFECTS ON BASE STRENGTH OF AMINES

NAME	STRUCTURE	K_b
aminomethane (methylamine)	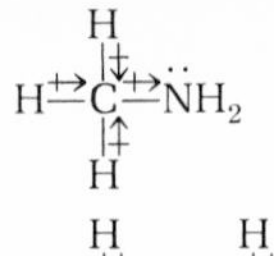	5.0×10^{-4}
N-methylaminomethane (dimethylamine)	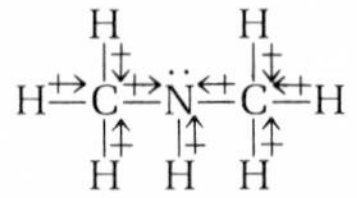	5.1×10^{-4}
aminoethane (ethylamine)	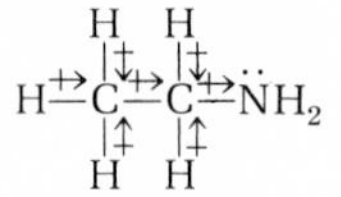	5.6×10^{-4}
N,N-dimethylaminomethane (trimethylamine)	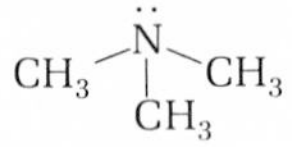	5.3×10^{-5a}
aniline	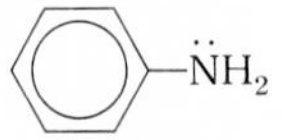	3.8×10^{-10}
p-chloroaniline	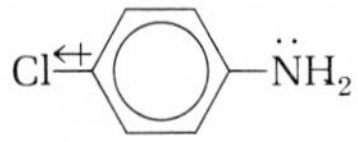	1.5×10^{-10}
o-chloroaniline	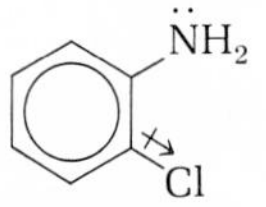	5.0×10^{-12}
p-nitroaniline	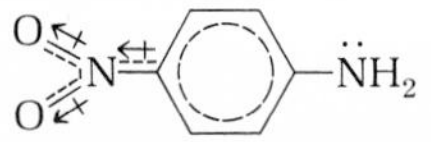	1.0×10^{-13b}

[a] Note the *decreased* base strength resulting from the steric ("crowding") effect of three methyl groups on the lone-pair orbital region.
[b] Note the additional effect of participation of the nitro group in the electron delocalization system.

31.3 INTRAMOLECULAR REACTIONS

In addition to inductive influences, functional groups may react directly with each other. Such reactions may, in fact, occur within the same molecule, provided a geometry is possible that brings the groups into close proximity.

Cyclization reactions

Hydroxy acids may undergo esterification reactions like those between simple alcohols and carboxylic acids. Again, if the molecular geometry is just right, an intramolecular reaction may occur. Such internal esters are called *lactones*.

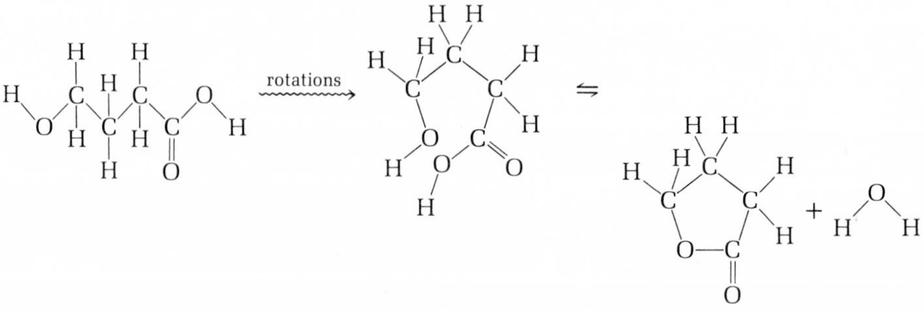

(a lactone)

Table 31.4 FORMATION OF ACETALS AND HEMIACETALS

REACTANTS	PRODUCT[a]	COMMENTS
aldehyde + alcohol e.g.,	aldehyde hemiacetal	an equilibrium process, involving addition of the alcohol to the carbonyl group

(benzaldehyde)   (methanol)   (benzaldehyde methyl hemiacetal)

| ketone + alcohol e.g., | ketone hemiacetal | |

(acetone)   (ethanol)   (acetone ethyl hemiacetal)

| hemiacetal + alcohol e.g., | acetal | catalyzed by acid or, in biological systems, by specific enzymes; easier for aldehydes than for ketones because of *steric hindrance* |

(benzaldehyde methyl hemiacetal)   (benzaldehyde dimethyl acetal)

[a] The hemiacetal functional group consists of a hydroxyl and an alkoxyl group on the same carbon $\begin{bmatrix} & | & \\ -C & -O-R \\ & | & \\ & O-H & \end{bmatrix}$, whereas the acetal group has two alkoxyl groups on the same carbon $\begin{bmatrix} & C-O-R \\ & | \\ & O-R' \end{bmatrix}$, R and R' may be the same or different alkyl groups.

Lactones are relatively rare in nature, but the burning pain from a *fire ant* sting has been attributed partly to a compound called *iridolactone* in the venom.

A cyclization reaction of particular importance in carbohydrate chemistry involves the formation of *acetals* and *hemiacetals*. These are functional groups we have not yet discussed, formed by the reaction between alcohols and the carbonyl group of an aldehyde or ketone (Table 31.4). Since simple sugars contain hydroxyl groups and either aldehyde or ketone groups, intramolecular reactions are expected.

Since alcohols may add to the π-bond system of an aldehyde or ketone carbonyl group, hydroxy aldehydes or ketones may also form ring systems by intramolecular reaction.

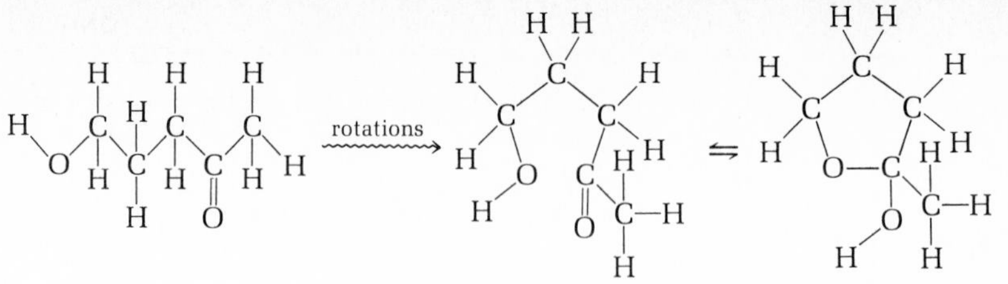

(a cyclic hemiacetal)

Reactions of the latter type are important in understanding the ring structures of simple sugars.

In most cases, cyclization reactions occur when either five-membered or six-membered rings can be formed. Smaller rings involve "bond-angle strain" so that their formation is energetically unfavorable.

Example 1

Which of the following compounds would be expected to undergo an intramolecular cyclization reaction? Draw structural formulas for the probable products.

a. lactic acid CH_3CHCO_2H
$\quad\quad\quad\quad\quad\quad\quad\quad|$
$\quad\quad\quad\quad\quad\quad\quad\quad OH$

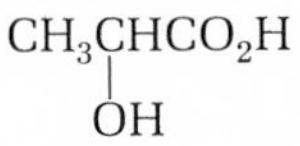

b. 5-hydroxyhexanoic acid
c. 4-hydroxybutanal

Solution

a. The two functional groups are too close together for an intramolecular reaction. A lactone in this case would involve a "strained" three-member ring:

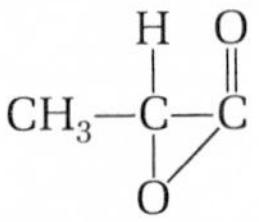

b. Cyclization is feasible, resulting in a six-member lactone:

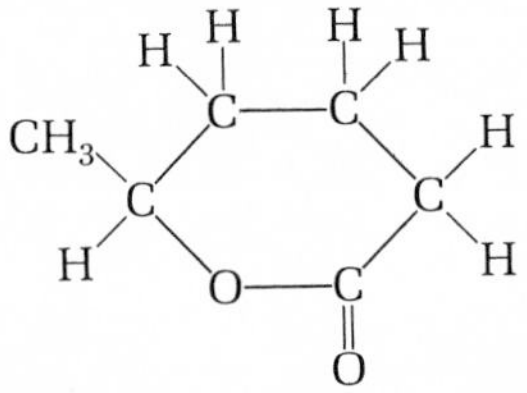

c. A five-member cyclic hemiacetal is expected:

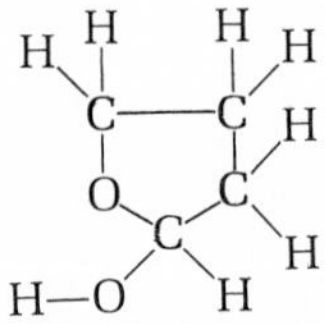

31.4 INTERMOLECULAR REACTIONS

A *polymer* is a large molecule consisting of many small molecules (*monomer units*) connected by covalent bonds.

Compounds with more than one functional group may react with other molecules in similar ways to those illustrated for monofunctional substances. Frequently, however, reactions in polyfunctional cases may result in the formation of polymers, both natural and synthetic (Unit 33). Simple sugars may polymerize via formation of acetal groups. Biochemists refer to these groups as *glycoside* links. Both starch and cellulose, for examples, are polymers of glucose. The polymerization of amino acids (Unit 32) results in the formation of polypeptides and proteins. In this case the "connecting group" is of the amide type (biochemical term, *peptide bond*), similar to that in the synthetic polymer *nylon*:

Note the use of an acid chloride (Unit 29) in the formation of *nylon-66*.

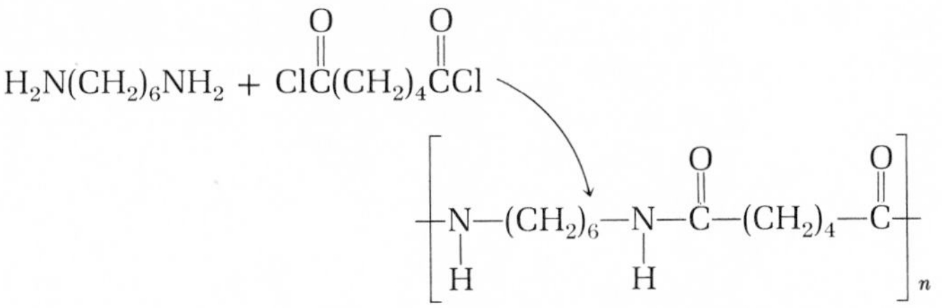

nylon-66
(a polyamide)

31.5 CARBOHYDRATES: DIFUNCTIONAL MOLECULES

Most substances classified as carbohydrates are polyhydroxy aldehydes or ketones or compounds that yield such molecules after hydrolysis. The chemistry of carbohydrates is a far from simple subject, but it may be conveniently introduced by considering various individual aspects of their overall pattern of behavior.

Independent reactions

The hydroxyl groups of carbohydrates may be esterified in the same ways as those employed for simple alcohols, for example,

Note that the hemiacetal hydroxyl (boldface symbols) is also esterified.

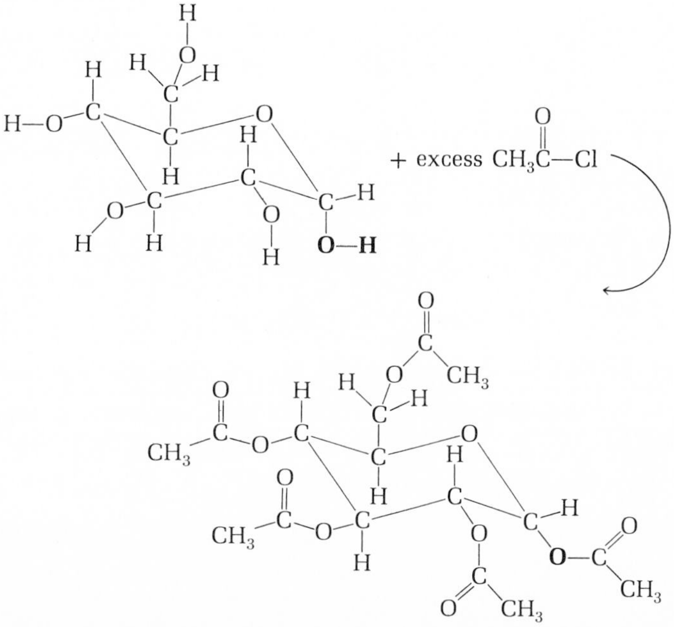

Esterification reactions are particularly useful with macromolecular carbohydrates such as starches and celluloses. Treatment of celluloses with acetyl chloride or acetic anhydride produces "cellulose acetate," a transparent polymeric material used to make clear plastic for overhead projectors, product packages, and so on. If an inorganic rather than an organic acid is used, the esters may have quite different properties. Nitric acid, for example, converts starches and celluloses to "nitrostarch" and "cellulose nitrate," respectively, both of which find uses as explosives.

Intramolecular reactions　Since carbohydrates may contain both carbonyl and hydroxyl groups, formation of cyclic hemiacetals is to be expected. Indeed, evidence indicates that most carbohydrates exist principally in this form.

Simple sugars (*monosaccharides*) exist in aqueous solution as equilibrium mixtures of cyclic hemiacetals and open-chain hydroxyaldehydes or hydroxyketones, although crystallographic studies indicate that the cyclic forms occur alone in "perfect" crystals of the sugars. With glucose, an *aldohexose* (6-carbon "aldehyde" sugar), the aqueous equilibrium system may be represented as

β-cyclic form　　　　　open-chain form　　　　　α-cyclic form

A similar equilibrium occurs with fructose ("fruit sugar," a *ketohexose*):

β-cyclic form　　　　　open-chain form　　　　　α-cyclic form

Note that these sugars contain several asymmetric centers, each contributing to optical activity (Unit 28). The distinction between α and β forms lies in the different geometries of the hemiacetal groups. In the α form, the hemiacetal oxygens are farther apart than in the β form. A similar distinction is made for the full acetals.

Intermolecular reactions: the glycoside link

In the presence of acid catalyst (or with appropriate enzymes), hemiacetals and alcohols react to form acetals:

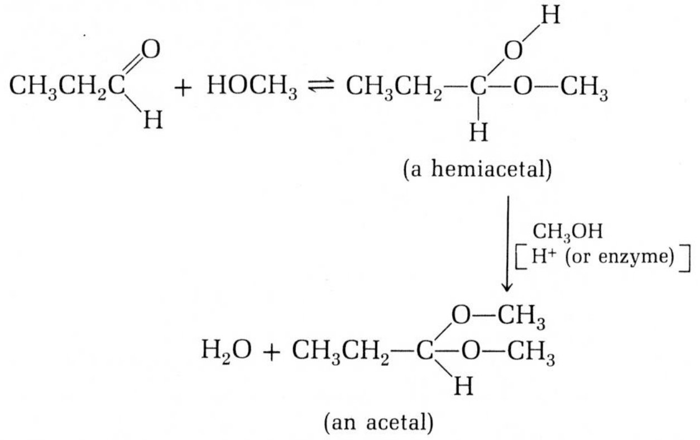

$$CH_3CH_2C\underset{\diagdown H}{\overset{\diagup O}{}} + HOCH_3 \rightleftharpoons CH_3CH_2-\underset{\underset{H}{|}}{\overset{\overset{\diagup O-H}{}}{C}}-O-CH_3$$

(a hemiacetal)

$$\downarrow \begin{bmatrix} CH_3OH \\ H^+ \text{ (or enzyme)} \end{bmatrix}$$

$$H_2O + CH_3CH_2-\underset{\underset{H}{|}}{\overset{\overset{\diagup O-CH_3}{}}{C}}-O-CH_3$$

(an acetal)

The acetal may be converted back to aldehyde and alcohols by hydrolysis.

Simple carbohydrates may react with each other to form acetals. The bond formed is commonly referred to as a *glycoside* link. The formation and hydrolysis of these glycoside links are important features of the metabolism of carbohydrates in living organisms.

Sucrose (table sugar) is a disaccharide (made from *two* simple sugars) formed by the combination of a glucose and a fructose molecule.

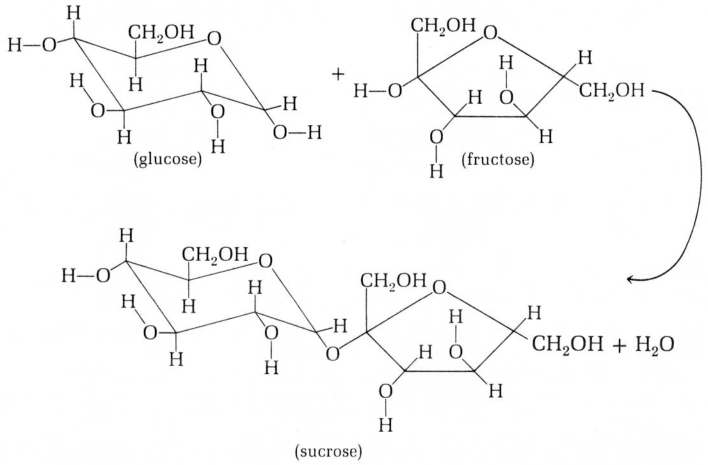

(glucose) + (fructose)

(sucrose)

Hydrolysis: breaking the glycoside link

Acetal formation involves the "release" of water as a product. Water may, in turn, act as a reactant in reversing this process by a hydrolytic splitting of the acetal group. In biological systems, some specific enzymes catalyze the combination of small

Figure 31.3 α- and β- Glycosides (skeletal structure). Note the different relative positions of the oxygen atoms in the α- and β- glycoside links.

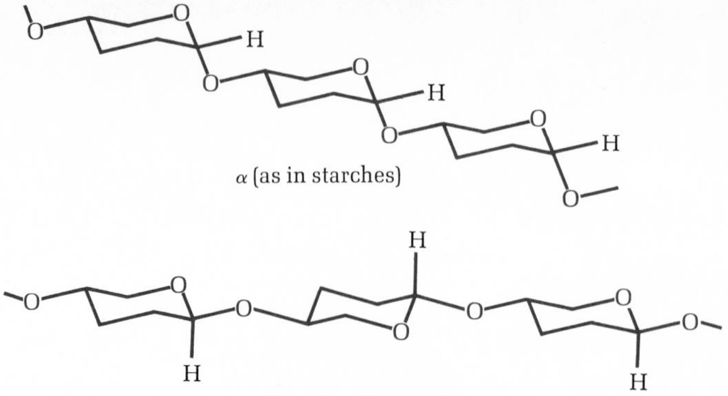

α (as in starches)

β (as in celluloses)

saccharides into larger ones, also producing one molecule of water for each glycoside link formed, whereas other enzymes catalyze the hydrolysis of large saccharides to smaller ones. In simple solutions both the formation and hydrolysis of acetals are acid catalyzed. When acetal formation is desired, anhydrous conditions are used to optimize yields.

Glucose is the monomer unit from which a number of biologically important macromolecular carbohydrates (polysaccharides) are formed. Among these are two major families of compounds, the starches and the celluloses. Both of these materials are hydrolyzed by aqueous acids to form smaller fragments, eventually glucose units. However, the starches are enzymatically hydrolyzed by enzymes found in human saliva and the celluloses are not. The explanation for this apparent anomaly lies in the stereoselectivity of enzyme catalysts (Unit 34) and the difference in the stereochemistry of the glycoside links involved. Starches contain α-glycoside links and celluloses contain β-glycoside links (Figure 31.3). Only the α geometry will fit the site at which hydrolysis is catalyzed on the enzyme system in human saliva.

There are many other important polyfunctional compounds in nature. The amino acids and proteins are discussed in Unit 32 and the nucleic acids in Unit 34. In every case the molecules must be considered in terms of the functional groups they contain. These groups may react independently or, when inductive effects are operative, they may be mutually influenced. Both intramolecular and intermolecular processes may be of importance in the behavior of polyfunctional compounds.

SUGGESTED READING

Barker, Robert. 1971. *Organic Chemistry of Biological Compounds.* Englewood Cliffs, N.J.: Prentice-Hall.

DeBey, H. J. 1968. *Introduction to the Chemistry of Life: Biochemistry.* Reading, Mass.: Addison-Wesley.

EXERCISES

1. Strychnine is an alkaloid first isolated by Pelletier and Caventou in 1819 from the bark and seeds of *Strychnos nux vomica*. It is a potent poison recently publicized for its use in killing coyotes. Its structure is formulated as

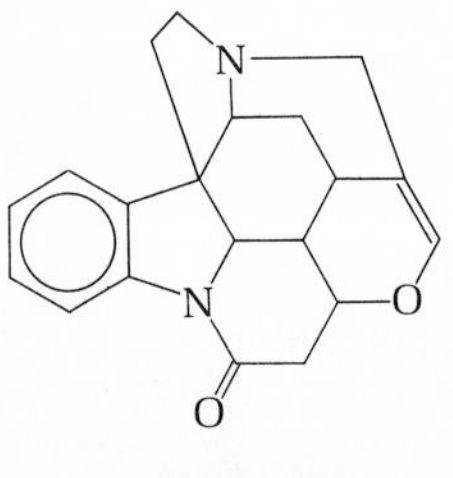

a. Copy this structural formula and label each functional group.

(e.g., 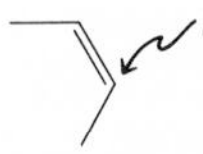 alkene)

b. Draw the structural formula of the product expected from reaction of strychnine with bromine.

c. Draw the structural formula of the product expected when strychnine is dissolved in excess cold hydrochloric acid.

d. Draw the structural formula of the product expected from boiling strychnine with aqueous sodium hydroxide.

2. Number the following in order of their relative acidities, listing the least acidic first:

acetic acid; chloroacetic acid; γ-chlorobutyric acid; fluoroacetic acid; isobutyric acid

3. Draw the structures of the major products expected from the intramolecular reactions of

a. 4-hydroxyhexanoic acid; **b.** 2,5-dihydroxypentanal

4. Referring to the structural formula for sucrose (Figure 31.1) draw a structure for the product expected from reaction of sucrose with excess acetyl chloride.

5. Draw the structures of three different forms of each of the monosaccharides resulting from the hydrolysis of sucrose.

PRACTICE FOR PROFICIENCY

Objective (a)

1. Methadone, used in therapy for drug addiction, may be formulated as

a. Label each functional group on this formula.

b. Draw the structural formula of the product expected from treatment of methadone with HCl.

c. Draw the formula of the product expected from treatment of methadone with lithium aluminum hydride.

2. For each of the following compounds

a. Identify all functional groups.

b. Draw the structure for the product expected (if any) when the compound is treated with anhydrous HCl.

c. Draw the structure(s) for the product(s) expected (if any) when the compound is boiled with aqueous sodium hydroxide.

d. Draw the structure of the product expected (if any) when the compound is treated with lithium aluminum hydride.

nicotine, the tobacco alkaloid

novocaine, a local anesthetic

norlutin, one of the first synthetic birth control drugs

tetracycline, a "broad spectrum" antibiotic

methyl anthranilate, a grape flavoring agent

Objective (b)

3. Arrange the following in order of acid strength, listing the strongest first:

benzoic acid; p-chlorobenzoic acid; 2,4-dinitrobenzoic acid

4. Arrange the following in order of base strength, listing the strongest first:

aniline; methylamine; p-nitroaniline

5. Arrange the following in order of base strength, listing the strongest first:

$ClCH_2CO_2^-$; $Cl_3CCO_2^-$; $(CH_3)_3CCO_2^-$; $CH_3CO_2^-$

Objective (c)

6. Draw structural formulas for the products expected from intramolecular reactions of

a. 4-hydroxypentanoic acid

b. 5-hydroxy-2-hexanone

d. $HOCH_2CH_2CHCO_2H$
            $\underset{OH}{|}$

e. $CH_3CH_2CHCH_2CH_2CH_2CCH_2CH_3$
            with $\underset{OH}{|}$ and $\overset{O}{\overset{||}{C}}$

Objective (d)

7. The structure of *maltose* (malt sugar) is

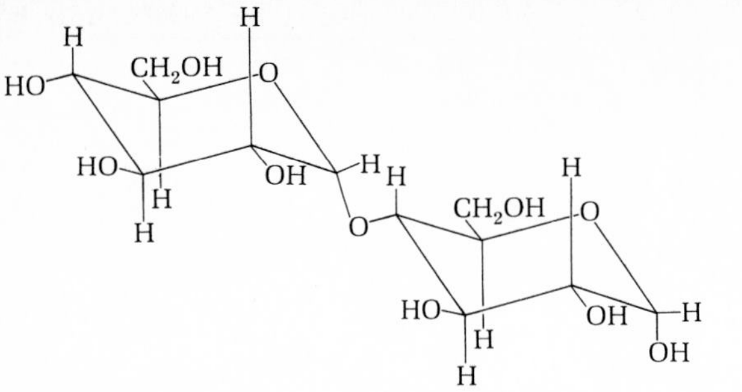

a. What monosaccharide would be expected as the hydrolysis product of *maltose*?

b. Draw the structures of two cyclic and one open-chain form of the hydrolysis product.

c. Draw the structure of the compound expected when *maltose* is treated with excess acetyl chloride.

d. Explain why maltose is expected to be quite soluble in water.

e. Draw the structure of the methyl acetal formed when *maltose* reacts with methanol containing a trace of anhydrous HCl.

8. When cotton fibers under go partial hydrolysis, a disaccharide is obtained called *cellobiose*:

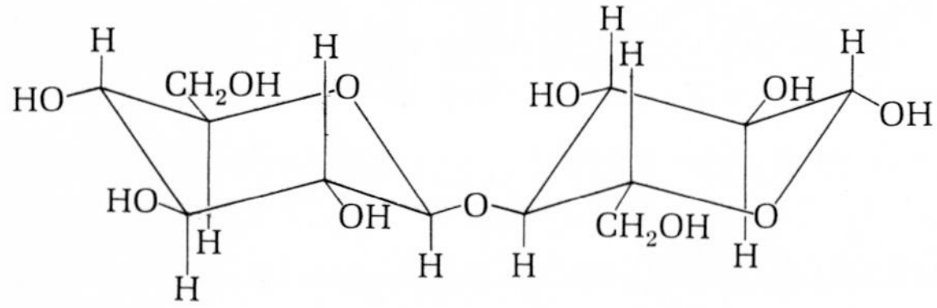

Further hydrolysis of *cellobiose* yields the same product as that obtained by hydrolysis of *maltose* (question 7). How-

ever, the enzyme *maltase* that catalyzes the hydrolysis of *maltose* has no effect on *cellobiose*. The hydrolysis is catalyzed by an enzyme found in certain almonds, but this enzyme has no effect on *maltose*. Account for these observations.

9. Glucose is sometimes classified as a *reducing sugar* because in aqueous solution it is a good reducing agent (i.e., it is easily oxidized).

a. Draw the structure of glucose in a form indicating a functional group expected to be easily oxidized (Unit 29).

b. Explain why *sucrose* is not a *reducing sugar*.

BRAIN TINGLER

10. A sugar found in honey has the structure:

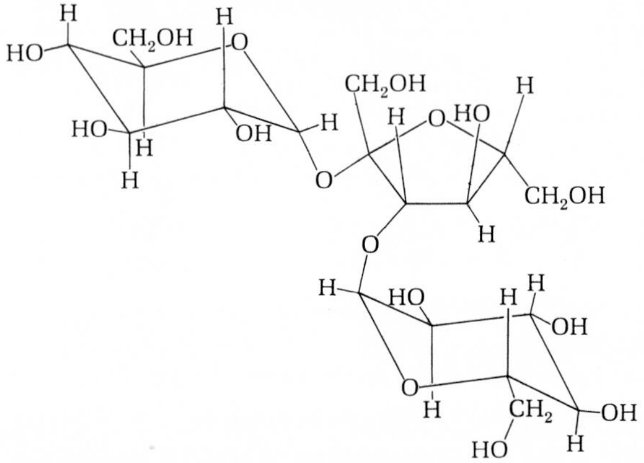

Partial hydrolysis of this *trisaccharide*, catalyzed by a specific enzyme, yields a monosaccharide and a disaccharide, both of which have been described in this unit. What are these partial hydrolysis products?

Unit 31 Self-Test

1. Label each functional group in the structural formula for *reserpine*, a potent tranquilizer:

2. Draw the structural formulas for all organic products expected as the result of boiling reserpine with aqueous sodium hydroxide. (Refer to the reactions of functional groups tabulated in Unit 29).

3. Indicate the relative acidities of the following compounds by numbering them 1–5 (least acidic to be labeled 5):
acetic acid; dimethylpropionic acid; methoxyacetic acid; β-methoxypropionic acid; trifluoroacetic acid

4. Draw structural formulas for the major products expected from intramolecular reactions of
a. γ-hydroxybutyric acid; **b.** 3,6-dihydroxy-2-hexanone

5. Draw the structural formula for the product formed by treatment of ribose with excess acetyl chloride.

6. Draw the structure of the product formed when ribose is treated with methanol containing a small amount of anhydrous HCl.

7. Draw an open-chain form of ribose.

Amino acids, peptides, and proteins

> . . . Over a period of thirty years I have witnessed the change from almost complete ignorance to a significant amount of knowledge and understanding of the structure of proteins.
> *Daedalus* (Fall 1970)
> LINUS PAULING

Objectives

(a) Be able to identify acidic and basic groups in common amino acids (Tables 32.1 and 32.3).

(b) Be able to apply information on functional group properties (Unit 29) to reactions involving amino acids and peptides (Section 32.4).

(c) Be able to calculate and use R_f values from chromatography patterns to identify amino acids (Section 32.6).

(d) Be able to deduce the amino acid sequences of simple peptides from identification of products from partial hydrolysis (Section 32.7).

(e) Be able to describe the types of bonds determining protein geometry and the structural changes expected from reactions involving these bonds (Section 32.8).

When protein foods are metabolized (Unit 34), the complex proteins are first broken down by hydrolysis of amide bonds to simpler species, the peptides, and these are eventually hydrolyzed to amino acids. The different amino acids (Table 32.1) undergo a variety of reactions in normal metabolic processes. One of the common amino acids, present in a relatively large amount in milk proteins, is phenylalanine:

$$\bigcirc-CH_2-\overset{\overset{\displaystyle H}{|}}{\underset{\underset{\displaystyle NH_3^+}{|}}{C}}-CO_2^-$$

In the normal metabolic pathways, phenylalanine undergoes an oxidative degradation. A small fraction is converted to the keto acid, phenylpyruvic acid,

$$\bigcirc-CH_2-\overset{\overset{\displaystyle O}{\|}}{C}-CO_2H$$

but most of the phenylalanine is changed to tyrosine,

$$HO-\bigcirc-CH_2-\overset{\overset{\displaystyle H}{|}}{\underset{\underset{\displaystyle NH_3^+}{|}}{C}}-CO_2^-$$

by a reaction catalyzed by the enzyme phenylalanine hydroxylase. The tyrosine formed undergoes further degradations, eventually being converted to simple aliphatic acids or, by an alternative pathway, into a component of the melanin pigments of skin or hair.

In the 1930s, Dr. Asbjorn Folling, A Swedish physician, discovered that the urine of two mentally retarded children contained abnormally high amounts of phenyl-

Table 32.1 SOME COMMON AMINO ACIDS $\left(R-\overset{\overset{\displaystyle H}{|}}{\underset{\underset{\displaystyle NH_3^+}{|}}{C}}-COO^- \right)$

STRUCTURE	R CHARACTER	COMMENTS
$H-\overset{\overset{\displaystyle H}{\|}}{\underset{\underset{\displaystyle ^+NH_3}{\|}}{C}}-COO^-$ glycine	hydrogen	neutral; *no asymmetric carbon*
$CH_3-\overset{\overset{\displaystyle H}{\|}}{\underset{\underset{\displaystyle ^+NH_3}{\|}}{C}}-COO^-$ alanine	alkyl	neutral
$\bigcirc-CH_2-\overset{\overset{\displaystyle H}{\|}}{\underset{\underset{\displaystyle ^+NH_3}{\|}}{C}}-COO^-$ phenylalanine	aromatic	neutral
$HOCH_2-\overset{\overset{\displaystyle H}{\|}}{\underset{\underset{\displaystyle ^+NH_3}{\|}}{C}}-COO^-$ serine	hydroxyl	neutral

Table 32.1 (Continued)

STRUCTURE	R CHARACTER	COMMENTS
$CH_3SCH_2CH_2\overset{\overset{\text{H}}{|}}{\underset{\underset{+NH_3}{|}}{C}}-COO^-$ methionine	thioether	neutral
$H_3\overset{+}{N}(CH_2)_4\overset{\overset{\text{H}}{|}}{\underset{\underset{NH_2}{|}}{C}}-COO^-$ lysine	amino	basic
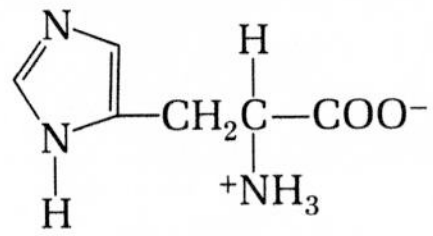 histidine	imidazole	basic
$HOOC(CH_2)_2\overset{\overset{\text{H}}{|}}{\underset{\underset{+NH_3}{|}}{C}}-COO^-$ glutamic acid	carboxyl	acidic
$HO-\langle\bigcirc\rangle-CH_2-\overset{\overset{\text{H}}{|}}{\underset{\underset{+NH_3}{|}}{C}}-COO^-$ tyrosine	phenol	acidic
$HSCH_2\overset{\overset{\text{H}}{|}}{\underset{\underset{+NH_3}{|}}{C}}-COO^-$ cysteine	sulfhydryl	acidic

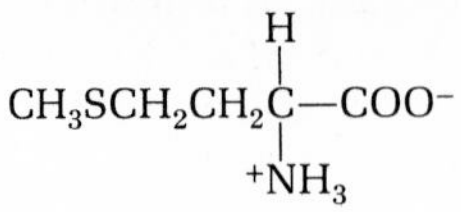

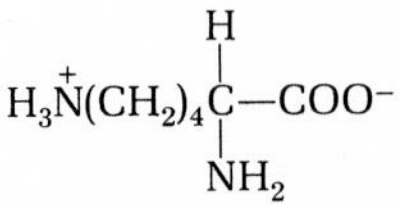

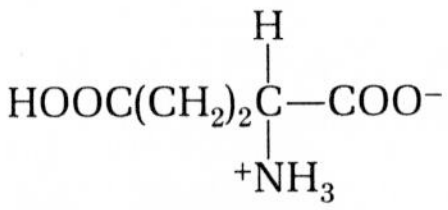

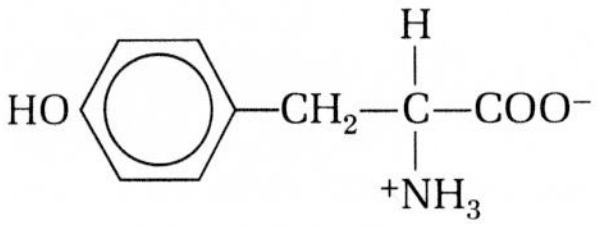

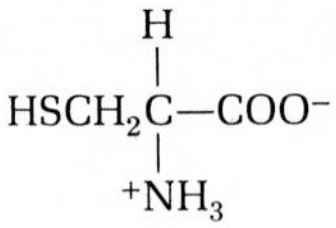

pyruvic acid. Subsequent research showed that the mental retardation was the result of a metabolism error caused by a deficiency in the phenylalanine hydroxylase enzyme system. The disease, called "phenylketonuria" (PKU), could be detected by a simple color test applied to the wet diaper of an infant. Like phenols, some keto acids react with iron(III) chloride to form a colored product. The reaction apparently detects the enol form of the keto acid:

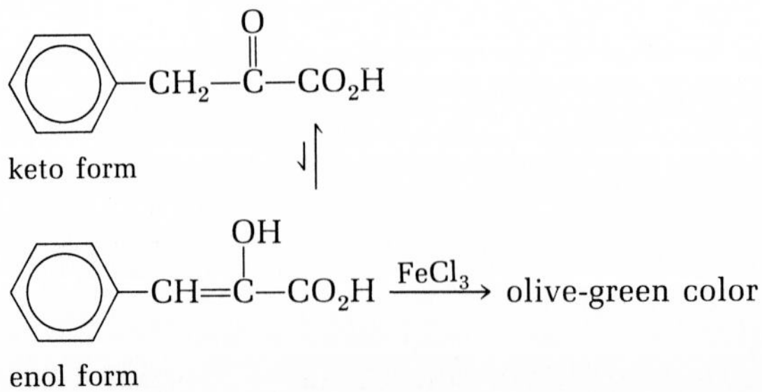

$$\langle\bigcirc\rangle-CH_2-\overset{\overset{\text{O}}{\|}}{C}-CO_2H$$

keto form

$$\langle\bigcirc\rangle-CH=\overset{\overset{\text{OH}}{|}}{C}-CO_2H \xrightarrow{\text{FeCl}_3} \text{olive-green color}$$

enol form

The Guthrie test, developed by Dr. Robert Guthrie at the State University of New York at Buffalo.

This simple test has now been largely replaced by a more sensitive, and generally more reliable, test using a few drops of blood to detect abnormally high concentrations of phenylalanine in the blood serum.

If the disease is detected in early infancy, the treatment is quite simple. The diet of the baby is altered to reduce the intake of proteins rich in phenylalanine, and supplementary tyrosine is added to make up for that which would normally have been formed from phenylalanine.

PKU is only one of a number of molecular diseases, many of which involve errors in amino acid metabolism.[1]

In this unit we shall study some of the fundamental aspects of the chemistry of amino acids, and we shall see how the structures of peptides may be determined and how various bonds contribute to the geometries of natural proteins. Amino acid sequences and geometric features of proteins are the keys to their functions in living systems.

32.1 AMINO ACIDS: IONIC FUNCTIONAL GROUPS

The simplest amino acids are bifunctional, containing a single basic group and a single acidic group. They are commonly represented by the general formula

$$R-\underset{\underset{H}{|}}{\overset{\overset{\ddot{N}H_2}{|}}{C}}-CO_2H$$

Now, we know enough chemistry to suspect that an amine and a carboxylic acid should not coexist without a proton exchange taking place. We would not, for example, expect a mixture of methylamine and acetic acid to remain stable. Rather, they should react to form the *salt* methylammonium acetate. Logic, then, suggests that amino acids are probably not very stable in the *molecular* form, and we might propose a more probable formulation:

$$R-\underset{\underset{H}{|}}{\overset{\overset{{}^+NH_3}{|}}{C}}-CO_2{}^-$$

Such a structure represents a species containing both a positive and a negative charge. Ions of this type are called *dipolarions* (or, in the older literature, *zwitterions*).

That is a good logical conclusion, but we have learned by now that we should not be satisfied with "logic" in science. Molecules are relatively immune to human thought processes and if we are to determine the correct formulation for amino acids, we must find experimental methods for testing our hypotheses. Such tests have been applied exhaustively to a variety of amino acids, and they confirm beyond any reasonable doubt that the dipolarionic formulation is the best description. Let us use glycine to illustrate the types of evidence supporting this conclusion.

Solution characteristics of glycine

a. K_a and K_b Values (see also Unit 19)

Simple carboxylic acids have K_a values around 10^{-5}, and simple alkylammonium ions have K_a values around 10^{-11}. Aliphatic amines have K_b values around 10^{-3}–10^{-4}, and aliphatic carboxylate salts have K_b values around 10^{-9}–10^{-10}.

[1] *The Metabolic Basis of Inherited Diseases*, edited by J. B. Stanbury et. al. (New York: McGraw-Hill, 1966).

This solvent is useful because it does not participate in proton exchange (as in the case of D_2O, for example) and it contains no "ordinary" hydrogen. Deuterium does not absorb radio-frequency radiation in the same region as ordinary hydrogen. The d_7 notation refers to the replacement of all seven hydrogens in dimethylformamide by deuterium:

$$D-\overset{\overset{O}{\|}}{C}-N\overset{CD_3}{\underset{CD_3}{\Big\langle}}$$

Crystalline glycine

For glycine, $K_a = 1.6 \times 10^{-10}$ and $K_b = 2.5 \times 10^{-12}$, indicating that the acidity is due primarily to the $-NH_3^+$ group and the basicity to the $-CO_2^-$ group.

b. nmr Spectrum

Glycine, dissolved in dimethylformamide-d_7, shows only two absorptions in the nmr spectrum, one characteristic of $-CH_2-$ and the other similar to the $-NH_3^+$ group in ethylammonium chloride. Only a trace of O—H absorption is observed, showing that the carboxyl group is nearly 100 percent ionized.

Glycine is not as soluble in this solvent as one would like for an optimum spectrum, but the data are consistent with the dipolarionic formulation.

The preceding evidence strongly supports the dipolarionic structure of an amino acid in solution, but we know that some molecular substances (e.g., HCl) do not become ionic until dissolved in a polar solvent.

What about the structure of amino acids in the pure crystalline form?

a. Vapor Pressure and Melting Behavior

Molecular substances having about the same formula weight as glycine would probably have fairly high vapor pressures at temperatures above room temperature. For example, ethyl acetate (molecular weight, 88) has a vapor pressure of 100 torr at 27°C, and aniline (molecular weight, 93) has a vapor pressure of 1 torr at 35°C and 100 torr at 120°C. The hydrogen-bonding characteristics of aniline account for the differences between it and ethyl acetate. Acetamide, also having hydrogen bonding, is isomeric with glycine, and acetamide has a vapor pressure of 1 torr as the crystalline solid (at 65°C), and liquid acetamide has a vapor pressure of 100 torr at 158°C.

Glycine has a very low vapor pressure at temperatures below 200°C and cannot be sublimed. Glycine also has a high melting point (~230°C), like an ionic salt. Again, the evidence seems to indicate a dipolarionic structure.

b. Infrared Spectrum

The best evidence for the structure of glycine is its infrared spectrum, using a solid sample. Although infrared spectra are often determined for liquids or solutions, it is fairly easy to obtain spectra of solids. A small amount of the crystalline solid is ground with pure crystalline potassium bromide. The mixture is placed in a special pellet press and compressed into a thin, translucent wafer. This is used as the sample for the spectrum determination.

Infrared spectra of amino acids (Table 32.2) furnish convincing proof that the structures are best represented by dipolarionic formulations even in the crystalline state.

We shall use this formulation, then, in our subsequent discussions. When we have more than one acidic or basic group in the amino acid (Table 32.1), the selection of the "correct" groups to draw in ionic forms is a bit tricky. Actually, there will be several different species as components of an equilibrium system, and a single formulation will represent only the major component. The formulation is drawn on the basis of relative K_a and K_b values, considering inductive effects. Table 32.1 shows examples of such compounds.

32.2 COMMON FEATURES OF MOST AMINO ACIDS

Aqueous solutions of amino acids can absorb added acid or base with less of a pH change than that shown by pure water. The amino acids are therefore rather effective buffers and since they may either donate or accept protons, they are *amphoteric* (Units 19 and 20).

Amphoterism

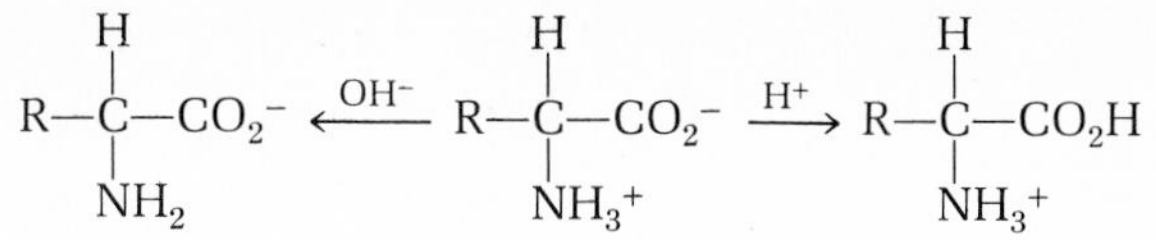

$$R-\underset{\underset{NH_2}{|}}{\overset{\overset{H}{|}}{C}}-CO_2^- \xleftarrow{\;OH^-\;} R-\underset{\underset{NH_3^+}{|}}{\overset{\overset{H}{|}}{C}}-CO_2^- \xrightarrow{\;H^+\;} R-\underset{\underset{NH_3^+}{|}}{\overset{\overset{H}{|}}{C}}-CO_2H$$

Geometry

Most natural amino acids are found to contain a carboxylate group and an ammonium group attached to the same carbon. These relative positions are called *alpha, α* (not to be confused with the "α geometry" of glycoside links). In addition, most of the amino acids contain an asymmetric carbon (the α carbon) in a particular configuration called L (Figure 32.1).

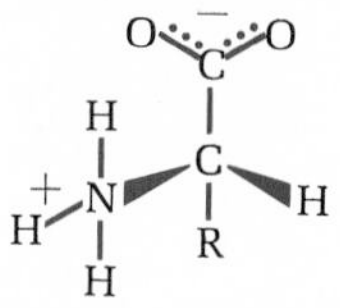

Figure 32.1 An L-α-amino acid. By convention, the side groups project above the plane of the paper and the top and bottom groups project below the plane of the paper. If the —COO⁻ group is in the position shown, then the —NH₃⁺ group is on the *left* for the L form and on the right for the mirror-image D isomer.

Table 32.2 INFRARED SPECTRAL EVIDENCE FOR DIPOLARIONIC STRUCTURES

INFRARED ABSORPTION (cm^{-1})	ORIGIN	FREE AMINES (R—NH₂)	"AMMONIUM" SALTS (R—NH₃⁺)	CARBOXYLIC ACIDS (RCO₂H)	CARBOXYLATE SALTS (RCO₂⁻)	AMINO ACIDS
3400–3300	—N—H asymmetric stretch	×	—	—	—	—
3300–3250	—N—H symmetric stretch	×	—	—	—	—
3300–2500 (broad)	—O—H stretch	—	—	×	—	—
3100–2600 (broad)	—N⁺—H stretch	—	×	—	—	×
2220–2000	—N⁺—H asymmetric bend	—	×	—	—	×
1755–1730	C=O stretch (in CO₂H)	—	—	×	—	—
1650–1580	—N—H bend	×	—	—	—	—
1600–1590	—C(=O)(O) asymmetric stretch	—	—	—	×	×
1550–1490	—N⁺—H symmetric bend	—	×	—	—	×
1410–1380	—C(=O)(O) symmetric stretch	—	—	—	×	×

32.3 STRUCTURAL VARIATIONS

Twenty-three different amino acids have been found in hydrolysis mixtures from common natural proteins, and a number of additional amino acids have been detected in some rather unusual compounds. There is little point in listing all of the known amino acids, but the examples in Table 32.1 indicate some of the structural differences found.

32.4 REACTIONS OF AMINO ACIDS

All amino acids participate in proton-exchange reactions. The acidic group in the simple bifunctional amino acids is the relatively weak alkylammonium ion. Polyfunctional amino acids may contain other acidic groups.

Other reactions typical of amines (or alkylammonium salts), carboxylic acids (or carboxylate salts), and other functional groups commonly encountered are to be expected. In addition, the amino acids may undergo reactions such as decarboxylation (loss of CO_2) or oxidative deamination (conversion of $H-\overset{|}{\underset{|}{C}}-NH_3^+$ to $\overset{|}{C}=O$).

These and other special reactions are typical of the enzyme-catalyzed processes of metabolism systems.

Some of the more common reactions of amino acids are listed, with specific examples, in Table 32.3.

Amino acids are the monomer units from which proteins are formed. The linkages connecting these monomers are amide bonds involving nitrogen from one amino acid and the carbon from the carboxylate group of another. Usually, other groups in polyfunctional amino acids remain free to react in their own characteristic ways.

Such bonds in proteins and smaller amino acid polymers are more commonly referred to as peptide bonds.

32.5 PEPTIDES AND PROTEINS

Although it might seem difficult to believe, the chemistry studied thus far is really quite simple—at least by comparison with the chemistry associated with even the most primitive living system. Consider the number of different reactions possible (Table 32.3) for a single amino acid such as tyrosine:

$$H-O-\underset{}{\bigcirc}-CH_2-\overset{\overset{\displaystyle H}{|}}{\underset{\underset{\displaystyle {}^+NH_3}{|}}{C}}-COO^-$$

Now multiply the complexity of this case by a few hundred and you will have a general idea of the variety of reactions that might be encountered with a fairly large protein molecule. Finally, multiply this by a few thousand and you will see that it is not surprising that man does not yet know all of the chemistry of a single living cell. Indeed, the surprising thing is that chemists have succeeded in unraveling as many aspects of molecular biology as they have so far.

Peptides and proteins are of interest not only because of their chemical properties but, more important, because of the tremendous variety of functions they fulfill in a living organism. A few examples will illustrate both the complex chemistry of these substances and the variable roles they play in the life process.

Table 32.3 COMMON REACTIONS OF AMINO ACIDS

1. "INDEPENDENT" GROUP REACTIONS

   a. ACIDITY

carboxyl

$$HOOC(CH_2)_2\overset{\overset{\displaystyle H}{|}}{\underset{\underset{\displaystyle {}^+NH_3}{|}}{C}}-COO^- \rightleftharpoons {}^-OOC(CH_2)_2\overset{\overset{\displaystyle H}{|}}{\underset{\underset{\displaystyle {}^+NH_3}{|}}{C}}-COO^- + H^+ \qquad K_a = 10^{-4}\text{--}10^{-5}$$

phenol

$$H-O-\underset{}{\bigcirc}-CH_2-\overset{\overset{\displaystyle H}{|}}{\underset{\underset{\displaystyle NH_2}{|}}{C}}-COO^- \rightleftharpoons {}^-O-\underset{}{\bigcirc}-CH_2-\overset{\overset{\displaystyle H}{|}}{\underset{\underset{\displaystyle NH_2}{|}}{C}}-COO^- + H^+ \qquad K_a = \sim 10^{-10}$$

Table 32.3 (Continued)

ammonium

$$CH_3-\overset{\overset{\displaystyle H}{|}}{\underset{\underset{\displaystyle H}{\overset{+}{N}}{\underset{\displaystyle H}{|}}}{C}}-COO^- \rightleftharpoons CH_3-\overset{\overset{\displaystyle H}{|}}{\underset{\underset{\displaystyle H}{:N-H}}{C}}-COO^- + H^+ \qquad K_a = {\sim}10^{-9}\text{-}10^{-10}$$

sulfhydryl

$$H-S-CH_2-\overset{\overset{\displaystyle H}{|}}{\underset{\underset{\displaystyle NH_2}{|}}{C}}-COO^- \rightleftharpoons {}^-S-CH_2-\overset{\overset{\displaystyle H}{|}}{\underset{\underset{\displaystyle NH_2}{|}}{C}}-COO^- + H^+ \qquad K_a = {\sim}10^{-11}$$

b. BASICITY

amino

$$H_3\overset{+}{N}(CH_2)_4\overset{\overset{\displaystyle H}{|}}{\underset{\underset{\displaystyle NH_2}{|}}{C}}-COO^- + H-O-H \rightleftharpoons H_3\overset{+}{N}(CH_2)_4\overset{\overset{\displaystyle H}{|}}{\underset{\underset{\overset{+}{N}}{|}}{C}}-COO^- + OH^- \qquad K_b = 10^{-4}\text{-}10^{-5}$$

imidazole

$$\text{(imidazole ring)}-CH_2-\overset{\overset{\displaystyle H}{|}}{\underset{\underset{\displaystyle {}^+NH_3}{|}}{C}}-COO^- + H-O-H \rightleftharpoons \text{(imidazole ring)}-CH_2\overset{\overset{\displaystyle H}{|}}{\underset{\underset{\displaystyle {}^+NH_3}{|}}{C}}-COO^- + OH^- \qquad K_b = {\sim}10^{-8}$$

carboxylate

$$CH_3-\overset{\overset{\displaystyle H}{|}}{\underset{\underset{\displaystyle {}^+NH_3}{|}}{C}}-COO^- + H-O-H \rightleftharpoons CH_3-\overset{\overset{\displaystyle H}{|}}{\underset{\underset{\displaystyle {}^+NH_3}{|}}{C}}-COOH + OH^- \qquad K_b = {\sim}10^{-9}\text{-}10^{-10}$$

c. ESTERIFICATION

$$R-\overset{\overset{\displaystyle H}{|}}{\underset{\underset{\displaystyle {}^+NH_3}{|}}{C}}-COO^- + HOR' \overset{H^+}{\rightleftharpoons} R-\overset{\overset{\displaystyle H}{|}}{\underset{\underset{\displaystyle {}^+NH_3}{|}}{C}}-\overset{\overset{\displaystyle O}{\|}}{C}-O-R' + HOH$$

d. DECARBOXYLATION

$$\text{(imidazole)}-CH_2-\overset{\overset{\displaystyle H}{|}}{\underset{\underset{\displaystyle {}^+NH_3}{|}}{C}}-COO^- \xrightarrow{\text{histidine decarboxylase (enzyme)}} \text{(imidazole)}-CH_2-\overset{\overset{\displaystyle H}{|}}{\underset{\underset{\displaystyle NH_2}{|}}{C}}-H + CO_2$$

histamine (toxic)

e. OXIDATIVE DEAMINATION

phenylpyruvic acid

$$\text{(benzene)}-CH_2-\overset{\overset{\displaystyle H}{|}}{\underset{\underset{\displaystyle {}^+NH_3}{|}}{C}}-COO^- \xrightarrow{\text{an oxidative enzyme}} \text{(benzene)}-CH_2-\overset{\displaystyle C}{\underset{\underset{\displaystyle O}{\|}}{}}-COOH$$

(occurs in abnormal amounts as a result of the molecular disease phenylketonuria, PKU)

2. INTERMOLECULAR REACTION

a. AMIDE FORMATION

$$R-\overset{\overset{\displaystyle H}{|}}{\underset{\underset{\displaystyle {}^+NH_3}{|}}{C}}-COO^- \xrightarrow{\text{heat}} \left[\begin{array}{c} \overset{\displaystyle H}{|}\quad\overset{\displaystyle O}{\|}\quad\;\;\overset{\displaystyle H}{|}\quad\overset{\displaystyle O}{\|} \\ R-C-C-N-C-C \\ |\quad\;\;\;\;|\;\;\;| \\ N-H\;\;H\;\;R \\ | \\ O=C \\ | \\ N-C-H \\ |\;\;\;| \\ H\;\;R \end{array}\right]_n$$

(a polyamide)

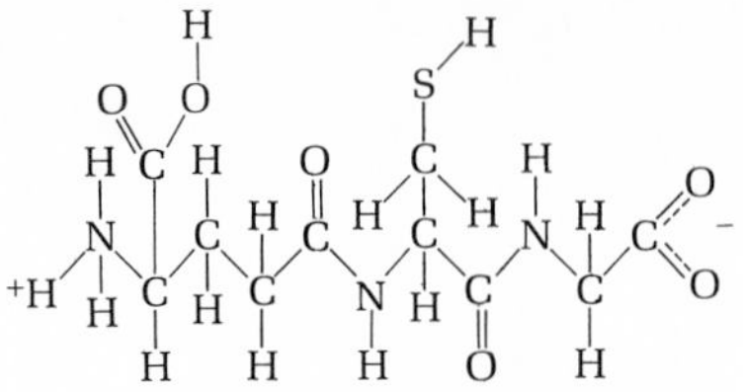

Example 1 (Glutathione)

γ-L-glutamyl-L-cysteinylglycine

This relatively simple tripeptide (made from three amino acids) occurs universally in all living matter. It apparently performs many different functions, sometimes acting as a coenzyme (a reversible aid to enzyme activity). One of its most important roles is to keep cysteine thiols on proteins in the reduced form (Figure 32.2).

Figure 32.2 Glutathione in a redox role (G = glutathione).

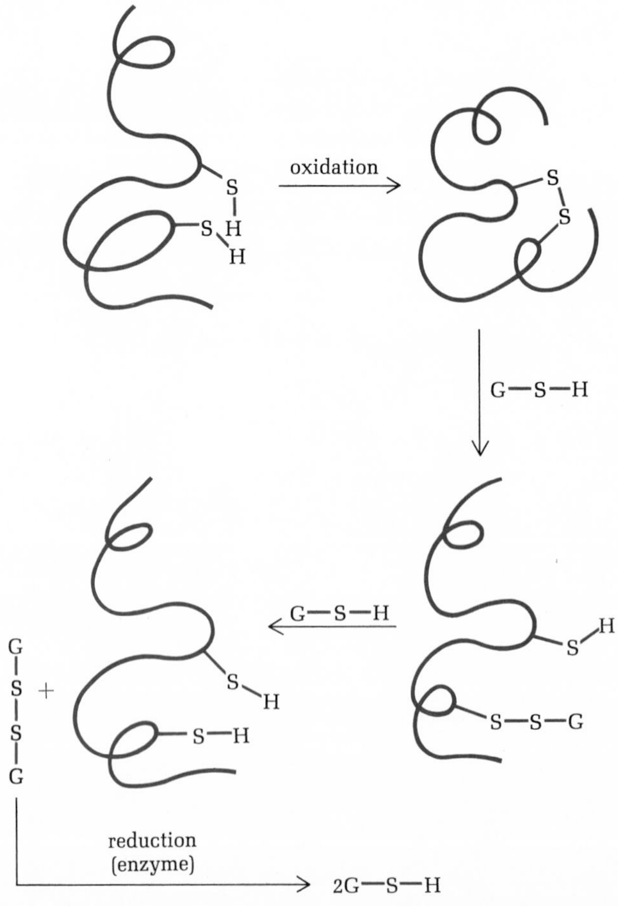

Example 2 (Procamine)

PROCamine is named for its codiscoverers Larry Peck and Rod O'Connor, of Texas A & M University.

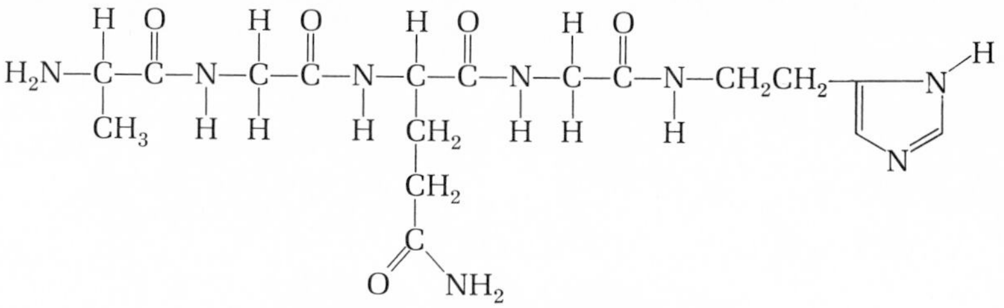

L-alanylglycyl-L-glutaminylglycylhistamine

This compound is found in the venom of the common honeybee and represents an entirely new class of natural products, the *histamine*-terminal peptides. It is not yet known what role such substances play in the complex insect venom, but studies on simpler histamine-terminal peptides indicate that they may be strong chelating agents (Unit 27) for ions such as Cu^{2+} and that they may slowly hydrolyze at physiological pH to release free histamine.

Example 3 (Oxytocin)

This *polypeptide* (contains "many" amino acids) is cyclic in character (Figure 32.3). It acts to stimulate contraction of smooth muscle, causing milk ejection in the lactating female.

Figure 32.3 Oxytocin.

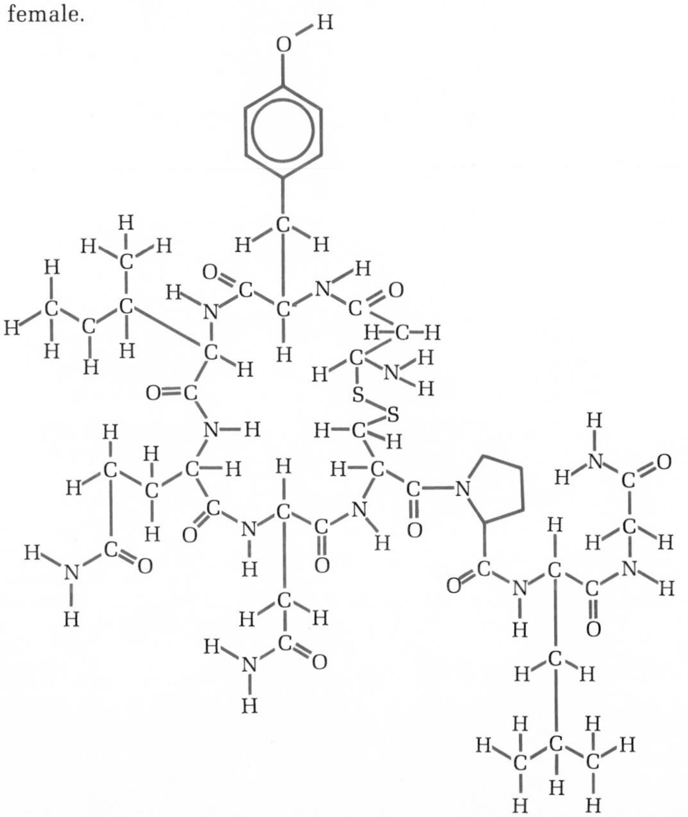

Example 4 (Gramicidin-S)

Another cyclic polypeptide, this substance has totally different properties from those of oxytocin. Gramicidin-S is a naturally occurring antibiotic. Note that the ring in oxytocin involves a disulfide (S—S) link, which can be broken by a mild reducing agent, whereas the ring system in gramicidin-S is held together only by peptide (amide) links (Figure 32.4).

Figure 32.4 Gramicidin-S.

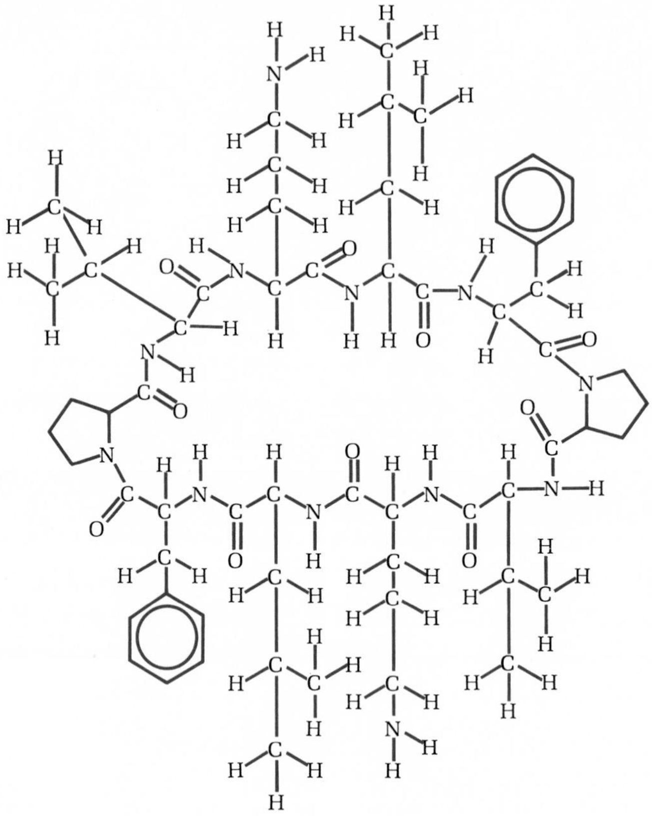

Example 5 (Bradykinin)

This linear *nonapeptide* (made from *nine* amino acid units) is an extremely potent smooth-muscle hypotensive agent. It is broken free from certain plasma proteins by the action of some of the snake venoms. Similar peptides have been found in some of the wasp venoms themselves. Bradykinin (Figure 32.5) is also a pain-producing compound.

Figure 32.5 Bradykinin.

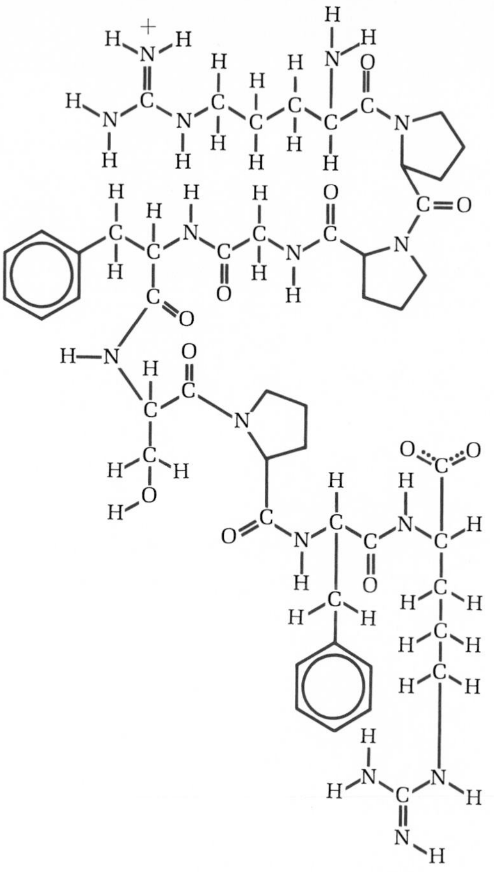

An alternative way of indicating peptide structure is the use of shorthand symbols; for example, for bradykinin,

$$
\begin{array}{c}
\text{arg---pro} \\
| \\
\text{phe---gly---pro} \\
| \\
\text{ser---pro---phe---arg}
\end{array}
$$

(arg = arginine, pro = proline, gly = glycine, phe = phenylalanine, and ser = serine.)

Example 6 (Myoglobin (a complex protein))

Myoglobin (Figure 32.6) consists of 153 amino acid units in a long polypeptide chain curled into a highly complex geometry. In one portion of the molecule is an Fe^{2+} ion, coordinated in a specific way such that it is able to complex reversibly with an oxygen molecule. Myoglobin is a component of muscle tissue, where it functions as an oxygen carrier.

Figure 32.6 Myoglobin.

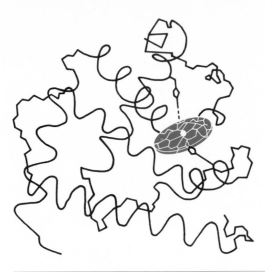

From the preceding examples it is obvious that peptides and proteins may be very complicated molecules, playing a broad spectrum of roles in living systems. The description of their molecular structures should be a challenging problem.

32.6 AMINO ACID COMPOSITION

One of the necessary stages in the description of any macromolecule is the identification of the monomer units from which it is made. Starches, for example, can be hydrolyzed into smaller molecules, eventually glucose. Peptides and proteins, on the other hand, seldom yield a single hydrolysis product. Rather, a mixture of several amino acids is usually obtained.

The most common hydrolysis of peptide bonds uses aqueous acid (Figure 32.7). Certain of the amino acids may, however, undergo structural changes under normal acid hydrolysis conditions so that other procedures (e.g., alkaline hydrolysis or hydrolysis catalyzed by appropriate enzymes) may be necessary to ensure that all component amino acids have been identified.

In order to identify the components of such a hydrolysis mixture, separation techniques must be employed that take advantage of the relatively minor differences in the structures of the amino acids. Two-dimensional chromatography (Figure 32.8) is one such technique. By developing a chromatogram in one dimension with a given solvent system and then in another direction with a different solvent, a fairly effective separation of an amino acid mixture can be obtained. Comparison of a two-dimensional chromatogram, after detection with a color-producing reagent such as ninhydrin, with patterns obtained from known amino acids may be sufficient in some instances to identify the amino acids in the mixture. Quantitative estimation is, however, difficult and this method is usually limited to simpler peptides.

For a discussion of chromatography, see Excursion 4.

Two-dimensional chromatography most commonly employs either a paper sheet or some powdered adsorbent adhering to a glass or plastic backing as the support medium. The latter technique is called "thin-layer chromatography." Usually a mixed solvent system is used so that separation by partitioning is the principal factor involved. For two-dimensional chromatography, the mixture is applied as a small spot near a lower corner of the support medium. The paper or thin-layer plate is then

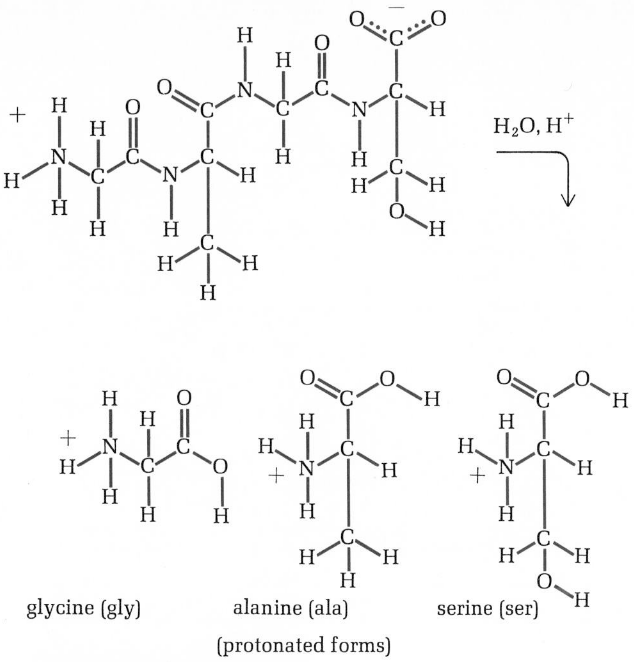

Figure 32.7 Acid hydrolysis of a tetrapeptide. [Hydrolysis involves breaking the peptide links (amide bonds)].

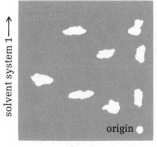

H_2O, H^+

glycine (gly)   alanine (ala)   serine (ser)

(protonated forms)

solvent system 1 ——→

origin

Figure 32.8 Two-dimensional chromatogram.

←—— solvent system 2

dipped into the solvent mixture to a line just below the applied spot, and the solvent moves up through the support, which acts as a wick. After the solvent front reaches a desirable height, the chromatogram is removed, dried, turned 90°, and dipped into a second solvent system. The second development is followed by drying and spraying with some detection reagent to produce color at the locations of the separated amino acids.

Measurement is made from the *center* of the spot to the origin.

Spots may be characterized by so-called R_f values, which measure the distance the component has moved from the origin relative to the distance the solvent front has moved (as measured from the origin).

$$R_f = \frac{\text{distance of spot center from origin}}{\text{distance of solvent front from origin}}$$

(32.1)

R_f values may vary with a number of experimental conditions, such as thickness of the support or temperature. As a result, chromatography patterns usually require confirmation by other analytical methods before unambiguous results can be expected, except in relatively simple cases.

In two-dimensional chromatography, R_f values are measured separately for each of the two solvent systems employed. A comparison with R_f values for known amino acids in the same chromatography system (Table 32.4) permits a tentative identification of each amino acid.

Table 32.4 R_f **VALUES FOR SOME COMMON AMINO ACIDS (USING THIN-LAYER CHROMATOGRAPHY ON SILICA GEL G)**

	R_f VALUE[a]	
AMINO ACID	60 PERCENT ETHANOL (aq) (BY WEIGHT)	1-BUTANOL/ACETIC ACID/WATER (60:20:20 PARTS BY WEIGHT)
alanine	0.47	0.27
glutamic acid	0.63	0.27
glycine	0.43	0.23
histidine	0.32	0.05
lysine	0.03	0.05
phenylalanine	0.63	0.51
serine	0.48	0.19
tyrosine	0.66	0.48

[a] Based on values as determined by William Rosenbrook, Jr., at Montana State University, 1963. For two-dimensional chromatograms, R_f values depend somewhat on which solvent is used first.

Example 7

Hydrolysis of a small peptide resulted in a mixture of amino acids which yielded the two-dimensional chromatograph shown. Assuming that the mixture contained only amino acids listed in Table 32.4, what amino acids were present in the peptide?

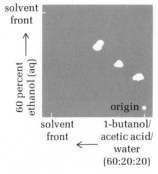

solvent front

60 percent ethanol (aq) →

origin

solvent front

1-butanol/ acetic acid/ water (60:20:20) ←

Solution

Using a ruler to measure the distances involved, we find the following approximate R_f values, beginning with the spot shown nearest the top in the drawing:

60 percent ethanol

$$\text{first } R_f = \frac{6.8 \text{ units}}{10.0 \text{ units}} = 0.68$$

$$\text{second } R_f = \frac{6.4 \text{ units}}{10.0 \text{ units}} = 0.64$$

$$\text{third } R_f = \frac{4.6 \text{ units}}{10.0 \text{ units}} = 0.46$$

$$\text{fourth } R_f = \frac{3.2 \text{ units}}{10.0 \text{ units}} = 0.32$$

1-butanol, etc.

$$\frac{4.8 \text{ units}}{10.0 \text{ units}} = 0.48$$

$$\frac{5.1 \text{ units}}{10.0 \text{ units}} = 0.51$$

$$\frac{2.7 \text{ units}}{10.0 \text{ units}} = 0.27$$

$$\frac{0.6 \text{ units}}{10.0 \text{ units}} = 0.06$$

Checking these values against those in Table 32.4 suggests the following most probable assignment:

first: tyrosine
second: phenylalanine
third: alanine
fourth: histidine

Within recent years, instruments have been employed for the automatic analysis of amino acid mixtures, as well as mixtures of other types of compounds. Separation is performed on ion-exchange resins by elution with buffers of variable composition. Eluted fractions are automatically mixed with a color-producing reagent (e.g., ninhydrin) and analyzed by a recording colorimeter system. Figure 32.9 shows the Beckman Model 116 Amino Acid Analyzer, and Figure 32.10 is an example of the type of printout from analysis of an amino acid mixture. The elution time, under standardized conditions, is used to identify each amino acid, and the peak intensities are descriptive of the relative amounts of each amino acid after corrections for individual characteristics.

Automatic analysis by gas chromatography has also been developed. Since amino acids contain ionic groups, they are not volatile enough for such vapor-phase

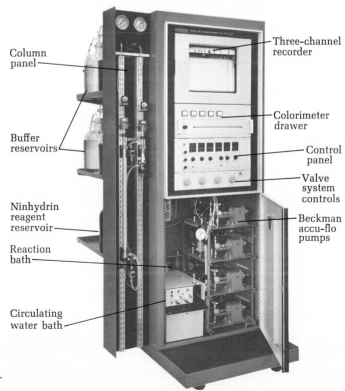

Figure 32.9 Automatic amino acid analyzer.

A: Column panel for mounting ion-exchange columns.
B: (Behind panel) Reservoirs for buffer, ninhydrin, and NaOH.
C: Strip-chart recorder.
D: Switches to activate pumps, reaction bath, and circulating water bath.
E: (Behind switch panel) Colorimeters.
F: Flow meter.
G: Control switches.
H: Valve panel for circulation of buffers and ninhydrin.
I: Circulating pumps.
(Photo courtesy of Beckman Instruments, Inc.)

Column panel

Buffer reservoirs

Ninhydrin reagent reservoir

Reaction bath

Circulating water bath

Three-channel recorder

Colorimeter drawer

Control panel

Valve system controls

Beckman accu-flo pumps

Figure 32.10 Segment of analysis recording. Under the conditions used, these peaks correspond to (1) tryptophan, (2) lysine, (3) histidine, (4) ammonia, (5) arginine, (6) threonine, (7) serine, (8) glutamic acid, (9) proline, (10) glycine, (11) alanine, (12) cystine.

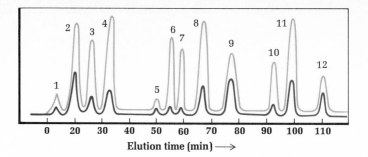

separations and must be converted into more volatile forms for separation and analysis by this technique. For example,

nonvolatile   volatile

32.7 AMINO ACID SEQUENCE

A knowledge of the amino acid composition of a peptide or protein is by no means a sufficient description of the molecule. Even a simple dipeptide made from two different amino acids could have either of two isomeric structures, for example,

By convention, the shorthand representation for a peptide indicates that the amino acid represented first (at left) is the N-terminal end and the one represented last (at right) is the C-terminal end.

glycylalanine
(gly—ala)

alanylglycine

(ala—gly)

The different isomers possible for a protein containing 20 different amino acids in sufficient numbers to give 150 peptide bonds is truly an astronomical figure. It is essential, then, that one be able to determine the sequence in which the different monomer units are connected in polypeptides.

At present, this procedure is a tedious one, often requiring several years of careful research. In some cases it is possible to add a reagent to the N terminal (free $^+NH_3$) or C terminal (free COO^-) end of a peptide chain, hydrolyze the adjacent amide bond preferentially, and identify the "tagged" amino acid. A sequential repetition of this procedure may gradually elucidate the amino acid sequence of the molecule. In large polypeptides and proteins this procedure is more difficult. A promising development is the analysis of sequence by fragmentation patterns in the mass spectrum, an exercise made feasible by preferential fragmentation at peptide bonds. However, computer programs for handling the very complex data from large polypeptides are still in their infancy, and rapid solutions to the puzzles of protein structure are still to be developed.

In the cases of relatively small peptides, it is often possible to deduce the amino acid sequence from analysis of the composition of mixtures of amino acids and peptide fragments resulting from incomplete hydrolysis of the original peptide. Many dipeptides and tripeptides are commercially available and can be used for chromatographic comparison with such partial-hydrolysis mixtures. The key to the deductive process is the fact that the amino acid sequence is retained in the smaller peptides produced.

Example 8

A tetrapeptide is partially hydrolyzed, and the products are identified by comparative chromatography as

    ala—gly
    phe—ala—gly
    gly—gly

along with the free amino acids and other peptides not identified. What was the amino acid sequence of the original peptide?

Solution

Since the original compound was a tetrapeptide (containing four amino acids) and the sequence is preserved in the partial-hydrolysis products, the last two peptides identified indicate that the original sequence was

$$H_3\overset{+}{N}-\underset{\underset{\displaystyle C_6H_5}{\overset{|}{CH_2}}}{\overset{\overset{\displaystyle H}{|}}{C}}-\overset{\overset{\displaystyle O}{\|}}{C}-\underset{\overset{|}{H}}{N}-\underset{\underset{CH_3}{\overset{|}{C}}}{\overset{\overset{\displaystyle H}{|}}{C}}-\overset{\overset{\displaystyle O}{\|}}{C}-\underset{\overset{|}{H}}{N}-CH_2-\overset{\overset{\displaystyle O}{\|}}{C}-\underset{\overset{|}{H}}{N}-CH_2-CO_2^-$$

phe—ala—gly—gly

32.8 MACROMOLECULAR GEOMETRY

Many, indeed most, proteins perform their natural functions only when in a specific shape. X-ray investigations of proteins and metallic derivatives of proteins have determined the structures of a large number of these macromolecules in the crystalline form. Changes in geometry are to be expected, however, when such molecules are dissolved or suspended in an aqueous medium, since there are many regions within the molecule at which hydrogen bonding with water can occur. Direct study of structures in solution or suspension is difficult, although certain measurements may be interpreted as evidence of a general molecular shape, such as fibrous or globular. A rather novel attempt at the description of such systems is using computer-generated oscilloscope displays of macromolecules as influenced by such factors as hydration and change of pH.

There are a number of factors contributing to the shape of a polypeptide or protein. Many of these can be altered by simple reactions, and such reactions may produce either reversible or irreversible *denaturation* (change in natural geometry and function) of the macromolecule. Examples of such factors are outlined in Figure 32.11.

All of the types of bonds described may contribute to the characteristic geometry of a polypeptide or protein. Because of the large number of amide (peptide) links in

Figure 32.11 Bonding factors influencing protein geometry. (*a*) Disulfide bonds. (*b*) Ionic attractions. (*c*) Hydrophobic "bonds." (*d*) Internal hydrogen bonding.

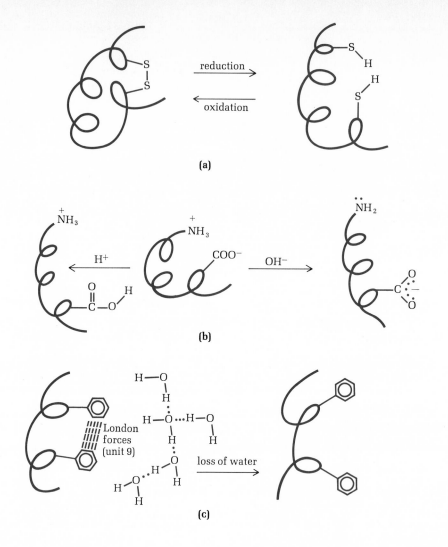

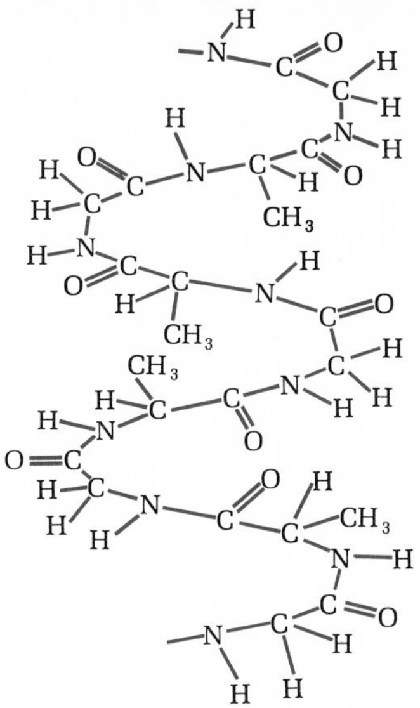

Figure 32.12 (left) Segment of a typical silk fibrous protein. (Hydrogen bonding not indicated)

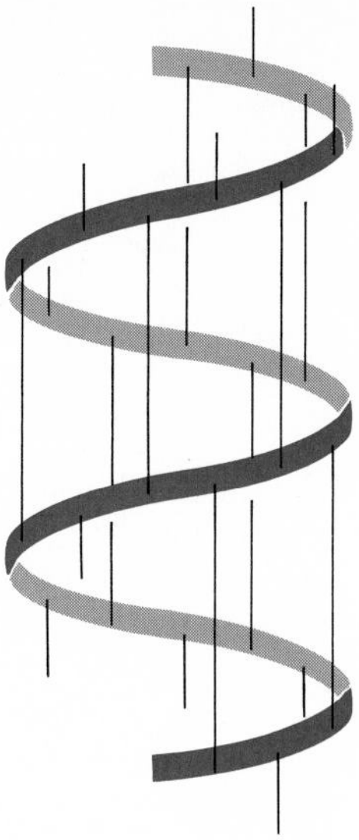

Figure 32.13 (right) Segment of α helix (showing hydrogen bonds).

these molecules, hydrogen bonding is one of the most important factors determining macromolecular shape. Although a detailed consideration of proteins is beyond the scope of this text, it may be of interest to mention briefly a few examples of their structures and functions.

Some fibrous proteins Some of the simplest naturally occurring proteins are the linear polymers of silk from such sources as the silkworm and the spider. These appear, from x-ray diffraction patterns, to be "linear" (zigzag) molecules made from only a few amino acids. For example, ordinary commercial silk is primarily alternating sequences of glycine and L-alanine (Figure 32.12). Other fibrous proteins occur in such materials as muscle tendon, hair, horn, fingernail, claw, and porcupine quill. Unlike the linear silk strands, these latter proteins occur in a more compressed structure, the α *helix* (Figure 32.13), in which the principal feature is formation of internal hydrogen bonds involving nearly all of the amide groupings in the molecule.

Some globular proteins Among the more important nonfibrous proteins are two that act as oxygen carriers, hemoglobin and myoglobin (Figure 32.6 and Unit 27). Both molecules are large polypeptides curled into compact globular forms. In the myoglobin a single Fe^{2+} ion acts as Lewis acid for coordination with molecular oxygen, and there are four such ions in the hemoglobin molecule.

Antibodies Organisms may combat the invasion of foreign bacteria, viruses, or toxins in various ways. One way is by formation of proteins that combine with the foreign substance to remove it from circulation. The mechanisms of antibody formation and activity are not yet entirely known, but the study of this aspect of protein biochemistry is one of the "hottest" areas in current research.

Enzymes
(See also Unit 34) Most of the reactions in living systems would occur spontaneously under appropriate laboratory conditions. You may recall, however, that reactions involving uncharged

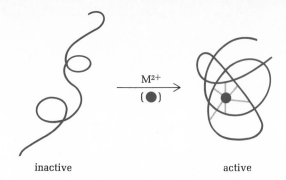

Figure 32.14 Configurational changes resulting from coordination with a metal ion (Unit 27).

inactive   active

molecules are usually slow and frequently inefficient (poor yields). The organism cannot afford too many slow or inefficient reactions if it is to remain alive. Highly specific (and frequently stereoselective) catalysts are used to speed up reactions and control product formation. These catalysts, called *enzymes,* are composed mainly of large protein molecules. Their behavior depends on their ability to adsorb selectively certain reactants (or portions thereof) in regions called *active sites.* Here the reactions can be controlled by factors that orient the reactants for effctive collision and, in some cases, activate the functional groups by electronic interactions. The active site must have the correct geometry for this process and must be surrounded by functional groups providing the proper characteristics for adsorption of reactants, activation of reaction, and facile desorption of products.

Enzyme molecules may assume configurations in which their catalytic function ceases. Such inactive forms may be converted into active catalysts by reaction with small organic molecules (coenzymes) or coordination with specific metal ions in ways that form the protein into the proper shape for its necessary function (Figure 32.14).

32.9 SOME UNUSUAL AMINO ACIDS

Although most amino acids that occur free or in polypeptides in nature have the L structure, some rare cases have been noted in which the D isomers are found (Figure 32.15).

It is possible to determine whether D amino acids are present in a hydrolysis mixture by subjecting the mixture to some reaction catalyzed by a stereoselective enzyme. For example, hydrolysis of a natural peptide indicated the presence of three alanine, one serine, one histidine, and two glutamic acid units. An aliquot of the mixture was treated with an L amino acid oxidase system and was then subjected to analysis again. This time only serine and histidine were found. A second aliquot was treated with a D amino acid oxidase and analyzed, revealing only three alanine and two glutamic acid units. From this it could be concluded that the original peptide contained D-histidine, D-serine, L-alanine, and L-glutamic acid. Care must be exercised to determine that no stereochemical changes have been produced by any hydrolysis procedures used.

We shall discuss some further aspects of protein chemistry in Unit 34.

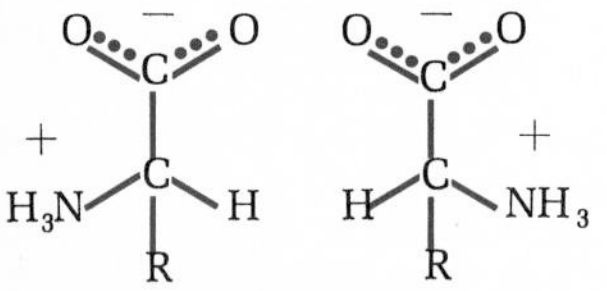

L
(most common)

D
(rare)

Figure 32.15 Optical isomers of amino acids.

SUGGESTED READING

Dickerson, R., and I. Geis. 1969. *The Structure and Action of Proteins.* New York: Harper & Row.
Kopple, K. D. 1966. *Peptides and Amino Acids.* Menlo Park, Calif.: W. A. Benjamin.

BEE STING

The sting of the common honeybee is never a pleasant sensation. For some, it could be fatal. Yet we may someday recognize the lowly bee as the source of valuable medications.

Death from the sting of a single bee, often within less than 20 minutes after the sting, is not uncommon. In most cases death appears to be the result of a severe allergic reaction to proteins in the bee venom. Many persons are allergic to bee sting and must take precautions to avoid serious reactions. The emergency treatment most widely recommended for a severe allergic reaction to the sting is an injection of epinephrine (adrenalin), administered within minutes of the onset of the "shock" reaction.

Bee venom may have significant medical applications. For many years the venom has been widely used in the treatment of rheumatoid arthritis, athough this is not currently endorsed by the American Medical Association. More recently it has been discovered that test animals injected with bee venom can withstand otherwise lethal doses of radiation, suggesting that the venom or some of its components might prove useful as a *radioprotective* drug.

The chemical investigation of an insect venom is quite different from the study of a single product from a natural source. Insect venoms are complex mixtures; 39 different compounds have been identified in honeybee venom and a small fraction of the mixture has yet to be analyzed.

First we must have a *pure* sample of the natural venom. This was a major stumbling block in the early research work in the field, but we can now obtain the pure venom by electrical excitation of individual bees.[2] Then special microchemical techniques must be employed for separation and analysis of the venom components, since it is quite expensive to procure pure venom. (The commercially available "crude" venom sells for $80 to $100 per gram; it is estimated that a gram of ultrapure natural bee venom would cost in excess of $4000.)

Modern techniques of chromatography (Excursion 4) do permit microscale separations, and microspectrophotometric analyses are helpful. In many cases, however, characterization of specific components requires very careful use of chemical degradations on a tiny scale.

We now know that bee venom is composed principally of amino acids, small and large peptides, and some enzymes. We also know a great deal about the physiological properties of many of these components, but many questions still remain to be answered.

The major component (approximately 50 percent of the venom solids) is *mellitin* (actually a "family" of closely related compounds). The mellitins are polypeptides containing 26 amino acid units. They are very toxic and extremely active in rupturing red blood cells. Two other large peptides have been identified, *apamin* and MCD-*peptide,* each constituting about 2 percent of the venom solids. Apamin is the smallest known neurotoxic ("nerve-poisoning") peptide and the MCD-peptide has the unique activity of releasing histamine by rupture of certain skin cells.

Two enzyme systems have been definitely identified and evidence suggests that others may be present. *Hyaluronidase* catalyzes the fragmentation of the hyaluronic acid polymers acting as "intercellular cement" in tissue, thus enabling the venom to

[2] See O'Connor, Rosenbrook, and Erickson in the Volume 139, 1963, issue of *Science* (p. 139).

spread rapidly through the tissue around the area of the sting. *Phospholipase*-A, like mellitin, acts to rupture red blood cells, in this case by catalyzing the destruction of certain cell membrane lipids.

The venom also contains 19 free amino acids, a small amount of free histamine, and a number of small peptides. Among these are two histamine-terminal peptides representing a previously unknown type of natural product. One of these is *procamine*, described in Example 2, and now available in a synthetic form.[3]

Some of the components responsible for the allergic reaction to bee sting have been identified; the hyaluronidase and phospholipase-A systems are definitely allergenic to humans. The radioprotective properties of the whole venom are also exhibited by mellitin and phospholipase-A. Neither of these would be suitable for therapeutic use in humans because of their toxicities, but recent studies of histamine-terminal peptides suggest that some compounds of this type may be both radioprotective and relatively nontoxic.

There is still a great deal to learn about the chemistry and biological activity of bee venom. In the meantime, if a bee should give you a little sample some day, here are some things to do.

1. *Flick* out the stinger with a knife blade or fingernail. Don't "grab and pull." The bee leaves behind the barbed stinger with a sac of venom attached. If you squeeze this while trying to pull out the stinger, you will inject more venom into the wound.

2. Use an ice cube or cold compress on the sting to reduce pain and swelling (and spreading action, by slowing down the activity of the hyaluronidase system).

3. If you should experience symptoms such as skin blotching, generalized itching, nausea, widespread swelling, or difficulty in breathing, consult a doctor right away. These symptoms may indicate a venom allergy, in which case an allergist familiar with insect sting problems should be consulted.

[3] See Peck and O'Connor in the Volume 22, 1974, issue of the *Journal of Agricultural and Food Chemistry* (pp. 51–53).

EXERCISES

1. Threonine is one of the essential amino acids. ("Essential" amino acids are those that must be present in the diet, in protein foods, to maintain normal health and growth.) Its chemical name is 2-amino-3-hydroxybutanoic acid. Draw the structures of major organic products expected from treatment of threonine with
a. excess dilute NaOH; **b.** excess dilute HCl; **c.** methanol, with H_2SO_4 catalyst (assume no polymer formation); **d.** excess acetyl chloride; **e.** Draw the structure of the dipolarionic form of threonine.

2. An equimolar mixture of alanine, serine, lysine, glutamic acid, and tyrosine is dissolved in dilute HCl such that the pH of the final solution is 1.0 and the concentration of each amino acid is 0.01 M. Draw the structures of the principal organic species in the solution.

3. Very recently, DOPA (3,4-dihydroxyphenylalanine) has been found to be effective in the treatment of Parkinson's disease.
a. Draw the structure of DOPA (principal form). DOPA can be converted to the hormone epinephrine (adrenaline) in the body. The first step in the reaction is the removal of CO_2 by an enzyme (similar to the decarboxylation of histidine to histamine). The resulting product is hydroxytyramine, or dopamine.
b. Draw the structures of dopamine (as an equilibrium mixture).

4. **a.** Draw structures (in dipolarionic forms) of all the possible

tripeptides that would give only alanine and serine on hydrolysis.

b. Draw structures (in dipolarionic forms) of all the possible tripeptides that would give only alanine and serine on hydrolysis, in the molar ratios of 2:1 (alanine:serine).

c. A tripeptide is only partially hydrolyzed. The hydrolysis mixture is chromatographed and found to contain (in addition to some unreacted tripeptide) some free alanine, some free serine, some serylserine, and some alanylserine. What was the structure of the tripeptide? Explain.

5. Hair is a fibrous protein containing polypeptide chains bound together by —S—S— linkages. In the hair-waving process, some of these linkages are first broken by treatment with a reducing agent such as ammonium thioglycollate:

$$\{-CH_2-S-S-CH_2-\} + 2HS-CH_2-COONH_4 \longrightarrow$$

$$\{-CH_2-SH + HS-CH_2-\} + \begin{array}{l} S-CH_2-COONH_4 \\ | \\ S-CH_2-COONH_4 \end{array}$$

The hair is now set around curlers, thereby bringing different —SH groups opposite each other. A mild oxidizing agent is now applied, resulting in the formation of new —S—S— linkages, and a "permanent" curl. When potassium bromate is the oxidizing agent, the other products are potassium bromide and water. Using R—SH as the symbol for the reduced fibrous strand, write a balanced equation for the oxidation reaction.

PRACTICE FOR PROFICIENCY

Objectives (a) and (b)

1. Aspartic acid is sometimes called 2-aminobutanedioic acid. Draw the structures (a) of the principal form of aspartic acid, and (b) of the major organic products expected from treatment of aspartic acid with
a. excess dilute NaOH; **b.** excess dilute HCl; **c.** excess acetyl chloride; **d.** excess methanol, with H_2SO_4 catalyst

2. An equimolar mixture of aspartic acid, phenylalanine, and lysine is dissolved in dilute NaOH such that the pH of the solution is 13.0 and the concentration of each amino acid is 0.05 M. Draw the structures of the principal organic species in the solution.

3. Draw the structural formulas of the organic products expected from
a. enzymatic decarboxylation of histidine; **b.** oxidative deamination of glutamic acid; **c.** mild oxidation of cysteine

4. Briefly summarize the evidence for the dipolarionic structure assigned to glycine.

5. The dipolarionic form of histidine is formulated as

When a 1.0-mg sample of histidine is dissolved in 10 ml of 1.0 M HCl, the net charge on the principal organic solute probably is
a. −1; **b.** −2; **c.** +2; **d.** +1; **e.** zero

6. When a 1.0-mg sample of histidine is dissolved in 10 ml of 1.0 M NaOH, the net charge on the principal organic solute probably is
a. +1; **b.** +2; **c.** −2; **d.** −1; **e.** zero

7. The dipolarionic form of lysine is formulated as

$$H_3\overset{+}{N}(CH_2)_4\overset{\overset{\displaystyle H}{|}}{\underset{\underset{\displaystyle NH_2}{|}}{C}}-CO_2^-$$

When a 1.0-mg sample of lysine is dissolved in 10 ml of 0.10 M HCl, the net charge on the principal organic solute probably is
a. −1; **b.** +1; **c.** −2; **d.** +2; **e.** zero

8. When a 1.0-mg sample of lysine is dissolved in 10 ml of a solution buffered at the pH of its isoelectric point, the net charge on the principal organic solute probably is
a. +1; **b.** +2; **c.** −2; **d.** −1; **e.** zero

9. The dipolarionic form of tyrosine is formulated as

$$HO-\underset{}{\bigcirc}-CH_2-\overset{\overset{\displaystyle H}{|}}{\underset{\underset{\displaystyle +NH_3}{|}}{C}}-CO_2^-$$

When a 1.0-mg sample of tyrosine is dissolved in 10 ml of 0.10 M HCl, the net charge on the principal organic solute probably is
a. −1; **b.** −2; **c.** +2; **d.** +1; **e.** zero

10. When a 1.0-mg sample of tyrosine is dissolved in 10 ml of 0.10 M NaOH, the net charge on the principal organic solute probably is
a. +1; **b.** +2; **c.** −2; **d.** −1; **e.** zero

11. What is the maximum number of moles of acetyl chloride that could react with 1 mole of tyrosine?
a. 6; **b.** 5; **c.** 4; **d.** 3; **e.** 2

12. What is the maximum number of moles of acetyl chloride that could react with 1 mole of lysine?
a. 5; **b.** 4; **c.** 3; **d.** 2; **e.** 1

Objective (c)

13. By referring to the chromatogram and table of R_f values, specify the probable amino acid composition of the mixture chromatographed.

AMINO ACID	R_f
ala	0.47
gly	0.43
his	0.32
lys	0.03
phe	0.63

a. his, ala; **b.** gly, ala; **c.** lys, his; **d.** gly, phe; **e.** phe, lys

14. By referring to the chromatogram and table of R_f values, specify the probable amino acid composition of the mixture chromatographed.

AMINO ACID	R_f
ala	0.47
glu	0.63
gly	0.43
his	0.32
lys	0.03

a. gly, ala; **b.** lys, his; **c.** gly, glu; **d.** glu, lys; **e.** his, ala

15. By referring to the chromatogram and table of R_f values, specify the probable amino acid composition of the mixture chromatographed.

AMINO ACID	R_f
ala	0.47
glu	0.63
gly	0.43
his	0.32
lys	0.03

a. ala, gly; **b.** lys, his; **c.** gly, glu; **d.** his, glu; **e.** ala, lys

16. By referring to the chromatogram and table of R_f values, specify the probable amino acid composition of the mixture chromatographed.

AMINO ACID	R_f
ala	0.47
glu	0.63
gly	0.43
his	0.32
lys	0.03
phe	0.63

a. phe, lys; **b.** his, glu; **c.** ala, gly; **d.** lys, his; **e.** phe, glu

Objective (d)

17. Give the dipolarionic structure of the tripeptide that can be partially hydrolyzed to a mixture of phenylalanine, methionine, phenylalanylphenylalanine, and phenylalanylmethionine.

18. Give the probable dipolarionic structures of the tripeptides for which partial hydrolysis yields a mixture of
   a. glycine, serine, glycylglycine, and serylglycine
   b. phenylalanine, serine, phenylalanylphenylalanine, and serylphenylalanine
   c. alanine, serine, alanylserine, and alanylalanine
   d. gly, phe, phe-gly, phe-phe
   e. gly, phe, ser, phe-gly, ser-phe

Objective (e)

19. Explain briefly how a protein might be denatured by addition of a strong acid.

20. There are a number of types of bonding contributing to the natural geometry of a complex protein. For each of the following factors known to cause *denaturation* of various proteins, indicate the type(s) of bonding most likely to be affected and draw simple schematic representations of the structural changes involved.

   a. heating the protein to a temperature just below that required for rupture of "full" covalent bonds
   b. addition to an aqueous suspension of a protein of cold dilute HCl
   c. dilution of a natural aqueous suspension of the protein with 95% ethanol
   d. addition of a mild reducing agent

BRAIN TINGLER

21. Draw structural formulas for all of the products expected from acid hydrolysis (assuming no structural changes in amino acids themselves) and indicate molar ratios of the products from
   a. glutathione; **b.** procamine; **c.** bradykinin; **d.** oxytocin; **e.** gramicidin-S

Unit 32 Self-Test

(Consult Table 32.1 when necessary.)

1. Draw the best structural formulations for the amino acids indicated and label acidic and basic groups as shown in the example.

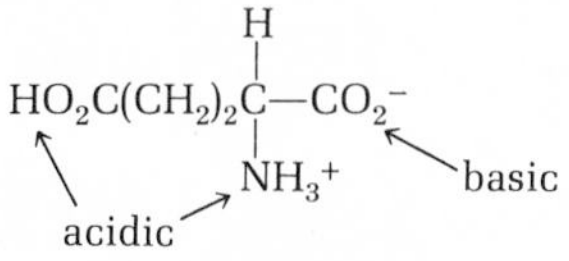

$$HO_2C(CH_2)_2\overset{\overset{\displaystyle H}{|}}{C}-CO_2^-$$

acidic → NH$_3^+$ ← basic

glutamic acid

a. serine; **b.** lysine; **c.** tyrosine

2. Draw the structures of the major organic products expected when serine is treated with the reagents indicated.

a. excess aqueous NaOH
b. excess aqueous HCl
c. excess CH$_3$CCl
$\overset{\parallel}{O}$
d. excess ethanol, with H$_2$SO$_4$ catalyst (assume no polymer formation)
e. H$_2$SO$_4$ catalyst (polymer formation)

3. Indicate the net charge on each of the major organic products expected when the amino acids shown are dissolved in excess aqueous sodium hydroxide:

a. phenylalanine; **b.** lysine; **c.** glutamic acid; **d.** cysteine
e. 3,4-dihydroxyphenylalanine (DOPA)

4. Draw the structural formulas of the organic products expected from:

a. enzymatic decarboxylation of tyrosine
b. oxidative deamination of alanine

5. Briefly summarize the evidence suggesting that amino acids are best represented as dipolarionic species.

6. By referring to Table 32.4 and measuring the appropriate distances on the chromatogram shown below, specify the probable amino acid composition of the mixture chromatographed.

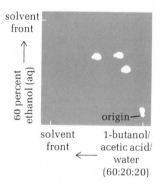

solvent front

60 percent ethanol (aq) →

origin

solvent front / 1-butanol/ acetic acid/ water (60:20:20)

7. a. Give the structures of all possible tripeptides that would yield alanine, tyrosine, and glycine on hydrolysis. (Draw dipolarionic formulations.)

b. Give the structure of the tripeptide that is partially hydrolyzed to a mixture of glycine, alanine, tyrosine, glycylalanine, and alanyltyrosine. (Draw the dipolarionic formulation.)

8. Give the structure of the organic product that would result from treatment of the peptide in question 7b with excess acetyl chloride.

9. Explain briefly why a protein might be denatured by a significant change in the pH of the solution in which it is dispersed.

Unit 33

COKE, PEPSI choose plastic bot-
tles.
Chemical & Engineering News,
June 9, 1975

Synthetic polymers

Objectives

(a) Given the structural formula of a representative segment of a polymeric molecule, or the name of one of the familiar polymers, be able to classify the polymer (Section 33.2).
  (1) as a *homopolymer* or a *copolymer*
  (2) as *linear, cross-linked,* or *space network*
(b) Be able to associate the names of some of the more familiar synthetic polymers with structural characteristics (Table 33.1).
(c) Be able to write equations to represent *condensation* polymerization, *ionic* polymerization, and *free-radical addition* polymerization (Section 33.1).

It has been said that "necessity is the mother of invention," and the vast industrial complex involved in polymer chemistry is a testimony to this. Polymers are enormous molecules made by connecting small molecules in repetitive sequences. There are many examples of polymers in nature—proteins (Unit 32), starches and celluloses (Unit 31), and nucleic acids (Unit 34) being the most common. Chemists have been studying these macromolecules for a long time, seeking clues to their structures, functions, and syntheses. In addition, they have investigated ways of making a variety of polymers not found in nature. Polymer chemistry did not come into its own, however, until the emergency conditions of World War II forced the United States to turn to the polymer chemists with some serious problems.

The silkworm and the rubber tree had been doing some polymer chemistry for a long time. The polyamides from the silkworm and the hydrocarbon polymers from the rubber tree found ready markets in the pre-War world. With the advent of war, the demand for these products increased—suddenly silk was needed for parachutes and rubber for the tires of countless jeeps and trucks. Unfortunately, silkworms and rubber trees were largely owned by the Japanese by early 1942, and they were not particularly interested in the U.S. market at that time.

The polymer chemists were ready and in an amazingly short time, small-scale operations were expanded to a massive production operation: synthetic rubber for the wheels of war and nylon as a substitute for silk.

Today, synthetic polymers are an integral part of our lives. Polyamide (e.g., nylon) and polyester (e.g., Dacron) fibers are more common than cotton or wool. We store our food in polyethylene refrigerator bowls, fry our eggs in Teflon-coated pans, pack our picnic beer in polystyrene ice chests, and line our garbage cans with plastic bags decorated with flowery prints. Today's chemists have contributed some marvelous artifacts for the archeologists of the future.

33.1 TYPES OF POLYMERIZATION REACTIONS

Polymerization reactions fall generally into one of the two main types, condensation or addition. *Condensation polymerizations* involve the splitting out of small molecules (NH_3, H_2O, HCl) by formation of linkages such as ester or amide. They are characterized by a rapid disappearance of the small molecules (monomers) used as initial reactants and a gradual increase in size of the macromolecular products as intermediate species combine. As an example, nylon-66 (hexamethylene adipamide) may be formed by condensation of 1,6-diaminohexane (hexamethylenediamine) with adipyl chloride:

A methylene unit is CH_2.

Initial condensation (amide formation)

$$
\overset{O}{\overset{\|}{Cl C}}(CH_2)_4\overset{O}{\overset{\|}{C}}Cl + H_2\ddot{N}(CH_2)_6\ddot{N}H_2 \longrightarrow
$$

$$
\overset{O}{\overset{\|}{Cl C}}(CH_2)\overset{O}{\overset{\|}{C}}\underset{H}{N}(CH_2)_6\ddot{N}H_2 + HCl
$$

Chain growth

$$
2\overset{O}{\overset{\|}{Cl C}}(CH_2)_4\overset{O}{\overset{\|}{C}}(CH_2)_6\ddot{N}H_2 \longrightarrow
$$

$$
\overset{O}{\overset{\|}{Cl C}}(CH_2)_4\overset{O}{\overset{\|}{C}}\underset{H}{N}(CH_2)_6\overset{O}{\overset{\|}{N C}}(CH_2)_4\overset{O}{\overset{\|}{C}}\underset{H}{N}(CH_2)_6\ddot{N}H_2 + HCl
$$

Continued growth

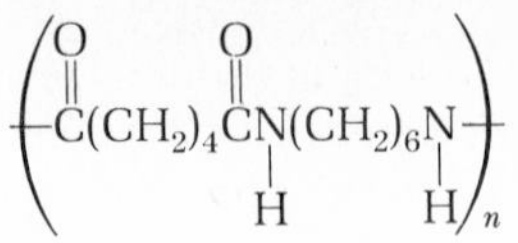

$$\left(\!-\!\overset{\displaystyle O}{\overset{\|}{C}}(CH_2)_4\overset{\displaystyle O}{\overset{\|}{C}}\underset{\underset{H}{|}}{N}(CH_2)_6\underset{\underset{H}{|}}{N}\!-\!\right)_{\!n}$$

nylon-66 polymer

As the name suggests, *addition polymerizations* involve the typical conversion of π-bond systems to σ-bond systems (Unit 29). Addition polymerizations are typically chain reactions, either ionic or free radical, and are characterized by relatively slow consumption of monomer and fairly rapid increase in polymer size. As an example, Teflon (polytetrafluoroethylene) is prepared by a free-radical polymerization. Very pure tetrafluoroethylene will polymerize so rapidly that the uncontrolled process may cause an explosion. To dissipate the heat from this highly exothermic reaction, polymerization may be performed in aqueous dispersions, using some water-soluble peroxide as the source of free radicals for chain initiation.

Chain initiation

$$H_2O_2 \longrightarrow 2(\cdot OH) \qquad \text{(other peroxides may also be used)}$$

Polymer initiation

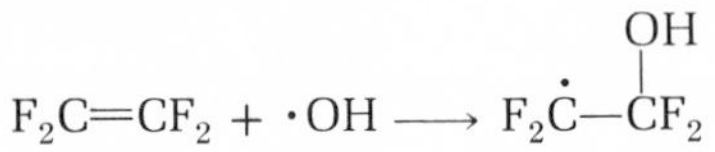

$$F_2C{=}CF_2 + \cdot OH \longrightarrow F_2\overset{\cdot}{C}\!-\!\overset{\overset{\displaystyle OH}{|}}{C}F_2$$

Polymer growth

$$F_2C{=}CF_2 + F_2\overset{\cdot}{C}\!-\!\overset{\overset{\displaystyle OH}{|}}{C}F_2 \longrightarrow F_2\overset{\cdot}{C}\!-\!CF_2\!-\!CF_2\!-\!\overset{\overset{\displaystyle OH}{|}}{C}F_2 \longrightarrow$$

$$F_2\overset{\cdot}{C}\!\left(\!CF_2CF_2\!\right)_{\!\overline{n}}\!CF_2OH$$

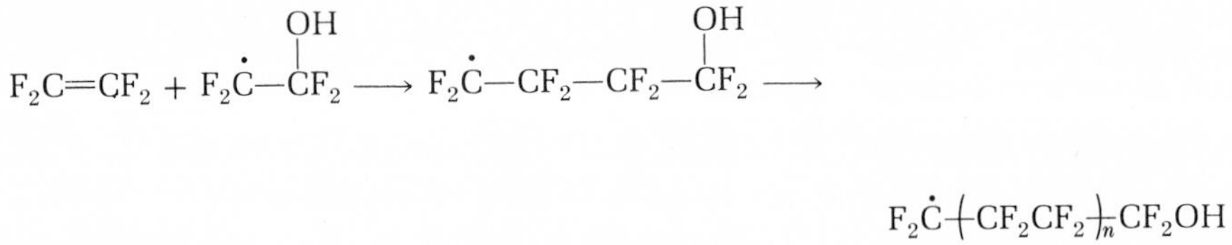

continued growth
by reaction
with monomer

Termination may involve coupling of two polymeric radicals or other processes involving polymer radicals themselves or with added chain-terminating species.

Ionic polymerization also involves an addition-type reaction, but reactants are either "simple" ions or radical ions (charged species containing one or more unpaired valence electrons). When these reactants are used in relatively nonpolar solvents, the species are primarily closely associated *ion pairs*, but mobile "free" ions are found when water or another polar liquid serves as the solvent. Either cationic or anionic monomer units may be employed, with the ion of opposite charge usually present as a small inorganic ion (e.g., Li^+ or NH_2^-). Two examples will serve to illustrate *ionic* polymerization.

Potassium amide in polystyrene formation (in liquid ammonia)

Chain initiation

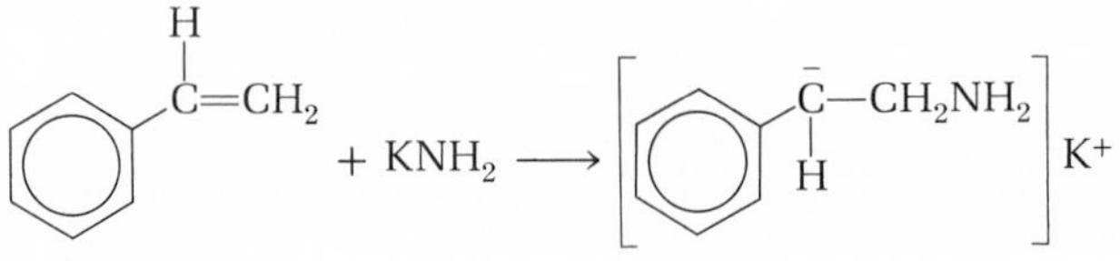

(styrene)   (ion stabilized by charge delocalization)

Polymer growth

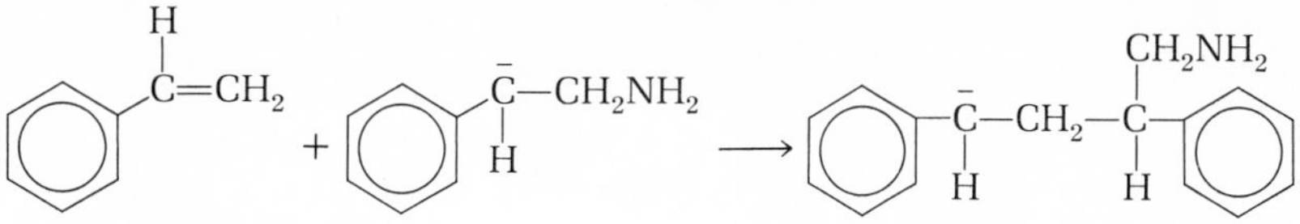

Continued growth

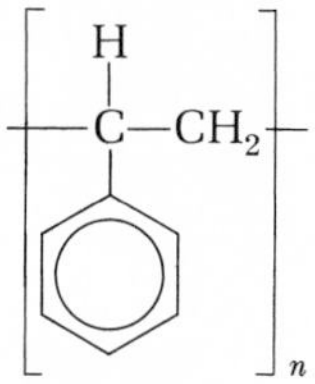

polystyrene

Chain termination typically occurs in this particular case by transfer of a proton of an ammonia molecule to neutralize the charge on the polymeric ion. This process is quite competitive with chain propagation, so polystyrene formed by this process consists mainly of relatively short polymers.

Radical-ions in butadiene polymerization

Chain initiation

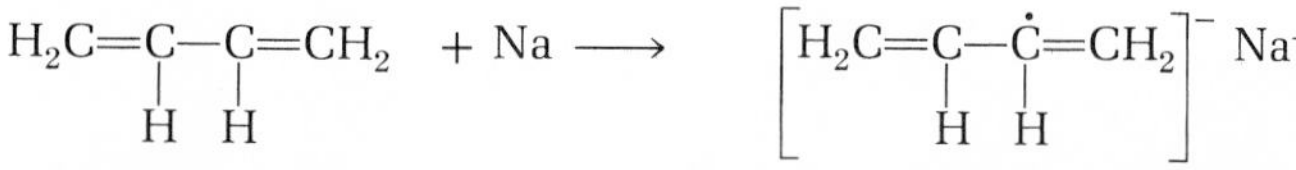

(1,4-butadiene)
(better represented by delocalized π bonding)

(the "extra" electron considered as being in part of the total delocalized molecular orbital system)

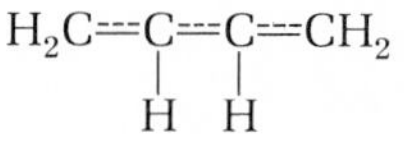

Polymer growth

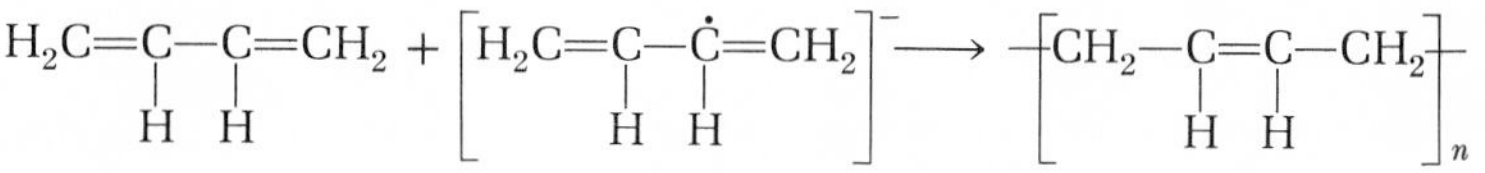

polybutadiene

Note that the actual polymer obtained is more complex than our simple representation would indicate. About half of the incorporated monomer units are of the general type indicated above, with the hydrogens in the *trans* geometry (Unit 28). The remaining units are isomeric. Polybutadiene prepared in this way was one of the earliest types of synthetic rubber, called *BUNA-rubber,* from *bu*tadiene and *na*trium (Latin name for sodium).

33.2 CLASSIFICATION OF POLYMERS

There are a number of different ways used to classify polymers. We may, for example, describe the type of functional group involved in linking together the monomer units. Thus *nylon* is a *polyamide* and *Dacron* is a *polyester* (Table 33.1). Alternatively, we may use two more general categories: *homopolymers* are those prepared from a single monomer compound, and two or more different monomer

Table 33.1 SOME FAMILIAR SYNTHETIC POLYMERS

CLASSIFICATION	EXAMPLES[a]	MONOMER(S)	COMMENTS
linear[b] homopolymers	"high density" polyethylene [structure]	[structure of ethene] (ethene, "ethylene")	prepared from petroleum products (annual U.S. production 4×10^9 kg); a *thermoplastic* (capable of being heat softened and molded into various shapes)
	polyvinyl chloride (PVC) [structure]	[structure of chloroethene] (chloroethene, "vinyl chloride")	a *thermoplastic* used in making various plastic wrapping materials, bottles, "artificial leathers", etc. (now considered a potential environmental problem)
	polystyrene [structure]	[structure of styrene] (styrene)	a common plastic for many household items (When the semiliquid plastic is whipped with air and cooled, the product is *Styrofoam*.)
	polyvinylidene chloride [structure]	[structure of 1,1-dichloroethene] (1,1-dichloroethene, "vinylidene chloride")	used in making *Saran* wrap
	polytetrafluoroethylene [structure]	[structure of tetrafluoroethene] (tetrafluoroethene, "tetrafluoroethylene")	Teflon
	polyvinyl acetate [structure]	[structure of vinyl acetate] ("vinyl acetate")	used in adhesives and in many "chewing gums"
	polymethyl methacrylate [structure]	[structure of methyl methacrylate] ("methyl methacrylate")	used to make transparent plastic solids (almost unbreakable) such as *Perspex*, *Lucite*, *Plexiglass*.

Table 33.1 (Continued)

CLASSIFICATION	EXAMPLES[a]	MONOMER(S)	COMMENTS
linear[b] homopolymers (continued)	*polybutadiene* some and some $-[CH_2CH=CHCH_2]_n-$	$H_2C=C-C=CH_2$ $\quad\quad$ H$\;$H (1,4-butadiene)	some of the earliest "synthetic rubber" (not nearly as good as "natural rubber" for automobile tires)
linear copolymers	*styrene-butadiene copolymer*	styrene and 1,4-butadiene	usually employs 3:1 mole ratio of butadiene-to-styrene (better in some respects than "natural rubber")
	polyhexamethylene adipamide	$H_2N(CH_2)_6NH_2$ (1,6-diaminohexane, "hexamethylenediamine") + $ClC(CH_2)_4CCl$ ("adipyl chloride")	*nylon-66*
	polyethylene terephthalate	$HOCH_2CH_2OH$ (1,2-ethanediol, "ethylene glycol") + $HO_2C-\bigcirc-CO_2H$ (terephthalic acid)	used for synthetic fibers (*Dacron*) and plastic film (*Mylar*)
cross-linked homo-polymers	*"low-density" polyethylene* (see Figure 33.1)	ethylene, polymerized under conditions different from those used for linear ("high-density") polyethylene	used for electrical insulation, plastic dishes, etc.
cross-linked copoly-mers	*polycyclohexanedimethyl terephthalate* (see Figure 33.1)	$HOCH_2-\bigcirc-CH_2OH$ (1,4-cyclohexanedimethanol) + $HO_2C-\bigcirc-CO_2H$ (terephthalic acid)	used for certain fibers (e.g., *Kodel*)

Table 33.1 (Continued)

CLASSIFICATION	EXAMPLES[a]	MONOMER(S)	COMMENTS
space network homo-polymers	diamond (see Figure 33.1)	carbon atoms in 3-D tetrahedral pattern	very hard crystals, found in nature, but now also man made
	quartz (see Figure 33.1)	$[SiO_4]$ tetrahedral units	
space network co-polymer	*phenol-formaldehyde resin* (see Figure 33.1)	(phenol) + $H_2C{=}O$ (methanal, "formaldehyde")	first patented in 1909 as *Bakelite*

[a]Any particular polymeric substance will typically include a variety of molecules differing in chain length, branching, and cross-linking. The structural representations shown are necessarily oversimplified.
[b]Most "linear" polymers contain some "branched chains" (Unit 28) and some degree of cross-linking.

compounds are used to make *copolymers*. Polyethylene, for example, is a *homopolymer* and nylon is a *copolymer*.

The examples we have considered so far have illustrated cases in which long fibrous polymers are formed. Such macromolecules are referred to as *linear* polymers. When segments of the same or different polymer chains are interconnected by covalent bonds, the polymers are said to be *cross-linked* (Figure 33.1). *Space network* polymers involve a high degree of cross-linking, often with a three-dimensional order approaching that of perfect crystals (Unit 11). Diamond and quartz may be thought of

Figure 33.1 Cross-linking in polymers.

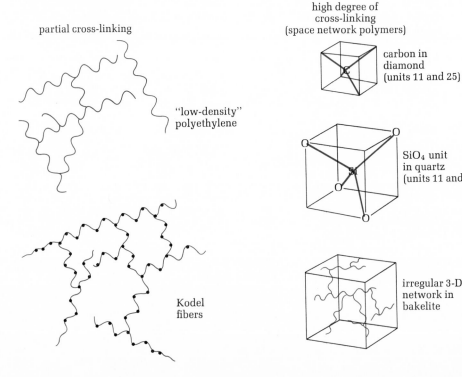

as highly ordered *space network polymers*, the former involving the "tetrahedral carbon atom" as the monomer unit and the latter as a regular array of tetrahedral (SiO_4) units.

33.3 POLYMER PROPERTIES

The characteristics of a particular polymer may depend on different reaction conditions employed to influence such factors as average chain length and degree of cross-linking, as in the case of "high-density" and "low-density" polyethylene (Table 33.1). In many cases, however, the properties of a polymeric material may reflect the *physical* treatment used in manufacturing a particular product. For example, the polyester polyethylene terephthalate (Table 33.1) may be heated until it is a viscous liquid and then extruded as Dacron fibers (Figure 33.2) or rolled into thin plastic sheets (*Mylar* film).

liquified polyester

Figure 33.2 Fiber production by *extrusion*. The device used is called a *spinneret* and resembles a shower head. Fibers are typically extruded into an inert liquid.

dacron fibers

In many cases the polymers alone do not possess all of the properties desired for a final consumer product. Automobile tires of synthetic rubber, for example, also contain charcoal and various chemical "antioxidants," along with strengthening fibers such as rayon, nylon, fiberglass, or even steel. In other cases, *plasticizers* are added to make otherwise brittle polymers softer and more flexible. One of the most commonly used *plasticizers* until the mid-1970s employed a mixture of chlorinated biphenyls (Figure 33.3). These have now been banned from most uses because of

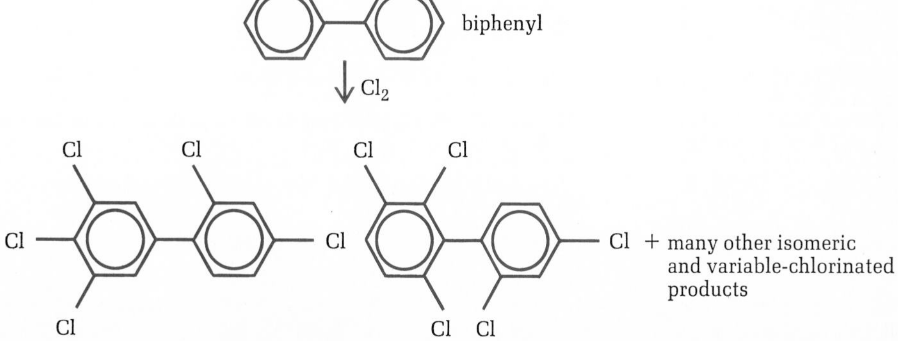

Figure 33.3 Polychlorinated biphenyls (PCB's), now banned as plasticizers in most cases.

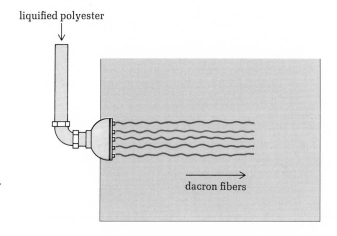

biphenyl

Cl_2

+ many other isomeric and variable-chlorinated products

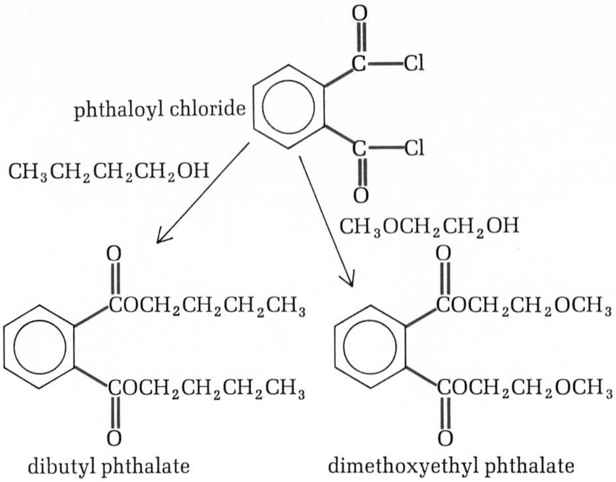

Figure 33.4 Phthalate esters, common plasticizers. Specific esters are used to impart specific properties to the polymer products.

evidence that they are long-range environmental hazards. Phthalate esters (Figure 33.4) are now the most frequently employed plasticizers, although some preliminary studies suggest that these chemicals, too, may be less innocuous than we would like them to be.

33.4 POLYMERS AND THE ENVIRONMENT

The chemical industry, government agencies, and individual consumers face some difficult problems stemming from our massive use of synthetic polymers.

One of the problems is related to the source of the monomer units required for many of our more common polymer products. Ethylene and propylene (and compounds, such as vinyl chloride, prepared from ethylene) are now obtained primarily from the petroleum industry. With the increasing pressure of the "energy crisis," it becomes more and more difficult to justify use of a significant fraction of our petroleum resources for polymer manufacture. Also, spiraling costs of petroleum are reflected in the economic factors of petroleum-based polymer production.

In some cases, alternative sources of petroleum-based monomers may be developed. The most promising of these appears to be in the area of hydrocarbon production from coal, but the large-scale development of such processes is probably many years in the future. It will undoubtedly be necessary to find alternatives to some of our current applications of petroleum-based polymers and the recycling of appropriate plastics, often entirely feasible, must certainly be encouraged.

The impact of synthetic polymers on the environment is no small problem. Although some of the newer plastics are *biodegradable* (i.e., converted to simple products by natural microorganisms), most of the synthetic polymers now in use are not. Biodegradability alone is not a sufficient criterion for environmental safety. Wastes may be dumped into the environment at a rate exceeding that at which they can be degraded, and aerobic (oxygen-requiring) microorganisms feeding on our waste products may consume oxygen needed by other members of the ecosystem or proliferate to the extent that the microorganisms themselves pose a problem.

Some plastics form hazardous gases, such as HCl or HCN, when they are burned, so incineration cannot be employed safely for the disposal of some types of polymers. We are, at the same time, running short of spaces for "dumps" and "land fills," particularly near large metropolitan areas.

Synthetic polymers illustrate very clearly the socioeconomic impact of science and technology. Although offering us many benefits, synthetic polymers also present us with some extremely difficult problems. To solve some of the problems, we may very well have to give up many of the "benefits."

SUGGESTED   **Commoner, Barry.** 1971. *The Closing Circle.* New York: Bantam Books.
READING   **Mandelkern, L.** 1972. *An Introduction to Macromolecules.* New York: Springer-Verlag.
            Margerison, D., and G. C. East. 1967. *Introduction to Polymer Chemistry.* New York: Pergamon Press.

EXERCISES

1. Indicate whether each of the following is a *homopolymer* or a *copolymer:*

a.

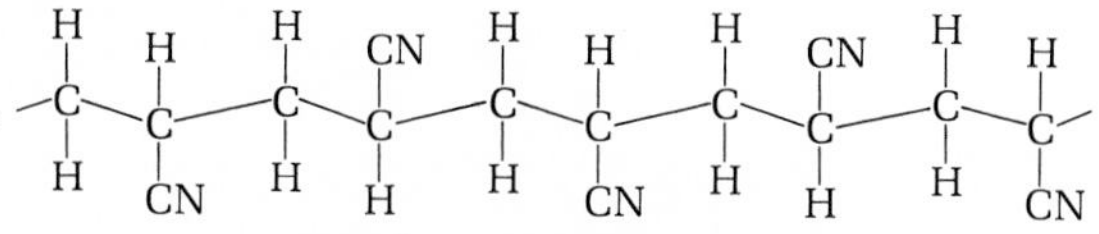

(a polyacrylonitrile used for synthetic fibers, such as Orlon)

b.

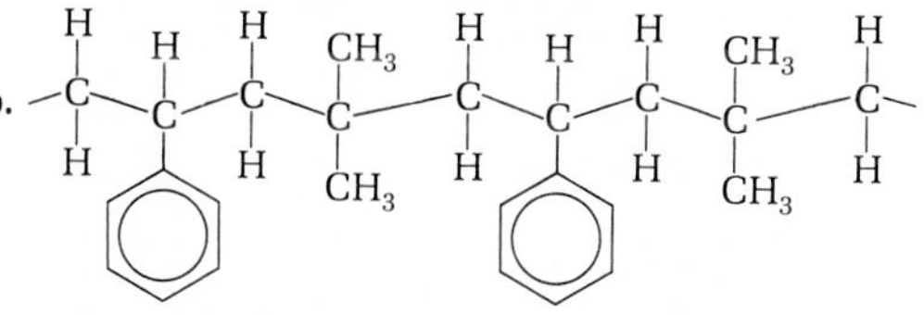

(Buna-S, the synthetic rubber used extensively in modern automobile tires)

c. nylon-66
d. teflon
e. Dacron

2. Indicate whether each of the following is considered a *linear* polymer, a *cross-linked* polymer, or a *space network* polymer.
a. polyvinyl chloride; **b.** "low-density" polyethylene; **c.** "high-density" polyethylene; **d.** Bakelite; **e.** Teflon

3. Match the polymer "product" names with the corresponding structural features.
NAMES: (A) polystyrene, (B) Lucite, (C) nylon-66; (D) Dacron, (E) Saran

a. 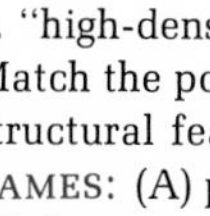 **b.**

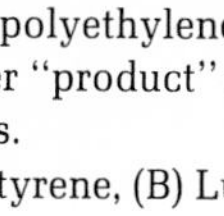

c. 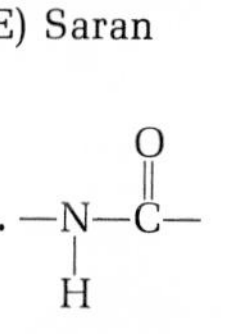 **d.**

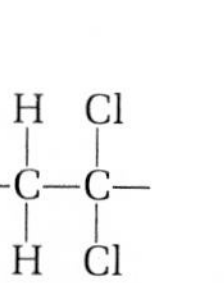

e.

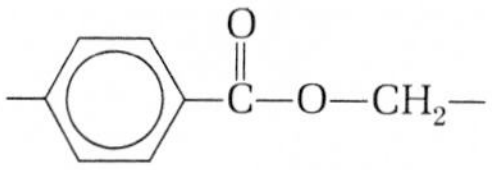

4. For each of the following processes, give a series of equations up to the point at which four monomer units have combined.
a. the condensation polymerization leading to Kodel fibers

from terephthalic acid $\left(HO_2C-\bigcirc-CO_2H \right)$ and

1,4-cyclohexanedimethanol $\left(HOCH_2-\bigcirc-CH_2OH \right)$

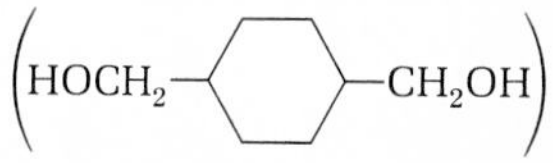

b. the free-radical addition of propylene (propene) to form polypropylene, used as fibers for "indoor–outdoor" carpets
c. the ionic polymerization of styrene, initiated by addition of a proton (from a strong acid) to the "ethylenic" π bond of styrene

PRACTICE FOR PROFICIENCY

Read and be prepared to discuss (or report on) one or more of the following articles. (Your instructor may wish to suggest others.)

1. "The Early Days of Polymer Science," by H. Mark, in the November, 1973, issue of the *Journal of Chemical Education* (pp. 757–760).
2. "Sulfur–Nitrogen Polymers Behave Like Metals," by Chris Murray, in the May 26, 1975, issue of *Chemical and Engineering News* (pp. 18–19).
3. "How Giant Molecules Are Measured," by P. J. W. Debye, in the September, 1957, issue of *Scientific American* (pp. 90–97).
4. "Some Stereochemical Principles from Polymers," by Charles Price, in the November, 1973, issue of the *Journal of Chemical Education* (pp. 744–747).

BRAIN TINGLER

5. Survey issues since 1971 of any two of the following journals for articles dealing with environmental problems associated with polymer biodegradability, plasticizers, or polymer incineration. Prepare a summary report of your findings, with a bibliography.

a. *Chemistry;* b. *Chemical and Engineering News;* c. *Scientific American;* d. *Science;* e. *Environmental Science and Technology*

Unit 33 Self-Test

Your comprehension of the material in this unit, and of related material from earlier units, is best demonstrated by your ability to read and discuss one or more of the articles listed in *Practice for Proficiency.*

1. Which of the following statements concerning epinephrine (adrenalin) is incorrect?

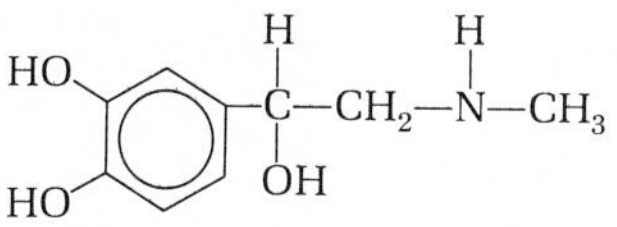

epinephrine

a. Epinephrine should react readily with aqueous HCl to form a salt.

b. Epinephrine contains both alcohol and phenol-type hydroxyl groups.

c. Neither epinephrine nor any of its simple salts is likely to be water soluble.

d. One mole of epinephrine will react with 2 moles of sodium hydroxide.

e. One mole of epinephrine will react with 4 moles of acetyl chloride.

2. Which is probably the weakest base?

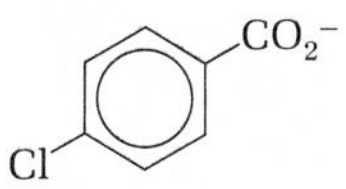

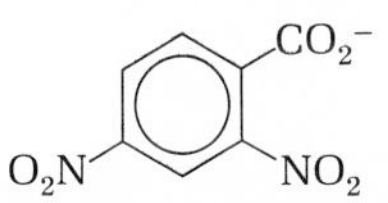

3. Which is the least likely to undergo a cyclization reaction?

a. $HOCH_2CH_2CH_2CO_2H$

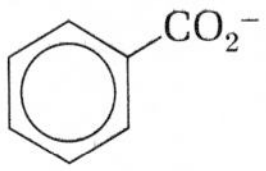

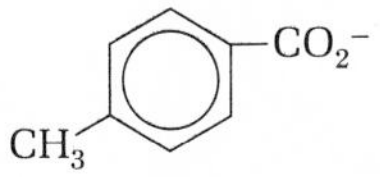

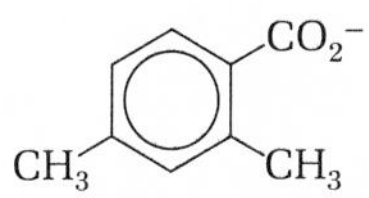

b. 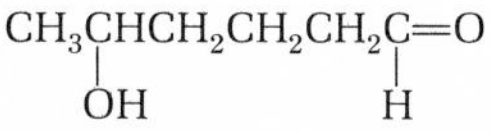 $CH_3CHCH_2CH_2CH_2C{=}O$ with OH and H

c. 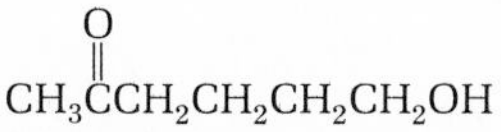 $CH_3CCH_2CH_2CH_2CH_2OH$ (O)

d. 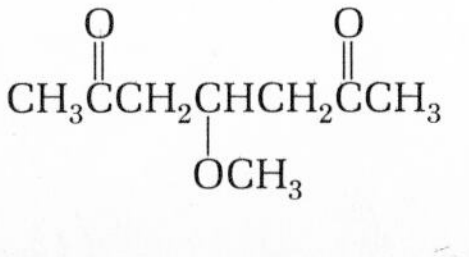 $CH_3CCH_2CHCH_2CCH_3$ with O, OCH_3, O

e. 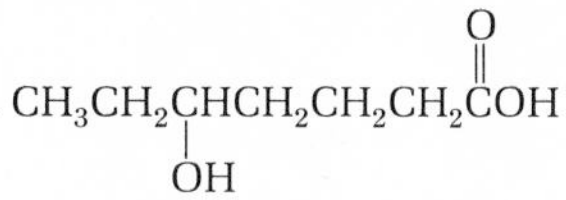 $CH_3CH_2CHCH_2CH_2CH_2COH$ with OH and O

4. How many moles of acetyl chloride are expected to react with a single mole of the compound shown?

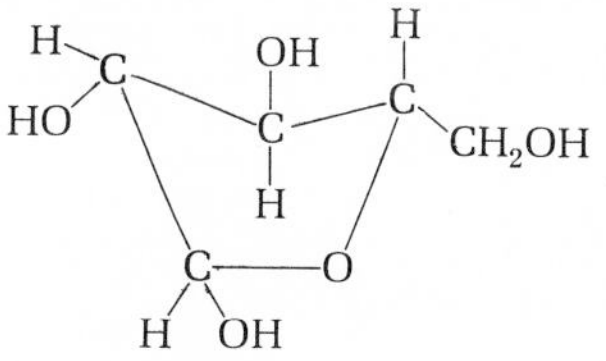

a. 1; **b.** 2; **c.** 3; **d.** 4; **e.** 5

5. The monosaccharide shown in question 4 is
  a. the α-cyclic form of glucose
  b. the β-cyclic form of glucose
  c. the cyclic form of an aldopentose
  d. the cyclic form of a ketopentose
  e. none of these

6. The dipolarionic form of histidine is formulated as

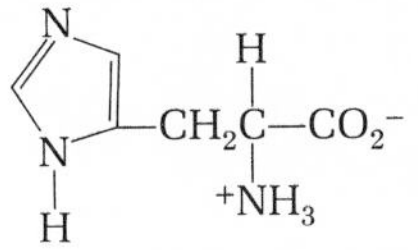

When a 1.0-mg sample of histidine is dissolved in 10 ml of 1.0 M HCl, the net charge on the principal organic solute probably is
  a. $+1$; **b.** $+2$; **c.** -2; **d.** -1; **e.** zero

7. What is the maximum number of moles of acetyl chloride that could react with 1 mole of tyrosine?

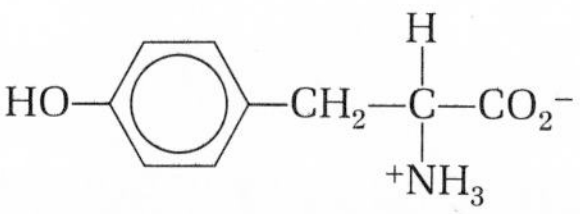

tyrosine

a. 2; **b.** 3; **c.** 4; **d.** 5; **e.** 6

8. The partial hydrolysis of a tripeptide yielded a mixture of glycine, serine, glyclyglycine, and serylglycine. Which of the following formulations may represent the original tripeptide? (Remember that the conventional formulation shows the N-terminal unit on the left.)
   a. gly-ser-gly; b. ser-gly-ser; c. gly-gly-ser; d. gly-ser-ser; e. ser-gly-gly

9. There are a number of types of bonding responsible for the natural geometry of a complex protein. Among these, the type most affected by altering the pH of the aqueous environment of the protein (without significant dilution or heating) is the

a. C—C covalent bond; b. ionic bond; c. hydrophobic bond; d. peptide bond; e. disulfide bond

10. Which of the following statements concerning nylon-66 is incorrect?
   a. It is a copolymer.
   b. It is a polyamide.
   c. It is considered as a *linear* polymer.
   d. It is typically formed by a process of *free-radical addition* polymerization.
   e. A given sample contains molecules of different chain lengths.

Excursion 5

Structure Determination and Synthesis

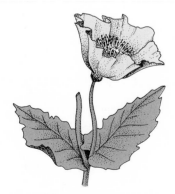

Figure E5.1 The opium poppy (Papaver somniferum).

The motives behind the work of research chemists are many and varied, but surely one of the most interesting is the desire to duplicate, or even excel, nature in the preparation of biologically active chemicals. Even a hint in the folklore of some isolated tribe about a natural product of unusual activity is enough to set chemists on a search for the structure of the essential compound in the material. Once the structure is known, chemists will not be satisfied until the compound can be prepared synthetically and structural modifications have been attempted to secure even more useful materials than nature originally provided.

In this excursion we shall use a single example to illustrate how the modern natural product chemist combines a variety of clues from spectroscopic and chemical investigations to deduce the structure of a biologically active molecule. The compound selected for discussion is papaverine, an alkaloid from the opium poppy (Figure E5.1). We shall trace the investigation from the initial isolation to the final structure proof, including the laboratory synthesis of the alkaloid.

1 ISOLATION AND PURIFICATION

The opium poppy contains a number of *alkaloids* (organic compounds containing nitrogen, usually basic) that have potent biological activities. Morphine (Unit 29) is the best known of these opium alkaloids, but another important product is *papaverine*. Papaverine is a white crystalline solid constituting about 1 percent of the dry weight of the poppy seeds. It is useful as an antispasmodic agent and, unlike morphine, is not habit forming.

Initial extraction of macerated poppy seed pods with ethanol, using a continuous

859

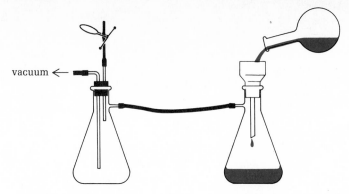

vacuum ←

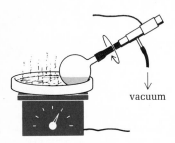

vacuum

Figure E5.3   Rotary thin-film evaporation.

extraction procedure (Excursion 4), yields a solution that contains several different alkaloids, along with pigments, polymeric materials, and other ethanol-soluble compounds. When this solution is chilled, many of the less soluble materials precipitate, but the papaverine remains in solution if the volume of ethanol used for the extraction was chosen properly. Filtration of the chilled mixture (Figure E5.2) then gives a crude solution of papaverine and other soluble products.

This mixture may be separated by column chromatography on alumina (Excursion 4) after evaporation of the ethanol. Typically, the original solution is placed in a round-bottom flask and this is attached to a rotary thin-film evaporator (Figure E5.3). As the flask rotates in a heating bath, a thin film of solution is spread over the interior of the flask and the solvent is removed by vacuum, usually from a water aspirator. The residual solids are chromatographed on an alumina column, and the fraction of interest may be detected by biological testing for antispasmodic activity or by suitable spectroscopic investigations.

The chromatographed papaverine is not generally pure enough for structural study, so additional purifications by fractional crystallization and sublimation (Excursion 4) are required. The final product is a white solid melting at 147°–148°C and subliming at 135°–140°C at 11 torr.

2 THE MOLECULAR FORMULA

Elementary analysis indicates the presence of carbon, hydrogen, and nitrogen and the absence of sulfur, phosphorus, the halogens, and metals. A quantitative analysis reveals 4.13 percent nitrogen and combustion analysis (Unit 28) shows 70.78 percent carbon and 6.24 percent hydrogen. By difference, then, the papaverine molecule contains 18.85 percent oxygen.

From these data one could calculate an empirical formula. However, a molecular weight determination allows a direct calculation of the molecular formula. A cryoscopic determination (Unit 15) yields the value of 339 ± 0.8 g mole^{-1}. The molecular formula is then calculated as follows:

Carbon

$$\frac{339 \text{ g}}{\text{mole}} \times \frac{0.7078 \text{ g C}}{\text{g(cpd.)}} \times \frac{1 \text{ g-atom C}}{12.01 \text{ g C}} = 20 \frac{\text{g-atoms C}}{\text{mole}}$$

Hydrogen

$$\frac{339 \text{ g}}{\text{mole}} \times \frac{0.0624 \text{ g H}}{\text{g(cpd.)}} \times \frac{1 \text{ g-atom H}}{1.008 \text{ g H}} = 21 \frac{\text{g-atoms H}}{\text{mole}}$$

Nitrogen

$$\frac{339 \text{ g}}{\text{mole}} \times \frac{0.0413 \text{ g N}}{\text{g(cpd.)}} \times \frac{1 \text{ g-atom N}}{14.01 \text{ g N}} = 1 \frac{\text{g-atom N}}{\text{mole}}$$

Oxygen

$$\frac{339\,g}{mole} \times \frac{0.1885\,g\,O}{g(cpd.)} \times \frac{1\,g\text{-atom}\,O}{16.00\,g\,O} = 4\,\frac{g\text{-atoms}\,O}{mole}$$

Hence, the molecular formula must be $C_{20}H_{21}NO_4$.

3 NUMBER OF RINGS AND PI BONDS

Papaverine is found to be basic ($K_b = 8.5 \times 10^{-9}$), so the nitrogen must be present in an amine group.

The relationship between hydrogen and carbon in an alkane was expressed as C_nH_{2n+2} (Unit 20). If we worked out a similar generalization for organic amines, allowing for "loss" of two hydrogen atoms for each ring or π bond, we would find that, for a molecule such as papaverine, the carbon-hydrogen relationship is expressed by

$$C_nH_{(2n+3-2x)}$$

in which x is the number of rings or π bonds (or both). Applying this to our molecular formula, $C_{20}H_{21}NO_4$, shows that

$$2(20) + 3 - 2x = 21$$
$$x = 11$$

Thus, there are 11 π bonds or rings (total) in the papaverine molecule.

Papaverine can be hydrogenated, under rather strenuous conditions. Quantitative hydrogenation shows that papaverine reacts with 8 moles of H_2 per mole of papaverine. Assuming that all π bonds have added hydrogen, we may then conclude that the papaverine molecule contains three rings (11 − 8).

4 STRUCTURAL INFORMATION FROM NONDESTRUCTIVE METHODS

The ultraviolet spectrum of a papaverine solution contains several absorption maxima between 2000 and 3300 Å. This suggests the presence of delocalized π-bond systems, usually associated with aromatic rings. Careful study of the ultraviolet spectrum by an experienced spectroscopist might suggest other structural features, but additional information is more readily obtained from the infrared and nmr spectra (Unit 30).

The infrared spectrum (Figure E5.4) reveals several important structural clues. The transparency of papaverine above 3050 cm^{-1} shows the absence of O—H and N—H bonds. From this we know that there are no hydroxyl groups (alcohol, phenol, carboxylic acid, hemiacetal) in the molecule and can conclude that the basic nitrogen is tertiary (three C—N, no H—N bonds). The absence of absorption in the region

Figure E5.4 Infrared spectrum of papaverine.

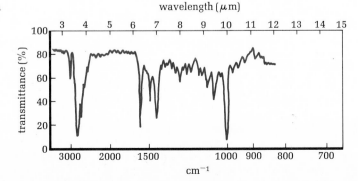

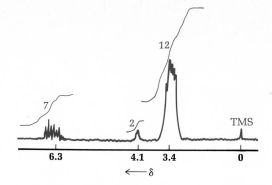

1750–1610 cm^{-1} indicates no carbonyl (aldehyde, ketone, ester, amide, carboxyl). Absorption bands at ~3050 (aromatic C—H), 1600, 1500, and 1450 cm^{-1} indicate aromatic C—H. Since all other common oxygen-containing functional groups are eliminated, it is likely that the four oxygens in the molecule are involved in ether linkages. From the infrared spectrum, then, we suggest the following features for the papaverine molecule:

one or more ⬡ (or aromatic ring systems other than benzene-type)

four
$$\begin{array}{c} C \quad\quad C \\ \diagdown \quad \diagup \\ O \end{array}$$

one
$$\begin{array}{c} C \quad\quad C \\ \diagdown \quad \diagup \\ N \\ | \\ C \end{array}$$
(no N—H)

Again, interpretation by an experienced analyst might add more details.

The nmr spectrum (Figure E5.5) furnishes some further useful data. Peaks in the vicinity of $\delta = 6.3$ probably correspond to hydrogens on aromatic rings (seven such hydrogens), the absorption near $\delta = 4.1$ is typical of a methylene group between two aromatic rings (two hydrogens, therefore one methylene group), and the peaks near $\delta = 3.4$ probably result from hydrogens on four methoxyl groups (OCH$_3$).

There are now a limited number of structures possible that are consistent with the spectroscopic data. Since four of the carbons must be in —OCH$_3$ groups and one in a —CH$_2$— group, the remaining 15 must be in three aromatic ring systems. Two benzene rings would require 12 carbons. It seems likely, therefore that some fused-ring system, possibly containing the nitrogen, is present. Nondestructive methods have not, however, indicated a unique structure. It is possible that the fragmentation pattern in the mass spectrum might solve the problem, but in the absence of sufficient comparison data it is probably better to rely on a more conventional chemical degradation.

Ring systems containing atoms in addition to carbon are said to be heterocyclic.

5 STRUCTURAL INFORMATION FROM DESTRUCTIVE METHODS

There are many ways in which complex molecules may be broken up into simpler fragments for identification by comparison with known compounds. In this case one would select the method best suited to finding intact rings with minimum loss of substituent groups.

Chemical oxidations are possible that convert alkyl groups on aromatic rings to a carboxyl group directly on the ring without affecting many other groups, for example,

When papaverine is subjected to such an oxidation, a mixture of products is obtained. Certain types of products are sought on the basis of structural features known to be present. Two products of particular interest are found in reasonably large percentages in the reaction mixture, and these are identified by comparative chromatography with known compounds. The products are shown to be

It is important to note that an oxidative degradation of this type is not a "clean" quantitative reaction. A mixture of products is always obtained. In this particular case, the two products of interest must have originated from *different* papaverine molecules.

These products contain structural features identified in the papaverine molecule except for the absence of a methylene group and the presence of carboxyl groups. It seems reasonable to assume that the methylene was oxidized to carboxyl, so that the identified fragments were connected originally through a methylene group. The structure suggested for papaverine, then, is

This structure is consistent with all data obtained for papaverine and is therefore presumed to be correct, subject to further confirmatory investigation.

6 STEREOCHEMISTRY

In many natural products, such as morphine, the two-dimensional structural formula is not a complete description of the molecule. If the molecule contains asymmetric

Figure E5.6 Comparison of papaverine and vanillin.

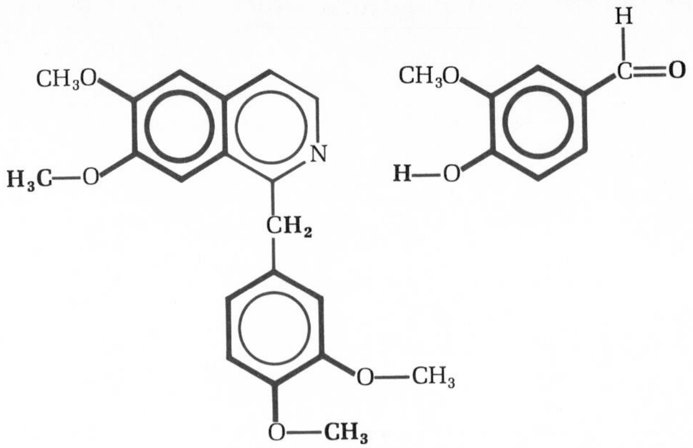

centers (Unit 28), it is necessary to determine which of the possible optical isomers is the one found in the natural source. In other cases, carbon–carbon double bonds in chains give rise to various possibilities for cis-trans isomers (Unit 28). X-ray crystallography (Excursion 1) is usually the best method for determining the actual three-dimensional structure when isomeric problems occur.

Papaverine contains neither asymmetric centers nor π-bond systems capable of cis–trans isomerism. As a result, the structure proposed is a completely satisfactory description of the papaverine molecule.

7 SELECTION OF A SYNTHETIC APPROACH

In any synthesis it is desirable to find inexpensive, readily available starting materials with as many structural features as possible in common with the desired product. Then reactions must be found that can convert these starting materials into the product sought.

An examination of the structure of papaverine (Figure E5.6) shows that vanillin (a natural flavoring agent from the vanilla bean, now available synthetically) has many common features. In addition, most of the synthetic steps required to convert vanillin to papaverine are relatively simple ones. The approach selected uses vanillin to produce the two dimethoxy-ring systems in papaverine and a series of additional reactions to complete the assembly of the total papaverine molecule. Several of the reactions involved have been described (Unit 29) in our discussion of simple monofunctional compounds. Some of the steps, however, require more complex reactions. We shall explore each step of a synthesis scheme summarized in Figure E5.7.

8 A STEPWISE APPROACH

1. METHYLATION OF VANILLIN The phenolic hydroxyl group in vanillin is a reactive center that must be protected before additional steps can be performed. Fortunately, a methyl ether is desired at this position (Figure E5.6), so this can serve as the protective group and at the same time get us one step closer to the final product. A general method for preparing ethers involves replacement of halogen by an alkoxide or phenoxide ion:

R, R' = alkyl groups; Ar = aromatic group; X = Cl, Br, I (See Unit 29).

$$R—X + {}^-OR' \longrightarrow R—O—R' + X^-$$

$$R—X + {}^-OAr \longrightarrow R—O—Ar + X^-$$

One might, then, treat vanillin with sodium hydroxide to obtain

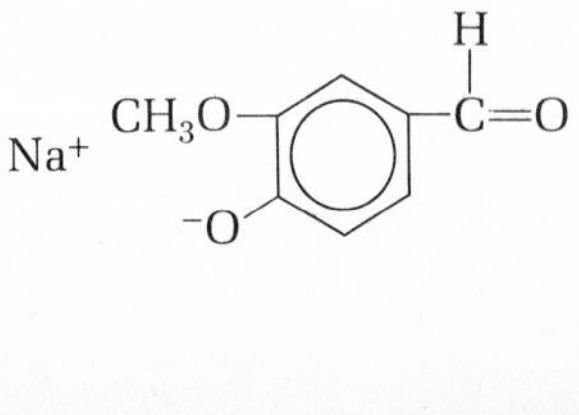

Figure E5.7 Summary of a synthetic scheme.

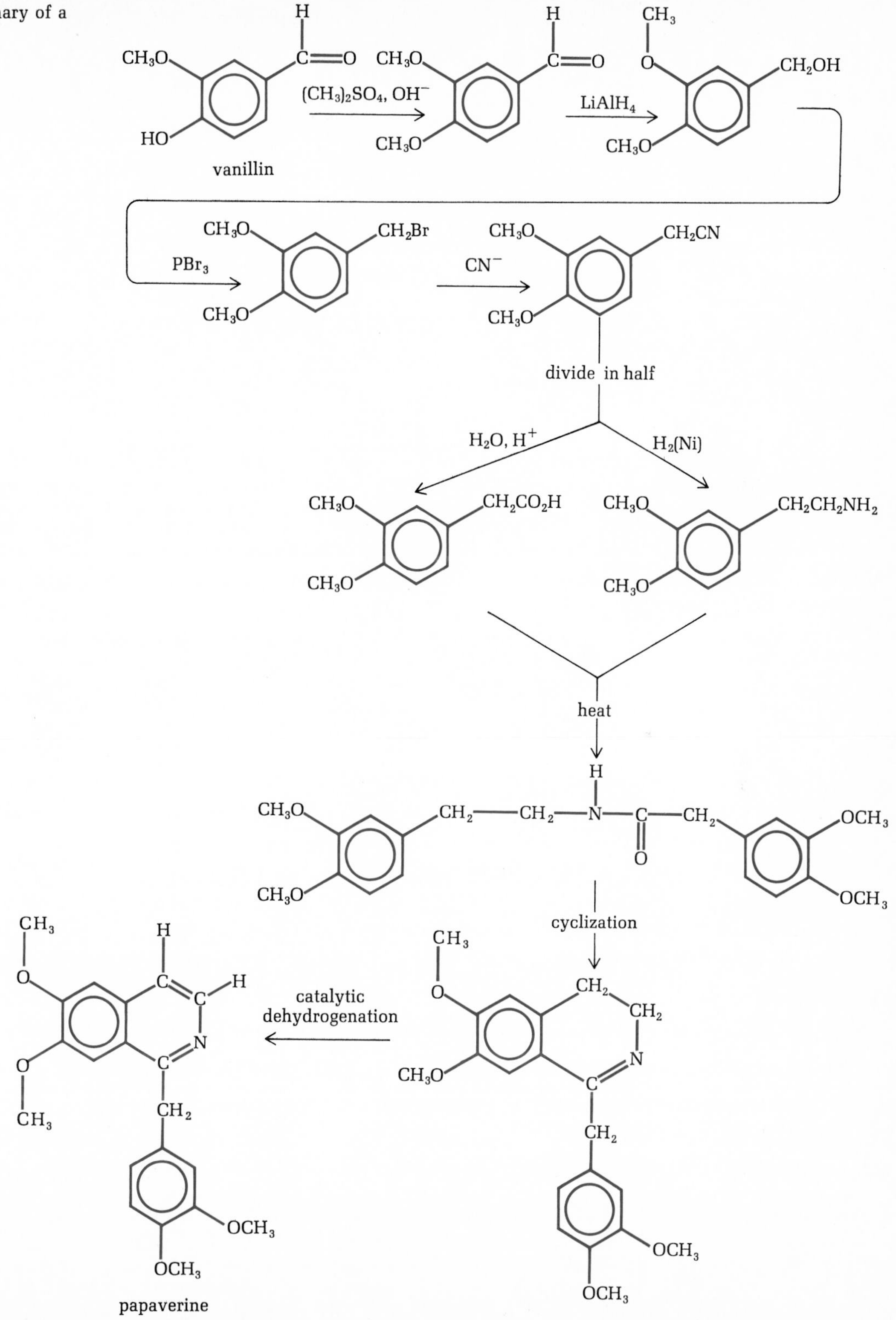

vanillin
$(CH_3)_2SO_4$, OH^-
$LiAlH_4$
PBr_3
CN^-
divide in half
H_2O, H^+
$H_2(Ni)$
heat
cyclization
catalytic dehydrogenation
papaverine

and allow this to react with methyl iodide. In practice, dimethyl sulfate,

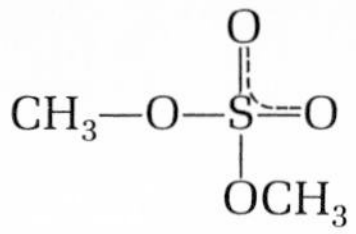

$$CH_3-O-\underset{\underset{OCH_3}{|}}{\overset{\overset{O}{\|}}{S}}\!=\!O$$

is used rather than methyl iodide because of its greater ease of handling and storage. This reagent behaves in essentially the same way as methyl iodide; that is, it acts as an alkylating agent. The first step in the synthesis, then, is

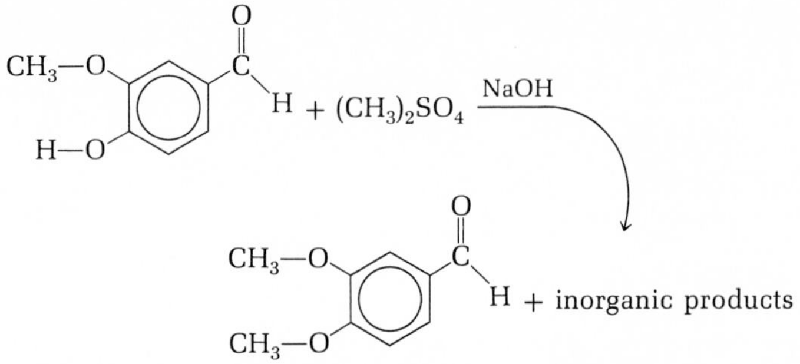

Liquid–liquid extraction is used to separate the 3,4-dimethoxybenzaldehyde ("veratraldehyde") from the water-soluble inorganic products. After removal of solvent, the extracted aldehyde is pure enough for the next step.

2. REDUCTION OF CARBONYL Aldehydes may be reduced to primary alcohols in a number of ways. One of the simplest uses lithium aluminum hydride (Unit 29). This is our second step:

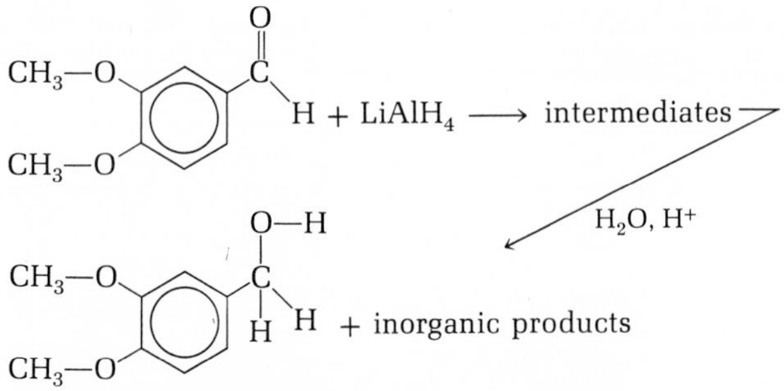

Again, liquid–liquid extraction separates the alcohol (note the large nonpolar region) from water-soluble inorganic materials.

3. CONVERSION OF ALCOHOL TO HALIDE Primary alcohols are best converted to halides by reagents such as phosphorus tribromide (Unit 29):

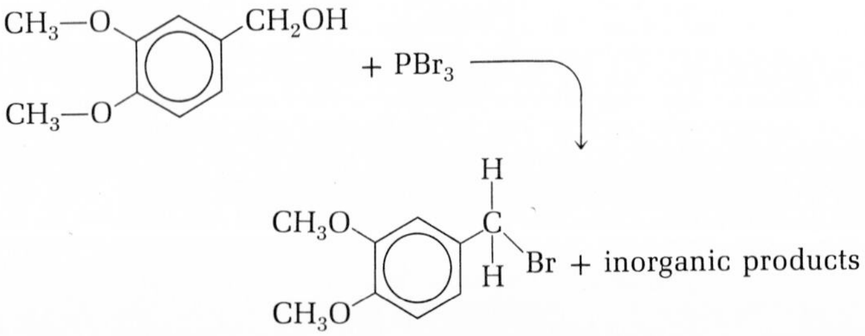

After extraction of the water-soluble inorganic products, the bromide may be purified by distillation.

4. HALOGEN REPLACEMENT BY CN⁻ Organic halides typically undergo halogen replacement by negative groups. When cyanide ion is used, a class of compounds called *nitriles* is obtained. *This is an effective way of introducing an additional carbon atom into a molecule.*

(a nitrile)

Again, purification by extraction, followed by distillation, is employed.

5. CONVERSION OF NITRILE TO AMINE The carbon–nitrogen triple bond, like other π-bond systems, will undergo addition reactions. Hydrogen can be added by a mild metal-catalyzed reaction under conditions such that the nitrile group is reduced, but the delocalized π-bond system of the aromatic ring remains unchanged.

Approximately half of the nitrile is used for this reaction, the other half being reserved for step 6.

The powdered metal catalyst is removed by filtration, and the amine is purified by fractional distillation.

6. HYDROLYSIS OF NITRILE Unsymmetrical reagents, such as H_2O, may also add to π bonds. Since the carbon–nitrogen triple bond of the nitrile group is polar ($-\overset{+\ \ \ }{C}\equiv N$,) the addition may be thought of as proceeding as follows:

Except in rare cases, two hydroxyls on the same carbon are unstable and eliminate the elements of water, so

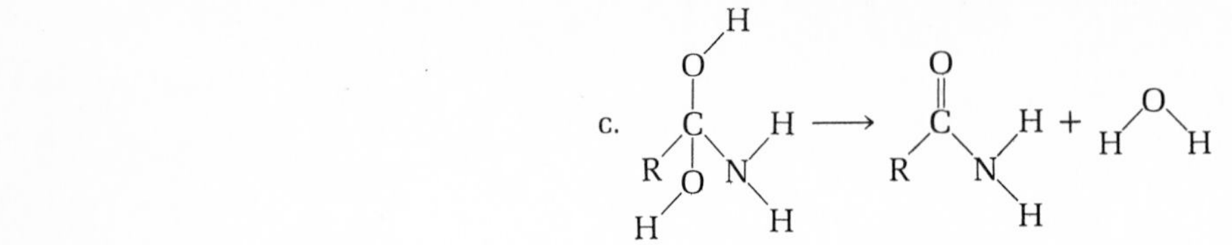

c.

$$\underset{\underset{H}{\overset{\overset{\displaystyle O-H}{|}}{\underset{\overset{|}{O}}{\overset{|}{C}}}}{R}\overset{\displaystyle O-H}{\underset{\overset{|}{\underset{H}{N}}}{}} \longrightarrow \underset{\underset{\overset{|}{\underset{H}{N}}}{R}}{\overset{\overset{\displaystyle O}{\|}}{C}}\overset{\displaystyle H}{} + \underset{H}{\overset{\displaystyle O}{}}\overset{}{H}$$

Thus, nitriles may be converted to amides by hydrolysis, but amides are themselves readily hydrolyzed (Unit 29):

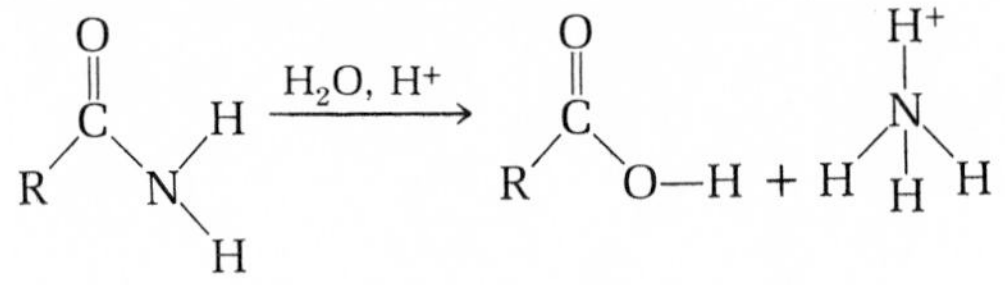

$$\underset{R}{\overset{\overset{\displaystyle O}{\|}}{C}}\underset{\overset{|}{\underset{H}{N}}}{\overset{\displaystyle H}{}} \xrightarrow{\text{H}_2\text{O, H}^+} \underset{R}{\overset{\overset{\displaystyle O}{\|}}{C}}\overset{\displaystyle O{-}H}{} + H\underset{H}{\overset{\overset{\displaystyle H^+}{|}}{N}}H$$

The acid-catalyzed hydrolysis of nitriles may then be summarized as

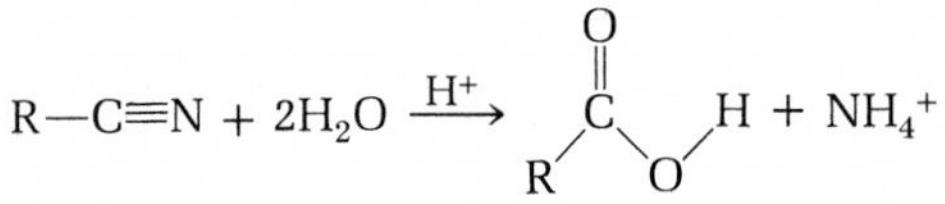

$$R{-}C{\equiv}N + 2H_2O \xrightarrow{\text{H}^+} \underset{R}{\overset{\overset{\displaystyle O}{\|}}{C}}\overset{\displaystyle H}{\underset{O}{}} + NH_4^+$$

The remainder of the nitrile prepared in step 4 is then hydrolyzed in acid:

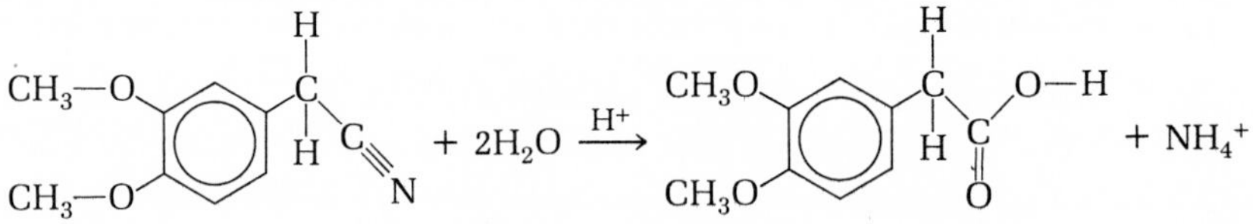

The resultant acid is a crystalline solid only sparingly soluble in water. It is isolated by filtration and purified by recrystallization.

Note that an alternative procedure could have used conversion of the carboxylic acid to the acid chloride, followed by addition of the amine (Unit 29).

7. AMIDE FORMATION Amines and carboxylic acids react by proton exchange to form salts similar to ammonium acetate. Such salts may form amides when heated, losing the elements of water.

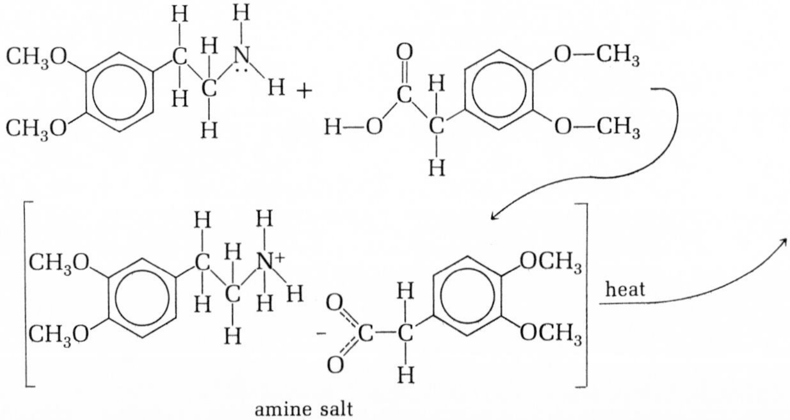

amine salt

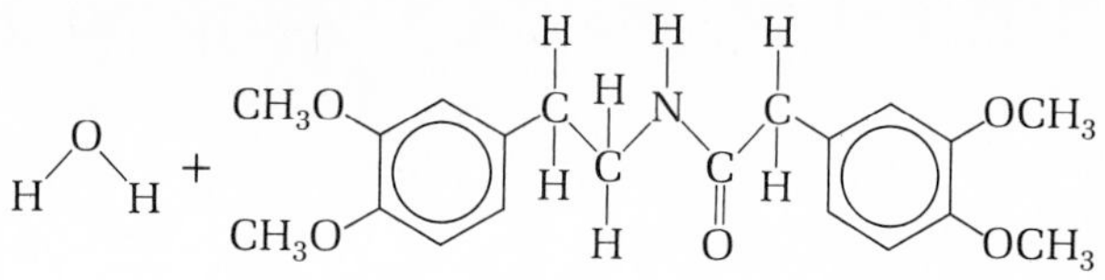

amide

Both the salt and the amide are white solids that may be purified by recrystallization.

8. RING CLOSURE The amide formed in step 7 appears quite unlike the structure of papaverine. However, if one remembers that rotations around σ bonds occur continuously in the molecule, then it is reasonable to assume that the molecular shape will occasionally resemble that of papaverine (conformation c below).

a.

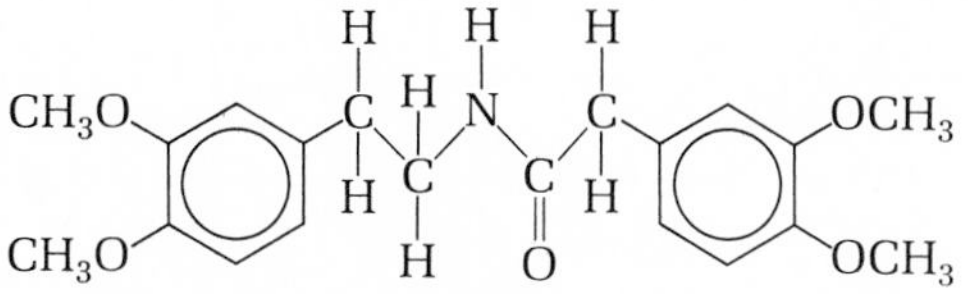

$\longrightarrow$ rotational motions

b.

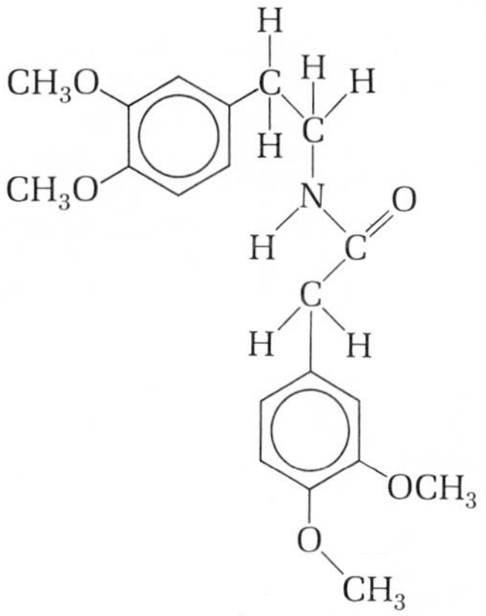

rotational motions

c.

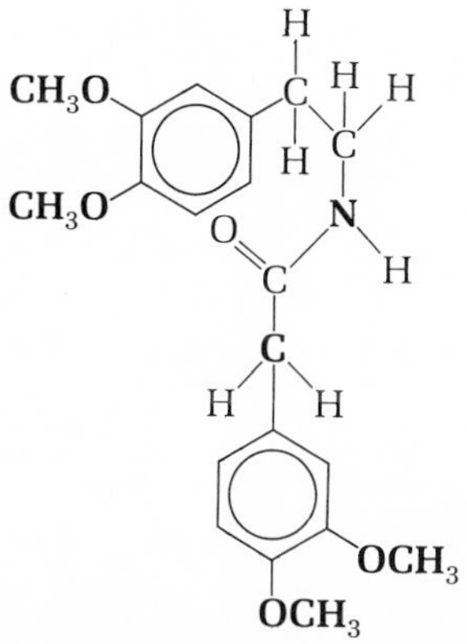

rotational motions

(other shapes)

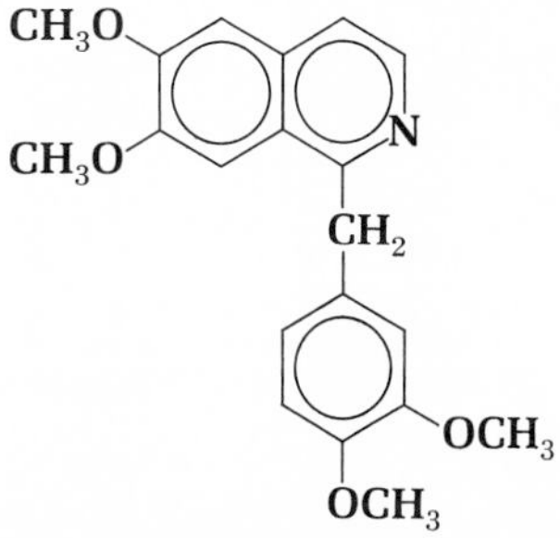

papaverine

Without specifying the reaction conditions, which are beyond the scope of our present interests, it may be indicated that a ring closure (cyclization) can be effected by loss of water (C=O oxygen, N—H proton, and one aromatic ring hydrogen) when the molecular geometry approaches that of conformation (c) under appropriate circumstances. The cyclized product has the structure

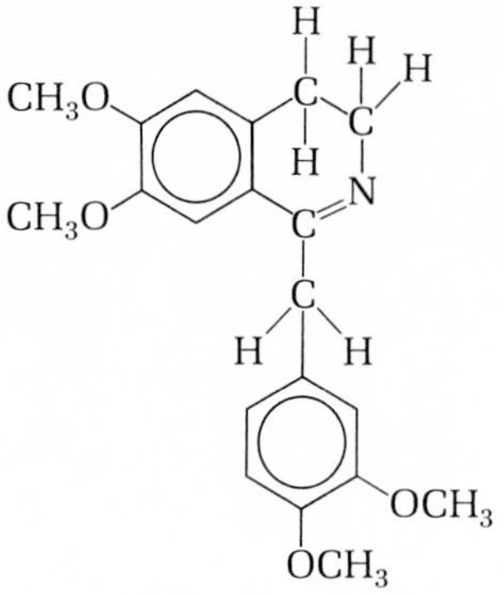

9. AROMATIZATION Hydrogens may be removed catalytically in a reversal of the hydrogenation of a π bond. The driving force for such a *dehydrogenation* reaction in this case is the formation of a delocalized π-bond (aromatic) system.

The product from step 8 is warmed with finely powdered palladium to effect the dehydrogenation to the desired product, papaverine:

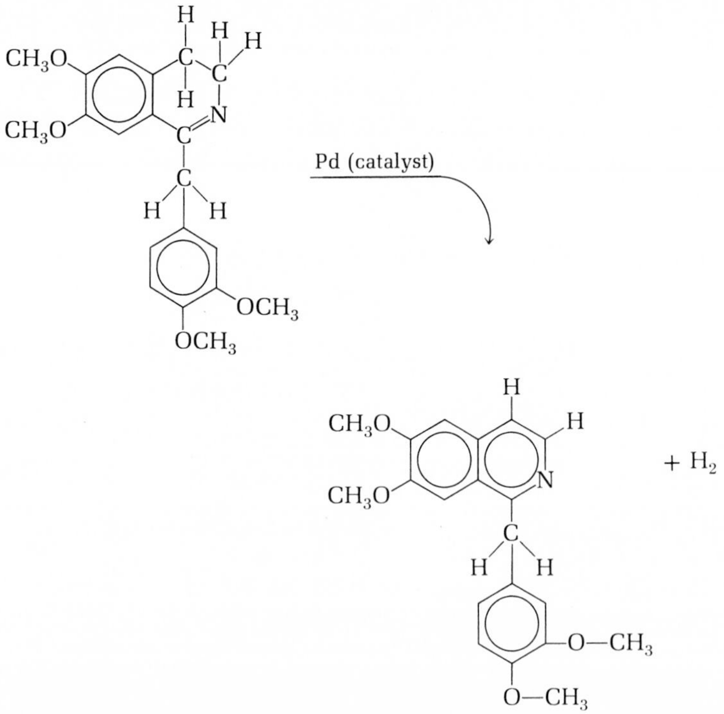

Pd (catalyst)

$+ H_2$

The synthetic material is found to be identical in all respects, including biological activity, with papaverine from the opium poppy. The synthesis thus confirms the structure suggested for papaverine. In addition, it utilizes simple and inexpensive procedures, so the compound can now be manufactured on a large scale for medical uses.

The ability to follow a multistep synthesis such as the scheme outlined for papaverine is considerably easier than the devising of such a synthesis originally. A great deal of knowledge, experience, and serendipity ("blind luck") are required for successful research involving structural determination and synthesis.

9 STOICHIOMETRY IN MULTISTEP SYNTHESES

The *balanced* equations for each step in a synthetic scheme show the stoichiometric ratios of reactants required. Since organic compounds rarely behave as simply as ionic species, the yields for various steps are almost always less than quantitative. In many cases, an excess of the least expensive reactant is employed to improve product yields.

The *net* yield for a multistep synthesis is generally poor, so the number of steps should be minimized and alternative reactions should be considered whenever they offer a more economical route to the final product.

When net yields have been found in "pilot-plant" trials of a synthesis, stoichiometric principles, combined with yield data from the various steps, permit calculation of the quantities of reactants required at each stage for formation of the desired quantity of final product.

SUGGESTED READING

Bates, Robert B., and John P. Schaefer. 1971. *Research Techniques in Organic Chemistry.* Englewood Cliffs, N.J.: Prentice-Hall.
Ferguson, Lloyd N. 1972. *Organic Chemistry: A Science and an Art.* Boston: Willard-Grant Press.
Ireland, R. 1969. *Organic Synthesis.* Englewood Cliffs, N.J.: Prentice-Hall.

If you find natural product chemistry of interest, you may wish to read the very interesting paperback; *Narcotics: Nature's Dangerous Gifts,* by Norman Taylor (New York, Dell, 1970).

SPECIAL TOPICS

The sum total of chemical knowledge is so vast that no individual can hope to be expert in all areas of the science. Practicing chemists have to be well versed in their areas of specialization and reasonably proficient in peripheral fields. This requires a considerable background in training and experience and a major continuing effort to keep abreast of new developments. One of the things that all scientists must learn during their academic preparation is how to use the scientific literature (Appendix H) so that they can find and use information when they need it.

The material covered in this textbook so far has been designed to give you a general overview of some of the more important areas of chemistry and an adequate background for further study. Our "overview" would be incomplete if we neglected two major specialty areas: biochemistry and radiochemistry.

Unit 34 describes just a few of the topics included in the broad field of biochemistry. This is one of the largest and most productive areas of current scientific investigation—involving chemists, physicists, biologists, biochemists, plant and animal scientists, and researchers in a variety of medical fields.

The topic of radiochemistry is discussed in Unit 35. Radioactive isotopes find applications ranging from the dating of archeological discoveries to the nuclear power industry. Some level of understanding of radiochemistry is essential to the appreciation of both the potential benefits and the potential hazards we face in the present "nuclear age."

The final textual material is a brief *epilogue*, containing a little "homespun philosophy" from the author.

CONTENTS

Some aspects of biochemistry

Happy is the man that findeth wisdom, and the man that getteth understanding.

Proverbs 3:13

Objectives

(a) Be able to identify systematic names of enzyme systems with the general type of enzyme activity involved (Table 34.1).

(b) Be able to describe *general* characteristics of human metabolic processes involving carbohydrates and proteins (Sections 34.6 and 34.8).

(c) Be able to identify monomer units characteristic of ribonucleic acids and deoxyribonucleic acids (Section 34.10).

(d) Be able to read with understanding and to discuss meaningfully some relatively nontechnical articles concerning various aspects of biochemistry.

Biochemists have learned a great deal about the life process and the chemical changes that take place within living systems, but we have only seen the "tip of the iceberg." Every answer obtained seems to pose more questions. We still do not know with certainty that "life" can be explained solely by the known laws of chemistry and physics. It is probable that "life" is more complex than we imagine; it is possible that it is more complex than we can imagine.

We can prevent conception by chemical means, and we have developed chemical ways of killing every life form we know. We do not know very much about extending life beyond its normal span or of regenerating life in dead tissue. Science, like other human endeavors, often seems to tackle the easier problems first. It is, after all, much easier to bomb a city than to build one worth living in.

The excitement of modern biochemistry is due to the courage now shown in attacking the "difficult" problems. Living systems are unbelievably complex, but biochemists are daring to challenge this complexity in a wide variety of research areas. Biochemists are vital members of research teams studying ways of improving nutrition for the millions of undernourished people throughout the world; of solving the problems of tissue rejection in medical transplants; of improving the environment by finding alternatives to chemical insecticides and fertilizers, or increasing the biodegradability of waste products; of developing chemical treatments for the scourges of drug addiction or cancer. Truly biochemistry is one of the most challenging and rewarding fields of modern research.

This unit is designed to offer some glimpses of the fascination and complexity of modern biochemistry. Virtually every aspect of the chemistry discussed in this text plays some role in biochemical research. We shall consider briefly three areas of biochemistry—the enzymes, some aspects of metabolism, and the function of nucleic acids in the genetic code.

34.1 ENZYMES

Most of the reactions involving organic compounds in living organisms require catalysis in order for them to proceed at rates sufficient for normal function of the organism. The most common catalytic agents for such reactions are the *enzymes*. These macromolecules are primarily proteins (Figure 34.1) formed into a particular

Figure 34.1 Lysozyme, an enzyme. Lysozyme is an enzyme capable of dissolving the cell walls of certain bacteria. The enzyme occurs in tears, saliva, blood serum, and a number of other animal and plant fluids. Its recent synthesis has involved collaborative efforts between Dr. Arthur Robinson's group at the University of California at San Diego and Dr. John Rupley's group at the University of Arizona. Photograph courtesy of Dr. Leon Barstow, University of Arizona.

geometry best suited to their specific catalytic requirements. Every living system contains a wide variety of enzyme catalysts, each with its own specific role in the overall chemical scheme of the organism. Like other catalysts (Unit 16), the enzymes serve to lower the activation energy for a reaction, but—unlike most inorganic catalysts—the enzymes are typically stereoselective. Thus an enzyme that can catalyze the oxidative deamination of L-alanine would probably have no effect on the mirror-image isomer, D-alanine (Unit 32). Any theory of enzyme activity, then, should account for both the catalytic function and the stereoselectivity of these complex molecules.

34.2 ENZYME STRUCTURE

Most enzymes can exist in inactive forms that can be easily converted to the catalytically active geometry when necessary. In many cases the major portion of the enzyme is a protein (called the *apoenzyme*) that requires either some metal ion (an *activator*) or some simpler organic molecule (a *coenzyme*) for conversion to the form

Figure 34.2 A model for enzyme action (a hydrolytic enzyme).

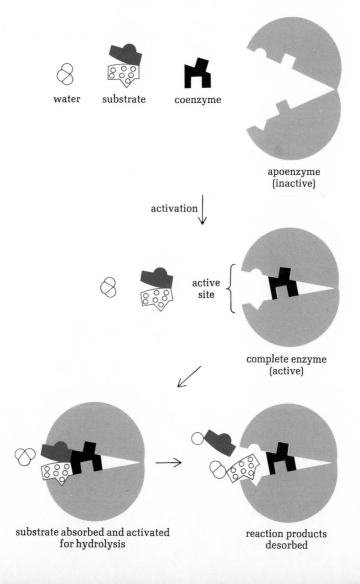

water   substrate   coenzyme

apoenzyme
(inactive)

activation

active
site

complete enzyme
(active)

substrate absorbed and activated
for hydrolysis

reaction products
desorbed

necessary for catalytic activity. Many of the molecules acting as coenzymes appear to involve structures derived from certain vitamins, so the importance of a proper vitamin balance in the diet may be directly related to the organism's requirements for maintenance of adequate enzyme activity.

A number of the more important enzyme systems have been well characterized. Evidence suggests that the catalytic function typically occurs at a small region of the macromolecule called the *active site*. This may be considered as an irregular depression in the surface of the tightly curled macromolecule, shaped so that a particular reactant molecule (or molecular segment) can fit in the specific alignment best suited for reaction to occur. The part of the enzyme molecule forming the active site contains functional groups capable of attracting and orienting the reactant species (the *substrate*). In addition, the electronic interactions of substrate bonds with active site groups may serve to activate the substrate for more facile reaction. After reaction the product molecules are desorbed from the enzyme, since their characteristics fail to meet the requirements for optimum attraction to the groups lining the active site. A schematic representation for this model of enzyme activity is shown in Figure 34.2.

In some cases evidence has accumulated that suggests an actual participation of enzyme groups in one or more elementary processes of the catalyzed reaction (Unit 16).

34.3 TYPES OF ENZYME SYSTEMS

Proton-transfer reactions in aqueous media are very rapid, so such processes do not require catalysis. The biochemical reactions most dependent on enzymes are those involving molecular species, whose reactions are typically slow at the relatively low temperatures of living systems. Since most organic compounds of biological importance contain oxygen or nitrogen (or both), reactions at C—O or C—N bonds are frequently involved.

The biochemical reaction sequences referred to generally as *metabolism* often include hydrolytic processes involving ester, amide (peptide), or acetal (glycoside) bonds. A variety of enzymes participate in these reactions. Others may act as catalysts for the reverse processes, that is, *formation* of ester, peptide, or glycoside groups. Oxidation–reduction reactions are also important in many phases of metabolic events, and these utilize still other enzyme systems. There is, indeed, some enzyme system known for almost every type of organic reaction that might occur with the chemical substances found in living systems. Modern terminology has attempted to use names for enzymes that are descriptive of the type of substrate they require (Table 34.1), but many enzymes discussed in the older literature have less

Table 34.1   SOME COMMON ENZYME SYSTEMS[a]

ENZYME NAME AND (TYPICAL SOURCE)	SUBSTRATE	PROCESS CATALYZED
HYDROLASES		
salivary α-amylase (saliva)	starches	hydrolysis of α-glycoside links to yield smaller polysaccharides
pancreatic α-amylase (pancreas)	starches and smaller polysaccharides	hydrolysis of α-glycoside links to yield mainly maltose (a disaccharide)
maltase (intestinal secretions)	maltose	hydrolysis of α-glycoside link to yield glucose units
pepsin (stomach fluids)	proteins	hydrolysis of peptide bonds to yield polypeptides
chymotrypsin (pancreas)	proteins or polypeptides	hydrolysis of peptide bonds to yield small peptides
dipeptidase (intestinal secretions)	dipeptides	hydrolysis of peptide bond to yield amino acids
gastric lipase (stomach fluids)	fats	hydrolysis of ester bonds to yield glycerin, carboxylic acids, and glycerides (partially esterified glycerin)
pancreatic lipase (pancreas)	fats	hydrolysis of ester bonds to form glycerin, carboxylic acids, and glycerides
acetylcholine esterase (nervous system)	acetylcholine	hydrolysis of ester bond to yield choline and acetic acid
lysozyme (tears, saliva, mucous)	cell wall polysaccharides in bacteria	hydrolysis of certain types of glycoside units

Table 34.1 (Continued)

ENZYME NAME AND (TYPICAL SOURCE)	SUBSTRATE	PROCESS CATALYZED
HYDROLASES		
urease (kidney)	urea	$H_2N-\overset{\overset{O}{\|}}{C}-NH_2 + H_2O \longrightarrow 2NH_3 + CO_2$
ATPase (cellular fluids)	ATP (Figure 34.9)	$ATP + H_2O \longrightarrow ADP + PO_4^{3-}$
SYNTHETASES		
RNA polymerase (cellular fluids)	nucleotides (Section 34.10)	formation of phosphate ester links in nucleic acids
alanyl-sRNA synthetase (cellular fluids)	alanine, sRNA, and ATP	formation of ester bond between alanine and sRNA as precursor to protein synthesis
OXIDOREDUCTASES		
ethanol dehydrogenase (liver)	ethanol	$CH_3CH_2OH \longrightarrow CH_3\overset{\overset{}{\underset{\underset{H}{\|}}{}}}{C}{=}O$
L-amino acid oxidase (kidneys)	L-amino acids	L-amino acid $+ H_2O + O_2 \atop \downarrow$ ketoacid $+ NH_3 + H_2O_2$
D-amino acid oxidase (liver and kidneys)	D-amino acids	D-amino acid $+ H_2O + O_2 \atop \downarrow$ ketoacid $+ NH_3 + H_2O_2$
catalase (blood)	H_2O_2	$2H_2O_2 \longrightarrow O_2 + 2H_2O$
phenylalanine 4-hydroxylase (liver)	phenylalanine	phenylalanine $\longrightarrow$ tyrosine
TRANSFERASES		
maltose phosphorylase (intestinal secretions)	maltose	maltose $+ PO_4^{3-} \atop \downarrow$ glucose $+$ glucose-1-phosphate
alanine aminotransferase (cellular fluids)	alanine	alanine $+$ 2-oxoglutaric acid $\atop \downarrow$ pyruvic acid $+$ glutamic acid
glucokinase (cellular fluids)	glucose	glucose $+ ATP \longrightarrow ADP +$ glucose-6-phosphate
LYASES		
histidine decarboxylase (liver, kidneys)	histidine	histidine $\longrightarrow$ histamine $+ CO_2$
serine dehydratase (cellular fluids)	serine	serine $\longrightarrow$ pyruvic acid $+ NH_3$
ISOMERASES		
alanine racemase (certain bacteria)	L-alanine	L-alanine $\longrightarrow$ D-alanine
lactate racemase (certain bacteria)	L-lactic acid	L-lactic acid $\longrightarrow$ D-lactic acid

[a] Enzyme "systems" are not necessarily unique molecular species.

revealing names. It is important to note that enzyme names refer to species having a particular catalytic activity and not, necessarily, to a unique chemical compound. For example, phospholipase-A in bee venom is a different molecular substance from that in rattlesnake venom, but both species have the same type of function.

34.4 ENZYME MALFUNCTION   Without proper levels of enzyme activity, living systems cannot function normally. We have already seen how a deficiency in a single enzyme, phenylalanine hydroxylase (Unit 32), can result in major abnormalities. Three other types of enzyme-related problems may be mentioned.

Coenzyme deficiency   Many of the vitamins are converted in the body to essential coenzymes. Although the human body is capable of synthesizing several of the vitamins, this capability is limited, and the diet must contain certain levels of particular vitamins. As an example, vitamin B_1 (thiamine) is converted in the body to thiamine pyrophosphate (Figure 34.3), a coenzyme for certain vital decarboxylase systems. Failure to maintain a sufficient thiamine level in the diet results in the malfunction of these enzyme systems and leads to the molecular disease called *beriberi*.

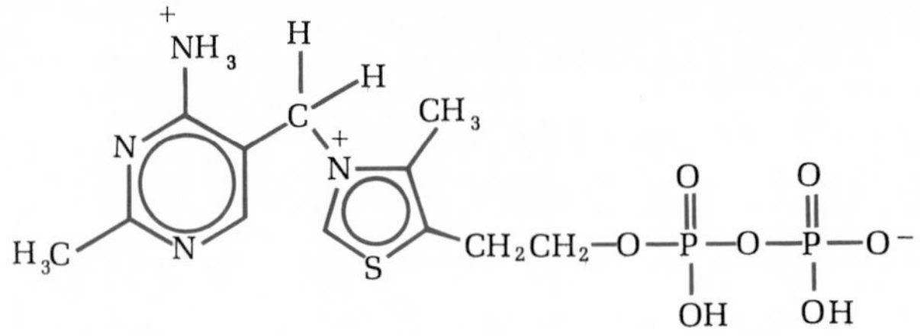

Figure 34.3 Thiamine pyro-phosphate (TPP).

Metal ion competition   Many enzymes are activated by specific metal ions that frequently coordinate at the active site to facilitate binding of the substrate as an additional ligand (Unit 27). Problems may arise if there is serious competition for the metal ion needed. For example, cyanide ion forms strong complexes with certain metal ions, and one of the results of cyanide poisoning may be the coordination of vital activating metal ions with CN^-.

A different problem may arise in the case of poisoning by "heavy metal" ions such as Pb^{2+} or Hg^{2+}. These may replace the normal metal ion in the enzyme system, especially when the foreign ion is present in high concentrations, to tie up the enzyme in a nonfunctional form.

Enzyme inhibitors   Metal ions may also deactivate (inhibit) enzymes by forming salts or complexes that convert the enzyme to a noncatalytic form. The metabolism of glucose depends, in one indispensable step, on an enzyme called *glyceraldehyde 3-phosphate dehydrogenase*. This enzyme contains sulfhydryl groups, which form salts with metal ions. Mercury(II) ion, for instance, forms a very stable salt with this enzyme, completely inhibiting its catalytic function and disrupting the normal pathway for glucose metabolism.

Another type of inhibition occurs when a foreign organic molecule is bound tightly in the active site of the enzyme, preventing the entrance of the normal substrate. Many of the "nerve gases" manufactured as chemical warfare agents act by inhibiting various enzymes in this way. In particular, inhibition of the acetylcholine esterase system prevents the hydrolysis of acetylcholine (Figure 34.4) necessary for the transmission of nerve impulses.

Figure 34.4 Hydrolysis of acetylcholine.

$$\underset{\text{(acetylcholine)}}{(CH_3)_3\overset{+}{N}CH_2CH_2O\overset{\displaystyle O}{\overset{\|}{C}}CH_3} + H_2O \xrightarrow{\text{(acetylcholine esterase)}} \underset{\text{(choline)}}{(CH_3)_3\overset{+}{N}CH_2CH_2OH} + CH_3CO_2H$$

34.5 SOME ASPECTS OF METABOLISM   The human animal is a fantastically complex system of chemical factories, capable of manufacturing literally thousands of different compounds. Most of these are synthesized by enzyme-catalyzed processes occurring inside individual cells. The raw materials for these processes include water, oxygen, and a variety of substances taken in as "food." Most of the cellular processes require molecules small enough to be transferred through the cell membrane, but most of the "food" is macromolecular—proteins or polysaccharides. The transportation system for delivery of the "starting materials" to the cellular "factories" is the aqueous medium of the bloodstream. The initial processes of the body, then, must break down the proteins and carbohydrates into small molecules capable of being transported through the circulatory system and absorbed through cell membranes. Fats, also present in the diet as useful sources of chemicals for cellular processes, must be converted to water-soluble (or dispersible, Unit 14) species for transportation and absorption.

The chemical sequences of metabolism, then, begin with *digestive* processes, which are essentially hydrolytic degradations that convert foods into simpler compounds. After these products are transported to the appropriate cellular "factories," they are used as energy sources or as reactants (or as both) for production of other chemicals needed by the body. As in the case of industrial factories, biochemical processes have waste products, and these are eliminated in various ways. The details of the many chemical reactions involved in metabolism are beyond the scope of this text and, indeed, our present knowledge is limited in this area. We shall consider only a general overview of the types of processes involved.

34.6 CARBOHYDRATES

The carbohydrates in foods are generally of three types—celluloses, starches, and sugars. Humans have no enzymes capable of breaking the β-glycoside links of the celluloses, so these molecules are not utilized and pass through the body essentially intact. Starches, on the other hand, contain the α-glycoside link, and the amylase enzyme systems can catalyze the hydrolysis of these bonds. This process begins in the mouth as the food is mixed with saliva containing an amylase enzyme system. The general scheme of carbohydrate metabolism is outlined in Figure 34.5. The simpler sugars from the food or from the degradation of starches may undergo a variety of interrelated reactions, many of which are involved in the vital processes of energy release. Carbohydrates and fats are the principal "fuels" for the body.

34.7 FATS (LIPIDS)

Lipid is a general term referring to biochemical species that are insoluble in water, but soluble in solvents such as chloroform or ether. The fats (esters of glycerin) are only one of the kinds of lipids found in living organisms.

See the Suggested Reading at the end of this unit for details of lipid metabolism.

The fats are primarily esters of glycerin (Unit 29). Their initial degradation occurs in the stomach and intestines, where *gastric lipase, pancreatic lipase,* and *intestinal lipase* enzyme systems catalyze their hydrolysis to glycerin and long-chain carboxylic acids. Since fats are not soluble in water, bile salts from the liver are used to disperse the fats in micelle units like those in aqueous dispersions of oil in soapy water (Unit 14).

Like the carbohydrates, fats may be eventually utilized in a number of different ways, often as important energy sources. They may also be stored for future utilization, a procedure of utmost importance to hibernating animals, which depend on stored fats almost exclusively for their energy source.

The details of various metabolic pathways for fats are beyond the scope of this text, but we may note that their oxidation as fuels requires a sequence of reactions, each of which removes a two-carbon fragment from the carbon chain of the carboxylic acids formed by hydrolysis of the fats.

34.8 PROTEINS

Like the fats, proteins are not affected by the salivary enzyme system, so their hydrolysis begins in the stomach. Here the *pepsin* begins the hydrolytic cleavage of peptide bonds, converting the large protein molecules into smaller polypeptide fragments. The process continues in the small intestine, where a variety of enzymes

Figure 34.5 Carbohydrates. Notes: (1) If blood sugar concentration is too high, excess glucose is converted in the liver to the polysaccharide glycogen for storage. This is hydrolyzed back to glucose units as necessary to maintain proper blood sugar concentration. (2) Anaerobic glycolysis requires no oxygen and is a rapid way of producing ~16 kcal of energy per mole of glucose consumed. Each glucose molecule is converted to two molecules of lactic acid. This process is used to produce brief bursts of energy for muscle action when the more efficient aerobic mechanism, depending on oxygen, is too slow for the immediate requirements. (3) Aerobic glycolysis yields ~48 kcal per mole of glucose consumed. Glucose is converted to pyruvic acid

$$\underset{(\text{CH}_3\overset{\displaystyle O}{\overset{\displaystyle \|}{\text{C}}}\text{CO}_2\text{H})}{}$$

(CH$_3$CCO$_2$H), which is utilized in the Krebs cycle, being eventually converted to CO$_2$. (4) For details of this pathway or further information on the alternative metabolic routes, see the Suggested Reading at the end of this unit.

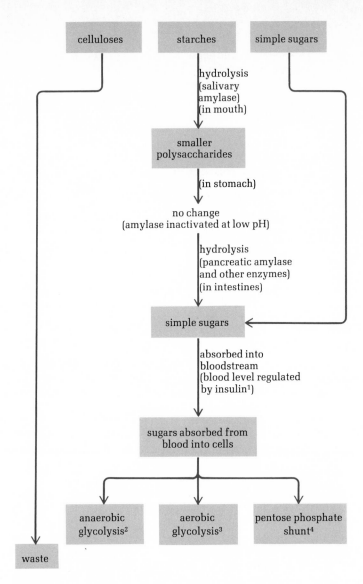

are involved in the degradation to small peptides and amino acids suitable for absorption through the intestinal membranes into the bloodstream.

The protein hydrolysis products are significantly different from those of the fats and carbohydrates. Neither amino acids nor small peptides are significant energy sources, and the body has no mechanism for "storing" protein as it does fat or carbohydrate materials.

Since natural proteins have so many different monomer units (Units 31 and 32), protein metabolism is exceedingly complex; each amino acid may participate in any of a variety of reactions. The general possibilities are indicated in Figure 34.6. Many of the steps in amino acid metabolism are still only partially understood. Since proteins are the chief source of nitrogen for cellular processes, mechanisms must be available for converting amino acids into many different kinds of nitrogen-containing molecules.

Figure 34.6   Proteins.

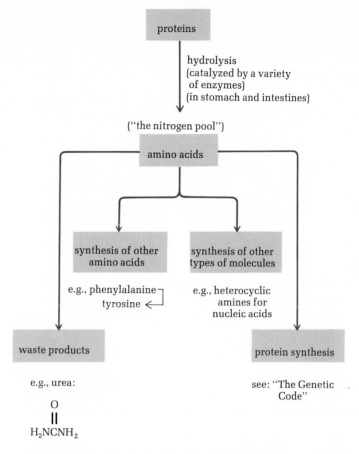

34.9 THE GENETIC CODE What determines "inherited" characteristics? Why do biological species reproduce "after their own kind"? How is the process of protein synthesis controlled in living systems?

The key to these and many other questions in molecular biology now appears to lie with substances called *nucleic acids*. These complex macromolecules contain the necessary "information" for their own self-replication, for the synthesis of other types of similar materials, for the determination of protein composition and structure, and for the transmission of hereditary traits. Surely the nucleic acids are the most fascinating molecules yet to be studied by man. First isolated from the nuclei of white blood cells and fish sperm by Miescher in 1869, the nucleic acids remained largely a mystery until 1953. Although it had long been recognized that these substances were intimately involved in the genetic mechanism, it remained for James Watson[1] and Francis Crick to propose a satisfactory structural description of the macromolecules that could explain their biological functions. Many aspects of nucleic acid behavior are still unknown. Many others are largely speculative. The

---

[1] For a truly absorbing account of the very typical mixture of genius, perspiration, fumbling, and luck inherent in most major advances in science, read the description by Watson of how he and Crick worked out the problem that culminated in their Nobel Prize in 1962 (*The Double Helix*, by James D. Watson, New York: New American Library, 1968; paperback).

general picture of polymers whose chemical structures control the genetic mechanisms is, however, now generally accepted.

The nucleic acids are enormous molecules, frequently associated with proteins ("nucleoproteins") in the cell. Once biologists had established the involvement of such compounds in cell division and protein biosynthesis, it remained for chemists to determine the molecular descriptions consistent with such functions. Since these compounds are polymeric, like the polysaccharides (e.g., starches and celluloses) and the proteins, it might be possible to break down the macromolecules into "monomer" units by hydrolysis. Hydrolysis of polysaccharides yields simple carbohydrates, principally glucose. The proteins, on the other hand, yield a rather diverse mixture of hydrolysis products, the amino acids. It should be interesting to see the results of hydrolysis of the nucleic acids.

34.10 MONOMER UNITS: THE NUCLEOTIDES

RNA and DNA are "class" names, like "protein" or "starch," and do not refer to specific compounds.

Hydrolysis of a nucleic acid yields three different types of substances—a simple sugar, inorganic phosphate, and some heterocyclic amines. Early investigations of these products revealed that there were two distinct types of nucleic acids. One yielded ribose (Figure 34.7) as the only carbohydrate product, and the other hydrolyzed to form a similar sugar, deoxyribose, having one less oxygen than ribose. The two types were, consequently, called RNAs (ribonucleic acids) and DNAs (deoxyribonucleic acids), respectively.

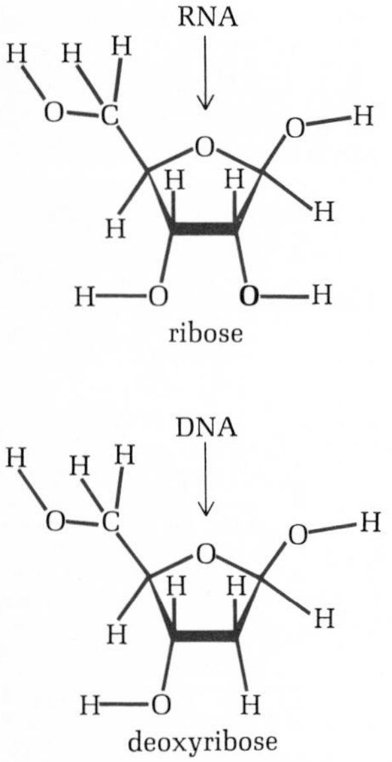

Figure 34.7 Carbohydrate products from hydrolysis of nucleic acids.

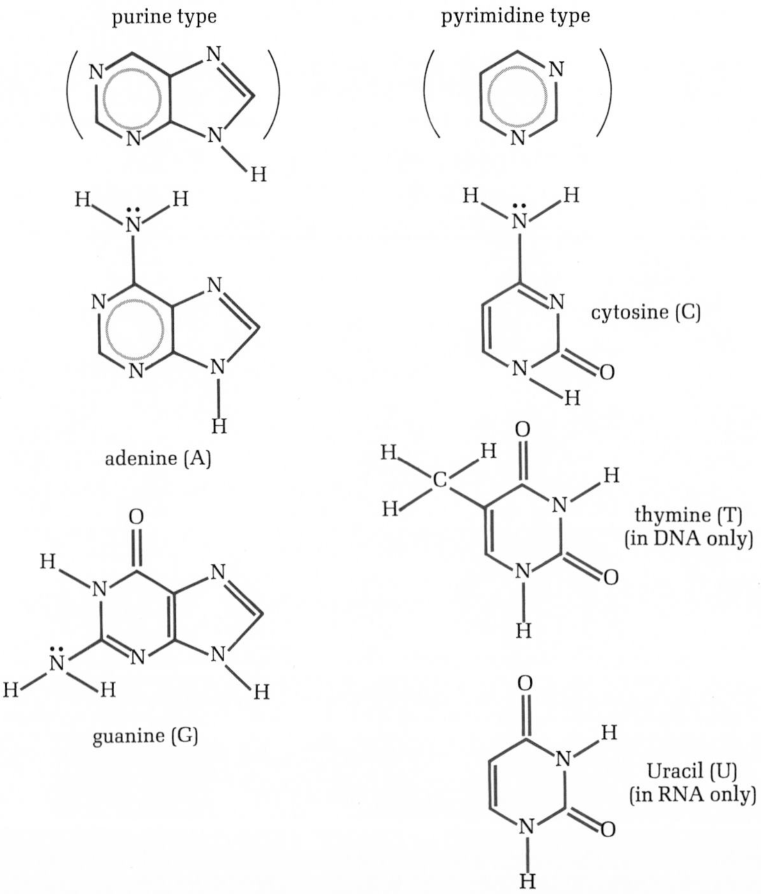

Figure 34.8 Heterocyclic amines from hydrolysis of nucleic acids. The bases shown are the ones most commonly found. Less common amines have structures very similar to these.

Table 34.2 RELATIVE MOLAR RATIOS OF HYDROLYSIS PRODUCTS OF NUCLEIC ACIDS

DNA FROM	DEOXYRIBOSE	PHOSPHATE	HETEROCYCLIC AMINES[a]			
			A	G	C	T
calf thymus cells	1.00	1.00	0.28	0.21	0.21	0.28
gypsy moth	1.00	1.00	0.21	0.31	0.33	0.20
silkworm virus	1.00	1.00	0.29	0.22	0.21	0.28

[a] Based on data reported by G. R. Wyatt, *Chemistry and Physiology of the Nucleus* (New York: Academic, 1952).

Both RNA and DNA yield orthophosphate (Unit 26) in a protonated or non-protonated form (actually as some equilibrium mixture) as determined by the pH of the hydrolysis conditions. In addition, a group of heterocyclic amines (Figure 34.8) are found. Although some individual characteristics distinguish each amine, all belong to either the group called *purines* (two fused heterocyclic rings) or the group called *pyrimidines* (one heterocyclic ring).

The qualitative analysis of these hydrolysis products was an important first step, but revealed little useful information about macromolecular structure. It was more difficult to obtain quantitative data, partly because of the problems of obtaining a highly purified nucleic acid and partly because of the similarities among the heterocyclic amines, which required quite sophisticated separation procedures. As quantitative data began to accumulate, it became apparent that the components of the polymeric molecules were present in certain specific ratios (Table 34.2). Any proposed structures must be consistent with these ratios.

These hydrolysis data suggest that there is a 1:1:1 ratio of sugar to phosphate to amine, with variation in which amine is involved. This strongly indicated a monomer unit of the type

phosphate
/   \
sugar——amine

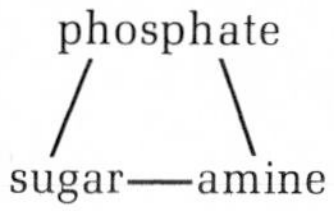

(i.e., some compound containing these three components in linkages that could be broken by hydrolysis). Such compounds were not unknown in biological systems. Indeed, removal of two phosphates from ATP (Figure 34.9) would produce exactly the type of monomer required. The definitive description of the actual monomer unit is

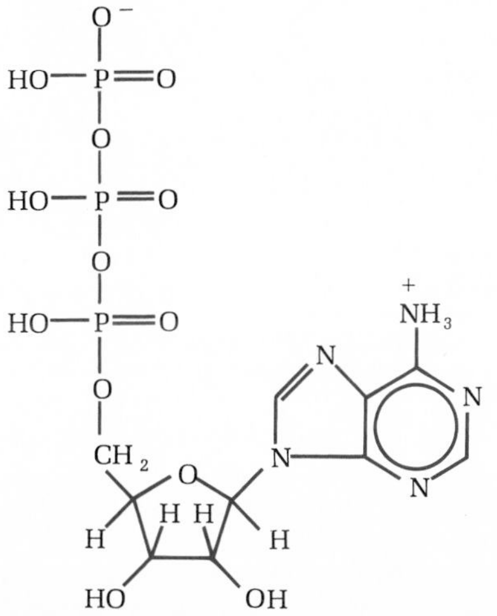

Heterocyclic refers to ring systems containing other atoms (in this case nitrogen) in addition to carbon.

It is also very important to note that adenine and thymine occur in equimolar amounts, whereas guanine and cytosine occur in different (but still equimolar) amounts.

Figure 34.9 Adenosine triphosphate (ATP), an important reactant in many biochemical processes. The P—O—P bonds are referred to as "high-energy phosphate" bonds because the hydrolysis of such anhydride bonds is quite exothermic (Unit 26).

Figure 34.10 A nucleotide unit. The unit shown is called *cytidylic acid*. When a different amine is involved, the nucleotide may be adenylic acid (from adenine), guanydilic acid (from guanine), uridylic acid (from uracil), and so on. Strictly speaking, these should be indicated as, for example, ribocytidylic acid or deoxyribocytidylic acid to show which sugar is involved.

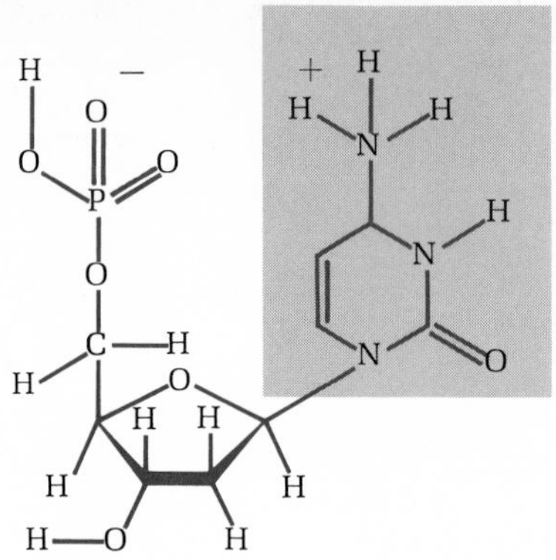

a fitting subject for a chapter of its own. A few of the difficulties included the proof of which positions in the sugar molecule and the various amines were involved in bonding together the segments of the monomer, the determination of the stereochemistry of the monomers, and the isolation of comparison monomers from natural nucleic acids by careful partial hydrolysis procedures. In the end all evidence supported structures of monomer units like that shown in Figure 34.10. These compounds, called nucleotides, have been found free in biological systems.

34.11 THE STRUCTURE OF DNA

The description of the monomer units was a long way from the determination of total molecular structure and function. By the early 1950s, x-ray data had been obtained for fibrous nucleic acids that strongly suggested they were spirals of uniform diameter. The final piece of the structural puzzle was contributed by Watson and Crick in 1953. Interestingly enough, their "research" did not involve either chemical degradations or x-ray crystallography, as such. They used scale models of nucleic acid components to try various ways of piecing these together in a pattern consistent with the chemical and x-ray data contributed by others. The key to the structure lay with the arrangements of the heterocyclic amines.

Evidence had accumulated that the molecules were double helices of sugar–phosphate–sugar alternating units, with the two strands of the helix held by hydrogen bonds between amine moieties. However, all attempts to construct models of identical strands (i.e., hydrogen bonding between two amines of the same structure) were unsuccessful. Regions containing two paired pyrimidine bases were smaller than regions having paired purine bases, and this was inconsistent with the x-ray data supporting a uniform diameter of the helix. Finally, by accident, Watson found that models of adenine paired best with those of thymine, while optimum pairing occurred for guanine with cytosine (Figure 34.11). Uracil could readily be substituted for the structurally similar thymine. The puzzle was solved. The genetic material had been described, and the "code" that determined the specific behavior of each type of nucleic acid was simply the unique sequence of heterocyclic amines in each particular molecule.

Like most puzzle solutions in science, the description of DNA raised more new questions than it answered. Some of the new research generated by the models of Watson and Crick has given at least partial explanations of the various steps in replication of nucleic acids, protein biosynthesis, and other related processes in cells and viruses. The interested student should consult some of the excellent articles and paperbacks available on these subjects.[2]

[2] See, for example, V. G. Allfrey and A. E. Mirsky, "How Cells Make Molecules," in the September, 1961, issue of *Scientific American;* Isaac Asimov, *The Genetic Code* (New York: New American Library, 1962, paperback); and J. Horwitz and J. J. Furth, "Messenger RNA," in the February, 1962, issue of *Scientific American.*

Figure 34.11 Base pairing by optimum hydrogen bonding.

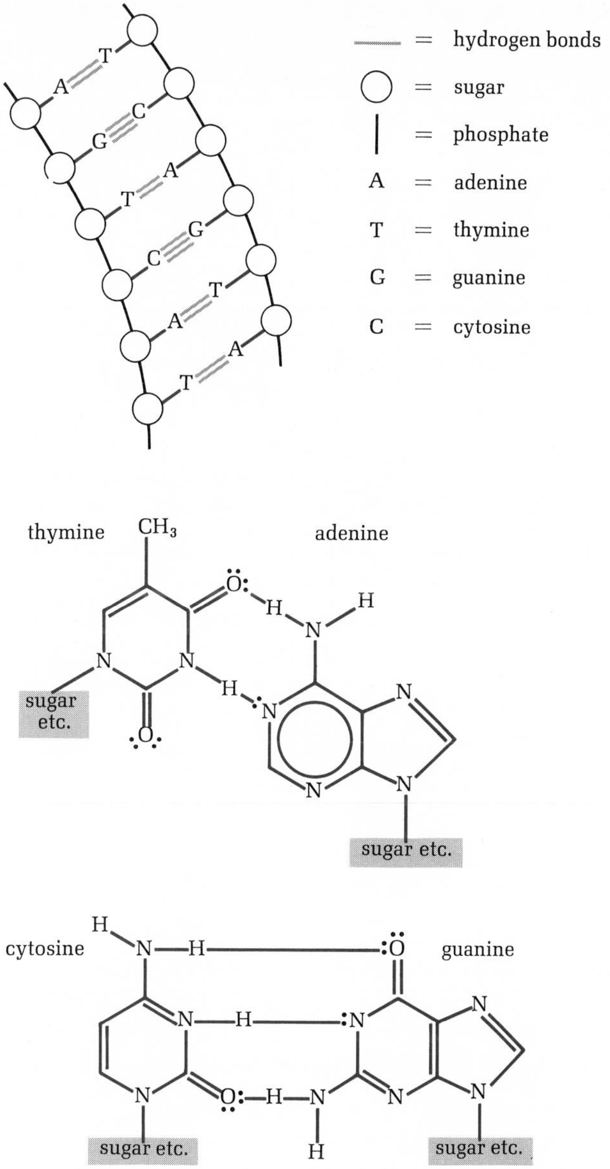

—— = hydrogen bonds

◯ = sugar

| = phosphate

A = adenine

T = thymine

G = guanine

C = cytosine

34.12 ARTIFICIAL SYNTHESIS OF LIFE

Biochemists estimate that a simple bacterial cell contains some 6000 different kinds of molecules. They believe that we understand now about one-third of the reactions occurring in the cell. It appears, then, that the synthetic production of a living system—even the most simple one—is several years away. Nonetheless, the manufacture of a living organism from nonliving materials must be considered a real possibility. The problems are enormous, but possibly solvable. The real key lies in whether the life process involves only changes that can be understood in terms of the laws of chemistry and physics. Most scientists are currently inclined to believe that such is the case, but it will be some time before direct investigations will be really feasible.

LIVING SYSTEMS: ACCIDENT OR CREATION?

As research in the field of molecular biology progresses, the temptation for scientists to be overly impressed by their own knowledge is almost overwhelming. It has become possible within the last 25 years to produce common organic molecules by electric discharge (simulating lightning) through gas mixtures like those believed to constitute the atmosphere of the primitive earth. Mixtures of amino acids heated on powdered rock and quenched by cooling rain (simulated) have formed polypeptides whose macrostructures are similar to those of many proteins. Synthetic nucleic acids have demonstrated the self-replicating properties of their natural relatives. Do these and other similar developments *prove* the theories of the accidental origin of life? One might believe so from many current texts and classes.

However, it is important to realize the limitations of scientific investigation. By its nature, science requires observation and investigation. It is possible to determine one or more routes by which an event may now occur. It is not possible to determine through science *the unique way* in which an event took place at a time before observations were made. No reputable scientist would dream of picking up a strange sample of alanine and, from his knowledge of synthetic chemistry, stating *as fact* exactly how that particular sample was made. The same scientist might find it difficult to resist the impulse to state rather dogmatically the route by which complex life forms evolved on the primitive earth.

There is no reason for scientists to discard a belief in God. Indeed, there are many who feel the need for a faith that suggests human life is more than a series of chemical changes and that it need not end when some of these reactions cease. The belief in the existence of a soul, a way in which a life ought to be lived, and a life beyond this transient one cannot be subjected to scientific tests. At worst, such a belief is a harmless fantasy. If true, and many of us are convinced that it is, it is far more important than anything science, technology, or this world has to offer.

SUGGESTED READING

Barrington, E. J. W. 1968. *The Chemical Basis of Physiological Regulation.* Glenview, Ill.: Scott, Foresman.

DeBey, H. J. 1968. *Introduction to the Chemistry of Life: Biochemistry.* Reading, Mass.: Addison-Wesley.

Locke, D. M. 1969. *Enzymes: The Agents of Life.* New York: Crown.

Scientific American. 1968. *Bio-Organic Chemistry.* Readings from *Scientific American,* San Francisco: Freeman.

Woese, Carl. 1967. *The Genetic Code: The Molecular Basis for Genetic Expression.* New York: Harper & Row.

EXERCISES

1. Match the enzyme system names with the corresponding enzyme-catalyzed processes.

ENZYME SYSTEMS: (A) alanine aminotransferase, (B) alanine racemase, (C) ATPase, (D) catalase, (E) histidine decarboxylase, (F) L-amino acid oxidase, (G) maltase, (H) methanol dehydrogenase, (I) phenylalanine 4-hydroxylase, (J) urease

a. $CH_3OH \longrightarrow H_2C{=}O$ _____

b. $2H_2O_2 \longrightarrow 2H_2O + O_2$ _____

c. $H_2N\overset{O}{\overset{\|}{C}}NH_2 + H_2O \longrightarrow 2NH_3 + CO_2$ _____

d. $H_3\overset{+}{N}{-}\underset{\underset{CH_3}{|}}{\overset{\overset{CO_2^-}{|}}{C}}{-}H \longrightarrow H{-}\underset{\underset{CH_3}{|}}{\overset{\overset{CO_2^-}{|}}{C}}{-}\overset{+}{N}H_3$ _____

e. $H_3\overset{+}{N}{-}\underset{\underset{CH_3}{|}}{\overset{\overset{CO_2^-}{|}}{C}}{-}H + H_2O + O_2 \longrightarrow$

$CH_3\overset{O}{\overset{\|}{C}}CO_2H + NH_3 + H_2O_2$ _____

f. $H_3\overset{+}{N}{-}\overset{\overset{CO_2^-}{|}}{\underset{|}{C}}{-}H \longrightarrow H_3\overset{+}{N}{-}\overset{\overset{CO_2^-}{|}}{\underset{|}{C}}{-}H$ _____

(phenyl group below left C; below right C a phenyl with OH)

g. $H_3\overset{+}{N}{-}\underset{\underset{CH_3}{|}}{\overset{\overset{CO_2^-}{|}}{C}}{-}H + HO_2C\overset{O}{\overset{\|}{C}}(CH_2)_2CO_2H \longrightarrow$

$CH_3\overset{O}{\overset{\|}{C}}CO_2H + H_3\overset{+}{N}{-}\underset{\underset{\underset{\underset{CO_2H}{|}}{CH_2}}{|}}{\overset{\overset{CO_2^-}{|}}{C}}{-}H$ _____

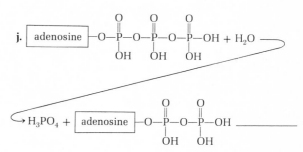

h. (histidine structure) ${-}CH_2\overset{+}{\underset{}{C}}CO_2^- \longrightarrow$ $+NH_3$

(imidazole) ${-}CH_2CH_2NH_2 + CO_2$ _____

i. $C_6H_{22}O_{11} + H_2O \longrightarrow 2C_6H_{12}O_6$ _____

j. $\boxed{adenosine}{-}O{-}\overset{O}{\overset{\|}{P}}{-}O{-}\overset{O}{\overset{\|}{P}}{-}O{-}\overset{O}{\overset{\|}{P}}{-}OH + H_2O \longrightarrow$
(below P's: OH, OH, OH)

$\longrightarrow H_3PO_4 + \boxed{adenosine}{-}O{-}\overset{O}{\overset{\|}{P}}{-}O{-}\overset{O}{\overset{\|}{P}}{-}OH$ _____
(below P's: OH, OH)

2. Indicate whether each of the following statements is true or false:

a. The hydrolysis of starches *begins* with enzymes found in the mouth. _____

b. The hydrolysis of proteins *begins* with enzymes found in the lower intestinal tract. _____

c. Humans contain salivary enzymes that hydrolyze both starches and celluloses. _____

d. Humans contain enzyme systems capable of converting some polysaccharides to glucose and also enzymes capable of polymerizing glucose molecules. _____

e. Some amino acids required by humans must be included in their diets. _____

f. The principal waste product of human metabolism of proteins is uric acid. _____

g. Insulin is required in humans to permit the digestion of polysaccharides containing β-glycoside links. _____

h. Most of the nitrogen required by humans for the biosynthesis of heterocyclic amines is obtained from protein foods. _____

i. Some of the amino acids present in human tissue proteins are synthesized in the body from other amino acids. _____

j. Surplus protein in the body is stored in the liver as the polypeptide *glycogen*. _____

3. Indicate whether each of the following compounds is an expected hydrolysis product of DNA, of RNA, or of both:
a. PO_4^{3-}; **b.** adenine; **c.** uracil; **d.** thymine; **e.** a simple sugar of molecular formula $C_5H_{10}O_4$

PRACTICE FOR PROFICIENCY

Read and be prepared to discuss (or write about) one or more of the following articles. (Your instructor may wish to suggest others.)

1. "Non-Covalent Interactions: Key to Biological Flexibility and Specificity," by Earl Frieden, in the December, 1975, issue of the *Journal of Chemical Education* (pp. 754–761).
2. "How the Liver Metabolizes Foreign Substances," by Attallah Kappas and Alvito P. Alvares, in the June, 1975, issue of *Scientific American* (pp. 22–31).
3. "Cyanate and Sickle-Cell Disease," by Anthony Cerami and Charles M. Peterson, in the April, 1975, issue of *Scientific American* (pp. 44–50).
4. "The Manipulation of Genes," by Stanley N. Cohen, in the July, 1975, issue of *Scientific American* (pp. 24–34).
5. "The Natural Origin of Optically Active Compounds," by W. E. Elias, in the July, 1972, issue of the *Journal of Chemical Education* (pp. 448–454).
6. "Insect Hormones and Insect Control: Or, Sex and the Single *Pyrrhocoris Apterus*," by Charles Berkoff, in the September, 1971, issue of the *Journal of Chemical Education* (pp. 577–581).
7. "Hereditary Fat-Metabolism Diseases," by Roscoe O. Brady, in the August, 1973, issue of *Scientific American* (pp. 88–97).

BRAIN TINGLER

8. Find a recent article discussing the plans for the possible worldwide moratorium on "genetic engineering." Prepare a case either in favor of or in opposition to such a proposal. (*Hint:* The ban was first suggested, in July, 1974, by a committee of molecular biologists sponsored by the National Academy of Sciences. Dr. Paul Berg of the Stanford Biochemistry Department was the committee chairman.)

Unit 34 Self-Test

Your understanding of this unit and of the pertinent background material from previous units is best measured by your comprehension of one or more of the articles suggested in "Practice for Proficiency."

Unit 35

It seems possible that the uranium nucleus . . . may, after neutron capture, divide itself. . . .

LISA MEITNER

(1939; commenting on the discovery by Otto Hahn in Germany of the fission of uranium after neutron bombardment)

Radiochemistry

Objectives

(a) Be able to use the "nuclear belt of stability" (Figure 35.1) to predict probable types of nuclear decay, expressing these predictions by nuclear equations (Section 35.3).

(b) Be able to calculate energy changes associated with nuclear transformations (Section 35.4).

(c) Given the half-life of a radioisotope and information related to the fraction decayed, be able to calculate the decay time involved (Section 35.5).

(d) Be able to write balanced equations representing artificial transmutations and nuclear fission (Sections 35.7 and 35.8).

The demand for more energy sources, coupled with the desire to reduce atmospheric pollution from fossil fuel power plants, is focusing increasing attention on the use of nuclear reactors for production of electric energy. It has been estimated[1] that nuclear power plants will supply one-fourth of the nation's electricity by 1980 and more than half by the year 2000. If, as the proponents of nuclear power contend, these installations are virtually accident proof and much "cleaner" than most conventional power plants, then the principal concern of environmentalists must be the disposal of the by-products of nuclear reactor function—"radioactive wastes."

Radioactive materials pose a unique problem in waste disposal. They must be stored in such a way that neither the radiation nor the actual radioactive substances can contact living systems. Many of these materials continue their decay processes for very long periods of time, so their storage cannot be considered a short-term problem. The energy released during radioactive decay is often sufficient to maintain the wastes at a very high temperature. Plans have been initiated for storing reactor wastes in vaults 500 to 2000 feet underground in deep salt beds in Kansas. Theoretically, the heat from these wastes will eventually melt their containers and surrounding salt, trapping the materials in molten pockets of salt far underground. Whether or not this storage will prove satisfactory is still the subject of considerable research and speculation.

To understand the problems of dealing with radioactive materials, we need to know something about the ways in which radioactive decay takes place, the characteristics of various types of radiation, and the effects of radiation on living systems. These topics form the basis for this unit.

35.1 THE ATOMIC NUCLEUS

The experiments culminating in Rutherford's suggestion of the nuclear model for the atom permitted estimations of the nuclear size. Modern investigations have refined these measurements, and we now believe the hydrogen nucleus (a single proton) to have a diameter of about 2.7×10^{-5} Å (2.7×10^{-13} cm). The nucleus of a uranium atom is about six times larger. By contrast, the atomic diameter for helium, based on the van der Waals radius, is about 1 Å. This means that the *nuclear* diameter is less than 0.01 percent of the effective diameter of a complete atom.

The development of the nuclear model posed two significant problems. No force ever identified could possibly be strong enough to hold positive protons together in the tiny volume of the nucleus—electrostatic repulsions should send these flying apart. In addition, atomic weights increase more rapidly than atomic numbers, suggesting that something other than simple protons must be present in the nucleus.

Rutherford postulated in 1920 the existence of some neutral particle in the atomic nucleus, but it was not until 1932 that Chadwick demonstrated experimentally the presence of the *neutron*. This neutral particle, of about the same mass as a proton, could account for the difference between atomic weight and atomic number, and for the existence of isotopes.

In addition, evidence was accumulating that certain proton–neutron combinations were uniquely stable (Figure 35.1), suggesting that the neutron in some way provided a "nuclear glue" to bind together the protons of the atomic nucleus.

Today we know a great deal more about the nuclei of atoms. Sophisticated experiments have shown that many nuclei are essentially spherical, whereas others are shaped more like a football. A variety of additional subatomic particles have been identified, and some of these have been related to the mechanism by which nuclei are stabilized. However, no single theory has proved fully satisfactory in

[1] *John Holdren and Philip Herrera,* Energy *(San Francisco: Sierra Club, 1971).*

explaining nuclear properties, and we shall leave a detailed treatment of these theories to more advanced texts.

Of concern to us, then, are the questions:

Which nuclei are stable and which radioactive?
What order of magnitude is the energy binding together the nuclear particles?
What are the implications of nuclear instability?

35.2 MAGIC NUMBERS

The "nuclear belt of stability" shown in Figure 35.1 indicates those isotopes that do not undergo spontaneous radioactive decay. This means that these nuclei contain a neutron-to-proton ratio that imposes some particular stability. If we add to this information the relative natural abundances of the isotopes (Unit 2), some peculiar features are noted. Isotopes in which the number of neutrons or the number of protons is 2, 8, 20, 28, 50, 82, or 126 are both stable and relatively abundant compared to other isotopes. In fact, some special stability seems to be associated with species in which both the protons and neutrons are present in one of these "magic numbers."

A particular exception is tritium ($^3_1$H), which is radioactive.

Figure 35.1 "Nuclear belt of stability."

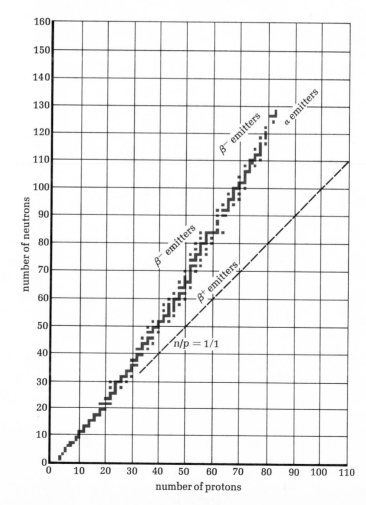

Lead-208, for example (82 protons, 126 neutrons), is unusually stable and quite abundant.

These numbers are reminiscent of the 2, 10, 18, . . . electrons associated with noble gas electron stability and suggest that some quantum mechanical model may eventually be applied to nuclear stability in a way similar to that now used to describe electron behavior.

In the absence of a simple theoretical model for nuclear behavior, we shall be forced to rely on the empirical tabulation of Figure 35.1, noting that the neutron-to-proton ratio associated with stability gradually increases from 1:1 for the lighter elements to about 1.5:1 for the heavy elements. Beyond $^{209}_{83}\text{Bi}$, no stable nuclei are known, although some (such as uranium-238) decay very slowly. The "magic numbers" are, then, useful at present for only limited prediction of stability and relative abundance.

The half-life for $^{238}_{92}\text{U}$ is 4.5×10^9 years.

35.3 RADIOACTIVITY

In 1896, Henri Becquerel was studying the effect of sunlight on various phosphorescent minerals, among them a uranium ore. During a period of several cloudy days, he left the uranium sample in a drawer along with some photographic paper. Much to his surprise, he discovered that the plates had been fogged by exposure to some invisible radiation from the uranium. He called this mysterious property of the ore *radioactivity*, and within a short time of his serendipitous discovery, a whole new field of scientific research had been launched.

We know now that radioactivity consists of either subatomic particles or very high energy radiation. Radioactivity occurs as an unstable nucleus undergoes a change to a more stable product.

We may characterize the more common types of radioactivity as

1. "penetrating" radiation (x-rays or γ rays)
2. subatomic particles (α particles, β particles, positrons, or neutrons)

Let us consider each of these in some detail.

Penetrating radiation

The true distinction between γ rays and x-rays is the *origin* of the radiation. The γ rays originate from the nucleus, whereas x-rays are formed as the result of electronic transitions outside the nucleus. Their wavelength regions overlap to some extent.

Sometimes referred to as *K* capture, since the electrons in the *K* shell (Unit 1) are closest to the nucleus and, hence, most readily captured.

This is, according to modern evidence, something of a simplification, but it does represent the net result.

Gamma (γ) *rays* are electromagnetic radiation in the wavelength region 0.005–1 Å. This is shorter wavelength radiation, and thus higher energy (Unit 30), than that of the more familiar x-ray region. Almost all nuclear transformations are accompanied by γ radiation representing part of the energy difference between the original nucleus and the more stable product of its decay. This phenomenon is similar to the emission of light when electrons drop from higher to lower energy states (Unit 30).

X-rays result from a particular type of nuclear change called *electron capture*, which is one way in which a nucleus can gain stability if its original neutron-to-proton ratio was too low. We may represent electron capture by a *nuclear equation*, which requires both mass and charge balance in the same way as an equation for a chemical process. To indicate balance, a complete isotope symbolism (Unit 2) is written, indicating an electron as $_{-1}^{0}e$. Both subscript and superscript numbers must add to the same net sum for reactants and products; for example,

$$^{15}_{8}\text{O} + _{-1}^{0}e \longrightarrow {}^{15}_{7}\text{N}$$

indicates that oxygen-15 undergoes electron capture to form the stable nitrogen-15. The electron captured is usually from the orbital closest to the nucleus (1s or K shell). When this is replaced by an electron "dropping down" from a higher orbital, the energy of the *electronic* change is emitted as x-radiation. The captured electron converts one of the nuclear protons to a neutron.

Subatomic particle emission

An alternative decay path for a "neutron-poor" radioactive isotope is *positron emission* (symbolized as β^+ or $_{+1}^{0}e$). A positron is an electron of *positive* charge. It may be thought of as originating by conversion of a proton to a neutron and a positron. The resulting isotope, like that produced by electron capture, has the same mass number as the "parent" isotope, but an atomic number one smaller, for example,

$$^{11}_{6}C \longrightarrow {}^{11}_{5}B + {}^{0}_{+1}e$$

It is observed that electron capture is more common with heavy isotopes, whereas lighter isotopes seem to prefer positron emission.

Two paths might be suggested for the decay of "neutron-rich" isotopes (neutron-to-proton ratio too high for stability). One possibility would be for the unstable nucleus to eject one of the surplus neutrons in order to gain stability. Neutron emission is observed, but only in rather special circumstances. For example, carbon-13 is considered a stable isotope (that is, "ordinary" $^{13}_{6}C$ is not radioactive), but the *high-energy $^{13}_{6}C$ species resulting from α-bombardment of beryllium-9* does emit a neutron. It was this reaction that Chadwick utilized in the experiments leading to characterization of the neutron:

$$^{9}_{4}Be + {}^{4}_{2}He \longrightarrow \quad [{}^{13}_{6}C] \quad \longrightarrow {}^{12}_{6}C + {}^{1}_{0}n$$
$$\text{(high energy}$$
$$\text{intermediate)}$$

Except under special conditions, neutron-rich nuclei gain stability by the alternative path of ejecting an electron *from the nucleus*. Such electrons are commonly referred to as β particles (β^{-}, $_{-1}^{0}e$) to distinguish them from the orbital electrons outside the nucleus. They may be thought of as originating from the conversion of a neutron to a proton, although the actual mechanism is not that simple. Beta (β) decay is a very common form of radioactivity, occurring with many of the species familiar from regular news stories of our "nuclear age"; for example,

$$^{14}_{6}C \longrightarrow {}^{14}_{7}N + {}^{0}_{-1}e$$
$$^{90}_{38}Sr \longrightarrow {}^{90}_{39}Y + {}^{0}_{-1}e$$

The α particle is the nucleus of a helium atom ($^{4}_{2}He$), so α emission provides a way for an unstable nucleus to lose two protons and two neutrons simultaneously. Although α decay is sometimes observed with lighter isotopes, this type of radioactivity is most common with the heavier radioactive isotopes, particularly those with a mass number greater than 200, for example,

$$^{238}_{92}U \longrightarrow {}^{234}_{90}Th + {}^{4}_{2}He$$

Two other types of nuclear change will be discussed later. In nuclear *fission*, the atomic nucleus is split into a set of new nuclei. This is the basis of the reaction in the "atomic" bomb and the nuclear power plant. The other situation occurs when subatomic particles are combined to form new species, as in nuclear *fusion* or the artificial formation of new elements of higher atomic number.

Example 1

On the basis of the preceding discussion, predict the most probable type of radioactive decay for the following isotopes, indicating the predictions by balanced nuclear equations: $^{29}_{15}P$, $^{17}_{7}N$, $^{218}_{84}Po$, $^{81}_{37}Rb$.

Solution

First, each isotope should be compared to the "nuclear belt of stability" (Figure 35.1).

This shows that

$^{29}_{15}P$ has a neutron-to-proton ratio too low for stability ($14:15 < 1:1$).
$^{17}_{7}N$ is too "neutron-rich" ($10:7 > 1.2:1$).
$^{218}_{84}Po$ is *beyond* the "belt of stability" (nucleus too "large").
$^{81}_{37}Rb$ is "neutron-poor" ($44:37 < 1.2:1$).

Next, we must recall how stability can be increased and, when two alternatives are possible, the decay most frequently observed. This can then be expressed by an appropriate equation. $^{29}_{15}P$ needs to increase the $n:p$ ratio. Lighter elements tend to use positron emission to accomplish this:

$$^{29}_{15}P \longrightarrow {}^{0}_{+1}e + {}^{29}_{14}Si$$

(Note that the product isomer was determined by the requirement of atomic number 14 for charge balance $15 = 1 + 14$.)

$^{17}_{7}N$ could become more stable by either neutron emission or β decay. The latter is more common:

$$^{17}_{7}N \longrightarrow {}^{0}_{-1}e + {}^{17}_{8}O$$

$^{218}_{84}Po$ can best become "lighter" by α decay:

$$^{218}_{84}Po \longrightarrow {}^{4}_{2}He + {}^{214}_{82}Pb$$

$^{81}_{37}Rb$ is a "heavier" isotope, so we might expect electron capture:

$$^{81}_{37}Rb + {}^{0}_{-1}e \longrightarrow {}^{81}_{36}Kr$$

Note that the product isotopes predicted may, in some cases, be unstable themselves. Our predictions applied only to *initial* decay, and successive transformations often occur until a stable isotope results.

35.4 ENERGY AND NUCLEAR TRANSFORMATION

The mass number for an isotope represents the sum of the number of protons and neutrons in the nucleus. One might expect to be able to calculate the isotopic mass (Unit 2) from the atomic number, the mass number, and the experimentally determined masses of the proton, neutron, and electron (Table 35.1). Let us see what happens if we try this for $^{200}_{80}Hg$:

$$120 \text{ neutrons} \times 1.0087 \text{ amu}/n = 121.04 \text{ amu}$$
$$80 \text{ protons} \times 1.0073 \text{ amu}/p = 80.58 \text{ amu}$$
$$80 \text{ electrons} \times 0.0005 \text{ amu}/e = \underline{0.04 \text{ amu}}$$
$$201.66 \text{ amu}$$

However, the rest mass of $^{200}_{80}Hg$ has been determined from sophisticated experimental measurements to be only 199.97 amu. The difference between the calculated and observed isotopic masses represents a mass–energy conversion from which the *binding energy* of this nucleus can be calculated, using the Einstein equation

$$E = mc^2$$
$$\text{mass difference} = 201.66 - 199.97 = 1.69 \text{ amu}$$

Using appropriate conversion factors (Appendix A):

$$E = \frac{1.69 \text{ amu}}{1} \times \frac{(3.00 \times 10^{10} \text{ cm sec}^{-1})^2}{1} \times \frac{1 \text{ g}}{6.02 \times 10^{23} \text{ amu}}$$

$$\times \frac{1 \text{ erg}}{1 \text{ g cm}^2 \text{ sec}^{-2}} \times \frac{2.39 \times 10^{-8} \text{ cal}}{1.00 \text{ erg}} \times \frac{1 \text{ kcal}}{10^3 \text{ cal}}$$

$$= 6.0 \times 10^{-14} \text{ kcal (per atom)}$$

The relationship between *amu* and grams is based on the definition of gram-mole and Avogadro's number (Unit 4):

$$1 \text{ g} = 6.02 \times 10^{23} \text{ amu}$$

Table 35.1 REST MASSES[a] FOR SOME SUBATOMIC PARTICLES

SPECIES	REST MASS (amu)
$_{-1}^{0}e$	0.00055
$_{0}^{1}n$	1.0087
$_{1}^{1}H$	1.0073
$_{1}^{2}H$	2.0136
$_{1}^{3}H$	3.0155
$_{2}^{3}He$	3.0149
$_{2}^{4}He$	4.0015
$_{6}^{12}C$	11.9967
$_{6}^{13}C$	13.0001
$_{6}^{14}C$	14.0000
$_{82}^{206}Pb$	205.929
$_{82}^{207}Pb$	206.931
$_{82}^{208}Pb$	207.932
$_{92}^{233}U$	232.989
$_{92}^{235}U$	234.993
$_{92}^{238}U$	238.000

[a] Rest masses other than those for $_{-1}^{0}e$ and $_{0}^{1}n$ are tabulated for *nuclei*, so masses of appropriate numbers of electrons must be added to obtain *isotopic* masses of neutral atoms, e.g., carbon-12 (Unit 2) 11.9967 + 6(0.00055) = 12.0000.

Note that mass change may be calculated either from isotopic masses or from nuclear masses (Table 35.1) as long as allowance is made for all particles involved.

Now, to gain some appreciation for the order of magnitude of this nuclear binding energy, let us compare it on a *molar* scale with the energy of typical chemical processes:

$$\frac{6.0 \times 10^{-14}\ \text{kcal}}{1\ \text{atom}} \times \frac{6.0 \times 10^{23}\ \text{atoms}}{1\ \text{mole}} = 3.6 \times 10^{10}\ \text{kcal mole}^{-1}$$

For comparison, the standard heat of combustion for methane is only 213 kcal mole^{-1}. The binding energy of a moderately heavy nucleus, then, is on the order of *100-million times* greater than the heat energy associated with typical chemical reactions.

If we have available the masses of "parent" and "product" isotopes, we can use the mass differences to estimate the energy accompanying radioactive decay. This procedure can yield a valuable insight into the energetics of nuclear change.

Example 2

What energy change (in kilocalories per mole) accompanies the β decay of carbon-14? The isotopic masses involved have been found to be 14.00307 for nitrogen-14 and 14.00324 for carbon-14.

Solution
The β decay of carbon-14 is represented by

$$_{6}^{14}C \longrightarrow _{-1}^{0}e + _{7}^{14}N$$

Since a carbon-14 atom has six electrons and six protons, β emission yields nitrogen-14 with seven protons, but only six electrons (i.e., N^{+} ion). Charge balance is maintained by the β particle (e^{-}). The mass change, then, is concerned only with the difference between the isotopic masses:

$$_{6}^{14}C:\ 14.00324\ \text{amu}$$
$$_{-1}^{0}e + _{7}^{14}N^{+}:\ \underline{14.00307\ \text{amu}}$$
$$\text{mass loss:}\ \ 0.00017\ \text{amu}$$

Then

$$E = mc^2$$

$$= \frac{1.7 \times 10^{-4}\ \text{amu}}{1\ \text{atom}} \times \frac{(3.00 \times 10^{10}\ \text{cm sec}^{-1})^2}{1}$$

$$\times \frac{1\ \text{g}}{6.02 \times 10^{23}\ \text{amu}} \times \frac{1\ \text{erg}}{1\ \text{g cm}^2\ \text{sec}^{-2}} \times \frac{2.39 \times 10^{-8}\ \text{cal}}{1.00\ \text{erg}}$$

$$\times \frac{1\ \text{kcal}}{10^3\ \text{cal}} \times \frac{6.02 \times 10^{23}\ \text{atoms}}{1\ \text{mole}}$$

$$= 3.7 \times 10^6\ \text{kcal mole}^{-1}$$

Conversely, the small energy changes of typical chemical reactions correspond to mass changes far too small for detection with even the most sensitive balances.

Now, in what way is this energy manifested? Some may appear as γ radiation and some undoubtedly as heat. Most, in the case of carbon-14 decay, is accounted for by the kinetic energy of the expelled β particle. Since the β particle is so light, it must be emitted with a tremendous velocity. Some of the energy is used in "recoil" of the disintegrating nucleus (like the "kick" of a shotgun). The important features of energy change accompanying radioactive decay may be summarized as follows:

1. A very small change in mass is associated with an enormous energy.

2. The energy of nuclear change imparts a very high velocity to emitted particles, thus making them capable of penetrating rather deeply into matter.
3. Some of the energy of nuclear change may, and usually does, appear as γ radiation.
4. Some of the energy of nuclear change appears as heat.

All of these factors are important to an understanding of the hazards of radioactive materials. These are not, however, the *only* factors. The nuclear decay rate must also be considered.

35.5 RATES OF RADIOACTIVE DECAY

Recent evidence has shown that radioactive decay rates are *not* constant. They may vary with changes in temperature, magnetic fields, or chemical environment. Although it is possible that these factors, or others, may have altered decay rates *drastically* over geologic time periods, only variations of a few percent have been directly observed so far. In the absence of more definitive information, we will proceed on the assumption that decay rates are "approximately" constant.[2]

The fact that a particular isotope is "unstable" says nothing about how quickly it will disintegrate, in the same way that thermodynamics (Unit 5) says nothing about the speed of a chemical process. The rate of decay is an inherent characteristic of each unique isotope and is unaffected by the chemical or physical environment of the isotope. Since this rate depends only on the single species, we may express the rate of decay by an equation for a first-order process (Unit 16):

$$\text{decay rate} = kC_R$$

in which R refers to the radioactive isotope.

Decay rates are most usefully compared on the basis of half-life times which, for first-order processes (Unit 16), are expressed by

$$t_{1/2} = \frac{0.693}{k} \tag{35.1}$$

If the half-life time is known, the rate constant (k) can be calculated and used to determine the relationship between decay time and concentration of radioactive isotope (or of product isotope formed). For first-order processes,

$$\log \frac{C_{R(t)}}{C_{R(0)}} = \frac{-kt}{2.303} \tag{35.2}$$

in which $C_{R(0)}$ represents the initial concentration and $C_{R(t)}$ the concentration of radioisotope remaining at time t.

Half-life times vary greatly, as exemplified by the successive decay scheme for uranium-238 (Figure 35.2). Radioisotopes with very long half-lives, such as $^{238}_{92}\text{U}$ ($t_{1/2} = 4.5 \times 10^9$ years), are not environmental hazards since they decay so slowly as to represent no significant radiation hazard to humans. At the other end of the scale, species with very short half-lives obviously represent no long-range hazard, but exposure to significant amounts of such isotopes can be very dangerous in terms of high radiation dosage. The major environmental problems are associated with radioisotopes of "intermediate" half-lives. In particular, those with half-lives of a few years to a few thousand years represent a combination of long term and continuous radiation that could prove very hazardous.

Example 3

How long would it take for 10 percent of a radium salt containing pure radium-226 ($t_{1/2} = 1600$ years) to decay?

[2]For details, see "Radioactivity reexamined," by H. C. Dudley, in the April 7, 1975 issue of *Chemical and Engineering News* (p. 2).

Figure 35.2 Successive decay of uranium-238.

1. $^{238}_{92}U \xrightarrow[t_{1/2} = 4.5 \times 10^9 \text{ years}]{} {}^{4}_{2}He + {}^{234}_{90}Th$

2. $^{234}_{90}Th \xrightarrow[t_{1/2} = 24 \text{ days}]{} {}^{0}_{-1}e + {}^{234}_{91}Pa$

3. $^{234}_{91}Pa \xrightarrow[t_{1/2} = 68 \text{ sec}]{} {}^{0}_{-1}e + {}^{234}_{92}U$

4. $^{234}_{92}U \xrightarrow[t_{1/2} = 2.4 \times 10^5 \text{ years}]{} {}^{4}_{2}He + {}^{230}_{90}Th$

5. $^{230}_{90}Th \xrightarrow[t_{1/2} = 8.0 \times 10^4 \text{ years}]{} {}^{4}_{2}He + {}^{226}_{88}Ra$

6. $^{226}_{88}Ra \xrightarrow[t_{1/2} = 1600 \text{ years}]{} {}^{4}_{2}He + {}^{222}_{86}Rn$

7. $^{222}_{86}Rn \xrightarrow[t_{1/2} = 3.8 \text{ days}]{} {}^{4}_{2}He + {}^{218}_{84}Po$

8. $^{218}_{84}Po \xrightarrow[t_{1/2} = 3.0 \text{ min}]{}$ or $\begin{cases} {}^{0}_{-1}e + {}^{218}_{85}At \\ {}^{4}_{2}He + {}^{214}_{82}Pb \end{cases}$

9. $^{218}_{85}At$ — α decay
   $^{214}_{82}Pb$ — β decay $(t_{1/2} = 27 \text{ min})$
   $\longrightarrow {}^{214}_{83}Bi$

10. $^{214}_{83}Bi \xrightarrow[t_{1/2} = 20 \text{ min}]{}$ or $\begin{cases} {}^{0}_{-1}e + {}^{214}_{84}Po \\ {}^{4}_{2}He + {}^{210}_{81}Tl \end{cases}$

11. $^{210}_{81}Tl$ — β decay
    $^{214}_{84}Po$ — α decay $(t_{1/2} = 10^{-4} \text{ sec})$
    $\longrightarrow {}^{210}_{82}Pb$

12. $^{210}_{82}Pb \xrightarrow[t_{1/2} = 22 \text{ years}]{} {}^{0}_{-1}e + {}^{210}_{83}Bi$

13. $^{210}_{83}Bi \xrightarrow[t_{1/2} = 5.0 \text{ days}]{}$ or $\begin{cases} {}^{0}_{-1}e + {}^{210}_{84}Po \\ {}^{4}_{2}He + {}^{206}_{81}Tl \end{cases}$

14. $^{206}_{81}Tl$ — β decay
    $^{210}_{84}Po$ — α decay $(t_{1/2} = 140 \text{ days})$
    $\longrightarrow {}^{206}_{82}Pb$ stable

Solution

If 10 percent decays, then

$$\frac{C_{R(t)}}{C_{R(0)}} = 0.90$$

so

$$\log 0.90 = \frac{-kt}{2.303}$$

and

$$k = \frac{0.693}{t_{1/2}} = \frac{0.693}{1600 \text{ years}} = 4.3 \times 10^{-4} \text{ year}^{-1}$$

$$t = \frac{-2.303 \log 0.90}{4.3 \times 10^{-4} \text{ year}^{-1}} = 250 \text{ years}$$

The assumption of a constant decay rate is essential to the technique of radioactive dating used to establish the age of fossils, rock samples, and so on. It is important to realize that this *is* an assumption and that we have no way of proving that some cosmic event might not have altered this process. However, we accept the assumption because of the vast accumulation of experimental evidence presently supporting it. It is in excellent agreement with historically dated events, although we obviously cannot apply this check to prehistoric dating.

Radiocarbon dating is, perhaps, of chief interest. If we add the assumption that the $^{14}_6\text{C} : {}^{12}_6\text{C}$ ratio of the atmosphere has remained constant, then at the time a plant dies, its $^{14}_6\text{C} : {}^{12}_6\text{C}$ ratio is the same as that of the atmosphere. The ratio is altered as carbon-14 decays, so the $^{14}_6\text{C} : {}^{12}_6\text{C}$ ratio in the dead plant drops at a rate determined by the decay rate of carbon-14. Determination of the $^{14}_6\text{C} : {}^{12}_6\text{C}$ ratio in a piece of petrified wood, for example, then permits a calculation of the age of the sample.

Correcting for new $^{14}_6\text{C}$ introduced by explosion of nuclear weapons during the past few years is required.

Example 4

A charcoal sample from a tree burned during the volcanic eruption that formed Crater Lake, Oregon, was found to have a $^{14}_6\text{C} : {}^{12}_6\text{C}$ ratio that was about 46 percent of that found in a standard wood sample from a tree cut down in 1940 (prior to nuclear weapons testing). The half-life of carbon-14 is 5570 years. What was the approximate date of formation of Crater Lake?

Solution

From equation (35.1), we can find the rate constant for decay of carbon-14:

$$k = \frac{0.693}{t_{1/2}} = \frac{0.693}{5570 \text{ years}} = 1.245 \times 10^{-4} \text{ year}^{-1}$$

Comparison of isotope ratios is equivalent to comparison of radioisotope concentrations, so in equation (35.2)

$$\log \frac{46}{100} = \frac{-(1.245 \times 10^{-4} \text{ year}^{-1})t}{2.303}$$

$$\log 0.46 = 0.663 - 1 = -0.337$$

$$t = \frac{2.303 \times (-0.337)}{-(1.245 \times 10^{-4} \text{ year}^{-1})} = 6.2 \times 10^3 \text{ years}$$

This indicates that the eruption killed the tree ∼6200 years earlier than the 1940 tree died. The date of eruption then, was 1940 − 6200 = −4260 (ca. 4300 B.C., to two significant figures).

Other isotopes may also be used in radioactive dating. Obviously, carbon-14 is limited both to samples containing carbon and to samples rather "youthful" on a geological time scale (remember that $t_{1/2} = 5570$ years). The $^{238}_{92}U:^{206}_{82}Pb$ ratio is sometimes employed with the assumption that the lead-206 content of a rock sample originated entirely from decay of uranium-238 (Figure 35.2). Both potassium-to-argon and rubidium-to-strontium decay schemes are also applicable to ancient rocks. Radioisotope dating was used on "moon rock" samples retrieved by the Apollo astronauts, revealing a rather surprising age of around 3.5×10^9 years for lunar surface samples.

$$^{87}_{37}Rb \longrightarrow ^{0}_{-1}e + ^{87}_{38}Sr,$$
$$t_{1/2} = 5.7 \times 10^{10} \text{ years}$$

35.6 RADIATION DETECTION

As α particles, positrons, or β particles penetrate matter, they transfer energy by collision with particles of the material. Many of these collisions are inelastic (Unit 9). Some such collisions may act to excite orbital electrons, yielding light when the electrons return to their ground states. This is the basis of the simple fluorescent screen used to detect α particles in the famous Rutherford experiments, as well as the fluorescent dials of watches.

The dials are painted with a substance such as zinc sulfide containing a trace of a radium salt. The dangers of radioactive materials were dramatized many years ago by the tragic results to watch dial painters who dampened their brushes periodically by inserting the tips into their mouths.

In other cases, collisions result in ionization of atoms or molecules encountered by the radioactive particles. High-energy radiation (γ rays or x-rays) also produces this effect. The ionization of a low-pressure gas is the phenomenon detected by the familiar Geiger–Müller counter (Figure 35.3).

Other, more sophisticated, techniques are also available to detect radiation. Most, however, depend on ionization or electronic excitation for their function.

Film badges and dosimeters (Figure 35.4) are used to maintain a record of expo-

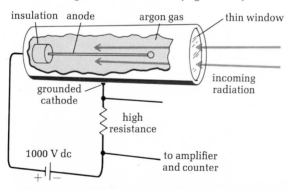

Figure 35.3 Geiger-Müller counter.

Figure 35.4 Film badge and dosimeter.

sure to penetrating radiation and are important safety devices for persons working with radiation or radioactive materials.

Elements of atomic numbers 43, 61, 85, 87, and >92.

35.7 ARTIFICIAL NUCLEAR TRANSFORMATION

A number of elements in the periodic table have never been found in nature. Most of these have been prepared in instruments called *high-energy accelerators* capable of speeding subatomic particles by applied electric fields to a "target" containing the element to be transformed. Others were made by neutron "bombardment."

As examples, the first transuranium elements were produced at the University of California at Berkeley in 1940:

$$^{238}_{92}U + ^{1}_{0}n \longrightarrow ^{239}_{92}U$$

$$^{239}_{92}U \longrightarrow ^{0}_{-1}e + ^{239}_{93}Np \qquad \text{(neptunium)}$$

$$^{239}_{93}Np \longrightarrow ^{0}_{-1}e + ^{239}_{94}Pu \qquad \text{(plutonium)}$$

In general, neutron bombardment is less effective than bombardment with high-energy (accelerated) positive ions, and this is now usually the method employed, for example (net processes),

$$^{244}_{96}Cm + ^{4}_{2}He \longrightarrow ^{1}_{1}H + 2^{1}_{0}n + ^{245}_{97}Bk \qquad \text{(berkelium)}$$

$$^{238}_{92}U + ^{12}_{6}C \longrightarrow 4^{1}_{0}n + ^{246}_{98}Cf \qquad \text{(californium)}$$

See the Suggested Reading at the end of the unit.

Most of the transuranium isotopes have very short half-lives, and the stories of their preparation and characterization are fascinating reading. New elements are still being made,[3] and artificial transmutations offer a fertile field for research. Many synthetic isotopes are valuable in medicine and in industrial applications.

35.8 NUCLEAR FISSION

The Franck Report (June 11, 1945) was signed by a number of eminent scientists, among them Glenn Seaborg, later Chairman of the U.S. Atomic Energy Commission.

The first atomic bomb was exploded in a New Mexico desert on July 16, 1945. Less than a month later, against the advice of a select panel of scientists, Hiroshima and Nagasaki, Japan, were obliterated by the most awesome weapon then known to man. Many feel that man has developed the weapon of his own ultimate destruction.[4]

Whether the fantastic energy of the nucleus will prove too much for man to handle is a question for which no answer is apparent. That this energy *can* be used for man's benefit, rather than for his extinction, has been demonstrated by science. Nuclear power plants offer a way of harnessing fission processes for much-needed production of electric energy.

A number of heavy isotopes undergo *fission* (splitting) after bombardment with neutrons. Of these, uranium-235 is the most widely employed:

$$^{1}_{0}n + ^{235}_{92}U \longrightarrow \quad ^{236}_{92}U$$
$$\text{very unstable}$$

$$^{236}_{92}U \longrightarrow \text{smaller nuclei} + \text{neutrons} + \text{energy}$$

The energy associated with this process, representing the difference between the large binding energy of the uranium nucleus and the sum of the smaller binding energies of the lighter nuclei, is enormous. For each gram of uranium-235, around 2×10^7 kcal of energy is released. Less than 100 pounds of $^{235}_{92}U$ is required for a "simple" atomic bomb. The fission of heavy nuclei is not a "clean" process. Different nuclei split to form different products—more than 200 isotopes of some 30-odd elements have been identified among the fission products of uranium-235. Since these are all lighter elements, the neutron-to-proton ratios in the isotopes formed from the

---

[3] Element 106 was first described by Ghiorso and co-workers in December, 1974 issue of *Physical Review Letters* (pp. 1490–1493).

[4] *The Late Great Planet Earth,* by Hal Lindsey, (Grand Rapids, Mich.: Zondervan Publishing House, 1970) offers some freightening predictions.

heavy uranium are too high for stability; the fission products are thus radioactive.

The most important feature of nuclear fission is that the process that requires only a single neutron for initiation forms a number of neutrons as products. For example, a typical sequence is

$$\,^1_0n + \,^{235}_{92}U \longrightarrow \,^{236}_{92}U \longrightarrow 3\,^1_0n + \,^{87}_{35}Br + \,^{146}_{57}La$$

If the sample is small, most of these neutrons escape, but in a large sample most are absorbed within the sample to produce additional fission reactions. This *chain reaction*, self-perpetuating, occurs in any sample sufficiently large. The minimum sample size for chain reaction is called the *critical mass*. Substances such as graphite or cadmium are unusually efficient in absorbing neutrons and may be used to control the rate of fission for a slow energy release rather than a nuclear explosion. The heat of a controlled fission process is the energy source for the nuclear reactor (Figure 35.5).

The combination (fusion) of light nuclei is a greater energy source than fission. Fusion of ^2_1H with ^3_1H to form ^4_2He and a neutron, for example, releases about four times as much energy per gram as in the fission of $^{235}_{92}U$. In addition, such fusion processes do not produce the radioactive wastes (fission products) typical of fission reactions. Although uncontrolled fusion processes have been developed (the "hydrogen" bomb), much research is still necessary before controlled fusion becomes available as a useful energy source. The "fuels" for fusion reactions are much more abundant than those required for fission reactors. It is not unlikely that controlled fusion will some day help relieve the world's energy crisis.[5]

Hydrogen fusion is the energy source of our sun and other stars.

35.9 RADIOACTIVITY AND MAN

The use of radioactive iodine for the study of possible malfunction of the thyroid gland, once a "routine" procedure, has now been largely abandoned for young people after the discovery of thyroid tumors in a high percentage of younger patients subjected to the analysis.

Alpha (α) particles are "heavy" with respect to β particles and thus travel more slowly. Their usual energies do not permit them to penetrate deeply into living tissue. This, of course, applies to α-emitters *outside* the body. If α-emitters are *ingested* they can be extremely damaging because, although the heavy 2+ particle cannot travel far through tissue, it is very effective as an ionizing agent in the region penetrated. On the other hand, β particles are much more penetrating and, once within a living

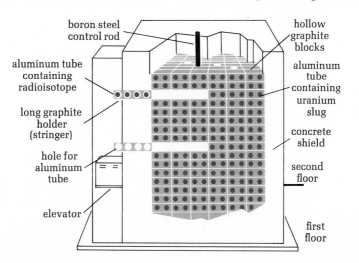

Figure 35.5 Schematic drawing of the first nuclear reactor at Oak Ridge, Tennessee. (From J. H. Wood, C. W. Keenan, and W. E. Bull, *Fundamentals of College Chemistry,* New York, Harper & Row, 1972.)

[5]For some cautions against placing too much hope in controlled fusion, see the editorial by Philip H. Abelson in the July 23, 1976 issue of *Science* (p. 279).

system, their ionizing action may disrupt vital biochemical processes. Like x-radiation, γ radiation is even more hazardous in its ionizing effect on living matter.

Studies of the survivors of atomic explosions in Japan and of their children have confirmed that radiation produces significant genetic damage. Even a small amount of radiation can be linked to an increased probability of mutation.

Larger exposures produce more immediate damage, although long-term effects cannot be discounted. Studies have confirmed, for example, the development of cancerous growths and leukemia many years after exposure to high radiation dosages. That serious harm can result from ionizing radiation is beyond question, with effects ranging from minor loss of white blood cells, from low-level radiation, to death within a short time of massive radiation exposure.

The most serious consequences accompany ingestion or inhalation of radioactive substances that become "trapped" within the body. Strontium-90 from fallout of atmospheric explosion of nuclear weapons has been found in milk. This isotope may, in the body, replace some calcium in the bones, where it becomes a permanent radiation source to the bone marrow. The half-life of strontium-90 is 29 years.

One of the concerns over the use of nuclear power plants is the problem of disposal of radioactive waste products, many of which have "intermediate" half-lives. If buried wastes were ever to permeate underground water supplies, the results could be disastrous.

The discovery of radioactivity, like so many advances in science or technology, must be considered a mixed blessing to man. Although scientists can present reasonably well-defined alternatives for the uses of nuclear reactions and reliable evaluations of the effects of their misuse, the ultimate choices will most likely be political. In a free society, all of us bear some responsibility for these decisions.

SUGGESTED READING

Choppin, Gregory. 1964. *Nuclei and Radioactivity.* Menlo Park, Calif.: W. A. Benjamin.
Kendall H. W., and W. K. H. Panofsky. 1971. "The Structure of the Proton and the Neutron", *Scientific American,* **224,** 61–77.
Seaborg, G. T. 1963. *Man-Made Transuranium Elements.* Englewood Cliffs, N.J.: Prentice-Hall.
Wiegand, C. E. 1972. "Exotic Atoms", *Scientific American,* November, 102–110.
Zafiratos, C. D. 1972. "The Texture of the Nuclear Surface", *Scientific American,* October, 100–108.
(Additional articles of interest may be found in later issues of *Scientific American.*)

EXERCISES

1. On the basis of the "nuclear belt of stability" (Figure 35.1), write balanced nuclear equations indicating a likely type of initial decay for
   a. $^{226}_{88}Ra$; b. $^{7}_{4}Be$; c. $^{35}_{16}S$

2. From rest masses given in Table 35.1, calculate the energy of the nuclear fusion process (in kilocalories per mole):

   $$^{2}_{1}H + {}^{3}_{1}H \longrightarrow {}^{4}_{2}He + {}^{1}_{0}n$$

3. A sample of petrified wood purchased at a curio shop was found to have a $^{14}_{6}C : {}^{12}_{6}C$ ratio about 29% of that of a standard 1940 wood sample. The half-life of carbon-14 is 5570 years. What was the approximate date of death of the tree from which the petrified wood was obtained?

4. Complete and balance the nuclear equations given:
   a. preparation of radioactive cobalt for use in radiation therapy

   $$^{59}_{27}Co + {}^{1}_{0}n \longrightarrow \underline{\hspace{2cm}}$$

   b. the first nuclear transformation produced by man (Lord Rutherford, 1919)

   $$^{14}_{7}N + {}^{4}_{2}He \longrightarrow {}^{1}_{1}H + \underline{\hspace{2cm}}$$

   c. β decay of iodine-131 used in radiation therapy of malignant goiter

   $$^{131}_{53}I \longrightarrow \underline{\hspace{2cm}} + \underline{\hspace{2cm}}$$

   d. the first observation of nuclear fission

   $$^{235}_{92}U + {}^{1}_{0}n \longrightarrow 3{}^{1}_{0}n + {}^{94}_{36}Kr + \underline{\hspace{2cm}}$$

OPTIONAL

5. Write a brief explanation of why nuclear reactor waste materials are "hot" in both a thermal and radioactive sense.

PRACTICE FOR PROFICIENCY

Objectives (a) and (d)

1. On the basis of the "nuclear belt of stability" (Figure 35.1), write balanced nuclear equations indicating a likely type of *initial* decay for

   a. $^{14}_{6}C$; **b.** $^{11}_{6}C$; **c.** $^{232}_{92}U$; **d.** $^{90}_{38}Sr$; **e.** $^{3}_{1}H$; **f.** $^{60}_{27}Co$; **g.** $^{14}_{8}O$; **h.** $^{211}_{83}Bi$; **i.** $^{13}_{7}N$; **j.** $^{40}_{19}K$.

2. Complete and balance

   a. $^{211}_{85}At + {}^{0}_{-1}e \longrightarrow$ _____

   b. $^{44}_{20}Ca + {}^{2}_{1}H \longrightarrow {}^{1}_{0}n +$ _____

   c. $^{7}_{3}Li + {}^{1}_{1}H \longrightarrow 2$ _____

   d. $^{14}_{7}N + {}^{4}_{2}He \longrightarrow {}^{17}_{8}O +$ _____

   e. $^{2}_{1}H + {}^{3}_{1}H \longrightarrow {}^{1}_{0}n +$ _____

   f. $^{59}_{27}Co +$ _____ $\longrightarrow {}^{60}_{27}Co$

   g. $^{81}_{37}Rb + {}^{0}_{-1}e \longrightarrow$ _____

   h. $^{29}_{15}P \longrightarrow {}^{29}_{14}Si +$ _____

   i. $^{17}_{7}N \longrightarrow {}^{0}_{-1}e +$ _____

   j. $^{1}_{0}n + {}^{235}_{92}U \longrightarrow {}^{87}_{35}Br +$ _____ $+ 3$ ____

Objective (c)

3. A sample of rock from a geological expedition was found to have a $^{87}_{38}Sr : {}^{87}_{37}Rb$ mole ratio of 0.031. The half-life for β-decay of $^{87}_{37}Rb$ is 5.7×10^{10} years. Using appropriate assumptions, estimate the age of the sample.

4. A sample of petrified wood found near Strawberry Lake in the Rockies was found to have a $^{14}_{6}C : {}^{12}_{6}C$ ratio about 20% of that of a standard 1940 wood sample. What was the approximate date of death of the tree from which the petrified wood was obtained? (The half-life of carbon-14 is 5570 years.)

5. A wooden coffin was removed from an Egyptian tomb that was sealed 3600 years ago. The half-life of carbon-14 is 5570 years. What portion of the original amount of carbon-14 remains in the coffin today?

6. The isotope $^{210}_{84}Po$ undergoes α decay to $^{206}_{82}Pb$, with a half-life of 140 days. How long would it take for 85% of a pure $^{210}_{84}Po$ sample to change to lead-206?

7. The isotope $^{234}_{90}Th$ undergoes β decay to $^{234}_{91}Pa$, with a half-life of 24 days. How long would it take for 60% of a pure sample of $^{234}_{90}Th$ to change to $^{234}_{91}Pa$?

8. The isotope $^{210}_{82}Pb$ has a half-life of 22 years. What percentage of a pure $^{210}_{82}Pb$ sample prepared in May, 1958, would have decayed by May, 1975?

9. The isotope $^{226}_{88}Ra$ has a half-life of 1600 years. What percentage of a pure radium-226 sample prepared in 1925 should remain as $^{226}_{88}Ra$ exactly 50 years later?

10. It was found that the radioactive gas radon undergoes a rapid α decay. If 67% of a purified $^{222}_{86}Rn$ sample had decayed in 6.08 days, what is the half-life of this isotope?

    a. 1.7 days; **b.** 2.3 days; **c.** 3.0 days; **d.** 3.8 days; **e.** 4.2 days

Objective (b)

11. Calculate the energy change (using Table 35.1), in kcal mole^{-1} for the process represented by

    $$^{235}_{92}U \longrightarrow {}^{233}_{92}U + 2{}^{1}_{0}n$$

12. What energy change would accompany the conversion of lead-206 to lead-207 by neutron bombardment, in kilocalories per mole? (See Table 35.1.)

13. Calculate the energy change for conversion of 1.00 g of $^{14}_{6}C$ to $^{14}_{7}N$ by β decay.

SPECIES	REST MASS (amu)
$^{0}_{-1}e$	0.00055
$^{14}_{6}C$	14.00324
$^{14}_{7}N$	14.00252

14. Calculate the energy change for formation of 1.00 g of α particles by the process

    $$^{3}_{2}He + {}^{1}_{0}n \longrightarrow {}^{4}_{2}He$$

SPECIES	REST MASS (amu)
$^{1}_{0}n$	1.0087
$^{3}_{2}He$	3.0149
$^{4}_{2}He$	4.0015

BRAIN TINGLERS

15. Han Van Meegeran (1889–1947) is considered by many to have been the master of art forgery, having amassed an estimated $3,000,000 from his "art." Perhaps his most famous work was *Christ and His Disciples at Emmaus,* "discovered" as an "original Vermeer" (1632–1675) in 1937 and sold for $280,000. In 1945 Van Meegeran confessed to the forgery of six "Vermeers," including the *Emmaus,* but art experts examined the paintings and declared them to be true "originals." The only support to Van Meegeran's confession came from a Belgian chemist, Paul Coremans, but it was not until 1968 that the *Emmaus* was proved a forgery with radiochemical evidence obtained by Bernard Keisch of the Mellon Institute.

    The evidence hinged on two nuclear transformations involving lead in white pigments used by both Vermeer and Van Meegeran in their paintings. The lead used in making the pigments contained some of the isotope $^{210}_{82}Pb$, which undergoes β decay with a half-life of 22 years, followed by successive decay steps (Figure 35.2) rapidly leading to stable lead-206. Also present, as an impurity in the lead used, was a trace of radium-226, which very slowly decays by a series of steps (Figure 35.2) to lead-210. The difference between the $^{210}_{82}Pb : {}^{226}_{88}Ra$ ratios in the white pigments in "real" and "forged" Vermeers was the vital clue to proving the *Emmaus* a forgery. What *kind* of difference must have been found?

    To check your answer, see "Chemistry in Art: Radiochemistry and Forgery," by F. E. Rogers, in the June, 1972, issue of the *Journal of Chemical Education* (pp. 418–419).

You might also like to read Mark Twain's short story "Is He Living or Is He Dead?," which, for all we know, may well have been "inspiration" for Van Meegeren.

16. Using information from newspapers, news magazines, and scientific journals (such as *Science, Chemistry, Scientific American, Chemical & Engineering News*) prepare a summary report on "the pros and cons of the nuclear power plant issue." Cite references used.

Unit 35 Self-Test

1. Complete and balance (consulting Figure 35.1 when necessary):

   a. $^{40}_{19}K \longrightarrow$ _____

   b. $^{29}_{15}P \longrightarrow$ _____

   c. $^{218}_{84}Po \longrightarrow$ _____

2. Calculate the net energy change (using data from Table 35.1), in kilocalories per mole, for the overall process represented by the following. (For necessary conversion factors, see Appendix A.)

   $$^{238}_{92}U \longrightarrow {}^{206}_{82}Pb + 8{}^{4}_{2}He + 6{}^{0}_{-1}e$$

3. One of the methods used for dating the rock samples brought from the moon by the Apollo 11 flight used the potassium-40 to argon-40 ratio based on positron decay of $^{40}_{19}K$ ($t_{1/2} = 1.3 \times 10^9$ years). One of the samples tested gave a $^{40}_{19}K : {}^{40}_{18}Ar$ mole ratio of 0.19. Using appropriate assumptions, estimate the age of the sample.

4. Complete and balance

   a. $^{54}_{26}Fe + {}^{4}_{2}He \longrightarrow {}^{1}_{0}n +$ _____

   b. $^{6}_{3}Li + {}^{1}_{0}n \longrightarrow {}^{4}_{2}He +$ _____

   c. $^{246}_{96}Cm + {}^{13}_{6}C \longrightarrow {}^{254}_{102}No +$ _____

Epilogue

Where Do We Go From Here?

This text was designed to lay the groundwork for further study of science, either through additional formal course work or simply through learning what science is doing as reported in the daily news. In either case, it is hoped that this background will provide some knowledge of chemistry, some appreciation for how the chemist looks at the world in investigative ways, and a feeling for the limitations of science.

If you do plan a career requiring additional study of chemistry, you will be exploring many of the topics we have covered in greater depth, and your understanding will increase with familiarity and practice. Typically, the chemistry student will go on from here to organic chemistry, a detailed study of carbon compounds and their behavior. Following, or perhaps accompanying, this study will be a course in analytical chemistry, where you will learn more about making and interpreting experimental investigations. At a more advanced stage, physical chemistry will use an accumulated background in mathematics and physics to develop theoretical treatment of many of the concepts we have had to discuss on a rather elementary level. Special courses offer opportunities to delve into areas such as inorganic or biochemistry and, of course, graduate work can add new dimensions to your understanding, chances to specialize in topics of particular interest and—through participation in actual research—the possibility of pushing forward the frontiers of science.

An academic program in chemistry can open the doors to many exciting careers. Chemists are involved in a fantastic variety of activities. Many, of course, do the things we typically expect—industrial or academic research or teaching. But chemists also work in medical research, in oceanography, in agriculture, or in environmental science. There are Ph.D. chemists who are lawyers, corporation executives, legislators, and college administrators.

Chemistry can be useful, of course, in many careers not requiring extensive study of the subject. Chemical information, concepts, and techniques are important to biologists, engineers, nurses, nutritionists, and many others. In fact, some knowledge of chemistry is almost essential to an understanding of our modern technological society, and certainly an appreciation of the laws of science and the limitations of science is necessary if we are to make intelligent choices about many of the problems we all must face.

We must understand, however, that science alone is not enough. It cannot find us real "truth." It cannot tell us why we are here, what we should do, or how we should live. Science and the discoveries it provides are neither good nor bad, but the things humanity does with these may ultimately salvage the earth . . . or utterly destroy it.

Where do we go from here? Now that is up to you.

Appendix A

Some Common Units and Conversion Factors

1. Metric System Prefixes[1]

giga (G) $= 10^9$
mega (M) $= 10^6$
kilo (k) $= 10^3$
deci (d) $= 10^{-1}$
centi (c) $= 10^{-2}$
milli (m) $= 10^{-3}$
micro (μ) $= 10^{-6}$
nano (n) $= 10^{-9}$
pico (p) $= 10^{-12}$

2. Metric/English Conversions

a. *Length*
2.54 cm = 1 in. (exactly defined)
1.0000 m = 39.370 in.
1.0000 km = 0.62137 mile

b. *Volume*
(Metric "bulk"/volume)
1.0000 liter = 1.0567 qt (liquid)
1.0000 m^3 = 35.314 ft^3

c. *Mass and weight* (Compared at Standard Earth Gravity)
1.0000 kg = 2.2046 lb (avoirdupois)

[1] Compound prefixes are to be avoided. For example, 10^{-8} meter could be expressed as 10 nanometers, but not as 10 millimicrometers. Note also that attaching a prefix to a unit constitutes a *new* unit. For example, 1 mm^3 = $(10^{-3}$ m$)^3$ = 10^{-9} m^3, and not 1 m (m)^3 = 10^{-3} m^3.

1.0000 ton (av.) = 907.18 kg
1.0000 lb = 453.59 g

d. *Pressure*
1.0000 atmosphere (atm) = 760.00 torr = 760.00 mm Hg

e. *Temperature*
°C (centigrade or Celsius) $= \frac{5}{9}$(°F − 32)
°K (absolute or Kelvin) = °C + 273.15°

3. International System of Units (SI)

In 1960, an agreement was reached on a system of international units (the *Système International,* SI). In many respects, the new system is based on the metric units, first used in 1790. Although many scientific journals and textbooks have adopted the SI units, they have not yet gained universal acceptance. Throughout this book, we have used, for the most part, units consistent with SI, with three major exceptions. In referring to very small dimensions, we have sometimes employed the *angstrom* (Å) $= 10^{-10}$ meter, in deference to common practice in the bulk of the chemical literature prior to 1960. For pressures, we have employed either *atmospheres* or *torr*, which are more convenient than is the SI unit of *pascal* (Pa) = newton meter^{-2}. Finally, we have chosen to use *calories* as the basic unit of energy, rather than the SI unit *joule* (J). This particular choice is, perhaps, the most difficult to defend because it is the most strongly contested by those scientists concerned with precision of terminology. The choice was made on the basis of the fact that the majority of the currently available chemical literature employs the calorie as an energy unit. There is, beyond question, a definite trend toward the SI units, and you should be aware of at least the most common of these.[2]

Some Basic SI Units

Physical quantity	SI unit (symbol)[3]
length	meter (m)
mass	kilogram (kg)
time	second (s)
electric current	ampere (A)
thermodynamic temperature	Kelvin (K)

Some Derived SI Units

Physical quantity	SI unit (symbol)[3]	Definition of unit[3]
energy	joule (J)	$kg\ m^2\ s^{-2}$
force	newton (N)	$kg\ m\ s^{-2}$
electric charge	coulomb (C)	$A\ s$
electric potential difference	volt (V)	$kg\ m^2\ s^{-3}\ A^{-1}$
frequency	hertz (Hz)	$cycles\ s^{-1}$
pressure	pascal (Pa)	$N\ m^{-2}$
temperature (customary)	degree Celcius (°C)	$K - 273.15°$

Some Units Accepted by SI (but not recommended)

Physical quantity	Unit (symbol)	SI relationship
volume	liter (l)	$10^{-3}\ m^3$
energy	electron volt (eV)	$1.6021 \times 10^{-19}\ J$
mass	tonne [metric ton](t)	$10^3\ kg$

[2] For further details, see "International System of Units (SI)," by Martin A. Paul in *Chemistry*, Vol. 45, p. 14 (1972) and "The International System of Units (SI)," *National Bureau of Standards Special Publication 330*, 1972, U.S. Government Printing Office, Washington, D.C.

[3] Note that we have used more generally employed abbreviations for time in seconds (sec) and current in amperes (amp) throughout this book, and we have used *deg K* (or °K), rather than just K alone.

Some Units Not Accepted by SI (but employed in this textbook)

Physical quantity	Unit (symbol)	SI equivalent
length	angstrom (Å)	$10^{-10}\ m$
volume	milliliter (ml)	cm^3 ($10^{-6}\ m^3$)
force	dyne (dyn)	$10^{-5}\ N$
pressure	atmosphere (atm)	$101.325\ kN\ m^{-2}$
	torr (torr)	$133.322\ N\ m^{-2}$
energy	erg (erg)	$10^{-7}\ J$
	calorie (cal)	$4.184\ J$

4. Mathematical Formulas

area of a rectangle = length $\times$ width
area of a triangle = $\frac{1}{2}$(length of base $\times$ height)
area of a circle = $\pi \times$ (radius)^2
surface area of a sphere = $\pi \times$ (diameter)^2
volume of an orthoganal box = length $\times$ width $\times$ height
volume of a sphere = $\frac{4}{3}\pi$ (radius)^3

5. Logarithmic Manipulation (base 10)

$\log(m \times n) = \log m + \log n$
$\log(m/n) = \log m - \log n$
$\log(1/m) = -\log m$
$\log(m)^x = x \log m$
$\log(\sqrt[x]{m}) = (\log m)/x$

6. Physical constants

absolute zero = $0°K = -273.15°C$
atomic mass unit (amu) = $1.6602 \times 10^{-24}\ g$
Avogadro's number = 6.0225×10^{23} unit particles mole^{-1}
Boltzmann's constant = 1.3805×10^{-16} erg °K^{-1} per molecule
charge on the electron = $1.6021 \times 10^{-19}\ C$
Faraday's constant (F) = 9.6487×10^4 C mole^{-1} (electrons)
gravitational acceleration = 980.66 cm sec^{-2}
molar gas constant (R) = 0.08206 liter atm mole^{-1} °K^{-1}
$= 1.987$ cal mole^{-1} °K^{-1}
molar volume = 22.414 liters mole^{-1} (at STP)
Planck's constant (h) = 6.6257×10^{-27} erg sec
(1.00 erg = 2.39×10^{-8} cal)
speed of light (c) = 2.9979×10^{10} cm sec^{-1}
standard temperature and pressure (STP) = $273.15°$K, 1.000 atm

Appendix B

Four-Place Logarithms (base 10)

Note: Numbers are in form required by scientific notation. Mantissas are decimals. Decimal points are omitted from the table.

No.	0	1	2	3	4	5	6	7	8	9
10	0000	0043	0086	0128	0170	0212	0253	0294	0334	0374
11	0414	0453	0492	0531	0569	0607	0645	0682	0719	0755
12	0792	0828	0864	0899	0934	0969	1004	1038	1072	1106
13	1139	1173	1206	1239	1271	1303	1335	1367	1399	1430
14	1461	1492	1523	1553	1584	1614	1644	1673	1703	1732
15	1761	1790	1818	1847	1875	1903	1931	1959	1987	2014
16	2041	2068	2095	2122	2148	2175	2201	2227	2253	2279
17	2304	2330	2355	2380	2405	2430	2455	2480	2504	2529
18	2553	2577	2601	2625	2648	2672	2695	2718	2742	2765
19	2788	2810	2833	2856	2878	2900	2923	2945	2967	2989
20	3010	3032	3054	3075	3096	3118	3139	3160	3181	3201
21	3222	3243	3263	3284	3304	3324	3345	3365	3385	3404
22	3424	3444	3464	3483	3502	3522	3541	3560	3579	3598
23	3617	3636	3655	3674	3692	3711	3729	3747	3766	3784
24	3802	3820	3838	3856	3874	3892	3909	3927	3945	3962
25	3979	3997	4014	4031	4048	4065	4082	4099	4116	4133
26	4150	4166	4183	4200	4216	4232	4249	4265	4281	4298
27	4314	4330	4346	4362	4378	4393	4409	4425	4440	4456
28	4472	4487	4502	4518	4533	4548	4564	4579	4594	4609
29	4624	4639	4654	4669	4683	4698	4713	4728	4742	4757
30	4771	4786	4800	4814	4829	4843	4857	4871	4886	4900
31	4914	4928	4942	4955	4969	4983	4997	5011	5024	5038
32	5051	5065	5079	5092	5105	5119	5132	5145	5159	5172
33	5185	5198	5211	5224	5237	5250	5263	5276	5289	5302
34	5315	5328	5340	5353	5366	5378	5391	5403	5416	5428
35	5441	5453	5465	5478	5490	5502	5514	5527	5539	5551
36	5563	5575	5587	5599	5611	5623	5635	5647	5658	5670
37	5682	5694	5705	5717	5729	5740	5752	5763	5775	5786

No.	0	1	2	3	4	5	6	7	8	9
38	5798	5809	5821	5832	5843	5855	5866	5877	5888	5899
39	5911	5922	5933	5944	5955	5966	5977	5988	5999	6010
40	6021	6031	6042	6053	6064	6075	6085	6096	6107	6117
41	6128	6138	6149	6160	6170	6180	6191	6201	6212	6222
42	6232	6243	6253	6263	6274	6284	6294	6304	6314	6325
43	6335	6345	6355	6365	6375	6386	6395	6405	6415	6425
44	6435	6444	6454	6464	6474	6484	6493	6503	6513	6522
45	6532	6542	6551	6561	6571	6580	6590	6599	6609	6618
46	6628	6637	6646	6656	6665	6675	6684	6693	6702	6712
47	6721	6730	6739	6749	6758	6767	6776	6785	6794	6803
48	6812	6821	6830	6839	6848	6857	6866	6875	6884	6893
49	6902	6911	6920	6928	6937	6946	6955	6964	6972	6981
50	6990	6998	7007	7016	7024	7033	7042	7050	7059	7067
51	7076	7084	7093	7101	7110	7118	7126	7135	7143	7152
52	7160	7168	7177	7185	7193	7202	7210	7218	7226	7235
53	7243	7251	7259	7267	7275	7284	7292	7300	7308	7316
54	7324	7332	7340	7348	7356	7364	7372	7380	7388	7396
55	7404	7412	7419	7427	7435	7443	7451	7459	7466	7474
56	7482	7490	7497	7505	7513	7520	7528	7536	7543	7551
57	7559	7566	7574	7582	7589	7597	7604	7612	7619	7627
58	7634	7642	7649	7657	7664	7672	7679	7686	7694	7701
59	7709	7716	7723	7731	7738	7745	7752	7760	7767	7774
60	7782	7789	7796	7803	7810	7818	7825	7832	7839	7846
61	7853	7860	7868	7875	7882	7889	7896	7903	7910	7917
62	7924	7931	7938	7945	7952	7959	7966	7973	7980	7987
63	7992	8000	8007	8014	8021	8028	8035	8041	8048	8055
64	8062	8069	8075	8082	8089	8096	8102	8109	8116	8122

No.	0	1	2	3	4	5	6	7	8	9
65	8129	8136	8142	8149	8156	8162	8169	8176	8182	8189
66	8195	8202	8209	8215	8222	8228	8235	8241	8248	8254
67	8261	8267	8274	8280	8287	8293	8299	8306	8312	8319
68	8325	8331	8338	8344	8351	8357	8363	8370	8376	8382
69	8388	8395	8401	8407	8414	8420	8426	8432	8439	8445
70	8451	8457	8463	8470	8476	8482	8488	8494	8500	8506
71	8513	8519	8525	8531	8537	8543	8549	8555	8561	8567
72	8573	8579	8585	8591	8597	8603	8609	8615	8621	8627
73	8633	8639	8645	8651	8657	8663	8669	8675	8681	8686
74	8692	8698	8704	8710	8716	8722	8727	8733	8739	8745
75	8751	8756	8762	8768	8774	8779	8785	8791	8797	8802
76	8808	8814	8820	8825	8831	8837	8842	8848	8854	8859
77	8865	8871	8876	8882	8887	8893	8899	8904	8910	8915
78	8921	8927	8932	8938	8943	8949	8954	8960	8965	8971
79	8976	8982	8987	8993	8998	9004	9009	9015	9020	9025
80	9031	9036	9042	9047	9053	9058	9063	9069	9074	9079
81	9085	9090	9096	9101	9106	9112	9117	9122	9128	9133
82	9138	9143	9149	9154	9159	9165	9170	9175	9180	9186

No.	0	1	2	3	4	5	6	7	8	9
83	9191	9196	9201	9206	9212	9217	9222	9227	9232	9238
84	9243	9248	9253	9258	9263	9269	9274	9279	9284	9289
85	9294	9299	9304	9309	9315	9320	9325	9330	9335	9340
86	9345	9350	9355	9360	9365	9370	9375	9380	9385	9390
87	9395	9400	9405	9410	9415	9420	9425	9430	9435	9440
88	9445	9450	9455	9460	9465	9469	9474	9479	9484	9489
89	9494	9499	9504	9509	9513	9518	9523	9528	9533	9538
90	9542	9547	9552	9557	9562	9566	9571	9576	9581	9586
91	9590	9595	9600	9605	9609	9614	9619	9624	9628	9633
92	9638	9643	9647	9652	9657	9661	9666	9671	9675	9680
93	9685	9689	9694	9699	9703	9708	9713	9717	9722	9727
94	9731	9736	9741	9745	9750	9754	9759	9763	9768	9773
95	9777	9782	9786	9791	9795	9800	9805	9809	9814	9818
96	9823	9827	9832	9836	9841	9845	9850	9854	9859	9863
97	9868	9872	9877	9881	9886	9890	9894	9899	9903	9908
98	9912	9917	9921	9926	9930	9934	9939	9943	9948	9952
99	9956	9961	9965	9969	9974	9978	9983	9987	9991	9996

Appendix C

Equilibrium Constants At 20-25°C

Table 1 ACID DISSOCIATION CONSTANTS

ACID	K_a
acetic (CH_3CO_2H)	1.8×10^{-5}
benzoic ($C_6H_5CO_2H$)	6.6×10^{-5}
boric (ortho) (H_3BO_3)	6.0×10^{-10} (K_{a_1})
carbonic ($H_2O + CO_2$)	4.2×10^{-7} (K_{a_1})
	4.8×10^{-11} (K_{a_2})
formic (HCO_2H)	2.0×10^{-4}
hydrobromic (HBr)	$\sim 10^9$
hydrochloric (HCl)	$\sim 10^6$
hydrocyanic (HCN)	4.8×10^{-10}
hydrofluoric (HF)	6.9×10^{-4}
hydroiodic (HI)	$\sim 10^9$
hydrosulfuric (H_2S)	1.0×10^{-7} (K_{a_1})
	1.3×10^{-13} (K_{a_2})
hypochlorous (HOCl)	3.2×10^{-8}
nitric (HNO_3)	~ 30
nitrous (HNO_2)	4.5×10^{-4}
oxalic ($H_2C_2O_4$)	6.3×10^{-2} (K_{a_1})
	6.3×10^{-5} (K_{a_2})
perchloric ($HClO_4$)	$\sim 10^{10}$
phosphoric (ortho) (H_3PO_4)	7.5×10^{-3} (K_{a_1})
	6.2×10^{-8} (K_{a_2})
	2.0×10^{-13} (K_{a_3})
phosphorous (H_3PO_3)	1.6×10^{-2} (K_{a_1})
	6.9×10^{-7} (K_{a_2})
propionic ($CH_3CH_2CO_2H$)	1.4×10^{-5}
sulfuric (H_2SO_4)	$\sim 10^3$ (K_{a_1})
	1.3×10^{-2} (K_{a_2})
sulfurous (H_2SO_3)	1.6×10^{-2} (K_{a_1})
	1.3×10^{-7} (K_{a_2})

Table 2 BASE IONIZATION CONSTANTS

BASE	K_b
ammonia (NH_3)	1.8×10^{-5}
aniline ($C_6H_5NH_2$)	3.8×10^{-10}
dimethylamine [$(CH_3)_2NH$]	5.1×10^{-4}
ethylamine ($CH_3CH_2NH_2$)	5.6×10^{-4}
methoxide ion (CH_3O^-)	$\sim 10^3$
methylamine (CH_3NH_2)	5.0×10^{-4}
trimethylamine [$(CH_3)_3N$]	5.3×10^{-5}

Table 3 SOLUBILITY PRODUCTS

SALT	K_{sp}	SALT	K_{sp}
Acetates		Hydroxides	
$Ag(CH_3CO_2)$	4.1×10^{-3}	(for amphoteric hydroxides, see	
$Hg_2(CH_3CO_2)_2$	3.6×10^{-10}	Table 5)	
Arsenates		$Ca(OH)_2$	4.0×10^{-6}
Ag_3AsO_4	1.0×10^{-22}	$Cd(OH)_2$	2.0×10^{-14}
		$Co(OH)_2$	2.0×10^{-16}
Bromates		$Cu(OH)_2$	1.8×10^{-19}
$AgBrO_3$	6.0×10^{-5}	$Fe(OH)_2$	1.8×10^{-15}
$Ba(BrO_3)_2$	5.6×10^{-6}	$Fe(OH)_3$	6.0×10^{-38}
Bromides		$Hg(OH)_2$	3.2×10^{-26}
$AgBr$	5.0×10^{-13}	$Mg(OH)_2$	1.2×10^{-11}
Hg_2Br_2	5.0×10^{-23}	$Mn(OH)_2$	2.0×10^{-13}
$PbBr_2$	1.0×10^{-6}	Iodides	
Carbonates		AgI	8.5×10^{-17}
Ag_2CO_3	8.2×10^{-12}	Hg_2I_2	4.5×10^{-29}
$BaCO_3$	1.6×10^{-9}	HgI_2	2.5×10^{-26}
$CaCO_3$	4.7×10^{-9}	PbI_2	8.3×10^{-9}
$CuCO_3$	2.5×10^{-10}	Phosphates	
$FeCO_3$	2.0×10^{-11}	Ag_3PO_4	1.6×10^{-18}
$MgCO_3$	4.0×10^{-5}	$Ba_3(PO_4)_2$	3.2×10^{-23}
$NiCO_3$	1.4×10^{-7}	$Ca_3(PO_4)_2$	1.3×10^{-32}
$PbCO_3$	1.5×10^{-13}	Li_3PO_4	3.2×10^{-13}
$SrCO_3$	7.0×10^{-10}	$Sr_3(PO_4)_2$	4.0×10^{-28}
$ZnCO_3$	3.0×10^{-11}	Sulfates	
Chlorides		Ag_2SO_4	6.4×10^{-5}
$AgCl$	2.4×10^{-10}	$BaSO_4$	7.9×10^{-11}
Hg_2Cl_2	1.2×10^{-18}	$CaSO_4$	2.5×10^{-5}
$PbCl_2$	1.6×10^{-5}	$PbSO_4$	1.3×10^{-8}
Chromates		$SrSO_4$	3.2×10^{-7}
Ag_2CrO_4	1.9×10^{-12}	Sulfides	
$BaCrO_4$	8.5×10^{-11}	Ag_2S	5.5×10^{-51}
$CaCrO_4$	7.1×10^{-4}	CdS	1.0×10^{-28}
$PbCrO_4$	1.8×10^{-14}	CoS	5.0×10^{-22}
$SrCrO_4$	5.7×10^{-5}	CuS	8.0×10^{-37}
Fluorides		FeS	4.0×10^{-19}
BaF_2	1.0×10^{-7}	Fe_2S_3	$\sim 10^{-88}$
CaF_2	1.7×10^{-10}	HgS	1.6×10^{-54}
LiF	$\sim 10^{-2}$	MnS	7.0×10^{-16}
MgF_2	6.8×10^{-9}	NiS	3.0×10^{-21}
SrF_2	2.5×10^{-9}	PbS	7.0×10^{-29}
		SnS	1.1×10^{-26}
		ZnS	1.6×10^{-23}

Most simple salts *not* listed in this table are soluble in water.

Table 4 FORMATION CONSTANTS OF COMPLEXES

$$(K_{net} = K_1 \times K_2 \times K_3 \cdots)$$

EQUILIBRIUM	K
Ammine complexes	
$[Ag(H_2O)_2]^+ + NH_3 \rightleftharpoons [Ag(H_2O)(NH_3)]^+ + H_2O$	$2.0 \times 10^3 \ (K_1)$
$[Ag(H_2O)(NH_3)]^+ + NH_3 \rightleftharpoons [Ag(NH_3)_2]^+ + H_2O$	$8.0 \times 10^3 \ (K_2)$
$[Cd(H_2O)_6]^{2+} + NH_3 \rightleftharpoons [Cd(H_2O)_5(NH_3)]^{2+} + H_2O$	$4.5 \times 10^2 \ (K_1)$
$[Cd(H_2O)_5(NH_3)]^{2+} + NH_3 \rightleftharpoons [Cd(H_2O)_4(NH_3)_2]^{2+} + H_2O$	$1.3 \times 10^2 \ (K_2)$
$[Cd(H_2O)_4(NH_3)_2]^{2+} + NH_3 \rightleftharpoons [Cd(H_2O)_3(NH_3)_3]^{2+} + H_2O$	$2.8 \times 10^1 \ (K_3)$
$[Cd(H_2O)_3(NH_3)_3]^{2+} + NH_3 \rightleftharpoons [Cd(H_2O)_2(NH_3)_4]^{2+} + H_2O$	$8.5 \times 10^0 \ (K_4)$
$[Co(H_2O)_6]^{2+} + NH_3 \rightleftharpoons [Co(H_2O)_5(NH_3)]^{2+} + H_2O$	$4.5 \times 10^2 \ (K_1)$
$[Co(H_2O)_5(NH_3)]^{2+} + NH_3 \rightleftharpoons [Co(H_2O)_4(NH_3)_2]^{2+} + H_2O$	$3.0 \times 10^1 \ (K_2)$
$[Co(H_2O)_4(NH_3)_2]^{2+} + NH_3 \rightleftharpoons [Co(H_2O)_3(NH_3)_3]^{2+} + H_2O$	$8.0 \times 10^0 \ (K_3)$
$[Co(H_2O)_3(NH_3)_3]^{2+} + NH_3 \rightleftharpoons [Co(H_2O)_2(NH_3)_4]^{2+} + H_2O$	$4.0 \times 10^0 \ (K_4)$
$[Co(H_2O)_2(NH_3)_4]^{2+} + NH_3 \rightleftharpoons [Co(H_2O)(NH_3)_5]^{2+} + H_2O$	$1.3 \times 10^0 \ (K_5)$
$[Co(H_2O)(NH_3)_5]^{2+} + NH_3 \rightleftharpoons [Co(NH_3)_6]^{2+} + H_2O$	$2.0 \times 10^{-1} \ (K_6)$
$[Cu(H_2O)_6]^{2+} + NH_3 \rightleftharpoons [Cu(H_2O)_5(NH_3)]^{2+} + H_2O$	$1.4 \times 10^4 \ (K_1)$
$[Cu(H_2O)_5(NH_3)]^{2+} + NH_3 \rightleftharpoons [Cu(H_2O)_4(NH_3)_2]^{2+} + H_2O$	$3.2 \times 10^3 \ (K_2)$
$[Cu(H_2O)_4(NH_3)_2]^{2+} + NH_3 \rightleftharpoons [Cu(H_2O)_3(NH_3)_3]^{2+} + H_2O$	$7.8 \times 10^2 \ (K_3)$
$[Cu(H_2O)_3(NH_3)_3]^{2+} + NH_3 \rightleftharpoons [Cu(H_2O)_2(NH_3)_4]^{2+} + H_2O$	$1.4 \times 10^2 \ (K_4)$
net $[Zn(H_2O)_4]^{2+} + 4NH_3 \rightleftharpoons [Zn(NH_3)_4]^{2+} + 4H_2O$	$2.8 \times 10^9 \ (K_{net})$
Chloro complexes	
$[Cd(H_2O)_6]^{2+} + Cl^- \rightleftharpoons [CdCl(H_2O)_5]^+ + H_2O$	$1.0 \times 10^2 \ (K_1)$
$[CdCl(H_2O)_5]^+ + Cl^- \rightleftharpoons [CdCl_2(H_2O)_4] + H_2O$	$5.0 \times 10^0 \ (K_2)$
$[CdCl_2(H_2O)_4] + Cl^- \rightleftharpoons [CdCl_3(H_2O)_3]^- + H_2O$	$4.0 \times 10^{-1} \ (K_3)$
$[CdCl_3(H_2O)_3]^- + Cl^- \rightleftharpoons [CdCl_4(H_2O)_2]^{2-} + H_2O$	$2.0 \times 10^{-1} \ (K_4)$
$[Hg(H_2O)_4]^{2+} + Cl^- \rightleftharpoons [HgCl(H_2O)_3]^+ + H_2O$	$5.5 \times 10^6 \ (K_1)$
$[HgCl(H_2O)_3]^+ + Cl^- \rightleftharpoons [HgCl_2(H_2O)_2] + H_2O$	$3.0 \times 10^6 \ (K_2)$
$[HgCl_2(H_2O)_2] + Cl^- \rightleftharpoons [HgCl_3(H_2O)]^- + H_2O$	$7.1 \times 10^0 \ (K_3)$
$[HgCl_3(H_2O)]^- + Cl^- \rightleftharpoons [HgCl_4]^{2-} + H_2O$	$1.0 \times 10^1 \ (K_4)$
net $[Pb(H_2O)_4]^{2+} + 3Cl^- \rightleftharpoons [PbCl_3(H_2O)]^- + 3H_2O$	$2.5 \times 10^1 \ (K_{net})$
net $[Sn(H_2O)_4]^{2+} + 4Cl^- \rightleftharpoons [SnCl_4]^{2-} + 4H_2O$	$1.8 \times 10^1 \ (K_{net})$
net $[Sn(H_2O)_6]^{4+} + 6Cl^- \rightleftharpoons [SnCl_6]^{2-} + 6H_2O$	$6.6 \times 10^0 \ (K_{net})$
Cyano complexes	
$[Cd(H_2O)_6]^{2+} + CN^- \rightleftharpoons [Cd(CN)(H_2O)_5]^+ + H_2O$	$3.0 \times 10^5 \ (K_1)$
$[Cd(CN)(H_2O)_5]^+ + CN^- \rightleftharpoons [Cd(CN)_2(H_2O)_4] + H_2O$	$1.3 \times 10^5 \ (K_2)$
$[Cd(CN)_2(H_2O)_4] + CN^- \rightleftharpoons [Cd(CN)_3(H_2O)_3]^- + H_2O$	$4.3 \times 10^4 \ (K_3)$
$[Cd(CN)_3(H_2O)_3]^- + CN^- \rightleftharpoons [Cd(CN)_4(H_2O)_2]^{2-} + H_2O$	$3.5 \times 10^3 \ (K_4)$
net $[Fe(H_2O)_6]^{2+} + 6CN^- \rightleftharpoons [Fe(CN)_6]^{4-} + 6H_2O$	$1.0 \times 10^{24} \ (K_{net})$
net $[Fe(H_2O)_6]^{3+} + 6CN^- \rightleftharpoons [Fe(CN)_6]^{3-} + 6H_2O$	$1.0 \times 10^{31} \ (K_{net})$
Fluoro complexes	
net $[Al(H_2O)_6]^{3+} + 6F^- \rightleftharpoons [AlF_6]^{3-} + 6H_2O$	$6.1 \times 10^{19} \ (K_{net})$
net $[Fe(H_2O)_6]^{3+} + 5F^- \rightleftharpoons [FeF_5(H_2O)]^{2-} + 5H_2O$	$2.0 \times 10^{15} \ (K_{net})$
Thiocyanato complexes	
$[Co(H_2O)_6]^{2+} + NCS^- \rightleftharpoons [Co(NCS)(H_2O)_5]^+ + H_2O$	$1.0 \times 10^3 \ (K_1)$
$[Co(NCS)(H_2O)_5]^+ + NCS^- \rightleftharpoons [Co(NCS)_2(H_2O)_4] + H_2O$	$1.0 \times 10^0 \ (K_2)$
$[Co(NCS)_2(H_2O)_4] + NCS^- \rightleftharpoons [Co(NCS)_3(H_2O)_3]^- + H_2O$	$2.0 \times 10^{-1} \ (K_3)$
$[Co(NCS)_3(H_2O)_3]^- + NCS^- \rightleftharpoons [Co(NCS)_4(H_2O)_2]^{2-} + H_2O$	$1.0 \times 10^0 \ (K_4)$
$[Fe(H_2O)_6]^{3+} + NCS^- \rightleftharpoons [Fe(NCS)(H_2O)_5]^{2+} + H_2O$	$1.1 \times 10^3 \ (K_1)$
Thiosulfato complexes	
net $[Ag(H_2O)_2]^+ + 2S_2O_3^{2-} \rightleftharpoons [Ag(S_2O_3)_2]^{3-} + 2H_2O$	$2.9 \times 10^{13} \ (K_{net})$

Table 5 SOME COMMON AMPHOTERIC HYDROXIDES

EQUILIBRIUM	K
<u>Aluminum</u>	
$[Al(H_2O)_6]^{3+}(aq) + H_2O(l) \rightleftharpoons [Al(OH)(H_2O)_5]^{2+}(aq) + H_3O^+(aq)$	$K_{a_1} = 1.4 \times 10^{-5}$
net $Al(OH)_3(s) \rightleftharpoons Al^{3+}(aq) + 3OH^-(aq)$	$K_{sp} = 5.0 \times 10^{-33}$
$Al(OH)_3(s) + OH^-(aq) + 2H_2O(l) \rightleftharpoons [Al(OH)_4(H_2O)_2]^-(aq)$	$K_4 = 4.0 \times 10$
<u>Chromium</u>	
$[Cr(H_2O)_6]^{3+}(aq) + H_2O(l) \rightleftharpoons [Cr(OH)H_2O)_5]^{2+}(aq) + H_3O^+(aq)$	$K_{a_1} = 1.0 \times 10^{-10}$
net $Cr(OH)_3(s) \rightleftharpoons Cr^{3+}(aq) + 3OH^-(aq)$	$K_{sp} = 7.0 \times 10^{-31}$
$Cr(OH)_3(s) + OH^-(aq) + 2H_2O(l) \rightleftharpoons [Cr(OH)_4(H_2O)_2]^-(aq)$	$K_4 = 9.0 \times 10^4$
<u>Zinc</u>	
$[Zn(H_2O)_4]^{2+}(aq) + H_2O(l) \rightleftharpoons [Zn(OH)(H_2O)_3]^+(aq) + H_3O^+(aq)$	$K_{a_1} = 2.5 \times 10^{-10}$
net $Zn(OH)_2(s) \rightleftharpoons Zn^{2+}(aq) + 2OH^-(aq)$	$K_{sp} = 4.5 \times 10^{-17}$
net $Zn(OH)_2(s) + 2OH^-(aq) \rightleftharpoons [Zn(OH)_4]^{2-}(aq)$	$K_{3,4} = 1.0 \times 10^{-15}$

Appendix D

Standard Electrode Potentials

Temperature = 298°K
All solids in most stable form
All gases at 1.00 atm pressure
All solutes at unit activity (approx. 1.0 M)

COUPLE	$\mathcal{E}°$ (volts)
A. Solutions of pH $\leq$ 7.0	
$Li^+(aq) + e^- \rightleftharpoons Li(s)$	-3.05
$K^+(aq) + e^- \rightleftharpoons K(s)$	-2.93
$Cs^+(aq) + e^- \rightleftharpoons Cs(s)$	-2.92
$Ra^{2+}(aq) + 2e^- \rightleftharpoons Ra(s)$	-2.92
$Ba^{2+}(aq) + 2e^- \rightleftharpoons Ba(s)$	-2.90
$Sr^{2+}(aq) + 2e^- \rightleftharpoons Sr(s)$	-2.89
$Ca^{2+}(aq) + 2e^- \rightleftharpoons Ca(s)$	-2.87
$Na^+(aq) + e^- \rightleftharpoons Na(s)$	-2.71
$Mg^{2+}(aq) + 2e^- \rightleftharpoons Mg(s)$	-2.37
$Be^{2+}(aq) + 2e^- \rightleftharpoons Be(s)$	-1.87
$Al^{3+}(aq) + 3e^- \rightleftharpoons Al(s)$	-1.66
$Ti^{2+}(aq) + 2e^- \rightleftharpoons Ti(s)$	-1.63
$Mn^{2+}(aq) + 2e^- \rightleftharpoons Mn(s)$	-1.18
$V^{2+}(aq) + 2e^- \rightleftharpoons V(s)$	-1.16
$TiO^{2+}(aq) + 2H^+(aq) + 4e^- \rightleftharpoons Ti(s) + H_2O(l)$	-0.89
$B(OH)_3(aq) + 3H^+(aq) + 3e^- \rightleftharpoons B(s) + 3H_2O(l)$	-0.87
$(SiO_2)_n(s) + 4nH^+(aq) + 4ne^- \rightleftharpoons nSi(s) + 2nH_2O(l)$	-0.86
$Zn^{2+}(aq) + 2e^- \rightleftharpoons Zn(s)$	-0.76
$Cr^{3+} + 3e^- \rightleftharpoons Cr(s)$	-0.74
$As(s) + 3H^+(aq) + 3e^- \rightleftharpoons AsH_3(g)$	-0.60
$Ga^{3+}(aq) + 3e^- \rightleftharpoons Ga(s)$	-0.53

STANDARD ELECTRODE POTENTIALS (Continued)

COUPLE	$\mathcal{E}°$ (volts)
$H_3PO_2(aq) + H^+(aq) + e^- \rightleftharpoons P(s) + 2H_2O(l)$	-0.51
$H_3PO_3(aq) + 2H^+(aq) + 2e^- \rightleftharpoons H_3PO_2(aq)$ $+ H_2O(l)$	-0.50
$Fe^{2+}(aq) + 2e^- \rightleftharpoons Fe(s)$	-0.44
$Cr^{3+}(aq) + e^- \rightleftharpoons Cr^{2+}(aq)$	-0.41
$Cd^{2+}(aq) + 2e^- \rightleftharpoons Cd(s)$	-0.40
$Se(s) + 2H^+(aq) + 2e^- \rightleftharpoons H_2Se(g)$	-0.40
$Ti^{3+}(aq) + e^- \rightleftharpoons Ti^{2+}(aq)$	-0.37
$PbSO_4(s) + 2e^- \rightleftharpoons Pb(s) + SO_4{}^{2-}(aq)$	-0.36
$Co^{2+}(aq) + 2e^- \rightleftharpoons Co(s)$	-0.28
$H_3PO_4(aq) + 2H^+ + 2e^- \rightleftharpoons H_3PO_3(aq)$ $+ H_2O(l)$	-0.28
$V^{3+}(aq) + e^- \rightleftharpoons V^{2+}(aq)$	-0.26
$Ni^{2+}(aq) + 2e^- \rightleftharpoons Ni(s)$	-0.25
$Sn^{2+}(aq) + 2e^- \rightleftharpoons Sn(s)$	-0.14
$Pb^{2+}(aq) + 2e^- \rightleftharpoons Pb(s)$	-0.13
$2H^+(aq) + 2e^- \rightleftharpoons H_2(g)$	ZERO Reference Standard
$P(s) + 3H^+(aq) + 3e^- \rightleftharpoons PH_3(g)$	$+0.06$
$Sn^{4+} + 2e^- \rightleftharpoons Sn^{2+}(aq)$	$+0.15$
$Cu^{2+}(aq) + e^- \rightleftharpoons Cu^+(aq)$	$+0.15$
$SO_4{}^{2-}(aq) + 3H^+(aq) + 2e^- \rightleftharpoons HSO_3{}^-(aq)$ $+ H_2O(l)$	$+0.17$
$Cu^{2+}(aq) + 2e^- \rightleftharpoons Cu(s)$	$+0.34$
$Cu^+(aq) + e^- \rightleftharpoons Cu(s)$	$+0.52$
$I_2(aq) + 2e^- \rightleftharpoons 2I^-(aq)$	$+0.54$
$MnO_4{}^-(aq) + e^- \rightleftharpoons MnO_4{}^{2-}(aq)$	$+0.56$
$O_2(g) + 2H^+(aq) + 2e^- \rightleftharpoons H_2O_2(aq)$	$+0.68$
$Fe^{3+}(aq) + e^- \rightleftharpoons Fe^{2+}(aq)$	$+0.77$
$Hg_2{}^{2+}(aq) + 2e^- \rightleftharpoons 2Hg(l)$	$+0.79$
$Ag^+(aq) + e^- \rightleftharpoons Ag(s)$	$+0.80$
$2Hg^{2+}(aq) + 2e^- \rightleftharpoons Hg_2{}^{2+}(aq)$	$+0.92$
$HNO_2(aq) + H^+(aq) + e^- \rightleftharpoons NO(g) + H_2O(l)$	$+1.00$
$Br_2(l) + 2e^- \rightleftharpoons 2Br^-(aq)$	$+1.07$
$ClO_4{}^-(aq) + 2H^+(aq) + 2e^- \rightleftharpoons ClO_3{}^-(aq)$ $+ H_2O(l)$	$+1.19$
$ClO_3{}^-(aq) + 3H^+(aq) + 2e^- \rightleftharpoons HClO_2(aq)$ $+ H_2O(l)$	$+1.21$
$O_2(g) + 4H^+(aq) + 4e^- \rightleftharpoons 2H_2O(l)$	$+1.23$
$MnO_2(s) + 4H^+(aq) + 2e^- \rightleftharpoons Mn^{2+}(aq)$ $+ 2H_2O(l)$	$+1.23$
$Cr_2O_7{}^{2-}(aq) + 14H^+(aq) + 6e^- \rightleftharpoons 2Cr^{3+}(aq)$ $+ 7H_2O(l)$	$+1.33$
$Cl_2(aq) + 2e^- \rightleftharpoons 2Cl^-(aq)$	$+1.36$
$Au^{3+}(aq) + 3e^- \rightleftharpoons Au(s)$	$+1.50$
$MnO_4{}^-(aq) + 8H^+(aq) + 5e^- \rightleftharpoons Mn^{2+}(aq)$ $+ 4H_2O(l)$	$+1.51$
$2HOCl(aq) + 2H^+(aq) + 2e^- \rightleftharpoons Cl_2(aq) + 2H_2O(l)$	$+1.63$
$HClO_2(aq) + 2H^+(aq) + 2e^- \rightleftharpoons HOCl(aq)$ $+ H_2O(l)$	$+1.64$
$PbO_2(s) + SO_4{}^{2-}(aq)$ $+ 4H^+(aq) + 2e^- \rightleftharpoons PbSO_4(s)$ $+ 2H_2O(l)$	$+1.68$
$H_2O_2(aq) + 2H^+(aq) + 2e^- \rightleftharpoons 2H_2O(l)$	$+1.77$
$Co^{3+}(aq) + e^- \rightleftharpoons Co^{2+}(aq)$	$+1.82$
$O_3(g) + 2H^+(aq) + 2e^- \rightleftharpoons O_2(g) + H_2O(l)$	$+2.07$
$F_2(g) + 2e^- \rightleftharpoons 2F^-(aq)$	$+2.87$

COUPLE	$\mathcal{E}°$ (volts)
B. Solutions of pH > 7.0	
$Ca(OH)_2(s) + 2e^- \rightleftharpoons Ca(s) + 2OH^-(aq)$	-3.03
$Mg(OH)_2(s) + 2e^- \rightleftharpoons Mg(s) + 2OH^-(aq)$	-2.69
$[Al(OH)_4(H_2O)_2]^-(aq) + 3e^- \rightleftharpoons Al(s) + 4OH^-(aq)$	
$+ 2H_2O(l)$	-2.35
$Zn(OH)_2(s) + 2e^- \rightleftharpoons Zn(s) + 2OH^-(aq)$	-1.24
$Se(s) + 2e^- \rightleftharpoons Se^{2-}$	-0.93
$Fe(OH)_2(s) + 2e^- \rightleftharpoons Fe(s) + 2OH^-(aq)$	-0.88
$S(s) + 2e^- \rightleftharpoons S^{2-}$	-0.48
$NO_3^-(aq) + H_2O(l) + 2e^- \rightleftharpoons NO_2^-(aq)$	
$+ 2OH^-(aq)$	$+0.01$
$IO_3^-(aq) + 3H_2O(l) + 6e^- \rightleftharpoons I^-(aq) + 6OH^-(aq)$	$+0.26$
$ClO_2(g) + e^- \rightleftharpoons ClO_2^-(aq)$	$+1.16$

USE OF STANDARD POTENTIALS

1. Sign conventions

The so-called European (or IUPAC) convention used in the preceding tables refers to the experimentally determined *electrostatic potential* of an electrode. As a result, the sign is *invariant;* that is, the same sign is always used for the potential, whether the half-reaction involved is oxidation or reduction in a particular electrochemical cell. The sign used is that found experimentally by comparison to the universal reference, the standard hydrogen electrode (Unit 23).

The American system, as originally used by Lewis, Randall, and Latimer, is more easily related to thermodynamic considerations. In this system, the sign of the potential is *bivariant,* being identical with that of the standard potential in the preceding tables only when the half-reaction in question is actually reduction in the particular cell involved. Then the *oxidation* potential (used by many texts) is of opposite sign from that of the *reduction* potential shown in our tables.

To minimize confusion, we shall use the standard electrode potentials as recommended by the International Union of Pure and Applied Chemists (IUPAC), so that $\mathcal{E}^0$ values will be invariant in sign. To use these potentials for calculation of voltage of galvanic or electrolytic cells, we shall identify half-reactions as *cathode* or *anode* half-reactions. The (internal)[1] *cathode* half-reaction is that actually occurring as reduction in the particular cell, and the *anode* half-reaction is that occurring as oxidation (electron loss). On this basis:

$$\mathcal{E}^0_{cell} = \mathcal{E}^0_{cathode} - \mathcal{E}^0_{anode}$$

2. "Unknown" electrode potentials

The preceding tables include only some of the electrode potentials that have been determined. In certain cases, it is possible to calculate an electrode potential from those for other half-reactions involving the species of interest. For cases in which equations for electrode couples must be *added,* the number of electrons shown in each half-reaction equation must be considered:

$$(n_a + n_b)\mathcal{E}^0_{(net)} = n_a\mathcal{E}_a{}^0 + n_b\mathcal{E}_b{}^0$$

Example

Given the data

$$Fe^{2+} + 2e^- \rightleftharpoons Fe \qquad \mathcal{E}^0 = -0.44$$
$$Fe^{3+} + e^- \rightleftharpoons Fe^{2+} \qquad \mathcal{E}^0 = +0.77$$

find the standard electrode potential for

$$Fe^{3+} + 3e^- \rightleftharpoons Fe$$

Solution[2]
The equation for the couple in question will result from addition of the two equations given:

a. $Fe^{2+} + 2e^- \rightleftharpoons Fe$
b. $\underline{Fe^{3+} + e^- \rightleftharpoons Fe^{2+}}$
 $Fe^{3+} + 3e^- \rightleftharpoons Fe$

The electrode potential in question, then, is given by

$$(2 + 1)\mathcal{E}^0_{net} = (2)(-0.44) + (1)(+0.77)$$

from which

$$\mathcal{E}^0_{net} = \frac{-0.88 + 0.77}{3} = -0.037$$

[1] Note that the chemist is concerned primarily with processes *within* the cell, that is, with the *internal* cathode and anode. The internal cathode of a galvanic cell is the *external* anode (labeled + on a cell) and the internal anode is the external *cathode* (labeled − on a cell). The distinction is easily remembered by observing that the *internal anode* is the electrode from which electrons leave the cell.

[2] An alternative approach, useful for both added and subtracted couples, involves application of free energy relationships, as described in Units 23 and 24.

3. Potentials at other than standard conditions

The Nernst equation may be used to calculate electrode or cell potentials at other than standard conditions. In its most common form, the equation is expressed as

$$\mathcal{E} = \mathcal{E}^0 - \frac{0.0592}{n} \log Q \qquad \text{at } 298°K$$

The use of this equation is described in detail in Unit 24. It is important to note that the equation permits only an approximation of potential when concentrations, rather than activities, are used in the Q term.

Potentials at cell temperatures other than 298°K cannot be calculated directly from $\mathcal{E}^0$ values tabulated for 298°K. Although the complete form of the Nernst equation contains a temperature term, this does not permit temperature correction of electrode or cell potentials, because $\mathcal{E}^0$ itself varies with temperature. Empirical corrections are possible using experimentally determined thermodynamic data or $\mathcal{E}^0$ values at temperatures other than 298°K.

4. References

Anson, F. C. 1959. *J. Chem. Ed.*, **36**, 394.

Licht, T. S., and A. J. Bethune. 1957. *J. Chem. Ed.*, **34**, 433.

Lyons, E. H. 1967. *Introduction to Electrochemistry*. Lexington, Mass.: Raytheon.

(See also Units 23 and 24).

Glossary of Technical Terms

abundance indication of the amount of one species in a mixture (e.g., the percentage of silicon in the earth's crust)

acid in most common usage, a chemical species acting as a source of hydrogen ion (H^+) (*see also* Unit 19.)

activation energy internal energy difference (ΔE) between reactants and transition state

active site the location in an enzyme at which catalytic activity is exhibited

activity in relationship to concentration, the *apparent* concentration of a chemical species (e.g., the concentration of a solute as determined from a solution property such as vapor pressure)

actual yield the amount of product obtained from a chemical process

adsorption adherence of a particle to the surface of a material

aliphatic type of organic compound or molecular segment that does not contain aromatic (e.g., benzene) groupings

alkali metal one of the group IA elements (e.g., sodium or potassium)

alkaline earth metal one of the group IIA elements (e.g., magnesium or calcium)

allotropes crystalline or molecular modifications of a pure substance (e.g., rhombic and monoclinic sulfur, or O_3 and O_2)

amalgam alloy in which mercury is a major component

amorphous noncrystalline solid

amphoteric capable of acting both as an acid and as a base

anhydrous not associated with water molecules

anion negatively charged particle

anode the internal electrode of an electrochemical cell at which oxidation occurs

aqueous solution in which the solvent is water

atomic mass unit (amu) defined as exactly one-twelfth the mass of the neutral atom $^{12}_{6}C$

atomic number the number of protons in the nucleus of an atom (equal to the number of electrons in the *neutral* atom)

atomic weight the weighted average of masses of natural isotopes of an element

autocatalysis chemical process for which a product acts to increase the reaction rate

Avogadro's number the number of unit particles (approximately 6.02×10^{23}) per mole

base in most common usage, a chemical species acting to combine with a hydrogen ion (H^+) (*see also* Unit 19)

biodegradable capable of being converted to simpler chemical species by processes involving one or more living organisms

Bohr atom atomic model employing quantized orbits for electrons moving around the nucleus of an atom

boiling conversion of a liquid to a gas (vapor) under the conditions that the vapor pressure of the liquid just exceeds the external pressure

Boltzmann distribution statistical distribution of particle speeds or translational kinetic energies in a collection of particles

buffer substance, or solution, that resists pH change on dilution or on addition of acid or base

calorimeter device used in measurement of heat change associated with a chemical or physical process

catalyst species that acts to increase the rate of a chemical reaction

cathode the internal electrode of an electrochemical cell at which reduction occurs

cation positively charged particle

chain reaction process in which consumption of reactants serves to furnish new species capable of participation in further reaction

chalcogen one of the group VIA elements (e.g., oxygen or sulfur)

charge density particle charge divided by particle volume

chemical family elements of similar chemical properties, typically members of the same group in the periodic table

coefficients numbers placed before chemical symbols or formulas in a chemical equation

colligative properties that depend primarily on the relative numbers of particles in a system, rather than on the nature of the particles.

combustion chemical combination with oxygen, usually during burning

complete formula combination of chemical symbols and subscripts indicating actual numbers of atoms combined in a molecule or poly-atomic ion

compound a pure substance that can be decomposed into two or more simpler substances

Compton scattering experimental evidence that x-ray photons have particle characteristics as exhibited by momentum transfer in collisions of x-rays with electrons

conjugate pair two chemical species differing by only a hydrogen ion (e.g., NH_3 and NH_4^+)

constitutive property determined by the nature of the chemical species involved

continuous spectrum spectral region in which no wavelengths are missing

coordinate covalent bond the bond resulting from donation of both bonding electrons by a single atom or ion

covalent chemical bonding in which a pair of electrons is shared by two or more atoms

covalent radius one-half the distance between the nuclei of two covalently bonded atoms

crenation shriveling of biological cells by loss of water through the cell membrane

critical temperature the temperature above which only a single-phase fluid can exist

crystal habit external appearance of a well-formed crystal

crystal structure complete description of the three-dimensional arrangement of atoms (or ions) in a crystal

crystal system one of seven fundamental three-dimensional patterns found in crystals

crystalline solid substance consisting of a repetitive, orderly three-dimensional array of unit particles

degenerate of equal energy

delocalized charge dissipated over more than one atom, or electrons shared among more than two atoms

density ratio of mass to volume

diamagnetic repelled by a magnetic field (a property associated with the absence of unpaired electrons)

diffusion movement of one or more particles through a collection of other particles

dipole unsymmetrical distribution of valence electrons within a molecule

dipole-dipole forces electrostatic attractions between oppositely oriented polar molecules

dipole-induced dipole forces electrostatic attractions between a polar molecule and a temporarily polarized molecule

dipole-ion forces electrostatic attractions between ions and polar molecules

discontinuous spectrum spectral region having one or more missing wavelengths or wavelength regions

disproportionation simultaneous oxidation and reduction of the same chemical species

ductility capability of being drawn out into wire

dynamic equilibrium characterized by opposing processes occurring at the same rate, within a closed system

effective collision collision of reactant species such that reaction occurs

efficiency (thermodynamic) relationship expressing the fraction of available heat that may be converted to work

effusion movement of gaseous particles into a vacuum or near-vacuum

elastic collision collision with conservation of translational kinetic energy

electrolyte substance that, in solution or in the liquid state, can conduct an electric current by movement of ions

electron subatomic particle having a negative charge and a mass about 0.05% of that of a proton

electron affinity energy difference between a neutral gaseous atom and its corresponding −1 anion

electron configuration complete assignment of quantum numbers (or related symbolisms) for all electrons of an atom or monatomic ion

electron diffraction interference phenomena indicating wave characteristics of electrons

electronegativity attraction of an atom in a covalent bond for the bond-pair electrons

element pure substance that cannot be decomposed into simpler substances by chemical or physical methods

elementary process one of the steps in the overall conversion of initial reactants to final products in a chemical reaction

empirical based on experimental observation rather than on theory

empirical formula simplest chemical formula for a molecule or poly-atomic ion, showing only the *ratio* of combined atoms

endothermic process requiring net heat input

endpoint the stage in a titration at which the indicator changes color

energy level indication of the energy of one state in comparison with another

enthalpy heat content (thermodynamic symbol, H)

entropy measure of energy content not available for useful work, related to disorder (thermodynamic symbol, S)

equivalence point stage of reaction at which chemically equivalent amounts of reactants have been combined

equivalent weight the weight (in atomic mass units) of a substance associated with one unit charge (*see also Index listings*)

excited state condition corresponding to higher than "normal" energy

exothermic process releasing heat

flood to add a very large excess of one reactant

fluid substance capable of flow involving independent movement of unit particles

formality concentration unit expressing number of gram-formula weights of solute per liter of solution prepared

formation in thermodynamic terms, combination of free elements to produce a compound or polyatomic ion

formula weight summation of all atomic weights represented by symbols and their subscripts in a chemical formula

free energy measure of the energy theoretically available for maximum useful work (thermodynamic symbol, G at constant pressure or A at constant volume)

free radical highly reactive species having one or more unpaired electrons

functional group within a molecule or polyatomic ion, the particular arrangement of atoms associated with characteristic chemical properties

galvanic cell an electrochemical cell acting to produce current flow by a spontaneous oxidation–reduction reaction

gram-formula weight formula weight expressed in units of grams

ground state lowest energy state

group a vertical row in the periodic table

half-life the time required for consumption of half the original quantity of material

halogen one of the group VIIA elements (e.g., chlorine or bromine)

heat capacity enthalpy change associated with temperature change of a one-phase system (typically expressed as calories per gram, per degree)

heat of crystallization enthalpy change associated with conversion of a specified amount of a liquid to a crystalline solid, at the freezing point

heat of fusion enthalpy change associated with conversion of a specified amount of a solid to liquid, at the melting point

heat of liquefaction enthalpy change associated with conversion of a specified amount of a gas (vapor) to liquid, at the boiling point of the liquid

heat of vaporization enthalpy change associated with conversion of a specified amount of a liquid to gas (vapor), at the boiling point

hemolysis the bursting of biological cells caused by rupture of the cell membrane

heterogeneous composition varies from one region to another within a sample

heterogeneous catalysis process in which the catalyst is in a different phase from that of the reactants

homogeneous uniform composition throughout the substance

homogeneous catalysis process in which the reactants and catalyst are in the same phase

hybrid orbitals orbitals possessing some characteristics of different simple atomic orbitals

hydration association with water molecules

hydrogen bond electrostatic attraction between the hydrogen of a highly polarized bond (typically O—H, N—H, or F—H) and the lone-pair electrons of a small electronegative atom (typically fluorine, oxygen, or nitrogen) in a neighboring molecule or functional group

hydrolysis reaction in which water participates as a reactant

hydrophilic attractive to water ("water-loving")

hydrophobic not attractive to water ("water-fearing")

immiscible not mutually soluble

inelastic collision collision involving net loss of translational kinetic energy

inhibitor chemical substance added to retard a reaction

internal energy total energy of a system (thermodynamic symbol, E)

ion a charged particle

ionic bond electrostatic attraction of ions of opposite charge

ionic radius one-half the *effective* diameter of a monatomic ion, typically measured in a crystal lattice

ionization potential energy required for removal of an electron from a gaseous atom or cation

isoelectronic containing the same number of electrons

isomers compounds having the same molecular formula but different structures

isotopes atoms of the same atomic number, but with different atomic weights (i.e., same number of protons, different numbers of neutrons in their nuclei)

IUPAC International Union of Pure and Applied Chemistry

kernel atomic nucleus and all electrons except those in the last principal energy level

kinetic energy (KE) energy of motion, defined as one-half the mass times the square of the speed

kinetics study of factors affecting rate of chemical reaction

lattice point center of mass of a unit particle in a crystal lattice network

Le Chatelier's principle a system at equilibrium will react to an applied stress in such a way as to counteract the effect of the stress

ligand species furnishing an electron pair for formation of a coordinate covalent bond to a metal ion or atom

limiting reagent reactant whose complete consumption by a chemical reaction determines the maximum amount of product that can be formed

liquid crystal substance or mixture characterized by fluidity approaching that of a liquid, while retaining a degree of order in molecular orientations approaching that of a crystalline solid

London dispersion forces interactions among neighboring particles attributed, in simplified terms, to temporary polarizations of "electron clouds" with respect to their associated nuclei

lone pair a pair of electrons in a valence orbital not involved in a chemical bond

lyophilization removal of solvent from a frozen solution by use of a vacuum ("freeze-drying")

macromolecules very large molecules such as those found in natural proteins or in synthetic fibers

malleability capability of being pressed into thin sheets or foils

mass number sum of number of protons and neutrons in the nucleus

mean free path average distance between successive collisions of a particle moving through a group of other particles

mechanism the steps by which a chemical reaction proceeds from initial reactants to final products

meniscus curved surface at a liquid-gas interface

metabolism chemical reactions within a living system

metal element or alloy characterized by high electrical conductivity that typically decreases with increasing temperature

metallic bond electrostatic force between metal cations in a crystal lattice and the mobile valence electrons that permeate the lattice

metallic radius distance from the nucleus of a metal cation in a metallic crystal to the edge of the nearest neighboring metal cation

metalloid element characterized by increasing electrical conductivity with increasing temperature

metastable having only temporary stability, like a precariously balanced rock

metathetical a chemical reaction in which no electrons are transferred (all oxidation states remain unchanged)

micelle colloidal aggregate in which nonpolar groups are clustered in the center, with ionic or polar groups protruding into the surrounding solvent

millimole one-thousandth (10^{-3}) mole, corresponding to approximately 6.02×10^{20} unit particles

miscible mutually soluble

molality concentration unit defined as the number of moles of solute per kilogram of solvent

molarity concentration unit defined as the number of moles of solute per liter of solution

mole 6.02×10^{23} unit particles of a substance

molecularity number of reactant species involved in an elementary process

molecule group of atoms connected by covalent bonding

momentum product of mass times velocity, a vector quantity

net equation chemical equation showing only those species actually participating in chemical change

neutralization reaction between equivalent amounts of acid and base

neutron electrically neutral subatomic particle having about the same mass as a proton

noble gas one of the group ZERO elements (e.g., helium or neon)

node in orbital terminology, a region of zero electron density

nonmetal element characterized by relatively high ionization potential and by tendency to form monatomic anions rather than cations

nonpolar having symmetrical distribution of electron density

normality concentration unit expressing number of gram-equivalent weights of solute per liter of solution

octahedral an eight-sided figure, having *six* corners

orbit fixed path (trajectory)

orbital quantum number specification of component of energy state of an electron as determined by the type of orbital occupied (commonly employs letter designations s, p, d, or f, as described in Unit 1)

orbitals three-dimensional regions occupied by moving electrons in atoms, ions, or molecules

orientation quantum number specification of component of energy state of an electron as determined by the *relative* spatial orientation of the orbital occupied (commonly employs subscript designation, such as p_x, p_y, p_z, as described in Unit 1)

osmosis process whereby ions or molecules pass through the pores of a semipermeable membrane

oxidation electron loss (increase in oxidation number)

oxidation number (oxidation state) number assigned to an element symbol in the formula of a molecule or ion on the basis of the *net charge* resulting from algebraic summation of the kernel charge and the charges of valence electrons assigned to the atom represented (*see also Index listings*)

oxidizing agent species acting as electron acceptor

paramagnetic attracted by a magnetic field (associated with the presence of unpaired electrons)

partial pressure pressure that would have been exerted by one component of a gaseous mixture if that component occupied the container alone

parts per million (ppm) concentration unit expressing the number of parts (by mass or volume) of solute per million parts of solution, usually as milligrams of solute per kilogram of solution

Pauli exclusion principle no orbital can be occupied by more than two electrons, and then only if the electrons are of opposite spins

percentage yield ratio of actual yield to theoretical yield, expressed as percentage

period a horizontal row in the periodic table

pH a notation used for expressing acidity or alkalinity of a solution (pH $= - \log C_{H^+}$); pH < 7 acidic, pH 7 neutral, pH > 7 alkaline

physical property property, such as melting point or hardness, that can be measured without chemical interaction

polar having unsymmetrical distribution of electron density

polymorphism existence of allotropic forms

potential energy the energy of a system *in comparison with* some other system to which it could be changed

precipitation from solution, formation of an insoluble solid

pressure force per unit area

principal quantum number specification of component of energy state of electron related, in simplified terms, to average distance of electron–nucleus separation

probability plot graphic display of probability of finding electrons at particular points in space

product species formed by a chemical reaction

proton positively charged subatomic particle (symbolized as $^1_1H^+$ or just H^+)

quantum mechanics applications of modern mathematical theory of wave phenomena to electrons and other small particles

quench sudden termination of a reaction (e.g., as by chilling)

rate speed of a chemical process

rate law mathematical expression relating concentration dependence of reaction speed

reactant species consumed by a chemical reaction

reaction order the sum of the exponents in the rate law for a reaction

reducing agent species donating electrons in an electron-exchange process

reduction gain of electrons (decrease in oxidation number)

representative element one of the *A* group elements, characterized in most cases by noninvolvement of *d*-orbital electrons in chemical bonding

resonance hybrid set of electron-dot or line-bond formulas showing alternative locations of pi bonds

reversible process for which it is experimentally feasible to reconvert the final state (or products) to the original state (or reactants)

rotation spinning motion

salt ionic compound of cation other than H^+ with anion

saturated solution solution of maximum stable concentration

semipermeable having a pore size that restricts the size of particles capable of passing through the barrier (e.g., a membrane in osmosis)

solubility indication of the quantity of solute that can be dissolved by a specified amount of solvent

solute substance that is dissolved

solution homogeneous mixture

solvent substance acting to dissolve another

specific heat ratio of the heat capacity of a substance to that of water

spectator ion ion present, but not participating in a chemical change

spin quantum number specification of component of energy state of electron associated with the direction of electron spin relative to that of another electron

spontaneous occurring naturally without aid from outside the system

standard state specified criteria and conditions for reference purposes

state function property, such as enthalpy, determined only by initial and final states, not by specific path

steady state characterized by equal input and output rates

stoichiometry application of mathematics to chemical relationships

strong (acid or base) essentially completely ionized in aqueous solution

sublevel subdivision of principal energy level, corresponding in common usage to energy of a particular orbital

substrate species acted upon by an enzyme

supercooling lowering the temperature of a gas to below the condensation temperature without liquefaction, or of a liquid to below the freezing point without causing crystallization (solidification)

superheating raising the temperature of a liquid to above the boiling point without causing boiling

supersaturated containing more solute per unit volume of solution than would the stable saturated solution

surface tension cohesive character of a liquid surface acting to resist penetration or rupture of the surface.

surroundings in thermodynamic usage, the environment externally encompassing a thermodynamic system

temperature a measurement of the average translational kinetic energy of the particles in a system

theoretical yield the maximum amount of product that could be obtained by complete conversion from the limiting reagent in a chemical process

titration analytical procedure in which the required volume of a reactant solution is measured to determine the quantity of reactant consumed

transition element one of the B group elements (e.g., copper or silver), usually characterized by *d*-orbital involvement in bonding

transition state high-energy intermediate between reactants and products

translational kinetic energy the component of total kinetic energy associated with translational motion

translational motion movement of a particle from one location to another

transuranium element one of the elements having an atomic number larger than 92 (e.g., plutonium or berkelium)

triple point the condition of equilibrium among solid, liquid, and gaseous phases of a substance

unit cell fundamental segment of a crystal lattice that describes the repetitive three-dimensional pattern of the lattice

universe in thermodynamic usage, the combination of a thermodynamic system plus its surroundings

valence electrons those, in the highest principal energy level of the atom, which may participate in chemical bonding

van der Waals forces general term for attractions between electrically neutral particles

van der Waals radius distance from the center of an atom to the effective edge of the nearest *nonbonded* adjacent atom

vapor pressure the pressure exerted by a vapor in equilibrium with its liquid phase

vibration motion involving the stretching or bending of covalent bonds

viscosity resistance to flow of a liquid or gas

weak (acid or base) only partially ionized in aqueous solution

work the product of the force exerted and the distance through which the force acts (expressed in joules, J)

yield the amount of product formed by a chemical reaction

Zeeman effect the splitting of spectral lines by a magnetic field

Inorganic Nomenclature

1. NAMING CATIONS

a. Cations of metals forming only one stable ion

Elements in groups IA and IIA form only one type of stable ion. The alkali metals (IA) form +1 ions, and the alkaline earth metals (IIA) form +2 ions. In addition, certain other elements commonly form only one type of stable ion. These include

aluminum (Al^{3+}), scandium (Sc^{3+}), zinc (Zn^{2+}), cadmium (Cd^{2+}), and silver (Ag^+)

The cations of these elements are given the same name as the elements themselves. For example,

Ag^+ is silver (ion)
Na^+ is sodium (ion)
Al^{3+} is aluminum (ion)

Compounds of these elements are named, except for complexes (Sections 1d and 2d), by first giving the name of the cation. For example,

NaCl is *sodium* chloride
$Al(NO_3)_3$ is *aluminum* nitrate

b. Cations of metals forming more than one stable ion

Transition metals (the B groups in the periodic table) often form more than one common stable ion. In the modern nomenclature system, these cations are given the ordinary English name of the element followed, in parentheses, by a Roman numeral indicating the oxidation state (Unit 3). For example,

Au^+ is gold(I) (ion)
Cu^{2+} is copper(II) (ion)
Fe^{3+} is iron(III) (ion)

One special case is the diatomic mercury(I) ion, Hg_2^{2+}.

Compounds of these elements, except for complexes (Sections 1d and 2d), are named like those of the simpler cations. For example,

CuBr is *copper(I)* bromide
$CuBr_2$ is *copper(II)* bromide
FeS is *iron(II)* sulfide
Fe_2S_3 is *iron(III)* sulfide
$MnCl_2$ is *manganese(II)* chloride

In the older naming system, a method was used to distinguish between the two most common oxidation states. The higher oxidation state was indicated by the suffix *ic* and the lower by the suffix *ous*. A segment of the Latin name of some elements (Table 1) was used as the first part of the name. For other elements, a segment of the ordinary English name was used. For example,

CuBr was *cuprous* bromide
$CuBr_2$ was *cupric* bromide
FeS was *ferrous* sulfide
Fe_2S_3 was *ferric* sulfide

but

$HgCl_2$ was *mercuric* chloride

Table 1 LATIN NAMES FOR SOME COMMON ELEMENTS

ENGLISH NAME	LATIN NAME	LATIN SEGMENT USED IN NAMING COMPOUNDS
iron	ferrum	*ferr*
copper	cuprum	*cupr*
tin	stannum	*stann*
gold	aurum	*aur*
lead	plumbum	*plumb*
silver[a]	argentum	*argent*
mercury	hydrargyrum	—[b]
antimony	stibnum	—[b]

[a] "Silver" is the name used except in negative complexes.
[b] English name segments are used (*mercur* and *antimon*).

c. Covalent cations containing nitrogen

Several cations are commonly encountered in salts of amines. Some of the more common are as follows:

Amine	Cation
NH_3 (ammonia)	NH_4^+ (ammonium)
CH_3NH_2 (methylamine)	$CH_3NH_3^+$ (methylammonium)
$C_2H_5NH_2$ (ethylamine)	$C_2H_5NH_3^+$ (ethylammonium)
$(CH_3)_2NH$ (dimethylamine)	$(CH_3)_2NH_2^+$ (dimethylammonium)
$(CH_3)_3N$ (trimethylamine)	$(CH_3)_3NH^+$ (trimethylammonium)
$C_6H_5NH_2$ (aniline)	$C_6H_5NH_3^+$ (anilinium)

d. Complex cations

The name of a complex cation ends with the common English name of the metal followed by a Roman numeral (in parentheses) indicating the oxidation state of the metal. For example,

$[HgCl]^+$ is chloro*mercury*(II) (ion)
$[Cu(NH_3)_4]^{2+}$ is tetraammine*copper*(II) (ion)

The names of *ligands* (ions or molecules bonded to the metal ion) are combined to form a prefix to the name of the metal. Negative ligands are named before neutral ligands and, within these groups, less complex species are listed before more complex species. If more than one of the same ligand is present, the number is indicated by di (2), tri (3), tetra (4), penta (5), or hexa (6). Some common ligands are listed in Table 2. For example,

Table 2 SOME COMMON LIGANDS

SPECIES	COMMON NAME	NAME AS A LIGAND
F^-	fluoride ion	fluoro
Cl^-	chloride ion	chloro
Br^-	bromide ion	bromo
I^-	iodide ion	iodo
S^{2-}	sulfide ion	sulfo
OH^-	hydroxide ion	hydroxo
CN^-	cyanide ion	cyano
NCS^-	thiocyanate ion	thiocyanato[a]
NO_2^-	nitrite ion	nitro
$S_2O_3^{2-}$	thiosulfate ion	thiosulfato
CO	carbon monoxide	carbonyl
NO	nitrogen(II) oxide (nitric oxide)	nitrosyl
H_2O	water	aquo
NH_3	ammonia	ammine

[a] When bonding is known, sulfur-metal bonding is indicated as S-thiocyanato and nitrogen-metal bonding is indicated as N-thiocyanato.

$[PbCl(H_2O)_3]^+$ is chlorotriaquolead(II) (ion)
$[Fe(NCS)(H_2O)_5]^{2+}$ is thiocyanatopentaaquoiron(III) (ion)

2. NAMING ANIONS

a. Monatomic anions

Monatomic anions are named by adding the suffix *ide* to a segment of the common English name of the element. For example,

F^- is fluor*ide*
I^- is iod*ide*
O^{2-} is ox*ide*
S^{2-} is sulf*ide*
N^{3-} is nitr*ide*

Compounds of these ions, except for complexes (Sections 1d and 2d), are named by listing the cation name first and, as a separate word, the anion name last. For example,

KCl is potassium *chloride*
Na_2O is sodium *oxide*
Fe_2S_3 is iron(III) *sulfide*
Li_3N is lithium *nitride*

b. Polyatomic anions (inorganic)

1. Some general rules are indicated for certain oxyanions in Table 3.
2. Two oxyanions, NCO^- (cyanate) and SO_4^{2-} (sulfate), have sulfur analogs that are often encountered. These may be considered as having an oxygen replaced by sulfur, and they are given the names of the corresponding oxyanion with the prefix *thio*:

NCS^- is *thio*cyanate
$S_2O_3^{2-}$ is *thio*sulfate

Table 3 NAMING OXYANIONS OF GROUPS VA–VIIA

ELEMENT GROUP	OXIDA-TION STATE	NAMING	EXAMPLE
VA	+1	hypo____ite	$H_2PO_2^-$ *hypophosphite*
	+3	____ite	NO_2^- *nitrite*
	+5	____ate	NO_3^- *nitrate*
VIA	+4	____ite	SO_3^{2-} *sulfite*
	+6	____ate	SO_4^{2-} *sulfate*
VIIA	+1	hypo____ite	ClO^- *hypochlorite*
	+3	____ite	ClO_2^- *chlorite*
	+5	____ate	BrO_3^- *bromate*
	+7	per____ate	IO_4^- *periodate*

3. Current IUPAC practice is to name *hydrogen* as a separate word when hydrogen is part of a polyatomic anion. This practice has not yet been fully adopted, and we have used the older, more common, system throughout this text. For example,

Anion	Older name	IUPAC-preferred name
SH^-	hydrosulfide	hydrogen sulfide (ion)
HCO_3^-	bicarbonate	hydrogen carbonate
HSO_4^-	bisulfate	hydrogen sulfate
HSO_3^-	bisulfite	hydrogen sulfite

4. Some common anions must simply be memorized. These include the following:

OH^- is hydroxide
SH^- is hydrosulfide (IUPAC-preferred: hydrogen sulfide)
MnO_4^{2-} is manganate
MnO_4^- is permanganate

CrO_4^{2-} is chromate
$Cr_2O_7^{2-}$ is dichromate

Others will be encountered from time to time.

c. Polyatomic anions (organic)

There are a very large number of anions that may be formed from the organic acids. Two naming systems are used, *with the trivial names most commonly employed for the simpler acids*. Table 4 lists a typical group of anions that are frequently encountered. Note that the anion name uses a segment of the acid's name, with the suffix *ate*. For complete details of naming organic acids, see Appendix G.

d. Complex anions

Complex anions are named in much the same way as complex cations (Section 1d), with two important differences. The base name uses a segment of the metal name with the suffix *ate*, followed by a Roman numeral (in parentheses) showing the oxidation state of the metal. *Latin names are used for some metals, as shown in Table 1*. The prefix combination of ligand names is exactly like that used for the complex cations (Section 1d). For example,

$[Ag(CN)_2]^-$ is dicyanoargentate(I) (ion)
$[FeF_4(H_2O)_2]^-$ is tetrafluorodiaquoferrate(III) (ion)
$[Cu(CN)_6]^{4-}$ is hexacyanocuprate(II) (ion)
$[HgCl_4]^{2-}$ is tetrachloromercurate(II) (ion)
$[PtF_6]^{2-}$ is hexafluoroplatinate(IV) (ion)

3. NAMING SALTS AND MOLECULAR COMPOUNDS

a. Salts

Salts are named by listing cations first and anions last. Each ion is named as a separate word, with no indication

Table 4 SOME ORGANIC ANIONS

PARENT ACID	IUPAC[a] NAME	TRIVIAL NAME	ANION	IUPAC[a] NAME	TRIVIAL NAME
HCO_2H	methanoic acid	formic acid	HCO_2^-	methanoate	formate
CH_3CO_2H	ethanoic acid	acetic acid	$CH_3CO_2^-$	ethanoate	acetate
$CH_3CH_2CO_2H$	propanoic acid	propionic acid	$CH_3CH_2CO_2^-$	propanoate	propionate
$HO_2C—CO_2H$	ethanedioic acid	oxalic acid	$^-O_2C—CO_2^-$ $C_2O_4^{2-}$	ethanedioate	oxalate

benzoic acid
($C_6H_5CO_2H$)

benzoate
($C_6H_5CO_2^-$)

phthalic acid

phthalate

salicylic acid

salicylate

[a] Name approved by International Union of Pure and Applied Chemistry.

of the relative numbers of ions (except in some special cases to avoid ambiguity). For example,

$BaCl_2$ is barium chloride

K_2HPO_4 is potassium hydrogen phosphate

NaH_2PO_4 is sodium dihydrogen phosphate

$Mg(NH_4)PO_4$ is magnesium ammonium phosphate

Na_3PO_4 is sodium phosphate (sometimes called trisodium phosphate)

$[Cu(NH_3)_4]SO_4$ is tetraamminecopper(II) sulfate

$K_4[Fe(CN)_6]$ is potassium hexacyanoferrate(II) (older name: potassium ferrocyanide)

$K_3[Fe(CN)_6]$ is potassium hexacyanoferrate(III) (older name: potassium ferricyanide)

$[Co(NH_3)_6][AuCl_4]_2$ is hexaamminecobalt(II) tetrachloroaurate(III)

b. Molecular compounds

Common molecular compounds (neutral, acidic, and basic) are listed in Table 3.2, Unit 3. In addition, there are molecular complexes (called coordination compounds). These are named exactly like the complex cations, but the metal has an oxidation state of zero. For example,

$[Fe(CO)_5]$ is pentacarbonyliron(0)

The naming rules given in this appendix will be sufficient for most simple species. For more details of modern inorganic nomenclature, see *J. Amer. Chem. Soc.* 82, 5523 (1960).

Appendix G

Organic Nomenclature

1. ALKANES

IUPAC[1] nomenclature for alkanes (condensed rules[2])

1. The base name is that of the alkane (Table 1) having the same carbon number as that of the longest continuous carbon chain in the compound being named.

2. Each carbon in the basic chain is numbered sequentially, starting at the end that will result in the "smallest numbers" for those carbon atoms attached to *branch groups*. To determine the "smallest" location numbers, number the longest chain from both ends and select two sets of location numbers for branch groups. The set to be used is that having the smallest number at the *first* difference encountered. For example,

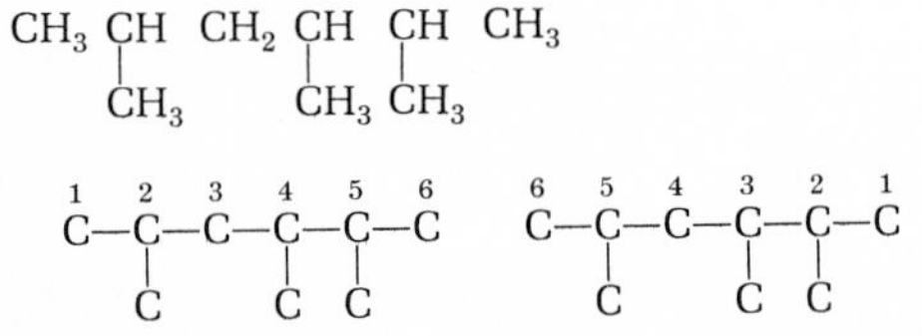

$$CH_3 \; CH \; CH_2 \; CH \; CH \; CH_3$$
$$\quad\;\; CH_3 \qquad CH_3 \; CH_3$$

$$
\begin{array}{cc}
\overset{1}{C}-\overset{2}{C}-\overset{3}{C}-\overset{4}{C}-\overset{5}{C}-\overset{6}{C} & \overset{6}{C}-\overset{5}{C}-\overset{4}{C}-\overset{3}{C}-\overset{2}{C}-\overset{1}{C} \\
\;\;\;| \quad\; |\;\; | & \quad\;\; | \quad\; |\;\; | \\
\;\;\;C \quad\; C\;\; C & \quad\;\; C \quad\; C\;\; C \\
2,4,5- & \quad 2,3,5-
\end{array}
$$

$$3 < 4, \text{ so:}$$

2,3,5-trimethylhexane

3. Hydrocarbon branch groups are named as *alkyl* groups, using the name of the corresponding *alkane* with the *ane* ending changed to *yl* (e.g., $-CH_3$ is methyl, $-CH_2CH_3$ is ethyl).

4. *Each* alkyl group is given the number of the carbon atom to which it is attached in the basic chain.

5. If there is more than one branch group of the same type, this is indicated by use of a prefix on the alkyl group name.

Table 1   THE FIRST FIFTEEN ALKANES

NUMBER OF CARBON ATOMS	MOLECULAR FORMULA	NAME
1	CH_4	methane
2	C_2H_6	ethane
3	C_3H_8	propane
4	C_4H_{10}	butane
5	C_5H_{12}	pentane
6	C_6H_{14}	hexane
7	C_7H_{16}	heptane
8	C_8H_{18}	octane
9	C_9H_{20}	nonane
10	$C_{10}H_{22}$	decane
11	$C_{11}H_{24}$	hendecane[a]
12	$C_{12}H_{26}$	dodecane
13	$C_{13}H_{28}$	tridecane
14	$C_{14}H_{30}$	tetradecane
15	$C_{15}H_{32}$	pentadecane

[a] Older name: undecane.

[1] International Union of Pure and Applied Chemistry.
[2] The rules given here are sufficient for most cases encountered. For a more detailed treatment of the IUPAC nomenclature system, see the Chemical Rubber Company's *Handbook of Chemistry and Physics*.

(di = 2, tri = 3, tetra = 4, penta = 5, hexa = 6, hepta = 7, octa = 8, nona = 9, deca = 10.) *The location numbers of all identical groups are retained.*

6. The combined branch group names (either alphabetically or in order of increasing complexity) and location numbers serve as a prefix to the *base name.* Separate numbers from numbers by commas, numbers from words by hyphens, and do not separate words from other words.

7. If branch groups are in *equivalent* positions,[3] the one written first in the name (whether alphabetically or in order of complexity) is given the smaller number. For example,

$$CH_3CH_2CHCH_2CHCH_2CH_3$$
$$\quad\quad\ \ |\quad\quad\ \ |$$
$$\quad\quad CH_3\quad CH_2CH_3$$

alphabetically	*order of complexity*
3-ethyl-5-methylheptane	3-methyl-5-ethylheptane

either name is correct, but

5-ethyl-3-methylheptane and
5-methyl-3-ethylheptane are incorrect

Example 1

Give the IUPAC name for

$$CH_3CH_2CHCH_2CHCH_2C(CH_3)_3$$
$$\quad\quad\ \ |\quad\quad\ \ |$$
$$\quad\quad CH_3\quad CH_2CH_3$$

Solution

First, redraw the structural formula in expanded form, keeping only branch groups "compressed" (until some experience has been gained, it may be useful to mark the basic chain):

$$\begin{array}{c} H\ \ H\ \ H\ \ H\ \ H\ \ H\ \ \ CH_3\ H \\ |\ \ \ |\ \ \ |\ \ \ |\ \ \ |\ \ \ |\ \ \ \ \ |\ \ \ | \\ H-C-C-C-C-C-C-C——C—H \\ |\ \ \ |\ \ \ |\ \ \ \ \ \ \ |\ \ \ CH_3\ H \\ H\ \ H\ \ \ \ \ \ CH_3\ \ \ CH_2CH_3 \end{array}$$

Note that two —CH_3 groups are part of the basic chain at each end.

Now apply naming rules, in order:

1. There are eight carbons in the longest continuous chain, so the base name (Table 1) is *octane.*
2. We have two choices for numbering. If we start at the left end, as drawn, there are groups attached to carbons 3, 5, 7, 7. From the right, groups are numbered 2, 2, 4, 6. As the latter yields the smallest numbers ($2 < 3$), this is the sequence selected:

$$\begin{array}{c} \quad\quad\quad\quad\quad\quad CH_3 \\ \quad\quad\quad\quad\quad\quad\ |2 \\ \overset{8}{C}-\overset{7}{C}-\overset{6}{C}-\overset{5}{C}-\overset{4}{C}-\overset{3}{C}——\overset{2}{C}-\overset{1}{C} \\ \quad\quad |\quad\quad\ |\quad\quad\ | \\ \quad\quad CH_3\quad CH_2CH_3\ CH_3 \end{array}$$

3. There are three methyl groups and one ethyl group attached to the basic chain.

[3] Note that rule 7 applies *only* to groups in *equivalent* positions. Otherwise, rules 2 and 6 are used.

4. The methyl group location numbers are 2, 2, 6, and that for the ethyl group is 4.
5. The three methyl groups are indicated as *2,2,6-trimethyl.*
6. The combined name, then, is either

    4-ethyl-2,2,6-trimethyloctane (alphabetically)

or

    2,2,6-trimethyl-4-ethyloctane (in order of complexity)

2. CYCLOALKANES

1. The base name is that of the alkane (Table 1) having the same number of carbons as the *ring* in the cycloalkane, with the added prefix *cyclo.* For example,

$$\begin{array}{c} \quad\ CH_2 \\ CH_2—CH_2 \end{array} \equiv \triangle \quad \text{is cyclopropane}$$

$$\begin{array}{c} CH_2—CH_2 \\ |\quad\quad\ | \\ CH_2—CH_2 \end{array} \equiv \square \quad \text{is cyclobutane}$$

$$\begin{array}{c} \quad CH_2 \\ CH_2\quad CH_2 \\ |\quad\quad\ \ | \\ CH_2\quad CH_2 \\ \quad CH_2 \end{array} \equiv \bigcirc \quad \text{is cyclohexane}$$

2. In monosubstituted cycloalkanes, no group location number is required. For example,

$$\square_{CH_3}\quad \text{is methylcyclobutane}$$

$$\bigcirc^{CH_2CH_3}\quad \text{is ethylcyclohexane}$$

3. In polysubstituted cycloalkanes, the first group written in the name is given the location number 1, and other location numbers are selected by numbering the ring carbons sequentially to give the smallest location numbers. For example,

$$\begin{array}{c} CH_2CH_3 \\ \bigcirc—CH_3 \\ CH_2CH_3 \end{array}$$

1-methyl-2,5-diethylcyclohexane (in order of complexity)

or

1,4-diethyl-2-methylcyclohexane (alphabetically)

3. ALKENES

1. The base name is formed from that of the alkane (Table 1) having the number of carbons found in the longest carbon chain in the alkene *that contains the double bond.* The name ending for alkenes is *ene* (rather than *ane,* as for alkanes).
2. The carbon atoms of the basic chain are numbered se-

quentially, *beginning at the end nearest*[4] *the double bond*. The location number of the first atom of the double bond is inserted just before the base name. (For a cycloalkene, the double bond is assumed to be between carbons 1 and 2, so no double-bond location number is needed for the name.)

3. If the compound can exist as a pair of geometric isomers, indicate the specific isomer (if known) by -*cis*- or -*trans*- (Unit 28) just preceding the location number of the double bond. For example,

CH_3 structures

3-methyl-*cis*-2-pentene

3-methylcyclopentene

4. AROMATIC HYDROCARBONS

1. The common names shown in Table 2 are used as base names.

Table 2   SOME COMMON AROMATIC HYDROCARBONS

benzene   toluene   o-xylene   m-xylene   p-xylene

cumene   mesitylene   styrene

naphthalene

anthracene

phenanthrene

2. In substituted benzenes, ring positions may be indicated either by numbers (as in cyclohexane) or by the lowercase letters *o*- (read "ortho"), *m*- (read "meta"), or *p*- (read "para"). (The letter designation is used only for disubstituted benzenes. Ortho groups correspond to 1,2 positions; meta to 1,3; and para to 1,4.) If one group on the ring (e.g., the —CH_3 group in toluene) determines the basic name, its point of attachment is understood to be number 1. (In disubstituted cases, the *o*-, *m*-, or *p*- designation refers to positions relative to such a name-determining group.) For example,

m-diethylbenzene   1,2,4-triethylbenzene   o-ethyltoluene
or   or
1,3-diethylbenzene   2-ethyltoluene

3. Position numbers in some other aromatic ring systems are indicated in Table 2.

5. MONOFUNCTIONAL COMPOUNDS

1. Certain aromatic compounds use common names (Table 3).
2. Some functional groups do not affect the name endings used in the IUPAC nomenclature system. These groups are treated in the same way as alkyl groups, as illustrated in Table 4.
3. *In other cases, such as the alkenes, the longest chain is always numbered beginning at the end nearest the functional group that determines the name ending.* The procedures used are illustrated in Table 5.

6. POLYFUNCTIONAL COMPOUNDS

1. Certain aromatic compounds use common names (Table 6).
2. In polyfunctional compounds having only one group that affects the name ending, nomenclature is based on rules illustrated in Table 5. For example,

$$\underset{\overset{|}{Cl}}{CH_3CHCH_2}\overset{\overset{O}{\|}}{C}CH_2CH_3$$

5-chloro-3-hexanone

3. In polyfunctional compounds having a carbon–carbon double bond and one other functional group that affects name ending, both groups are indicated in the name ending, using numbering from the end nearest the C=C bonding (except with aldehydes, acids, or acid derivatives). For example,

$$\underset{\overset{|}{CH_3}}{CH_3}\overset{\overset{O}{\|}}{C}CHCH_2CH_2CH=CH_2$$

5-methyl-1-heptene-6-one

CH_3
    C=CHCH_2CH_2CO_2H
CH_3

5-methyl-4-hexenoic acid

---

[4] If the double bond is in the center of the chain, number from the end giving the "smallest" group location numbers (see rule 2, under Alkanes).

Table 3 SOME COMMON AROMATIC COMPOUNDS (MONOFUNCTIONAL)

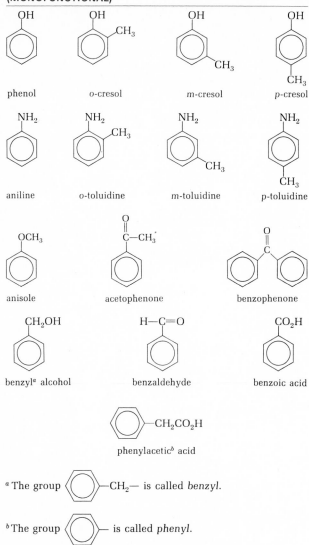

phenol   o-cresol   m-cresol   p-cresol

aniline   o-toluidine   m-toluidine   p-toluidine

anisole   acetophenone   benzophenone

benzyl[a] alcohol   benzaldehyde   benzoic acid

phenylacetic[b] acid

[a] The group C₆H₅—CH₂— is called *benzyl*.

[b] The group C₆H₅— is called *phenyl*.

---

4. In other cases, the functional group determining the name ending is selected in the order of preference:

a. —CO_2H (or derivatives)

b. 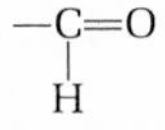 —C=O
    |
    H

c. —C—C=O
    |
   —C— 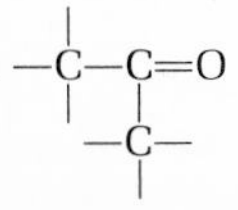
    |

d. —OH

Table 4 COMMON FUNCTIONAL GROUPS NOT AFFECTING NAME ENDINGS

CLASS	FUNCTIONAL GROUP	EXAMPLE
halide	—F (fluoro)	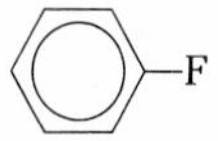 fluorobenzene
	—Cl (chloro)	$CHCl_3$ trichloromethane (chloroform)
	—Br (bromo)	$CH_3CH_2CH_2Br$ 1-bromopropane
	—I (iodo)	$(CH_3)_2CHCH_2CH_2CHCH_2CH_3$ with I on carbon 5-iodo-2-methylheptane
ether	—O—C— (alkoxy)	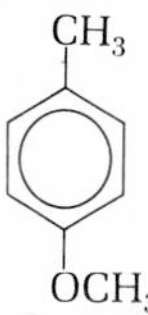 p-methoxytoluene
nitro compound	—NO_2 (nitro)	CH_3NO_2 nitromethane; 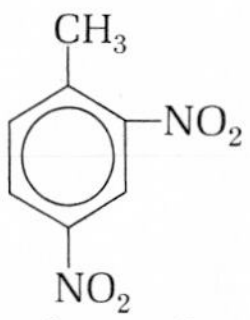 2,4-dinitrotoluene
amine[a]	—N— (amino)	$CH_3\ddot{N}H_2$ aminomethane (methylamine)
primary	—NH_2 (amino)	CH_3—C—CH_3 with :NH_2; 2-aminopropane (isopropylamine)
secondary	—N—C— (N-alkylamino) with H	$CH_3CH_2CH_2$C—CH_3 with :N—CH_3; 2-(N-methylamino) pentane[b]
tertiary	—N—C— (N,N-dialkylamino) with —C—	CH_3CH_2C—$CH_2CH_2CH_3$ with :$N(CH_3)_2$; 3-(N,N-dimethylamino) hexane[b]

[a] Remember that certain aromatic amines have specific trivial names (e.g., aniline, or p-toluidine).
[b] Note use of parentheses.

Table 5 COMMON FUNCTIONAL GROUPS DETERMINING NAME ENDING[a]

CLASS	FUNCTIONAL GROUP	NAME ENDING	EXAMPLE
alkene	$\diagdown$C=C$\diagup$	. . . ene	CH_3—$\overset{\overset{H}{\mid}}{\underset{\underset{CH_3}{\mid}}{C}}$—$CH_2CH_2C\overset{\overset{H}{\mid}}{=}$$CH_2$ 5-methyl-1-hexene
alkyne	—C≡C—	. . . yne	CH_3—$\overset{\overset{H}{\mid}}{\underset{\underset{CH_3}{\mid}}{C}}$—$CH_2CH_2$C≡CH 5-methyl-1-hexyne
alcohol	—OH (hydroxyl group)	. . . ol	CH_3—$\overset{\overset{H}{\mid}}{\underset{\underset{CH_3}{\mid}}{C}}$—$CH_2CH_2$$\overset{\overset{H}{\mid}}{\underset{\underset{OH}{\mid}}{C}}$—$CH_3$ 5-methyl-2-hexanol

aldehyde	—C=O
	  $\mid$
	  H
	(—C=O is called
	a carbonyl group)
		. . . al	CH_3—$\overset{\overset{H}{\mid}}{\underset{\underset{CH_3}{\mid}}{C}}$—$CH_2CH_2CH_2C\overset{\overset{H}{\mid}}{=}$O
			5-methylhexanal[b]

ketone	$\overset{\mid}{\underset{\underset{\mid}{\overset{\mid}{C}}}{C}}$—C=O (carbonyl group)	. . . one	CH_3—$\overset{\overset{H}{\mid}}{\underset{\underset{CH_3}{\mid}}{C}}$—$CH_2\overset{\overset{O}{\parallel}}{C}CH_2CH_3$ 5-methyl-3-hexanone
carboxylic acid	$\overset{\overset{O}{\parallel}}{C}$—OH (carboxyl group)	. . . oic acid	CH_3—$\overset{\underset{\underset{CH_3}{\mid}}{C}}{}CH_2CH_2CH_2CO_2H$ 5-methylhexanoic[b] acid
salt of carboxylic acid	$\overset{\overset{O}{\parallel}}{C}$—O$^-$ (carboxylate group)	metal (space) . . . oate	CH_3—$\overset{\underset{\underset{CH_3}{\mid}}{C}}{}CH_2CH_2CH_2CO_2^-$ Na^+ sodium 5-methylhexanoate[b]
ester	$\overset{\overset{O}{\parallel}}{C}$—O—$\overset{\mid}{\underset{\mid}{C}}$	alkyl (space) . . . oate	CH_3—$\overset{\underset{\underset{CH_3}{\mid}}{C}}{}CH_2CH_2CH_2\overset{\overset{O}{\parallel}}{C}OC_2H_5$ ethyl 5-methylhexanoate[b]
amide	$\overset{\overset{O}{\parallel}}{C}$—N— $\mid$ (Note: N- sub- stituted amides are indicated as in the cases of N- sub- stituted amines)	. . . amide	CH_3—$\overset{\underset{\underset{CH_3}{\mid}}{C}}{}CH_2CH_2CH_2\overset{\overset{O}{\parallel}}{C}NH_2$ 5-methylhexanamide[b] $CH_3\overset{\underset{\underset{CH_3}{\mid}}{C}}{}CH_2CH_2CH_2\overset{\overset{O}{\parallel}}{C}N(CH_3)_2$ N,N-dimethyl-5-methylhexanamide[b]

Table 5 (Continued)

CLASS	FUNCTIONAL GROUP	NAME ENDING	EXAMPLE
acid chloride	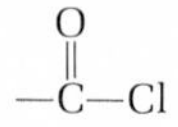 O∥ —C—Cl	. . . *oyl* (*space*) *chloride*	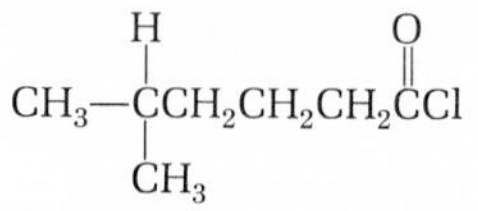 CH_3—CCH$_2$CH$_2$CH$_2$CCl (with H and CH$_3$) 5-methylhexanoyl[b] chloride
acid anhydride	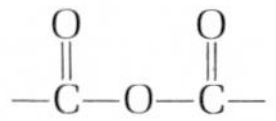 O O ∥ ∥ —C—O—C—	. . . *oic* (*space*) *anhydride*	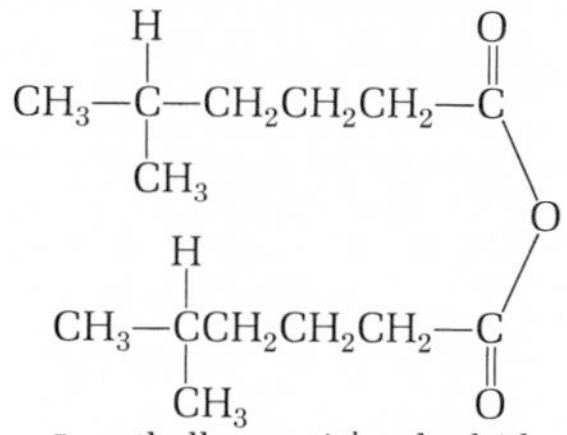 5-methylhexanoic[b] anhydride

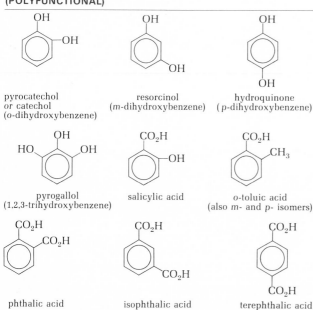

[a] For further details of nomenclature, see the Chemical Rubber Company's *Handbook of Chemistry and Physics.*

[b] Note that for functional groups in which the group carbon is, of necessity, the end carbon in the basic chain, no group location number is used. This carbon is always the number 1 carbon, so no ambiguity is involved.

Table 6 SOME COMMON AROMATIC COMPOUNDS (POLYFUNCTIONAL)

pyrocatechol or catechol (o-dihydroxybenzene)

resorcinol (*m*-dihydroxybenzene)

hydroquinone (*p*-dihydroxybenzene)

pyrogallol (1,2,3-trihydroxybenzene)

salicylic acid

o-toluic acid (also *m*- and *p*- isomers)

phthalic acid

isophthalic acid

terephthalic acid

The functional groups b–d above are then named and numbered as with groups not affecting the name ending:

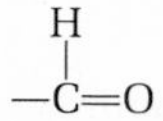

H | —C=O

is called *oxo* when part of the basic chain or *formyl* when it serves as a branch. For example,

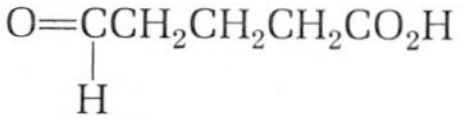

O=CCH$_2$CH$_2$CH$_2$CO$_2$H (with H)
5-oxopentanoic acid

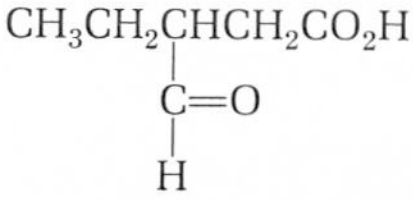

CH$_3$CH$_2$CHCH$_2$CO$_2$H (with C=O, H)
3-formylpentanoic acid

The ketone carbonyl is indicated by *oxo*. For example,

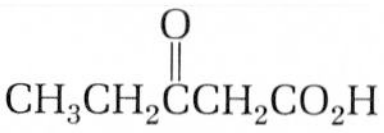

O ∥
CH$_3$CH$_2$CCH$_2$CO$_2$H
3-oxopentanoic acid

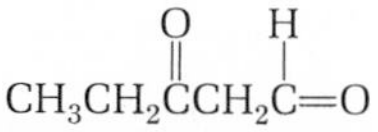

O H ∥
CH$_3$CH$_2$CCH$_2$C=O
3-oxopentanal

The alcohol or phenol —OH group is named as *hydroxy*. For example,

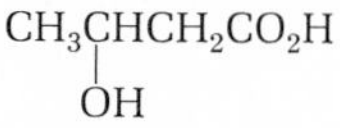

CH$_3$CHCH$_2$CO$_2$H | OH
3-hydroxybutanoic acid

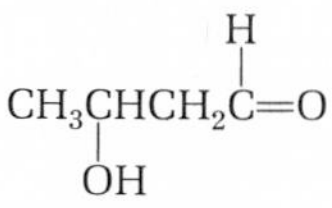

H
CH$_3$CHCH$_2$C=O | OH
3-hydroxybutanal

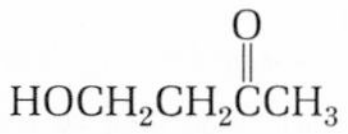

O ∥
HOCH$_2$CH$_2$CCH$_3$
4-hydroxy-2-butanone

For additional nomenclature rules, see the Chemical Rubber Company's *Handbook of Chemistry and Physics.*

Appendix H

The Chemical Literature

No single textbook can possibly introduce all of the basic ideas of chemistry or pursue any one idea in really exhaustive detail. A textbook, instead, is like the *Reader's Digest* in that it presents some rather brief discussions of many different ideas. If any one of these particularly appeals to the reader, he or she will want to pursue the subject in greater depth.

The *primary literature* is the collection of articles written by persons directly involved in the work being described. This represents the best source of information as to how the experimental or theoretical scientist views his or her own work and presents original ideas. Publications such as the *Journal of the American Chemical Society,* the *Journal of Organic Chemistry,* the *Journal of the Chemical Society* (England), or *Analytical Chemistry* are examples of sources of original articles. Professional chemists will scan as many of these journals in their own specialty area as time permits on a regular basis. So many papers are published annually now that no chemist can hope to read more than a fraction of the articles available. Even then, waiting lists for publication are so long that the articles may appear from six months to two years after the reported research was completed. The chemist must attend meetings and seminars and maintain an active correspondence with other scientists in the field in an effort to keep abreast of current developments.

Of greater general interest to students are the publications that summarize or review areas of special interest. At the introductory level, four journals are of particular importance. *Chem-istry* and *Scientific American* contain articles of interest to those having a limited background in chemistry. *Science,* like *Scientific American,* contains articles in many other fields, but the former journal is at a more advanced level. The *Journal of Chemical Education,* although designed primarily for teachers, has many excellent review articles of value to students. At the upper division and graduate level, students will wish to become familiar with such publications as *Chemical Reviews* and *Accounts of Chemical Research.*

Abstract journals present brief synopses of the primary literature so that detailed literature searches can be made and original articles of interest identified for further study. The most important abstract journals are *Beilstein's Handbuch*[1] *der Organischen Chemie* (German) and the English language *Chemical Abstracts.* The former is rather tricky to use until one has gained some practice and, of course, is useful only to those able to read German. *Chemical Abstracts* may be used to find information on particular chemical compounds or on specific subject areas. Excellent index issues are available by author, subject, chemical formula, and patent number. Abstract journals are always several months behind the primary literature.

When extensive information is not essential, *handbooks*

---

[1] *Beilstein's* articles are more in the nature of collections of data and references, so these volumes are really a bridge between the abbreviated handbooks and the more extensive abstract journals.

provide the most convenient source for brief pieces of information. The single best handbook for chemical data, as well as an extensive collection of other useful items, is the *Handbook of Chemistry and Physics* (Chemical Rubber Company, Cleveland, Ohio). Every student planning on advanced study in chemistry should own one of these books and become familiar with its use. In addition, the *Merck Index* (Merck and Co., Inc., Rahway, New Jersey) has much of value concerning the more than 10,000 chemicals of interest to those in the medical sciences and related areas of chemistry.

Hours spent in the chemistry section of the library investigating the types of literature available will be a wise investment for those planning careers in chemistry and related fields. But it may sometimes appear that "if all the journals were laid end-to-end they would not reach a definite conclusion."

Answers to Exercises, Practice for Proficiency, and Self-Tests

UNIT 1

Exercises

1. (a) D **(b)** C **(c)** I **(d)** B **(e)** G **(f)** A **(g)** E **(h)** F **(i)** H

2. (b) (A) **(c)** (A, B) **(e)** (A, B) **3. (a)** (see Fig. 1.10 through 5p sublevel) **(b, c, d)** (see Fig. 1.5) **4. (a)** $1s\,\updownarrow\,2s\,\updownarrow\,2p_x\,\updownarrow\,2p_y\,\updownarrow\,2p_z\,\updownarrow\,3s\,\updownarrow\,3p_x\,\updownarrow\,3p_y\,\updownarrow\,3p_z\,\updownarrow\,4s\,\updownarrow$ or   [Ar] $4s^2$ **(b)** $1s\,\updownarrow\,2s\,\updownarrow\,2p_x\,\updownarrow\,2p_y\,\updownarrow\,2p_z\,\updownarrow\,3s\,\updownarrow\,3p_x\,\uparrow\,3p_y\,\uparrow\,3p_z\,\uparrow$ or   [Ne] $3s^23p^3$
(c) $1s\,\updownarrow\,2s\,\updownarrow\,2p_x\,\updownarrow\,2p_y\,\updownarrow\,2p_z\,\updownarrow\,3s\,\updownarrow\,3p_x\,\updownarrow\,3p_y\,\updownarrow\,3p_z\,\updownarrow\,4s\,\updownarrow\,3d_{xy}\,\updownarrow\,3d_{yz}\,\updownarrow$ $3d_{xz}\,\updownarrow\,3d_{x^2-y^2}\,\updownarrow\,3d_{z^2}\,\updownarrow\,4p_x\,\updownarrow\,4p_y\,\updownarrow\,4p_z\,\updownarrow\,5s\,\updownarrow\,4d_{xy}\,\updownarrow\,4d_{yz}\,\updownarrow\,4d_{xz}\,\updownarrow$ $4d_{x^2-y^2}\,\updownarrow\,4d_{z^2}\,\updownarrow\,5p_x\,\updownarrow\,5p_y\,\updownarrow\,5p_z\,\updownarrow$ or $1s^22s^22p^63s^23p^64s^23d^{10}4p^65s^2$ $4d^{10}5p^6$ or   [Xe] **(d)** same as **c**, except $5p_z\,\uparrow$ or   [Kr] $5s^24d^{10}5p^5$ **(e)** same as **c** **5. (a)** $1s\,\updownarrow\,2s\,\uparrow\,2p_x\,\uparrow\,2p_y\,\uparrow$ or   [He] $2s^12p^2$ **(b)** $1s\,\updownarrow\,2s\,\updownarrow\,2p_x\,\updownarrow\,2p_y\,\updownarrow\,2p_z\,\updownarrow\,3s\,\uparrow\,3p_x\,\uparrow$ or   [Ne] $3s^13p^1$
6. (a) 1.28×10^4 Å **(b)** 4.34×10^3 Å **(c)** 9.49×10^2 Å
(d) 5.2×10^{-19} cal (13.7 eV) per atom

Practice for Proficiency

1. (a) D **(b)** N **(c)** F **(d)** H **(e)** I **(f)** J **(g)** A **(h)** M **(i)** B
(j) L **(k)** K **(l)** G **(m)** C **(n)** E **2. (a)** (A) **(b)** (A, B)
(c) (A, B) **(e)** (A, B) **3. (a)** (see Fig. 1.10), **(b, c)** (see Fig. 1.5) **4. (a)** $1s\,\updownarrow\,2s\,\updownarrow\,2p_x\,\updownarrow\,2p_y\,\updownarrow\,2p_z\,\updownarrow\,3s\,\updownarrow$ or   [Ne] $3s^2$
(b) same as **a,** with addition of $3p_x\,\uparrow\,3p_y\,\uparrow$ or   [Ne] $3s^23p^2$

(c) same as **a,** with addition of $3p_x\,\updownarrow\,3p_y\,\updownarrow\,3p_z\,\updownarrow$ or   [Ne] $3s^23p^6$ **(d)** $1s\,\updownarrow\,2s\,\updownarrow\,2p_x\,\updownarrow\,2p_y\,\updownarrow\,2p_z\,\updownarrow\,3s\,\updownarrow\,3p_x\,\updownarrow\,3p_y\,\updownarrow\,3p_z\,\updownarrow\,4s\,\updownarrow\,3d_{xy}\,\updownarrow$ $3d_{yz}\,\updownarrow\,3d_{xz}\,\updownarrow\,3d_{x^2-y^2}\,\updownarrow\,3d_{z^2}\,\updownarrow\,4p_x\,\uparrow\,4p_y\,\uparrow$ or   [Ar] $4s^23d^{10}4p^2$
(e) $1s\,\updownarrow\,2s\,\updownarrow\,2p_x\,\updownarrow\,2p_y\,\updownarrow\,2p_z\,\updownarrow\,3s\,\updownarrow\,3p_x\,\updownarrow\,3p_y\,\updownarrow\,3p_z\,\updownarrow\,4s\,\updownarrow\,3d_{xy}\,\updownarrow$ $3d_{yz}\,\updownarrow\,3d_{xz}\,\updownarrow\,3d_{x^2-y^2}\,\updownarrow\,3d_{z^2}\,\updownarrow\,4p_x\,\updownarrow\,4p_y\,\updownarrow\,4p_z\,\updownarrow\,5s\,\updownarrow\,4d_{xy}\,\updownarrow\,4d_{yz}\,\updownarrow\,4d_{xz}\,\updownarrow$ $4d_{x^2-y^2}\,\updownarrow\,4d_{z^2}\,\updownarrow$ or   [Kr] $5s^23d^{10}$ **(f)** $1s\,\updownarrow\,2s\,\updownarrow\,2p_x\,\updownarrow\,2p_y\,\updownarrow\,2p_z\,\updownarrow$ $3s\,\updownarrow\,3p_x\,\updownarrow\,3p_y\,\updownarrow\,3p_z\,\updownarrow$ or   [Ar]   or $1s^22s^22p^63s^23p^6$ **(g)** same as **f**
(h) same as **e** **5. (a)** $1s\,\updownarrow\,2s\,\uparrow\,2p_x\,\uparrow$ or   [He] $2s^12p^1$ **(b)** $1s\,\updownarrow\,2s\,\updownarrow$ $2p_x\,\updownarrow\,2p_y\,\updownarrow\,2p_z\,\updownarrow\,3s\,\uparrow\,3p_x\,\uparrow\,3p_y\,\uparrow\,3p_z\,\uparrow$ or   [Ne] $3s^13p^3$ **(c)** $1s\,\updownarrow\,2s\,\updownarrow$ $2p_x\,\updownarrow\,2p_y\,\updownarrow\,2p_z\,\updownarrow\,3s\,\updownarrow\,3p_x\,\updownarrow\,3p_y\,\updownarrow\,3p_z\,\updownarrow\,4s\,\uparrow\,3d_{xy}\,\updownarrow\,3d_{yz}\,\updownarrow\,3d_{xz}\,\updownarrow\,3d_{x^2-y^2}\,\updownarrow$ $3d_{z^2}\,\updownarrow\,4p_x\,\uparrow\,4p_y\,\uparrow$ or   [Ar] $4s^13d^{10}4p^2$ **(d)** $1s\,\updownarrow\,2s\,\updownarrow\,2p_x\,\updownarrow\,2p_y\,\updownarrow\,2p_z\,\updownarrow$ $3s\,\updownarrow\,3p_x\,\updownarrow\,3p_y\,\updownarrow\,3p_z\,\updownarrow\,4s\,\updownarrow\,3d_{xy}\,\updownarrow\,3d_{yz}\,\updownarrow\,3d_{xz}\,\updownarrow\,3d_{x^2-y^2}\,\updownarrow\,3d_{z^2}\,\updownarrow\,4p_x\,\updownarrow$ $4p_y\,\updownarrow\,4p_z\,\updownarrow\,5s\,\updownarrow\,4d_{xy}\,\updownarrow\,4d_{yz}\,\updownarrow\,4d_{xz}\,\updownarrow\,4d_{x^2-y^2}\,\updownarrow\,4d_{z^2}\,\updownarrow\,5p_x\,\uparrow\,5p_y\,\uparrow\,5p_z\,\uparrow$ or [Kr] $5s^14d^{10}5p^3$ **(e)** $1s\,\updownarrow\,2s\,\updownarrow\,2p_x\,\updownarrow\,2p_y\,\updownarrow\,2p_z\,\uparrow\,3s\,\uparrow$ or   [He] $2s^22p^53s^1$ (no sublevel transitions, principal level transitions of higher energy) **6. (a)** 4.86×10^3 Å (visible)
(b) 9.8×10^{-20} cal atom^{-1} **7.** K$^+$, Cl$^-$ and Cd, Sn^{2+} (Ca^{2+}, for example, is isoelectronic with K$^+$, Cl$^-$) **8.** elements 7, 15, 33, 51 in group VA; elements 12, 20, 38, 56 in group IIA; radium uses s-type orbital, bismuth uses s- and p-type orbitals, (and d) **9.** Rb$^+$, Br$^-$, Kr isoelectronic; Sr^{2+}, Se^{2-}

10. Cr: [Ar] $4s^1 3d^5$, Cu: [Ar] $4s^1 3d^{10}$, filled or half-filled d-orbital set particularly stable

Self-Test

1. (a) C **(b)** B **(c)** A **(d)** G **(e)** D **(f)** F **(g)** E   **2.** c, e, f **3.** (see Fig. 1.10 through $3p$ sublevel; b, c (see Fig. 1.5) **4. (a)** $1s \uparrow\downarrow 2s \uparrow\downarrow 2p_x \uparrow\downarrow 2p_y \uparrow\downarrow 2p_z \uparrow\downarrow 3s \uparrow\downarrow 3p_x \uparrow\downarrow 3p_y \uparrow\downarrow 3p_z \uparrow\downarrow 4s \uparrow\downarrow 3d_{xy} \uparrow\downarrow 3d_{yz} \uparrow\downarrow 3d_{xz} \uparrow\downarrow 3d_{x^2-y^2} \uparrow\downarrow 3d_{z^2} \uparrow\downarrow 4p_x \uparrow 4p_y \uparrow 4p_z \uparrow$ **(b)** [Xe] $6s^1$ **5.** $1s \uparrow\downarrow 2s \uparrow\downarrow 2p_x \uparrow\downarrow 2p_y \uparrow\downarrow 2p_z \uparrow\downarrow 3s \uparrow 3p_x \uparrow 3p_y \uparrow$ **6. (a)** 3.97×10^3 Å **(b)** 1.20×10^{-19} cal atom^{-1}

UNIT 2
Exercises

2. (a) H **(b)** C **(c)** A **(d)** I **(e)** F **(f)** J **(g)** B **(h)** E **(i)** G **(j)** D   **3. (a)** $6p$, $6n$, $6e$, **(b)** $8p$, $10n$, $10e$ **(c)** $11p$, $12n$, $10e$ **4.** b, d, and e are incorrect   **5. (a)** Al ($Z = 13$, at. wt. $= 26.9815$), Na ($Z = 11$, at. wt. $= 22.9898$), As ($Z = 33$, at. wt. $= 74.9216$), B ($Z = 5$, at. wt. $= 10.811$), Pb ($Z = 82$, at. wt. $= 207.2$) **(b)** S^{2-}, K^+, F^- (no stable monatomic ion of carbon), Sr^{2+} **(c)** P $<$ Al $<$ Na $<$ K **(d)** $Al^{3+} < Mg^{2+} < Cl^- < S^{2-}$ **(e)** highest first ionization potentials: F, Mg, Al, N; largest *negative* value of electron affinities: F, Na, Al, O **6.** 32.064 (sulfur)

Practice for Proficiency

1. (a) aluminum **(b)** Sb **(c)** cadmium **(d)** Cu **(e)** silver **(f)** Fe **(g)** tin **(h)** K   **2. (a)** E **(b)** N **(c)** I **(d)** H **(e)** A **(f)** F **(g)** C **(h)** P **(i)** O **(j)** B **(k)** G **(l)** L **(m)** D **(n)** M **(o)** K **(p)** J   **3. (a)** $12p$, $12n$, $12e$ **(b)** $7p$, $8n$, $10e$ **(c)** $1p$, $2n$, $0e$ **(d)** $15p$, $16n$, $15e$ **(e)** $17p$, $19n$, $18e$ **(f)** $13p$, $13n$, $10e$ **(g)** $92p$, $146n$, $92e$ **(h)** $53p$, $74n$, $54e$ **(i)** $38p$, $52n$, $36e$ **(j)** $101p$, $157n$, $101e$ **4.** a (only) is incorrect **5. (a)** As ($Z = 33$, at. wt. $= 74.9216$), Hg ($Z = 80$, at. wt. $= 200.59$), Na ($Z = 11$, at. wt. $= 22.9898$), Au ($Z = 79$, at. wt. $= 196.97$), Be ($Z = 4$, at. wt. $= 9.012$), Cu ($Z = 29$, at. wt. $= 63.546$), Sn ($Z = 50$, at. wt. $= 118.69$), Ni ($Z = 28$, at. wt. $= 58.71$), He ($Z = 2$, at. wt. $= 4.0026$), Fe ($Z = 26$, at. wt. $= 55.847$) **(b)** Al^{3+}, Br^-, Ca^{2+}, Kr, and Si (no stable monatomic ions) **(c)** (1) Br $<$ Se $<$ Ca $<$ K; (2) Ne $<$ B $<$ Li $<$ Na; (3) Cl $<$ S $<$ Se $<$ Te **(d)** (1) $H^+ \ll Be^{2+} < F^- < N^{3-}$; (2) Ca^{2+}, K^+, Br^-, Se^{2-}; (3) $Li^+ < Na^+ < F^- < Cl^-$; **(e)** (1) Be (electron closer to nucleus); (2) He (electron closer to nucleus *and* stability of $1s^2$ configuration); (3) Cl (stability of [Ar]); (4) Be (high energy of $2s^2 2p^1$ compared to $2s^2 2p^0$ for the effective nuclear charge involved); (5) K (high-energy difference between $3p^6 4s^0$ and $3p^5 4s^0$ compared to that between $3p^6 4s^1$ and $3p^6 4s^0$) **6.** 35.453 (chlorine)   **7.** zinc   **8.** cadmium   **9.** Cobalt ([Ar] $4s^2 3d^7$) and nickel ([Ar] $4s^2 3d^8$) are "anomalies," but chemical differences between these species should be much less than those between argon and potassium ([Ar] $4s^1$)   **10.** Li^+ (1.8×10^{-19}), Na^+ (4.5×10^{-20}), K^+ (1.6×10^{-20}), Be^{2+} (2.6×10^{-18}), Mg^{2+} (2.8×10^{-19}), Ca^{2+} (7.9×10^{-20}), B^{3+} (1.4×10^{-17}), Al^{3+} (9.2×10^{-19}), C^{4+} (4.5×10^{-17}), Si^{4+} (2.2×10^{-18}), Ge^{4+} (1.0×10^{-18}); (Note, for example, that Li^+ is closer to Mg^{2+} than to Na^+, Be^{2+} is closer to Al^{3+} than to Mg^{2+}, and B^{3+} is closer to Si^{4+} than to Al^{3+}.)

Self-Test

1. (a) Ar **(b)** tin **(c)** Au **(d)** Br **(e)** potassium **(f)** Zn **(g)** magnesium **(h)** Pb **(i)** Na **(j)** chlorine **(k)** Cu **(l)** mercury   **2. (a)** H **(b)** F **(c)** A **(d)** B **(e)** D **(f)** E **(g)** J **(h)** G **(i)** C **(j)** I   **3. (a)** 35 **(b)** 45 **(c)** 36   **4.** c   **5.** 26, 55.847   **6.** d   **7.** d   **8.** b   **9.** b   **10.** 39.102 (potassium)

UNIT 3
Exercises

2. (a) potassium bisulfate (IUPAC: potassium hydrogen sulfate) **(b)** $NH_4CH_3CO_2$ **(c)** copper(I) cyanide (old: cuprous cyanide) **(d)** Ag_2CrO_4 **(e)** chromium(III) oxide (old: chromic oxide) **(f)** Na_2S **(g)** aluminum carbonate **(h)** $Fe_3(PO_4)_2$ **(i)** cobalt(II) hydroxide (old: cobaltous hydroxide) **(j)** $Hg_2(NO_3)_2$ [mercury(I) ion is Hg_2^{2+}]   **3. (a)** E **(b)** D **(c)** B **(d)** A **(e)** C **4.** (shown balanced, as answers to 5.)

(a) *complete* $CaCO_3(s) + 2HCl(aq) \longrightarrow$
$$H_2O(l) + CO_2(g) + CaCl_2(aq)$$
net $CaCO_3(s) + 2H^+(aq) \longrightarrow$
$$H_2O(l) + CO_2(g) + Ca^{2+}(aq)$$
(b) *complete* $2NH_3(aq) + H_2SO_4(aq) \longrightarrow (NH_4)_2SO_4(aq)$
net $NH_3(aq) + H^+(aq) \longrightarrow NH_4^+(aq)$
(c) *complete* $4HCl(aq) + Ca(OCl)_2(aq) \longrightarrow$
$$Cl_2(g) + 2H_2O(l) + CaCl_2(aq)$$
net $2H^+(aq) + Cl^-(aq) + OCl^-(aq) \longrightarrow$
$$H_2O(l) + Cl_2(g)$$

5. (see **4.**)   **6. (a)** M **(b)** O—R **(c)** O—R **(d)** M **(e)** O—R

Practice for Proficiency

1. (a) CH_3CO_2H **(b)** phosphine **(c)** OH^- **(d)** nitric acid **(e)** C_4H_{10} **(f)** bisulfite (IUPAC: hydrogen sulfite ion) **(g)** $CHCl_3$ **(h)** phosphoric acid **(i)** H_2S **(j)** dichromate **(k)** H_2SO_4 **(l)** benzene **(m)** HCN **(n)** carbonate **(o)** CH_3OH **(p)** methylamine **(q)** $HClO_4$ **(r)** ammonium **(s)** $S_2O_3^{2-}$ **(t)** hypochlorous acid   **2. (a)** $NaHCO_3$ **(b)** potassium permanganate **(c)** Al_2O_3, **(d)** magnesium nitride **(e)** $AgCH_3CO_2$ **(f)** iron(III) nitrate (old: ferric nitrate) **(g)** $Ca(ClO_3)_2$ **(h)** manganese(II) sulfate (old: manganous sulfate) **(i)** $(NH_4)_2Cr_2O_7$ **(j)** copper(I) sulfide (old: cuprous sulfide) **(k)** $Zn_3(PO_4)_2$ **(l)** methylammonium bromide **(m)** SnF_2 **(n)** lead oxalate   **3. (a)** D **(b)** A **(c)** B **(d)** H **(e)** E **(f)** I **(g)** J **(h)** C **(i)** G **(j)** F **4.** (shown balanced, as answers to **5.**)

(a) *complete* $2NH_3(aq) + Ca(OCl)_2(aq) \longrightarrow$
$$2ClNH_2(g) + Ca(OH)_2(aq)$$
net $NH_3(aq) + OCl^-(aq) \longrightarrow ClNH_2(g) + OH^-(aq)$
(b) *complete* $Na_2CrO_4(aq) + Pb(NO_3)_2(aq) \longrightarrow$
$$PbCrO_4(s) + 2NaNO_3(aq)$$
net $CrO_4^{2-}(aq) + Pb^{2+}(aq) \longrightarrow PbCrO_4(s)$
(c) *complete* $2NaOH(aq) + 6H_2O(l) + 2Al(s) \longrightarrow$
$$3H_2(g) + 2Na[Al(OH)_4](aq)$$
net $2OH^-(aq) + 6H_2O(l) + 2Al(s) \longrightarrow$
$$3H_2(g) + 2[Al(OH)_4]^-(aq)$$
(d) *complete* $2HCl(aq) + Na_2SO_3(aq) \longrightarrow$
$$SO_2(g) + H_2O(l) + 2NaCl(aq)$$
net $2H^+(aq) + SO_3^{2-}(aq) \longrightarrow SO_2(g) + H_2O(l)$

(e) *complete* $5CaCO_3(s) + 3(NH_4)_2HPO_4(aq) + H_2O(l) \longrightarrow$
         $Ca_5(OH)(PO_4)_3(s) + 4NH_4(HCO_3)(aq) + (NH_4)_2CO_3(aq)$
net $5CaCO_3(s) + 3HPO_4{}^{2-}(aq) + H_2O(l) \longrightarrow$
         $Ca_5(OH)(PO_4)_3(s) + 4HCO_3{}^{-}(aq) + CO_3{}^{2-}(aq)$

5. (See **4.**)   **6. (a)** $\overset{+I}{K}\overset{+VII}{Mn}\overset{-II}{O_4}$ **(b)** $\overset{+II}{Ca}\overset{+III}{C_2}\overset{-II}{O_4}$ **(c)** $\overset{+1}{K}\overset{+III}{Al}\overset{+VI}{(S}\overset{-II}{O_4)_2}$
(d) $\overset{+I}{Cu_2}\overset{+III}{C_2}\overset{-II}{O_4}$ **(e)** $\overset{+II}{Fe}\overset{+V}{(N}\overset{-II}{O_3)_2}$

7. (a) (O—R) $Zn + 2H^+ \longrightarrow Zn^{2+} + H_2$
(b) (M) $Zn^{2+} + H_2S \longrightarrow ZnS + 2H^+$
(c) (M) $3CaCO_3 + 2H_3PO_4 \longrightarrow Ca_3(PO_4)_2 + 3CO_2 + 3H_2O$
(d) (O—R) $4NH_3 + 7O_2 \longrightarrow 4NO_2 + 6H_2O$

8. (a) *complete* $NaOH(aq) + HCl(aq) \longrightarrow$
         $NaCl(aq) + H_2O(l)$
net $OH^-(aq) + H^+(aq) \longrightarrow H_2O(l)$
(b) *complete* $KOH(aq) + HCN(aq) \longrightarrow$
         $KCN(aq) + H_2O(l)$
net $OH^-(aq) + HCN(aq) \longrightarrow CN^-(aq) + H_2O(l)$
(c) *complete* (assuming sufficient KOH)
   $2KOH(aq) + H_2SO_4(aq) \longrightarrow K_2SO_4(aq) + 2H_2O(l)$
net $2OH^-(aq) + H^+(aq) + HSO_4{}^-(aq) \longrightarrow$
         $SO_4{}^{2-}(aq) + 2H_2O(l)$
(d) *complete* $NH_3(aq) + HNO_3(aq) \longrightarrow NH_4NO_3(aq)$
net $NH_3(aq) + H^+(aq) \longrightarrow NH_4{}^+(aq)$
(e) *complete* $Al(OH)_3(s) + 3HClO_4(aq) \longrightarrow$
         $Al(ClO_4)_3(aq) + 3H_2O(l)$
net $Al(OH)_3(s) + 3H^+(aq) \longrightarrow Al^{3+}(aq) + 3H_2O(l)$

9. (a) $\underline{HCl}(aq) + NH_3(aq) \longrightarrow NH_4Cl(aq)$
(b) $\underline{HNO_3}(aq) + NH_3(aq) \longrightarrow NH_4NO_3(aq)$
(c) $\underline{H_2SO_4}(aq) + 2NH_3(aq) \longrightarrow (NH_4)_2SO_4(aq)$
(d) $\underline{H_3PO_4}(aq) + 3NH_3(aq) \longrightarrow (NH_4)_3PO_4(aq)$

10. (All are oxidation–reduction reactions.)
(a) $Pb(s) + PbO_2(s) + 2H_2SO_4(aq) \longrightarrow 2PbSO_4(s) + 2H_2O(l)$
(b) $Zn(s) + 2MnO_2(s) + 2NH_4Cl(aq) \longrightarrow$
         $ZnCl_2(aq) + Mn_2O_3(s) + 2NH_3(aq) + H_2O(l)$
(c) $Cd(s) + NiO_2(s) + 2H_2O(l) \longrightarrow Cd(OH)_2(s) + Ni(OH)_2(s)$
(d) $CH_4(g) + 2O_2(g) + 2KOH(aq) \longrightarrow K_2CO_3(aq) + 3H_2O(l)$

Self-Test
1. (a) acetic acid   **(b)** HNO_3 **(c)** ethanol   **(d)** CH_3NH_2
(e) phosphoric acid   **2. (a)** $(NH_4)_2SO_4$ **(b)** $Cu(NO_3)_2$
(c) Na_3PO_4 **(d)** $AgBr$ **(e)** $KClO_3$ **(f)** sodium dichromate
(g) lead sulfate   **(h)** potassium bicarbonate (IUPAC: potassium
hydrogen carbonate)   **(i)** calcium oxide   **(j)** iron(III) nitrate
(old: ferric nitrate)   **3. (a)** B   **(b)** F   **(c)** E   **(d)** D   **(e)** A
(f) C   **(g)** G   **(h)** I   **(i)** H
4. (Shown balanced, as answers for **5.**)
(a) *complete* $CO_2(g) + NH_3(aq) + NaCl(aq) + H_2O(l) \longrightarrow$
         $NaHCO_3(s) + NH_4Cl(aq)$
net $CO_2(g) + NH_3(aq) + Na^+(aq) + H_2O(l) \longrightarrow$
         $NaHCO_3(s) + NH_4{}^+(aq)$
(b) *complete* $AgNO_3(aq) + HBr(aq) \longrightarrow$
         $AgBr(s) + HNO_3(aq)$
net $Ag^+(aq) + Br^-(aq) \longrightarrow AgBr(s)$
(c) *complete* $Zn(s) + 2HCl(aq) \longrightarrow H_2(g) + ZnCl_2(aq)$
net $Zn(s) + 2H^+(aq) \longrightarrow H_2(g) + Zn^{2+}(aq)$
(d) *complete* $Al(OH)_3(s) + 3HCl(aq) \longrightarrow$
         $AlCl_3(aq) + 3H_2O(l)$
net $Al(OH)_3(s) + 3H^+(aq) \longrightarrow Al^{3+}(aq) + 3H_2O(l)$

(e) *complete* $CO_2(g) + Ca(OH)_2(aq) \longrightarrow$
         $CaCO_3(s) + H_2O(l)$
net $CO_2(g) + Ca^{2+}(aq) + 2OH^-(aq) \longrightarrow$
         $CaCO_3(s) + H_2O(l)$

5. (See **4.**)
6. (a) $(M) + \cdots + 2NaCl \longrightarrow \cdots + 2AgCl$
(b) $(M) + \cdots + 6H^+ \longrightarrow 2Al^{3+} + 3H_2O$
(c) (O—R) $5CO + \cdots \longrightarrow 5CO_2 + \cdots$
(d) (O—R), . . . , $2Cu$
(e) (O—R) $2NH_4NO_3 \longrightarrow 2N_2 + O_2 + 4H_2O$

UNIT 4
Exercises
1. (a) 47 g   **(b)** 138   **(c)** 138 g   **(d)** 66 g   **2. (a)** 60   **(b)** 132
(c) 102   **(d)** 95   **(e)** 63   **3.** 6.64 g (CO_2), 2.72 g (H_2O)
4. (a) acetic acid   **(b)** 93%   **5. (a)** 68 kg   **(b)** approx. \$4.74

Practice for Proficiency
1. (a) N_2 **(b)** 17.0   **(c)** 28.0 g   **(d)** 60.7 g   **(e)** 2.02 g
(f) 1.79×10^3 **(g)** 85.0%   **(h)** 42 g   **(i)** 2.1×10^{24} **2. (a)** 96
(b) 82   **(c)** 188   **(d)** 116   **(e)** 46   **(f)** 400   **(g)** 78   **(h)** 119   **(i)** 98
(j) 249   **3.** 121 g   **4. (a)** 152 kg   **(b)** 87 kg   **5. (a)** 308 kg
(b) 205 kg   **6. (a)** 9.2×10^8 kg   **(b)** 1.3×10^9 kg
7. (a) 118 kg   **(b)** 97 kg   **(c)** 73 kg (NH_4NO_3 has largest percentage nitrogen, 35%)   **8. (a)** 5.4 kg   **(b)** 4.4 kg   **9. (a)** 185 g
(b) 213 g   **10. (a)** 136 kg   **(b)** 210 kg   **11. (a)** NaCl
(b) 67%   **12. (a)** HNO_3 **(b)** 93%   **13. (a)** 4.3×10^3 kg
(b) 4.6×10^3 kg   **14. (a)** 96 lb   **(b)** \$1.71   **15.** 171 kg

Self-Test
1. (a) 46   **(b)** 98   **(c)** 61   **(d)** 358   **(e)** 80   **2.** 44.2 g
3. 58.0%   **4. (a)** C_6H_5Br **(b)** 28.0 g   **(c)** 78.1   **(d)** 0.157 g
(e) C_6H_6 **5.** approx. \$53.40

UNIT 5
Exercises
1. (a) E   **(b)** C   **(c)** D   **(d)** B   **(e)** A   **2. (a)** F   **(b)** T   **(c)** F
(d) F   **(e)** T   **3. (a)** -488.4 **(b)** -30.5 **(c)** -189.8 (The difference represents use of H_2S rather than the free elements.)
(d) -8.6 **(e)** -28.2 ("Violence" is a *rate* phenomenon.)
4. 376 kcal   **5.** 62%   **6.** No. If ΔH is positive and ΔS is
negative, ΔG must be positive ($\Delta G = \Delta H - T\Delta S$). Hence, synthetic diamonds could only be manufactured from powdered
charcoal under exothermic conditions such that negative ΔH
term outweighs positive value of $(-T\Delta S)$ term.

Practice for Proficiency
1. (a) H   **(b)** C   **(c)** J   **(d)** D   **(e)** E   **(f)** K   **(g)** B   **(h)** F   **(i)** I
(j) G   **(k)** A   **2. (a)** T   **(b)** T   **(c)** F   **(d)** T   **(e)** T   **(f)** T   **(g)** T
(h) T   **(i)** F (*solids* in most stable forms)   **(j)** F   **(k)** T   **(l)** F
(only true for entropy of the *universe*)   **(m)** F   **(n)** T   **(o)** T
3. (a) -326.5 kcal mole^{-1} (ethanol)   **(b)** -118.3 kcal mole^{-1}
(ethanol)   **(c)** -134.4 kcal mole^{-1} (H_2S)   **(d)** -46.6 kcal mole^{-1}
(Na_2CO_3)   **(e)** complete combustion is more exothermic by
405.6 kcal mole^{-1} (benzene)   **(f)** -34.8 kcal mole^{-1} (NH_3)
(g) -38.0 kcal mole^{-1} (NaF)   **(h)** -69.8 kcal mole^{-1} (NH_3)
(i) $+36.1$ kcal mole^{-1} (glycine)   **(j)** $+1023$ kcal mole^{-1} (sulfanilamide)   **4.** 5.7×10^3 kcal   **5.** $+9.9 \times 10^2$ kcal
6. -11.9 kcal g^{-1} (acetylene) versus -10.0 kcal g^{-1} (benzene)   **7.** 62%   **8.** 10%   **9.** d   **10.** -229 kcal mole^{-1}
(cyclamic acid), "food calorie value" $= -4.6$ kcal g^{-1}

Self-Test

1. (a) B **(b)** E **(c)** D **(d)** F **(e)** A **(f)** C **2. (a)** F **(b)** F
(c) T **(d)** T **(e)** F **3.** -42.6 kcal mole^{-1} (HCl) **4.** approx.
1000 kcal **5.** 48% **6.** c

UNIT 6
Exercises

2. (a) HF (acid), OH$^-$ (base) **(b)** NH$_3$ (base), HCl (acid)
(c) NH$_4^+$ (acid), CN$^-$ (base) **(d)** CO$_3^{2-}$ (base),
CH$_3$CO$_2$H (acid) **(e)** HSO$_4^-$ (acid), HPO$_4^{2-}$ (base)
3. (a) (HCO$_3^-$ amphoteric)
$$HCO_3^-(aq) + H^+(aq) \longrightarrow CO_2(g) + H_2O(l)$$
$$HCO_3^-(aq) + OH^-(aq) \longrightarrow CO_3^{2-}(aq) + H_2O(l)$$
(b) NH$_4^+$(aq) + H$^+$(aq) (N.R.)
$$NH_4^+(aq) + OH^-(aq) \longrightarrow NH_3(aq) + H_2O(l)$$
(c) (H$_2$PO$_4^-$ amphoteric)
$$H_2PO_4^-(aq) + H^+(aq) \longrightarrow H_3PO_4(aq)$$
$$H_2PO_4^-(aq) + OH^-(aq) \longrightarrow HPO_4^{2-}(aq) + H_2O(l)$$
$$[H_2PO_4^-(aq) + 2OH^-(aq) \longrightarrow PO_4^{3-}(aq) + 2H_2O(l)]$$
(d) (H$_2$NNH$_3^+$ amphoteric)
$$H_2NNH_3^+(aq) + H^+(aq) \longrightarrow H_3NNH_3^{2+}(aq)$$
$$H_2NNH_3^+(aq) + OH^-(aq) \longrightarrow H_2NNH_2(aq) + H_2O(l)$$
4. (a) CH$_3$CO$_2$H(aq) + OH$^-$(aq) $\longrightarrow$ CH$_3$CO$_2^-$(aq) + H$_2$O(l)
(b) HCO$_3^-$(aq) + H$^+$(aq) $\longrightarrow$ CO$_2$(g) + H$_2$O(l)
(c) Ba^{2+}(aq) + HSO$_4^-$(aq) $\longrightarrow$ BaSO$_4$(s) + H$^+$(aq)
(d) NH$_3$(aq) + H$^+$(aq) $\longrightarrow$ NH$_4^+$(aq)
(e) 2Ag$^+$(aq) + SO$_4^{2-}$(aq) $\longrightarrow$ Ag$_2$SO$_4$(s) (N.R. if sufficiently
dilute, as Ag$_2$SO$_4$ is moderately soluble.)
(f) HCO$_3^-$(aq) + OH$^-$(aq) $\longrightarrow$ CO$_3^{2-}$(aq) + H$_2$O(l)
(g) Ag$^+$(aq) + Br$^-$(aq)
(h) N.R.
(i) Sr^{2+}(aq) + SO$_4^{2-}$(aq) $\longrightarrow$ SrSO$_4$(s)
(j) BaSO$_3$(s) + H$^+$(aq) + HSO$_4^-$(aq) $\longrightarrow$
$$BaSO_4(s) + SO_2(g) + H_2O(l)$$
5. 2.8 tons **6.** $+135$ kcal mole^{-1} (The LiF bond is
"stronger.")

Practice for Proficiency

1. (a) CH$_3$NH$_2$(base), HBr (acid) **(b)** H$_2$PO$_4^-$ (acid),
NH$_3$ (base) **(c)** H$_2$PO$_4^-$ (base), H$_2$SO$_3$ (acid) **(d)** H$_2$O (acid),
F$^-$ (base) **(e)** H$_2$O (base), HF (acid) **(f)** CO$_3^{2-}$ (base),
HCN (acid) **(g)** HCO$_3^-$ (base), HSO$_4^-$ (acid) **(h)** HSO$_3^-$ (base),
HSO$_4^-$ (acid) **(i)** HSO$_3^-$ (acid), NH$_3$ (base) **(j)** NH$_4^+$ (acid),
CO$_3^{2-}$ (base) **2.** H$_2$PO$_4^-$, H$_2$O, HSO$_3^-$
3. (a) complete Ba(OH)$_2$(aq) + 2HNO$_3$(aq) $\longrightarrow$
$$Ba(NO_3)_2(aq) + 2H_2O(l)$$
net OH$^-$(aq) + H$^+$(aq) $\longrightarrow$ H$_2$O(l)
(b) complete K$_2$HPO$_4$(aq) + NaOH(aq) $\longrightarrow$
$$K_2NaPO_4(aq) + H_2O(l)$$
net HPO$_4^{2-}$(aq) + OH$^-$(aq) $\longrightarrow$ PO$_4^{3-}$(aq) + H$_2$O(l)
(c) N.R.
(d) complete MgSO$_4$(aq) + HBr(aq) $\longrightarrow$
$$MgBr(HSO_4)(aq)$$
net SO$_4^{2-}$(aq) + H$^+$(aq) $\longrightarrow$ HSO$_4^-$(aq)

(e) complete (NH$_4$)$_2$CO$_3$(aq) + H$_2$SO$_4$(aq) $\longrightarrow$
$$(NH_4)_2SO_4(aq) + CO_2(g) + H_2O(l)$$
net CO$_3^{2-}$(aq) + H$^+$(aq) + HSO$_4^-$(aq) $\longrightarrow$
$$CO_2(g) + SO_4^{2-}(aq) + H_2O(l)$$
(f) complete NH$_4$NO$_3$(aq) + LiOH(aq) $\longrightarrow$
$$LiNO_3(aq) + NH_3(aq) + H_2O(l)$$
net NH$_4^+$(aq) + OH$^-$(aq) $\longrightarrow$ NH$_3$(aq) + H$_2$O(l)
(g) complete NH$_3$(aq) + CH$_3$CO$_2$H(aq) $\longrightarrow$
$$NH_4CH_3CO_2(aq)$$
net NH$_3$(aq) + CH$_3$CO$_2$H(aq) $\longrightarrow$
$$NH_4^+(aq) + CH_3CO_2^-(aq)$$
(h) N.R.
(i) complete BaSO$_3$(s) + 2HCl(aq) $\longrightarrow$
$$BaCl_2(aq) + SO_2(g) + H_2O(l)$$
net BaSO$_3$(s) + 2H$^+$(aq) $\longrightarrow$
$$Ba^{2+}(aq) + SO_2(g) + H_2O(l)$$
(j) complete CuO(s) + 2HNO$_3$(aq) $\longrightarrow$
$$Cu(NO_3)_2(aq) + H_2O(l)$$
net CuO(s) + 2H$^+$(aq) $\longrightarrow$ Cu^{2+}(aq) + H$_2$O(l)
4. (a) complete CdCl$_2$(aq) + H$_2$S(aq) $\longrightarrow$
$$CdS(s) + 2HCl(aq)$$
net Cd^{2+}(aq) + H$_2$S(aq) $\longrightarrow$ CdS(s) + 2H$^+$(aq)
(b) complete Na$_2$SO$_4$(aq) + Pb(NO$_3$)$_2$(aq) $\longrightarrow$
$$PbSO_4(s) + 2NaNO_3(aq)$$
net SO$_4^{2-}$(aq) + Pb^{2+}(aq) $\longrightarrow$ PbSO$_4$(s)
(c) complete FeSO$_4$(aq) + Ba(OH)$_2$(aq) $\longrightarrow$
$$Fe(OH)_2(s) + BaSO_4(s)$$
net Fe^{2+}(aq) + SO$_4^{2-}$(aq) + Ba^{2+}(aq) + 2OH$^-$(aq) $\longrightarrow$
$$Fe(OH)_2(s) + BaSO_4(s)$$
(or, write as two separate equations)
(d) complete CuBr$_2$(aq) + 2AgNO$_3$(aq) $\longrightarrow$
$$Cu(NO_3)_2(aq) + 2AgBr(s)$$
net Br$^-$(aq) + Ag$^+$(aq) $\longrightarrow$ AgBr(s)
(e) N.R.
(f) complete CaCl$_2$(aq) + Na$_2$CO$_3$(aq) $\longrightarrow$
$$CaCO_3(s) + 2NaCl(aq)$$
net Ca^{2+}(aq) + CO$_3^{2-}$(aq) $\longrightarrow$ CaCO$_3$(s)
(g) complete NH$_4$NO$_3$(aq) + NaOH(aq) $\longrightarrow$
$$NH_3(aq) + H_2O(l) + NaNO_3(aq)$$
net NH$_4^+$(aq) + OH$^-$(aq) $\longrightarrow$ NH$_3$(aq) + H$_2$O(l)
(h) complete MgCO$_3$(s) + H$_2$SO$_4$(aq) $\longrightarrow$
$$MgSO_4(aq) + CO_2(g) + H_2O(l)$$
net MgCO$_3$(s) + H$^+$(aq) + HSO$_4^-$(aq) $\longrightarrow$
$$Mg^{2+}(aq) + SO_4^{2-}(aq) + CO_2(g) + H_2O(l)$$
(i) complete Fe$_2$O$_3$(aq) + 6HCl(aq) $\longrightarrow$
$$2FeCl_3(aq) + 3H_2O(aq)$$
net Fe$_2$O$_3$(aq) + 6H$^+$(aq) $\longrightarrow$ 2Fe^{3+}(aq) + 3H$_2$O(l)
(j) N.R. (No reaction with "vinegar" and "hard water".)
5. 32 kg **6.** 29 g **7.** 7.4 g **8.** $+120$ kcal mole^{-1}
9. -17.0 kcal mole^{-1} (NaCl) **10. (a)** (consult instructor)
(b) electrolytes: NaI, HI, FeCl$_3$, ionic compounds: only NaI
and HI [Note: HI is *not* ionic by usual criteria. (See Unit 7.)
Apparent ionic character is the result of proton exchange
with SO$_2$.]

Self-Test

1. (a) HPO_4^{2-} (acid), OH^- (base) **(b)** HPO_4^{2-} (base), HSO_4^- (acid) **(c)** NH_3 (base), HCO_3^- (acid) **(d)** CH_3CO_2H (acid), HCO_3^- (base) **(e)** CH_3CO_2H (acid), NH_3 (base) **2.** HPO_4^{2-} and HCO_3^-

3. (a) *complete* $NH_3(aq) + KHSO_4(aq) \longrightarrow K(NH_4)SO_4(aq)$

net $NH_3(aq) + HSO_4^-(aq) \longrightarrow$
$$NH_4^+(aq) + SO_4^{2-}(aq)$$

(b) *complete* $Al_2O_3(s) + 6HNO_3(aq) \longrightarrow$
$$2Al(NO_3)_2(aq) + 3H_2O(l)$$

net $Al_2O_3(s) + 6H^+(aq) \longrightarrow 2Al^{3+}(aq) + 3H_2O(l)$

(c) *complete* $Na_2CO_3(aq) + SrCl_2(aq) \longrightarrow$
$$2NaCl(aq) + SrCO_3(s)$$

net $CO_3^{2-}(aq) + Sr^{2+}(aq) \longrightarrow SrCO_3(s)$

(d) N.R.

(e) *complete* $PbCO_3(s) + H_2SO_4(aq) \longrightarrow$
$$PbSO_4(s) + H_2O(l) + CO_2(g)$$

net $PbCO_3(s) + H^+(aq) + HSO_4^-(aq) \longrightarrow$
$$PbSO_4(s) + H_2O(l) + CO_2(g)$$

4. 61 g **5.** -56.9 kcal mole^{-1} (Al_2O_3)

UNIT 7
Exercises

1. (a)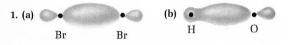
Br Br **(b)** H O

2. (a)

$$H{:}\overset{\displaystyle H}{\underset{\displaystyle H}{C}}{:}N{:}H \qquad H-\overset{\displaystyle H}{\underset{\displaystyle H}{C}}-N\overset{\displaystyle H}{\underset{\displaystyle H}{}}$$

(b)

$$H{:}\overset{\displaystyle H}{\underset{\displaystyle H}{C}}{:}\ddot{O}{:}H \qquad H-\overset{\displaystyle H}{\underset{\displaystyle H}{C}}-\ddot{O}-H$$

(c) $H{:}\ddot{\ddot{O}}{:}\ddot{\ddot{Cl}}{:} \qquad H-\ddot{\ddot{O}}-\ddot{\ddot{Cl}}{:}$

(d) $H{:}\overset{\displaystyle \cdot\cdot}{\underset{\displaystyle \cdot\cdot}{S}}{:}H \qquad H-\overset{\displaystyle \cdot\cdot}{\underset{\displaystyle \cdot\cdot}{S}}-H$

(e) $H{:}C{:}{:}{:}N{:} \qquad H-C{\equiv}N{:}$

3. (a) polar covalent **(b)** polar covalent **(c)** nonpolar covalent **(d)** ionic **(e)** polar covalent **4. (a)** 132 g **(b)** 138 g **(c)** 90 g **5. (a)** -1.2×10^2 kcal **(b)** -3.3×10^2 kcal **(c)** -5.2×10^2 kcal

6.

$$\left[:\ddot{O}\diagdown_{\underset{\displaystyle :\ddot{O}:}{C}}\diagup\ddot{O}:\right]^{2-} \longleftrightarrow \left[:\ddot{O}\diagdown_{\underset{\displaystyle :\ddot{O}:}{C}}\diagup\overset{\cdot\cdot}{O}:\right]^{2-} \longleftrightarrow \left[:\ddot{O}\diagdown_{\underset{\displaystyle :\ddot{O}:}{C}}\diagup\ddot{O}:\right]^{2-}$$

Practice for Proficiency

1. (a)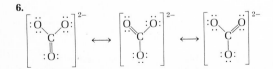
H H **(b)**
S H

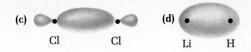

(c) Cl Cl **(d)** Li H

(e)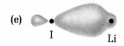
I Li

2. (a) $H{:}\overset{\cdot\cdot}{N}{:}N \qquad H-\overset{\displaystyle H}{\underset{\displaystyle H}{N}}-H$

(b) $H{:}\overset{\displaystyle :\ddot{O}:}{\underset{\displaystyle :\ddot{O}:}{\ddot{O}{:}\ddot{Cl}{:}\ddot{O}}}{:} \qquad H-\overset{\displaystyle :\ddot{O}:}{\underset{\displaystyle :\ddot{O}:}{\ddot{O}-Cl-\ddot{O}}}$

(c) $H{:}\overset{\displaystyle H\ H}{\underset{\displaystyle H\ H}{C}}{:}C{:}\ddot{O}H \qquad H-\overset{\displaystyle H\ H}{\underset{\displaystyle H\ H}{C}}-C-\ddot{O}-H$

(d) $H{:}\ddot{O}{:}H \qquad H-\ddot{O}-H$

(e) $\ddot{O}{:}{:}C{:}{:}\ddot{O} \qquad \ddot{O}{=}C{=}\ddot{O}$

(f) $:\ddot{F}{:}Be{:}\ddot{F}: \qquad :\ddot{F}-Be-\ddot{F}:$

(g) $:N{:}{:}{:}N{:} \qquad :N{\equiv}N:$

(h) $H{:}\overset{\displaystyle :O:}{C}{:}\ddot{O}{:}H \qquad H-\overset{\displaystyle :O:}{\overset{\|}{C}}-\ddot{O}-H$

(i) $\cdot\ddot{N}{:}{:}\ddot{O} \qquad \cdot\ddot{N}{=}\ddot{O}$

(j) $:C{:}{:}{:}\ddot{O} \qquad :C{\equiv}O:$

3. b **4.** b **5. (a)** polar covalent **(b)** nonpolar covalent **(c)** ionic **(d)** polar covalent **(e)** polar covalent **6.** c **7.** b **8.** a

9. (a)

$$\overset{\displaystyle :O:}{\cdot N}\underset{\displaystyle :\ddot{O}:}{} \longleftrightarrow \overset{\displaystyle :\ddot{O}:}{\cdot N}\underset{\displaystyle :O:}{}$$

(Additional contributors could be drawn showing a lone pair on N and the unpaired electron on O.)

(b) $\left[H-\ddot{O}-\overset{\displaystyle :\ddot{O}:}{C}-\ddot{O}:\right]^- \longleftrightarrow \left[H-\ddot{O}-\overset{\displaystyle :\ddot{O}:}{C}{=}\ddot{O}\right]^-$

(c) $:\ddot{N}-N{=}\ddot{O} \longleftrightarrow :\ddot{N}{=}N-\ddot{O}:$ (Other possible contributors are
$:\ddot{N}{=}N{=}\ddot{O}:$ and $:N{\equiv}N-\ddot{O}:$)

(d) $\left[H-C-\ddot{\underset{..}{O}}: \right]^{-} \longleftrightarrow \left[H-C=\ddot{O} \right]^{-}$
$\qquad \overset{|}{:\underset{..}{O}:} \qquad\qquad \overset{|}{:\underset{..}{O}:}$

6. $\left[H-\overset{\overset{H}{|}}{\underset{\underset{H}{|}}{C}}-C-\ddot{\underset{..}{O}}: \right]^{-} \longleftrightarrow \left[H-\overset{\overset{H}{|}}{\underset{\underset{H}{|}}{C}}-C=\ddot{O} \right]^{-}$
$\qquad\qquad\qquad\quad :\underset{..}{O}: \qquad\qquad\qquad :\underset{..}{O}:$

UNIT 8
Exercises

10. (a) $H:\overset{H}{\underset{H}{\overset{..}{C}}}:\overset{H}{\underset{H}{\overset{..}{C}}}:\overset{H}{\underset{H}{\overset{..}{C}}}:\ddot{\underset{..}{O}}:H$ $\left(H-\overset{\overset{H}{|}}{\underset{\underset{H}{|}}{C}}-\overset{\overset{H}{|}}{\underset{\underset{H}{|}}{C}}-\overset{\overset{H}{|}}{\underset{\underset{H}{|}}{C}}-\ddot{\underset{..}{O}}-H \right)$

$H:\overset{H}{\underset{H}{\overset{..}{C}}}:\overset{H}{\underset{H}{\overset{..}{C}}}:\ddot{\underset{..}{O}}:\overset{H}{\underset{H}{\overset{..}{C}}}:H$ $\left(H-\overset{\overset{H}{|}}{\underset{\underset{H}{|}}{C}}-\overset{\overset{H}{|}}{\underset{\underset{H}{|}}{C}}-\ddot{\underset{..}{O}}-\overset{\overset{H}{|}}{\underset{\underset{H}{|}}{C}}-H \right)$

$H:\overset{H}{\underset{H}{\overset{..}{C}}}:\overset{H}{\underset{:\underset{..}{O}:}{C}}:\overset{H}{\underset{H}{\overset{..}{C}}}:H$ $\left(H-\overset{\overset{H}{|}}{\underset{\underset{H}{|}}{C}}-\overset{\overset{H}{|}}{\underset{\underset{\overset{|}{H}}{:\underset{..}{O}:}}{C}}-\overset{\overset{H}{|}}{\underset{\underset{H}{|}}{C}}-H \right)$

(b) $H-\overset{\overset{H}{|}}{\underset{\underset{H}{|}}{C}}-\overset{\overset{H}{|}}{\underset{\underset{\overset{|}{H}}{:\underset{..}{O}:}}{C}}-\overset{\overset{H}{|}}{\underset{\underset{H}{|}}{C}}-H$

Self-Test

1. (a)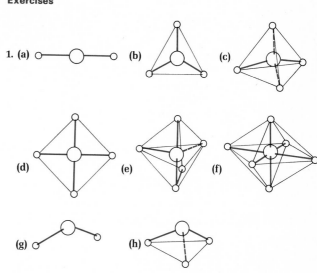
Br   H **(b)** O   Cl

2. (a) $\left[H:\overset{H}{\underset{H}{\overset{H}{\overset{..}{C}}}}:\overset{H}{\underset{H}{\overset{..}{N}}}:H \right]^{+}$ $\left[H-\overset{\overset{H}{|}}{\underset{\underset{H}{|}}{C}}-\overset{\overset{H}{|}}{\underset{\underset{H}{|}}{N}}-H \right]^{+}$

(b) $H:\overset{H}{\underset{H}{\overset{..}{C}}}:\overset{H}{\underset{H}{\overset{..}{C}}}:H$ $H-\overset{\overset{H}{|}}{\underset{\underset{H}{|}}{C}}-\overset{\overset{H}{|}}{\underset{\underset{H}{|}}{C}}-H$

(c) $H:\overset{..}{\underset{H}{N}}:H$ $H-\overset{..}{\underset{\underset{H}{|}}{N}}-H$

(d) $\left[:\overset{:\ddot{O}:}{\underset{:\ddot{O}:}{\ddot{O}:\ddot{Cl}:\ddot{O}:}} \right]^{-}$ $\left[:\overset{:\ddot{O}:}{\underset{:\ddot{O}:}{\ddot{O}-Cl-\ddot{O}:}} \right]^{-}$

(e) $H:\overset{H}{\underset{H}{\overset{:\ddot{O}:}{C}}}:\overset{:\ddot{O}:}{C}:\ddot{O}:H$ $H-\overset{\overset{H}{|}}{\underset{\underset{H}{|}}{C}}-\overset{\overset{:\ddot{O}:}{||}}{C}-\ddot{O}-H$

3. (a) I **(b)** PC **(c)** NP **(d)** PC **(e)** I   **4.** 4.8 tons   **5.** d

1. (a) o——◯——o **(b)** **(c)**

(d) **(e)** **(f)**

(g) o——◯——o **(h)**

2. (a) trigonal planar **(b)** linear **(c)** pyramidal (angles $<$ 109.5°) **(d)** trigonal planar **(e)** trigonal planar **(f)** tetrahedral **(g)** octahedral **3. (a)** sp^2 from boron with p from fluorine **(b)** p from arsenic with s from hydrogen **(c)** sp^3 from oxygen with s from hydrogen **(d)** sp^3 from carbon with p from chlorine **(e)** sp^2 from phosphorus with p from chlorine for three bonds; pd from phosphorus with p from chlorine for two bonds **(f)** sp^3d^2 from selenium with p from fluorine **(g)** sp from beryllium with s from hydrogen **(h)** sp^3 from nitrogen with s from hydrogen **4. (a)** side view

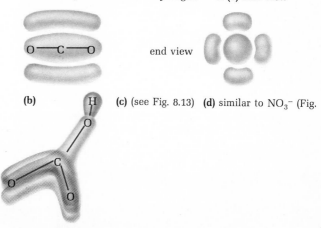

O——C——O   end view

(b) **(c)** (see Fig. 8.13) **(d)** similar to NO_3^- (Fig. 8.12) **(e)** (see Fig. 8.11)

5. (a) 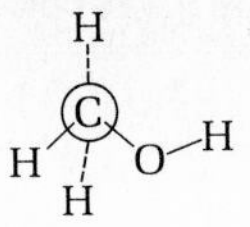 (polar)

(b) H—C≡N (polar)
(linear)

(c) 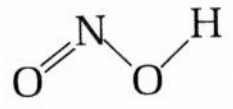 (polar)

(d) 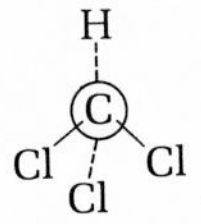 (polar)

(e) 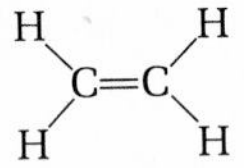 (nonpolar)
(coplanar)

Practice for Proficiency

1. (see Fig. 8.1)

2. (a) 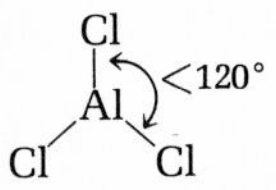 <120°

(b) 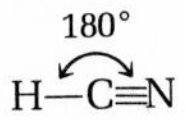 H—C≡N 180°   **(c)** :—S⟨ <120° (with O's)

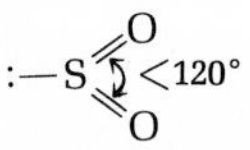

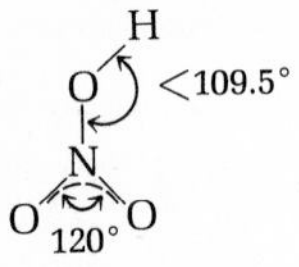

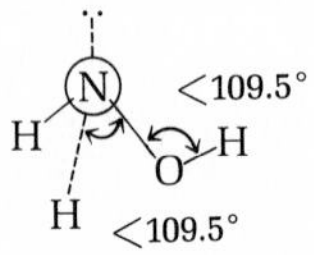

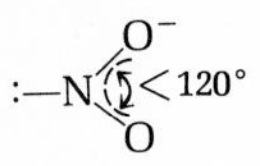

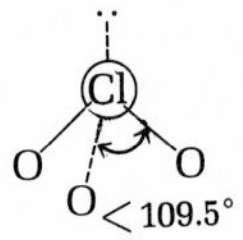

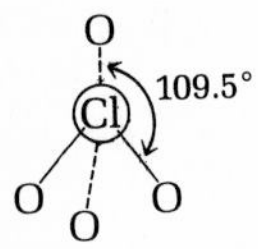

(d) <109.5°, 120°

(e) <109.5°, <109.5°

(f) :—N⟨ <120°   **(g)** <109.5°

(h) 109.5°   **(i)** 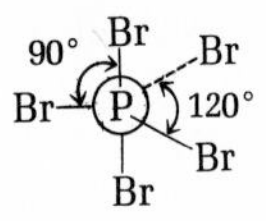 90°, 120°

(j) 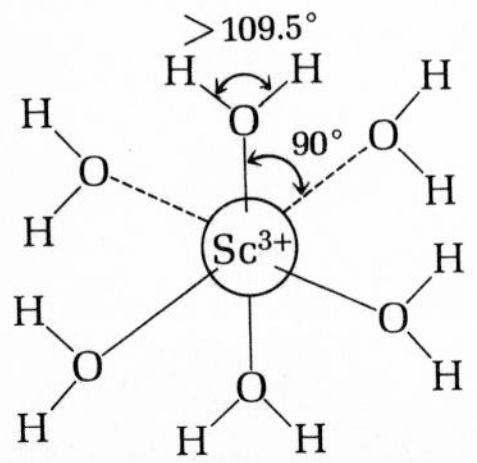 >109.5°, 90°

3. e **4.** c **5.** (check with instructor regarding "pictorial" representations) **(a)** p orbital from selenium with s orbital from hydrogen **(b)** sp orbitals from carbon with s orbital from hydrogen and sp from nitrogen **(c)** sp^2 from carbon with s from hydrogens and sp^2 from oxygen

(d)*

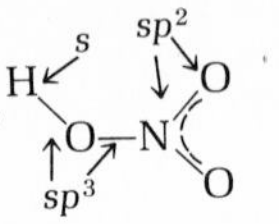

(e)* 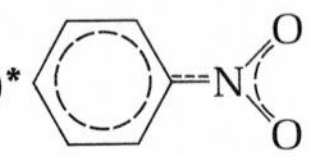
sp^2 throughout

(f)* 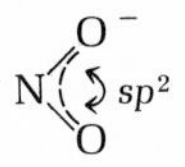 sp^2

(g) sp^3 from nitrogen with s from hydrogen

(h)*

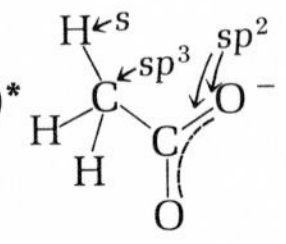

(i)* 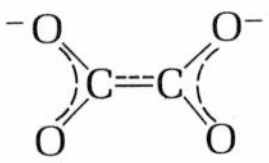
sp^2 throughout

(j) 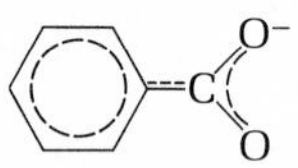
sp^2 throughout

6. c **7.** d

8. (a)

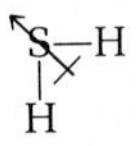

(b) S=C=S (zero net dipole)
linear

(c)

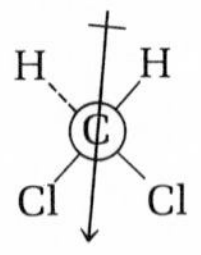

(d)

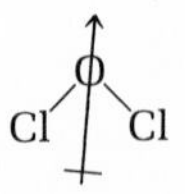

(e) 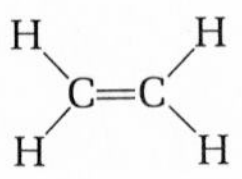 (zero net dipole)
coplanar

* Asterisk (*) denotes delocalized π bonding.

(f)

H—B—H
|
H

(g) N=O
|
Cl

(h)

Cl—Al—Cl
|
Cl

(zero net dipole)

trigonal
planar

(i) Cl—⬡—Cl   (zero net dipole)

(j) ⬡—O—H

9. a   **10.** a, c, d

Self-Test

1. (a) C **(b)** A **(c)** B **(d)** D **(e)** F **(f)** E   **2. (a)** $>90°$
(b) 120° **(c)** 120° **(d)** 109.5° **(e)** 120°   **3.** b   **4.** c   **5.** b
and **c** are polar:

(b) :S with O above and O below (dipole arrows)   **(c)** H—C≡N

UNIT 9
Exercises

1. (a) C **(b)** G **(c)** J **(d)** H **(e)** D **(f)** I **(g)** B **(h)** E **(i)** A
(j) F

2. (a) —◯→   **(b)** ◯—◯—◯   **(c)** (bent molecule with dipole arrows)

(d) (bent molecule with dipole arrows)   **(e)** (bent molecule with dipole arrows)

3. (a) 3T, 3V, 3R **(b)** 3T, 1V, 2R   **(c)** 3T, 4V, 2R **(d)** 3T, 6V,
3R **(e)** 3T, 7V, 2R **(f)** 3T, 12V, 3R   **4. (a)** ionic bonds
(b) metallic bonding **(c)** London forces **(d)** covalent bonds
(e) hydrogen bonds   **5. (a)** (1) CH_3OH [65°], CH_3SH [7°]; (2)

CHF_3 [−82°], $CHCl_3$ [61°]; (3) (furan ring with O) [33°], (thiophene ring with S) [84°] **(b)** (1)

benzene, (2) water, (3) water, (4) chloroform, and (5) chloroform

Practice for Proficiency

1. (a) F **(b)** C **(c)** B **(d)** E **(e)** D **(f)** G **(g)** A **(h)** H **(i)** I **(j)** M
(k) H **(l)** K **(m)** J   **2. (a)** F **(b)** T **(c)** F **(d)** T **(e)** T **(f)** F
(g) T **(h)** T **(i)** F   **3. (a)** 1 **(b)** 4 **(c)** 3 **(d)** 9 **(e)** 12 **(f)** 4
(g) 6 **(h)** 30 **(i)** 6 **(j)** 7   **4. (a)** (see Fig. 9.1f) **(b)** (see Fig.
9.1d) **(c)** (see Fig. 9.1e) **(d)** (see Fig. 9.1c) **(e)** (see Fig. 9.6)
(f) (similar to NH_3, Example 1)   **5. (a)** ionic bond
(b) dipole–ion force **(c)** hydrogen bond **(d)** dipole–dipole
force **(e)** London dispersion forces **(f)** metallic bond
(g) dipole-induced dipole force **(h)** nonpolar covalent bond
(i) polar covalent bond   **6. (a)** 4 **(b)** 4 **(c)** 3 **(d)** 2 **(e)** 4
(f) 6 **(g)** 2 **(h)** 9 **(i)** 5 **(j)** 1   **7. (a)** $C_3H_7CH_2OH$ [117°],
$C_3H_7C=O$ [74°] **(b)** $C_2H_5OC_2H_5$ [34°], CCl_4 [76°]

(c) ⬡—NO_2 [211°], ⬡—CH_3 [108°] **(d)** HF [20°],

HCl [−85°] **(e)** Cl—⬡—Cl [174°], ⬡(Cl)(Cl) [181°]

(f) $CH_3CH_2CH_2NH_2$ [48°], $(CH_3)_3N$ [3°] **(g)** HI [−35°], ICl [97°]

(h) ⬡—F [85°], ⬡—I [188°] **(i)** CH_3CCH_3 (with O double bond) [56°],

CH_3COH [118°], **(j)** CF_4 [−132°], SiF_4 [−86°]
8. (a) water **(b)** chloroform **(c)** water **(d)** benzene
(e) chloroform **(f)** water **(g)** benzene **(h)** chloroform
(i) water **(j)** chloroform   **9.** b. (1-octanol)   **10.** The most
important characteristic is hydrogen bonding, resulting in
strong forces among water molecules (contrary to the "ideal"
assumption of negligible interparticle forces). Of considerably
less importance, vibrational excitation may occur on collision
so that there is a net loss of *translational* kinetic energy.
Self-Test

1. (a) I **(b)** E **(c)** F **(d)** C **(e)** B **(f)** H **(g)** D **(h)** A **(i)** J
(j) G

2. (a) —◯→   **(b)** (structure with dipole arrows)

(c) ◯=◯=◯ (with dipole arrows)   **(d)** (bent molecule with dipole arrows)

3. (a) 1 **(b)** 3 **(c)** 6 **(d)** 9

4. CH_3(C=O)CH_3 ⋯ H—O—CH_3 (hydrogen bond to acetone)

5. (a) [benzene-OH structure] [182°], [benzene-SH structure] [170°]

(b) $CH_3CH_2OCH_2CH_3$ [35°], $CH_3CH_2SCH_2CH_3$ [92°]
(c) CH_3CH_2F [−38°], CH_3CH_2I [72°]   **6. (a)** B **(b)** W **(c)** C

UNIT 10
Exercises

1. (a) Kelvin **(b)** 273°K; 760 (1.00) **(c)** PV **(d)** atm; liters
2. (a) 48 ml **(b)** 1.13×10^{-3} g cm^{-3} **(c)** 1.37 atm **(d)** 37 g mole^{-1} **3. (a)** 5.9×10^6 liters **(b)** 4.2×10^4 liters
4. (a) 0.026 **(b)** 633 torr **(c)** 0.073 g   **5. (a)** 40 atm **(b)** 39 atm (Attractions among water molecules favor inelastic collisions and reduce number of particles colliding with container walls per unit time.)

Practice for Proficiency

1. (a) 22.4 **(b)** 0.082 **(c)** 273 **(d)** nRT, P_2V_2/T_2 **2. (a)** 3.4 liters **(b)** 34 ml **(c)** 0.564 g liter^{-1} **(d)** 65 g mole^{-1}
3. 84 ml   **4.** 0.89 g liter^{-1} **5.** 5.9×10^3 torr   **6.** 44 g mole^{-1};   **7.** 19 ml   **8.** 2.8 g liter^{-1} **9.** 4.6×10^3 torr (6.0 atm)   **10.** 34 g mole^{-1} **11.** 72 liters   **12.** 633 liters
13. 211 g   **14.** 676 liters   **15.** 27 g   **16.** 112 g   **17.** 36 atm (2.7×10^4 torr)   **18.** 18.2%   **19.** 76 g   **20.** 15 g
21. 34 g   **22.** 28 g   **23.** 21 g   **24. (a)** 676 torr **(b)** 0.95 **(c)** 0.143 g   **25. (a)** 9.8 atm (ideal); 9.9 atm (v. d. W.)*
(b) 9.8 atm (ideal); 9.0 (v. d. W.) **(c)** 1.6 atm (ideal); 1.7 (v. d. W.) (Helium is "nearly ideal" at both 300° and 50°K, but strong intermolecular forces among Cl_2 molecules have a significant effect at 300°K.)   **26.** $^{237}_{92}U$

Self-Test

1. (a) 273°K; 1.00 (760) **(b)** 22.4 **(c)** nR   **2. (a)** 44.8 liters
(b) 6.5×10^{-4} g cm^{-3} **(c)** 150 ml **(d)** 800 torr   **3.** 72 g mole^{-1} **4. (a)** 0.051 **(b)** 598 torr **(c)** 3.2×10^{-3} g   **5.** 53 liters   **6.** 12.1 atm (ideal: 12.3 atm)

UNIT 11
Exercises

1. (a) C **(b)** D **(c)** H **(d)** G **(e)** F **(f)** E **(g)** A **(h)** B
2. (a) diagonal line through back left corner of the top face to front right corner of bottom face **(b)** any plane that cuts the cube into equal halves, **(c)** exact center of cube
3. (a) molecular, might melt below 0°C (actually melts at 114°C), probably nonconductor **(b)** metallic, probably melts well above 0°C, good conductor **(c)** ionic, melts above 0°C, nonconductor **(d)** covalent (network), *very* high melting point, good conductor because of mobile electrons between planar network sheets **(e)** covalent (network), *very* high melting point, nonconductor   **4. (a)** ethylene glycol **(b)** −10°C
(c) 40°C   **5. (a)** miscible **(b)** predicted immiscible (actually miscible, as described in Unit 9) **(c)** miscible   **6. (a)** (see Fig. 11.21), CaO **(b)** (see Fig. 11.20), HfO₂ **(c)** (see Fig. 11.20), CuI   **7.** KIO₃, potassium iodate   **8. (a)** 1.3 g cm^{-3}
(b) 2.29 g cm^{-3} **(c)** 1.75 g cm^{-3} **9.** 5.25 millipoise
10. 32 dyn cm^{-1}

* V. d. W. = van der Waal's.

Practice for Proficiency

1. (a) C **(b)** B **(c)** D **(d)** H **(e)** A **(f)** G **(g)** E **(h)** F
2. (see Fig. 11.10)   **3. (a)** (see Fig. 11.20) **(b)** (see Fig. 11.21)   **4. (a)** (see Fig. 11.21) **(b)** (see Fig. 11.20)
5. (a, b) (see Fig. 11.11) **(c)** plane bisecting both carbon atoms and perpendicular to the plane of the molecule **(d)** any plane cutting the unit cell into equal halves **(e)** exact center of cube   **6.** a   **7.** e   **8. (a)** glycerin **(b)** 10°C **(c)** 100°C
9. b and c   **10.** d   **11.** a   **12. (a)** HfO₂ **(b)** MnO
(c) SnS **(d)** FeS **(e)** SnCl₂   **13.** BaO   **14. (a)** 3.4 g cm^{-3}
(b) 2.7 g cm^{-3} **(c)** 1.42 g cm^{-3} **15.** 14.2 millipoise
16. 51 dyn cm^{-1} **17.** 16.0 millipoise   **18.** 5.3 Å

Self-Test

1. (a) A **(b)** G **(c)** H **(d)** C **(e)** B **(f)** D **(g)** I **(h)** E

2. (a) (see Fig. 11.10) **(b)**

3. c   **4. (a)** metallic (best conductivity) **(b)** ionic
(c) molecular **(d)** atomic (lowest melting point) **(e)** covalent [network]   **5. (a)** glycerin **(b)** 0° **(c)** 50°   **6.** a and b
7. SnO   **8.** Cu₂O   **9.** 1.54 g cm^{-3} **10. (a)** 14.6 millipoise
(b) 47 dyn cm^{-1}

UNIT 12
Exercises

1. (a) C **(b)** A **(c)** E **(d)** B **(e)** D   **2. (a)** approx. 250 torr
(b) approx. 4 atm **(c)** approx. 50°C **(d)** spontaneous vaporization **(e)** spontaneous vaporization   **3.** 197.0, gold
4. 6.3 kcal   **5. (a)** ~150 torr **(b)** 79°C

Practice for Proficiency

1. (a) D **(b)** C **(c)** B **(d)** A **(e)** H **(f)** I **(g)** E **(h)** G **(i)** F
2. b   **3.** e   **4.** c   **5. (a)** approx. 40 torr **(b)** approx. −75°C **(c)** approx. 80°C **(d)** approx. 3300°C, 130 atm
6. a   **7.** a   **8.** c   **9.** b   **10.** chromium   **11.** cadmium   **12.** −3.9 kcal   **13.** 26 kcal   **14.** 2.5 kcal
15. −1.46 kcal   **16.** 67°C   **17.** 65°C   **18. (a)** 0.2 torr
(b) 0.5 torr **(c)** 0.8 torr   **19. (a)** 220°C **(b)** 115°C
(c) 6.8 torr   **20.** −18.7 kcal

Self-Test

1. (a) G **(b)** D **(c)** H **(d)** B **(e)** C **(f)** F   **2.** e   **3.** 195 (platinum)   **4.** 4.0 kcal   **5. (a)** 61 torr **(b)** 7°C

UNIT 13
Exercises

1. (a) no. gram-formula weights solute **(b)** no. kilograms solvent **(c)** no. gram-equivalents solute **(d)** (no. g solute)/[no. g (solute + solvent)] **(e)** no. kilograms solution   **2. (a)** Add 9.3 g of the salt to <125 ml distilled water; mix thoroughly; dilute to 125 ml; mix thoroughly **(b)** Pour about 420 ml concentrated H_2SO_4 carefully (with stirring) into <24.5 liters distilled water and mix thoroughly; when cool, dilute to 25.0 liters; mix thoroughly **(c)** Transfer 1.1 ml concentrated HNO_3 carefully (with stirring) into <98 ml distilled water; mix thoroughly; dilute to 100 ml (when cool); mix thoroughly
3. (a) 16 M **(b)** 37 m **(c)** 16 N   **4. (a)** 100.5 **(b)** 111.6
(c) 25.1   **5.** 0.328 M   **6.** 6.0 ml

Practice for Proficiency

1. (See Table 13.2.)   **2. (a)** Add 6.1 g of the salt to $<$250 ml distilled water; mix thoroughly; when cool, dilute to 250 ml; mix thoroughly   **(b)** Transfer 150 ml of the concentrated acid carefully (with stirring) into $<$2.8 liters of distilled water; mix thoroughly; when cool, dilute to 3.0 liters   **(c)** Add 1.3 g $KMnO_4$ to $<$125 ml distilled water; mix thoroughly; dilute to 125 ml; mix thoroughly   **(d)** Add 4.3 g $MgSO_4$ to $<$25 liters distilled water; dilute to 25 liters; mix thoroughly   **(e)** A mixture of 8.6 g NaCl and 991.4 g H_2O would give the correct composition, but probably less than 1.00 liter, so about a 10% excess should be used (9.5 g NaCl and 1090 g H_2O), mixing thoroughly   **3. (a)** 15 M **(b)** 58 m **(c)** 44 N **4. (a)** 63.02   **(b)** 21.01   **(c)** 71.00   **(d)** 8.74   **(e)** 32.67   **(f)** 41.0 (H_3PO_3 only has two "acidic" hydrogens.)   **(g)** 16.02 (capable of accepting two protons)   **5. (a)** 7.93 M, 10.6 m, 7.93 N **(b)** 11 M, 25 m, 11 N **(c)** 3.01 M, 3.40 m, 6.02 N **(d)** 3.61 M, 4.37 m, 10.8 N **(e)** 10.8 M, 14.7 m, 10.8 N **6. (a)** 22 g   **(b)** 43 g   **(c)** 11 g   **(d)** 3.6 g   **(e)** 22 g   **(f)** 1.6 g   **(g)** 13 g   **(h)** 100 g   **(i)** 0.29 g   **(j)** 75 g   **7. (a)** 8.3 ml   **(b)** 90 ml   **(c)** 13 ml   **(d)** 2.9 ml   **(e)** 9.5 ml   **8.** 3.3 ml   **9.** 36 ml   **10.** 102 g   **11.** 1.2 g   **12.** 28.6 ml   **13.** 1.7 liters   **14.** 27 liters   **15.** 13 liters   **16.** 714 ml   **17.** 78 ml   **18.** 14 liters   **19.** 16 g   **20.** 25 g

Self-Test

1. (a) no. gram-equivalents solute   **(b)** no. kilograms solvent   **(c)** (no. moles solute)/(no. liters solution)   **2.** 37 m **3.** 9.7 g   **4.** 0.30 liters (300 ml)   **5. (a)** 30.6   **(b)** 32.7   **(c)** 57   **6. (a)** 6.02 M, 8.02 m, 12.0 N **7.** 0.158 M **8.** 0.66 g

UNIT 14

Exercises

(Discuss the examples you selected with your instructor.)

Practice for Proficiency

1. (a) foam (solid)   **(b)** aerosol (smoke)   **(c)** emulsion   **(d)** sol   **(e)** foam (liquid)   **2.** lyophobic   **3. (a)** rigidity   **(b)** thermal convection currents (and probably electrostatic repulsions)   **(c)** thermal convection currents   **(d)** electrostatic repulsions of micelles   **(e)** semirigidity   **4, 5.** (Discuss your choices with your instructor.)

Self-Test

1. (a) D   **(b)** A   **(c)** G   **(d)** F   **(e)** C   **(f)** B   **(g)** E   **2.** philic, phobic   **3. (a)** rigidity   **(b)** rigidity   **(c)** surface charge   **(d)** surface charge   **(e)** "other factors"   **4. (a)** shaking very finely powdered $CaCO_3$ with water, or carefully mixing appropriate aqueous solutions of Ca^{2+} and CO_3^{2-} **(b)** (See discussion of Fig. 14.6.)

UNIT 15

Exercises

1. only **c** and **d**   **2.** 30.7 torr   **3. (a)** 107.0°C   **(b)** -22.7°C   **(c)** 205 atm   **4.** $\sim$160 g mole^{-1} **5.** $\sim$92,000 g mole^{-1}

Practice for Proficiency

1. only **a** and **c**   **2.** 231 torr   **3.** -10.4°C   **4.** 15 atm   **5.** 100.3°C   **6.** -2.2°C   **7.** 8.2 atm   **8.** 100.7°C   **9.** 22 atm   **10.** -50°C   **11.** 126 g mole^{-1} **12.** 206 g

mole^{-1} **13.** 168 g mole^{-1} **14.** $\sim$39,000 g mole^{-1} **15.** $\sim$56,000 g mole^{-1} **16. (a)** 703 torr   **(b)** 0.92

Self-Test

1. 89 torr   **2.** 91.5°C   **3.** -2.2°C   **4.** 7.7 atm   **5.** $\sim$260 g mole^{-1}

UNIT 16

Exercises

1. (a) D   **(b)** A   **(c)** E   **(d)** B   **(e)** C   **2. (a)** bimolecular   **(b)** $N_2H_2O_2$ **(c)** N_2O_2, N_2O **(d)** third order   **3. (a)** $Ag^+(aq) + Cl^-(aq) \longrightarrow AgCl(s)$ (particle size and electrostatic attractions)   **(b)** at 250°C (more frequent collisions and higher percentage particles of high energy)   **(c)** with Fe^{3+} catalyst (alternate pathway of lower activation energy)   **4. (a)** more frequent collisions and (of major importance) higher percentage of high-energy collisions   **(b)** alternate pathway of lower activation energy, possibly involving weakening of bonds through chemisorption effects   **5. (a)** rate $= k \times C_{N_2O_5}$ **(b)** rate $= k \times C_{Cl_2} \times C_{Fe^{2+}}$ **6.** rate $= k \times C_{ClO_2}^2 \times C_{OH^-}$, $k = 1.2 \times 10^2$ liter^2 mole^{-2} sec^{-1} **7.** (others are possible)

   (a) *slow* $N_2O_5 \longrightarrow N_2O_4 + O$
        fast $N_2O_5 + O \longrightarrow N_2O_4 + O_2$
   (b) *slow* $Cl_2 + Fe^{2+} \longrightarrow Cl + Cl^- + Fe^{3+}$
        fast $Cl + Fe^{2+} \longrightarrow Cl^- + Fe^{3+}$

8. (others are possible)

    fast $ClO_2 + OH^- \rightleftharpoons HClO_3^-$ [steady state]
    slow $HClO_3^- + ClO_2 \longrightarrow ClO_3^- + HClO_2$
    fast $HClO_2 + OH^- \longrightarrow ClO_2^- + H_2O$
    ALTERNATIVE:
    fast $2ClO_2 \rightleftharpoons Cl_2O_4$ [steady state]
    slow $Cl_2O_4 + OH^- \longrightarrow ClO_3^- + HClO_2$
    fast $HClO_2 + OH^- \longrightarrow ClO_2^- + H_2O$

9. *fast* $ClO_2 + {}^{18}_8OH^- \rightleftharpoons HO_2Cl^{18}_8O^-$ [steady state]
      slow $HO_2Cl^{18}_8O^- + ClO_2 \longrightarrow O_2Cl^{18}_8O^- + HClO_2$
      fast $HClO_2 + {}^{18}_8OH^- \longrightarrow ClO_2^- + H_2{}^{18}_8O$

10. (a) 38 kcal mole^{-1} **(b)** 3.5×10^{-4}

Practice for Proficiency

1. (a) D   **(b)** A   **(c)** B   **(d)** C   **(e)** F   **(f)** E   **(g)** G   **(h)** H   **2. (a)** The aqueous process is more rapid because of better collision possibilities and because of electrostatic interactions among "free" Ag^+ and Br^- ions.   **(b)** 250°C (more frequent collisions and greater percentage of high-energy collisions)   **3. (a)** more frequent collisions and greater percentage of high-energy collisions   **(b)** alternate pathway of lower activation energy (probably involving adsorption of CO and O_2 molecules)   **(c)** proper collision geometries (probably infrequent with large molecules, such as cyclohexene)   **(d)** rate law is first order in both $S_2O_8^{2-}$ and I^- (slow step depends on effective collision of I^- with $S_2O_8^{2-}$)   **4.** rate $= k \times C_{CH_3CHO}^2$ **5.** rate $= k \times C_{S_2O_8^{2-}} \times C_{I^-}$ **6. (a)** first order   **(b)** 0.29 hr^{-1} **(c)** 2.9×10^{-2} mole liter^{-1} hr^{-1} **7. (a)** rate $= k \times C_{H_2} \times C_{NO}^2$ **(b)** third order   **(c)** 1.4 liter^2 mole^{-2} hr^{-1} **8.** rate $= k \times C_{ClO_2} \times C_{F_2}$

9. (others possible)

FOR QUESTION 4:

slow $2CH_3CHO \longrightarrow [C_4H_8O_2]$

fast $[C_4H_8O_2] \longrightarrow CH_4 + CO + CH_3CHO$

FOR QUESTION 5:

slow $S_2O_8{}^{2-} + I^- \longrightarrow SO_4{}^{2-} + ISO_4{}^-$

fast $ISO_4{}^- + I^- \longrightarrow SO_4{}^{2-} + I_2$

fast $I_2 + I^- \longrightarrow I_3{}^-$

FOR QUESTION 6:

slow $N_2O_5 \longrightarrow NO_2 + NO_3$

fast $N_2O_5 + NO_3 \longrightarrow NO_2 + N_2O_6$

fast $N_2O_6 \longrightarrow 2NO_2 + O_2$

10. (others possible)

fast $H_2 + NO \rightleftharpoons H_2NO$ [steady state]

slow $H_2NO + NO \longrightarrow N_2 + H_2O_2$

fast $H_2O_2 + H_2 \longrightarrow 2H_2O$

ALTERNATIVE:

fast $2NO \rightleftharpoons N_2O_2$ [steady state]

slow $N_2O_2 + H_2 \longrightarrow [H_2N_2O_2]$

fast $[H_2N_2O_2] + H_2 \longrightarrow N_2 + 2H_2O$

(Intermediate identification favors second alternative.)

11. (others possible)

slow $ClO_2 + F_2 \longrightarrow F_2ClO_2$

fast $F_2ClO_2 + ClO_2 \longrightarrow 2FClO_2$

(transition state formula: F_2ClO_2)

12. *initiating* $Cl_2 \longrightarrow 2Cl\cdot$

propagating $Cl\cdot + CH_4 \longrightarrow CH_3Cl + H\cdot$

propagating $Cl\cdot + CH_4 \longrightarrow \cdot CH_3 + HCl$

terminating $2Cl\cdot \longrightarrow Cl_2$

terminating $H\cdot + Cl\cdot \longrightarrow HCl$

terminating $2\cdot CH_3 \longrightarrow C_2H_6$

(Free radicals proposed were $Cl\cdot$, $H\cdot$, and $\cdot CH_3$.)

13. (a) 52 kcal mole^{-1} **(b)** 0.65 sec^{-1} **(c)** 6×10^{12} sec^{-1}

14. (a) 6.3×10^{-1} sec^{-1} **(b)** 1.1 sec **(c)** 9.5×10^{-2} mole liter^{-1} sec^{-1} **15.** The OH^- is a catalyst. (Discuss details with your instructor.)

Self-Test

1. (a) C **(b)** B **(c)** A **(d)** D   **2.** b   **3.** a

4. (a) rate $= k \times C_{NO}^2 = C_{Cl_2}$ **(b)** rate $= k \times C_{NO_2} \times C_{F_2}$

(c) rate $= k \times C_{H_2}^2 \times C_{NO}$

5. (others possible)

(a) *fast* $2NO \rightleftharpoons N_2O_2$ [steady state]

    slow $N_2O_2 + Cl_2 \longrightarrow 2NOCl$

(b) *slow* $NO_2 + F_2 \longrightarrow [NO_2F_2]$

    fast $[NO_2F_2] + NO_2 \longrightarrow 2NO_2F$

(c) *fast* $H_2 + NO \rightleftharpoons H_2NO$ [steady state]

    slow $H_2 + H_2NO \longrightarrow H_2O + H_2N$

    fast $H_2N + NO \longrightarrow N_2O + H_2$

6. (a) 24 kcal mole^{-1} **(b)** 1.7×10^{-4} sec^{-1} **(c)** $\sim 10^{13}$ sec^{-1}

UNIT 17

Exercises

1. (a) $K = \dfrac{C_{NO_2}^2}{C_{N_2O_4}}$ **(b)** $K = \dfrac{C_{H^+}^2 \times C_{S^{2-}}}{C_{H_2S}}$

(c) $K = \dfrac{C_{CH_3NH_3^+} \times C_{OH^-}}{C_{CH_3NH_2}}$ **(d)** $K = \dfrac{C_{H_2BO_3^-}^2 \times C_{H_3BO_3}^2}{C_{B_4O_7^{2-}}}$

(e) $K = \dfrac{C_{[Co(NH_3)_6]^{2+}}}{C_{Co} \times C_{NH_3}^6}$ **(f)** $K = C_{Ag^+}^2 \times C_{S^{2-}}$

2. 2.2×10^{-5} **3.** $C_{NO} = 2.8 \times 10^{-5}$ mole liter^{-1}, $C_{O_2} = 0.79$ mole liter^{-1}, $C_{NO_2} = 2.0 \times 10^{-2}$ mole liter^{-1}

4. $C_{HCO_2H} \approx 2$ M, $C_{H^+} \approx 0.02$ M, $C_{HCO_2^-} \approx 0.02$ M, $C_{OH^-} \approx 5 \times 10^{-13}$ M **5. (a)** decrease **(b)** increase **(c)** decrease **(d)** no change

Practice for Proficiency

1. (a) $K = \dfrac{C_{NO_2}^2}{C_{NO}^2 \times C_{O_2}}$ **(b)** $K = \dfrac{C_{H^+}^2 \times C_{SO_3^{2-}}}{C_{H_2SO_3}}$

(c) $K = \dfrac{C_{HF} \times C_{OH^-}}{C_{F^-}}$ **(d)** $K = C_{Ba^{2+}} \times C_{SO_4^{2-}}$

(e) $K = \dfrac{P_{CO_2}^2}{P_{CO}^2 \times P_{O_2}}$ **(f)** $K = \dfrac{C_{[Fe(CN)_6]^{3-}}}{C_{Fe^{3+}} \times C_{CN^-}^6}$

(g) $K = \dfrac{C_{H^+} \times C_{HCO_3^-}}{C_{CO_2}}$ **(h)** $K = \dfrac{1}{C_{Al^{3+}} \times C_{OH^-}^3}$

(i) $K = C_{Mg^{2+}} \times C_{F^-}^2$ **(j)** $K = C_{H^+} \times C_{OH^-}$

2. (a) the ones for question **1(d)** and **(i)** **(b)** the one for question **1(j)** **3.** 2.2×10^{-10} mole liter^{-1} **4.** 4.5×10^{-2} atm

5. 5.6×10^{-10} **6.** $C_{CO} = 1.9 \times 10^{-13}$ mole liter^{-1}, $C_{O_2} = 0.29$ mole liter^{-1}, $C_{CO_2} = 1.5 \times 10^{-2}$ mole liter^{-1}

7. 3.0×10^{-3} M **8.** 6.0×10^{-6} M **9.** 8.0×10^{-3} M

10. 8.0×10^{-5} M **11.** $\sim 6 \times 10^{-3}$ M **12. (a)** decrease **(b)** decrease **(c)** no change **(d)** decrease **13.** a **14.** d

15. e **16.** 2.5×10^9 liters

Self-Test

1. (a) $K = \dfrac{C_{NO}^2}{C_{N_2} \times C_{O_2}}$ **(b)** $K = C_{Ag^+}^2 \times C_{S^{2-}}$

(c) $K = \dfrac{C_{HCO_3^-} \times C_{OH^-}}{C_{CO_3^{2-}}}$ **(d)** $K = \dfrac{C_{H^+}^2 \times C_{C_2O_4^{2-}}}{C_{H_2C_2O_4}}$

(e) $K = \dfrac{C_{H_2PO_4^-} \times C_{OH^-}^2}{C_{PO_4^{3-}}}$

2. 5.5×10^{-10} **3.** 1.9×10^{-3} mole liter^{-1}

4. 6.6×10^{-3} M **5. (a)** decrease **(b)** no change **(c)** increase **(d)** no change

UNIT 18

Exercises

1. (a) $MgCO_3(s) \rightleftharpoons Mg^{2+}(aq) + CO_3{}^{2-}(aq)$

(b) $Al(OH)_3(s) \rightleftharpoons Al^{3+}(aq) + 3OH^-(aq)$

(c) $Cu_2S(s) \rightleftharpoons 2Cu^+(aq) + S^{2-}(aq)$

2. (a) $K_{sp} = C_{Ba^{2+}} \times C_{CrO_4^{2-}} = 2.2 \times 10^{-10}$

(b) $K_{sp} = C_{Ag^+}^2 \times C_{C_2O_4^{2-}} = 5.3 \times 10^{-12}$

(c) $K_{sp} = C_{Pb^{2+}}^3 \times C_{PO_4^{3-}}^2 = 1.5 \times 10^{-32}$

3. only a **4. (a)** 1.4×10^{-5} g $(100$ ml$)^{-1}$ **(b)** 0.44 g $(100$ ml$)^{-1}$ **(c)** 0.023 g $(100$ ml$)^{-1}$ **5. (a)** 3.5×10^{-3} g $(100$ ml$)^{-1}$ **(b)** 5.6×10^{-5} g $(100$ ml$)^{-1}$ **(c)** 5.7×10^{-11} g $(100$ ml$)^{-1}$

6. (a) 3.5 g **(b)** three

Practice for Proficiency

1. (a) $BaSO_4(s) \rightleftharpoons Ba^{2+}(aq) + SO_4{}^{2-}(aq)$

(b) $Fe(OH)_3(s) \rightleftharpoons Fe^{3+}(aq) + 3OH^-(aq)$

(c) $Ag_2CrO_4(s) \rightleftharpoons 2Ag^+(aq) + CrO_4{}^{2-}(aq)$

(d) $PbCl_2(s) \rightleftharpoons Pb^{2+}(aq) + 2Cl^-(aq)$

(e) $Li_2CO_3(s) \rightleftharpoons 2Li^+(aq) + CO_3^{2-}(aq)$
(f) $Ca_3(PO_4)_2(s) \rightleftharpoons 3Ca^{2+}(aq) + 2PO_4^{3-}(aq)$
2. (a) $K_{sp} = C_{Ba^{2+}} \times C_{SO_4^{2-}}$ (b) $K_{sp} = C_{Fe^{3+}} \times C_{OH^-}^3$
(c) $K_{sp} = C_{Ag^+}^2 \times C_{CrO_4^{2-}}$ (d) $K_{sp} = C_{Pb^{2+}} \times C_{Cl^-}^2$
(e) $K_{sp} = C_{Li^+}^2 \times C_{CO_3^{2-}}$ (f) $K_{sp} = C_{Ca^{2+}}^3 \times C_{PO_4^{3-}}^2$
3. (a) $K_{sp} = C_{Ag^+} \times C_{CH_3CO_2^-}$ (b) $K_{sp} = C_{Pb^{2+}} \times C_{Br^-}^2$
(c) $K_{sp} = C_{Ag^+}^3 \times C_{PO_4^{3-}}$ (d) $K_{sp} = C_{Cr^{3+}} \times C_{OH^-}^3$
(e) $K_{sp} = C_{Pb^{2+}}^3 \times C_{AsO_4^{3-}}^2$
4. (a) $K_{sp} = C_{Mg^{2+}} \times C_{F^-}^2 = 6.9 \times 10^{-9}$
(b) $K_{sp} = C_{Ba^{2+}}^3 \times C_{PO_4^{3-}}^2 = 1.7 \times 10^{-23}$
5. (a) 5.0×10^{-13} (b) 1.0×10^{-6} (c) 1.6×10^{-18}
(d) 1.2×10^{-11} (e) 1.3×10^{-29} 6. only **b**
7. (a) 6.9×10^{-4} g $(100 \text{ ml})^{-1}$ (b) 3.1×10^{-9} g $(100 \text{ ml})^{-1}$
8. (a) 2.4×10^{-12} g $(100 \text{ ml})^{-1}$ (b) 8.0×10^{-35} g $(100 \text{ ml})^{-1}$
9. c   10. c   11. e   12. b   13. d   14. e   15. d
16. c   17. e   18. d   19. d   20. c   21. a   22. a
23. c   24. (a) 4.8   (b) 5.4 g   25. 3.3×10^{-3} ppm
Self-Test
1. $1.5 \times 10^{-3} M$ 2. 9.8×10^{-8} 3. no   4. 3.5×10^{-6}
g $(100 \text{ ml})^{-1}$ 5. 4.9×10^{-20} g $(100 \text{ ml})^{-1}$ 6. ~46   7. two

UNIT 19
Exercises

1. (a) e.g., $Al(OH)_3$ (b)

conjugate pair

$F^- \; + \; H_2O \; \rightleftharpoons \; HF \; + \; OH^-$
(base)   (acid)   (acid)   (base)

conjugate pair

2. (a) 3.0   (b) 13.0   (c) 2.89   (d) 10.98   3. (a) 8.74   (b) 5.70
(c) 8.73   4. (a) chloric > sulfurous > boric, (b) $K_{a_2} \approx 10^{-7}$;
$K_{a_3} \approx 10^{-12}$ 5. pH = 10.9 (alizarin yellow indicator)
Practice for Proficiency

conjugate pair

1. acid, no   2. $CH_3CO_2^- + H_2O \rightleftharpoons CH_3CO_2H + OH^-$
(base)   (acid)   (acid)   (base)

conjugate pair

3. a   4. d   5. b   6. c   7. d   8. (a) 2.70   (b) 11.60
(c) 2.02   (d) 12.0   9. (a) 8.32   (b) 5.7   10. (a) 3.10   (b) 10.60
(c) 3.89   (d) 10.78   (e) 2.45   11. (a) 10.98   (b) 2.69   (c) 11.60
(d) 4.10   (e) 1.12   12. a   13. e   14. (a) 4.82   (b) 9.00
(c) 4.97   (d) 11.5   (e) 10.4   15. chloric > nitrous > boric
16. d   17. b   18. b   19. a   20. (a) HNO_3 > HNO_2
(b) H_2SO_4 > H_2SO_3 (c) $H_3PO_3 \approx H_3PO_2$ (same formal charge
on phosphorus)   21. $K_{a_2} \approx 10^{-7}$; $K_{a_3} \approx 10^{-12}$ 22. 8.17
23. 6.1   24. 10.9   25. phenol red or thymol blue (basic) for
22, methyl red or bromthymol blue for 23, alizarin yellow for
24   26. 6.23% (N)
Self-Test

conjugate pair

1. $NH_4^+ + H_2O \rightleftharpoons NH_3 + H_3O^+$
(acid)   (base)   (base)   (acid)

conjugate pair

2. (see Table 19.1)   3. (a) 2.70   (b) 12.70   (c) 1.73
(d) 11.06   4. (a) 11.31   (b) 5.70
5. $HClO_4 > H_2SO_4 > HNO_2 > H_4SiO_4$ 6. 5.2, methyl red

UNIT 20
Exercises
1. (a) 4.18   (b) 4.34   (c) 6.72   2. (a) 5.0 pH units   (b) 0.10 pH
unit   3. (a) 0.09 pH unit   (b) 3.76 pH units   (c) 7.2 pH
units   4. (a) 0.19 mole (H^+) (b) 0.13 mole (OH^-) (c) 0.39
mole (H^+) 5. 6.4
Practice for Proficiency
1. (a) 4.44   (b) 4.44   2. 4.34   3. 3.7   4. 8.3   5. 10.2
6. 8.65   7. 5.11   8. 4.27   9. 7.4   10. (a) 5.3 pH units
(b) 0.01 pH unit   11. 4.92   12. 9.95   13. ~13
14. 4.64   15. (a) 0.088 mole (H^+) (b) 0.41 mole (H^+), 0.41
mole (OH^-) (c) 0.018 mole (H^+) 16. (a) 0.10 pH unit
(b) 0.04 pH unit   (c) 0.41 pH unit   17. (a) 0.11 pH unit
(b) 0.03 pH unit   (c) ~1.0 pH unit   18. (a) 0.04 pH unit
(b) 0.4 pH unit   (c) 7.4 pH units   19. (a) 0.30 pH unit
(b) 0.30 pH unit   (c) 0.09 pH unit   20. (Consult your instructor.)
Self-Test
1. 3.16   2. 8.35   3. 4.99   4. 2.64   5. 0.24 mole (H^+); 0.17
mole (OH^-) 6. 2.73

UNIT 21
Exercises
1. $C_{H_3PO_3} = 0.15 M$; $C_{H_2PO_3^-} = 4.9 \times 10^{-2} M$; $C_{H^+} = 4.9 \times$
$10^{-2} M$; $C_{HPO_3^{2-}}$ and C_{OH^-} are negligible   2. ≥ 7.91
3. 2.82 to 7.21   4. ≥ 3.13 5. (a) $7.2 \times 10^{-20} M$ (b) The
concentrations of "intermediate" complexes are *not* negligible.
Practice for Proficiency
1. $C_{H_2SO_3} = 0.42 M$; $C_{HSO_3^-} \approx 8 \times 10^{-2} M$; $C_{H^+} \approx 8 \times 10^{-2} M$;
$C_{SO_3^{2-}} \approx 1.3 \times 10^{-7} M$; $C_{OH^-} \approx 1.2 \times 10^{-13} M$;   2. (Consult
your instructor.)   3. (a) $C_{H_2SO_4}$ negligible; $C_{HSO_4^-} \approx 0.29 M$;
$C_{H^+} \approx 0.31 M$; $C_{SO_4^{2-}} \approx 0.012 M$; C_{OH^-} negligible   (b) $C_{SO_4^{2-}}$ only
slightly less than 0.3 M (simplifications inadequate because
autoprotolysis of water *not* negligible)   (c) $C_{H_3PO_4} \approx 0.12 M$;
$C_{H^+} \approx 3 \times 10^{-2} M$; $C_{H_2PO_4^-} \approx 3 \times 10^{-2} M$; other species negligible   (d) simplifications inadequate   (e) simplifications inade-
quate   4. 7.0   5. 1.1 (incomplete precipitation)   6. 7.9
7. d   8. e   9. 2.2   10. d   11. d   12. e   13. d
14. b   15. e   16. $6.4 \times 10^{-13} M$ 17. b   18. $6 \times 10^{-11} M$
19. $3 \times 10^{-20} M$ 20. $6 \times 10^{-19} M$
Self-Test
1. 1.09   2. 7.41   3. $SrCO_3$ 4. ZnS; PbS
5. $\sim 2 \times 10^{-7} M$

EXCURSION 2
Exercises
1. 13.55%   2. 35.46%   3. 6.51%   4. 0.32%

UNIT 22

Exercises

1. (a) $\overset{+I\ +III\ -II}{H\ N\ O_2}$ **(b)** $\overset{+I\ +IV\ -II}{H\ S\ O_3^-}$ **(c)** $\overset{\boxed{-4/3}\ +I\ -II}{C_3\ H_6O}$ **(d)** $\overset{0\ +I\ -II}{C_2H_4O_2}$

2. (a) $4H^+ + Cu + 2NO_3^- \longrightarrow Cu^{2+} + 2NO_2 + 2H_2O$

(b) $42H^+ + 8Al + 3\underline{Cr_2O_7^{2-}} \longrightarrow 8Al^{3+} + 6Cr^{2+} + 21H_2O$

(c) $3(CH_3)_2CHOH + 2\underline{MnO_4^-} \longrightarrow$

$$3CH_3\overset{O}{\underset{\parallel}{C}}CH_3 + 2MnO_2 + 2H_2O + 2OH^-$$

(d) $8OH^- + Mn^{2+} + 4\underline{MnO_4^-} \longrightarrow 5MnO_4^{2-} + 4H_2O$

(e) $\underline{KOCl} + KCl + 2HNO_3 \longrightarrow 2KNO_3 + Cl_2 + H_2O$

3. (a) 71% **(b)** 38% **4.** 0.210 $\underline{N}$

Practice for Proficiency

1. (a) $\overset{+V\ -II}{N\ O_3^-}$ **(b)** $\overset{+II\ +I\ +IV\ -II}{Ni\ (H\ C\ O_3)_2}$ **(c)** $\overset{-II\ +I\ -II}{C\ H_4O}$ **(d)** $K_2\overset{+I\ +VI\ -II}{Mn\ O_4}$

(e) $\overset{+II\ \ +V\ -II}{Cu_3(As\ O_4)_2}$ **(f)** $\overset{+I\ +VI\ -II}{K_2Cr_2\ O_7}$ **(g)** $\overset{-I\ +I}{C_6H_6}$ **(h)** $\overset{+I\ +III\ +VI\ -II}{K\ Al\ (S\ O_4)_2}$

(i) $\overset{+I\ +III\ -II}{Na_2C_2\ O_4}$ **(j)** $\overset{+I\ +V\ -II}{H_4P_2\ O_7}$ **2.** Cl_2 **3.** O_2 **4.** $H_2C_2O_4$

5. sulfide ion

6. (a) $2H^+ + \underline{Ag} + NO_3^- \longrightarrow Ag^+ + NO_2 + H_2O$

(b) $H^+ + 2MnO_4^- + 5\underline{HSO_3^-} \longrightarrow 2Mn^{2+} + 5SO_4^{2-} + 3H_2O$

(c) $\underline{C_6H_5CH_3} + 2MnO_4^- \longrightarrow$

$$C_6H_5CO_2^- + 2MnO_2 + OH^- + H_2O$$

(d) $2OH^- + 2\underline{Cr(OH)_4^-} + 3H_2O_2 \longrightarrow 2CrO_4^{2-} + 8H_2O$

(e) $2\underline{Al} + 2NaOH + 6H_2O \longrightarrow 2NaAl(OH)_4 + 3H_2$

7. (a) (1) Ag; (2) $\overset{\text{(or)}}{\underset{\longleftarrow}{HSO_3^-}}$; (3) $C_6H_5CH_3$; (4) $Cr(OH)_4^-$; (5) Al
(b) water

8. (a) $8OH^- + \underline{\underline{I^-}} + 4Cl_2 \longrightarrow IO_4^- + 8Cl^- + 4H_2O$

(b) $H^+ + \underline{Co^{2+}} + HNO_2 \longrightarrow Co^{3+} + NO + H_2O$

(c) $\underline{Zn} + 2Cr^{3+} \longrightarrow Zn^{2+} + 2Cr^{2+}$

(d) $\underline{\underline{I_2}} + 3H_2O \longrightarrow 5I^- + IO_3^- + 6H^+$

(e) $\underline{\underline{I_2}} + 2S_2O_3^{2-} \longrightarrow 2I^- + S_4O_6^{2-}$

(f) $2OH^- + 2\underline{Cu^{2+}} + \underline{\underline{CN^-}} \longrightarrow 2Cu^+ + CNO^- + H_2O$

(g) $2\underline{MnO_2} + 3NaBiO_3 + 10HNO_3 \longrightarrow$

$$2NaMnO_4 + 3Bi(NO_3)_3 + NaNO_3 + 5H_2O$$

(h) $4H^+ + 3\underline{MnO_4^{2-}} \longrightarrow MnO_2 + 2MnO_4^- + 2H_2O$

(i) $3\underline{\underline{H_3AsO_3}} + \underline{BrO_3^-} \longrightarrow 3H_3AsO_4 + Br^-$

(j) $2\underline{\underline{H_3AsO_4}} + 5\underline{\underline{H_2SnCl_4}} + 10HCl \longrightarrow$

$$2As + 5H_2SnCl_6 + 8H_2O$$

9. (a) 71% **(b)** 74% **10.** 3.8 liters **11.** 9.43 liters
12. 23 ml **13.** 53 g **14.** 4.22 M **15.** 42.8 ml
16. 11 ml **17.** 9.4 ml **18.** 23 ml **19.** 2.00 liters
20. 5.6×10^{-3} M; rate of Cl_2 oxidation significantly less than rate of $H_2C_2O_4$ oxidation; stomach flush with dilute MnO_4^- solution with monitoring for color indication of excess MnO_4^-

Self-Test

1. (a) redox; $\boxed{SO_3^{2-}}$; $\boxed{S}$ **(b)** metathetical **(c)** redox; $\boxed{ClO_2}$; $\boxed{ClO_2}$ (disproportionation) **(d)** metathetical **(e)** redox; $\boxed{O_3}$;

$\boxed{XeO_3}$ **2. (a)** $\overset{+I\ \ -I}{Na_2O_2}$ **(b)** $\overset{+II\ -II\ +I}{Ca(OCl)_2}$

3. (a) $Al + 3NO_3^- + 6H^+ \longrightarrow Al^{3+} + 3NO_2 + 3H_2O$

(b) $3CH_3OH + 4MnO_4^- \longrightarrow$

$$3HCO_2^- + 4MnO_2 + 4H_2O + OH^-$$

(c) $C_6H_5CH_3 + K_2Cr_2O_7 + 4H_2SO_4 \longrightarrow$

$$C_6H_5CO_2H + K_2SO_4 + Cr_2(SO_4)_3 + 5H_2O$$

4. 79% **5.** 285 ml

UNIT 23

Exercises

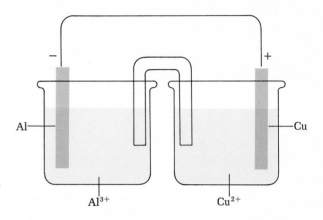

2. 37 g **3. (a)** $\sim$190 hr **(b)** 1.4 hr **4. (a)** (1) 1.20 V, Cd; (2) 1.36 V, H_2; (3) 2.04 V, Pb; (4) 5.79 V, Cs **(b)** +0.42 V
5. $\Delta G_f° = -25.6$ kcal mole^{-1}; $\Delta S_f° = -16$ cal mole^{-1} deg^{-1}

Practice for Proficiency

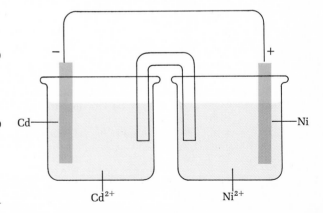

2. (Representations for **a–d** would be similar to Fig. 23.4. Both electrodes in **e** involve gas bubbled over platinum.) The *positive* external electrodes are: **(a)** hydrogen electrode **(b)** Ag **(c)** hydrogen electrode **(d)** Cu **(e)** hydrogen electrode

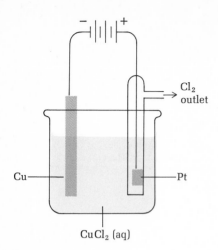

Cl₂ outlet

Cu — — Pt

CuCl₂ (aq)

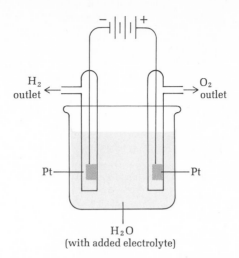

H₂ outlet

O₂ outlet

Pt — — Pt

H₂O
(with added electrolyte)

4. e **5.** d **6.** c **7.** e **8.** 81 g **9. (a)** 534 hr **(b)** 264 liters **10.** 7.1 g **11.** 126 liters **12.** 45 hr
13. 2.6 amps **14.** 0.21 g hr⁻¹ **15.** 67 hr **16.** 11 amps
17. 52 ml **18.** 83% **19.** 0.17 g hr⁻¹ **20.** 24 ml
21. 1.3 hr **22.** ~1200 amps **23.** 94 ml **24.** a **25.** a
26. e **27.** b **28.** a **29.** d **30. (a)** (1) 2.00 V, Al;
(2) 0.71 V, Mg; (3) 2.71 V, Mg **(b)** +0.16 V **31.** $\Delta G_f° =$
-31.4 kcal mole⁻¹; $\Delta S_f° = -48.0$ cal mole⁻¹ deg⁻¹
32. (a) -117 kcal mole⁻¹ **(b)** -125 kcal mole⁻¹ **(c)** -172 kcal mole⁻¹ **(d)** -313 kcal mole⁻¹ **(e)** -98 kcal mole⁻¹
33. (a) approx. $86,000 **(b)** approx. $24,000 **(c)** approx. $7200. Differences in electric costs reflect differences both in atomic weights *and* ionic charges. Market prices do not follow electric bill trends, as costs of ores and other plant costs are major factors.

Self-Test

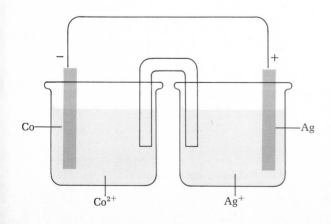

Co

Ag

Co²⁺

Ag⁺

2. 0.076 g **3.** ~550 sec **4. (a)** (1) 3.73 V; Mg, Mg²⁺;
(2) 1.51 V; Cl₂, Cl⁻; (3) 2.46 V; Al, Al³⁺ **(b)** +0.91 V
5. $\Delta G_f° = -31.3$ kcal mole⁻¹; $\Delta S_f° = -29.2$ cal mole⁻¹ deg⁻¹

UNIT 24
Exercises
1. (a) -0.85 V **(b)** -0.015 V **2. (a)** $+0.84$ V **(b)** $+1.34$ V
  (c) (a) Zn|Zn(1.0 × 10⁻³ M)||H⁺(0.50 M); H₂(0.80 atm)|Pt
  (b) Al|Al³⁺(2.0 × 10⁻⁵ M)||Cd²⁺(0.30 M)|Cd
3. (a) 3 × 10⁻¹² **(b)** 5 × 10²⁵ **(c)** 5.6 × 10³ **(d)** 1 × 10⁻⁷³
4. 3 × 10⁻⁸

Practice for Proficiency
1. (a) Cu|Cu²⁺(0.10 M)||Ag⁺(0.10 M)|Ag
  (b) Pb|Pb²⁺(0.05 M)||Cu²⁺(0.05 M)|Cu
  (c) Pt|H₂(1 atm), H⁺(0.0010 M)||Hg²⁺(0.10 M)|Hg
  (d) Pt|H₂(0.8 atm), H⁺(0.005 M)||Cl₂(0.8 atm), Cl⁻(0.30 M)|Pt
2. (a) 0.44 V; Ni + 2Cu⁺ ⟶ Ni²⁺ + 2Cu;
    Ni|Ni²⁺(0.10 M)||Cu⁺(1.0 × 10⁻⁶ M)|Cu
  (b) 0.10 V; 2Ag + Cu²⁺ ⟶ 2Ag⁺ + Cu;
    Ag|Ag⁺(1.0 × 10⁻¹⁰ M)||Cu²⁺(0.10 M)|Cu
  (c) 0.61 V; Zn + 2H⁺ ⟶ Zn²⁺ + H₂;
    Zn|Zn²⁺(0.10 M)||H⁺(1.0 × 10⁻³ M); H₂(1.0 atm)|Pt
  (d) 0.49 V; Zn + 2H⁺ ⟶ Zn²⁺ + H₂;
    Zn|Zn²⁺(0.10 M)||H⁺(1.0 × 10⁻¹⁰ M); H₂(1.0 atm)|Pt
  (e) 1.63 V; 2Al + 6H⁺ ⟶ 2Al³⁺ + 3H₂;
    Al|Al³⁺(0.050 M)||H⁺(0.010 M); H₂(1.0 atm)|Pt
3. (a) 1.53 V **(b)** 0.69 V **(c)** 0.40 V **(d)** 0.18 V **(e)** 0.97 V
4. (a) ~10²⁰² **(b)** ~3 × 10⁻⁴⁵ **(c)** ~3 × 10³⁵ **(d)** ~10⁻²³⁷
(e) ~10³⁷¹ **5. (a)** 1.2 × 10⁷ **(b)** 8 × 10¹⁸ **(c)** 8 × 10⁶
(d) 1.3 × 10³¹ **(e)** 1 × 10¹⁵ **6.** 2 × 10⁻¹² **7.** 2.2
8. 2 × 10⁻⁵ **9.** 7 × 10⁻¹² **10.** 2.4 **11.** 9 × 10⁻⁵
12. 1 × 10⁻¹⁶ **13.** 3.7 **14.** c **15.** d

Self-Test
1. -1.81 V **2.** 2.92 V
3. Mg|Mg²⁺(1.0 × 10⁻³ M)||Ag⁺(2.0 × 10⁻⁶ M)|Ag
4. ~10²⁰ **5.** ~10⁻⁸

UNIT 25

Exercises

1. (a) K **(b)** K **(c)** Na^+ **(d)** Al **(e)** LiF **(f)** Si **(g)** $B(OH)_3$
(h) Si^{4+} **(i)** $Be(OH)_2$ and $Al(OH)_3$ **(j)** Ca
2. (a) $2Li(s) + H_2(g) \longrightarrow 2LiH(s)$
(b) $2Na(s) + 2H_2O(l) \longrightarrow 2NaOH(aq) + H_2(g)$
(c) $Ca(s) + Cl_2(g) \longrightarrow CaCl_2(s)$
(d) $2Mg(s) + O_2(g) \longrightarrow 2MgO(s)$
(e) $2Na(s) + O_2(g) \longrightarrow Na_2O_2(s)$
(f) $K(s) + O_2(g) \longrightarrow KO_2(s)$
(g) $3Sr(s) + N_2(g) \longrightarrow Sr_3N_2(s)$
(h) $6Li(s) + N_2(g) \longrightarrow 2Li_3N(s)$
(i) $Sr(s) + 2H_2O(l) \longrightarrow Sr(OH)_2(aq) + H_2(g)$
(j) $Mg(s) + F_2(g) \longrightarrow MgF_2(s)$
3. (a) 51 kg **(b)** 56% **(c)** 6.35 g **4. (a)** 12.9 g (Na) hr^{-1}
(b) 6.80 g (Mg) hr^{-1} **(c)** 5.04 g (Al) hr^{-1} **5.** $\sim 10^{-4}$
6. 42 kg **7.** 4.9 **8.** 77% **9.** 17 g **10.** 0.75 V

Practice for Proficiency

1. (a) Ca **(b)** Mg **(c)** Na **(d)** Ba **(e)** $CaSO_4$ **(f)** Be and Al
(g) Ge **(h)** K **(i)** C **(j)** K
2. (a) *amphoteric* $Al(OH)_3(s) + 3H^+(aq) \longrightarrow$
$$Al^{3+}(aq) + 3H_2O(l)$$
$$Al(OH)_3(s) + OH^-(aq) \longrightarrow$$
$$[Al(OH)_4]^-(aq)$$
(b) *acidic* $B(OH)_3(s) + OH^-(aq) \longrightarrow [B(OH)_4]^-(aq)$
(c) *basic* $Ba(OH)_2(s) + 2H^+(aq) \longrightarrow Ba^{2+}(aq) + 2H_2O(l)$
(d) *basic* $LiOH(s) + H^+(aq) \longrightarrow Li^+(aq) + H_2O(l)$
(e) *amphoteric* $Be(OH)_2(s) + 2H^+(aq) \longrightarrow$
$$Be^{2+}(aq) + 2H_2O(l)$$
$$Be(OH)_2(s) + OH^-(aq) \longrightarrow$$
$$[Be(OH)_3]^-(aq)$$
(f) *acidic* $Si(OH)_4(s) + OH^-(aq) \longrightarrow$
$$[SiO(OH)_3]^-(aq) + H_2O(l)$$
(g) *basic* $Mg(OH)_2(s) + 2H^+(aq) \longrightarrow Mg^{2+}(aq) + 2H_2O(l)$
(h) *amphoteric* $Ga(OH)_3(s) + 3H^+(aq) \longrightarrow$
$$Ga^{3+}(aq) + 3H_2O(l)$$
$$Ga(OH)_3(s) + OH^-(aq) \longrightarrow$$
$$[Ga(OH)_4]^-(aq)$$
(i) *basic* $Ca(OH)_2(s) + 2H^+(aq) \longrightarrow Ca^{2+}(aq) + 2H_2O(l)$
3. (a) $Ca(s) + Cl_2(g) \longrightarrow CaCl_2(s)$
(b) $2Na(s) + O_2(g) \longrightarrow Na_2O_2(s)$
(c) $2Li(s) + H_2(g) \longrightarrow 2LiH(s)$
(d) $2Cs(s) + 2H_2O(l) \longrightarrow 2CsOH(aq) + H_2(g)$
(e) $Mg(s) + Br_2(l) \longrightarrow MgBr_2(s)$
(f) $2Ca(s) + O_2(g) \longrightarrow 2CaO(s)$
(g) $Cs(s) + O_2(g) \longrightarrow CsO_2(s)$
(h) $3Mg(s) + N_2(g) \longrightarrow Mg_3N_2(s)$
(i) $Ba(s) + O_2(g) \longrightarrow BaO_2(s)$ (and $2Ba(s) + O_2(g) \longrightarrow$
$$2BaO(s))$$
(j) $Sr(s) + 2H_2O(l) \longrightarrow Sr(OH)_2(aq) + H_2(g)$
4. (a) 44.6 g **(b)** 1.0 liter **(c)** 1.6 liters **5.** 40 metric tons
6. 5.0 **7.** 2.6 hr **8.** 87.5% **9.** 197 hr **10. (a)** K^+ **(b)** Ba^{2+}

Self-Test

1. (a) Mg **(b)** Li **(c)** K **(d)** B **(e)** MgF_2
2. (a) $Mg + H_2 \longrightarrow MgH_2$
(b) $2Cs + 2H_2O \longrightarrow 2CsOH + H_2$
(c) $2Na + Cl_2 \longrightarrow 2NaCl$
(d) $4Li + O_2 \longrightarrow 2Li_2O$

(e) $2Na + O_2 \longrightarrow Na_2O_2$
(f) $Rb + O_2 \longrightarrow RbO_2$
(g) $6Li + N_2 \longrightarrow 2Li_3N$
(h) $B(OH)_3 + H_2O \rightleftharpoons [B(OH)_4]^- + H^+$
(i) $Al(OH)_3 + OH^- \rightleftharpoons [Al(OH)_4]^-$
(j) $Al(OH)_3 + 3H^+ \longrightarrow Al^{3+} + 3H_2O$
3. 71% **4.** 88 metric tons **5.** 4.8 **6.** 8.0 amps **7.** 1.93 V

UNIT 26

Exercises

1. (a) potassium nitrite **(b)** H_3PO_4 **(c)** sodium thiosulfate
(d) AsH_3 **(e)** ozone **(f)** xenon tetrafluoride **(g)** $HClO_3$
(h) potassium periodate **(i)** NaOCl **(j)** $HBrO_2$

2. (a) $\ddot{S} :: C :: \ddot{S}$ (linear)

(b) $:\ddot{C}l \cdot \ddot{N} \cdot \ddot{C}l:$ (orbitals approximately tetrahedral,
$\quad \ddot{C}l:$ atoms pyramidal)

(c) $:\ddot{C}l : \overset{\ddot{C}l \cdot \ddot{C}l}{P} : \ddot{C}l:$ (trigonal bipyramidal)
$\quad :\ddot{C}l:$

(d) $:\ddot{O} \cdot \ddot{N} :: \ddot{O}^- \longleftrightarrow \ddot{O} :: \ddot{N} \cdot \ddot{O}:^-$ (orbitals approximately
$\qquad\qquad\qquad\qquad$ trigonal planar, atomic
$\qquad\qquad\qquad\qquad$ geometry "bent")

(e) $:\ddot{O} : \ddot{C}l:^-$ (orbitals approximately tetrahedral, atoms
$\quad :\ddot{O}:$ pyramidal)

(f) $H : \ddot{O} : \overset{:\ddot{O}:}{N} :: \ddot{O} \longleftrightarrow H : \ddot{O} : \overset{:O:}{N} : \ddot{O}:$ (trigonal planar around N)

(g) $\overset{:\ddot{F}:}{\underset{:\ddot{F}:}{:\ddot{F}: Xe :\ddot{F}:}}$ (octahedral)

3. (a) White phosphorus consists of P_4 molecules, readily soluble in a nonpolar solvent such as CS_2, whereas red phosphorus is polymeric. **(b)** There are more solute *particles* (P_4) in
the phosphorus solution than in the sulfur (S_8) solution.
4. (a) $NH_4^+(aq) + NO_2^-(aq) \longrightarrow 2H_2O(l) + N_2(g)$
(b) $3SiO_2(s) + 5C(s) + Ca_3(PO_4)_2(s) \longrightarrow$
$$3CaSiO_3(s) + 5CO(g) + P_2(g);$$
$$2P_2(g) \xrightarrow{\text{(cool)}} P_4(g) \xrightarrow{\text{(bubble thru } H_2O)} P_4(s)$$
(c) $As_4O_6(s) + 6C(s) \longrightarrow As_4(g) + 6CO(g)$
(d) $2KClO_3(s) \xrightarrow{\text{(heat)}} 2KCl(s) + 3O_2(g)$
(e) $2Br^-(aq) + Cl_2(g) \longrightarrow Br_2(l) + 2Cl^-(aq)$
(f) $2Al(s) + 2NaOH(aq) + 6H_2O(l) \longrightarrow$
$$2Na[Al(OH)_4](aq) + 3H_2(g)$$
5. (a) $2NO(g) + O_2(g) \longrightarrow 2NO_2(g)$
(b) $H_3PO_3(aq) + 2OH^-(aq) \longrightarrow HPO_3^{2-}(aq) + 2H_2O(l)$
(c) $CH_3NH_2(aq) + H_2O(l) \rightleftharpoons CH_3NH_3^+(aq) + OH^-(aq)$
(d) $H_2S(aq) + 2OH^-(aq) \longrightarrow S^{2-}(aq) + 2H_2O(l)$

(e) $P_4(g) + 6Cl_2(g) \longrightarrow 4PCl_3(g)$
(f) $P_4(g) + 10Cl_2(g) \longrightarrow 4PCl_5(g)$
(g) $2IO_3^-(aq) + 5HSO_3^-(aq) \longrightarrow$
$$I_2(s) + 5SO_4^{2-}(aq) + 3H^+(aq) + H_2O(l)$$
(h) $N_2(g) + 3H_2(g) \rightleftharpoons 2NH_3(g)$
(i) $CaH_2(s) + 2H_2O(l) \longrightarrow Ca(OH)_2(s) + 2H_2(g)$
(j) $4LiH + AlCl_3 \longrightarrow LiAlH_4 + 3LiCl$
6. (a) 0.5 **(b)** 9.6 **(c)** 7.4 **(d)** 10.2 **7.** 14 liters **8.** 37 ml
9. 0.89 kg (C_6H_6); 2.7 kg (Cl_2) **10.** 78 ml
Practice for Proficiency
1. (a) nitrous acid **(b)** H_3PO_2 **(c)** potassium bisulfate (IUPAC; potassium hydrogen sulfate) **(d)** Na_3AsO_4 **(e)** sodium selenide **(f)** hypochlorous acid **(g)** $KClO_4$ **(h)** magnesium chlorate **(i)** $HClO_2$ **(j)** lithium aluminum hydride **(k)** $H_4P_2O_7$ **(l)** silver nitrate **(m)** $SbCl_3$ **(n)** PH_3 **(o)** ozone **(p)** $NaBH_4$

2. (a)  :F:P:F: (pyramidal)
　　:F:

(b) 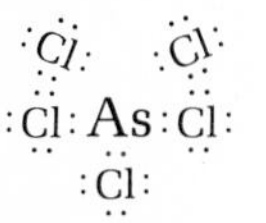 :Cl:As:Cl: (trigonal bipyramidal)
　　:Cl:
　　:Cl:

(c)  O::S::O (orbitals approximately trigonal planar, atomic geometry "bent")

(d) 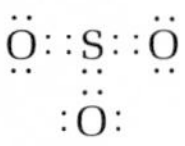 O::S::O (trigonal planar)
　　:O:

(e) 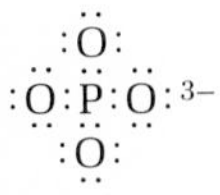 :O:P:O: ³⁻ (probably should be represented as resonance hybrid involving some P=O double bonding, tetrahedral)
　　:O:

(f) 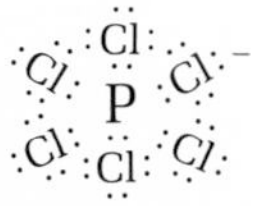 :Cl: P :Cl: ⁻ (octahedral)
　　:Cl: :Cl:

(g) (See Fig. 26.12.)

(h) 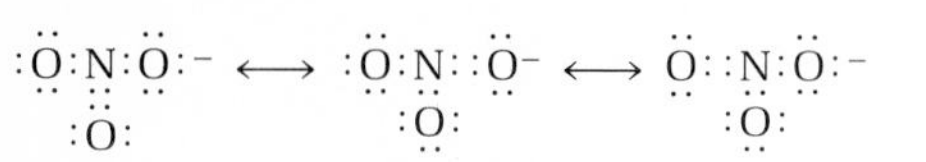 :O:N:O:⁻ ⟷ :O:N::O⁻ ⟷ O::N:O:⁻ (trigonal planar)
　　:O: 　　:O: 　　:O:

(i)  :F: S :F: (octahedral)
　　:F: :F:

(j) 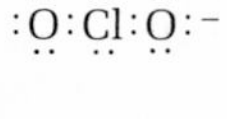 :O:Cl:O:⁻ (orbitals approximately tetrahedral, atomic geometry "bent")

3. (a) The P_4 molecules of white phosphorus are held in the crystal lattice only by London forces, whereas covalent bonds must be broken to liberate gaseous phosphorus from the polymeric "red" allotrope. **(b)** In both crystalline forms of sulfur, molecular S_8 units are available to dissolve, whereas "plastic" sulfur is polymeric. **(c)** The O_2 molecule contains unpaired electrons, but O_3 does not. **(d)** The α-monoclinic form is of the molecular crystal type, as Se_8 units. **(e)** White phosphorus consists of P_4 molecules.

4. (a) $2NH_4NO_3(s) \longrightarrow 2N_2(g) + O_2(g) + 4H_2O(g)$
(b) $H_3PO_4(aq) + 3NaOH(aq) \longrightarrow Na_3PO_4(aq) + 3H_2O(l)$
(c) $H_2NOH(l) + HCl(g) \longrightarrow [H_3NOH]Cl(s)$
(d) $Ca(OH)_2(aq) + H_2SO_3(aq) \longrightarrow CaSO_3(s) + 2H_2O(l)$
(e) $3NO_2(g) + H_2O(l) \longrightarrow 2HNO_3(aq) + NO(g)$
(f) $SO_3(g) + H_2O(l) \longrightarrow H_2SO_4(aq)$
(g) $NH_4^+(aq) + NO_2^-(aq) \longrightarrow N_2(g) + 2H_2O(l)$
(h) $2KClO_3 \xrightarrow{\text{heat}} 2KCl + 3O_2$
(i) $H_3PO_2(aq) + NaOH(aq) \longrightarrow NaH_2PO_2(aq) + H_2O(l)$
(j) $4NaH + B(OCH_3)_3 \longrightarrow NaBH_4 + 3NaOCH_3$
5. (a) (See answer to Exercise 4b, Unit 26.)
(b) $Sb_4O_6(s) + 6C(s) \longrightarrow Sb_4(g) + 6CO(g)$
(c) $2H_2O(l) \xrightarrow{\text{elec.}} 2H_2(g) + O_2(g)$
(d) $2NaCl(aq) + 2H_2O(l) \xrightarrow{\text{elec.}}$
$$2NaOH(aq) + Cl_2(aq) + H_2(aq)$$
(e) (See answer to Exercise 5g, Unit 26.)
6. (a) 12.2 **(b)** 1.23 **(c)** 11.5 **7.** ~20 ml **8.** 45 g
9. 10.6 **10.** 174 g **11.** 250 liters **12. (a)** 7.2×10^4 liters
(b) 9.9 kg
Self-Test
1. (a) HIO_4 **(b)** calcium hypochlorite **(c)** $KClO_3$
(d) $Mg(BrO_2)_2$ **(e)** $Na_2S_2O_3$ **(f)** arsine **(g)** $H_4P_2O_7$
(h) bismuth(V) oxide (or bismuth pentoxide) **(i)** N_2O
(j) hydrogen selenide **(k)** magnesium sulfite
2. (a) 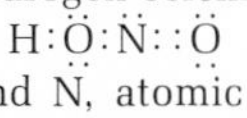 H:O:N:O (orbitals approximately trigonal planar around N, atomic geometry around N "bent") **(b)** (See Exercise 2e, Unit 26.) **(c)** (See Exercise 2c, Unit 26.) **(d)** (See Practice for Proficiency 2c, Unit 26.) **(e)** (See Exercise 2g, Unit 26.) **3.** a **4.** b
5. (a) $NH_4NO_3(s) \xrightarrow{200°C} N_2O(g) + 2H_2O(g)$
(b) $3NO_2(g) + H_2O(l) \longrightarrow 2HNO_3(aq) + NO(g)$
(c) $H_3PO_2(aq) + OH^-(aq) \longrightarrow H_2PO_2^-(aq) + H_2O(l)$
(d) $3Ag(s) + 4H^+(aq) + NO_3^-(aq) \longrightarrow$
$$3Ag^+(aq) + NO(g) + 2H_2O(l)$$
(e) $NH_4^+(aq) + NO_2^-(aq) \longrightarrow N_2(g) + 2H_2O(l)$
(f) $Cl_2(g) + 2I^-(aq) \longrightarrow 2Cl^-(aq) + I_2(s)$
6. (a) 1.8 **(b)** 8.5 **(c)** 0.28 **7.** 50 liters **8.** 2.3 liters
9. 10.5 **10.** 63 liters

UNIT 27
Exercises
1. (a) $+II, +IV$ **(b)** $+II, +IV, +VII$ (also $+III, +V, +VI$)
(c) $+II, +III$ (also $+I, +IV$) **(d)** $+I, +II$ **(e)** only $+II$

2. (a) [Cr]: (Ar) $4s\uparrow$, $3d\uparrow$, $\uparrow$, $\uparrow$, $\uparrow$, $\uparrow$ [Cr^{2+}]: (Ar) $3d\uparrow$, $\uparrow$, $\uparrow$, $\uparrow$,
[Mn^{2+}]: (Ar) $3d\uparrow$, $\uparrow$, $\uparrow$, $\uparrow$, $\uparrow$; [Fe]: (Ar) $4s\uparrow$, $3d\uparrow\downarrow$, $\uparrow$, $\uparrow$, $\uparrow$, [Ni^{2+}]:
(Ar) $3d\uparrow\downarrow$, $\uparrow\downarrow$, $\uparrow\downarrow$, $\uparrow$, $\uparrow$ **(b)** Cr: 6.93, Cr^{2+}: 4.90; Mn^{2+}: 5.92; Fe:
3.88; Ni^{2+}: 2.83 **3. (a)** [Cl—Au—Cl]$^-$

(b)
$$\left[\begin{array}{c} \text{Cl} \\ \text{Cl} \quad \overset{\displaystyle \text{Ni}}{} \quad \text{Cl} \\ \text{Cl} \end{array}\right]^{2-}$$

(c)
$$\left[\begin{array}{c} \text{CN} \\ \text{NC}-\text{Ni}-\text{CN} \\ \text{CN} \end{array}\right]^{2-}$$

(d)
$$\left[\begin{array}{c} \text{NH}_3 \\ \text{H}_3\text{N} \cdots \cdots \text{Co} \cdots \cdots \text{NH}_3 \\ \text{H}_3\text{N} \cdots \cdots \quad \cdots \cdots \text{NH}_3 \\ \text{NH}_3 \end{array}\right]^{3+}$$

4. (a) [Cu(NH$_3$)$_4$(H$_2$O)$_2$]$^{2+}$ (Zn^{2+} has no unpaired electrons;
Cu^{2+} has an unpaired electron.) **(b)** [FeF$_6$]$^{4-}$ (CN$^-$ is a
"stronger" ligand than F$^-$, so is more likely to "force" electrons to pair in three of the d orbitals.) **5. (a)** sp^3 **(b)** dsp^2
(c) d^2sp^3 **(d)** sp^3d^2
6. Cu^{2+}, Cd^{2+}, and Zn^{2+}, equations:

AT STEP 2
$$\text{Cu}^{2+}(\text{aq}) + \text{H}_2\text{S}(\text{aq}) \longrightarrow \text{CuS}(\text{s}) + 2\text{H}^+(\text{aq})$$
$$\text{Cd}^{2+}(\text{aq}) + \text{H}_2\text{S}(\text{aq}) \longrightarrow \text{CdS}(\text{s}) + 2\text{H}^+(\text{aq})$$

AT STEP 3
$$3\text{CuS}(\text{s}) + 8\text{H}^+(\text{aq}) + 2\text{NO}_3{}^-(\text{aq}) \longrightarrow$$
$$3\text{Cu}^{2+}(\text{aq}) + \text{S}(\text{s}) + 4\text{H}_2\text{O}(\text{l}) + 2\text{NO}(\text{g})$$
$$3\text{CdS}(\text{s}) + 8\text{H}^+(\text{aq}) + 2\text{NO}_3{}^-(\text{aq}) \longrightarrow$$
$$3\text{Cd}^{2+}(\text{aq}) + 3\text{S}(\text{s}) + 4\text{H}_2\text{O}(\text{l}) + 2\text{NO}(\text{g})$$
$$\text{Cu}^{2+}(\text{aq}) + 4\text{NH}_3(\text{aq}) \rightleftharpoons [\text{Cu}(\text{NH}_3)_4]^{2+}(\text{aq})$$
$$\text{Cd}^{2+}(\text{aq}) + 4\text{NH}_3(\text{aq}) \rightleftharpoons [\text{Cd}(\text{NH}_3)_4]^{2+}(\text{aq})$$

AT STEP 4
$$[\text{Cu}(\text{NH}_3)_4]^{2+}(\text{aq}) + \text{S}_2\text{O}_4{}^{2-}(\text{aq}) + 4\text{OH}^-(\text{aq}) \longrightarrow$$
$$\text{Cu}(\text{s}) + 2\text{SO}_3{}^{2-}(\text{aq}) + 4\text{NH}_3(\text{aq}) + 2\text{H}_2\text{O}(\text{l})$$

AT STEP 5
$$[\text{Cd}(\text{NH}_3)_4]^{2+}(\text{aq}) + \text{H}_2\text{S}(\text{aq}) \longrightarrow$$
$$\text{CdS}(\text{s}) + 2\text{NH}_4{}^+(\text{aq}) + 2\text{NH}_3(\text{aq})$$

AT STEP 8
$$[\text{Zn}(\text{OH})_4]^{2-}(\text{aq}) + \text{H}_2\text{S}(\text{aq}) \longrightarrow$$
(from step 6)
$$\text{ZnS}(\text{s}) + 2\text{H}_2\text{O}(\text{l}) + 2\text{OH}^-(\text{aq})$$

7. 1.35 kg **8.** 0.15 M **9.** 0.28 V **10.** 32 amps

Practice for Proficiency

1. Sc: +III (only); Cr: +III, +VI (also +II, +IV, +V); Ni:
+II (also +I, +III, +IV); Au: +I, +III (also +IV); Ag: +I
(also +II, +III); V: +II thru +V; Fe: +II, +III (also +I,
+IV, +V, +VI) **2. (a)** [Fe]: (Ar) $4s\uparrow\downarrow$, $3d\uparrow\downarrow$, $\uparrow$, $\uparrow$, $\uparrow$, $\uparrow$; [Fe^{2+}]:
(Ar) $3d\uparrow\downarrow$, $\uparrow$, $\uparrow$, $\uparrow$, $\uparrow$; [Fe^{3+}]: (Ar) $3d\uparrow$, $\uparrow$, $\uparrow$, $\uparrow$, $\uparrow$; [Sc]: (Ar) $4s\uparrow\downarrow$,

$3d\uparrow$; [Sc^{3+}]: (Ar): [V]: (Ar) $4s\uparrow$, $3d\uparrow$, $\uparrow$, $\uparrow$; [V^{3+}]: (Ar) $3d\uparrow$, $\uparrow$;
[Cr^{3+}]: (Ar) $3d\uparrow$, $\uparrow$, $\uparrow$; [Mn]: (Ar) $4s\uparrow\downarrow$, $3d\uparrow$, $\uparrow$, $\uparrow$, $\uparrow$, $\uparrow$; [Co] (Ar)
$4s\uparrow\downarrow$, $3d\uparrow\downarrow$, $\uparrow\downarrow$, $\uparrow$, $\uparrow$, $\uparrow$; [Co^{2+}]: (Ar) $3d\uparrow\downarrow$, $\uparrow\downarrow$, $\uparrow$, $\uparrow$, $\uparrow$; [Co^{3+}]: (Ar) $3d\uparrow\downarrow$,
$\uparrow$, $\uparrow$, $\uparrow$, $\uparrow$; [Ni]: (Ar) $4s\uparrow\downarrow$, $3d\uparrow\downarrow$, $\uparrow\downarrow$, $\uparrow$, $\uparrow$; [Cu]: (Ar) $4s\uparrow$, $3d\uparrow\downarrow$, $\uparrow\downarrow$,
$\uparrow\downarrow$, $\uparrow\downarrow$, $\uparrow\downarrow$; [Cu$^+$]: (Ar) $3d\uparrow\downarrow$, $\uparrow\downarrow$, $\uparrow\downarrow$, $\uparrow\downarrow$, $\uparrow\downarrow$; [Cu^{2+}]: (Ar) $3d\uparrow\downarrow$, $\uparrow\downarrow$, $\uparrow\downarrow$, $\uparrow\downarrow$, $\uparrow$;
[Zn]: (Ar) $4s\uparrow\downarrow$, $3d\uparrow\downarrow$, $\uparrow\downarrow$, $\uparrow\downarrow$, $\uparrow\downarrow$, $\uparrow\downarrow$; [Zn^{2+}]: (Ar) $3d\uparrow\downarrow$, $\uparrow\downarrow$, $\uparrow\downarrow$, $\uparrow\downarrow$, $\uparrow\downarrow$
(b) Fe: 4.90; Fe^{2+}: 4.90; Fe^{3+}: 5.92; Sc: 1.73; Sc^{3+}: 0; V: 3.88; V^{3+}:
2.83; Cr^{3+}: 3.88; Mn: 5.92; Co: 3.88; Co^{2+}: 3.88; Co^{3+}: 4.90; Ni:
2.83; Cu: 1.73; Cu$^+$: 0; Cu^{2+}: 1.73; Zn: 0; Zn^{2+}: zero
3. (a) [Ag(NH$_3$)$_2$]$^+$ **(b)** tetracyanodiaquocuprate(II)
(c) [AuCl$_4$(H$_2$O)$_2$]$^-$ **(d)** hexaamminecobalt(II) **(e)** [Fe(CN)$_6$]$^{4-}$
(f) potassium hexanitrocobaltate(III) **(g)** [Cu(NH$_3$)$_4$]SO$_4$
(h) sodium dicyanoargentate(I) **(i)** [Sc(H$_2$O)$_6$](NO$_3$)$_3$
(j) copper(II) hexacyanoferrate(III) **4. (a)** [NC—Ag—CN]$^-$

(b)
$$\left[\begin{array}{c} \text{Cl} \\ \text{Cl} \quad \overset{\displaystyle \text{Pb}}{} \quad \text{Cl} \\ \text{Cl} \end{array}\right]^{2-}$$

(c)
$$\left[\begin{array}{c} \text{Cl} \\ \text{Cl}-\text{Pt}-\text{Cl} \\ \text{Cl} \end{array}\right]^{2-}$$

(d)
$$\left[\begin{array}{c} \text{H}_2\text{O} \\ \text{H}_2\text{O} \cdots \cdots \quad \text{OH}_2 \\ \text{Cu} \\ \text{H}_2\text{O} \cdots \cdots \text{OH}_2 \\ \text{H}_2\text{O} \end{array}\right]^{2+}$$

(e)
$$\begin{array}{c} \text{CO} \\ \text{OC} \cdots \text{Fe} \cdots \text{CO} \\ \text{CO} \\ \text{CO} \end{array}$$

5. e **6.** c **7.** d **8.** e **9.** c **10.** d **11.** present: Ag$^+$,
Fe^{3+}, Ni^{2+}; absent: Cd^{2+}, Cu^{2+}, Hg^{2+}, Cr^{3+}, Zn^{2+}, Co^{2+},
Mn^{2+} **12.** possible presence: Cd^{2+}, Cu^{2+}, Hg^{2+}; absent: Ag$^+$,
Co^{2+}, Fe^{3+}, Mn^{2+}, Ni^{2+} **13. (a)** 24 kg **(b)** ~280 hr
14. 2.6×10^6 liters **15.** ~870 liters **16.** 8.5 g
17. ~10 amps **18.** 6.7 hr **19.** −1.15 V **20. (a)** 2×10^7
(b) ~10^{35} **21.** (Consult your instructor.)
Self-Test
1. (a) +III (only) **(b)** +III, +VI (also +II, +IV, +V)
(c) +II, +IV, +VII (also +III, +V, +VI) **(d)** +II, +III (also
+I, +IV, +V, +VI) **(e)** +II (only) **2.** Ti^{2+} : (Ar) $3d\uparrow$, $\uparrow$;
[Zn^{2+}]: (Ar) $3d\uparrow\downarrow$, $\uparrow\downarrow$, $\uparrow\downarrow$, $\uparrow\downarrow$, $\uparrow\downarrow$; Mn^{2+} : (Ar) $3d\uparrow$, $\uparrow$, $\uparrow$, $\uparrow$, $\uparrow$; [Cu$^+$]:
(Ar) $3d\uparrow\downarrow$, $\uparrow\downarrow$, $\uparrow\downarrow$, $\uparrow\downarrow$, $\uparrow\downarrow$; Cu^{2+} : (Ar) $3d\uparrow\downarrow$, $\uparrow\downarrow$, $\uparrow\downarrow$, $\uparrow\downarrow$, $\uparrow$ **3. (a)** penta-
aquoamminecopper(II) **(b)** trichlorotriaquocuprate(II)
(c) potassium hexanitrocobaltate(III) **(d)** (NH$_4$)$_2$[PtCl$_6$]

(e) [Fe(EDTA)]⁻   **4. (a)** [CuCl$_4$(H$_2$O)$_2$]$^{2-}$ **(b)** [CoCl$_6$]$^{3-}$
5. (a) dsp^2 **(b)** sp **(c)** sp^3 **(d)** d^2sp^3 **(e)** sp^3d^2 **6. (a)** Ag$^+$
(b) Cd^{2+}, Co^{2+}, Cr^{3+}, Cu^{2+}, Fe^{3+}, Hg^{2+}, Mn^{2+}, Ni^{2+}, Zn^{2+} (assuming no complex formation)   **7.** 3.5×10^3 liters
8. 0.11 M **9.** 34 hr   **10.** ~1.3×10^9

11. (a)

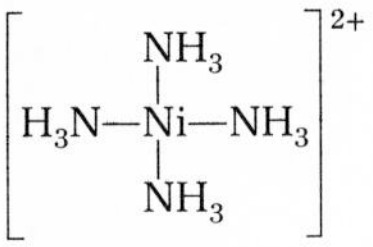

$$\left[\begin{array}{c} NH_3 \\ | \\ H_3N-Ni-NH_3 \\ | \\ NH_3 \end{array} \right]^{2+}$$

(b) [NC—Ag—CN]⁻

(c)

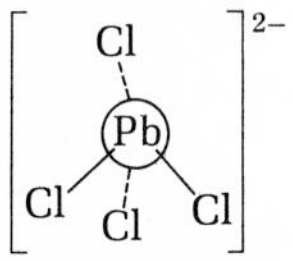

$$\left[\begin{array}{c} Cl \\ \diagup \\ Pb \\ Cl \quad Cl \\ Cl \end{array} \right]^{2-}$$

(d)

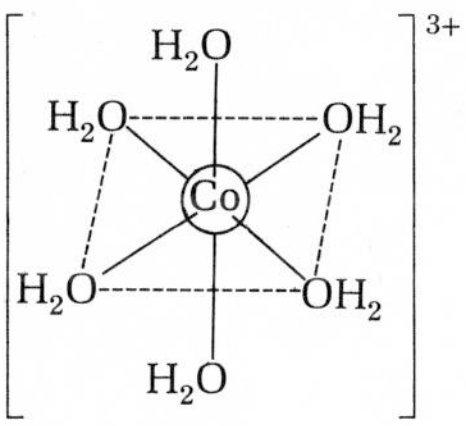

$$\left[\begin{array}{c} H_2O \\ H_2O \cdots\cdots OH_2 \\ Co \\ H_2O \cdots\cdots OH_2 \\ H_2O \end{array} \right]^{3+}$$

(e)

$$\left[\begin{array}{c} Cl \\ Cl\cdots\cdots Cl \\ Co \\ Cl\cdots\cdots Cl \\ Cl \end{array} \right]^{3-}$$

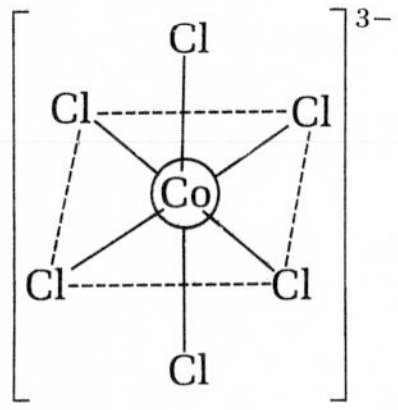

UNIT 28
Exercises

2. (a) halide **(b)** aldehyde **(c)** alkene (cycloalkene)
(d) phenol **(e)** ketone **(f)** alcohol **(g)** ester **(h)** amide
(i) aromatic hydrocarbon **(j)** ether   **3. (a)** 2-methyl-4-ethylhexane

(b) (CH$_3$)$_2$CHCH$_2$$\overset{\displaystyle CH_2CH_3}{\underset{\displaystyle CH_2CH_3}{C}}CH_2CH_2CH_2CH_2CH_2CH_3$

(c) 3-methyl-3-ethylheptane (or 3-ethyl-3-methylheptane)
(d) (CH$_3$)$_4$C **(e)** 4,6-dimethyl-6-ethyldecane

4. (a)

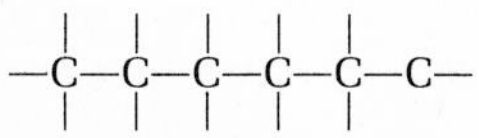

—C—C—C—C—C—C— (hexane);

—C—C—C—C—C— (2-methylpentane);
   |
  —C—

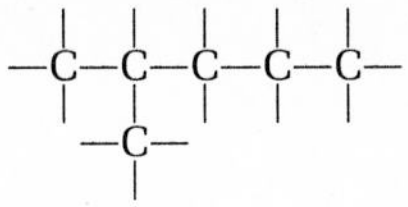

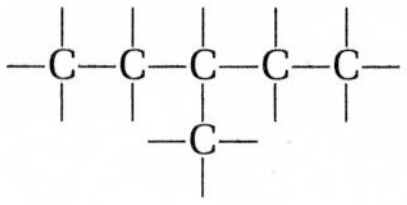

—C—C—C—C—C— (3-methylpentane);
       |
      —C—

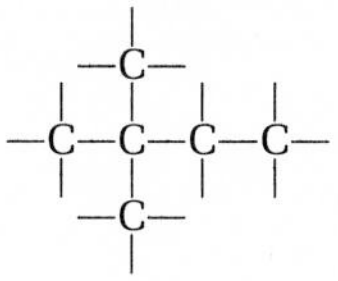

—C—C—C—C— (2,2-dimethylbutane);

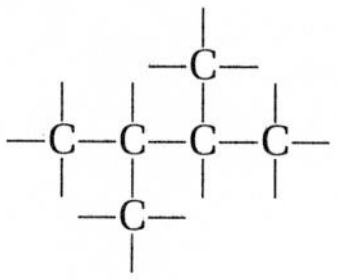

—C—C—C—C— (2,3-dimethylbutane)

(b) C=C—C—C—C—C— (1-hexene);

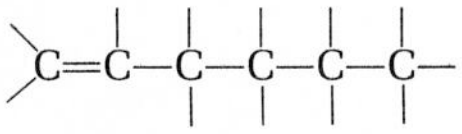

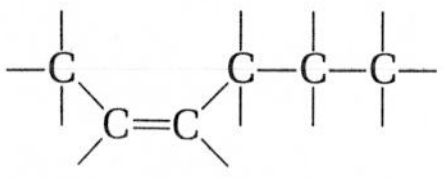

C=C (cis-2-hexene);

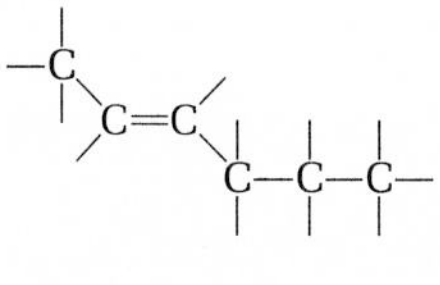

C=C (trans-2-hexene);

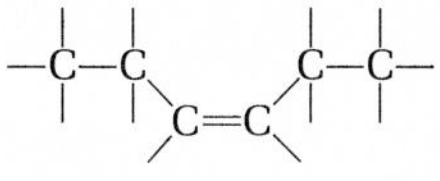

C=C (cis-3-hexene);

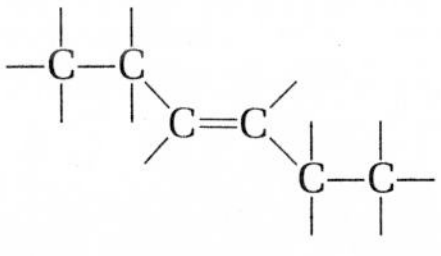

C=C (trans-3-hexene);

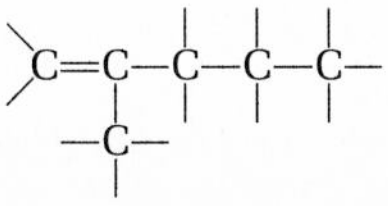

C=C—C—C—C— (2-methyl-1-pentene);

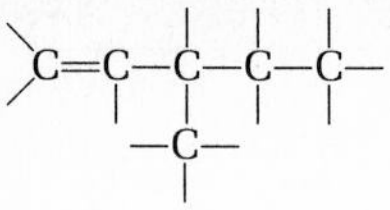

C=C—C—C—C— (3-methyl-1-pentene); 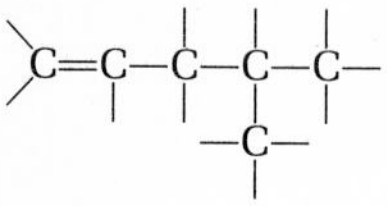
  |
 —C—
  |

C=C—C—C—C— (4-methyl-1-pentene); 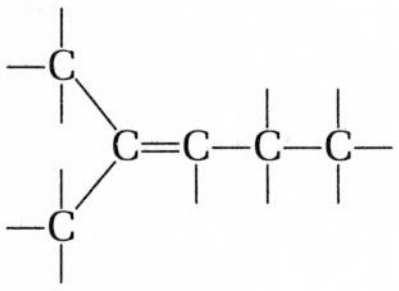
      |
     —C—
      |

—C
  \
   C=C—C—C—C— (2-methyl-2-pentene); 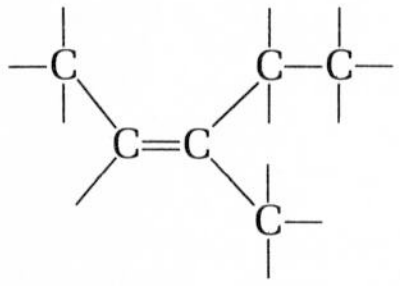
  /
—C

—C   C—C—
  \   /
   C=C—C   (3-methyl-cis-2-pentene); 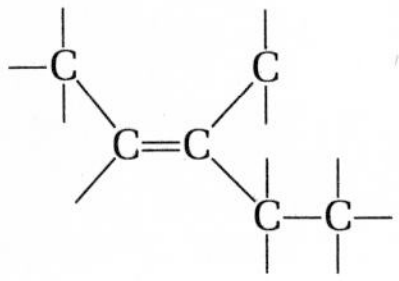
  /   \
         C—

—C   C
  \   / \
   C=C—C   (3-methyl-trans-2-pentene); 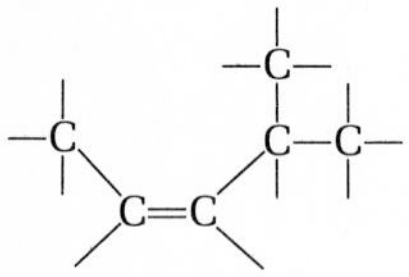
  /   \
         C—C—

          —C—
         / \
—C   C—C—
  \   /
   C=C   (4-methyl-cis-2-pentene); 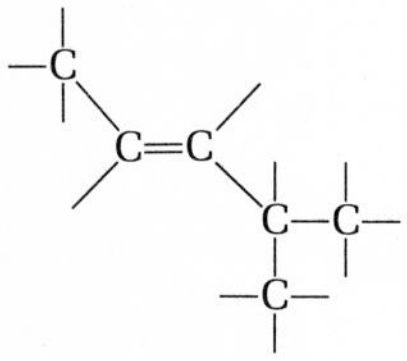
  /
—C

—C
  \   /
   C=C—C   (4-methyl-trans-2-pentene); 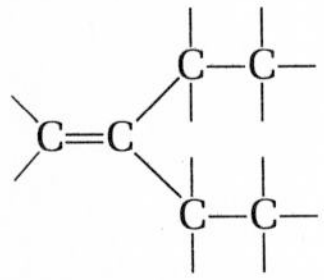
  /   \
         C—C—
          |
         —C—
          |

         C—C—
        /
C=C   (2-ethyl-1-butene); 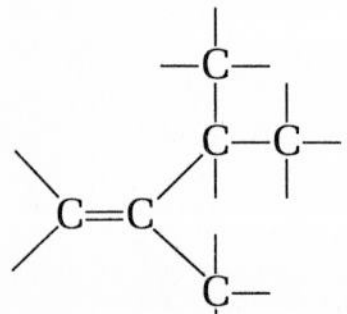
        \
         C—C—
          |

         —C—
        /
       C—C—
C=C   (2,3-dimethyl-1-butene); 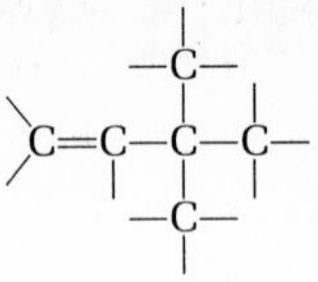
        \
         C—
          |

          —C—
         |
C=C—C—C— (3,3-dimethyl-1-butene); 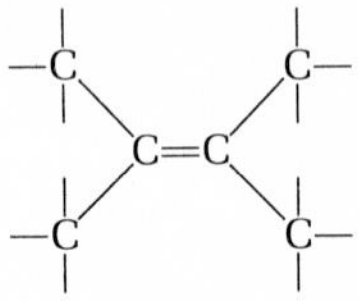
       |
      —C—

—C   C—
  \   /
   C=C   (2,3-dimethyl-2-butene)
  /   \
—C   C—

5. (a) o-xylene (b) m-toluidine (c) p-cresol
(d) 1,3,5-trihydroxybenzene (e) 2,4,6-trinitrotoluene

6. (a)

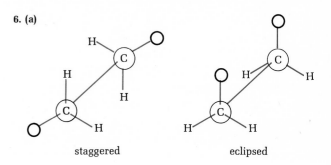

staggered   eclipsed

(b)

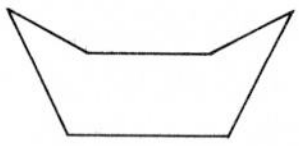

boat 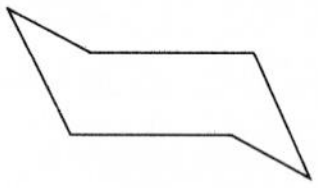   chair

(c)
  —C—   —C—
   | |
  —C—   —C—
   | |
H—C—CH₃   H₃C—C—H
   | |
  —C—   —C—
   | |
  —C—   —C—
   | |
  —C—   —C—

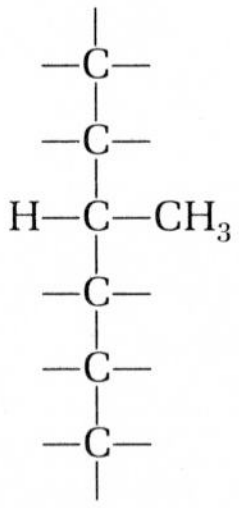

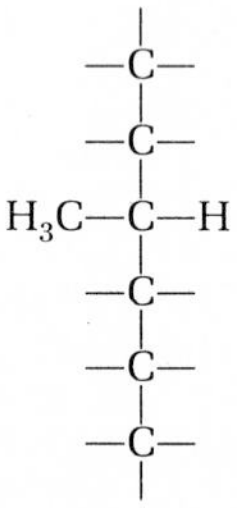

7. (a) cis-3-hexene (more polar) (b) 1,2,3-trimethylbenzene
(more polar) (c) p-cresol (hydrogen bonding) 8. (a) yes
(b) no (c) the alkenes (d) cis-2-butene (permanent
asymmetry)
Practice for Proficiency
1. (a) ketone (b) ether (c) aldehyde (d) carboxylic acid
(e) ester (f) amine (g) alkyne (h) alcohol (i) amide
(j) aromatic hydrocarbon   2. (a) amine (b) alkane (cycloalkane) (c) alkene (d) aromatic hydrocarbon (e) phenol
(f) carboxylic acid (g) alcohol (h) amine (i) ester
(j) aldehyde

3. (a) 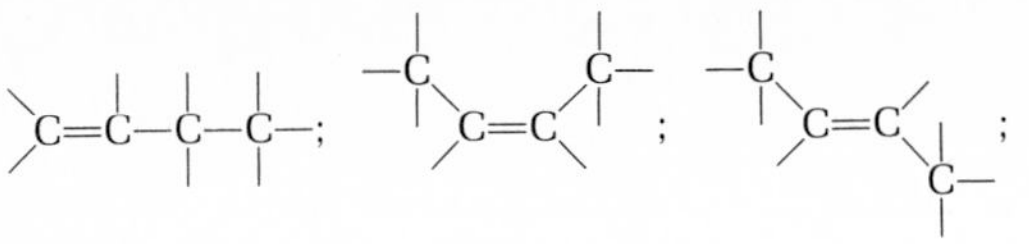 [structural formulas of alkenes]

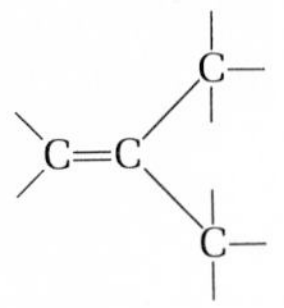

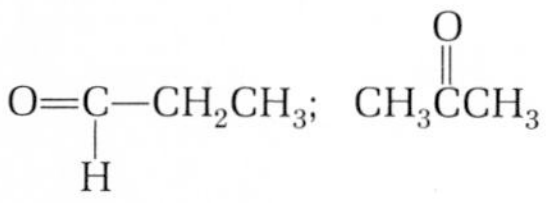

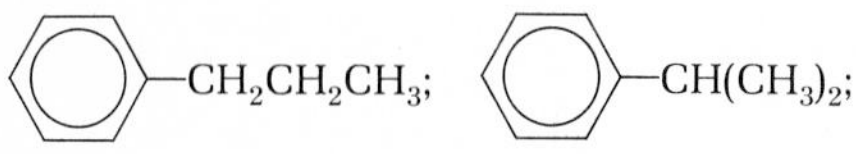

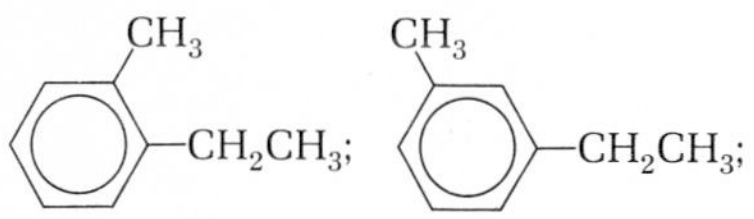

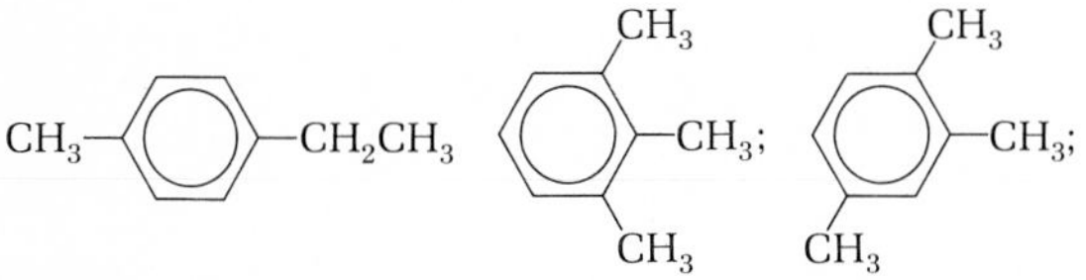

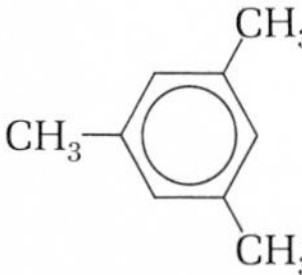

(h) [structural formulas] and [structural formula]

(i) HO—$\overset{CH_3}{\underset{\underset{CH_3}{CH_2}}{C}}$—H and H—$\overset{CH_3}{\underset{\underset{CH_3}{CH_2}}{C}}$—OH

(j) [see Exercise **6(c)**]

(b) O=$\overset{}{\underset{H}{C}}$—CH₂CH₃; CH₃$\overset{O}{C}$CH₃

4. (a)

[structural diagrams] staggered   eclipsed

(c) [benzene]—CH₂CH₂CH₃; [benzene]—CH(CH₃)₂;

[benzene with CH₃]—CH₂CH₃; [benzene with CH₃]—CH₂CH₃;

CH₃—[benzene]—CH₂CH₃; [benzene with CH₃]—CH₃; [benzene with CH₃]—CH₃;

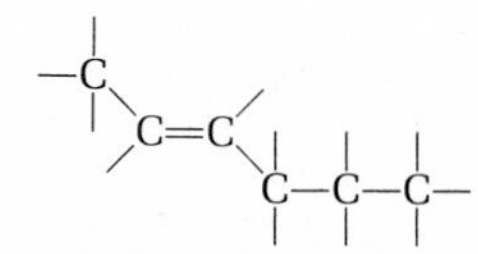

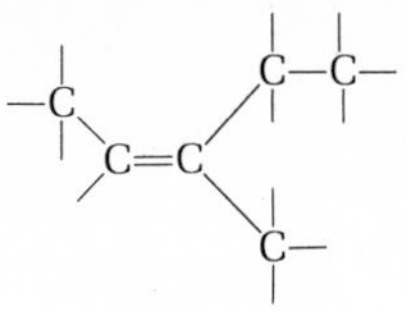

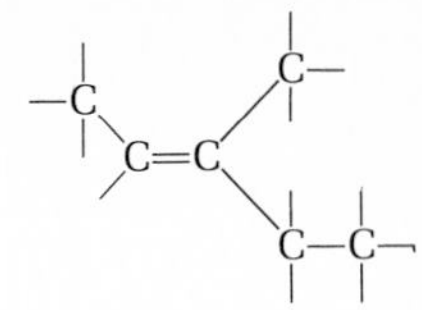

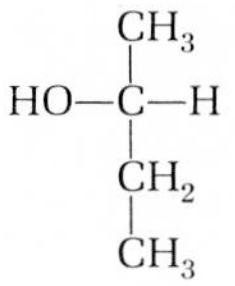

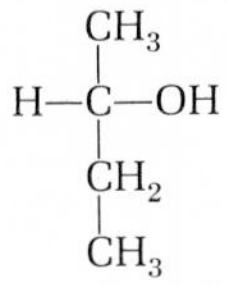

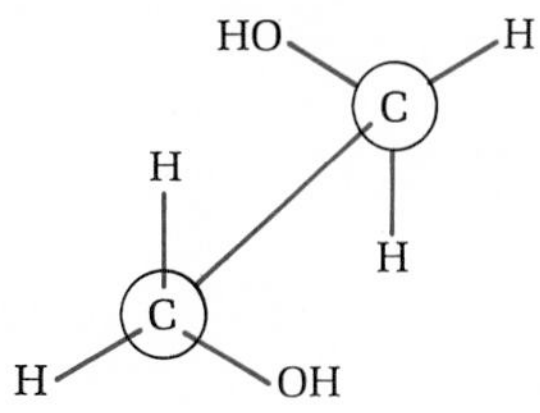

[benzene with CH₃ groups]

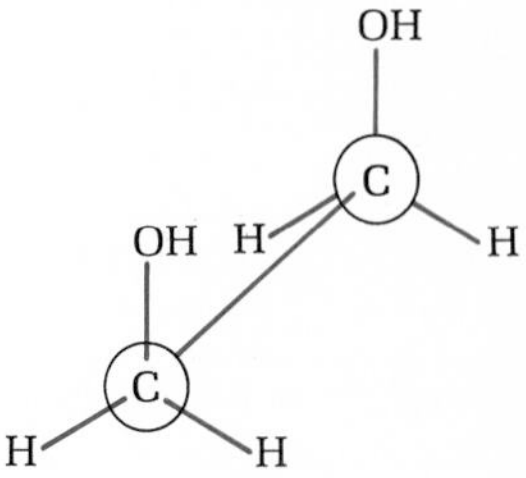

(b) HO—$\overset{Br}{\underset{\underset{CH_3}{CH_2}}{\underset{CH_2}{C}}}$—H and H—$\overset{Br}{\underset{\underset{CH_3}{CH_2}}{\underset{CH_2}{C}}}$—OH

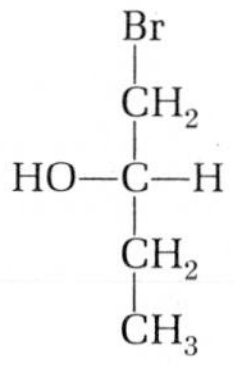

(d) [benzene]—CH₂OH; [benzene]—OCH₃;

[benzene with CH₃]—OH; [benzene with CH₃]—OH; CH₃—[benzene]—OH

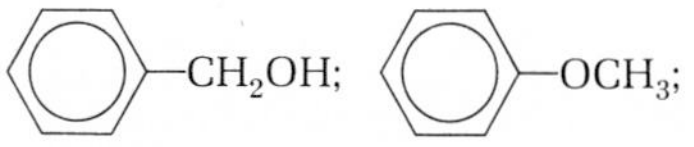

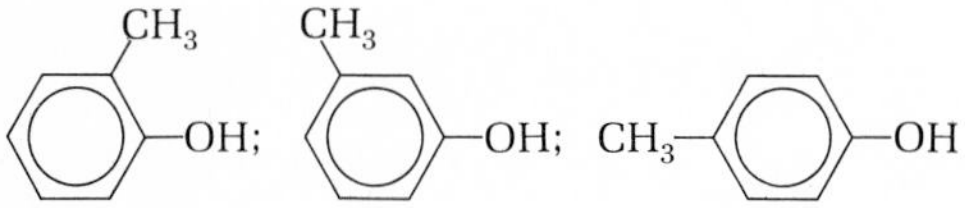

(e) CH₃OCH₂CH₂CH₃; CH₃OCH(CH₃)₂; CH₃CH₂OCH₂CH₃

(f) Any four of these possibilities: an aldehyde, a ketone, a cyclic compound having an ether or alcohol group, or an alkene having an ether or alcohol group

(g) 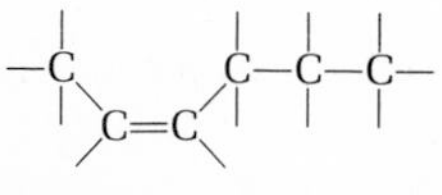 [structural formulas] and [structural formula]

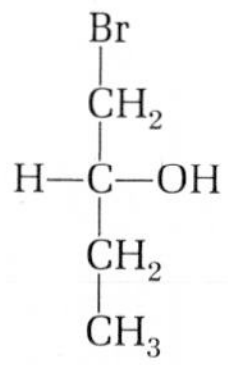

5. (named in order of structures given for question 3)—[3a]: 1-butene; *cis*-2-butene; *trans*-2-butene; 2-methyl-1-propene. [3c]: propylbenzene; isopropylbenzene; *o*-ethyltoluene; *m*-ethyltoluene; *p*-ethyltoluene; 1,2,3-trimethylbenzene; 1,2,4-trimethylbenzene; 1,3,5-trimethylbenzene. [3g]: *cis*-2-hexene; *trans*-2-hexene. [3h]: 3-methyl-*cis*-2-pentene; 3-methyl-*trans*-2-pentene   **6. (a)** decane **(b)** 2-methylbutane **(c)** 2,4-dimethyl-hexane **(d)** cyclopropane **(e)** methylcyclobutane **(f)** cyclo-pentene **(g)** 4-ethylcyclohexene **(h)** 3,4-dimethyl-*cis*-3-hep-tene **(i)** 2-methyl-2-pentene **(j)** 4-ethyl-*cis*-3-octene
7. (a) *p*-xylene **(b)** *m*-cresol **(c)** *o*-bromotoluene **(d)** *m*-nitro-benzoic acid **(e)** *o*-chloroaniline   **8. (a)** heptane
(b) CH₃(CH₂)₃CH₃ **(c)** butane **(d)** CH₃(CH₂)₈CH₃
(e) 3-methyl-*trans*-2-pentene   **9. (a)** 2,2-dimethylbutane
(b) CH₃CH₂CHCH₂CH(CH₂)₃CH₃ **(c)** 2,5-dimethylheptane
           CH₃   CH₃
(d) 1,1-dimethylcyclopentane **(e)** *cis*-1,2-dimethylcyclopentane
(f) *trans*-1,2-dimethylcyclohexane

(g) 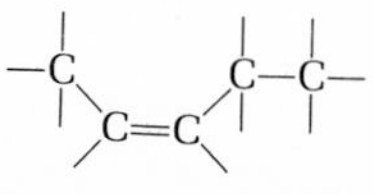 [structural formula]

(h)

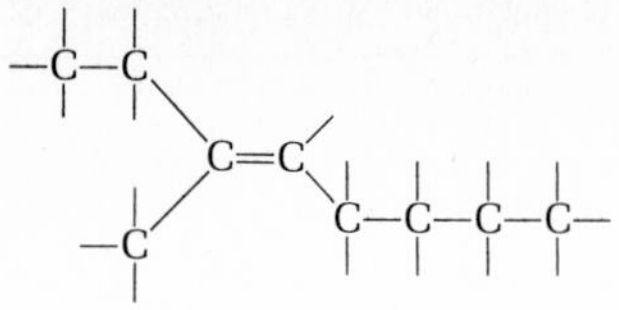

(i)

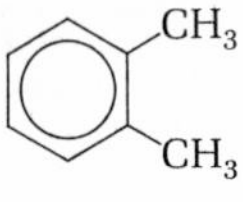

o-xylene structure with two CH₃ groups

(j)

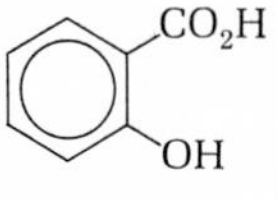

salicylic acid structure with CO₂H and OH

10. (a) 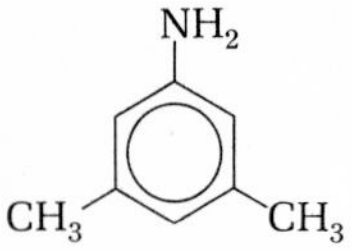 (hydrogen bonding)

structure with NH₂ and two CH₃ groups

(b) o-dichlorobenzene (polar)

(c)  (hydrogen bonding)

phenol structure with OH and CH₃

(d) cis-2-butene (more polar)

(e) 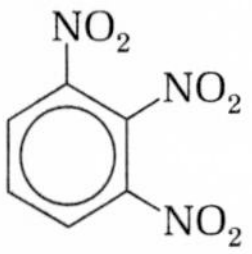 (more polar)

structure with three NO₂ groups

11. a and e (Structural similarity of optical isomers results in identical boiling points.) **12. (a)** C_6H_{14} **(b)** [See Exercise 4a.]

Self-Test

1. (a) alkyne **(b)** aldehyde **(c)** ether **(d)** amide **(e)** alcohol **(f)** ester **(g)** phenol **(h)** amine **(i)** ketone **(j)** aromatic hydrocarbon **2. (a)** 2,2,4-trimethylhexane **(b)** 2,2-dimethyl-4-ethyloctane

3. (a) $(CH_3)_3CCH_2CH_2CH(CH_2)_3CH_3$
$\qquad\qquad\qquad\qquad\quad |$
$\qquad\qquad\qquad\qquad CH_2CH_3$

(b) $CH_3CH_2CHCH_2CH_2CH_2CH_3$
$\qquad\qquad\quad |$
$\qquad\qquad CH_2CH_3$

4. (a)

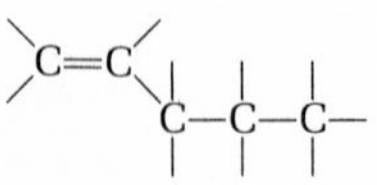

(b)

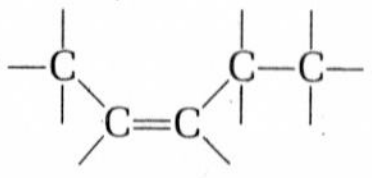

(c)

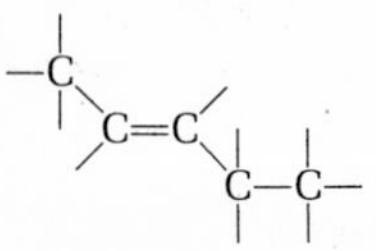

(d)

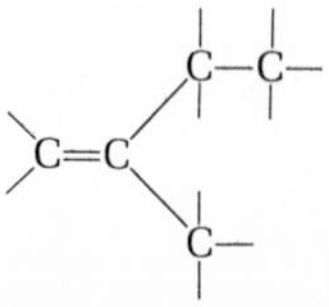

(e)

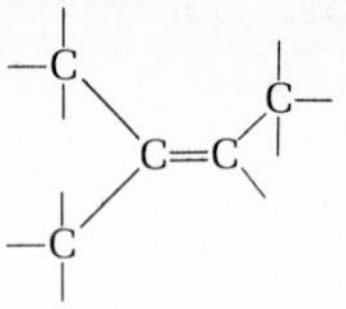

(f)

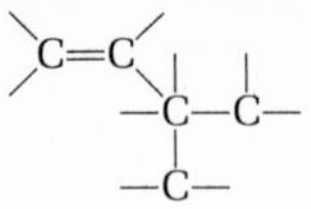

5. (a) 1-pentene **(b)** cis-2-pentene **(c)** trans-2-pentene **(d)** 2-methyl-1-butene **(e)** 2-methyl-2-butene **(f)** 3-methyl-1-butene **6. (a)** (1) m-xylene; (2) o-cresol; (3) p-toluidine; (4) methylcyclopentane; (5) 4-ethyl-cis-3-heptene

(b) (1) 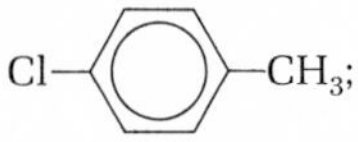 Cl—⟨benzene⟩—CH₃; (2) 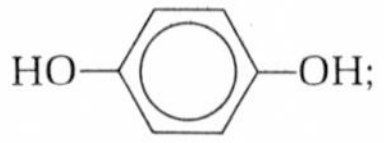 HO—⟨benzene⟩—OH;

(3)

structure with OH, two O₂N, NO₂

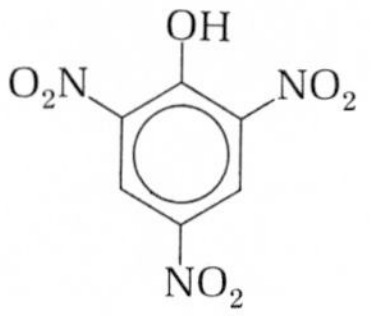

7. (a)

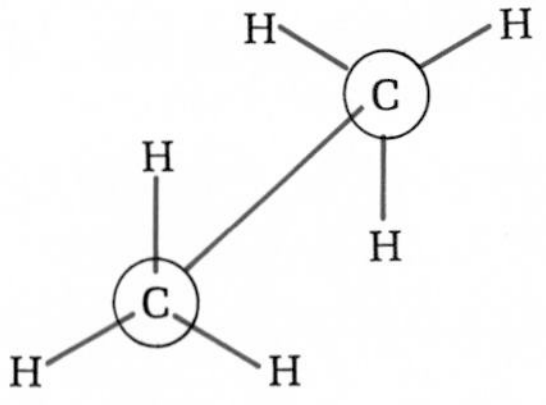

staggered 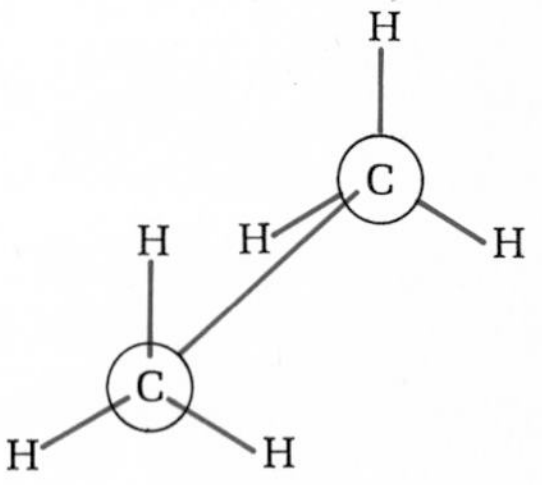   eclipsed

(b) 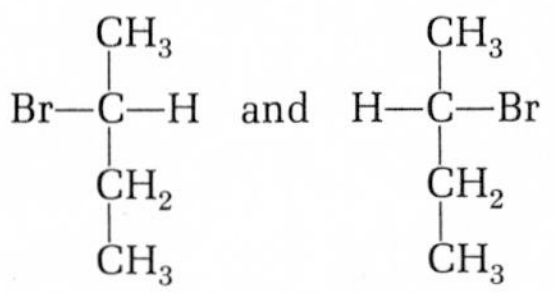
$\qquad\quad CH_3 \qquad\qquad CH_3$
$\qquad\qquad |\qquad\qquad\qquad |$
$Br—C—H \quad and \quad H—C—Br$
$\qquad\qquad |\qquad\qquad\qquad |$
$\qquad\quad CH_2 \qquad\qquad CH_2$
$\qquad\qquad |\qquad\qquad\qquad |$
$\qquad\quad CH_3 \qquad\qquad CH_3$

8. (a) cis-2-butene **(b)** o-xylene **(c)** m-cresol

UNIT 29

Exercises

1. (a) 5-bromo-2-methylheptane **(b)** o-methoxyphenol **(c)** 6-methyl-3-heptanol **(d)** 3,6-dimethylheptanal **(e)** 6-methyl-3-heptanone **(f)** 3,6-dimethylheptanoic acid **(g)** 5-amino-2-methylheptane **(h)** ethyl 2,5-dimethylhexanoate **(i)** 3,6-dimethylheptanamide **2. (a)** formic acid **(b)** potassium acetate **(c)** propionaldehyde **3. (a)** 2-propanol (hydrogen bonding) **(b)** 2-propanol (smaller hydrophobic region) **4.** a and c (charge delocalization of anion) **5. (a)** $HOCH_2CH_2OH$
(b) $BrCH_2CH_2Br + 2OH^- \longrightarrow HOCH_2CH_2OH + 2Br^-$

6. (a) $CH_3CH_2OH \xrightarrow{Cr_2O_7{}^{2-}} CH_3CO_2H \xrightleftharpoons[\quad]{\overset{CH_3CH_2OH}{(H^+)}}$

$$CH_3\overset{O}{\overset{\|}{C}}OCH_2CH_3 \ (+ H_2O)$$

(b)

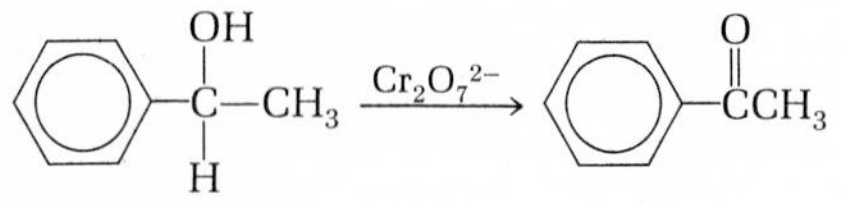

(c)

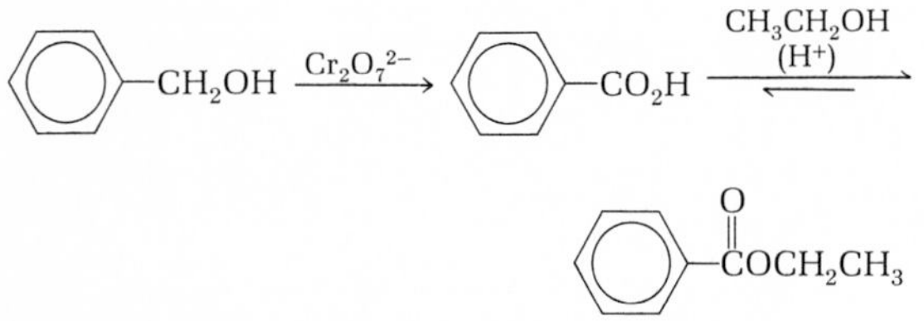

Practice for Proficiency

1. (a) 7-methyl-4-nonanol   **(b)** 6-amino-3-methylnonane
(c) 6,6-dimethyloctanoic acid   **(d)** ethyl 4-bromohexanoate
(e) o-methoxytoluene   **2. (a)** $CH_3CH_2CH_2C{=}O$ with H

(b) $Na(HCO_2)$ **(c)** CH_3OCCH_3 **3.** c   **4.** a   **5.** d
6. b   **7.** a   **8.** e   **9.** c   **10.** a   **11.** acetone (smaller hydrophobic region)   **12.** a   **13.** d   **14.** c   **15.** a
16. c   **17.** e   **18.** 2   **19.** 2   **20. (a)** carboxylic acid
(b) aromatic hydrocarbon   **21.** c   **22.** e   **23.** a
24. c   **25.** b

26. (a) $(CH_3)_2CHOH \xrightarrow{Cr_2O_7{}^{2-}} CH_3\overset{O}{\overset{\|}{C}}CH_3$

(b)

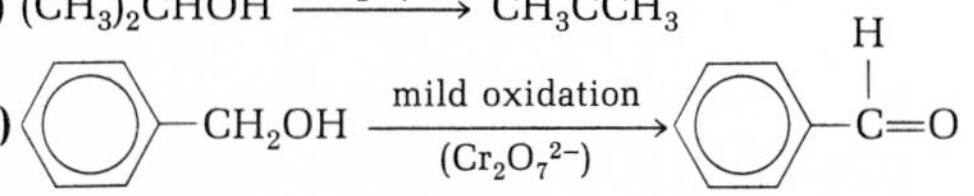

(c)

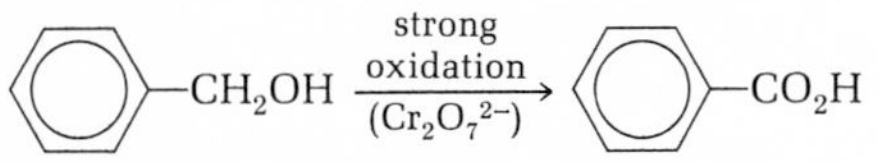

(d)

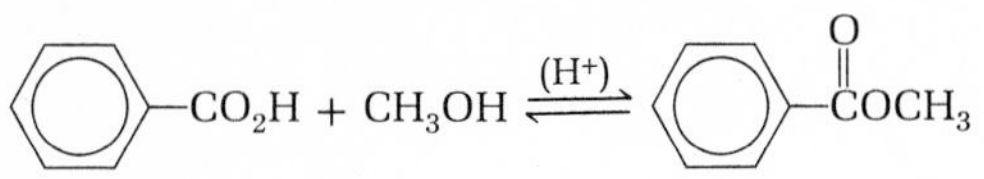

(from **c**)

(e)

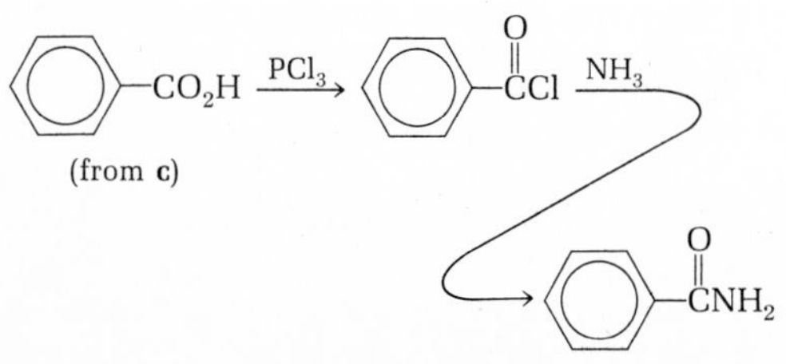

(from **c**)

(f) $CH_3CH_2OH \xrightarrow{Cr_2O_7{}^{2-}} CH_3CO_2H \xrightleftharpoons[\quad]{(H^+)}$

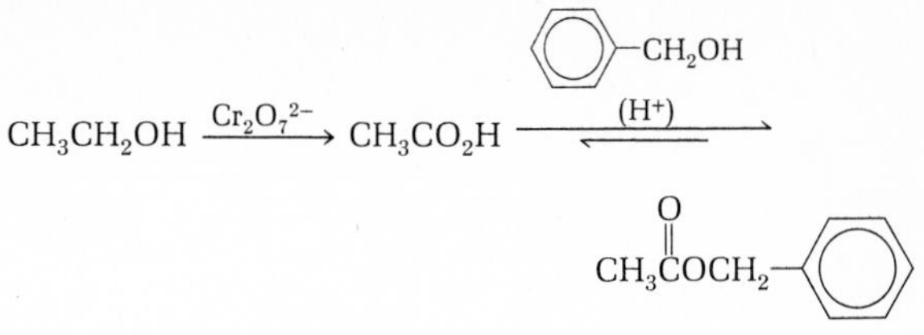

(g) $CH_3CH_2CH_2OH \xrightarrow{PCl_3} CH_3CH_2CH_2Cl \xrightarrow{NH_3}$

$$CH_3CH_2CH_2NH_2$$

$$CH_3CH_2CH_2CH_2OH \xrightarrow{Cr_2O_7{}^{2-}} CH_3CH_2CH_2\overset{O}{\overset{\|}{C}}OH \xrightarrow{PCl_3}$$

$$CH_3CH_2CH_2\overset{O}{\overset{\|}{C}}Cl;$$

$$CH_3CH_2CH_2\overset{O}{\overset{\|}{C}}Cl \xrightarrow{CH_3CH_2CH_2NH_2}$$

$$CH_3CH_2CH_2\overset{O}{\overset{\|}{C}}\underset{H}{N}CH_2CH_2CH_3$$

27. 4×10^{-8} **28.** 2.5×10^{-6} **29.** 3.1×10^{-5}
30. 4×10^{-9} **31.** 2.8×10^{-7}

32. (A) (B) (C) (D)

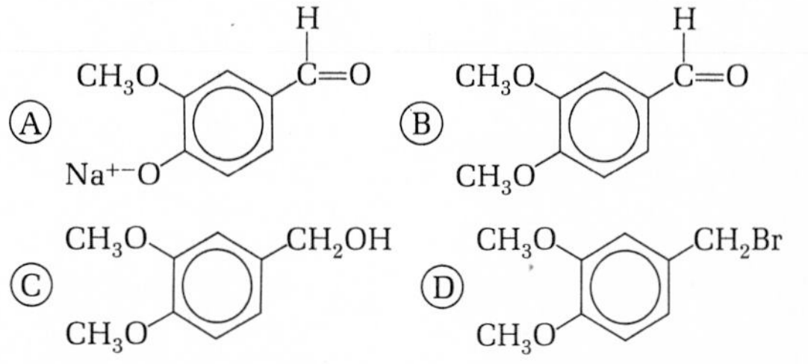

Self-Test
1. (a) 1-chloro-3,3-dimethylbutane   **(b)** 3,3-dimethyl-1-butanol
(c) 3,3-dimethylbutanoic acid   **(d)** methyl 3,3-dimethyl-
butanoate   **(e)** p-methoxyaniline   **2. (a)** butyric acid
(b) ethyl propionate   **(c)** formaldehyde   **3. (a)** $(CH_3)_3COH$

(b)

       4. (a) amine   **(b)** alkene   **(c)** amide

5. (a) $CH_3CH_2CH_2\overset{H}{\underset{}{C}}{=}O \xrightarrow{LiAlH_4} CH_3CH_2CH_2CH_2OH$

(b) $CH_3\overset{H}{\underset{}{C}}{=}O \xrightarrow{Cr_2O_7{}^{2-}} CH_3CO_2H$

(c) 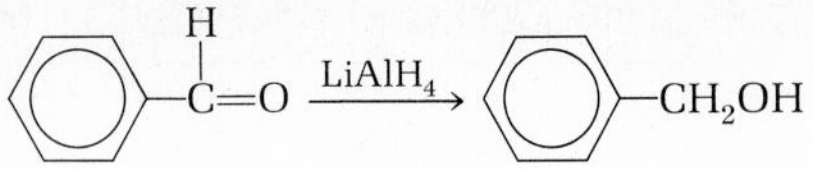 [benzaldehyde structure with H—C=O] $\xrightarrow{\text{LiAlH}_4}$ [benzene with —CH$_2$OH]

(d) $CH_3CH_2\overset{\text{H}}{\underset{}{C}}=O \xrightarrow{\text{LiAlH}_4} CH_3CH_2CH_2OH \xrightarrow{\text{PCl}_3} CH_3CH_2CH_2Cl$

(e) $CH_3(CH_2)_4\overset{\text{H}}{\underset{}{C}}=O \xrightarrow{\text{Cr}_2\text{O}_7^{2-}} CH_3(CH_2)_4CO_2H \xrightarrow{\text{PCl}_3}$

$\rightarrow CH_3(CH_2)_4\overset{\text{O}}{\overset{\|}{C}}Cl \xrightarrow{\text{NH}_3} CH_3(CH_2)_4\overset{\text{O}}{\overset{\|}{C}}NH_2$

6. $\sim 7 \times 10^{-3}$

UNIT 30
Exercises
1. benzophenone (most extensive delocalized π-bond system)　　**2. (a)** $6.2 \times 10^{-5}\,M$　**(b)** $1.7 \times 10^{-3}\,M$　　**3.** 87%
4. (a) carboxylic acid or amide　**(b)** propylbenzene; isopropylbenzene; o-ethyltoluene; m-ethyltoluene; p-ethyltoluene; 1,2,3-trimethylbenzene; 1,2,4-trimethylbenzene; 1,3,5-trimethylbenzene　**5. (a)** $(CH_3)_3COH$　**(b)** 3-methyl-2-butanone

(c) $CH_3\overset{\text{O}}{\overset{\|}{C}}OCH_2CH_2CH(CH_3)_2$ (Discuss with your instructor.)

Practice for Proficiency
1. (a) $1.6 \times 10^{-2}\,M$　**(b)** $6.2 \times 10^{-5}\,M$　　**2.** $7.2 \times 10^{-5}\,M$
3. $6.0 \times 10^{-4}\,M$　　**4.** $2.7 \times 10^{-5}\,M$　　**5.** $3.0 \times 10^{-3}\,M$
6. 14 mg liter^{-1}　　**7.** 70%　　**8.** 50%　　**9.** p-nitrobenzaldehyde (most extensive delocalized π-bond system)　　**10.** azobenzene (most extensive π-bond system)　　**11. (a)** yes　**(b)** no　**(c)** yes (useful only if mixture components have no significant absorption overlap)　　**12.** d　　**13.** b　　**14.** d　　**15.** c
16. b　　**17.** b　　**18.** b　　**19.** d　　**20.** b

21. $CH_3CH_2\overset{\text{O}}{\overset{\|}{C}}OCH_3$

Self-Test
1. nitrobenzene　　**2.** $1.0 \times 10^{-3}\,M$　　**3.** 93%
4. (a) carboxylic acid or amide

(b) actual compound: [benzene ring]—CH_2CO_2H;　other plausible

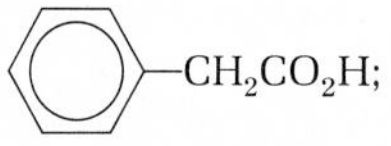

structures: CH_3—[benzene ring]—CO_2H;　[benzene ring with CO_2H and CH_3 ortho]　;　[benzene ring with CO_2H and CH_3 meta]

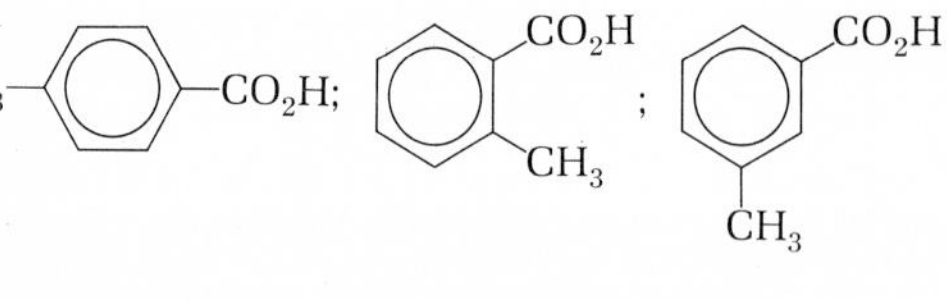

5. (a) $(CH_3)_2CHOCH(CH_3)_2$　**(b)** 2-methylbutanoic acid

EXCURSION 4
Exercises
1. first: cyclohexene; second: cyclohexanone; last: cyclohexanol　　**2.** first: p-xylene; second: N,N-dimethylaniline; last: p-chloroaniline　　**3.** first: p-dichlorobenzene; second:

p-chloronitrobenzene; last: p-chloroaniline　　**4. (a)** 670 torr
(b) 0.85　　**5. (a)** 703 torr　**(b)** 0.92　　**6. (a)** 648 torr　**(b)** 0.88
7. Dissolve mixture in ether and extract with aqueous NaOH to remove the phenol.　　**8.** Dissolve in ether and extract with aqueous HCl to remove the aniline. Then extract remaining ether solution with aqueous NaOH to remove the phenol.
9. Dissolve in ether and extract with aqueous NaOH to remove the salicylic acid.

UNIT 31
Exercises

1. (a)

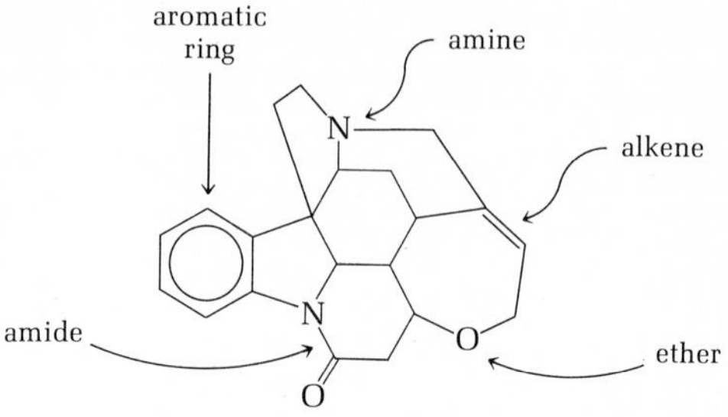

(b) (Same as **a**, but alkene group converted to dibromide)

(c) (same as **a**, but amine group converted to [structure: N^+—H])

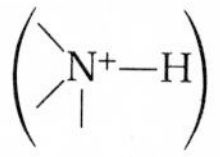

(d) (Same as **a**, but ring opened by hydrolysis of amide group, leaving cyclic amine and sodium salt of "opened" carboxylate group)　　**2.** ① isobutyric; ② acetic; ③ γ-chlorobutyric; ④ chloroacetic; ⑤ fluoroacetic (most acidic)

3. (a) 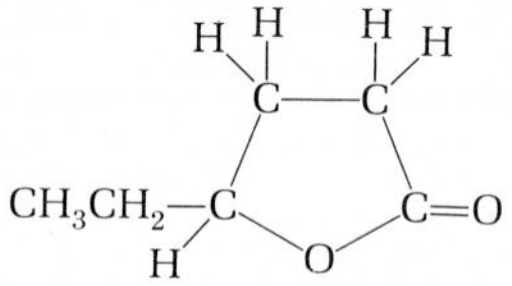 [cyclic structure: CH_3CH_2—C with H H, C with H H, C=O, O in ring]　　**(b)** 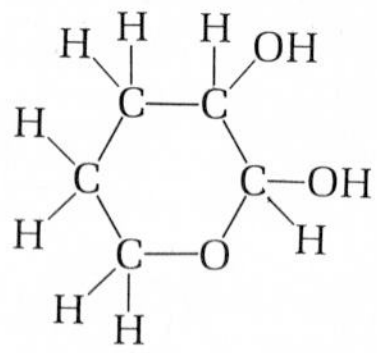 [cyclic structure with OH, C—OH, O, H's]

4. (all O—H groups converted to O—$\overset{\text{O}}{\overset{\|}{C}}CH_3$ groups)　　**5.** (See structures of glucose and fructose, Section 31.5.)

Practice for Proficiency

1. (a)

aromatic ring;　amine;　aromatic ring;　ketone;　CH_3CH_2

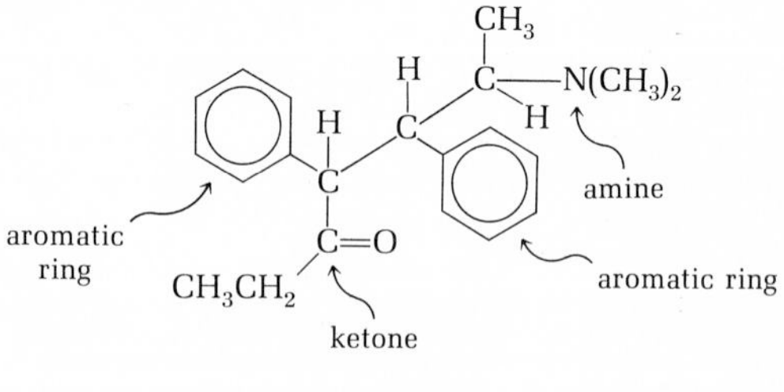

(b)

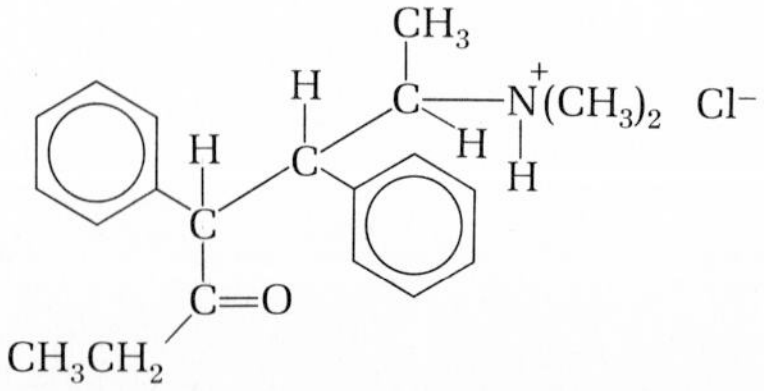

(c)

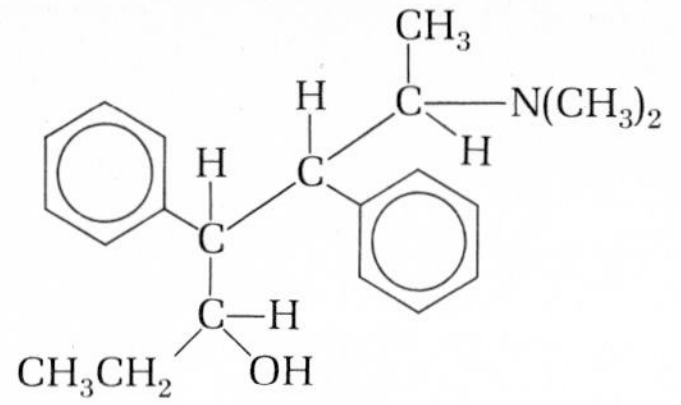

2. (a)

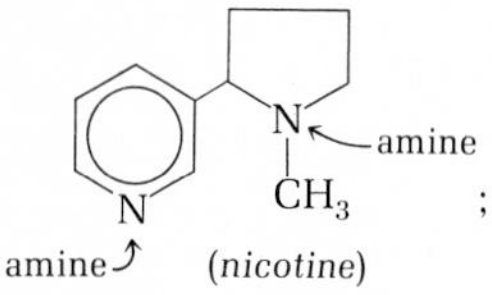

amine ⌐ (*nicotine*)

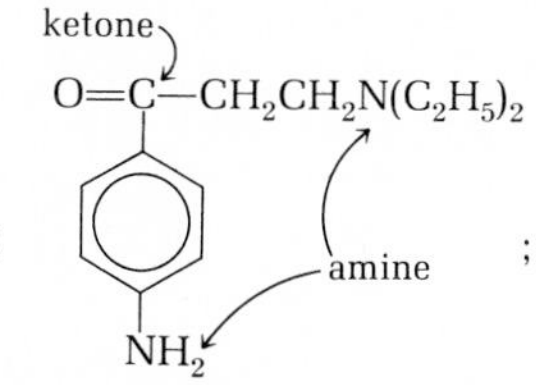

(*novocaine*)

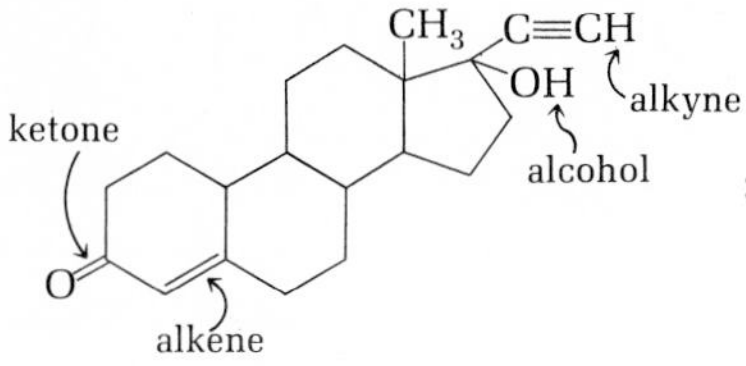

(*Norlutin®*)

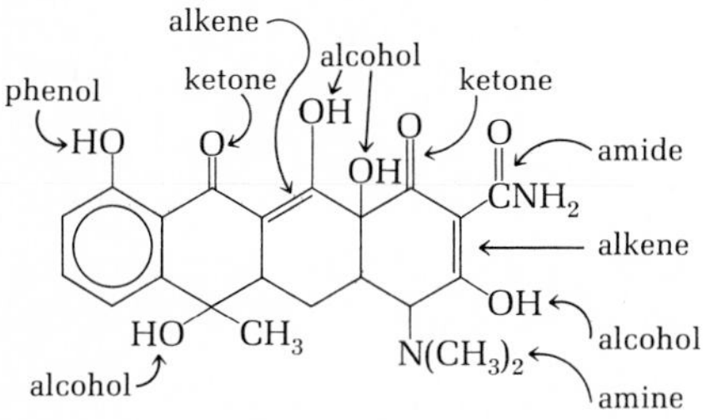

(*tetracycline*)

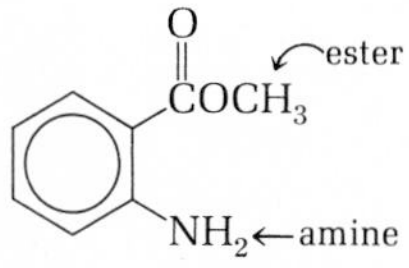

(*methyl anthranilate*)

(b) nicotine $\xrightarrow{\text{excess HCl}}$

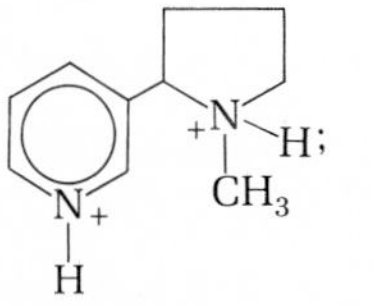

novocaine $\xrightarrow{\text{excess HCl}}$

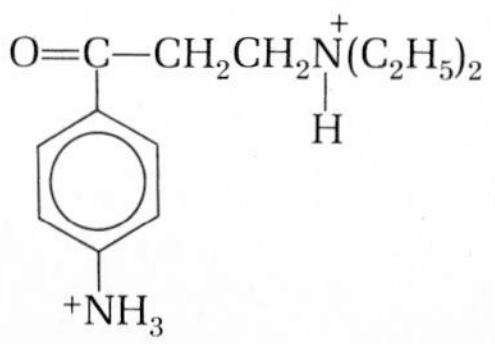

Norlutin® $\xrightarrow{\text{HCl}}$ (no reaction);

tetracycline $\xrightarrow{\text{HCl}}$

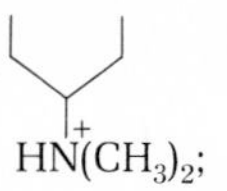

methyl anthranilate $\xrightarrow{\text{HCl}}$

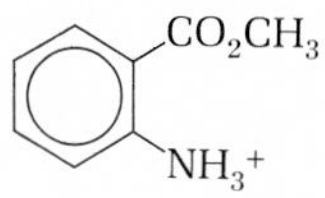

(c) nicotine $\xrightarrow{\text{H}_2\text{O, OH}^-}$ (no reaction);

novocaine $\xrightarrow{\text{H}_2\text{O; OH}^-}$ (no reaction);

Norlutin® $\xrightarrow{\text{H}_2\text{O, OH}^-}$ (no reaction);

tetracycline $\xrightarrow{\text{H}_2\text{O, OH}^-}$

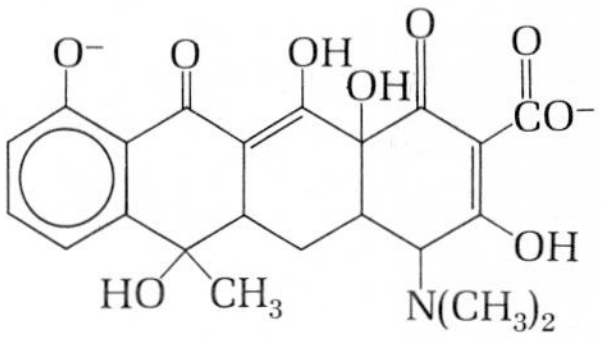

$(+\text{NH}_3)$;

methyl anthranilate $\xrightarrow{\text{H}_2\text{O, OH}^-}$

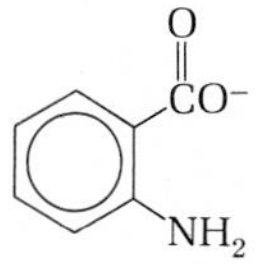

$(+\text{CH}_3\text{OH})$

(d) nicotine $\xrightarrow{\text{LiAlH}_4}$ (no reaction);

novocaine $\xrightarrow{\text{LiAlH}_4}$

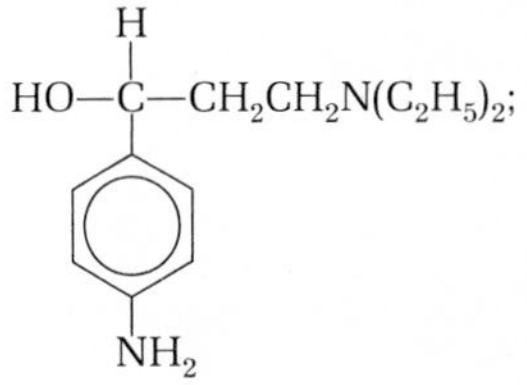

Norlutin® $\xrightarrow{\text{LiAlH}_4}$

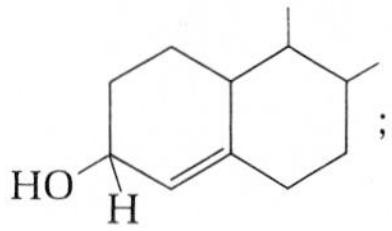

;

tetracycline $\xrightarrow{\text{LiAlH}_4}$

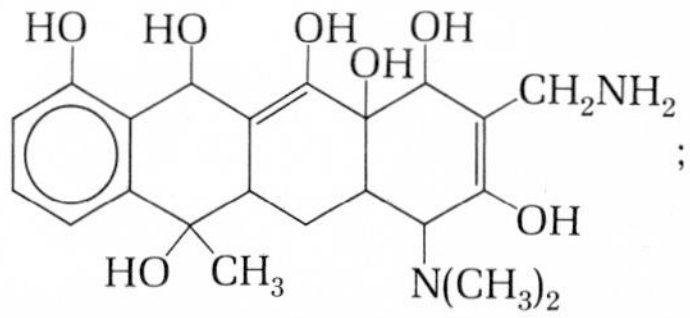

;

(amide reduced)

methyl anthranilate $\xrightarrow{\text{LiAlH}_4}$

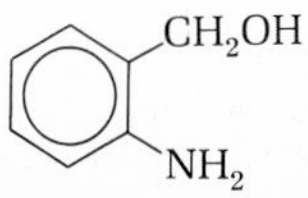

3. 2,4-dinitrobenzoic $>$ p-chlorobenzoic $>$ benzoic
4. methylamine $>$ aniline $>$ p-nitroaniline
5. $(CH_3)_3CCO_2^- > CH_3CO_2^- > ClCH_2CO_2^- > Cl_3CCO_2^-$

6. (a)

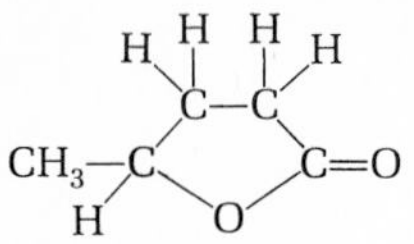

(b)

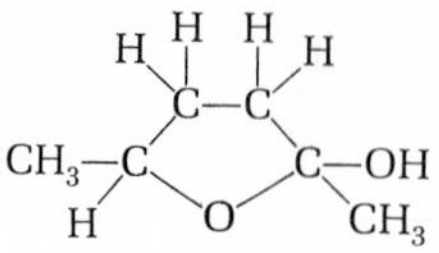

(c)

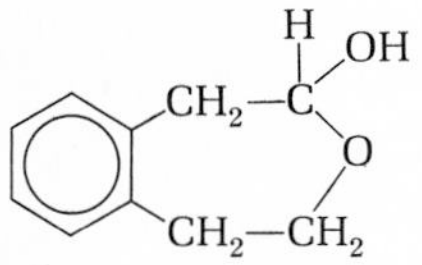

(d)

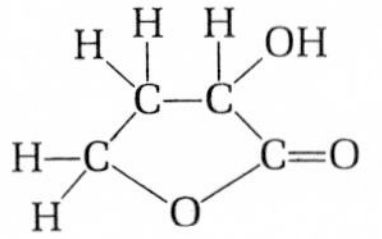

(e)

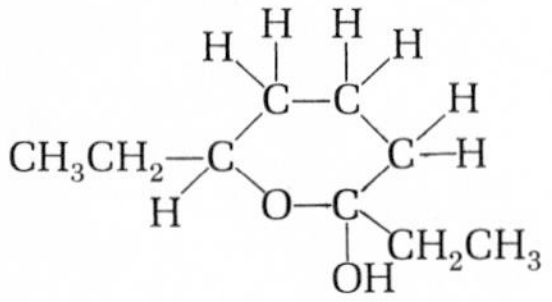

7. (a) glucose **(b)** (see glucose structures, Section 31.5) **(c)** (all

O—H groups converted to $O—\overset{\overset{O}{\|}}{C}CH_3$ groups) **(d)** many OH groups for hydrogen bonding with water

(e)

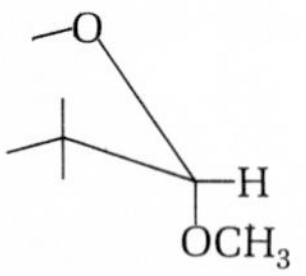

8. Cellobiose and maltose differ in the geometry of the glycoside links, and this difference is reflected in hydrolysis catalyzed by stereospecific enzymes. **9. (a)** open-chain form (see Section 31.5) indicates aldehyde group **(b)** no open-chain form possible, because no *hemiacetal* linkages **10.** glucose and sucrose

Self-Test

1.

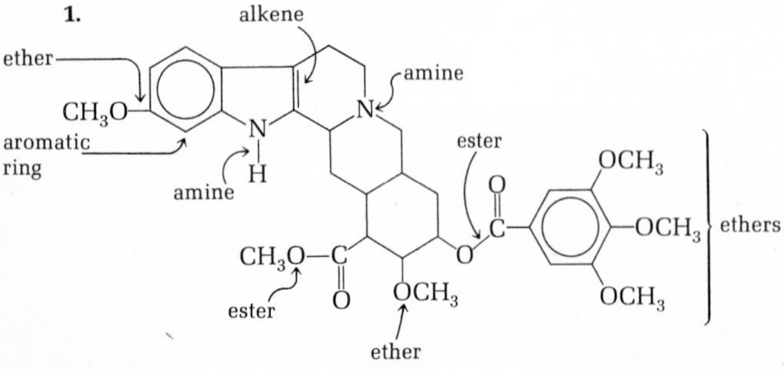

2.

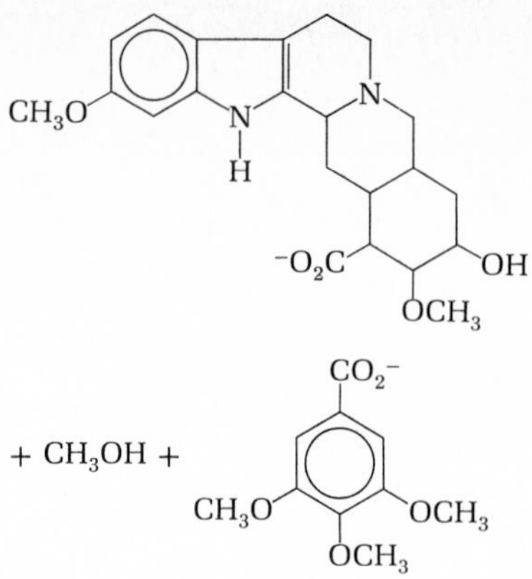

$+ CH_3OH +$

3. ① trifluoroacetic; ② methoxyacetic; ③ β-methoxypropionic; ④ acetic; ⑤ dimethylpropionic

4. (a)

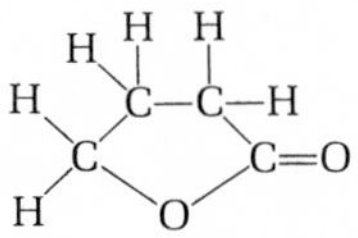

(b)

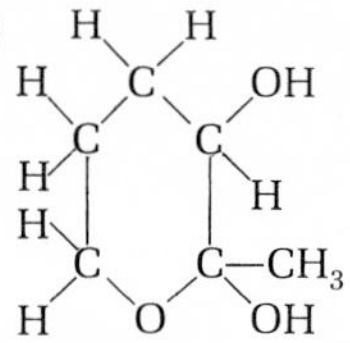

5.

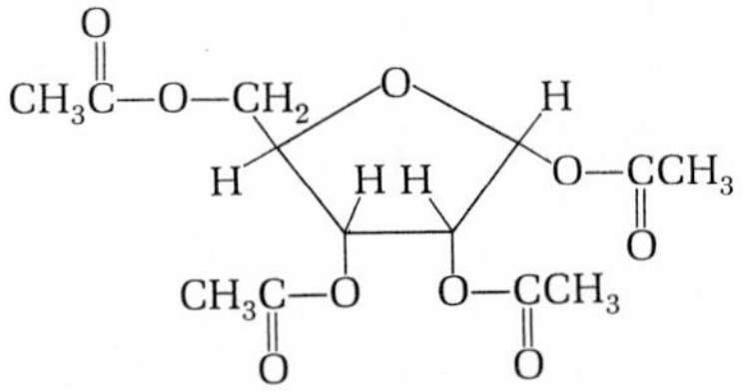

6.

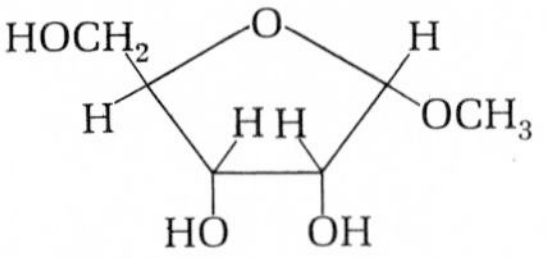

7.

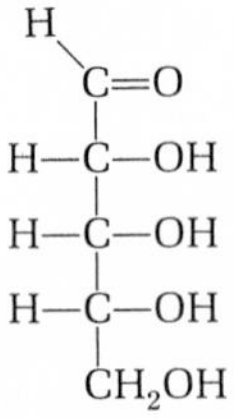

1. (a) $CH_3\overset{\overset{HO}{|}}{C}\overset{\overset{H}{|}}{\underset{\underset{NH_2}{|}}{C}}-CO_2^-$ **(b)** $CH_3\overset{\overset{HO}{|}}{C}\overset{\overset{H}{|}}{\underset{\underset{^+NH_3}{|}}{C}}-CO_2H$

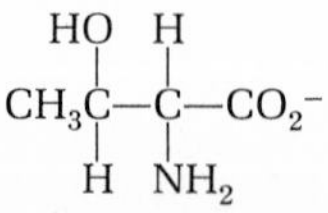

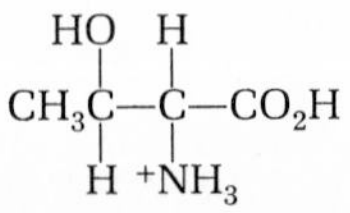

Practice for Proficiency

(c) $CH_3\overset{HO}{\underset{H}{\overset{|}{C}}}-\overset{H}{\underset{+NH_3}{\overset{|}{C}}}-\overset{O}{\overset{||}{C}}OCH_3$ (d) $CH_3\overset{CH_3\overset{O}{\overset{||}{C}}}{\underset{H}{\overset{O\ \ \ H}{\overset{|\ \ \ |}{C}}}}-\overset{H}{\underset{NH}{\overset{|}{C}}}-CO_2H$
$O{=}\overset{}{C}CH_3$

1. (a) $HO_2CCH_2\overset{H}{\underset{+NH_3}{\overset{|}{C}}}-CO_2^-$

(e) $CH_3\overset{HO}{\underset{H}{\overset{|}{C}}}-\overset{H}{\underset{+NH_3}{\overset{|}{C}}}-CO_2^-$

(b) aspartic acid $\xrightarrow{OH^-}$ $^-O_2CCH_2\overset{H}{\underset{NH_2}{\overset{|}{C}}}-CO_2^-$;

2. (ala) $CH_3\overset{H}{\underset{+NH_3}{\overset{|}{C}}}-CO_2H$; (ser) $HOCH_2\overset{H}{\underset{+NH_3}{\overset{|}{C}}}-CO_2H$;

aspartic acid $\xrightarrow{H^+}$ $HO_2CCH_2\overset{H}{\underset{+NH_3}{\overset{|}{C}}}-CO_2H$;

(lys) $H_3\overset{+}{N}(CH_2)_4\overset{H}{\underset{+NH_3}{\overset{|}{C}}}-CO_2H$; (glu) $HO_2C(CH_2)_2\overset{H}{\underset{+NH_3}{\overset{|}{C}}}-CO_2H$;

aspartic acid $\xrightarrow[\text{excess}]{\overset{O}{\overset{||}{CH_3CCl}}}$ $HO_2CCH_2\overset{H}{\underset{NH}{\overset{|}{C}}}-CO_2H$;
$O{=}\overset{}{C}CH_3$

(tyr) $HO-\bigcirc-CH_2\overset{H}{\underset{+NH_3}{\overset{|}{C}}}-CO_2H$

aspartic acid $\xrightleftharpoons[(H^+)]{CH_3OH}$ $CH_3O\overset{O}{\overset{||}{C}}CH_2\overset{H}{\underset{+NH_3}{\overset{|}{C}}}-\overset{O}{\overset{||}{C}}OCH_3$

3. (a) $HO-\underset{HO}{\bigcirc}-CH_2\overset{H}{\underset{+NH_3}{\overset{|}{C}}}-CO_2^-$

2. (asp) $^-O_2CCH_2\overset{H}{\underset{NH_2}{\overset{|}{C}}}-CO_2^-$; (phe) $\bigcirc-CH_2\overset{H}{\underset{NH_2}{\overset{|}{C}}}-CO_2^-$;

(b) $HO-\underset{HO}{\bigcirc}-CH_2CH_2NH_2 \rightleftharpoons$

(lys) $H_2N(CH_2)_4\overset{H}{\underset{NH_2}{\overset{|}{C}}}-CO_2^-$

$^-O-\underset{HO}{\bigcirc}-CH_2CH_2NH_3^+ \rightleftharpoons$

3. (a) $\underset{N}{\overset{N}{\rceil}}-CH_2CH_2NH_2$ **(b)** $HO_2C(CH_2)_2\overset{O}{\overset{||}{C}}CO_2H$;

$HO-\underset{-O}{\bigcirc}-CH_2CH_2NH_3^+$

(c) $^-O_2C\overset{H}{\underset{+NH_3}{\overset{|}{C}}}CH_2S{-}SCH_2\overset{H}{\underset{+NH_3}{\overset{|}{C}}}CO_2^-$

4. (a) $H_3\overset{+}{N}\overset{H\ \ O}{\underset{CH_3}{\overset{|\ \ ||}{C}}}-\overset{}{C}-N-\overset{H\ \ O}{\underset{CH_3}{\overset{|\ \ ||}{C}}}-\overset{}{C}-N-\overset{H}{\underset{CH_2OH}{\overset{|}{C}}}-CO_2^-$ *and* ala-ser-ala;
$[ala]{-}{-}{-}[ala]{-}{-}{-}[ser]$

4. physical properties indicating ionic character, K_a and K_b values typical of $-NH_3^+$ and $-CO_2^-$, IR and nmr spectra (see Section 32.1) **5.** c ($+2$) **6.** d (-1) **7.** d ($+2$)
8. e (0) **9.** d ($+1$) **10.** c (-2) **11.** e (2) **12.** d (2)
13. d (gly, phe) **14.** c (gly, glu) **15.** d (his, glu)
16. d (lys, his)

ser-ala-ala; ser-ser-ala; ser-ala-ser, ala-ser-ser **(b)** ala-ala-ser, ala-ser-ala, ser-ala-ala **(c)** ala-ser-ser
5. $6RSH + BrO_3^- \longrightarrow 3R{-}S{-}S{-}R + Br^- + 3H_2O$

17. $H_3\overset{+}{N}C-C-N-C-C-N-C-CO_2^-$ (structure with H, O groups; side chains CH_2—phenyl, CH_2—phenyl, $CH_2CH_2SCH_3$)

18. (a) $H_3\overset{+}{N}C-C-NCH_2C-NCH_2CO_2^-$ (side chain CH_2OH)

(b) $H_3\overset{+}{N}C-C-NCHC-NCHCO_2^-$ (side chains CH_2OH-phenol, CH_2-phenyl, CH_2-phenyl)

(c) $H_3\overset{+}{N}C-C-NCHC-NCHCO_2^-$ (side chains CH_3, CH_3, CH_2OH)

(d) $H_3\overset{+}{N}C-C-NCHC-NCH_2CO_2^-$ (side chains CH_2-phenyl, CH_2-phenyl)

(e) $H_3\overset{+}{N}C-C-NCHC-NCH_2CO_2^-$ (side chains CH_2-phenol, CH_2-phenyl)

19. Addition of H^+ will neutralize charges of $-CO_2^-$ groups, removing their contribution to ionic bonding regions of the protein. **20.** (For representations of structural changes, see Fig. 32.11.) **(a)** hydrogen bonds, London forces (and probably ionic bonds) **(b)** ionic bonds, **(c)** hydrogen bonds (and probably London forces), **(d)** disulfide bonds
21. (a) (one each: glu, cys, gly)

$HO_2C(CH_2)_2\overset{H}{\underset{+NH_3}{C}}CO_2H;$ $HSCH_2\overset{H}{\underset{+NH_3}{C}}CO_2H;$ $H_3\overset{+}{N}CH_2CO_2H$

(b) (two gly and one each: ala, glu, histamine, NH_4^+)

$H_3\overset{+}{N}CH_2CO_2H;$ $CH_3\overset{H}{\underset{+NH_3}{C}}CO_2H;$ $HO_2C(CH_2)_2\overset{H}{\underset{+NH_3}{C}}CO_2H;$ $NH_4^+;$

(imidazole ring) $-CH_2CH_2NH_3^+$

(c) (three pro, two each: arg and phe, and one each: gly and ser)

(d) (3NH_4^+; one each of others) $NH_4^+;$

$H_3\overset{+}{N}CH_2CO_2H;$ $(CH_3)_2CHCH_2\overset{H}{\underset{+NH_3}{C}}CO_2H;$ (pyrrolidine ring)$-CO_2H;$

$HO_2C\overset{H}{C}CH_2S-SCH_2\overset{H}{C}CO_2H;$ (side chains $+NH_3$, $+NH_3$)   (phenyl)$-CH_2\overset{H}{\underset{+NH_3}{C}}CO_2H;$

$CH_3CH_2\overset{H}{\underset{CH_3}{C}}-\overset{H}{\underset{+NH_3}{C}}-CO_2H;$ $HO_2C(CH_2)_2\overset{H}{\underset{+NH_3}{C}}CO_2H;$

$HO_2CCH_2\overset{H}{\underset{+NH_3}{C}}CO_2H$

(e) (two of each) (pyrrolidine ring)$-CO_2H;$ (phenyl)$-CH_2\overset{H}{\underset{+NH_3}{C}}CO_2H;$

$(CH_3)_2CHCH_2\overset{H}{\underset{+NH_3}{C}}CO_2H;$ $(CH_3)_2CH\overset{H}{\underset{+NH_3}{C}}CO_2H;$

$H_3\overset{+}{N}(CH_2)_3\overset{H}{\underset{+NH_3}{C}}CO_2H$

Self-Test

1. (a) $HOCH_2\overset{H}{\underset{+NH_3}{C}}CO_2^-$ basic / acidic   **(b)** $H_3\overset{+}{N}(CH_2)_4\overset{H}{\underset{NH_2}{C}}CO_2^-$ acidic / basic

(c) HO—⟨benzene ring⟩—CH$_2$CCO$_2^-$ ⟿ basic

acidic ↗ $^+$NH$_3$

(C has H above and $^+$NH$_3$ below)

H$_3$$^+$N—CH$_2$—C(=O)—N(H)—C(CH$_3$)(H)—C(=O)—N(H)—C(H)(CH$_2$—⟨benzene ring⟩—OH)—CO$_2^-$

2. (a) HOCH$_2$CCO$_2^-$ (H above C, NH$_2$ below) **(b)** HOCH$_2$CCO$_2$H (H above C, $^+$NH$_3$ below)

(c) CH$_3$COCH$_2$CCO$_2$H (O above first C; H above second C, HN—CCH$_3$ with O below) **(d)** HOCH$_2$C—COCH$_2$CH$_3$ (H and O above, $^+$NH$_3$ below)

(e) [—OCH$_2$CCOCH$_2$CC—]$_x$ (HO above, $^+$NH$_3$ below each, repeating)

3. (a) (-1) **(b)** (-1) **(c)** (-2) **(d)** (-2) **(e)** (-3)

4. (a) HO—⟨benzene ring⟩—CH$_2$CH$_2$NH$_2$, **(b)** CH$_3$CCO$_2$H (O above C)

5. physical properties consistent with ionic character and K_a, K_b, IR, nmr data consistent with —NH$_3^+$ and —CO$_2^-$ groups   **6.** glu, lys, phe, and ser

7. (a) H$_3$$^+$N—C(H)(CH$_3$)—C(=O)—N(H)—CH$_2$—C(=O)—N(H)—C(H)(CH$_2$—⟨benzene ring⟩—OH)—CO$_2^-$

(b) H$_3$$^+$N—CH$_2$—C(=O)—N(H)—C(CH$_3$)(H)—C(=O)—N(H)—C(H)(CH$_2$—⟨benzene ring⟩—OH)—CO$_2^-$

H$_3$$^+$N—C(H)(CH$_3$)—C(=O)—N(H)—C(H)(CH$_2$—⟨benzene ring⟩—OH)—C(=O)—N(H)—CH$_2$—CO$_2^-$

H$_3$$^+$N—C(CH$_2$—⟨benzene ring⟩—OH)(H)—C(=O)—N(H)—CH$_2$—C(=O)—N(H)—C(H)(CH$_3$)—CO$_2^-$

H$_3$$^+$N—C(CH$_2$—⟨benzene ring⟩—OH)(H)—C(=O)—N(H)—CH$_2$—C(=O)—N(H)—C(H)(CH$_3$)—CO$_2^-$

H$_3$$^+$N—CH$_2$—C(=O)—N(H)—C(CH$_2$—⟨benzene ring⟩—OH)(H)—C(=O)—N(H)—C(H)(CH$_3$)—CO$_2^-$

8. CH$_3$—C(=O)—N(H)—CH$_2$—C(=O)—N(H)—C(CH$_3$)(H)—C(=O)—N(H)—C(H)(CH$_2$—⟨benzene ring⟩—O—C(=O)—CH$_3$)—CO$_2$H

9. A pH change might change ionic groups so that geometric features depending on ionic attractions are altered.

UNIT 33
Exercises

1. (a) homopolymer **(b)** copolymer **(c)** copolymer **(d)** homopolymer **(e)** copolymer **2. (a)** linear **(b)** cross-linked **(c)** linear **(d)** space network **(e)** linear **3. (a)** C **(b)** E **(c)** A **(d)** B **(e)** D

4. (a)

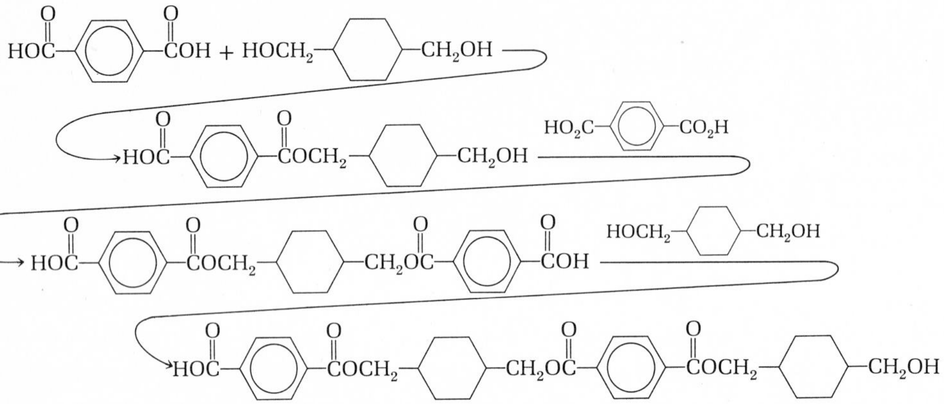

(b)

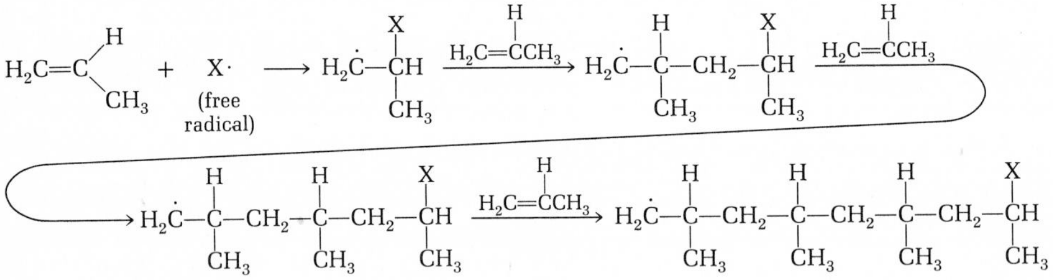

(c)

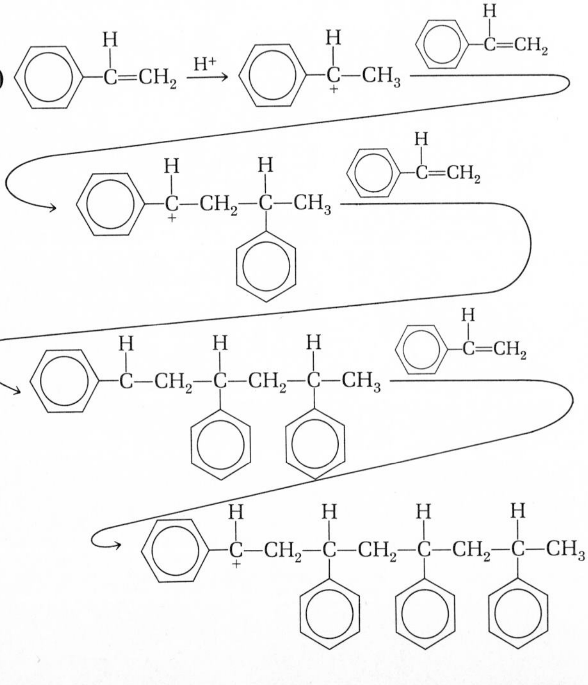

Practice for Proficiency *and* **Self-Test**
(Consult with your instructor.)

UNIT 34
Exercises
1. (a) H **(b)** D **(c)** J **(d)** B **(e)** F **(f)** I **(g)** A **(h)** E **(i)** G
(j) C   **2. (a)** T **(b)** F **(c)** F **(d)** T **(e)** T **(f)** F **(g)** F **(h)** T
(i) T **(j)** F   **3. (a)** both **(b)** both **(c)** RNA **(d)** DNA
(e) DNA

Practice for Proficiency *and* **Self-Test**
(Consult with your instructor.)

UNIT 35
Exercises
1. (a) $^{226}_{88}\text{Ra} \longrightarrow {}^{4}_{2}\text{He} + {}^{222}_{86}\text{Rn}$
  (b) $^{7}_{4}\text{Be} \longrightarrow {}^{0}_{+1}e + {}^{7}_{3}\text{Li}$
  (c) $^{35}_{16}\text{S} \longrightarrow {}^{0}_{-1}e + {}^{35}_{17}\text{Cl}$
2. 3.4×10^8 kcal mole^{-1} (He)   **3.** $\sim$8200 B.C.
4. (a) $^{59}_{27}\text{Co} + {}^{1}_{0}n \longrightarrow {}^{60}_{27}\text{Co}$
  (b) $^{14}_{7}\text{N} + {}^{4}_{2}\text{He} \longrightarrow {}^{1}_{1}\text{H} + {}^{17}_{8}\text{O}$
  (c) $^{131}_{53}\text{I} \longrightarrow {}^{0}_{-1}e + {}^{131}_{54}\text{Xe}$
  (d) $^{235}_{92}\text{U} + {}^{1}_{0}n \longrightarrow 3{}^{1}_{0}n + {}^{94}_{36}\text{Kr} + {}^{139}_{56}\text{Ba}$
5. (Discuss with your instructor.)

Practice for Proficiency
1. (a) $^{14}_{6}\text{C} \longrightarrow {}^{0}_{-1}e + {}^{14}_{7}\text{N}$ **(b)** $^{11}_{6}\text{C} \longrightarrow {}^{0}_{+1}e + {}^{11}_{5}\text{B}$
  (c) $^{232}_{92}\text{U} \longrightarrow {}^{4}_{2}\text{He} + {}^{228}_{90}\text{Th}$ **(d)** $^{90}_{38}\text{Sr} \longrightarrow {}^{0}_{-1}e + {}^{90}_{39}\text{Y}$
  (e) $^{3}_{1}\text{H} \longrightarrow {}^{0}_{-1}e + {}^{3}_{2}\text{He}$ **(f)** $^{60}_{27}\text{Co} \longrightarrow {}^{0}_{-1}e + {}^{60}_{28}\text{Ni}$
  (g) $^{14}_{8}\text{O} \longrightarrow {}^{0}_{+1}e + {}^{14}_{7}\text{N}$ **(h)** $^{211}_{83}\text{Bi} \longrightarrow {}^{4}_{2}\text{He} + {}^{207}_{81}\text{Tl}$
  (i) $^{13}_{7}\text{N} \longrightarrow {}^{0}_{+1}e + {}^{13}_{6}\text{C}$ **(j)** $^{40}_{19}\text{K} + {}^{0}_{-1}e \longrightarrow {}^{40}_{18}\text{Ar}$
2. (a) $^{211}_{85}\text{At} + {}^{0}_{-1}e \longrightarrow {}^{211}_{84}\text{Po}$ **(b)** $^{44}_{20}\text{Ca} + {}^{2}_{1}\text{H} \longrightarrow {}^{1}_{0}n + {}^{45}_{21}\text{Sc}$
  (c) $^{7}_{3}\text{Li} + {}^{1}_{1}\text{H} \longrightarrow 2{}^{4}_{2}\text{He}$ **(d)** $^{14}_{7}\text{N} + {}^{4}_{2}\text{He} \longrightarrow {}^{17}_{8}\text{O} + {}^{1}_{1}\text{H}$
  (e) $^{2}_{1}\text{H} + {}^{3}_{1}\text{H} \longrightarrow {}^{1}_{0}n + {}^{4}_{2}\text{He}$ **(f)** $^{59}_{27}\text{Co} + {}^{1}_{0}n \longrightarrow {}^{60}_{27}\text{Co}$
  (g) $^{81}_{37}\text{Rb} + {}^{0}_{-1}e \longrightarrow {}^{81}_{36}\text{Kr}$ **(h)** $^{29}_{15}\text{P} \longrightarrow {}^{29}_{14}\text{Si} + {}^{0}_{+1}e$
  (i) $^{17}_{7}\text{N} \longrightarrow {}^{0}_{-1}e + {}^{17}_{8}\text{O}$ **(j)** $^{1}_{0}n + {}^{235}_{92}\text{U} \longrightarrow$

$$^{87}_{35}\text{Br} + {}^{146}_{57}\text{La} + 3{}^{1}_{0}n$$

3. 2.6×10^9 years   **4.** $\sim$11,000 B.C.   **5.** 64%   **6.** $\sim$380
days   **7.** $\sim$32 days   **8.** $\sim$42%   **9.** $\sim$98%   **10.** 3.8
days   **11.** 2.9×10^8 kcal mole^{-1} (U)   **12.** 1.4×10^8 kcal
mole^{-1} (Pb)   **13.** 2.6×10^5 kcal   **14.** 1.2×10^8 kcal
15. (See the June 1972 issue of the *Journal of Chemical Education*, pp. 418–419.)

Self-Test
1. (a) $^{40}_{19}\text{K} \longrightarrow {}^{0}_{+1}e + {}^{40}_{18}\text{Ar}$ **(b)** $^{29}_{15}\text{P} \longrightarrow {}^{0}_{-1}e + {}^{29}_{14}\text{Si}$
  (c) $^{218}_{84}\text{Po} \longrightarrow {}^{4}_{2}\text{He} + {}^{214}_{82}\text{Pb}$
2. 1.1×10^9 kcal mole^{-1} (U)   **3.** $\sim 3 \times 10^9$ years
4. (a) $^{54}_{26}\text{Fe} + {}^{4}_{2}\text{He} \longrightarrow {}^{1}_{0}n + {}^{57}_{28}\text{Ni}$ **(b)** $^{6}_{3}\text{Li} + {}^{1}_{0}n \longrightarrow {}^{4}_{2}\text{He} + {}^{3}_{1}\text{H}$
  (c) $^{246}_{96}\text{Cm} + {}^{13}_{6}\text{C} \longrightarrow {}^{254}_{102}\text{No} + 5{}^{1}_{0}n$

Appendix J

Answers to Section Overview Tests

SECTION ONE

1. (a) J **(b)** D **(c)** I **(d)** H **(e)** K **(f)** M **(g)** A **(h)** B **(i)** G
(j) F **(k)** E **(l)** N **(m)** C **(n)** L **2.** (d) **3.** (c) **4.** (a)
5. (c) **6.** (c) **7.** (b) **8.** (d) **9.** (e) **10.** (c)
11. (e) **12.** (d) **13.** (b) **14.** (d) **15.** (a)

SECTION TWO

1. (a) K **(b)** J **(c)** F **(d)** G **(e)** O **(f)** M **(g)** C **(h)** N **(i)** A
(j) B **(k)** L **(l)** D **(m)** I **(n)** H **(o)** E **2. (a)** F **(b)** F **(c)** T
(d) T **(e)** F **(f)** F **(g)** T **(h)** T **(i)** F **(j)** T **(k)** T **(l)** F
(m) F **(n)** T **(o)** T **3. (a)** CH_3NH_2 **(b)** chloroform
(c) $HClO_4$ **(d)** ammonium nitrate **(e)** $FeSO_4$ **(f)** phosphoric
acid **(g)** KOH **(h)** ethanol **(i)** NO **(j)** sodium dichromate
4. (a) $Na_2CO_3(s) + HNO_3(aq) \longrightarrow$
$$CO_2(g) + NaNO_3(aq) + H_2O(l)$$
net $\quad Na_2CO_3(s) + H^+(aq) \longrightarrow$
$$CO_2(g) + Na^+(aq) + H_2O(l)$$
(b) $CO_2(g) + Ca(OH)_2(aq) \longrightarrow CaCO_3(s) + H_2O(l)$
net $\quad CO_2(g) + Ca^{2+}(aq) + OH^-(aq) \longrightarrow$
$$CaCO_3(s) + H_2O(l)$$
(c) $Al(s) + H_2SO_4(aq) \longrightarrow Al(HSO_4)_3(aq) + H_2(g)$
net $\quad Al(s) + H^+(aq) \longrightarrow Al^{3+}(aq) + H_2(g)$
5. (a) $Na_2CO_3 + 2HNO_3 \longrightarrow CO_2 + 2NaNO_3 + H_2O$
net $\quad Na_2CO_3 + 2H^+ \longrightarrow CO_2 + 2Na^+ + H_2O$
(b) same as **4(b)**
net $\quad CO_2 + Ca^{2+} + 2OH^- \longrightarrow CaCO_3 + H_2O$

(c) $2Al + 6H_2SO_4 \longrightarrow 2Al(HSO_4)_3 + 3H_2$
net $\quad 2Al + 6H^+ \longrightarrow 2Al^{3+} + 3H_2$
6. (a) 72 g **(b)** 23.8 mg **(c)** 274 g (270 to 2 places) **7. (a)** CO
(b) 80% **(c)** 54 g **8.** -173.6 kcal mole^{-1} (CH_3OH)
9. -1980 kcal **10.** 57% **11. (a)** **12.** 15¢/g (CH_3OH)
13. (c)

SECTION THREE

1. (a) HSO_3^-, $\boxed{OH^-}$ **(b)** $\boxed{HSO_3^-}$, HSO_4^- **(c)** $\underline{NH_4^+}$, $\boxed{CN^-}$
(d) $\boxed{H_2PO_4^-}$, $\underline{HSO_4^-}$ **(e)** $\underline{H_2PO_4^-}$, $\boxed{HCO_3^-}$ **2.** HSO_3^- and
$H_2PO_4^-$
3. (a) *complete* $\quad CH_3CO_2H(aq) + NaOH(aq) \longrightarrow$
$$NaCH_3CO_2(aq) + H_2O(l)$$
net $\quad CH_3CO_2H(aq) + OH^-(aq) \longrightarrow$
$$CH_3CO_2^-(aq) + H_2O(l)$$
(b) *complete* $\quad H_2SO_4(aq) + Ba(OH)_2(aq) \longrightarrow$
$$BaSO_4(s) + 2H_2O(l)$$
net $\quad H^+(aq) + HSO_4^-(aq) + Ba^{2+}(aq) + 2OH^-(aq) \longrightarrow$
$$BaSO_4(s) + 2H_2O(l)$$
(c) N.R.
(d) *complete* $\quad Cu(NO_3)_2(aq) + Na_2S(aq) \longrightarrow$
$$CuS(s) + 2NaNO_3(aq)$$
net $\quad Cu^{2+}(aq) + S^{2-}(aq) \longrightarrow CuS(s)$
(e) *complete* $\quad 2AgNO_3(aq) + MgCl_2(aq) \longrightarrow$
$$2AgCl(s) + Mg(NO_3)_2(aq)$$
net $\quad Ag^+(aq) + Cl^-(aq) \longrightarrow AgCl(s)$

4. b   **5.** d   **6. (a)**

H   F

F   F

(s-p σ bond)   (p-p σ bond)

7. (a) H:$\ddot{\text{O}}$:$\ddot{\text{N}}$::$\ddot{\text{O}}$ H—$\ddot{\text{O}}$—N=$\ddot{\text{O}}$

(b) H:C:::N: H—C≡N:

(c) H:$\ddot{\text{N}}$:H   H—N—H

H   H

(d) $\ddot{\text{O}}$::C::$\ddot{\text{O}}$ $\ddot{\text{O}}$=C=$\ddot{\text{O}}$

(e) H:C::$\ddot{\text{O}}$ H—C=$\ddot{\text{O}}$

:$\ddot{\text{O}}$:   :$\ddot{\text{O}}$:

H   H

8. (a) PC   **(b)** PC   **(c)** I   **(d)** NP   **(e)** PC   **9.** (See Fig. 8.1, Unit 8.)   **10. (a)** pyramidal ($<$109.5°)   **(b)** tetrahedral (109.5°) **(c)** trigonal planar (120°)   **(d)** linear (180°)   **(e)** octahedral (90°) **11.** e   **12.** d   **13.** b

SECTION FOUR

1. (a) C   **(b)** P   **(c)** M   **(d)** D   **(e)** B   **(f)** J   **(g)** N   **(h)** H   **(i)** O **(j)** E   **(k)** I   **(l)** A   **(m)** L   **(n)** F   **(o)** K   **(p)** G   **2.** c **3. (a)** hydrogen bonding   **(b)** metallic bond   **(c)** ionic bond **(d)** London dispersion forces   **(e)** dipole–dipole forces **4.** e   **5.** a   **6.** e   **7.** b   **8.** d   **9.** e   **10.** b   **11.** e **12.** e   **13.** e   **14.** e   **15.** c   **16.** d   **17.** a   **18.** c **19.** a   **20.** a

SECTION FIVE

1. (a) $M = \dfrac{\text{no. moles solute}}{\text{no. liters solution}}$ **(b)** $F = $ (same as M)

(c) $N = \dfrac{\text{no. gram-eqs solute}}{\text{no. liters solution}}$ **(d)** $m = \dfrac{\text{no. moles solute}}{\text{no. kg solvent}}$

(e) $\text{ppm} = \dfrac{\text{no. mg solute}}{\text{no. liters solution}}$ or $\text{ppm} = \dfrac{\text{no. mg solute}}{\text{no. kg solution}}$

(if dilute aqueous solution)   (general case)

2. c   **3.** a   **4.** b   **5.** e   **6.** d   **7.** a, c constitutive; b, d, e colligative   **8. (a)** 2.55 m **(b)** 2.29 M **(c)** 6.86 N **9.** a   **10.** d   **11.** b   **12.** d   **13.** a   **14.** c   **15.** a **16.** a   **17.** c   **18.** b

SECTION SIX

1. e   **2.** c   **3.** c   **4.** d   **5.** b   **6. (a)** bimolecular **(b)** third order   **(c)** H_2NO and N_2O **(d)** $H_2N_2O_2$ **(e)** yes **7.** d   **8.** a   **9.** c   **10.** c   **11.** c   **12.** a   **13.** b **14.** c   **15.** d   **16.** c   **17.** c   **18.** d   **19.** b   **20.** b **21.** d   **22.** d   **23.** a   **24.** b   **25.** b

SECTION SEVEN

1. e   **2.** e   **3.** a   **4.** b   **5.** b   **6.** a   **7.** e   **8.** d **9.** e   **10.** b   **11.** d   **12.** a   **13.** b (methyl red) **14.** e   **15.** c   **16.** c   **17.** a   **18.** e   **19.** b   **20.** d **21.** c   **22.** d

SECTION EIGHT

1. d   **2.** b   **3.** d   **4.** c   **5.** d   **6.** a   **7.** d   **8.** d **9.** d   **10.** a   **11.** c   **12.** c   **13.** a   **14.** d   **15.** c

SECTION NINE

1. b   **2.** e   **3.** d   **4.** a   **5.** b   **6.** d   **7.** a   **8.** d **9.** b   **10.** b   **11.** e   **12.** b   **13.** c   **14.** b   **15.** d **16.** e   **17.** d   **18.** e   **19.** c   **20.** d   **21.** c   **22.** a **23.** d   **24.** e   **25.** b   **26.** a   **27.** a   **28.** d   **29.** b **30.** e

SECTION TEN

1. e   **2.** c   **3.** a   **4.** c   **5.** d   **6.** b   **7.** c   **8.** c **9.** c   **10.** a   **11.** e   **12.** a   **13.** a   **14.** c   **15.** b **16.** b   **17.** b   **18.** a   **19.** d   **20.** c   **21.** d   **22.** a **23.** a   **24.** c   **25.** d   **26.** c   **27.** c   **28.** c   **29.** e **30.** $C_6H_4Cl_2$

SECTION ELEVEN

1. c   **2.** e   **3.** d   **4.** d   **5.** c   **6.** b   **7.** a   **8.** e **9.** b   **10.** d.

Index

77 78 79 80 9 8 7 6 5 4 3 2 1